Proceedings of the Eighth International Conference

Electrorheological Fluids and Magnetorheological Suspensions

Proceedings of the Eighth International Conference

Electrorheological Fluids and Magnetorheological Suspensions

Nice, France 9 – 13 July, 2001

G. Bossis

Nice University, France

World Scientific
New Jersey • London • Singapore • Hong Kong

Published by

World Scientific Publishing Co. Pte. Ltd.

P O Box 128, Farrer Road, Singapore 912805

USA office: Suite 1B, 1060 Main Street, River Edge, NJ 07661

UK office: 57 Shelton Street, Covent Garden, London WC2H 9HE

British Library Cataloguing-in-Publication Data
A catalogue record for this book is available from the British Library.

ELECTRORHEOLOGICAL FLUIDS AND MAGNETORHEOLOGICAL SUSPENSIONS
Proceedings of the Eighth International Conference

ISBN 981-02-4937-3

Printed in Singapore by Uto-Print

PREFACE

The Eighth International Conference on Electrorheological (ER) Fluids and Magnetorheological (MR) Suspensions and their Applications was held in the city of Nice, France, July 9 to 13, 2001.

The Conference started with two memorial lectures in the honor of the pioneers of electrorheology, W. Winslow and of magnetorheology, J.Rabinow, who both died recently. I would like, on behalf of our community, to thank Pr. R. Tao (Temple University) and Dr D. Carlson (Lord Corporation) for giving these memorial lectures and writing the obituaries which are placed hereafter.

Two prizes, sponsored by Lord Corporation and Winslow's daughter: Nancy Winslow Keyes, in order to help young scientists to attend the conference, were awarded to Sonia Melle and Ji Woo Kim for their very promising work, respectively in the field of magnetorheology and electrorheology.

Since the 1st International Conference on ER and MR Suspensions held in Boston in 1987 a lot of progress has been done and smart fluids are now on the market, mainly in applications related to controllable damping. The fundamental mechanisms responsible for the change of viscosity, as well as the potentiality of new classes of field responsive materials like polymer blends or filled elastomers are now better understood. Some important issues for the development of future applications, as for instance the factors governing the long term stability of ER and MR fuids or the complex interdependence between structure and rheology, have also been addressed during this conference. A total of 145 papers consisting of 85 oral and 60 poster presentations, including 19 invited talks by the world's leading authorities, were given , covering all the topics related to ER and MR fluids. The Proceedings contain 130 papers giving the up to date information on the state of the art of ER/MR research.

I thank all the participants who contributed to the success of the conference and our sponsors: National Center for Scientific Research, University of Nice Sophia-Antipolis, City of Nice, Department of Alpes-Maritimes, Region Provence Côte d'Azur, Lord Corporation, Metravib, Hutchinson-Paulstra, Fludicon, Thermo Rheo and Thomson Marconi Sonar. My special appreciation is also due to our secretary Mrs Scatena for her outstanding support in organizing the conference.

Georges Bossis
Nice University
November, 2001

On April 17, 2000, the noted scientist and engineer Willins M. Winslow passed away in Denver, Colorado, USA at the age of 96. His discovery of the electrorheological (ER) effect made more than half a century ago continues to inspire us today.

Winslow was born in the Denver area on January 1, 1904. After earning his engineering degree from University of Colorado at Boulder in 1926, he went to work for Western Electric Company in Chicago. Eighteen months later he returned to Denver to marry Vera Vail and started a new job at Public Service Company, where he served for 43 years and retired in 1972.

His first interest in electrorheology began after reading an article in 1937 describing the "Johnsen Rahbeck Effect." The article discussed large forces of attraction between electrodes separated by certain poor insulator as slate. Winslow immediately began experimenting with different materials to try to find a practical use for this theory. In the process, in his basement-laboratory he found that some liquids become solid when subjected to an electric field.

He received a patent for his discovery in 1947 after demonstrating the operability of an invention for which there was no theoretical explanation. The patent office examiner termed the phenomenon the "Winslow effect", which is now commonly called the ER effect.

Widely published, Winslow was an internationally recognized authority in the field he helped found. At 90, he remained as a consultant for the Lubrizol Corporation of Ohio and continued to do experiment in his basement-laboratory.

Winslow left behind his permanent contributions to the area of ER fluids, which will last forever. ER industrial applications, such as walking machines for handicapped, began to appear on the market and are well received. Future ER applications may be on automobile, helicopter, aircraft, and earthquake-proofing building and bridges.... As gathering here to disseminate the advance of ER technology, we honor the man who first discovered the phenomenon.

Jacob Rabinow was born in Kharkov, Russia on January 8, 1910. After what he called a too eventful life as a child in European Russia and Siberia during the revolution, his family moved to China in 1919 and eventually settled in Brooklyn, New York in 1921. In 1933 he earned a BS in Engineering from the City College of New York and a year later a graduate degree in Electrical Engineering.

In 1938 Rabinow began his inventing career at the National Bureau of Standards, now the National Institute of Standards and Technology (NIST). Here he worked on many ordnance devices during the war and rose to become Chief of the Electro-Mechanical Ordnance Division. In 1947 Rabinow applied for a patent on a "Magnetic Fluid Torque and Force Transmitting Device" which was later granted as US Patent 2,575,360.

Rabinow become one of the most prolific inventors of the 20th century. He was awarded 230 US patents on a wide variety of mechanical and electrical devices. Among these was the first magnetic disc memory, a straight-line phonograph, magnetic character reading machines, automatic letter sorting machines and the magnetic particle clutch.

In 1954 he formed his own engineering company which in 1964 became part of Control Data Corporation where he served as Vice President and Head of the Rabinow Advanced Development Laboratory until 1972. In 1968 he formed the RABCO Company to manufacture straight-line phonographs. The Harmon Kardon Corporation later acquired this company. In 1972 he rejoined the National Bureau of Standards where he held several positions including Chief Research Engineer, National Engineering Laboratory. Although he retired from government service in 1989 he continued as a consultant with NIST in the evaluation of inventions in the Office of Energy-Related Inventions.

Rabinow received many awards and honors throughout his career. These included the President's Certificate of Merit in 1948 and the Industrial R&D Scientist of the Year Award 1960. In 1998 Jacob Rabinow was awarded the prestigious Lemelson-MIT Lifetime Achievement Award.

Jacob Rabinow died on September 11, 1999.

CONTENTS

III. MATERIAL PROPERTIES

IV. PHYSICAL MECHANISMS

I. APPLICATIONS

MULTIPLE APPLICATION OF MAGNETORHEOLOGICAL EFFECT IN HIGH PRECISION FINISHING

WILLIAM KORDONSKI AND DON GOLINI

QED Technologies, 1040 University Avenue, Rochester NY 14607 USA

E-mail: kordonski@qedmrf.com

Magnetorheological finishing (MRF) is enabling technology that may produce surface accuracy of the order of 10 nm peak to valley and surface micro-roughness less than 10Å. In this technique, magnetorheological fluid performs the primary function, which is deterministic material removal. There are several auxiliary devices that also function on a "smartness" of magnetorheological fluid. A non-intrusive magnetorheological valve is an integral part of the finishing machine "circulatory system" and provides the pump "stiffness" and computer control of the flow rate. To provide MR polishing fluid continuous circulation, a special means for fluid removal from the polishing wheel and returning to the delivery system has been designed. Once the fluid leaves the polishing zone, it is removed from the wheel surface by a suction cup. The cup has a magnetic system that forms a dynamic magnetorheological fluid seal in the gap between the cup and the wheel surface. Finite element analysis software has been used for magnetic systems and flow design and optimization.

1. Introduction

In the most common polishing method, a glass work piece is pressed with a constant force against a moving polishing lap covered with slurry containing small abrasive particles (~1-3μm in diameter). The relative motion of the work piece and the lap, combined with an applied pressure and the chemical activity of the slurry, provide conditions for polishing which is considered as a removal of a soft hydrated layer by a scratching that is performed by the abrasive particles embedded in the polisher. In this technique, each geometry of the part requires a dedicated, appropriately shaped polishing lap. In addition, the polishing lap shape changes with time and must be periodically reconditioned to obtain consistent removal efficiency. The traditional method therefore does not offer a flexible, cost-effective option for finishing, particularly for advanced optical shapes such as aspheres [1] or conformal optics [2].

A new precision polishing method called magnetorheological finishing (MRF) has been developed to overcome fundamental limitations of traditional finishing techniques [3]. MRF may produce surface accuracy of the order of 10 nm peak to valley and surface micro-roughness less than 10Å on optical glasses, single crystals (calcium fluoride, silicon) and ceramics. High precision spheres and flats, nearly any asphere, square and rectangular aperture optics, prisms and cylindrical optics may be corrected and finished with MRF. Currently manufactured by QED Technologies, MRF equipment is being employed in the production of high precision optics in North America, Europe, and Asia. In this paper, emphasis is given to the various aspects of utilization of unique properties of magnetorheological fluid in MRF. The MR fluid primary function (deterministic material removal), as well as several auxiliary devices that also function on a "smartness" of magnetorheological fluid, are discussed.

2. Process of Material Removal in MRF

The process of material removal in MRF is based on latest advances in non-Newtonian fluid dynamics, physical chemistry, physics of magnetic phenomena, material science, and has no analogy in the practice of polishing.

The basic idea of MRF lies in the fact that, in certain conditions, a chemical mechanical process, wherein material removal occurs by a sorption detachment and capture of molecularly small fragments of a hydrated silicon layer by polishing particles rather than by mechanical scratching, may prevail in glass polishing with a water based slurry. It assumes that in contrast to conventional polishing, the surface normal stress and the surface indentation are not dominant in the process of material removal. In this case, certain mechanical work must be done at the surface to supply energy via surface shear stress to form chemical bonds between lattice defects in polishing crystals (active centers) and the silicon atoms of the surface layer. An important benefit of the process is lack of the subsurface damaged layer even at high material removal rates. The surface shear stress can be conveniently generated, for example, by boundary viscous fluid flow.

In MRF, this is arranged utilizing properties of magnetorheological fluid [4]. The magnetorheological polishing fluid is deposited by a nozzle on the rim of a rotating wheel, which transports the fluid to the work piece surface. The wheel rim and the surface to be polished form a converging gap exposed to a magnetic field. The moving wall, which is the rim surface, generates a flow of magnetically stiffened MR polishing fluid through the converging gap. It is commonly accepted that magnetically stiffened MR fluid can be considered as a Bingham material with a yield stress controlled by a magnetic field [5]. The flow of Bingham material generates a unique pressure distribution in the gap that is associated with an unsheared fluid (core), which is attached to the moving wall [6] (Fig.1). This means that a quasi-solid moving boundary is effectively formed very close to the surface of the work piece resulting in the high shear stress in the contact zone and material removal over a portion of the work piece surface. This area is designated as a polishing spot. The material removal is enhanced by nonmagnetic abrasive particles, which are constituents of the slurry that are forced out to the polishing interface by a magnetic field gradient.

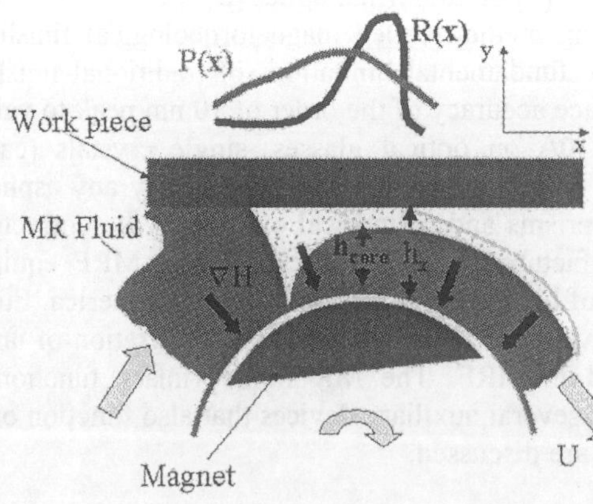

Fig 1. Schematic of the flow in polishing interface
P(x) and R(x) are experimental pressure and removal rate distribution

Unfortunately, a more sophisticated treatment of the flow is not possible within the limits of known analytical models. Nevertheless some results in conjunction with experimental data can be used for a more detailed flow analysis.
It was shown in [6] that

$$h_{core} = h_x - \sqrt{\frac{2\eta_0 U}{dp/dx}} \tag{1}$$

where η_0 is plastic viscosity and dp/dx is pressure gradient, which is depended on fluid yield stress (magnetic field intensity) and other flow parameters. Parameters h_x and U are identified in Fig.1.
Equation (1) enables an effective geometry of polishing interface to be found using experimental data for pressure gradient distribution. Thereafter a standard FEA Newtonian code can be used for the shear stress calculation. As soon as the shear stress distribution along the flow centerline is determined, the results can be related to the experimentally derived pressure distribution and material removal profile.

Some results of the analysis are shown in Fig. 2a,b. Measurements of pressure and material removal distribution were taken at two different magnetic fields with other conditions remaining equal.

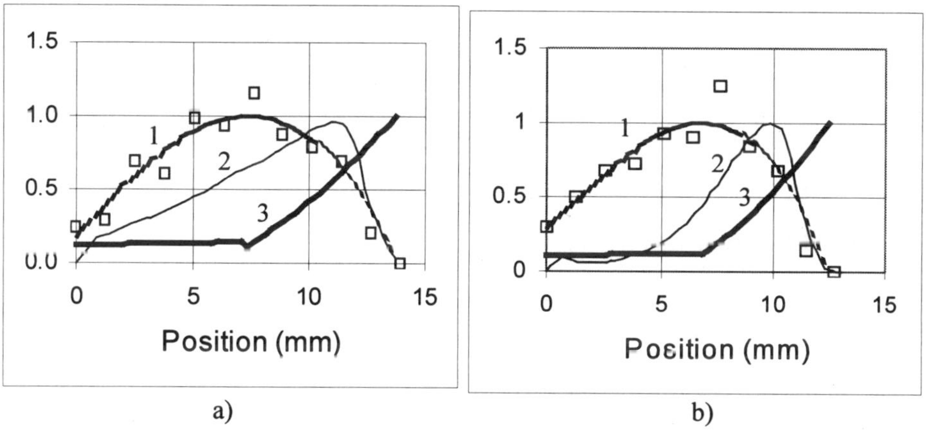

Fig. 2. Pressure, removal and shear stress distribution in polishing spot
a) magnetic field 170 kA/m; b) magnetic field 260 kA/m
1- pressure; 2 – removal; 3- shear stress

The distribution of the pressure is typical for lubrication flow and does not coincide with the shape of removal, whereas the shear stress distribution does have characteristics similar to the removal profile. There is appreciable evolution of all profiles with the magnetic field. Of course, this simple approach has many limitations. It does not give an exact match between the shear stress and removal rate profile, but it does give support to the hypothesis that MRF is a shear stress controlled removal process.
Some conclusions supporting the above basic assumptions may also be drawn from experiments where the drag force produced by MR polishing fluid flow against glass work piece surface was measured [7]. The experiment was designed to provide only one variable parameter, namely, glass/abrasive chemical activity. Chosen abrasives were

cerium oxide, aluminum oxide and diamond, and their ability to chemically bond with the glass surface decreases as they are listed [8]. All other conditions remained equal. Standard rheological measurements were made to be certain that the chosen range of abrasive particles concentration did not change bulk mechanical properties of tested MR fluids.

As shown from Fig. 3, the addition of abrasive particles reduces the drag force suggesting that mechanical properties of the fluid near the contact polishing zone are significantly changed. This is due to the fact that the magnetic field gradient forces out nonmagnetic abrasive particles to the polishing interface contributing to an increase in concentration of nonmagnetic component at the interface. At the same time, the cerium oxide maintains a highest drag force, which may result from stronger particle/surface interaction due to more pronounced chemical bonding.

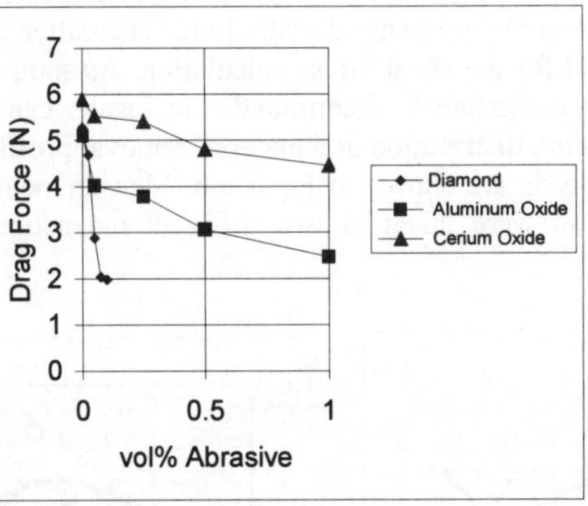

Fig. 3. Effect of nonmagnetic abrasives on the drag force in MRF

3. MR Fluid Handling and Delivery

The important part of the MR finishing process is proper functioning of the MR polishing machines "circulatory system", schematically shown in Fig. 4.

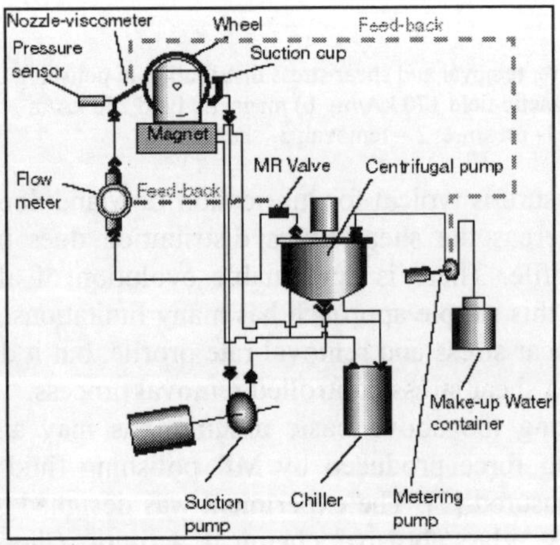

Fig. 4. MR fluid delivery system schematic

It delivers the MR polishing fluid to the polishing wheel, takes the fluid off of the wheel, and performs the fluid conditioning.

The "heart" of the system is a specially designed vertical centrifugal pump. The pump provides smooth (pulse less) flow, as well as the MR slurry mixing and agitation. A non-intrusive, fast-response magnetorheological valve is an integral part of the pump and provides the pump "stiffness" as well as feedback-assisted control of the flow rate. The valve magnetic circuit optimization was done with 3D magnetic software [9]. An example of the result of valve magnetic circuit transient process modeling is shown in Fig. 5. The obtained time constant of ~ 20ms coincides well with the experimental value.

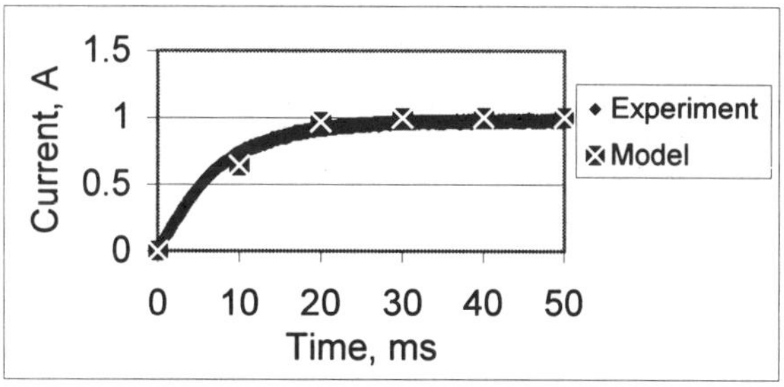

Fig.5. MR valve transient characteristic

The nozzle, together with a pressure sensor and a flow meter, is an in-line viscometer. The viscometer picks up changes in the slurry concentration due to water evaporation or solids sedimentation and activates an appropriate response of the fluid conditioner. In the viscometer design, provision has been made for correct MR fluid initial viscosity measurements, including nozzle shielding from the magnetic field and temperature control.

To provide MR polishing fluid continuous circulation, a special means for fluid removal from the wheel and returning to the system has been designed. Once the fluid leaves the polishing zone, it is removed from the wheel surface by a suction cup.

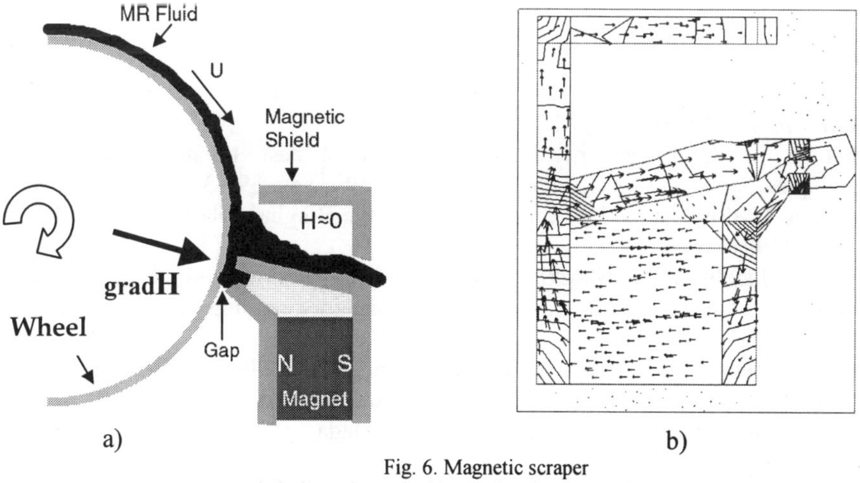

Fig. 6. Magnetic scraper
a) design schematic; b) calculated magnetic field map

The cup (wiper) schematically shown in Fig.6a has a permanent magnetic system that forms a dynamic magnetorheological fluid seal in the gap between the cup and the wheel surface. In operation scraped from the wheel by the magnetic seal, the MR fluid is collected in the magnetically shielded chamber. Thereafter the demagnetized low viscous fluid is drawn off and returned to the delivery system by a suction pump. The condition

$$\mu_o M \text{grad} H V > \rho U^2 S, \tag{2}$$

which assumes that the fluid is kept in the gap by a magnetic volume force that exceeds the hydrodynamic force of inflowing fluid, was used as a basis for the seal design with 3D magnetic software [9]. In (2) M is a magnetorheological fluid magnetization, grad$H$ is a magnetic field gradient, V is a volume of the MR fluid in the gap, ρ is the MR fluid density, U is the MR fluid velocity, S is a contact area and μ_o is the magnetic constant. The result of modeling is shown in Fig. 6b. As is seen from Fig.6b, there is a gradient fringing field at the wheel/scraper contact zone and there is no magnetic field in the scraper chamber. The magnetic scraper offers few advantages over the mechanical one. There is no need for periodical adjustment and the wear of the wheel is significantly reduced.

4. Acknowledgements

The authors gratefully acknowledge the extensive contribution of the following individuals: colleagues at the University of Rochester – A. Shorey, S. Jacobs; and co-workers at QED Technologies – S. Hogan, J. Carapella and A. Sekeres.

References

1. Golini, D., Improve your image with aspheres. *Optics and Photonics News*, August (1997) pp. 41 - 45.
2. Thompson, K. P and Rodgers, J. M., Conformal optics - key issues in a developing technology". *Optics and Photonics News* October (1997) pp. 24 - 30.
3. Golini, D., Jacobs, S., Kordonski, W., and Dumas, P.. Precision optics fabrication using magnetorheological finishing. *SPIE Proceedings: Advanced materials for optics and precision structures. Critical Reviews of Optical Science and Technology* **CR67**-16 (1997) pp. 251-273.
4. Kordonski W. and Jacobs S.. Magnetorheological Finishing. *Inter. Journal of Modern Physics,* **10** (1996) pp. 2837-2848.
5. Kordonski W. I.. Magnetorheological effect as a base of new devices and technologies, *J. Magnetism &Magnetic Mat.* **122** (1993) pp. 395-398.
6. Tichy J. A. Hydrodynamic lubrication theory for the Bihgham plastic flow model. *J. Rheol.* **34** (1991) pp. 477-495.
7. Shorey A., Jacobs S., Kordonski W., and Gans R., Experiments and observations regarding the mechanism of glass removal in magnetorheological finishing. *Apl. Optics* **40** (2001) pp. 20-33.
8. Shorey A., mechanism of material removal in magnetorheological finishing (MRF) of glass. Ph.D. dissertation (University of Rochester, Rochester, N.Y.) (2000).
9. MagNet 6 software. Infolytica Corp., Montreal, Canada

NEW MICROSCOPIC POLISHING WITH MAGNETIC COMPOUND FLUID (MCF)

KUNIO SHIMADA

Akita Prefectural University
84-4 Aza-ebinokuchi, Tsuchiya, Honjyo, 015-0055, Japan
E-mail: shimada@akita-pu.ac.jp

YOICHI AKAGAMI

Akita Prefectural Industrial Technology Center
4-11 Aza-Sanuki, Arayamachi,Akita 010-1623, Japan
E-mail: akagami@akita-iri.pref.akita.jp

SHINICHI KAMIYAMA

Akita Prefectural University
84-4 Aza-ebinokuchi, Tsuchiya, Honjyo, 015-0055, Japan
E-mail: kamiyama@akita-pu.ac.jp

TOYOHISA FUJITA, TOSHIO. MIYAZAKI AND ASTUSHI SHIBAYAMA

Akita University
1-1 Tegatagakuenmachi, Akita 010-8502, Japan
E-mail: fujita@akita-pu.ac.jp , miyazaki@akita-pu.ac.jp , sibayama@akita-pu.ac.jp

We propose a magnetic compound fluid (MCF) as a new smart fluid. This fluid includes nm size magnetite and μ m size iron particles in a solvent. The magnetic effects of this fluid can be expected to lie in the range between those of magnetic fluid (MF) and magneto-rheological fluid (MRF). Moreover, the magnitude of the shear stress to the shear rate under a steady magnetic field can be larger than that of MRF by varying the compound rate of the MF and MRF. This report shows an experimental application of microscopic polishing using MCF. We examined the surface roughness of flat titanium material attached to a rotating plate at 15 rpm under a magnetic field and found that MCF produced a more uniform surface polish than did MR. As well, the roughness of R_y and R_a are larger under a fluctuating magnetic field than under a steady magnetic field. This more effective polishing can be explained in terms of the cluster model of particles.

1 Introduction

MF is a nano-fluid containing nano-sized magnetite particles. MF's magnetic pressure and apparent viscosity under a magnetic field are smaller than those of MRF. However, the stability of the particle distribution in a MF solvent is greater than that of MRF. On the other hand, MRF has a yield stress. Especially in the case of high magnetic fields and mass concentrations, MRF differs from MF in regard to its hydrodynamic characteristics. Specifically, MRF can be appropriately dealt with as a powder rather than a fluid [1]. Because MRF has μ m size iron particles, the stability of its particle distribution is lesser than MF.

Due to these differences, the engineering applications are different between MF and MRF. However, engineering applications require larger magnetic pressure and apparent viscosity under a magnetic field, and more stable distribution of particles while maintaining its behavior as a fluid. In magnetic polishing using a magnetic field, greater magnetic pressure is especially required.

In the present paper, we propose a magnetic fluid compounded by nm-size magnetite and μ m-size iron particles in a solvent as a new magnetic intelligent or "smart" fluid. This report describes an experimental application of microscopic polishing using MCF. We measured the relation of shear stress to shear rate of MCF including abrasive particles using a rotating rheometer. We examined the surface roughness of flat titanium material attached to a rotating plate under a magnetic field and compared the results of roughness between the use of MCF and MR.

2 MCF and hydrodynamic characteristics

2.1 MCF and experimental method

Our proposed MCF contains 10 nm magnetite coated with oleic acid and 1.2-1.6 μ m iron (HQ) in kerosene. Namely, it compound of MF and MRF.

We examined the shear stress to a given shear rate using cone and concentric cylinder types of rotating rheometers under a uniform magnetic field. In the case of the cone type, the magnetic field is applied transversely to the lower plate. In the case of the concentric cylinder type, the magnetic fields are applied transversely and longitudinally to the cylinder axis [2]. The temperature of the fluid is constant at room temperature.

2.2 Results and discussion

Figure 1 shows the comparison of stress-strain curves of MCF with Fe_3O_4: 26.4 vol.%, HQ: 30.8 vol.% and Kerosene: 42.8 vol.% with and without the application of a magnetic field. The shear stress to a shear rate varies in terms of the type of rheometer and direction of the applied magnetic field.

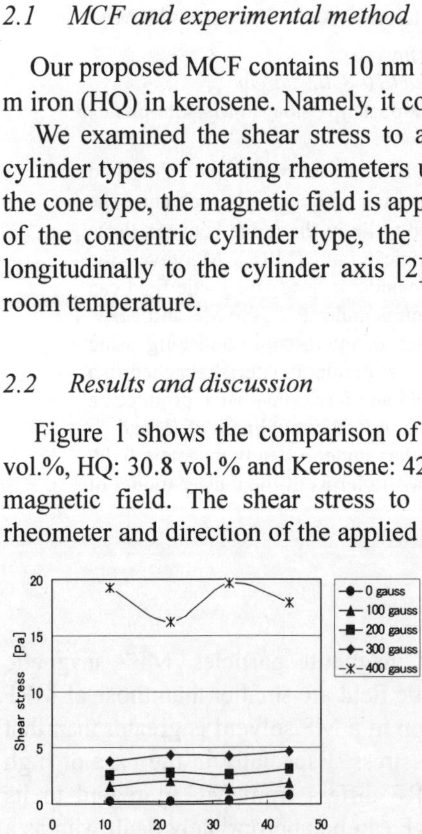

(a) Case with using a cone-type rheometer.

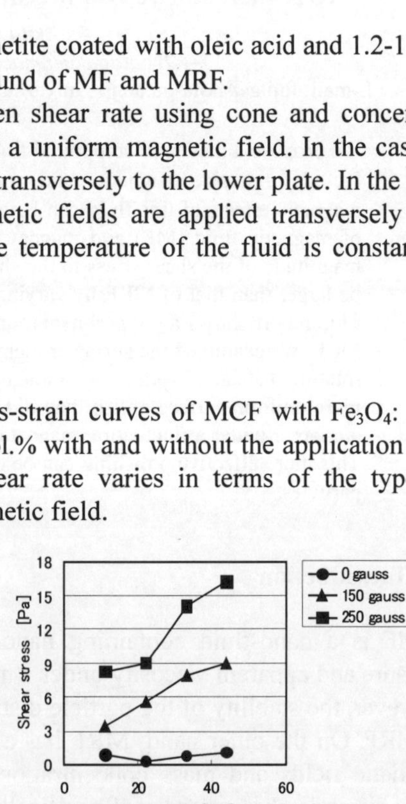

(b) Case with using a concentric cylinder-type rheometer under a transverse magnetic field.

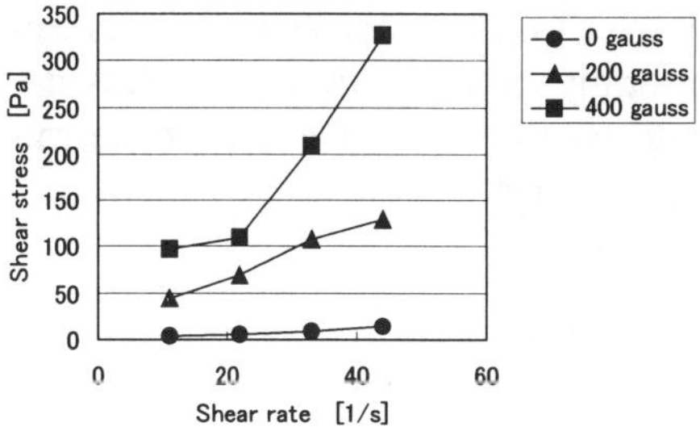

(c) Case with using a concentric cylinder-type rheometer under a longitudinal magnetic field.
Figure 1. Relation between shear stress and shear rate of MCF.

In the case of microscopic polishing using MCF, the relation between the shear stress and shear rate of MCF which includes abrasive particles must be investigated. Figure 2 shows a comparison of stress-strain curves of MCF with magnetite: 10.6 vol.%, HQ: 7.0 vol.%, Al_2O_3: 11.3 vol.% and Kerosene: 71.1 vol.%. This fluid is compounded by the weight rates of 4.2g from MCF with 50wt%HC-50 30ml + HQ60g, Al_2O_3 2g and kerosene 2g.

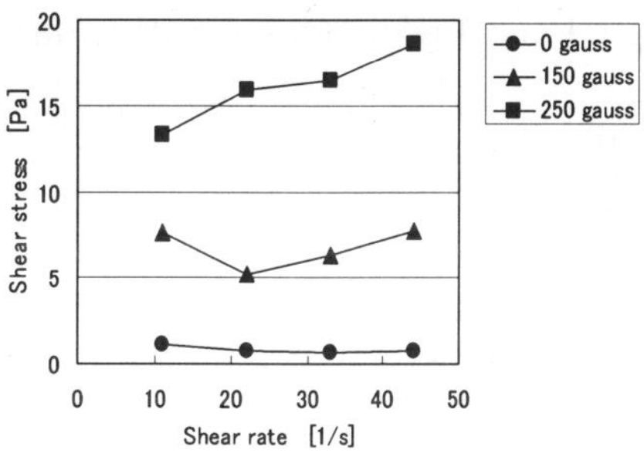

(a) Case with using a cone-type rheometer.

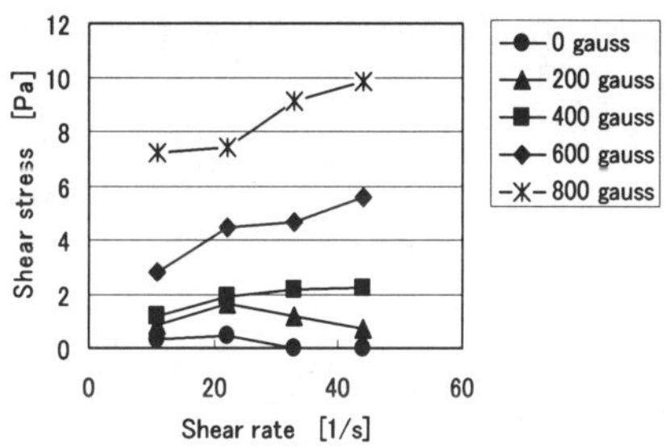

(b) Case with using a concentric cylinder-type rheometer under a transverse magnetic field.

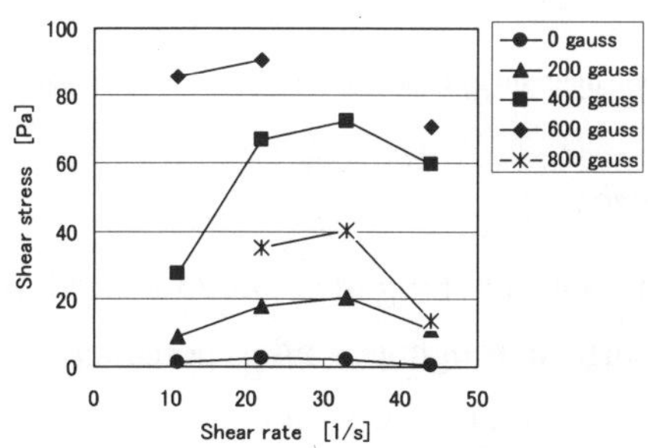

(c) Case with using a concentric cylinder-type rheometer under a longitudinal magnetic field.
Figure 2. Relation between shear stress and shear rate of MCF with abrasive particles.

As shown in Fig. 1, shear stress and shear rate show a consistent correlation in any type of rheometer or direction in which a magnetic field is applied. Therefore, polishing with the MCF which include abrasive particles can be expected to polish a rotating plate more uniformly because the shear stress do not change due to the strength of the shear rate on the plate. We expect the same result in the case of polishing using a rotating cylinder.

3 Microscopic polishing using MCF

3.1 Experimental method

Figure 3 shows the experimental apparatus for polishing using MCF. We used 99.5 % pure titanium (50mm×50mm×3mm) which is equally as strong as steel, non-corrosive, and heat resistant. Under room temperature, the fluid compounded by the weight rates of 4.2g from the MCF with magnetite (26.4 vol.%, HQ: 30.8 vol.% and kerosene: 42.8 vol.%, Al_2O_3 2g and kerosene 2g) is inserted between the polishing pad and the titanium specimen. The magnetic pole with the polisher pad fixed at the surface is rotated at a constant rotating speed of 280 rpm while a pressure of 5kgf is presented on the specimen. The polishing pad is for hanging over the workpiece. The magnetic field is applied to the MCF containing abrasive by an electromagnet. The duration of each polishing period was five minutes. The surface roughness of the specimen was measured using a Rank Taylor Hobson S4C surface analyzer [3].

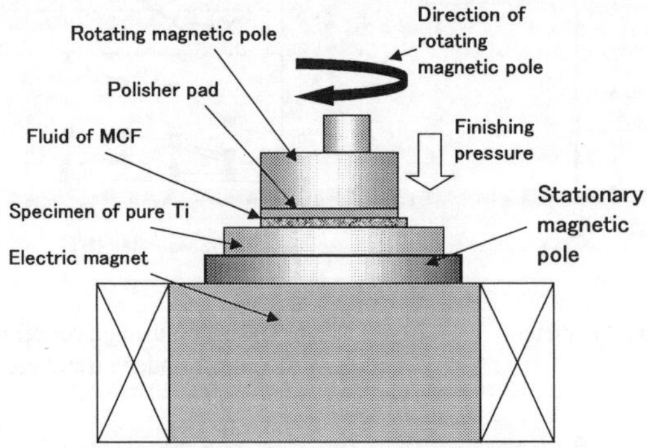

Figure 3. Schematic diagram of experimental apparatus for magnetic fluid polishing

3.2 Results and discussion

Figure 4(a) shows the surface roughness as evaluated by the surface analyzer before polishing. The surface roughness after polishing is shown in Fig. 4(b-1), having used a fluctuating magnetic field with a rectangular pulse with a duty of 30%, frequency 0.1 Hz and amplitude 1300 gauss which is the measured value at the position of the used MCF. Figure 4(b-2) shows the results when using a steady magnetic field with amplitude 1300 gauss. Figure 4(b-3) shows the results when using no using magnetic. Figure 4(b-4) shows the results when using a fluctuating magnetic field with a rectangular pulse with a duty of 30%, frequency 0.1 Hz and amplitude 2100 gauss. Figure 4(b-5) shows the results when using MRF 2g with iron particles (CL) of diameter 7～8 μ m, Al_2O_3 1g and silicon oil 2g, and fluctuating magnetic field with a rectangular pulse with a duty of 30%, frequency 0.1 Hz, amplitude 2400 gauss. Only in the case shown in Fig. 4(b-5) did the polished surface contain many scratches. As demonstrated by the use of MCF on the rotating disk under uniform shear stress as shown in Fig. 2, MCF can effectively and uniformly polish a rotating disk.

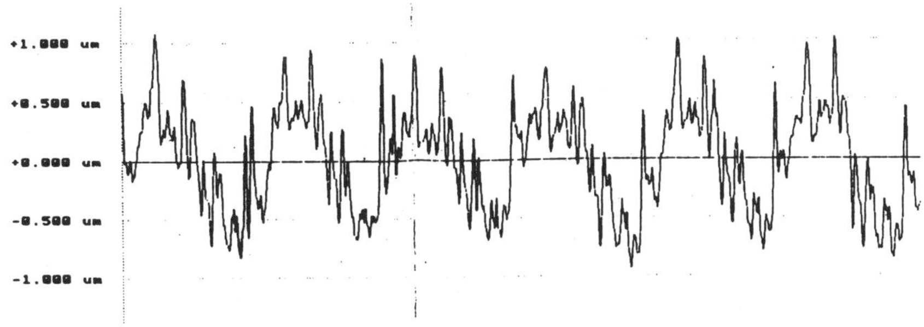

(a) Measured roughness of the plate before polishing

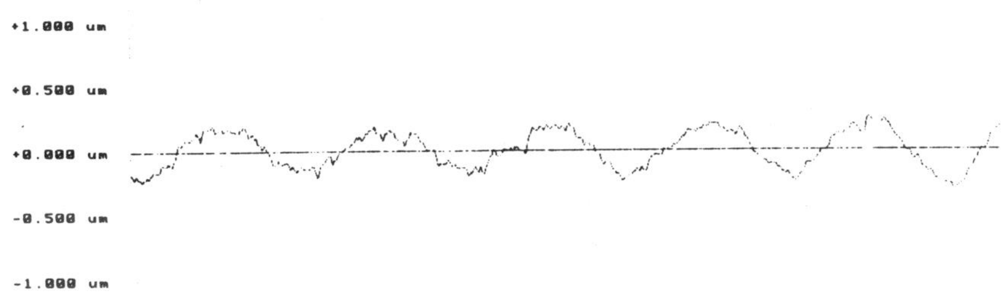

(b-1) In the case of a fluctuating magnetic field with an amplitude of 1300 Gauss using MCF

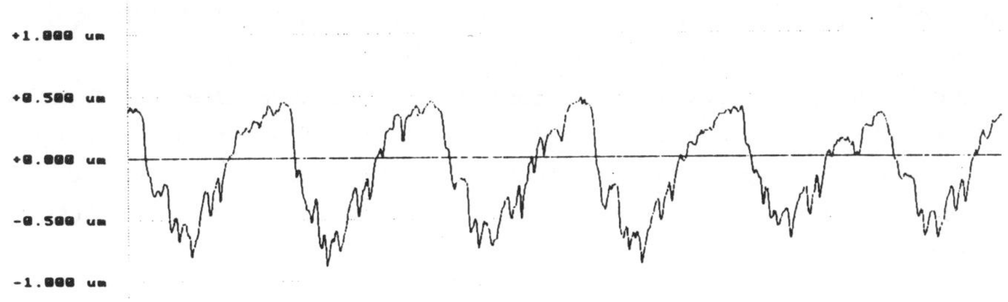

(b-2) In the case of a steady magnetic field with a strength of 1300 Gauss using MCF

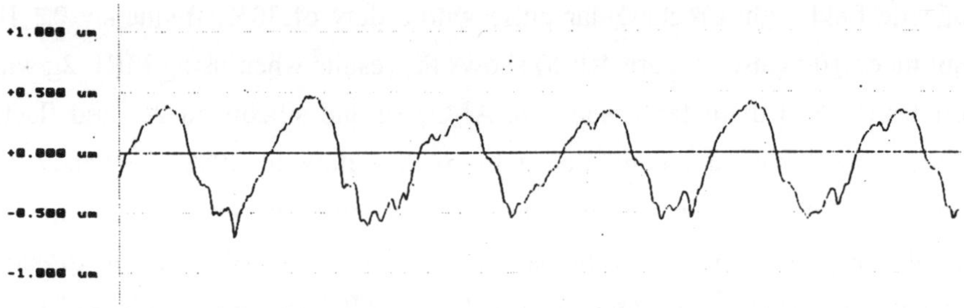

(b-3) In the case without a magnetic field using MCF

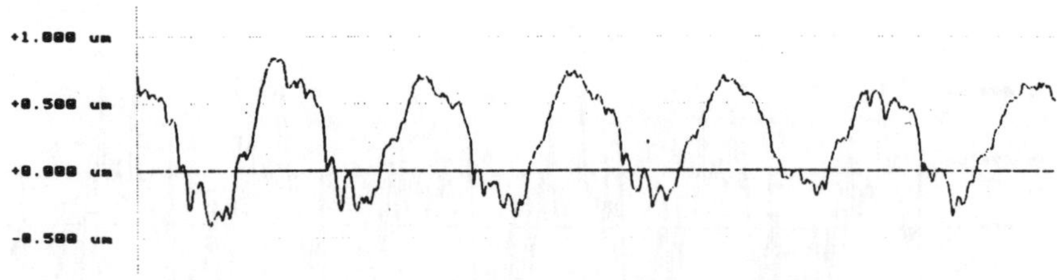

(b-4) In the case of a fluctuating magnetic field with an amplitude of 2100 Gauss using MCF

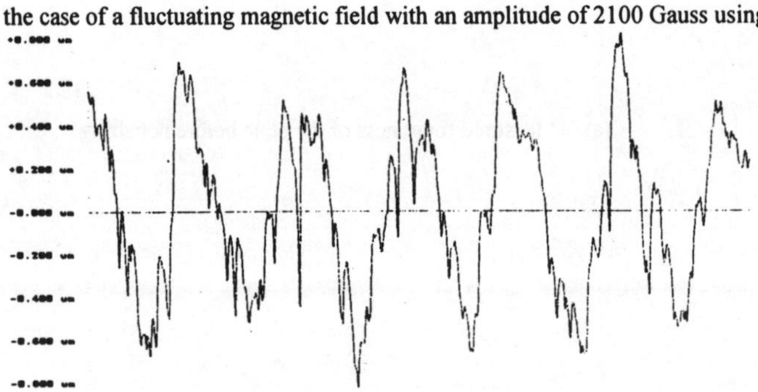

(b-5) In the case of a fluctuating magnetic field with an amplitude of 2400 Gauss using MRF

(b) Measured roughness of the plate after polishing

Figure 4. Test roughness of the plate

Table 1 shows the surface roughness of R_a and R_y. The use of a fluctuating magnetic field is more suitable than a steady one for polishing, as seen from Figs. 4(b-1)-4(b-3). However, the use of a fluctuating magnetic field requires an appropriate magnetic amplitude, as seen in Figs. 4(b-1) and 4(b-4).

Figure 5 shows the schematic diagram of the particle aggregates of the MCF containing abrasive. Many long and large-diameter magnetite and iron chains form in clusters and push the abrasive particle on the attached polished material surface. As in other experimental investigation, these chains could be easily observed by microscope.

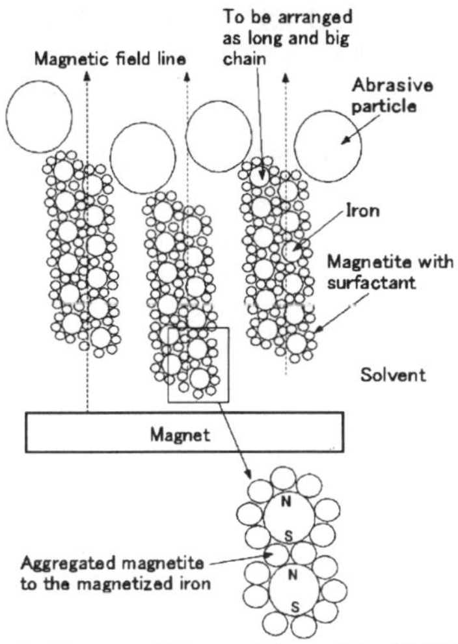

Figure 5. Schematic diagram of the particle model of MCF contatining abrasive.

Table 1. Test results regarding roughness shown in Fig. 4

Rougness	R_a (μ m)	R_y (μ m)
Before polishing	0.3673	1.9418
After ploshing in Fig. 4		
(b-1)	0.1319	0.6301
(b-2)	0.3333	1.3396
(b-3)	0.2979	1.0574
(b-4)	0.3350	1.2653
(b-5)	0.3222	1.5070

4.Conclusions

Micro-polishing using MCF under a fluctuating magnetic field is more suitable than MRF for uniformly polishing a rotating disk surface.

References

1. G. L. Gulley and R. Tao, Structures of amagnetorheological fluid, Proceedings of 7th International Conference ERF & MRS, (1999), pp.331-335.
2. T. Fujita, J. Mochizuki and I. J. Lin, Viscosity of electrorheological magneto-dielectric fluid under electric and magnetic fields, J. Magn. Magn. Mat., 122, (1993), pp.29-33.
3. Y. Akagami, K. Asari, B. Jeyadevan, T. Fujita and N. Umehara, ER fluid finishing using rotating electrode, Proceedings of 7th International Conference ERF & MRS, (1999), pp.821-826.

IMPACT OF DROPLETS OF MAGNETO-RHEOLOGICAL SUSPENSION UNDER APPLIED MAGNETIC FIELDS

S. SUDO AND M. FUNAOKA

College of Science and Engineering, Iwaki Meisei University, Iino 5-5-1, Chuohdai, Iwaki 970-8551, Japan

E-mail: sudou@iwakimu.ac.jp

H. NISHIYAMA

Institute of Fluid Science, Tohoku University, Katahira 2-1-1, Aoba-ku, Sendai 980-8577, Japan

E-mail: nishiyama@ifs.tohoku.ac.jp

An experimental study is presented of the dynamic behavior of the impacting droplet of magneto-rheological suspension under applied magnetic fields. Effects of magnetic fields on the impact phenomena between a MR fluid droplet and paper on the hard rubber mat were revealed with a three-dimensional motion analysis system. In case of perpendicular magnetic field to a target surface, the MR fluid droplet did not collapse completely by its impact, but MR fluid spikes were formed by the magnetic field. In case of parallel magnetic field to a target surface, the MR fluid droplet showed complicated deformation. In case of strong parallel field, the MR fluid droplet showed breakup into few droplets.

1 Introduction

The study of various phenomena of magneto-rheological suspensions is of fundamental interest and importance with respect to a variety of applications. MR suspensions are a typical representative of 'intelligent' fluids, i.e., these suspensions show the dramatic but reversible changes in the rheological characteristics between Newtonian fluid without magnetic field and solid-like state with magnetic fields[1]. MR suspensions are fluids of micro-sized magnetizable particles dispersed in carrier liquids. These suspensions respond readily to magnetic fields. Therefore, controllable magnetorheological fluid devices have been developed in commercial production[2].

On the other hand, the impact of a liquid droplet onto the surfaces of another liquid or solid has been studied for many years, and it is still currently of interest in such applications as spray painting, spray cooling, spray coating, ink jet printing, fire suppression, and soil erosion by rain. For example, Aziz et al.[3] studied the impact and solidification of molten tin droplets on a stainless steel surface. Kang and Lee[4] showed the dynamic behavior of a water droplet impinging upon an inclined heated surface. Bergeron et al.[5] showed controlling the impact of drops onto smooth, non-wetting hydrophobic solid surfaces by the addition of small amounts of a flexible polymer. Bussmann et al.[6] presented a numerical methodology for examining droplet fingering and splashing, by initiating a perturbation of the flow shortly after the impact of a droplet. Kim et al.[7] developed a linear perturbation theory to investigate the interface instabilities of a radially-expanding, liquid jet in cylindrical geometries when a liquid droplet impinges on a solid surface. Authors also studied the impact phenomena between a magnetic fluid drop and paper on the hard rubber mat, subject to magnetic field[8].

In spite of many investigations, however, no study has been published on the drop impact of MR suspension on the solid surface under an applied magnetic field.

In the present study, the impact phenomena between a droplet of magneto-rheological suspension and paper on the hard rubber mat subjected to applied magnetic fields were investigated. Effects of applied magnetic fields on the spread area of MR droplet were studied with a three-dimensional motion analysis system under a perpendicular field to the target surface. The process of forming spikes of MR suspension after the drop impact was revealed. It was also revealed that a parallel field to the target surface may lead to the droplet into breakup.

2 Experimental Apparatus and Procedures

The experimental method is similar to that presented by authors in our previous paper[8]. Figure 1 shows the experimental apparatus. The main components are composed of (1) a 25ml burette to form and release the MR drop, (2) a flat target surface on which the droplet fell, (3) a three-dimensional motion analysis system to record and analyze the impact phenomena of MR droplet, and (4) a permanent magnet installed under the rubber mat to magnetize the MR drop (the direction of an applied magnetic field is perpendicular to the target plane), or an electromagnet (the direction of an applied magnetic field is parallel to the target plane).

The drops of MR suspension fell on a piece of paper that was bonded to the hard rubber mat. The impact velocity of the drop was varied by the change of the height from which it fell. The permanent magnet used to magnetize the MR drop was a disk type with 70mm diameter and 10mm thick. The strength of the magnetic field

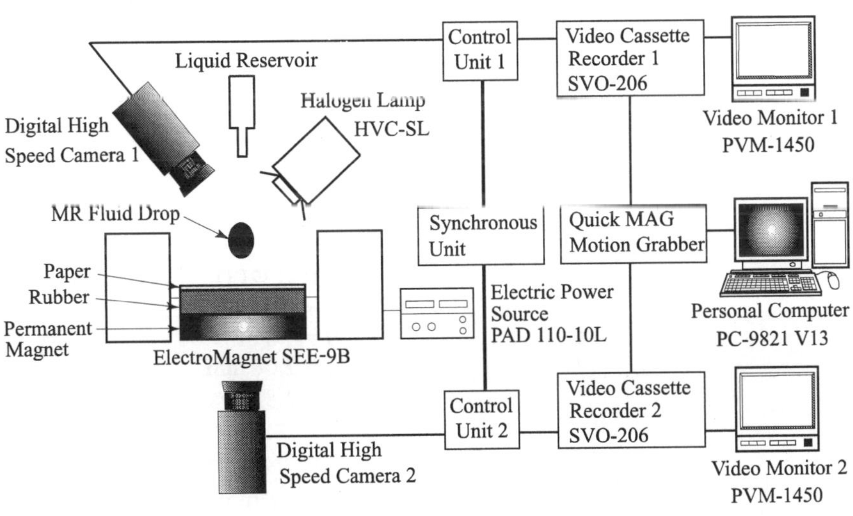

Figure 1. Schematic diagram of experimental apparatus

was changed with the distance between the magnet and the target surface. In this case, the direction of an applied magnetic field is perpendicular to the target plane. The water-cooled electromagnet was used to apply the parallel magnetic field to the target plane. The field strength of the electromagnet was changed by the electric current. The average diameter of the MR drops was 2.08mm. The drops of MR suspension showed the deformations under the action of magnetic fields. The impact phenomena of MR drops were analyzed by the three-dimensional motion analysis system. The system was composed of two high speed video cameras, synchronous unit, two video cassette recorders, a motion grabber, two video monitors, and a personal computer. The high speed video camera was generally operated at a speed of 13500 frames per second. Sample MR suspension used in the experiment was MR fluid MRF-132LD made by Lord Corporation.

3 Experimental results and discussion

3.1 Impact of a MR drop under a non-magnetic field

The impact phenomena of a MR droplet under a non-magnetic field showed the well-known process as the impact of a water droplet. After drop impact the MR fluid flowed radially outward from the underside of the drop to form a thin liquid layer on the paper surface. As the drop collapses, the MR fluid layer increased in area. Figure 2 shows how the radius of the base of the MR fluid spread varies with time measured from the moment of impact. The maximum spread area of MR fluid depends on the impact velocity of droplet. In this experiment, the impact velocity of the drop was varied by the change of the height from which it fell. If we neglect air resistance and assume that the free-fall acceleration does not vary with altitude, then the vertical motion of a freely falling object is equivalent to motion

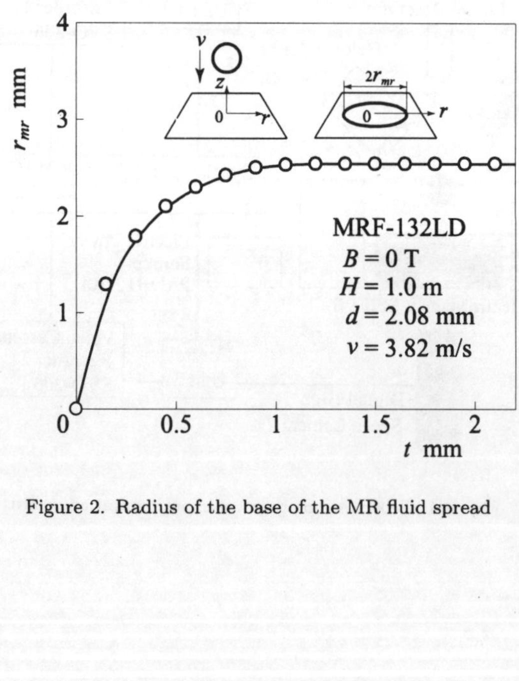

Figure 2. Radius of the base of the MR fluid spread

in one dimension under constant acceleration. Therefore, the impact velocity is described as follows :

$$v = (2gH)^{\frac{1}{2}} \tag{1}$$

where g is the gravitational accelration and H the falling height. Due to air resistance, actual impact velocity is smaller then the value of Eq.(1).

3.2 Impact of a MR fluid droplet under a perpendicular magnetic field

The impact phenomena of a MR fluid droplet under a perpendicular magnetic field are described in this paragraph. In this case, a permanent magnet was installed under the rubber mat to magnetize the MR fluid drop. The permanent magnet forms magnetic field gradient in its circumference. The impact velocity may replace Eq.(1) with Eq.(2) [9].

$$v = \left\{2(g - \frac{\mu_0}{\rho}\bar{M}\frac{dH_m}{dz})H\right\}^{\frac{1}{2}} \tag{2}$$

$$\bar{M} = \frac{1}{H_m}\int_0^{H_m} M dH_m \tag{3}$$

where ρ is the density of MR fluid, μ_0 the permeability of free space, z the vertical coordinate to be the z axis and call z positive upward, and H_m the strength of magnetic field. The origin of the z coordinate is located on the target surface. Therefore, in this case, the impact velocity at the constant H is higher compared with the case of non-magnetic field. In spite of higher impact velocities, the spread area of MR fluid decreased with the applied magnetic field. Figure 3 shows a sequence of photpgraphs of the impact phenomena produced by the MR fluid drop after it had fallen through a vertical height H = 0.5 m and colloided with the paper pasted on the hard rubber mat under a perpendicular magnetic field to target surface. It can be seen that the MR fluid droplet is deformed before the impact. The deformation of MR drop depended on the strength of magnetic field and the

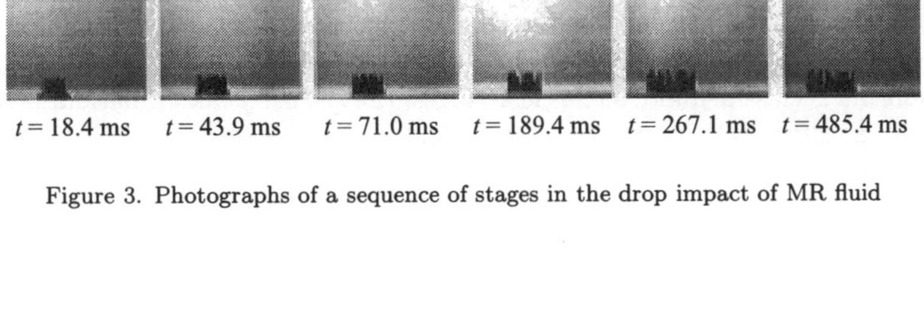

Figure 3. Photographs of a sequence of stages in the drop impact of MR fluid

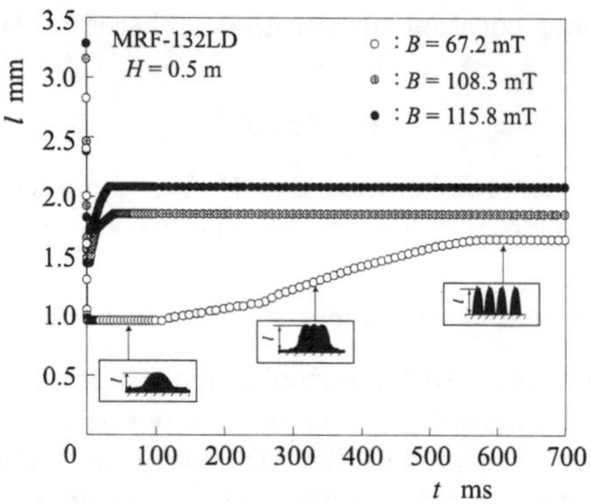

Figure 4. Effects of perpendicular magnetic field on the drop impact

impact velocity. In general, MR fluids are non-colloidal suspensions of polarizable particles having a size on the order of a few microns[2]. Considering the particle size in the carrier liquid, the magnetization of MR fluid might be characterized by a Brownian rotational diffusion time τ_B.

$$\tau_B = \frac{3V\eta_0}{kT} \tag{4}$$

where V is the particle volume, η_0 the viscosity of the carrier liquid, k Boltzmann's constant, and T the absolute temperature in degrees Kelvin. We can estimate as $\tau_B \approx 10^{-2}$s for the MR fluid. The relaxation time of surface deformation of MR suspension, τ_s, is also estimated the order of 10^{-2} s. These relaxation times, τ_B and τ_s, characterize the impact phenomena of MR fluid droplet. It can be seen also from Fig.3 that the MR fluid droplet shows the incomplete collapse. After drop impact, the MR suspension flows slightly and radially outward from the underside of the drop with an applied magnetic field. This peak formation can be attributed to the following rheological properties[10] for MR fluid.

$$\tau = \eta_0\dot{\gamma} + \mu_0 H_m^2 \phi_a \alpha^{-1} \chi_a^2/(2+\chi_a) \tag{5}$$

where τ is the shear stress, $\dot{\gamma}$ the shear rate, ϕ_a the concentration of the dispersed phase, α a constant, and χ_a the susceptibility. After peak formation, the peak of MR fluid is cracked as if a flower bud may expand. This phenomenon is attributable to the surface instability produced by the magnetic field. Furthermore, the interaction between the growth of surface instability and magnetic charge (magnetic dipole) formes many MR fluid spikes. Figure 4 shows effects of perpendicular magnetic field on the impact process of the MR fluid drop. The process of impact phenomena is classified into some clear stages, that is, the peak formation, breakup of peak, and the spike formation of MR fluid. It can be seen from Fig.4 that the progress time of each stage in the impact process depends on the strength of applied magnetic field. The increase of magnetic field accelerates the spike formation.

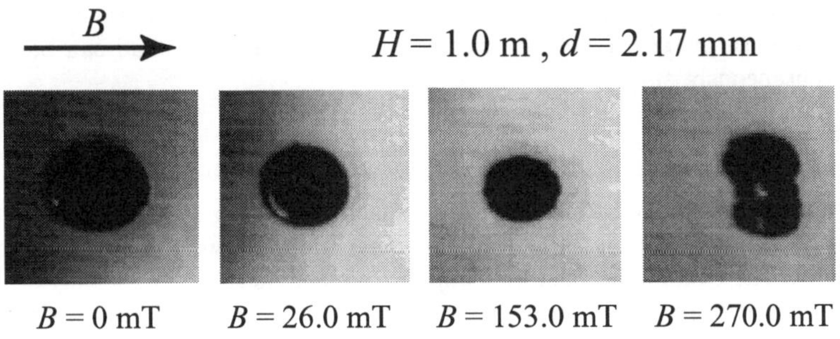

Figure 5. Photographs of impact patterns left by drops of MR suspension

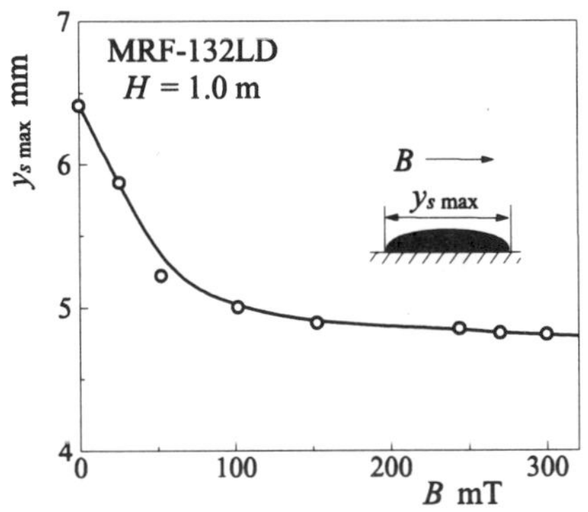

Figure 6. Relation between fluid spread along the magnetic field and the field strength

3.3 Impact of a MR fluid droplet under a horizontal magnetic field

In this paragraph, the impact phenomena of a MR fluid droplet under a horizontal magnetic field are discribed. The impact phenomena in the horizontal magnetic field were very different from that of the perpendicular magnetic field. Before drop impact, the MR drop is deformed by the applied magnetic field. The drop deformation of MR suspension was formed by the interaction of the surface instability produced by the magnetic field with air resistance during the free fall. Figure 5 shows the patterns left by MR fluid drops after they had fallen onto horizontal paper under a parallel field to the target plane. In general, a magnetizable drop in a magnetic field is expected to elongate along the magnetic field. However, the impact pattern of MR drop showed the different result. Figure 6 shows the fluid spread along the magnetic field. The maximum spread of MR suspension along the magnetic field, y_s max, decreases with the strength of magnetic field. This phenomenon may be attributed to the interaction between the surface instability and air resistance. This kind of instability for MR fluid drop are basically the same phenomenon observed on a magnetic fluid drop between horizontal flat surfaces when a uniform magnetic field is oriented along these surfaces and can be turned on instantaneously[11].

4 Conclusions

The impact phenomena between a drop of magneto-rheological suspension and paper on the hard rubber mat under applied magnetic fields were investigated experimentally. The results obtained are summarized as follows.
(1) In the case of perpendicular magnetic field to the target surface, the process of impact phenomena is classified into some clear stages, such as the peak formation, breakup of peak, and the spike formation of MR suspension. The progress time of each stage in the impact process depends on the strength of applied magnetic field.
(2) In the case of parallel magnetic field to the target surface, fluid spread of MR suspension along the magnetic field decreases with the magnetic field strength. In case of strong parallel field, especially, the MR fluid droplet shows the growth of surface instability and breakup into few droplets.

References

1. W.I. Kordonsky, *J. Magn. Magn. Mater.* **122**, 395 (1993).
2. J.D. Carlson, D.M. Catenzarite, and K.A.St. Clair, *Int. J. Modern Phys. B* **10**, 2857 (1996).
3. S. D. Aziz and S.Chandra, *Int. J. Heat Mass Transfer* **43**, 2841 (2000).
4. B. S. Kang and D. H. Lee, *Exp. Fluids* **29**, 380 (2000).
5. V. Bergeron, D. Bonn, J. Y. Martin and L. Vovelle, *Nature* **405**, 772 (2000).
6. M. Bussmann, S. Chandra and J. Mostaghimi, *Phys. Fluids* **12**, 3121 (2000).
7. H. Y. Kim, Z. C. Feng and J. H. Chun, *Phys. Fluids* **12**, 531 (2000).
8. S. Sudo, N. Wakamatsu, T. Ikohagi, H. Nishiyama, M. Ohabe and K. Katagiri, *J. Magn. Magn. Mater.* **201**, 285 (1999).
9. S. Sudo, M. Ohaba, K. Katagiri and H. Hashimoto, *Magnetohydrodynamics* **33**, 480 (1997).
10. V.I. Kordonsky, Z.P. Shulman, S.R. Gorodkin, S.A.Demchuk, I.V. Prokhorov, E.A. Zaltsgendler and B.M. Khusid, *J. Magn. Magn. Mater.* **85**, 114 (1990).
11. V. Bashtovoi, S. Pogirnitskaya and A. Reks, *J. Magn. Magn. Mater.* **201**, 300 (1999).

Caster-walker with intelligent brakes employing ER fluid composed of liquid crystalline polysiloxane

Akio INOUE[1], Ushio RYU[2], Syohji NISHIMURA[2]

[1]Asahi Kasei Corporation,
[2]Asahi Engineering Co., Ltd.

Abstract

A caster-walker with ER fluid brakes installed in the rear wheels has been developed and tested. The brake actuation is "intelligent", in response to signals from a speed sensor and microprocessor, permitting it to prevent stumbling by the patient in mounting or dismounting the walker and on downslopes due to walker acceleration. The walker also serves effectively as a walk trainer. The ER brake is advantageous for small-wheeled vehicles because it is compact and simple, and consumes very little power. In this report, we describe the structure and characteristics of the brake, and key points in the brake design and development.

Key words: ER fluids, caster-walker, brake, intelligent

1. Introduction

ER (electroreological) fluids, with their fast, reversible change in viscosity on exposure to an electric field, have for some time been the subject of widespread interest because of their potentials for use in automotive applications. Although many types of ER fluids and applications have been proposed, however, practical use has been elusive. The ER fluids have generally been of the "dispersion type", consisting of dielectric particles in an insulating oil, and problems impeding their practical utilization have centered on variations in performance characteristics, insufficient durability, and particle sedimentation or abrasion.

We have resolved these problems, through the development of "homogeneous-type" ER fluids based on liquid crystalline polymers. These ER fluids resemble a grease in consistency. They are composed of liquid crystalline polysiloxane (Fig. 1) and a diluent. They exhibit a large viscosity increase (generated shear stress) under application of an electric field, and excellent stability and durability because they contain no particles and are therefore inherently free from the problems of sedimentation and abrasion.[1].

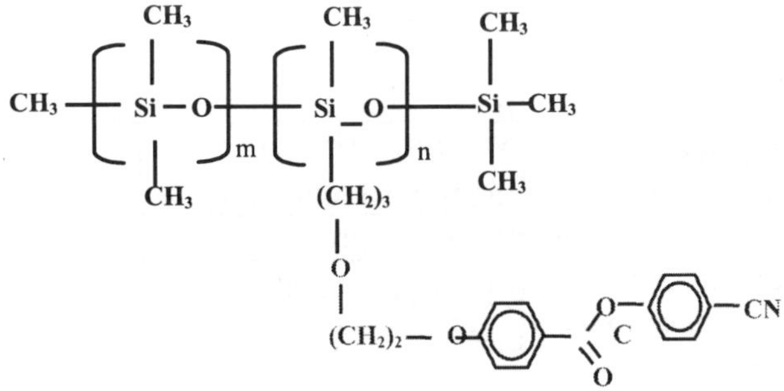

Figure 1. Structure of liquid crystalline polysiloxane

In comparison with the extremely fast response shown by dispersion-type ER fluids, homogeneous-type ER fluids generally show slower response and are therefore less suitable for oscillation absorption or other applications requiring very high-speed response. For applications such as brakes and clutches, however, we found that their response times are well within the required range, and determined that their far larger viscosity increase ("ER effect") on application of an electric field would be highly advantageous for the development of simple, compact, effective brake and clutch designs.[2].

In our development work, extending over some years, we encountered and were able to overcome a number of difficulties related to the homogeneous ER fluid and the brake design, and obtain highly efficient, extremely stable brake performance.

In our investigations of potential applications for such brakes, we noted the tendency for stumbling to occur among patients with ataxic gait or other disorders when using conventional caster-walkers, due to sudden walker movement during mounting or dismounting and to walker acceleration on down-slopes.

In collaboration with a Hyogo Prefecture rehabilitation hospital, we developed a caster-walker with ER brakes, as shown in Photograph 1, which arrest the movement of the wheels automatically by sensing abrupt or unusual walker motions.[3].

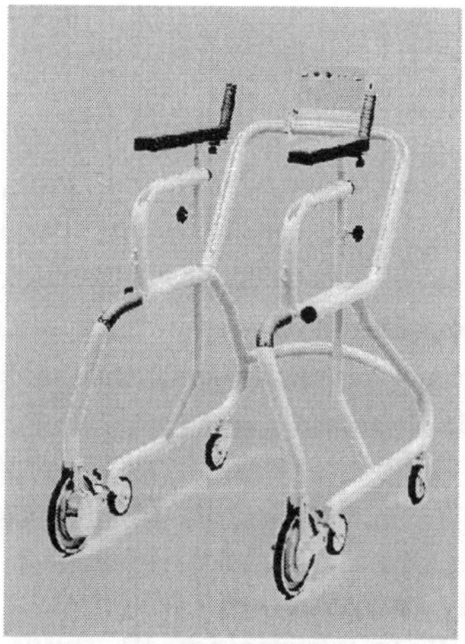

Photograph 1

In this paper, we describe the relation between the structure and the power of the ER brake, the key points in the brake design and manufacture, and the performance of the brake and the caster-walker.

2. Homogeneous ER fluid composed of liquid crystalline polymer

2.1 Characteristics and performance

Table 1. Characteristics of ER fluids made at Asahi Kasei

ER fluid type	Dispersion	Homogeneous
Rheological behavior	Bingham flow	Newtonian flow
Generated shear stress (Pa / 300sec-1)	600	8000
Viscosity (Poise) (under no electric field)	1	10
Response (msec)	<3	20~80 (rise)
Operating temperature (℃)	(0)~60	10(0)~60
Current density (μA/cm2)	<1	<1

ER fluids may be classified as homogeneous-type and dispersion-type. Under no electric field, both types show Newtonian flow in which the shear stress axis is proportional to the shear rate, in a straight-line relationship. Under an electric field, however, they exhibit quite different behaviors. By convention, we plot the shear stress along the Y-axis and the shear rate along the X-axis, and the slope of the line thus represents the viscosity. With a dispersion-type ER fluid, the line shifts toward the Y-axis with no change in its slope with increasing electric field strength, and it thus exhibits Bingham flow. With a homogenous-type ER fluid, in contrast, the slope increases with increasing electric

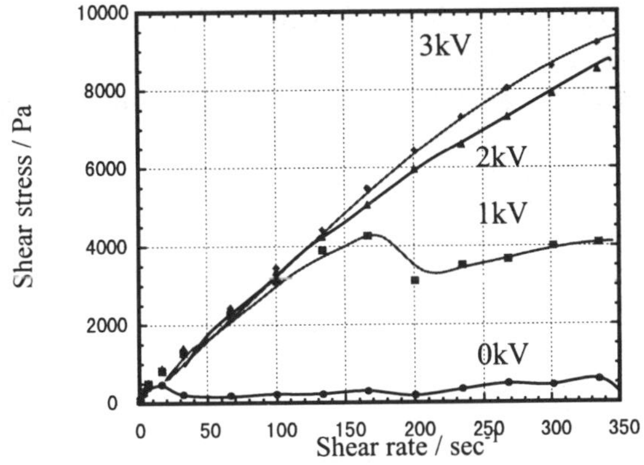

Fig. 2. Shear stress and electric field strength, with homogeneous ER fluid.

field strength, and it thus exhibits Newtonian flow. Table 1 summarizes the differing characteristics of the two fluid types.

Homogeneous ER fluids show a much larger ER effect than do dispersion ER fluids, despite their high viscosity with no electric field. Their ER effect is about 8000 Pa on application of 2 kV/mm at a shear rate of 300 per second. Their response is comparatively slow, but they show low current density and they are highly stable at 10 to 60 degrees Celsius. Homogeneous ER fluids, with their liquid crystalline polymers, are generally susceptible to crystallization or freezing below 10 degrees C and therefore lose their ER effect, but we have also developed homogeneous ER fluids that work at lower temperatures, down to about zero degrees.

3. ER brakes

3.1 Basic configurations

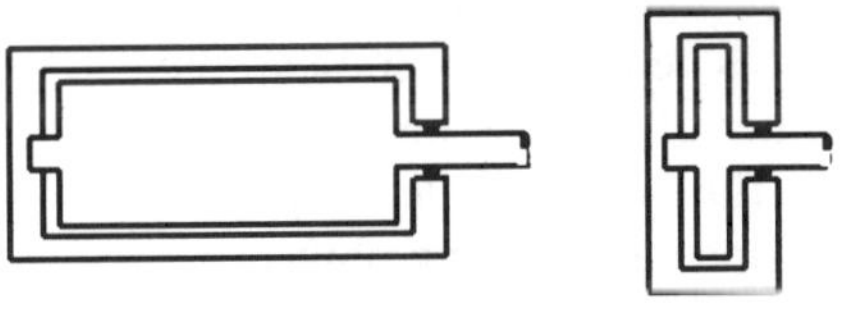

Fig. 3. Basic configurations of ER fluid brakes: cylinder- and disk-type.

The rotary brakes employing ER fluid can be broadly classified as cylinder-type and disk-type, as shown in Fig. 3. The clearance (usually less than 1 mm) between the outer and inner elements is filled with ER fluid. The elements in contacting with the ER fluid serve as positive and negative electrode plates. The braking force generated by the ER effect depends on the strength of the voltage between these electrodes.

3.2 Braking power

The braking power of the two brake types can be calculated as follows.

・Disk-type

$$Td = \int 2\pi r \cdot F \cdot r \cdot dr$$
$$= 2\pi F \int r2dr$$
$$= 2\pi Fr^3 + C \qquad (1)$$

・Cylinder-type

$$Tc = 2\pi r \cdot F \cdot r$$
$$= 2\pi Fr^2 \qquad (2)$$

T:brake power[N-m] γ :shear rate[1/s]
F:shear stress[Pa] δ :rotation speed[1/s]
r:radius[cm] a, b: factors
L:cylinder wide[cm] C:constant

The following applies, for a homogeneous ER fluid employing liquid crystalline polymer, which shows Newtonian flow.

$$F = a\gamma = br \qquad (3)$$

Therefore, we find the following relations between the brake power and the brake design.

a) Disk-type

$$Td = 2\pi Fr^3 + C$$
$$= 2\pi br^4 + C \qquad (4)$$

b) Cylinder-type

$$Tc = 2\pi Fr^2$$
$$= 2\pi br^3 \qquad (5)$$

Thus, in brakes using homogenous-type ER fluid, the braking power is generally proportional to the third power of the cylinder radius in the cylinder type, and to the fourth power of the disk radius in the disk type.

3.3 Compact brake design

To obtain a highly compact brake employing a homogenous ER fluid, it is generally advantageous to use a multi-element configuration, which provides a large element surface area in a small brake volume.

In a disk-type brake, the use of thin disks enables the packing of a large number of disks in a small brake volume. However, such disks may tend to deform under the forces exerted on them by the ER fluid during braking. Strong forces tend to occur near the center of the disk because of the Wisenberg effect under rotation, as the homogenous ER fluid is a polymer solution.

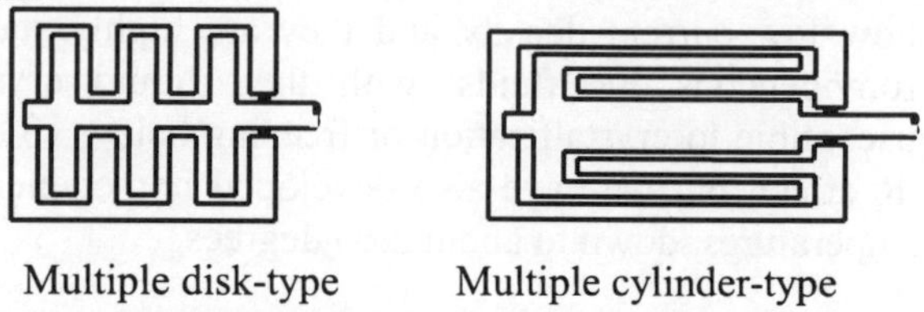

Fig. 4. Multi-disk and multi-cylinder configurations, for compact brake design.

Since the braking power is proportional the fourth power of the disk diameter and to the third power of the cylinder diameter, the contribution of the central part of the disk or cylinder to the braking power is relatively small. The effect of the central part of the disk or cylinder is one of increasing the current rather than the braking power. It is therefore desirable to use a design in which the central region of the brake is unoccupied by either disk or cylinder surfaces.

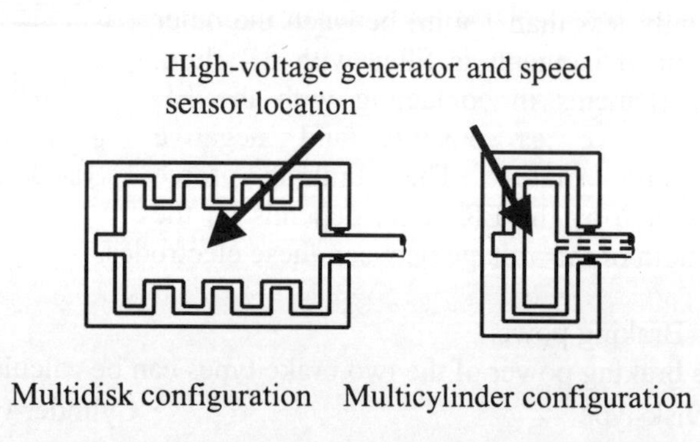

Fig. 5. Brakes with built-in high-voltage supply and speed sensor.

The problem of the disk deformation with a disk-type brake, in short, can be solved most effectively by using multiple washer-shaped disks, as shown in Fig. 5. With either the disk- or the cylinder-type, moreover, it is possible to install a voltage generator and a speed sensor in the central cavity, thus avoiding the need for relatively long high-voltage lines from an outside generator and facilitating the safety of the brake design and operation. In this way, it is possible to build a very compact wheel with an integrated brake unit, in which the outer cylinder serves as the wheel and the inner cylinder serves as the stationary brake board.

4. Cylinder-type brake with multicylinder electrodes

4.1 Brake structure

The structure of the cylinder-type brake with two double-walled cylindrical electrodes, one rotary and the other stationary, is shown in Fig. 6. The rotary electrode, with concentric cylinder diameters of 60 mm and 46 mm and a cylinder length of 46 mm, is connected to the negative outer terminal through a bearing on the shaft. The stationary electrode is built into the plastic housing, and connected to the positive terminal via a bolt. The gap between two electrodes (0.3 mm) is filled with ER fluid (about 14 ml).

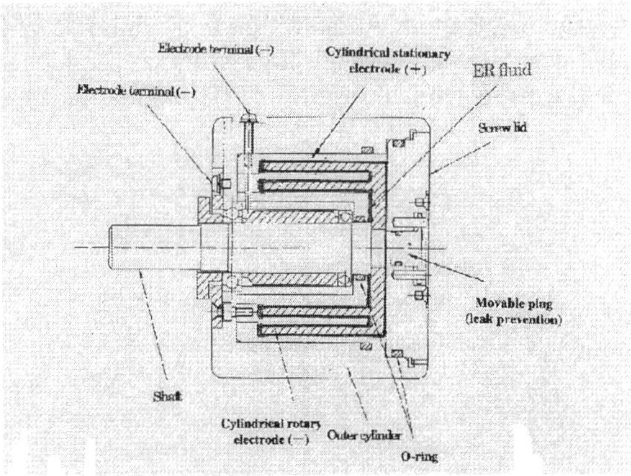

Fig. 6. Cross section of ER brake

4.2 Key aspects of ER brake design and development

It is necessary to maintain an electrode gap of precisely 0.3 mm. Any air trapped in the ER fluid will tend to cause an instability of brake performance and an electric break down. It is therefore essential to fill the very narrow gap completely with the ER fluid, which has a visconsity similar to that of grease, without permitting any air entry into the fluid. It is also necessary to prevent leakage around the shaft seal due to expansion of the fluid under temperature change.

4.3 Performance of the ER brake

Fig. 7 shows the braking power (generated torque) and current density versus the strength of the electric field applied to the brake shown in Fig. 7. The braking power is generally proportional to the applied field strength and to the current. A torque of 8 N-m is generated by application of 500 V (at a current of 200 μA), which translates to a very low energy consumption of 0.1 W by the brake.

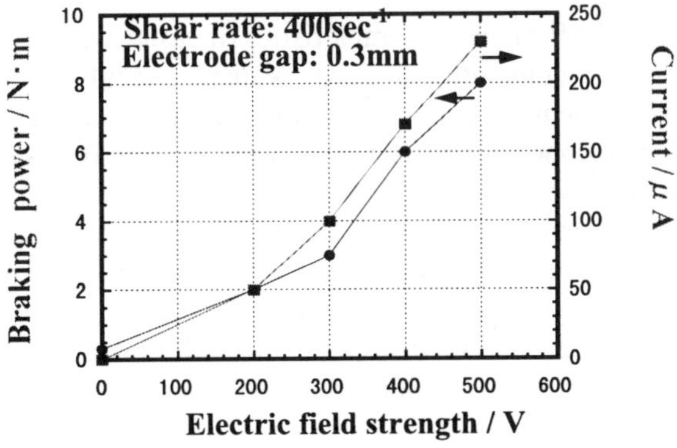

Fig. 7. ER brake performance

5. Caster walker

5-1. Wheel with built-in ER brake

The structure of the cylinder-type ER brake, built into the wheel of the caster-walker, is shown in Fig. 8. In this configuration, the stationary cylinder has an outer diameter of 10.0 cm and a cylinder length of 1.5 cm and serves as the positive electrode. The rotary cylinder, with an inner diameter of 10.1 cm, is inside the stationary cylinder and serves as the negative electrode. The electrode gap is 0.05 cm and is filled with 13 ml of ER fluid. A gap of 0.05 cm, rather than 0.03 mm, has been chosen to obtain the cost advantages of large-volume production, facilitate ER fluid filling, and enhance mechanical precision. The diameter of the wheel is 20 cm.

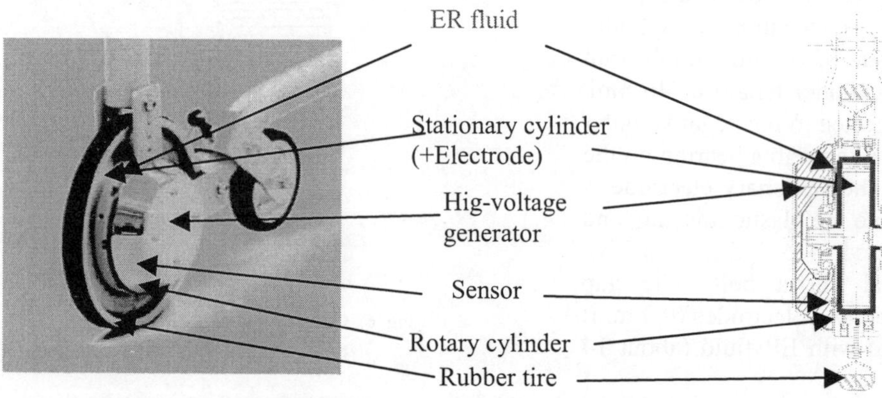

Figure 8 Structure of wheel with built-in ER brake

5.2 Composition of caster walker

The caster-walker, shown in Fig. 2, is equipped with a complete ER brake system. An ER brake is installed on each of the two rear wheels, and contains a speed sensor (Hall device) and a high-voltage generator (max., 1000 V). A mode and speed adjuster and a controller (CPU and 6 V battery) are installed in the front of the armrest. The electric field from the high-voltage generator is applied to the ER brake. The appropriate field strength is calculated and maintained by the controller, in response to signal inputs from the sensor. The ER brake system can be easily added on to commercial caster-walkers.

5.3 Performance characteristics of caster-walker

With a conventional caster walker, patient stumbling tends to occur due to walker movement during mounting or dismounting, and to walker acceleration on downslopes. The ER brake caster-walker is designed to prevent such stumbles, and also to serve as a walk trainer. It is operational in three modes. In the first mode, the speed sensor can detect abrupt wheel rotation during mounting or dismounting, and actuate the ER brake, thus arresting the movement of the walker and preventing patient stumbling. In the second mode, the sensor and the ER brake serve as a speed controller on downslopes. The ER brake provides moderate braking, to prevent acceleration of the walker beyond a pre-set speed, and thus prevent patient stumbling.

In the third mode, the walker is also effective as a walk trainer, as it can be set to provide the desired level of resistance to walker movement by the ER brake.

This caster-walker is equipped with the cylinder–type ER brakes installed in its rear wheels, as shown in Fig. 8. The braking power is about 8 N-m under application of 1000 V at each wheel, giving a total of 16 N-m for the two brakes. The measured value of 8 N-m is near the calculated value. It is sufficient to stop the movement of a walker bearing the full weight of a 60-kg patient on a 15-degree downslope.

Under these conditions, the current through the ER brakes is less than 200 μA and their energy consumption is about 0.2 watt. That of the whole system is about 1 watt. Under continuous rotation with an applied voltage of 1,000 V for 30 minutes at a room temperature of 25 degrees C, the temperature elevation of the brake assembly is only about 15 degrees C.

As shown by this application, the ER brake system is simple and compact, consumes very little energy, and is highly advantageous for small-wheeled vehicles.

5.4 Safety devices

Safety measures built into the system include protection from contact with the high-voltage generator despite the very small operating current, a self-diagnosis system for detection of low battery voltage, electric line disruption, and misoperation, a warning lamp for abnormalities, and a mechanical emergency lock to arrest caster movement.

5.5 Clinical evaluations

The ER walker has performed well in clinical evaluations at rehabilitation centers. The participating doctors and therapists have all found the walker effective for patients with ataxic gait and post-surgical patients, and especially for patients with Parkinson's disease. The only unfavorable response was a note by some of them, that the residual load sometimes slowed down the handling and stowage of the walker.

6. Further applications development and additional studies

6.1 New clinical applications

Because of its low energy consumption, intelligent-system performance characteristics, compact size and light weight, and reliability and safety, the ER brake will bring important advances in health care applications ranging from wheelchairs and gurneys to guide robots for the blind and small vehicles for the aged.

6.2 Future objectives

Safety is a key element for the full, long-range success of the ER brake walker, and proof of its performance in the prevention of stumbling is a long and continuing process, because of the many and varied environments in which it may be used and the physical condition and behavior of the patients it protects.

References

1.Inoue A., Ide Y., Maniwa S.,Oda H. and Yamada H., MRS Bulletin, **23** (1998) pp43-49
2.Inoue A., J.of SICE, **34** (1995) pp698-701
3.Inoue A., Ryu U., Miyabe Y. and Nakagawa A., Proceeding of JMSE Autumn Meeting 2000 in Wakayama Univ., 1322

LYAPUNOV-BASED CONTROL OF A BRIDGE USING MAGNETO-RHEOLOGICAL FLUID DAMPERS

XIAOJIE WANG AND FARAMARZ GORDANINEJAD

Composite and Intelligent Materials Laboratory
Department of Mechanical Engineering
University of Nevada, Reno, NV 89557, USA
e-mail: faramarz@unr.edu

This study focuses on the effect of semi-active magneto-rheological fluid (MRF) dampers in reducing the response of a scaled bridge structure subjected to a random loading. A fluid-mechanics based model that can characterize the nonlinear dynamic behavior of MRF dampers is employed. A state-variable model for an integrated system of a 1/12 scaled, two-span bridge and two MRF dampers is established. A feedback on-off control law is employed based on Lyapunov approach that guarantees the system stability for uncertain bounded input disturbances. An output feedback Lyapunov controller is implemented in the experimental study to verify the presented method. Both the theoretical and experimental studies show that the Lyapunov based control systems can effectively reduce the relative displacement between the deck and the abutment of the bridge when subjected to various input motions. In addition, when comparing to experiment results, it is demonstrated that the proposed analytical model can predict the response of the system, accurately.

1 Introduction

Semi-active control of bridges under random loading such as earthquakes, strong storms, and vehicular traffic has received attention in recent years[1-5]. Comparing to active and passive systems, semi-active control can offer the combined advantages of both systems with its control flexibility and low energy requirement. A variety of semi-active control devices have been developed for the purposes of structural vibration control. Examples of such devices include variable orifice fluid dampers, variable stiffness devices, friction controllable isolators and braces, and controllable MRF fluid dampers. Specifically for MRF devices, activation of a MRF damper's built-in electro-magnetic field causes a fast and dramatic change in the apparent viscosity of the MR fluids contained in the damper. These fluids can reversibly change from liquid to semisolid in milliseconds. The result is an infinitely variable, controllable damper capable of very large damping forces. Several semi-active control strategies for MRF dampers have been applied to vibration control of civil structures[4-10].

This paper aims at the vibration control of a scaled bridge model for a proof-of-concept evaluation utilizing MRF dampers as the semi-active control devices. The structure considered in this study is a two-span scaled bridge. The MRF dampers are developed at the University of Nevada, Reno (UNR). A bang-bang control strategy based on Lyapunov stability theory is proposed for controlling MRF dampers to reduce bridge responses due to random vibration. A theoretical study, incorporating a newly developed model that can accurately capture the dynamic behavior of MRF dampers, is proposed.

2 Experimental Set up

A 1/12-scaled, two-span bridge equipped with two UNR MRD-001 dampers is constructed. A schematic of the experimental set up is presented in Figure 1. The abutments are made from 1.27cm thick steel plate and have adjustable mounts that extend under the deck. Ball bearings are placed between the extended mounts and the deck. The deck is a solid steel plate with 30.5

cm width and 61cm in length. A steel box filled with lead is mounted on top of the deck and has a total combined mass of 680kg. A graphite/epoxy-concrete (G/E-C) composite column is joined via solid circular steel inserts to the base at the center of the deck allowing the deck to move horizontally between the two abutments. The length, and the outer and inner radii of the G/E-C column are 54.8cm, 8.1cm and 7.6cm, respectively. The base of one solid circular insert bolts to the magnesium base plate between the abutments while the solid circular protrusion is pressed into the end of the column. The base of the other solid circular insert bolts directly to the underside of the deck while its circular protrusion is pressed into the other end of the column. The MRF dampers are mounted on the underside of the bridge deck and attached to one of the abutments. A Ling Dynamic Systems shaking table with vibrator (V830-335T) capable of generating a peak load of 9.8kN at frequencies 2-3000 Hz, is used to simulate random base excitations.

3 Theoretical Modeling

The bridge system can be modeled as a single-degree-of-freedom system. The equation of motion subjected to a ground excitation can be expressed as:

$$M\ddot{x}(t) + C_m\dot{x}(t) + Kx(t) + F(t) = -M\ddot{x}_0(t) \tag{1}$$

where M is the mass of the bridge deck, C_m is the structural damping coefficient, K is the column stiffness, $x, \dot{x}, \ddot{x}$ are relative displacement, velocity and acceleration of deck with respect to ground, respectively. $\ddot{x}_0(t)$ is the input acceleration, and $F(t)$ is the controllable damping force of the MRF dampers. System identification of the bridge model was conducted with a serieis of sinosoidal motions from 2Hz ~12Hz, and logarithmic decrement method along with the snap tests. The structural damping ratio and the natural frequency of the bridge system (with no dampers attched) are determined to be 0.087 and 4.5Hz, respectively.

A dynamic model based on fluid mechanics approach utilizing a Herschel-Bulkley constitutive equation has been demonstrated to accurately predict the dynamic response of the UNR MRD-001 damper over a wide range of operating conditions[11]. The model includes dissipated as well as stored energies. Combining the dynamic modeling of the MRF dampers with the Bingham plastic equation, the model for a controllable bridge system can be expressed as:

$$M\ddot{x}(t) + C\dot{x}(t) + Kx(t) + A_pP = -M\ddot{x}_0(t) \tag{2}$$

where the rate of the pressure drop across the piston can be expressed, as follows[11]:

$$\dot{P} = C_p\dot{x}(t) - \frac{C_p}{A_p}\left(\frac{\pi D_1^4 N}{128\mu L_{mrf}}\right)P + \frac{C_p}{A_p}\left(\frac{1}{24}\frac{\pi D_1^3 N}{\mu} - \frac{2}{3}\frac{\pi L_{mrf}^3 N}{\mu}\left(\frac{\tau_y}{P}\right)^3\right)\tau_y sign(p) \tag{3}$$

where

$$C = C_m + \frac{128\mu L_{vis} A_p{}^2}{\pi D_2{}^4} \text{ and } C_p = \beta\left(\frac{1}{L_1 - x(t)} + \frac{1}{L_2 + x(t)}\right) \tag{4}$$

Here, $P = P_1 - P_2$ is the pressure drop between MRF damper chamber one and chamber two, A_p is the area of the piston, L_1, L_2: the length of chamber one and two, L_{vis} and D_2 are the length and diameter of channels not effected by MRF action, L_{mrf} is the length of MRF valve, D_1 is the diameter of MRF valve, β is the effective bulk modulus of the fluid in the chambers, N is the number of designed MRF valves, τ_y is the fluid yield stress which is a function of the external field, and μ is the plastic viscosity of the MR fluid. Defining state variable $Z = [z_1, \ z_2, \ z_3]^T$, where $z_1 = x$ as relative displacement of bridge deck and $z_2 = \dot{x}$ as relative velocity of bridge deck, and $z_3 = P$ as the pressure drop across the piston, and letting $u = \tau_y$ as control input, then, a nonlinear state space model of the controllable bridge system is obtained as:

$$\dot{Z} = A(Z)Z + B(Z)sign(z_2)\cdot u + E\ddot{x}_0 \tag{5}$$

where $A = \begin{bmatrix} 0 & 1 & 0 \\ -\dfrac{K}{M} & -\dfrac{C}{M} & -\dfrac{A_p}{M} \\ 0 & C_p & -\dfrac{C_p}{A_p}\left(\dfrac{1}{128}\dfrac{\pi D_1{}^4 N}{\mu L_{mrf}}\right) \end{bmatrix}$,

$$B = \begin{bmatrix} 0 \\ 0 \\ \dfrac{C_p}{A_p}\left(\dfrac{1}{24}\dfrac{\pi D_1{}^3 N}{\mu} - \dfrac{2}{3}\dfrac{\pi L_{mrf}{}^3 N}{\mu}\left(\dfrac{u}{P}\right)^3\right) \end{bmatrix}, \text{and } E = \begin{bmatrix} 0 \\ -1 \\ 0 \end{bmatrix}.$$

(6)

If the relative displacement $x(t)$ is much smaller than L_1 and L_2, then, $C_p \cong \beta(\frac{1}{L_1} + \frac{1}{L_2})$.

4 Controller Design

One of the main advantages of the Lyapunov method is that it is applicable to nonlinear and time-varying systems as well as linear and time-invariant ones. This is important since the MRF dampers have a nonlinear behavior. This strategy requires selection of a Lyapunov function, a positive definite function V. Stability is guarantied if the time derivative of this function $\dot{V}$ is negative semi definite. A control law can be designed by minimizing the time derivative of a Lyapunov function. Since the Lyapunov function can be selected arbitrarily, therefore, the controller design is not unique. This could be regarded as an advantage of this method, since its flexibility allows designing a controller that satisfies a range of goals. In this study the Lyapunov direct method is used to design the controller. Consider a Layapunov function candidate, as follows:

$$V(Z) = Z^T RZ \tag{7}$$

where R is the positive definite solution of the algebraic Layapunov equation:

$$PR + A^T R + Q = 0 \tag{8}$$

for a given symmetric positive definite Q. Corresponding to the state-space Equations (5) and (6), the derivative of Lyapunov function for a solution of Equation (7) is

$$\dot{V}(Z) = -Z^T QZ + 2Z^T RB\,\mathrm{sgn}(z_2)\cdot u + 2Z^T RE\ddot{x}_0(t) \tag{9}$$

In order to minimize $\dot{V}(z)$ for all possible $\ddot{x}_0(t)$, one needs to only minimize $Z^T RB\,\mathrm{sgn}(P)\cdot u$ for $0 \le u \le u_{\max}$. If R expressed by:

R=r_{ij}; where i and j=1,2,3 (10)

then, $Z^T RB\,\mathrm{sgn}(z_2)u = (r_{13}z_1 + r_{23}z_2 + r_{33}z_3)b_3\,\mathrm{sgn}(z_2)\cdot u$ (11)

where $b_3 = \dfrac{C_p}{A_p}\left(\dfrac{1}{24}\dfrac{\pi D_1^{\,3} N}{\mu} - \dfrac{2}{3}\dfrac{\pi L_{mrf}^{\,3} N}{\mu}\left(\dfrac{u}{P}\right)^3\right)$

Since $b_3 < 0$ for $0 \le u \le u_{\max}$, Thus, a control scheme is obtained, as follows:

$$u = \begin{cases} u_{\max} & if\left(r_{13}z_1 + r_{23}z_2 + r_{33}z_3\right)\mathrm{sgn}(z_2) > 0 \\ 0 & if\left(r_{13}z_1 + r_{23}z_2 + r_{33}z_3\right)\mathrm{sgn}(z_2) \le 0 \end{cases} \tag{12}$$

The semi-active control system for the proposed controller has been proved to be stable[12]. This control law requires measurements of relative displacement, relative velocity and internal pressure drop. A block diagram of the Lyapunov based semi-active control system is shown in Figure 2.

5 Theoretical and Experimental Results

Experimental and theoretical studies were conducted to investigate the coupled relationship of the bridge and the MRF dampers. The system was excited by an input random ground motion. The geometric and materials properties used on this study are: A_p L_1, L_2, L_{vis}, D_1, L_{mrf}, D_2, β, N, μ, τ_y with magnitudes of 1.9×10^{-3}m^2, 0.057m, 0.057m, 0.00939m, 0.00635m, 0.00292m, 0.00635m, $6.5\times10^7\,pa$, 4, $0.4\,pa\cdot s$, and $3.0\times10^4\,pa$ respectively.

A comparison of the experimental and analytical results in the passive-off (*i.e.*, damper is not activated) and passive-on (*i.e.*, damper is fully activated) cases are shown in Figures 3(a) and 3(b), respectively. It is evident that the analytical model is capable of capturing the system response. Next, the Lyapunov-based control scheme was examined experimentally. The proposed control rule requires all the state outputs and the control gains depends on the choice of positive definite matrix Q. A judicious selection of the gains, for example, setting $r_{33} = 0$ that can help avoiding the MRF damper internal pressure drop measurement. Figure 3(c) illustrates a time history of the bridge deck dynamic response under random ground excitation with feedback control. For comparison, a simulation of the feedback-controlled response of the bridge using the same ground motion is also plotted in this Figure. It can be seen that the predicted responses agree well with the experimental results.

6 Conclusions

This study was focused on the application of semi-active MRF dampers for controlling the response of a bridge structure system subjected to random vibration. Experimental study was performed on an integrated system consisting of both 1/12 scaled, two-span bridge and two MRF dampers. Theoretical modeling of the dampers was performed utilizing a fluid mechanics-based model that includes the nonlinear dynamic behavior of MRF dampers. A feedback bang-bang control law was developed based on Lyapunov approach that guarantees the stability of the coupled structure/damper system. Both the theoretical and experimental studies show that the Lyapunov based control systems can effectively reduce the bridge relative displacement between the deck and the abutment when subjected to various input motions.

7 Acknowledgments

This study is funded by the U. S. National Science Foundation, Dr. S. C. Liu, Program Director. The authors wish to thank Gregory H. Hitchcock for helpful comments.

Reference

1. Patten, W. N., Sack R. L., and He, Q., "Controlled semi-active hydraulic vibration absorber for bridge," *Journal of Structural Engineering*, Vol.122, No.2, 1996, pp.187-192.
2. Symans, M. D., and Kelly, S. W., "Fuzzy logic control of bridge structures using intelligent semi-active seismic isolation," *Earthquake Engineering and Structural Dynamics*, Vol.28, No.1, 1999, pp.37-60.
3. Yang, J. N., and Wu, J. C., "Hybrid control of seismic-excited bridge structures," *Earthquake Engineering and Structural Dynamics*, Vol.24, 1995, pp.1437-1451.
4. Gordaninejad, F., Saiidi, M., Hansen, B. C., and Chang, F-K, "Control of bridges using magneto-rheological fluid dampers and fiber-reinforced, Composite-Material Column," *Smart Systems for Bridges, Structures, and Highways, Proceedings of the 1998 SPIE Conference on smart materials and structures*, Vol. 3325, pp. 2-11, 1998
5. Liu, Y., Gordaninejad, F., Evrensel, C. A., Wang, X., Hitchcock, G., " Vibration control of a bridge using failsafe magneto-rheological fluid dampers: a comparative study," *submitted*
6. Jansen, L.M., and Dyke, S.J. "Semi-active strategies for MR dampers: a comparative study", *ASCE Journal of Engineering Mechanics*, 2000, Vol.126, No.8, pp.795-803.
7. Zhou, L., and Chang C. "Adaptive fuzzy control for a structure-MR damper system", *Proceedings of SPIE's 7th Int. Symposium on Smart Materials and Structures, Newport Beach, California, March,* 2000, Vol.3988, pp.105-115.
8. Hiemenz, G. J., Choi, Y. T., and Wereley, N. M, "Seismic control of civil structures utilizing semi-active MR bracing systems", *Proceedings of SPIE's 7th Int. Symposium on Smart Materials and Structures, Newport Beach, California, March,* 2000, Vol.3988, pp.217-228.
9. Xu, Y. L., Qu, W. L., and Ko, J. M., "Seismic response control of frame structures using magnetorheological/electrorheological dampers", *Earthquake Engineering and Structural Dynamics*, Vol.29, 2000, pp.557-575.

10. Gavin, H. P., "Control of Seismically Excited Vibration Using Electrorheological Materials and Lyapunov Methods", *IEEE Trans. on Control Systems Technology,* Vol.9, No.1, 2000, pp.27-36.
11. X.Wang and F.Gordaninejad. "Dynamic Modeling of Semi-Active ER/MR Fluid Dampers," *8th Annual SPIE Symposium on Smart Structures , Conference on Damping and Isolation*, Newport Beach, CA, 4-8 March, Paper No.4331-09.
12. Leitman, G., "Semi-active Control for Vibration Attenuation", *J. of Intell. Mat. Syst. and Struct.*, pp.841-846, 1994.

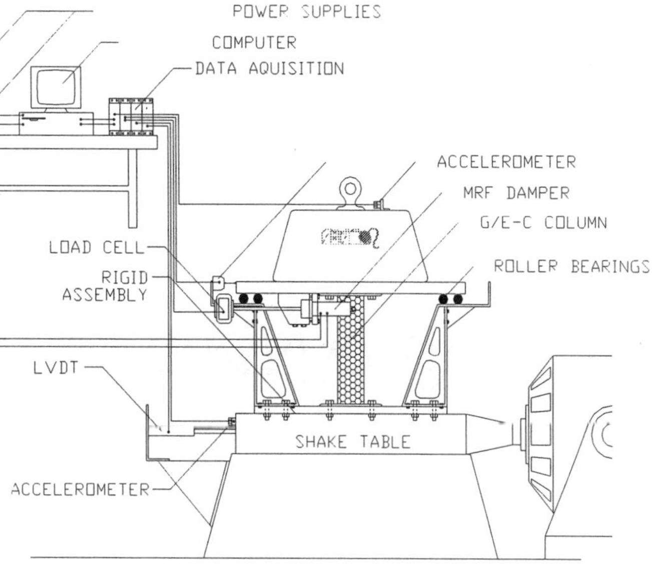

Figure 1. Schematic of the experimental setup.

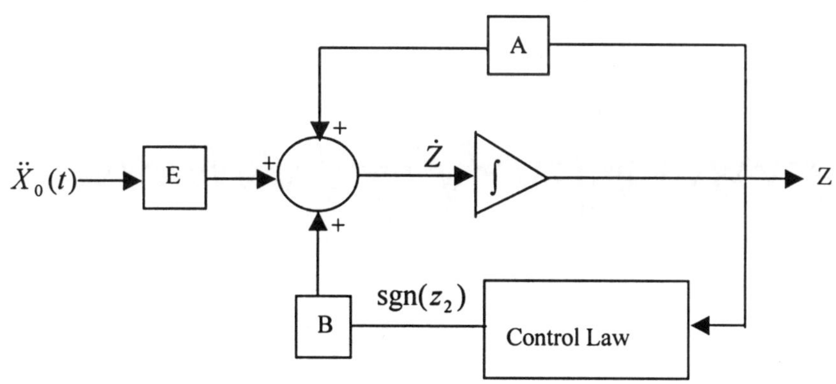

Figure 2. Block diagrams of the Lyapunov-based feedback controller.

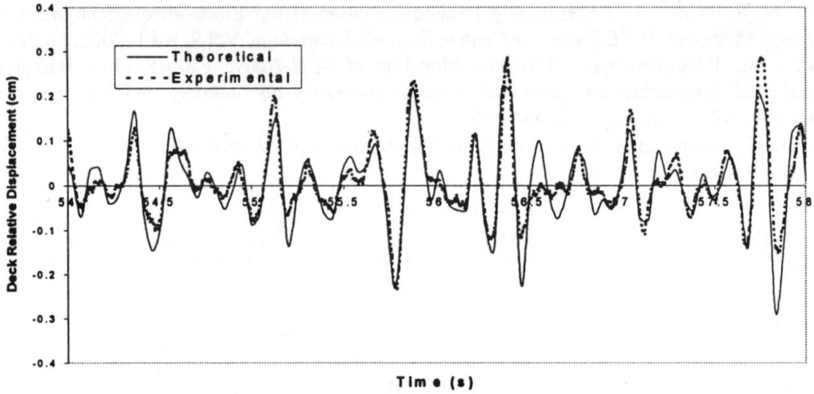

Figure 3 (a)

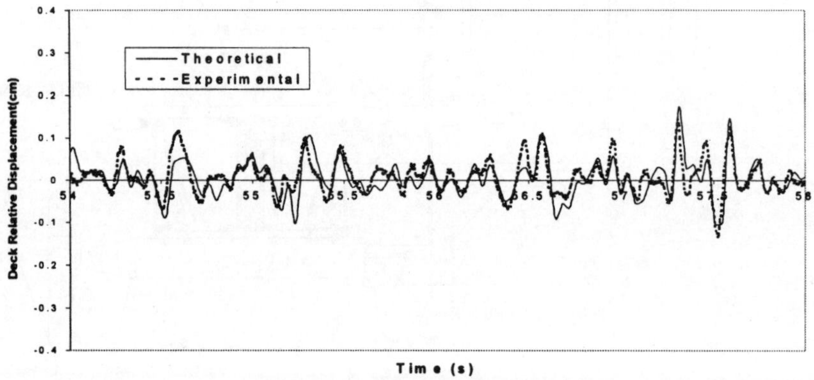

Figure 3 (b)

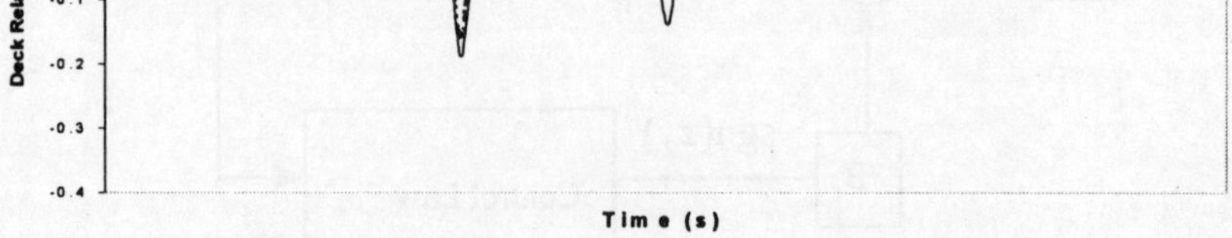

Figure 3 (c)

Figure 3. Comparisons of measured and predicted responses of bridge deck relative displacement in passive-off (control input u=0, 3(a)), passive-on (control input u=Maximum, 3 (b)) and Lyapunov Feedback Control (3(c)).

INDUSTRIAL BENEFITS OF ERF-TECHNOLOGY (A DEVELOPMENT REPORT FROM THE INDUSTRY)

DOROTHEA ADAMS, LUCIEN JOHNSTON

Fludicon / Schenck Pegasus GmbH, Landwehrstr. 55, 64293 Darmstadt, Germany

To demonstrate the industrial application and benefits of the ERF-technology, a German Ministry of Research backed project was initiated with the working title "Adaptronic Transport Systems with electrorheological fluids (ERFs) for the transport of sensitive goods". This project was successfully completed by Schenck Pegasus GmbH and partners and resulted in a fully functional ERF-hydraulic transporter system. The general system design including the specific ERF-actuators with integrated ring-segment ER-valves is discussed. Parallel to fully active systems, research work is being carried out on ERF-damper design for commercial automobiles and on the development of an ERF-servovalve.

1 Introduction

Since the development of ER-fluids which are non-abrasive and can be used in combination with standard hydraulic components, there have been concentrated efforts to bring this technology out of the laboratories and into the industry. The high dynamic response of ER-fluids, their ability to be progressively adjusted and the simple design of effective ER-valves allowing for large pressure drop spread are three key features which need to be implemented when exploiting the industrial benefits of this technology. The sum of the key features of the ERF technology is the additional functionality which allows innovative solutions to problems which cannot be solved with "classical" technologies. To demonstrate this additional functionality, customers must be given not only an ERF application which encompasses the key features, but a complete system consisting of fluid, mechanical application, high voltage controllers, sensoring and the control system/algorithms which, working as a unit, make this functionality available.

2 The Adaptronic Transport System (ADT)

Utilising the key features of ERFs, a German Ministry of Research backed project was initiated to develop a system for the transport of sensitive goods. Figure [1] shows the realised transport system at the end of the three-year project period.

Fig. 1: Adaptronic Transport System

The hydraulically driven system, mounted on the loading area of a truck, uncouples sensitive goods from road unevenness (vibration isolation) and aligns forces resulting from motion dynamics (cornering, accelerating) orthogonal to the load surface. Situated between vehicle chassis and load platform are three standard hydraulic cylinders (carrying the base load) and three ERF-cylinders (for dynamic platform motion) powered by the common hydraulic unit. The sensoring system provides information (displacement, force, position) to the self-developed electronic hardware which delivers the control signals via the high voltage controllers to the ER-valves integrated within the ERF-cylinders.

This system was completed at the end of 1999. First tests on a four-poster hydraulic test bed showed an increased vibration isolation at low frequencies. Further tests on an open road showed the active compensation of acceleration and deceleration (platform leaned forward or backward) and also the improved cornering capabilities of the platform. In this case, the platform leaned into the curve like a motorcyclist. The control system was thus able to align forces resulting from motion dynamics orthogonal to the load surface.

This new system can find application in the transport of injured people or sensitive goods and also in shipping and military domains. In general, active vibration control is considered the objective of this application.

3 RheAct cylinders

The ERF-cylinders at the heart of the ADT system have been designed as double-acting, single-rod actuators with integrated ER-valves. The four ER-valves are arranged as a full hydraulic bridge, allowing for maximum control possibilities. To minimize the cylinder dimensions and improve their hydraulic stiffness, the ER-valves were arranged as four electrically and hydraulically isolated positive electrode ring segments inserted between cylinder wall and outer casing. A schematic representation of the arrangement and a close-up view of the cylinder is given in figure [2].

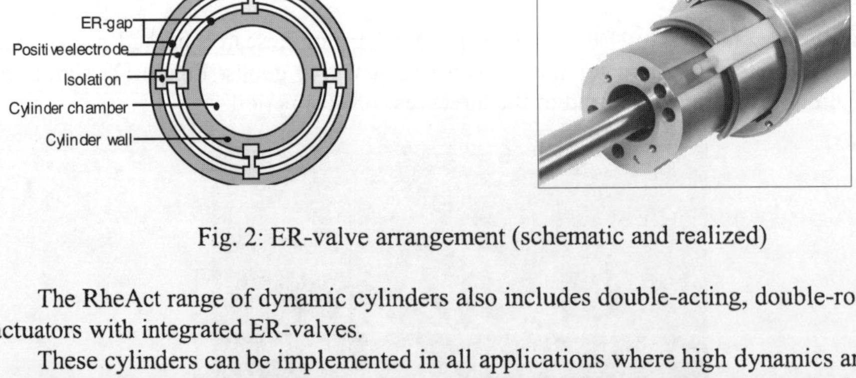

Fig. 2: ER-valve arrangement (schematic and realized)

The RheAct range of dynamic cylinders also includes double-acting, double-rod actuators with integrated ER-valves.

These cylinders can be implemented in all applications where high dynamics and frequencies in combination with the power density of hydraulic systems is required. The first application presently being tested is an elastomer testing machine for frequencies in excess of 1000Hz. Open loop control has been achieved beyond this value. The next step is to realise control hardware which will allow closed loop control up to this frequency. To this effect, a 10kHz loop update rate is a minimum requirement.

4 RheDamp vibration damper

One of the applications which can enormously benefit from the added functionality of ER-fluids is the automotive shock absorber/damper. The patented ERF-damper is a modified automotive monotube shock absorber which makes use of a single spiral channel as an ER-valve, eliminating the need for a complicated plunger valve. Figure [3] shows a cutaway section of the shock absorber. The positive electrode in the plunger is separated from the tube wall by a spiral plastic strip. This strip provides the necessary guidance, maintains the correct ER gap (1mm) and artificially increases the length of the ER valve. With a straight ER valve, the plunger would have been three times the length – drastically restricting the available stroke. The positive electrode in the plunger is isolated from the rod by a thermally stable plastic and the shielded control cable led through the hollow rod for connection to the control unit mounted in the vehicle.

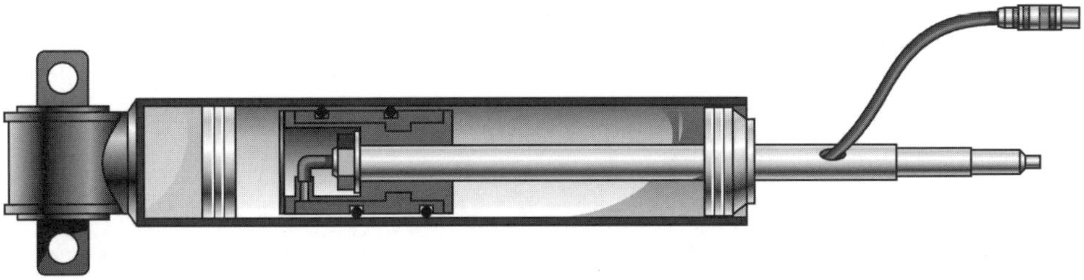

Fig. 3: Section through RheDamp shock absorber

Tests at the TU Darmstadt in Germany have confirmed the adjustable damping characteristics and an open-loop system response time of 6ms has been determined.

The damping characteristics of this damper have been integrated into a complete vehicle simulation system and a suitable control algorithm designed and implemented. A simulation of the vehicle undergoing a double lane change at 100km/h on an uneven road was enacted without and then with the ERF damper. Figure [4] shows two still frames of the simulation. The film of the simulation impressively demonstrates the potential which such a continuously variable shock absorber offers with regard to comfort and safety.

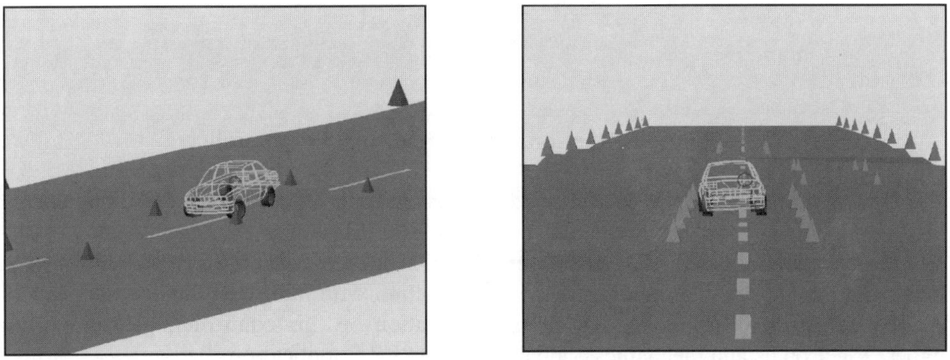

Fig. 4: Simulation results of vehicle without and with RheDamp shock absorber

5 RheCon ER controller

To realize the benefits of the ERF technology regarding improved control conditions, it is necessary to have a high voltage supply unit which fulfills the following criteria:

- Highly dynamic field generation and discharge
- Stable field control independent of load
- High efficiency
- Output voltage 0- 6kV
- Sufficient output current

Having been unable to acquire a suitable unit externally, Fludicon has developed an ERF controller which fulfills all these criteria. With rise and fall times of 0.3ms, the true dynamic response of ER fluids can now be enhanced. Accurate modulation of the unit can be realised up to 1000Hz for a sine wave with an amplitude of 6kV. The load independent stability allows for correspondingly higher frequencies at lower amplitudes (e.g. 1500Hz modulation at 4kV output voltage amplitude). Figure [5] shows the RheCon controller as a stand-alone unit.

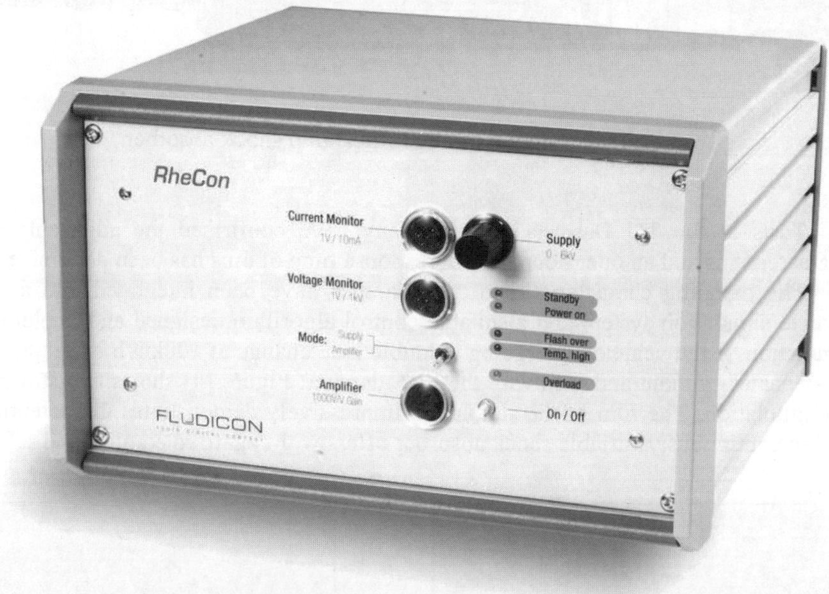

Figure 5: RheCon, a highly dynamic ER controller

Taking the needs of the ER fluid research community into consideration, this unit is also equipped with an accurate load current sensor, flashover monitoring and further features. Compliancy with CE norms ensures that all necessary safety standards have been adhered to. The unit can be used in conjunction with industrial applications and/or rheological testing equipment. For more information on this controller, please see the paper "Electronics and mechanics – a balance act in ERF applications".

6 RheValve ER servovalve

Given the available shear stresses of modern ER fluids (~ 8000Pa at 5.5kV/mm and 20000s^{-1} shear rate) with a moderate zero field viscosity of 30mPas, application dimensions can be quickly estimated using a simple Bingham model. Despite calculation inaccuracies, the feasibility of an ERF solution for a specific application can be deduced. Where application size is critical, given a desired on-off ratio, it is sometimes the case that no presently available fluids possess the necessary shear strength under electric field to allow for the implementation of a direct ER gap valve.

To overcome this problem, Fludicon has developed an ER servovalve. This servovalve combines the advantages of classical servohydraulics (large gain from a small signal) with those of ER fluid technology (adaptive control). Figure [6] shows a cutaway CAD model of this new servovalve.

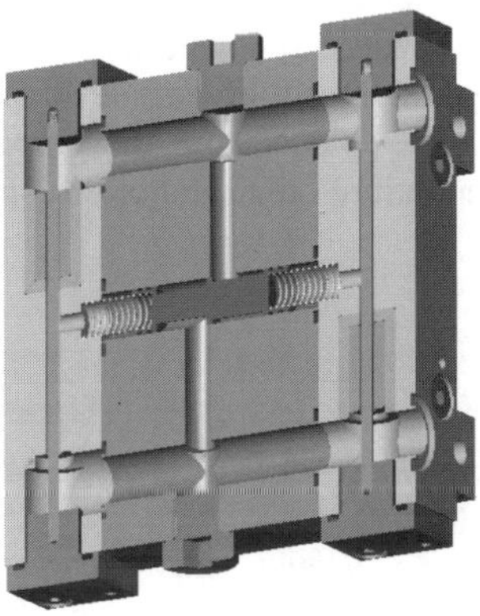

Fig. 6: RheValve servovalve (cutaway CAD model)

Given control channels with ER gaps of approximately 0.5mm, a servohydraulic pressure gain of factor eight can be achieved at volume flows in excess of 50litre/min. – a realistic value for standard hydraulic applications.

By activating the ERF in the control channel only, a reduction in the required voltage by factor two can be achieved. This brings enormous advantages with regard to the size of the HV controller and also the required power consumption.

Tests of the RheValve in combination with an automotive shock absorber will be carried out this year in collaboration with an industrial partner.

7 Industrial applications

A successful transfer of ERF benefits into commercial applications requires the joint effort of ERF specialists and industrial partners. Given the advanced state of prototype applications as discussed above, the industry is starting to take a new look at the ERF technology with regard to their own products. We are working together with several companies in different fields on various ERF based applications. Examples include rotational dampers, combined gas-spring and damper units and automotive shock absorbers as well as the testing machine mentioned in chapter 3.

8 Summary

An industrially viable ER fluid, combined with competence in mechanical hardware, electronic controllers and control methods have allowed us to successfully demonstrate the ability of the ERF technology to solve real-world problems. This success has given customers the confidence to integrate ERF technology into their development projects.

From a mechanical hardware point of view, focus is now being placed on actuator optimization, servovalve technology and the improvement of damping characteristics for various shock absorber applications.

The newest ERF controllers cover all requirements regarding energy, signal form, dynamics and load current feedback for a whole range of applications, so that the selective miniaturization for specific applications can be realized.

To ensure the commercial feasibility of development projects in the pipelines of partners, focus must now be placed on the improvement of ER fluid regarding sedimentation stability, temperature range and on-off shear stress gain.

References

1. Wolff-Jesse, C., *Untersuchung des Einsatzes elektrorheologischer Fluessigkeiten in der Hydraulik*, Verlag Mainz, 1997
2. Fees, G.; Rosenfeldt, H.; Wendt, E.; Technical report "Hochdynamischer ER-Aktuator RheAct" , Proceedings, 1IFK Aachen, 1998
3. Adams, D.; Rosenfeldt, H,; Final report "Adaptronisches Transportsystem mit Elektrorheologischen Fluessigkeiten (ERFs) zur Befoerderung sensibler Gueter" of the same titled BMBF project, Schenck Pegasus GmbH, Darmstadt, 2000
4. Barz, M.; Breuer, B.; Internal report "Experimentelle Untersuchungen an einem ERF-Schwingungsdaempfer", TU Darmstadt, 1999

VIBRATION ISOLATION OF STRUCTURAL SYSTEMS USING SQUEEZE MODE ER MOUNT

S. R. HONG AND S. B. CHOI

Smart Structures and Systems Laboratory, Department of Mechanical Engineering, Inha University, Incheon 402-751, Korea

W. J. JUNG

Agency for Defense Development, Chinhae 645-600, Korea

W. B. JEONG

Vibration and Noise Laboratory, Department of Mechanical Engineering, Pusan National University, Pusan 609-735, Korea

This paper presents vibration control of flexible structures using squeeze mode ER mounts. After devising the squeeze mode ER mount, its field-dependent damping forces are evaluated at various exciting frequencies. The ER mounts are then incorporated with a flexible beam structure. The governing equation of motion is obtained and an optimal controller which represents controllable damping force is designed to attenuate unwanted structural vibration. The controller is experimentally realized and control responses are presented in time and frequency domains. In addition, vibration control responses of a flexible frame structure integrating with the proposed ER mounts are investigated.

1 Introduction

There are basically three methodologies to attenuate unwanted vibration of flexible structures; passive, active and semiactive. The passive method utilizes rubber mounts or viscoelastic constrained layer in order to suppress vibration in the relatively high frequencies[1]. This method is effective for the large-sized flexible structures, but its performance limitation is inevitable in the relative low frequency in which a large vibration magnitude occurs. Active vibration control method is usually integrated with smart material actuators such as piezoelectric materials and shape memory alloys[2,3]. This method is effective for the small-sized flexible structures and its performance is fairly good in a wide frequency range. However, the power consumption becomes more serious as the size of the structural system becomes bigger. Recently, the semiactive vibration control method featured by electrorheological(ER) or magnetorheological(MR) fluid has been initiated and investigated by many researchers. ER or MR fluids which undergo significant instantaneous reversible changes in material characteristics when subjected to electric or magnetic fields have several advantages in the sense of vibration control such as fast response. The vibration control of flexible structures using ER or MR fluid can be achieved from two different methods. The first approach is to replace conventional viscoelastic materials by the ER or MR fluid layer[4,5]. This method is very effective when the size is small and the thickness is thin. The second approach is to utilize ER or MR mount[6~9] in which the structure is to be supported by the mount at discrete positions[10~13]. This method is effective for the large size of the flexible structures.

This research work presents vibration control of flexible beam and frame structures utilizing ER mounts. After devising the squeeze-mode ER mount, its field-dependent damping force and controllability are experimentally evaluated. The ER mounts are then incorporated with beam and frame structures and the governing equations of motion are derived. An optimal controller is formulated and empirically realized to attenuate unwanted vibration, and control responses are presented in time and frequency domains.

2 Squeeze-mode ER Mount

The schematic configuration of the ER mount proposed in this work is shown in Fig.1. The lower electrode is fixed to the base, while the upper electrode is to be moved up and down. Thus, the squeeze-mode motion of the ER fluid occurs in the housing. The coil spring is attached to provide a static mass which in turn is the mass of a flexible beam or frame structure. The total force of the proposed ER mount can be obtained by

$$F(t) = kh(t) + c_f(t)\dot{h}(t) + F_{er}(t) \tag{1}$$

where,

$$\begin{aligned} c_f(t) &= \frac{3}{2}\frac{\pi\eta R^4}{(h_o + h(t))^3} \\ F_{er}(t) &= \frac{4}{3}\frac{\pi R^3}{h_o + h(t)}\tau_y(E)\,\mathrm{sgn}(\dot{h}(t)) \end{aligned} \tag{2}$$

In the above, k is stiffness constant of the coil spring, $c_f(t)$ damping coefficient of the ER fluid in the absence of the electric field, $h(t)$ exciting displacement, h_o initial gap between lower and upper electrodes, and R radius of the circular electrode. $F_{er}(t)$ is controllable damping force owing to the electric field of E, and $\tau_y(E)$ is the field-dependent yield stress which is given by αE^{β}. Here, α and β are intrinsic values of the ER fluid to be experimentally determined. By considering the mass of the beam and frame structures, the radius of the circular electrode is designed to be $15mm$ and the initial gap is fixed by $3mm$. On the other hand, for the ER fluid chemical statch and silicone oil whose viscosity is 30cSt are used for particle and base liquid, respectively.

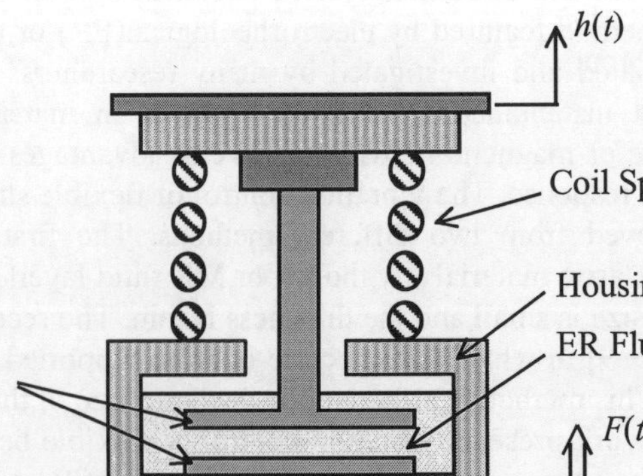

Figure 1. Configuration of the proposed ER mount.

3 Vibration Control of Beam Structure

A steel beam structure(length × width × height = $1500mm \times 60mm \times 15mm$) incorporated with the proposed ER mount is shown in Fig.2. The spring mounts are suported at l_1 and l_5, while the ER mounts at l_2 and l_4. The governing equation of motin of the proposed system is derived using Hamilton's principle, and decoupled modal equation is given by

$$\ddot{q}_i(t) + 2\varsigma_i\omega_i\dot{q}_i(t) + \omega_i^2 q_i(t) = \frac{Q_i(t)}{I_i} + \frac{Q_{di}(t)}{I_i} \tag{3}$$

where,

$$\begin{aligned} Q_i(t) &= -\phi_i\,(l_2)c_{f2}(t)\dot{y}(l_2,t) - \phi_i\,(l_4)c_{f4}(t)\dot{y}(l_4,t) \\ &\quad - \phi_i\,(l_2)F_{er2}(t) - \phi_i\,(l_4)F_{er4}(t) \\ Q_{di}(t) &= \phi_i\,(l_{ex})F_{ex}(t) \\ y(x,t) &= \sum_{i=1}^{\infty}\phi_i(x)q_i(t) \end{aligned} \tag{4}$$

In the above, $q_i(t)$ is generalized modal coordinate, $\phi_i(x)$ is mode shape, ω_i and ζ_i are natural frequency and damping ratio of the i th vibration mode. I_i is generalized mass, Q generalized force and $F_{ex}(t)$ external force. Considering the boundary conditions, the first and second modes are chosen to be dominant for vibration characteristic. Thus, by defining the state variable $x(t)(=[q_1(t)\;\dot{q}_1(t)\;q_2(t)\;\dot{q}_2(t)]^T)$, the following system model is obtained.

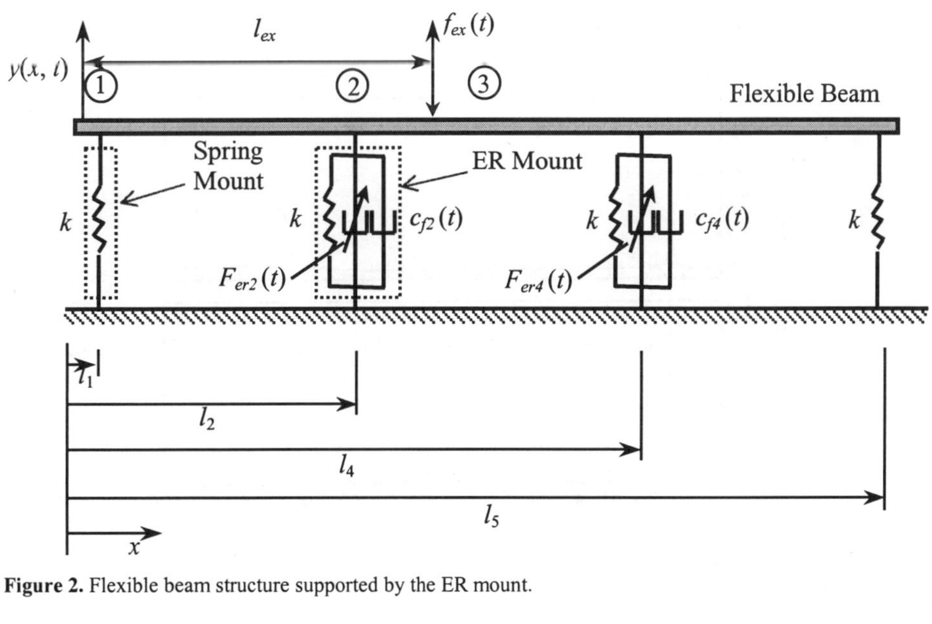

Figure 2. Flexible beam structure supported by the ER mount.

$$\dot{x}(t) = Ax(t) + Bu(t) + \Gamma d(t)$$
$$y(t) = Cx(t) \tag{5}$$

In the above, $u(t)(=[F_{er2}(t)\ F_{er4}(t)]^T)$ represents control damping force and $d(t)$ is external disturbance. Among many potential controller candidates, an optimal controller to minimize the following performance measure is adopted.

$$J = \int_0^\infty \{x(t)^T Qx(t) + u(t)^T Ru(t)\}dt \tag{6}$$

where Q and R weighting matrices. The controller is then obtained from the Riccati equation as follows.

$$u(t) = -R^{-1}B^T Kx(t) \tag{7}$$

where K is the feedback gain. Since the proposed ER mount is semiactive, the following actuating condition should be imposed to increase energy dissipation.

$$u_j(t) = \begin{bmatrix} u_j(t) & for\ u_j(t)\dot{y}(l_j,t) > 0 \\ 0 & for\ u_j(t)\dot{y}(l_j,t) \le 0 \end{bmatrix},\ j = 2,4 \tag{8}$$

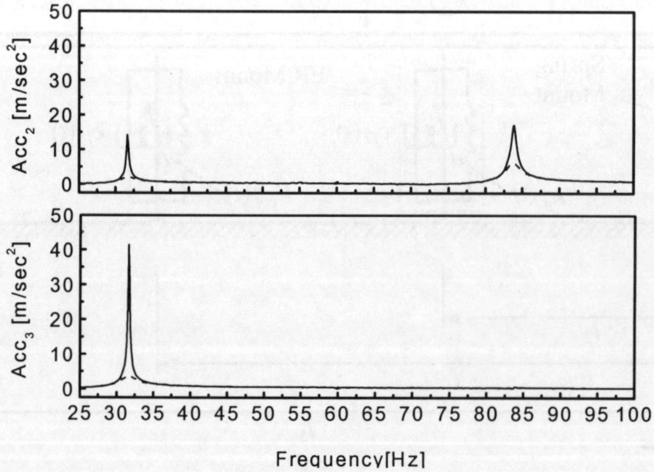

Figure 3. Acceleration response of the beam structure.

Once the control damping force is determined, the control voltage to be applied to the ER mount is given by

$$V_j(t) = E_j(t)(h_o + y(l_j,t)) = \left[\frac{3}{4} \frac{h_o + y(l_j,t)}{\pi R^3 \alpha} u_j(t) \right]^{\frac{1}{\beta}} (h_o + y(l_j,t)),\ j = 2,4 \quad (9)$$

It is noted that when we implement the optimal controller a full-order Luenberger observer is used to estimate state variable $x(t)$. This is because we directly measure only displacement at the location l_2.

Fig.3 presents the acceleration measured at the posion 1, 2, and 3, respectively. At the position 1 and 2 the first and second mode vibrations are remarkably reduced by controlling the ER mount. More specifically, at the position 1 the aceleration of $39 m/\sec^2$ without control is reduced to $5 m/\sec^2$ by activating the controller at the first mode natural frequency. The first mode vibration at the middle position(3) is also substantially reduded by activating the optimal controller associated with the ER mounts.

4 Vibration Control of Frame Structure

The schematic diagram of the frame structure supported by the ER mount is shown in Fig.4. The frame structure consisting of the steel L angles has the length of $1200mm$ and width of $800mm$. Four spring mounts are fixed at four corners($L1$, $R1$, $L3$, $R3$), and two ER mounts are attached at $L2$ and $R2$ positions. In order to formulate the decoupled modal equation, the mode shape of the frame structure is experimentally identified. In fact, we find out that there is no difference in mode shape between without and with the electric field. It is also noted that the measured first and second bending mode natural frequencies agree well to the results analyzed from the finite element method. After identifying modal parameters, the governing equation of motion is established and the same optimal controller as the beam structure is designed to attenuate unwanted vibration. Figure 5 presents the measured acceleration responses at three different positions. It is clearly observed that the vibrations at the resonance frequencies are substantially reduced by activating the controller associated with the ER mounts. For instance, at the first mode frequency, the acceleration of $60 m/\sec^2$ at $R1$ is reduced to $12 m/\sec^2$ by controlling the damping force of the ER mount. In addition, it is found out that the displacement at each position can be remarkably reduced at the resonance frequencies by controlling the damping force of the ER mount.

5 Conclusion

Vibration control of beam and frame structures were investigated by applying the squeeze-mode ER mounts. After verifying the field-dependent damping force and damping force controllability of the ER mount, both beam and frame structures supported by spring mounts and ER mounts have been constructed. It has been demonstrated through realization of the optimal control scheme that unwanted vibration levels such as acceleration can be substantially reduced at the resonance frequencies by activating the that the squeeze-mode ER mount can be effectively employed to distributed parameter

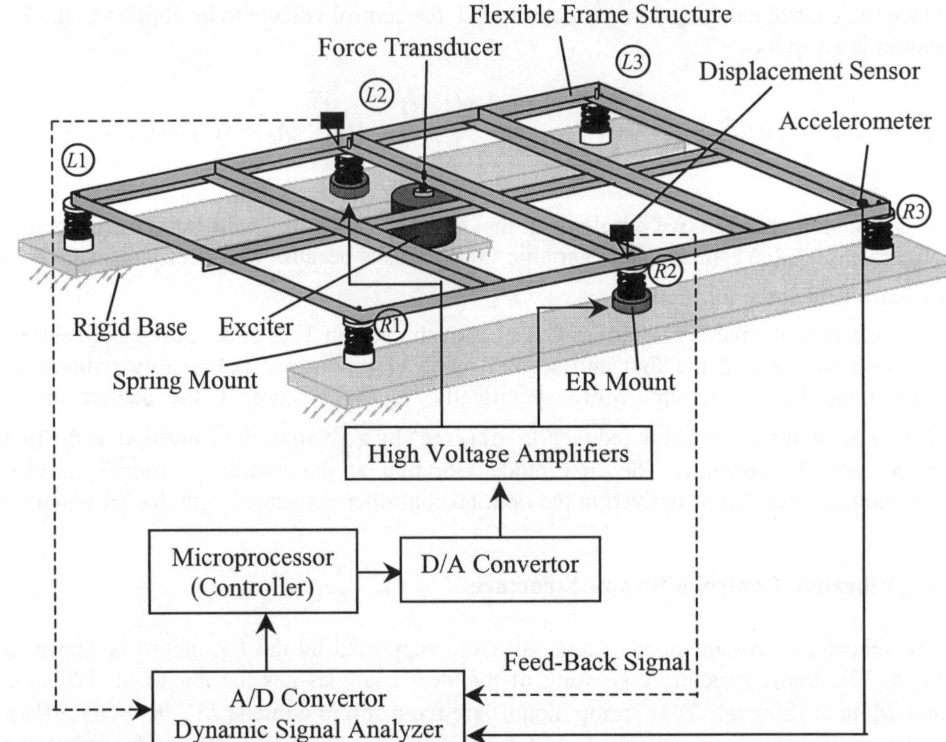

Figure 4. Schematic diagram of the frame structure test.

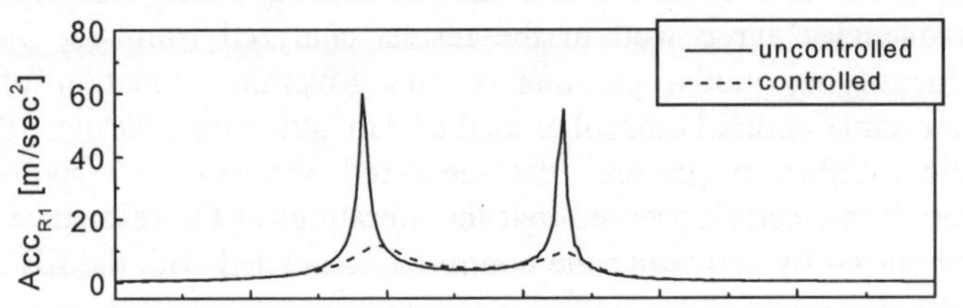

Figure 5. Acceleration response of the frame structure.

flexible structures in order to suppress unwanted vibration in the relatively high frequency range.

Acknowledgement

This work was supported by Korea Agency for Defense Development under contact No. UD990006BD. This financial support is gratefully acknowledged.

References

1. Chang, K. C., Soong, T. T., Oh, S. T. and Lai, M. L., Seismic behavior of steel frame with added viscoelastic dampers. *Journal of structural engineering* **121** (1995) pp.1418–1426.
2. Choi, S. B., Cheong, C. C. and Kim, S. H., Control of flexible structures by distributed piezofilm actuators and sensors. *Journal of intelligent material systems and structures* **6** (1995) pp.430–435.
3. Choi, S. B. and Cheong, C. C., Vibration control of a flexible beam using sma actuators. *AIAA Journal of guidance, control, and dynamics* **19** (1996) pp.1178–1180.
4. Choi, S. B., Thomson, B. S. and Gandhi, M. V., Experimental control of a single-link flexible arm incorporating electro-rheological fluids. *AIAA Journal of guidance, control, and dynamics* **18** (1995) pp.916–919.
5. Yalcintas, M. and Dai, H., Magnetorheological and electrorheological materials in adaptive structures and their performance comparison. *Smart materials and structures* **8** (1999) pp.560–573.
6. Choi, S. B. and Choi, Y. T., Sliding mode control of shear-mode type ER engine mount. *KSME International Journal* **13** (1999) pp.26–33.
7. Hong, S. R., Choi, S. B., Jung. W. J., Ham, I. B. and Kim, D. K., Vibration control of an ER mount subjected to high static loads. *Journal of sound and vibration* **242** (2001) pp.740–748.
8. Choi, S. B., Choi, Y. T., Cheong, C. C. and Jeon, Y. S., Performance evaluation of a mixed mode ER engine mount via hardware-in-the-loop simulation. *Journal of intelligent material systems and structures* **10** (1999) pp.671–677.
9. Williams, E. W., Rigby, S. G., Sproston, J. L. and Stanway, R., Electrorheological fluids applied to an automotive engine mount. *Journal of non-newtonian fluid mechanics* **47** (1993) pp.221–238.
10. Wang, K. W., Kim, Y. S. and Shea D. B., Structural Vibration control via electrorheological-fluid-based actuators with adaptive viscous and frictional damping. *Journal of Sound and Vibration* **177** (1994) pp.227–237.
11. Choi, S. B., Vibration control of a flexible structure using ER dampers. *ASME Journal of Dynamic Systems, Measurement and Control* **121** (1999) pp.134–138.
12. Dyke, S. J., Spencer, B. F., Sain, M. K. and Carlson, J. D., Seismic response reduction using magnetorheological dampers. *Proceedings of the IFAC world congress* (1996) pp.145–150.
13. Hiemenz, G. J. and Wereley, N. M., Seismic response of civil structures utilizing semi-active MR and ER bracing systems. *Journal of Intelligent material systems and structures* **10** (1999) pp.646–651.

THE RHEOLOGICAL CHARACTERISTICS OF ELECTRORHEOLOGICAL FLUIDS IN DYNAMIC SQUEEZE

ALI K. EL WAHED AND JOHN L. SPROSTON
Department of Engineering, University of Liverpool, Liverpool L69, 3GH, UK
E-mail: em50@liverpool.ac.uk *and* j.l.sproston@liverpool.ac.uk

This paper is concerned with an experimental and theoretical determination of the rheological performance of an electrorheological (ER) fluid when subjected to time-dependent applied loads. The experimental facility was built as a squeeze cell in which the fluid is sandwiched between two electrodes, one fixed and the other moving, which permits the instantaneous measurement of the mechanical and electrical responses of the fluid. The transient rheological characteristics of the fluid were assessed for various mechanical force levels and for constant voltage excitation of the fluid. Input and output stress levels across the fluid were monitored enabling the dynamic response of the fluid to be determined using a combination of displacement, force, velocity and acceleration transducers. The experimental results were compared with the results from a modified theoretical analysis, which employs a bi-viscous shear stress/shear strain characteristic of the electrically stressed fluid together with a fluid yield stress, which has a strain-direction dependence on the electrical field.

1 Introduction

Vibration control is one of the most promising areas for the commercial exploitation of new ER fluids, which can provide damping forces within industrial requirements. Bullough and Foxon [3] reported the first practical application of ER fluids in a valve operated vibration damper. As a result, the area of vibration control has seen a surge in the number of investigations particularly in engine mounts [15], primary shock absorbers [13,20] and rotor support systems [23] as well as in adaptive structures [5].

The majority of these investigations have employed the ER fluids in simple shear mode or flow mode. An alternative arrangement, squeeze mode, in which the fluid is subjected to compressive and tensile stresses has been identified and investigated by Stanway et al [18]. In this mode of operation, the fluid would provide a yield stress that is an order of magnitude higher than that in shear mode [11]. As a result, subsequent investigations, by the authors and other researchers, have focused on the performance of ER fluids in squeeze subjected to various mechanical and electrical inputs (see for example, Sproston et al [16,17], El Wahed et al [6,7,8], Jung and Choi [9], Morishita and An [12] and Khusid et al [10]. Stanway et al [19] provide comprehensive reviews of available literature. In addition, the numerical simulation of ER fluid behaviour has received considerable attention and has been investigated using different approaches. Peel et al [14] and Coulter et al [4] reported numerical investigations in which the ER fluid in flow mode and shear mode was modelled respectively. Theoretical modelling of an ER fluid in squeeze flow has been investigated thoroughly (see for example, Williams et al [21], Sproston et al [16], El Wahed et al [8] and Yang [22]).

This paper is concerned with the dynamic performance of an ER fluid in a squeeze device in which the fluid is subjected to compressive and tensile forces. The experimental results are compared with the predictions of a theoretical model, which assumes a bi-viscous fluid characteristic. Interest is also focused on a new approach in which the yield stress of an ER fluid in squeeze is estimated.

2 Experimental Arrangement

The experimental rig (Figure. 1) consists of a Ling Dynamic Systems electromagnetic shaker (Model No. V450) which is capable of providing vertical oscillatory motion with maximum amplitude of 19 mm (peak-peak) over a frequency in the range DC to 7.5 kHz. The shaker head is attached rigidly to a Kistler (Model No. 9311A) piezoelectric force link and an earthed electrode having a recessed cavity of diameter 95 mm which provides the reservoir for the ER fluid. The high voltage upper electrode is a circular disc of diameter 56 mm, its circumferential edge and rear face surrounded and supported by a PTFE collar. This is rigidly attached to a second identical force link and positioning assembly to the supporting frame.

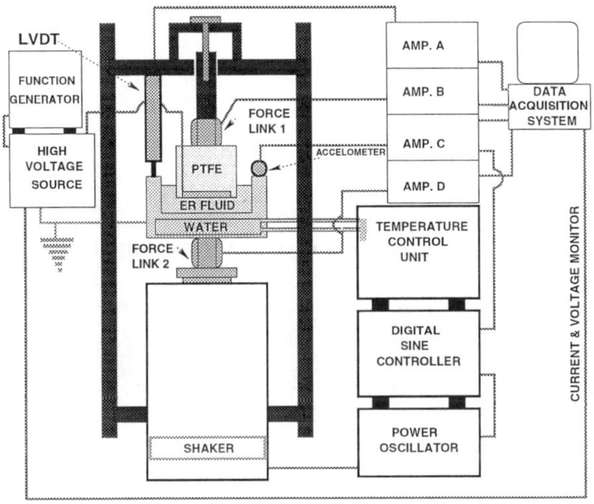

Figure 1. Experimental Arrangement

The instantaneous displacement of the lower electrode is determined using an RDP (type GTX 2500) LVDT, the velocity using an RDP (Type 240A0500) self-induced velocity sensor and the acceleration using an Endevco (Type 7254-100) accelerometer, all three devices attached to the upper surface of the lower electrode. Electrical excitation of the ER fluid is achieved by means of a Trek (Model 664) high voltage amplifier, driven by a Thander (Model TG102) function generator. Data acquisition and processing was achieved using a Measurement Group (Type ESAM) analogue to digital converter that is controlled by a Pascal program running on an IBM compatible personal computer. The ER fluid temperature was controlled at 30 $^\circ C$ by re-circulating water through a second closed cavity in the lower electrode using a Grant Instruments (Model LTD6) temperature controller. The ER fluid used in this investigation is a suspension of agglomerated corn flour in silicone oil. The solid phase was in the range 10 to 28 μm and the weight fraction was 57%.

3 Theoretical Model

The device in Figure 1 is modelled as an electrorheological fluid contained in the space between two parallel circular electrodes of radius a. The lower electrode oscillates about its mean position ($z=0$) with a displacement $y\ (t)$ whilst the top electrode is stationary.

The theoretical modelling is directed towards the determination of the input force $F_I(t)$ and the transmitted force $F_2(t)$ from a set of experimental data for $y(t)$.

The analysis is based on that reported by Williams et al [21] who solved the quasi-steady equations for the fluid flow where the constitutive model is the bi-viscous model in which the ratio γ represents the ratio of the viscosities in the 'yielded' to the 'un-yielded' regions ($\gamma = \eta / \eta_r$).

The non-dimensional pressure gradient G given by

$$G = -\frac{(h - y(t))}{2\tau_1}\left(\frac{dp}{dr}\right) \tag{1}$$

is determined as a function of the non-dimensional radius S, which is given by

$$S = -\frac{\dot{y}(t)\,\eta\, r}{(h - y(t))^2\,\tau_1} \tag{2}$$

where h is the mean separation of the electrodes, $\dot{y}(t)$ is the velocity of the lower electrode and r is the radial co-ordinate. Integration over the upper electrode then gives the transmitted force $F_2(t)$ in the form

$$F_2 = \frac{2\pi\tau_1 a^3 \phi(\chi)}{h - y(t)} \tag{3}$$

where χ is the value of S at $r = a$, and the non-dimensional output force ϕ is

$$\phi(\chi) = \frac{1}{108}\left(\frac{\gamma}{\chi}\right)^3 + \frac{1}{\chi^3}\int_{s=\frac{\gamma}{3}}^{\chi} S^2\, G\, dS \tag{4}$$

When F_2 is obtained from equation (3) the input force F_I is then calculated from

$$F_1 - F_2 = M\,\ddot{y} \tag{5}$$

where M and $\ddot{y}$ are the mass and acceleration of the lower electrode assembly.

4 Results and Discussion

The tests carried out consisted of the simultaneous measurement of the instantaneous values of the input force, the transmitted force across the ER fluid, the displacement, velocity and acceleration of the lower electrode. These measurements were collected at a sampling frequency of 5 kHz for a range of mechanical frequencies, for DC excitation.

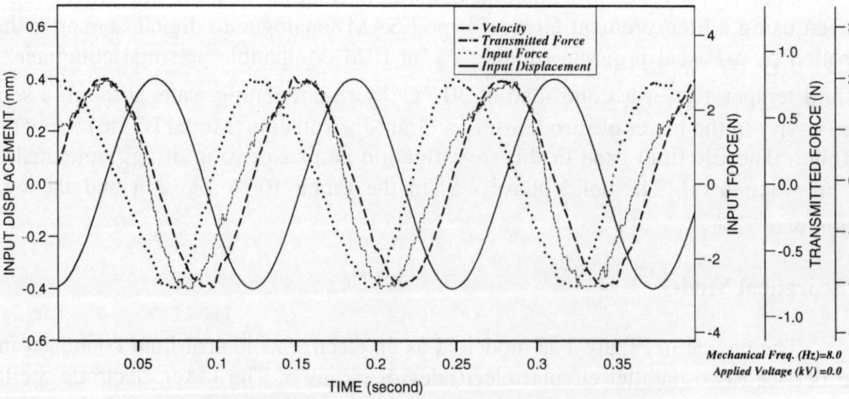

Figure 2. ER Fluid in Squeeze under Constant Voltage

For a mechanical frequency of 8 Hz, the variation in input displacement and velocity of the lower assembly together with the input force delivered by the shaker and the transmitted force across the ER fluid for zero voltage is presented in Figure 2. It is seen that the maximum value of the transmitted force occurs when the lower assembly passes through its mean position and travels with its maximum velocity.

For 4 kV/mm, Figure 3, the maximum input displacement is reduced by a factor of 15, in comparison with the zero-voltage case. This is accompanied by a maximum velocity reduction of about 6. It is also seen that the transmitted force coincides with the input force, Figure 3, confirming the development of a high yield stress when the fluid is energised by sufficiently high voltage in contrast to 18 ms lag time that appears to lapse before the input force is transmitted across the fluid (zero-voltage case), Figure 2.

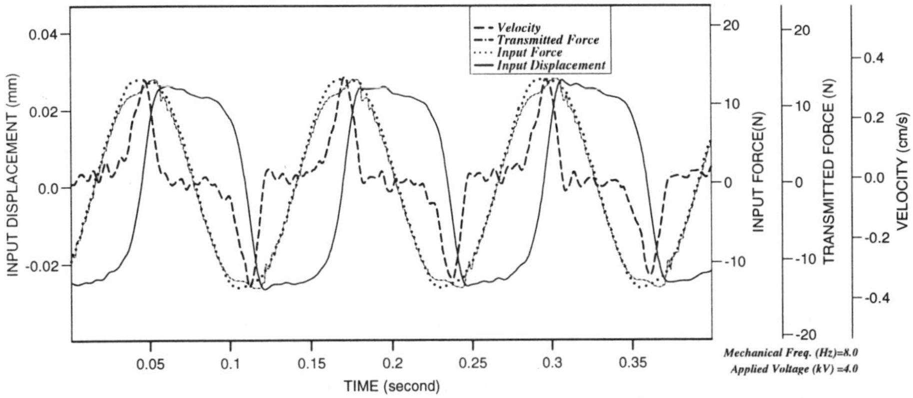

Figure 3. ER Fluid in Squeeze under Constant Voltage

The behaviour of the ER fluid cell was compared with that predicted by the bi-viscous model, and the results are shown in Figure 4 for one force cycle for 1 kV. In this figure, the experimental and theoretical data for the transmitted force seem to compare well when the lower assembly approaches its stationary position but the agreement becomes poor particularly when the lower assembly travels with its maximum velocity.

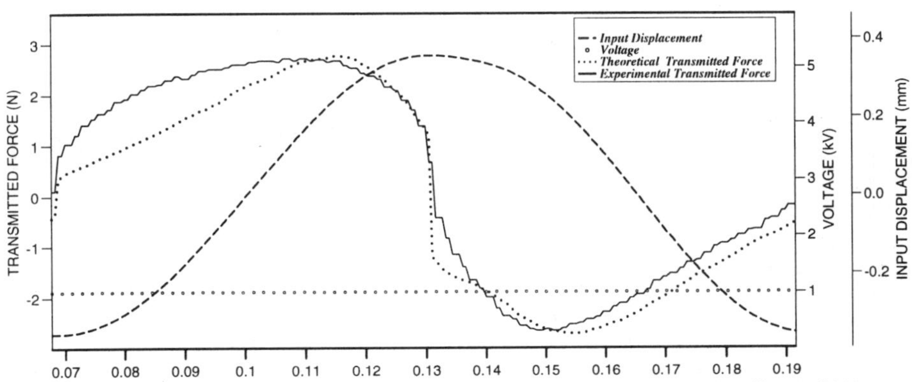

Figure 4. Variation of Transmitted Force with Time

This was also the case when the performance of a similar ER cell, driven by a smaller shaker, was simulated using the present theory [6,17]. The above parameters have been predicted when the yield stress, τ_1, was determined using a power-law in the form

$$\tau_1 = AE^b \tag{6}$$

where the constants A and b are system dependent, and E is the electric field strength. This law was derived for an ER fluid in shear; the situation here is more complex as the fluid is subjected to compressive and tensile loading in which the gap and the electric field continuously change. Moreover, Bonnecaze and Brady [1] have demonstrated the existence of more than one kind of yield stress whose value depends on whether the fluid is in static or dynamic stress.

As a result it was thought prudent to consider modifying the way in which the yield stress of the ER fluid in squeeze is determined. Recently, Bose et al [2] reported a new technique in which the yield stress is determined using two different formulae, namely the power-law Eq. (6) and a linear-law, each of which is applicable over a specific limited range of applied fields. The piece-meal representation of the yield stress adopted by Bose et al [2] is applied in the present investigation and the linear equation takes the form

$$\tau_1 = \tau_{eo} + \frac{\tau_{eo}}{E_o} E \tag{7}$$

where τ_{eo} is an offset, which represents the lowest value of the yield stress calculated over an electric field range, the start of which is marked by E_0, using Eq. (7) and the results showed no improvement in agreement.

As a result, it was decided to estimate the yield stress using a combination of the present algorithm as well as the experimental values of input displacement and transmitted force. Firstly, the present theory is used to estimate the instantaneous values of the yield stress using the power-law, which are then used to determine the transmitted force. Secondly, successive iterations are then performed in order to achieve agreement between the predicted and experimental values of the transmitted force. During each of these iterations, a new χ value is calculated using the new yield stress determined using equation (3), in which the experimental results are substituted for F_2.

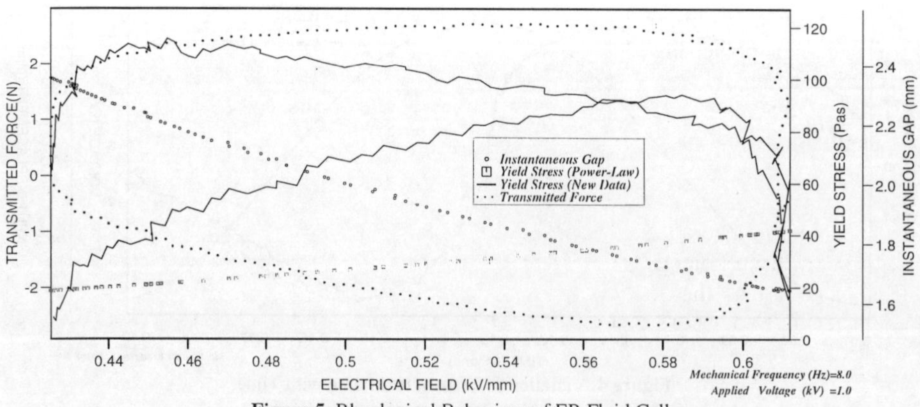

Figure 5. Rheological Behaviour of ER Fluid Cell

The instantaneous yield stress values that satisfied the above condition are shown in Figure 5 as a function of applied field. The yield stress values determined using the power-law is also shown. It can be seen that the yield stress reaches its minimum value when the lower electrode approaches its stationary position. However, the yield stress results which are based on the power-law reach their maximum value about the above electrode position and this would explain the relatively poor agreement between the predicted and the experimental results.

A program was written to determine the best equation to simulate the new yield stress results. From this, a quadratic equation was found which was of the form

$$\tau_1 = c + \frac{d}{E} + \frac{e}{E^2} \qquad (8)$$

where c, d and e are constants that are dependent on the direction of strain (compressive or tensile). The variation of the transmitted force with instantaneous inter-electrode gap for 1 kV (using the present theory in conjunction with the newly derived yield stress formula) is shown in Figures 6, for the cases when the yield stress is determined using the power-law, the linear/power-law and the quadratic equation.

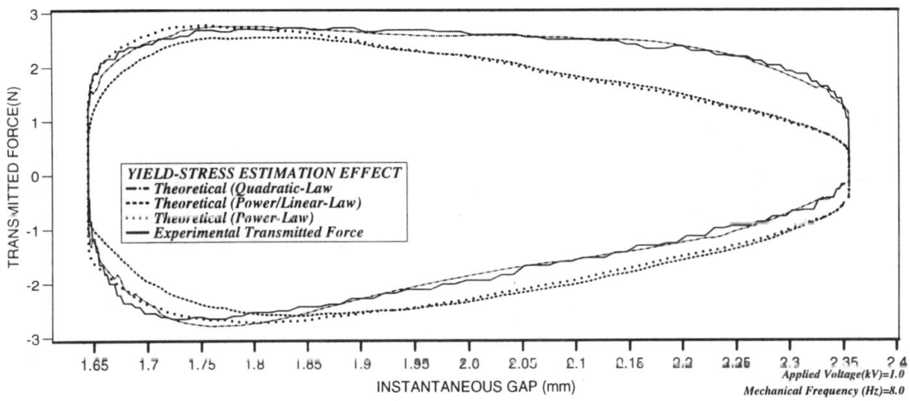

Figure 6. Variation of Transmitted Force with Instantaneous Gap

It is seen that the data predicted using the latter equation compared very well with the experimental results whilst the former laws produce predictions that agree with the experimental results only when the lower assembly approaches its stationary position. This suggests that the power-law equation might only be applicable when the fluid is subjected to near-static loading.

5 Conclusions

This paper has shown that an ER fluid in dynamic squeeze was seen that the fluid was capable of generating a force transmissibility with a value close to unity when energised by a sufficiently high electric field and, as a result, the damping force was sufficient to arrest the motion of the shaker operating over a range of mechanical frequencies.

The response of the cell under DC excitation was satisfactorily predicted by a theoretical model, which assumed a bi-viscous fluid characteristic in conjunction with an estimation of the fluid yield stress using a quadratic equation in E^{-1} instead of the more common power-law formula. This model is considered useful for optimising the fluid performance when employed in a vibration control device.

6 Acknowledgement

This work is supported by EPSRC research grant GR/M25407.

References

1. Bonnecaze, R T and Brady, J F, 1992, *J. Rheology,* Vol. 36, pp. 73-115.
2. Bose, H, Hoppe, R H W and Mazurkevitch, G, 2001, *J. Non-Newtonian Fluid Flow, to be published (Personal Communication).*
3. Bullough, W. A. and Foxon, M. B., 1978, *J. Sound and Vibration,* Vol. 56, pp. 35-44.
4. Coulter, J P, Weiss, K D and Carlson, J D, 1993, *J. Intelligent Material Systems and Structures,* Vol. 4, pp. 248-259.
5. Ehrgott, R. and Masri, S., 1993, *3rd Int. Conf.Adaptive Str.,* pp. 161-175.
6. El Wahed A K, Sproston J L and Stanway R, 1998, *J. Physics D: Applied Physics,* Vol. 31, pp. 2964-2974.
7. El Wahed A K, Sproston J L and Stanway R, 1999, *J. Colloid and Interface Science,* Vol. 211, pp. 264-280.
8. El Wahed A K, Sproston J L and Williams, E. W., 2000, *J. Physics D: Applied Physics,* Vol. 33, pp. 2995-3003.
9. Jung S Y and Choi S B, 1995, *Tribology Transactions,* Vol. 38, No. 4, pp.857-862.
10. Khusid B, Acrivos A, Khodorkovsky Y and Beltran M, 1999, *International Journal of Modern Physics B,* Vol. 13, Nos. 15&16, pp. 2143-2150.
11. Monkman, G J, 1995, *J. Physics D: Applied Physics,* Vol. 28, pp. 588-593.
12. Morishita, S and An, Y K, 1996, *JSME Int. J., Series C,* Vol. 39, No. 4, pp. 702-707.
13. Nakano, M. and Yonekawa, T., 1997, *Noise & Vib.Worldwide,* Vol. 28, No. 8, pp. 21-29.
14. Peel, D J, Stanway, R, Sproston, J L and Bullough, W A, 1995, *ASME Design Eng. DE: Vibration Control, Analysis and Identification,* Vol. 84-3, Part C, pp. 737-744.
15. Petek, N. K., Goudie, R. J. and Boyle, F. P., 1988, *SAE Technical Paper No. 881785.*
16. Sproston, J. L., Stanway, R., Williams, E. W. and Rigby, S. G., 1994, *J. Electrostatics,* Vol. 32, pp. 253-259.
17. Sproston, J. L., El Wahed, A. K., Williams, E W and Stanway, R., 1996, *Int. J. Modern Physics B,* Vol. 10, Nos. 23 & 24, pp. 3081-3091.
18. Stanway R, Sproston J L, Prendergast M J, Case J R and Wilne C E, 1992, *J. Electrostatics,* Vol. 28, pp. 89-94.
19. Stanway R, Sproston J L and El Wahed A K, 1996, *J. Smart Materials and Structures,* Vol. 5, No. 4, pp. 464-482.
20. Wereley, N. M. and Lindler, J., 1999, *ASME, Aerospace Division Publication AD, "Adaptive Structures and Material Systems",* Nashville, USA, Vol. 59, pp. 67-75.
21. Williams E W, Rigby SG, Sproston J L and Stanway R, 1993, *J. Non-Newtonian Fluid Mechanics,* Vol. 47, pp. 221-238.
22. Yang, F, 1997, Tension and compression of electrorheological fluid, *J. Colloid and Interface Science,* Vol. 192, pp. 162-165.
23. Yao, G Z, Qiu, Y, Meng, G, Fang, T and Fan, Y B, 1999, Vibration control of a rotor system by disk type electrorheological damper, *J. of Sound and Vibration,* Vol. 219, pp. 175-188.

DEVELOPMENT OF ER BRAKE AND ITS APPLICATION TO PASSIVE FORCE DISPLAY

JUNJI FURUSHO, MASAMICHI SAKAGUCHI, NAOYUKI TAKESUE AND KEN'ICHI KOYANAGI

Osaka University, 2-1 Yamadaoka, Suita, Osaka 565-0871, Japan
E-mail: furusho@mech.eng.osaka-u.ac.jp

Force display systems are important in virtual reality and other applications. Conventional force displays, however, are active systems with actuators, so that it may become danger inherently. Consequently, passive force display is effective method for assuring safety. In this paper, we developed a brake using ER fluid and passive force display using ER brakes. Then we discuss two degree of freedom passive force display and basic control experiments.

1 Introduction

Force information is often required for tele-operation system and virtual reality. Many types of force display systems are now available, but most of them are active systems using servomotors or other actuators. While these active systems can provide operators with a wide variety of force senses, they inherently involve a potential hazard in that they could uncontrollably move and hurt operators if something really goes wrong. On the other hand, passive force displays using brakes or clutches to present resistance to an operator's force or movement is quire safe [1]. The authors have developed an ER brake using an electrorheological fluid as its functional material. They have also developed a passive force display using such ER brakes. In this paper they will outline the system and discuss the basic experiments on it.

2 ER brake

Electrorheological (ER) fluid is a type of functional fluid, and attracting much attention these days [2,3]. The shear stress of This type ER fluid is almost independent of the shear rate, but changes according to the magnitude of the applied electric field (as shown in Fig.1). Making use of the unique characteristics of particle-type ER fluid, the authors developed a clutch-type actuator[4,5,6] and studying their applications[7,8].

Figure 2 shows the basic structure of an ER brake consisting of a fixed cylinder and a rotating cylinder with an ER fluid between them. The two cylinders also serve as a pair of electrodes. The rotating cylinder is fitted with an output shaft and is driven by an external force through the shaft. When a voltage is applied between the cylinders, or electrodes, an electric field is produced in the ER fluid, then the viscosity of the fluid increases. Thus the rotating cylinder is forced to slow down, or braked, providing the rotating output shaft with resistance (braking torque). The specification of the developed ER brake are shown in Table 1. The maximum braking torque also depends on the characteristics of the ER fluid used. Figure 1 is the characteristic of the ER Fluid used in this paper, ARP–06, which is provided for research purposes by Asahi Chemical Industry Co., Ltd.

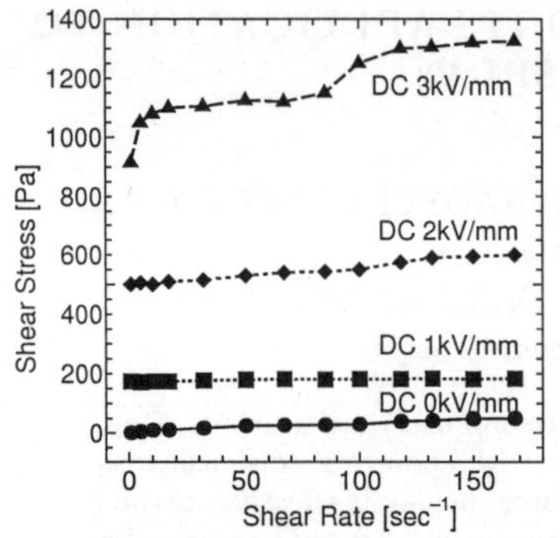

Figure 1. Characteristics of the ER fluid used in this paper.

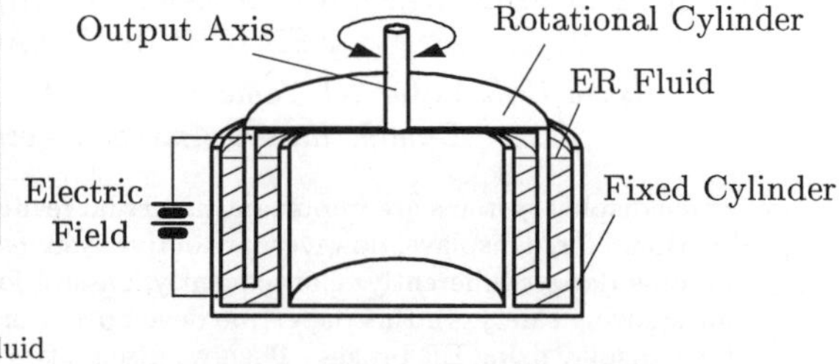

Figure 2. Basic structure of ER brake.

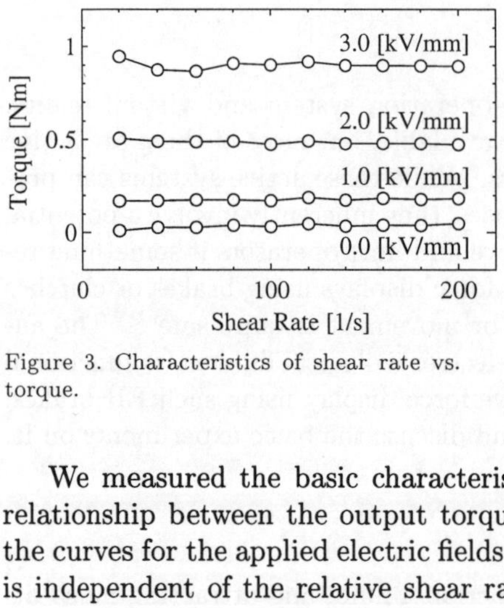

Figure 3. Characteristics of shear rate vs. torque.

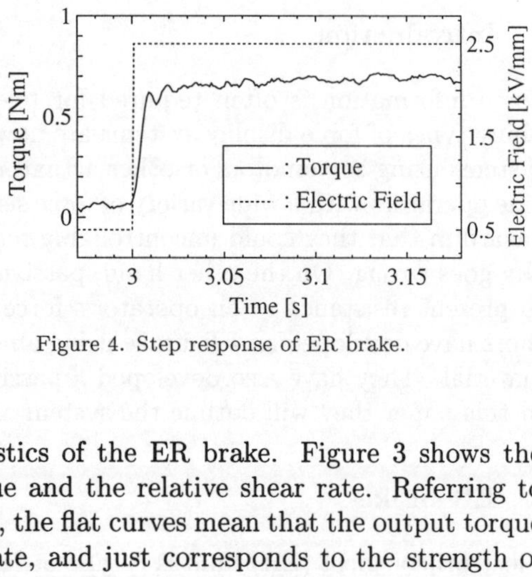

Figure 4. Step response of ER brake.

We measured the basic characteristics of the ER brake. Figure 3 shows the relationship between the output torque and the relative shear rate. Referring to the curves for the applied electric fields, the flat curves mean that the output torque is independent of the relative shear rate, and just corresponds to the strength of the applied electric field. Figure 4 shows the output torque response as measured in a step-response experiment. Note that the electric field was increased from 0.5 to 2.5 [kV/mm] instantly at time 3.0 [s]. As the output torque has changed in several milliseconds. In conclusion, unlike the conventional friction brake, the ER brake can produce an accurately controlled dynamic braking torque, and the braking torque of the ER brake can rise faster than the powder clutch.

Table 1. Specifications of developed ER brake.

Diameter	156	[mm]
Height	135	[mm]
Weight	2.6	[kg]
Electrode-to-Electrode Distance	1.0	[mm]
Brake Torque		
Shera Stress of ER Fluid: 1[kPa]	1.3	[Nm]
Shera Stress of ER Fluid: 5[kPa]	6.3	[Nm]

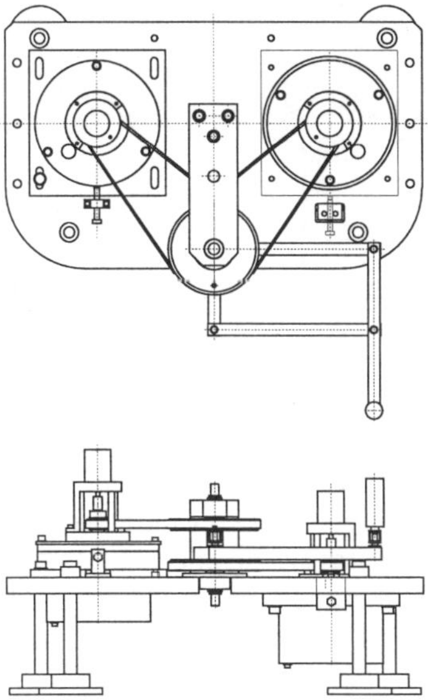

Figure 5. Top view and front view of 2 DOF passive force display using ER brakes.

Figure 6. Picture of passive force display system with computer display.

3 Passive force display using ER brakes

The authors have developed a 2 DOF passive force display using ER brakes. Figure 5 shows the front view and top view of the deveoped system. Figure 6 is a picture of the force display system with computer display. The system uses two ER brakes. They are connected through a belt-pulley mechanism to a parallel-link mechanism. And each ER brake control the movement of the first or second link independently. An operator holds the handle at the tip of the parallel-link mechanism. The reduction ratio of the belt-pulley mechanism is 1:4. The first and second links of the parallel-link mechanism are 20 [cm] long each. Each ER brake is provided with a high-resolution encoder to measure its rotational position and thereby the position of the handle. A force sensor is setting in the handle. The operator manipulate the handle while watching the computer display.

4 Basic experiments of force presentation

4.1 Virtual Walls

The authors set up the system so that it represents virtual walls on a two-dimensional plane, and made preliminary experiments using it.

Figure 7 shows the sketch of top view of the passive force display. In the figure, the coordinates of the handle is (x, y) and the line of the virtual wall is $Y = k$. When the handle touches a virtual wall, that is to say, y is less than or equal to k, electric field is applied to the ER brakes, then operator felt a reaction force from the wall. Since the ER brakes feature fast response, a sensation of touching a hard

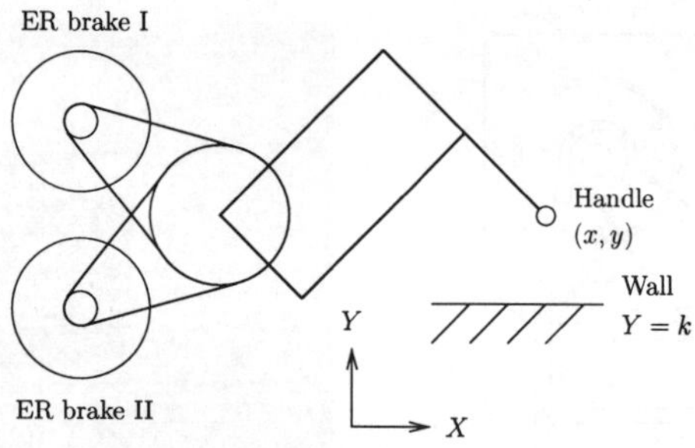

Figure 7. Sketch of top view of passive force display.

surface was well represented. However, the energized ER brakes provided resistance force equally to the detaching movement of the handle, representing a sticky wall. To prevent this phenomenon, the ER brakes were instantly de-energized when the detaching force to the handle was detected, that is to say, F_y, the y directional force of handle, is greater than zero. The control algorisms of the ER brakes is shown in Table 2.

Figures 8 show the experimental results for touch and detach of the virtual wall. The horizontal axes show the time, and vertical axes show the y directional velocity and reaction force measured at the handle. When the handle touched the wall, at the both type of the walls, the velocity of the handle decreased and reaction force increased quickly (Fig. 8 (a), shown only sticky wall). When the handle detached the wall, large force was needed to detach the sticky wall, but handle was detached by small force at the less sticky wall (Fig. 8 (b) & (c)).

Table 2. Control algorisms of ER brakes.

	Conditions	State of ERBs
Sticky wall	$y > k$	OFF
	$y \leq k$	ON
Less sticky wall	$y > k$	OFF
	$y \leq k$, $F_y > 0$	OFF
	$y \leq k$, $F_y \leq 0$	ON

Table 3. Operating Patern When Tracing Over Wall

State	ERB I	ERB II
$0 < \theta < \theta_1$	ON	OFF
$\theta_1 \leq \theta \leq \theta_2$	ON	ON
$\theta_2 < \theta < 180$	OFF	ON

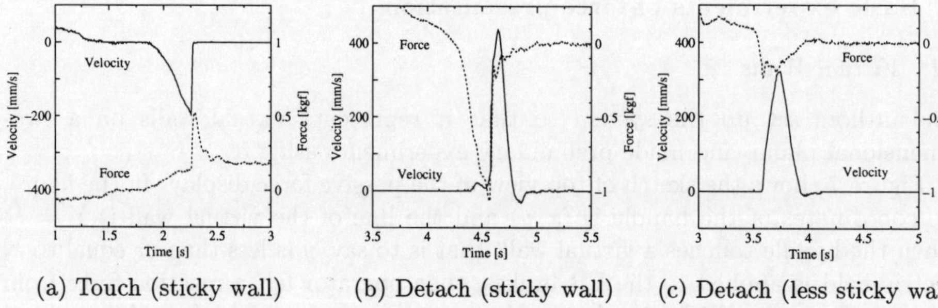

(a) Touch (sticky wall) (b) Detach (sticky wall) (c) Detach (less sticky wall)

Figure 8. Experimental results (sticky wall & less sticky wall).

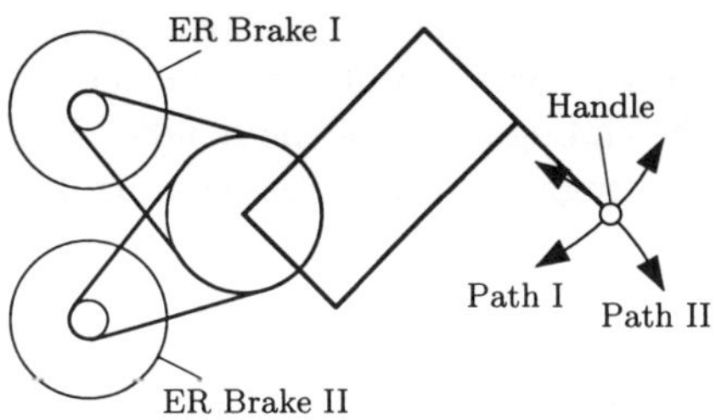

Figure 9. Handle track when one brake was operated.

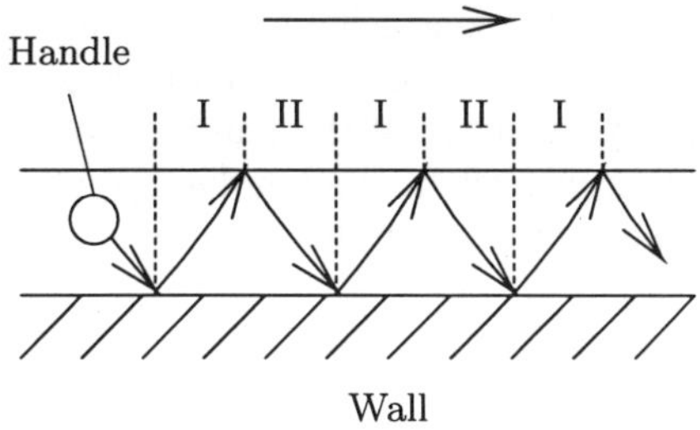

Figure 10. Narrow width area on virtual wall.

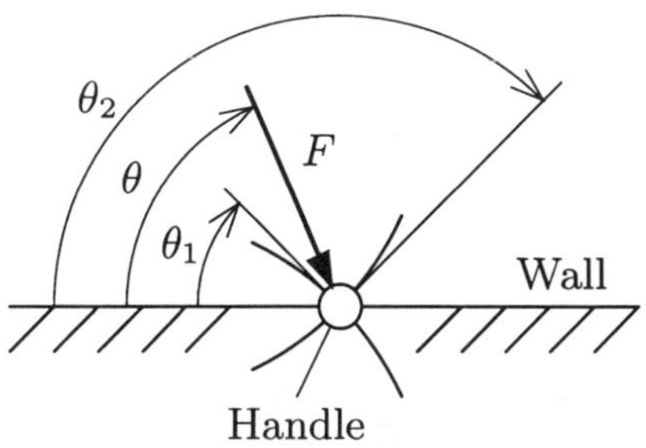

Figure 11. Relation between Wall and Direction of Operation Force

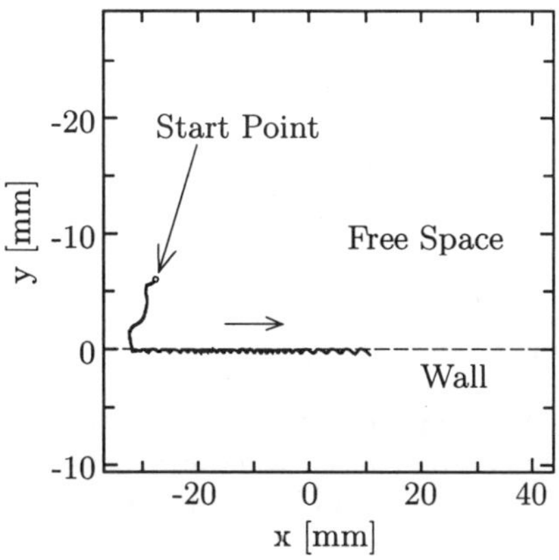

Figure 12. Experimental Result

4.2 Sensation of tracing over a virtual wall

The authors also studied how to represent a sensation of tracing over a virtual wall. With the passive force display system, an operator could feel a virtual wall because the operator felt a reaction force from the wall when the handle touched the wall. Unlike with a real wall, however, once the handle touched the virtual wall, the operator could neither move the handle into the wall nor slide the handle over it, because both the ER brakes were operated at the same time to represent a reaction from the wall. Thus, it became clear that representing a sensation of sliding over a surface requires clever switching of the two ER brakes such that only either of them is operated at a time.

In this system, if only one ER brake was operated, the handle was restricted to move along the arc track as path I or path II shown in Fig.9. Then, the handle can trace over the virtual wall with saw-teeth movement in the narrow width area as shown in Fig. 10. Here, force direction information of the handle is introdced to switch the ER brakes smoothly. Figure 11 shows relation between the wall and a direction of operation force, where θ is the angle between the force direction and the wall, and θ_1 and θ_2 are the angles between the wall and tangent lines of each arc tracks which are drawn by the handle when only one ER brake is ON. When the handle touched the wall and the force direction was within $0° < \theta < \theta_1$ or

$\theta_2 < \theta < 180°$, one ER brake was released but the other operated instead so that the handle would be prevented from getting into the virtual wall. By this method, vibratory behavior of the handle depends on the sliding speed of the handle. Fine vibration was felt if the handle moved slow, but rough vibration was felt with fast movement of the handle.

5 Conclusion

(1) We have developed the ER brakes using ER fluid. We conducted the basic experiments with them and ascertained there characteristics as follows: (a) Quick response, (b) Low moment of inertia, (c) Good controllability of brake torque.

(2) We have developed a 2 DOF passive force display using ER brakes.

(3) We made some force presentation experiments using passive force display as follows: (a) Touch and detach the virtual wall, (b) Tracing over the virtual wall.

References

1. H. Davis and W. Book, "Torque Control of a Redundantly Actuated Passive Manipulator", *Proc. American Control Conf.*, pp.959–963, 1997.
2. M. Nakano and K. Koyama ed, *Proc. 6th Int. Conf. on ER Fluids, MR Suspensions and Their Applications*, World Scientific, 1998.
3. J. Furusho, "Control of Mechatronics Systems Using Electrorheological Fluids", *J. of the Society of Instrument and Control Engineers*, Vol. 34, No. 9, pp. 687–725, 1995 *(in Japanese)*.
4. J. Furusho and M. Sakaguchi, "Active Control Using Electrorheological Fluids", *Japanese J. of Tribology*, Vol. 41, No. 6, pp.645–654, 1996.
5. M. Sakaguchi, J. Furusho, G. Zhang, and Z. Wei, "Development of ER Actuator and Basic Study on its Force Control System", *J. of Robotics Society of Japan*, Vol. 16, No. 8, pp.1108–1114, 1998 *(In Japanese)*.
6. M. Sakaguchi, G. Zhang, and J. Furusho, "Modeling and Motion Control of an Actuator Unit Using ER Clutches", *Proc. 2000 IEEE Int. Conf. on Robotics and Automation*, pp.1347–1353, 2000.
7. J. Furusho and M. Sakaguchi, "New Actuators Using ER Fluid and Their Applications to Force Display Devices in Virtual Reality and Medical Treatments", *Int. J. of Modern Physics B*, Vol.13, Nos.14,15&16, pp.2151–2159, 1999.
8. M. Sakaguchi, J. Furusho, and E. Genda, "Basic Study in Rehabilitation Training System Using ER Actuators", *Proc. 1999 IEEE Int. Conf. on Systems, Man, and Cybernetics*, pp.135–140, 1999.

WHAT MAKES A GOOD MR FLUID?

J. DAVID CARLSON
Lord Corporation, 110 Lord Drive, Cary, North Carolina 27511, USA
E-mail: jdcarlson@lord.com

Experience in manufacturing MR fluids for commercial application has shown that some of the greatest barriers to commercial success are not factors or conditions normally considered in the laboratory. The present paper looks at conditions found in MR fluid devices operating in real-world applications where shear rates may exceed 10^5 sec^{-1} and devices are called upon to operate for very long periods of time. The problem of "In-Use-Thickening" wherein a MR fluid subjected to long-term use progressively thickens until it eventually becomes an unworkable paste is presented. The search for a solution to this heretofore unrecognized problem delayed commercial introduction of the Lord truck seat damper system for several years. Today, good fluids are able to operate for long periods with minimum in-use-thickening.

1 Introduction

The most common response to the question of what makes a good MR fluid is likely to be "high yield strength" or "non-settling". However, those particular features are perhaps not the most critical when it comes to ultimate success of a magnetorheological fluid. The most challenging barriers to the successful commercialization of MR fluids and devices have actually been less academic concerns.

Figure 1 - Rabinow's magnetic fluid demo.[1]

As anyone who has made MR fluids knows, it is not hard to make a strong MR fluid. Over fifty years ago both Rabinow and Winslow described basic MR fluid formulations that were every bit as strong as fluids today. A typical MR fluid used by Rabinow consisted of 9 parts by weight of carbonyl iron to one part of silicone oil, petroleum oil or kerosene.[1] To this suspension he would optionally add grease or other thixotropic additive to improve settling stability. The strength of Rabinow's MR fluid can be estimated from the result of a simple demonstration that he performed. As shown in Figure 1, Rabinow was able to suspend the weight of a young woman from a simple direct shear MR fluid device. He described the device as having a total shear area of 8 square inches and the weight of the woman as 117 pounds. For this demonstration to be successful it was thus necessary for the MR fluid to have a yield strength of at least 100 kPa.

MR fluids made by Winslow were likely to have been equally as strong. A typical fluid described by Winslow consisted of 10 parts by weight of carbonyl iron suspended in mineral oil.[2] To this suspension Winslow would add ferrous naphthenate or ferrous oleate as a dispersant and a metal soap such as lithium stearate or sodium stearate as a thixotropic additive.

The formulations described by Rabinow and Winslow are relatively easy to make. The yield strength of the resulting MR fluids is entirely adequate for most applications. Additionally, the stability of these suspensions is remarkably good. It is certainly adequate for most common types of MR fluid application. As early as 1950 Rabinow pointed out that complete suspension stability, i.e. no supernatant clear layer formation, was not necessary for most MR fluid devices. MR fluid dampers and rotary brakes are in general highly efficient mixing devices. When the piston in a MR fluid damper moves, the MR fluid jets through the orifices quite rapidly causing it to swirl and eddy vigorously even for low piston speed. Similarly, the shear motion that occurs in a MR brake causes vigorous fluid motion. As long as the MR fluid does not settle into a hard sediment, normal motion of the device is generally sufficient to cause sufficient flow to quickly remix any stratified MR fluid back to a homogeneous state. For a small MR fluid damper such as the Lord Motion Master RD-1005, two or three strokes of a damper that has sat motionless for several months are sufficient to return it to a completely remixed condition.

Except for very special cases such as seismic dampers, lack of complete suspension stability is not a necessity. It is sufficient for most applications to have a MR fluid that soft settles – upon standing a clear layer may form at the top of the fluid but the sediment remains soft and easily remixed. Attempting to make these MR fluids absolutely stable may actually compromise their performance in a device. One of the areas where MR fluids find their greatest application is in linear dampers that effect semi-active control. These include small MR fluid dampers for controlling the motion of suspended seats in heavy-duty trucks, larger MR fluid dampers for use as primary suspension shock absorbers and struts in passenger automobiles and special purpose MR fluid dampers for use in prosthetic devices.

In all of these devices one of the most important fluid properties is a low-off state viscosity. While in all of these examples having a MR fluid with a high yield strength in the on-state is important, it is equally important that the fluid also have a very low off-state. The very ability of an MR fluid device to be effective at enabling a semi-active control strategy such as "sky-hook" damping depends on being able to achieve a sufficiently low off-state. Care must be taken in choosing fluid stabilizing additives so that they do not adversely affect the off-state viscosity.

Earthquake dampers and other some other special applications in which the device will sit quiescent for very long periods of time represent special cases where fluid stability issues may have overriding importance. Because of the transient nature of seismic events these dampers never see regular motion, which can be relied on to keep the fluid mixed. This lack of motion also has it benefit. Unlike dampers used in highly dynamic environments, seismic dampers do not need to sustain millions of cycles. The fact that durability and wear are not issues gives the fluid designer grater latitude to formulate a highly stable fluid. MR fluids for these applications are typically formulated as shearing-thinning thixotropic gels.

2 The MR Fluid Environment Outside the Laboratory

In exploring the question of "what makes a good MR fluid", it is instructive to consider the factors and conditions seen by an MR fluid, not in the laboratory, but in devices in actual service. Even under normal device operation these conditions can be quite extreme. The specific environment to which a MR fluid is exposed inside a MR device is quite different from the regime that is normally measured in the laboratory. Consider the conditions found inside a MotionMaster™ RD-1005-3 damper manufactured by Lord Corporation for use in heavy-duty truck and bus seating.[3,4,5] This damper is designed to generate a nominal resistive force of about 1200 N at an input current of 1 amp. The shear rate inside the MR fluid valve under normal operating conditions when the speed across the damper is in the range of .05-0.2 m/s is 1-4 x 10^4 sec^{-1}. Under extreme conditions this damper may experience speeds in excess of 1 m/s or shear rates greater than $2x10^5$ sec^{-1}. An automotive primary suspension MR fluid shock absorber may experience shear rates that are even larger, perhaps as much as 106 sec^{-1}. The peak mechanical power dissipation in the RD-1005-3 damper is normally in the range of 60-240 watts. Since the damper contains about 70 cm^3 of fluid this corresponds to about 0.9 to 3.4 watts/cm^3 of fluid in the damper. Under extreme conditions the peak power dissipation may exceed 1200 watts or 17 watts/cm^3.

Conditions in a rotary brake are, in general, not as extreme until one gets to the power per unit volume level. The Rheonetic™ MRB-2107 brake is designed to produce a maximum dissipative torque of about 6 Nm at an input of 1 amp.[6] Under normal operating speeds of 100-1000 RPM this corresponds to shear rates of 10^3-10^4 sec^{-1}. Power dissipation ranges from 63-630 watts. Since the fluid volume in the brake is only 5 cm^3, this corresponds to 13-130 watts/cm^3.

Before a MR fluid is placed into a device is must be made. The manufacturing process for commercial MR fluids is quite different from the mixing of MR fluids in a laboratory. A MR fluid that is to be commercially useful outside of the laboratory must be amenable to volume production. As such the process for making the fluid must be scalable well beyond a liter-sized batch by many orders of magnitude. The process must also yield a result that is repeatable and reproducible. Today's highly competitive, global market, particularly any market related to the automotive industry, demands that MR fluids be made with a process that is QS-9000 compliant and certified.[5,6]

The scale of fluid production necessary for even a modest automotive application is large. A single MR fluid device such as a damper on a single automotive platform model can easily require a total fluid volume production on the order of 10^5 liters/year. Depending on the iron particle loading, that amount corresponds to 2-4 x 10^5 kg/year or 1-2 tons of MR fluid per day. A 50 gallon (185 liter) barrel of MR fluid weighs between 1/2 ton and 1 ton. Thus, one must be able to reliably and reproducibly make several barrels of MR fluid per day on a continuous basis to support even a single, modest automotive application. Being able to cope with the logistics of handling and mixing large quantities of heavy, pyrophoric materials is an absolute necessity.

3 The Problem of In-Use-Thickening

During the development and commercialization phases of several recent MR fluid devices, problems with the MR fluid were discovered that were not apparent in the early research phases of these projects. An example of one such problem is a phenomenon called "In-Use-Thickening" or IUT. If an ordinary MR fluid is subjected high stress and high shear rate over a long period of time, the fluid will thicken. Superficially, this process appears much like the process of churning cream to make butter. An originally low-viscosity, i.e. low off-state, MR fluid progressively becomes thicker and thicker until it eventually becomes an unmanageable paste having the consistency of shoe polish.

An example of the "in-use-thickening" or IUT problem is illustrated in Figure 2. In this example, the off-state force of an early version of the MotionMaster™ RD-1005 truck seat damper is shown over the course of a life-cycle test. The damper was operated at 1 Hz with a 25.4 mm stroke in the "on-state" with 1 amp applied continuously such that the damper force was approximately 1200 N. Under these conditions, the RMS mechanical power dissipation was approximately 68 watts. Aluminum cooling fins were fastened to the damper body to augment convection cooling and keep the damper temperature from exceeding 100C. Periodically, the current applied to the damper was removed and the off-state force was measured. These periodic off-state forces over the course of about 600,000 on-state cycles or about 167 hours of operation are plotted in Figure 2.

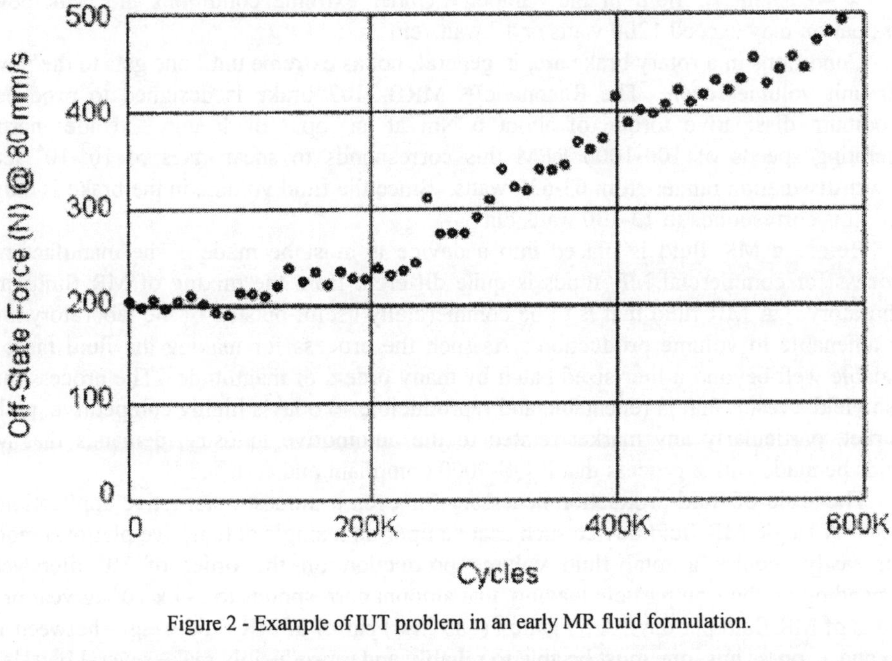

Figure 2 - Example of IUT problem in an early MR fluid formulation.

The IUT phenomenon appears as a progressive increase in the off-state force. By the time the damper had experienced 600,000 on-state cycles, the off-state force has increased from 200 N to 500 N. This factor of 2.5 increase in off-state force is sufficient to render the damper unsuitable for implementing effective semi-active vibration control. For such, a low off-state force is just as important as a high on-state force. While this fluid provided very good performance when the damper was first tested, the fluid thickening that developed eventually rendered the damper unacceptable.

The off-state force increase that is the hallmark of the IUT problem is due to an increase if the off-state viscosity of MR fluid subjected to long-term stress. The cause of this viscosity increase is believed to be due to spalling of the friable surface layer from the surface of the carbonyl iron particles that typically comprise the particle component of MR fluid. This surface layer composed of iron oxides, carbides and nitrides is rather brittle. When subjected to high inter-particle stresses this surface layer fractures and breaks into small pieces that separate from the primary particle. These very small nanometer-sized secondary particles have a very large surface area to weight ratio. As such even a very weight small amount of these secondary particles is capable of significantly affecting the rheology of the overall MR fluid.

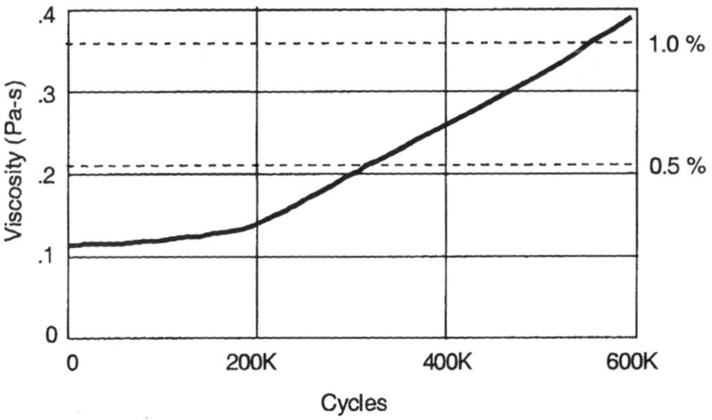

Figure 3 - Viscosity increase due to simulated surface spalling compared to observed IUT.

The increase in MR fluid viscosity that results from the addition of a volume of nano-sized particulates that may have spalled from the surface of the carbonyl iron particles is illustrated in Figure 3. In this graph the observed progressive off-state viscosity increase due to IUT is compared to the viscosity increase caused by replacing a small part of the carbonyl iron mass with nano-sized ferrite particles.. The 0.5% and 1% dashed lines in this graph correspond to the increased viscosity that results if that amount of the carbonyl iron is replaced by nano-sized ferrite. Replacing 0.5% of the mass of carbonyl iron with the same mass of nano-sized ferrite particles results in almost a factor of two increase in MR fluid of-state viscosity. Replacing 1% of the carbonyl iron mass with ferrite results in approximately a factor of three increase in viscosity. Thus, even if only a small portion of the carbonyl iron particles is shed secondary nano-sized debris, a very large increase in off-state viscosity can occur.

IUT was not identified as a serious problem until long-term life testing to qualify commercial MR fluid truck seat dampers was well underway. Finding and implementing a viable solution to the IUT problem required the focused efforts of a six-person team working over a period of about two years. The IUT problem delayed commercial introduction of the MotionMaster truck seat damper for over two years. Figure 4 shows the progressive progress that was made in dealing with the IUT problem Starting with the original MR fluid formulation, MR fluid formulations that showed progressively less IUT were developed over the course of several years intensive development effort. Eventually, a fluid formulation that showed no significant IUT after two million cycles was found. Over two years of intensive fluid development were required to progress from Curve 1 to Curve 4, a fluid that could survive the minimum 2 million cycles required for a commercial MR fluid seat damper. Today the "in-use-thickening" problem has not only been identified but has been solved. Today, good MR fluids show no measurable in-use-thickening after more than 10 million cycles in the MotionMaster™ RD-10005 damper.

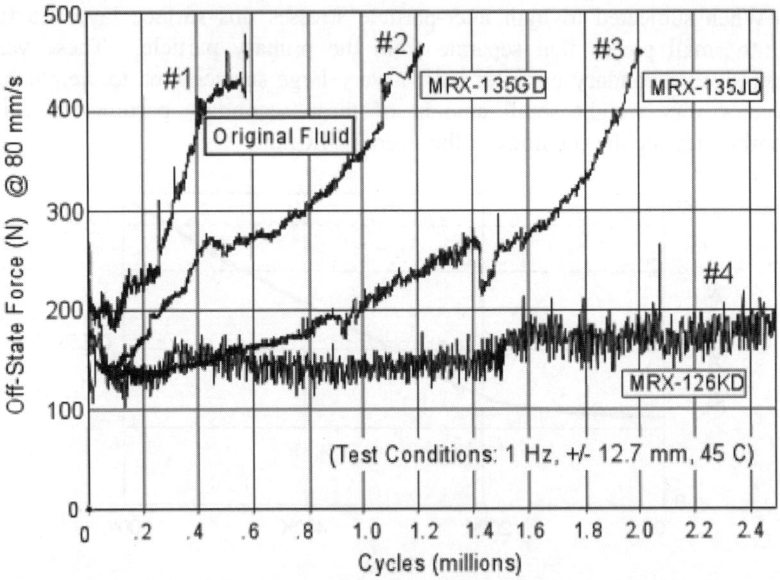

Figure 4 - Progressive development of IUT resistant MR fluid formulations.

4 MR Fluid Life

Depending on the conditions of the specific application, all MR fluids will eventually show some degree of deterioration. Such deterioration is usually manifested as a thickening of the fluid as described above although problems may occur as well. Silicone oil based fluids, for instance, are prone to cross-linking if exposed to high temperatures for extended periods or to ionizing radiation. The amount of deterioration generally depends on the shear rate, temperature and duration.

A measure that is useful in predicting the expected life of a MR fluid is the lifetime dissipated energy or LDE defined in Equation 1.

$$LDE = \frac{1}{V} \int_0^{life} P \cdot dt \tag{1}$$

where P is the instantaneous mechanical power being converted to heat in the MR device. Thus, LDE is simply the total mechanical energy dissipated per unit volume of MR fluid over the life of a device.

In the case of MR Fluid #4 in Figure 3 the damper was operated at an amplitude of +/-12.7 mm at 1 Hz with 1 amp applied to generate 1200 N for 2.5 million cycles. The RMS power dissipation was 68 watts and the damper contained 70 cm3 of fluid. Thus, over the course of approximately 29 days of continuous operation the fluid in this damper accumulated a LDE of $2.4\text{x}10^6$ J/cm^3. It is our experience that the best MR fluids today can sustain approximately 10^7 J/cm^3 before they have thickened to the point where they are no longer useful in a controllable MR fluid device.

5 Conclusion

The answer to the question "What makes a good MR fluid?" is "It depends". It depends on the type of device in which the MR fluid is used, the conditions to which the fluid is exposed and the duration of that exposure. MR fluids that are considered good in one application may fail miserably in another type of device. MR fluid development is of course a balancing act that is highly coupled with MR device design. In evaluating the quality of an MR fluid it is important to consider the actual conditions to which it will be exposed and not just the rheological behavior measured under normal laboratory conditions. MR fluid durability and life have been found to be more significant barriers to commercial success than yield strength or stability. Amenability of a particular MR fluid formulation to being scaled to volume production must also be considered. Challenges for future MR fluid development are fluids that operate in the high shear regime of 10^4 to 10^6 sec^{-1} and fluids able to sustain high LDE greater than 10^7 J/cm^3.

References

1. *Magnetic Fluid Clutch*, Technical News Bulletin, National Bureau of Standards, 32/4 (1948) 54-60.
2. W. Winslow, *Field Responsive Fluid Couplings*, US Patent No. 2,886,151, (1959).
3. *Motion Master Ride Management System*, Pub No. PB8008a, Lord Corp., Cary (1998).
4. J.D.Carlson, and M.J.Chrzan, *Magnetorheological Fluid Dampers*, US Patent No. 5,277,281 (1994).
5. J.D.Carlson, K.A.St.Clair, M.J.Chrzan and D.R.Prindle, *Controllable Vibration Apparatus*, US Patent No. 5,878,851 (1999).
6. J.D.Carlson, D.M.Catanzarite and K.A.St.Clair, *Commercial Magneto-Rheological Fluid Devices*, Proc. 5th Int. Conf. on ER Fluids, MR Fluids and Assoc. Tech., July 1995, W.A.Bullough, Ed., World Scientific, Singapore (1996) 20-28.
7. International Automotive Sector Group, http://www.qs-9000.org/ (1999).
8. *Quality Systems Requirements QS-9000*, AIAG-Automotive Industry Action Group, Southfield, MI (1998).

A Study on The Vibration Attenuation of a Driver Seat Using an MR Fluid Damper

YEONGROK LEE AND DOYOUNG JEON

Dept. of Mechanical Engineering, Sogang University, Sinsudong, Mapoku

Seoul, 121-742, Korea

A seat suspension system with a controlled MR(Magneto Rheological) fluid damper is introduced to improve the ride quality and prevent the health risk of a driver compared to conventional seats. The system locates between a seat cushion and base, and is composed of a spring, MR fluid damper and controller. The MR fluid damper designed in valve mode is capable of producing a wide range of damping force according to applied currents. In experiments, a person was sitting on the controlled seat excited by a hydraulic system. The skyhook control, continuous skyhook control and relative displacement control were applied and the continuous skyhook control improved the vibration suppression by 36.6%.

1. Introduction

A newspaper survey in Korea reveals that 96% of bus drivers suffer from job related diseases such as backbone pain, neck pain and stomach disease. They are mainly caused by high vibration during drive especially under the rough road condition.

The seat vibration should be reduced at the seat level in addition to vehicle suspension level which can not be soft enough for the driving stability of vehicle.[6] Currently the passive methods such as soft foam and air suspended seat are used and may work in low frequencies and may not work in the frequency range between 4 and 8 Hz where human body is very sensitive according to ISO.[1] Therefore, an advanced system is necessary to effectively reduce vibration in a wide range of frequency.

The seat damper system in fig. 1 has been developed and the system works without an air suspension.[2] The MR fluid damper is controlled in real time and changes its damping force accordingly. Several control algorithms including the skyhook control, continuous skyhook control and relative displacement control have been applied.[11]

In the following sections, the details of seat system and experimental results are shown.

2. A controlled seat and an MR fluid damper

Fig. 1 depicts the MR fluid damper used in experiments.[4, 5] It has a stroke of 60mm and its diameter is 45mm. For the fluids, MRF-132LD from the Lord corp. is used. To measure the damping force, one end of the damper was connected to a loadcell and the other end was attached to a CNC machine which could generate constant and various velocities. The position and velocity of piston and the loadcell measurement were stored in a PC through a data acquisition board. Fig. 2 shows the relation between the measured damping force and velocity according to applied currents.

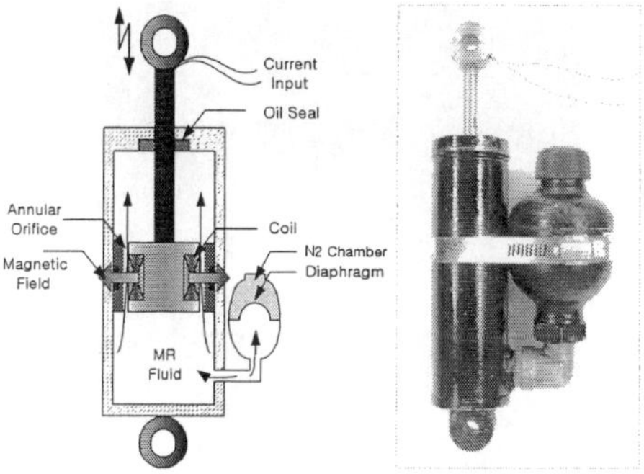

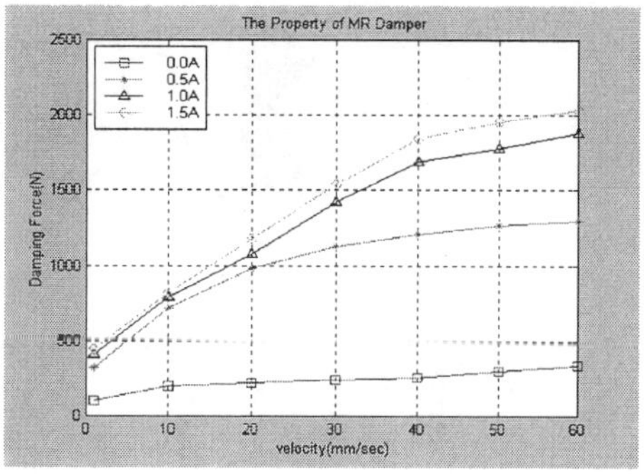

Fig. 1. The MR fluid damper in experiment Fig. 2. The relation between damping force and velocity

Fig. 3 shows a relation between the damping force and stroke at the velocity of 40mm/s. The slope in fig. 3 shows that the gas chamber plays as an air spring. From the least square method, the relation between the damping constant and applied voltage is approximated as

$$volt = 2.0\times10^{-4} c_d - 0.5093 \tag{1}$$

where c_d is the damping constant.

Fig. 4 shows the seat in experiment. A commercialized bus driver seat was modified to generate the up and down motion, i.e., the fixed mechanism between cushion and base was replaced by a flexible mechanism composed of a spring and MR fluid damper. The seat may have a vertical motion up to 60mm. As the cushion has a vertical motion, the mechanism works as a four bar linkage with a slider and the slider is constrained to have 60mm vertical motion. The spring has a nonlinearity such that the seat locates closely to the neutral position with some variation of driver's weight.

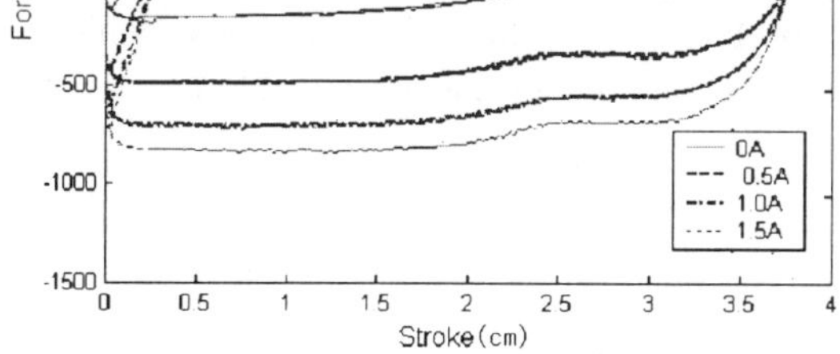

Fig. 3. The relation between stroke and force

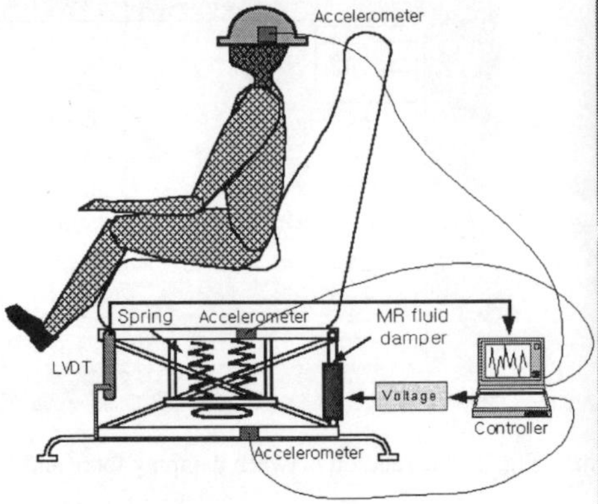

(a) Schematic plot

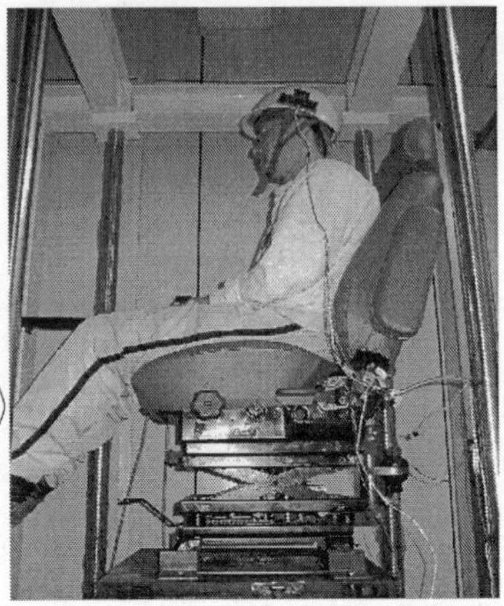

(b) Photo

Fig. 4. Experimental setup for seat

The accelerometer at the base measures the vibration input from bus chassis and its measurement was compared with that of the accelerometer attached to the driver's hat. The LVDT measures the position of cushion and base.

The Lab-PC+ of National Instrument Co. was used as a microprocessor. It had 2 D/As and 8 A/Ds with 2 MHz clock. The sampling time for control was 4msec.

3. Design of controller

The three control algorithms were implemented; sky-hook control, continuous sky-hook control and relative displacement control.

The sky-hook control has the maximum and minimum values according to the following condition; the damping constant c is

$$c = \begin{cases} c_{\max} & if\ \dot{z}_s(\dot{z}_s - \dot{z}_b) > 0 \\ c_{\min} & if\ \dot{z}_s(\dot{z}_s - \dot{z}_b) \le 0 \end{cases} \tag{2}$$

where $\dot{z}_s$ is a driver seat velocity and $\dot{z}_b$ is the velocity of base, i.e., the bus chassis.[12] The drawback of this algorithm is that the control output has discontinuity and its performance deteriorates as the actuator has a time delay.

To overcome the problem, the continuous sky-hook control was applied. It has continuous control outputs by introducing a new variable c_d ;

$$c_d = c_{sh} \frac{\dot{z}_s}{\dot{z}_s - \dot{z}_b} \tag{3}$$

where c_{sh} is the sky-hook coefficient determined by a designer, $\dot{z}_s$ is the velocity of the seat cushion and $\dot{z}_b$ is the velocity of the base. In addition to the algorithm in eq. (2), the control becomes;[13]

$$c = \begin{cases} c_{\max} & \text{if } c_{sh}\dfrac{\dot{z}_s}{\dot{z}_s - \dot{z}_b} > c_{\max} \\ c_{sh}\dfrac{\dot{z}_s}{\dot{z}_s - \dot{z}_b} & \text{if } c_{\min} \le c_{sh}\dfrac{\dot{z}_s}{\dot{z}_s - \dot{z}_b} \le c_{\max} \\ c_{\min} & \text{if } c_{sh}\dfrac{\dot{z}_s}{\dot{z}_s - \dot{z}_b} < c_{\min} \end{cases} \tag{4}$$

The relative displacement control is a heuristic algorithm to reduce the shock at both limits of vertical motion and is given as

$$c = \begin{cases} c_{\max} & \text{if } |d| > d_{th} \\ c_{\min} & \text{if } |d| < d_{th} \end{cases} \tag{5}$$

where $|d|$ is $|\, z_s - z_b \,|$ and the d_{th} is the threshold value.[3,9,10]

4. Experimental results

For experiments, a sweep sine vibration of 5mm amplitude and up to 5Hz was applied to the base of driver seat.[7] Fig. 5 shows the experimental result when a person was sitting on the seat when the sky-hook control was applied, the vertical acceleration of the driver's head reduced 10% at the peak compared to the case when a passive damper was used.

Fig. 6 shows the time domain response for random vibration of base. When a passive damper was installed, the amplitude of vibration may be amplified. With controls, however, the vibration clearly reduced for random inputs as shown in fig. 6 (b)

In Fig. 7, the RMS(Root Mean Square) of the acceleration of the driver's head was compared under the ISO 2631 which tells that human body is very sensitive in the frequency range of 4 and 8Hz.[1, 8] The continuous sky-hook reduces the vibration transmission during the frequency range.

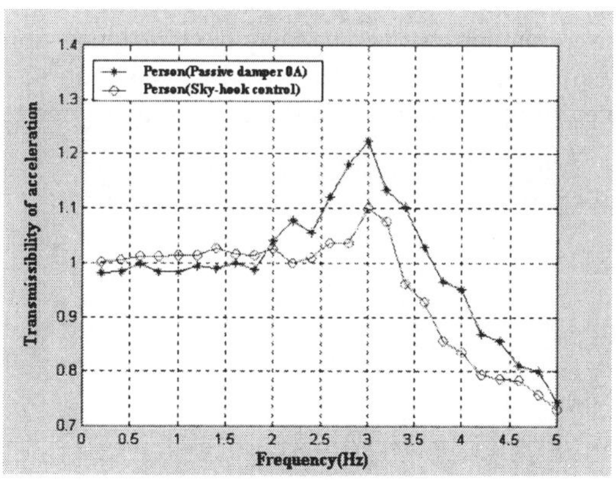

Fig. 5. Acceleration transmissibility in experiment

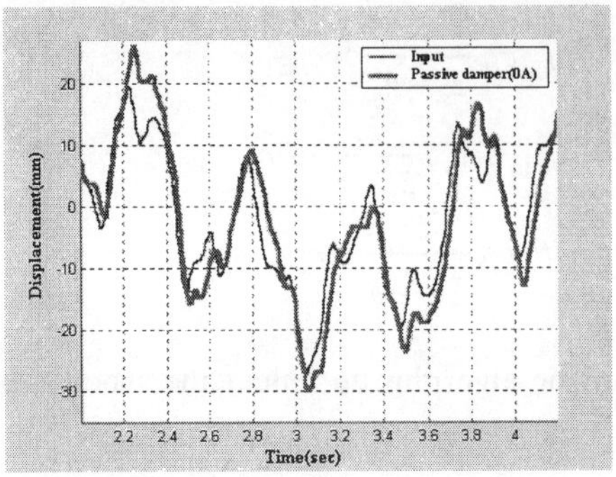

(a) Passive damper for random input

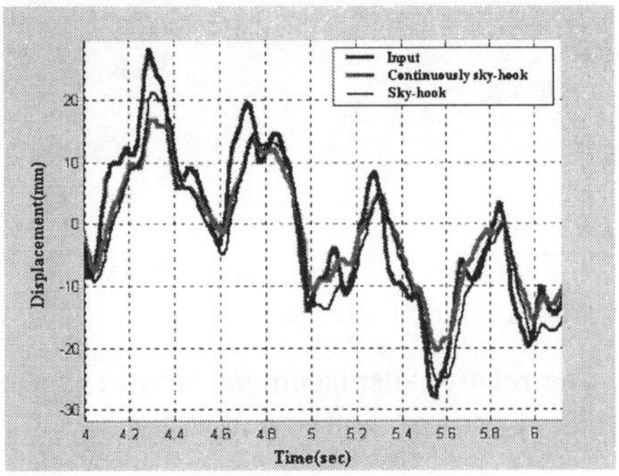

(b) Controlled seat for random input

Fig. 6. Displacement response in the time domain

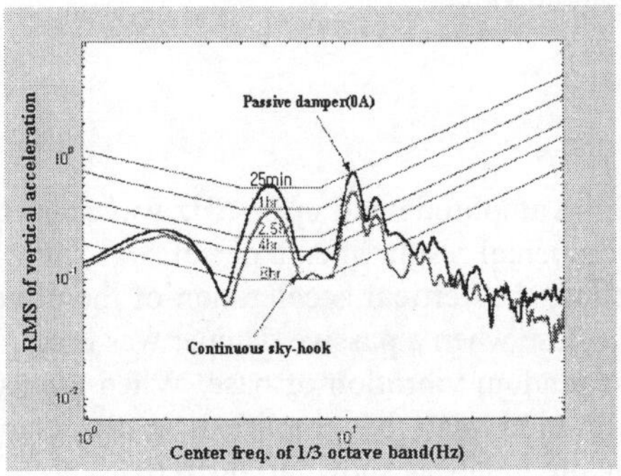

Fig. 7. RMS of acceleration of the driver's head

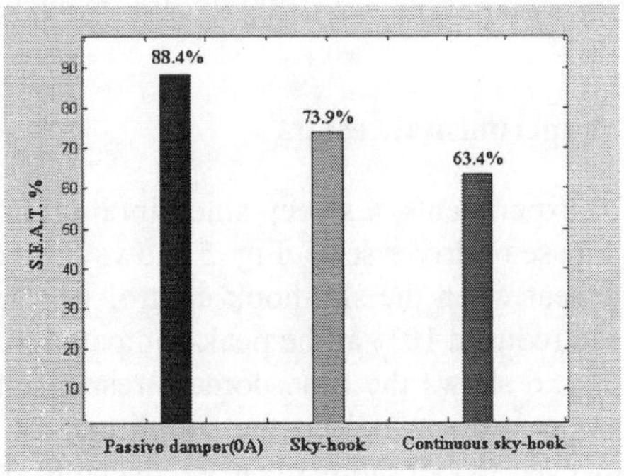

Fig. 8. Comparison of S.E.A.T values

Fig. 8 shows the comparison of the SEAT(Seat Effective Amplitude Transmissibility) For various controllers.[1] The continuous sky-hook shows the best result among them. Since the time delay of the MR fluid damper is about 20msec, the continuous control output performs better than the discontinuous control such as the on/off control.

When there is a bump, the passive system may have residual vibration as shown in Fig. 9 (a). The relative displacement control removes the overshoot motion as shown in Fig. 9 (b). In terms of acceleration, the continuous sky-hook shows good results as shown in Fig. 10.

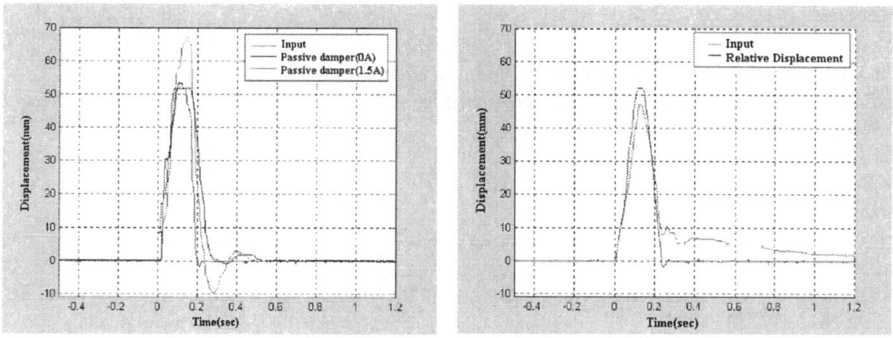

(a) Result of passive damper (b) Result of relative displacement control

Fig. 9. Displacement response for a bumped input

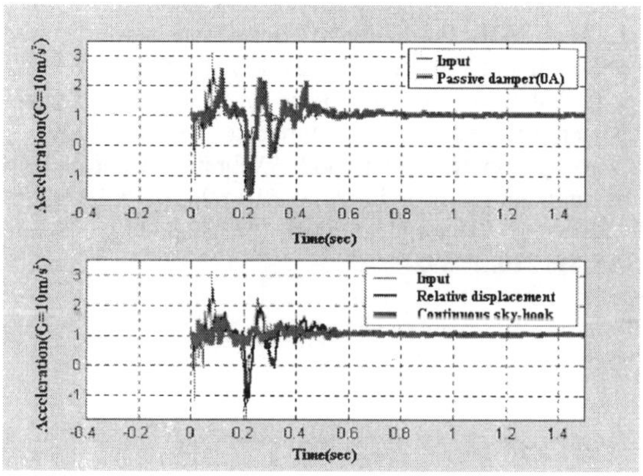

Figure. 10. Acceleration response for a bumped input

5. Conclusion

A controllable seat was developed to reduce the vibration transmitted from vehicle chassis to the driver. The MR fluid damper was controlled by several control algorithms such as the sky-hook control, continuous sky-hook control and relative displacement control. Among them, the continuous sky-hook shows the best result for random road condition and bump input. In terms of SEAT, the continuous sky-hook reduces vibration by 36.6%.

Acknowledgements

This study was supported by research grants from the Korea Science and Engineering Foundation (KOSEF) through the Applied Rheology Center (ARC), an official KOSEF-created engineering research center (ERC) at Korea University, Seoul, Korea.

Reference

1. M.J. Griffin, *Handbook of Human Vibration*, (Academic Press, London, 1990).
2. G.J. Stein , "A Driver's Seat with Suspension of Electropneumatic Type", Transactions of the ASME, p. 230-235, 1997.
3. M.J. Griffin and X. Wu, "A Semi-Active Control Policy to Reduce the Occurrence and Severity of End-stop Impacts in a suspension Seat with an Electrorhelogical Fluid Damper", Journal of Sound and Vibration, p. 781-793, 1997.
4. P.J. Wahlers, "Method for Controlling a Seat Suspension System", United States Patent No.5908456, 1999.
5. J.D. Carlson, M.J. Chrzan and N.C. Raleigh, "Magnetorheological Fluid Damper", United States Patent No.5277281, 1994.
6. J.Y. Wong, *Theory of Ground Vehicles*, (John-Wiley, 1993).
7. Y. Wan and J.M. Schimmels, "Optimal Seat Suspension Design Based on Minimum Simulated Subjective Response", Journal of Biomechanical Engineering, p. 409-416, 1997.
8. C.M. Harris, *Shock and vibration Handbook* (McGraw-Hill, New York, 1991).
9. M.J. Griffin and X. Wu, "The Influence of End-stop buffer characteristics on the Severity of Suspension Seat End-stop Impacts", Journal of Sound and Vibration, p. 989-996 1998.
10. M.J. Griffin and L. Wei, "The Prediction of Seat Transmissibility from Measures of Seat Impedance", Journal of Sound and Vibration, p. 121-137, 1998.
11. K. Kim and D. Jeon, "Vibration Suppression in an MR Fluid Damper Suspension System", Journal of Intelligent Material Systems and Structures, vol. 10, Oct, 1999.
12. D. Karnopp, M.J. Crosby and R.A. Harwood, "Vibration Control Using Semi-active Force Generators", Journal of Engineering for Industry, p. 619-626, May, 1974.
13. M. Nagai and T. Shioneri, "Theoretical Study on Semi-active Suspensions by Nonlinear Feedback Control", SAE 891210, 1989.

ELECTRORHEOLOGICAL DAMPER ANALYSIS USING AN EYRING CONSTITUTIVE RELATIONSHIP

LIONEL BITMAN, YOUNG-TAI CHOI, AND NORMAN M. WERELEY

Smart Structures Laboratory, Dept. of Aerospace Engineering, University of Maryland, College Park, MD 20742, U.S.A.

E-mail: wereley@eng.umd.edu

We validate and assess the applicability of an Eyring constitutive model. The Bingham plastic model has a zero shear rate discontinuity, which leads to inaccuracies in modeling and simulation, and is characterized by two rheological constants for a constant field: yield stress and postyield viscosity. The Eyring model has a smooth transition through the zero shear rate condition, and also has two rheological constants for a constant field. An ER damper having a damping level comparable to a shock absorber for a small-sized passenger car damper is manufactured, and its performance is predicted using the Eyring model. To accurately identify the rheological parameters of the Eyring model, a parameter identification method was used. The daming force versus piston displacement and velocity behaviors of the ER damper are experimentally measured with respect to applied electric field and excitation frequency. To validate the Eyring model, the experimental results are compared with predictions in the damping force versus piston displacement and velocity behaviors.

1 Introduction

ER (electrorheological) or MR (magnetorheological) fluids are known to produce yield shear stresses, when external field is applied to the fluid. These yield stresses can be adjusted by changing the intensity of applied field and its response is rapid. Advantages such as continuous control of yield stress and fast response have led to substantial applied research of ER and MR fluids to various systems such as dampers, valves, clutches and brakes [1-4]. In developing ER or MR applications, it is important to accurately model device characteristics, so as to save developing time and increase cost-effectiveness. So far, Bingham plastic model has been widely used in the evaluation and prediction of ER or MR devices. Even though Bingham plastic model is mathematically simple, it has some deficiencies particularly in the representation of preyield behavior. However, since the Bingham plastic model exhibits discontinuous characteristics in shear stress versus shear rate behavior at the zero shear rate condition. Thus, the Bingham plastic model cannot accurately account for the practical behavior of the progressive and continuous yield shear stress that is usually observed as the shear rate increases in the preyield region. Consequently, this paper explores an alternative constitutive model to improve modeling of ER or MR applications in a practical sense.

We validate and assess the applicability of an Eyring constitutive model. The Bingham plastic model has a zero shear rate discontinuity, which leads to inaccuracies in modeling and simulation, and is characterized by two rheological constants for a constant field: yield stress and postyield viscosity. The Eyring model has a smooth transition through the zero shear rate condition, and also has two rheological constants for a constant field. An ER damper having a damping level comparable to a shock absorber for a small-sized passenger car damper is manufactured, and its performance is predicted using the Eyring model. To accurately identify the rheological parameters of the Eyring model, a parameter identification method was used. The force versus piston displacement and velocity behavior of the ER damper is experimentally measured with respect to applied electric field and excitation frequency. To validate the Eyring model, the experimental results are compared with predictions in the damping force versus piston displacement and velocity behaviors.

2 Eyring model

When external field is applied to ER or MR fluids, typical behavior is shown in Figure 1. The Bingham plastic model exhibits a discontinuity in preyield (low shear rate) region. However, the Eyring model has a smooth transition behavior through the preyield region that is much closer to the practical situation [3,4]. The key difference between the alternative models is the low shear rate behavior – the high shear rate behavior is similar. Maximum shear stress at maximum shear rate is the same in both models.

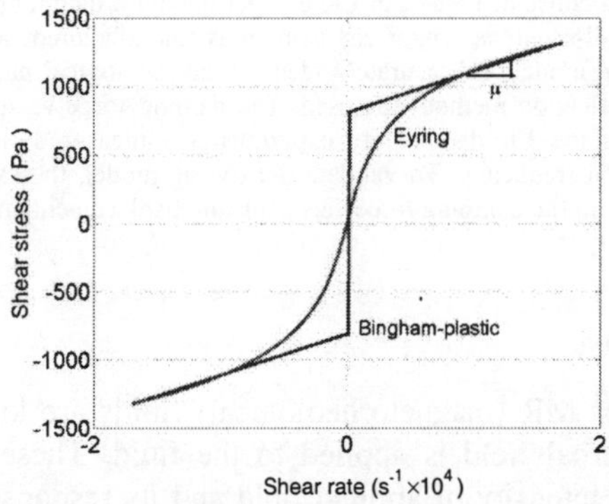

Figure 1. Comparison of Eyring to Bingham plastic model in terms of shear stress versus shear rate.

The constitutive equation of Eyring model for the relationship between the shear stress τ and the shear rate (du/dy) can be represented as follows:

$$\tau = \frac{1}{K}\operatorname{asinh}\left(\frac{1}{\xi}\frac{du}{dy}\right) \tag{1}$$

where u is the fluid velocity, y is the lateral coordinate, and K and ξ are the two rheological parameters of the Eyring model. The parameter K accounts for the magnitude of the shear stress and ξ for the steepness of the shear rate gradient through the preyield region. It is noted that the parameter ξ partially affects the magnitude of the shear stress. In addition, as shown in Eq. (1), the mathematical form of the Eyring model is simpler than the Bingham plastic model, and it does not contain the notions of yield shear stress τ_y and postyield viscosity μ.

3 Parameter identification

To establish the Eyring model structure, the parameters K and ξ are identified by using experimental data. In this study, for more accurate parameter identification, an optimization method is used on the basis of the measured damping force characteristics of an ER damper. It is noted that MR dampers exhibit qualitatively similar damping force characteristics upon application of magnetic field, so that the Eyring model would be equally applicable. For ER dampers, electric voltages are used to create external fields to the fluid domain. In this study, experimental data from an ER damper is used to identify the parameters K and ξ.

3.1 Manufacture of flow-mode ER dampers

A flow-mode ER damper with the damping force level comparable to that needed for a small-sized passenger car damper was designed and manufactured as shown in Figure 2. The ER damper used in this study is a monotube damper with single rod. The damper consists of hydraulic and pneumatic reservoirs separated by a floating piston. Inside the hydraulic reservoir, the piston rod is attached to a piston head holding two concentric tubular electrode lengths. During piston rod motion, ER fluids pass through the electrode gap between two concentric tubes in the piston head and can be energized by applied electric fields. The plastic spacer electrically insulates the two concentric tube electrodes to create electrical voltage potential in the gap. The important design parameters of the ER damper are as follows: electrode length is 43 mm and electrode gap is 0.65 mm. The ER fluid, made in the laboratory, consisted of corn starch and peanut oil. The composition ratio of the ER fluid is chosen by 40% corn starch and 60% peanut oil by weight.

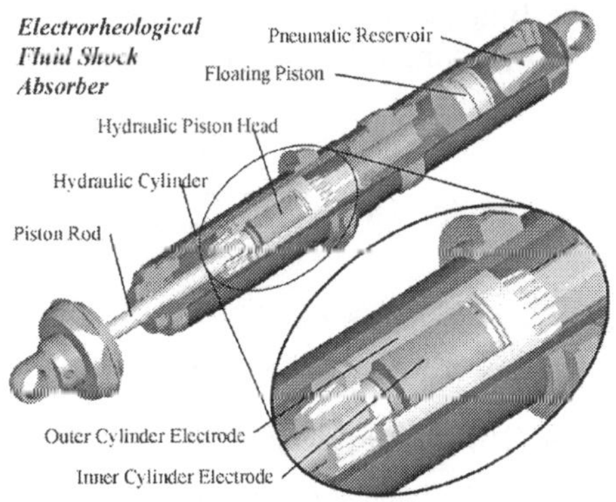

Figure 2. Schematic diagram of the ER damper.

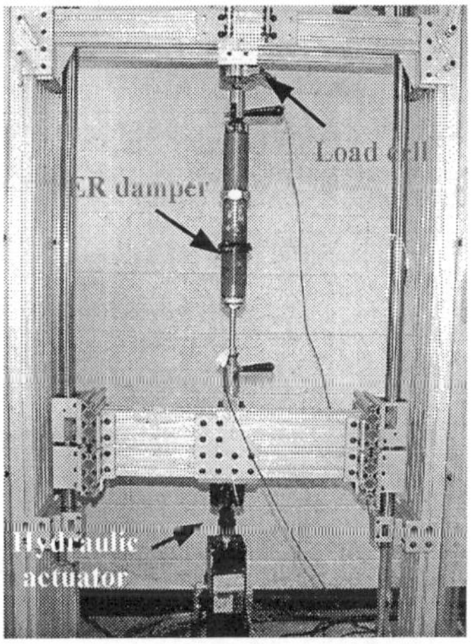

Figure 3. Experimental setup for the damping force test of the ER damper.

3.2 Damping force test of flow-mode ER dampers

Experimental setup for the damping force test of the ER damper is shown in Figure 3. The ER damper is placed between the load cell and hydraulic excitation actuator. The load cell is fixed and one end of the ER damper is connected to the hydraulic actuator. When the hydraulic actuator moves up and down by a command signal, the ER damper produces the damping force and it is measured by the load cell [3]. In this study, the excitation used is a sinusoidal displacement with constant amplitude of 0.5 inches. Four different excitation frequencies were tested: 0.25, 0.50, 0.75 and 1.0 Hz. The various electric field ranged from 0 to 7.0 kV/mm.

3.3 Dynamic equation of flow-mode ER dampers

In order to derive dynamic equation of the ER damper based on Eyring model, it is assumed that the electrode gap between two cylinders is approximated by parallel plates. It implies that the fluid velocity profile in the gap is symmetric [4]. In addition, the fluid flow in the gap is assumed to be quasi-steady, so that dynamic effects, such as inertia, are not considered in this study. It is noted that above assumptions are not valid at high excitation frequency.

The governing equation in electrode gap of the ER damper is given by

$$\frac{\partial \tau}{\partial y} = \frac{\Delta P}{L} \tag{2}$$

where ΔP is the pressure drop between both ends of the electrode and L is the electrode length. If substituting Eq.(1) into Eq.(2) and integrating twice with respect to y, then, the fluid velocity profile in the gap can be obtained on the basis of Eyring model as follows:

$$u(y) = \left(\frac{L\xi}{\Delta PK}\right)\left\{\cosh\left(\frac{\Delta P}{L}Ky\right) - \cosh\left(\frac{\Delta P}{2L}Kd\right)\right\} \tag{3}$$

where d is the electrode gap. Typical fluid velocity profile of Eyring model in the electrode gap is shown in Figure 4. As observed in this figure, Eyring model exhibits a smooth and continuous velocity profile, unlike Bingham plastic model where there is a large discontinuity called the "plug" region over which the velocity gradient is zero.

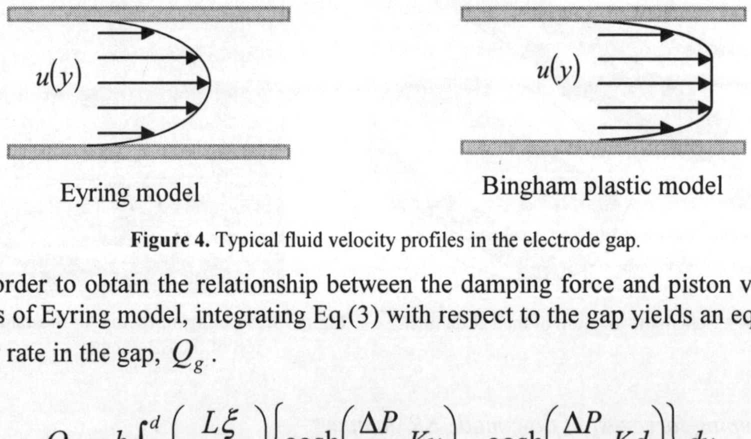

Figure 4. Typical fluid velocity profiles in the electrode gap.

In order to obtain the relationship between the damping force and piston velocity on the basis of Eyring model, integrating Eq.(3) with respect to the gap yields an equation for the flow rate in the gap, Q_g.

$$Q_g = b\int_0^d \left(\frac{L\xi}{\Delta PK}\right)\left\{\cosh\left(\frac{\Delta P}{L}Ky\right) - \cosh\left(\frac{\Delta P}{2L}Kd\right)\right\}dy \tag{4}$$

where b is electrode width. Using conservation of flow rate, $Q_g = A_p V_p$, and $\Delta P = -F/A_p$, relates damping force F to piston velocity V_p.

$$V_p = \frac{b\xi Ld}{FK}\left\{\cosh\left(\frac{FKd}{2LA_p}\right) - \left(\sinh\left(\frac{FKd}{2LA_p}\right)\right)\cdot\left(\frac{2LA_p}{FKd}\right)\right\} \tag{5}$$

Here, A_p is the effective cross-sectional piston area. The expression of the piston velocity V_p is obtained as a function of the damping force, F in Eq. (5). Consequently, in theoretical analysis, the damping force, F, is the independent variable. However, in the experiments, the independent variable is the piston velocity V_p. Given the damping force, F, geometry properties of the ER damper, and the parameters K and ξ, the piston velocity is easily calculated from Eq. (5). However, the parameters K and ξ are not directly found from the experimental data. Consequently, parameter identification of K and ξ is needed.

3.4 *Optimization method*

In this study, for more accurate identifications of the parameters K and ξ, an optimization method is used. On the other hand, the shear stresses of ER or MR fluids are known to be dependent on applied field and shear rate. For the ER damper, the damping force depends on applied field and excitation frequency. Therefore, the parameters K and ξ can be expressed as a function of applied field and excitation frequency.

Based on preliminary work, the parameters K and ξ are functions of electric field and frequency and behave according to the functions below:

$$K(E,f)=\frac{a_K}{\sqrt{1+\left(b_K\ (E-c_K)\right)^2}} \tag{6}$$

$$\xi(E,f)=\frac{a_\xi}{\sqrt{1+\left(b_\xi\ (E-1)\right)^2}} \tag{7}$$

where E is the applied electric field, f is the excitation frequency, and a_K, b_K, c_K, a_ξ, and b_ξ are coefficients of functions for the parameters K and ξ. It is noted that these coefficients are dependent on only excitation frequency and can be expressed as second degree polynomials of excitation frequency, f. The coefficients $a_K, b_K, c_K, a_\xi, b_\xi$ can be identified by minimizing the following cost function, J.

$$J(a_K,b_K,c_K,a_\xi,b_\xi)=\frac{\sum_f\sum_E (G_1+G_2+2G_3+3G_4)/7}{N_f * N_E} \tag{8}$$

with

$$G_1=\frac{1}{N_i}\sqrt{\sum_i\left(\frac{V_p(t_i)-V_{pm}(t_i)}{\overline{V}_{pm}}\right)^2},\quad G_2=\frac{1}{N_i}\sqrt{\sum_i\left(\frac{X_p(t_i)-X_{pm}(t_i)}{\overline{X}_{pm}}\right)^2}$$

$$G_3 = \left|1 - \frac{\overline{V}_p}{\overline{V}_{pm}}\right|, \qquad G_4 = \left|1 - \frac{\overline{X}_p}{\overline{X}_{pm}}\right|$$

where, V_{pm} and X_{pm} are the measured piston velocity and displacement, respectively. And $\overline{V}_{pm}$ and $\overline{X}_{pm}$ are the maximums of the measured piston velocity and displacement, respectively. In addition, N_f is the number of excitation frequencies tested, N_E is the number of electric fields tested, and N_i is the number of sampling points over one period of time. In this study, N_f=4, N_E=28, and N_i is nominally 4000.

The minimization of the cost function given by Eq.(8) will give the polynomial expressions for $a_K, b_K, c_K, a_\xi, b_\xi$ coefficients as functions of excitation frequency. These optimal $a_K, b_K, c_K, a_\xi, b_\xi$ coefficients obtained in this study are shown in Figure 5. As observed, the coefficients are strong functions of excitation frequency. Given an excitation frequency, we get the value of these coefficients. Then, the parameters K and ξ are obtained by substituting the coefficients into Eqs. (6) and (7) and shown in Figure 6.

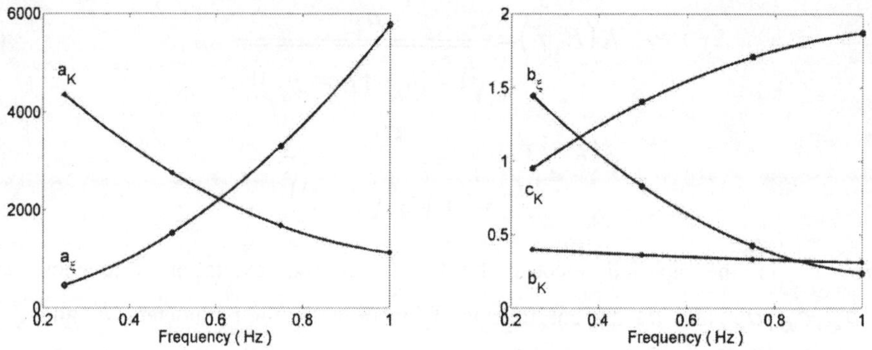

Figure 5. The $a_K, b_K, c_K, a_\xi, b_\xi$ coefficients identified by the optimization method.

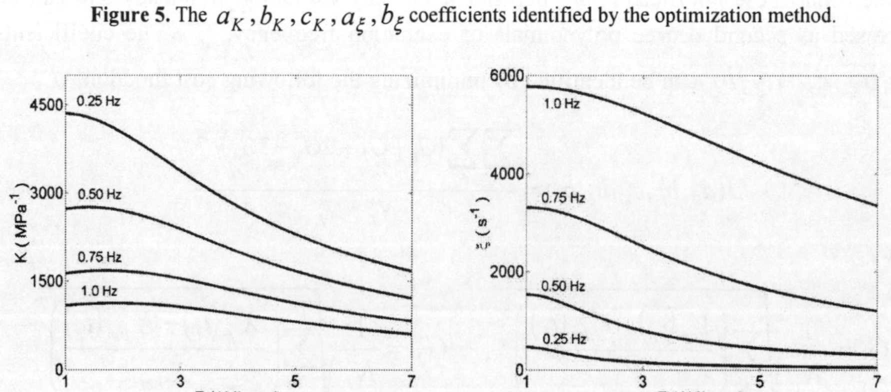

Figure 6. The parameters K and ξ obtained by the optimization method.

4 Experimental validation

In order to prove the validity of the Eyring model, the experimental data are compared with theoretical predictions. Figure 7 presents experimental and theoretical results for the damping force characteristics of the ER damper. In this case, the excitation frequency is 0.75 Hz and the applied electric fields shown are 1.0, 3.5, 5.0 and 6.5 kV/mm. In addition, the solid line represents the experimental data and the dashed line represents theoretical data. As clearly observed, the theoretical data obtained using Eyring model are in a good accordance with the experimental data.

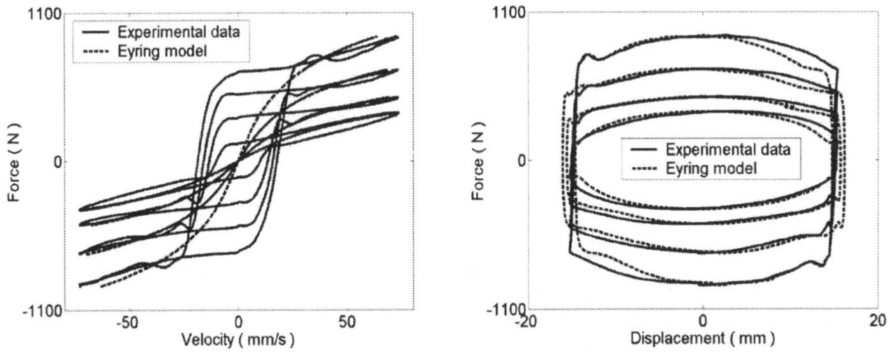

Figure 7. Experimental and theoretical results for the damping force characteristics of the ER damper.

5 Conclusion

The validation and applicability of the Eyring model, as an alternative constitutive model to Bingham plastic model, was evaluated using experimental and theoretical results comparing the performance of an ER damper. We demonstrated that the Eyring model provided a good correlation between the experimental and theoretical results.

References

1. Williams, S, Rigby, S.G., Sproston, J.L., and Stanway, R., Electrorheological fluids applied to as automotive engine mount, *J. Non-Newtonian Fluid Mechanics*, **47** (1993) pp.221-238.
2. Han, S.S., Choi, S.B., Cheong, C.C., Position control X-Y table mechanism using electro-rheological clutches, *Mechanism and Machinery Theory*, **35** (2000) pp.1563-1577.
3. Kamath, G., Hurt, M. and Wereley, N.M., Analysis and testing of Bingham plastic behavior in semi-active electrorheological fluid dampers, *Smart Materials and Structures*, **5** (1996) pp.576-590.
4. Wereley, N.M. and Pang, L., Nondimensional analysis of semi-active electrorheological and magnetorheological dampers using approximate parallel plate models, *Smart Materials and Structures*, **7** (1998) pp.732-743.

INVESTIGATION OF DYNAMIC PROPERTIES AND CONTROL METHOD INFLUENCES ON MR FLUID DAMPERS PERFORMANCE

ANDRZEJ MILECKI, ADAM MYSZKOWSKI, DARIUSZ SĘDZIAK

Poznan University of Technology,Insitute of Mechanical Technology
ul. Piotrowo 3, 61-138 Poznan, Poland
e-mail: Andrzej.Milecki@put.poznan.pl

A semi-active, controllable fluid dampers and their control devices are described and studied in this paper. Such dampers can be used to eliminate oscillations in different servo drive robots and machine tools systems, where their good dynamic parameters are required. In this paper a different control circuit and its influence on MR damper properties are investigated. The experimental results of coil current transients response during switching on and off with different circuits and elements are shown. Construction of the above mentioned damper is described and finally experimental results of control circuits influence on dampers dynamic performance are presented.

1 Introduction

In the last few years, the benefits of electronically controlled servo drives have been amply demonstrated. The new generation of electronic control systems enables their easy connection with host controllers like Numerical Controllers, Programmable Logic Controllers or Robot Controllers and their application in machine tools and robots. A main requirement of such applications is operational safety, flexibility, high dynamic and high accuracy [1]. To get the high dynamics required of the system the drive should be sufficiently fast and adequately damped. Usually in such applications oscillation can lead to tool breakage or sub-standard surface finish. They also reduce the positioning accuracy and can lead to long unstable cycles at low velocities ("stick-slip" effect) or during positioning processes. The main reasons for oscillation formation are the high dynamic forces and the high gain coefficient on the one hand, and stiffness and many linear variables like static friction or backlash between moving elements on the other.

If, in the application, oscillations cannot be tolerated, the value of gain coefficient should be decreased. The oscillations can be reduced using a passive damper. This element generates an opposite velocity direction forces, which depend only on kind of damper and velocity value. Unfortunately there are no possibilities to change force value by using any control signal. The usage of a passive damper results in increasing of losses and static friction. It is especially disadvantageous in high dynamic systems. The possibility of usage an electrically controlled damper can solve mentioned above problems. Its damping force can be controlled in dependence of velocity, position, position error or acceleration. It would allow for the matching of machine dynamics to the variables of a treatment process and it would increase operational safety. Good examples of electrically controlled dampers are semi-active magneto-rheological fluid dampers that could realize such a task [2, 3]. Their application can increase the overall system-damping ratio, correct the stability conditions and reduce or even eliminate oscillations. The most important is the fact, that a damper should be sufficiently fast and should generate forces adequate to requirements. The aim of this paper is to investigate the influence of a magneto-rheological damper controlling circuits elements on dampers dynamic and static properties.

2 Theoretical model of MRF damper dynamics

The properties of magneto-rheological fluids allow their use as dampers controlled by an electric current. The damping force can act only in a direction opposite to the direction of the velocity and can only absorb mechanical energy. The model consists of Coulomb friction placed in parallel with a viscous damper and a damper controlled by magnetic field. The force generated by the device is given by following equation [4]

$$F_d(s) = \left\{ F_{st} + \frac{b_H}{T_m s + 1} H(s) \right\} \cdot \text{sign}[s \cdot y(s)] + b_v [s \cdot y(s)] \tag{1}$$

where: F_{sf} – static friction force, b_v – viscous damping coefficient, b_{H} – gain coefficient of the force generated by magnetic field, H – magnetic field strength, T_m – time constant.

The time constant T_m describes the dynamics of the viscosity changes of magneto-rheological fluid. For simple analysis the change of a current and the corresponding change of magnetic field as a function of voltage on the coil can be described by following equation

$$H(s) = \frac{k_H}{T_e s + 1} U(s) \tag{2}$$

where: $k_H = z \cdot R$ – coefficient, U – voltage, $T_e = L/R$ – electrical time constant.

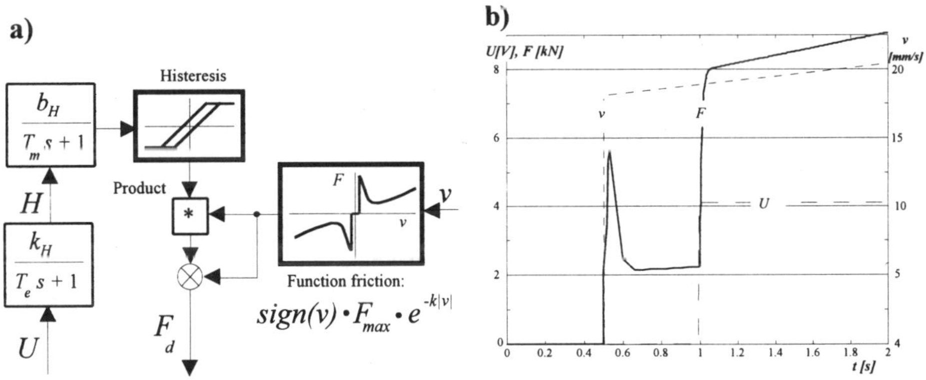

Fig. 1. Simple dynamic model of a magneto-rheological fluid damper (a), and result of simulation in Matlab-Simulink (b)

The magnetic circuit is characterized by residual flux density and force hysteresis and they should also be added into the damper model. The velocity of the piston has also an important influence on the dynamic properties of the magneto-rheological fluid damper. If the velocity is high, the time during which the particles are in magnetic field is short. This results in the saturation of the damping force at high velocities. The model of a damper prepared in Matlab-Simulink program is shown on figure 1a. The force generated by the model is formed by the sum of the Coulomb friction and the current-varying force. The switch and Matlab function are controlled by absolute value of velocity. They guarantee that the force is equal to zero at zero velocity. The current varying force depends on the velocity value and its sign. The model has two inputs: control voltage and piston velocity. In figure 1b examples of simulation results are shown.

3 MR damper usage in servo drive oscillation damping

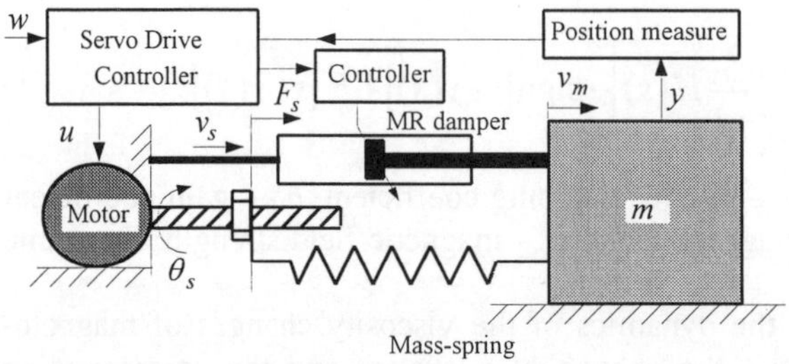

Fig. 2. Model of the servo drive with mass-spring system and a damper

In figure 2, as an example, a simple linear model of a servo drive unit is shown. It is assumed that a high-gain, servo-controlled motor is used to produce the force and displacement. This implies the motor friction to be hardly comparable to the rest of the system. The motor is coupled to a mass, sliding on a frictional surface. The spring represents the stiffness of the transmission and is placed in parallel to the semi-active damper.

The described above application indicates the necessity to use special control rules for a magneto-rheological damper to reduce the oscillations [1, 4]. The simple method is to compare velocity on the input v_s and velocity of the output v_m. The damper should be switched off only when v_s is approximately equal to v_m. In another solution if $(v_m - v_s) \cdot v_m > 0$, the damper is switched on, otherwise switched off. This control method needs to measure velocities, which in practice is not simple. There is another possibility of using only the position error signal to control. A voltage signal to the damper is switched on when the position error is lower then the assumed value (for example 5 mm). The main advantage of this method is its simplicity. Additionally, the controller can generate a forcing signal (at the moment of switching on), which reduces the electrical time constant. A further possibility is to build a controller, which generates an output signal proportional to the position error. The resulting force is more equivalent to the position error. If this error is small, the regulator gives a voltage output signal equal to zero and the damping force is also equal to zero. The damper does not work and drive does not lose its power.

4 MR damper control devices and its investigations

As described above the MR damper used in servo drive unit works in two states beeing switched on and off. Only in poor applications there is a need to change the voltage value given to the coils. It is very important that the obtained damper reaction time was as short as possible. In the simplest case the ordinary mechanical, operating contact (relay contact) can be used. Howsoever its switching on and off time is far too long for high dynamic applications. The better solution is to use Darlington transistor as shown in fig. 3a (for example TIP 142). Nowadays there are different PWM solenoid drivers on a market (for example DRV 101). They enable to build an easy circuit, which can be operated conveniently (fig. 3b). Adjustable PWM allows to a fine control of the power delivered to the coil. The circuit can provide a strong initial current to the load by switching to the constant DC source during an adjustable time. The delay time can be set from 15 μs to 100 ms. This solution enables the abbreviation of the coil time constant. In both circuits it is necessary to use the diode, which protects the switching elements against high switching on and off overvoltages.

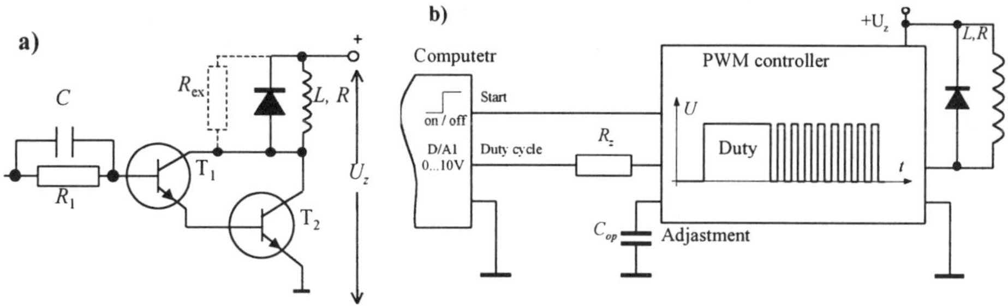

Fig. 3. The control circuits: a) Darlington transistor, b) PWM integrated controller

For the reaction time abbreviation task it is very interesting how the switching elements influence on coil current dynamic properties. During the investigation the ordinary contact switch, Darlington transistor and PWM controller to switch on/off a linear damper with a coil L = 0,21 H and R = 10 Ohm were used.

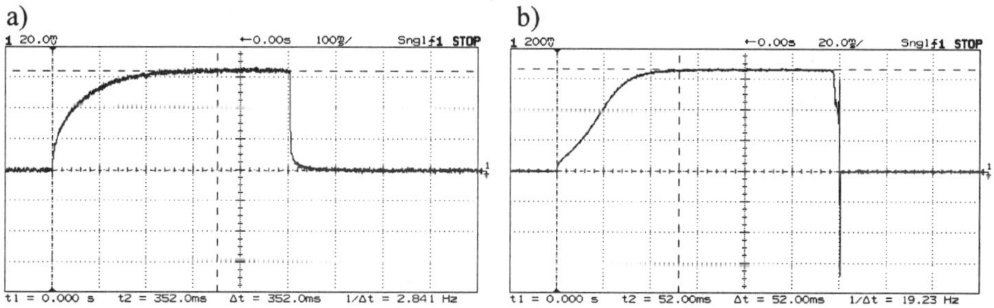

Fig. 4. The coil current transient: a) with a diode (1 V), b) without a diode (10 V)

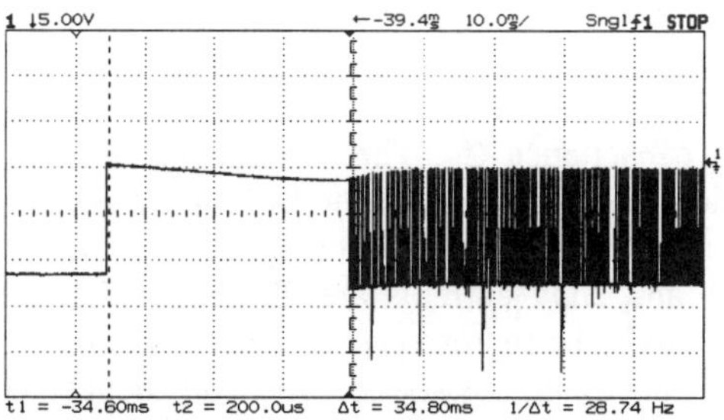

Fig. 5. The voltage with PWM controller and 10 V supply voltage

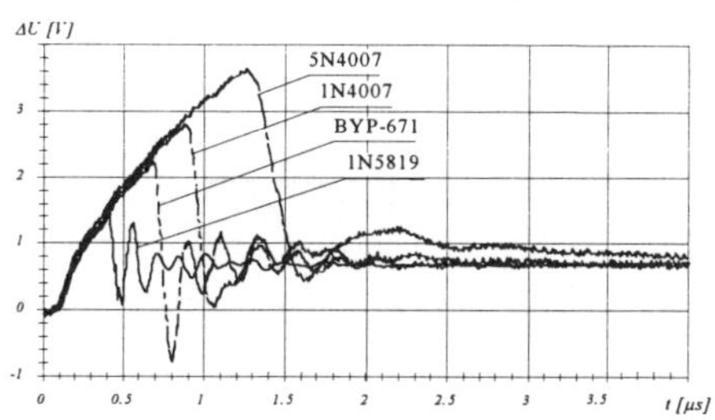

Fig 6. Diode type influence on current increasing after switching on

On a fig. 4 a coil changes by using a contact after switching on and of are shown. The increasing time when applying 1 V supply voltage is 250 ms, but by using 10 V supply it is much shorter (25 ms). The usage of the diode eliminates the overvoltage. The circuit is strongly non-linear. The investigations have shown that the decreasing time varies from 50 to 200 ms and it depends of diode type.

At the beginning of operation with PWM controller the supply voltage is temporarly switched to the coil (fig. 5). After the delay the circuit goes to normal PWM operation. Fig. 6 shows the diode type influence on current reaction time. The best results can be obtained by using fast Schotky diode. Apparently oscillations evidence the presence of the capacity in the circuit. At fig 7 the coil current transients during switching on with that circuit are shown. In that case the current increasing time is from 5,5 to 7,5 ms.

a) b)

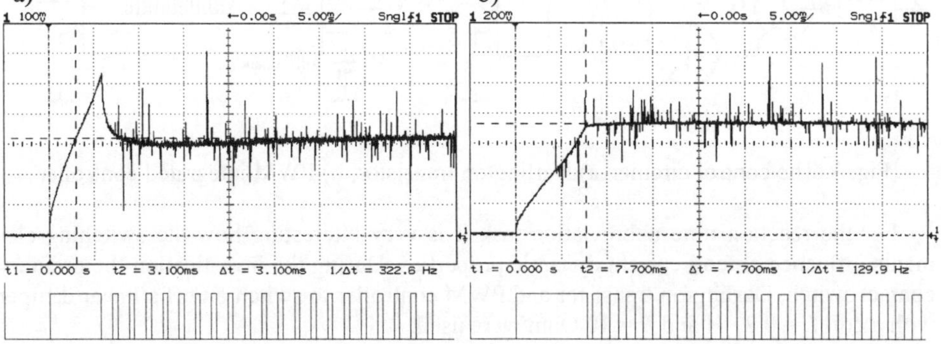

Fig. 7. Coil current transients with PWM controller with supply voltage: a) 5 V, b) 10 V

5 Theoretical model of MR damper electrical model circuit in PWM mode

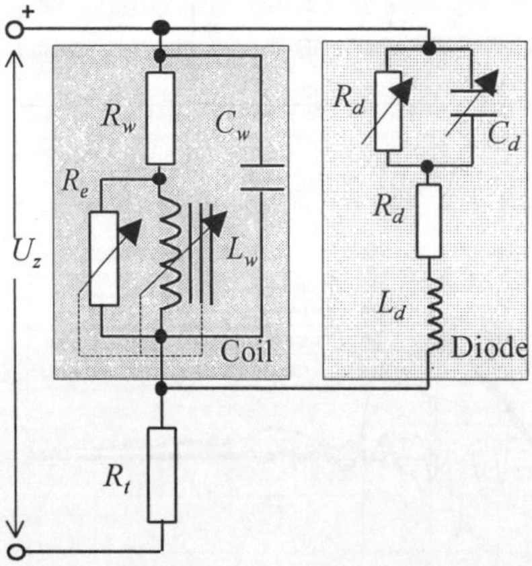

Fig. 8. Equivalent schematic diagram

During switching on/off and in PWM mode the currents contain different variable components. Their description should contain eddy currents and losses in conducts. The transient currents in the damper windings are not known before the field calculation. Therefore the equations describing electric circuit must be coupled to the field describing equations and feed system [5]. In general the coil electric circuit may contain winding resistance R_w linked to the flux Ψ and coil capacitance C_w. The experimental investigations have shown that the MR fluid has similar interacting as an electromagnet core (fig. 6). This and eddy currents are the reason that the flux has to be represented by parallel connected inductance and resistance. The coil inductance L_w and edgy current resistance R_e depend of transient magnetic induction and current frequency.

The transistor model is simplified and represented only by resistance R_t (almost equal to zero or to infinity). The diode is represented by non-linear resistance computed on the basis of current-voltage curve u_d versus i_d, non-linear capacitance C_d parallel to resistance R_d, series resistance R_{ds} and inductance L_d representing diode leads. The whole scheme diagram is shown on fig. 8. It is so complicated that for simulation purposes only finite element implementation must be used [5].

6 Experimental MR damper investigations

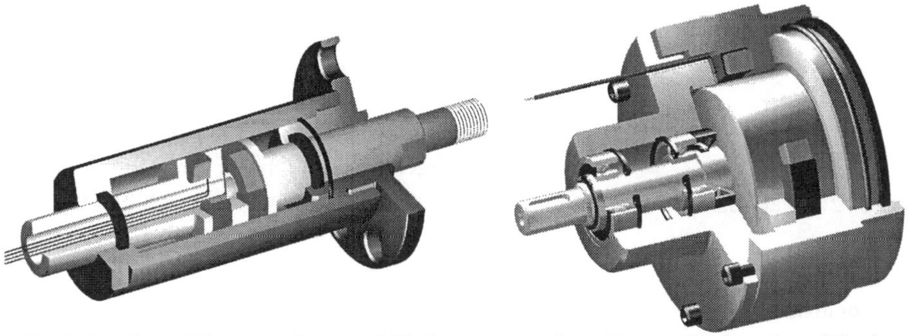

Fig. 9. Design of linear and rotary MR dampers made at Poznan University of Technology

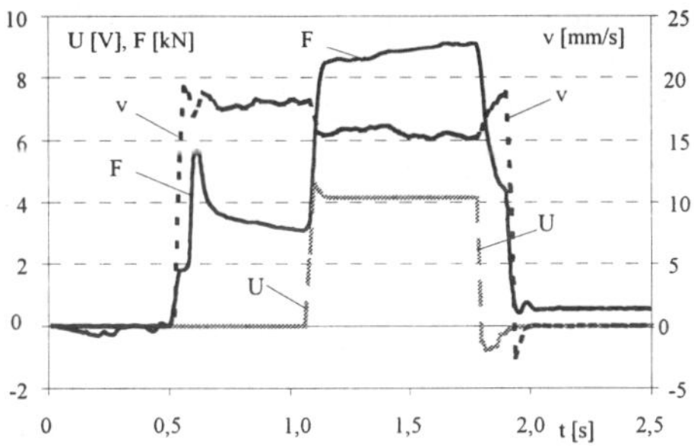

Fig. 10. Measured velocity and force after voltage switching on and off

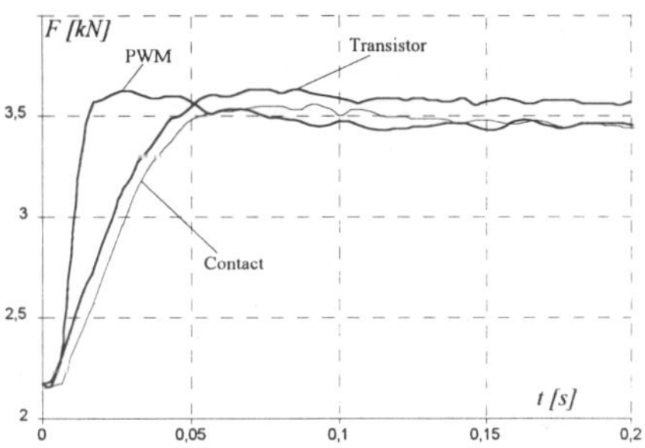

Fig. 11. Linear damper step responses

At Poznan University of Technology different linear and rotary MR dampers (fig. 8) were designed, built and investigated on a research rig [1, 4]. The electrohydraulic servo drive unit was used to generate the moving force. The linear damper was attached to a 100 kN force transducer. The displacement of the damper was measured using a linear variable different transducer (LVDT) connected to the actuator. All sensors were connected to a computer, which was used to collect all the data.

During investigations the electrohydraulic servo drive piston was moved with a constant velocity on a way of 50 mm. After overstepping equal to 12 mm the damper was switched on. It was switched off after overstepping of 24 mm. The investigations have shown that the damper reaction time by supply voltage equal to 1,5 V was about 180 ms and by supply voltage 8,8 V it was 30 ms.

Figure 10 shows the investigation results of linear MR damper dynamic properties. The voltage switching on results as damping force increasing and decreasing the piston velocity. The recorded transient is well related to the curves obtained from simulation (fig 1).

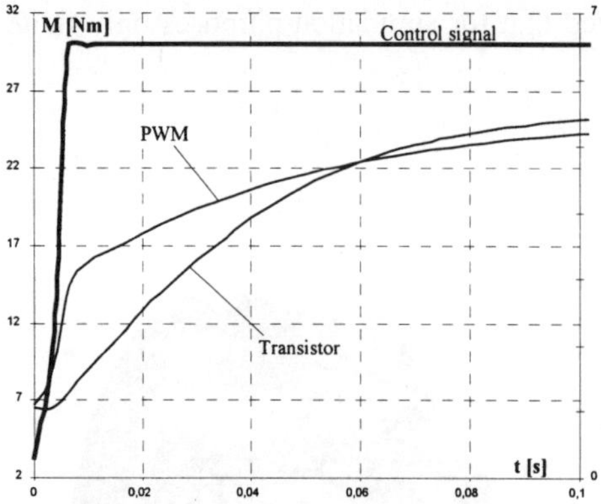

Fig. 12. Rotary damper step response

On fig. 11 damper step responses with different control circuit elements are shown. The fastest reaction time is obtained by using PWM controller and is equal to 15 ms. The reaction time depends also on supply voltage value. It shortens with increasing of the supply voltage value.

On fig. 12 rotary damper step responses are shown. The investigations revealed that the rotary damper reaction time is comparably with the linear on by low velocity (12 rpm). By higher velocity it is considerably longer than the linear one and its value rather poorly depends on revolution velocity.

7 Conclusion

Magnetorheological fluid dampers have provided technology that has enabled effective application in different servo drive systems used in modern machines and robots. Their main destination is the reduction of oscillation (without decreasing the gain coefficient), control of velocity and positioning accuracy improvement. Because of high dynamic parameters of the servo drive, the damper should be equivalently fast and should generate the necessary force or moment. The damper reaction time depends on switching circuits and the type of diode used. Especially good results can be obtained by using a special PWM controller. The investigations have also shown that the MR damper coil circuit is very complicated and can be analyzed only using the finite element methods. The main disadvantage of ME damper usage is the cost and complication of the servo drive and the control systems. The linear and rotary dampers, designed and constructed in Poznan University of Technology have good parameters and can be employed in many applications. In future investigation the authors will work on dampers optimal circuit design. They will improve the control methods of MR dampers used for oscillation and vibration damping.

Bibliography

[1] Milecki A., Analysis of magneto-rheological fluid damper application in electrohydraulic servo drive, Proceedings of Conference „Engineering Mechanics '98", 11.05-14.05.98, Srvatka, Czech Republic, 1998.

[2] Carlson J.D., Low-cost MR fluid sponge devices, Proceedings of the 7th International Conference on ER Fluids and MR Suspensions, published by World Scientific Publishing Co. Pte. Ltd. 2000.

[3] Dyke S.J., Spencer B.F. Jr., Sain M.K., Carlson J.D.: Seismic response reduction using magnetorheological dampers, Proceedings of the IFAC World Congress, San Francisco, California, June 30 – July 5, 1996.

[4] Milecki A., Investigation and control of magneto-rheological fluid dampers, Machine Tools & Manufacture 41, 2001.

[5] Szelag W., Milecki A., Electromagnetic field analysis in MR damper, Conference: Computer Application in Electrotechnics, Poznan, 1999.

DESIGN OF THE ELECTRORHEOLOGICAL SHOCK ABSORBER*

ZHANG SHAOHUA LI MEIYAN
School of Vehicle & Traffic Engineering, Beijing Institute of Technology,

Beijing 100081, P.R.China

Abstract The apparent viscosity of the electrorheological(ER) fluids can be changed with the variation of applied electric field strength. The damping force offered by ER shock absorbers can then be controlled by using the characteristics. It is found that, besides the performance of the ER fluids, the construction design of the shock absorber also strongly affects the characteristics of ER shock absorbers. In this paper, this phenomenon is discussed in detail with different construction styles of ER shock absorbers and mechanical analysis is done with hydro-mechanical and rheological theories.
Key words Electrorheology; shock absorber; construction design; apparent viscosity

1 Introduction

The shock absorbers with variable damping force are mostly liquid ones with double sleeves so far and can be divided into two different styles according to the principle of damping force variation. The system in which damping force adjustment is concentrated on piston valve is called "intercontrolled systems"; In another system, the adjustments of the damping force(include the press damp generated by the bottom valve) are executed between piston and reservoirs, and this kind of system is called "external control system"[1]. From the view of construction, ER shock absorbers are of new type of variable damping system which belongs to the second control fashion discussed above. The ER shock absorber uses ER fluids as its working media and provides variable damp through changing the shearing stress and apparent viscosity of the ER fluids under different applied electric field. Research on ER shock absorber includes construction design, performance of ER fluids and control technique of the damping force. In allusion to the distorted curves of damping force *vs.* displacement obtained by performance test, influence of construction design on the performance of ER shock absorber is analyzed in the following parts.

2 Construction and working principle of ER shock absorber

The ER shock absorber used in this test consists of piston and three concentric cylinders(inner cylinder 1, middle cylinder 2 and outer cylinder 3). The piston is laid in the inner cylinder which is separated into two cavities, ER fluids flow in the loop comprises of these two cavities and the round gap composed of inner cylinder and middle cylinder which are the two electrodes of the electric field. The shearing stress and apparent viscosity of the ER fluids can be adjusted by changing electric field strength and thus the working damp of the shock absorber can be regulated.

* The project supported by the National Natural Science Foundation(K59872005)

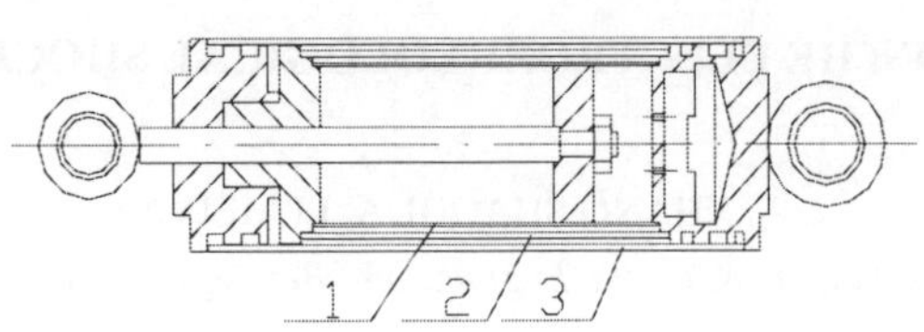

Fig.1 Sketch diagram of ER shock absorber

3 Performance of the ER fluids used in shock absorber

The ER fluids used in this test is a suspension composed of dispersed particles and common lubricating oil. This test was done with electrorheo-meter CV20 produced by HAAKE Corp., and the intention was to ascertain the diversification degree of apparent viscosity *vs*. applied electric field strength and temperature. Fig.2 and Fig.3 give their specific curves.

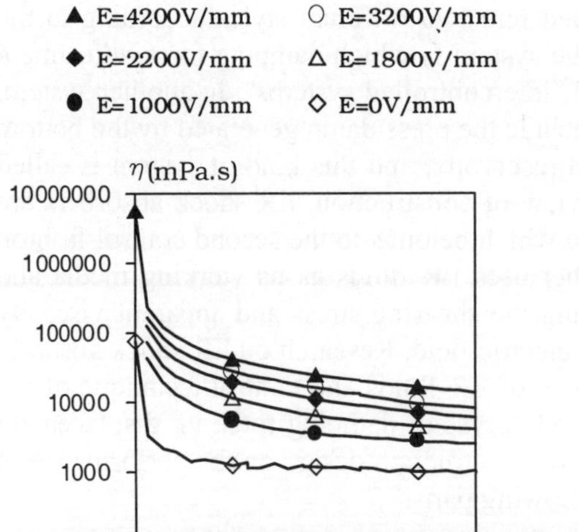

Fig.2 Curves of apparent viscosity vs. shear rate under different electric field strength

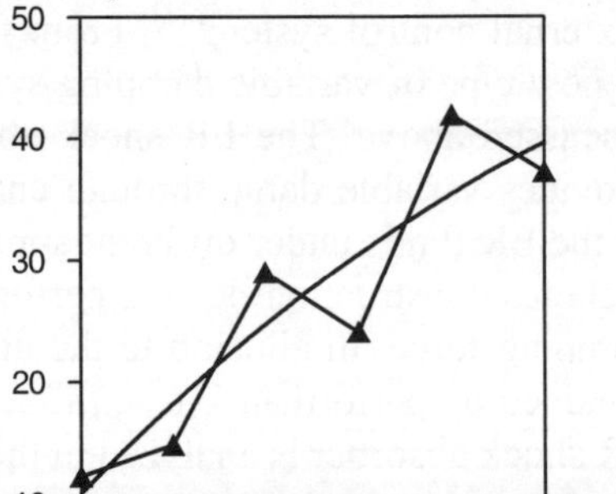

Fig.3 Curve of viscosity vs. temperature

It can be seen from Fig.2 and Fig.3 that the apparent viscosity of the ER fluids have an obvious change with the variation of the electric field strength, and the ratio of η/η_0 increases with the increase of temperature, where η and η_0 respectively represent the apparent viscosity of ER fluids with and without electric field. This feature is very important for ER fluids to compensate the reduction of viscosity caused by temperature rise.

4 Description of the test system and procedure

The test of ER shock absorber discussed in this paper was accomplished with EVH Road Simulator. The main instruments used include (1)EVH-80-250-15 electro-hydro vibration simulator, (2)BLR-1 pull-press sensor, (3)YD-15 resistance dynamic strain gauge, (4)PC Data acquisition & processing system and (5)0~10000V DC high voltage power supply. The sketch diagram of this test system is given in Fig.4

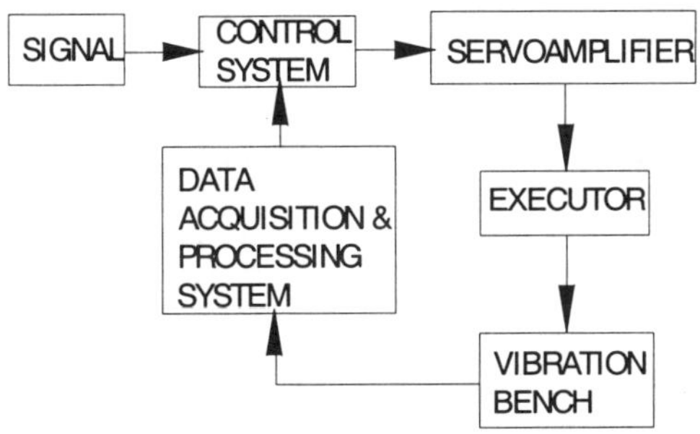

Fig.4 Sketch diagram of test system

The simulative frequency provided by the EVH Road Simulator is 0.05Hz,0.2Hz,0.5Hz,1Hz,2Hz, respectively, vibration amplitude adopted during the test procedure is 10mm,15mm,25mm,50mm in turn and the electric field strength is 0 and 2100V/mm.

5 Analysis of the test result

It can be seen clearly from Fig.5(a,b,c,d) that under the case of different frequency and amplitude, the Force-Displacement curve of the ER shock absorber show an "aberration" of empty procedure during its restore procedure. This phenomenon shows that when the shock absorber is compressed and the ER fluids flow through the uniform electric field between middle and inner cylinders, the cavity with piston rod is not filled with ER fluids sufficiently because of the increase of damp and the defect in construction design. In this shock absorber, the flow Q through the electric field can be expressed as the equation of

$$Q = \frac{\pi d}{\eta_p}\left(\frac{\Delta P \delta^3}{12l} - \frac{\tau_0(E)\delta^2}{2}\right)$$

where d—diameter of the inner cylinder□m

η_p—apparent viscosity of the ER fluids□Pa.s

ΔP —pressure differential, Pa

δ—gap between the inner cylinder and the middle cylinder□m

l—length of the inner cylinder,m

τ_0 —yield stress of the ER fluids,Pa

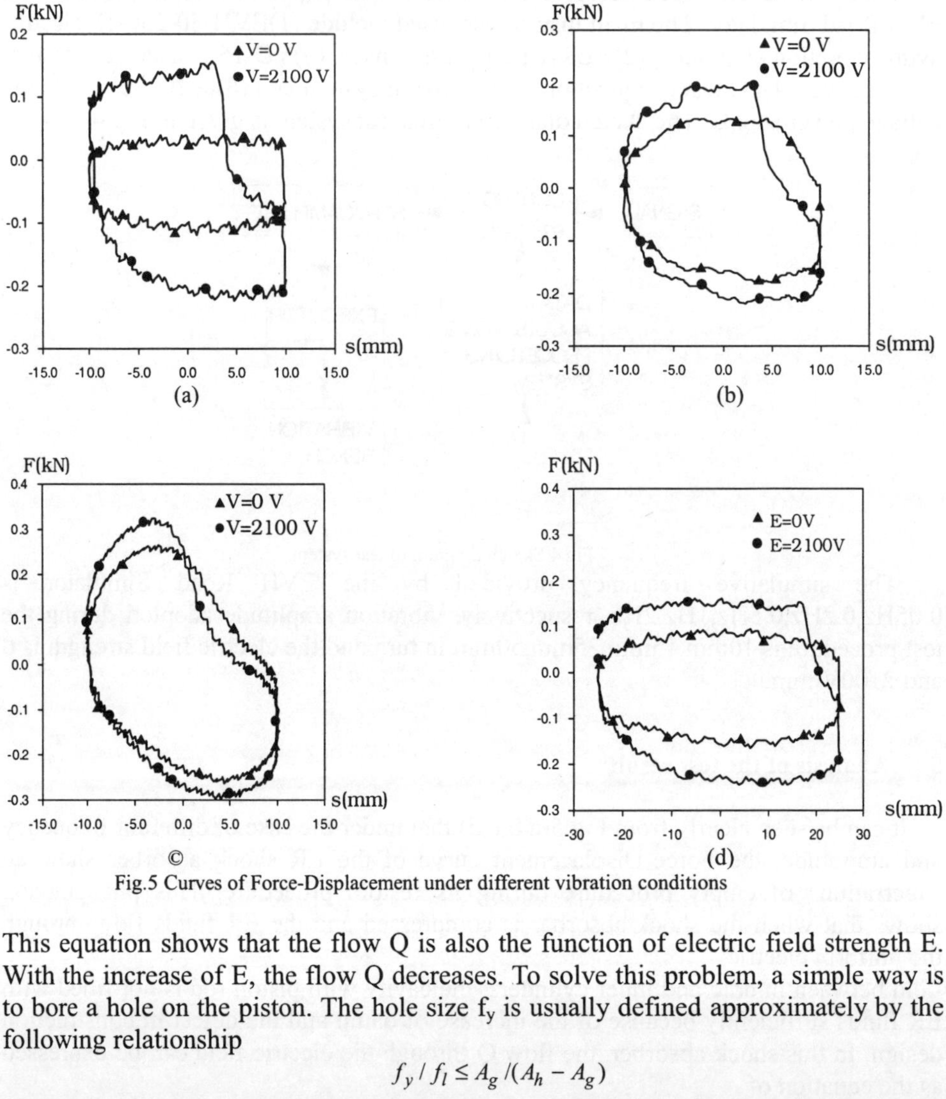

Fig.5 Curves of Force-Displacement under different vibration conditions

This equation shows that the flow Q is also the function of electric field strength E. With the increase of E, the flow Q decreases. To solve this problem, a simple way is to bore a hole on the piston. The hole size f_y is usually defined approximately by the following relationship

$$f_y / f_l \leq A_g /(A_h - A_g)$$

where f_y is the area of the hole on piston, f_l, area of the hole towards the electric field at the bottom of the inner cylinder, A_g, sectional area of the piston rod, and A_h, area of the piston.

In addition, it can be seen from Fig.5 that with the increase of the frequency and amplitude, this kind of "aberration" decreases gradually, but the variation of curves for damping force *vs.* displacement will also abate as the variation of applied electric field strength. It shows that the damping force of the absorber caused by electric field decrease. It is known from the performance of ER fluids given in Fig.2 that with the increase of the frequency and amplitude, velocity of the ER fluids flowing through electric field increases and makes the shear rate of ER fluids increase, so the apparent viscosity of the ER fluids decreases and the damping force decreases correspondingly. In Fig.5, (a) frequency 0.05Hz, amplitude 10mm; (b) frequency 0.2Hz, amplitude 10mm; (c) frequency 0.5Hz, amplitude 10mm; (d) frequency 0.05Hz, amplitude 25mm.

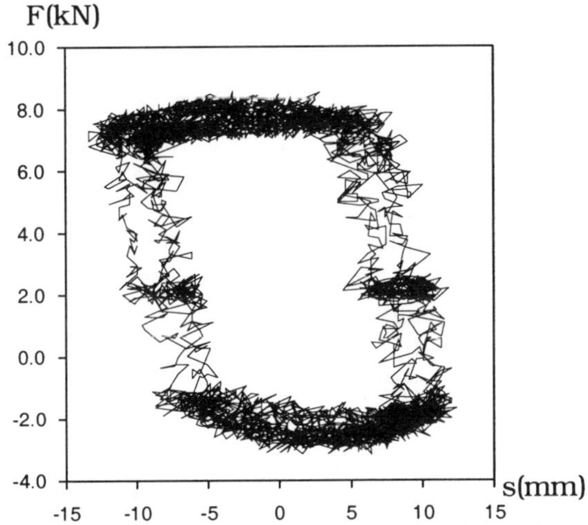

Fig.6 Curves of Force-Displacement after the construction was improved

Fig.6 shows the curves of damping force *vs.* displacement obtained after the construction wasimproved. Although the data have some obvious fluctuation compared with those of previous test due to changing of the PC Data acquisition & processing system, it also can be seen from the figure that the damping force during the restore procedure has a significant improvement due to the setting hole on the piston. Meanwhile the discharge caused by the hole makes the damping force decrease during the whole procedure. The inner loop in Fig.6 represents the case of zero electric field, and the outer loop is the case when the electric field strength is 2100V/mm.

The phenomenon mentioned above can also be explained by the following expressions[2]

$$F_y = \frac{\pi^3 h^6 \rho}{128\varepsilon^2 f^2} V^2 + \frac{3\pi l \eta_p (d^2 - h^2)^2}{4d\delta^3} V + \frac{3\pi l A(d^2 - h^2)}{2\delta} E^2$$

$$F_f = \frac{3\pi l \eta_p (d^2 - h^2)^2}{4d\delta^3} V + \frac{3\pi l A(d^2 - h^2)}{2\delta} E^2$$

where F_f—restore procedure damp, N

F_y—compress procedure damp□N
h—diameter of the piston rod□m
V—velocity of the piston□m/s
ρ—density of the ER fluids□kg/m^3
f—section area of the bottom valve, m^2
A—constant related to the characteristic of ER fluids
E—electric field strength, V/m
ε— flow coefficient

These two expressions show that the damping force generated by ER shock absorber is composed of two parts: background damp and electric induced damp. The background damp is determined mainly by the construction of absorber, and the electric induced damp is determined by the applied electric field. The rate of electric induced damp to background damp shows that the adjustability of ER shock absorber is decided by the electric field strength and the apparent viscosity of ER fluids. The larger the applied electric field strength is and the lower of the ER fluids' apparent viscosity is, the wider the adjustable range would be.

6 Conclusions

The test of ER shock absorber shows that the construction has a significant influence on the performance of ER shock absorber. The to-and –fro flow Q of ER fluids between two cavities of the shock absorber is also the function of electric field strength E. Problems in design can directly affect the damping efficacy as well as the working quality when it works together with a suspension. Adjustable range of damping force is mostly decided by the performance of ER fluids. ER fluids with lower apparent viscosity and higher ability of anti-breakdown are the essential requirement to its application in shock absorber.

Acknowledgement: The performance test of ER shock absorber in this paper was conducted under the help from Professor Li Xiaolei and Engineer Jiang Hong.

References

1. Yu Defu. Automobile random vibration & the suspension control. Beijing: Ordnance Industrial Pr.,1992

2. Gao Jingmin. Research on the ER shock absorber and its control system: [dissertation]. Beijing: Beijing Institute of Technology,1999

FUNDAMENTAL TRIBOLOGY: ESF

W A BULLOUGH**, P L WONG*, C FENG* AND W C LEUNG*

* *Department of Manufacturing Engineering and Engineering Management, City University of Hong Kong, Tat Chee Avenue, Hong Kong, P R China.*

** *SMMART Engineering Network, The University, Mappin St, Sheffield S1 3JD, UK*

Friction and wear phenomena associated with electro-structured fluids (ESF) are reviewed with respect to operation in the boundary lubricated region. Opportunities for fundamental research into "system effects" are revealed in the light of recent results from magneto-rheological fluid, steel on steel, block on ring, sliding contact tests. Consideration of characterisation means and of conditions for best performance are stimulated by the wide ranging nature of the information presented.

1 Introduction

The tribology of ESF is at first sight a specialised topic which involves pre-conditioned mixtures of complex base or carrier fluids and relatively high concentrations of particles or polymers. More conventional tribological situations operating in the boundary regime encounter much lower loadings of particulate (mainly wear debris) matter. In both cases the ultimate aim of research comprises the setting of a code of practice for the prediction and limitation of friction (coefficient μ) and wear on the bounding sliding surfaces. Taken altogether any advance on the present situation in either, implies a fundamental understanding of the mechanisms involved in both.

In the present paper only unexcited MRF mixtures are considered, principally within the conditions imposed by a rotating, hard and fine ground steel ring and a stationary, initially smooth steel block, the former being harder than the latter, with the MRF particles having a nominal intermediate hardness. Such studies produce only an indication of the general order and magnitude of friction and wear severity for different fluids, surfaces and operating parameters – rubbing speed, pressure in contact etc. More specific tests are required for individual applications, e.g. reciprocating motion in a seal. Nevertheless such factors as particle load sharing, coagulation and system geometry effects seen in one situation will occur to a greater or lesser extent in another. It is the essence of this text to illustrate key fundamental physical facts as they are thrown up in such tribology and to relate the findings to boundary lubrication (BL) in general.

2 Previous Studies

The initial paper on BL in ERF [1] was meant as a warning for fluid developers at the time not to ignore lubricity when considering the design of fluids. Little concern seemed apparent with respect to potentially catastrophic bearing and seal wear and friction. The results were interesting and surprising since particles larger than the nominal film size seemed to be passing through the contact zone and, in some cases, friction was reduced by their introduction. The problem of blending tribo-chemistry and rheology with the electro-physics of the dielectric mismatch phenomenon was amply demonstrated by comparisons of tribo performance with control test results on a well known Shell lubricating oil: wear using ERF was one order greater, at least. Studies on MRF have shown [2] that magnetic polarisation is more compatible with surfactant chemistry with respect to sedimentation and anti-coagulation. Break down due to solid contaminant, air or water in a controller is not really a factor. Reasonable friction and wear performance can be established.

Electro-hydrodynamic lubrication work [3] has shown that in ball on disc tests with ESF, particles generally avoided the contact zone. When subject to line contact as in a roller bearing, where the avoidance of the conjunction was more difficult, quite violent unsatisfactory operation was witnessed: particles at the inlet agglomerated being occasionally nipped onto the roller/rotating plate interface and flushed through with a large periodic change in film thickness [4]. Hydrodynamic lubrication type phenomena are discussed in [3].

More recently the predominance of the effect of ploughing by debris particles attached to the soft block of a boundary lubricated pair, rather than that of adhesion has been suggested [5]. If this proves to be the case then system effects such as roughness of surface, micromachined surfaces, or hard coated implanted surfaces [6], all of which could provide relief channels for particles or debris, could be important aspects of investigation. Also, coagulation of particulate matter must be important. Thus, for the specific case of MRF experimental results have been arranged, and secondary tests carried out to consolidate them, in order to provide information relevant to the argument.

3 Schemes of Testing and Results

Descriptive matter appertaining to results, experimental conditions [7 and 8] and readings etc. are contained within the figure captions. All particles are centrifuged from MRF 132 cleaned and re-mixed. In all cases excepting figure 10 [9] the ring is the outer ring of a SKF tapered roller bearing #32206/62 mm diameter. The block is always ground BS4569 (1971)-1307 cold worked tool steel. The former has hardness 63 HRC and the latter 15 HRC, both have original surface finishes better than 0.2μ CLA.

All low (% C) concentrations of particles in test fluids use Shell Tellus 15 as the base oil (15 c St @ 40° C). All particles were cleaned by acetone before mixing. In general, for both high and low concentration tests, each data point is produced by a new alcohol wiped and acetone washed sliding pair: both block and ring are so cleaned and ultrasonically cleaned before weighing or microscopic examination. Tellus 32 has a viscosity close to MRF 132 mixture and Tellus 15 close to that of the MRF base fluid from MR 132. When syringe fed supply is utilised (mainly low % C) the fluid floods the contact but once, being wiped off by a sponge following its passage through the contact. When the bath is used (mainly for high % C), fluid and wear debris can pass through this contact on more than one occasion. Particle remnants on worn surfaces were identified by X-ray spectrum analyser, surface topology of block and ring were evaluated by Talysurf scanner. Wear scar width was measured with a X25 surface profile projector.

4 Discussion

It is evident from an inspection and assimilation of the above data that a full understanding and hence optimization of the performance of MRF in a sliding contact is some way off. The generalized, simple and arbitrary test method adopted for a single fluid has produced results which, extensive as they are, leave large gaps in the evidence on which to base a comprehensive physical mechanism. However, it does seem that tribochemistry alone is unlikely to produce a final answer with respect to friction and wear reduction. Though well supplied with a carrier fluid which needs to be "doped" to inhibit particle settlement and coagulation, and the knowledge that smooth surfaces can be separated by rolling particles, conditions in the contact seem best suited to BL type analysis.

It is evident that particles smaller than the nominal film thickness traverse the gap. On their way through they roll on the block in well defined "plough" tracks in which some of them stick. Subsequently the stationary particles then perhaps work harden and polish the ring surface. The effect of %C in all of this is not yet totally established save that it is not a simple matter: coagulation of particles in low %C mixtures in T15 can cause more wear than high %C, well dispersed particle fluids do.

Clearly there is a need to establish the effect of time of rubbing on the phenomena. If it is accepted that the soft block will be shown to be damaged very quickly but subsequently skates on the block via embedded particles therein, then perhaps there exists a case for introducing surface implantations or micromachining to produce escape routes for the MRF particles i.e. effective drainage ditches so that system effects are utilized as well as those from a choice of surface materials and tribo/rheo/chemical properties within the fluid mix. The base oil tests probably start in metal to metal contact with oil latterly entering the contact via wear scars to effect a hydrodynamic film. The use of a PxV characterization results in some scatter.

The wide range of data presented, and its form of presentation, is meant to stimulate and influence the planning of future investigations and to provide a means of linking ESF with more conventional tribological developments. Preliminary calculations have indicated that ploughing will probably constitute the majority basis of the friction magnitude. In choosing the fluid parameters in future tests it should be understood that strict limitations apply to %C and the base fluid from an MRF device operation and particle supply point of view. The use of T15 and MRF particles was imposed purely from the point of view of fluid husbandry - only one litre of MRF132 was available. In general only high concentrations of various %C need be investigated versus P and V to point to best wear and friction conditions. These tests, which are currently being carried out will also provide evidence which will support, or otherwise, the particle load and ploughing predominance that have been indicated in the simple calculations. Also, preliminary surface machining/system effects tests are imminent.

5 Conclusions.

The tribology of MRF in sliding contacts is a complex affair. Not only are many parameters under the investigators control but the effects of %C magnitude, surface(s) conditions, speed and pressure factors can be variable. A large amount of testing in the time domain is called for and system effects must be clearly demonstrated if progress is to be made. Different materials and topologies are to be investigated.

It would seem that whatever rationalization can be brought into the test plan (e.g. by use of the present apparatus) it must be remembered that this will bring with it a degree of arbitrariness in the interpretation of results. Specific tests on e.g. a sliding bearing for MRF cylinder applications may bring contra-indications of best speed and load than arise in say a rotary shaft seal. Indeed the entire geometry in the sense of seal before bearing or vice versa may be reversed should the manufactured topology of the rubbing surfaces prove to be an important factor in reducing friction and wear.

6 References

1. S Lingard, W A Bullough. Tribological Aspects of ER Fluid Behaviour with respect to Non Polar Base Fluids. Proc. 2nd Int. Conf. on ER Fluids, Raleigh, NC. August 1989. Ed. Conrad H and Carlson D, Technomic Press, Lancaster Pa. pp 158-175.
2. M R Jolly, J M Bender and J D Carlson, Properties and applications of commercial magnetorheological fluids. Proc. SPIE 5th An. Int. Symp. on Smart Structures and Materials, San Diego, CA. 15 March 1998, p1-5.
3. P L Wong and W A Bullough, Tribology of ESF, Proc. Minsk Int. Forum IV, Tom 7, Heat and Mass Transfer, Byelorussian Acad. Sic, 22-26 May 2000, pp 145-154.
4. X.Chen, Behaviour of a Magneto-Rheological Fluid on Elastohydrodynamic Lubricated Line Contact, - private research report, Bearing Research Institute, Shanghai University, China (2000)
5. Nan P Suh, Engineered Surfaces: Theoretical Basis and Applications, Plenary Lecture, 7th Int. Conf. Tribology, South African Inst. Tribology 21-23 March (2001).
6. K Steinhoff, A Kapoor and N Gullion, Controlled wear as a mechanism for the design of geometrically defined nanometric surface structures of forming tools. Proc. 6th Int. Conf. Tech. of Plasticity, Nuremberg, Sept. 1999.
7. P L Wong et al, Tribological Effect of a MRF on Steel/Steel Conjunction Under Boundary Lubrication Conditions. 7th Int. Conf. Tribology, South African Inst. Tribology 21-23 March (2001).
8. P L Wong et al, Tribological Performance of Magneto-rheological suspension. Wear 247 (2001) pp 33-40.
9. C Q Feng, P L Wong and W A Bullough, Observations of fatigue wear on contacts under boundary lubrication with magneto-rheological suspensions. 3rd Int. Symp. on Tribo. Fatigue, Beijing, China (2000).

7 Acknowledgement.

Thanks are due to Lord Corporation for supply of fluid, advice and information and to British Council/City University, Hong Kong for their support of the collaboration.

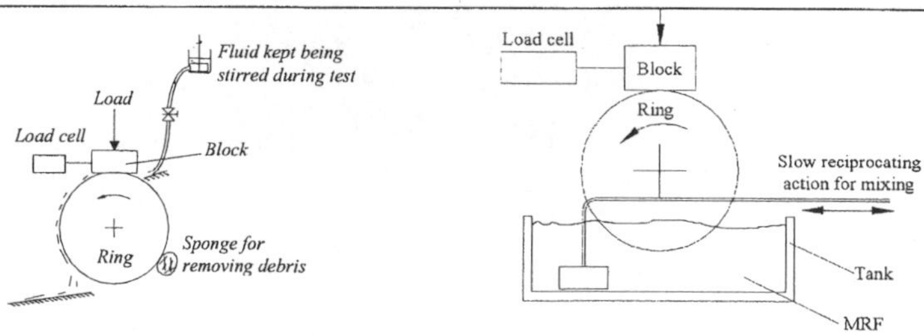

Fig.1 Schematic diagram of boundary lubrication test rig. (a) For low % particle concentration test; (b) For high % particle concentration

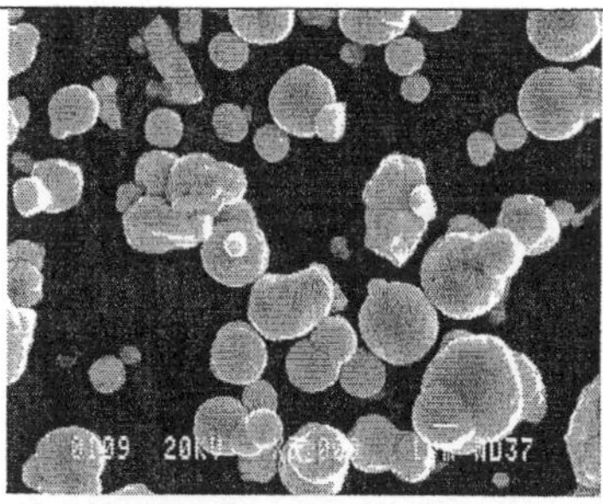

Fig.2 Morphology of iron particles as centrifuged out from Lord MRF 132. Scanning electron microscope (SEM) indicates a spherical shape of 1 to 5 μm diameter.

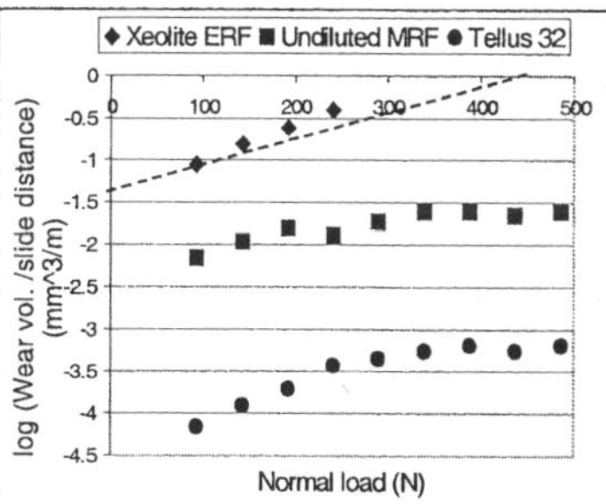

Fig.3 Comparative wear performance of various fluids. Speed is 500 rpm and duration of each test is 1 hour. The two ERF mixtures Lipol 30/w (dotted line) [1,10] and home-made 35% by volume zeolite/silicone oil mixture show highest wear rates. MRF 132 is one order better than ERF but worse than mineral based Shell Tellus (T)32 by one or two orders.

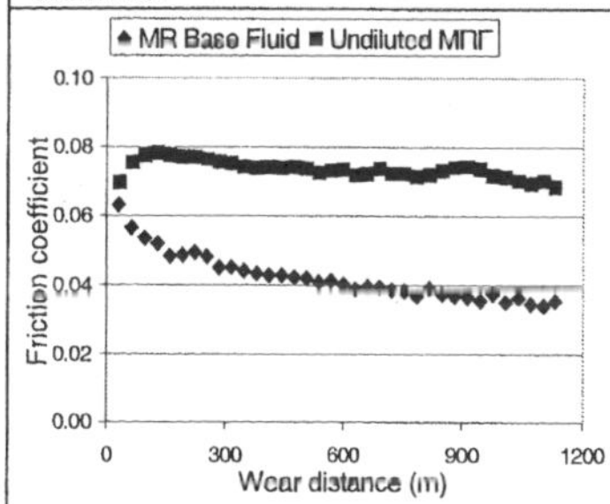

Fig.4 μ vs distance for 292 N load at 500 rpm. MRF is clearly worse than its base oil alone. Typical ERF (Lipol) has a nominal coefficient of friction (μ) of 0.4.

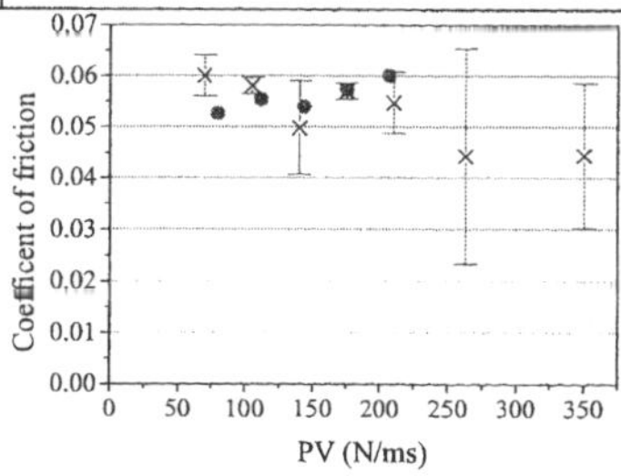

Fig. 5 μ vs load pressure P x sliding velocity V for MRF 132, in contact for constant load X and constant speed •. 4 hour test with readings taken during last 200′ of test, i.e. after 'run in'. μ is steady for constant speed results increasing only slightly with load. In constant load tests μ is steady at low speed becoming smaller at higher speeds (mixed lubrication encountered?)

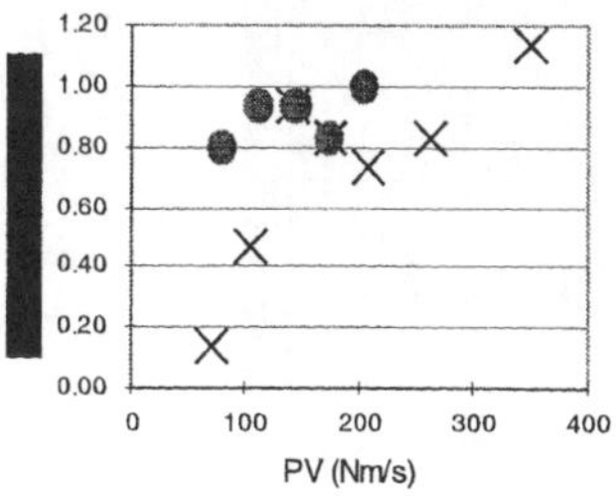

Fig. 6 Ring wear weight loss for different PV factors for constant load X and constant velocity •. Conditions as for Fig 5. Wear under constant load varies markedly with speed.

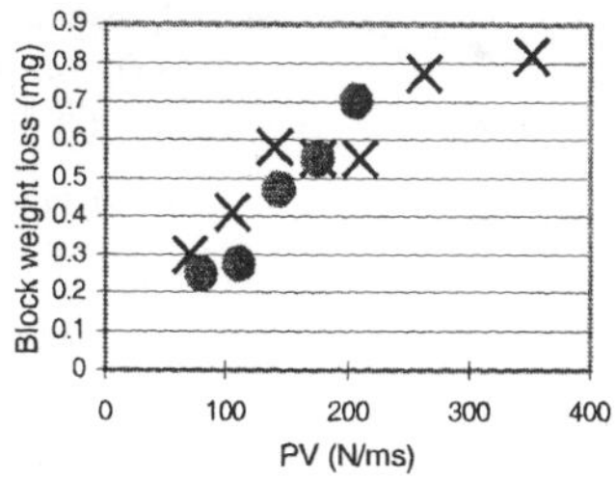

Fig 7 Block wear weight loss for different PV factors on same test as Fig 6 (• : constant speed; x: constant load). The ring wear is slightly greater than the block wear.

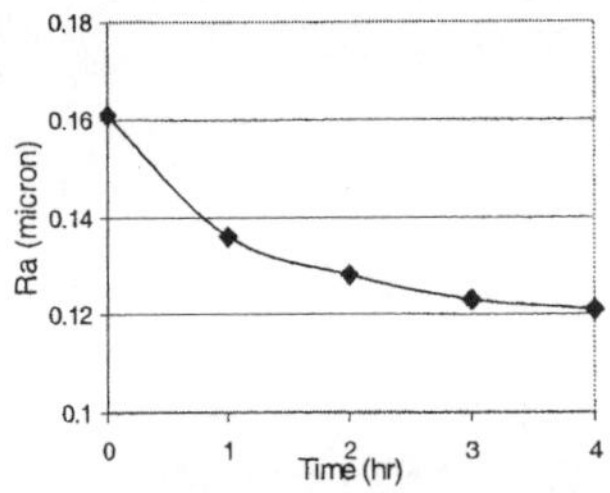

Fig.8 Change in ring roughness during a wear test. T15 with 10g/L of particles, load 532 N, speed 0.32 m/s. Particles evidently polish the ring.

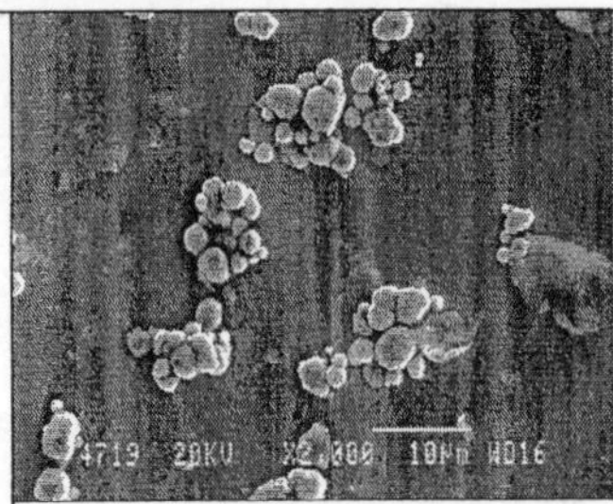

Fig.9 Agglomeration of particles on block surface. T15 with 40g/L of particles. Some particles are flattened, bunch and become attached to block surface - after once through test and subsequent flushing, but not ultrasonic cleaning of block surface. Sliding speed 0.32 m/s and load 532 N.

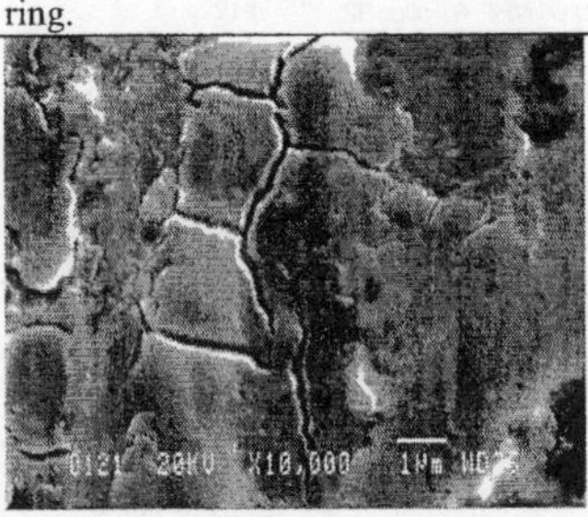

Fig.10 G Cr 15 heat treated ring surface after 4 hour test with T15 base fluid plus 20g/L particles. Surface inspection demonstrates high temperature oxydisation indicating transient high temperature frictional heating. Thermal cracks are formed in detached surfaces within oxydised region; the growth of these is subsequently expedited by action of particles to form pits in the swelling. SKF ring suffers similar problems but to lesser extent. Difference in performance derives from different microstructures/ring material.

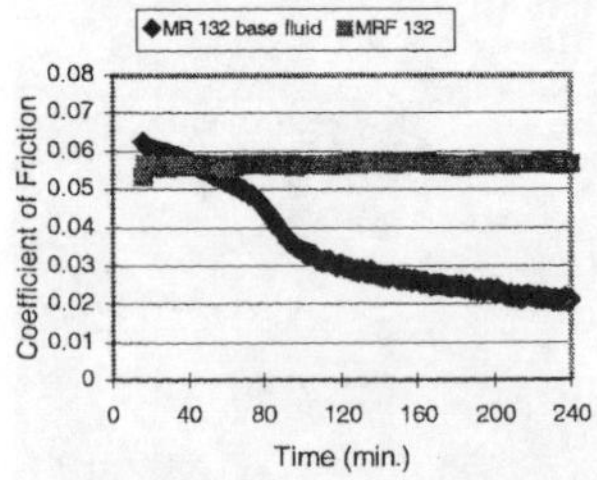

Fig. 11: μ vs running time with MRF 132 and MRF 132 base fluid. Base oil curves drop considerably after 30′ running. Steady μ of MRF particle fluids infers friction is due to continuous replenishment of particles.

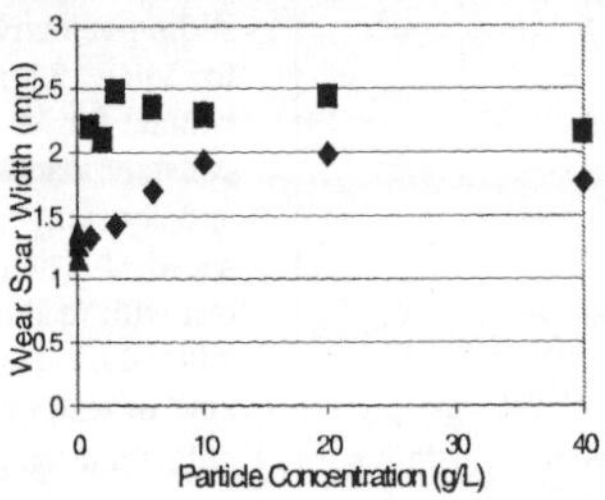

Fig.12 Block scar width for different low % C: results taken by two operators. Speed 0.32 m/s, load 532 N, 1 hour duration. At higher concentration, wear seems to fall – results seem contradictory to MRF 132 base fluid results: indicates effect of bunching of particles. Different base fluid involved.

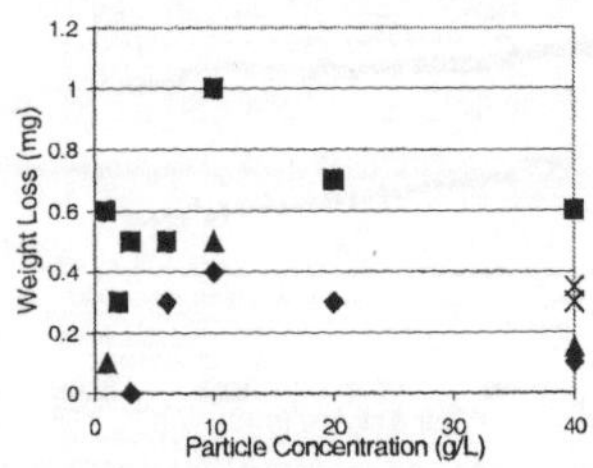

Fig.13 Ring weight loss for low % C of particles. Conditions as Fig 12. Ring loses less weight than block by a factor of two.

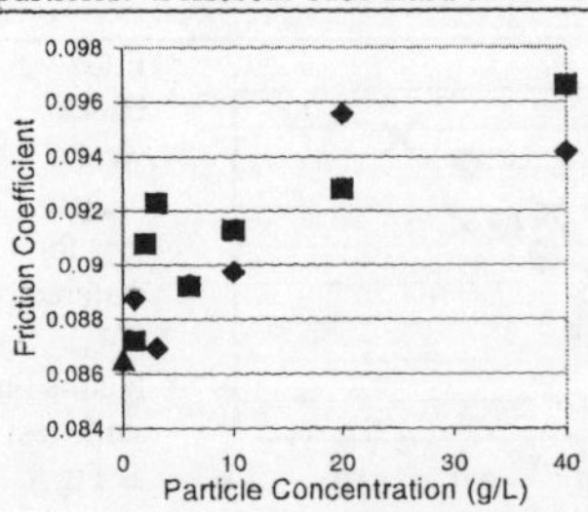

Fig.14 μ vs % C, conditions as Fig 12 : μ little affected at low % C but slightly increases with concentration

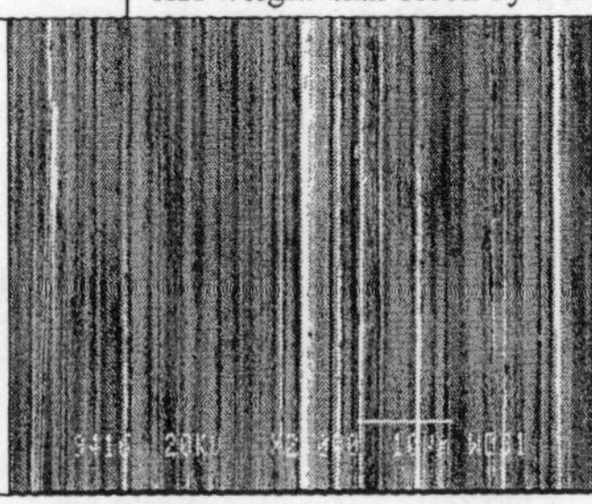

Fig. 15 SEM of ring original surface. 0.05 – 0.07 μm CLA determined by directional grinding grooves.

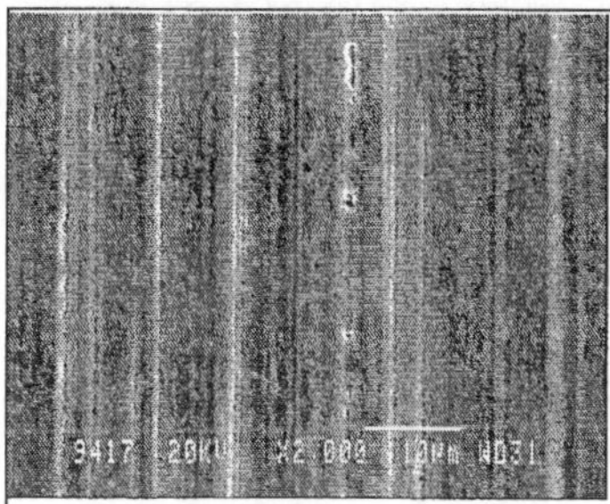

Fig. 16 Ring surface after 4 hours test with MRF 132, speed 0.32 m/s, load 638 N. (X2000) Sharp ridges on initial surface have been flattened.

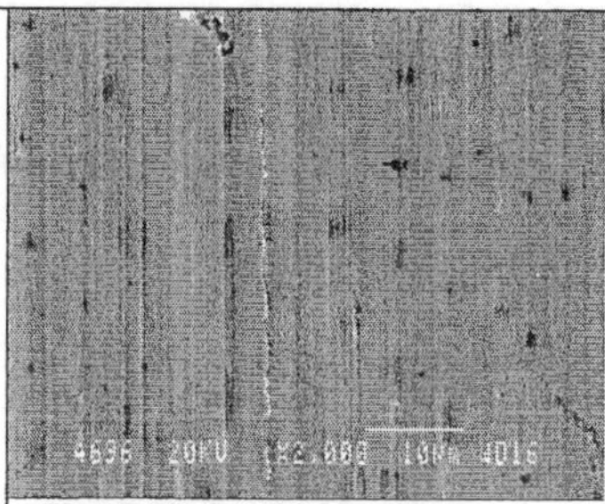

Fig.17 Block original surface

Fig.18 Optical microscopy of block wear scar, after 4 hour test with MR 132 fluid, speed 0.32 m/s (X30), load 532N. Wear scar after use with MRF 132 is rougher than with base fluid. Well dispersed particles embedded on inlet to flushed clean block.

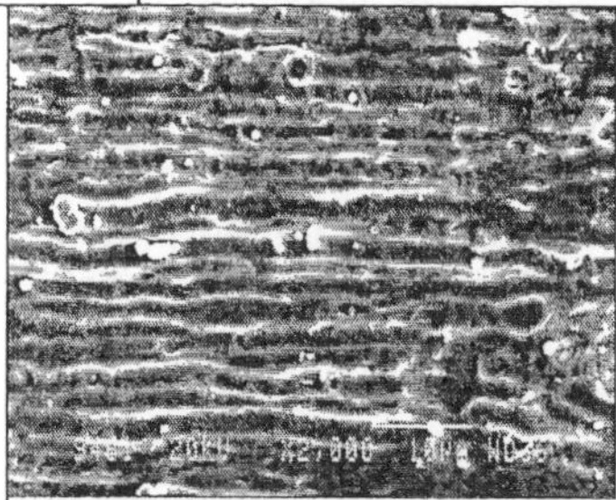

Fig. 19 Inlet region of block after 4 hour test. Plough grooves are of similar diameter to particles. Also, plough grooves are dense corresponding to high % C of MRF 132. (speed: 0.26 m/s: load: 532 N: 4 hours).

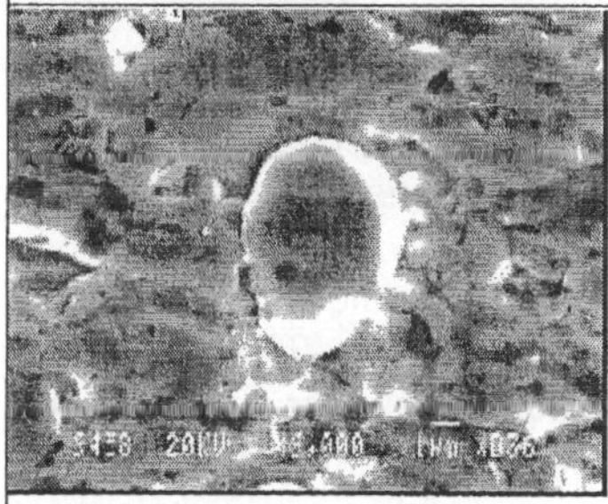

Fig. 20 Slightly deformed particle as found in block wear scar, MRF 132, speed 0.26 m/s, load 532 N, 4 hours duration (X500). Particle identified by Xray spectrum analysis.

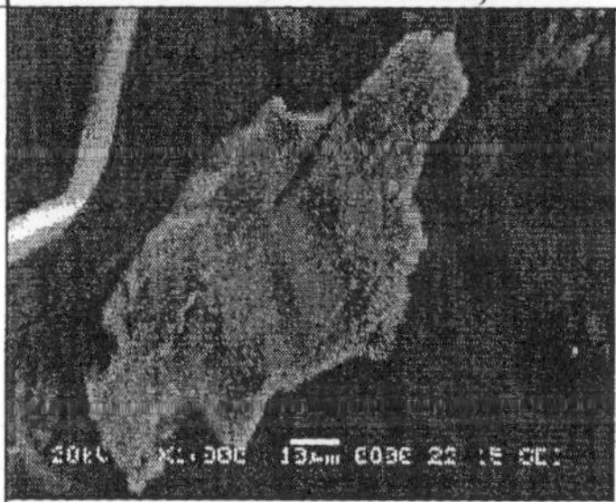

Fig. 21 Typical wear debris collected in one of MR 132 pure base oil tests. Its shape is very different to that of damaged MR particle. The plough grooves in the inlet region appear to be mainly produced by iron particles.

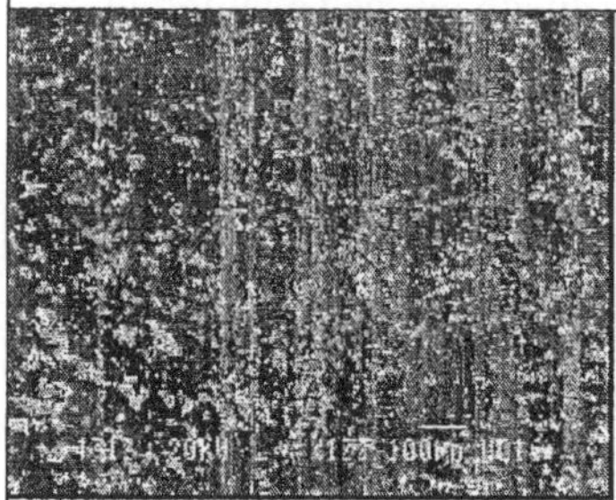

Fig.22 Block surface after 1 hour test, speed 0.32 m/s, load 532 N, % C = 6g/L. MR particles are embedded in block.

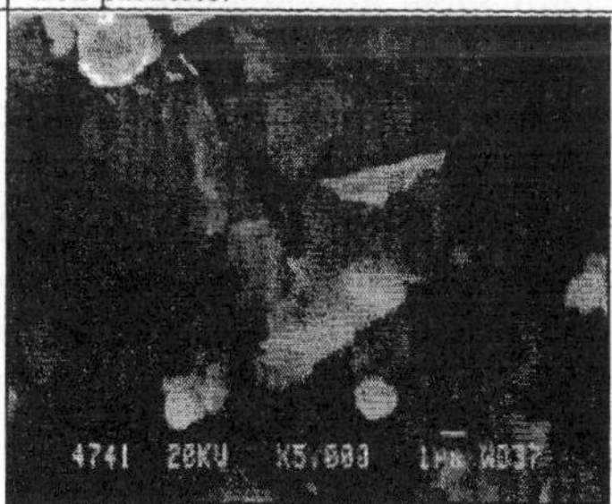

Fig.23 Particles collected after passing through contact once. Base oil T15, particles from MRF 132. Sliding speed 0.32 m/s, load 532 N, block G Cr 15 Chinese bearing steel. Particles are plastically deformed and even flattened

HEAT TRANSFER FROM AN ESF RADIAL PLATE CLUTCH SURFACE

D. J. ELLAM, W. A. BULLOUGH, V. ORAVSKY

SMMART Engineering Network, University, Dpt. of Mechanical. Engineering.
Mappin Street, Sheffield, S1 3JD, England

The aim of this work was to investigate the rate of heat transfer from a moving radial plate clutch surface. The paper presents experimental results over a range of angular speeds (Ω) and fluid gap widths (h), which are then favorably compared to analytical and Computational Fluid Dynamics (CFD) solutions for the same geometry and operating conditions. Verifying the heat transfer capabilities of the latter goes some way towards validating a CFD package as a viable virtual prototyping method. In light of the results assumptions were established to allow the comparison of a duel channel radial and concentric clutch designs on a heat transfer basis.

Introduction

An ESF clutch operates in the shear mode thereby requiring an adequate heat dissipation capability to keep the ESF at an acceptable operating temperature. This is particularly so when continually switching in order to achieve an electronically controlled variable configuration of output motion [1]. In some instances multi-plate clutches will be required to provide a large output torque in a small, low inertia device. It follows that sooner or later a means of predicting the temperature within, and heat transfer through, a device of complex geometry will be required and that the non-linear properties of the ERF/MRS will give rise to a problem best handled by CFD in steady and unsteady non-isothermal situations.

For the same outer radius dimension both radial and concentric, dual channel, clutch type designs offer similar torque control ratio (T_{total}/T_e) and acceleration (T_{total}/I) capabilities. It may well be, therefore, that the capability for heat transfer dictates the best type of design.

In a previous work [2], the heat transfer from the surface of a concentric clutch was modeled and subsequent developments [3] have validated the technique. This model was latterly tested by running the outer rotor at speed whilst holding the inner stationary and measuring the temperature on the outer surface by a remote infra red thermometer. The use of a well known viscous fluid allowed the slipping torque and hence the work input ΩT to be calculated. A balance between the heat transfer from essentially the rotating outer case would equal the calculated work rate when the clutch reached equilibrium test conditions.

In the present work, much the same approach is followed but this time on a radial plate clutch in which the upper plate is driven and the lower plate is held stationary and thermally insulated. Hence the heat transferred through the upper plate is equal to the power input ΩT. Unlike the predominately uniform shear field in the large diameter, small fluid gap concentric clutch the shear rate in the radial device varies with radius. This situation invites concern with respect to temperature gradients across and along the oil film and plates.

Radial clutch description

The clutch comprises two mild steel (EN3) radial discs (85 mm dia., ≈ 10 mm thick). The upper disc is attached to a motor by a solid shaft (20 mm dia.) of the same material. The shaft is partly insulated (shaft insulation 36 mm dia. x 60 mm) see Fig.1. Gauge blocks were used to set a gap between the discs, which contains hydraulic oil of known fluid properties (Shell Tellus 37).

The bottom plate was held stationary and was thermally insulated by the use of Tufnol and Kao Wool so that a zero heat flux boundary condition could be assumed. Five k-type thermocouples are implanted approximately 1.0 mm away from the wetted surface, located at radial positions of 0, 10, 20, 30 and 40 mm. To allow for a 1-d analysis the outer edges of the device and driving shaft are also insulated. Refer to Fig.1.

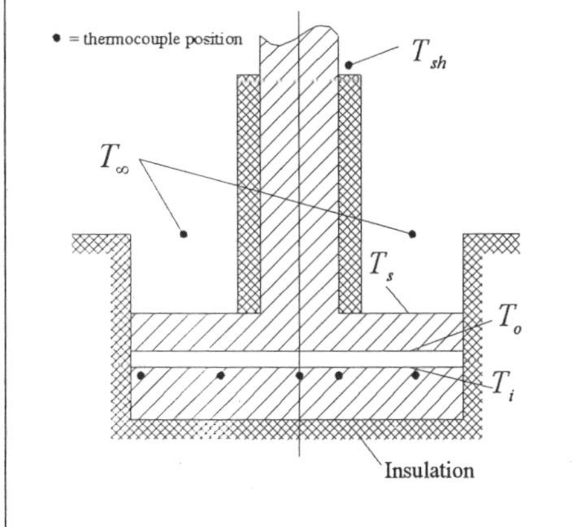

Fig.1: Schematic of experimental apparatus

To determine the amount of heat that is conducted up though the shaft, in the CFD model, it is important to accurately know the temperature T_{sh}. The accuracy of an infer red thermometer which could remotely measure this temperature was deemed inadequate for this situation. The temperature T_{sh} was therefore inferred from the near air temperature. To calculate the amount of heat convected from the disc surface the bulk air temperature T_∞ was also measured.

Experimental Method

All thermocouples were first checked to be working correctly and calibrated. The gap size was set at either 0.5, 0.75 or 1.0 mm. The motor was set to rotate at 500, 750, or 1000 rpm. Temperature readings were then taken every twenty minutes or so until the temperatures had reached their steady state values. The readings consisted of the five lower plate temperatures T_i, two readings above the plate allowing an average T_∞ to be determined and the shaft near air temperature T_{sh}.

Theoretical Model and Solution of Heat Transfer in the Experimental Device

The analysis described here is based in principle on an analysis described in significant detail in [4]. There is however one fundamental difference involving the modeling of heat transfer rate though the shaft, $\dot{Q}_{shaft}$. In this paper the experimental device geometry, material properties and measured temperatures (along with an assumption that is later proved to be valid) are used to determine $\dot{Q}_{shaft}$. This is in direct contrast to [4], where the experimental data is utilised to calculate a parameter to be associated with $\dot{Q}_{shaft}$. Such a model gives results close to experimental data as a consequence of absorbing all experimental error and other uncertainties in an empirical coefficient it is difficult to analyze the importance of the shaft with respect to the other parts.

Only the rate of steady state heat transfer at constant rotational speed is considered. Insulation on circumferential surfaces allows 1-d heat transfer to be assumed. Furthermore, it is assumed that all mentioned temperatures are uniform in the radial direction. The fluid is taken to be Newtonian with viscosity dependent on the temperature $(T_i/2 + T_o/2)$ – average temperature in the oil film.

Frictional tangential forces in the fluid generate an amount of heat equal to $T\Omega$. As a result, heat is transferred through the oil, upper disc and shaft (via conduction). A proportion of the heat is transferred from the rotating disc into the surrounding air via convection, at a rate approximated by the Newton's law of cooling. The heat transfer coefficient for this convection is given in a semi-empirical relationship expressed through averaged Nusselt, $\overline{Nu}$, Reynolds, Re, and Prandtl, Pr, numbers ([4] & [5]).

Determining $\dot{Q}_{shaft}$ involves assuming that the exact shaft temperature at the location where the insulation ends (i.e. at length L_{sh} = 60 mm above the lower disc) is equal to the air temperature T_{sh} at a location very close to the shaft at the same vertical position. This allows the conductivity equation to be applied for heat rate through the shaft. Above the end of the insulation a small length (35 mm) of shaft remains before one of two large bearings are encountered which are in turn attached to a large supporting steel structure. The geometry of the apparatus was therefore such that very little heat would be convected from the remaining shaft due to the supporting structure providing a more favorable means for the heat to travel. Since the flow is therefore highly one-dimensional the above assumption is expected to be correct. To be fully confident in using this assumption some of the experimental data was inputted into a CFD model to determine approximate temperature distributions. From the results over the full range of parameters the assumption is found to be sound.

Mutual comparison of the above mentioned heat flows provides a system of equations defining the three unknown temperatures T_s, T_i and T_o. The shaft, T_{sh} and air, T_∞ temperatures are known allowing the theoretical T_i to be determined, using a simple iterative procedure, which can then be compared to the experimentally measured value.

CFD Analysis (software package Fluent5)

The CFD model was set up in comparison to the above mentioned analytical theory. The exact geometry of the experimental device was used to produce an axis-symmetric grid required for the finite volume solution procedure. Table 1 outlines which relevant equations are applied and the boundary conditions that were used to close the system of equations. CFD applies the full set of Navier-Stokes and energy equations, thereby avoiding the requirement for the semi-empirical heat transfer coefficient for convection as used in theoretical analysis.

Flow	The flow of the oil and air is set to be laminar. Viscous heating terms are included. Furthermore, buoyancy terms in the air region are included.
Air(ρ,μ,k)	Ideal gas, properties as a function of cell temp.. (k = thermal conductivity)
Oil(μ,k)	Newtonian (Tellus 37), properties as a function of cell temp. [4].
Boundary Condition	Insulated walls = zero heat flux condition, rotating walls = correct angular velocity. The insulation end is set to temp. T_{sh}. The boundary of the air region is placed at a vertical position equal to the T_∞ location. This boundary is then set to be equal to atmospheric pressure and temp. T_∞.

Table 1: CFD parameters for analysis on Fluent5.

Convergence was verified by making sure that the residual history of each equation was sufficiently low and that the net heat flux equated to $T\Omega$. Grid sufficiency was met using a built in grid adaptation facility, which increases the grid density in regions where the derivatives of variables are high. The temperature T_i and resulting heat fluxes though the various parts can then recorded.

Results: Experimental vs. CFD vs. Numerical

The experimental and CFD data verified that the assumption of uniform radial temperature distribution in all metal and fluid parts was correct. No variation on the lower plate was recorded in all experiments. In most cases there was little difference between T_{sh} and T_{∞}. The CFD analysis clearly showed that although there is a radial distribution in shear rate the high conductivity of the metal parts distributes the temperature out evenly. The essential results are shown in Table 2. Theory gives results that are rather high (4.1 ^oC in one case). The CFD results fare better, the reason for which is discussed below.

Gap		Experimental Temp.		Lower Plate Temp. (*T_i*)			*Q_sh/Q_s*	
		Shaft	Air		diff. (T - T_exp.)			
Width	Speed	*T_sh*	*T_∞*	exp.	analy.	CFD	analy.	CFD
[mm]	[rpm]	[°C]	[°C]	[°C]	[°C]	[°C]		
0.5	500	24.90	25.40	28.10	**2.09**	**1.75**	1.5	3.1
		24.50	24.80	28.30	**1.54**	**1.19**	1.5	2.9
	750	25.80	25.90	31.80	**3.08**	**2.43**	2.7	2.2
		25.80	25.95	31.70	**3.19**	**2.54**	2.7	2.2
	1000	26.50	26.30	35.80	**3.52**	**2.45**	3.9	0.5
		26.50	26.35	35.60	**3.73**	**2.68**	3.9	0.5
0.75	500	26.60	26.95	29.10	**1.31**	**0.82**	1.0	3.1
		26.60	26.95	29.20	**1.21**	**0.72**	1.0	3.1
	750	26.60	26.50	31.60	**2.03**	**1.04**	1.9	2.2
		23.95	24.10	29.10	**2.66**	**1.60**	2.1	2.2
	1000	24.45	24.20	32.25	**3.58**	**1.71**	3.1	1.9
		24.50	24.45	31.70	**4.09**	**2.31**	3.1	1.9
1.00	500	27.40	27.55	29.00	**1.46**	**0.87**	0.7	0.3
		25.60	25.75	27.30	**1.56**	**0.97**	0.8	0.3
	750	25.40	25.35	30.10	**1.64**	**0.27**	1.6	0.5
		27.10	27.20	31.00	**2.05**	**0.82**	1.5	0.4
	1000	25.40	25.25	32.00	**2.99**	**0.82**	2.4	0.5
		25.30	25.20	31.90	**3.03**	**0.99**	2.4	0.5

Table 2: Results: measured experimental temperatures and resulting theoretical predictions

Discussion

The semi-empirical expression used to calculate the heat transfer coefficient for the heat flow rate into the air was formulated for a complete circular rotating disc. This does not take into account shaft/wall constraints present in the experimental set up (see Fig.1). It was however hypothesized [4] that the expression would remain true if the area for convection were based on the disc area minus the shaft insulation area. In light of the results, it was thought prudent to check whether the expression was valid for the experimental set-up. In a CFD simulation, a hypothetical disc was set to rotate in a large region of air (atmosphere), an amount of heat flux was set to leave the disc and enter the atmosphere. Setting a far field temperature T_{∞} closes the system of governing equations and allows T_s (surface temp. of disc) to be determined. Over a range of parameters, remarkably close coloration between disc theory and CFD was found. A shaft and wall constraint as seen in the experimental setup was then introduced. It can be seen from Fig. 2, which show the results of this case, that the semi-empirical expression over predicts the surface temperature. This would explain why the CFD results are closer to the

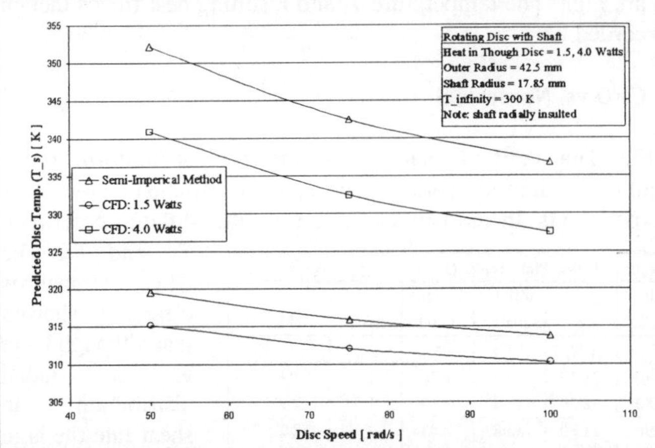

Fig. 2: Rotating disc in air with shaft and wall constraint (as in Fig.1) Considering convection only (testing semi-empirical formulae).

experimental data than the theoretical model, and why CFD gives lower values of $\dot{Q}_{shaft}/\dot{Q}_s$.

For theory and CFD all the temperatures predicted read high and the measured difference between *CFD* T_i to experimental T_i appears to be approximately proportional to $[T_i - T_{room}]$. This is likely to be to due to the theoretical models being *ideally* insulated on the lower and outer parts whereas experimentally it was not. Heat loss though insulated area would be a function of $[T_i - T_{room}]$. So that a 1-d assumption could be used in the theoretical analysis most of the apparatus was insulated. As a result the area assumed ideally insulated is relatively large in comparison to the non-ideally insulated surfaces. Furthermore, thermal expansion of supporting material is likely to be a factor. The Perspex casing which housed the lower disc and insulating fiber material was of a ring type construction. Although constrained to expand axially thermal expansion in the radial direction would be possible quite favorable and may have resulted in some lateral shrinkage, resulting in larger oil film gap and subsequent reduction in $T\Omega$, this error would also be a function of $[T_i - T_{room}]$.

These facts along with the difficulty in accurately setting a uniform gap would explain the 0.3-2.7 °C error in the *CFD* T_i predictions. In this case it is important to note the necessary inclusion of the driving shaft: it carries away a significant proportion of the generated heat.

Comparison of Clutch Designs

An attempt is now made to compare two different clutch designs. In order to produce a competitive commercial device it is important to consider acceleration (T_{totol}/I) and control ration (T_{total}/T_e) as well as T_{total}. A variable speed driven rotor is therefore likely to be made of a lightweight material, whereas the outer casing that rotates at constant speed is likely to be made from a heavier highly conducting material to maximize heat dissipation from the device and help maintain speed i.e. reduce regulation.

Dual channel, horizontal axis radial and concentric clutch designs are compared. The geometries are shown Fig. 3. It is likely that the outer casing and consequent surface temperature T_s would be approximately uniform. Shafts will have similar effects in each design and are so therefore not included. Having chosen a mean radius, dimension z and gap size h, the overall geometry of the radial clutch outer casing (R_r and L_r) are known. For set speed differences Ω, fluid properties and air temperature T_∞, a uniform surface temperature (T_s) can be calculated. It is assumed that,

$$T\Omega = (T_s - T_\infty)(A_r h_r + A_a h_a) \tag{1}$$

where h_r and h_a are the heat transfer coefficients for dissipation of heat by convection in the radial and axial directions which are calculated using the following semi-empirical formulas [5]:

$$Nu = 0.4(Re^2 + Gr)^{0.25} \quad Gr = \rho^2 \beta g R^3 \pi^{1.5} \Delta\theta / \mu^2 \tag{2}$$

$$Nu = 0.18[(0.5Re^{0.5} + Gr)Pr]^{0.315} \quad Gr = \rho^2 \beta g D^3 \Delta\theta / \mu^2 \tag{3}$$

The Bingham plastic equation is used to determine the value of torque T_{total}. A further parameter, length of the electrode L_e, is required for the concentric clutch. This length was chosen to make T_s equal to that of the radial clutch. The amount of T_{total}/I for the same parameters giving the same T_s can now be compared as shown in Fig. 4.

It is observed that the plastic viscosity plays a noticeable but relatively insignificant role when compared to the effects of fluid yield shear stress. It can be seen that the radial disc design gives the better results even though the concentric design was approximately twice as long in axial dimension. In all cases, T_{total}/T_e was approximately the same. Also coefficients h_r & h_a were similar.

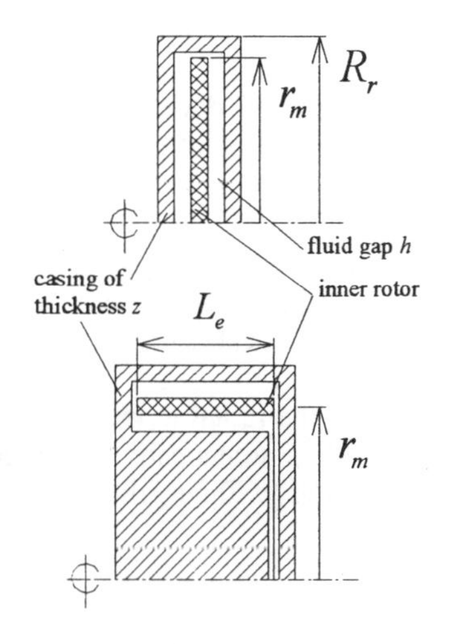

Fig. 3: Clutch models

It must be noted that L_e need not be constrained to being relatively small. Increasing this dimension has no direct effect on T_{total}/I but would give an improved T_{total}. However, the inner electrode being of a low conducting material will make it difficult to dissipate heat from the inner channel. To get good T_{total}/I a commercial device is likely to be of the form length > radius resulting in a large surface area for radial heat flow. A multi-plate radial clutch occupying similar dimensions would therefore be better since it would allow radial flow. Furthermore, a radial design does not exhibit plug flow behavior over

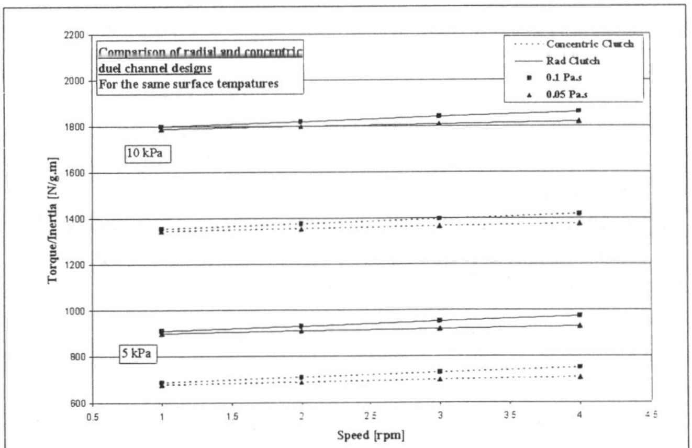

Fig. 4: Comparison of radial and concentric dual channel designs. T/I for the same case surface temperature. Note: inertia's (I) are similar.

its full range of operation generally leading to lower viscous heating.

Conclusion

As is evident from comparisons of experimental and theoretical data, of a viscous clutch device set-up, the correct magnitudes of heat transfer can be predicted effectively. For more complex set-ups, such as multi plate clutches, in an unsteady operation, CFD promises to be a simple and important tool for heat transfer evaluation purposes.

In the experimental study, the viscous medium was Newtonian oil. Heat transfer in non-Newtonian fluids is a complex subject. However, it is proposed that modeling heat transfer in ESFs need not be that complicated due to the following three reasons:

- Most ESF devices operate with ESF in a very narrow channel for which the distribution of temperature across the gap width is likely to be nominal as indicated in the present simulations.
- Using CFD on ESFs the most important variable (yield stress) can be entered as a function of temperature, excitation and shear rate.
- Furthermore, correctly determined non-Newtonian velocity profiles will give the rate of heat generation on cell-by-cell basis.

Should the radial model prove useful then through flow could be used to cool the sheared fluid automatically, i.e. by a centrifugal pump type action. The range of cooling flow required without disrupting the electrically applied stress has already been established [6].

[1] W. A. Bullough, "*Electrostructured fluids and smart machines*", one of a series of lectures given at Centre for Int. Mechanical Sciences, Udine, Italy, June 2000, (in press, Spring-Verlag-Berlin).

[2] R. Smyth, K. H. Tan, and W. A. Bullough, "*Heat transfer modeling of a cylindical ER catch*", Int. Jnl. Mod. Phys., **8,** p. 3033-3050 (1994).

[3] W. A. Bullough et. al, "*ESF clutch drive mechanisms and the ER linear reversing motion demonstrator*", lecture given at Centre for Int. Mechanical Sciences, Udine, Italy, June 2000, (in press, Spring-Verlag-Berlin).

[4] V. Oravsky, "*Heat transfer from an ESF radial plate clutch surface II*", Proc. 8th Int. Con. on ERF, MRS, Nice, France, 2001 (World Scientific Publishing).

[5] H. Y. Wong, Handbook of essential formulae and data on heat transfer for engineers, (Logman, London & New York, 1977).

[6] D. J. Ellam, R. J. Atkin, and W. A. Bullough, "*Appraisal of an ESF radial plate clutch with cooling flow using a NNF CFD solver*", ASME Int. Mech. Eng. Congress and Exposition, Orlando, Florida, 2000 (ASME), p. 117-125.

VIBRATION CONTROL OF A HELICOPTER TAIL LANDING GEAR SYSTEM USING ELECTRORHEOLOGICAL AND MAGNETORHEOLOGICAL SHOCK STRUTS

YOUNG-TAI CHOI AND NORMAN M. WERELEY

Dept. of Aerospace Engineering, University of Maryland, College Park, MD 20742, USA
E-mail: nicechoi@wam.umd.edu, wereley@eng.umd.edu

This study demonstrates the feasibility and effectiveness of ER (electrorheological) and MR (magnetorheological) fluid-based landing gear systems on attenuating dynamic load and vibration due to the landing impact. First, the theoretical model for ER/MR shock struts, which are the main components of the landing gear system, is developed based on experimental data. With this theoretical model, the telescopic type landing gear system using the ER/MR shock struts is constructed and its governing equation is derived. Then, the robust sliding mode controller against parameter variations and external disturbances is designed and controlled performance of the ER/MR landing gear system is evaluated during touchdown of the aircraft.

1 Introduction

During touchdown, an aircraft is exposed to a short duration of impulsive impact. The landing impact has been recognized as a significant factor in structural fatigue damage, dynamic stress on the aircraft airframe, and crew and passenger discomfort. In order to attenuate the landing impact transmitted to aircraft, several types of landing gear systems have been employed in commercial airplanes. One potential method to implement landing gear systems is to use smart fluids such as electrorheological (ER) and magnetorheological (MR) fluids. In general, ER and MR fluids are known to have outstanding advantages of continuously controllable rheological properties and fast response time. Therefore, the landing gear systems using ER or MR fluids have salient features such as continuously controllable mechanical damping characteristics and wide control bandwidth.

So far, of the research published, studies of the application of ER and MR fluids to the landing gear system are largely unexplored [1]. Consequently, the main contribution of this study demonstrates the feasibility and effectiveness of ER and MR fluid-based landing gear systems on attenuating dynamic load and vibration due to the landing impact. In order to achieve this goal, the theoretical model for ER/MR shock struts, which are the main components of the landing gear system, is developed based on experimental data obtained at different excitation velocities and fields. With the theoretical model, the landing gear system using ER/MR shock struts is constructed and its governing equation is derived. From the obtained governing equation, the robust sliding mode controller against parameter variations and external disturbances is formulated to attenuate the acceleration and displacement of the landing gear system. Controlled performances of the landing gear system under the robust sliding mode controller are evaluated during touchdown of the aircraft.

2. ER/MR shock struts

A schematic diagram of flow-mode type of controllable ER/MR shock struts proposed in this study is shown in Figure 1. The shock struts are composed of gas and hydraulic reservoirs. The hydraulic reservoir is divided into the upper and lower chambers by the piston head and fully filled with controllable fluids. There is a valve

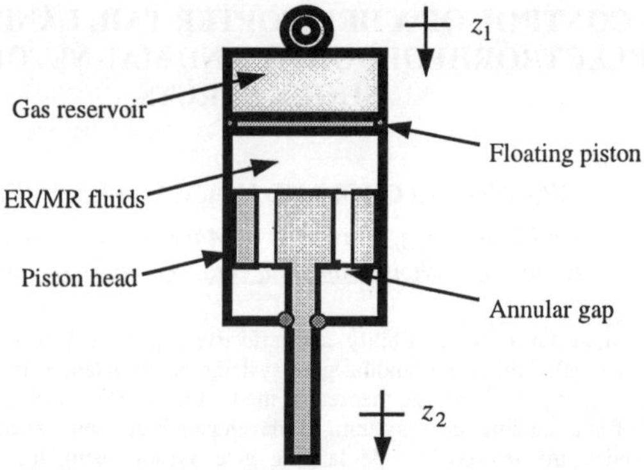

Figure 1. Schematic diagram of the controllable ER/MR shock strut.

through the piston head, typically an annular valve, where the fluid is activated through the application of field. As the piston moves, the resulting pressure moves fluids through the annular gap from the upper chamber to lower chamber. In order to compensate for the changing fluid volume occupied by the piston rod in the lower chamber, the gas reservoir is located above the upper chamber. In the absence of an external field, the ER/MR shock struts produce the same level of damping forces as passive shock struts comprised of viscous damping force, gas spring force and so forth. However, when field is applied to the controllable fluid shock strut, additional damping force due to field-induced yield stress is produced in the annular valve. This damping force can be continuously controlled by adjusting the intensity of applied field [1,2].

The damping force of the ER/MR shock strut (F_d) can be expressed as a combination of nonlinear viscous damping (F_v), gas spring (F_g) and yield (F_y) forces:

$$F_d = F_v + F_g + F_y \tag{1}$$

The nonlinear viscous damping force (F_v) can be given by

$$F_v = c(\dot{s}, G)\dot{s} \tag{2}$$

where $\dot{s}(=\dot{z}_1 - \dot{z}_2)$ is the relative velocity of shock strut and G is the applied field. In the case of ER fluids, $G\,(=E\,)$ is the electric field that is proportional to applied voltage. In the case of MR fluids, $G\,(=H\,)$ is the magnetic field that is proportional to applied current. $c(\dot{s}, G)$ is the nonlinear viscous damping function of the relative velocity and applied field. In the ideal Bingham model, the post-yield viscous damping of ER/MR damper is changed by neither velocity nor applied field. However, this assumption is valid at the conditions of low velocity and low field strength. Since the shock struts normally operate under the conditions of high velocity range and high field strength, a constant viscous damping is not sufficient to precisely predict the behavior of the strut. Therefore, in this study, a variable damping, $c(\dot{s}, G)$, is used to predict the behavior of the struts and the values of $c(\dot{s}, G)$ are identified through experiments. For doing so, first, the damping force of a shock strut is tested under different piston velocities and field strengths. Next, from the measured data, the viscous damping is estimated over each piston velocity and applied field through a curve fitting method. Then, a lookup table for the viscous damping is made from piston velocity and applied field. The viscous damping at the untested conditions between experimentally estimated data is obtained by an interpolation method.

Since the gas in the pneumatic reservoir undergoes the polytropic process during touchdown, the gas spring force (F_g) due to the gas pressure can be given by

$$F_g = A_r P_{g0} \left(\frac{v_{g0}}{v_{g0} - A_r s} \right)^n \tag{3}$$

where v_{g0} is the initial volume of the pneumatic chamber, P_{g0} is the initial gas pressure of the pneumatic chamber, and n is the polytropic exponent.

The yield force (F_y) due to the yield stresses of ER and MR fluids is given by

$$F_y = \frac{\dot{s}}{|\dot{s}|}(A_p - A_r)2\frac{L}{h}\tau_y(G) \tag{4}$$

where A_p is the cross-sectional area of piston head, A_r is the cross-sectional area of piston rod, L is the field activation length, h is the field activation gap and $\tau_y(G)$ is the yield stresses of the ER or MR fluid. It is well known that τ_y is an exponential function of applied field, $\tau_y = \alpha G^{\beta}$. Here, α and β are characteristic values determined by experiments.

In order to validate our theoretical model for the ER/MR shock struts, an ER shock strut was fabricated and tested. The requiring damping force level of the ER shock strut in this study is chosen on the basis of the conventional shock strut for a helicopter tail landing gear [2]. The electrode length (*L*) and gap (*h*) of the ER shock strut are 43 mm and 0.6 mm.

Figure 2 shows the maximum damping force performance of the ER shock strut at different excitation velocities. Below the piston velocity of 0.2 m/sec, it is clearly observed that the damping force increases as the applied electric field increases. However, over 0.2 m/sec, the increment of the damping force in response to applied field decreases. It arises from the fact that as the fluid velocity in the gap of the strut increases, the particle chains in the fluid are easier to break. Figure 3 presents the viscous damping estimated by a curve fitting method with the measured data. In this figure, the solid circle stands for the viscous damping directly estimated from the measured damping force characteristics. The viscous damping between the directly estimated values is obtained by the interpolation method. Figure 4 presents comparison between the theoretical and experimental data for the ER shock strut. As observed in this figure, the theoretical data show favorable agreement with the experimental data. Therefore, it is proved that the proposed theoretical model for the ER shock strut is appropriate for predicting its behavior with respect to applied field.

3. ER/MR landing gear system

In general, there are several types of landing gear systems. Our analysis in this study will focus on a telescopic type of landing gear system. The transient behavior of the landing gear system can be easily evaluated because the telescopic type of landing gear has no linkage mechanism. The schematic configuration of the landing gear system considered in this analysis is shown in Figure 5. It is noted that our analysis is based on the assumption that the airframe is a rigid body mass (the upper mass) and the wheel tire is a rigid body

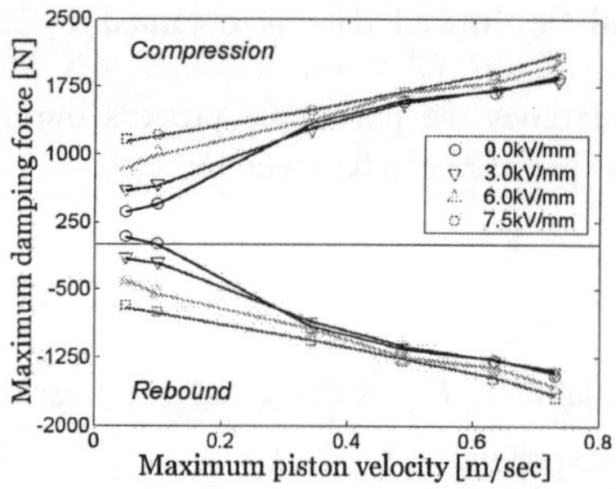

Figure 2. Maximum damping force versus piston velocity.

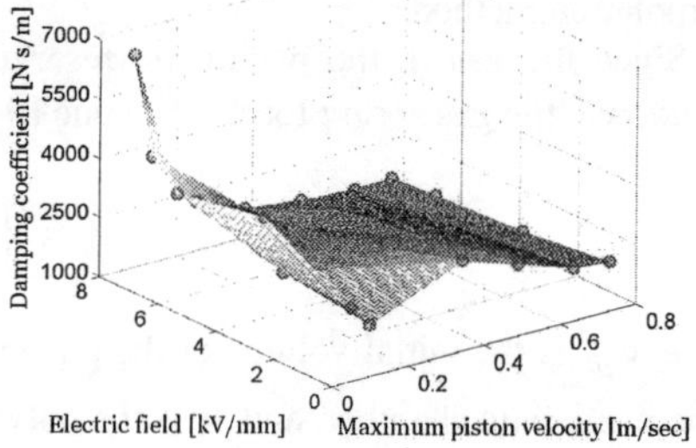

Figure 3. Estimated damping coefficient.

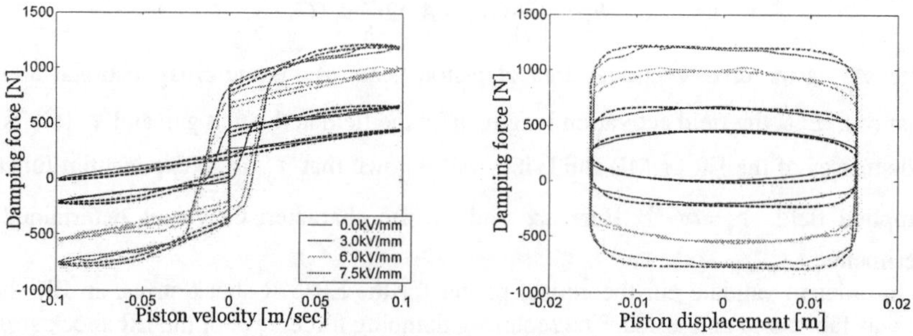

(a) damping force vs. piston velocity (b) damping force vs. piston displacement

Figure 4. Comparison between theoretical (dash line) and experimental (solid line) data.

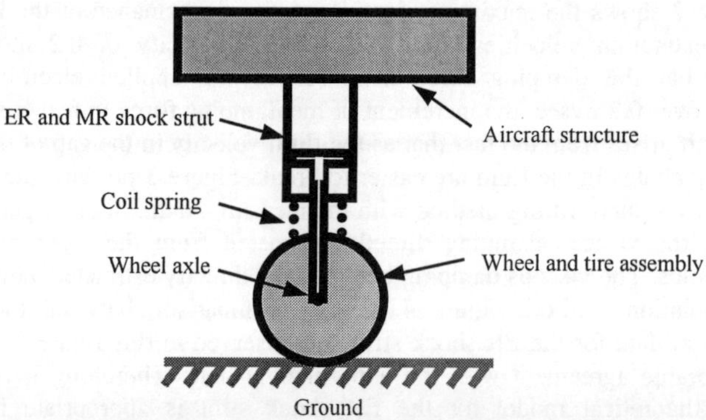

Figure 5. Schematic configuration of telescopic type of ER/MR landing gear system.

mass (the lower mass) and nonlinear spring.

The equations of motion for the landing gear system can be expressed as

$$\frac{W_1}{g}\ddot{z}_1 + F_{ds} + L_f - W_1 = 0$$
$$\frac{W_2}{g}\ddot{z}_2 - F_{ds} + F_t - W_2 = 0 \tag{6}$$

where W_1 is the weight of the airframe, W_2 is the weight of the wheel tire, g is the gravitation acceleration, L_f is the lift force, and $F_{ds}(=F_d+F_s)$ is the damping force of the ER/MR shock strut with coil spring. Also $F_s(=k_c(z_1-z_2))$ is the coil spring force and k_c is the stiffness coefficient of the coil spring. In addition, F_t is the wheel tire spring force and can be represented by

$$F_t = m z_2^r \tag{7}$$

where m and r are constants corresponding to various regimes of the tire deflection process. The regimes of the tire deflection process can be divided into two parts by the occurrence of tire bottoming. For a regime before tire bottoming, m=78.6×10^3 and r=1.34 [3]. For a regime after tire bottoming, the values of m and r become large numbers since a severe impact takes place when the tire bottoming occurs.

On the other hand, the ER/MR shock struts are inflated to some finite pressure in the fully extended position. Thus the shock strut does not begin to move under impact until sufficient force is developed to overcome the initial preloading imposed by the gas force [4,5]. Since the shock strut is effectively rigid in compression as well as in bending, the landing gear system may be considered to have only one degree of freedom during the initial stage of the impact. The equations of motion for the one-degree-of-freedom system are derived in order to permit determination of the initial conditions required for the analysis of the landing gear behavior subsequent to the beginning of shock strut motion.

Since $\ddot{z}_1=\ddot{z}_2=\ddot{z}$ and $\dot{z}_1=\dot{z}_2=\dot{z}$ during this first phase of the impact when the vertical force at the wheel axle, F_{dv}, is less than the initial gas force of the shock strut, F_{p0}, so that $F_{dv} \le | F_{p0} |$, the equation of motion can be given by

$$\frac{(W_1+W_2)}{g}\ddot{z}+F_{ds}+F_t+L_f-W_1-W_2=0 \tag{8}$$

where, $F_{dv}=F_t+(W_2/g)\ddot{z}-W_2$. It is noted that the states of Eq.(8) are used as initial conditions for solving Eq.(6).

In order to attenuate the dyanmic load and vibration of the landing gear system, a robust sliding mode control algorithm is adopted in this study. It is well known that the robust sliding mode controller has strong stability in the presence of parameter variations and external disturbances [6]. The dynamics for the ER/MR landing gear system can be rewritten in the form of a single input nonlinear system

$$\begin{aligned}\ddot{z}_1 &= \frac{g}{W_1}\left(-F_v-F_a-F_s+W_1-u-v\right)\\ \ddot{z}_2 &= \frac{g}{W_2}\left(F_v+F_a+F_s+F_t+W_2+u\right)\end{aligned} \tag{9}$$

where, u is the control input that physically means the yield force, and v are the external disturbances like lift force.

Then, a sliding surface function is defined as follows

$$s=(\frac{d}{dt}+\lambda)z_1 \tag{10}$$

where $\lambda>0$ is the slope of the sliding surface. For the practical ER/MR landing gear system, there are some variations in the parameters such as upper vehicle mass, viscous damping, and so forth. Therefore, in this study, the following parameter variations are considered to design the robust sliding mode controller.

$$500kg \le \frac{W_1}{g} \le 800kg \ , \qquad 1500N \cdot s/m \le c(\dot{s},G) \le 7000N \cdot s/m \, ,$$

$$70000N/m \le k_s \le 90000N/m \, , \quad 0 \le v \le 2(W_1 + W_2)$$

It is assumed that the estimation errors of the parameter variations and dynamics are bounded as follows

$$\overline{W}_1 \ge |W_1 - \hat{W}_1| , \quad \overline{c}(\dot{s},G) \ge |c(\dot{s},G) - \hat{c}(\dot{s},G)| , \quad \overline{k}_s \ge |k_s - \hat{k}_s| , \quad \overline{F}_a \ge |F_a - \hat{F}_a| \tag{11}$$

where, $\hat{W}_1, \hat{c}(\dot{s},G), \hat{k}_s, \hat{F}_a$ are the estimation and $\overline{W}_1, \overline{c}(\dot{s},G), \overline{k}_s, \overline{F}_a$ are the bounds on the estimation errors. Then we can formulate the robust sliding mode controller satisfying the sliding mode condition, $s\dot{s} < 0$.

$$u = \frac{\hat{W}_1}{g} \lambda \dot{z}_1 - \hat{F}_v - \hat{F}_a - \hat{F}_s + \hat{W}_1 + k_g \operatorname{sgn}(s) \tag{12}$$

where, $\hat{F}_v, \hat{F}_a, \hat{F}_s$ are the estimation dynamics, and k_g is the control gain given by

$$k_g = \frac{W_{1\max}}{g} \left[\overline{c}(\dot{s},G) |\dot{z}_1 - \dot{z}_2| + \overline{F}_a + \overline{k}_s |z_1 - z_2| + \overline{W}_1 + \xi_v + (1-\phi_w)\lambda |\dot{z}_1| \right] \tag{13}$$

where $\phi_w = \dfrac{W_{1\min}}{W_{1\max}}$, $\xi_v \ge 2(W_1 + W_2)$. On the other hand, the proposed robust sliding mode control input is designed in the active actuator manner. However, since the ER/MR shock strut is semi-active actuator, the following semi-active condition should be superposed on the Eq. (12).

$$u = \begin{bmatrix} u & \text{if } u \cdot (\dot{z}_1 - \dot{z}_2) \ge 0 \\ 0 & \text{if } u \cdot (\dot{z}_1 - \dot{z}_2) < 0 \end{bmatrix} \tag{14}$$

This semi-active condition constrains the control input to ensure that the ER/MR landing gear system is dissipating energy.

Figure 6 presents controlled performances of the ER/MR landing gear system under the robust sliding mode controller. In this case, the upper mass is chosen as 800kg and the sink velocity of the aircraft is set to 0.76m/sec. It considers the practical condition that the aircraft carrying payload is landing softly. As observed in this figure, the acceleration of the upper mass is significantly reduced by employing the robust sliding mode controller to the ER/MR landing gear system. In addition, the robust sliding mode controller gives good performance in displacement attenuation. Figure 7 shows controlled performances of the ER/MR landing gear system in case of zero payload. The acceleration magnitude of the upper mass is increased since the aircraft mass is lighter than that of Figure 6. It is observed that the robust sliding mode controller still give good performance on attenuating the acceleration and displacement of the landing gear system.

4. Conclusion

An analysis of a telescopic type of helicopter tail landing gear system using ER/MR shock struts was constructed and its feasibility and effectiveness on attenuating the dynamic load and vibration were demonstrated in this study. The theoretical model for ER/MR shock struts was developed based on experimental data obtained at different excitation velocities and fields for an ER shock strut constructed at Maryland. With the theoretical model, the ER/MR landing gear system was constructed and its governing equation was derived. From the obtained governing equation, the robust sliding mode

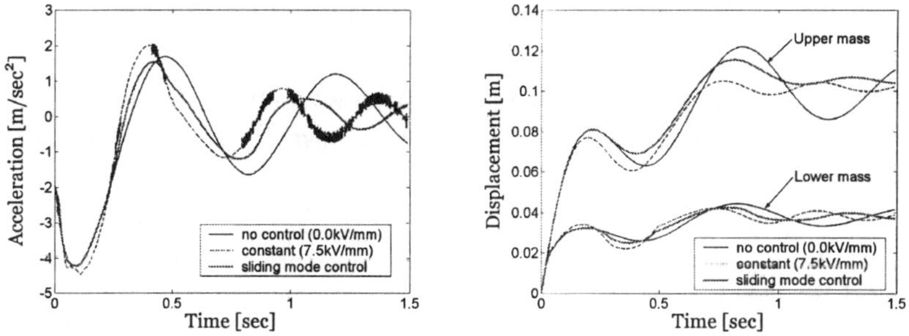

(a) acceleration (b) displacement

Figure 6. Controlled performance of the ER/MR landing gear system in case of carrying payload.

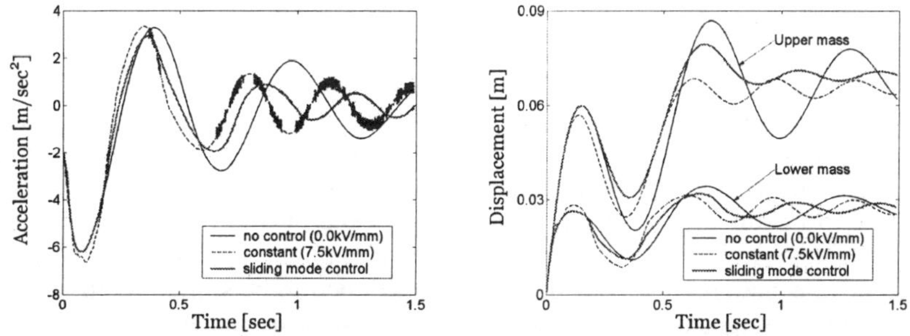

(a) acceleration (b) displacement

Figure 7. Controlled performance of the ER/MR landing gear system in case of zero payload.

controller against parameter variations and external disturbances was formulated to attenuate the acceleration and displacement of the landing gear system. It has been demonstrated that acceleration and displacement are significantly attenuated by employing the robust sliding mode controller to the ER/MR landing gear system regardless of parameter variations.

References

1. Lou, Z, Ervin, R.D., Filisko, F.E. and Winkler, C.B., An Electrorheologically controlled semi-active landing gear, *SAE Technical Paper Series* **931403** (1993).
2. Choi S.B., Choi Y.T., Chang, E.G., Han, S.J., and Kim, C.S., Control characteristics of a continuously variable ER damper, *Mechatronics* **8** (1998) pp.143-161.
3. Reddy P. J., Nagaraj V.T. and Ramamurti V., Analysis of a semi-levered suspension landing gear with some parametric study. ASME *J. Dynamic Systems, Measurement, and Control* **218** (1984) pp.218-224.
4. Yadav D. and Ramamoorthy R.P., Nonlinear landing gear behavior at touchdown. ASME *J. Dynamic Systems, Measurement, and Control* **113** (1991) pp.677-683.
5. Milwitzky B. and Cook F.E., Analysis of landing gear behavior. *NACA Report* **1154** (1953).
6. Slotine J.J. and Li W., Applied nonlinear control. (Prentice-Hall International Inc., New Jersey, 1991).

ELECTRO-STRUCTURED FLUID SEALS

R. J. ATKIN, D. J. ELLAM, AND W. A. BULLOUGH

SMMART Engineering Network, University of Sheffield, Sheffield, S1 3JD, UK

The aim of this paper is to bring together various works on concentrically located rotating shaft seals which utilize electro-structured fluid sealant, in order to produce a generalized approach to their design, thus resolving seemingly incompatible reports on the effects of rotational speed on sealing capabilities. In turn the results from these works are reviewed, a viscoplastic analysis of the two dimensional problem is presented and the effects on leakage rate of radius ratio, rotational speed and axial pressure gradient as they interact with fluid properties are illustrated.

Introduction

Experiments, and analysis based on steady, two-dimensional flow induced by a pressure gradient in the axial (shaft) direction and rotary motion in the other (circumferential) direction carried out on an ERF [1], indicates that the motion in the circumferential direction of one surface with respect to the stationary surface always causes flow to take place in a direction appropriate to that of the pressure gradient inducing it. In a second work [2] no equivalent theory is supplied but experimentally a MRF seal is found to withstand a significant axial pressure gradient without leakage occurring. Since the constitutive equation for ERF and MRF is understood to be similar, both materials being modelled as a Bingham plastic, these conflicting results are in need of examination.

Recently the use of a Computational Fluid Dynamics package (CFD) has been used to solve viscoplastic flows by use of an algorithm attached to a fundamental Newtonian fluid solver which is inclusive of centrifugal effects. Given that the viscoplastic continuum analysis, albeit in some cases approximated by the absence of centrifugal terms and limited by use of the lubrication assumption compares well with the CFD results in both radial plate and concentric cylindrical clutch cases [3], [4], it is now used to resolve the apparent paradox between the results reported in [1] and [2].

From the outset due caution needs to be exercised: at low shear rates when the exact properties of the various fluid types are not fully known. This can be due to the presence of surfactant and anti coagulant additives – see for example [5]. This position reinforces calls for a more comprehensive understanding and perhaps more accepted standard methods of fluid testing. CFD analysis is becoming very powerful yet the results of the analysis can vary significantly depending on the definition of yield stress and plastic viscosity. These difficulties may be overcome by the adaptation of the main basic effective viscosity vs. deformation rate vs. excitation test standard.

Theoretical Analysis: Bingham Plastic Flat Plate Theory

Since the publication of [1] the theory has been developed to include the plug regime, which occurs when the effect of relative speed of the seal surfaces compared to the pressure gradient on the flow rate is relatively low. A review of this extended analysis is presented below.

Within the gap width, where the field strength is assumed constant, flow occurs in both the circumferential and axial directions, the former caused by rotation of the inner electrode and the latter due to the difference in pressure between inlet and outlet along the shaft. Since the inter electrode gap width h is small compared to the mean radius r_m of the electrodes the flow is considered to be between two flat plates, $x = 0$ which moves with speed $V(=r_l\Omega)$ in the y-direction and a fixed plate $x = h$. The pressure gradient $G(=dp/dz = \Delta P/L)$ is in the z-direction. If $v(x)$ and $w(x)$ denote the velocity components in the y- and z- directions respectively the two rates of strain components ½(dv/dx) and ½(dw/dx) give an expression for the resultant 2-d shear rate,

$$\dot{\gamma} = \tfrac{1}{2}\{(dv/dx)^2 + (dw/dx)^2\}^{1/2}. \tag{1}$$

The non-zero shear stresses τ_{xy} and τ_{xz} and the magnitude of the resultant shear stress, τ, are given by:

$$\tau_{xy} = a, \quad \tau_{xz} = Gx + b, \quad \tau = \sqrt{a^2 + (Gx+b)^2}\,, \tag{2}$$

for which a and b are constants.

Flow occurs when $\tau > \tau_e$. Here the yield stress τ_e is a positive constant. Velocity components, which satisfy the no-slip condition on the boundaries $x = 0$ and $x = h$, can be determined. To satisfy the boundary conditions on w it follows that $b = -Gh/2$ and the resultant shear stress is symmetric about $x = h/2$. Using (2) (with $b = -Gh/2$) the condition $\tau > \tau_e$ can be recast in the form $a^2/\tau_e^2 + G^2/(2\tau_e/h)^2 > 1$. For a given material and known geometry this gives a criteria for the values of a and G for flow to occur. When the operating parameters are such that the values of a and G lie inside the ellipse $a^2/\tau_e^2 + G^2/(2\tau_e/h)^2 = 1$ no axial flow can occur and the shaft cannot rotate. The main results are given below (see [1] & [6] section 2, for further details).

Case (i) $|a| > \tau_e$

Here flow occurs across the whole gap with no plugs present. Both v and w depend upon G and V and so are affected by varying either the axial pressure gradient or the speed of the plate. To satisfy the condition $v(h) = 0$,

$$|a|\left[h - (2\tau_e/G)\sinh^{-1}(Gh/2|a|)\right] = \mu V. \tag{3}$$

The magnitude of the shear stress $|a|$ depends upon G and V as well as the fluid parameters τ_e and plastic viscosity μ and the gap width h. If the plate moves, for any finite pressure gradient flow occurs in the axial z-direction. The leakage flow rate is given by:

$$\dot{q}_z = -\frac{Gh^3\pi r_m}{6\mu} + \frac{2\pi r_m}{\mu}\left\{\frac{\tau_e h}{4G}\sqrt{G^2h^2 + 4a^2} - \frac{\tau_e a^2}{G^2}\sinh^{-1}(Gh/2|a|)\right\}. \tag{4}$$

$\dot{q}_z$ has Newtonian and Bingham contributions, the latter always trying to reduce the former. When $G = 0$, $\dot{q}_z = 0$ and at high values of G the Newtonian term dominates. For very low flow rates when $|Gh/2a| << 1$ (3) (4) reduce to

$$|a| = \mu(V/h) + \tau_e, \quad \dot{q}_z = -(Gh^3\pi r_m/6\mu)(1 - \tau_e/|a|), \tag{5}$$

so that in this approximation $|a|$ is independent of G but $\dot{q}_z$ depends upon G and V and is zero if either G or V vanish. Equation $(5)_1$ also arises from the analysis of shear flow when $w = 0$. For higher pressure gradients using (3) it is found that $|a|$ decreases as $|G|$ increases and eventually there is a transition from $|a| > \tau_e$ to $|a| < \tau_e$.

Case (ii) $|a| < \tau_e$

In this case there is a central plug symmetrically placed about x = *h/2* of thickness $2(\tau_e^2 - a^2)^{1/2} / |G|$, which is less than *h* provided $|G| > 2(\tau_e^2 - a^2)^{1/2} / h$. Again both *v* and *w* depend upon *G* and *V*. Results corresponding to (3) and (4) are obtained. The transition point from case (i) to (ii) is found to move to higher flow rates with decreasing τ_e and *V*. Typical behaviour for $\tau_e = 2$ and 10 kPa at 600 rpm is shown in Figs. 1 & 2 of ref [4]. Flow profiles are shown in Figs. 3 & 4.

Computational Fluid Dynamics Package

The same problem is solved under identical flow assumptions as those outlined above using a commercial CFD package (Fluent v4) for flow between two concentric cylinders but without making the flat plate approximation adopted in the theoretical analysis.
The governing Navier-Stokes equations are discretised using a finite volume method which involves splitting the continuum into a finite number of cells, allowing the partial differential equations to be integrated on a control volume basis. Each cell volume has a viscosity which is dependent on the local deformation rate $\dot{\gamma}$. This non-Newtonian viscosity μ_{nn} is determined using the bi-viscous model where μ_{max} is set at 1000 Pa.

$$\mu_{nn} = \frac{\tau_e}{\dot{\gamma}} + \mu \quad \leq \mu_{max} \tag{6}$$

Since the flow aligns with the grid a first order power law interpolation scheme is adequate to determine cell face values. Velocities are determined from the momentum equations, and the pressure is determined using the SIMPLE pressure-velocity coupling algorithm.

The seal is modelled using an inner and outer cylindrical wall; all derivatives are zero in the tangential direction allowing a 2-d axisymmetric grid to be used. A uniform pressure condition was set at the inlet and outlet allowing a pressure difference to be set over 100 mm of length. The inner cylindrical wall $r = r_1$ was set to rotate at a tangential speed $V = r_1\Omega$.

Review

The aim of this section is to consider why some experimental results [1] and the Bingham plastic flat plate theory suggest that it is impossible to have an effective ESF seal when the shaft is rotating whilst other experimental results reported in [2] for their MRF indicate that the opposite is true. A list of the values of the relevant parameters used by these authors is given in Table 1; those indicated by a (*) are estimates.

	Ref [1]	Ref [2]
Yield stress	ERF: 256 – 1500 Pa	MRF: ~ 5 - 10 kPa *
Shaft diameter	60 mm	20 mm
Gap width	1 – 2 mm	0.075 - 0.15 mm
Shaft speed	500-2000 rpm (~50 - 200 rad/s)	< 250 rpm (~25 rad/s)
Pressure gradient	1000Pa/10mm = 0.1 GPa/m	10 min, 300 max GPa/m *

Table 1: Seal parameters used in [1] and [2]

The experiments in [1] were conducted in the range $|a| > \tau_e$ but, from the magnitudes of the yield stress and speed, it appears that the results reported in [2] could be in the range $|a| < \tau_e$ when, in the parallel plate case, there is a central plug. It also seems pertinent at this point to consider any effects due to cylindrical geometry. For purely rotational flow ($w = 0$) between two concentric cylinders with radii r_1 and r_2 ($r_1 < r_2$), caused by either cylinder rotating with angular speed Ω with the other stationary, at small enough speeds it is possible to have flow in only part of the gap with a plug attached to outer cylinder. The angular speed at which the plug disappears for a Bingham plastic material is given by [7]

$$\Omega_{limit} = \frac{\tau_e}{2\mu}\left(\frac{r_2^2}{r_1^2} - 1 - \ln\frac{r_2^2}{r_1^2}\right). \tag{7}$$

For $\Omega > \Omega_{limit}$ there is flow across the whole gap. Estimates of Ω_{limit} are given in Table 2.

case	r_1	$r_2 - r_1$	τ_e	μ	Ω_{limit}
	mm	mm	Pa	Pa.s	rad/s
[2]	10	0.15	10000	0.1	22
	10	0.15	10000	0.5	11
[1]	30	2	1000	0.1	44
on smaller shaft	10	2	1000	0.1	4963

Table 2: Estimates of Ω_{limit}

The value of Ω_{limit} is significantly affected by the fluid properties. It can be concluded from line 3 of the above table that experimental work in [1] is likely to have been carried out at rotational speeds greater than Ω_{limit}, the limiting speed for plugs being present, when the theory predicts flow across the whole gap. This conclusion is further validated by the fact that the results (within experimental error) do match up to the theory. However with the same parameters and a smaller shaft radius very high speeds are needed. As illustrated in line 1 in the above table, for a smaller shaft (as used in [2]) the limiting speed is much higher than many of the experimental speeds used. This consequently means that plugs are likely to be present which goes some way to explaining why it may be possible to have full sealing.

CFD Analysis

Further insight into the effect of the relevant parameters when flow is present in both axial and circumferential directions is gained from the CFD analysis. Since shaft diameter is likely to be a significant factor, results are obtained for two sizes corresponding to the apparatus used in [1] and [2]. To highlight the important parameters an orthogonal array, which limits the total number of simulations to 16, is used (Tagucci Analysis). The table below shows the flow rate results for all 16 simulations, also indicated is the theoretical transition speed Ω_{limit}.

						Shaft radius 10 mm			Shaft radius 30 mm		
No.	Ω	Gap Size	Yield Stress	Vis.	dP/dx	Ω_limit eqn. (5)	Flow Rate [Lt/hr] *Diff = from theory*		Ω_limit eqn. (5)	Flow Rate [Lt/hr] *Diff = from theory*	
	rad/s	mm	kPa	Pa.s	GPa/m	rad/s	CFD	Diff [%]	rad/s	CFD	Diff [%]
1	10	0.5	10	0.05	10	492	0.20	84.8	55	4.03	60.7
2	30	0.5	10	0.10	20	246	4.71	40.3	28	39.85	30.8
3	10	0.2	10	0.10	10	40	0.05	47.1	4	0.64	12.4
4	10	0.5	5	0.05	20	246	7.47	38.8	28	60.71	26.5
5	30	0.2	10	0.05	20	79	0.41	46.7	9	4.21	14.5
6	30	0.5	5	0.10	10	123	4.51	36.4	14	35.56	23.0
7	10	0.2	5	0.10	20	20	0.44	37.5	2	3.52	20.8
8	30	0.2	5	0.05	10	40	0.43	28.6	4	4.06	6.7
					Average	161	2.28	45.0	18	19.07	24.4

Table 3: Chosen parameters and resulting flow rates

Here the values of Ω are close to the range used in [2]. For the chosen range of parameters the results from the CFD simulations differ from those of the flat plate theory between 7 - 85 %, the largest discrepancy occurring when $\Omega \ll \Omega_{limit}$ (simulation 1). The flow profiles corresponding to simulation 1 when r_I = 10 mm are shown in Fig. 1. The circumferential velocity is seen to exhibit plug-like behaviour, corresponding to high values of μ_{nn}, and the axial velocity is non-symmetric. The plug-like region is attached to the outer cylinder and extends over almost half the gap. The large discrepancy indicates that the flat plate approximation is inapplicable and that it is necessary to consider the cylindrical geometry. Other large discrepancies indicate that for this range of parameters care has to be taken when using the flat plate approximation.

Care in interpreting the results of Fig. 1 should be exercised on account of the concentric cylindrical arrangement and the selection of the value of μ_{max}. It will be noted that no constant velocity plug pen se is evident.

The CFD results are used in Fig. 2 to show the average effect of each parameter. A considerably lower flow rate on the smaller shaft can now be seen. At the relatively low speeds of 10 & 30 rad/s the shaft speed is found to be relatively unimportant. The yield stresses and pressure gradients are found to have an almost linear effect, i.e. double the yield stress halves the flow rate. As clearly represented in Fig. 2 the significant parameter of most interest appears to be the gap size. It must be noted that the resulting flow rates are quite small (litres/hour).

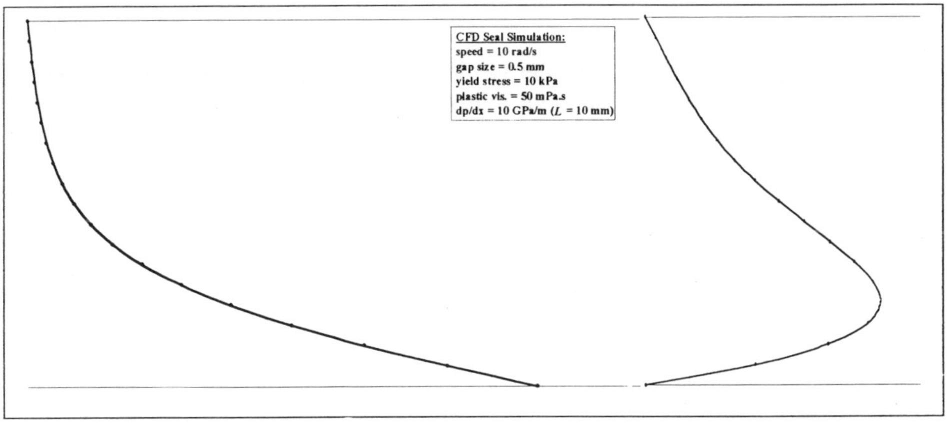

Fig. 1: CFD results, velocity profiles for case no. 1 on 10 mm shaft. Circ. (left) and Axial (right).

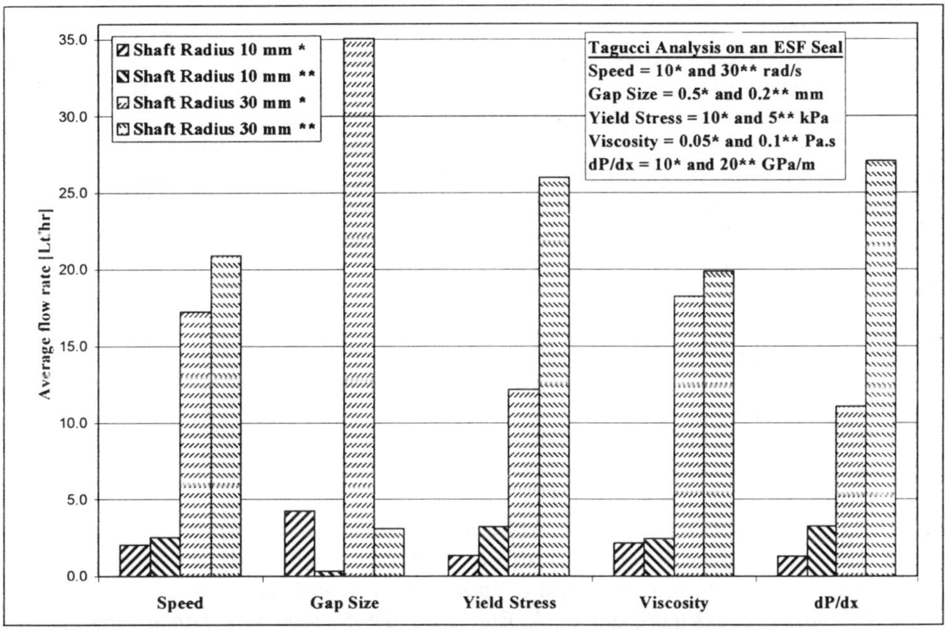

Fig. 2: Effect of each parameter on average flow rate using CFD simulation results.

It can also be shown that for smaller Ω_{limit}/Ω (less extensive plug behaviour) the results are generally closer. Also in all cases the larger *r/h* values (30 mm shaft) the results are considerably closer (Table 3). Over the range of parameters that were chosen the gap size is a more significant factor then yield stress and pressure gradient. Also it is found that the plastic viscosity and rotational speed are generally less important.

Conclusion

The present analyses indicate that, the results in [1] and [2] can both be reconciled. Although the analyses predict that there is leakage when rotation occurs, for some range of parameters this may be very small (see Fig. 2). For all parameter values taken in Table 3 it is found that smaller flow rates are obtained with a smaller shaft radius. Since this model only approximates the real fluid, the flow rate could be zero as found experimentally in [2] and [8].

Whilst formula (6) for Ω_{limit} used to predict the occurrence of plugs in cylindrical geometry is restricted to purely rotational flow, the CFD results, which allow for flow in both the circumferential and axial directions, substantiate the occurrence of a plug-like region attached to the outer cylinder. The CFD analysis also allows for the effect of the relevant parameters on flow rate to be investigated and indicates that the gap width and the shaft radius are the most significant.

For some parameter values in Table 3 there are large discrepancies between the values for flow rates predicted by the CFD and those which follow from the theoretical analysis in which the flow is taken to be between two parallel flat plates. Here the speeds where chosen to be within the range used in [2]. However in earlier work [1], [3], [4] and [6] where higher speeds are used very good agreement is found between this theory and the results from CFD. This indicates that care has to be taken when replacing the cylindrical geometry by two parallel flat plates.

Finally, in all the analysis presented the excitation field is assumed to be constant across the gap. The effect of a non-uniform field distribution can be included in both theoretical analysis and CFD and further work is in progress.

References

[1] R. J. Atkin, T. J. Corden, T. G. Kum, and W. A. Bullough, "*Two-dimensional Bingham plastic flow in a cylindical pressurised clutch*", Int. J. Mod. Phys. B, **8,** p. 2955-2965 (1994).

[2] W. I. Kordonski and S. R. Gorodkin, "*Magnetorheological fluid seal*", Proc. 5th Int. Con. on ERFs, MRSs and Associated Technology, Sheffield, UK, 1995 (World Scientific Publishing) p. 704-709.

[3] D. J. Ellam, R. J. Atkin, and W. A. Bullough, "*Appraisal of an ESF radial plate clutch with cooling flow using a NNF CFD solver*", 2000 ASME Int. Mech. Eng. Congress and Exposition, Orlando, Florida, 2000 (ASME), p. 117-125.

[4] R. J. Atkin, D. J. Ellam, and W. A. Bullough, "*Thoughflow Effect on Slip Torque in an ESF Concentric Clutch*", Proc. 13th Int. Congress on Rheology, Cambridge 2000, (British Soc. Of Rheology) p. 414-416.

[5] M. R. Tully, J. W. Binder, and a. J. D. Carlson, "*Properties and applications of commercial magnetorheological fluids*", 5th SPIE Symposium on Smart Structures and Materials, 1991, p. 202-214.

[6] W. A. Bullough, R. J. Atkin, S. Urang, T. G. Kum, C. Musch, and T. Rober, "*Two-dimensional flow of an ESF: experiment, CFD, Bingham plastic analysis*", Int. J. Mod. Phys. B, (2001).

[7] W. L. Wilkinson, *Non-Newtonian fluids* (Pergamon Press, Oxford, 1960).

[8] T. Fujita et. al., "*Characteristics for characterisation of magnetorheological suspensions for seals*", Proc. 7th Int. Con. on ERFs, MRSs and their applications, *Honolulu, Hawaii*, 1999 (World Scientific Publishing) p. 709-715.

EXPERIMENTAL INVESTIGATIONS OF THE EFFECTS OF ELECTRODE GEOMETRY ON THE PERFORMANCE OF ELECTRORHEOLOGICAL FLUID VALVES

G. GEORGIADES AND S. O. OYADIJI

Manchester School of Engineering, University of Manchester, Manchester M13 9PL, UK
E-mail: georgiadesg@fs1.eng.man.ac.uk, s.o.oyadiji@man.ac.uk

One of the key factors affecting the performance of ER fluid devices is the geometry of the electrodes, which also affects the geometry of the device itself. For an ERF valve, the pressure drop and the flow rate characteristics dictate the performance and efficiency of the device. The geometrical configurations considered in applications up to now mainly involve two parallel surfaces, most of these being parallel flat electrodes and annular electrode configurations. However, there are other electrode configurations that may be potentially more efficient than the traditional parallel flat electrode configurations. The static and dynamic yield stresses of different ER valve geometrical configurations were characterised. The efficiency considered is not the electrical efficiency of the structure, but rather the locking pressure efficiency of the geometry of the electrodes and the valve, closely linked with the pressure drops achieved at different applied electric fields. The data produced was processed, analysed and plotted to give the characteristic operation curves for the different electrode configurations. In this work, the performance of ERF valves based on the traditional parallel flat electrode configurations is compared with those of ERF valves based on square teeth and tapered electrode configurations.

1 Introduction

Electrorheological fluid devices are difficult to model due to the complexity of the ER effect. Modelling gets even more complex when the electrodes are not parallel electrodes [1]. Nevertheless, it is possible that more effective geometries do exist [2,3,4,5]. If modelling is not an option then other methods of getting information about their performance and subsequently their control must be investigated.

This paper looks into how other geometries could be tested experimentally, yielding information about their performance and efficiency, and determining their characteristics without going through the mathematical modelling of the electrodes. The obvious way is to link performance to the pressure drop characteristics.

2 Apparatus

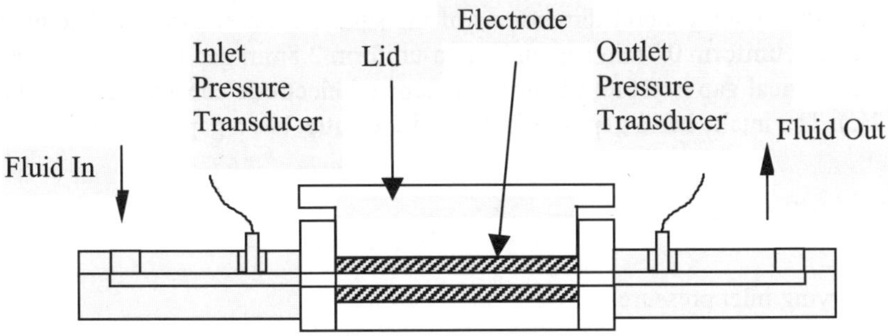

Figure 1: Apparatus used for testing

The apparatus in Figure 1 shows the rig constructed for the tests. It was constructed as a set of modular units to enable the easy replacement of one set of electrodes by another set of a different geometrical configuration. This means that in the tests only electrodes change and all other components of the rig stay the same. The electrodes have a general surface of 100x50 mm (see Figure 1), and thus give a duct flow area of 50xheight. The upper electrode is attached to the lid. Using spacers, various heights of the duct (distance between electrodes) can be achieved. Two pressure transducers were attached to measure the pressure at the inlet and the outlet of the valve. An air-line (up to 80psi) pressurises a cylinder containing the ER fluid that is forced through the valve. An air pressure regulator controls the supplied air (and fluid) pressure. A 15kV, 3mA power supply is used for the tests. A computer controls the supplied electric field and logs the two pressures, the applied electric field, the supplied voltage and the temperature of the negative electrode. The ER fluid used for the tests consists of silicon oil and starch at a concentration of 50% (by weight).

3 Experimental Testing

3.1 Electrode geometry

Three different electrode geometries were used for the tests. They are shown in Figure 2. Figure 2(a) shows the general dimensions of the flat parallel electrodes. The electrodes were tested at 1.1mm, 3.3mm and 5.5mm gaps.

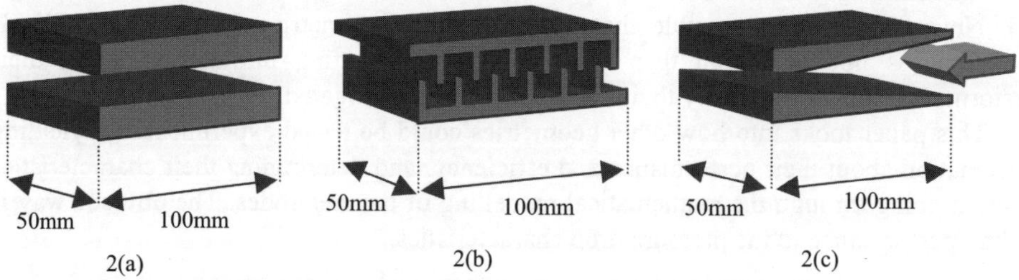

Figure 2: Flat parallel, square teeth and tapered electrode geometries

Figure 2(b) shows the general dimensions of the square teeth electrodes. The electrodes were tested at a uniform 0.7mm gap and a non-uniform 2.8mm gap (horizontal gap still at 0.7mm but vertical gap is set to 2.8mm). The tapered electrode configuration is shown in Figure 2(c). The intake has a gap of 5.7mm, and the outlet has a gap of 1.5mm.

3.2 Testing Procedures

3.2.1 Varying inlet pressure

In this series of tests, at a given applied electric field the input pressure was increased from 0psi to 70psi and back to 0psi (note: 1psi= 6895Pa). The electric field was then altered and the procedure repeated. The typical performance curves obtained from such a test are shown in Figure 3.

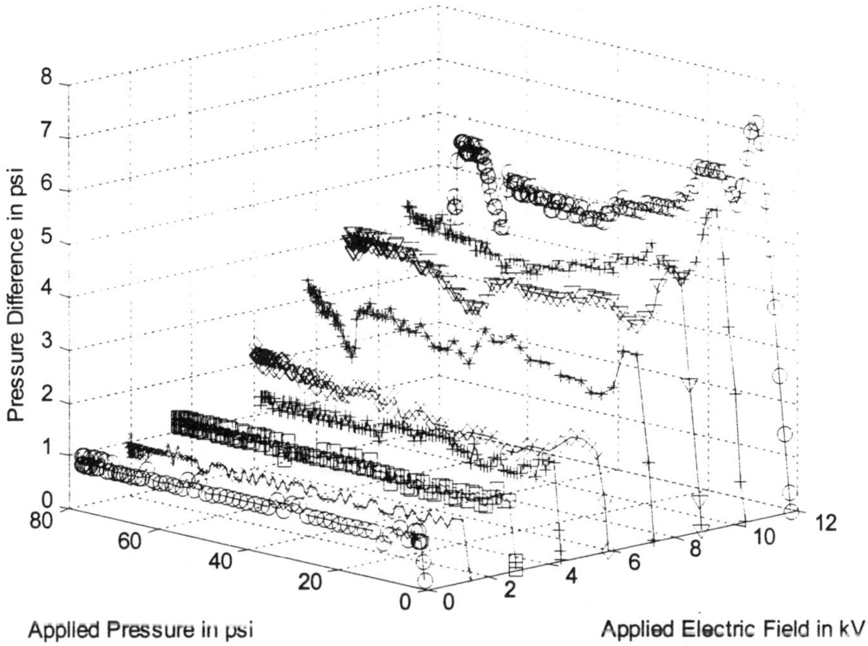

Figure 3: Typical performance curves obtained for varying inlet pressure for the flat electrode configuration at 5.5mm gap

Each curve represents a different applied electric field. All curves start with a straight line of a gradient of about one. This is the stage where the flow is completely stopped and the pressure difference (between input and output) equals input pressure. This straight line leads to a peak, showing the achieved static locking pressure. Increasing the applied pressure, the particle chains of fibrils, which were formed by the ER effect, break, flow is initiated and the pressure difference drops to a dynamic pressure that increases with input pressure.

3.2.2 Varying applied electric field

In this testing procedure, at a given inlet pressure, the applied electric field was increased and decreased. The inlet pressure was then changed and the procedure repeated. This test is more realistic in checking the valve mode of the device. When using ER valves the actuation pressure is approximately constant. The objective is then to control the flow of the ER fluid through the valve using a control variable (in this case the applied electric field). Figure 4 shows the characteristic curves produced from such a test. The curves show both, the increasing (loading) and the decreasing (unloading) of the electric field. As the applied electric field increases, the pressure drop increases until it reaches a limiting magnitude. This is when the achieved pressure difference equals the inlet pressure and there is no flow of the fluid. On reducing the applied electric field, the pressure drop curve follows a different route. This is due to the fact that the static yield stress [6,7] is higher than the dynamic one so that the particle chains hold longer. Thus a hysterisis loop is formed.

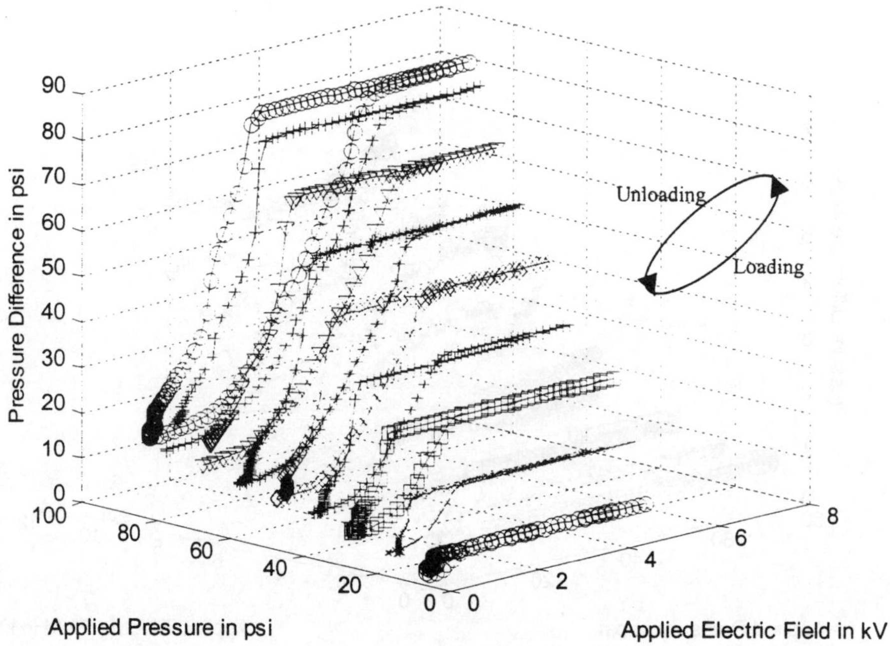

Figure 4: Typical performance curves obtained for varying electric field for the square teeth electrode configuration at 5.5mm gap.

When finally the particle chains break, the pressure does not return to the initial loading curve but instead follow a return path above the loading one.

4 Results and Discussion

From the varying pressure tests described in section 3.2.1 and their characteristic curves obtained, the static pressures (peaks) and the dynamic pressures at an applied pressure of 70psi are plotted for the different electrode geometry configurations on Figures 5 and 6. The pressure drop is shown to decrease as the electrode gap decreases. This implies that the ER effect is enhanced as the electrode gap decreases. The square teeth electrodes with the smallest gap and elongated path give the greater ER effect. The tapered electrodes with the decreasing gap (5.7mm to 1.5mm) in the direction of the flow, give an ER effect lower than that of the flat parallel electrodes at 1.2mm gap, but a higher ER effect than that of the flat parallel electrode at 3.3mm and 5.5mm gaps. Figure 6 also shows a comparison between the static and dynamic pressure drop curves for the flat parallel electrodes at a gap of 1.2mm. The static pressure drop curve is higher. The shaded area shows the difference between the static and dynamic pressures.

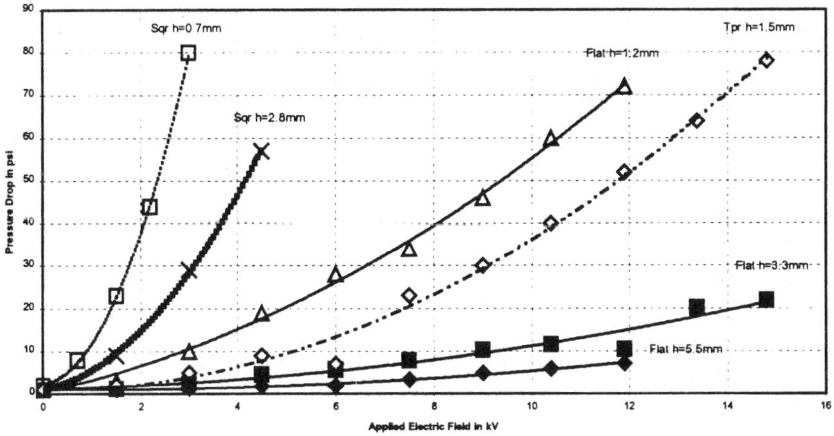

Figure 5: Comparison of the static pressure drop characteristics for the different electrode configurations.

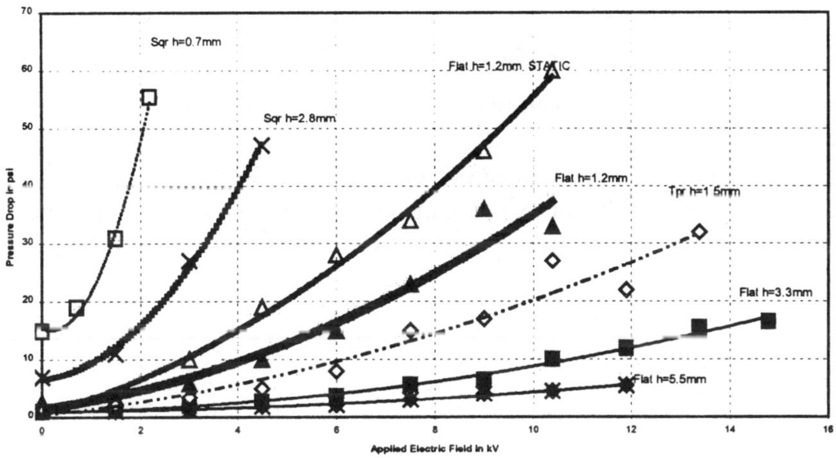

Figure 6: Comparison of the dynamic pressure drop characteristics for the different electrode configurations.

The varying electric field tests described in section 3.2.2 were conducted to test the valves in conditions similar to valve operation. That is having a fluid at a constant pressure and controlling the output pressure by varying the applied electric field. Figure 7 shows one of the cycles in Figure 4. The cycle is the one for the approximate inlet pressure of 49psi. The point at which the flow stops at a constant inlet pressure of 49psi is indicated in Figure 7 as point 1. The corresponding points at other inlet pressures were obtained from Figure 4 and were used to plot the "Dynamic Hold" curve in Figure 8. On reducing the electric field, the pressure difference is maintained until the chains break and flow initiates. This point is indicated as point 2 on Figure 7. The corresponding points from the loading curves in Figure 4 were used to plot the curve labelled 'Static Hold' in Figure 8. The third set of points labelled 'Static Break' in Figure 8 are the points where the pressure drop gradient starts reducing.

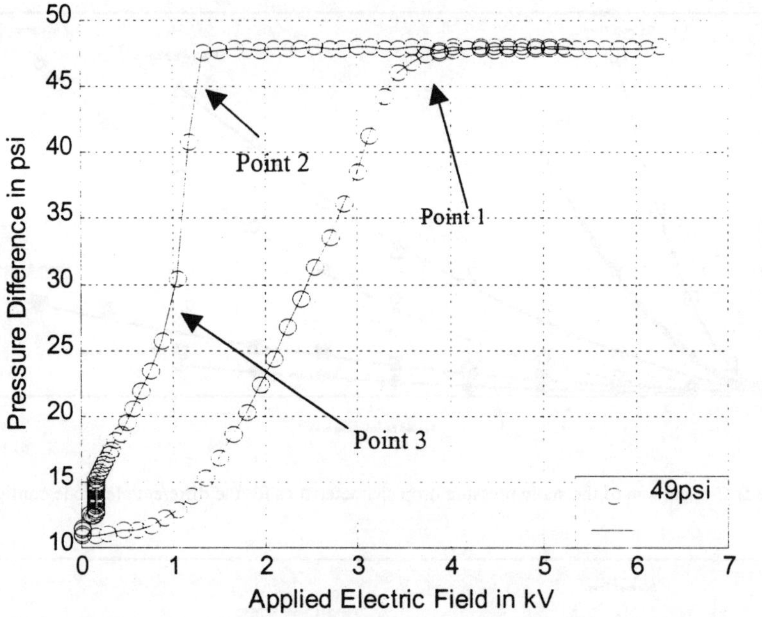

Figure 7: Typical Loading Curve for Varying Electric Field Tests

The three curves in Figure 8 show the operating areas of the particular valve. In order to close completely the valve at a particular working pressure, an electric field should be applied that is to the right of the "Dynamic Hold" curve. Applying an electric field that is

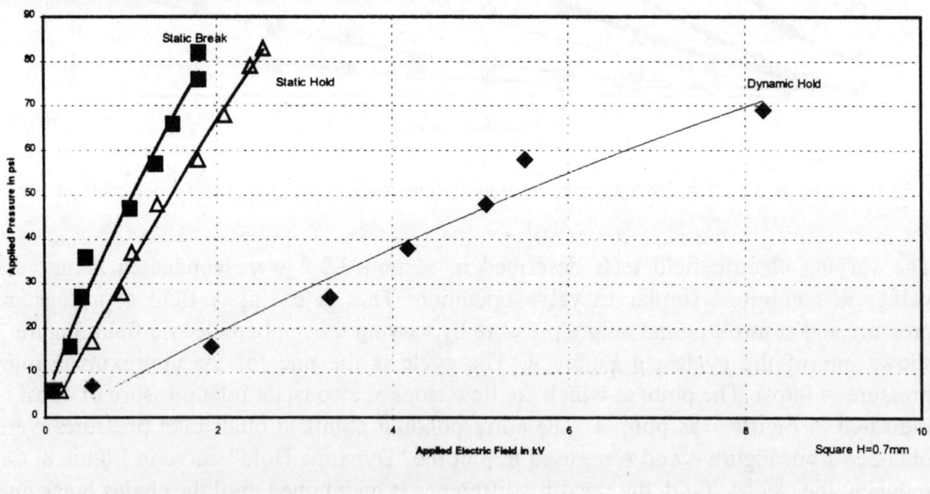

Figure 8: Varying electric field test on square teeth electrodes valve for a gap of 0.7mm

to the left of the "Static Break " curve, the valve should open. The loading curve shown in Figure 7 is a typical curve for the square teeth electrodes at a gap of 0.7mm. However, there are slightly different versions of the curve for the different electrode configurations.

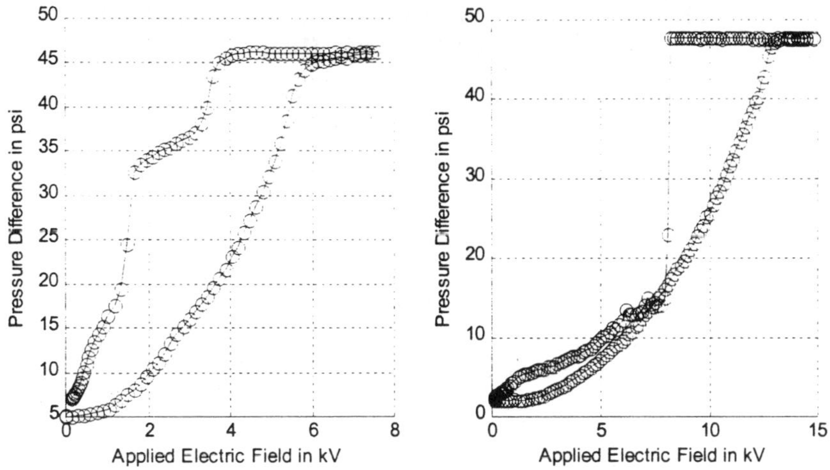

Figure 9(a & b): Variations in curves

Figure 9a shows the curves for the testing of the square teeth electrodes at a vertical gap of 2.8mm. The characteristic that stands out is the double break observed on reducing the electric field. This is probably due to the multiple gap size along the duct; the horizontal gap is 0.7mm whereas the vertical gap was 2.8mm. The curves for the tapered electrodes (Figure 9b) show a very sharp drop on unloading (reducing the electric field), taking the pressure to the level of the loading (increasing electric field) part of the curve.

5 Conclusions

The characterization of the valves using the above tests shows the different distinctive characteristics of the different electrode configurations. It also shows how the operating curves of the valves can be obtained. Further work should involve testing the valves in half open mode and building knowledge-based controllers for the control of the valves.

References

1. Coulter P. J., Weiss K. D., and Carlson J. D., Engineering Applications of Electrorheological Materials, *J Intell. Mater. Syst. Struc*, **4**, (1993)
2. Monkman G. J., Addition of Solid Structures to Electrorheological Fluids, *J Rheology*, (1991)
3. Katsikopoulos P. and Zukoski C., Effects of Electrode Morphology on the Electrorheological Response, *Proc. 5th Int Conf Electro-Rheological Fluids,* (1994)
4. Abu-Jdayil B. and Brun O., Effects of Electrode Morphology on the Slit Flow of an Electrorheological Fluid, *J Non-Newtonian Fluid Mech.*, **63** (1996)
5. Abu-Jdayil B. and Brun O., Study of the Flow Behaviour of Electrorheological Fluids at Shear and Flow-Mode, *Chemical Engineering and Processing* **36** *(1997)*
6. Weiss K. D., Carlson J. D., and Coulter P. J., Material Aspects of Electrorheological Systems , *J Intell. Mater. Syst. Struc,* **5**, (1993)
7. Weiss K. D., Carlson J. D. and Nixon D. A., Viscoelastic properties of Magneto- and Electro- Rheological Fluids, *J Intell. Mater. Syst. Struc,* **5**, (1994)

CONTROLLING TUBE VIBRATIONS USING ELECTRO-RHEOLOGICAL FLUIDS

S.O. OYADIJI

Dynamics and Aeroelasticity Research Group, Manchester School of Engineering, University of Manchester, Manchester M13 9PL, U.K.

The vibration transmissibility characteristics of a cantilevered brass tube are controlled by means of an electro-rheological fluid which is employed as a constrained layer damping medium. The ER fluid, which is a 50 wt % mixture of silicone oil and starch, was contained in the annular space formed between the outer surface of the tube and the inner surface of a bigger brass tube. It is shown that the modal characteristics of the tube depend on the electric field strength and on the height of the ER fluid. The modal frequencies of the tube increase as the electric field potential is increased and as the height of the ER fluid is increased. Also, the loss factors generally increase as the electric field is increased, but decrease as the height of ER fluid in the annular space is increased. When the annular space is 100% filled with ER fluid, the fluid effectively couples the inner and outer tubes into one vibrating system. Overall, the modal transmissibility of the tube is least when the annular space is filled with the ER fluid to a level of 12.5% of the maximum and an electric field potential of 3 kV is applied.

1 Introduction

Passive control of the vibrations of structures such as beams, plates, panels, tubes and shells using the constrained layer damping (CLD) treatment technique is well established.[1] The technique, which employs solid viscoelastic materials for the CLD treatment, is a relatively inexpensive and convenient method for controlling the vibrational response characteristics of structures compared to the more expensive alternative of local stiffening of flexible sections of the structure or of complete re-design of the structure.

Using the CLD technique, existing structures can be readily retrofitted with CLD treatment in order to suppress unacceptably large vibration amplitudes in particular regions of the structure. However, the successful application of the CLD technique sometimes suffers a drawback in terms of the unavailability of viscoelastic materials with dynamic properties to match those predicted to be optimal on the basis of theoretical analysis of the viscoelastically-damped structure. Thus, there exists the need for smart materials whose properties can be tuned fairly readily and easily in order to match the predicted optimal dynamic properties required.

Since the complex dynamic stiffness/modulus properties of electro-rheological (ER) fluids vary between those of viscoelastic liquids (at low levels of applied electric field strengths) and those of viscoelastic solids (at high levels of applied electric field strengths), ER fluids can be adapted to meet this need. In a previous paper,[2] an ER fluid layer was used as a CLD treatment for a clamped-clamped plate. Several investigators[3-10] have also applied ER fluids to damp the vibrations of beams, plates and panels.

In this paper, the measured flexural vibrational response characteristics of a circular-section tube, which are damped by means of partial and total ER fluid-based CLD treatments, are presented. It is shown that the modal characteristics of the treated tube are strongly dependent on the electric field strength and on the location and size of the ER fluid CLD treatment. Examples of potential applications of such a composite tube include robotic arms and artificial fingers with variable stiffness.

2 Experimental Procedure

The experimental apparatus consists of a cantilevered brass tube of dimensions 15.7 mm ID x 19.1 mm OD x 820 mm long whose vibrations are to be damped. The tube is surrounded concentrically by a larger brass tube of dimensions 25.1 mm ID x 28.8 mm OD x 800 mm long. Thus, a radial gap of 3.0 mm is created between the inner and outer tubes. The tube assembly is mounted on a linear motion table as shown in Figure 1, which also shows the associated experimental apparatus. With the tubes empty, their cantilevered ends are subjected to random excitation by an electromagnetic vibrator, which is driven from a spectrum analyser via a power amplifier as shown in Figure 1. By means of accelerometers, which are fixed to the tube assembly via plastic insulators, the acceleration responses of the base and of the tips of the inner and outer tubes are measured. From these measured time histories, the spectrum analyser computes the acceleration transmissibility frequency response functions (FRFs) of the inner and outer tubes as the ratios of the Fourier transforms of the tip responses to the Fourier transform of the base response.

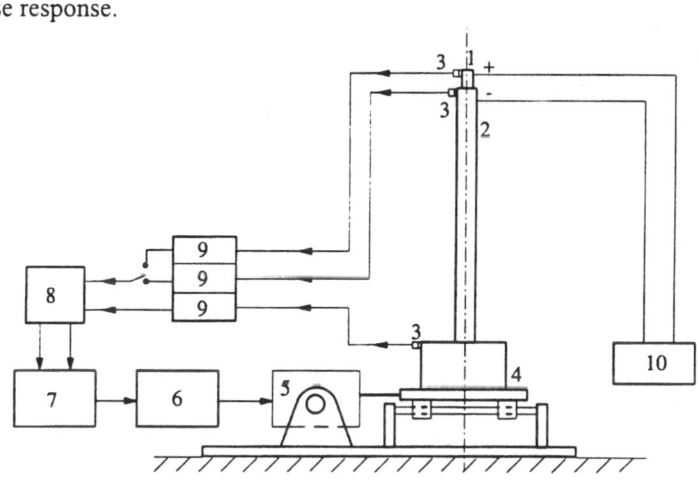

Figure 1: Experimental Set-up

The annular space between the two tubes is now filled to different levels with an ER fluid, which is a 50 wt % mixture of silicone oil and starch powder. The height of ER fluid in the annular space is varied from 0 to 800 mm. A positive electrical potential is applied to the inner cylinder while the outer cylinder is maintained at zero (earth) potential. The potential difference applied across the annular gap is varied in steps from

0 to 9kV. Thus, the dynamic properties of the ER fluid are varied. The acceleration transmissibility FRF of the inner tube is then determined as described above.

3 Results and Discussion

The comparison of the acceleration transmissibility FRFs of the inner and outer tubes is shown in Figure 2. By means of a curve fitting technique the modal properties of the first three bending modes of vibration of the tubes were derived. These properties are tabulated in Table 1. It can be seen that the modal loss factor of the tubes is approximately 0.002.

Table 1: Modal Properties of the Inner and Outer Brass Tubes

Mode No.	Modal Properties			
	Inner Tube		Outer Tube	
	f (Hz)	η	f (Hz)	η
1	20.0	0.0019	28.5	0.0023
2	112	0.0023	178	0.0025
3	312	0.0012	489	0.0011

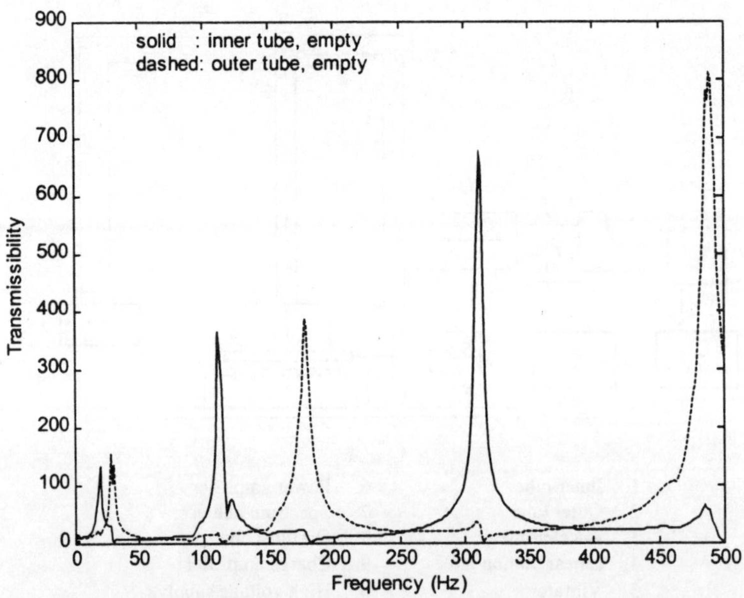

Figure 2: FRF of Inner and Outer Tubes when Annular Gap Empty

Figure 3 shows the comparison of the transmissibility FRF of the tube when the annular gap was empty and when it was filled with the ER fluid to a level of 12.5% of the maximum height and electric field potentials of 0, 3 and 6 kV were applied. It is seen that the vibrations of the tube are considerably suppressed by filling it with the ER fluid. The best reduction of the modal transmissibility of the tube is achieved when an electric field potential of 3 kV is applied to the ER fluid. It is also seen from Figure 3 that the modal frequencies of the tube increase as the electric field potential is increased. This is due to the increase in stiffness of the ER fluid as the electric field strength is increased.

Table 2: Modal Properties of the Inner Tubewhen Annular Gap Empty and Filled with ERF (h/H=0.125)

Mode No.	Modal Frequencies of Inner Tube (Hz)			
	Empty	V= 0kV	V=3kV	V=6kV
1	20.0	18.2	18.8	20.0
2	112	110	114	118
3	312	304	332	342
Mode No.	Modal Loss Factors of Inner Tube			
	Empty	V= 0kV	V=3kV	V=6kV
1	0.0019	0.0052	0.0059	0.0065
2	0.0023	0.0053	0.0098	0.0085
3	0.0012	0.0068	0.0115	0.0067

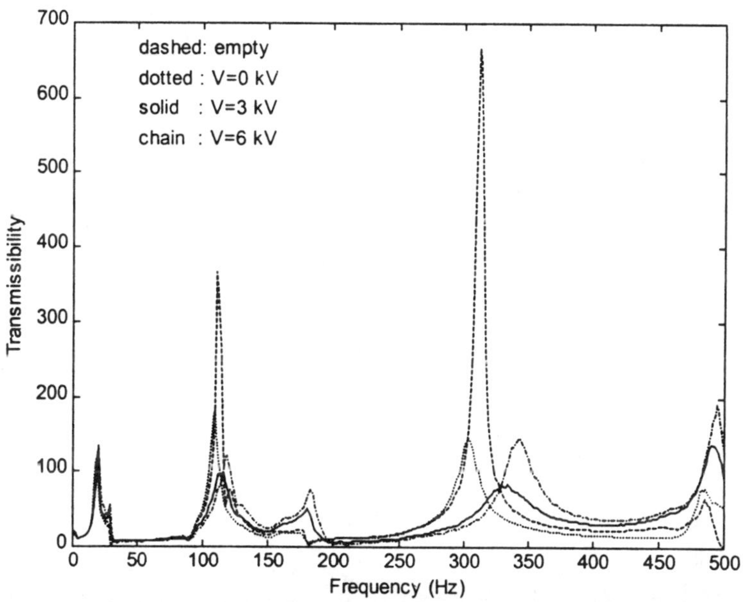

Figure 3: FRF of Inner Tube when Annular Gap Empty and Filled with ERF (h/H=0.125, V=0, 3, 6 kV)

Numerical values of the modal properties of the inner tube derived from the FRFs shown in Figure 3 are tabulated in Table 2. It is seen that when the annular gap is filled with ER fluid to a level of 12.5% but with no applied electric field (V= 0kV), the modal frequencies are slightly less than the corresponding modal frequencies for the empty case. This is due to the added mass effect of the fluid coupled with no increase in stiffness since the ER fluid is not activated. However, the table shows that the loss factor is increased by between 2 and 5 times.

When an electric field of 3 kV is applied, modal frequencies increase beyond the corresponding values of modal frequencies of the empty case. This is due to the stiffening effect provided by the activated ER fluid. Also, the loss factors are further increased. In particular, the loss factor of the third mode is now about 10 times greater than the loss factor of the tube when the annular gap is empty. Further increase in the electric field to 6kV causes a further stiffening effect and a corresponding further increase in the modal frequencies of the composite tube. But at 6kV, the loss factor is slightly less than the loss factor at 3kV electric field.

Table 3: Modal Properties of the Imer Tube when Annular Gap Empty and Filled with ERF (V=3 kV)

Mode No.	Modal Frequencies of Inner Tube (Hz)			
	Empty	h/H=0.125	h/H=0.500	h/H=1.000
1	20.0	18.2	23.8	22.5
2	112	110	151	145
3	312	304	355	398

Mode No.	Modal Loss Factors of Inner Tube			
	Empty	h/H=0.125	h/H=0.500	h/H=1.000
1	0.0019	0.0052	0.0058	0.0049
2	0.0023	0.0053	0.0068	0.0046
3	0.0012	0.0068	0.0070	0.0024

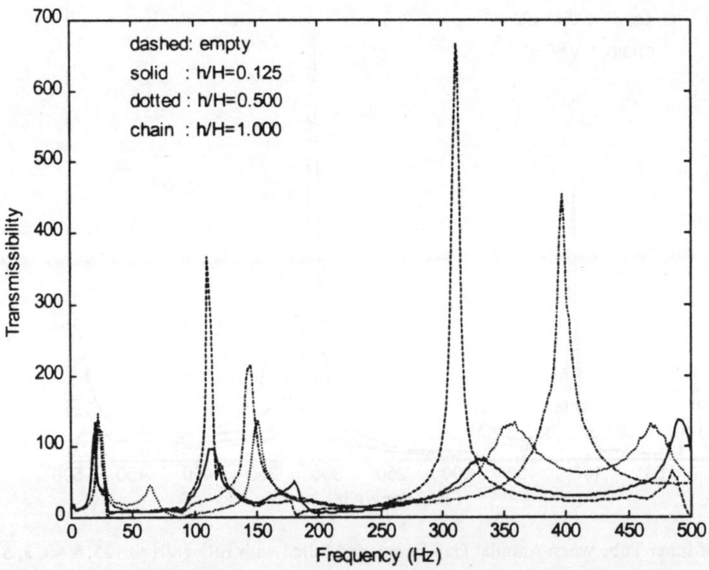

Figure 4: FRF of Inner Tube when Annular Gap Empty and Filled with ERF (V=3 kV, h/H=0.125, 0.5, 1.0)

Similarly, Figure 4 shows the transmissibility FRF of the tube when the annular gap was empty and when it was filled with the ER fluid to levels of 12.5%, 50% and 100% of the maximum height and an electric field potential of 3 kV was applied. The figure shows that the resonance peaks of the tube vibrations are least when the ER fluid level is 12.5% of the maximum. As the ER fluid level increases, it is seen that the magnitudes of the modal transmissibilities increase. Furthermore, the modal frequencies of the tube increase as the level of the ER fluid is increased. This is due to the increase in the combined stiffness of the tube-ERF-tube composite structure. Overall, the modal transmissibility of the tube is least when the annular space is filled with the ER fluid to a level of 12.5% of the maximum and an electric field potential of 3 kV is applied.

Table 4: Modal Properties of Inner and Outer Tubeswhen Annular Gap Empty and Filled with ERF

Mode No.	Modal Frequencies (Hz)			
	Inner Tube, Empty	h/H=1.000, V=0kV	h/H=1.000, V=6kV	Outer Tube, Empty
1	20.0	23.0	23.0	28.5
2	112	144	144	178
3	312	400	397	489
Mode No.	Modal Loss Factors of Inner Tube			
	Inner Tube, Empty	h/H=1.000, V=0kV	h/H=1.000, V=6kV	Outer Tube, Empty
1	0.0019	0.0051	0.0050	0.0023
2	0.0023	0.0054	0.0048	0.0025
3	0.0012	0.0041	0.0023	0.0011

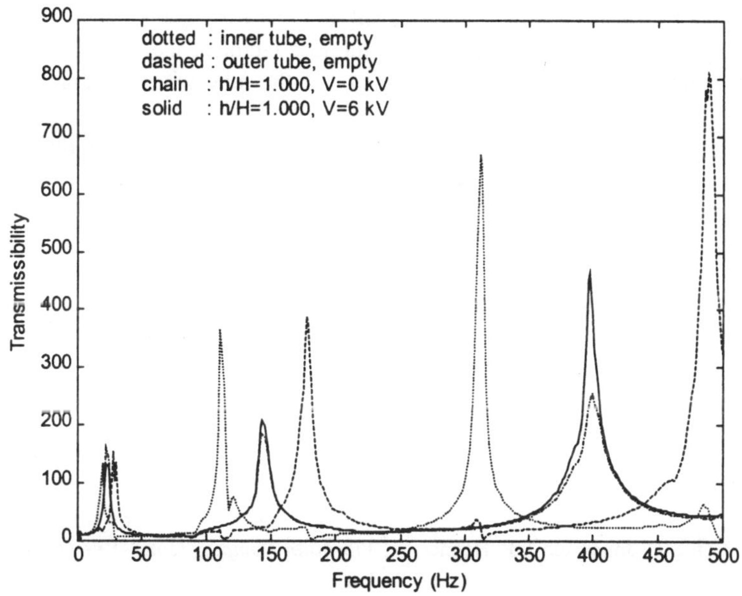

Figure 5: FRFs of Inner and Outer Tubes when Annular Gap Empty and Filled with ERF (h/H=1.0, V=0, 6 kV)

These facts are more clearly shown by the numerical values of the modal parameters tabulated in Table 3. It is seen that the modal frequencies of modes 1 and 2 are highest when an ER fluid level of 50% is employed, while the modal frequency of mode 3 is highest when the ER fluid level used is 100%. The table also shows that the highest relative increase in modal loss factor is achieved when an ER fluid level of 50% is used.

The acceleration transmissibility FRFs of the inner and outer tubes when the annular gap is empty and when it is 100% filled with ER fluid are compared in Figure 5. The figure shows that the fluid effectively couples the inner and outer tubes into one vibrating system. Table 4 shows that the modal frequencies of the fluid-filled case are approximately the arithmetic average of the modal frequencies of the inner and outer tubes when the annular gap is empty. Conversely, the loss factors of the fluid-filled case are approximately the arithmetic sum of the modal loss factors of the inner and outer tubes when the annular gap is empty.

4 Conclusions

An ER fluid, which comprises of 50 wt % mixture of silicone oil and starch, has been used successfully as a constrained layer damping medium to control the vibration transmissibility characteristics of a cantilevered brass tube. The ER fluid was contained in the annular space formed between the outer surface of the tube and the inner surface of a bigger brass tube which was placed concentrically around the first tube. The results show that the modal characteristics of the tube depend on the electric field strength and on the height or volume of the ER fluid CLD treatment.

The modal frequencies of the tube increase as the electric field potential is increased and as the height of ER fluid in the annular space is increased. This is due to the increase in stiffness of the ER fluid as the electric field strength is increased. Also, the loss factors generally increase as the electric field is increased, but decrease as the height of ER fluid in the annular space is increased. In particular, the loss factor of the third mode is about 10 times greater than the corresponding loss factor of the tube when the annular gap is empty.

When the annular space is 100% filled with ER fluid, the results show that the fluid effectively couples the inner and outer tubes into one vibrating system. Under this condition, the modal frequencies of the fluid-filled case are approximately the arithmetic average of the modal frequencies of the inner and outer tubes when the annular space is empty. Conversely, the loss factors of the fluid-filled case are approximately the arithmetic sum of the modal loss factors of the inner and outer tubes when the annular gap is empty. Overall, the modal transmissibility of the tube is least when the annular space is filled with the ER fluid to a level of 12.5% of the maximum and an electric field potential of 3 kV is applied.

References

1. A D. Nashif, D. I. G. Jones and J. P. Henderson, *Vibration Damping* (John Wiley & Sons, New York 1985).
2. S. O. Oyadiji, *J Intell. Mater. Syst. Struc.* **7**, 541 (1996).
3. G. Haiqing and L. M. King, *J Intell. Mater. Syst. Struc* **8**, 401 (1997)
4. C. Y. Lee and C. C. Cheng, *J Intell. Mater. Syst. Struc* **9**, 608 (1998)
5. M. Yalcintas and J. P. Coulter, *J Intell. Mater. Syst. Struc* **7**, 128 (1998)
6. C. D. Rahn and S. Joshi, *Trans. ASME. J Vib. Acoustics.* **120**, 221 (1998)
7. J. Leng, A. Asundi and Y. Liu *Trans. ASME. J Vib. Acoustics* **121**, 508 (1999)
8. Y. K. Park and S. B. Choi, *J Sound Vib.* **225**, 391 (1999)
9. T. Takawa, T. Fukuda and K. Nakashima, *Smart Mat. Struc.* **9**, 215 (2000)
10. N. R. Harland, B. R. Mace and Jones, *IEEE Trans. Cont. Sys. Tech.* **9**, 209 (2001)

NONTRADITIONAL ERF-BASED DEVICES AND TECHNOLOGIES

KOROBKO E.V.

Heat and Mass Transfer Institute, Belarussian Academy of Sciences ,
220276,, Minsk, P.Brovky 15, Belarus

Stractural rearrangements of an electrorheological fluid (ERF) in an electric field cause a change not only in its mechanical characteristics (viscosity, elasticity, plasticity) but also in such physical properties as thermophysical, optical, electrical, heat-transfer, etc. We have considered the possibility of control of these parameters by an electric field on the basis of our experimental results and an analysis of the relevant data available in the scientific literature and in the patents.

1. The State-of-the Art of ERF Applications in Practice

Engineering and technological aspects of the use of electrorheological fluids (ERF) are extremely diverse and numerous. The electrorheological effect (ER-effect) allows low power electric control signals to be simply and efficiently transferred to load-bearing hydraulic actuators. The important distinctive features of such devices are the absence of movable mechanical elements, the low consumption of energy, and the high speed of response.

A review of numerous publications devoted to the practical applications of ERFs has revealed several stages in the history of implementation of their advantages in concrete technologies and devices which are distinguished sometimes by the keenest interest of researchers, engineers and of a market as a whole but sometimes by the loss of this interest. In their duration and scope of embracing the branches of economy these stages correspond to the level of evolution of our knowledge of the nature and specificity of manifestation of the ER-effect. Moreover, on frequent occasions science and practice have stimulated one another by posing new problems, focussing attention on particular problems.

Thus, in the initial stage (in the 50-ies) investigation of the flow regularities in plane and coaxial cylindrical channels has opened up fresh opportunities for creation of controlled hydraulic systems. Just in this stage measurements were made of the flow rate-head characteristics, whose change in an electric field was related to the increase of the ERF viscosity and initially was described by simple linear dependence and explained by internal structural rearrangements of particles of the solid phase, which are visible to the naked eye in the low-concentration electrorheological fluids. The structural nature of the effect has impelled rheologists to measure the shear stress versus shear rate and to plot flow curves. These dependencies have shown the pseudoplastic nature of ERF flow, i.e. the decrease of viscosity with an increase of the shear rate which has resulted in development of rheological models of the type of the power-law and exponential ones. At a later time these models have allowed engineers to design the basic traditional hydraulic elements, i.e. . ER-valves, to create their improved alternate designs of small size in the form of stacks of plane (disk), coaxial cylindrical or spiral electrodes. On their basis shut-off units(throttling valves) of hydraulic systems, controlled shock-absorbers, hydraulic pumps, brake pushers, and the other devices of a new generation working in the «open-shut» regime were developed . An attempt to apply valves-pumps for tracking and damping systems, oscillators of fluid pressure, in which the change of an electric field and the corresponding change in the fluid viscosity must follow an impact load, has necessitated investigation of time-dependent regularities of the ERF response.

More profound rheological studies with the aid of vibroviscosimetric measurements, results of the experiments on determination of thixotropy and creep have revealed, in the first place, the viscoelastic behavior of ERFs at small strains, second, have shown the necessity of accounting for the prehistory, when the ERFs preliminary have experienced particular mechanical shearing within some time interval. It has been established that ERFs represent the complex fluids , for which rheological properties depend on the time of field action. These results have urged investigators to develop viscoelastic rheological models and have provided a basis for describing the unsteady-state processes. An understanding of kinetics of the structurization process and of the rheological behavior of ERFs has made it possible to explain the shift of resonance curves on the amplitude-frequency dependencies and to optimize control of ER dampers in vibratory systems, to create impact dampers, anti-shock and brake devices.

As an outgrowth of the traditional hydrodynamic applications at the same time (in the 60-70ies) the facilities have appeared which implement the quasi-solid state of ERFs at high intensities of an electric field in the range of small shearing loads, in particular, a clutch .

The attempts of their application are known, for instance, in fans that provides the possibility of increasing their efficiency. However, for the basic spectrum of type items designed to transfer large forces or used at considerable shear rates and due to temperature limitations no acceptable electrorheological clutch has been designed yet.

The most successful example of application of the quasi-solid mode of ERF operation is ER fixing devices. These are facilities of the type of cramps-straps with electrorheological elements for fixing of movable objects under cutting (for instance, tubes of a large diameter), the fixing devices intended for objects of both the two-dimensional and complex geometry (turbine blades) used in mechanical treatment of their surface or gripping devices in the form of a robot arm used for transfer and positioning of work pieces. Recently we have implemented the same principle in vibration tests of radioelectronic devices [1, 2]. Now we will dwell on this example in more detail.

Competitiveness of manufactured radioelectronic devices at the home and foreign markets depends on the quality of their functioning under conditions of strong mechanical disturbances.

Vibration tests based on vibration simulation pertain to the fundamental experimental methods used for evaluation of the serviceability of produced items. In the radioelectronic industry such items as integrated circuits, capacitors, transistors are tested on a mass scale for vibration strength and vibration resistance. As a rule, bench tests are carried out with the use of special-purpose facilities which allow reproduction of real vibrations or their equivalents within the ranges of frequencies from 0.1 to 5.000 Hz, of displacements from a few tens of micrometers to a few tens of millimeters, of accelerations from units to a few hundreds of m/sec^2, of spectral density of accelerations up to 10 m^2 / sec^3 with a reproduction error of 10-20 % (of acceleration, displacement) and 1.5 to 3.0 dB (spectral density) and their transfer to different items, including from elements, for instance, resistors to whole structures.

An important link in a system providing the quality of vibrations transferred to an object is the fixing devices used to fasten the object to a vibration source, in particular, to an electrodynamical vibration-testing unit. In industry different methods and fixing devices are known and everywhere used to fix items in vibration tests and during their treatment. The major part of those are facilities employing bolts, clamps, cramps. The essential feature required of any fastener is maximum complete transfer of the mechanical energy from the table of a vibration - testing unit to a tested item which corresponds to the transfer factor equal to unity in the entire working range of vibration tests.

The majority of the known mechanical fixing devices not only fail to meet this requirement but moreover, their use creates additional difficulties associated with a technology of their manufacture and application to multipurpose objects of different dimensions and shape. In the other cases for these purposes adhesives, resins, compounds serve which calls for potting and laborious operations to make items free of adhesive joints. We have suggested a new and efficient method of fixing of radioelectronic items which is based on the electrorheological effect and does not suffer from the drawbacks mentioned above. The effect lies in the fast reversible increase of the effective viscosity and elasticity of a material in the presence of a strong electric field. The material represents a nonaqueous dispersed system, namely, a paste. The main components of such an ER-adhesive are a nonpolar dispersed phase with high resistively and a solid dispersed phase with rather high permittivity. The ER-adhesive is applied to the contact surface of a fixing device afterwards a tested item is pressed by some effort against it.

Dictated by goals, the ER-fixing devices can have rather various designs. For instance, if voltage is inapplicable to a tested item for initiation of an electric field,
then for fixing purposes the use can be made of a facility of the planar type. For the tests of cylindrical capacitors the facility depicted in Figure 1 has been

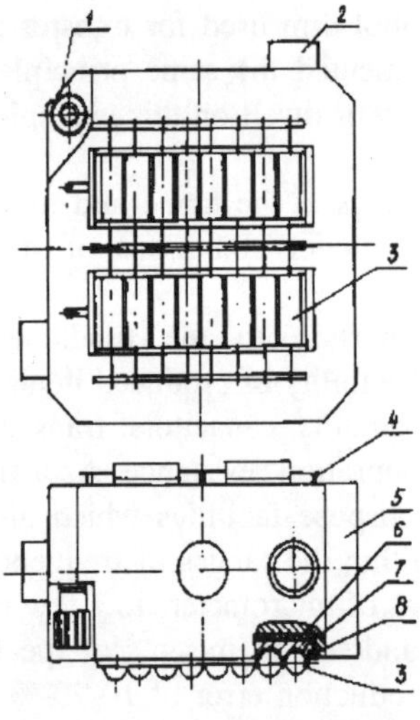

manufactured. High voltage is applied to terminal block 1. Tested capacitors 3 are placed in recesses of the cylindrical form which follow the capacitor surface. Between metallic insert 7 and capacitors 3 there is ER-adhesive 8 which hardens upon applying high voltage and makes the capacitors monolithic. The facility was installed on a working table of the vibration-testing unit.

Fig.1 Device for fixing the capacitors. 1, terminal block; 2, current protection unit; 3, capacitors; 4, plate for making of solderent terminals of capacitors; 5, body, 6, insulator; 7, insert; 8, ER-adhesive

We have studied the frequency dependence of the transfer coefficient of mechanical loads which is determined as a ratio of the output to input value of acceleration $K = g_1 / g_2$. As samples, we used the aluminum and steel ones with the thickness h = 15 mm, the contact area $S = 12.55\ cm^2$, and with m = 51.5 g and m = 135g, respectively.

Figure 2 represents the transfer coefficient of the system at the vibration frequency ranged from 1 to 5.000 Hz. The transfer coefficient retains the value K = 1 in the range up to 3,800 Hz which corresponds to reliable fixing of an object in vibration tests and meets the requirements of state Standards. Thus, the new method to attain K = 1 by creating the conditions of effective transfer of the mechanical energy from the table of the vibration-testing unit to a tested item by means of the ER-adhesive ensures the high quality of vibration tests of radioelectronic items. In addition to providing the stable output characteristics, the suggested fixing devices possess the required reliability and safety in service. Among the advantages are also their universality, simplicity of design, low energy intensity of rigging (currents - of about 10^{-5} A), and the efficiency determined by the time of response of the ER-adhesive quicker than 0.1 sec.

In the last two decades the works of researchers have been mainly aimed at a study of the nature of the electrorheological effect, the reasons which cause and determine the peculiar features of structurization in the presence of an electric field. The accumulated new knowledge has allowed us to achieve marked progress in the field of improvement of the composition and properties of l fluids [3, 4]. Thus, to improve the electrorheological response and to widen the temperature range of ERF applications, we have carried out target-oriented modification of the properties of components of the known ERFs by introducing additives by subjecting the surface of particles of the solid phase to physical or chemical treatment, by creation of two-layer particles. Moreover, radically new formulas of ERF compositions, for instance, of the homogeneous type based on liquid crystals, have been developed, the use of which will allow one to remove limitations on the sedimentation factor. The ERF compositions were created which extended the spheres of application of the electrorheological effect. Thus, the invention of the polymer-containing quick-setting ERFs has permitted us to begin development of a technology of moulding of films with parameters and structure formed by an electric field . The emergence of the ERFs with improved characteristics has provided the basis to inventors for

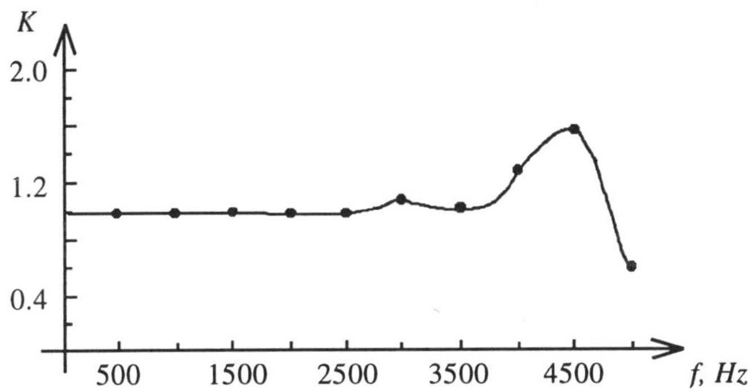

Fig.2. Transfer characteristic of the resonance table disk.

development of original proposals aimed at improvement and creation of controlled mechanical devices such as fishing-rods, facilities for rehabilitation of recovering patients, geophones. The potentialities of application of electrorheological fluids for chemical technologies have also increased for fragmentation of dispersed material by increasing ERF viscosity in a gap between colloid mill surfaces under friction; metering different fluid-dispersed materials by discrete de-energization of electrodes and ejection of the material portion to a production line; filtration by vibroelectrocontrollable filters; precipitation by coagulation of an electrorheological filler at electric field with material particles in sumps; control of flow rate characteristics of hydraulic systems; in robotics systems, i.e., manipulator grabs, positioning systems; for voltage control during gas electrocleaning and in electric filters..

In the 80-90ies the curiosity of investigators has led them to a study not only of the mechanical, but also of the other physical properties of electrorheological fluids and new non traditional applications of ERFs.

In particular, investigations were made of the thermophysical properties of electrorheological fluids, i.e. thermal conductivity and thermal diffusivity, in the presence of electric fields. Their dependencies on the electric field intensity have served as the basis for development of a differential technique of determination of thermal conductivity of media. This method is as follows. A heated medium and the reference medium are heated by equal heat fluxes and then a temperature is measured at several points. In so doing, for increasing the accuracy and for decreasing the labor input, a constant electric field is applied to the reference medium, which represents an electrorheological fluid, thus attaining the equality of temperatures at the chosen points equidistant from the heat sources. Next, the electric field intensity is measured and the sought thermal conductivity is determined from the calibration curves. The results of experimental studies of convective heat transfer of ERFs produced the dependencies of the heat transfer coefficient on the electric field intensity which made it possible to recommend them for cooling of the objects with a high potential on their surface, for instance, of radioelectronic devices calling for the use of dielectric heat-transfer agents due to this reason. Application of the electrorheological effect shows promise for control of a thermal boundary layer and can be efficiently used to act on temperature fields and the characteristics of heat-transfer apparatuses.

Measurement of the acoustic characteristics of electrorheological fluids in electric fields, namely, of the sound velocity and the absorption coefficient, has allowed inventors to propose the designs of controlled acoustic devices with tunable parameters, such as sonars, hydrophones, delay lines, filters, lenses. Moreover, electrorheological fluids were applied as matching interlayers, for instance in non-destructive test instruments to perform control of the strength of a workpiece with the aim of improving the quality of acoustic matching of the material impedance of a converter, a contact medium, and a tested item . Similar practical applications are also possible in the field of development of optical devices.

Electrorheological fluids can find application not only in actuators as electric -to-mechanical energy converters but also can serve as indicators, transducers in various technological processes. Thus, a change in electric conductivity allows one to judge about a change of the parameters of an electrical medium or an object, for instance, of a vibration level, temperature, or voltage. It was used the device for control of surface roughness of a workpiece , in which for increasing the accuracy of measurements a potential difference between a controlled element and a surface coated with an ERF layer is induced and the surface roughness is evaluated from the force of separation of the element from the fluid.

The rheological characteristics, which in an electric field have the unambiguous dependence on the concentration of the solid phase, the presence of water and of the other additives for an electrorheological fluid can be employed to determine characteristics of a wide spectrum of unknown materials by the comparison method of measurements and by specific laboratory devices of the type of viscosimeters.

Of considerable interest in sight is the opportunity of control of the surface tension [5] and, therefore, of the process of spreading of an electrorheological fluid layer over a surface and control of the form of the ERF drops. Recently, we have experimentally showed the emergence of normal stresses/ in an ERF upon switching an electric field which along with the phenomenon of dilatancy can be employed for creation of the mechanisms of micromechanics.

2. Concluding remarks

The accumulated knowledge of the nature and the regularities of manifestation of the ER-effect allows us to make an optimistic forecast of the future of ER technologies by having prognosticated a break-through in creation of extraordinary new designs and extension of the function of the existing devices and improvement of the traditional technologies. Thus, it is very interesting to study the possibility of control of the processes of mixing of fluids which has not been investigated yet. Naturally, the undoubted merits of ERFs, their inexpensive cost, the ready availability of the ER technologies must not shut consumer's eyes to such their drawbacks as the limited stability of the properties, the limited temperature range of activity, the special requirements for personal and so on which are essential for development of the ER devices on a mass scale.

Aknowledgments

The present work was carried out with the financial support of the Fundamental Research Foundation of Belarus, grant N F99R-124 RF.

References

1. A.V. Luikov , Electrorheological Effect (in Russian), Minsk, Nauka i Tekhnika, (1972), p. 176.
2. E.V. Korobko Electrostructurized (Electrorheological) Fluids: Some Features of Fluid Mechanics and Potentialities of Their Use (in Russian), Minsk, ASC HMTI, (1996), p.189.
3. Proc. of the 6th Int. Conf. on Electro-Rheological Fluids , Magneto-Rheological Suspensions and Their Applications. ed. M. Nakano, K. Koyawa., July 22-25, Yonezawa, Japan, (1997).
4. Proc. of the 7th Int. Conf. on Electro-Rheological Fluids , Magneto-Rheological Fluids and MagnetopRheological Suspensions. ed. R. Tao., July 19-23, Honolulu, Hawaii (1999).
5. E.V. Korobko , R.G. Gorodkin, N.A. Goncharova and V.I.Baykov ERF's Surferce Tension in Electric Field. in IHMT Proc. 2001. in press.

EMBOLIZATION OF BLOOD VESSELS AS A CANCER THERAPY USING MAGNETORHEOLOGICAL FLUIDS

G. A. FLORES AND J. LIU

Department of Physics/Astronomy, California State University
1250 Bellflower Boulevard, Long Beach, CA 90840, USA

This paper demonstrates a possible *in-vitro* cancer therapy by mechanically blocking the blood vessels to a tumor through injecting a model magnetorheological (MR) fluid and applying an external magnetic field. The biocompatible MR fluid is made of magnetite particles coated with starch and suspended in either water or sheep blood. Two particle sizes of 1.0 and 0.25 μm are used. A simple "blood" network consists of 4 branches of "blood" vessels made of silicone tubes of 0.4 mm in diameter. One of the branches contains a "tumor" made of a cylindrical cavity with a diameter of either 3.0 mm or 5.0 mm to simulate stage II or III tumors. The cavity is connected with either two or four vessels to the rest of the branch and is placed between two magnetic poles. By measuring the weight of the leaking fluid downstream from the magnets, the sealing quality is monitored. A dilute MR fluid (1.0% particle volume fraction) is pumped slowly through the network. As the fluid goes through the cavity, seals form within seconds after a magnetic field (0.43 - 0.62 T) is turned on. They block the fluid flow to the "tumor". With sheep blood as the suspension medium, both size of magnetite particles forms a good quality seal for the 3 mm cavity. Thus, a mechanical method for stopping "blood" flow is demonstrated in a multi-tube "blood" network for stage II tumor.

1. Introduction

Human cells have a definite timetable when to divide and when not to divide. For example, skin cells will divide to heal a cut, or the lining of a woman's uterus will rebuild every month during her child-bearing years [1]. Cancerous tumors start out as a group of mutated cells that divide uncontrollably [2]. Once the mass reaches a size of a few cubic millimeters, the tumor cannot survive until it receives a new blood supply through a process called *Angiogenesis* [3]. The tumor will continue to feed from this new supply and spread, or metastasize. Multi-modal techniques may work against most types, but are relatively ineffective if the tumor begins to mutate [4].

Surgical removal is the most basic therapy to eliminate cancer from the body, but the risk of not removing the entire mass is very high resulting in metastasis. Angiogenesis inhibitors called Angiostatin and Endostatin have proved successful in treating mice with tumors, but it is still years, if at all, that they can be applied to human use due to the scarcity of producing such drugs [3-5]. Chemotherapy and radiation are effective treatments, but their toxicity to the patient is so high that the balance between killing the tumor and the human body is a constant battle. Here we propose a new technique in cancer therapy with less toxicity than chemotherapy and less expensive than angiogenesis inhibitors. By injecting a magnetorheological (MR) fluid into the blood vessel and applying an external magnetic field on top of the tumor, a seal forms and mechanically blocks the blood vessels to a tumor that causes the tumor death. Magnetizable particles are tested for delivering drugs to tumors in both animal and human subjects with no adverse side effects [6, 7]. During the control study, a positive effect is found without using drugs [6]. This inspired our cancer treatment research.

Our work focuses on simulation experiments to demonstrate the cancer treatment principle and formulate important controlling parameters. By placing MR fluids in a silicone tube and pumping it through a pair of magnetic poles, an embolism forms within the tube at the site nearest to the poles. Our early work showed that blocking the fluid flow is possible within a single-tube system, simulating one blood vessel. We found that the seal remains even at pressures exceeding those found inside human arterioles and capillaries [8, 10]. We also found the dependency upon particle concentration, flow rate and magnetic field strength in the seal formation: The lower the particle volume fraction, the slower the flow rate and the weaker the magnetic field are, the weaker the formed seal is and the longer the sealing time is needed [9-10]. A seal can be formed within one hour under high pressure after the fluid has reached the field site. This makes the technique possible for clinic application.

Here we report the recent progress using a multiple-tube system with a simulated tumor cell inside. These condition changes – although still a simulation – makes our work a step closer to reality than before. We will show *in-vitro* that the sealing ability of MR fluids under an applied magnetic field still remains in a multiple-tube system with a larger cavity. It is our hope that this technique will become a valuable tool against cancer.

2. Experimental Set up and Procedures

The fluid used throughout the experiments is a suspension of pure magnetite particles (MAG) from Chemicell GmbH, Germany in either sheep blood or water. The particles are coated with starch to make it biocompatible. In our past experiments [10] using the same type of particles, the sealing effect was found to reduce after human blood was used as the suspending medium. We found that the particle coating, which was negatively charged, interacted electro statically with the human red blood cells (hRBC). To reduce electrostatic repulsive interactions with the RBC, the supplier has modified this new MAG sample with a hydrophilic starch coating. This starch coating also helps make the particles "appear" invisible to the body.

Two different diameters (***2a***) of particles are used for different injection methods. The 0.25-μm particles are designed for the drug delivery application and are suited for intra-venous applications. They have been tested in both animals and humans before without adverse effects [7]. The 1.0 μm particles are intended for increasing the efficiency of mechanical blockage of the fluid flow and are suited for an intra-arterial application. We would have a greater particle loss if they were placed intravenously. Based on our earlier experiments [9-10], a particle volume fraction (**Φ**) of 1.0% is chosen for the sealing experiments since it can adequately seal the tubes and has minimum interference with normal blood chemistries.

Figure 1 shows the setup used in this work. A silicone tube is placed inside the MR source beaker to draw the sample toward a peristaltic pump from Manostat ("SARAH" model, Barnant Company, USA) that will send the fluid through the system. The initial inner diameter (***D***) of this tube is 1.6 mm, much larger than that found in a human arteriole (~0.1 mm) but smaller than human arteries [12-14]. To simulate a human arterial network, this tube is subdivided into two - 0.8 mm tubes first and then to four - 0.4 mm tubes. The fluid flow rate (***Q*** ~ 0.008 ml/min) is similar to that in human arteriole. However, the flow velocity of the fluid (1.17 mm/s) in a 0.4 mm tube is slower than that found in human arteriole (7.5 mm/s for an 0.05 mm arteriole) due to the larger cross-sectional area used in our tube.

Cavity cells of diameters (***S***) 3.0 mm and 5.0 mm are used to "simulate" a tumor cell. These sizes were chosen because of its similarity to a Stage II and III breast cancer or melanoma [8, 15]. Either one or two tubes are connected to either side of the cavity cell to simulate a tumor with either two or four blood vessels. A pair of electromagnetic coils with two pole pieces (1018 carbon steel with pole diameter of 16.0 mm) generates and focuses a magnetic field to the cavity cell. Each electromagnet has $\boldsymbol{N}$ = 1150 turns. The flux density ($\boldsymbol{B}$) generated at $\boldsymbol{I}$ = 5.0 A is as follows: for $\boldsymbol{S}$ = 5.0 mm $\boldsymbol{B}$ = 0.43 T at the cavity center and $\boldsymbol{B}$ = 0.48 T at the tube-cell interfaces for 4 branch configuration; for $\boldsymbol{S}$ = 3.0 mm $\boldsymbol{B}$ = 0.62 T at the cavity center and $\boldsymbol{B}$ = 0.63 T at the tube-cell interfaces for 4 branch configuration. The 3-mm cavity cell is so small that the 4-branch configuration can be connected to the cavity via only two leads. Thus, it is essentially a 2-branch configuration.

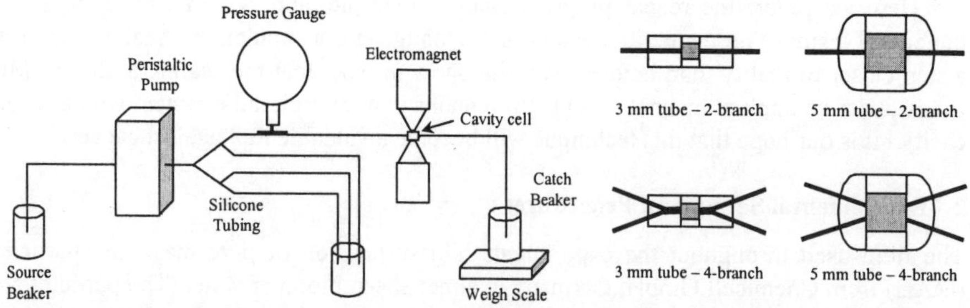

Figure 1: Experimental "simulation" of a tumor within a cardiovascular system. Also seen here is a close-up diagram of a 5 mm and a 3 mm cavity with 2-branch and 4-branch configurations.

One tube is connected to a DSF26 digital pressure gauge, manufactured by Kolbold USA, Inc. to measure the resistive pressure ($\boldsymbol{P}$) on the seal. The gauge is connected to a PC computer through a Data Acquisition (DAQ) card from National Instruments, Inc. The pressure gauge sends the data to the DAQ card as a voltage from 0 → 10 V as set by the factory. To translate this voltage range into 0 → 200 mmHg, we use a graphical program called LabVIEW, also by National Instruments, to record and save the data. We can measure five data points every second, or a maximum of 9,000 data points over 30 minutes for the entire experiment. To measure the fluid weight ($\boldsymbol{W}$), we use a Mettler-Toledo PB-153 balance connected to the computer via a RS-232 port. We can read and save the values through LabVIEW also at the same rate.

3. Results and Discussion

3.1. *Sheep Blood as the suspension medium*

Sheep blood consists sheep red blood cells (RBC) with an equal volume of Alsever's solution (a red blood cell preservative) within distilled water. The sheep RBC (~4.5 μm) is almost half the diameter of the human RBC (~8.0 μm), and the packed RBC volume fraction in sheep (~35%) is also lower than that in human (~45%) [11].

3.1.1 *Magnetite particles: 2a = 1.0 μm (intra-arterial simulation)*

Experiments to create a blockage were performed using two size cavity cells and tube configurations to simulate tumor blood vessel distribution. Figure 2 shows the weight that passes through the cavity as a function of time. Figure 2a shows the data measured in the 3 mm cell where $\boldsymbol{B}$ = 0.62 – 0.63 T. Before the magnetic field is applied, $\boldsymbol{t}$ < 15 min, $\boldsymbol{W}$

increases stepwise as the fluid dripping down the tube drop by drop. Here the two lines are for two connection configurations. But once the magnetic field is turned on at t = 15 min, the fluid weight stays constant and a stable seal is formed within seconds. In the case of single branch blood network as in our earlier work [8], the pressure in the tube builds up quite high and the seal formation time is long. Here the pressure in the tube remained low throughout because the excess fluid is diverted toward the other three branches of the blood network, as is expected in human physiology. The decrease in W after the seal is formed comes from evaporation of the liquid from the catch beaker. The evaporation effect is more obvious in this experiment than before due to small amount of sample used in the narrow tubes.

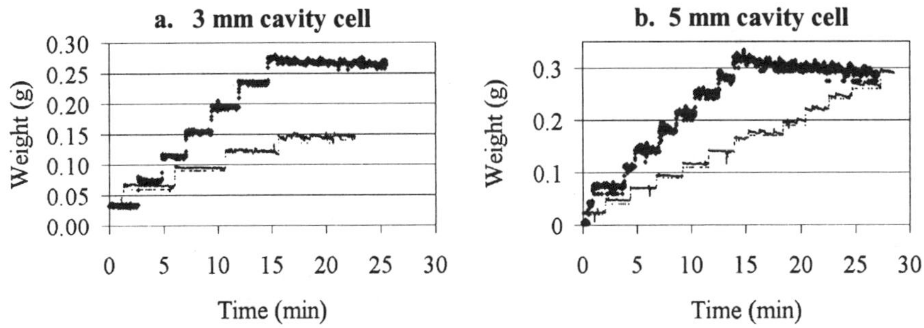

Figure 2. Fluid weight (W) as a function of time. Thin lines show 2-branch cavity configuration, while thick lines show 4-branch cavity configuration. Φ = 1.0%. Q = 0.008 ml/min. Magnetic field is applied at t = 15 min.

Figure 2b shows data obtained in the 5 mm cavity cell where B = 0.43 for 2-branch configuration and B = 0.48 T for 4-branch configuration. Here, the 4-branch tumor blood vessel distribution forms a stable seal, but the 2-branch configuration does not. In the 4-branch configuration, the tubes at the cell interface are closer to the magnetic poles than the tubes in the 2-branch configuration and thus experience a higher magnetic field. The data suggests that the sealing does not occur inside the cavities, but in the tubes leading to and from the cavities. This is what we have expected. It is easier to block a river with sand at a narrow place than a wide area.

3.1.2 *Magnetite particles: 2a = 0.25 μm (intra-venous simulation)*

Repeat the same experiments as above with a 0.25 μm magnetite particles in MAG. We have previously seen that smaller particles (1 μm) can form a tighter seal than larger ones (3 μm) because of the particles gap size, forming a "mesh" [8, 10]. Figure 3a shows the sealing for a 3 mm cavity. This cell formed a seal with almost no pressure. That is to say, P was practically nil throughout the experiment and a stable seal is seen. However, in the case of the 5 mm cavities, the results are much different. Figure 3b shows an uneven weight increase with time (t < 15 min). Here we also observed tube pressure increase. This may be due to the formation of clusters in the fluid that travels unevenly through the tube. When the field is turned on, a stable seal is formed for the 2-branch cavity. In the case of the 4-branch cavity, a seal is initially formed, but collapses shortly thereafter. The seal later regroups and reforms itself for the duration of the experiments. Even with the increase in P, the seal has held. We have documented the maximum seal strength of MR particles in past experiments [8-10].

Notice that without magnetic field, the step size in W is smaller now (Fig. 3a) than before (Fig. 2a) and the number of drops is about a factor three greater now than before. This indicates that the drop size at the end of tube before the weighing scale is smaller using small particles than using larger particles. This may be due to the surface tension of the fluid droplets that is affected by the particle size: shear alignment of the particles in a flow results in lower fluid viscosity. This shear thinning effect is greater with larger number of particles for the same particle volume fraction. Thus a colloid fluid with smaller particles may have lower fluid viscosity than that with larger particles.

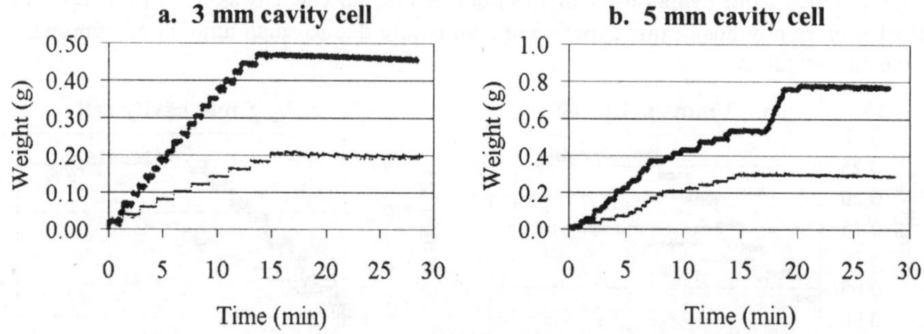

Figure 3. Fluid weight (*W*) as a function of time. Thin lines show 2-branch cavity configuration, while thick lines show 4-branch cavity configuration. Φ = 1.0%. ***Q*** = 0.008 ml/min. Magnetic field is applied at t = 15 min.

3.2. *Water as the suspension medium*

Figure 4 shows the same measurements as in section 3.1 repeated using de-ionized water as the suspension fluid. Stable seals are formed for all the eight different parameters in cavity size and configuration, and particle size. It is clear that the seal formed with water as the suspension fluid is stronger than that formed with sheep blood as the suspension fluid. There are two possible explanations: 1) the higher viscosity of the sheep blood raises the tube pressure that breaks the seal; 2) the electrostatic repulsion between coating of the particle and red blood cells weakens the formed seal.

From the pressure measurements, we find that the pressure increases with time sometimes for the sample using sheep blood as suspension medium. Whereas no pressure increases is observed using water as the suspension medium. This is due to the viscosity difference between the MR fluids using the two different suspension media. With water as the suspension medium, the fluid is able to run through the system quickly and divert to other branches of the network once one branch is blocked. Thus, fluid flows without the hindrance of any noticeable resistance. With sheep blood as the suspension medium, the fluid may stuck at one place and flow becomes irregular if there is a cluster of particles or blood formed. The increased viscosity diverts the fluid to other branches slowly. This increases the pressure exerted on the seal and breaks the seal as shown in Fig. 3b. Even before magnetic field is applied, the fluid flow is different comparing Fig. 3b with 4d. The interaction between particles and blood may be responsible for this.

The resistance ***R*** of the fluid flow within a tube is defined as the change of pressure (**Δ*P***) over ***Q***.

$$R = \Delta P/Q = 8l\eta/\pi r^4 \qquad \text{(Poiseuille's Law for a streamlined flow).}$$

Where ***l*** is the length of the blood vessel, **η** is the blood viscosity, and ***r*** is the radius of the blood vessel in question [13]. The unit of ***R*** is usually given as dyne sec/cm^3 or mmHg min/ml. Thus, with the same flow rate, tube length and radius, the resistance to fluid is different from the viscosity of the two suspension media alone. The estimated sheep blood viscosity is about a factor of 2.5 higher than water based on the human blood viscosity that is about 3 times higher than water.

Is this viscosity difference enough to explain the differences in the seal strength between these two suspension media? It is necessary to measure the tube pressure in a single-branch experiment for the seal strength as we did earlier [10] to conclude this and to determine the degree of the contribution from the electrostatic repulsion between particles and red blood cells.

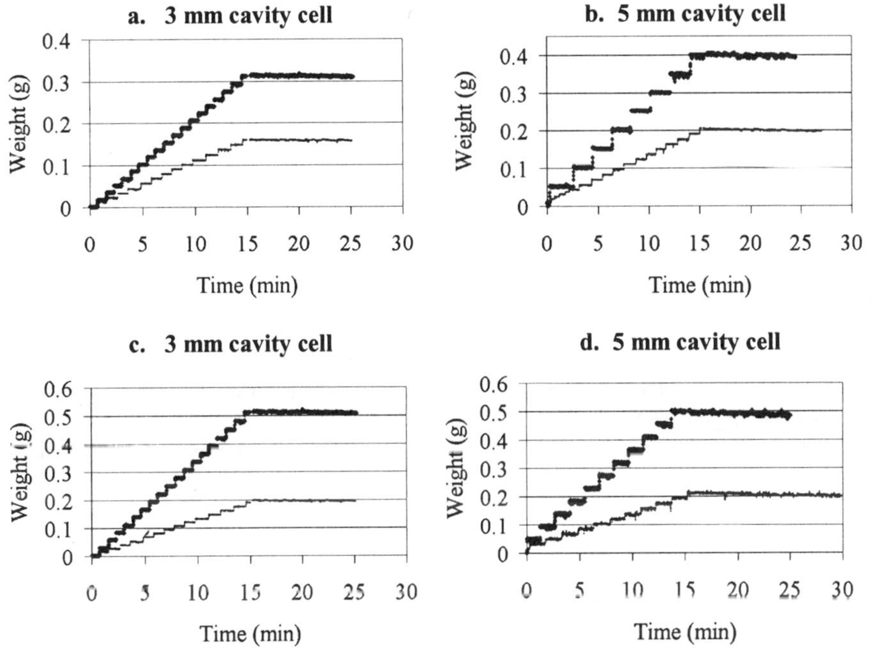

Figure 4. Fluid weight (***W***) of MAG sample as a function of time. Thin lines show 2-branch configuration, while thick lines show 4-branch configuration. Particle diameter $2a$ = 1.0 μm for a) and b), and $2a$ = 0.25 μm for c) and d). Φ = 1.0% and Q = 0.008 ml/min. Magnetic field is applied at t = 15 min.

Using water as the suspension medium, the step size before applying magnetic field decreases and the number of drops increases in most of the data shown in Fig. 4 as compared to Figs. 2 and 3. Since viscosity affects the surface tension of the droplet formed by the fluid, lower viscosity form smaller size droplets and more number of drops measured in the weight. Thus the step size in ***W*** confirms the viscosity differences between the two suspension fluids.

4. Conclusion

In a 4-branch "blood" network, blockage of water and sheep blood flow is simulated. 3 mm and 5 mm cavity cells are connected in one branch to simulate a Stage II and III tumor with different arteriole distribution (either 2 or 4 vessels). Two size magnetite particles are tested: 1.0 μm for intra-arterial treatments, and 0.25 μm for intravenous treatments. Using water as the suspension medium, MR fluid forms a stable seal under all the testing parameters. While using sheep blood as the suspension fluid, the 3 mm cell forms a stable seal for both particle sizes within seconds after the field is turned, but the 5 mm cell fails for the larger particles in a two-vessel configuration. We believe this is due to the greater magnetic field exerted on the particles in a smaller cavity cell. Thus, a magnetic field of 0.63 T is sufficient to treat a 3-mm "tumor". It requires more than 0.43 T to treat a 5 mm tumor. One needs to balance the availability of the magnetic field strength verses the tumor size that can be treated. Further work is needed to understand the details of the reduced seal strength when using blood as suspension medium as compared to using water.

We hope that this work will stimulate and help design clinical trials in the near future for cancer therapy.

Acknowledgements

We would like to gratefully acknowledge the support of Research Corporation (CC4801), NASA (NAG3-1830) and California State University Long Beach. We would also like to acknowledge Dr. Christian Bergemann of Chemicell GmbH, Germany for the magnetite samples, and Dr. Carol Itatani, CSULB, for her help on sheep physiology.

References

1. "Angiogenesis Inhibitors In Cancer Research", National Cancer Institute Cancer Trials Online, www.nci.nih.gov, 31 March 1999.
2. S. Brownlee, "When Will We Cure Cancer?" Time Magazine. November 8, 1999. Chicago, 80.
3. S. Brownlee and N. Shute, U.S. News and World Report, 18 May 1998, p62.
4. J. Metz, Univ. of Pennsylvania Cancer Center. 27 August, 1999.
5. J. Folkman, Scientific American. September 1996, New York, 150.
6. A.A. Kuznetsov, A.R. Harutyunyan, E.K. Dobrinsky, V.I. Filippov, A.G. Malenkov, A.F. Vanin, O.A. Kuznetsov. *Scientific and Clinical Applications of Magnetic Carriers*, U. Häfeli, W. Schütt, J. Teller, M. Zborowski (Eds.), Plenum Press, New York, (1997) 379.
7. A.S. Lübbe, C. Bergemann, J. Brock, D.G. McClure, J. Magnetism and Mag. Mat. 194 (1999) 149-155.
8. G.A. Flores, R. Sheng, J. Liu, *the 7th International Conference on Electrorheological Fluids, Magnetorheological Suspensions, and their Related Technologies*, R. Tao (Ed.), World Scientific, Singapore, (2000) 716.
9. G.A. Flores R. Sheng, J. Liu, J. Magnetism and Mag. Mat. 194 (1999) 167-175.
10. J. Liu, R. Sheng, and G. A. Flores, J. Magnetism and Mag. Mat. 225 (2001) 209-217.
11. J.F. Hecker, *The Sheep as an Experimental Animal*, Academic Press, London, 1983.
12. H.V. Sparks, T.W. Rooke, *Essentials of Cardiovascular Physiology*, University of Minnesota Press, 1987.
13. R.C. Little, *Physiology of the Heart and Circulation*, Year Book Medical Publication, Chicago, 1977.
14. R.M. Berne, M.N. Levy, *Cardiovascular Physiology*, 5th Ed. C. V. Mosby, Co., St. Louis, 1986.
15. National Cancer Institute NCI/PDQ Patient Statement: "Melanoma" and "Breast Cancer". University of Pennsylvania Cancer Center, www.oncolink.com, February 2001.

IMPEDANCE CONTROL OF A ONE SHAFT-TYPE CLUTCH USING HOMOGENEOUS ELECTRORHEOLOGICAL FLUID

TARO NAKAMURA AND NORIHIKO SAGA

Department of Machine Intelligence and System Engineering, Akita Prefectural University, 84-4 Tsuchiya Honjo, Akita 015-0055, JAPAN
E-mail: taro@akita-pu.ac.jp

MASARU NAKAZAWA

Department of Functional Machinery and Mechanics, Shinshu University, 3-5-1 Tokida Ueda, NAGANO 386-8567, JAPAN
E-mail: nakazawa@kappa.shinshu-u.ac.jp

This study demonstrates the application of a precise and variable impedance control to the homogeneous ER clutch. The viscous coefficient of a homogeneous ER fluid can be changed by the application an electric field. However, the ER effect of the fluid is easily changed by shear rate variation or temperature variation. Therefore, some feedback compensations are necessary for these systems to maintain a more precise viscosity.

In this paper, we apply a force-based impedance control to the homogeneous ER clutch in order to more effectively use ER fluid. First, we study the fundamental characteristics of the homogeneous ER clutch. Next, we apply the viscous control based on the force control in order to maintain a more precise viscosity. The results of our experiment clearly show that the viscosity of the clutch has been accurately and stably controlled by the proposed control method.

1 Introduction

Most mechanical systems are modeled in consideration of mechanical impedance composed of stiffness, mass, and damping. Further, these systems are widely applied to vibration control and the impedance control [1]. Recently, a variable viscous control has been applied to this type of system. For example, Satoh *et al.* [2] applied a semi-active hydraulic oil damper, which can continuously control viscosity, to a quarter-size automobile model, and discussed the resulting ride quality using numerical simulation. Ikeura *et al.* [3] showed that the damping factor is important in the co-operation of two people carrying an object. They further applied variable damping control to a robot to enable its cooperation with human.

On the other hand, ER (electro rheological) fluids have the inherent ability of generating passive resistance forces due to their drastic increase of shear stress in strong electric fields. Particular, homogeneous ER fluid [4] can change its viscous coefficient under the application of an electric field. Therefore, the application of this fluid to a damping device can achieve variable damping control in a mechanical system. Takesue *et al.* [5] proposed a direct drive motor attached an ER damper using homogeneous ER fluid. They showed that response performance of the direct drive motor with high gain feedback was improved when the ER damper was given a constant electric field. However, it is preferable in a mechanical system that the viscosity be continuously controlled. Further, in the case of applying this fluid to a damping device in a machine system, it is possible that the characteristics of the ER damper change, because the ER effect of the fluid is easily changed by shear rate variation or temperature variation [6]. Therefore, it is necessary to apply feedback control to ER fluid.

In this paper, we apply the continuous and variable viscous controller to the ER clutch whose parameter variations are robustly controlled. This paper is organized in the following way: First of all, we explain the developed one shaft-type ER clutch [7] and describe about its fundamental characteristics. Next, we propose a variable viscous controller which utilizes a force-based impedance control for a semi-active device, and apply this controller to the ER clutch which is controlled by feedback compensations.

2 One shaft-type ER clutch

We have developed a one shaft-type ER clutch for installations at the joint part of a robot. A detailed diagram of the developed ER clutch is shown in Figure 1. The following describes the primary features of this clutch.

- A gap distance between electrodes is stable for a large radial load by installing the shaft in the clutch.
- The clutch's electrode shapes combined with multiplex cylinders and disks. Therefore, an efficient output torque can be expected.

Next, we discuss the fundamental characteristics of the clutch. The relationship between the clutch's rotational speed and torque is shown in Figure 2. From this figure, we can see that the viscosity tends to increase with the increase in the electric field. However, notice that the torque

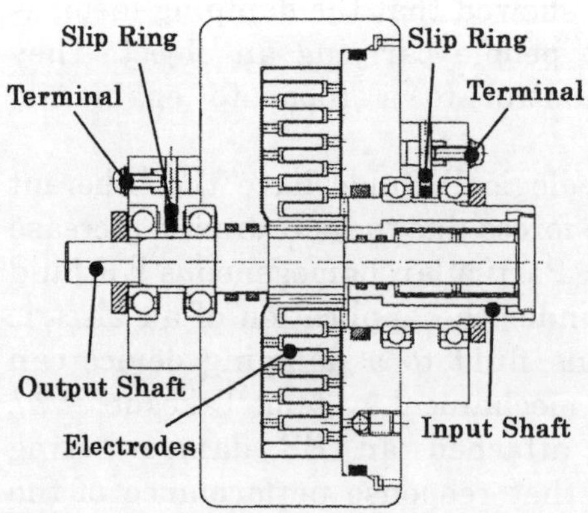

Figure 1. The Structure of the ER Clutch.

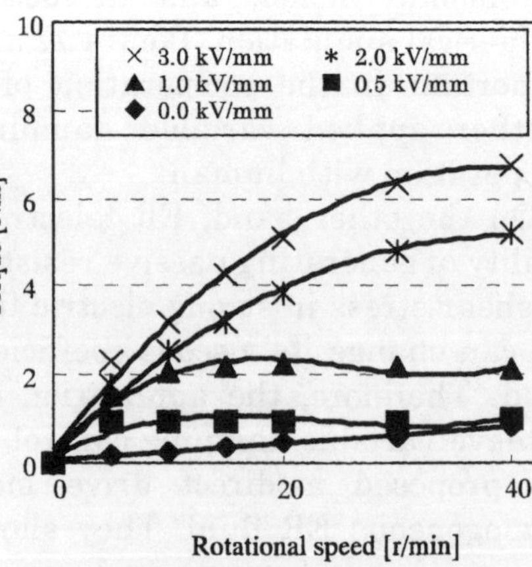

Figure 2. Relationship between the rotational speed and the torque.

tends to saturate with increasing rotational speed, although the output torque appears proportional to the rotational speed.

3 Variable viscous control

3.1 Desirable impedance model

We use an impedance control method proposed by Hogan [1], in order to control the viscosity of the ER clutch. The impedance control is the control method which controls the dynamic relation between a position and a force using an appropriate mechanical impedance model. Hence, the equation for the desirable impedance model is defined as follows:

$$f_d = m_d \dot{\omega} + d_d \omega \ , \tag{1}$$

where f_d is a desirable force calculated by the desirable impedance model, and ω is a rotational speed applied to the ER clutch. m_d and d_d are desirable impedance parameters for the inertia and the viscosity, respectively. Further, m_d is almost ignored due to the fact that inertia of the shaft and electrodes is much smaller than viscosity of ER fluid. Therefore, the desirable impedance Z_d of the clutch is as follows:

$$Z_d = d_d \ . \tag{2}$$

The desirable impedance is provided by only the viscosity parameter. Hence, in this paper, the impedance control is used for viscosity control.

3.2 Variable viscous control for semi-active mechanism

In this paper, we propose a semi-active variable viscous control method which as shown in Figure 3. This control system uses both a control signal indicating viscosity and rotational speed as an external input factor to the ER clutch. Therefore, the torque of the ER clutch is controlled by the control signal only when the ER clutch is rotated by a rotary actuator. Equation (2) is applied to variable viscous control.

$$Z_d(t) = d_d(t) = \beta(t) Q_0 \, , \tag{3}$$

where $\beta(t)$ is an adjustable parameter ($\beta(t) \geq 1$) and Q_0 is the basic viscosity of the ER clutch (Q_0=0.2Nms/r). Hence, $f_d(t)$ is shown in equation (4) when the rotational speed is ω.

$$f_d(t) = \beta(t) Q_0 \omega \ . \tag{4}$$

A controlled force $f(t)$ is shown as follows by using a convolution integral:

$$f(t) = Q_0 \omega \int_0^t p_s(\tau) \beta(t-\tau) d\tau \quad , \quad 0 \leq \tau \leq t \ , \tag{5}$$

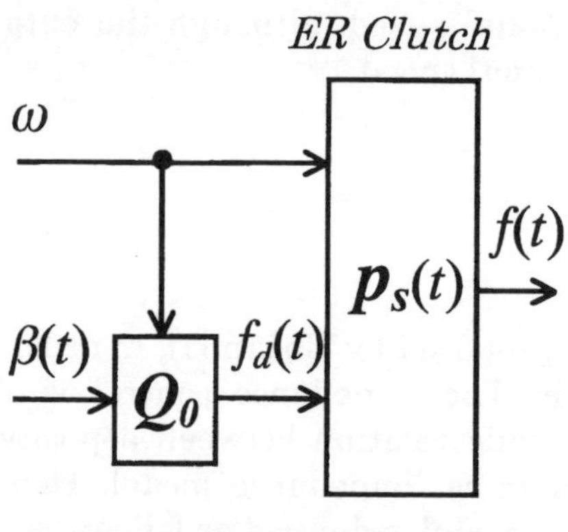

Figure 3. The block diagram of the semi-active variable viscous control.

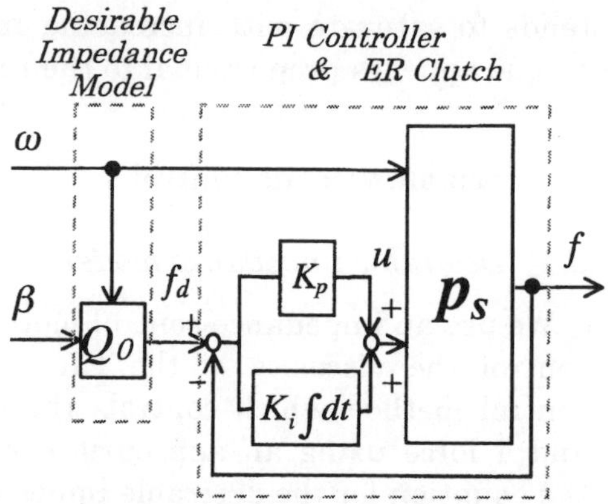

Figure 4. The block diagram of the force-based variable viscous control.

where $p_s(t)$ is a transfer function from $f_d(t)$ to $f(t)$. Further, $f(t)$ can also be simply expressed as follows:

$$f(t) = C(t)\omega \quad , \tag{6}$$

where $C(t)$ is the viscosity of the ER clutch when t [s] has passed.
$C(t)$ is also described as follows by equation (5) and the equation (6).

$$C(t) = Q_0 \cdot \int_0^t p_s(\tau)\beta(t-\tau)d\tau \ . \tag{7}$$

Therefore, if $p_s(t)$ is stably and precisely controlled, the viscosity of the ER clutch is determined in terms of the adjustable parameter $\beta(t)$ alone.

3.3 Force-based variable viscous control using PI control

A block diagram of the force-based variable control system is shown in Figure 4. From this figure, we apply a PI control to the force control of the ER clutch in order to stably and precisely control $p_s(t)$. Here, a control input $u(t)$ is shown in as follows:

$$u(t) = K_p e(t) + K_i \int e(t)dt \ , \tag{8}$$

where $e(t) = f_d(t) - f(t)$ is a force tracking error, and K_p and K_i are proportional and integral gain, respectively. Therefore, the force-based variable viscous control is operated as follows:

1. The desirable viscosity $d_d(t)$ is determined by the adjustable parameter $\beta(t)$. [Equation (3)]
2. When a rotational speed ω is applied to the ER clutch, the desirable force $f_d(t)$ is decided by using the rotational speed. [Equation(4)]
3. In order to achieve desirable force, the controlled force $f(t)$ is controlled by the PI controller. [Equation (8)]
4. When the output $f(t)$ converges to the desirable force $f_d(t)$, the viscosity of the ER clutch $C(t)$ is controlled to the desirable viscosity $D_d(t)$.

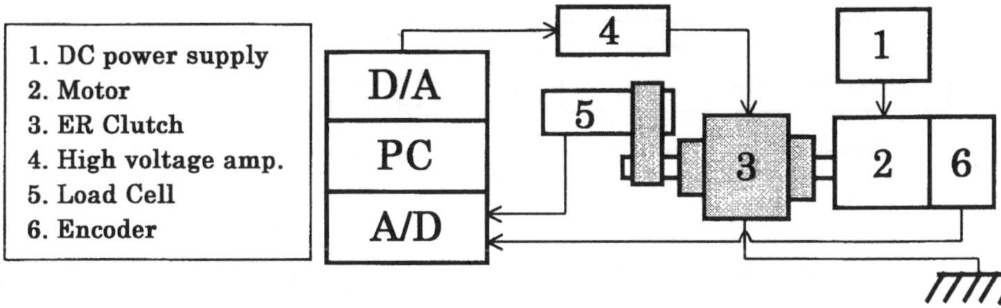

Figure 5. The Configuration of the experimental system.

4 Experiments

4.1 Experimental setup

A schematic diagram of the experimental setup is shown in Figure 5. The experiments are carried out as follows:

1. When a motor is driven by a reference signal from a DC power supply, the input shaft of the ER clutch rotates. Here, in order to use the rotational speed as an external input, notice that the motor is not controlled by a personal computer.
2. When the control input which is amplified by a high voltage amplifier is applied to the electrodes of the ER clutch, its viscosity is changed.
3. An arm is set at the output shaft of the ER clutch, and its torque is measured by a load cell. Then, an encoder which is attached to the motor also measures the rotational speed. Here, notice that the arm always makes contact with the load cell in order to be able to ignore the inertia of the ER clutch.
4. A personal computer calculates the control input based on torque and rotational speed. The control input then controls the viscosity of the ER clutch. Here, the sampling time for the calculation and the viscous control is 2ms.

4.2 Experimental results and discussion

First of all, we discuss the force controls which are constituted by an inner loop of the force-based variable viscous control. As one example, the step responses of the output torque are shown in Figure 6 (β=6, ω=40r/min). This figure shows that because the open loop control is not applied to the PI controller, the output torque has a steady-state error and cannot attain the desirable force. Whereas in the case of applied PI controller, the output torque more quickly converges to the desirable force $f_d(t)$. Therefore, we see that the PI controller accurately and quickly controls the output torque.

Next, we discuss the variable viscous control. The relationship between rotational speeds and output torques when using the viscous control is shown in Figure 7. The all of output torque is precisely proportional to

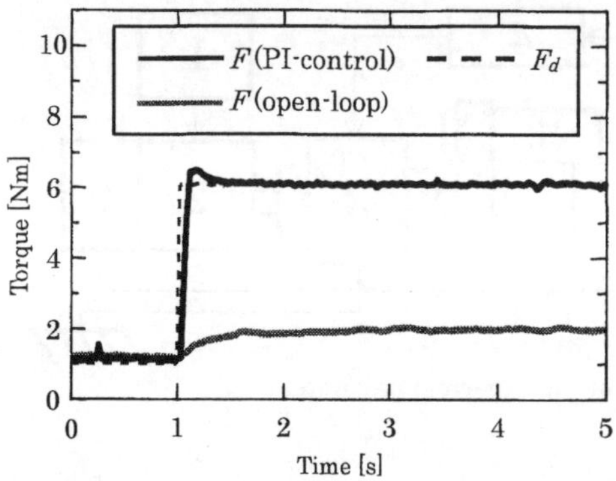

Figure 6. Step responses of the ER clutch (β=6,40r/min).

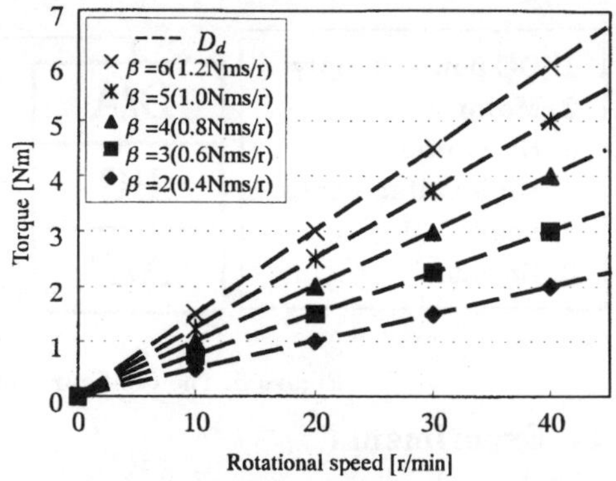

Figure 7. Relationship between rotational speed and torque using viscous control.

the rotational speed. Here, it should be emphasized that the saturation of the output torque as shown in Figure 2 is not confirmed by the high rotational speed. Therefore, we see from this figure that the viscosity is accurately controlled by the PI controller.

We next discuss the continuous variable viscous control. Figure 8 describes the viscous responses of the ER clutch when the desirable viscosity is changed by β in real time. This figure illustrates that in the case of using this control method, the viscosity is accurately and quickly tracked in real time in order to reach the desirable viscosity.

Finally, we examine the viscous response in regard to variable rotational speeds. Figure 9 shows the viscous response of the ER clutch when the rotational speed is continually changed in real time. In the case of open loop responses, the viscous response cannot reach the desirable viscosity, and is also decreased with increasing rotational speed due to the saturation of the viscosity. Whereas in the case of using the PI control, the trajectory maintained the desirable viscosity for changing the rotational speed, it also does not cause the saturation of the viscosity.

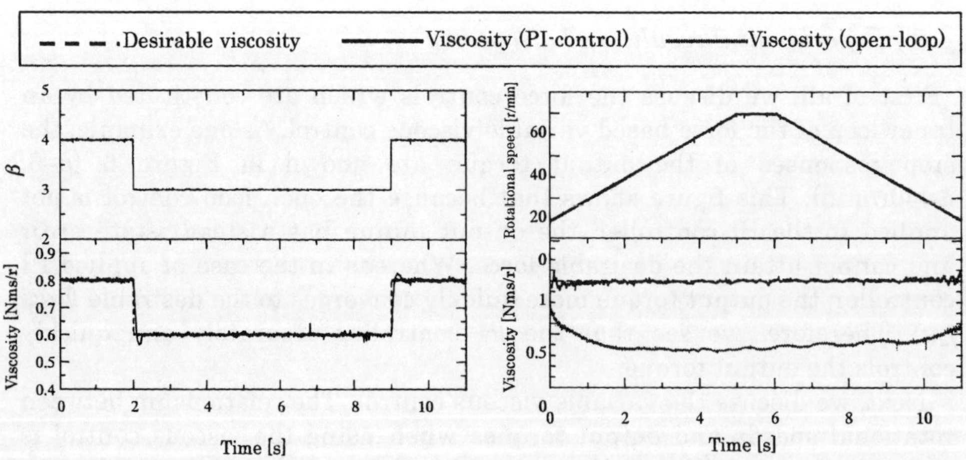

Figure 8. Viscous responses of the ER clutch (40r/min). **Figure 9.** Viscous responses of the ER clutch (β=5).

5 Conclusions

In order to accurately and stably control the viscosity of the homogeneous ER clutch, we have proposed a force-based variable viscous control which includes a force controller as the inner loop, and have applied it to the ER clutch. These results are shown as follows:

1. The PI controller has been applied as the force control which is comprised of the inner loop of the proposed variable viscous controller. As a result, the PI controller can accurately and quickly control the output torque to the desirable force.
2. In order to demonstrate variable viscous control, we have applied it to the ER clutch. The results shows the variable viscous control using PI controller accurately and quickly controls the viscosity in terms of the variation of β or the rotational speed.

6 Acknowledgements

The authors want to thanks the Asahi chemical industry Co. Ltd. For providing the ER fluid.

References

1. N.Hogan, Stable Execution of Contact Tasks Using Impedance Control, *IEEE International conference Robots and Manipulators* **2**, (1987) pp. 1047-1054.
2. N. Satoh, S. Hayashi, T. Hayase and I. Iimura, Vehicle Vibration Control Using Hydraulic Semi-Active Damper, *Transaction of The Japan Hydraulics & Pneumatics Society* **30**, (1999) pp.10-19.
3. R. Ikeura and H. Inooka, Variable Impedance Control of a Robot for Cooperation with Human, *Proceedings. IEEE International Conference on Robotics and Automation* **3**, (1995) pp.3097-3102.
4. A. Inoue and S. Misawa, Electrorheological Effect of Liquid Crystalline Polymers, *Journal of Applied Polymer Science* **55**, (1995) pp.113-118.
5. N. Takesue, G. Zhang, J. Furusho and M. Sakaguchi, Motion Control of Direct-Drive Motor By A Homogeneous ER Fluid, *Journal of Intelligent material Systems and Structures* **10**, (1999) pp.728-731.
6. N. Saito, T. Nakamura, N. Saga et al., Development of the robot Arm with soft touch-Fundamental Characteristics of ER Clutch for a Joint of the robot Arm-, *Proceedings of the 18th Annual Conference of the Robotics Society of Japan*, (2000) pp525-526.
7. N. Saga, T. Nakamura et al., Development and Experiment of ER clutch for robot Arms, *Proceedings The Fifth International Conference on Motion and Vibration Control*, (2000) pp.649-654.

EVALUATION OF A MAGNETORHEOLOGICAL SUSPENSION IN A SHOCK ABSORBER AND IN A MAGNETORHEOMETER

ANTONIO J. F. BOMBARD

Escola Federal de Engenharia de Itajubá, Av. BPS, 1303, Pinheirinho, Itajubá, MG, CEP 37.500-903, BRAZIL
E-mail: bombard@efei.br

MARCELO KNOBEL

Instituto de Física Gleb Wataghin, Unicamp, C.P. 6165, Campinas, SP, CEP 13.083-970, BRAZIL
E-mail: knobel@ifi.unicamp.br

INES JOEKES

Instituto de Química, Unicamp, C.P. 6154, Campinas, SP, CEP 13.083-970, BRAZIL
E-mail: ines@iqm.unicamp.br

The particle size distribution and magnetic susceptibility of some commercial carbonyl iron powders were measured. The polydispersity of the powders increased as follows: OX ≈ CS < HQ < SM << CC. The magnetic susceptibility increased in the order: CC ≈ HQ < OX < SM. Three grades were chosen to prepare magnetorheological suspensions (MRS). Yield stress without magnetic field and at 100 and 200 Gauss were measured. The yield stress under field increased in the order: CC < SM < OX. This result suggests that particle size distribution plays a role at least as important as magnetic susceptibility on the performance of MRS´s. Besides, a damper (Lord RD-1005-3) slightly modified and filled with an alternative MRS formulated with one of these powders was tested in a MTS 850 test system. Applying an electric current to the damper, ranging between 0 and 1 A, we measured a force of about 250 N between 0 and 100 mA, 500 N at 300 mA, 1000 N at 600 mA and 1300 N at 1 A. The behavior of the suspension is reproducible and reversible in this range of current, which suggests that it may be applied to manufacture controllable devices.

1 Introduction

There are two kinds of magnetorheological suspensions (MRS): micrometer-sized magnetic particles dispersed in an inert chemically and non-volatile liquid; and non-magnetic particles (silica, polystyrene beads) dispersed in a magnetic liquid medium, typically a ferrofluid (or magnetic colloid). MRS applications were first reported by *Rabinow* at the end of the forties.[1]

Rosensweig developed a model for the dependence of the yield stress σ_y on the magnetic susceptibility χ, volumetric fraction ϕ and applied magnetic field H_0, for MRS subjected to demagnetisation according to:[2]

$$\sigma_y = \frac{\mu_0 H_0^2}{4} \frac{\phi(1-\phi)\chi^2}{\{(1+\chi)(1+\phi\chi)[1+\chi(1-\phi)]\}^{1/2}} \tag{1}$$

Popplewell and Rosensweig have made a comparison between experimental values of yield stress and theoretical values calculated with the equation 1, and obtained a good agreement.[3] *Shulman et al* also report that yield stress has a quadratic dependence on the applied magnetic field and with the square of the susceptibility.[4] The magnetic susceptibility is defined by the ratio between magnetisation M and the field strength H and, as well as saturation magnetisation [5] play an important role in the performance of MRS.

Other important properties of magnetic dispersed phases are their particle size and particle size distribution. *Ota and Miyamoto*, using a cubic particle chain model with two particle sizes, calculated the static yield stress for a theoretical electrorheological fluid (ERF), and concluded that: "the ERF consisting of only the same particles gives the largest static yield stress." [6] Using another approach, *Lemaire et al* reported on the influence of particle size on MR: "it is better to have monodisperse sample in order to optimise the magnetorheological effect." [7] In other words, we can expect that the broader the dispersity of a given particle distribution, the lower the yield stress. *Foister*, [8] *Weiss et al*, [9] and others have used bimodal formulations to increase the MR effect, since this can reduce the viscosity of any concentrated suspension, as predicted by *Farris*.[10]

The objective of this work was to investigate how the magnetic susceptibility and the particle size distribution of some commercial carbonyl iron powders influence the yield stress of MRS within the same volume fractions.

Besides, we show the results of a mechanical test of a shock absorber prototype purchased from Lord Corporation, disassembled and reassembled with an alternative MRS formulated using one of these powders.

2 Experimental

Our magnetorheological suspensions were formulated with a naphtenic oil base and carbonyl iron powders (kindly donated by BASF AG). Thixotropic (Cab-O-Sil TS610, from Cabot) and stabilizer additives (Silwet L-7500, from Witco Corp. or Oleic acid) were also used in the formulations.

The mechanical tests were accomplished at room temperature, using a machine test system MTS 850. The monotube shock absorber tested has an electromagnet coil built-in with the shaft. The amplitude of the shaft of the shock was of 40 mm, and the frequency of 95 cycles/minute. A Minipa 3006D power source was used to supply DC current to the electromagnet.

The magnetization and susceptibility curves were measured with a Quantum Design MPMS SQUID (superconductor quantum interference device) magnetometer.

The viscosity and strain x shear stress curves were obtained with a controlled-stress rheometer (Paar Physica model MCR 300), using the plate-plate geometry with 50 mm diameter, a gap of 1 mm and room temperature. The magnetic field was applied through one pair of Helmholtz coils, and the magnetic induction **B**, was measured with a Hall effect probe and a gaussmeter (FW Bell model 5070).

3 Results and discussion

3.1 *Tests of the shock absorber prototype based on the alternative magnetorheological suspension*

In figure 1, it can be observed the effect of the current on the force measured using a Lord damper RD-1005 modified and filled with our MRS. Increasing the current increases the traction and compression forces, without saturation, until 1 A. Besides, as pointed out by *Li et al,* the amount of damping represented by the area of each cycle also increases with magnetic field.[11] The carbonyl iron powder BASF grade 'CC' was used in this formulation, showing to be efficient as MRS, and that the modification and change of the original MRS does not affected the damper.

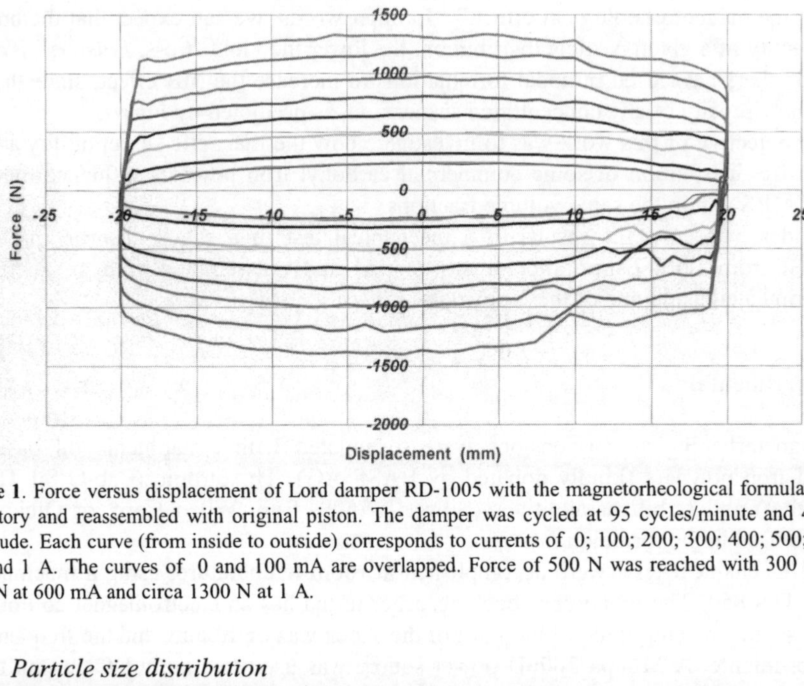

Figure 1. Force versus displacement of Lord damper RD-1005 with the magnetorheological formulated in our laboratory and reassembled with original piston. The damper was cycled at 95 cycles/minute and 40 mm of amplitude. Each curve (from inside to outside) corresponds to currents of 0; 100; 200; 300; 400; 500; 600; 750 mA and 1 A. The curves of 0 and 100 mA are overlapped. Force of 500 N was reached with 300 mA, circa 1000 N at 600 mA and circa 1300 N at 1 A.

3.2 *Particle size distribution*

The particle size distributions of the carbonyl iron powders as measured experimentally by laser light scattering, using a Malvern Mastersizer Micro instrument, are summarised on table 1 and the distributions profiles of the 5 grades are showed in the figure 2.

Table 1. Particle size distribution measured experimentally, using laser light scattering. Distribution number based.

Carbonyl iron powder grade	HQ	SM	OX	CC	CS
10% less than	0.57 μm	0.91 μm	1.64 μm	0.40 μm	2.45 μm
50% less than	0.88 μm	1.44 μm	2.64 μm	0.81 μm	3.91 μm
90% less than	1.67 μm	2.91 μm	4.62 μm	3.90 μm	6.98 μm
Span = $(D_{0.9}-D_{0.1})/D_{0.5}$	1.25	1.39	1.13	4.33	1.16

Span is the measurement of the width of the distribution, the smaller the value, narrower the distribution. The order of dispersity is: OX ≈ CS < HQ < SM << CC. Grade CC seems to have a bimodal size distribution. The finest powder is grade HQ, while CS grade has the largest particles. So, considering only dispersity, OX and CS should be the better of these 5 grades to formulate MRS.

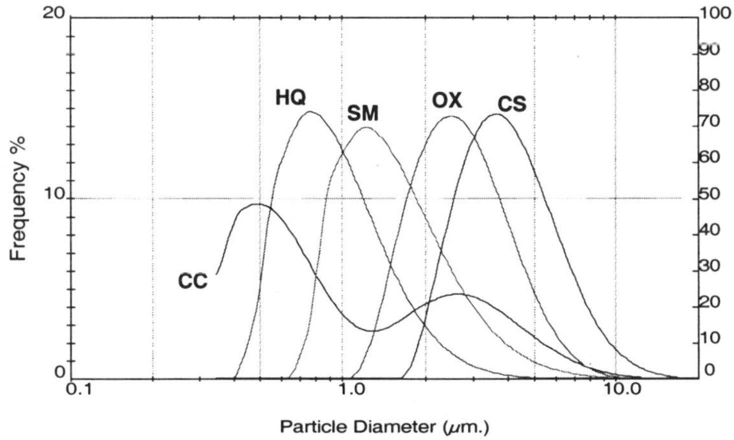

Figure 2. Particle size distribution (number based) of 5 different grades of carbonyl iron powder of BASF AG. The lower limit of sensibility of the Malvern Mastersizer was 0.3 μm.

3.3 Magnetization measurements

Figure 3 shows the magnetic susceptibility χ of the powders as a function of the magnetic field strength H. The curves were measured with the BASF powders as received. It can be seen that the magnetic susceptibility of the powders increases in the following order: CC ~ HQ (low field) < OX < SM. The mass fraction can be coarsely converted to volume fraction ($\phi \approx 19\%$ v/v), if all powders have the same density.

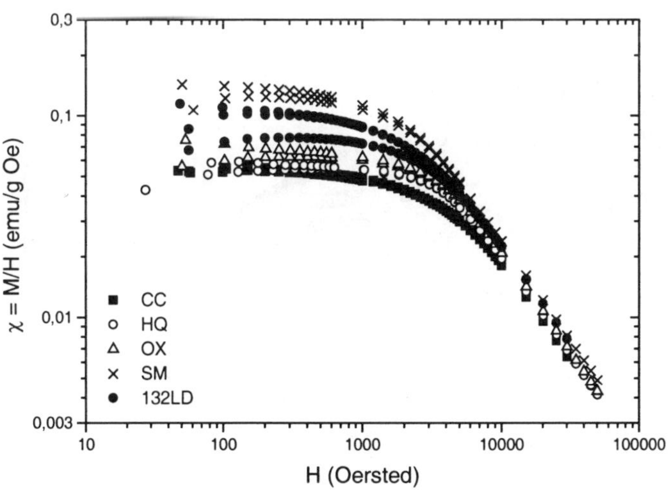

Figure 3. Magnetic susceptibility χ of 4 carbonyl iron powders from BASF AG as a function of the applied magnetic field strength H. Above 1000 Oersted χ drops quickly.

Then, considering equation 1, and the results in figure 3, could be expected that the MR effect, or the apparent viscosity increase, will follow the magnetisation order. That means: a MR suspension based on SM powder, should have the largest MR effect among the 5 grades; and another MR suspension based on CC powder, must be the worst, in terms of the viscosity increase, if magnetisation and susceptibility were the only difference among the powders.

3.4 Rheological measurements

Magnetorheological formulations employing BASF iron powders grades CC, OX and SM were prepared, as showed in table 2. According BASF, CC is a SiO_2 coated iron powder; OX has a coating of Fe_2O_3 and SM has just the oxygen level recommended for safety reasons. Firstly, we prepare a gel adding to the base oil (45 ml), oleic acid as surfactant (8 ml) and Cab-O-Sil TS610 (2.3515 g) as thixotropic additive. After homogenisation of this gel, the iron powders were weighted separately, added to the gel and mixed thoroughly.

Table 2: Composition of the MRS and chemical composition of powders

Formulation	Gel base mass	Carbonyl iron powder mass	% w/w Iron	Fe content % w/w	Others % w/w
MRS 'CC'	10.1384 g	20.0014 g	66.4	99.6	0.1 SiO_2
MRS 'OX'	10.4073 g	20.0007 g	65.8	96.4	2 Fe_2O_3
MRS 'SM'	10.2576 g	20.0006 g	66.1	99.3	0.6 oxygen

Figure 4 shows that the viscosity of MRS without magnetic field follow the order: OX < CC < SM; while at **B** = 200 Gauss, viscosity data of the three MRS were quite similar. It can suggest, perhaps, that the viscosity of MRS is influenced by the particle size and surface properties of the dispersed phase, in the absence of field.

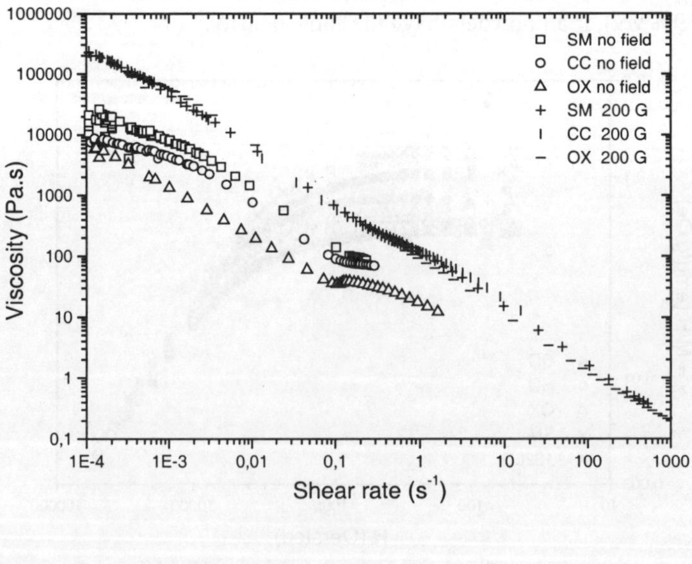

Figure 4. Viscosity curves of MRS based on SM, CC and OX carbonyl iron powders at 66% w/w. No magnetic field and with **B** = 200 Gauss.

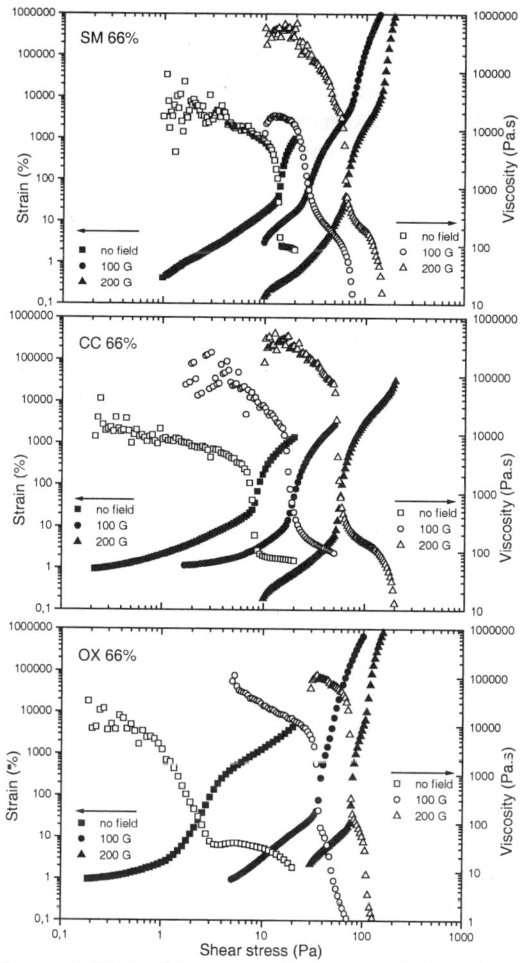

Figure 5. Strain (left axis) and viscosity (right axis) as function of shear stress. From top to bottom: MRS 'SM', 'CC' and 'OX'. The stress was increased logarithmically; strain and viscosity were measured to determine the yield stresses.

According several controlled-stress rheometer manufacturers (Paar, Haake, Bohlin, etc.), a good method to determine yield stress is to programme a shear stress ramp, and measure the resulting strain. *Barnes*, in a recent (and excellent) review about yield stress, recommended that a plot log-log of viscosity as a function of shear stress should be used to detect the yield stress.[12] The shear stress value associated with the sharp drop in viscosity is related to the yield stress value. We measured the MRS from the table 2 using this technique, with the yield values being calculated through the software US200 of the Paar Physica rheometer MCR300. The results are shown on figure 5, and the yield stress values are resumed in the table 3. It can be seen at figure 5, that MRS 'OX', present the more pronounced inflexion on strain-stress curves, at 100 and 200 Gauss. Besides, the yield stress increase of 'OX', from 0 (no field) to 100 G, is larger than to the other two MRS, 'SM' and 'CC'. By other hand, the viscosity drop of 'OX', at no field, is less sharp than others. Looking on table 3, the order of yield stress under magnetic field (100 or 200 G) is: CC < SM < OX. This is the inverse order of dispersity, as shown by span, on table 1. Since 'SM' has the highest magnetic susceptibility, this result suggests that particle size distribution plays a role at least as important as magnetic susceptibility, on the performance of MRS.

Table 3: Yield stress (Pa) of the MRS calculated from log (strain) x log (shear stress) plots.

MRS	OX	CC	SM
σ_y (no field)	1.3	7.2	11.6
σ_y (100 G)	31.2	16.0	23.5
σ_y (200 G)	72.8	49.9	58.0

4 Conclusion

The particle size distribution shows to be as important as magnetic susceptibility χ, influencing the performance of a magnetorheological suspension. Besides, it seems that the particle size influence the viscosity of the MRS, in the absence of the magnetic field.

5 Acknowledgements

We thank Professor Elizabeth P. G. Arêas and Professor Maria Regina Alcântara, from IQ-USP, for the use of Paar Physica MCR 300 rheometer. We also thank Mr. José Roberto Turini, from Magneti Marelli – Cofap (Brazil) for the tests with the Lord damper. The grant TEC-2271/96, from FAPEMIG is gratefully acknowledged.

References

1. Rabinow J.; "The magnetic fluid clutch"; *AIEE Trans.*; **67** (part II); 1948; pp. 1308 – 1315, and US Patent 2,575,360 – Nov. 20, 1951.
2. Rosensweig R. E.; "On magnetorheology and electrorheology as states of unsymmetric stress"; *J. Rheology*; **39**; 1995; pp. 179 – 192.
3. Popplewell J. and Rosensweig R. E.; "Magnetorheological fluid composites"; *J. Phys. D: Appl. Phys.*; **29**; 1996; pp. 2297 – 2303.
4. Shulman Z. P.; Kordonsky V. I.; Zaltsgendler E. A.; Prokhorov I. V.; Khusid B. M. and Demchuk S. A., "Structure, physical properties and dynamics of magnetorheological suspensions"; *Int. J. Multiphase Flow*; **12**(6); 1986; pp. 935 – 955.
5. Ginder J. M., Davis L. C.; "Shear stresses in magnetorheological fluids: Role of magnetic saturation"; *Appl. Phys. Lett.*; **65**(26); 1994; pp. 3410 – 3412.
6. Ota M. and Miyamoto T.; "Optimum particle size distribution of an electrorheological fluid"; *J. Appl. Phys.*; **76**(9); 1994; pp. 5528 – 5532.
7. Lemaire E.; Meunier A.; Bossis G.; Liu J.; Felt D.; Bashtovoi P. and Matoussevitch N.; "Influence of the particle size on the rheology of magnetorheological fluids"; *J. Rheology*; **39**(5); 1995; pp. 1011 – 1020.
8. Foister R. T.; "Magnetorheological fluids"; US Patent 5,667,715 – Sep. 16, 1997.
9. Weiss K. D.; Carlson J. D. and Nixon D. A.; US Patent 6,027,664 – Feb. 22, 2000.
10. Farris R. J.; "Prediction of the Viscosity of Multimodal Suspensions from Unimodal Viscosity Data"; *Trans. Soc. Rheol.*; **12**(2); 1968; pp. 281 – 301.
11. Li W. H.; Yao G. Z.; Chen G.; Yeo S. H. and Yap F. F.; "Testing and steady state modelling of a linear MR damper under sinusoidal loading"; *Smart Mat. Struct.*; **9**(1); 2000; pp. 95 – 102.
12. Barnes H. A.; "The yield stress – a review or 'παντα ρει' – everything flows?"; *J. Non-Newtonian Fluid Mech.*; **81**; 1999; pp. 133 – 178.

ELECTROSTRICTION EFFECT IN SOLID-STATE ELECTRORHEOLOGICAL COMPOSITES FOR SENSING APPLICATIONS

G. H. KIM AND Y. M. SHKEL

Dept of Mechanical Engineering and Polymer Engineering Center, University of Wisconsin - Madison, 1513 University Ave, WI 53706, USA
E-mail: yshkel@engr.wisc.edu

Solid electrorheological (ER) composites are prepared by arranging suspended particles in polymer melts with an electric field and cross-linking the obtained structure. Electrostriction phenomenon of ER composites is defined as a deformation effect on the dielectric properties. Composite materials with a chain-like structure demonstrate large electrostriction response to normal and shear deformations. High sensitivity of anisotropic composites to shear electrostriction can be utilized for vibration control and for shear stress/strain sensor applications. We model an electrostriction behavior and provide experimental data characterizing electromechanical properties of the synthesized ER composites.

1 Introduction

Anisotropic solid composites are called "smart materials" due to their electric field-controllable mechanical response. Two phenomena, *electrostriction*, defined as the field-induced deformation of a material, occurs in all materials[1] and *electrorheology* (ER), defined as the field-induced change in rheological properties mostly observed in anisotropic materials, are both manifestations of field-induced electrostatic stresses.[2] The same electrostriction material parameters[1,3] determining deformation effect on dielectric properties of the material define an electric field-induced stress in the material, *e.g.*, its electrorheology. Solid electrorheological (ER) composites are prepared by arranging suspended particles in polymer with an electric field and cross-linking the obtained structure.[4] A chain-like structure of ER composites enhance their electrostriction response to normal and shear deformations. This makes them feasible for fully integrated and embedded in ER device sensor solutions as well as for stand-alone sensing systems. High sensitivity of anisotropic composites to shear electrostriction can be utilized for vibration control in various "smart" bushing systems and for shear stress/strain sensor applications such as robot's grasping. A similar design can be formulated by combining shear electrostriction sensing within MR systems. We present continuous descriptions of the electrostriction phenomena and provide experimental data characterizing the electrorheology and electrostriction properties of the synthesized ER composites.

2 Solid-State Capacitance Sensor

A solid-state capacitor sensor is composed of a dielectric material of thickness, h, sandwiched between two parallel electrodes of area A (Fig. 1). Such a design ensures robustness of the sensor—high overload deformations elicit a non-linear response without collapsing the sensor, and the sensor geometry can vary with bending,

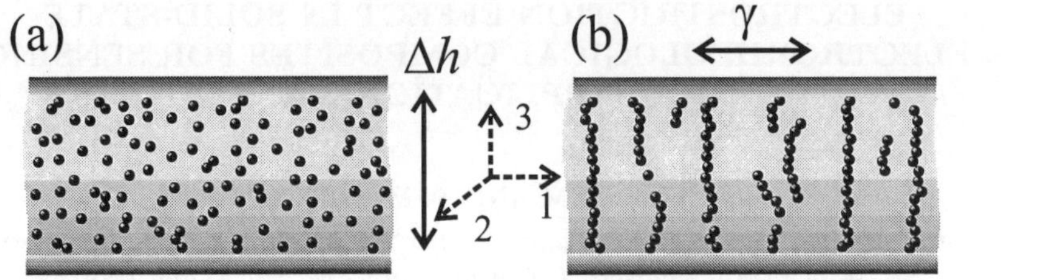

Figure 1. (a) Sensor with isotropic material between electrodes is sensitive to normal deformations. (b) Anisotropic ER composite enhances sensing capabilities for both normal and shear deformations.

---

twisting, etc. For some applications it can be very helpful to embed a sensor within existing material making it a part of the sensor itself. A chain-like structure of solid ER composites increases their electrostriction response to normal and shear deformation relative to similar composites with random distribution of inclusions. Utilizing the same material for actuator and sensing functions is advantageous for application of ER composites in various transducers and actuators systems.

2.1 Theoretical Analysis

Capacitance C of a capacitor composed of dielectric material having thickness h and dielectric constant ϵ sandwiched between parallel electrodes of area A is[3]

$$C = \frac{\epsilon_0 \epsilon A}{h}\left[1 + O\left(\frac{\log(\sqrt{A}/h)}{2\sqrt{A}/h}\right)\right], h \ll \sqrt{A}. \tag{1}$$

where $\epsilon_0 = 8.85 \times 10^{-12}$ F/m is permittivity of a free space. A fringe effect estimated by the second term in the brackets is less than 2% for reported geometry and, thus, will be ignored. A variation of capacitance $\Delta C/C$ in Eq. (1) is composed by variation of the gap $\Delta h/h$ and a variation of the dielectric constant $\Delta\epsilon/\epsilon$, while for rigid electrodes $\Delta A/A = 0$,

$$\frac{\Delta C}{C} = -\frac{\Delta h}{h} + \frac{\Delta \epsilon}{\epsilon}. \tag{2}$$

The term $\Delta\epsilon/\epsilon$ for isotropic materials: A normal deformation u_{33} of constrained isotropic dielectric affects its dielectric constant[5,6]

$$\Delta\epsilon = (\alpha_1 + \alpha_2)u_{33}. \tag{3}$$

The parameters α_1 and α_2, first introduced by Stratton[1] and Landau&Lifshitz[3] define deformation effect on the dielectric constant as well as electric field induced stress in dielectric materials and are called *electrostriction parameters.* Theoretical model for isotropic composites[6] predicts electrostriction parameters through dielectric constant of the continuous phase ϵ_c

$$\alpha_1 = -\frac{2}{5}\frac{(\epsilon - \epsilon_c)^2}{\epsilon_c}, \text{ and } \alpha_2 = -\frac{(\epsilon - \epsilon_c)(\epsilon + 2\epsilon_c)}{3\epsilon_c} + \frac{2}{15}\frac{(\epsilon - \epsilon_c)^2}{\epsilon_c}. \tag{4}$$

The term $\Delta\epsilon/\epsilon$ for anisotropic materials: The dielectric properties of transverse isotropic materials differ in direction defined by the unit vector **d** and in the plane perpendicular to **d**. These materials have a second order dielectric tensor of the form

$$\epsilon_{ik} = \epsilon^{\perp}\delta_{ik} + (\epsilon^{\|} - \epsilon^{\perp})d_i d_k, \tag{5}$$

where $\epsilon^{\|}$ and $\epsilon^{\perp}$ are the relative dielectric constants for probing fields applied parallel and perpendicular to **d**, respectively. The strain-dependent dielectric tensor requires 6 phenomenological *electrostriction coefficients* α_1–α_6.[2] These material properties are scalar functions of thermodynamic variables, such as the temperature, the electric field magnitude, and the density. Effect of normal deformation on $\Delta\epsilon = \epsilon_{33} - \epsilon^{\|}$ and the normal stress of transverse isotropic materials is[2]

$$\Delta\epsilon_{33} = K_n u_{33}, \text{ and } \sigma_{33}^{elect.} = \epsilon_0 \frac{E^2}{2}\left\{\epsilon^{\|} - K_n\right\}, \tag{6}$$

$$K_n = \alpha_1 + \alpha_2 + \alpha_3 + 2\alpha_4 + \alpha_5 + \alpha_6, \tag{7}$$

where deformation effect on polarization and normal stress are both determined by same parameter K_n. Given that strain tensor is defined as[7]

$$u_{ik} = \frac{1}{2}\left(\frac{\partial u_i}{\partial x_k} + \frac{\partial u_k}{\partial x_i} + \frac{\partial u_l}{\partial x_i}\frac{\partial u_l}{\partial x_k}\right), \tag{8}$$

deformation caused by shear displacement $U_2 = \gamma x_3$ yields for 3 nonzero strain components $u_{13} = u_{31} = \gamma/2$ and $u_{33} = \gamma^2/2$. Thus, $\Delta\epsilon$ and shear stress σ_{13} of transverse isotropic materials are

$$\Delta\epsilon = (K_\tau + K_n)\frac{\gamma^2}{2}, \text{ and } \sigma_{13}^{elect.} = \epsilon_0 K_\tau \frac{E^2}{2}\gamma, \tag{9}$$

$$K_\tau = \epsilon^{\|} - \epsilon^{\perp} - \alpha_3 - \alpha_4 - \alpha_6. \tag{10}$$

A theoretical model of isolated chains[2] predicts parameters K_n and K_τ as

$$K_n = -(\epsilon^{\|} - \epsilon_c)\left[1 + \frac{3\xi}{1-\xi}\right], \; K_\tau = 2(\epsilon^{\|} - \epsilon^{\perp}) + (\epsilon^{\|} - \epsilon_c)\frac{\xi}{1-\xi} \tag{11}$$

$$\xi = \frac{1}{2}\frac{\epsilon_p - \epsilon_c}{\epsilon_p + 2\epsilon_c}\zeta(3)\left(\frac{a}{l}\right)^3, \; \zeta(3) = 1.202\ldots \tag{12}$$

where ϵ_p and ϵ_c are dielectric constants of inclusions and continuous phase, a is the particle diameter and l is the distance between centers of particle in the chain.

3 Experiment

3.1 Sample Preparation

Either iron particles diameter $< 45\mu m$ ($\phi \approx 1.6$ vol%) or alumina particles diameter $< 100\mu m$ ($\phi \approx 2.6$vol%) were mixed with base prepolymer of silicone elastomer (from Silicones, Inc.) and de-gazed in a vacuum chamber. With added cross-linking agent and catalyst, the mixture was evacuated for 10 *min* and transferred into a set of casting cells resembling parallel plate capacitors with gap in 1 *mm* each. A composite with aligned particles was formed in cells by maintaining an

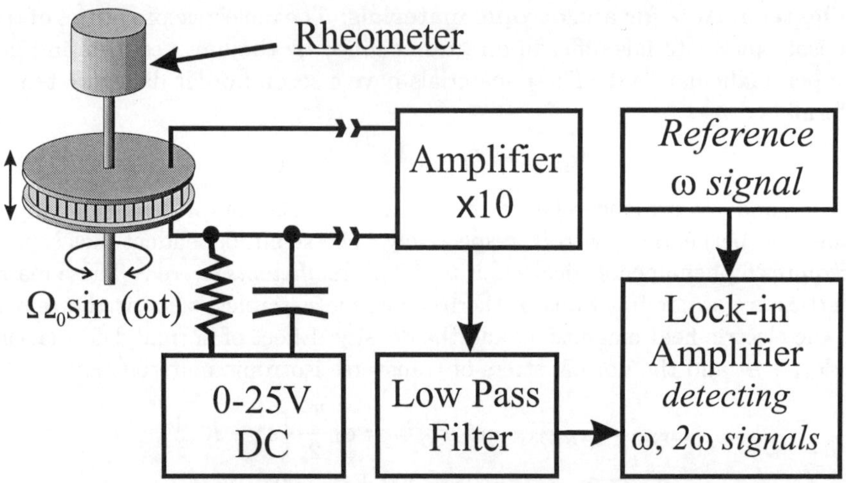

Figure 2. Power supply provides bias DC electric field; high impedance AC signal due to sample deformation is measured by Lock-in-Amplifier and compared with rheometric data.

electric field of 1.5 kV/mm at $f = 100\ Hz$ during the curing time. 1 $M\Omega$ blocking resistor connected in serious with the casting cells and a high voltage source was required for aligning composites with iron particles. Similar cells with the field off were assigned for samples with random distribution of particles.

3.2 Experimental Setup

AR1000 rheometer from TA Instruments with parallel plate configuration was modified for shear testing and employed for electrostriction and electrorheological characterization of samples with aligned and random structures. To study an ER effect, a high voltage potential in range $0 - 3500V$ was generated by a Trek HV amplifier. Rheological and ER properties of the samples where acquired at various frequencies, deformation magnitudes, and electric field strengths. In addition, the rheometer provided controllable displacements for an electrostriction study. A voltage variation due to shear electrostriction was registered with an experimental setup on Fig. 2. A voltage produced by the sample has enormously high source impedance—typically greater than 400 $M\Omega$—which cannot be measured by conventional instrumentation having typical input impedance $1M\Omega$. Electrostriction measurement of the sample resembles *Dynamic potential* technique implemented in capacitance microphones for amplifying of deformation-induced potentials.[8] High impedance DC power supply is providing a constant dielectric displacement **D** across the sample. AC signal due to the sample deformation is buffered and amplified by a high impedance amplifier and measured by Stanford Research Systems Lock-in-Amplifier.

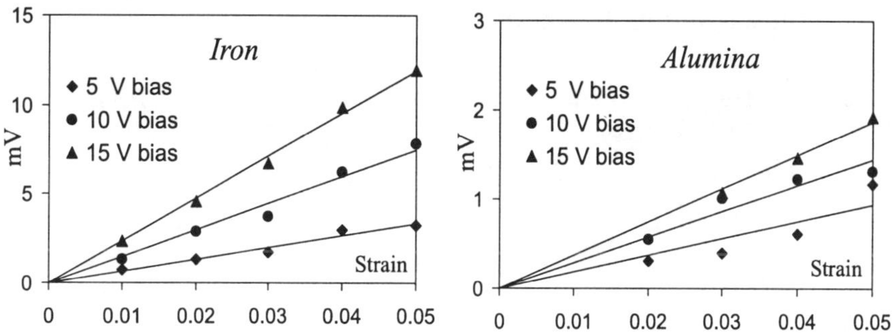

Figure 3. Shear strain/voltage response characteristics for iron and alumina composites. Data are adjusted for normal strain effect, the samples conductivity, and amplifier gain.

3.3 Experimental Results

Capacitance C of the sample and voltage V due to charge Q are related as $Q = VC$. Variation in the capacitance ΔC with deformation of the charged sample causes variation of the voltage ΔV across the sample

$$\frac{\Delta V}{V} = -\frac{\Delta C}{C}. \tag{13}$$

Voltage variations with respect to shear deformation of aligned iron and alumina samples are plotted on Fig. 3. These plots represent important design parameters for shear and normal strain sensors. Both iron and alumina samples with aligned structure demonstrate strong ER response, while samples with unaligned structures have no ER response. An increase in elastic modulus G' due to field is much larger for the iron than for the alumina composites, which is illustrated by Fig. 4.

4 Discussion

Displacement of the rheometer plate $U = \gamma h$, where $h = const$ is the sample thickness, causes shear $u_{13} = u_{31} = \gamma/2$ and normal $u_{33} = \gamma^2/2$ deformations, Eq. (8). The sample capacitance change $\Delta C/C$ and, therefore, the voltage variation $\Delta V/V$, are determined by Eq. (2) retaining only the term $\Delta\epsilon/\epsilon$

$$\begin{aligned}\frac{\Delta V}{V} = -\frac{\Delta\epsilon_{eff}}{\epsilon_{eff}} &= -\left(\frac{\partial\epsilon_{eff}}{\partial u_{33}}\frac{\partial u_{33}}{\partial\gamma} + \frac{\partial\epsilon_{eff}}{\partial u_{13}}\frac{\partial u_{13}}{\partial\gamma}\right)\frac{\gamma}{\epsilon_{eff}} \\ &= -\frac{1}{\epsilon_{eff}}(K_\tau + K_n)\frac{\gamma^2}{2}, \end{aligned} \tag{14}$$

where K_n and K_τ are material parameters introduced in Eqs. (7) and (10). Here, these parameters are introduced without references to a particular composite model. Following others,[9,10] same general consideration relates parameter K_τ to field-induced shear stress σ_{13} and parameter K_n to field-induced normal stress σ_{33}

$$\sigma_{13} = -\frac{1}{2}\frac{\partial\mathbf{E}\cdot\mathbf{D}}{\partial\gamma} = -\epsilon_0\gamma K_\tau E^2, \tag{15}$$

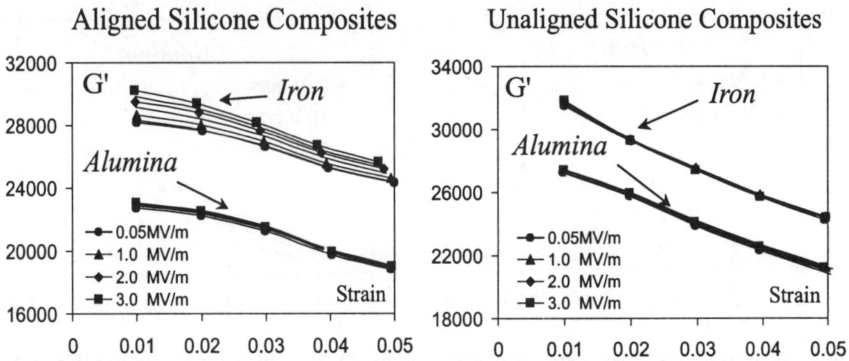

Figure 4. ER effect: increase of elastic modulus G'due to field for aligned and unaligned samples with iron and alumina particles.

$$\sigma_{33} = \frac{\mathbf{E}\cdot\mathbf{D}}{2} - \frac{1}{2}\frac{\partial D_l}{\partial u_{33}}E_l = \epsilon_0\frac{E^2}{2}\Big(\epsilon^{||} - K_n\Big). \tag{16}$$

Eq. 14 provides important parameters for a sensor design—a relationship between the voltage response ΔV, the deformation intensity γ and electrostriction properties of the composite. Tailoring a material structure for optimization of parameters K_n and K_τ is required for increasing a sensor efficiency. Eqs. (11) and (12) relate these parameters to material microstructure. Presented here experimental technique can be also applied to study electrostriction and electrorheology of solid composites. Shear electrostriction test determines a combination of two parameters $K_n + K_\tau$, thus, an additional, normal electrostriction test should be conducted. In according to definition of the parameter K_n by Eq. (6)), essentially same test is required for its measurement. This test would utilize CP20 Compressional Rheometer from TA Instruments for compressional oscillatory deformations instead of shear deformations discussed above; electronic measuring setup will stay the same. Planes are to conduct these measurements in near future.

Eq. 14 predicts that voltage response to shear deformations is quadratic in the shear strain magnitude. Thus, the voltage signal should be measured on double frequency of the rheometer oscillation. Unfortunately, a driver motor of the rheometer has a noticeable electric noise on the same frequency, which is masking a comparatively weak testing signal. This noise would be shielded in future developments of this measuring technique, while at present time, we employed another approach—oscillatory shearing with an offset: $\gamma = x_0 + \gamma_0 \sin\omega t$. Due to a simple algebraic relation, $\gamma^2 = x_0^2 + 2\gamma_0 x_0 \sin\omega t + \gamma_0^2 \sin^2\omega t$, such an offset allows us to conduct measurements using either the first or the second harmonics of the rheometer oscillation frequency. For the first harmonic, Eq. 14 can be modified as

$$\frac{\Delta V}{V} = -\big(K_\tau + K_n\big)\frac{2\gamma_0 x_0 \sin\omega t}{\epsilon_{eff}}.$$

Data presented on Fig. 3 are taken with the frequency of oscillatory shear deformations ($f = 10\ Hz$) and behave according to the above equation: they are linear with the strain magnitude, γ_0, and applied voltage, V. While these measurements reveal a combination of two parameters $K_n + K_\tau$, only the parameter K_τ is responsible for ER effect. Still, we can compare experimental values predicted from electrosrtiction and electrorheology measurements. Given that the offset deformation x_0 was produced by the offset stress of 3 Pa, shear modulus of the iron/silicone composite is $G' = 30000\ Pa$, strain $\gamma = 0.02$, and a gain of the high impedance amplifier is 10, we can estimate $K_n + K_\tau \approx 58$, for $\epsilon_{eff} = 5$. ER measurements, for $G'(3\ MV/m) = 29420Pa$ and $G'(0) = 27840Pa$, reveal that parameter $K_\tau = 2[G'(E) - G'(0)]/\epsilon_0 E^2 \approx 40$ has same order of magnitude.

5 Conclusion

We have demonstrated a feasibility of application ER composites for shear stress/strain sensing. Such easy manufacturing sensors can be applied in many different areas. There is a potential to combine ER devices with shear and normal strain/stress sensing, making truly "smart" structures. However, for better performance of stand alone sensors better ER composites with low conductivity and high anisotropy should be engineered. Presented in this article novel shear electrostriction phenomenon requires future investigation for designing materials with optimal sensing response. There is a big potential to apply developed sensing technique to analytical study of material microstructure and ER phenomenon.

Acknowledgments

The authors thanks Torrey Filanc-Bowen for discussion and help with measurements and TA Instruments for providing CP20 Compressional Rheometer. This work was supported in part by the University of Wisconsin I&EDR grant #2002038.

References

1. J. A. Stratton, *Electromagnetic Theory* (McGraw-Hill, New York, 1941).
2. Y. M. Shkel and D. J. Klingenberg, J.Rheol. **45(5)**, 1322 (1999).
3. L. D. Landau and E. M. Lifshitz, *Electrodynamics of Continuous Media*, (Pergamon, 1984).
4. B. Liu and M. T. Shaw, J.Rheol. **45(3)**, 641 (2001).
5. Y. M. Shkel and D. J. Klingenberg, J. Appl. Phys. **80**, 4566 (1996).
6. Y. M. Shkel and D. J. Klingenberg, J. Appl. Phys. **83(12)**, 7834 (1998).
7. Landau, L. D. and E. M. Lifshitz, *Theory of Elasticity*, (Pergamon, 1959.)
8. P. Horowitz and W. Hill, *The Art of Electronics* (Cambr. Univ. Press, 1998).
9. R. Tao, J.T. Woestman, and N.K. Jaggi, J. Appl.Phys.Lett. **55**, 1844 (1989).
10. L.C. Davis, J. Appl. Phys. **72**, 1334 (1992).

ER EFFECTS OF LIQUID CRYSTALLINE POLYMERS ON ONE-SIDED PATTERN ELECTRODES

NAOYUKI TAKESUE, JUNJI FURUSHO, AND MASAMICHI SAKAGUCHI
Graduate School of Engineering, Osaka University,
2-1 Yamadaoka, Suita, Osaka, 565-0871 Japan
E-mail: furusho@mech.eng.osaka-u.ac.jp

AKIO INOUE
Asahi Chemical Industry Co., Ltd.,
2-1 Samejima, Fuji, Shizuoka, 416-8501 Japan

Various devices and equipment using the ER effect have been developed and studied. When the ER fluid is used in shear mode, the ER fluid is filled between a pair of opposing parallel electrodes, and the rheological characteristics are controlled by applying an electric field across the electrodes. In general, a high voltage of a few kV is required to obtain an adequate ER effect. This causes problems associated with safty or cost. When a pair of opposing parallel electrodes is employed, use of a slip-ring or a brush is needed for the wiring to the moving part, again leading to problems associated with reliability or cost. In this paper, use of one-sided pattern electrodes is proposed as a solution to the above problems. A one-sided pattern electrode is an electrode in the form of a pattern arranged on only one of the two surfaces of a pair of disks, with an insulator arranged on the other surface. We developed two kinds of one-sided pattern electrodes and experimentally verified the manifestation of ER effects in liquid crystalline polymers.

1 Introduction

An ER (electro-rheological) fluid can change its rheological characteristics by the application of an electric field[1,2]. Various devices and equipment such as dampers, clutches and actuators using the ER effect have been developed[3,4,5]. We too have showed the effectiveness of ER fluids for robotic applications[6,7,8,9,10]. An ER damper can change the damping properties of the robot mechanically according to its operating situation.

In most of the studies, ER fluid fills between a pair of opposing parallel electrodes, and an electric field is applied across the electrodes. In general, an electric field strength of a few kV/mm is required to obtain an adequate ER effect. Thus, with a gap distance between the electrodes of 1 mm, although only a small current is required, a power source with an output voltage of a few kV is necessary, giving rise to possible problems associated with safety as well as cost.

Reduction of the gap distance may be a possible solution to the problems associated with high voltage. In order to maintain such a reduced gap distance, a precise mechanism that can center the rotation axis or a precision bearing is required. If a mechanism for maintaining gap distance fails, short-circuiting of the electrodes would result, leading to a breakdown or inadequate performance of the equipment.

When a pair of opposing parallel electrodes is employed, use of a slip-ring or a brush needs to be considered for wiring to the moving part, again leading to problems associated with reliability and cost.

In this study, use of one-sided pattern electrodes is proposed as a solution to the above problems. By using the proposed pattern electrodes, the gap between electrodes can be firmly controlled, and short-circuiting between electrodes can be eliminated so that operation at low voltage is possible. Wiring to the moving part is conventionally made using brushes or the like. However, with the one-sided pattern electrode, the need for this wiring to moving parts is completely eliminated, and has made maintenance easier. The cost of fabrication is expected to be reduced greatly by employing photomasks as used in the fabrication of semiconductors.

We developed two kinds of one-sided pattern electrodes and experimentally verified the manifestation of ER effects in a homogeneous ER fluid. The effect of disk gap upon viscosity change was also investigated.

2 One-sided pattern electrodes and the experimental apparatus

2.1 Concept of one-sided pattern electrodes

Conventionally, when an electric field is applied to an ER fluid, the electric field is applied across a pair of vertically opposing disk electrodes, as shown in Fig. 1 (a). In this type of application, in addition to the problem of wiring to the rotating part, control of the gap between the electrodes is important.

We examined the possibility of eliminating the electrode on the upper disk and applying the electric field only on the pattern electrodes formed on the lower disk, as shown in Fig. 1 (b). An arrangement called one-sided pattern electrodes, the use of which eliminates the need to wire to the moving part, and also enables the gap between electrodes to be better controlled. It is, in addition, considered to be advantageous for maintenance.

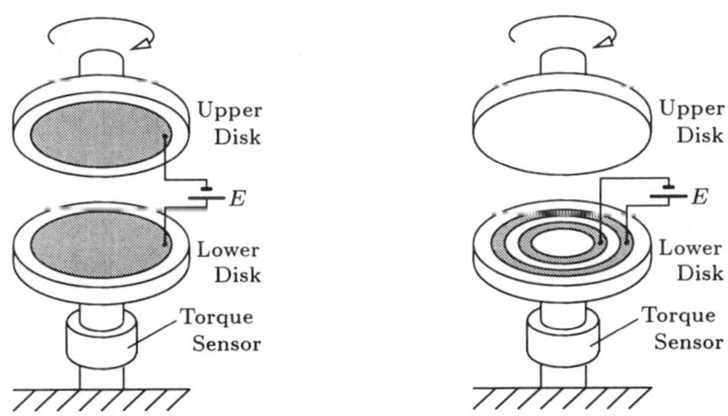

(a) Conventional electrodes (b) Pattern electrodes

Figure 1. Concept chart of pattern electrodes

2.2 Composition of electrodes and electric field

For the upper disk, disks made of stainless steel and of resin were prepared. For the lower disk, conventional disk electrodes and pattern electrodes were prepared. Composition of electrodes and electric fields as shown in Fig. 2 can be realized by changing the combination of the upper and lower disks.

A both-sided whole-surface electrode as shown in Fig. 2 (a) is the conventional composition, with both upper and lower electrodes being flat plane form. A one-sided pattern electrode as shown in Fig. 2 (b) is the combination proposed in this study. An insulator disk made of resin is used for the upper disk, and a pattern electrode is set for the lower disk and an electric field is applied.

(a) both-sided whole-surface electrodes (b) one-sided pattern electrodes

Figure 2. Types of conbinations of electrodes

2.3 Shapes of the pattern electrodes

Two shapes were considered for the one-sided pattern electrode: a concentric circles form of pattern electrodes model as shown in Fig. 3, and a radial form of pattern electrode model as shown in Fig. 4. Using these pattern electrodes, the ER effect can be measured in cases where the direction of an electric field is normal to and parallel with the direction of rotation.

These pattern electrodes were fabricated by etching photosensitive substrates for electronic circuits. The electrode has an effective outer radius of 60 mm and an inner radius of 30 mm. The gap distance between electrodes is 0.2 mm.

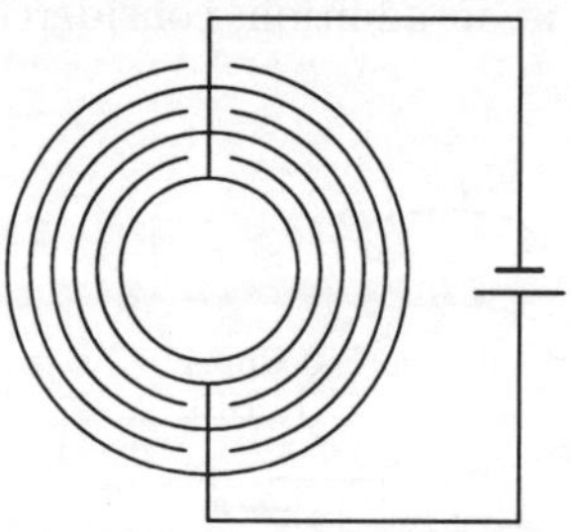

Figure 3. Model of concentric circles form of pattern electrodes

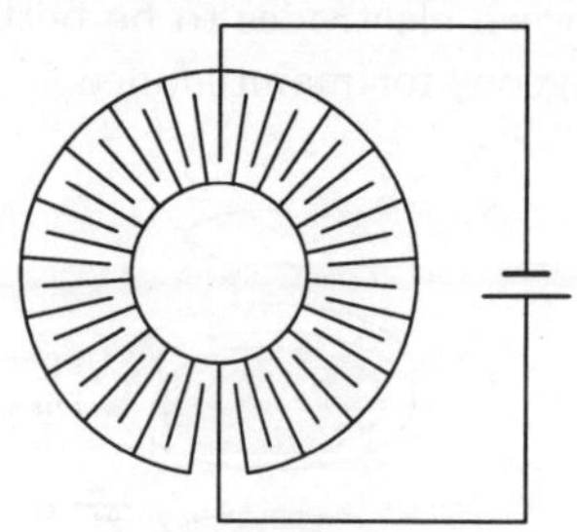

Figure 4. Model of radial form of pattern electrodes

2.4 Experimental apparatus

Figure 5 shows the transmitted torque measuring apparatus developed in this study. In order to investigate the ER effect, after an ER fluid is placed between two parallel disks, the upper disk sandwiching the fluid is rotated with an electric motor with speed reduction gear. Torque transmitted via the ER fluid is then measured with a torque sensor fixed on the lower disk. The distance between disks can be varied using a ball thread.

Figure 6 shows a photograph of the experimental apparatus on which the one-sided pattern electrodes were set. A fluid was then poured on the pattern electrodes, the upper resin disk was fixed so as to maintain a predetermined distance between disks, and the measurement performed.

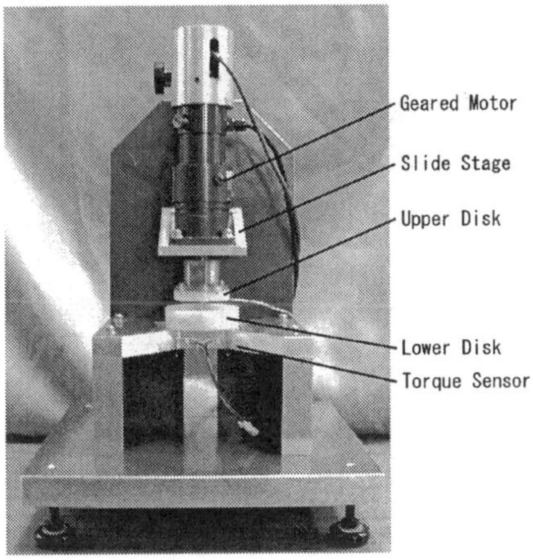

Figure 5. Overview of experimental apparatus

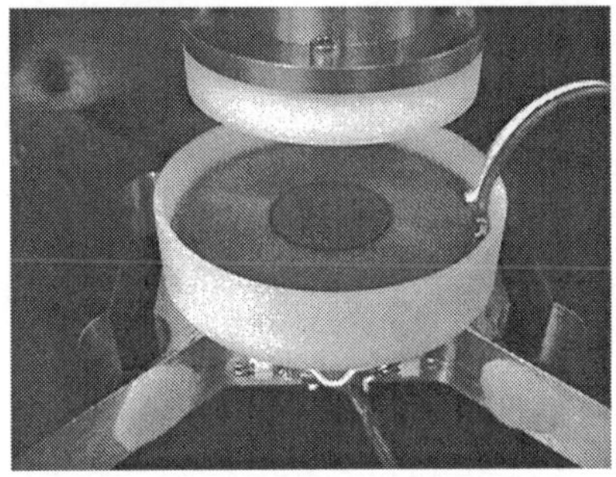

Figure 6. Picture of one-sided pattern electrodes

3 Measurement of the viscosity of ER fluids

The homogeneous ER fluid used in this study is a liquid crystalline polymer[2] developed by Asahi Chemical Industry Co., Ltd.

3.1 Experiments for both-sided whole-surface electrodes

Experiments were performed with both-sided whole-surface electrodes (Fig. 2 (a)). The gap distance between disks (electrodes) was set to 0.2 mm. Figure 7 shows the experimental results for the applied electric field strengths of 0, 1 and 2 kV/mm. The horizontal axis represents shear rates at the periphery of the disk, and the vertical axis represents shear stress calculated from the measured torque.

3.2 Experiments for one-sided pattern electrodes

Experimental results for the concentric circles pattern and for the radial pattern are shown in Fig. 8 and 9, respectively. The distance between disks in both cases was 0.2 mm, and applied electric fields were 0, 0.5, 1, 1.5, 2, 3 and 4 kV/mm.

As can be seen in Fig. 8, the same changes in viscosity are observed in the concentric circles pattern as in the both-sided electrodes, although the magnitude is about three to four tenths of that in the both-sided electrodes.

On the other hand, in the case of the radial form pattern, although the same change of characteristics as in the concentric circles pattern can be confirmed at low shear rates, the shear stress drops sharply at relatively high shear rates.

The above-described results show that the characteristics of viscosity change are different depending upon the direction of the electric field and the shear direction. It can also be seen that in the concentric circles form, the pattern electrode is better suited to applications such as dampers or torque control systems.

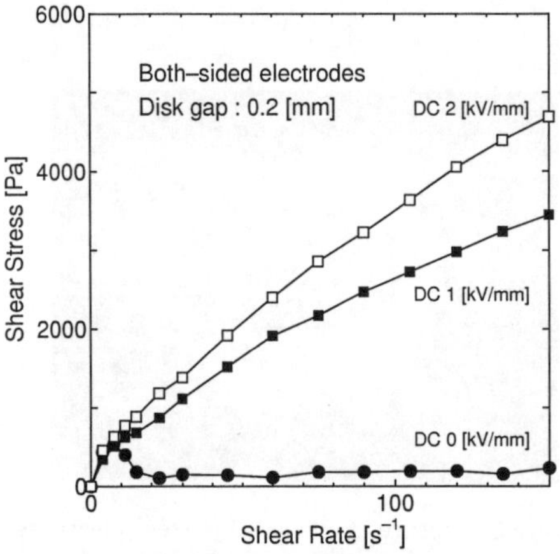

Figure 7. Characteristics of both-sided whole-surface electrodes system

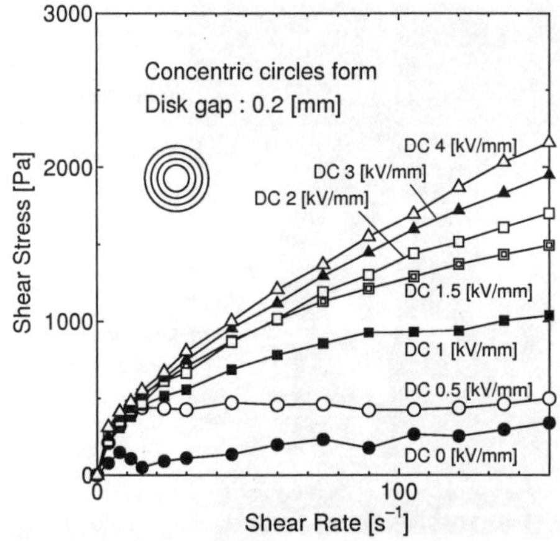

Figure 8. Characteristics of one-sided pattern electrodes system (concentric circles form)

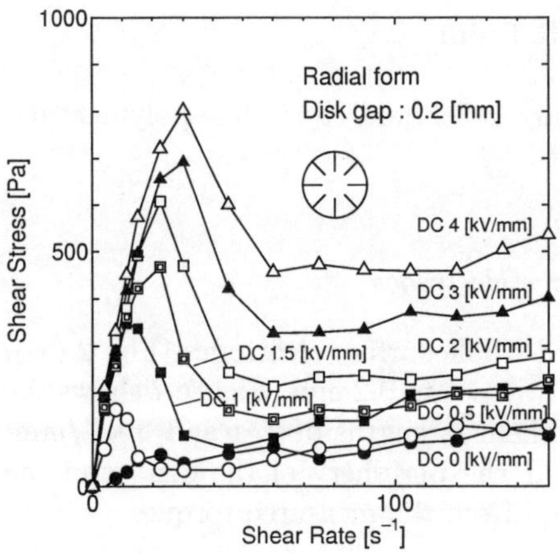

Figure 9. Characteristics of one-sided pattern electrodes system (radial form)

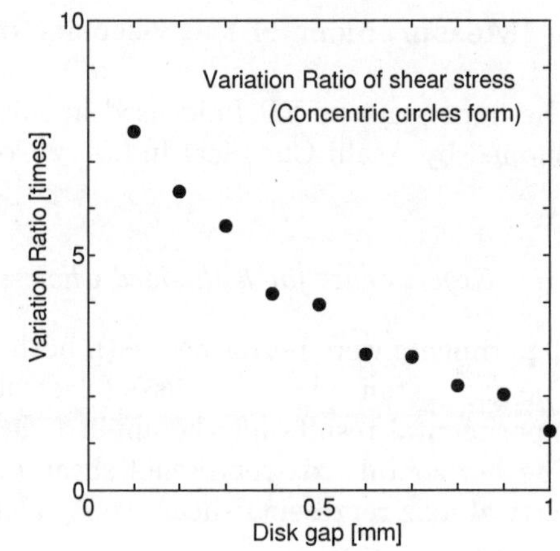

Figure 10. Variation ratio of shear stress for disk gap

4 Influence of the distance between disks

The experiments of the previous sections were performed with a constant disk gap. Since disk gap is considered to have a large influence upon the characteristics of one-sided electrodes, the effect of the disk gap was next investigated.

For the concentric circles form pattern, the disk gap was set to increments of 0.1 mm from 0.1 to 1.0mm and the experiment performed. Figure 10 shows the results for the variation ratio of shear stress for an electric field of 4 kV/mm compared to one of 0 kV/mm in each disk gap. In each case, the rotation velocity of the upper disk is 1 rad/s. It can be seen in the figure that the variation ratio of shear stress increases with decreasing disk gap.

5 Summary

This paper proposed one-sided pattern electrodes as a novel form for applying an electric field, and came to the following conclusions:

(1) We proposed one-sided pattern electrodes, and investigated the ER effects with this electrode composition. It has been found that the ER effects are exhibited also in the proposed one-sided pattern electrodes.

(2) The concentric circles form and radial form pattern electrodes have been developed and their characteristics have been investigated. In the case of the concentric circles form, a similar change in viscosity as in the case of both-sided electrodes takes place. On the other hand, in the case of the radial form, the same viscous effect as in the concentric circles form was observed at low shear rates, but at higher shear rates, shear stress dropped sharply.

(3) Effect of disk gap between upper and lower disks upon viscosity change in the case of one-sided electrodes has been studied. That shear stress and the rate of change in viscosity increase with decreasing disk gap.

References

1. W. M. Winslow, Induced Fibration of Suspensions, J. of Applied Physics, Vol.20, pp.1137-1140 (1949).
2. A. Inoue, and S. Maniwa, "Electrorheological Effect of Liquid Crystalline Polymers", J. Applied Polymer Science, Vol.55, pp.113-118 (1995).
3. Proceedings of the 5th International Conference on ER Fluids, MR Suspensions and Associated Technology, edited by W. A. Bullough, (World Scientific, 1995).
4. Proceedings of the 6th International Conference on ER Fluids, MR Suspensions and Their Applications, edited by M. Nakano, and K. Koyama, (World Scientific, 1997).
5. Proceedings of the 7th International Conference on ER Fluids and MR Suspensions, edited by R. Tao, (World Scientific, 1999).
6. M. Sakaguchi, and J. Furusho, "Force Display System Using Particle-Type Electrorheological Fluids", in Proceedings of the 1998 IEEE International Conference on Robotics and Automation, Vol.3, pp.2586-2591 (1998).
7. N. Takesue, G. Zhang, J. Furusho and M. Sakaguchi, "High Stiffness Control of Direct-Drive Motor System by a Homogeneous ER Fluid", IEEE Proc. Int. Conf. on Robotics and Automation, Vol.1, pp.188-192 (1999).
8. N. Takesue, G. Zhang, J. Furusho, and M. Sakaguchi, "Precise Position Control of Robot Arms Using a Homogeneous ER Fluid", IEEE Control Systems Magazine, Vol.19, No.2, pp.55-61 (1999).
9. M. Sakaguchi and J. Furusho, "Development of High-Performance Actuators Using ER Fluids", Journal of Intelligent Material Systems and Structures, Vol.10, No.8, pp.666-670 (1999).
10. G. Zhang, J. Furusho and M. Sakaguchi: "Vibration Suppression Control of Robot Arms Using a Homogeneous-Type Electrorheological Fluid", IEEE/ASME Transactions on Mechatronics, Vol.5, No.3, pp.302-309 (2000).

A STUDY ON THE PERFORMANCE ESTIMATION OF SEMI-ACTIVE SUSPENSION SYSTEM CONSIDERING THE RESPONSE TIME OF ELECTRO-RHEOLOGICAL FLUID

M. S. YEO
Department of Automotive Engineering, Inha Technical College
Incheon, KOREA
E-mail: msyeo@true.inhatc.ac.kr

H. G. LEE
Smart Structures and Systems Laboratory, Department of Mechanical Engineering
Inha University, Incheon 402-751, KOREA
E-mail: leehg@hananet.net

M. C. Kim
Department of Mechanical Engineering, Inha Technical College
Incheon, KOREA
E-mail: mckim@true.inhatc.ac.kr

This paper deals with the performance estimation of the semi-active suspension system considering the response time of Electro-Rheological fluid-based damper. The Bingham characteristics and response time of ER fluids are obtained experimentally. Then the ER damper is applied to the quarter car model and its performance using sky-hook control algorithm is compared to the system using conventional damper through computer simulation. As simulation results, the gain is improved above 50% but it is deteriorated rapidly with the increase of response time. So it is proved that the response time as well as the yield stress of ER fluids is the dominant factor to design the semi-active suspension system using the ER damper.

1 Introduction

Since the invention of vehicle, automotive engineering has been developed rapidly as a main field of mechanical engineering. Especially in order to achieve a comfortable ride, research on the suspension system has been focused. As known, suspension system supporting entire vehicle at axes improves ride comfort through absorption of vibration and impulse[1]. Also, it provides stability of steering, acceleration, and braking. However, the solution of the design contradiction between ride comfort and stability can't be achieved by passive suspension elements. With aid of hydraulic and electric actuators, some fully-active suspension systems to handle the compromise have been developed[2][3]. But these active systems require high cost and power, large package size, and poor operating economy. So, D.C. Karnopp[4] suggested a novel semi-active suspension system, which combines the advantages of both passive and active suspension systems. Variable damper employed in the semi-active suspension system should produce damping force independently with response time of several milliseconds via feedback control scheme. But it is impossible to implement continuously controlled damper due to complicated mechanical elements and slow response time. Recently a novel suspension system using electro-rheological fluids in order to overcome these shortcomings has been investigated by numerous researchers, which gives greater performances on structure design and response time rather than a conventional system.

Therefore, this study is to analyze an operating mechanism of ERF(Electro-Rheological Fluids) and to obtain experimentally Bingham behavior and response time of ERF. The required response time of ERF is suggested to apply the ER damper to semi-active suspension system with a real-time control scheme through simulation.

2 Bingham Property and Response Time of ER Fluid

2-1 Characteristics of ER fluid

ER fluids undergo instantaneous reversible changes in rheological properties when subjected to electric fields[5]. The fluids are dramatically changed from Newtonian flow to Bingham behavior which yield stress is controlled according to the intensity of the applied electric field. Numerous models[6][7] on the interaction between particles of the fluid under electric fields have been suggested, but there is no accepted theory to explain all of the ER effect. So it is difficult to adopt these models entirely into an application of various systems.

In this study Bingham properties such as yield stress and viscosity coefficient are experimentally obtained. On the basis of the measured Bingham properties, the

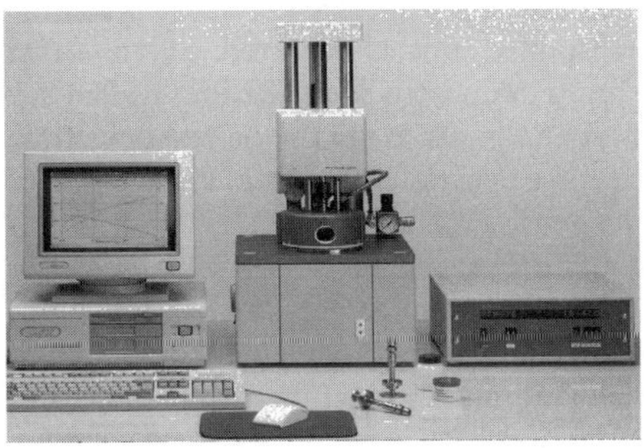

Figure 1. Photograph of measuring shear stress of ER fluid.

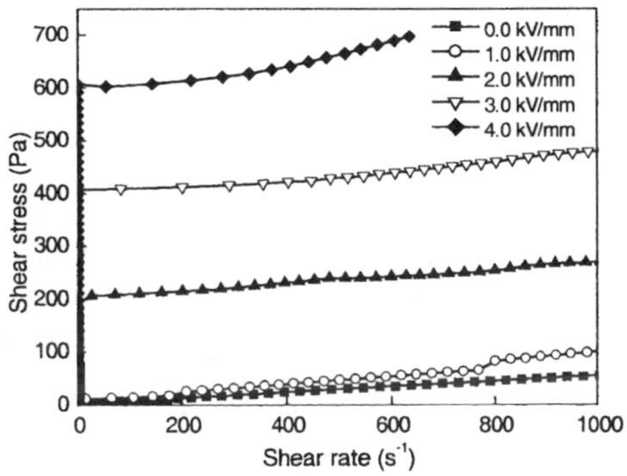

Figure 2. Bingham characteristics of ER fluid.

damping force of the ER damper is to be predicted. Figure 1 shows the viscometer employed to infer Bingham behavior of ER fluids. The measured results are represented in figure 2 and mathematically expressed by

$$\tau = \eta\dot{\gamma} + \tau_y(E) \tag{1}$$

where τ is shear stress, $\dot{\gamma}$ is shear rate, η is viscosity, and $\tau_y(E)$ is yield shear stress of the ER fluid. $\tau_y(E)$ is a function of the electric field E and exponentially increases with respect to electric field. Therefore eq. (1) can be represented by

$$\tau = \eta\dot{\gamma} + \alpha E^{\beta} \tag{2}$$

α and β are intrinsic values of the ER fluid, which are obtained by the experiment as this work. In this study, four types of ER fluids are employed and named as ERF 1, 2, 3 and 4. The first fluid of ERF1 is fabricated with gum arabic. The others come from Loctite, Bayer, and Shokubai, respectively. As well known in Table 1, the less the electrode gap, the higher the yield stress. It is estimated that the combining forces between the particles are increased, because the relative particle size to the gap distance becomes larger and the length of particle chain is shorter. ERF4 has the highest yield stress, so it seems to have a great advantage for application. But fluid performance such

Table 1. Yield stress of ER fluids.

	0.75mm	1.00mm	1.25mm
ERF1	$79.36E^{1.07}$	$66.04E^{1.11}$	$59.21E^{0.95}$
ERF2	$364.34E^{1.63}$	$374.53E^{1.68}$	$437.46E^{1.47}$
ERF3	$210.53E^{1.47}$	$190.39E^{1.48}$	$138.59E^{1.73}$
ERF4	$410.72E^{1.20}$	$379.79E^{1.25}$	$398.64E^{1.21}$

as stability on temperature, sediment, durability, and consuming power has also to be considered to employ ER fluids for various systems. The response time of ERF is one of major factors to be considered.

2-2 Response Time of ER fluid

Without electric fields, particles are randomly in ER fluid. But by applying electric fields, particle chains between electrodes are generated to resist the flow. The formulating and releasing processes of the particle chains with electric field take some time for the particles to move and a delay of response time is occurred. So this time delay has to be taken into consideration to design a semi-active suspension system for real-time control. Figure 3 shows the experimental apparatus to obtain the response time of ER fluids with respect to electric fields. By measuring the torque change under electric fields with aid of torque transducer, the response time is obtained.

Figure 4 represent results for the response time of ERF 1 and ERF 3, in which electric fields are applied at 0.99sec. In these fluids, the response time increases only a little with respect to the intensity of the fields, which is possible to be adopted for application systems. ERF1 has low yield stress and fast response time, on the other hand ERF3 has high yield stress and slow response time. So ERF1 and ERF3, which are representative in yield stress and response time, are selected for surveying the effect of response time. The response time of ERF1 is almost 80 msec at electric field of 4kV/mm and that of ERF3 is 50 msec. This is related to the interference between particles during applying electric fields. It is estimated that ER fluid for semi-active suspension system has to be optimally selected through considering the response time and the yield stress.

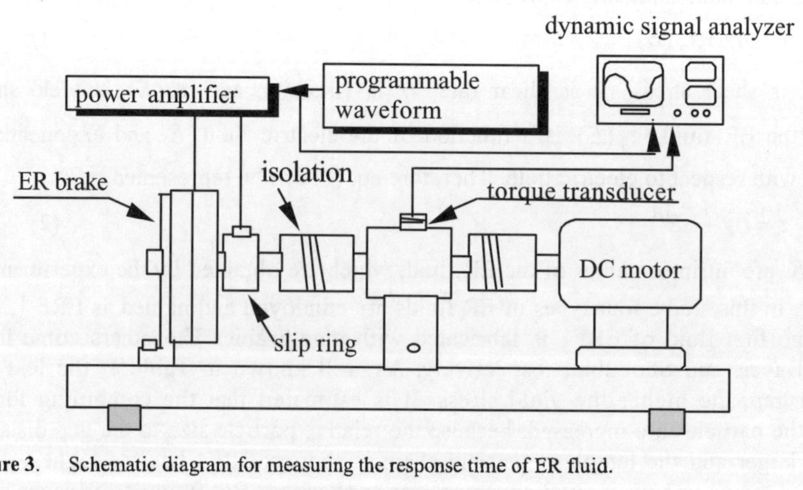

Figure 3. Schematic diagram for measuring the response time of ER fluid.

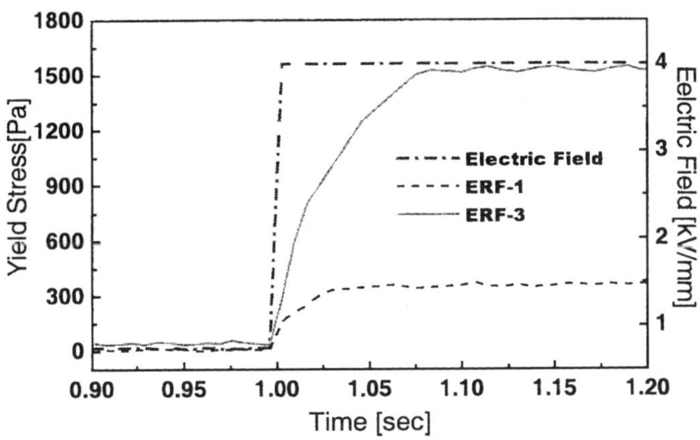

Figure 4. Response time of ERF 1 and ERF 3.

3 Performance Evaluations of Quarter Car Model

In order to predict the effect of response time for the semi-active suspension system with ERF damper, a quarter car model is adopted. Governing equations for sprung mass and unsprung mass are given by

$$M_s \frac{d^2 x_s}{dt^2} = F_D + k_s(x_u - x_s) \tag{3}$$

$$M_u \frac{d^2 x_u}{dt^2} = -F_D - k_s(x_u - x_s) + k_T(x_G - x_U) \tag{4}$$

Here M_s and M_u denote sprung and unsprung mass. k_s and k_T are spring constant of coil and tire, respectively. In a conventional damper, the damping force F_D is to be expressed as two types of compression and extension. On the other hand the damping force of ER damper as shown in equation (5) is given by summing F_N and F_{ER}. The former comes from viscous resistance of fluid and the latter from applied electric field. So the damping force[8] for ER damper is expressed as follows.

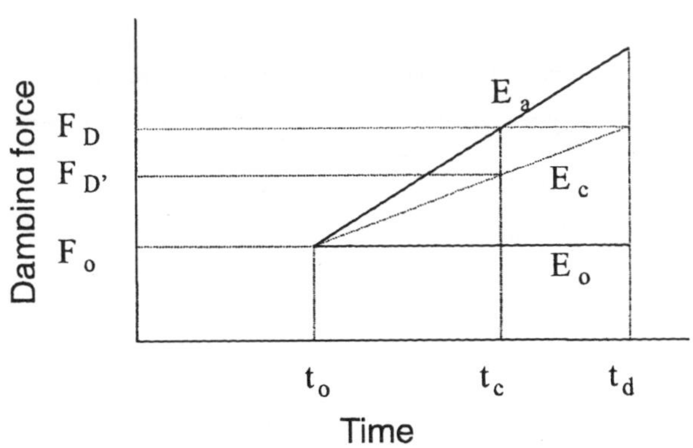

Figure 5. Effect of response time to damping force.

$$F_D = F_{ER} + F_N \tag{5-1}$$

$$F_{ER} = \frac{\pi}{2}\frac{L}{h}(d_2^2 - d_1^2)\tau_y , \qquad F_N = \frac{3\pi}{4}\frac{L}{d_3 h^3}(d_2^2 - d_1^2)\eta V_p \tag{5-2}$$

Where V_p, L and h represent damping coefficients of contraction and extension, speed of piston, length of duct, and gap between duct and wall, respectively. d_2 and d_3 are diameters of inner and outer cylinder and d_1 is piston rod diameter.

Sky-hook control scheme[9], which is convenient for the design of controller and input sensors, is used to control the damping force of ER damper.

$$F_D = C_{sky} \cdot V_s \cdot \lfloor V_s \cdot V_{(s-u)} \geq 0 \rfloor \tag{6}$$

Here, C_{sky} is sky-hook constant and the logical function $\lfloor V_s \cdot V_{(s-u)} \geq 0 \rfloor$ is 1 if TRUE and 0 if FALSE, in which V_s is absolute velocity of sprung mass and $V_{(s-u)}$ is relative velocity between sprung and unsprung mass. So electric field for damper control can be obtained by using the equations (2), (5) and (6).

In spite of the response time of ER fluid is very fast, the time delay is occurred due to sensors, signal processor and formation of high voltage in damper. As shown in figure 5, when the response time t_d of ER damper is shorter than control time interval t_c, the required damping force is supplied sufficiently to the suspension system. But if it is longer than that, the desired damping force F_D can't be obtained by the controlled electric field E_c. In order to achieve the desired damping force F_D within control time interval t_c, electric field E_a higher than E_c has to be applied. It can be expressed by

$$E_a = E_o + \frac{t_d}{t_c}(E_c - E_0) \tag{7}$$

Table 2 shows the input data employed to simulate the quarter car model under consideration of the response time of ER fluids. Assuming that road input is sinusoidal, dynamic characteristics are theoretically evaluated by sweeping the input frequency.

Figure 6 represents that the amplification ratio (gain) of sprung mass in the quarter model using ERF dampers with and without of sky-hook control scheme are compared with that using the conventional damper. The input frequency is swept from 0.1 to 25Hz

Table 2. Input data for quarter car simulation.

Description	Data	Description	Data
Electrode gap	1.0 mm	Unsprung mass	61 kg
Electrode length	241 mm	Spring stiffness	91.6 kN/m
Piston rod dia.	7 mm	Tire stiffness	162.2 kN/m
Outer dia.	38 mm	Compression damping coefficient	1,443 $N \cdot sec/m$
Inner dia	30 mm	Extension damping coefficient	5,181 $N \cdot sec/m$
Density	1,000 kg/m^3	Input Amplitude	10 mm
Sprung mass	339 kg		

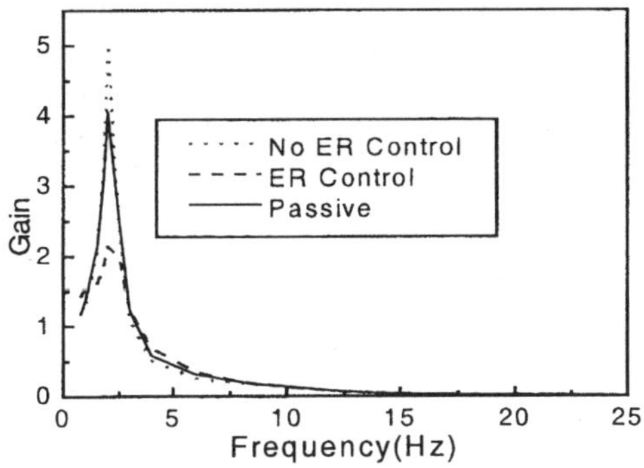

Figure 6. The frequency response of the sprung mass gain.

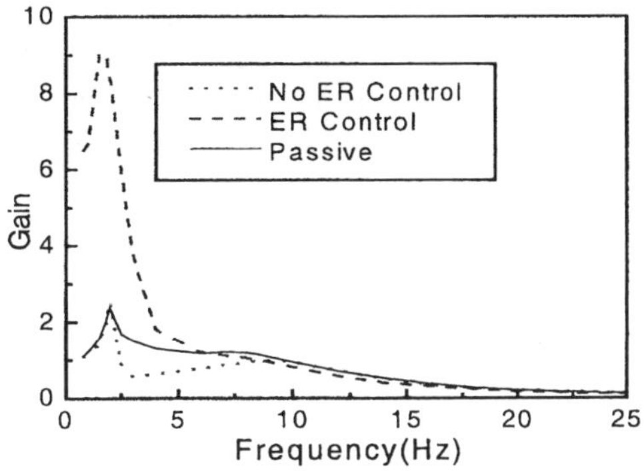

Figure 7. The frequency response of the unsprung mass gain.

which is sensitive to human beings and is distinct guide for defining human tolerance to whole-body vibration. The amplification ratio near 2Hz, which is natural frequency of sprung mass, is 2.15, 5.02, and 4.07. Vibration characteristics using sky-hook control scheme is improved up to about 50% and therefore it is conformed that the adoption of ERF damper to the real car is available. In the case without control scheme, the amplification ratio is increased about 25% because damping force, which is only due to viscous damping without ER effect, is not enough. So it is thought that the conventional passive damper is well designed. And the amplification ratio at the natural frequency ratio of unsprung mass, 9Hz, is nearly constant without any distinctive difference. Figure 7 denotes the amplification ratio of unsprung mass over the same frequency bandwidth. Vibration characteristics using sky-hook control scheme is overall worse than the others. This arises from the fact that the aim of sky-hook control scheme is to suppress the vibration of sprung mass and this suppression force is induced to the unsprung mass.

Figure 8 represents frequency response of sprung mass with respect to the time delay of ER damper using sky-hook algorithm. Amplification ratio at near 2Hz in the case of the response delay time of 5ms is 1.86. Its result is appreciated but the effect of vibration isolation decreases rapidly according to increase of the delay time. The amplification ratio with the delay time of 1,000ms is 4.08, which is similar to that using passive damper. So it is meaningless to use the ERF damper. Figure 9 shows the amplification ratio of sprung mass at natural frequency with respect to the response delay time. In the case of ERF 1, the ER control effect is not manifest because the damping force of ER damper is too less than that of passive damper. And the variation of vibration

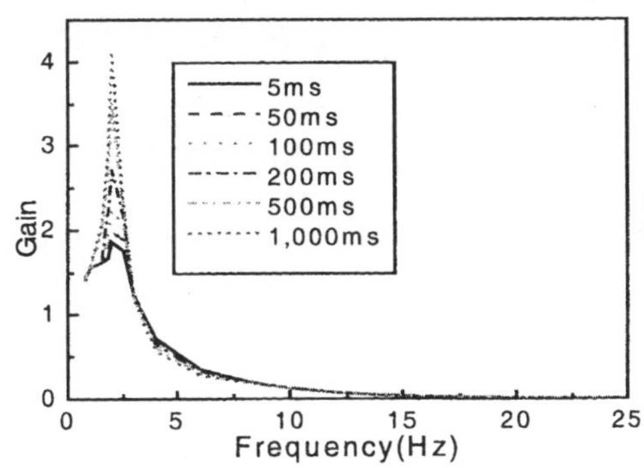

Figure 8. The sprung mass gain with the variation of time delay.

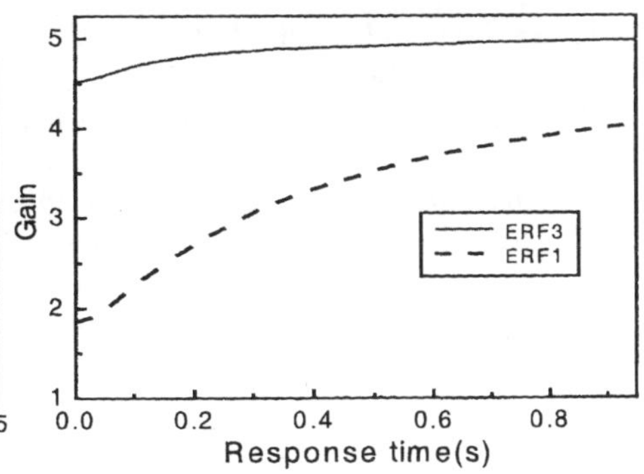

Figure 9. The sprung mass gain with the yield stress and time delay.

isolation effect is not large. In the case of ERF 3 with great ER effect, the control effect of ER damper is sufficient to improve the vibration characteristics, but the effect of vibration isolation decreases with the increase of response delay time, too. So the response time should be within 70msec for achieving vibration isolation up to 50% than passive suspension system in considering the cost of sensors, controller, etc., for semi-active suspension system. For the improvement up to 30%, the response time less than 236msec is required.

4 Conclusion

In this paper Bingham characteristics and response time of ER fluids are experimentally evaluated. Sky-hook control scheme is used to simulate a quarter car model for its convenient use. Amplification ratio to represent the effect of vibration insulation is improved up to about 50% than passive suspension system, which means that ERF damper is available to semi-active suspension system. But the effect of vibration isolation decreases with the increase of response delay time. In this ERF damper, the response time should be within 70ms for achieving vibration isolation up to 50% of the passive suspension system. For the improvement up to 30%, the response time less than 236ms is required.

Therefore, ER fluids to be utilized in the continuous controllable damper for semi-active suspension system should be simultaneously with high yield stress and fast response time.

References

1. Butsuen, T., The Design of Semi-Active Suspensions for Automotive Vehicles, *Ph. D. Thesis at MIT*, 1989.
2. Hedrick, J. K. and Wormley, D. N., Active Suspension for Ground Transport Vehicle, *Mechanics Transportation Suspension Systems, ASME, AMD* **15** (1975)
3. Goodall, R. M. and Kortum, W., Active Controls in Ground Transportation, *Vehicle System Dynamics*, **12** (1983)
4. Karnopp, D. C., Active Damping in Road Vehicle Suspension Systems, *Vehicle System Dynamics*, **11** (1982)
5. Winslow, W. M., Induced Fibration Suspension, *Journal of Applied Physics*, **43** (1949) pp. 119-133.
6. Klass, D. L. and Martinek, T. W., Electroviscous Fluids, *Journal of Applied Physics*, **38** (1967) pp. 67-74.
7. Pohl, H. A., Dielectrophoresis, *Cambridge Univ. Press*, (1978)
8. Petek, N. K., An Electronically Controlled Shock Absorber Using ER Fluid, *SAE Technical Paper*, No. 920275, (1992)
9. Suh, M.S. and Yeo, M.S., Development of Semi-Active Suspension Systems Using ER Fluids for the Wheeled Vehicles, *Journal of Intelligent Material Systems and Structures*, **10** (1999) pp. 743-747.

COMPARATIVE ANALYSIS OF THE TIME RESPONSE OF ELECTRORHEOLOGICAL AND MAGNETORHEOLOGICAL DAMPERS USING NONDIMENSIONAL PARAMETERS

YOUNG-TAI CHOI AND NORMAN M. WERELEY

Dept. of Aerospace Engineering, University of Maryland, College Park, MD 20742, U.S.A.

E-mail: nicechoi@wam.umd.edu, wereley@eng.umd.edu

This paper presents the characteristics of ER (electrorheological) and MR (magnetorheological) fluid-based systems through comparative analysis considering time response phenomena. In doing so, a nondimensional analysis, based on parallel plate geometry, is used to characterize the field-dependent properties of ER and MR dampers. Experimental tests are conducted for ER and MR dampers in order to validate our analysis.

1. Introduction

Electrorheological (ER) and magnetorheological (MR) fluids are known to exhibit similar field-dependent rheology from a bulk fluid perspective [1]. However, the implementation of ER and MR fluids in devices may require different strategies because of the inherent properties of ER and MR fluids such as density, viscosity and the magnitude of the yield stress. The difference between ER and MR fluid-based systems can be accurately understood by investigating the dynamic performance ranges and the time response characteristics of each controllable fluid system. The dynamic performance ranges of ER and MR fluid-based systems, e.g. the magnitudes of pressure drop or damping force, can be theoretically estimated by using quasi-steady fluid flow model. However, the quasi-steady fluid flow model does not give any information about the time response characteristics of the system. The reason is simple: it is assumed that the change of the fluid velocity occurs instantaneously when external fields are applied in the quasi-steady fluid flow model. Therefore, it is not possible to comprehensively assess the difference between ER and MR fluid-based systems. However, to date, most studies evaluating performance of ER and MR fluid-based systems in devices have made on the basis of the quasi-steady fluid flow model. Moreover, comparative analysis between ER and MR fluid-based systems are uncommon [1]. Almost no published research analytically compares ER and MR fluid-based systems by using nondimensional parameters that characterize the performance of ER and MR fluid-based systems. Therefore, the main contribution of our study is to explore the characteristics of ER and MR fluid-based systems on the basis of dynamic performance range and time response.

2. Nondimensional analysis for ER and MR valves

Let us consider a parallel-plate type of ER and MR valve to which the flow rate, Q, steadily passes through when an external field suddenly turns on. To develop the governing equation of the valve, it is assumed that ER and MR fluids are incompressible and in one-dimensional flow. The dynamic equation of motion is governed by [2]

$$\frac{\partial P(x,t)}{\partial x}=\frac{\partial \tau(y,t)}{\partial y}-\rho\frac{\partial u(y,t)}{\partial t} \tag{1}$$

Here, u is the velocity, τ is the shear stress, x is the longitudinal (valve axial) coordinate, y is the lateral coordinate, ρ is the density of the fluid, and P is the pressure. The solution of Eq. (1) can be obtained by solving two parts: homogeneous and particular partial differential equations [2]. The homogeneous equation gives the solution that physically corresponds to transient response and the particular equation to steady state response.

The homogeneous partial differential equation can be given by

$$\frac{\partial u_h(y,t)}{\partial t} = \frac{\mu}{\rho}\frac{\partial^2 u_h(y,t)}{\partial y^2} \tag{2}$$

Here, μ is the viscosity of the fluid. It is assumed that the time response of the high voltage and current amplifiers is not considered and that the yield stress of the fluids instantaneously occurs when electric or magnetic fields are applied. Using the method of separation of variables, the solution of the homogeneous partial differential equation is obtained by

$$u_h(y,t) = \sum_{n=1}^{\infty} \sin(\omega_n y)\left[C_n e^{-(\omega_n^2 \mu/\rho)t}\right] \tag{3}$$

Here, $C_n = \frac{2}{d}\int_0^d u_0(y)\sin(\omega_n y)dy$, $u_0(y) = \left(P'd^2/2\mu\right)\left[y/d - (y/d)^2\right]$, $\omega_n = (n\pi/d)$. In addition, $P' = -(\partial P/\partial x)$, is the pressure gradient and d is the gap between the parallel-plates in the valve.

When external fields are applied to the parallel-plate type of ER and MR valve, its steady state response can be described by the biviscous fluid flow model. The biviscous fluid flow model is known to give more accurate description than the Bingham plastic model in the pre-yield region where external shear stress is less than static yield stress. In the biviscous fluid flow model, the shear stress of the fluid is given by

$$\tau = \begin{cases} \mu_{pi}\dfrac{du}{dy} & at \quad |\tau| \le \tau_{ys} \\ \mu_{po}\dfrac{du}{dy} + \tau_{yd}\,\mathrm{sgn}(\dfrac{du}{dy}) & at \quad |\tau| > \tau_{ys} \end{cases} \tag{4}$$

Here, μ_{pi} is the pre-yield viscosity, $\mu_{po}(=\mu)$ is the post-yield viscosity, τ is the shear stress, τ_{ys} is the static yield stress, $\tau_{yd} = (1-\bar{\mu})\tau_{ys}$ is the dynamic yield stress of the fluid and $\bar{\mu} = \mu_{po}/\mu_{pi}$ is the nondimensional viscosity ratio. Finally, the solution of the particular partial differential equation can be obtained as

$$u_p = \begin{cases} u_{p1}(y,\infty) = \dfrac{P'd^2}{2\mu_{po}}\left[\left\{1-(1-\bar{\mu})\bar{\delta}\right\}\left(\dfrac{y}{d}\right) - \left(\dfrac{y}{d}\right)^2\right] & at \quad 0 \le y \le y_{pi} \\ u_{p2}(y,\infty) = \dfrac{P'd^2}{2\mu_{po}}\left[\bar{\mu}\left(\dfrac{y}{d}\right) - \bar{\mu}\left(\dfrac{y}{d}\right)^2 - \dfrac{(1-\bar{\delta})^2(\bar{\mu}-1)}{4}\right] & at \quad y_{pi} \le y \le y_{po} \\ u_{p3}(y,\infty) = \dfrac{P'd^2}{2\mu_{po}}\left[\left\{1-(1-\bar{\mu})\bar{\delta}\right\}\left(1-\dfrac{y}{d}\right) - \left(1-\dfrac{y}{d}\right)^2\right] & at \quad y_{po} \le y \le d \end{cases} \tag{5}$$

Here, $\bar{\delta} = \delta/d$ is the nondimensional plug thickness, $\delta = (y_{po} - y_{pi}) = 2\tau_{yd}/[(1-\bar{\mu})P']$ is the plug thickness, y_{po} is the top location of plug, and y_{pi} is the bottom location of plug. The derivation of Eq. (5) is omitted for conciseness, it is well described in reference [3,4].

Now, the complete solution for $u(y,t)$ is obtained by satisfying $u(y,t) = u_0(y)$ for $t = 0$

$$u_h(y,t) = u_h(y,t) + u_p(y,\infty) = \sum_{n=1}^{\infty} \sin(\omega_n y)\left[C_{nt} e^{-(\omega_n^2 \mu/\rho)t}\right] + u_p(y,\infty) \tag{6}$$

Here, the constant C_{nt} is found to be

$$C_{nt} = C_n - \frac{2}{d}\int_0^d u_p(y,\infty)\sin(\omega_n y)dy \tag{7}$$

Then integrating and simplifying Eqs. (6) and (7) yield the velocity profile for the parallel-plate type of ER and MR valve.

$$u(y,t) = \sum_{n=1}^{\infty} \sin(\omega_n y)\left[\frac{P'd^2\{2 - 2\cos(n\pi) - \Delta_1(\bar{\delta}) - \Delta_2(\bar{\delta}) - \Delta_3(\bar{\delta})\}e^{-(t/t_s)}}{2\mu_{po}(n\pi)^3}\right] + u_p(y,\infty) \tag{8}$$

where,

$$\begin{aligned}
\Delta_1(\bar{\delta}) &= \{-2 - [1-(1-\bar{\mu})\bar{\delta}](n\pi)^2 \bar{y}_{pi} + (n\pi\bar{y}_{pi})^2\}\cos(n\pi\bar{y}_{pi}) \\
&\quad + \{2 + (n\pi)[1-(1-\bar{\mu})\bar{\delta} - 2\bar{y}_{pi}]\sin(n\pi\bar{y}_{pi})\} \\
\Delta_2(\bar{\delta}) &= \{2\bar{\mu} - (1-\bar{\delta})^2(\bar{\mu}-1)(n\pi)^2/4 + \bar{\mu}(n\pi)^2(\bar{y}_{pi} - \bar{y}_{pi}^2)\}\cos(n\pi\bar{y}_{pi}) \\
&\quad + \{-2\bar{\mu} + (1-\bar{\delta})^2(\bar{\mu}-1)(n\pi)^2/4 - \bar{\mu}(n\pi)^2(\bar{y}_{po} - \bar{y}_{po}^2)\}\cos(n\pi\bar{y}_{po}) \\
&\quad + \{2\bar{y}_{pi}\sin(n\pi\bar{y}_{pi}) + \sin(n\pi\bar{y}_{po}) - 2\bar{y}_{po}\sin(n\pi\bar{y}_{po})\}\bar{\mu}n\pi \\
\Delta_3(\bar{\delta}) &= -2\cos(n\pi) + \{2 - (n\pi)^2(1-\bar{\mu})\bar{\delta} + (n\pi)^2\bar{y}_{po}(1 + (1-\bar{\mu})\bar{\delta} - \bar{y}_{po})\}\cos(n\pi\bar{y}_{po}) \\
&\quad - \{1 + (1-\bar{\mu})\bar{\delta} - 2\bar{y}_{po}\}(n\pi)\sin(n\pi\bar{y}_{po})
\end{aligned} \tag{9}$$

Here, $\bar{y}_{po} = y_{po}/d$, $\bar{y}_{pi} = y_{pi}/d$, and $t_s = \sum_{n=1}^{\infty} \rho d^2/((n\pi)^2 \mu_{po})$ is the time constant that characterizes the transient response of the fluid velocity.

To develop an expression for the pressure drop, ΔP, in terms of our nondimensional parameters, integrating and rearranging Eq. (8) yields

$$\Delta P = \left(\frac{12\mu_{po}L}{bd^3}\right)Q \Big/ \left[\sum_{n=1}^{\infty}\left(\frac{\cos(n\pi)-1}{(n\pi)^4}\right)\Delta_h(\bar{\delta})e^{-t/t_s} + \Delta_p(\bar{\delta})\right] \tag{10}$$

where, $\Delta_h(\bar{\delta}) = [12\Delta_1(\bar{\delta}) + 12\Delta_2(\bar{\delta}) + 12\Delta_3(\bar{\delta}) + 48\cos(n\pi) - 48]$

$\Delta_p(\bar{\delta}) = \left[(1-\bar{\delta})^2(1+\frac{\bar{\delta}}{2}) + \frac{3}{2}\bar{\mu}(1-\frac{\bar{\delta}^2}{3})\bar{\delta}\right]$, and L is the valve length.

Since ER and MR devices have different field inputs, a nondimensional input is required for comparative analysis between ER and MR devices. Thus, in this study, the Bingham number is introduced as nondimensional input.

$$Bi = \frac{\tau_{ys} b d^2}{\mu_{po} Q} = \frac{\Delta P \bar{\delta} d^2 A_d}{2\mu_{po} L Q} \tag{11}$$

It is noted that Bi is a measure of the relative importance of field-dependent yield stress and the viscous stress, and so can be used to compare ER/MR fluids. Here, $A_d = bd$ is the cross-sectional area of the valve. It is well known that the yield stress is an exponential function of applied field, $\tau_{ys} = \alpha G^{\beta}$. For ER fluids, G is the electric field, E, that is proportional to applied voltage. For MR fluids, G is the magnetic field, H, that is proportional to applied current. In addition, α and β are characteristic values determined by experiments. We substitute Eq. (10) in Eq. (11) and rewrite it as a function of $\bar{\delta}$ as

$$\sum_{n=1}^{\infty}\left(\frac{\cos(n\pi)-1}{(n\pi)^4}\right)\Delta_h(\bar{\delta})e^{-(t/t_s)} + \Delta_p(\bar{\delta}) - \frac{6\bar{\delta}}{Bi} = 0 \tag{12}$$

Given the flow rate, Q, and the Bingham number, Bi, the nondimensional plug thickness, $\bar{\delta}$, can be determined from Eq. (12) using a root finding algorithm [5]. The physically sensible root is the root that satisfies $0 < \bar{\delta} < 1$.

3. Analysis results for ER and MR valves

In order to accomplish comparative analysis between ER and MR valves, computer simulation is undertaken. All of the design parameters of the valves used in this study are the same other than smart fluids used. Geometrical dimensions are: gap (d =1 mm), width (b =10 mm) and length (L=10 mm). For the ER fluid, the lab-made ER fluid that consists of starch and peanut oil is employed. On the other hand, the product (MRF-132LD) from the Lord Corporation is used for the MR fluid. In this study, the yield stresses of ER and MR fluids are experimentally obtained as a function of applied field like.

$$\begin{aligned} \tau_{ys}(E) &= 168.69E^{1.30} \text{ [Pa]}, \quad \text{for the ER fluid} \\ \tau_{ys}(H) &= 1.91H^{1.73} \text{ [Pa]}, \quad \text{for the MR fluid} \end{aligned} \tag{13}$$

In the above, the unit of E is kV/mm, while the unit of H is A/mm. In addition, for the ER fluid, the post-yield viscosity (μ_{po}), nondimensional viscosity ratio ($\bar{\mu}$) and density (ρ) are found to be μ_{po} =0.174 Pa s, $\bar{\mu}$ =0.03 and ρ =1100 kg/m^3, respectively. For the MR fluid, μ_{po} =0.33 Pa s, $\bar{\mu}$ =0.09 and ρ =3055 kg/m^3, respectively.

Figure 1 represents the dynamic pressure ranges of ER and MR valves with respect to Bingham number and time. As stated earlier, since the type of applied field is different, Bingham number is used as the intensity of input for the same condition. In this study, since we focus on evaluating time response characteristics of the valve due to only application of field, flow rate (Q) passing through the valve is fixed at 1.5×10^{-4} m^3/s. As seen in this figure, the ER valve shows relatively higher dynamic pressure range than the

MR valve, for the same Bingham number of Bi=30. Here, the dynamic pressure range is the ratio of the magnitude of the pressure drop in presence of field to the magnitude of the pressure drop with no field. Therefore, this result implies that ER fluid-based systems can have higher controllable performance range than MR fluid-based systems, for the same Bingham number. In addition, ER valve reaches steady state faster than the MR valve, for the same Bingham number. This result implies that ER fluid-based systems can have faster time response than MR fluid-based systems for the same Bingham number. However, it is noted that the results in Figure 1 are obtained under the condition of the same Bingham number. In practice, ER and MR fluids are limited in the magnitude of the yield stress produced by applied field because of dielectric break down, saturation, and so on. Therefore, it is necessary to consider practical values of the Bingham number.

Figure 2 represents the dynamic pressure ranges of ER and MR valves with respect to practical values of the Bingham number and time. For the ER fluid, the maximum voltage input chosen is 8kV on the basis of the dielectric breakdown. The maximum voltage of 8kV corresponds to a nondimensional Bingham number of Bi=9.65. On the other hand, the maximum current input is chosen to be 1.5A for the MR fluid. Since the number of turns in the coil of the MR valve employed in this study is 150, the maximum current input of 1.5A corresponds to Bi=13.88 that is easily calculated by using Kirchoff's law for a magnetic circuit [6]. As clearly observed, the MR valve has higher dynamic pressure ranges than the ER valve, but its time response is still slower than the ER valve. Therefore it implies that ER fluid-based system is appropriate for practical applications demanding relatively low controllable performance range and fast time response, while the MR fluid-based system is suitable for applications with relatively high controllable performance and slow time response.

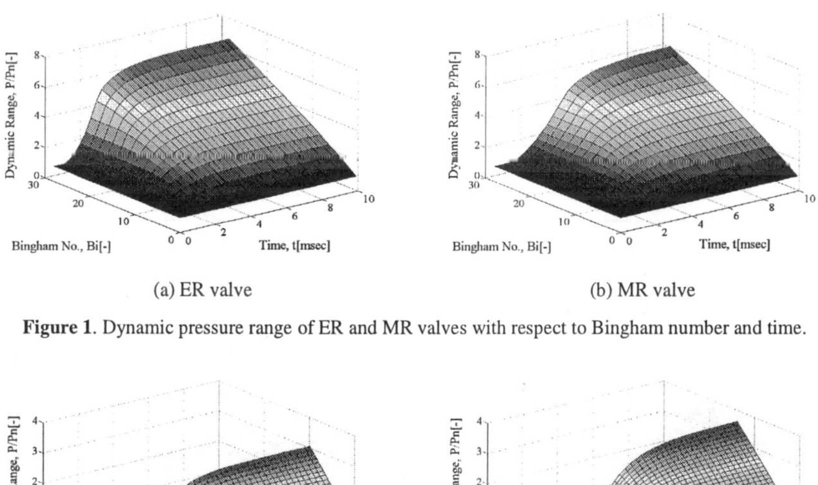

(a) ER valve (b) MR valve

Figure 1. Dynamic pressure range of ER and MR valves with respect to Bingham number and time.

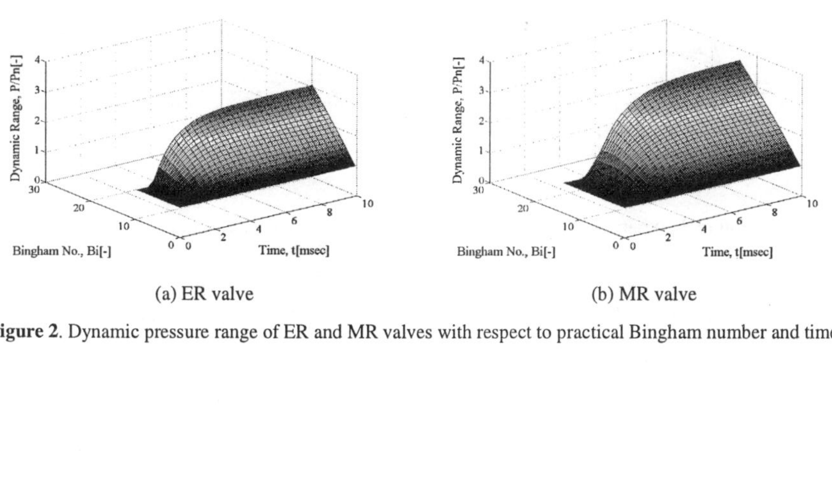

(a) ER valve (b) MR valve

Figure 2. Dynamic pressure range of ER and MR valves with respect to practical Bingham number and time.

4. Experimental results for ER and MR dampers

In order to validate our analysis, flow-mode, monotube type ER and MR dampers were constructed and evaluated via experiments and analysis. In this case, excitation piston velocity is constant so that the flow rate passing through the gap in the damper is also constant. The reason is that, as stated earlier, we focus on investigating time response characteristics of the ER and MR systems due to only application of field. Characterization of these time constants is essential when applying ER and MR dampers in control systems.

Figure 3 represents the experimental results for dynamic force ranges of ER and MR dampers with respect to applied fields. In this case, the damping force of the damper is measured when the damper is compressed and its piston velocity, V_p, is fixed at 0.041m/sec for the constant flow rate in the gap of the damper. The step input that turns on at the time of 0.015 sec is used as applied field. For the ER fluid, three different voltage inputs of 5.6kV (8kV/mm), 4.0kV (5.7kV/mm), and 2.4kV (3.4kV/mm) are applied and it is corresponding to nondimensional Bingham number, Bi=13.22, Bi=8.56, and Bi=4.36, respectively. On the other hand, for the MR damper, three current inputs of 1.2A (144A/mm), 0.9A (107A/mm), and 0.6A (72A/mm) are used corresponding to Bi=24.76, Bi=14.89, and Bi=7.45, respectively. In Figure 3, the solid line stands for theoretical data and the dashed lined for experimental data. To compute the damper time response, Bingham number is calculated using the flow rate, $Q=(A_p-A_r)V_p$, due to piston motion and the yield stress, $\tau_{ys}(G)$, due to applied field. The nondimensional plug thickness, $\bar{\delta}$, is then obtained by numerically solving Eq. (12) and the damping force of the damper at a certain field input is computed by $F=(A_p-A_r)\Delta P$. It is noted that the time response of the amplifier is considered in the calculation of the nondimensional plug thickness and the damping force of the damper as adding to its rise time in Eqs. (10) and (12). This additive delay is reasonable because the theoretical model in Eqs. (10) and (12) is derived under the assumption that the yield

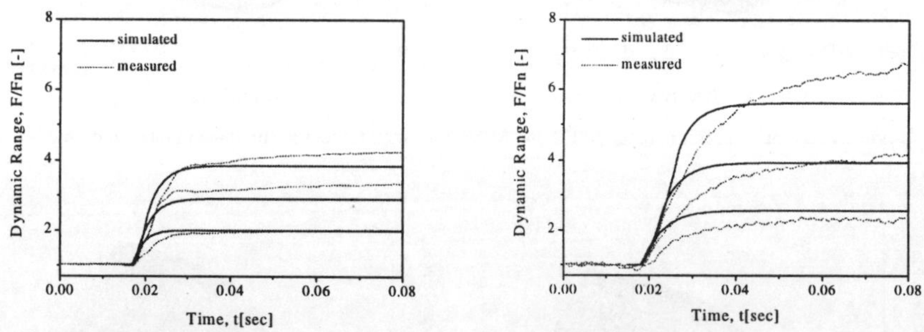

(a) ER damper (b) MR damper

Figure 3 Experimental results for dynamic force ranges of ER and MR dampers with respect to applied fields.

(Here, for the ER damper, 5.6kV→ Bi =13.22, 4.0kV→ Bi =8.56, and 2.4kV→ Bi =4.36 are applied. And, for the MR damper, 1.2A→ Bi =24.76, 0.9A→ Bi =14.89, and 0.6A→Bi=7.45 are used.)

stresses of the ER and MR fluids develop instantaneously when a field is applied. As seen in the figure, the MR damper shows higher dynamic force range, but its time response is slower than that of the ER damper. Here, the dynamic force range is the ratio of the magnitude of the damping force in the presence of field to the magnitude of the damping force with no field. On the other hand, for the experimental data, the damping force range still increases after reaching the steady state value. This is the reason why the spring force effect due to the pneumatic reservoir increases, as the damper is compressed. However, in this study, the spring force due to the pneumatic reservoir is not considered in order to simplify the analysis. Nevertheless, the theoretical data agrees favorably with the experimental data.

5. Conclusions

The dynamic performance range and time response of the ER and MR valves were evaluated through comparative work on the basis of considering the time response phenomena. It has been shown that the ER valve has higher controllable pressure range and faster time response than MR valve at the same Bingham number of Bi=30. However, for the practical values of the Bingham number limited by effects such as dielectric breakdown and saturation (that is, Bi=13.22 for the ER damper and Bi=24.76 for the MR damper), the ER valve shows faster response time, but does not have higher controllable pressure range than the MR valve. In order to validate our analysis, flow-mode type ER and MR dampers were constructed and evaluated via experiment and compared to theoretical calculations. Subsequently, it has been demonstrated that the theoretical model utilizing nondimensional parameters gives a favorable prediction for the dynamic force range and time response of the ER and MR dampers and the nondimensional expressions are considerably effective in the comparative analysis between ER and MR dampers.

References

1. Carlson J.D., Catanzarite D.N. and St. Clair K.A., Commercial magneto-rheological fluid devices, *Proc. Of 5th Int. Conf. On ER Fluids, MR Suspension and Associated Technology*, (1996) pp. 20-28.
2. Gavin H.P., Electrorheological dampers for structural vibration suppression, (PhD Dissertation, Dept. of Civil and Environmental Engineering, University of Michigan, 1997).
3. Wereley N.M. and Lindler J., Quasi-steady biviscous analysis of ER/MR dampers, *Proc. Of the 10th Int. Conf. On Adaptive Structures and Technologies*, (1999) pp.82-91.
4. Wereley N.M. and Pang L., Nondimensional analysis of semi-active electrorheological and magnetorheological dampers using approximate parallel plate models, *Smart Materials and Structures*, **7** (1998) pp. 732-743.
5. Nakamura S., Applied numerical methods in C. (Prentice-Hall International, Inc., London, 1993).
6. Cogdell J. R., Foundations of electric power. (Prentice Hall, New Jersey, 1999).

A BEARING APPLICATION USING MAGNETORHEOLOGICAL FLUIDS

ASHISH AGRAWAL, CONSTANTIN CIOCANEL, TONY MARTINEZ, SHEILA L. VIEIRA AND NAGI G. NAGANATHAN

M.I.M.E. Department, Nitschke Hall 4005
University of Toledo, Toledo, OH 43606, USA
e-mail: nagi.naganathan@eng.utoledo.edu

SCOTT ROBB AND JIM DUGGAN

Spicer Driveshaft Division – DANA Corporation
P.O. BOX 955, Toledo, OH 43697-0955, USA
e-mail: jim.duggan@dana.com

A conventional vehicle center bearing (CB) consists of a roller bearing that rests on a circular ″U″ shaped support, which includes a bladder formed by an elastomer, providing damping for radial vibrations between the roller bearing and the housing. In order to have a better control of the damping force of a CB, MR fluids can be used instead of the traditional elastomeric material. In this work we estimated the variation of the forces acting on a vehicle center bearing, performed the analysis of the magnetic field in a prototype bearing, and tested a MR CB prototype. We concluded that the major component of the reaction on a CB acts in the direction of the joint operating angle, and that MR fluids effectively can be used in vehicle center bearings to provide damping. Moreover, a design of a MR center bearing based in the shear mode is not the best option since weld of the plates may occur at high speeds of rotation of the shaft.

1 Introduction

Magnetorheological (MR) fluids belong to a family of smart materials whose rheological properties can be varied by means of an external applied magnetic field. This is interesting in providing simple, quiet, rapid response interfaces between electronic controls and mechanical systems. MR fluids have been actively studied with a focus on applications in mechanical components and devices. One of the newest uses for MR fluids is in driveline center bearings (CB) whose design intent to enable the fluid to provide a variable stiffness in the bearing.

In vehicle drivelines that consist of three or more universal joints, the driveline is supported by bearings called shaft support bearings or center bearings (CB). The CB in a vehicle supports the weight of the entire driveline assembly and also prevents the vibrations from being transmitted to the body of the vehicle. Conventional CBs consist of a roller bearing that rests on a ″U″ shaped support. This support includes a bladder made of an elastomeric material that provides damping for radial vibrations between the roller bearing and the housing.

Start-up shudder is a condition that exists when high torque inputs at initial vehicle launch cause suspension deflections, which allow universal joint angles to become large and uncancelled. This vibration condition has been observed in light truck design and some passenger car platforms, which use beam rear axles along with soft rate suspensions.

Launching a vehicle requires a relatively high torque input to overcome the vehicle's inertia. The axle ring gear, axle shafts, and tires are grounded producing zero rotational movement before the inertia of the vehicle is overcome. The large torque input at start up causes the suspension to deflect as the pinion rotates up the ring gear. With a stiff suspension, the rotational deflections of the axle under load are reduced but the ride quality becomes more "truck like". With a soft tuned suspension, the deflections under load cause the universal joint angles to increase at the axle end. The universal joint angles at the center bearing can decrease depending on drivetrain layout. The kinematic reaction of the uncancelled joint angle produces a secondary couple effect, which reacts at the center bearing in a plane, which is 90° to the plane of the true joint angle at the center bearing. The shaft and the bearing orbit is an ellipse. As the bearing accelerates through the allowable movement of the bracket and strikes the support at the end of the travel, a force equal to the mass of the shaft at the bearing times the acceleration is transmitted to the vehicle twice every shaft revolution.

As the vehicle's inertia is overcome and the vehicle begins to move, the torque required to maintain a constant acceleration is reduced. The deflected shape of the suspension returns to its design state and the powertrain angles once again return to their designed orientations. At this point the shudder is reduced or eliminated.

Tested solutions to this vibration problem have involved the use of additional suspension reaction points, i.e. "ladder bars" or "traction bars", to increase the rotational stiffness of the suspension. Alternative bracket designs to limit the movement of the shaft, i.e. pillow block bearings, have been effective in reducing the shudder. The solid bracket designs however tend to transmit all of the high frequency, low amplitude vibrations associated with higher speed shaft operation. One-piece shaft systems, like aluminum graphite and large diameter aluminum, eliminate the center bearing and thus the shudder mechanism. CV or double cardan joints at the axle may help, provided that the other joint angles in the driveline allow good cancellation when the axle and suspension are in their deflected shape.

Another solution is to make a CB with variable stiffness. At low frequencies of shaft rotation, it is desirable that the CB has a high stiffness, so the forces won't be transmitted to the vehicle. However, at high vehicle speeds, the frequency of the shaft vibration is high even though the displacement amplitude of the shaft is small. In this condition, it is desirable to have a bearing with low stiffness, so the vibration won't be transmitted to the vehicle body. This variable stiffness in a CB can be achieved by using MR fluids instead of the traditional elastomeric material, since their properties can be easily controlled upon the application of a magnetic field. Another advantage of this kind of bearing is that the MR fluid changes the frequency of CB out of the resonance.

There are some patents already developed regarding the use of MR fluids in center bearing applications. Pinkos and Shtarkman [1] proposed a design of a MR center bearing, shown in figure 1-(a). In this design, the MR fluid works in the shear flow mode to provide damping, and is located between the damper blades that move relative to each other. As the speed of the rotating shaft changes, the controller changes the coil current, which in turn varies the shear resistance offered by the MR fluid.

Duggan [2] proposed another design for a MR center bearing, in which the MR fluid operates in both shear and squeeze mode, as shown in figure 1-(b). The support member of this type of CB includes a bladder formed by an elastomeric material containing a MR fluid. By varying the power supplied to a coil located adjacent to the bladder, the vibration characteristics of the support member may be varied depending upon the operating conditions of the vehicle.

In this work we outline the basic steps in the design of the MR center bearing. The forces acting on the CB are estimated, the profile of the magnetic field in a prototype bearing is studied, three MR fluids are tested in shear mode in order to choose the one with better performance, and a MR CB prototype is tested.

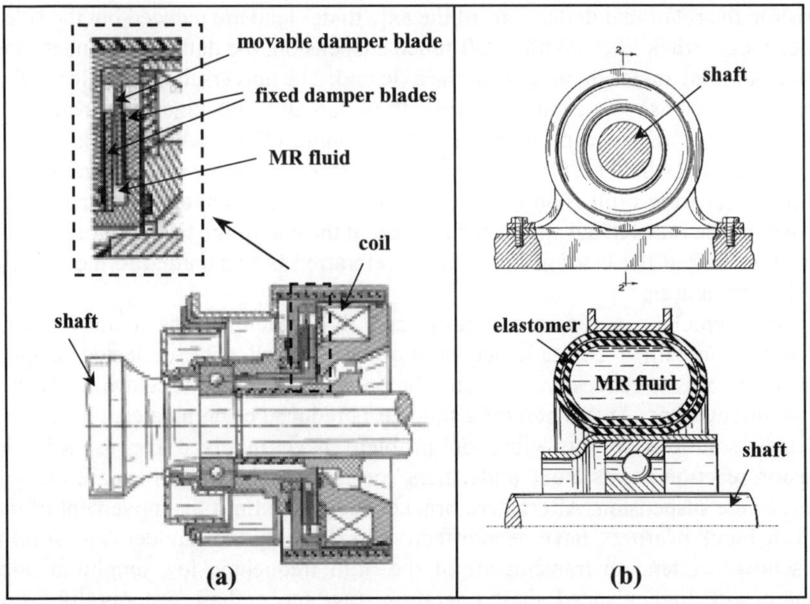

Figure 1. MR center bearing: (a) US Patent 5,730,531 (taken from ref. [1]); (b) US Patent Re.36,270 (taken from ref. [2]).

2 Forces Analysis in a Center Bearing

The forces acting on a CB can be categorized as a reaction on the bearing due to the weight of the driveline and due to the secondary couple. The weight of the driveline is a constant force that acts on the bearing. However, the secondary couple produced in the driveline depends on the torque transferred by the driveline and on the true joint angles of the universal joint that connects the driveline. This force component becomes very critical during the startup of the vehicle, when the driveline bends and the joint angles increase to about 20^0. The torque transferred by the driveline is high and hence the secondary couples are very high.

In this section, the effect of the secondary couples is studied from a basic free body approach. The equations for the secondary couple acting across a single joint are derived. The equations derived for a single joint can be extended to any driveline configuration, however, in this paper we will focus on a 2-joint driveline system.

The universal joint comprises of three rigid bodies: the driving yoke, the cross member and the driven yoke. The cross member consists of two arms at right angles, which connect the two yokes. As the torque is transferred across the joint, each of the yokes is loaded with a force. By making the cross axis being the global axis and using coordinate transformations, the yoke force matrices will be on the cross member.

By solving the equations of equilibrium of all the forces on the cross, we find the secondary couple component acting on each of the yokes, as shown in fig 2. Then the derived couple equations are applied to a 3-joint system in a driveline, as shown in figure 3. The calculated bearing reaction is plotted in figure 4-(a).

At square positions, when the shaft has rotated through 0 and 90^0, the reaction component is:

- $\theta = 0^0$

$$A = \frac{M_1 \sin\beta_1}{L_1} + \frac{M_1 \sin\beta_2 \cos\beta_1}{L_1 \cos\beta_2} + \frac{M_1 L_2 \sin\beta_2 \cos\beta_1}{L_1 L_3 \cos\beta_2} - \frac{M_1 L_2 \sin\beta_3 \cos\beta_1}{L_1 L_3 \cos\beta_2 \cos\beta_3} \quad (1)$$

- $\theta = 90^0$

$$A = \frac{M_1 \sin\beta_1}{L_1 \cos\beta_1} - \frac{M_1 \sin\beta_2}{L_1 \cos\beta_1} - \frac{M_1 L_2 \sin\beta_2}{L_1 L_3 \cos\beta_1} - \frac{M_1 L_2 \sin\beta_3 \cos\beta_2}{L_1 L_3 \cos\beta_1 \cos\beta_3} \quad (2)$$

where M_1 is the input torque, L_i are the dimensions defined in figure 3, β_1 and β_2 are the true joint angle of the 1st and 2nd joint, respectively.

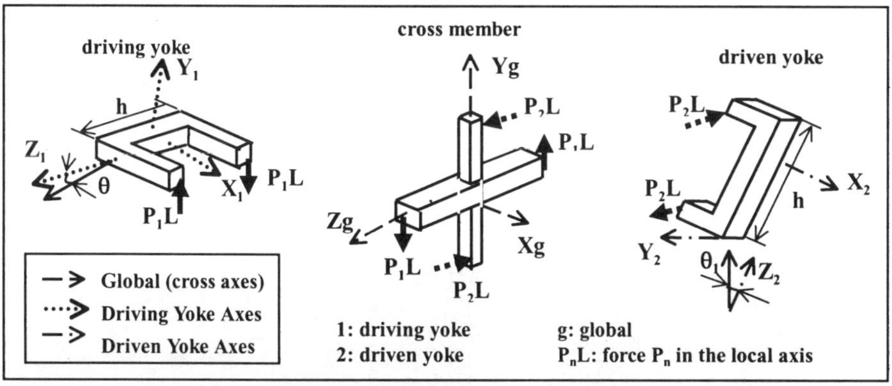

Figure 2. Equilibrium of a single joint.

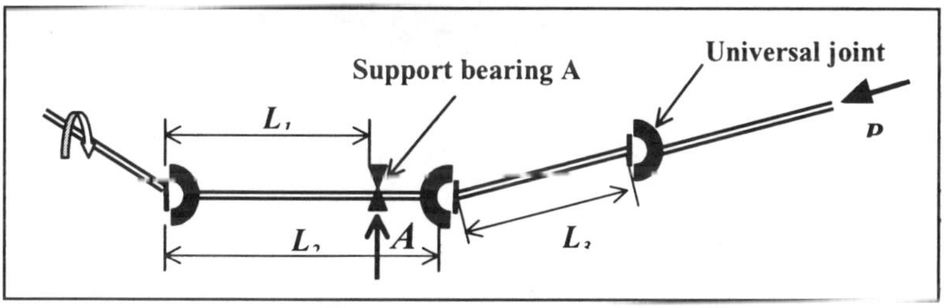

Figure 3. Three-joint driveline.

From figure 4-(a), we observe that in one complete rotation of the shaft, the bearing forces reach absolute maximum and zero forces twice. If the reaction forces are plotted radially, we obtain the figure 4-(b). This would be equivalent to look at the shaft from the end direction P. It can be seen that the bearing reaction is in the form of an ellipse with its major axis lying along an angle called the joint operating angle. The major component of the reaction acts in the direction of this angle and hence it is necessary to provide a higher stiffness to the bearing in this direction.

3 Magnetic field in a CB Prototype

The magnetic field inside a MR bearing prototype (figure 5-(a)) designed to operate in the shear flow mode was simulated. The modeling was made using FEA software Maxwell 3D and the B-H curve of the MR fluid. Some of the results are plotted in Figure 5-(b). It can be observed that the flux lines do not concentrate in the fluid region and hence the field inducted in the fluid is not very strong (around 0.1~0.3 T), indicating that this type of design is not the optimum one.

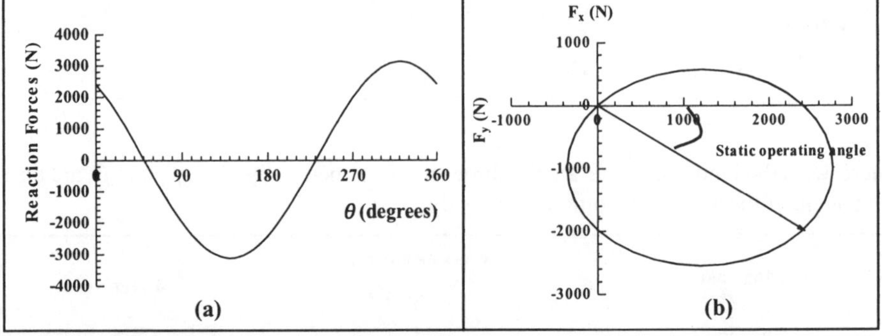

Figure 4. (a) Bearing Reactions and (b) Radial plot of the bearing reaction force for a 3-joint configuration (these are local axis bearing reactions).

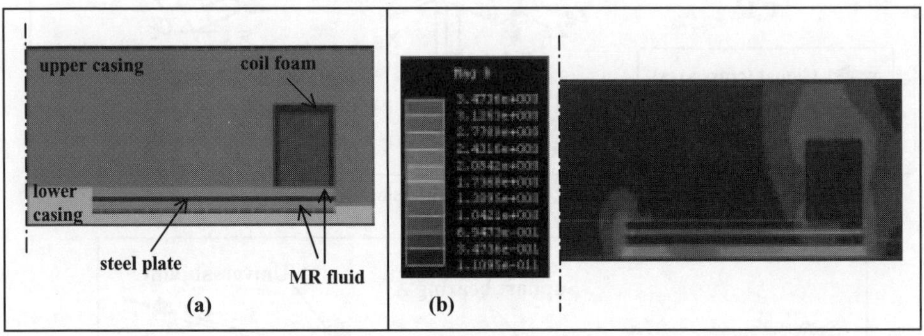

Figure 5. (a) FEA model of the MR prototype; (b) Flux distribution for the FEA model of the prototype for a field of 600 A-turn.

4 Experimental – Tests on a Center Bearing Prototype

In order to study the performance of MR fluids in a CB application, we conducted tests on a MR center bearing prototype, in which the MR fluid was subjected to shear mode. In the bearing design, the MR fluid was sandwiched between shear plates connected to the outer casing and the inner support. A coil wound inside the outer casing of the bearing assembly provided the magnetic field.

Figure 6 shows the experimental setup where the bearing was attached to a driveshaft. The motor assembly rotated the shaft from 0 to 4000 rpm, and an accelerometer was placed on top of the bearing to measure the acceleration of the shaft along the coordinate directions. The X-axis was on the plane of the shaft, and the Z-axis comes out of the plane of the paper. The bearing was filled with the fluid DC-2 (composed of nickel zinc ferrite and organic liquids) and mounted on the setup. The coil (resistance of 4.6 Ω) was energized by means of a variable voltage power source. The voltage was made to increase in steps of 1 volt and the shaft speed varied from the start position to 4000 rpm. The accelerations of the bearing produced in the Y and Z direction are the ones that interest the most. The axial acceleration is significant and the final bearing design should consider it.

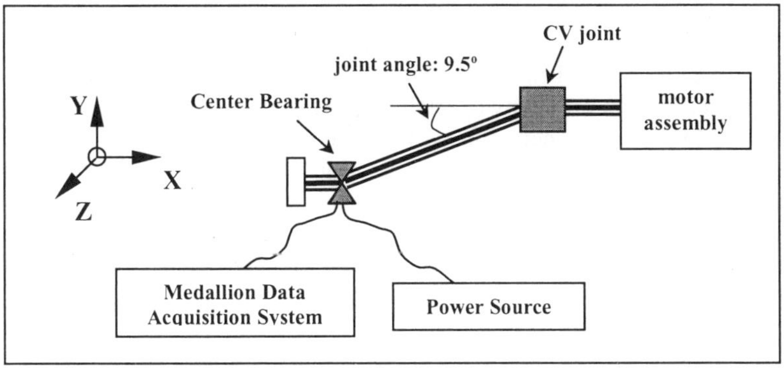

Figure 6. Esperimental setup for the MR center bearing.

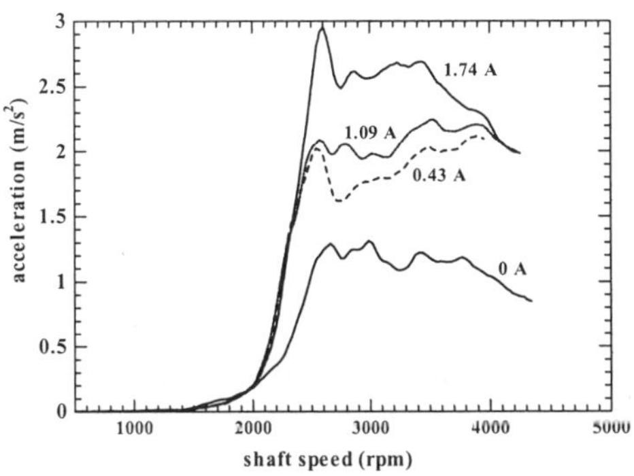

Figure 7. Acceleration of the shaft in the Z direction as a function of its speed.

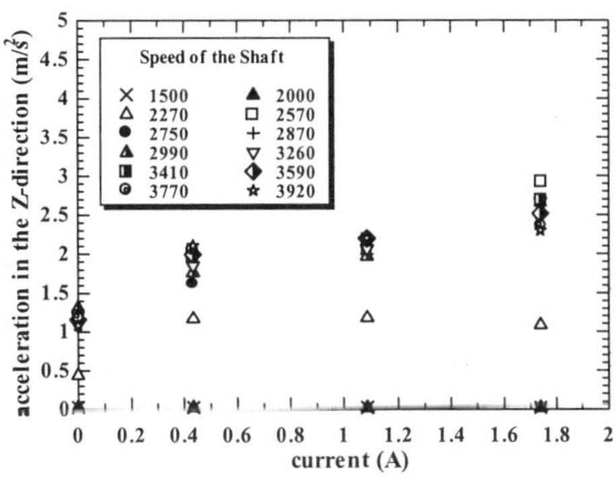

Figure 8. Acceleration of the shaft in the Z direction as a function of the voltage applied to the coil (magnetic field).

Figure 7 shows the acceleration of the shaft in the Z direction as a function of its speed. From this graph we observe that, increasing the magnetic field strength, the acceleration (i.e., the transmissibility) in the Z direction increases. The increase of transmissibility is due to the field-induced structure formed inside the MR fluid that produces an increase of the stiffness of the MR bearing. From this figure we also can see that, in the higher frequency range (speed higher than 2000 rpm), the increase of stiffness is much higher than that in the lower frequency range.

The increase of stiffness of the MR fluid in the lower frequency range is a desirable phenomenon. During the startup of the vehicle from the rest position, a large torque is transferred through the driveline, and, due to inertia effects, the entire driveline bends introducing large joint angles. These joint angles coupled with high torque results in large secondary couple[1] forces and bearing reactions. As a result, the bearing has to withstand large forces. In this condition, the displacement of the bearing is large, and the support bearing should preferably have a higher stiffness. This would prevent the bearing to transfer these high forces to the frame of the vehicle. Increasing the magnetic field (by increasing the coil current) would help in providing a higher stiffness for the bearing in the startup condition.

[1] Couples are product of the torque and the joint angle.

In the high frequency range, the bearing is subjected to lower reactions, since the displacement amplitude of the shaft is small. However, the vibration frequency is high. In this region of operation, the bearing should have a low stiffness to enable it to absorb the vibrations. This is accomplished by operating the bearing at the off-state condition, during which the MR fluid provides negligible stiffness to the bearing.

Figure 8 shows the behavior of the acceleration as a function of the current applied to the MR fluid. For all cases, increasing the current produced an increase of the acceleration, i.e. the transmissibility of the MR fluid increased as a result of the increase of its stiffness. Moreover, for speeds higher than 2500 rpm, the acceleration didn't change as much as for speeds below this value.

After we performed some tests with the MR center bearing, we found out that a metal-metal point contact in the bearing (which could be attributed to leakage problems) was occurring during the rotation at high speeds. As a result of the friction and consequent heating, the contact points welded making the continuation of tests impossible. So, it is clear that a proper bearing design is the foremost criteria in the development of a MR bearing, and that the use of shear plates is not a convenient configuration. A future design should consider other types of flow modes for the MR fluid.

5 Conclusions

In this work we estimated the forces acting on a center bearing, studied the profile of the magnetic flux in a MR CB, and tested an actual MR center bearing prototype. From the results we conclude:

- the bearing reaction is in the form of an ellipse with its major axis lying along the joint operating angle. The major component of the reaction acts in the direction of this angle, making necessary to provide a higher stiffness in this direction.
- MR fluids can be used in vehicle center bearings to provide damping;
- a design of a MR center bearing based in the shear mode is not the best option since weld of the plates may occur at high speeds of rotation of the shaft;
- the vehicle operating conditions (higher stiffness at lower speeds and lower stiffness at higher speeds) can be accomplished by varying the magnetic field applied to the MR fluid.

6 Acknowledgements

The authors thank Matt White and John Hickey for helping in the experiments with the center bearing.

References

1. Pinkos A. F. and Shtarkman E. M., Center Bearing Assembly with Rheological Fluid for Damping Vibrations. *US Patent 5,730,531* (1998).
2. Duggan J., Center Bearing Assembly Including Support Member Containing Rheological Fluid, *Patent US Re. 36,270* (1999).
3. SAE, *Universal Joint and Driveshaft Design Manual - Advances in Engineering Series No. 7*, published by The Society of Automotive Engineers, Inc. (1979).
4. Schmelz F., Seherr-Thoss C. H. and Aucktor E., *Universal Joints and Driveshafts - Analysis, Design, Applications*, translated by S. J. Hill and J. A. Tipper, published by Springer-Verlag.

STUDY OF THE BEHAVIOR OF MR FLUIDS IN SQUEEZE, TORSIONAL AND VALVE MODES

PRASHANT KULKARNI, CONSTANTIN CIOCANEL, SHEILA L. VIEIRA AND NAGI NAGANATHAN

M.I.M.E. Department, Nitschke Hall 4005, University of Toledo
Toledo, OH 43606, USA
E-mail: nagi.naganathan@utoledo.edu

In order to understandthe effectiveness of MR fluids insome applications , such as bearings, this work focuses on the properties of MR fluids under squeeze mode, torsional mode and squeeze combined with torsional mode. A special device was designed and fabricated to performthese experiments, which were carried out on an Instron Multiaxial Testing Machine. The hysteresis loops of the MR fluids were studied for different intensities of magnetic field, frequency,strain amplitude and angle amplitude. The results showed that the damping force and the area of the hysteresis loop of MR fluids increase with the magnetic field andstrain amplitude. Changes on the frequency did not influence the stress values significantly,within the range of frequencies used in these experiments. These results were compared to the ones obtained in valve mode by means of a RD1005 damper manufactured by Lord Corporation. The shape of the hysteresis loop curves and the MR effect were found to be different for the squeeze film damper and the RD1005 damper.

1 Introduction

Devices that use MR fluids can be classified in valve mode, direct shear mode, squeeze film mode and a combination of these three modes [1]. Of all the modes above, the shear mode is probably the most widely studied, while the squeeze mode is less understood.

Sproston *et al.* [2,3] studied ER fluids in squeeze mode. The forces transmitted through the ER fluid in the AC field case were found to be smaller than those observed in the DC case, and to be a function of the applied voltage and of the shape of the input waveform and electrical excitation frequency. In a later work, they compared the dynamic performance of an ER and a MR fluid when subjected to impulsive loads [4]. Vieira and Arruda [5,6] submitted ER fluids to mechanical testing (compression and tensile) in a MTS. The results showed that ER fluids are more resistant to compressive forces than to tensile forces. They also studied the fatigue life of ER fluids and concluded that it is possible to treat data on fatigue life of ER fluids by means of Weibull distribution and that it is possible to determine the time the "snap" starts by monitoring the electric current [7]. A similar work was done by Vieira *et al.* [8]. They studied the mechanical properties of an anhydrous ER fluid made of carbonaceous particles dispersed in silicone oil in tensile, compression and oscillatory squeeze tests. The tests were carried out with two different configurations: (a) the fluid was placed between the electrodes and (b) the electrodes were immersed inside the ER fluid. The results showed that the mechanical properties were dependent on the field strength, initial gap, upper electrode velocity and configuration. In oscillatory tests they observed that increasing the oscillation frequency f and the number of cycles N produced a decrease of the damping performance of the ER fluid. The electric field also played an important role on the shape of the hysteresis loop for both configurations. Tang *et al.* [9] did experiments dealing with the rheopexy of MR fluids and developed a theoretical model for the behavior of MR fluids in squeeze mode. Wang *et al.* [10] conducted experiments on a cone-shaped squeeze-film mode ER damper and highlighted the synergetic advantages of squeeze and shear modes working together.

MR fluids might be used in applications where both modes, squeeze and shear, are present, such as a center bearing from a driveshaft line [11]. So, in this work we focus on the properties of MR fluids under squeeze mode, torsional mode and squeeze combined with torsional mode. Moreover, the behavior of the MR fluid in a developed squeeze film damper studied and compared to the results obtained in a commercial linear damper.

2 Experimental

In this section the experimental setup used to perform the squeeze, torsional, squeeze combined with torsional, and valve mode is presented.

2.1 Squeeze film damper (Squeeze and Torsional modes)

This device was used to perform the squeeze and torsional tests and consists of two parallel plates surrounded by a rubber enclosure and has design features to apply cyclic loads, as shown in figure 1. Steel clamps were used to prevent any leakage. A coil with 2200 turns and 143 Ω, rated to work at 110 V DC, was wound on the lower steel plate to provide the magnetic field. On the upper plate there was a hole in order to fill the device with the MR fluid. The current was limited at 0.76 A by the coil design.

The fluid used in these tests was MRF 132-LD manufactured by Lord Corporation. After the apparatus was filled with a measured amount of MR fluid, it was mounted on the Instron Multiaxial Testing Machine, gripped and a 2mm gap was set before starting the loading. The tests were performed in a displacement control mode.

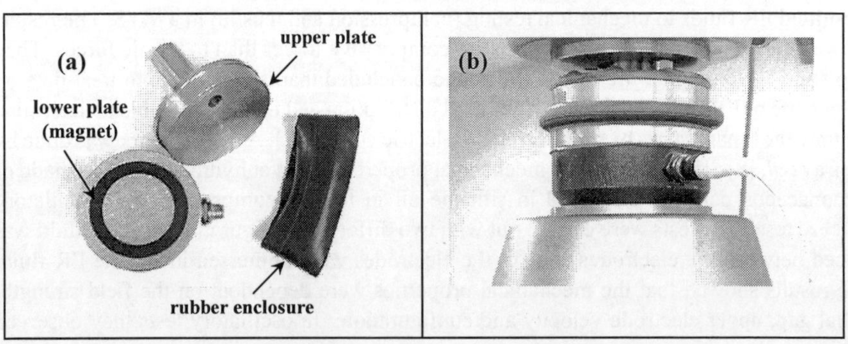

Figure 1. (a) Various parts of the squeeze film damper; (b) squeeze film damper mounted on the Instron Machine.

This setup was used to carry out three different types of experiments:

(i) *Squeeze mode:* where the lower plate was made to oscillate up and down at a given frequency (f) and axial displacement amplitude (x_0).
(ii) *Torsional mode (shear mode):* where the lower plate was made to undergo rotary oscillations about a mean position at a given frequency (f) and angle amplitude (θ_0).
(iii) *Torsional and Squeeze modes:* where the lower plate was forced to undergo simultaneous axial and torsional oscillations, at a given frequency, angle amplitude (θ_0), and axial displacement amplitude (x_0).

All these tests were conducted for different intensities of magnetic field, frequency, displacement amplitude and angle amplitude. The hysteresis loops were plotted for all the studied cases and the results are given and discussed in the following sections.

2.2 Linear damper (Valve mode)

A linear damper manufactured by Lord Corporation (model RD1005) was used to study the behavior of MR fluids in valve mode. The nominal number of turns of the magnetic coil in the damper is 207. The results obtained for this damper were compared with the ones for the squeeze film damper.

3 Results and Discussions

In this section, the results obtained with the developed squeeze film damper are presented and compared with the ones obtained with the Lord RD1005 damper.

3.1 Squeeze film damper

3.1.1 Squeeze Mode

As the magnetic field increased, we observed that the maximum compressive and tensile stresses along with the area of the loop curve increased, indicating the increase of the damping performance of the fluid (see figure 2). Also the values of stress attained in compression were greater than those obtained in tension. The compressive side of the hysteresis loop was highly non-linear and a peak on the left side of the curve was observed. Tang et al. [9] obtained similar results. From figure 2, we can also notice that the maximum compressive/tensile stresses attained for displacement amplitude x_0 = 1mm were higher than those obtained for the case of x_0 = 0.5mm.

The effect of frequency on the behavior of the loop curves was studied for x_0 = 0.5mm, i = 0.76 A, and $1 \leq f \leq 8$Hz. We observed that on the tension side of the loop curves, there was virtually no change with the frequency. However, on the compression side, the maximum stress attained dropped with the increase in frequency: comparing the results obtained for 1 and 8 Hz, a decrease of 31% was observed. This could be attributed to the fact that, as the fluid is compressed the distance between the two plates decreases and the magnetic field inside the fluid increases. In this way, the chains of MR fluid tend to become thicker, increasing the strength of the fluid. For smaller frequencies (1Hz) the chains have more time to thicken, forming more stable structures, than for the case of higher frequencies. This would reflect on the stress response of the fluid, which was found to be smaller for 8 Hz.

3.1.2 *Torsional Mode*

Figure 3 shows the hysteresis loop of torque as a function of angle of oscillation. The increase of the field produced an increase of the transmitted torque and the area of the hysteresis curve. The vertical portion of the curves is due to the yielding effect, and the corresponding torque, which was found to increase with the increase of the magnetic field, is called yield torque. From figure 3 we can observe that the yield torque is almost the same for different values of angle amplitude (θ_0).

Regarding changes in frequency, we observed that it did not affect the transmitted torque, at least in the range $1 \leq f \leq 8$ Hz.

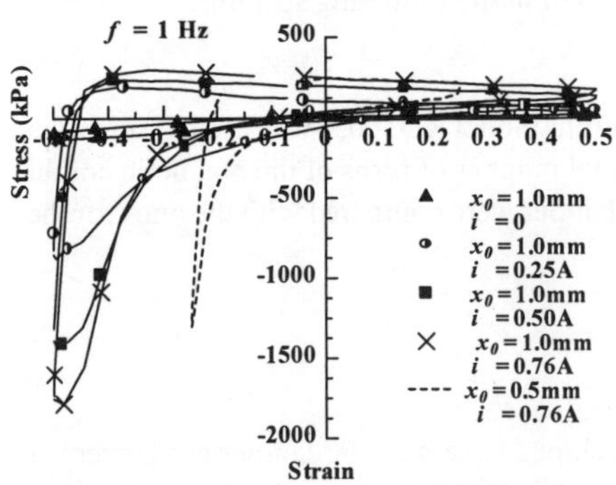

Figure 2. Effect of the field and displacement amplitude x_0 on the hysteresis loop of stress as a function of strain (squeeze mode)

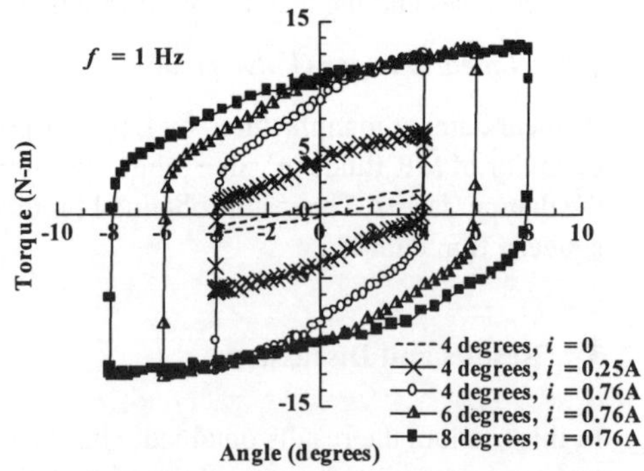

Figure 3. Effect of the field and angle amplitude θ_0 on the hysteresis loop of torque as a function of angle (torsional mode).

3.1.3 *Squeeze and Torsional Mode*

Figure 4 shows the behavior of the torque as a function of the angle. The non-linearity in the hysteresis loop is evident from the non-uniform widening of the curve towards left. The shape of the loop curve in this kind of mode was different from that obtained in pure torsional mode, since two peaks appeared for $\theta \sim -\theta_0$ on the left side of the curves. This was attributed to changes of the microstructure inside the MR fluid. Moreover, the area of the loop didn't exceed that obtained by pure torsional mode, meaning that the overall behavior of the MR fluid in pure torsional mode was better than that when squeeze was introduced.

For positive angles[1] and positive axial displacements (tension), the torque was stronger for the case of $x_0 = 0.5$ mm than for $x_0 = 1.0$ mm ($i = 0.76$A in Fig.4). This is due to the fact that the velocity of the lower plate and the gap between the plates are smaller (higher magnetic field) in the first case. So, the MR microstructure developed might be more stable, leading to higher values of torque. However, for negative values of angles and axial displacements (compression)[1] the peak torque was higher for $x_0 = 1.0$ mm ($i = 0.76$A), since its final gap was smaller and, consequently, the magnetic field was stronger. Probably, the appearance of the peak torque in the third quadrant for $x_0 = 1.0$ mm is related to fast

[1] In all the experiments where squeeze was combined with torsional mode, positives angles were always obtained when the MR fluid was subjected to tension, and negative angles when it was under compression.

changes on the microstructure due to the higher axial velocity of the lower plate. A similar behavior was observed for $i = 0.25$ A.

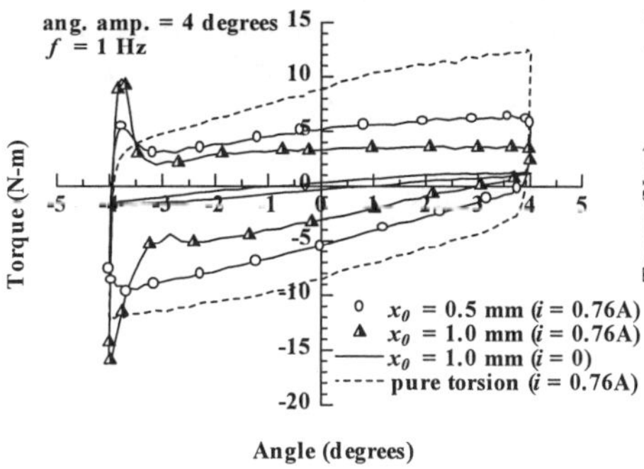

Figure 4 Effect of the fieldand axial displacement amplitude x_0 on the hysteresis curve of torque as a function of angle in squeeze and torsional mode.

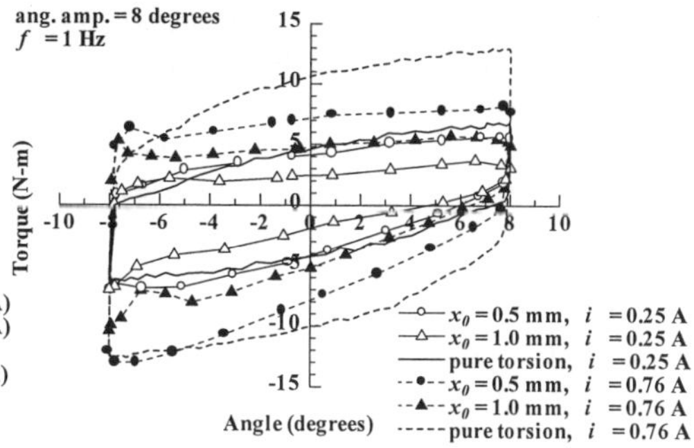

Figure 5 Effect of the axial displacement amplitude x_0 on the hysteresis curve of torque as a function of angle in squeeze and torsional mode.

For the case of $\theta_0 = 8°$, we obtained a different result. As shown in Fig.5, when the current was 0.25 A, the torque values at $\theta \sim -\theta_0$ for $x_0 = 0.5$mm and 1.0 mm were very close, and the loop for $x_0 = 0.5$mm was almost coincident with the loop of pure torsion. However, for $i = 0.76$ A, the torque at $\theta \sim -\theta_0$ was higher for the case of $x_0 = 0.5$mm, i.e., there was an inversion of the curves in the compression side for $\theta_0 = 8°$.

From the analysis of figures 4 and 5, we can say that, depending on the equilibrium between the hydrodynamic and magnetic forces, the introduction of squeeze can intensify or weaken the MR effect produced by the pure torsional mode. This contradicts the thought that the introduction of squeeze always increases the strength of MR fluid.

Another interesting thing to observe from these figures is that, the shape of both curves ($x_0 = 0.5$ mm and $x_0 = 1.0$ mm) is different, and the peak for $x_0 = 1.0$ mm is sharper. Probably this is due to the effect of changes of the relative axial velocity of the plates and changes of the magnetic field strength on the MR structure formed inside the fluid.

In order to check the effect of the introduction of shear on the pure squeeze mode, we conducted experiments for a fixed x_0 and electric current i_0 while θ_0 was changed from 0° (pure torsional mode) to 4°, 6° and 8°. For almost all the cases we studied, the peak stress, as well as the area of the loop curve, was greater for the case of squeeze combined with torsional mode when compared to the pure squeeze mode. This indicates that the introduction of shear ($\theta_0 \neq 0$) caused an increase of the squeeze strength of the MR fluid (see fig. 6). The exception was the case of $x_0 = 0.5$mm and $i = 0.76$ A, for which all the curves for different θ_0 were coincident with the loop of pure squeeze mode. These results indicate that shear changes the MR microstructure, sometimes helping the MR effect. It is possible that the thickness of the chains increases with the torsional motion of the lower plate, increasing the strength of the MR fluid. This means that the changes on the MR microstructure is governed not only by the magnetic field strength but also by the hydrodynamic forces caused by shear, which increase as the angle amplitude θ_0 increases. It is important to point out that there is a limit for which this happens. Our experiments showed that the higher peak stress was obtained for the case of $\theta_0 = 4°$. However, as the angle displacement continued to increase, the peak stress tended to the value obtained for pure squeeze mode.

We observed another interesting result. For $\theta_0 = 8°$ and $x_0 = 1$mm, the peak torque was slightly higher for $i = 0.25$ A than for 0.76 A and no significant change of the loop area was detected. This means that this lower value of current gives not only a higher damping force, but also saves energy. Moreover, the coil does not heat so much as in the case of 0.76 A. So, one can take advantage of a lower magnetic field strength associated with different kinds of modes in order to obtain almost the same damping performance.

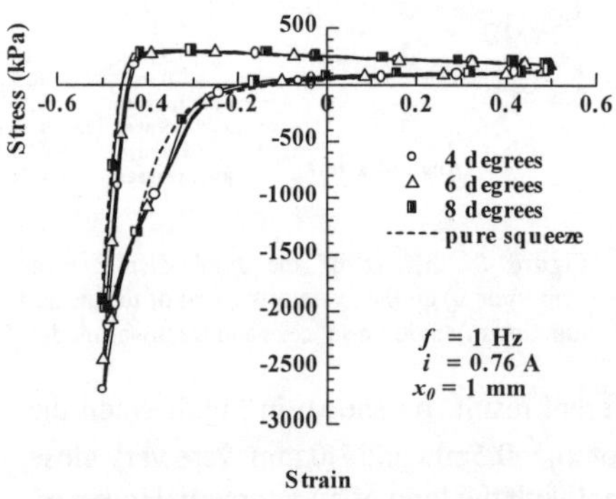

Figure 6 Hysteresis loop of stress as a function of strain for different angle amplitude, i= 0.76 A and axial displacement amplitude of 1 mm insqueeze and torsional mode.

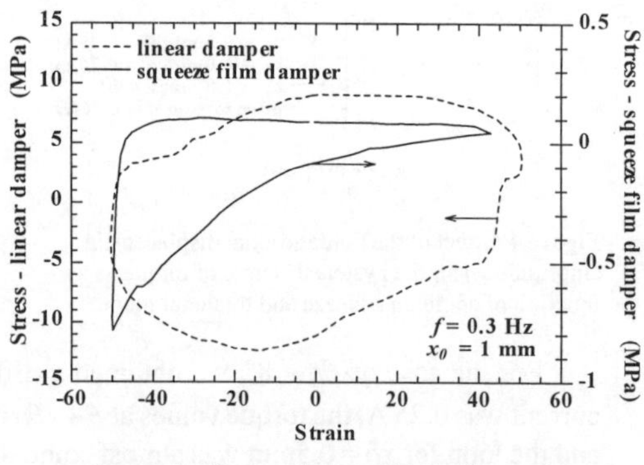

Figure 7 Hysteresis loop for the linear damper (field strength = 362 A-turns, 1.75A, 207 turns) and for the squeeze film damper (field strength =352 A-turns, 0.16A, 2200 turns).

3.2 Comparison of the linear damper and the squeeze film damper

In order to compare the behavior of MR fluids in both types of dampers, the tests were made under similar conditions of magnetic field (approximately same number of Ampere-turns), similar value of fluid strain and same frequency. As Lord Corporation does not sell the fluid inside the RD-1005 commercially, a fluid MRF-132LD, that is about 25% stronger than the fluid contained in the RD-1005, was used for tests with the squeeze film damper.

Figure 7 shows the comparison between the RD-1005 damper and the squeeze film damper. It can be seen from this figure that, even considering that the linear damper uses a weaker MR fluid, its pressure drop values are higher than the stress values obtained for the squeeze film damper (on the order of 10 to 100 times more, depending if the dampers are under compression or tension). It is known that part of the pressure drop (or stress) in a damper is due to the MR fluid characteristics and part is due to the damper design characteristics. In the case of the linear damper, the design characteristics play an important role on the intensity of pressure drop even in the absence of field. Moreover, the shapes of the curves are different, not only because the fluids are different, but also due to differences in the damper design and test modes (valve and squeeze modes).

4 Conclusions

Contrary to what is believed, the introduction of squeeze in the shear mode (torsional mode) not always increases the strength of the MR fluid. Moreover, shear can change the

MR microstructure, sometimes helping the MR effect: by introducing oscillatory shear (torsional mode) in the pure squeeze mode, it is possible to achieve a similar damping force and damping performance in the axial direction even for a smaller field strength. This saves energy and decreases the heating of the coil.

5 Acknowledgements

The authors thank Robert Csortos (Duramag Corporation) for his time and generous donation of the electromagnet; Dr. Carlson (Lord Corporation) for his help in providing the design details of the RD-1005 Damper; John Jaegley, Lawrence Cousino and Randall Reihing for their great support in the fabrication of the squeeze film damper.

References

1. Carlson J. D. and Jolly M. R., MR Fluid Foam and Elastomer Devices. *Mechatronics* **10** (2000) pp. 555-569.
2. Sproston J. L., El Wahed A. K., Williams E. W. and Stanway R., A Comparison of the Performance of ER Fluids in Squeeze. *5th Int. Conference on Electro-Rheological Fluids, Magneto-Rheological Suspensions and Associated Technology* (1996) pp. 259-269.
3. Sproston J. L., El Wahed A. K. and Stanway R., Electrorheological Fluids in Squeeze Under AC and DC Excitation. *6th Int. Conference on Electro-Rheological Fluids, Magneto-Rheological Suspensions and Their Applications* (1998) pp. 223-231.
4. El Wahed A. K., Sproston J. L. and Schleyer G. K., A Comparison Between Electrorheological and Magnetorheological Fluids Subjected to Impulsive Loads. *7th Int. Conference on Electro-Rheological Fluids and Magneto-Rheological Suspensions* (2000) pp. 401-410.
5. Vieira S. L. and Arruda A. C. F., Electrorheological Fluids Response Under Mechanical Testing. *J. Int. Mat. Syst. and Struct.* **9** (1) (1998) pp.44–52.
6. Vieira S. L. and Arruda A. C. F., Characterization of the Mechanical Properties of Electrorheological Fluids made of Starch and Silicone Fluid. *Journal of Modern Physics B*, V. 13, Nos. 14, 15 & 16 (1999) pp. 1908-1916.
7. Vieira S. L. and Arruda A. C. F., Study of Electrorheological Behavior Under Fatigue Tests. *J. Int. Mat. Syst. and Struct.* **9** (1) (1998) pp.53–59.
8. Vieira S. L., Nakano M., Oke R. and Nagata T., Mechanical Properties of an ER Fluid in Tensile, Compression and Oscillatory Squeeze Tests. *7th Int. Conference on Electro-Rrheological Fluids and Magneto-Rheological Suspensions* (2000) pp. 171-179.
9. Tang X., Wang X. J., Li W. H. and Zhang P. Q. 1998. Testing and Modeling of an MR Damper in the Squeeze Flow Mode. *6th Int. Conference on Electro-Rheological Fluids, Magneto-Rheological Suspensions and Their Applications*, pp. 870-878.
10. Wang X. J., Zhang P. Q., Tang X. and Tao R. 2000. Testing and Modeling a Cone-Shaped Squeeze-Film Mode Electrorheological Damper. *7th Int. Conference on Electro-Rheological Fluids and Magneto-Rheological Suspensions*, pp. 804-811.
11. Patent US Re: 36,270

ELECTRIC POWER AND EFFICIENCY UNDER A. C. ELECTRIC FIELD ON A ROTATING DEVICE OF DISK TYPE WITH SUSPENSION-TYPE ELECTRORHEOLOGICAL FLUID

KUNIO SHIMADA
Faculty of System Science And Technology, Akita Prefectural University, 84-4 Aza-ebinokuchi, Tsuchiya, Honjyo, 015-0055, Japan
E-mail: shimada@akita-pu.ac.jp

HITOSHI NISHIDA
Hokuriku Polytechnic College, 1289-1 Kawabuchi, Uozu, 937-0856, Japan
E-mail: nishida@hokuriku-pc.ac.jp

YOICHI AKAGAMI
Akita Prefectural Industrial Technology Center 4-11 Aza-Sanuki, Arayamachi, Akita 010-1623, Japan
E-mail: akagami@akita-iri.pref.akita.jp

The amount of electric power and the efficiency of a rotating disk under an A.C. electric field surrounded by suspension-type electrorheological fluid (ERF) were investigated. This report also dealt with the time lags of torque and current density from the applied electric field. Especially, the time lag of the current density was arranged by the power factor. The time lag of the torque becomes larger than that of current density. On the other hand, the data regarding torque and the current density have steady and varying characteristics as amplitude components. Therefore, four aspects of electric power require investigation: apparent power, effective power, steady power from the steady component of the current density, and total power. From the four electric powers, we can deal with four areas of efficiency regarding electric power in relation to torque. They depend on the angular velocity of the disk,,and frequency and intensity of the electric field. The results are explained using the particle aggregation model.

1 Introduction

In our previous paper [1], we experimentally investigated the torque and current density of a disk-type rotating device employing suspension-type ERF under an A.C. electric field. We then compared the results regarding the A.C. electric field with those achieved using a D.C. electric field. investigated on suspension-type ERF. This investigation seeks to determine the most appropriate electric fields for clutch and braking devices employing ERF.

Experimental data regarding the use of an A.C. electric field shows a lag in torque and current density from the time point when the electric field is applied. Furthermore, the data regarding torque and the current density have steady and varying characteristics as amplitude components. Therefore, we need to deal with four types of electric power: apparent power, effective power, steady power from the steady component of the current density, and total power. In contrast to our attention to electric power alone, we must deal with four forms of efficiency in regard to the rotating disk-type device both mechanically and electrically.

In the present study, we experimentally investigated the time lags of torque and current density, and four areas of power and efficiency under an A.C. electric field. We then compared the results regarding the A.C. electric field with those achieved using a D.C. electric field.

2 Experimental procedure

We used the same experimental apparatus employing a disk-type rotating device presented in previous experimental investigations [1-3]. The rotating disk in ERF has a diameter d_d=90.0mm with angular velocity ω. The electrode gap between the rotating disk and the casing is δ_d=1.0 mm, 1.5 mm. An A.C. electric field is supplied between the electrodes and has a frequency f from 1 Hz to 100 Hz.

We used four areas of power and efficiency under an A.C. electric field as follows:

$$W_{,eff}=E_{,A}I_{,A}(\cos\theta)_{,I} \quad (1),$$

$$\eta_{,eff}=T_{,A}\,\omega\,(\cos\theta)_{,T}/W_{,eff} \quad (2),$$

$$W_{,app}=E_{,A}I_{,A} \quad (3),$$

$$\eta_{,app}=T_{,A}\,\omega/W_{,app} \quad (4),$$

$$W_{,c}=E_{,A}\,I_{,c} \quad (5),$$

$$\eta_{,c}=T_{,c}\,\omega/W_{,c} \quad (6),$$

$$T\omega_{,AC}=(T_{,A}(\cos\theta)_{,T}+T_{,c})\omega \quad (7),$$

$$W_{,AC}=W_{,eff}+W_{,c} \quad (8),$$

$$\eta_{,AC}=T\omega_{,AC}/W_{,app} \quad (9),$$

where $E_{,A}$, $I_{,A}$ and $T_{,A}$ are the amplitude of the electric field, current density and torque, respectively, $(\cos\theta)_{,I}$ and $(\cos\theta)_{,T}$ are the power factor of current density and torque , and $I_{,c}$ and $T_{,c}$ are the steady component of current density and torque.

The ERF used in the present study is composed of many spherical particles in a solvent, and was made by Nippon Shokubai co. ltd. in Japan as TXER 7 [4].

3 Results and Discussion

3.1 *A. C. electric field*

Figure 1 shows the power factors of torque and current density, where E is the effective value of the electric field. The change of phase lag of the torque under an electric field is larger than the one of current density because of particle aggregates as shown in Fig. 13 [5] in the **Appendix**. However, the phase lag of current density is smaller than that of torque.

Figures 2 and 3 show effective and apparent powers, respectively. Figures 4 and 5 show effective and apparent efficiency, respectively. Because torque increases as ω increases, η increases as ω increases. However, $\eta_{,eff}$ has a peak at smaller E because of the torque phase lag. Because the increase of current density with increasing E is larger than torque, η decreases as E increases. Figures 6 and 7 show the steady component of power and efficiency, respectively. $W_{,eff}$, $W_{,app}$, $W_{,c}$ are almost constant at larger ω. On the other hand, as ω increases, $T_{,c}$ increases. And then $\eta_{,c}$ increases as

ω increases. $\eta_{,c}$ has a large value because $W_{,c}$ is much smaller than $T_{,c}$.

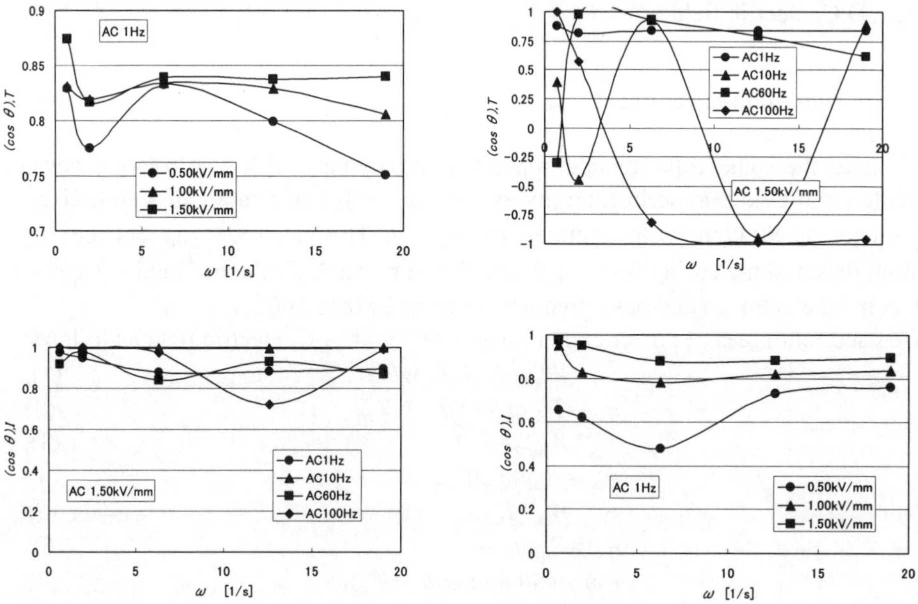

Figure 1. Power factors of torque and current density under A.C. electric field.

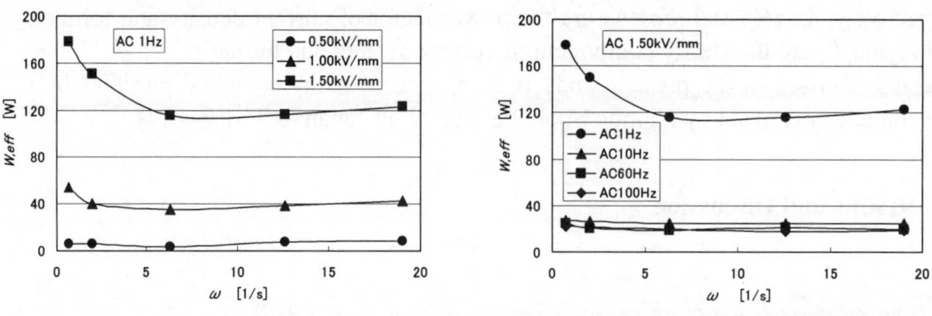

Figure 2. Effective power under A.C. electric field.

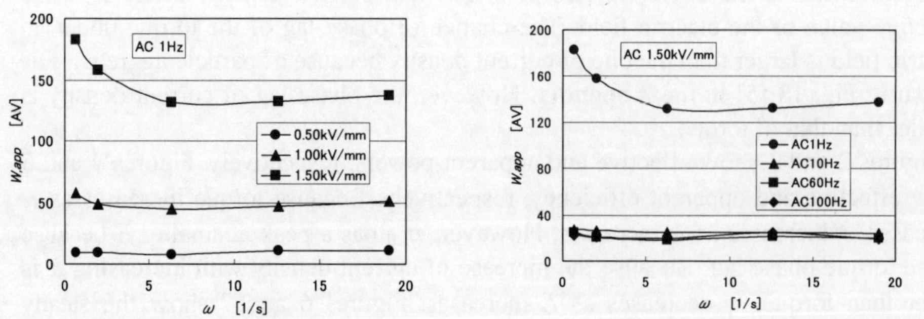

Figure 3. Apparent power under A.C. electric field.

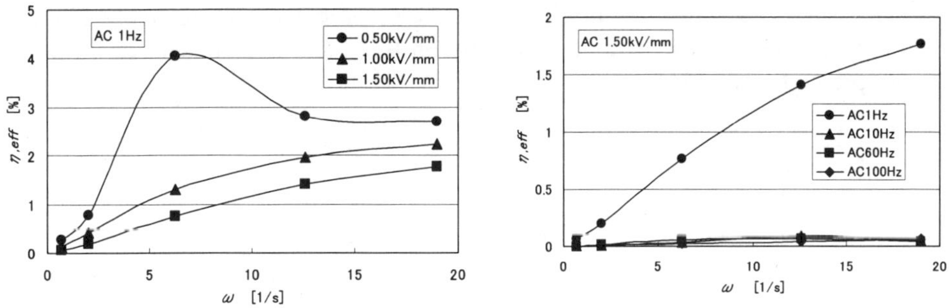

Figure 4. Effective efficiency under A.C. electric field.

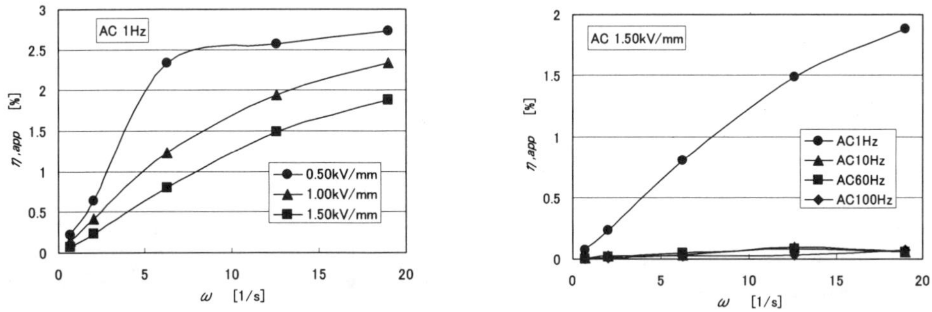

Figure 5. Apparent efficiency under A.C. electric field.

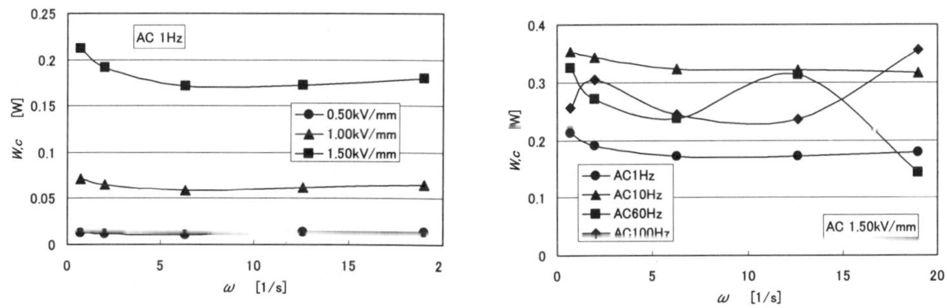

Figure 6. Steady component of power under A.C. electric field.

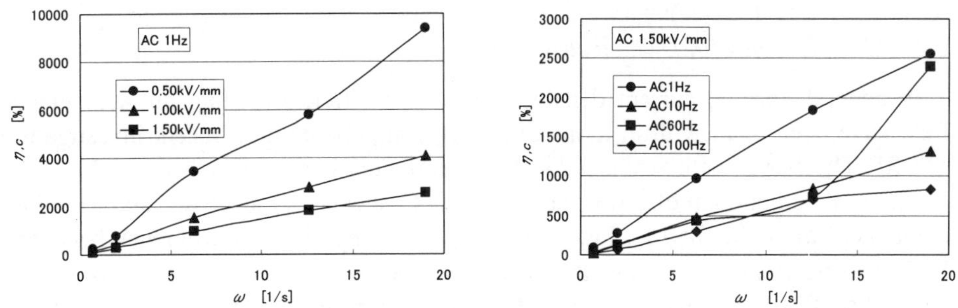

Figure 7. Steady component of efficiency under A.C. electric field.

Figures 8 and 9 show total mechanical and electric powers, respectively. Figure 10 shows total efficiency. $W_{,AC}$ depends mainly on $W_{,eff}$, and $T_{,AC}$ depends on $T_{,c}$ as seen in our previous report [6].

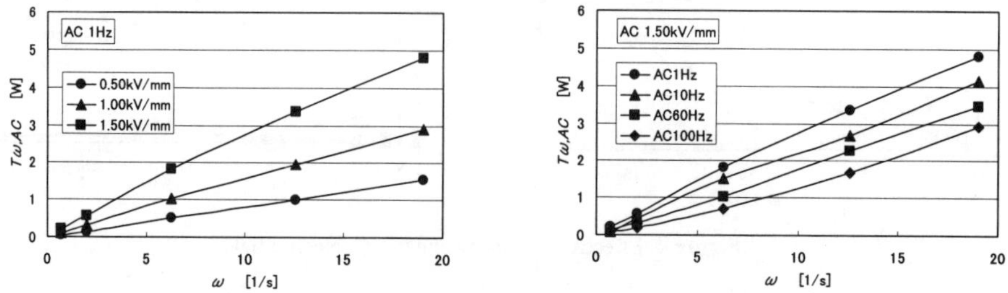

Figure 8. Total mechanical power under A.C. electric field.

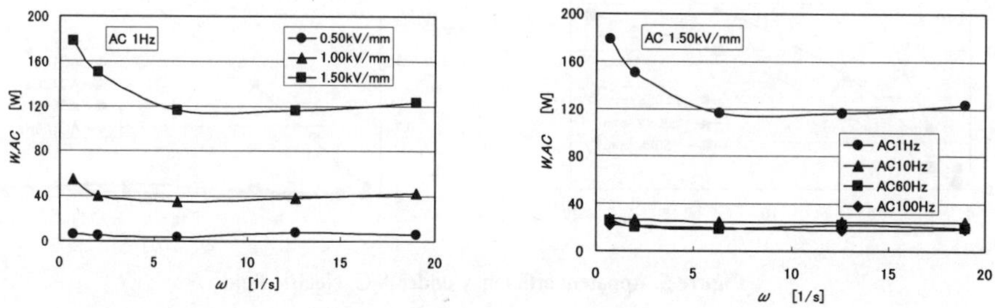

Figure 9. Total electric power under A.C. electric field.

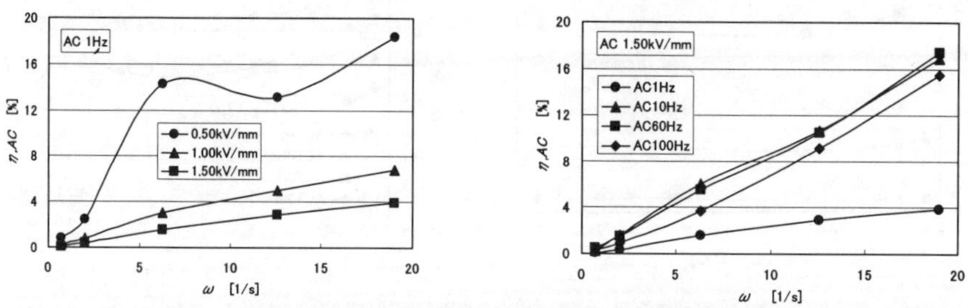

Figure 10. Total efficiency under A.C. electric field.

3.2 Comparison with D.C. electric field

In general, torque, current density, power and efficiency under an A.C. electric field have been dealt with in terms of steady components until recent investigations of ERF. Therefore, we investigated steady component in an A.C. electric field. Figure 11 shows a comparison of power between a D.C. electric field, $W_{,DC}$, and the steady component of A.C., $(W_{,c})_{,AC}$. Figure 12 shows a comparison of efficiency between a D.C. electric field, $\eta_{,DC}$, and the steady component of A.C., $(\eta_{,c})_{,AC}$. $(W_{,c})_{,AC}$ is larger than $W_{,DC}$. However, $T_{,c}$ is smaller than the torque of a D.C. electric field $T_{,DC}$. Therefore, $(\eta_{,c})_{,AC}$ is smaller $\eta_{,DC}$. As seen in the same qualitative results described in our previous paper [6], the relationship depends on frequency.

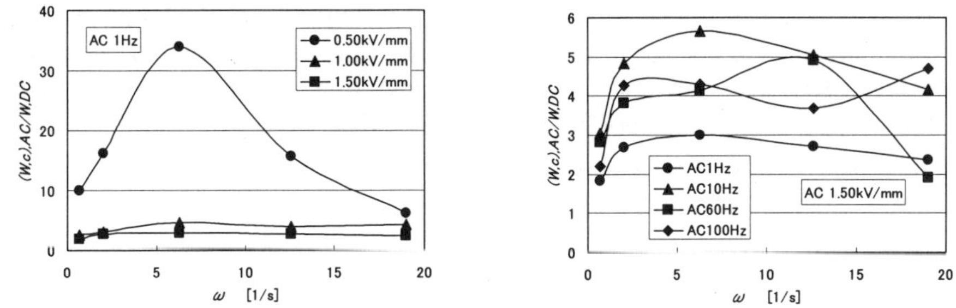

Figure 11. Total electric power under A.C. electric field.

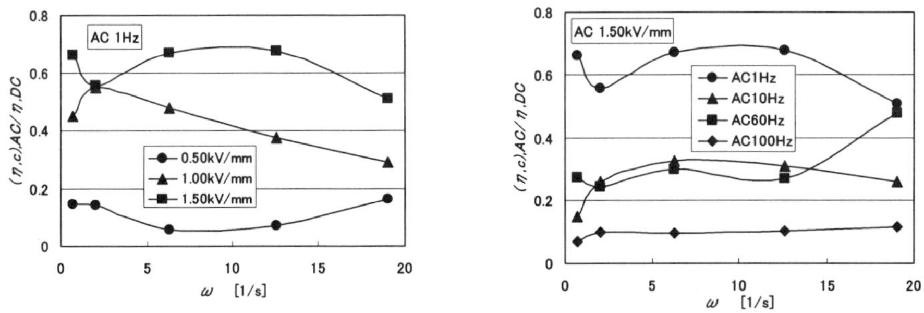

Figure 12. Total efficiency under A.C. electric field.

4 Conclusion

Under an A.C. electric field, we must four areas of power and efficiency because the torque and the current density have steady, varying characteristics as an amplitude components and time lag. These parameters depend on the angular velocity of the disk, and the frequency and intensity of the electric field.

5 Acknowledgements

The ERF used in the present study was provided by a grant from Nippon Shokubai co. ltd. of Japan.

References

1. K. Shimada and H. Nishida, Torque and Current Density on A Rotating Device of Disk Type under A. C. Electric Field with Suspension-type Electrorheological Fluid, (to be submitted).
2. H. Nishida, K. Shimada, M. Iwabuchi, T. Fujita and K. Okui, A study of steady torque characteristics of a rotating disk in ERF (1st report, comparison of flow curve between rotating concentric cylinder and rotating disk), Transactions of the Japan Society of Mechanical Engineers, **65**(633) B (1999), pp.1703-1709 (in Japanese).

3. H. Nishida, K. Shimada, M. Iwabuchi, T. Fujita and K. Okui, Effects of experimental factors on steady characteristics on devices of rotating disk type using ERF with smectite particles, Transactions of the Japan Society of Mechanical Engineers, **65**(629) C (1999), pp.269-276 (in Japanese).
4. T. Kawakami and Y. Asako, Electrorheological suspensions, Transactions of the Japan Society of Mechanical Engineers, №.97-1 (1997), pp.128-129 (in Japanese).
5. D. J. Klingenberg and C. F. Zukoski IV, Studies on the steady-shear behavior of electrorheological suspensions, Langmuir, **6** (1990), pp.15-24.
6. K. Shimada and H. Nishida, Torque and Current Density on A Rotating Device of Disk Type under A. C. Electric Field with Suspension-type Electrorheological Fluid, (to be submitted).

Appendix

Particles are moved according to an A.C. electric field between electrodes as shown in Fig. 13 as seen in our previous report [6]. The behavior of the other ERF with 6μm spherical diamond particles in silicone oil could be confirmed by microscopic observation as shown in photograph of Fig. 14. The fluid can be regarded to have the same behavior as particles. The first and second upper photographs in Fig. 14 show the movement of particles under an A. C. electric field and its sine curve, while the third photograph shows its pulse curve. At low frequency the particles are moved between electrodes in a circular motion, while at high frequency this movement could not be observed. Thus, in the case of a pulsed electric A.C. field, the particles move to one side of an electrode.

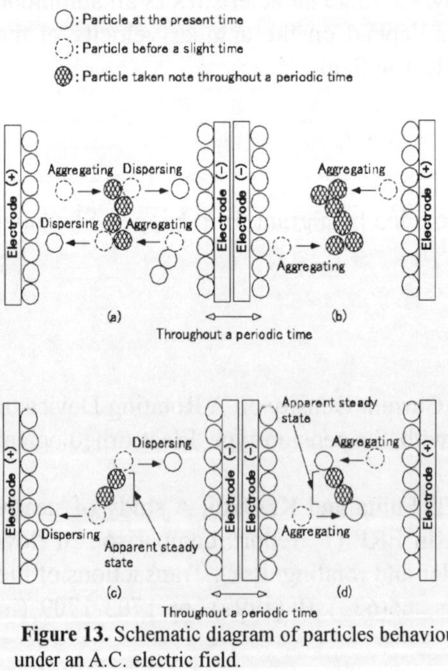

Figure 13. Schematic diagram of particles behaviour under an A.C. electric field.

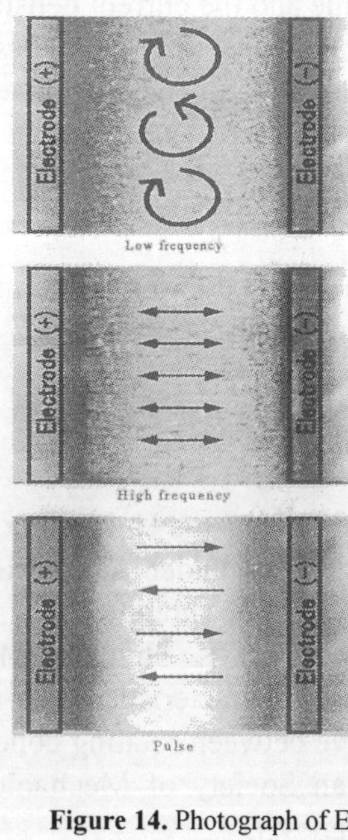

Figure 14. Photograph of ERF unde an A.C. electric field.

TORQUE AND CURRENT DENSITY ON A ROTATING DEVICE OF DISK TYPE UNDER A. C. ELECTRIC FIELD WITH SUSPENSION-TYPE ELECTRORHEOLOGICAL FLUID

KUNIO SHIMADA
Faculty of System Science And Technology, Akita Prefectural University, 84-4 Aza-ebinokuchi, Tsuchiya, Honjyo, 015-0055, Japan
E-mail: shimada@akita-pu.ac.jp

HITOSHI NISHIDA
Hokuriku Polytechnic College, 1289-1 Kawabuchi, Uozu, 937-0856, Japan
E-mail: nishida@hokurikua-pc.ac.jp

Torque and current density on a rotating disk immersed in suspension-type electrorheological fluid (ERF) under A.C. electric field with were measured. They have steady and varying characteristics as an amplitude components which depend on the angular velocity of the disk, and the frequency and intensity of the electric field. The results were demonstrated by means of a particle aggregation model. The constant torque and current density under D.C. electric field were compared to those under A.C. field. The magnitudes of the differences strongly depend on the angular velocity of the disk and the frequency of the electric field. When an A.C. electric field is used, a smaller frequency should be applied for the sake of greaten efficiency.

1 Introduction

In the present study, we focus on suspension-type ERF. The determination of the most appropriate electric field is very important in applications in clutch and braking devices employing ERF. A comparison of A.C. and D.C. electric fields' effects on torque and current is of particular relevance. In case of an A.C. electric field, it is necessary to investigate the frequency and intensity of the electric field as well as the angular velocity of the device.

Many experimental investigations have been performed regarding a D.C. electric field in devices employing rotational flow. As determined by a formulation based on the rotating body in the ERF, two types of bodies can be considered: the disk [1] and the cylinder. The cylindrical type has been the subject of many experimental investigations, while relatively few studies have been performed on a rotating disk under an A.C. electric field.

On the other hand, for practical applications, we can make an educated guess regarding the disk type with taking into consideration issues of dynamic and economical efficiency. Therefore, in the present study, we experimentally investigated the torque and current density in the disk-type rotating device under an A.C. electric field. We then compared the results regarding the A.C. electric field with those achieved using a D.C. electric field.

2 Experimental procedure

We used the disk-type rotating device used previously in experimental investigations [2,3]. The rotating disk in ERF has a diameter d_d=90.0mm and a thickness 8.00mm. The electrode gap between the rotating disk and the casing is δ_d=1.0 mm, 1.5 mm, and the casing is at rest.

An A.C. electric field is supplied between the electrodes and has a frequency f from 1 Hz to 100 Hz. Angular velocity is applied to the connected rotating disk by means of an A.C. surbomoter. The angular velocity and the torque of the rotating disk are measured, respectively, by a speed meter and a torque meter connected to the disk. The current passing through the ERF is measured.

The characteristics of the measured data show both constant and varying characteristics as an amplitude in case of the A.C. electric field, as shown in Fig. 1.

The ERF used in the present study is composed of many spherical particles in a solvent, and was made by Nippon Shokubai co. ltd. in Japan as TXER-7 [4].

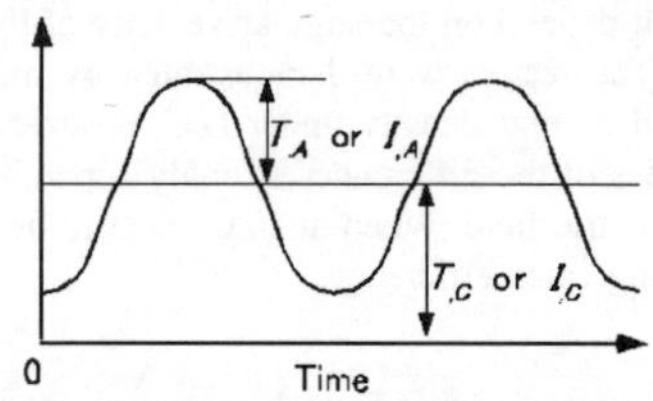

Figure 1. Schematic diagram of torque and current density components under an A.C. electric field.

3 Results and Discussion

3.1 Torque

Figure 2 shows the steady torque $T_{,DC}$ under a D.C. electric field. Figures 3 and 4 show the torque's varying as an amplitude $T_{,A}$ and constant $T_{,C}$ characteristics under an A.C. electric field, respectively. Steady torque under a D.C. electric field decreases and then increases with increasing angular velocity of the rotating disk ω. However, the changes in torque differ under a D.C. electric field differ from those under an A.C. electric field according to f. At lower f, $T_{,A}$ decreases with larger ω. However, at higher f, $T_{,A}$ increases and then decreases with increasing ω. At larger f levels, $T_{,A}$ is smaller. On the other hand, as f increases, the increase in $T_{,C}$ in accordance with increases in ω is larger. As f grows larger, $T_{,C}$ grows smaller.

In many previous experimental investigations regarding rotating cylinder in ERF, there has no been discussion concerning the clear differences between its constant and varying characteristics. Thus we found it appropriate to distinguish these characteristics under an A.C. electric field.

These results can be explained by using particle aggregates as shown in Fig. 5. Explanations in terms of particle behavior have been proposed by several researchers [5]. In the present study, we use the particles model as shown in Fig. 5. Particles move between the electrodes according to the A.C. electric field with repeats between

(a) and (b), or between (c) and (d), as shown in Fig. 5.

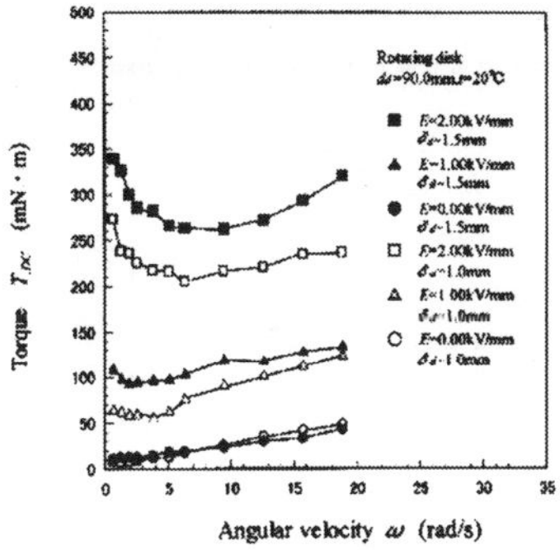

Figure 2. Steady torque under a D.C. electric field.

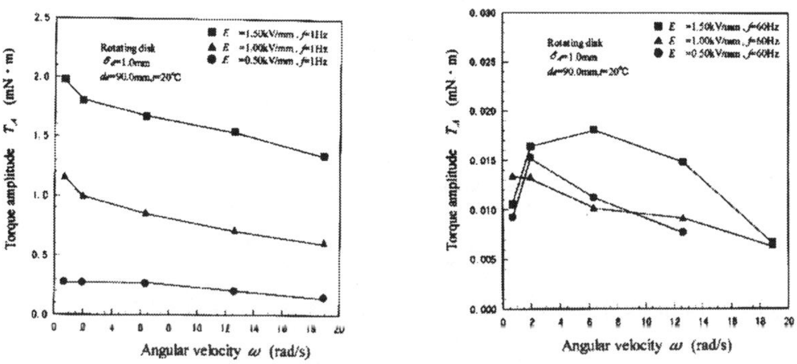

Figure 3. Amplitude component of torque under an A.C. electric field.

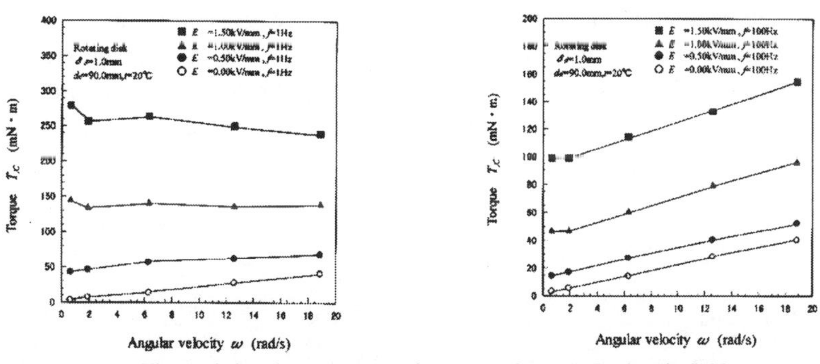

Figure 4. Steady component of torque under an A.C. electric field.

As Figs. 5(a) and (b), show an amplitude of moving particles between the electrodes; therefore, the amplitude of torque exists. At larger ω and f, the amplitude of moving particles between the electrodes becomes smaller because the number of particles which cannot move according to the changing electric field and shear flow is increasing. As a result, $T_{,A}$ changes according to the ω and f, as shown in Fig. 3.

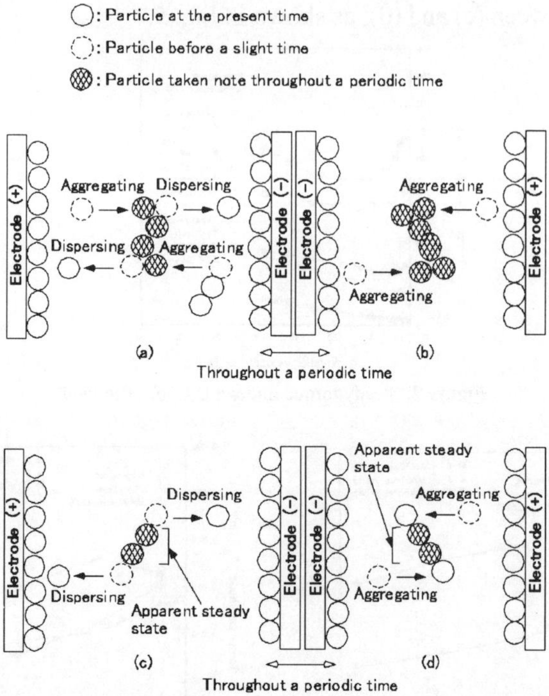

Figure 5. Schematic diagram of particles aggregation model under an A.C. electric field.

On the other hand, as seen in Figs. 5(c) and (d), some particles exist under an apparently static state; therefore, the steady component of torque exists. At larger ω, the particles under a static state apparently increase because the number of particles which cannot move according to the increasing shear flow is increasing. However, at larger f, the number of particles under a static state apparently decrease more than in the case of larger ω, because the particles are disturbed by the oscillating electric field. As a result, T_{C} changes according to ω and f, as shown in Fig. 4.

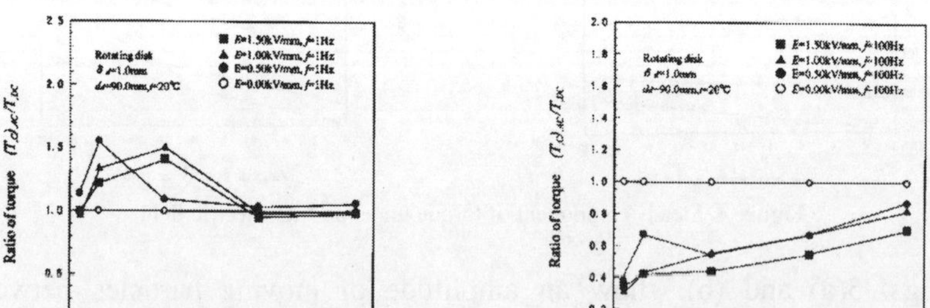

Figure 6. Comparison of steady torque betw een D.C. and A.C. electric field.

When we design an electric field to be applied in a rotating device, the question at to which type of electric field should be used, A.C. or D.C., becomes very important. Figure 6 can provide us with an answer. At lower f, with increasing ω, the ratio of steady torque between a D.C. and an A.C. electric field increases and then decreases to a saturated 1. However, at higher f, with increasing ω, the ratio increases to a saturated 1. When the ratio is larger than 1, an A.C. electric field is more. Thus, in the case of a device which employs an electric control method, the electric field for use in a disk-type device needs to be chosen according to f and ω, for the sake higher efficiency. As a matter of course, we must consider not only $T_{,C}$ and $T_{,A}$ but also the efficiency of electric and mechanical powers. Then, this efficiency is dealt with in the next session again.

3.2 *Current density*

Figure 7 shows the steady current density $I_{,DC}$ under a D.C. electric field. Figures 8 and 9 show the varying as an amplitude $I_{,A}$ and constant $I_{,C}$ characteristics of current density under an A.C. electric field, respectively. In general, steady current density under a D.C. electric field decreases with increasing shear rate D, which is defined by the angular velocity on the outside radius. However, changes in current density in the D.C. electric field differ from those under an A.C. electric field according to f. At every f, $I_{,A}$ decreases according to the increase of D. At larger f, $I_{,A}$ decreases. On the other hand, in case of larger D, $I_{,C}$ is almost constant at every f. At larger f, $I_{,C}$ is larger.

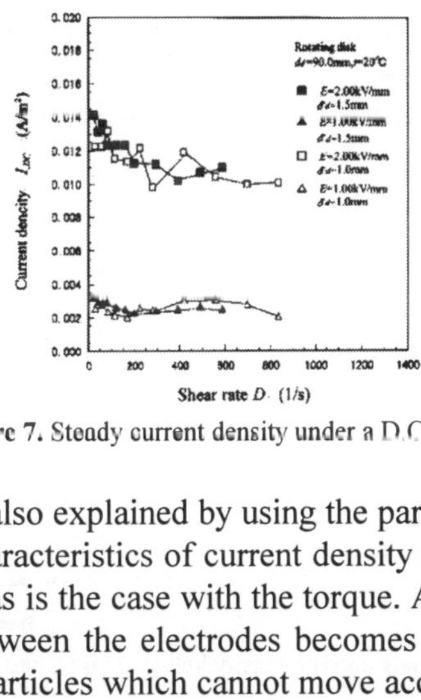

Figure 7. Steady current density under a D.C. electric field

These results can be also explained by using the particle aggregate model shown in Fig. 5. The constant characteristics of current density exists because the particles are in an apparently static, as is the case with the torque. At larger D and f, the amplitude of moving particles between the electrodes becomes smaller; therefore, the current decreases because the particles which cannot move according to the changing electric field and shear flow are increasing. As a result, $I_{,A}$ is changed according to D and f, as shown in Fig. 8.

At larger D, the number of particles which cannot move according to the changing electric field increases because the particles which cannot move according to the increasing shear flow are also increasing. The current then increases as well; however, the current density becomes saturated. On the other hand, at larger f, the particles under a static state apparently decrease more than in the case of larger D, because the particles are disturbed by the oscillating electric field. Therefore, the oscillating particles between the electrodes increase as the current increases. As a result, $I_{,A}$ is changed according to D and f, as shown in Fig. 8.

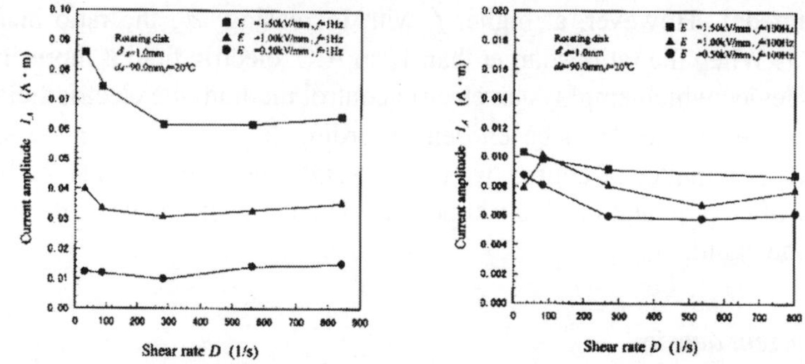

Figure 8. Amplitude component of current de nsity under an A.C. electric field.

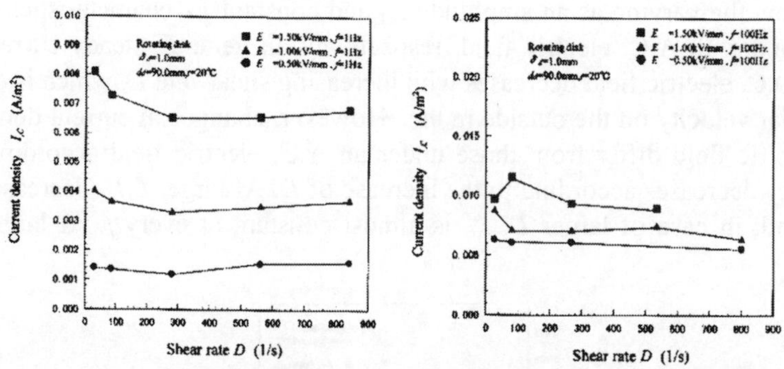

Figure 9. Constant characteristics of current density under an A.C . electric field.

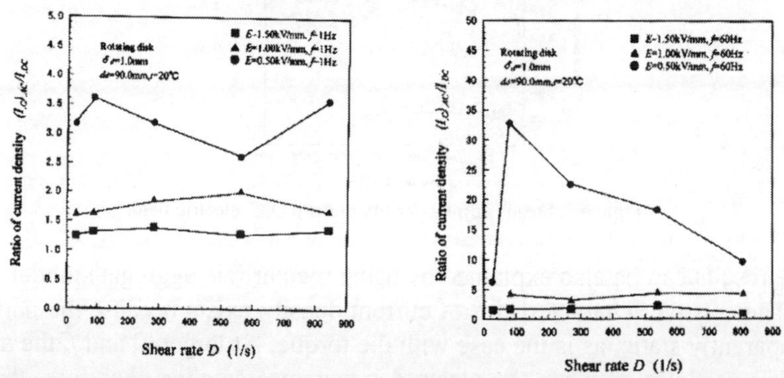

Figure 10. Comparison of steady current density between D. C. and A.C. electric fields.

At the same torque level, a comparison of current density between a D.C. electric field and an A.C. electric field is very important. Figure 10 shows the ratio of steady current density between a D.C. and an A.C. electric field. In contrast to the torque, at every f, the ratio is larger than 1, and at larger f, the ratio is greater.

From Figs.6 and 10, when we consider the efficiency defined by $\omega T/IE$, the efficiency is smaller at larger f in the case wherein an A.C. electric field rather than a D.C. electric field is used. However, more investigation of the efficiency is need to be made. Then, this subject is dealt with in the other present [6].

4 Conclusion

(1) We can distinguish constant and varying as an amplitude characteristics.
(2) At lower f, $T_{,A}$ decreases with larger ω. However, at higher f, $T_{,A}$ increases and then decreases with increasing ω. At larger f, $T_{,A}$ is smaller. On the other hand, at larger f, $T_{,C}$ increases according to increasing ω. At larger f, $T_{,C}$ is smaller.
(3) At every f, $I_{,A}$ decreases according to the increase of D. At larger f, $I_{,A}$ is smaller. On the other hand, in the case of larger D, $I_{,C}$ is almost constant at every f. At larger f, $I_{,C}$ is larger.
(4) At lower f, as ω increases, the steady torque ratio between the D.C. and A.C. electric fields increases and then decreases to a saturated 1. However, at higher f, with increasing ω, the ratio increases to a saturated 1.
(5) At every f, steady current ratio between the D.C. and A.C. electric fields is larger than 1. And at larger f, the ratio is greater.
(6) When we use an A.C. electric field, we should use smaller frequencies to achieve greater efficiency.

5 Acknowledgements

The ERF used in the present study was provided by a grant from Nippon Shokubai co. ltd. of Japan.

References

1. K. Shimada and T. Fujita, A device to regulate rotational speed using a rotating regulator with electrorheological fluid, Experimental Mechanics, **40**(2) (2000) pp.231-240.
2. H. Nishida, K. Shimada, M. Iwabuchi, T. Fujita and K. Okui, A study of steady torque characteristics of a rotating disk in ERF (1st report, comparison of flow curve between rotating concentric cylinder and rotating disk), Transactions of the Japan Society of Mechanical Engineers, **65**(633) B (1999), pp.1703-1709 (in Japanese).
3. H. Nishida, K. Shimada, M. Iwabuchi, T. Fujita and K. Okui, Effects of experimental factors on steady characteristics on devices of rotating disk type using ERF with smectite particles, Transactions of the Japan Society of Mechanical Engineers, **65**(629) C (1999), pp.269-276 (in Japanese).
4. T. Kawakami and Y. Asako, Electrorheological suspensions, Transactions of the Japan Society of Mechanical Engineers, №.97-1 (1997), pp.128-129 (in Japanese).
5. D. J. Klingenberg and C. F. Zukoski IV, Studies on the steady-shear behavior of electrorheological suspensions, Langmuir, **6** (1990), pp.15-24.
6. K. Shimada, H. Nishida and Y. akagami, Electric power and efficiency under A. C. Electric Field on A Rotating Device of Disk Type with Suspension-type Electrorheological Fluid, (to be submitted).

THREE-PORTS MICRO ER VALVE FOR ER SUSPENSION FABRICATED BY PHOTOLITHOGRAPHY

M. NAKANO, K. KATOU, A. SATOU, K. MIYATA AND K. MATSUSHITA

Faculty of Engineering, Yamagata University
4-3-16 Jonan, Yonezawa, Yamagata 992-8510, JAPAN
E-mail: nakano@mnaka.yz.yamagata-u.ac.jp

A three-ports micro ER valve has been fabricated by a photolithography, which is made of a patterned resist film between two-glass plates coated ITO electrode pattern. And one inlet channel (height h=0.1mm, width w=2mm) of the ER suspension diverges into two outlet channels having ER valves, so that the inflow of the ER suspension from the inlet port is divided into two outflows to the two outlet ports by synchronously controlling a pair of ER valves installed in the outlet channels. The PWM control strategy is applied to a pair of ER valves to change the flow rates, as proposed in our previous paper. The PWM control of each outlet micro-ER valve can realize the almost same changing characteristics of flow rate with the duty ratio as that of a conventional valve. The switching characteristics of a pair of ER valves have been investigated by measuring the flow rates from both outlet ports and visualizing the micro-flow behavior of ER suspension in the ER valves.

1 Introduction

Electrorheological (ER) suspensions are one family of the intelligent materials that behave like a Bingham fluid having yield stress, which can be rapidly changed in a reversible manner by applying electric fields [1]. This rheological change is caused by the formation of fibrous structure consisting of the dispersed particles, which are polarizable under the electric field. In a hydraulic circuit system using ER suspensions, when the ER suspension flows through the slit channel between two electrodes the dispersed particles are polarized and form many clusters, so that the flow rate and / or pressure drop across the channel can be controlled by changing the electric field strength applied to the electrodes because the flow resistance due to the clusters changes with the electric field strength [2]. That is, the valve consisting of only electrodes with no moving parts is realized using ER fluids. The ER valve has advantages such as a simple and compact construction and a high response rate, and therefore is anticipated some applications to micro systems such as a micro valve, a micro actuator and a micro pump. An extremely compact and simply constructed positioning unit has been developed by integrating the ER valves into the cylinder case [3]. And also, a miniature bellows actuator driven by a pair of PWM controlled ER valves has been developed, and has demonstrated the good control characteristics in our previous study [4].

In this study, a three-ports micro ER valve has been proposed and fabricated as the feasibility study for applying to a micro actuator. The electrodes and the flow channels are made by a photolithography technique used in the manufacturing process of semi-conductor. One inlet channel diverges into two outlet channels having ER valves, so that the inflow of the ER suspension from the inlet port is divided into two outflows to the two outlet ports by synchronously controlling a pair of ER valves installed in the outlet channels. The PWM or DC control strategy is applied to a pair of micro-ER valves to change the flow rates. The steady and switching characteristics of flow rates for a pair of micro-ER valves are investigated by measuring the flow rates from both outlet ports and visualizing the micro-flow behavior of ER suspension in the ER valves.

2 Experimental and Measurement

2.1 Fabrication of Three-ports Micro ER Valve

The three-ports micro ER valve is made of a patterned photo resist film between two-glass plates coated ITO electrode pattern. The negative resist for forming thick membrane is used to make the micro flow channel and micro ER valves of about 100 μm gap. Figure 1 shows the fabrication process of the three-ports micro ER valve. The fabrication procedure is as follows:

1. Two glass plates are coated with ITO (Indium-tin oxide) film.
2. The two ITO glass plates are coated with photo-resist, and then are irradiated by ultraviolet rays through an electrode-patterned photo mask to remain the resist of designed electrode shape.
3. Soaking them in a developer dissolves the part of the resist, which has not been irradiated, and only the resist of electrode shape remains on the ITO glass plate.
4. ITO membrane coated on the glass plates is etched with aqua regia to form the ITO membrane of electrode shape covered with the resist. After that, the resist on the ITO electrodes is removed.
5. One of two glass plates treated in the above processes is plastered with the resist of 100 μm thickness and irradiated through a patterned mask to form flow channels.
6. The resist corresponding to the part of the flow channels is dissolved by soaking them in a developer. The sidewall of the flow channels is formed with the remaining resist.
7. Finally, as shown in Fig.2, the glass plate with patterned flow channels made of the remaining resist, which is coated with a resist, is glued together with the other glass plate with patterned ITO electrodes by irradiating the ultraviolet rays.

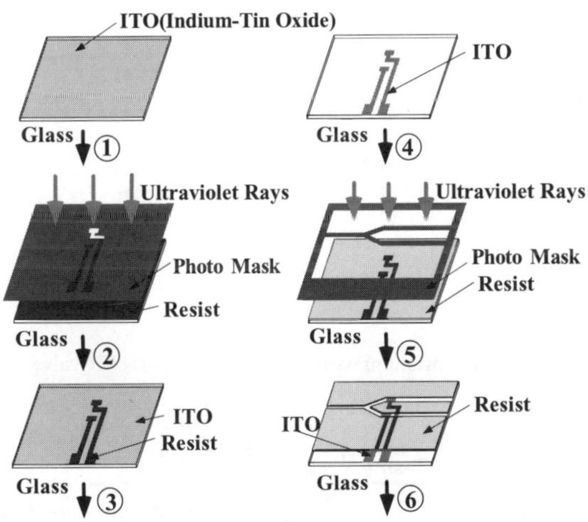

Figure 1. Fabrication process of three-ports ER valve by photolithography.

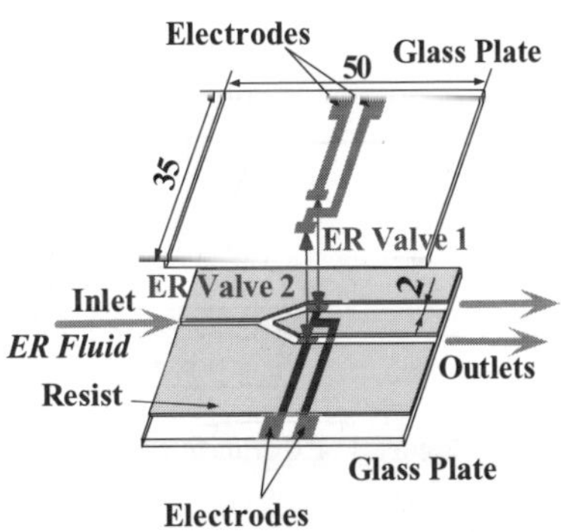

Figure 2. Micro three-ports ER valve.

2.2 Construction of Three-Ports Micro ER Valve

Figure 2 shows the construction of the three-ports micro ER valve. It has micro flow channels of 2 mm in width and about 100 μm in height, which consist of one inlet port and two outlet ports. The ER valves consisting of ITO electrodes of 4 mm in length are installed in two outlet channels. An ER fluid supplied to the inlet port at constant pressure is distributed to two outlet ports of which the flow rates are controlled by a pair of ER valves.

2.3 Experimental and Measuring System

The experimental system of constant pressure type for the micro three-ports ER valve and the measuring system are shown in Fig.3. An ER suspension is supplied to the ER valve from an ER fluid reservoir tank pressurized by compressed air of constant pressure. The pressure of supplied ER suspension is measured using a semi-conductor type pressure transducer at the inlet side of the ER valve, and is recorded on an analyzing recorder. Outlet flow rate **Q**, is obtained by estimating the volume of fluid moving in a measuring pipette per unit time duration of 0.1 sec from the video frames recorded on a CCD camera and video tape recorder. The time variation of flow rate is calculated by smoothing the time series data of the flow rate using the method of moving averages. DC voltage or PWM wave voltage is applied to the electrodes of the ER valve. The PWM wave is made with a control voltage input to a PWM circuit, which is applied to the ER valve through a high voltage amplifier. The flow behavior within the ER valve can be recorded on a CCD camera because the valve consists of transparent glass plates. An ER fluid (TX-ER6, Nippon Shokubai Co. Ltd.) containing sulfonated polymer particles of 5μm average diameter and 20 vol% is used.

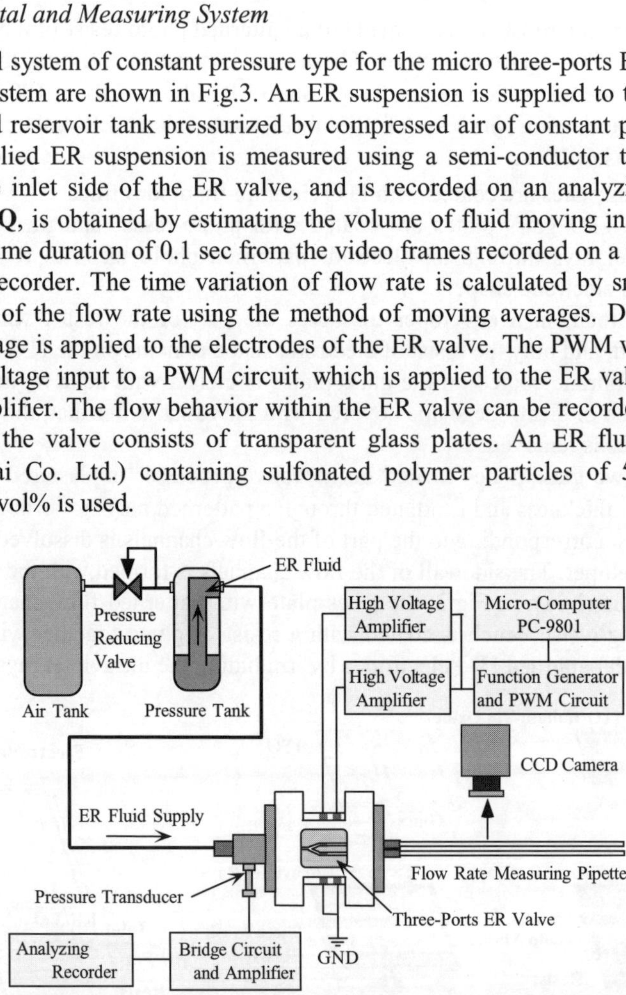

Figure 3. Experimental and measuring system of constant pressure type for three-ports ER valve.

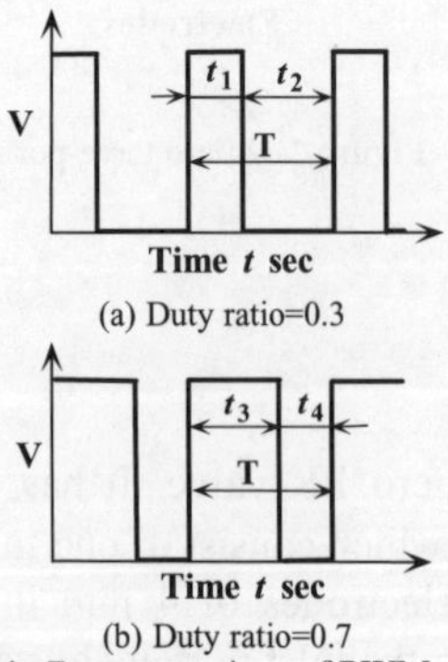

Figure 4. Representatives of PWM wave.

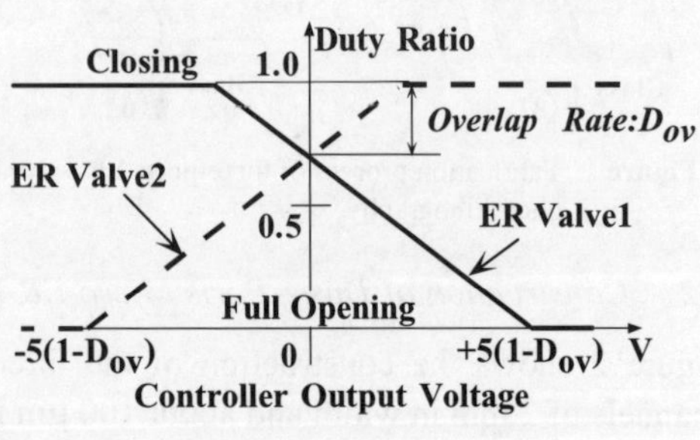

Figure 5. Relationship between controller output voltage and PWM duty ratio for a pair of ER valves.

2.4 PWM Flow Rate Control Strategy of a Pair of ER Valves

The PWM waves shown in Fig.4 are generated on a PWM circuit in which a saw wave of 100 Hz in carrier frequency is compared with a controller output voltage to determine a duty ratio. The electric field strength **E** of the PWM voltage amplitude is set for 1.3 kV/mm to stop flow at the maximum supply pressure of 30 kPa. The duty ratios of PWM waves applied to the ER valve 1 and 2 are varied in proportion to the controller output voltage according to the relationship shown in Fig.5. In this study, a sinusoidal wave of 0.25 Hz is used as the signal of the controller output voltage to vary the duty ratio. The switching timing between both ER valves can be controlled with an overlap rate $\mathbf{D_{ov}}$ between the duty ratios of both valves.

3 Steady Flow Rate Control Characteristics of Three-Ports Micro ER Valve

3.1 Steady Characteristics without Electric Field

Figure 6 shows the flow rate **Q** as a function of the supply pressure of ER fluid **ΔP** for a pair of ER valves under no electric field. The flow rates of both ER valves increase with the increase of the supply pressure nearly along a theoretical line. The slight difference between the flow rates of two ER valves seems to be caused by minute difference of flow channel shapes.

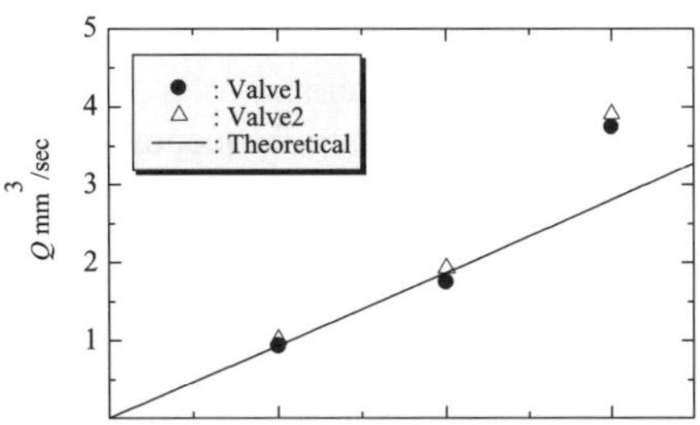

Figure 6. Flow rate against supply pressure of ER fluid for a pair of ER valves under no electric field strength.

3.2 Steady Characteristics with Applied Electric Field

The steady flow rate control characteristics of each valve are investigated at the supply pressure **ΔP** of 20 or 30 kPa, in the condition where one of the valves has the duty ratio of 1 to be completely closed, and the DC or the PWM wave voltage is applied to another valve. Figure 7 shows the case of DC voltage, and Fig.8 does the case of PWM wave voltage. When the DC voltage is applied, values of the closing electric field strength for two valves differ from each other, which seems to be caused by minute difference of flow

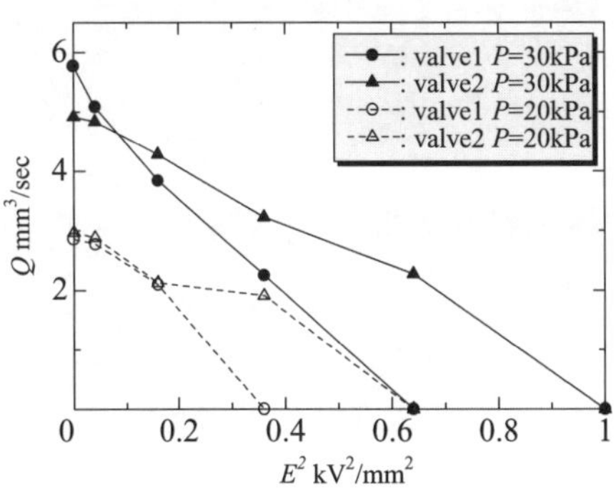

Figure 7. Steady characteristics of flow rate control when DC voltage is applied.

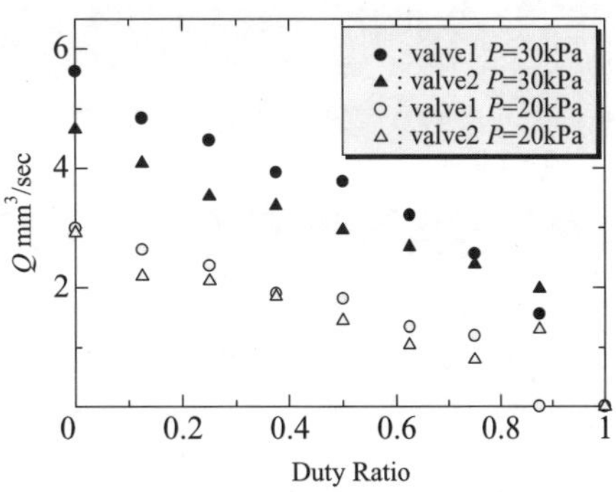

Figure 8. Steady characteristics of flow rate control when PWM control is applied.

channel shapes. Furthermore, a slight difference of supply pressures in each experiment can affect the flow rate because the flow channels are quite small. On the other hand, when the PWM wave voltage is applied, both valves can be closed at the duty ratio of 1 for each supply pressure, and almost equivalent flow rate control with the duty ratio is available for both valves, as seen in Fig.8. It means that to set the amplitude of the PWM wave voltage larger than the closing electric one at maximum supply pressure can crucially control the opening and closing of the valves, even if the pressure difference across the valve fluctuates. And also, the PWM control of the ER valves performs the almost same flow rate control characteristics, even if there is a slight difference between the valve closing electric fields of both valves.

4 Switching Characteristics of a Pair of ER Valves using PWM Control

4.1 Constant Duty Ratio at Valve Opening

The switching characteristics of a pair of ER valves were investigated using PWM control strategy at the switching frequency of one time per 2 sec. In this switching control, one valve is opened while another valve is completely closed. The opening rate of valves can

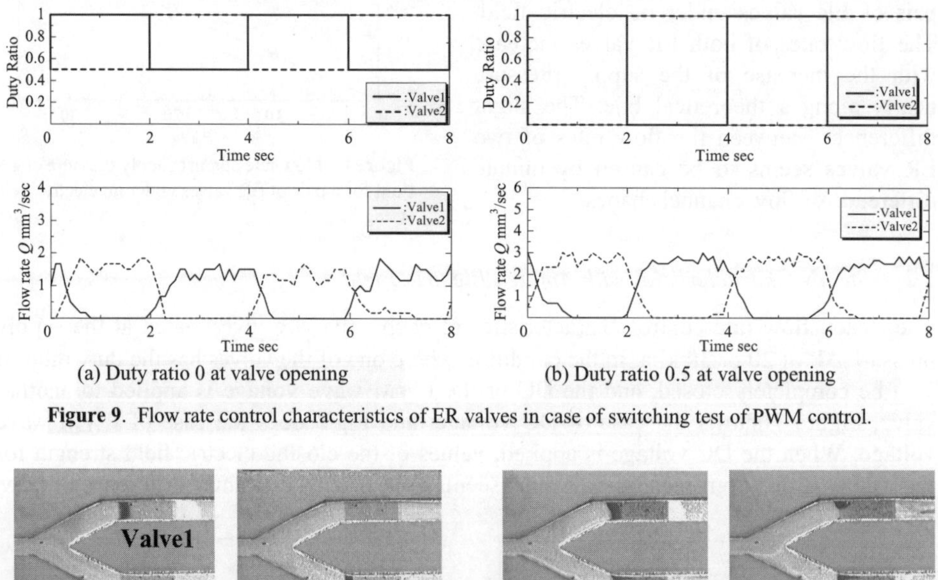

(a) Duty ratio 0 at valve opening (b) Duty ratio 0.5 at valve opening

Figure 9. Flow rate control characteristics of ER valves in case of switching test of PWM control.

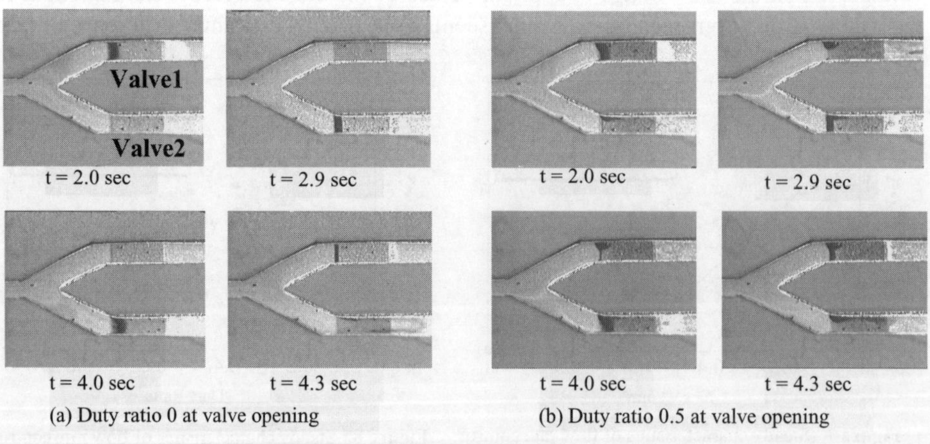

(a) Duty ratio 0 at valve opening (b) Duty ratio 0.5 at valve opening

Figure 10. Flow behavior of ER suspension within a pair of ER valves in case of switching test of PWM control be changed with the duty ratio of the PWM wave voltage. In this experiment, the supply pressure is 20 kPa, and the duty ratio at valve opening is set at a constant value, 0 for the case (a) and 0.5 for the case (b). Figure 9 shows the time variations of the duty ratio and the flow rates, and Fig.10 shows the visualized images of the flow behavior in the valves. For both cases of (a) and (b), the flow rates of the ER valves 1 and 2 at valve opening are almost constant and equivalent each other. And changing the duty ratio from (a) 0 to (b) 0.5 reduces the flow rate at valve opening to about half. Furthermore, the flow rate gradually increases or decreases just after every switching, representing some delays of valve responses in both cases.

The delay of the valve response at the switching from fully closed state to opened state depends on the time to wash out the stuffed particles between the electrodes, as seen in the valve 2 from t=4.0 sec to t=4.3 sec in Fig.10. At the switching from opened state to fully closed state, as seen in the valve 1 from t=4.0 sec to t=4.3 sec in Fig.10 the particles simultaneously form clusters to stop between the electrodes because the particle clusters are formed almost instantaneously by applying an electric field, and then some inflow particles are still gradually stuffed between the electrodes although only a little of carrier oil is flowing through among the particle clusters as observed in the valve 1 at t=4.3 sec, and finally the valve is completely closed, resulting in the delay of valve closing.

4.2 Continuously Changing Duty Ratio

Figure 11(a) shows the time variations of continuously changing duty ratio as a rectified half-sinusoidal wave and of the flow rates in the valve 1 and 2, and Fig 11(b) shows the visualized images of the flow behavior in the valves, in case of the supply pressure **ΔP**=20 kPa and the overlap duty ratio $\mathbf{D_{ov}}$=0 %. It can be seen from the lower figure in Fig. 11(a) that there is little overlap between the flow rates of both valves at every switching, i.e., no apparent delay at the switching is observed. This might be because forming or breaking of the particle clusters occurs gradually between the electrodes. It can be observed in Fig. 11(b) that both valves are closed at t=2.0 and 4.0 sec, the valve 1 is opened and the valve 2 is closed at t=2.9 sec, and the valve 1 is closed and the valve 2 is open at t=4.3 sec.

Figure 12 shows the case of **ΔP**=20 kPa and $\mathbf{D_{ov}}$=30 %. The overlaps between the flow rates of both valves are obtained at every switching as seen in the lower figure in Fig. 12(a), so that the switching at relatively high flow rate is realizable. It means that the

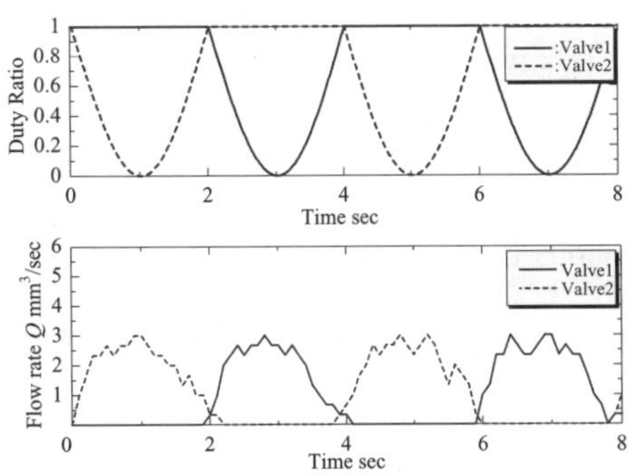

(a) Time histories of duty ratio and flow rates of a pair of ER valves

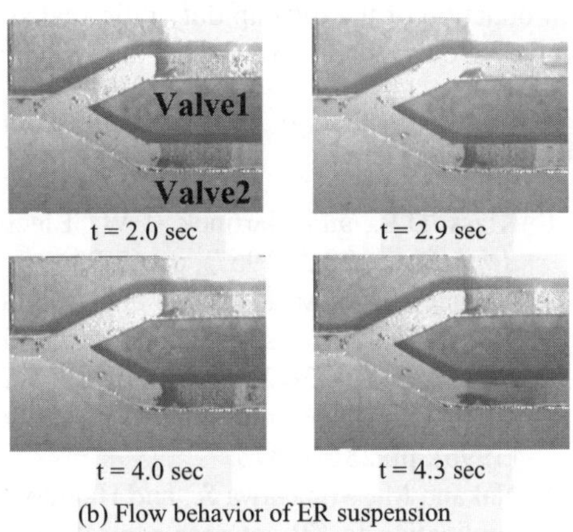

(b) Flow behavior of ER suspension

Figure 11. Flow rate control characteristic and flow behavior of a pair of micro ER valves when continuously changing duty ratio of PWM voltage (Overlap duty ratio D_{OV}=0 %).

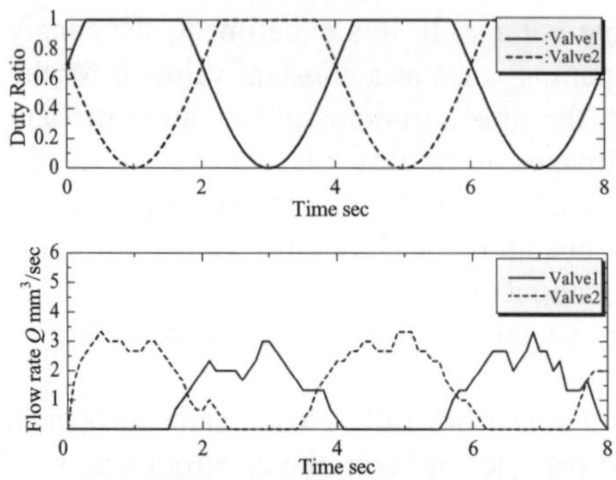

(a) Time histories of duty ratio and flow rates of a pair of ER valves

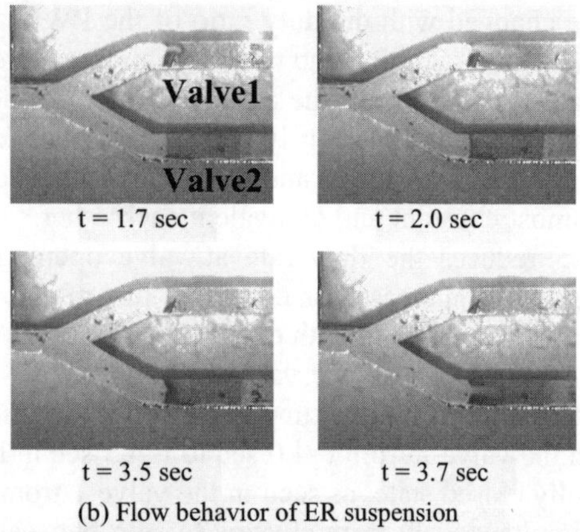

(b) Flow behavior of ER suspension

Figure 12. Flow rate control characteristic and flow behavior of a pair of micro ER valves when continuously changing duty ratio of PWM voltage (Overlap duty ratio D_{OV}=30 %.)

overlap duty ratio can change the timing of opening and closing of both valves. Maximum flow rate is almost the same as the case of $\mathbf{D_{ov}}$=0 %. Figure 12(b) exhibits that both valves are opened at t=2.0 and 3.7 sec, the valve 1 is just opened and the valve 2 has been already opened at t=1.7 sec, and the valve 1 has been already opened and the valve 2 is closed at t=3.5 sec.

5 Concluding Remarks

A three-ports micro ER valve with minute flow channels has been fabricated by a photolithography. It was found by the constant pressure experiments that the PWM control of a pair of micro ER valves could realize the continuous flow rate control from the closing to the full opening by changing the duty ratio, independent of the pressure difference, and the equivalent valve performance could be obtained even if there are some differences between the closing electric fields or the channel shapes. The continuous change of the duty ratio at the switching of a pair of ER valves could reduce the delay of the valve response observed in case of the discontinuous valve switching. The introduction of the overlap duty ratio realized the overlaps between the flow rates of both valves at every switching, that is, the switching at relatively high flow rate was realizable.

References

1. Klass, D.L., and Martinek, T.W., Electroviscous fluids. I: Rheological properties, *J. Appl. Phys.*, **38-1** (1967), pp.67-74.
2. Nakano M., Aizawa R., and Asako Y., Steady and transient responses of electrorheological suspension passing through a rectangular channel, *Int. J. of Mod. Physics B*, **10-23&24** (1996), pp.2965-2972.
3. Wolff C., Closed loop controlled ER-actuator, *Int. J. of Mod. Physics B*, **10-23&24** (1996), pp.2867-2876.
4. Nakano M., Minagawa S., and Hagino K., PWM flow rate control of ER valve and its application to ER actuator control, *Int. J. of Mod. Physics B*, **13-14, 15&16** (1999), pp.2168-2175.

ELECTRORHEOLOGICAL SEMI-ACTIVE DAMPER: POLYANILINE BASED ER SYSTEM

J. W. KIM, Y. H. CHO AND H. J. CHOI

Department of Polymer Science and Engineering, Inha University, Inchon, 402-751, Korea
E-mail: hjchoi@inha.ac.kr

H. G. LEE AND S. B. CHOI

Department of Mechanical Engineering, Inha University, Inchon, 402-751, Korea

Among various materials suitable for an electrorheological (ER) fluid, polyaniline and its many derivatives based upon modification of oxidation state, dopant and polymerization conditions are of technological interest for their adaptation into ER fluids. This is because they have better thermal stability and smaller density than other potential ER materials. Furthermore, polyaniline is easy to polymerize by oxidation polymerization at relatively low temperature and can be doped from a conducting emeraldine hydrochloride form to an insulating state using simple protonic acids. In this study, we adapted the polyaniline based ER fluids for semi-active cylindrical flow-mode type ER damper. A cylindrical flow-mode type ER damper was chosen in this study because it has similar geometrical configuration to the conventional damper normally used for passenger vehicle. The damping forces measured using a semi-active ER damper with polyaniline based ER fluids were found to be controlled by tuning an applied electric field for different piston velocity.

1 Introduction

Electrorheological (ER) fluids are those whose rheological properties can be changed by an imposed electric field. Commonly they are composed of a suspension of polarization solid particles dispersed in a non-conducting liquid. When an electric field is imposed, the rheological properties of the fluid vary, showing a characteristic fibrillation; the strings of the particles are oriented along the direction of the electric field. This structure is known to be induced by a mismatch of the dielectric constants of the suspended particles and the non-conducting oil [1,2]. Nevertheless this may not be the only mechanism, and the ER phenomenon can also be explained by a conductivity model [3]. Because of the controllable rheological properties, ER fluids can potentially be used as a smart material for active devices, which transform electric energy to mechanical energy.

Among various promising materials for ER fluids, semiconducting polymers, including poly(acene quinone) radicals [4,5], polyaniline [6,7], copolyaniline [8,9], poly(p-phenylene) [10,11], and polymer-clay nanocomposite [12-14], have been adopted as anhydrous ER fluids because of their ease in handling and superior physical properties, compared to various wet-base ER materials, including corn starch [15], silica gel [16], and a mesoporous molecular sieve [17]. Polyaniline and its many derivatives based upon modifying the oxidation state, dopant, and polymerization conditions [18] are of technological interest. This is because they have a better thermal stability and smaller density than the other polymers. Furthermore, polyaniline can be easily polymerized by an oxidation polymerization at a relatively low temperature and doped from a conducting emeraldine hydrochloride form to an insulating state using simple protonic acids.

On the other hand, ER fluids can solidify in the order of ms, and also they can fluidize under applied deformation that destroys the chain structure formed by the particles. The enormous change in the rheological properties of these fluids, coupled with the fast response time, offers many advantages in a variety of active devices including actuators, torque transducers and dampers [19]. Despite superior characteristics of the anhydrous ER fluids over wet-base ER fluids, little is known on their application into active devices.

In this study, polyaniline-based ER fluid was adapted as a smart material for a semi-active ER damper. Among the three types of ER damper, based on the working mode of the ER fluid: flow-mode type, shear-mode type, and squeeze-mode type [20], a cylindrical flow-mode type ER damper was selected because it has a similar geometrical configuration to the conventional damper normally used in passenger vehicles.

2 Experimental

2.1 *Materials*

The polyaniline (PANI) was synthesized through an oxidation polymerization. At first, 0.6 mol of aniline was added to 400 ml of 1M HCl, chilled, and stirred. A pre-chilled solution of ammonium peroxysulfate (0.36 mol in 200ml of 1M HCl) was then added dropwise to the reaction system, with continuous stirring for 1 hour. As a result, the solution color became dark-green within a few minutes of the addition. After adding drops for 2 hours, the reaction conditions were then maintained for 2 hours for the completion of the polymerization. The polymerization temperature was fixed at 0 °C. Recently, dodecylbenzenesulfonic acid-doped PANI particles were synthesized via emulsion polymerization, and their ER characteristics were investigated [21]. To obtain semiconducting polyaniline, the polyaniline particles were then dedoped by changing the pH of the aqueous medium containing the particles to pH 9.0 using an aqueous NaOH solution and HCl solution. The pH of the aqueous polyaniline suspension remained constant for one day. The pH-controlled particles were then filtered and washed using distilled water, ethanol, and cyclohexane in order to remove any oligomer and excess monomer and make the particle surface hydrophobic [22]. Finally, the products were dried in a vacuum oven at room temperature.

2.2 *Rheological measurement*

The ER fluid was prepared by dispersing the synthesized anhydrous polyaniline particles in silicone oil by 20 vol%. The density and kinematic viscosity of the silicone oil were $0.95 g/cm^3$ and 30cS at 25°C, respectively. The prepared ER fluids were then stored in desiccator prior to use. The rheological properties were determined using a rotational rheometer (Physica MC120) with Couette-type geometry, high-voltage generator, and oil bath for temperature control. The gaps of Z3-DIN are 1.06mm and its maximum stress is 1,141 Pa, respectively. Various DC fields (1~4kV/mm) are applied to the insulated bob by HVG. An electric field was applied for five minutes to obtain an equilibrium chain-like or columnar structure before applying the shear.

2.3 Damper test

A cylindrical flow-mode type ER damper was selected for the current study [20]. The ER damper was divided into an upper and lower chamber by the piston, and fully filled with ER fluid flows through a duct between the inner and outer cylinders from one chamber to the other (Fig. 1). The positive voltage produced by a high voltage generator was connected to the inner cylinder, and the negative voltage was connected to the outer cylinder. Meanwhile, the gas chamber positioned in the lower part acted as an accumulator of the ER fluid induced by the motion of the piston. The gas in the chamber was nitrogen and the gas pressure was 10 bars. The electrode gap (h) of the damper was 0.75 mm.

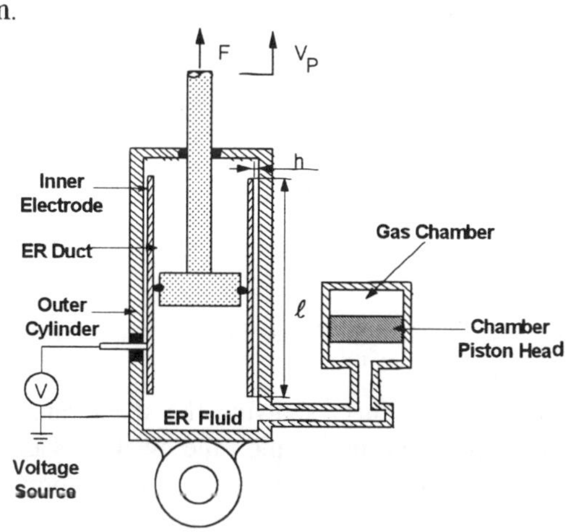

Figure 1. Schematic diagram of the ER damper.

The electric field was applied to the ER damper using a high voltage generator that had a gain of 1000 and output voltage range of ±10kV. Unlike a rheometer test, the measured ER properties, such as the damping force, were obtained without developing the electric field prior to the test. An electric heater surrounding the damper controlled the temperature of the ER damper. The excitation magnitude and frequency on the ER damper were ±20 mm and 1.89 Hz, respectively.

3 Results and Discussion

Polyaniline is regarded as being composed of structural repeating units formed by two aniline molecules. These repeating units can be formed in a reduce state or oxidized form. For a rigid main chain and the existence of dopants (e.g. Cl^-), metallic polyaniline (salt type) is infusible and insoluble in almost all organic solvents, except concentrated acids and N-methylpyrrolidone. These properties are a serious disadvantage in processability, yet good for ER fluids in which they produce thermal stability. The density of polyaniline was measured using a pycnometer at 25°C and determined as 1.30 g/cm^3. The conductivity of the polyanilines measured by the 2-probe method using compressed disks was found to be 9.0 x 10^{-12} S/cm [23].

The PANI-based ER fluid showed a general ER behavior under various conditions. Among these properties, the yield stress measured by a rotational rheometer [24] is a representative property for explaining the ER effect.

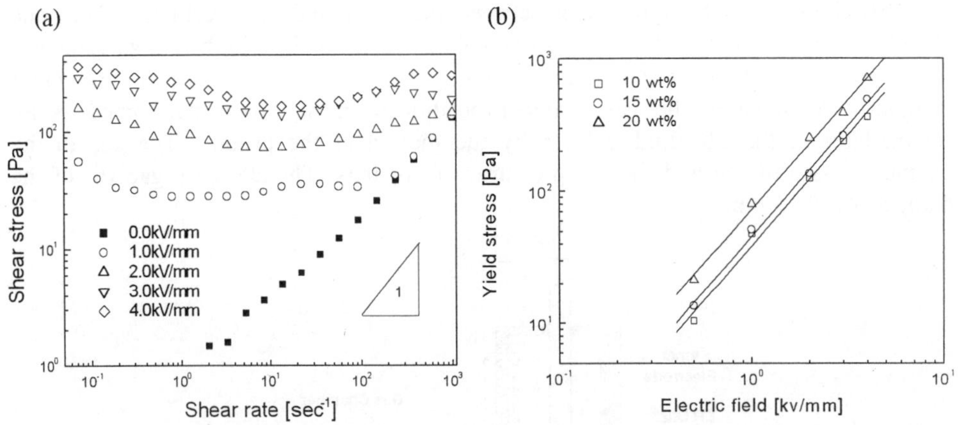

Figure 2. Shear rate *vs.* shear stress curve (a) and the change of yield stresses of PANI based ER fluids with various concentrations of suspensions (b) (20vol%, pH=9.0).

Figure 2 (a) represents a flow curve showing the shear rate *vs.* the shear stress at different electric field strengths. From this plot, the dynamic yield stress was obtained, as determined by an extrapolation from the constant stress region (plateau region). In the low shear rate region, the shear stress was sustained because the electrostatic force from polarization is considered to be larger than the hydrodynamic interaction. However, in the high shear rate region, the behavior of the flow curve became a Newtonian flow motion. This process has been previously explained based on a relaxation process from the dielectric spectrum [13]. Furthermore, the PANI based ER fluid exhibited a typical shear thinning behavior under an applied electric field.

Figure 2 (b) shows the static yield stress *vs.* electric field strength, which is representative value of common ER fluids, with different particle concentration at various electric fields. The value of yield stress is generally increased with both particle concentration and electric field strength. Recently, Choi et al. [25] suggested the scaling equation of universal yield stress for ER fluids. However, the limitation of both particle concentration and electric field strength exists. If the ER fluids possess many particles, its viscosity become too high. Increment of the initial viscosity decreases the controllable range of the ER suspension.

Figure 3 (a) presents the damping forces as a function of time at an operating temperature of 25 °C. As the electric field increased, the damping forces of the ER damper filled with the PANI-based ER fluid also increased, thereby requiring more force to move the piston. This was mainly attributed to the increment of the yield stress of the ER fluid. From this result, the damping force was determined as 210 N with a 0kV/mm electric field. In contrast, the damping force at 5kV/mm was 970 N and thus the increment of the damping force was 760 N. This result implies that tuning the electric field could continuously control a large range of the damping force of the ER damper. However, the curve shape of the PANI- based ER fluid for the damper was not well developed. It is assumed that the cause of this behavior was from a mismatch of the response time between the PANI-based ER fluid and the applied electric field.

(a) (b)

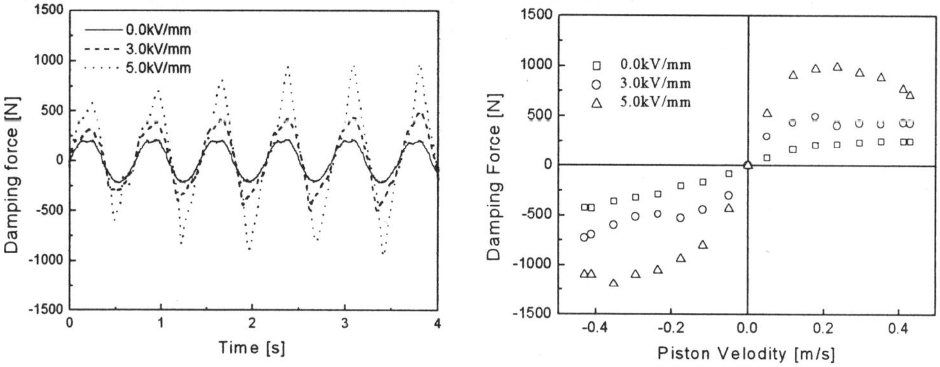

Figure 3. Damping force vs. time curves at 25°C (a) and damping force relative to piston velocity (b) (their volume fraction; 20vol/%).

Figure 3 (b) shows the electric field dependent damping force relative to the piston velocity, where increasing the excitation frequency from 0.4 to 3.0 Hz, while a fixed excitation amplitude value was maintained changing the piston velocity. The damping force increased with an increase in both the electric field and the piston velocity, except at 5kV/mm. At a higher frequency region, a decrease in the damping force occurred. This behavior was related to the chain structure of the ER fluid at a higher electric field, as the higher shear thinning occurs with a higher electric field.

Figures 4 shows the same damping force test of PANI-based ER fluid with small amount of an additive (surfactant, Span80) [26]. Nevertheless, it was much the same damping force at 5kV/mm. Besides it has shorter response time than the PANI-based ER fluid without surfactant. This is considered to be an increase of dielectric constant by surfactant, thereby, with faster polarization and faster response time.

(a) (b)

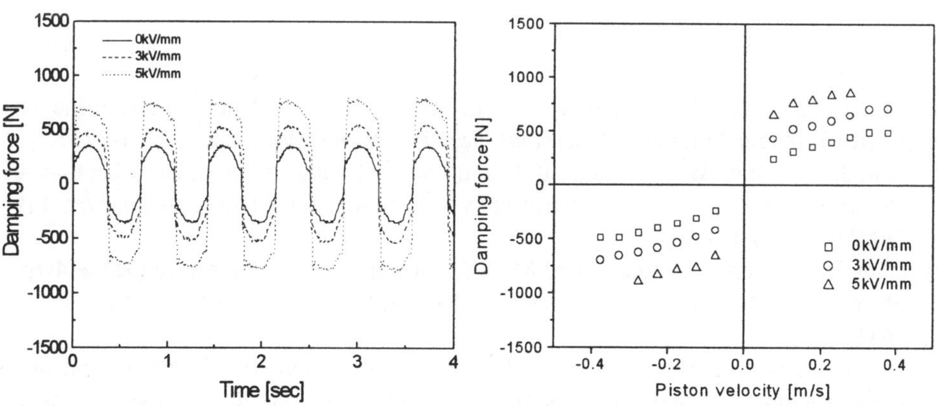

Figure 4. Damping force vs. time curves at 25°C (a) and damping force relative to piston velocity (b) with surfactant. (volume fraction; 17vol%)

4 Conclusion

A PANI-based ER fluid was synthesized and then adapted to a semi-active damper system. From both the ER properties and the damping force investigated, the synthesized PANI-based ER fluid exhibited a typical shear thinning behavior under an applied electric field. It was also found that tuning the applied electric field could control the damping force of the ER damper, and the damping force increased relative to the electric field and piston velocity, except at 5kV/mm.

5 Acknowledgements

This study was supported by research grants from the National Research Laboratory (NRL) Program (2001).

References

1. Parthasarathy M. and Klingenberg D. J., Electrorheology: Mechanisms and Models, *Mater. Sci. Eng.* **R17** (1996) pp.57–103.
2. Tao R., Zhang J., Shiroyanagi Y., Tang X. and Zhang X., Electrorheological Fluids under Shear, *Int. J. Mod. Phys. B* **15** (2001) pp.918-929.
3. Gonon P., Foulc J. N. and Atten P., A Conduction Model Describing Particle-Particle Interaction in the Case of Surface Conducting Particles, *Int. J. Mod. Phys. B* **15** (2001) pp.704-713.
4. Choi H. J., Cho M. S. and Jhon M. S., Hysteresis Behaviors of Poly(naphthalene quinone) radical Electrorheological fluid, *Int. J. Mod. Phys. B* **13** (1999) pp. 1901-1907.
5. Block H., Kelly J. P., Qin A. and Watson T., Materials and Mechanisms in Electrorheology, *Langmuir* **6** (1990) pp. 6-15.
6. Choi H. J., Cho M. S. and To K., Electrorheological and Dielectric Characteristics of Semiconductive Polyaniline-Silicone Oil Suspensions, *Physica A* **254** (1998) pp. 292-299.
7. Grow C. J. and Zukoski C. F., The Electrorheological Properties of Polyaniline Suspensions, *J. Colloid Interface Sci.* **136** (1989) pp. 175-188.
8. Cho M. S., Choi H. J. and To K., Effect of ionic pendent of polyaniline on electrorheological property, *Macromol. Rapid Commun.* **19** (1998) pp. 271-273.
9. Kim J. W., Jang W. H., Choi H. J. and Joo J., Synthesis and electrorheological characteristics of polyaniline derivatives with different substituents, *Synth. Met.* **119** (2001) pp. 173-174.
10. Choi H. J., Sim I. S. and Jhon M. S., Aqueous Ferric Chloride doped Poly(p-phenylene) for Electrorheological Material, *J. Mater. Sci. Lett.* **19** (2000) pp. 1629-1631.
11. Sim I. S., Kim J. W., Choi H. J., Kim C. A. and Jhon M. S., Preparation and Electrorheological Characteristics of Poly(p-phenylene) Based Suspensions, *Chem. Mater.* **13** (2001) pp. 1243-1247.

12. Kim J. W., Choi H. J. and Jhon M. S., Synthesis and Electrorheological Characterization of Emulsion Polymerized SAN-clay Nanocomposite Suspensions, *Macromol. Symp.* **155** (2000) pp. 229-237.
13. Kim B. H., Jung J. H., Kim J. W., Choi H. J. and Joo J., Physical Characterization of Polyaniline-Na+-montmorillonite Nanocomposite Intercalated by Emulsion Polymerization, *Synth. Met.* **117** (2001) pp. 115-118.
14. Choi H. J., Kim J. W., Joo J. and Kim B. H., Synthesis and electrorheology of emulsion intercalated PANI-clay nanocomposites, *Synth. Met.* **121** (2001) pp. 1325-1326.
15. Kobayashi K., Ogura J., Kodama R., Sakai T. and Sato M., The Flow Rate and Current Density of ERF (Corn Starch – Transformer Oil) Flowing Through an Annulus, *Int. J. Mod. Phys. B.* **13** (1999) pp. 1940-1948.
16. Wu C. W. and Conrad H., Electrorheology of Suspensions of Si Particles with an Oxide Film in Silicone Oil, *Int. J. Mod. Phys. B.* **13** (1999) pp. 1713-1720.
17. Choi H. J., Cho M. S., Kang K. K. and Ahn W. S., Electrorheological Characteristics of Mesoporous Molecular Sieve (MCM41) Suspension, *Micropor. Mesopor. Mater.* **39** (2000) pp.19-24.
18. Kang E. T., Neoh K. G. and Tan K. L., Polyaniline: A Polymer with Many Interesting Intrinsic Redox States, *Prog. Polym. Sci.* **23** (1998) pp. 277-324.
19. Shimada K., Nishida H. and Fujita T., Differences in Steady Characteristics and Response Time of ERF on Rotational Flow between Rotating Disk and Concentric Cylinder, *Int. J. Mod. Phys. B* **15** (2001) pp. 1050-1056.
20. Lee H. G., Choi S. B., Han S. S., Kim J. H. and Suh M. S., Bingham and Response Characteristics of ER Fluids in Shear and Flow Modes, *Int. J. Mod. Phys. B* **15** (2001) pp. 1017-1024.
21. Kim S. G., Kim J. W., Cho M. S., Choi H. J. and Jhon M. S., Viscoelastic Characterization of Semiconducting Dodecylbenzenesulfonic Acid Doped Polyaniline Electrorheological Suspensions, *J. Appl. Polym Sci.* **79** (2001) pp. 108-114.
22. Choi H. J., Kim T. W., Cho M. S., Kim S. G. and Jhon M. S., Electrorheological Characterization of Polyaniline Dispersions, *Eur. Polym. J.* **33** (1997) pp. 699-703.
23. Lee J. M. and Lim K. H., Electrochemical Synthesis of Conducting Polythiophene in an Ultrasonic Field, *J. Ind. Eng. Chem.* **6** (2000) pp. 157-162.
24. Kim S. G., Kim J. W., Jang W. H., Choi H. J. and Jhon M. S., Electrorheological Characteristics of Phosphate Cellulose-Based Suspensions, *Polymer* **42** (2001) pp. 5005-5012.
25. Choi H. J., Cho M. S., Kim J.W., Kim C. A. and Jhon M. S., A Yield Stress Scaling Function for Electrorheological fluids, *Appl. Phys. Lett.* **78** (2001) pp.3806-3808.
26. Lee H. J., Chin B. D., Yang S. M. and Park O. O., Surfactant Effect on the stability and Electrorheological Properties of Polyaniline Particle Suspension, *J. Colloid Int. Sci.* **206** (1998) pp.424-438.

USING OF FERROMAGNETIC PARTICLES SUSPENSION AS CONTROLLABLE DISSIPATIVE ENVIRONMENT

V.G. BASHTOVOI, D.N. KABACHNIKOV

Belarussian State Polytechnic Academy, Belarus, 220027, Minsk, 65 F.Skaryna Ave.
E-mail: bashv@itmo.by, kabachnikov@yahoo.com

G. BOSSIS

Universite de Nice, Parc Valrose, 06108 Nice, France
E-mail: bossis@unice.fr

Magneto-rheological (MR) suspensions are actively used in damping devices. As a rule suspensions are magnetic particles in a nonmagnetic environment. The control and field influence are carried out only at the expense of the MR effect. In this article the dissipative environment on a magnetic fluid basis is presented. This gives opportunity to use additional controlling mechanism by this environment at the expense (using) ponderomontive force. It will allow to create the local volume of magnetic environment in the given local place. The attention is given to account of the dissipation force at theoretical investigation of the plate oscillation and corresponding calculation of the damping decrement.

1 Introduction

It is well known, a MR effect with a large increase of the apparent viscosity in the presence of an external magnetic field is observed in suspensions of micronic magnetic particles [1]. Hence the use of MR suspensions gives the possibility of an effective control of the damper characteristics by the external magnetic field [2].

The influence of the external magnetic field and properties of the suspensions on the damping of the elastic plate free oscillation is studied theoretically and experimentally. The free end of the plate is placed between two drops of MR suspension as shown in Figure 1, which are retained and controlled by the external magnetic system and carry out the damping functions.

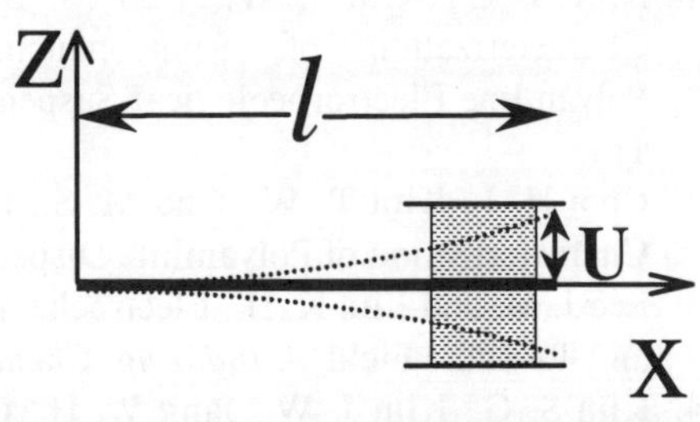

Figure 1. Model and geometry of the plate oscillations problem.

2 Theory

2.1 *Calculation of dissipation force in the suspension*

Calculate the viscous dissipation force F in the drop of the MR suspension with the radius R,

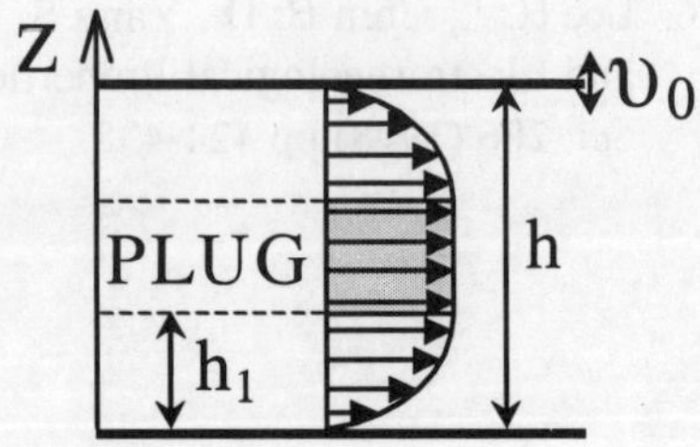

Figure 2. Geometry of the flow and velocity distribution.

submitted to the oscillations of the upper plate. Consider a layer of suspension with dynamic viscosity η and yield stress τ_0 between two flat parallel plates spaced by a distance h (see figure 2). One plate is stationary and the other one moves in the perpendicular direction with the velocity $|\upsilon_0|>0$. The Bingham plastic flow model is used to describe the shear stress-shear rate relation of MR fluid

$$\tau=\tau_0\text{sign}\left(\frac{d\upsilon}{dz}\right)+\eta\frac{d\upsilon}{dz} \qquad |\tau|>\tau_0 \tag{1.1}$$

$$\frac{d\upsilon}{dz}=0 \qquad |\tau|\le\tau_0 \tag{1.2}$$

Suspension is put in non uniform (10 Tl/m) along axes z magnetic field. In this case τ_{0zr} component yield stress much more then τ_{0zz} component yield stress ($\tau_{0zr}>>\tau_{0zz}$). Taking into consideration that the thickness of the sheared region on each side of the plug core (see figure 2) $h_1<<R$ [3], so τ_{0zr} component yield stress much more then τ_{0rr} component yield stress ($\tau_{0zr}>>\tau_{0rr}$). (Later everywhere $\tau_0=\tau_{0zr}$). The governing equation [4] is simplified then

$$\frac{dp(r)}{dr}=\frac{\partial\tau(r,z)}{\partial z} \tag{2}$$

The pressure gradient can be derived from Eqs. (1) and (2) with no-slip boundary conditions. A global mass balance is given by equating the radial mass flow with the rate of change of the fluid volume which leads to $\upsilon=\pm\frac{3\upsilon_0 rz(2h_1-z)}{2h_1^2(3h-2h_1)}$ (for the lower part of the sheared region) and

$$\frac{dp}{dr}=\mp\frac{3\upsilon_0\eta r}{h_1^2(3h-2h_1)} \tag{3}$$

(here and later the upper sign corresponds to compressibility of fluid drop, the lower sign corresponds to its extension) where h_1 is the thickness of the sheared region on each side of the unyielded core (see figure 2 4) is the solution of the equation

$$h_1^3-\frac{3}{2}hh_1^2-\frac{3}{2}\frac{\upsilon_0\eta r}{\tau_0}\left(h_1-\frac{h}{2}\right)=0. \tag{4}$$

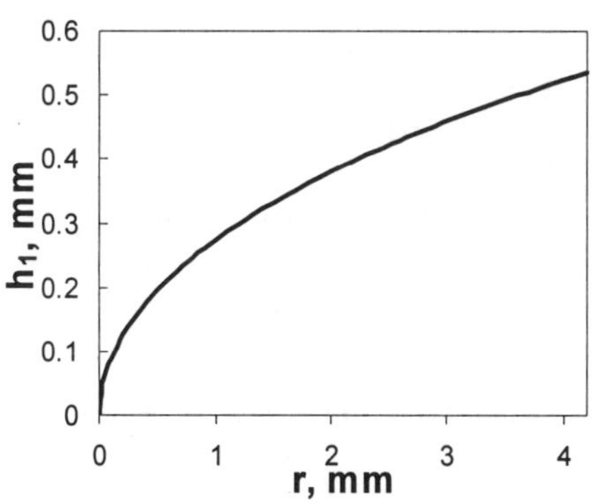

Figure 3. Diagram of the height of yield region h_1 on the radius of the MR suspension drop r (h=4.6mm, τ_0=100 Pa, η=0.28 Pa·s, υ_0=59 mm/s).

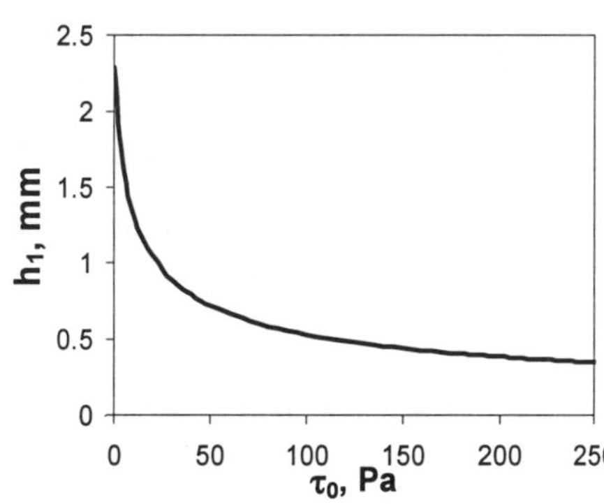

Figure 4. Dependence of the height of yield region h_1 on the yield stress τ_0 (h=4.6mm, r=4.2 mm, η=0.28 Pa·s, υ_0=59 mm/s).

Defining a dimensionless pressure drop $y=\frac{h}{2\tau_0}\frac{dp}{dr}$ and dimensionless parameter $W=\frac{|\upsilon_0|\eta r}{h^2\tau_0}$, Eq. (3) can be transformed into an equivalent third-order form, which is

$$\pm y^3+3\left(W+\frac{1}{2}\right)y^2-\frac{1}{2}=0. \tag{5}$$

Since in the experiment W<<1, so Eq. (5) can be solved empirically as [5]

$$y=\mp 1\mp(2W)^{0.5}. \tag{6}$$

The normal force on the plate placed between two fluid drops is obtained by integrating the pressure along the area of the plate

$$F=-\left(F_c+C\left|\frac{\partial J}{\partial t}\right|^{0.5}\right)\text{sign}\left(\frac{\partial J}{\partial t}\right) \tag{7}$$

where $F_c=\frac{8\pi\tau_0 R^3}{3h}$, $C=\frac{32\pi(2\tau_0\eta)^{0.5}R^{3.5}}{21h^2}$, J – the deviation of the plate.

2.2 *Calculation of decrement of the plate oscillation damping*

The minor free oscillations of a thin elastic plate with one end fixed and the other one free [6] with taking into account dissipation force (6) are described by the following equation:

$$\frac{\partial^2 U}{\partial t^2}=-\Omega^2\frac{\partial^4 U}{\partial x^4}-\frac{fl}{m}, \tag{8}$$

with the boundary conditions:

$$U\Big|_{x=0}=0,\quad \frac{\partial U}{\partial x}\Big|_{x=0}=0,\quad \frac{\partial^2 U}{\partial x^2}\Big|_{x=l}=0,\quad \frac{\partial^3 U}{\partial x^3}\Big|_{x=l}=0$$

and the initial conditions:

$$U\Big|_{t=0,x=l}=A,\quad \frac{\partial U}{\partial t}\Big|_{t=0}=0,$$

where, U(x,t) – the deviation of the plate, x – the longitudinal co-ordinate, t – the time, Ω – coefficient determining fundamental frequency of plate oscillation without damping, l – the plate length, m – the plate mass, A – the initial deviation of plate free end, f – distributed force connected with F as $\int_0^l f dx=F\left(\frac{\partial U}{\partial t}\Big|_{x=x_0}\right)$ (x_0 – the longitudinal co-ordinate of the plate corresponding to the center of the drop).

Expected $U(x,t)=U_0(x)U_1(t)$, and $U_0(x)$ is same that in absent of dissipation force in (8) and is equal to [7]

$$U_0(x) = -A\frac{(\cos\chi l + \mathrm{ch}\chi l)(\cos\chi x - \mathrm{ch}\chi x) + (\sin\chi l - \mathrm{sh}\chi l)(\sin\chi x - \mathrm{sh}\chi x)}{2\sin\chi l\,\mathrm{sh}\chi l}, \quad (9)$$

where $\chi=1.875/l$. So $f = U_0(x)f_0$ where $\frac{\partial f_0}{\partial x} = 0 \Rightarrow$

$$f = \frac{U_0(x)F\left(\left.\frac{\partial U}{\partial t}\right|_{x=x_0}\right)}{\int_0^l U_0(x)dx} \quad (10)$$

If K is the deviation of plate point $x=x_0$ so $K = U(x_0) = U_0(x_0)U_1$ then Eq. (8) can be transformed into equation of motion for this point

$$\frac{\partial^2 K}{\partial t^2} + \omega_0{}^2 K + \frac{l}{m}\frac{U_0(x_0)}{\int_0^l U_0(x)dx}\left(F_c + C\left|\frac{\partial K}{\partial t}\right|^{0.5}\right)\mathrm{sign}\left(\frac{\partial K}{\partial t}\right) = 0 \quad (11)$$

with the initial conditions:

$$K_{t=0} = U_0(x_0), \qquad \left.\frac{\partial K}{\partial t}\right|_{t=0} = 0,$$

where $\omega_0=3.52\cdot\Omega/l^2$ – the frequency of the plate free oscillations without dissipation.

The oscillations of the plate are damped due to the viscous damping inside the magnetic fluid. The decrement of oscillations D is obtained by taking the logarithm of the ratio of two oscillation amplitudes:

$$D = \frac{T}{t_{ij}}\ln\left(A_i/A_j\right), \quad (12)$$

where A_i and A_j are the amplitudes of the i-th and j-th oscillations correspondingly, t_{ij} – time between these amplitudes, T – period of oscillations.

As the equation of motion (11) includes the dependence on the yield stress τ_0 so decrement also depends on yield stress (figure 5).

Expected the yield stress τ_0 of the MR fluid obeys a power law with the magnetic field induction B, i.e.

$$\tau_0 = GB^n \quad (13)$$

where constants G and n can be derived from experiments. So dependence decrement D on the magnetic field induction B can be found.

3 Experiment

The elastic brass plate with dimensions 131x36.5x1.35 mm and mass 128 g (including mass of the sensor) was fixed on one end. The other free end was placed in the slot between two poles of a closed electromagnet system. The frequency of free oscillations of the plate ω_0 is equal to 143 s^{-1}. The slots between the plate and the poles of the electromagnet were filled with the MR suspension, which is retained in the slot by the magnetic field of the electromagnet. The field intensity inside the slot varied within the limits from 0 to 80 mTl. Two kind of MR suspensions were used. The both ones consisted of the magnetic fluid based on transformer oil (magnetization of saturation M_s=44.8 kA/m, viscosity η=0.10 Pa·s) and 19 % of mass of the magnetic powders. The first powder consisted of the Fe particles of size 5 – 10 μm and M_s = 815 kA/m, the second one – the ferric oxide particles of size 1–2 μm and M_s = 380 kA/m. Viscosity of the first suspension (S1) was equal 0.28 Pa·s and the second (S2) – 0.23 Pa·s. The total fluid volume as under and over the plate was equal to 0.5 cm^3. The slot size between any poles of the electromagnet and plate was fixed at 4.6 mm. The initial deviation of plate free end was equal to 1 mm. The piezoelectric acceleration sensor fixed on the vibrating plate allows to get the displacement. The electrical signal is amplified, digitized with the help of an analogy-digital card: National Instrument AT-MIO-64E and recorded on a PC.

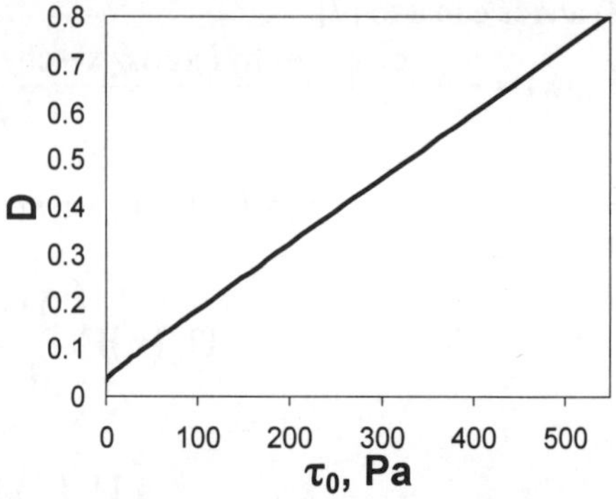

Figure 5. Dependence of the decrement D on the yield stress τ_0 (η_{S1}=0.28 Pa·s, A_i/A_j=10).

4 Discussion.

Under the action of the magnetic field these particles form elongated aggregates in the fluid which considerably increase the yield stress of the fluid. The experimental results which demonstrate this effect are presented in Figure 6. The decrement increases practically linearly with the magnetic field induction B. At the maximum field it is by an order of magnitude higher than the corresponding value obtained in the pure magnetic fluid. The lines are the theoretically dependencies calculated using Eqs. (11)-(13). Here, n chosen equal to be 1.3 for suspension S1 and 1 for suspension S2, A_i/A_j=10. From Eq. (13) the found value of G is $2.4 \cdot 10^4$ for suspension S1 and $4.5 \cdot 10^3$ for suspension S2. The calculation results under

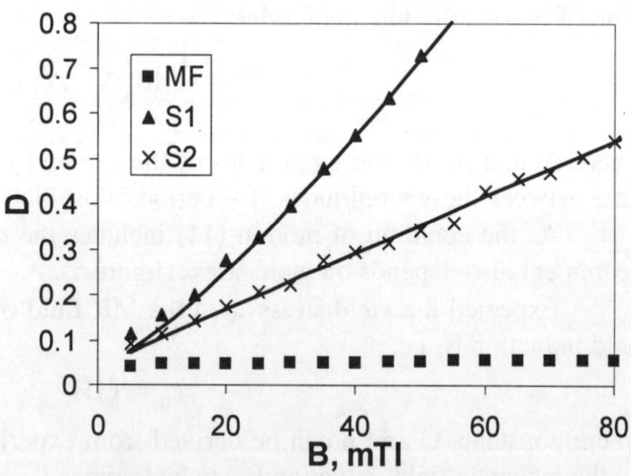

Figure 6. Dependence of the decrement D on magnetic field induction B .

such parameters fit the experiments pretty well in the large range of the magnetic field induction.

When plate velocity is equal to zero the dissipation force is equal to zero too ($F|_{\frac{\partial U}{\partial t}=0} = 0$). But when velocity change from zero to negligible quantity the dissipation force changes to F_c by jump. So motion of the plate is spasmodic. The break of velocity dependencies when velocity change sign is well seen on the Figure 7. This break is more evident for high yield stress and for low displacement. At these values the elastic force of the plate is comparable with F_c. The plate stop if displacement of the plate is less than critical $|U| \le U_c$ at the time moment when velocity is equal zero $\frac{\partial U}{\partial t} = 0$ (see figure 7). Calculate critical displacement of the plate at the point $x=x_0$ using (11)

$$K_c = \frac{l}{m\omega_0^2} \frac{U_0(x_0)}{\int_0^l U_0(x)dx} F_c \tag{14}$$

So for low velocity the maximum/minimum of this point velocity (and the maximum/minimum of plate velocity too) moves from the time moment when K=0 to $K = \mp K_c$. For high velocity its maximum/minimum is hardly earlier (at higher K) , in view of correction $C\left|\frac{\partial K}{\partial t}\right|^{0.5}$ (see (7)). But the maximum error differs from K_c less than 10%.

Dependence of the plate energy on time is presented on figure 8 where

$$E = E_k + E_p, \quad E_k = \frac{m}{2l} \frac{\int_0^l U_0^2(x)dx}{U_0^2(x_0)} \left(\frac{\partial K}{\partial t}\right)^2, \quad E_p = \frac{m\omega_0^2}{2l} \frac{\int_0^l U_0^2(x)dx}{U_0^2(x_0)} K^2. \tag{15}$$

Introduction in (8) distributed dissipation force (10) simplifies search of the problem solution. But evidently, it has error. At the experiment the dissipation force acts

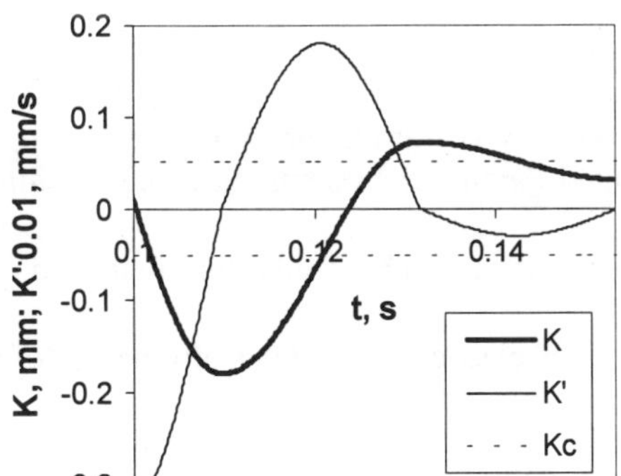

Figure 7. Dependence of the displacement K and velocity K′ of the plate at the point $x=x_0$ on time (τ_0=550 Pa).

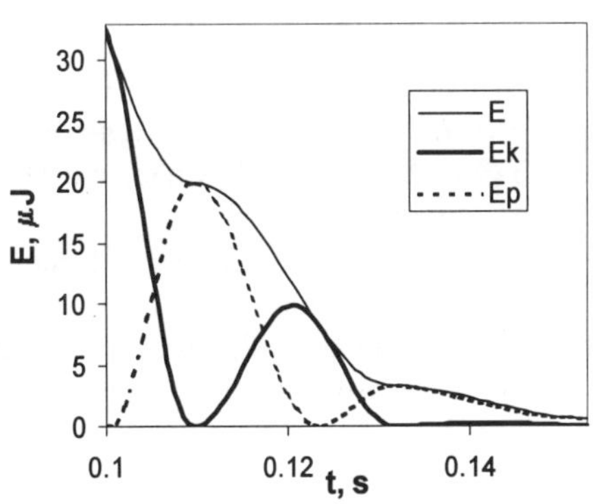

Figure 8. Dependence of the plate energy on time (τ_0=550 Pa).

locally at the point $x=x_0$. This moment changes distribution $U_0(x)$. For example, at the experiment for calculation critical displacement of the plate at the point $x=x_0$

$$K_c = \frac{(x_0/l)^3(\chi l)^4}{3m\omega_0^2}F_c \tag{16}$$

should be used [8]. The use of (16) increase of K_c by 14%. From the other point of view using local force, the calculation of the potential energy of the plate E_p leads to

$$E_p = \frac{3m\omega_0^2}{(x_0/l)^3(\chi l)^4}K^2, \tag{17}$$

which gives the same result as (15.3).

Thus, a very efficient control of the decrement value with a magnetic field which is very easy to obtain with a small coil is possible and explained theoretically.

5 Conclusion.

We have demonstrated the possibility to design a very efficient controllable damper by using drops of magnetic fluid containing micronic magnetic particles as damping element. Accounts of dissipation force at theoretical investigation of the plate oscillation and in corresponding calculation of the damping decrement provide well correspondence between theory and experiment.

References

1. Magnetic Fluids and Applications Handbook. Eds. B.M. Berkovski, V.G. Bashtovoi. Begell House Inc.Publishers, NY, 1996. – 851 p.
2. V. G. Bashtovoi, D. N. Kabachnikov, A. G. Reks, L. V. Suloeva, G. Bossis, O. Y. Volkova. *Magnetohydrodynamics* 36 (2000) 233-240.
3. J. P. Coulter, K. D. Weiss, J. D. Carlson, *J. of Intelligent Material Systems and Structures*. 4 (1993) 248.
4. L. D. Landau, E. M. Lifchitz. Physique theorique. T.6. Mecanique des fluides. Ed. 2, Moscow: Mir, 1989
5. X. Tang, X. J. Wang, W. H. Li, P. Q. Zhang. *Proceeding of 6th Intern. Conf. On Electrorheological fluids, Magnetorheological suspensions* (1998) 870-877.
6. – 329 p. (from book ?, Chapter 6 Distributed-Parameter Systems, Sec. 6.5 Bending vibration of a beam.)
7. L. D. Landau, E. M. Lifchitz. Physique theorique. T.7. Theorie de l'elasticite. Ed. 2, Moscow: Mir, 1990.
8. – 90 p. (from book ?, Chapter 2 Properties of elastic solids, 2.9 Deformation of a beam by distributed loads.)

SEMIACTIVE VIBRATION CONTROL OF A SMART SEAT WITH AN MR FLUID DAMPER CONSIDERING ITS TIME DELAY

CHANHO PARK AND DOYOUNG JEON

Dept. of Mechanical Engineering, Sogang University, Sinsudong, Mapoku
Seoul, 121-742, Korea

The Lyapunov type robust control is applied to the semiactive vibration control of a controllable seat system to accommodate the modeling uncertainty of system and time delay of the damper. The seat system has a flexible mechanism of a spring and MR fluid damper between a seat cushion and base. Simulation results show that the robust control considering the time delay of the damper suppresses the seat vibration effectively.

1. Introduction

In an effort to reduce the vibration of a bus driver's seat, the flexible mechanism composed of a spring and MR fluid damper was suggested. In case of buses and trucks, the vehicle suspension can not be soft enough to have a comfortable working environment because it may cause the instability of the vehicle [1,2].

The Lyapunov type robust control has been researched by Breinl and Leitmann [3] and Reithmeier and Leitmann [4]. It can reduce vibration under modeling uncertainties. In this research, the robust control is applied to effectively reduce the seat vibration.

In following sections, the smart seat system is introduced and the possible uncertainties are explained. The stability analysis and simulation results are given.

2. Smart Seat and MR Fluid Damper

Fig. 1 shows the schematic plot of the seat system. As the seat cushion moves up and down, the mechanism becomes a four-bar linkage with a slider. The slider has an end stop to have the vertical motion up to 60mm. The system is used in simulations.

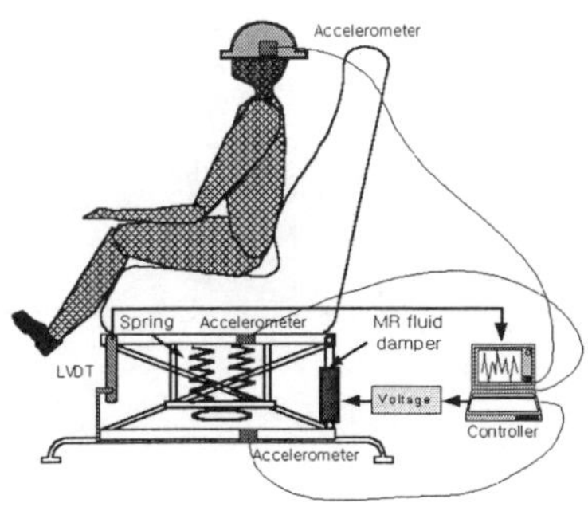

Fig. 1. Schematic plot of the smart seat

Fig. 2 shows the MR fluid damper in consideration. It has a stroke of 60mm and its diameter is 45mm. To measure the damping force, one end of the damper was connected to a loadcell and the other end was attached to a CNC machine which could generate constant and various velocities. The position and velocity of piston and the loadcell measurement were stored in a PC through a data acquisition board.

Fig. 3 shows the force measurement when the applied voltage changes in a step keeping the compression rate constant. When the voltage is from on to off, the time delay was 19msec. When the voltage is from off to on, the time delay was 28msec. The various time delay is considered as an uncertainty in the analysis.

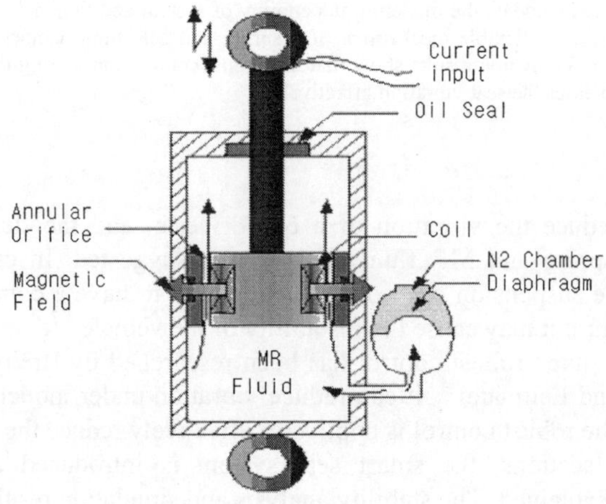

Fig. 2. The MR fluid damper

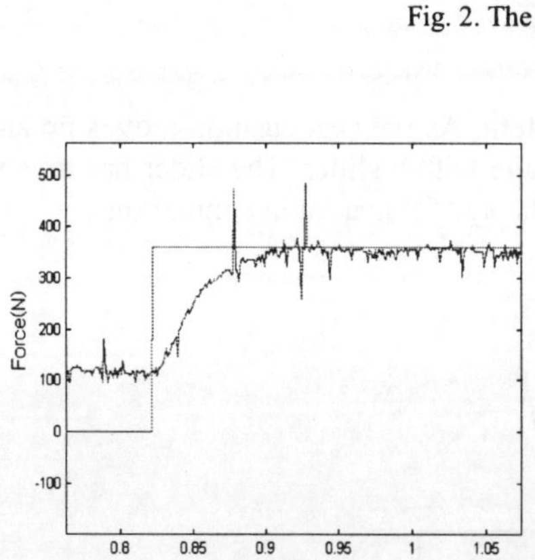

(a) From off to on

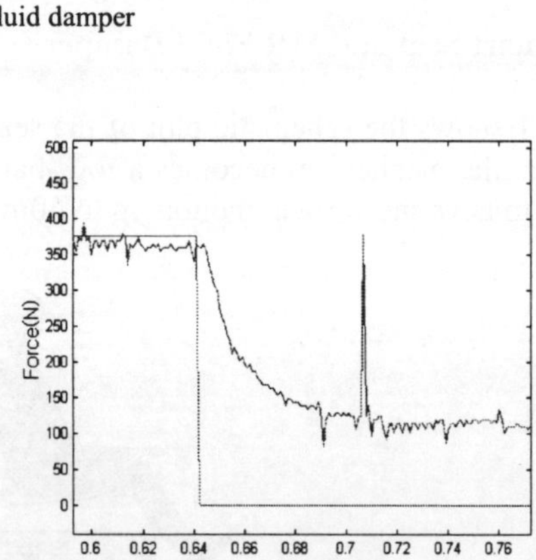

(b) From on to off

Fig. 3. Time delay of an MR fluid damper

3. Lyapunov Type Robust Control

The seat system in Fig. 1 may be represented as

$$m\ddot{x} + k(x - w) + c_p(\dot{x} - \dot{w}) + F = 0 \tag{1}$$

where m is the mass of seat and driver, k is the stiffness of spring and accumulator, c_p is the damping constant with zero current, x is the position of seat cushion, w is the position of base, i.e., vehicle chassis and F is the variable damping force. The time delay of the damper is represented as a first order differential equation as

$$\dot{F} = -\frac{1}{\tau}F + \frac{1}{\tau}F_c \tag{2}$$

where τ is the time delay.

When the state variable is defined as $X = [x \quad \dot{x} \quad F]^T$, the state equation of the smart seat system is described by

$$\dot{X} = AX + BF_c + E(x,w,t) \tag{3}$$

where

$$A = \begin{pmatrix} 0 & 1 & 0 \\ -\frac{k}{m} & -\frac{c_p}{m} & -\frac{1}{m} \\ 0 & 0 & -\frac{1}{\tau} \end{pmatrix}, \quad B = \begin{pmatrix} 0 \\ 0 \\ \frac{1}{\tau} \end{pmatrix}, \quad E(x,w,t) = \begin{pmatrix} 0 \\ \frac{k}{m}w + \frac{c_p}{m}\dot{w} \\ 0 \end{pmatrix}. \tag{4}$$

The linear part $A \in R^{3\times 3}$ is the constant stable matrix, $B \in R^{3\times 1}$ is the control input matrix and $E(x,w,t) \in R^{3\times 1}$ is the disturbance matrix of system. E is assumed to be bounded.

When the system is represented as in Eq. (3), a Lyapunov function candidate V for any symmetric positive definite matrix P is expressed as

$$V = \frac{1}{2}X^T PX. \tag{5}$$

The rate of change of energy can be regarded as

$$\dot{V} = \frac{1}{2}(X^T P\dot{X} + \dot{X}^T PX). \tag{6}$$

Substituting Eq. (3) into Eq. (6) results in

$$\dot{V} = -\frac{1}{2}X^T QX + X^T PBF_c + X^T PE \tag{7}$$

where $Q = -(PA + A^T P)$. When A is stable and Q positive definite, there is a unique symmetric positive definite matrix P [5].

When F_c is determined as

$$F_c = \begin{pmatrix} F_{\min} & if\ X^T PB > 0 \\ F_{\max} & if\ X^T PB < 0 \end{pmatrix} \tag{8}$$

it is proved by Reithmeier and Leitmann [4] that $\dot{V} < 0$, $\forall\ \|X\| > r$ where $r = \dfrac{2\varepsilon\lambda_{\max}(P)}{\lambda_{\min}(Q)}$ and $\|E\| \leq \varepsilon$. And the radius ρ of a ball of ultimate bound is therefore determined by

$$\rho = r\sqrt{\frac{\lambda_{\max}(P)}{\lambda_{\min}(Q)}}. \tag{9}$$

In case of the smart seat system, it is assumed that the uncertainties mainly come from the variation of driver's weight and time delay of the damper. Then the system equation in Eq. (3) can be represented as

$$\dot{X} = \hat{A}X + \hat{B}F_c + \hat{E} + \Delta E \tag{10}$$

where ^ defines the nominal values. Then the modeling uncertainty ΔE is expressed as

$$\begin{aligned} \Delta E &= (A - \hat{A})X + (B - \hat{B})F_c + (E - \hat{E}) \\ &= \{(\alpha - 1)A_1 + (\beta - 1)A_2\}X + (\beta - 1)\hat{B}F_c + (\alpha - 1)\hat{E} \end{aligned} \tag{11}$$

where $\alpha = \hat{m}/m$, $\beta = \hat{\tau}/\tau$, $A_1 = \begin{pmatrix} 0 & 0 & 0 \\ -\dfrac{k}{\hat{m}} & -\dfrac{c_p}{\hat{m}} & -\dfrac{1}{\hat{m}} \\ 0 & 0 & 0 \end{pmatrix}$ and $A_2 = \begin{pmatrix} 0 & 0 & 0 \\ 0 & 0 & 0 \\ 0 & 0 & -\dfrac{1}{\hat{\tau}} \end{pmatrix}$. (12)

Then the Eq. (10) can be rewritten as

$$\dot{X} = \hat{A}X + \beta\hat{B}F_c + \alpha\hat{E} + \{(\alpha - 1)A_1 + (\beta - 1)A_2\}X . \tag{13}$$

The third and fourth term is defined as a disturbance $\|E_{sys}\|$ and

$$\|E_{sys}\| < \alpha\varepsilon_0 + \{(\alpha - 1)\varepsilon_1 + (\beta - 1)\varepsilon_2\}\|X\| \tag{14}$$

where $\|\hat{E}\| < \varepsilon_0$, $\|A_1\| < \varepsilon_1$ and $\|A_2\| < \varepsilon_2$.

Then the system is stable for [4]

$$\|X\| > \frac{2\alpha\varepsilon_0\lambda_{\max}(P)}{\lambda_{\min}(Q) - 2\lambda_{\max}(P)\{(\alpha - 1)\varepsilon_1 + (\beta - 1)\varepsilon_2\}}. \tag{15}$$

4. Simulation Results

The performance of the robust controller was evaluated in simulations under the following conditions.

(i) For a realistic result, the driver is assumed to have four degree of freedom [6]; head, upper torso, viscera and lower torso.
(ii) The vehicle speed is 60km/hour.
(iii) The sampling time of the controller is 4msec and the integration interval of the Runge-Kutta algorithm is 40μsec.
(iv) The Hanning window is applied with 600 truncation points in order to smoothen the PSD graph.

The parameters for simulations are as follows

$$m = m_{seat} + m_{driver} = 14 + 70 = 84(kg)$$
$$k = k_{spring} + k_{accumulator} = 11000 + 6100 = 17100(N/m)$$
$$\tau = 0.028(\sec)$$
$$c_p = 500(Ns/m) . \qquad (16)$$

ISO 2631 [7] is the evaluation method of human exposure to whole body vibration in the frequency range of 1 to 80 Hz. In simulations, the root mean square of acceleration of the driver's head is plotted in the frequency domain. Fig. 4 shows the comparison of several control algorithms such as passive method, skyhook control, Lyapunov type robust control and Lyapunov type robust control considering the time delay of the damper. In the frequency range of 4 and 8 Hz where human is very sensitive, the Lyapunov type robust control considering the time delay suppresses the vibration most effectively.

Fig. 5 compares the SEAT [8] (Seat Effective Amplitude Transmissibility) values for these controls. SEAT is defined as the ratio of the power spectral density between the seat cushion and base in the frequency range of 0.5 and 80Hz. It is shown that the Lyapunov type robust control considering the time delay is most effective.

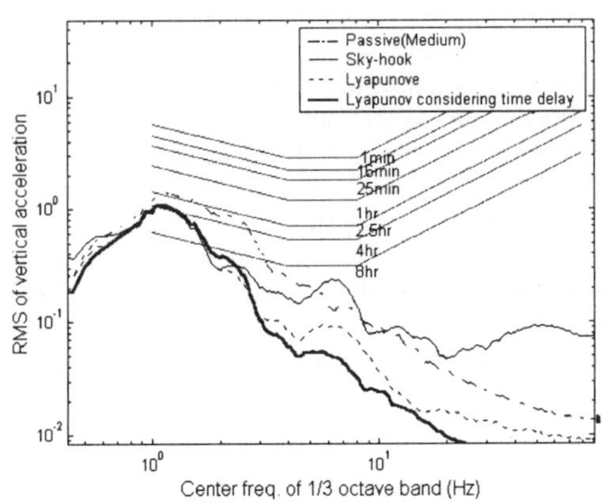

Fig. 4. The comparison of controls in ISO 2631 standard

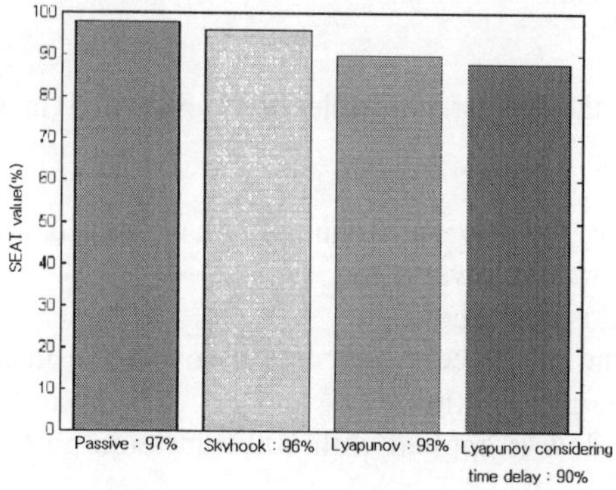

Fig. 5. The comparison of SEAT values

5. Conclusion

The Lyapunov type robust control is applied to suppress the vibration of a smart seat with an MR fluid damper. When the time delay of the damper is included in the design of the robust control, it provides the comfort of the seat most effectively in simulations.

Acknowledgements

This study was supported by research grants from the Korea Science and Engineering Foundation (KOSEF) through the Applied Rheology Center (ARC), an official KOSEF-created engineering research center (ERC) at Korea University, Seoul, Korea. The discussion with Prof. Leitmann at University of California, Berkeley is appreciated.

Reference

1. J. Y. Wong, *Theory of Ground Vehicles*, (John-Wiley, 1993).
2. K. Kim and D. Jeon, “Vibration suppression in an MR fluid damper suspension system”, Journal of Intelligent Material Systems and Structures, vol. 10, 1999.
3. W. Breinl and G. Leitmann, “State feedback for uncertain dynamical systems”, Applied Mathematics and Computation, pp. 65-87, 1987.
4. E. Reithmeier and G. Leitmann, "Structural vibration control", Journal of the Franklin Institute, pp. 1-24, 2000.
5. H. K. Khalil, *Nonlinear Systems*, (Prentice Hall, NJ, 2nd Ed.).
6. M. J. Griffin and X. Wu, "The influence of end-stop buffer characteristics on the severity of suspension seat end-stop impacts", Journal of Sound and Vibration, pp. 989-996, 1998.
7. C. M. Harris, *Shock and Vibration Handbook*, (McGraw-Hill, NY, 1991).
8. M. J. Griffin, *Handbook of Human Vibration* (Academic Press, London, 1990).

THERMAL EFFECTS OF A HOMOGENEOUS ER FLUID DEVICE

TARO NAKAMURA AND NORIHIKO SAGA

Department of Machine Intelligence and System Engineering, Akita Prefectural University, 84-4 Tsuchiya Honjo, Akita 015-0055, JAPAN
E-mail: taro@akita-pu.ac.jp

MASARU NAKAZAWA

Department of Functional Machinery and Mechanics, Shinshu University, 3-5-1 Tokida Ueda, NAGANO 386-8567, JAPAN
E-mail: nakazawa@kappa.shinshu-u.ac.jp

The viscous coefficient of a homogeneous ER fluid can be changed by the application of an electric field. However, the ER effect of the fluid is easily varied by temperature variation. Therefore, an envelopment and time in which the fluid is applied is limited.

In this paper, we examine the influence of the thermal effects induced on the fluid by the device, in order to construct more reliable homogeneous ER devices. We discuss two kinds of thermal effects caused by joule thermal and the dissipation of viscous energy.

The results of simulations and experiments suggest that the most of the exothermic elements of the homogeneous ER fluid are caused by the dissipation of viscosity energy in the fluid. Therefore, it seems reasonable to drive the clutch at a low rotating speed for short periods. We think that a robot arm for impedance control is more effective than a damper for vibration control.

1 Introduction

The viscous coefficient of a homogeneous ER fluid [1] can be changed by the application of an electric field. Therefore, it can be usefully applied to vibration control in a structural system or an impedance control of a robot [2][3].

We have previously developed a one-shaft type clutch using homogeneous ER fluid, and suggested its possible application in the clutch's impedance control [4]. In that study, we showed that the ER effect of the fluid is easily changed by temperature variation and the shear rate. However, few reports have addressed the thermal effect on the homogeneous ER clutch.

In the present paper, we examine the influence of the thermal effects induced on the fluid by the device, in order to construct more reliable homogeneous ER devices. First, we investigate the relationship between, viscosities, temperature, and shear rate. Next, we discuss the exothermic element of the homogeneous ER fluid induced by the ER clutch. Here, we calculate and experimentally verify the thermal effects caused by joule thermal and the dissipation of viscous energy.

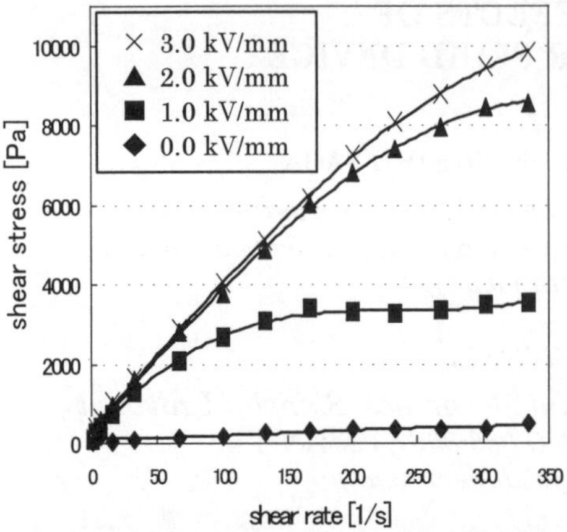

Figure 1. ER effects on homogeneous ER Fluid.

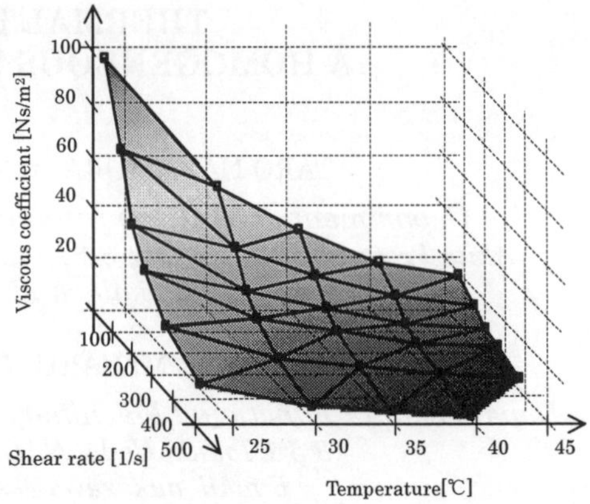

Figure 2. Viscosity vs. shear rates and temperatures (1kV/mm).

2 Characteristics of the homogeneous ER fluid

First of all, in order to discuss the characteristics of the homogenous ER fluid, the relationship between shear stresses and shear rates are shown in Figure 1. We see from this figure that the viscous coefficient of this fluid increases with increase in the application of the electric fields.

Next, we examine the relationship between viscosity and shear rate and temperature respectively. As one the example, the viscosity vs. shear rates and temperatures data with the application of an electric field (1kV/mm) is shown in Figure 2.

this figure shows that the viscosity tends to decrease with increasing temperature and shear rate. Here, using an equation of the relationship between viscosity and temperature proposed by Andrade [5], the following relation is obtained:

$$\eta(\nu,\theta) = \frac{A}{\nu^x}\exp\left(\frac{B}{\theta}\right) \quad , \tag{1}$$

where η, ν, and θ are the viscosity coefficient, the shear rate, and the temperature, respectively. A, B, x are a constant which is decided by the fluid. Here, these constants are obtained by approximating those data shown in Figure 2, which are A=31.87, B=84.05, and x=0.67, respectively.

3 Experimental setup

Studying about its thermal effects, we use the developed ER clutch as an ER device. The detailed structure of the ER clutch is shown in Figure 3. This figure shows that this ER clutch has electrode shapes which are combined with multiplex cylinders and a disk. Here, the outer diameter of the electrode is 95.4mm, the axial length of the electrode on an unilateral is 14mm, the effective electrode area is 7723mm^2, the distance between

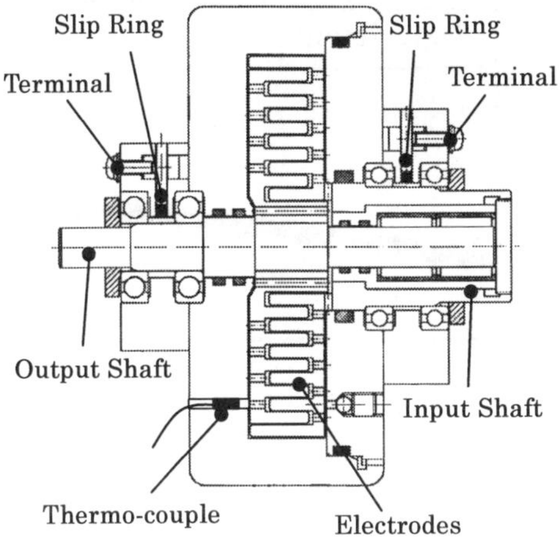

Figure 3. Structure of the ER Clutch.

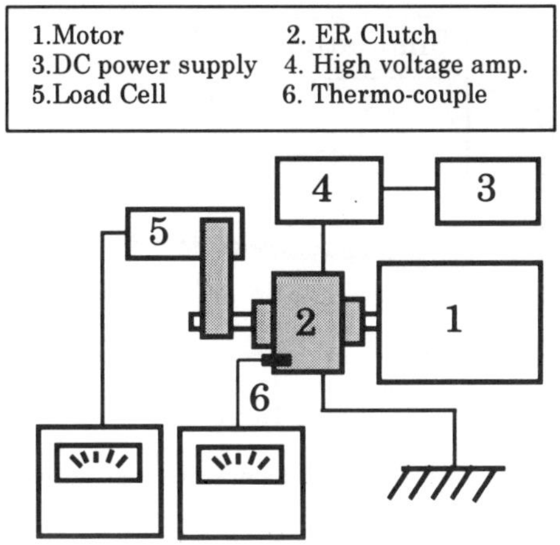

Figure 4. Configuration of the experimental system.

electrodes is 0.3mm, and the electrode material is SS400.

A schematic of the experimental setup is shown in Figure 9. The input shaft is driven by an AC motor (Oriental Motor Co.: KX PM6180 GD-AE). Output torque is then obtained by applying to the electrode a high voltage transformed by a high voltage amplifier (Matsusada Precision Co.: HOPS3B1). Here, the torque of the output shaft is measured by the load cell (KYOWA Electric Ins.Co.:LUB-20KB). Further, the temperature of the ER fluid is measured by a thermo-couple installed in the ER clutch.

4 Exothermic factors of the ER Fluid

The energy applied to the ER clutch is electric energy caused by joule thermal and mechanical energy caused by the dissipation of viscosity energy. Therefore, The thermal generated by each type of energy is verified by simulation and experiment.

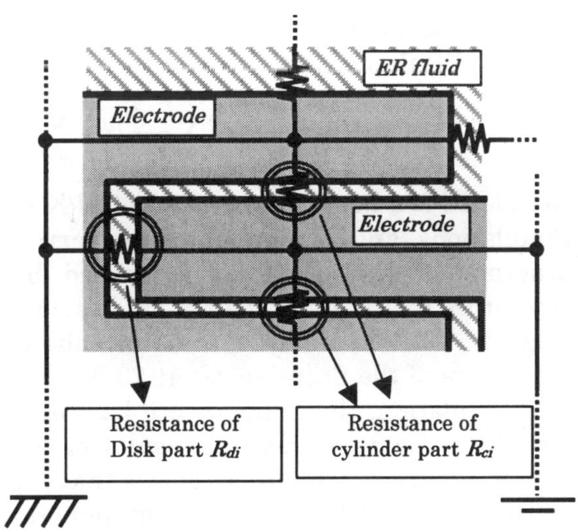

Figure 5. Enlargement of the electrode model.

4.1 *Joule thermal*

First of all, it is assumed that all electric power is converted into thermal. Thermal energy Q_J[W] per second based on the law of

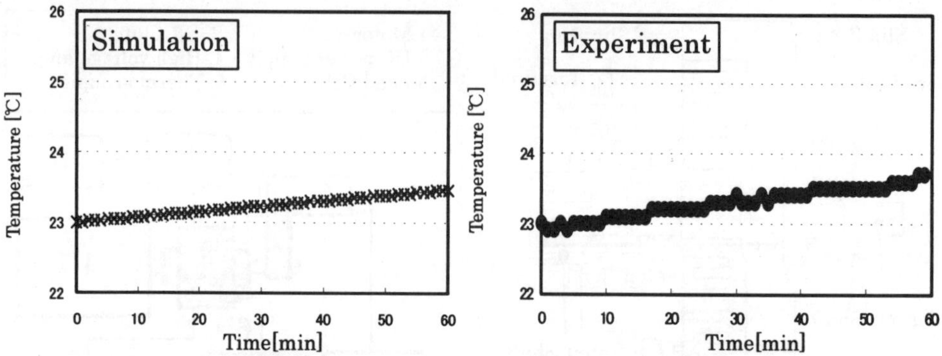

Figure 6. Temperature trajectories of the ER fluid caused by joule thermal.

the joule is written as follows:

$$Q_J = V^2/R_T \tag{2}$$

where V[V] is an impressed voltage, and R_T[Ω] is the synthesized electric resistance of the ER fluid. Here,' the electric resistance of the ER fluid 4.0 ×10^{-8}Ωm. Figure 5 shows the enlargement of the electrode model. The clutch's electrode shapes combined with multiplex cylinders and disks. Here, the electric resistance of each disk part is assumed to be $R_c(i)$ [Ω](i=1,2,⋯, n) and the electric resistance of each cylinder part is assumed to be $R_d(i)$ [Ω] (i=1,2,⋯,n-1). As a result, R_T[Ω] is written as follows:

$$R_T = \left(\sum_{i=1}^{n} \frac{1}{R_c(i)} + \sum_{i=1}^{n-1} \frac{1}{R_d(i)} \right)^{-1}, \tag{3}$$

where n shows a cylinder number of 11. Therefore, it is assumed that the heating value in the ER clutch is insulated, and the temperature θ_J[K] where the ER fluid rises t seconds later is written as follows:

$$\theta_J = \frac{V^2}{C} \cdot \left(\sum_{i=1}^{n} \frac{1}{R_c(i)} + \sum_{i=1}^{n-1} \frac{1}{R_d(i)} \right) \cdot t \quad . \tag{4}$$

Here, C[J/K] is a heat capacity. C=0.051J/K.

Simulations and experiments are carried out in a case where the exothermic of the ER fluid is caused by joule thermal. First, the temperature trajectory of the ER fluid calculated by equation (4) is shown in Figure 6 (2kV/mm). Here, the atmosphere temperature and the initial temperature of the fluid are set at 23.0-Celsius degrees. From the results of the simulation, the temperature has increased only 0.5 K 60 minutes later. Next, the experimental results of its temperature trajectory are shown in Figure 7. This figure shows that experimental results tend to be quantitatively similar to the simulation results. Therefore, from these results, the exothermic of the fluid which is caused by joule thermal appears to have little influence on the variation of the output torque.

4.2 *Dissipation of viscous energy*

We next examine the dissipation of viscous energy. As for the ER clutch, cylindrical electrodes are arranged between radius $r(i+1)$[m] from radius $r(i)$ [m](i=1,2,⋯,n) at equal intervals as shown in Figure 7, and the ER fluid of the viscous coefficient η is filled into the gap between the electrodes.

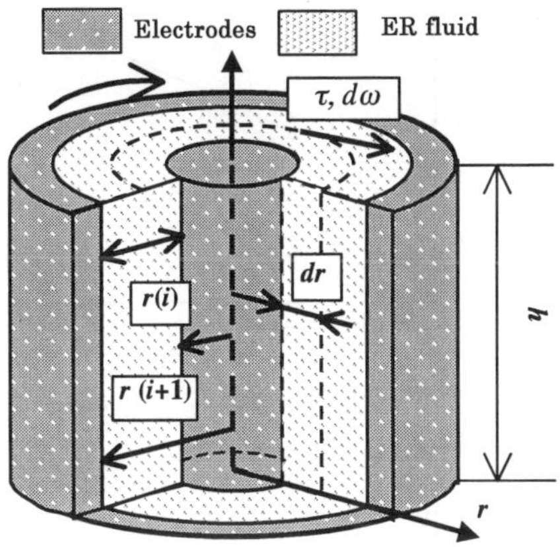

Figure 7. The dissipation of viscosity energy.

When the area, the stress, and the rotation speed are A[m²], τ [N/m²], and ω[r/s] in minute radius dr[m], respectively, the dissipation of viscous energy per second $dE(i)$[W] (i=1,2,⋯,n) is written as follows:

$$dE(i) = \frac{\partial(\tau A r\,\omega)}{\partial r}dr \quad . \tag{5}$$

Suppose cylinder length is h[m] and output torque is T[Nm], equation (5) is as follows:

$$dE(i) = \frac{T^2}{4\pi h\eta}\frac{\partial}{\partial r}\left(\frac{1}{r(i)^2} - \frac{1}{r^2}\right)dr \quad . \tag{6}$$

This equation is integrated from $r(i)$[m] to $r(i+1)$[m], and uses the relative rotation speed Ω[r/s] of both parties' electrodes. The simplified equation (7) is

$$\int_{r(i)}^{r(i+1)} dE(i) = 4\pi h\eta\Omega^2 \cdot \left(\frac{1}{r(i)^2} - \frac{1}{r(i+1)^2}\right)^{-1} \quad . \tag{7}$$

In addition, it is assumed that gap c[m] between electrodes is very small compared with radius $r(i)$[m] of the electrode. Thus, the heating value of the entire clutch induced by the rotation of the motor is shown by the following approximation equation.

$$Q_v = \sum_{i=1}^{n} dE(i) = \sum_{i=1}^{n} \frac{2\pi h\eta\Omega^2 r(i)^3}{c} \quad . \tag{8}$$

Next, the viscous coefficient η_E of the ER fluid when voltage E[V/mm] is applied to the ER clutch is shown with the approximation equation as follows by the use of the temperature of the shearing speed v[1/s] and function $\theta_v(t)$ from equation(1).

In addition, it is assumed that the heating value in the ER clutch is insulated, and the temperature θ_v[K] where the ER fluid rises t seconds later is written as follows:

$$\theta_v(t+\Delta t) = \frac{1}{C}\left\{\sum_{i=1}^{n}\frac{2\pi h\eta_E(v,\theta_v(t))\Omega^2 r(i)^3}{c}\cdot\Delta t\right\} + \theta_v(t) \quad . \tag{9}$$

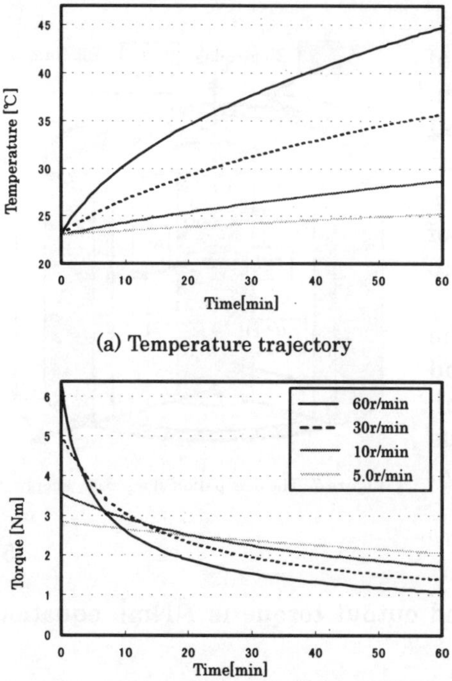

(a) Temperature trajectory

(b) Torque trajectory

Figure 8. Simulation of temperature characteristics caused by dissipation of viscous energy.

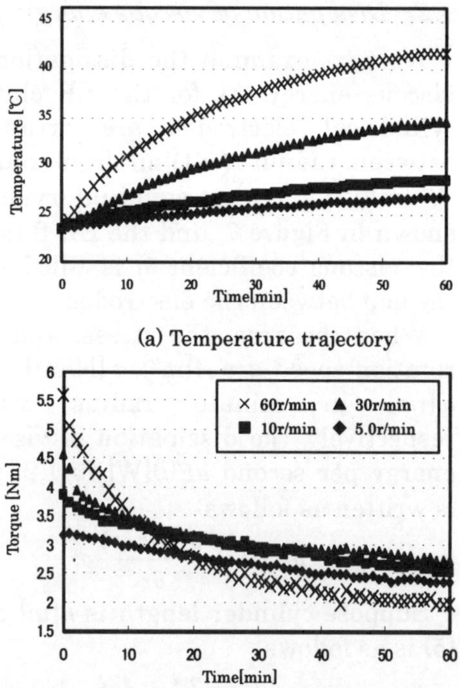

(a) Temperature trajectory

(b) Torque trajectory

Figure 9. Experiments of temperature characteristics caused by dissipation of viscous energy.

Simulations and experiments are carried out in a case in which the exothermic of the ER fluid is caused by the dissipation of the viscous energy (applying the electric fields 1kV/mm). First, the temperature trajectory of the ER fluid calculated by equation(9) is shown in Figure 6. The results of the simulation show that the increase of the temperature for each rotational speed is much larger than that caused by joule thermal. Further, the output torque then decreases with increasing time. Therefore, the exothermic of the fluid which is caused by the dissipation of the viscous energy greatly influences the variation of the output torque. Experimental results of the temperature trajectory are shown in Figure 10. This figure shows that the temperature trajectory tends to be saturated in the high temperature region, because the thermal is radiated. However, the experimental results tend to be very similar to simulation results qualitatively and quantitatively.

5 Discussion

The results of simulations and experiments indicate that most of the exothermic behavior of the homogeneous ER fluid is caused by the dissipation of viscous energy. Based on these results, a device using a homogeneous ER fluid has to be used taking into account the dissipation of viscous energy as shown in equation(9). Therefore, it is desirable that the device is used for a short time and is driven at a low rotational speed. For example, it is reasonable to suppose that a robot arm for impedance controls is more effective than a damper for vibration controls.

6 Conclusion

In order to construct more reliable a homogeneous ER device, we have examined the influence of the thermal effects induced on the ER fluid by the device. The results are as follows:

1. In regard to exothermic factors, simulations and experiments were carried out in the case in which an exothermic of the ER fluid is caused by joule thermal. As a result, the exothermic negligibly influences the variation of the output torque because the temperature increase is small.
2. Simulations and experiments have been carried out in a case in which the exothermic of the ER fluid is caused by the dissipation of the viscous energy. As a result, the increase of temperature at each rotational speed is much larger than that caused by joule thermal. Further, the output torque then decreases with time.

7 Acknowledgements

The authors want to thanks the Asahi chemical industry Co. Ltd. For providing the ER fluid.

References

1. A. Inoue and S. Misawa, Electrorheological Effect of Liquid Crystalline Polymers, *Journal of Applied Polymer Science* **55**, (1995) pp.113-118.
2. N. Takesue, G. Zhang, J. Furusho and M. Sakaguchi, Motion Control of Direct-Drive Motor By A Homogeneous ER Fluid, *Journal of Intelligent material Systems and Structures* **10**, (1999) pp.728-731.
3. N. Saito, T. Nakamura, N. Saga et al., Development of the robot Arm with soft touch-Fundamental Characteristics of ER Clutch for a Joint of the robot Arm-, *Proceedings of the 18th Annual Conference of the Robotics Society of Japan*, (2000) pp525-526.
4. N. Saga, T. Nakamura et al., Development and Experiment of ER clutch for robot Arms, *Proceedings The Fifth International Conference on Motion and Vibration Control*, (2000) pp.649-654.
5. E. N. da C. Andrade, *Nature 125*, (1930), 309

PERFORMANCE ANALYSIS OF SMART IMPACT DAMPER SYSTEM USING HERSCHEL-BULKLEY MODEL

DUG-YOUNG LEE

School of Automotive and Industrial Engineering, Taegu University
Kyungsan 712-714 KOREA
E-mail: dylee@taegu.ac.kr

YOUNG-TAI CHOI AND NORMAN M. WERELEY

Alfred Gessow Rotorcraft Center, Dept. of Aerospace Engineering, University of Maryland
College Park, MD 20742, USA
E-mail: nicechoi@wam.umd.edu, wereley@eng.umd.edu

Performance analysis for ER (electrorheological) and MR (magnetorheological) fluid-based impact damper systems is evaluated on the basis of the Herschel-Bulkley shear model. Generally, most ER/MR dampers have been analyzed based on simple Bingham-plastic shear model. However, the Bingham-plastic shear model cannot well describe the behavior of the damper in the condition of high velocity and high field input. Therefore, in this study, the Herschel-Bulkley shear model, in which the constant post-yield plastic viscosity in Bingham model is replaced with a power law model dependent on shear rate, is used to assess performance of ER/MR impact damper systems. In doing so, an ER/MR impact damper system is proposed and its governing equation of motion is derived on the basis of the Herschel-Bulkley shear model. Then, computer simulation is done to analytically evaluate characteristics of the ER/MR impact damper system under impact loads. As an important parameter in the Herschel-Bulkley shear model, effects of the flow behavior index, *n*, that is the degree of shear thinning or shear thickening, on impact damper system performance are shown.

1 Introduction

Smart fluids comprised of electrorheological (ER) and magnetorheological (MR) fluids, have the capability of controlling their rheological properties in response to applied field input. Many studies have been done on applications using smart fluids including ER and MR fluid-based impact damper systems that can be used in the joint mechanism of railroad vehicles, launch equipment of aircrafts, protection equipment of elevator's drop, and damping mechanism of boring tools or drills. Such impact damping systems are the focus of this study.

Generally, most ER/MR dampers have been analyzed based on Bingham-plastic shear model because the prediction of the damper characteristics is straightforward. However, the Bingham-plastic shear model cannot well describe the behavior of the damper in the conditions of high velocity and high field input. Shear thinning and thickening can manifest themselves at high shear rates and field input in post-yield region. Therefore, in this study, the Herschel-Bulkley shear model, in which the constant post-yield plastic viscosity in Bingham model is replaced with a power law model dependent on shear strain rate, is used to assess performance of ER/MR impact damper systems. In doing so, an ER/MR impact damper system is proposed and its governing equation of motion is derived on the basis of the Herschel-Bulkley shear model. Then, computer simulation is done to analytically evaluate characteristics of the ER/MR impact damper system under impact loads. As an important parameter in the Herschel-Bulkley shear model, we focus, in particular, on the effect of the flow behavior index, n, that is the degree of shear thinning or thickening, on impact damper system performance.

2 ER/MR impact damper systems

We consider that a load freely falls from a height h to an ER/MR impact damper, as shown in Figure 1. The load is described as a rigid mass m and a spring constant k_m. The source of the spring force is the elasticity of the load at the moment of impact

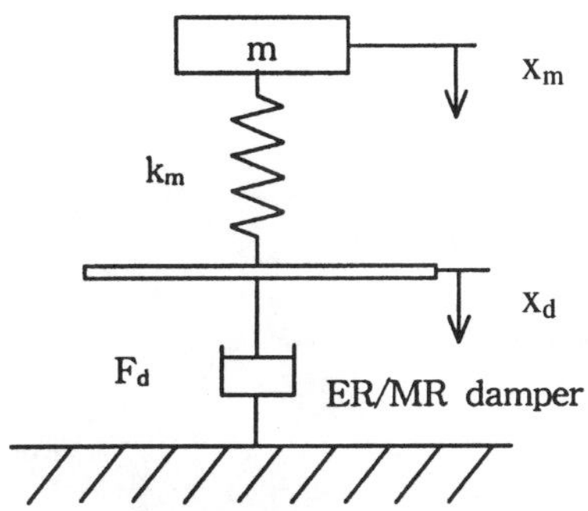

Figure 1. Schematic diagram of ER/MR impact damper systems.

For simplicity, we assume that the potential energy of the mass m at a height h is transferred to the ER/MR impact damper, and that the mass is in continuous contact with the impact damper after impact. In addition, the mass is falling vertically and the center of gravity point of the mass is aligned with the center of the ER/MR impact damper. The equation of motion of the ER/MR impact damper system is then

$$m\ddot{x}_m = -k_m(x_m - x_d) + mg \tag{1}$$

$$F_d = k_m(x_m - x_d) \tag{2}$$

with the following initial conditions

$$\dot{x}_d(0) = \dot{x}_m(0) = \sqrt{2gh} \tag{3}$$

where x_m is the displacement of the falling mass, x_d is the stroke of the impact damper, g is the acceleration due to gravity, and F_d is the damping force of the ER/MR impact damper. In next section, the damping force of the ER/MR impact damper, F_d is explicitly expressed based on the Herschel-Bulkley shear model.

3 Modeling of ER/MR impact damper systems

3.1 Configuration of ER/MR impact dampers

The configuration of the nominal MR impact damper is shown in Figure 2. It is noted that there is no difference in the configuration between ER/MR impact dampers other than how it produces an input field in the fluid path. For MR impact dampers, since magnetic field proportional to applied current should be produced in the fluid path, coil component is needed. However, for ER impact dampers, since electric field proportional to applied voltage should be applied, a simpler configuration is possible.

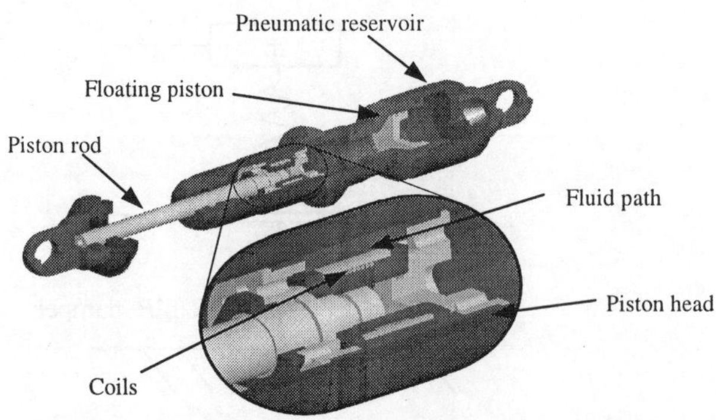

Figure 2. Configuration of the MR impact damper.

As shown in Figure 2, the MR impact damper is composed of hydraulic and pneumatic reservoirs. The hydraulic reservoir is fully filled with a MR fluid and again divided into two chambers by piston head: upper chamber and bottom chamber. In the piston head, there is a fluid path through which the MR fluid pumped by the piston passes between upper and bottom chamber. In addition, coils are wired around the piston head so that produces magnetic field in the fluid path. The pneumatic reservoir located at the bottom chamber below acts an accumulator to compensate the fluid volume of the piston rod in the upper chamber and the thermal expansion of the damper. In addition, the pneumatic reservoir is filled with nitrogen and highly pressurized in order to prevent cavitation during operation.

In the absence of field, the ER/MR impact damper has a viscous damping force due to the fluid path and gas spring force due to the gas pressure. However, when a field is applied to the ER/MR damper, additional damping force due to the yield stress of the ER/MR fluid is produced. This additional damping force can be continuously controlled by adjusting the intensity of the electric or magnetic fields.

3.2 Nonlinear Herschel-Bulkley model

The damping force of the ER and MR damper, F_d, can be expressed by

$$F_d = (A_p - A_r)\Delta P + A_r P_g \tag{4}$$

where, A_p and A_r are the cross-sectional areas of piston head and piston rod, respectively. ΔP is the pressure drop through the fluid path and P_g is the gas pressure in the pneumatic reservoir. If assuming that the gas in the pneumatic reservoir experiences adiabatic process under a motion, the gas pressure, P_g, can be expressed by

$$P_g = P_{go}\left(\frac{v_{go}}{v_{go} - A_r x_d}\right)^r \tag{5}$$

where P_{go} and v_{go} are the initial pressure and volume in the pneumatic reservoir and r is the specific heat ratio of nitrogen gas.

In order to get an explicit expression for the pressure drop, the Herschel-Bulkley shear model is used. Generally, ER/MR fluids exhibit non-Newtonian behavior when subjected to external electric or magnetic fields. Numerous empirical equations have been proposed to model the observed relations between shear stress, τ, and shear rate, *du/dz*, for time-independent non-Newtonian fluids. This may be adequately represented for many engineering applications by the yield stress, τ_y plus shear rate power law model, which is usually called the Herschel-Bulkley shear model. The constitutive shear flow relationship for the Herschel-Bulkley shear model for 1-dimensional flow can be expressed as follows [1]:

$$\tau = \tau_y + K\left(\frac{du}{dz}\right)^n \tag{6}$$

where $|\tau|$ must be greater than $|\tau_y|$ to initiate flow. The exponent, *n*, is called the flow behavior index and *K* is called the consistency index.

Depending on the choice of flow behavior index, *n*, it can be used to account for post-yield shear thickening or thinning behavior, as follows: (a) $n > 1$: dilatant or shear thickening behavior, (b) $n = 1$: Bingham behavior, (c) $n < 1$: pseudoplastic or shear thinning behavior.

3.3 Velocity profile and nondimensional plug thickness

A schematic of the velocity profile of ER/MR fluids fully developed in the gap of the fluid path is given in Figure 3. There are three flow regions in the gap in Figure 3. Regions 1 and 3 are post-yield regions where $|\tau|>|\tau_y|$, while region 2 is the pre-yield or plug flow region where $|\tau|<|\tau_y|$. The velocity profile must be symmetric about the center of the

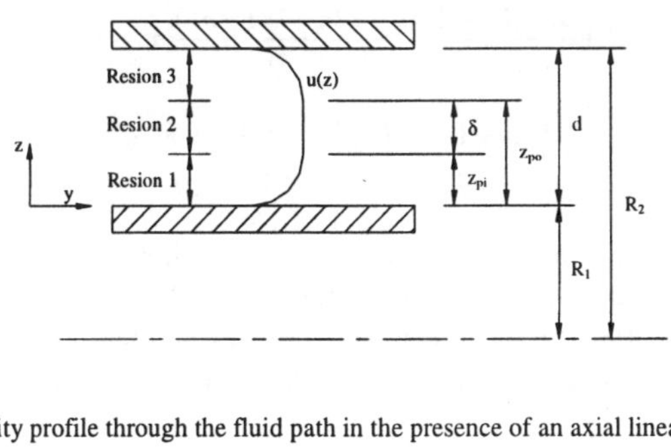

Figure 3. Typical velocity profile through the fluid path in the presence of an axial linear pressure gradient

fluid path because we assume that the pressure is uniform across the gap and that cylindrical or axis-symmetric geometry may be approximated by parallel plate geometry.

The velocity profile in each region can be decided from boundary conditions, compatibility conditions and symmetry conditions. The total fluid flow rate, Q_f, is given by the sum of the fluid flow rate in each region [2].

$$Q_f = bn\left(\frac{\Delta P}{KL}\right)^{1/n}\left(\frac{d-\delta}{2}\right)^{1+1/n}\frac{n(d+\delta)+d}{(2n+1)(n+1)} \tag{7}$$

where $b = 2\pi(R_1 + d)$ is the width of the MR fluid path approximated parallel-plate valve, L is the field-activated length and d is the field-activated gap. The total fluid flow rate can be also expressed in term of the stroke velocity of the impact damper as follows

$$Q_f = (A_p - A_r)\dot{x}_d \tag{8}$$

The boundary conditions, or shear stresses at z_{pi} and z_{po} in Figure 3 should yield an expression for nondimensional plug thickness, $\bar{\delta} = \delta / d$,

$$\bar{\delta} = \frac{2L\tau_y}{\Delta P d} \tag{9}$$

It is noted that the nondimensional plug thickness increases as the applied field increases due to the field dependence of the yield stress, τ_y. Also the yield shear stress of ER/MR fluids is a function of applied fields such as electric and magnetic fields and exponentially increase with respect to applied field, G. Therefore it can be represented by

$$\tau_y = \alpha G^{\beta} \tag{10}$$

where α and β are characteristic values of ER/MR fluids, which are usually determined by experiments.

From Equations (7) and (8), the pressure drop through the fluid path is

$$\Delta P = \left\{\frac{(A_p - A_r)\dot{x}_d}{bn}\right\}^n \frac{2^{n+1}}{(1-\bar{\delta})^{n+1}}\left\{\frac{(2n+1)(n+1)}{n(1+\bar{\delta})+1}\right\}^n \frac{KL}{d^{2n+1}} \tag{11}$$

Substituting Equations (11) into (9) and rearranging it with Taylor series expansion about $(1-\bar{\delta})^{1/n}$ yield the following nonlinear function for the nondimensional plug thickness.

$$\frac{n-1}{2n}\bar{\delta}^4 + \frac{2n^2+n-1}{2n^2}\bar{\delta}^3 + \frac{(2n+1)(1-n^2)}{2n^2}\bar{\delta}^2 - \frac{2n+1}{n}\bar{\delta}$$
$$-\frac{2(2n+1)(n+1)}{n}\frac{(A_p - A_r)\dot{x}_d}{bd^2}\left(\frac{K}{\tau_y}\bar{\delta}\right)^{1/n} + (n+1) = 0 \tag{12}$$

Given the yield stress and the stroke velocity, the nondimensional plug thickness at a certain of applied field can be obtained by numerically solving Equation (12). It is noted that the physically sensible root of Equation (12) is the root that satisfies $0 < \bar{\delta} < 1$.

On the other hand, the shear rate of the ER/MR fluid in the gap of the fluid path can be obtained by

$$\frac{du}{dz} = \left[\frac{\Delta P d}{KL} \left(\frac{1-\bar{\delta}}{2} - \frac{z}{d} \right) \right]^{1/n} \tag{13}$$

It is noted that the shear rate given by Equation (13) varies along with z-axis and that the maximum shear rate takes place at the wall, z=0, that is,

$$\left.\frac{du}{dz}\right|_{z=0} = \left\{ \frac{(A_p - A_r)\dot{x}_d}{bn} \right\}^n \frac{2^n}{(1-\bar{\delta})^n} \left\{ \frac{(2n+1)(n+1)}{n(1+\bar{\delta})+1} \right\}^n \frac{1}{d^{2n}} \tag{14}$$

4 Analytical results

In order to evaluate the performance characteristics of the ER/MR impact damper, computer simulation is done. The MR fluid used is a product (MRF-132LD) of Lord Cooperation and has characteristic values such as α =60 and β =1.3 is used in this study. To simplify the analysis, the MR fluid is assumed to have a nominal consistency index, K, of 0.33 Pa sec^n. Three impact dampers are installed to suppress the vibration of the mass at the condition that the mass is 200 kg and is fallen from 2 m height without initial velocity. The spring constant of mass, k_m, is chosen by 50 MN/m. The nominal design of the proposed ER/MR impact damper is as follows: the diameter of the piston is 76.2 mm, the field-activated gap of the fluid path is 0.75 mm, the field-activated length of the fluid path is 12.7 mm, the diameter of the piston rod is 50.8 mm, the maximum stroke of damper is 101.6 mm, and the initial pressure in the pneumatic reservoir is 2.0682 Mpa.

Figure 4 shows the damping forces of the ER/MR impact damper with respect to flow behavior indices, n, when the current of 1.5 A is applied. The higher the flow behavior index, n, the larger the damping force, F_d. It is the reason why as n increases, the magnitude of shear stress in post-yield region increases. The stroke of the impact damper is also shown in Figure 5. As clearly observed, the case of n=1.1 (shear thickening) reaches steady state value faster than other cases. One interesting thing is that although the difference of n values is small, its effects on the performance of the impact damper are great. It means that not only the yield stress but also the shear stress characteristics in post-yield region are important factors to choose the ER/MR fluids appropriate for applications. Especially, if the ER/MR fluid that has severe shear thinning characteristics (n<1) is chosen as working fluid of an application, careful consideration of these shear thinning/thickening characteristics should be made when implementing the device. It is the reason why the shear thickening effect (n>1) is likely to act as fail-safe device, but the shear thinning effect may be likely to damage the application, if the ER/MR fluid is chosen based on Bingham plastic model (n=1).

Figure 6 presents the nondimensional plug thickness after impact. It is shown that, at right after impact, the nondimensional plug thickness is almost the same as each other. However as time goes on, its time history is considerably different. The maximum shear rate at the gap of the ER/MR impact damper is shown in Figure 7. As observed, the magnitude of the maximum shear rate for n=1.1 is much larger than that for n=0.9. It arises from the pressure drop between the fluid paths increasing exponentially with respect to the fluid behavior index n.

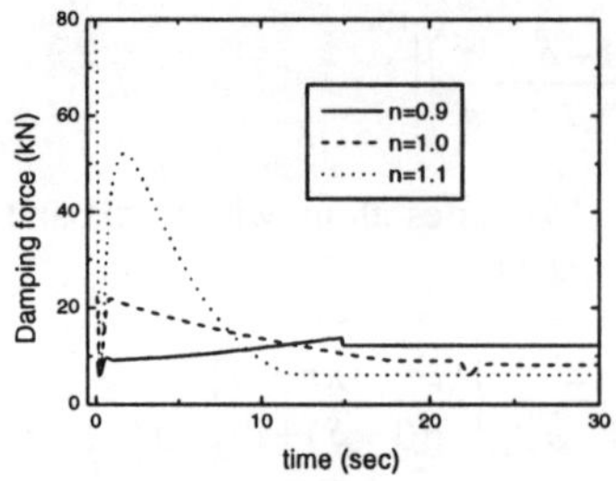

Figure 4. Effects of the fluid behavior index n on damping force.

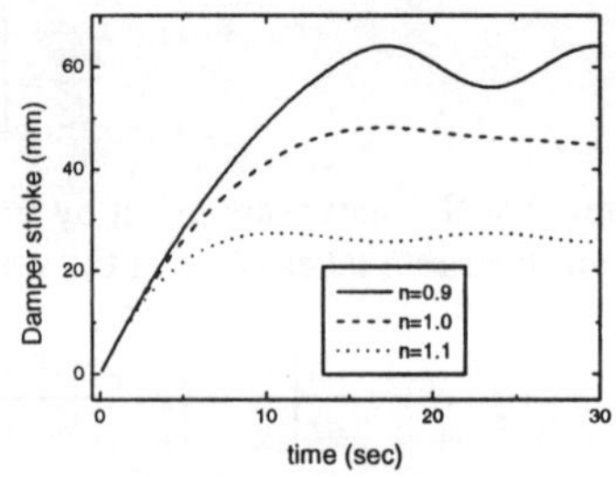

Figure 5. Effects of the fluid behavior index n on damper stroke.

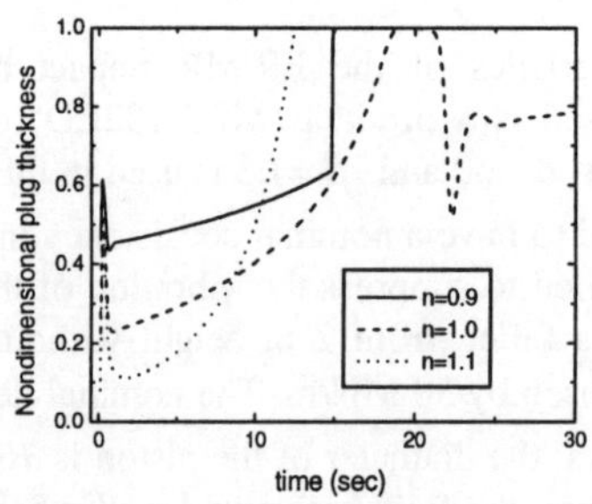

Figure 6. Effects of the fluid behavior index n on nondimensional plug thickness.

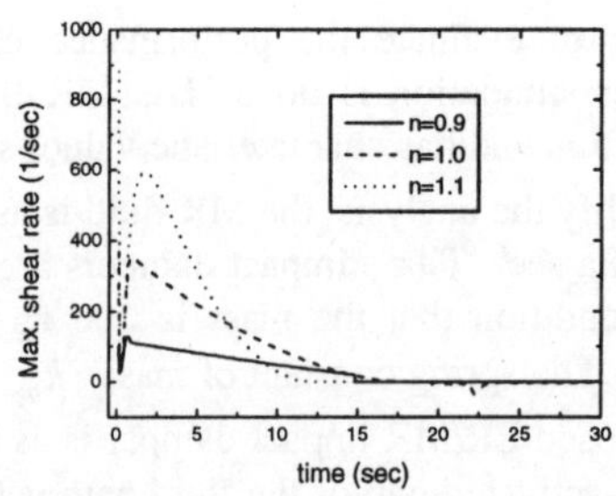

Figure 7. Effects of the fluid behavior index n on maximum shear rate.

5 Conclusion

In this study, performance analysis for ER and MR fluid-based impact damper systems was evaluated on the basis of the Herschel-Bulkley shear model. In doing so, an ER/MR impact damper system was proposed and its governing equation of motion was derived on the basis of the Herschel-Bulkley shear model. Then, computer simulation was done to analytically evaluate characteristics of the ER/MR impact damper under impact load. Effects of flow behavior index, n, on the impact damper performance were evaluated. It has been shown that, at the same field input, the higher the flow behavior index, n, the larger the damping force of the impact damper is. In addition, it has been observed that not only the yield stress but also the shear stress characteristics in post-yield region are important factors to choose the ER/MR fluids appropriate for applications.

References

1. Skelland A. H. P., Non-Newtonian Flow and Heat Transfer. (John Wiley and Sons, Inc., New York, 1967).
2. Lee D. Y. and Wereley N. M., Quasi-steady Herschel-Bulkley analysis of electro- and magneto-rheological flow mode dampers. *Journal of Intelligent Material Systems and Structures* **10** (1999) pp.761-769.

NUMERICAL SIMULATION OF MR FLUIDS DAMPING CHARACTERISTICS USING MODIFIED BINGHAM MODEL

H. NISHIYAMA, S. FUSHIMI
Institute of Fluid Science, Tohoku University,
2-1-1, Katahira, Aoba-ku, Sendai, 980-8577, Japan
E-mail: nishiyama@ifs.tohoku.ac.jp

M. NAKANO
Department of Mechanical Systems Engineering, Faculty of Engineering,
Yamagata University,
4-3-16, Jonan, Yonezawa, Yamagata, 992-8510, Japan
E-mail: nakano@mnaka.yz.yamagata-u.ac.jp

A numerical simulation is conducted to clarify the unsteady flow behavior and drag force around the oscillating flat plate immersed in MR fluids under the applied magnetic field. A modified Bingham model is adopted to take viscous flow behavior at the small shear rate and also yield stress into account. The experimental yield stress is given to the model as a function of magnetic field intensity. The governing equations for the oscillating flat plate and unsteady MR fluids flow are derived by modified Bingham model taking magnetic body force, yield stress and viscous force into account. The effects of magnetic field intensity, oscillating frequency and amplitude and further plate thickness on the fluid and damping characteristics around the oscillating flat plate are clarified by the numerical simulation.

1 Introduction

MR fluids [1] is one of the typical functional fluids [2], since it has advantages of controllable yield stress, large apparent viscosity, strong magnetic body force in the applied magnetic field. Furthermore, it can be changed easily from liquid to semi-solid by applying the strong magnetic field. These functions result from the robust cluster formation [3] and also from the high magnetization even in the weak applied magnetic field. Then it has been extensively utilized in the seat damper, shock absorber, brake, clutch and damping rubber so far [4]. Especially, MR damper is one of the most well developed products using the controllable functions of MR fluids compared with MF damper [5, 6, 7]. Then it is very important to clarify the correlation between the flow characteristics [8] and the damping characteristics taking the rheology properties [9] of MR fluids in the applied magnetic field into account.

In the present study, the governing equations for MR fluids unsteady flow and the oscillating flat plate are derived by the modified Bingham model taking the magnetically controllable yield stress into account. The effects of the magnetic field intensity, oscillating frequency and amplitude and further plate thickness on the unsteady velocity distribution, drag force and damping characteristics in the MR fluids damper are clarified by numerical simulation.

2 Rheological Properties and Modified Bingham Model

Figure 1 shows the flow curve of MR fluids (MRF-132LD) in the applied magnetic field using rotational viscometer. The apparent viscosity and the dynamic yield

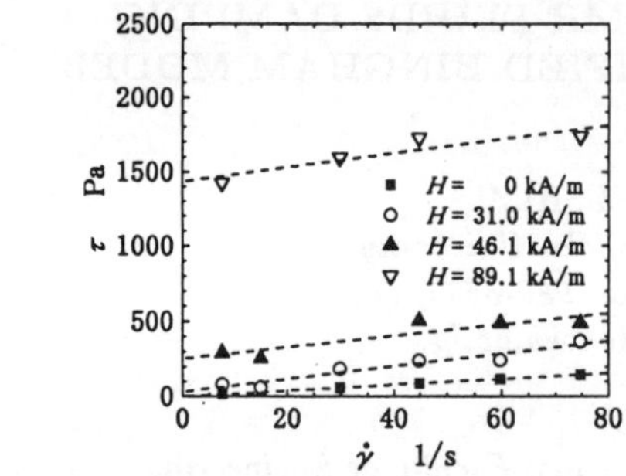

Figure 1: Flow curve in the applied magnetic field.

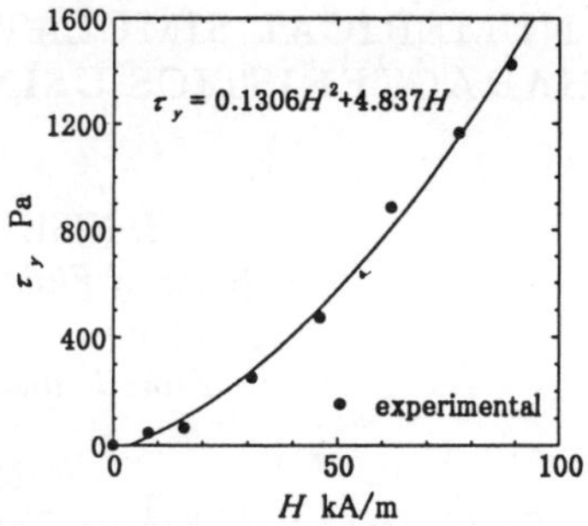

Figure 2: Yield stress with magnetic field intensity.

stress increase with the magnetic field intensity.

Figure 2 shows the dynamic yield stress as a function of magnetic field intensity as given by Eq. (1).

$$\tau_y = \alpha H^2 + \beta H \tag{1}$$

where $\alpha = 0.1306; \beta = 4.837$.

Referring to Fig. 1, the modified Bingham model [10] is proposed here to determine the shear stress explicitely at very small shear rate and also to obtain converged solution easily. The shear stress is given by Eq. (2).

$$\tau_{zx} = \left(\frac{\tau_y(H)}{\left|\dfrac{\partial u_z}{\partial x}\right| + \varepsilon(H)} + \eta \right) \frac{\partial u_z}{\partial x} \tag{2}$$

The first term in the right hand means the increased apparent viscosity due to the cluster formation. When the parameter $\varepsilon(H)$ equals to zero, the model becomes Bingham model. With increase of magnetic field intensity, ε decreases.

3 Theoretical Analysis

3.1 MR Fluids Flow Analysis Around the Oscillating Flat Plate

Figure 3 shows the schematic illustration of an oscillating flat plate immersed in a MR fluids damper. The following assumptions are introduced to derive the governing equations for MR fluids flow.

(1) MR fluids is incompressible and non-newtonial fluid.

(2) The flow is 1D and laminar.

$$u_z = u_z(t, x), \qquad p = p(t, z) \tag{3}$$

(3) The applied magentic field is given by

$$H_x = \frac{1}{2} H_{max} \left[1 + \cos\left(\frac{\pi}{l_z} z\right) \right], \qquad H_{max} = aI \tag{4}$$

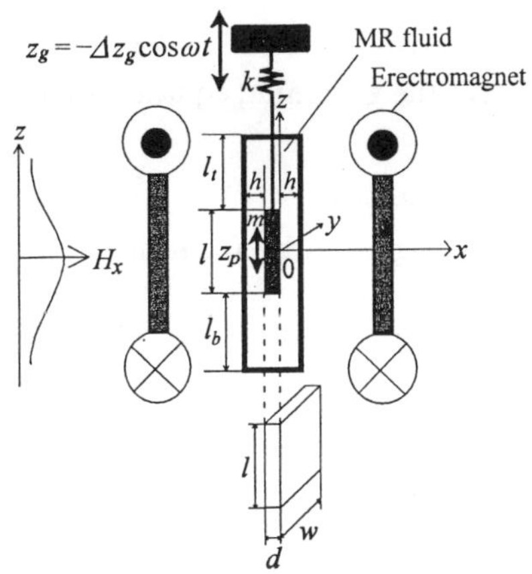

Figure 3: Schematic illustration of MR fluids damper.

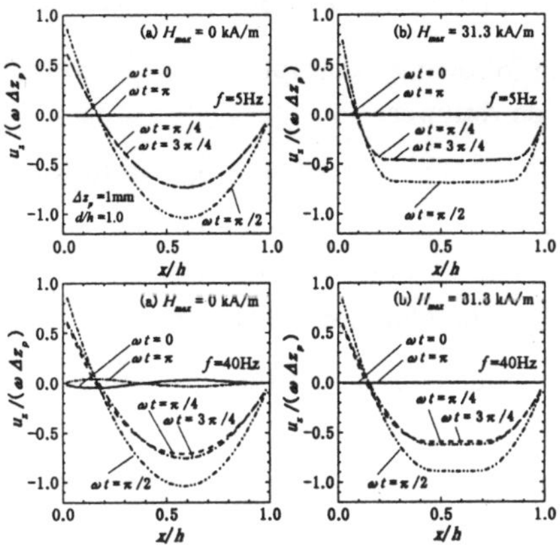

Figure 4: Velocity distribution with oscillation frequency.

where $l_z = 0.14$m, $a = 1.57 \times 10^4$.
(4) The magnetization is linear with magentic field intensity.

$$M_x = \bar{\chi} H_x \tag{5}$$

(5) The displacement of only forced oscillating flat plate is virtually given by

$$z_p = -\Delta z_p \cos\omega t \tag{6}$$

Governing Equations
Eq. of contrinuity:

$$u_z \mid_{x=0} d + 2\int_0^h u_z dx - 0 \tag{7}$$

Constitutive eq.:

$$\tau_{zx} = \left(\frac{\tau_y(H_x(z))}{\left|\frac{\partial u_z}{\partial x}\right| + \varepsilon} + \eta \right) \frac{\partial u_z}{\partial x} \tag{8}$$

where $0 < \varepsilon \ll V_0/h$.
Eq. of motion:

$$\rho\frac{\partial u_z}{\partial t} = -\frac{\partial p}{\partial z} + \frac{\partial}{\partial x}\left[\left(\frac{\tau_y(H(z))}{\left|\frac{\partial u_z}{\partial x}\right| + \varepsilon} + \eta \right) \frac{\partial u_z}{\partial x} \right] - \rho g + \mu_0 M_x \frac{\partial H_x}{\partial z} \tag{9}$$

Boundary condition

$$x = 0: \quad u_z = \frac{dz_p}{dt}, \qquad x = h: \quad u_z = 0 \tag{10}$$

3.2 MR Fluids Damping Analysis

The following assumptions are introduced to derive the governing equations for MR fluids damper.

(1) $\mu_0^2 M_n^2/2$ acted on upper and lower edges is neglected.

(2) The gravity force and buoyant force can be neglected by regarding the balancing point as the orgin of z axis.

(3) The Coulomb frictional force is neglected.

The equation of motion for oscillating flat plate is given by

$$m\frac{d^2 z_p}{dt^2} = -k(z_p - z_g) + F_p + F_\tau \tag{11}$$

where the pressure drag F_p and the frictional drag F_τ are given by

$$F_p(t, H_x(z)) = -dw\int_{z_p(t)-l/2}^{z_p(t)+l/2} \frac{\partial p}{\partial z} dz = -dwl\frac{\partial p'}{\partial z} - dw\mu_0\int_{H_x(z_p(t)-l/2)}^{H_x(z_p(t)+l/2)} M_x dH \tag{12}$$

$$F_\tau = 2\left[\int_{-w/2}^{w/2}\int_{z_p(t)-l/2}^{z_p(t)+l/2}\left(\left(\frac{\tau_y(H(z))}{\left|\frac{\partial u_z}{\partial x}\right| + \varepsilon} + \eta\right)\frac{\partial u_z}{\partial x}\right) dz dy\right]_{x=0} \tag{13}$$

4 Numerical Results and Discussion

4.1 Fluid Dynamic Characteristics

Figure 4 shows the effect of oscillation frequency on the velocity distribution with and without magnetic field. The larger plug flow is observed in the central region for the lower frequency f with the applied magentic field. This is because the pre-yield effect is dominant due to the robust cluster formation in the applied magnetic field. The flow behavior comes from post-yield effect near oscillating plate and wall.

Figure 5 shows the effect of the oscillation amplitude on the velocity distribution with and without magnetic field. Since the cluster is easily broken for the large amplitude oscillation resulting in the decrease in the effective viscous force, the plug flow region becomes narrower in the applied magnetic field.

Figure 6 shows the effect of plate thickness d/h on the velocity distribution with and without magnetic field. The shear flow mode is dominant against the pressure flow mode for the thin plate showing the wider plug flow region in the applied magnetic field.

Figure 7 shows the effect of the frequency on the drag force acted on the oscillating plate with and without magnetic field. F_p and F_τ denotes the pressure and the frictional drags respectively. When the magnetic field is applied, there is the nonlinear relation between the plate displacement and the drag force. Just when the plate stops in the applied magnetic field, there is the large and discontinuous jump of drag force changing its sign. This comes from the large frictional effect

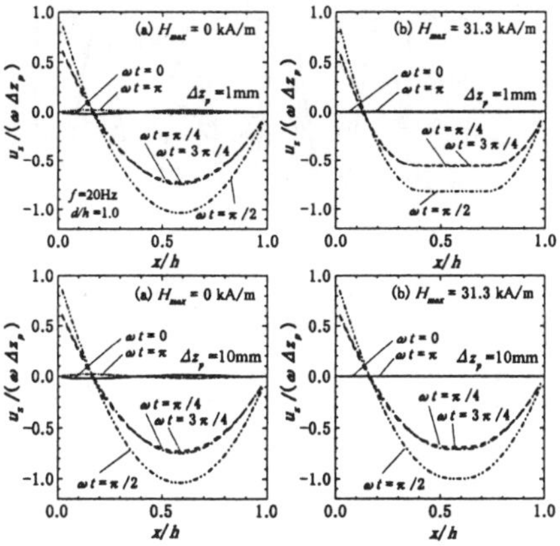

Figure 5: Velocity distribution with oscillation amplitude.

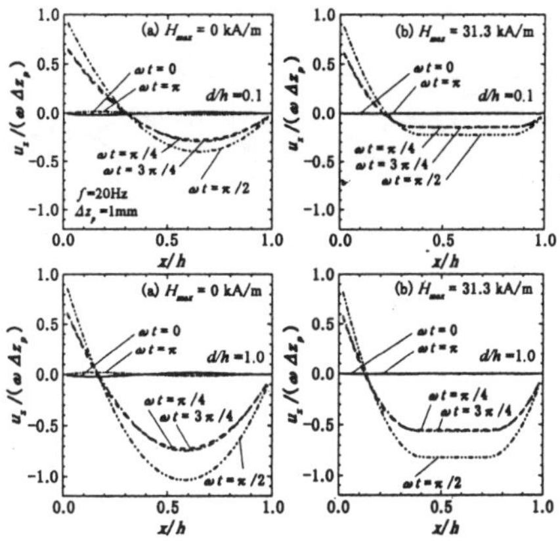

Figure 6: Velocity distribution with plate thickness.

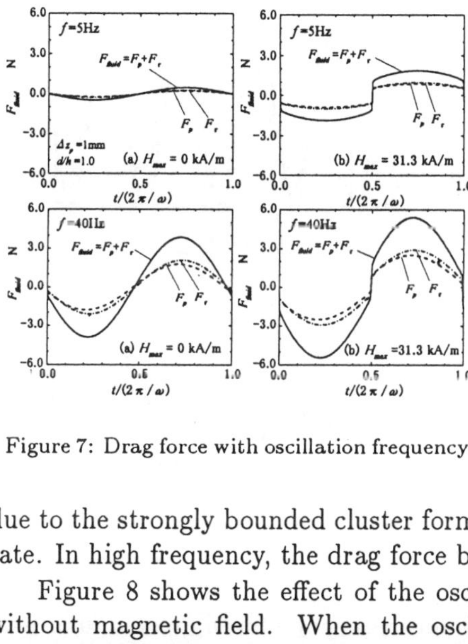

Figure 7: Drag force with oscillation frequency.

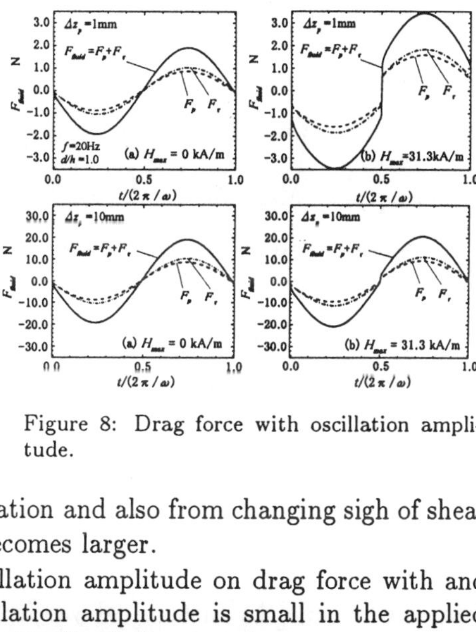

Figure 8: Drag force with oscillation amplitude.

due to the strongly bounded cluster formation and also from changing sigh of shear rate. In high frequency, the drag force becomes larger.

Figure 8 shows the effect of the oscillation amplitude on drag force with and without magnetic field. When the oscillation amplitude is small in the applied magnetic field, the relative drag force per amplitude becomes larger.

Figure 9 shows the effect of the plate thickness on the drag force with and without magnetic field. When the magnetic field is applied for thin plate, the frictional force is considerably larger than the pressure drag corresponding to the shear flow mode as shown in Fig. 6.

4.2 Damping Characteristics

Figure 10 shows the effect of frequency on the plate oscillation in the MR fluids damper. With increase of the magnetic field intensity, the oscillation is damped

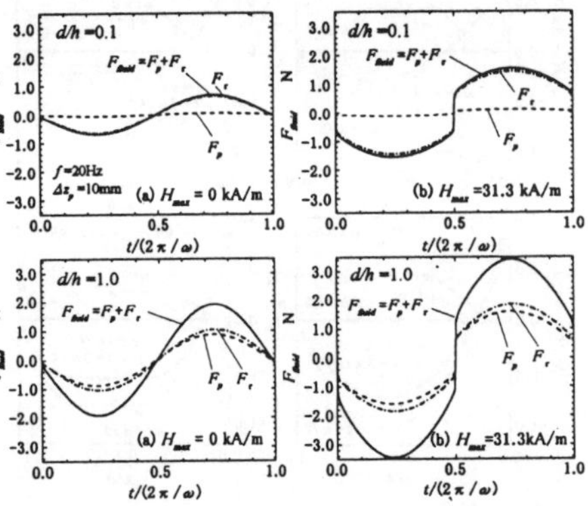

Figure 9: Drag force with plate thickness.

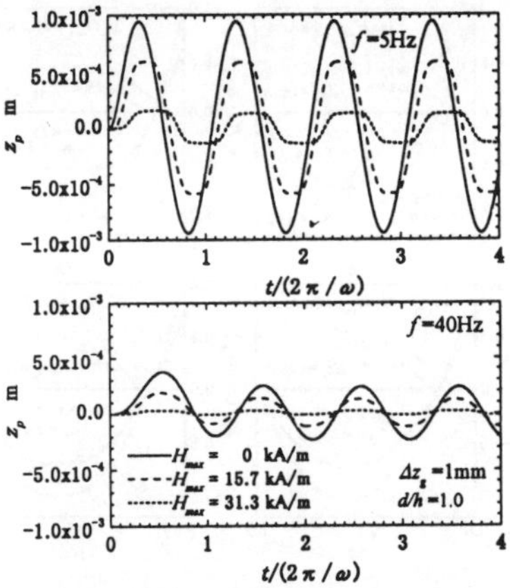

Figure 10: Plate displacement with oscillation frequency.

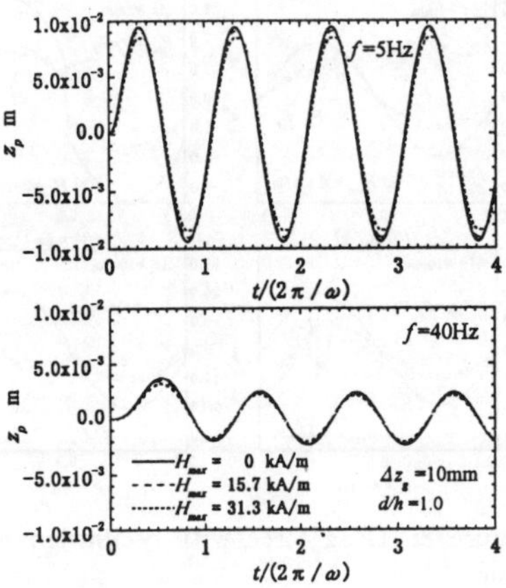

Figure 11: Plate displacement with large oscillation amplitude.

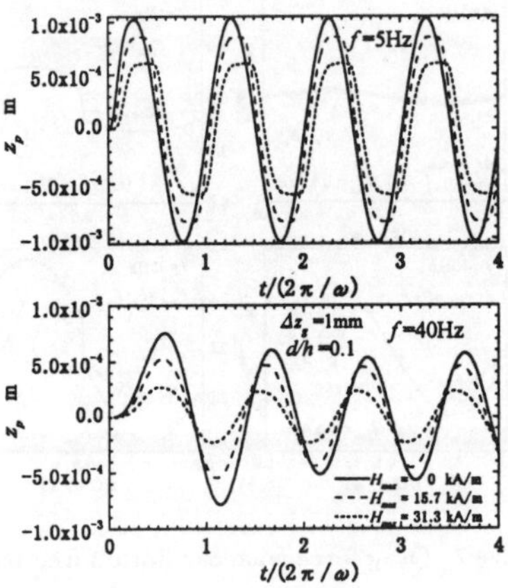

Figure 12: Plate displacement with thin plate.

considerably and the phase is delayed for lower frequency showing rectangular wave due to large apparent viscosity in pre-yield region. For larger frequency, after the transient oscillation, there is a periodic oscillation showing a smaller displacement of plate. Furthermore the phase seems to be advanced with increase of the magnetic field intensity.

Figure 11 shows the plate displacement characteristics with the large amplitude. There is small effect of applied magnetic field on the plate damping, compared with that for small amplitude as shown in Fig. 10.

Figure 12 shows the thin plate displacement characteristics. There is small damping effect even for small frequency by applying the magnetic field compared with that for thick plate as shown in Fig. 10.

5 Conclusions

The governing equations for the oscillating flat plate and MR fluids flow are derived by modified Bingham model taking magnetic force, yield stress and viscous force as a function of magnetic field intensity into account. The flow field, drag force and damping characteristics around the oscillaing flat plate in MR fluids are clarified numericallty. The results obtained in the pressent study are as follows.

1. There is linear relation between the plate oscillation and the drag force without magnetic field, however, nonlinear relation between them in the applied magnetic field.
2. There is large plug flow region for lower oscillation frequency in the applied magnetic field due to the cluster formation. There is the discontinuous jump of drag force acted on the stopping plate in the applied magnetic field due to the pre-yield. The drag force acted on the thin plate mainly comes from the frictional force as a function of magnetic field intensity. The large drag force acts on the plate for high frequency and large amplitude.
3. The smaller oscillation frequency and oscillation amplitude for thick plate, the better damping characteristics becomes in the applied magnetic field due to the contollable apparent viscosity.

Acknowledgements

We would like to give our sincere thanks to Associate Professor K. Ueno and Mr. T. Oyama in our numerical simulation and experimental study respectively.

References

1. R. G. Larson, in *The Structure and Rheology of Complex Fluids* (Oxford University Press, Oxford, 1999).
2. *Functional Fluids and Intelligent Fluids*, The Japan Society of Mechanical Engineers (Corona Pub. Co. Ltd., Tokyo, 2000), (in Japanese).
3. L. Zhou, W. Wen and P. Sheng, *Phys. Rev. Lett.* **81**, 1509 (1998).
4. *Proc. of the 7th Int. Conf. on ER Fluids and MR Suspensions*, Hawaii, (1999).
5. H. Nishiyama, T. Oyama and T. Fujita, *Int. J. Modern Phys.* B **15**, 829 (2001).
6. B. F. Jr. Spencer, S. J. Dyke, M. K. Sain and J. D. Carlson, *J. Eng. Mech.* **123**, 230 (1997).
7. S. Kamiyama, S. Kawabe, T. Yamane, T. Oyama and K. Ueno, *Non-linear Electromagnetic Systems* (IOS Press, 1998), p.875.
8. Y. J. Jang, M. S. Suh, M. S. Yeo and S. B. Choi, *J. Intelligent Mater. Systems and Structures* **7**, 604 (1996).
9. J. M. Ginder, *Encyclopedia Appl. Phys.*, **16**, 487 (1996).
10. M. Nakamura and T. Sawada, *J. Non-Newtonian Fluid Mech.*, **22**, 191 (1987).

ELECTROSENSITIVE FLUIDS IN ACOUSTIC DEVICES

ANATOLYI S. RUBANOV

Agricalture Technical University,
117 F.Scorina. Av., Minsk, 220072, Belarus

ALEXSANDRA O. KARABKO

Byelorussian State University,
Ya. Kolasa Str. 17, apt.3, 220050, Minsk, Belarus

The present work is aimed at elucidation of the specific features in propagation of ultrasound waves in an electrostructurized medium, of their dependence on an applied electric field and implementation of the effects obtained in practice.

1. Introduction

The unique combination of strong electric properties, fluidity and stability in electric fields provides the basis for using ER-fluids to solve a wide spectrum of problems in mechanical engineering, cosmic researches, power engineering, in rare-earth metal mining technology, medicine, and in the other branches of industry. A peculiar behavior of ER-fluids in external fields is of substantial interest for both improvement of traditional and development of new methods of control of informative (of weak intensity) and technological (of high power) fluxes of the energy of elastic waves as applied to non-destructive testing, ultrasound technology, underwater acoustics, metrology, and to the other fields of engineering. The main drawback of the traditional methods and means of engineering acoustics using a fluid as a sound-conducting (tracer) and technological medium lies in the absence of their controllability. This retards the further development and improvement of the devices of acoustic measurements and technologies to be used in hard-to-access places and in increased radiation zones, for work under vibration and weightlessness conditions, hinders their automatization. In particular, among the burning problems is creation of stable acoustic contact in both the acoustic control of workpieces and in the implementation of highly efficient technologies under conditions of local retention of liquid volumes with a developed free surface. A broad gamut of the unsolved problems exists which are concerned with formation of dynamical directional patterns of vibration sources, visualization of acoustic images, creation of the sources of elastic waves of the ultrasound and infrared ranges with controlled spectral characteristics.

Electrosensitive suspensions based on a dielectric disperse medium and a finely divided filler, whose internal structure changes reversibly in the presence of electric fields, opens up new promising avenues in physical acoustic measurements.

The available data on shearing hydromechanical characteristics of such suspensions show that due to an electric field the dielectric quasihomogeneous suspensions change their viscoelastic properties as a result of changes in the internal structure changes and interaction of the structure elements.

Applications of such media with their periodic structure rearranged by an electric field cover a rather wide sphere and include hydroautomation systems, electronic devices, fixture units, dampers, and the other facilities, in which these suspensions are subjected to the shearing constant (flow) or periodic (vibration) action. However, the acoustic properties of such suspensions have not been studied yet.

There are only scarce data on trial experiments[1] that have shown sensitivity of the acoustic characteristics of the medium to an electric field. However, the available information about the changes in the elastic and viscous properties of the disperse medium in an electric field[2] at the expense of the adequate response of the internal structure provides grounds to expect changes in the main characteristics of an acoustic signal, i.e. its velocity and the absorption coefficient, and open up·prospects for creation of returned acoustic devices.

Peculiarity of the system under consideration lies in interaction of the electrical and hydrodynamic types between individual elements, i.e. filler particles in the disperse dielectric medium. The first type is caused by an applied electric field that induces dipole moments in particles, whose values depend on the dielectric properties of materials and on the size of inclusions. The second type is associated with the fact that the structural formations, i.e. conglomerates and periodic bridge-like formations, scatter sound and, as a result, undergo changes in their sizes due to both the external acoustic wave and the waves rescattered by the neighboring structures. It is pertinent to note that the electrical interaction propagates in a dielectric with the velocity of light, i.e. practically momentarily, while the hydrodynamic type with the velocity of sound. The structural changes due to interaction of the second type generate in turn the changes in their dipole moments. As a result, the interaction of the first type also undergoes changes. Here the cross effects can occur that lead to anisotropy of the velocity of acoustic waves.

2. Experimental

The acoustic characteristics of the ERF in an electric field have been measured by two methods: pulse (direct) and interferometric (indirect). These methods are described in an the literature.[3, 4] In the case of longitudinal and transverse electric fields we used different acoustic cells. To study the transient processes with the field being switched and off on, the signal from the measuring cell was send to the amplitude detector through the amplifier, while the voltage from the output of the latter was fed to the storing oscillograph. The selected procedures to study of the influence of an electric field on the ERF acoustic parameters embody a considerable merit residing in the fact that the application of continuous acoustic waves of low amplitudes rules out their destructive influence on the medium structure thus allowing proper measurement of its viscosity and elasticity parameters.

In experiments, the use was made of electrosensitive liquid interlayers, which manifest the essential dependence of viscoelastic-plastic properties on a magnitude of an electric field applied to shear macroflows. The basis of a homogeneous suspention were the silicon fluid and the transforme oil, to which the finely divided (1-10 μm) oxide powder with a concentration from 2% wt. to 10% wt., the corresponding surfactant and activator were added.

To determine the velocity of ultrasound propagation and damping in liquid media, the time of ultrasound signal propagation $t_p$was compared with the delay time t_d of reference signal in the standard delay line. In this case, an electric signal is delayed to the time t_p = d_d which makes it possible to perform zero comparison and attain an accuracy in measurements of the time of ultrasound propagation up yo 0,5 nsec. The installation consistend of a series crystal oscillator-shaper of reference and delayed pulses, as such the I 2-17 commercial meter of time intervals was used, of an acoustic chamber, an amplifier-an amplitude detector, a follower, and a digital voltmeter.

Besides, between the amplifier and the oscillator there was the time comparison circuit connected via a driven sweep oscillator with an optical indicator, i.e. a light-emitting diode. In additional to generation of the reference probing pulses, the I 2-17 device also served for precise time delay of pulses, the value of which could be established relative to the reference pulse sequence with an accuracy of up to 0.5 nsec which was provided in a rough delay range by high stability of the frequency of the cristal oscillator, while in the precise range of delay by quantitative parameters of the tunable time delay line of signals. The delay value t_d was set equal to the time of ultrasound propagation in a medium which was controlled by the time comparison circuit that at $t_d < t_p$ generated a short triggering pulse for the driven sweep oscillator shaping a broader-band pulse and triggering the light-emitting diode. The measured time was $t_p = t' + t''$, where t' was the time of ultrasound propagation in a suspension, while t'' was the delay time in the structural elements of the chamber. In the course of investigations we obtained t_p = 7.9 msec, t' = 5 msec. Whence the relation resolving power of the installation was determined $\Delta\tau/\tau$ = 0.001%. The absolute value of the velocity is determined by the formula $V = h/\tau'$, where h is the dimension of the acoustic base recorded within the limits 0-10mm.

Ultrasound damping - α was determind by measuring a signal in the circuit of the amplitude detector, the follower, and of the digital volumetre. In so doing, we recorded the frequency and the amplitude of probing pulses.

Preliminary, we carried out experiments for transformer oil (in the delivery condition V = 1420 m/s), in which no change in the velocity of ultrasonic wavé attenuation was detected. The experimental results for ERF are presented in Fig.1 in the form of the dependence of the relative change of the ultrasound velocity on the electric field strength at converters' resonance frequency of 1 MHz . The ultrasound velocity was observed to increase with the electric potential. Here it was found that the unique solid phase concentration φ exerts an unambiguous influence on the dependence

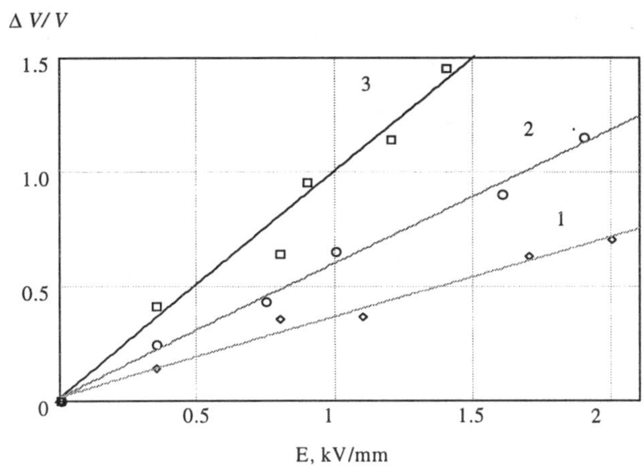

Fig 1. Relative change of the US velocity versus the electric field strength at different solid phase concentration.
1, φ=10 %; 2, 5; 3, 2.

$\Delta V/V \sim (E)$. The observed effect strengthens with the growth of φ.
Our experimental data show that the maximum change in $\alpha = \alpha(E)$ for the φ = 10% suspension of silica in the transforme oil at U = 8 kV made up 30%.
Measurement of the pulse delay time at the perpendicular orientation of the vectors and did not reveal any changes in the US velocity in ERF.
The experiments carried out with the electrorheological suspension based on transformer oil with diatomite, as a filler, at concentrations from 5% to 40% in longitudinal and transverse electrostatic fields with an intensity up to 30 kV/cm have shown that the characteristic time of the structural changes corresponding to a maximum increment of the absorption coefficient of acoustic vibrations in the tested medium is within the limits 0.3-4.0 sec.
For a more sophisticated treatment of kinetics of the processes that proceed in the ER-fluid when it is subjected to the action of an electrostatic field or when the latter is switched off, the following method of investigation is suggested. If the intensity of a varying electric field obeys the harmonic law $E(t) = \Delta E \cdot \cos \Omega t$,then the velocity of acoustic waves V(t) propagating in the investigated medium changes also according to the harmonic law as well

$$V(t) = V_0 + \Delta V \cos \Omega t \ , \qquad (1)$$

where v_0 is the velocity of propagation of acoustic vibrations in an undisturbed medium: $\Delta V = dV/dE \bullet \Delta E$ and Ω are the amplitude and the frequency of velocity modulation, respectivity.

If harmonic voltage of frequency ω is applied to the inlet of an interferometer with two linear acoustical-electrical transducers, a wave propagating in a resonator from a plane radiator can be described by the expression

$$\xi(x.t) = \xi_0 \exp(-\alpha x)\exp\{i\omega[t - \tau(x,t)]\}, \qquad (2)$$

where ξ_0 is the amplitude of vibrations in the plane of the radiating surface of the radiator; x is the coordinate reckoned off the radiator

$$\tau(x,t) = \frac{x}{< V(x,t) >}, \qquad (3) \qquad\qquad < V(x,t) >= \frac{1}{\tau(x,t)} \int_{t-\tau(x,t)}^{t} V(t')dt'. \qquad (4)$$

from relations (1), (3), (4) at $\Delta V << V_0$ (for definiteness sake we assume that $\Delta V > 0$) it is easy to find

$$\tau(x,t) = \frac{x}{V_0} - \frac{\Delta V}{V_0 \Omega}\left[\sin \Omega t - \sin\left(\Omega t - \frac{\Omega x}{V_0}\right)\right].$$

Then expression (2) acquires the form

$$\xi(x,t) = \xi_0 \exp\left[i\omega\left(t - \frac{x}{V_0}\right)\right] \exp\left\{i\theta\left[\sin \Omega t - \sin\left(\Omega t - \frac{\Omega x}{V_0}\right)\right]\right\}, \qquad (5)$$

where
$$\theta = \frac{\omega}{\Omega} \cdot \frac{\Delta V}{V_0}. \qquad (6)$$

An analysis of formula (5) shows that at smal values of modulation parameter θ ($\theta << 1$), provided the time of double pass of acoustic vibrations through a resonator differs from half the modulation period by the quantity multiple to the even number of modulation periods, i.e $\frac{2L}{V_0} \pm \frac{1}{2}\frac{2\pi}{\Omega_m} = 2m\frac{2\pi}{\Omega_m}$, $m = 1,2,...,$ (7)

the voltage at the interferometer outlet will be modulated only with respect to frequency and according to the harmonic law with the modulation frequency coinciding with that of the velocity Ω and with the frequency deviation equal to

$$\Delta\omega = \sqrt{1 + \left(\frac{1 - R^2}{1 + 2R\cos\gamma + R^2}\right)^2}\, \omega \frac{\Delta V}{V}$$

or with regard for the fact that $\Delta V = \frac{dV}{dE}\Delta E$

$$\Delta\omega = \sqrt{1 + \left(\frac{1 - R^2}{1 + 2R\cos\gamma + R^2}\right)^2}\, \frac{\omega}{V}\frac{dv}{dE}\Delta E \qquad (8)$$

Thus, with the frequency of modulation of the velocity of propagation of acoustic vibrations, which sutisfies relation (7), the frequency deviation of an electrical signal at the interferometer outlett is proportional to the amplitude of modulation of the velocity Δv. It should be noted that in the case of a weakly absorbing medium (a<<k) at the points of local minima of a slope of the phase characteristic(PC) curve (of maxima of the amplitude characteristic (AC)) the frequency deviation (5) acquires its maximum

$$\Delta\omega_{\max} = \frac{\sqrt{2(1+R^2)}}{1-R}\frac{\omega}{V}\frac{dV}{dE}\Delta E\,, \tag{9}$$

while its minimum at the points of local maxima of the slope of the AC minima) is

$$\Delta\omega_{\min} = \frac{\sqrt{2(1+R^2)}}{1+R}\frac{\omega}{V}\frac{dV}{dE}\Delta E \tag{10}$$

For measurements at the frequency $\omega/2\pi = 1$ MHz in water ($V_0 = 1.5\bullet10^3$ m sec^{-1}, a – $2.3 \bullet 10^{-2}$ m^{-1}), with a resonator length being L = 2 $\bullet10^{-2}$ m, the maximum and minimum values of frequency deviation (9) and (10) of voltage at the interferometer outlet equal, respectively, to $\Delta\omega_{max}/2\pi = 4.350$ Hz and $\Delta\omega_{min}/2\pi = 0.001$ Hz (at $r_1 = r_2 = 1$) will correspond to the amplitude of velocity modulation $\Delta V = 1.5\bullet 10^{-6}$ m sec^{-1} .Here, the minimum frequency of velocity modulation, provided condition (12) is fulfilled, for which relations (9) and (10) are valid, turns out to be equal to $\Omega/ 2\pi$ = 18.75 Hz.

Since the ER -fluid belongs to highly absorbing media (when R-0), the above arguments will be also valid and formula (10) will aquire the form

$$\Delta\omega = \sqrt{2}\,\frac{\omega}{\mathrm{v}}\frac{dV}{dE}\Delta E\,. \tag{11}$$

Proceeding form the data reported above, kinetics of the structural processes in an ER- fluid in an applied electric field varying according to the harmonic law can be investigated only at sufficiently low frequencies ($\Omega/ 2\pi$ = 0/1-30/0 Hz). Having

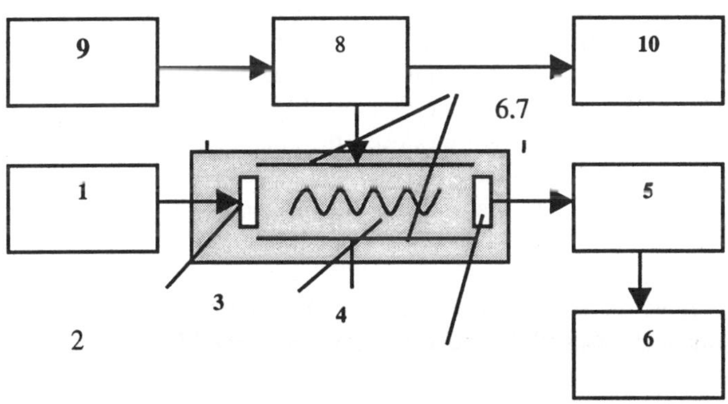

Fig. 2. Block-diagram of the experimental set-up.
1 , high-frequency generator; 2, radiating transducer; 3, investigated medium; 4, receiving transducer; 5, wide-band amplifier; 6, frequency deviation meter; 7, metallic plane parallel plates; 8, voltage multiplier; 9, law-frequency generator; 10, voltage meter.

the dependence of dv/dE on Ω which gives information on kinetics of the process.
For implementation of the suggested method we have made an experimental set-up, a block-diagram of which is given in Fig. 2.
A harmonic signal of the frequency 10^6 Hz from the generator outlet 0.1 is sent to a piezoelectric transducer emitting continuous vibrations into the investigated medium. The vibrations passed through the medium are transformed by a receiving acoustical - electrical transducer to an electrical signal practically linearly and are brought to the iamplifier inlet. The voltage-amplified signal arrives at the inlet of a frequency deviation meter.
Modulation of a measuring signal with respect to frequency is accomplished by changing in time the velocity of propagation of acoustic vibrations in the investigated medium in a varying electric field with its intensity up to 10^5 V/m. The field is initiatedby two plane-parallel plates perpendicularly to the direction of propagation of acoustic vibrations (a similar installation can be made for a longitudinal field). Voltage to the plates is applied from the outlet of a voltage multiplier, to the inlet of which a harmonic signal in the frequency range 5-30 Hz is sent from the outlet of generator G2.
The experiments have shown that the suggested method really «works' which is testified by the received frequency-modulated signal at the outlet of the receiving acoustical-electrical transducer. As a probing medium, the use was made of the electrorheological fluid based on transformer oil with diatomite, as a filler, with a concentration of 20%.

3. Conclusion

The results obtained indicate distinctly that ER-fluids in an electric field provide the possibility of creating various acoustic systems on their basis, namely, lenses, filters as well as allow their use in special arrangements, i.e. geophones, sound-absorbing interlayers which makes it possible to accomplish their remote control to improve accuracy and to extend the range of control of the main characteristics. In particular, we have created a model of the acoustic transducer, in which an ER suspension is used as a contact medium to increase the quality of acoustic control by improving the acoustic impedance matching of the transducer, the contact medium and a controllable item, that receives echo signals characterizing the acoustic contact and changes the electric intensity till a maximum amplitude of the received signal is attained.

Acknowledgments

The present work was carried out with the financial support of the Belorussian Fundament al Research Foundation, grant No. F99-281.

References

1. E.V. Korobko and I. A.Chernobay, *Inzenerno-Fisicheski Zhurnall.* **53,** N 2, (1985), p. 96.
2. E.V. Korobko V.D. Dreval and V.G. Kulichikhin, «Peculiar features in the rheological behavior of ERS», *Rheological Acta.* **33**, (1994), p.117.
3 A.V. Kolesnicov, *Ultrasound Measurements*, (Standards Press., Moskow, 1982), p. 248.
4. N.N. Brazhnikov, *Ultrasonic Phasemetry*, Energy Publ., Moskow, (1968), p.356.

ELECTRONICS AND MECHANICS – A BALANCE ACT IN ERF-APPLICATIONS

LUCIEN JOHNSTON, HORST ROSENFELDT
Fludicon GmbH, Landwehrstr. 55, 64293 Darmstadt, Germany

MANFRED KRÄMER
Schenck Fertigungs- GmbH, Landwehrstr. 55, 64293 Darmstadt, Germany

A new high voltage amplifier/supply unit specifically for the demands of ERF applications and ER-fluid testing has been designed, developed and tested. This unit is now commercially available. The heart of the amplifier is a 60kHz clocked blocking oscillator (B.O.) type high voltage transformer operating in current-control-mode. Main attributes of this patented transformer are a low leakage inductance coupled with a reduced influence of the internal capacitance on the natural frequency of the transformer. The B.O. type transformer is fed by the rectified mains voltage (230V AC). Controlled switching of the transformer on the primary side and high voltage dependent forced discharge of the charging capacitor secondary-side via high voltage sink allow externally variable HV (0-6kV) modulation up to 1kHz which is independent of the load. For detailed ER-fluid investigation and exact mechatronic control, accurate load current and load voltage monitoring, flashover indication and minimal signal distortion were design feature requirements.

1 Introduction

From an applications point of view, the ER-effect allows the design and development of truly mechatronic systems. Electrical signals result in a direct alteration of mechanical properties. The response time of commercially available ER fluids can only be extrapolated from system signal- and response-times. All data points to ER reaction times of less than 1ms. To make full use of this dynamic reaction, one of the key areas of application development is hence the availability of a supply unit for the necessary electrical field of several kilovolts which allows for accurate modulation in excess of 1kHz.

From an ER-fluid research and development standpoint, detailed knowledge of the actual load current and load voltage in any given test-rig setup will allow for a better understanding of the internal electrical field conditions and dielectric properties of the fluid under scrutiny. An electrical breakdown (flashover) counter with internal shutdown and automatic ramp-up, overload safety and accurate signal amplification are necessary design features of a state-of-the-art, compact high voltage amplifier for ERF technology.

2 HV Amplifier/Supply Unit Design

The newly designed HV amplifier/supply unit is based on a blocking oscillator type high voltage transformer. High voltage modulation is realised by closed loop control. Variance comparison of set point and instantaneous HV value is used, in combination with the primary side current-mode-control, as feedback signal for switching the transformer, the oscillator of which is clocked at 60kHz. Controlled on-off switching of the transformer primary-side and forced discharge of the charging capacitor via high voltage sink allow for a modulation of the output voltage which is to a large degree independent of the ERF-load. The transformer is fed by the rectified mains voltage. In comparison with linear power amplifiers using four quadrant output stages, this amplifier design allows for a very compact size and attractive cost-performance ratio. As the load current is a function of the output voltage, power loss is kept to a minimum and a higher load current is available at a reduced output voltage.

One of the main design features of the amplifier is the galvanic separation of primary and secondary sides of the transformer. By ensuring separation of earthing lines on the primary and secondary side, the actual load current flowing through the ER-fluid between the electrodes can now be accurately measured. The use of opto-couplers between primary and secondary sides avoids distortion of the measurements by any equalising currents.

3 Specifications

The following table gives a summary of the technical data of the amplifier.

Input signal (potentiometer or analogue)	0 – 6 V
Output voltage / current	0 – 6 kV / ≥ 20 mA
Mains supply	230 V AC
Max. output power	120 W
Monitors	Load current (1V/10mA) Load voltage (1V/kV)
Dimensions (width x height x depth)	257mm x 150mm x 256mm
Weight	3.8kg
Input impedance	40 kΩ
Response time (10%-90%)	0.3 ms (load 500kΩ//1nF)
Load capacitance	1 nF
Power consumption	150VA

Table 1: RheCon specifications

To allow for maximum flexibility in laboratory use or for field tests, the amplifier can be supplied as a benchtop unit (as shown in fig. [1]), or optionally inserted within a standard 19" rack. The compact dimensions allow for two RheCon amplifiers to be placed side-by-side within a single 19" rack. The size and also weight are inportant aspects to be taken into consideration when operating mobile ERF applications. For semi-active suspension systems utilizing four ERF dampers, for example, the units will fit comportably within the boot or trunk of the test automobile – hence allowing for accurate control and measurement of all relevant parameters (control voltage, output voltage, output current...) under field conditions.

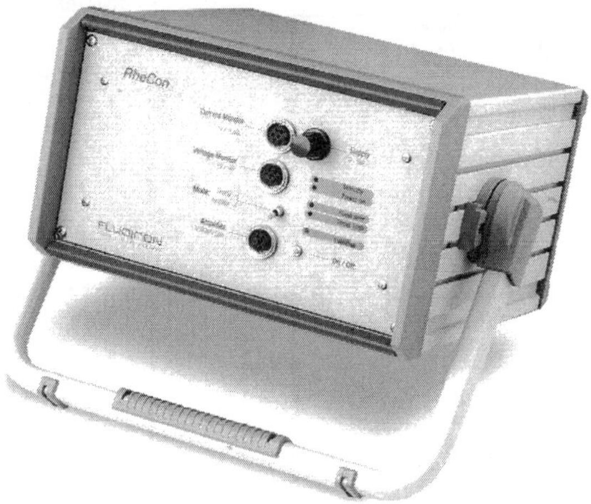

Fig.1: Photo of RheCon – benchtop unit

4 Sample measurement

The dynamic response of the amplifier has been thoroughly tested. Various wave forms (sine, triangle, step) at frequencies up to 1000Hz were carried out and the response documented.

In figure [2], the following settings were used:

- Load condition: 500kΩ//1nF.
- Step function: 0-6kV
- Frequency: 50Hz
- Scaling: 5ms/division (horizontal)
2kV/division (vertical)
- Upper curve: output voltage
Lower curve: control voltage

As can be seen from the curves in figure [2], a very high slew rate has been realised (in this case, 6kv/0.1ms). Even allowing for a conservative response time of 0.3ms, accurate modulation up to 1kHz for a sine wave can be realised.

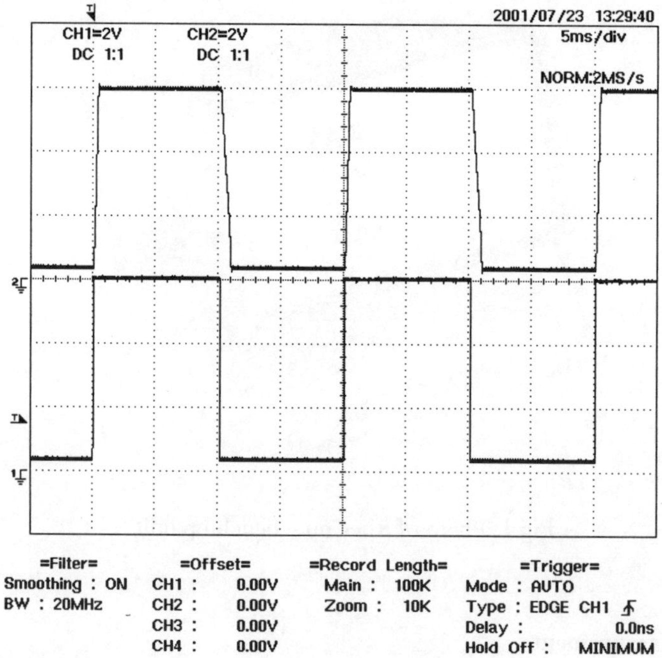

Fig. 2: Dynamic RheCon response.

5 Summary

Industrial application of the ERF technology requires the ability to combine the three basic components ER-Fluid – Mechanics – Electronics to a functioning unit. With the development of a compact, dynamic, cost-effective, CE-compliant and accurate ERF-controller, a further step towards this goal has been taken. At the same time, minimal nonlinear distortion, the accurate measurement of the load current and flashover registration will allow the implementation of the same amplifier in a laboratory environment for ER-fluid research activities.

One of the areas of ERF technology which can now be investigated more thoroughly is the volume flow dependence of load currents flowing through activated ER-fluids. This so-called sensor effect, although previously observed, has yet to be quantified to an extent which will allow the integration of this sensor function within the control algorithm of the ERF application. The use of an ER-fluid as power transfer-, control- and sensor-medium will further enhance the advantages of this technology.

DESIGN OF A HIGH-EFFICIENCY MAGNETORHEOLOGICAL VALVE

JIN-HYEONG YOO AND NORMAN M. WERELEY

Alfred Gessow Rotorcraft Center, Department of Aerospace Engineering University of Maryland, College Park, Maryland 20742 USA E-mail: wereley@eng.umd.edu

A high efficiency design was explored for meso-scale MR valves (<25 mm O.D.). The main design issues in the MR valve involve the magnetic circuit and nonlinear fluid mechanics. The performance of the MR valve will be limited by saturation phenomenon in the magnetic circuit and by the yield stress of the MR fluid. The non-dimensional plug thickness is evaluated as a basis for evaluating the valve efficiency. In this paper, design parameters of the MR valve were studied and an optimal performance was designed using these parameters. A maximum magnetic flux density at the gap was achieved in the optimized valve design based on a constraint on the outer diameter limitation. Valve performance was verified with simulation. A flow mode bypass damper system was fabricated and was used to experimentally validate the valve performance.

1 Introduction

The substantial field-induced yield stresses exhibited by MR fluids make possible numerous industrial applications. [1] Furthermore, there are several studies to develop active devices with electrorheological (ER) [2] or magnetorheological (MR) fluids [3] in hydraulic actuation systems. Using MR valves in hydraulic actuation systems accrues many advantages, including: valves have no moving parts, and mitigating complexity and durability issues in conventional mechanical valves. A Wheatstone bridge-based hydraulic actuator is being developed at the University of Maryland for compact actuation applications in unmanned air vehicles and helicopters. The MR valve is a key component of the actuation system. The actuator controls motion for a given pump flow rate, and avoids, to a large degree, fluid compliance. If a change in direction is required, the flow through each of the valves in the wheatstone bridge can be controlled smoothly via changing the applied magnetic field. Above all, most important advantages of MR valve will be the miniaturization and weight savings compared to a mechanical valve. This miniaturization can expand the application area to the aerospace industry, making it a feasible means of actuating trailing-edge flaps in helicopter blades [4] as an example. Two disadvantages are the block force and the cut-off frequency of this actuator. The block force depends on the yield stress of the MR fluid, and the cut-off frequency is a function of the response time of MR fluid.

In this paper, MR valves will be analyzed and evaluated experimentally to assess controllability of flow rate and pressure drop in the valve. Also, magnetic field analysis will be utilized to optimize electromagnetic performance with given material properties. Pressure drop achieved in the MR valve due to the applied current are also measured to validate the design methodology. The performance of the MR valve is evaluated using the non-dimensional plug thickness as a metric of the valve efficiency.

2 MR Valves

MR valves used in this study consist of a core and a flux return, and an annulus through which the MR fluid flows, as shown in Figure 1(a). The bobbin shaft is wound with insulated wire. A current applied through the wire coil around the bobbin creates a magnetic field in the gap between the core and the flux return. The magnetic field increases the yield stress of the MR fluid between the core and the flux return. This increase in yield stress alters the velocity profile of the fluid in the gap by creating a plug flow, decreases the volume flux, Q, and raises the pressure difference, ΔP, required for a given flow rate. We now consider the approximate parallel plate analysis of the flow mode valve system containing MR fluid, which follows the Bingham-plastic flow model. The typical velocity profile is illustrated in Figure 1(b). The total volume flux [5] is

$$Q = -\frac{bd^3}{12\mu L_a}(1-\bar{\delta})^2(1+\bar{\delta}/2)\Delta P \tag{1}$$

where the non-dimensional plug thickness is $\bar{\delta} = \delta / d$ and $\bar{\delta} = 0$ for Newtonian flow. Here, b is the circumference of the MR valve and μ is the differential post-yield viscosity. To verify the performance of the valve, an MR bypass damper was designed and fabricated. A schematic of a flow mode bypass damper is shown in Figure 2. The volume flux displaced by the hydraulic cylinder head is proportional to the cylinder head velocity, v_o, or $Q=A_p v_o$, where A_p is the area of the cylinder head minus the area of the shaft. Solving for the force acting at the shaft, the pressure drop becomes

$$\Delta P = \frac{F}{A_p} = \frac{12\mu L_a A_p}{bd^3(1-\bar{\delta})^2(1+\bar{\delta}/2)} v_0 \tag{2}$$

Measuring shaft force, F and velocity, v_o experimentally, we obtain $\bar{\delta}$ from

$$\frac{1}{2}\bar{\delta}^3 - \frac{3}{2}\bar{\delta} + 1 - \frac{12\mu L_a A_p^2}{bd^3}\left(\frac{v_0}{F}\right) = 0 \tag{3}$$

In evaluating the valve performance, the non-dimensional plug thickness in equation (1) plays a substantial role. Namely, in the case of $\bar{\delta} = 1$, there is no flow through the valve as in the case of an ideal valve (fully closed with infinite blocking pressure). Because the non-dimensional plug thickness is a function of flow rate and pressure difference in the valve, and has a value between 0 and 1, it is a good measure of valve efficiency.

2 Magnetic Circuit for the Valve

A meso-scaled MR valve design was explored. Its main parts are pictured in Figure 3. The yield stress of the MR fluid can be varied as a function of the applied magnetic field. Therefore, the magnetic field applied to the MR fluid must be correlated with prediction or measurement. The main design parameters of the magnetic circuit are gap distance, bobbin shaft diameter, bobbin core length, thickness of the flux return and number of windings in the coil, which is related to the length of the bobbin shaft. To achieve an efficient magnetic circuit, the area of the path of magnetic flux should be maintained constant. Theoretically, the smaller gap distance will be better because the permeability of the MR fluid at the gap is much less than the iron-based bobbin and flux return. Practical gaps typically range from 0.25 to 2 mm for ease of manufacture and assembly. A constant gap between the core and flux return should be maintained for uniformity of the magnetic flux in the gap. Furthermore, the distance of the gap should be considered in the concept of fluid flow. The gap determines the flow rate for a given pressure difference as shown in equation (1). In this study, the gap will be set to 0.5 mm. The flux return has two functions: one is as an element in the magnetic circuit and the other is as a connector to the tubing adapter with a seal as shown in Figure 3. In most cases of small valve design, the wall thickness of the flux return will be determined by the size of threads for the adapter.

In this paper, the magnetic field finite element analysis for the valve system is conducted using ANSYS/Emag 2D. The purpose of the analysis is to identify the effect of the saturation phenomenon in the magnetic circuit and to evaluate the effect of the design parameters at the valve. The magnetization data of Permalloy [6] steel and Hiperco 50-A material are used in this analysis. Throughout this analysis, two optimized models of MR valve are introduced. The first is a flux return with an outer diameter of 20 mm (Model 1), and second with an outer diameter of 25.4 mm (Model 2).

2.1 Bobbin diameter

The bobbin shaft radius is the most sensitive design parameter limiting the magnetic performance. In Figure 4, the magnetic flux density at the gap is plotted versus the bobbin shaft radius for a low permeability Permalloy steel, and a high permeability Hiperco 50A powder metallurgical alloy. It is desired to have as high a flux as possible. Both materials provide adequate flux above 4 mm bobbin shaft radius. However, the low permeability Permalloy rolls off below 4 mm much faster than the high Permeability Hiperco 50A, due to saturation. Therefore, to reduce the bobbin shaft and the valve diameter, higher permeability, more costly, magnetic materials must be used.

2.2 Core length

Figure 5 shows the trend of magnetic flux density at the gap as a function of core length. In the case of Permalloy, as the length of the core decreases, the magnetic flux density at the gap tends to increase because the magnetic flux density at the bobbin shaft decreases with decreasing core length. Decreasing the core length results in higher magnetic flux density at the gap by reducing saturation at the bobbin shaft. To verify this effect, the magnetic flux density at the gap was calculated for cases of 10, 4 and 2 mm core length, as shown in Figure 6. In the case of 2 mm core length has better performance. Figure 7 shows the magnetic flux density along the air gap with various active core lengths. In this narrow range of core lengths, as core length increases, the uniformity of the magnetic field across the active length improves, but the level of magnetic flux density also decreases.

2.3 Magnetic material

The maximum magnetic flux density at the gap is about 0.41 Tesla with Permalloy material in the case of Model 1. Considering the shear stress versus magnetic inductance of MRF-132LD (Lord Corp., 1999) and magnetization curve for Permalloy [6], 0.8 Tesla is the maximum magnetic flux density achievable with these materials. To achieve 0.8 Tesla at the gap, we should increase the outer diameter to 25.4 mm as Model 2. The maximum magnetic flux density of this valve will be about 0.8 Tesla with 1.6-Ampere input current, as shown in Figure 8(a). Table 1 shows the main design parameters of the each model. In the case of high permeability material (Hiperco 50-A) in Figures 4 and 5, the bobbin shaft can be reduced, and the core length increased, while still obtaining the same performance as the larger valve made of Permalloy steel.

3 Measurement of the Magnetic Flux density

In this experiment, a thin film (FH-301-060, F.W.BELL) hall sensor is used to measure the magnetic flux density at the small gap and 5080 hand-held gauss meter (F.W.BELL) is used for calibrating the hall sensor. Figure 8(b) compares the experimental data with the analytical prediction from ANSYS/Emag 2-D for Model 2. Taking the sensing error of the hall sensor into the consideration, the results in Figure 8(b) are in good agreement with each other. With these results, we conclude that the simulation result with ANSYS/Emag 2-D will be accurate enough to predict the performance of the magnetic field at the gap when filled with MR fluid.

4 Analytical results

Due to the non-linearity of the MR-fluid, determination of the performance of the valve requires numerical analysis. In the following analysis, we use a commercially available MR-fluid data, namely MRF-132LD (Lord Corp., 1999). To evaluate the valve performance, the pressure difference and flow rate will be calculated. For the Bingham-plastic model, the material properties of the dynamic yield stress, τ_y, and plastic viscosity, μ, of the MR-fluid will be required as a function of applied magnetic field. The dynamic yield stress for this fluid can be approximated by a cubic equation of the magnetic field, so that

$$\tau_y = a_3 B^3 + a_2 B^2 + a_1 B + a_0 .$$

The polynomial coefficients were determined by least-squares fit of the dynamic yield stress data as a function of magnetic field from data supplied by Lord Corporation (1999), and are: a_0=-0.877 kPa, a_1=17.42 kPa/Te, a_2=122.56 kPa/Te^2 and a_3=-86.51 kPa/Te^3. To simplify the analysis, the MR fluid is assumed to have a nominal plastic viscosity of 0.6 Pa sec.

The pressure difference versus flow rate diagram is shown in Figure 11(a) with experimental results for the case of Model 1. As applied current increases, the bobbin shaft saturates at low current, so that the yield stress is low. The shear stress of the MR fluid in the case of the Model 2 is shown in Figure 9. The maximum shear stress is about 47 kPa with 1.6-Ampere input current. Comparing the pressure differences for a fixed value of the flow rate in Figure 11(b) with Figure 11(a), we conclude that the valve performance of Model 2 is higher than that of Model 1.

5 Experimental results

To validate this non-dimensional analysis using the nonlinear Bingham-plastic shear flow and the magnetic circuit design with ANSYS/Emag 2-D, a high stroke ($\approx 20cm$) MR damper was constructed. The damper consists of four main parts: a hydraulic cylinder, industrial tube fittings, an accumulator and an MR bypass valve as pictured in Figure 10. The damper is charged with MR fluid, MRF-132LD (Lord Corporation). The accumulator connected to the cylinder increase the overall pressure inside the damper for the purpose of reducing the amount of air bubbles inside and preventing cavitation.

For the experimental validation of the flow mode bypass valve equations, force measurement from triangular displacement cycles on an MTS servo-hydraulic testing machine were recorded on a computer. The MR bypass valve damper shaft was mounted in the clevises, and the shaft of the damper was oscillated. The shaft displacement was measured using an LVDT in the hydraulic actuator and the applied load was measured using a 250 lb load cell.

In Figure 11, the flow characteristics of the Model 1 and Model 2 valve are plotted along with simulation results. The yield stress data of MR fluid is applied from the data sheet of Lord Corporation. Comparing Figure 11(a) with Figure 11(b), the enhancement of the flow characteristics is apparent. Figure 12 shows the test results of non-dimensional plug thickness of Model 1 and Model 2 valve with simulation results, respectively at a constant current input. From the valve performance point of view, we can say that the Model 2 valve has a higher efficiency than Model 1 because the input current and number of windings are same for the each model. Figure 13(a) shows the time responses of the each valve with same input step current. As the results of pressure difference are different, the response is not directly comparable. Figure 13(b) shows the normalized step responses for each valve. Comparing the step responses of each valve, the time response of the Model 2 is comparable to that of Model 1, in spite of the higher-pressure difference at the valve.

6 Conclusion

Considering the shear stress versus magnetic inductance of the MR-fluid (MRF-132LD, Lord Corp., 1999) and magnetization curve for Permalloy [6], a maximum magnetic flux density at the gap was achieved with optimized design and was verified with simulation and experiment. Based on these results, we conclude the following:

1. Using low permeability Permalloy steel material, 1500 kPa block pressure can be achieved with our MR valve design with a 25.4 mm outer diameter.
2. Using high permeability material, the size of the valve can be reduced and the active core length can be increased for a higher blocking pressure.
3. The non-dimensional plug thickness is a useful measure of valve efficiency, and as such, efficiency of the Model 2 valve is higher than Model 1.

References

1. Carlson, J. D. Catanzarite, D. M. Clair, K. A. S. “Commercial Magnetorheological Fluid Devices Technology,” International Journal of Modern Physics B, v.10, n.22-23, 1996, p.2857
2. Zheng Lou, Robert D. Ervin and Frank E. Filisko, “Behaviors of Electrorheological Valves and Bridges,” Proceedings of the International Conference on ER fluid, 15-16 October 1991, pp.398-423.
3. J. -H. Yoo, J. Sirohi and N. M. Wereley, “Design of a MR hydraulic power actuation system,” SPIE’s 8th Annual International Symposium on Smart Structures and Materials, 4-8 March 2001, Long Beach CA. Paper No. 4327-22.
4. Milgram, J. H. and Chopra, I., “Parametric design study for actively controlled trailing edge flaps,” Journal of the American Helicopter Society, v.43, n.2, 1998, p110-119.

5. Kamath, G. M., Hurt, M. K., and Wereley, N. M., "Analysis and Testing of Bingham Plastic Behavior in Semi-Active Electrorheological Fluid Dampers," Smart Materials and Structures, v. 5, n. 5, pp.576-590.
6. Herbert C. Roters, <u>Electromagnetic Devices</u>, John Wiley & Sons, Inc., 1941.

Table 1. The dimensions of the valves

	Outer Dia.	Bobbin Dia.	Core L/each	Air gap	No. of windings	Max. Te at the gap
Model 1	20.0 mm	7 mm	2 mm	0.5 mm	160 turns	0.41 Tesla
Model 2	25.4 mm	14 mm	3 mm	0.5 mm	160 turns	0.80 Tesla

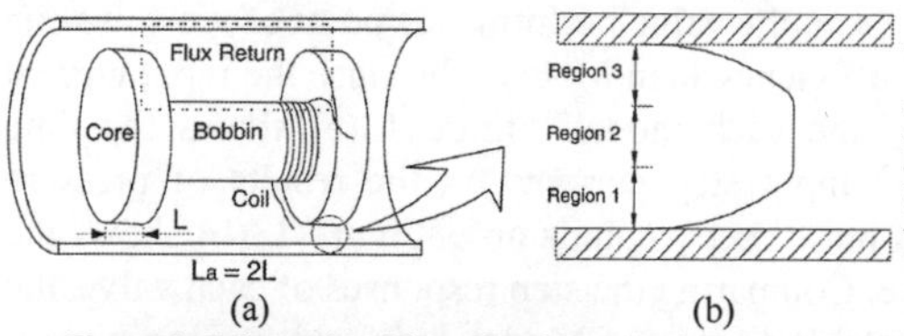

Figure 1. Schematic of the valve (a) and velocity profile (b).

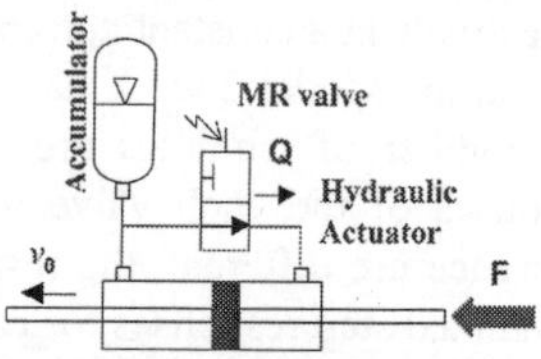

Figure 2. Test configuration for the MR valve

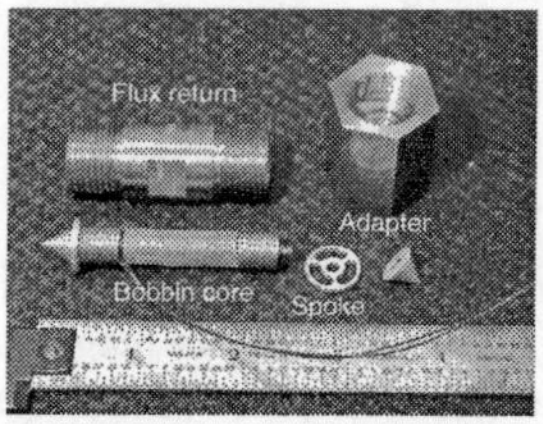

Figure 3. Photograph of a design of the MR valve parts.

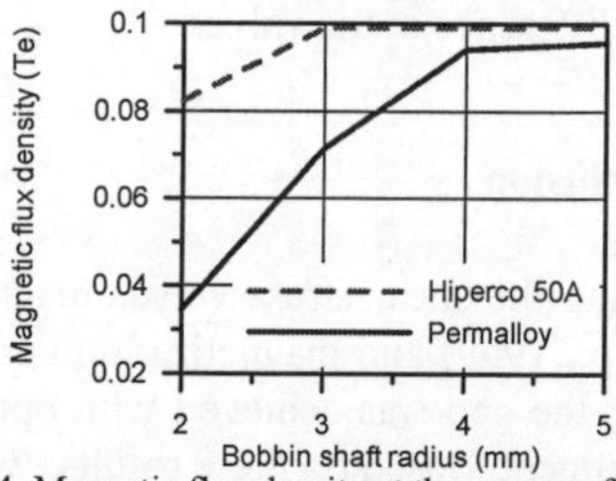

Figure 4. Magnetic flux density at the gap as a function of the bobbin shaft radius. (gap = 0.5 mm, air)

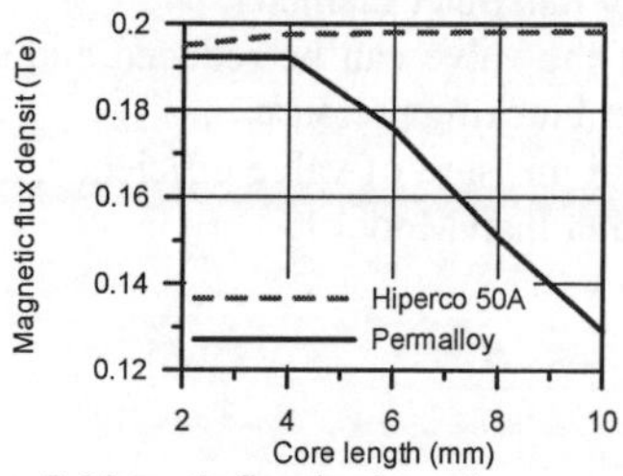

Figure 5. Magnetic flux density at the gap as a function of the core length. (gap=0.5 mm, air)

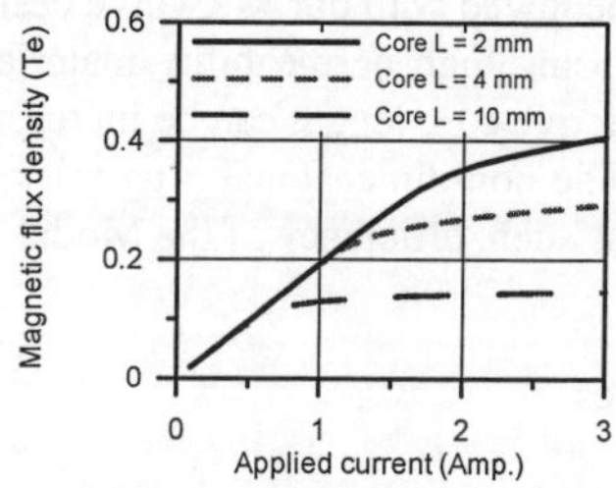

Figure 6. The magnetic flux density at the gap with various core lengths. (Permalloy, Model 1)

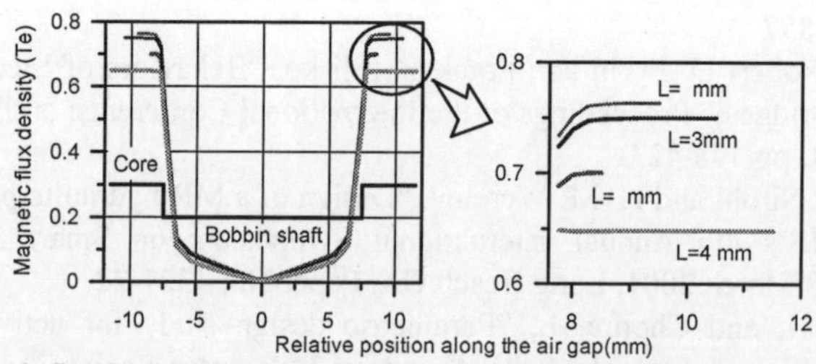

Figure 7. Magnetic flux density along the gap with various core lengths. (Permalloy, Model 2, i=1Amp)

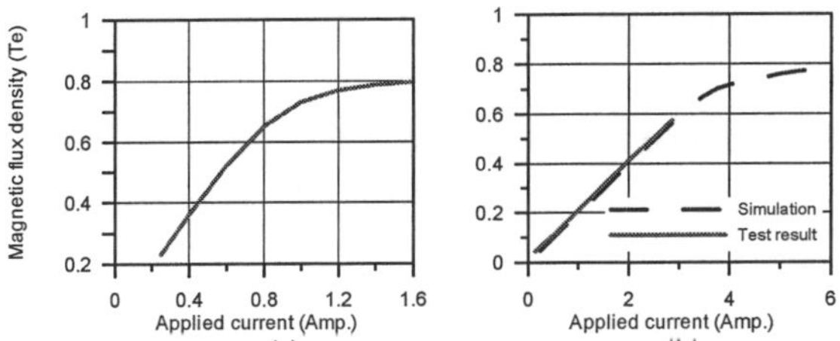

(a) (b)

Figure 8. Performance of the magnetic flux density at the gap (a) with MR fluid permeability. (b) comparison between test and simulation, air. (Permalloy, Model 2)

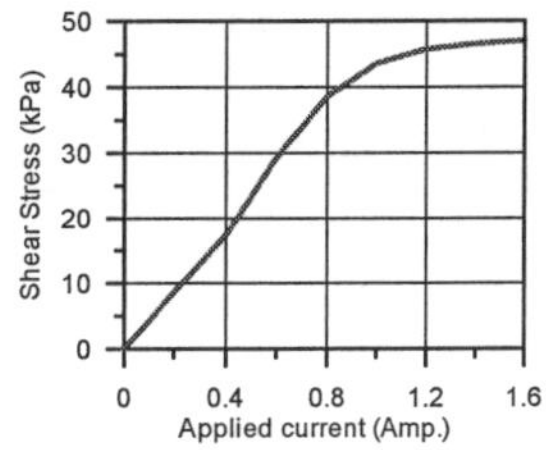

Figure 9. Shear stress of the MR fluid at the gap. (Permalloy, Model 2)

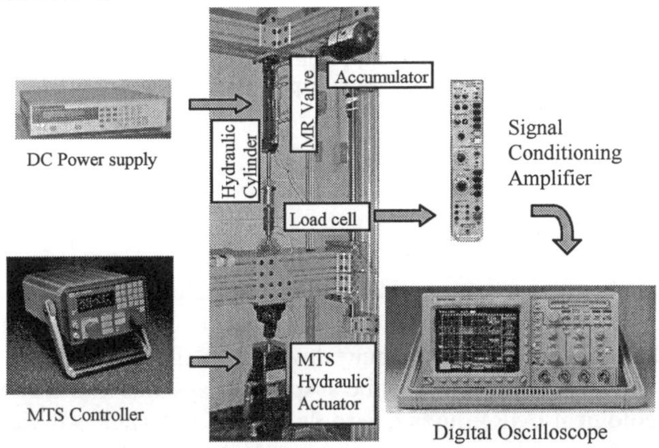

Figure 10. Experimental setup for performance measurement of the MR valve.

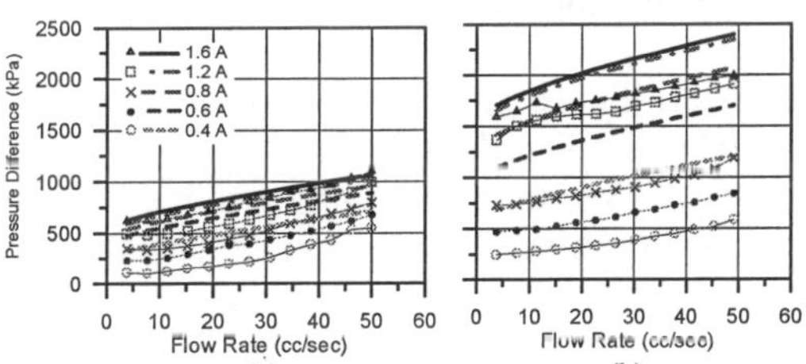

(a) (b)

Figure 11. Flow characteristics of the Model 1(a) and Model 2(b) MR valve. (Symbols: test and Lines: simulation)

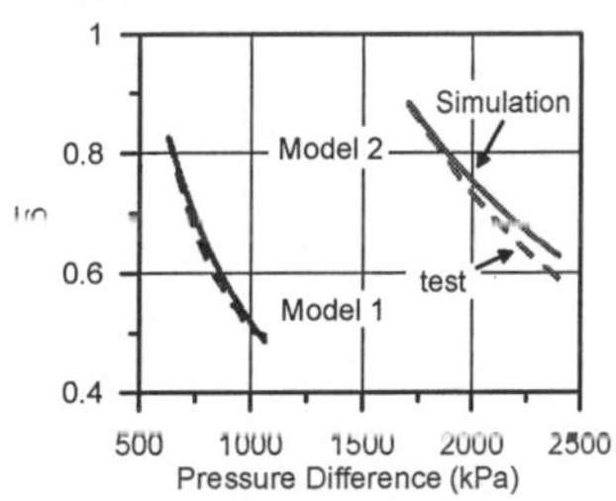

Figure 12. Comparison of the non-dim. δ between Model 1 and Model 2. ($i = 1.6\ Amp.$)

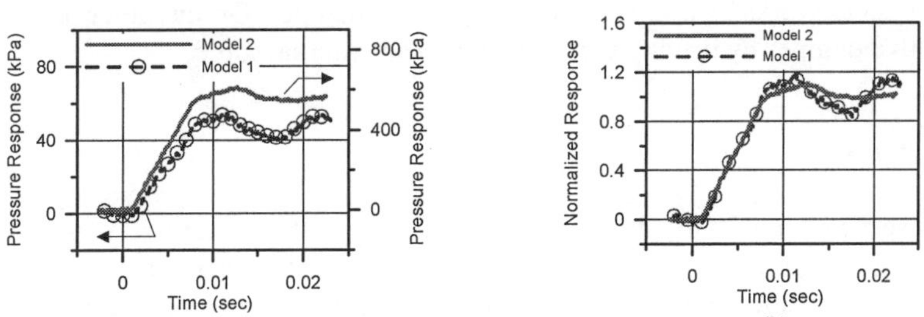

(a) (b)

Figure 13. Comparison of the time responses of (a) pressure difference and (b) normalized pressure difference between Model 1 and Model 2. (Test results)

ACTIVE CONTROL OF ROD VIBRATIONS, USING MAGNETIC FLUIDS

Z. WANG, G. BOSSIS, O.VOLKOVA

Laboratory of Condensed Matter Physics, Universite de Nice – Sophia Antipolis, Parc Valrose, 06108 Nice Cedex 2, France

V. BASHTOVOI, M. KRAKOV

Laboratory of Thermomechanics of Magnetic Fluids, Belarussian State Polytechnical Academy, 65 F.Scaryna Ave., Minsk, 220027 Belarus

The use of magnetic fluids in controlling rod vibrations is investigated. A prototype of ferrofluid vibration damper is designed and experimentally set up based on the principle of anti-resonance. The efficiency of this damping system is verified in experiments and well explained with classical equations of motion. The improvement of the present system towards active control of rod vibration is also discussed.

1 Introduction

One of the most prospective applications of the magnetic fluids is in the field of vibration controls. There have been several attempts in using the tunable viscosity of these fluids to develop the active damping devices. In the cases where high damping forces are needed, such as in the automobile suspension systems, more attentions are paid to the magnetorheological (MR) fluids due to their large field-induced yield stresses. The relatively 'softer' ferrofluids, on the other hand, have shown their advantages in constructing damping devices with more compact structure and less energy consuming[1,3].

In this paper, we present the use of ferrofluids in active control of rod vibrations based on the principle of anti-resonance. The damping device consists of two main parts: the primary mass, which is a closed container fixed to one end of the rod; the secondary mass is a magnet, or, as well, an electromagnet, which is floating on the ferrofluid inside the primary container. During the vibration of the system, the two masses oscillate out of phase with each other. The kinetic energy of the primary mass is transferred to the secondary one, and then dissipated by viscous friction in the ferrofluid. The vibration of the rod can also be controlled by tuning the viscosity of the ferrofluid with the magnetic field. Experiments and theoretical calculations have shown that when a permanent magnet is used, the system works as a Lanchester-like passive damper. On the other hand we can obtain an active damper by driving the magnet with an external force.

2 Experiments

A cross section view of the damper is shown in Fig.1. It is composed of a cylindrical box containing a cylindrical magnet magnetized along its symmetry axis. The box consists in a plastic tube with an inner diameter of 1.6cm and two plastic plugs that are used to block the two ends of the tube. The inner length of the box, L_l, is chosen to be 5cm, but it can be increased up to 7cm by varying the length of the plugs. To exclude the influence of compressing air, two small holes are opened on the top of the tube. An

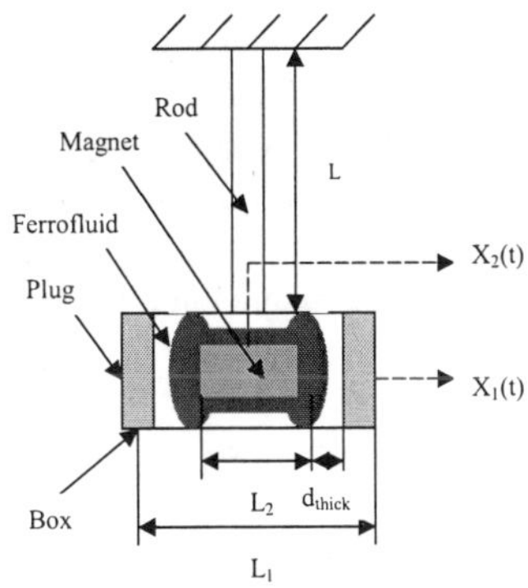

accelerometer is fixed at one end of the box and allows to measure the transient displacement. The total value of the primary mass, m_1, is 107.15g.

In a first step, a permanent magnet is used as the secondary mass, which has a diameter of 1.37cm and length, L_2, of 2cm.

Fig.1 Scheme of the experimental setup of the ferrofluid damper.

Most of the ferrofluid aggregates at the two ends of the magnet due to the magnetic field distribution and a dumbbell-like structure is formed as shown in Fig.1. When the viscosity of the ferrofluid is large enough, the magnet can be well centered in the box. Thus there will be no solid-friction to prevent its motion, and the viscous dissipation in the ferrofluid can improve the damping behavior of the system. Unlike the MR dampers where the device is generally filled with MR fluids, here only a small amount of ferrofluid is needed . The thickness of the ferrofluid layer at the two ends of the magnet is around 3mm in the axis direction. When the distance between the magnet and one plug is smaller than this thickness, the repulsive force resulting from the compression of the ferrofluid will bounce the magnet back. We have measured this repulsive force, F_{rep}, experimentally with a special device and found that it can be expressed by an exponential function. The magnet and the attached ferrofluid form the secondary mass, m_2, with a value of 27.3g. The steel rod connecting the device with a fixed table has a diameter of 5mm and length of 60cm. Giving an initial displacement to the rod in the horizontal direction, the system begins to oscillate.

The vibration behavior of the system is studied in two cases. One where the motion of the magnet is blocked, the only damping factor comes from internal energy dissipation inside the rod. In the other case the magnet is released, then the viscous friction due to the ferrofluid will play an important role. The experimental results are shown in Fig.2, where the initial displacement of the box is 2.0cm. In the blocked case, the device works as a normal single degree of freedom system under free vibration. The frequency of damped vibration ω_d and the logarithmic decrement δ are given by

$$\omega_d = \sqrt{1-\zeta^2}\,\omega_n, \qquad \delta = \frac{2\pi\zeta}{\sqrt{1-\zeta^2}}, \qquad (1)$$

where the damping ratio ζ is defined as the ratio of the damping constant of the rod C_1 and the critical damping constant $Cc=2(m_1+m_2)\omega_n$ with the natural frequency

$\omega_n=[K_1/(m_1+m_2)]^{1/2}$. K_1 is the stiffness of the steel rod. From the experimental results in Fig.2(a), ω_d and δ are found to be 21.4 s^{-1} and 0.0247, respectively. Then the values of C_1 and K_1 are respectively 2.2×10^{-2} Kg/s and 61.7 N/m, respectively.

When the magnet is released, it oscillates out of phase with the box. The viscous dissipation in the ferrofluid begins to play a dominant role. The viscous friction coefficient, C_2, of the ferrofluid between the magnet and the plastic tube is found equal to

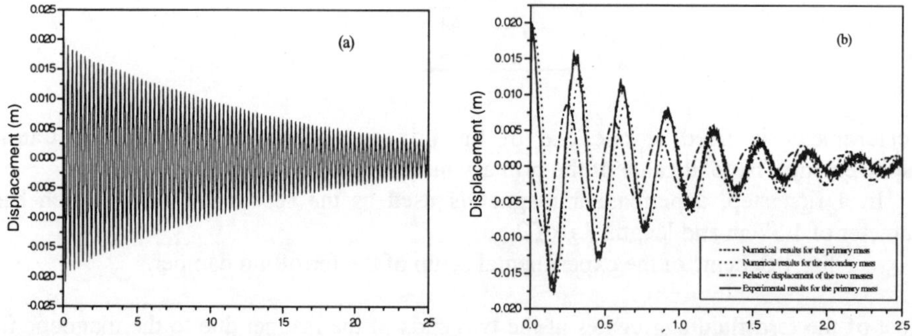

Fig.2 Experimental results for the transient displacement of the primary mass in the cases of (a) blocked magnet and (b) unblocked magnet. Numerical simulation results for the transient displacement of the primary and secondary masses also given in (b).

1.0 Kg/s. In this case, the system works as a damped dynamic absorber with two degrees of freedom. Since there is no spring force between the primary and secondary masses, the device is a Lanchester-like damper, except for the short-range repulsive force which intervenes only when the magnet is touching the plug. Fig.2(b) shows the high efficiency of this damper. The logarithmic decrement δ is increased to about 0.32. Thus the oscillation of the primary mass is stopped after a few periods. Similar results have been obtained for different initial displacements of the box. The results are found to be insensitive to the initial position of the magnet in the box. Because the viscous friction of the ferrofluid is relatively large, the maximum relative displacement between the centers of the magnet and of the box is about 1cm which is smaller than LL = 1.5cm. It means that the magnet mainly oscillates in the middle part of the box. The influence of the short-range repulsive force is thus not important in this regime.

3 Theoretical Analysis

3.1 *Optimum damping conditions*

For the damping system with two degrees of freedom in free vibration, the equations of motion of the two masses are :

$$m_1\ddot{x}_1 + C_1\dot{x}_1 + C_2(\dot{x}_1 - \dot{x}_2) + K_1x_1 + F_{rep}(xx) = 0, \tag{2}$$

$$m_2\ddot{x}_2 - C_2(\dot{x}_1 - \dot{x}_2) - F_{rep}(xx) = 0, \tag{3}$$

where x_1 and x_2 are the transient positions of the centers of the primary and secondary masses, respectively, and $xx = x_1 - x_2$. Numerical simulations are carried out to study the dynamics of this system.

As an example, the simulation results of the transient displacements of the two masses are shown in Fig.2(b) for the case of unblocked magnet. The experimental parameters are used in the simulation and the initial position of the magnet is supposed to

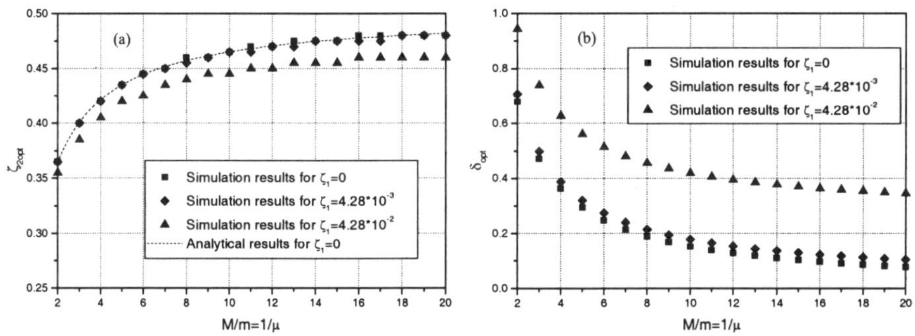

Fig.3 Simulation results for the optimum damping constant ζ_{2opt} (a) required for optimal operation of the device in free vibration and the corresponding logarithmic decrement δ_{opt} (b). The analytical result of ζ_{2opt} for a Lanchester damper under harmonic external force is also given in (a).

be in the middle of the box. It can be seen that good consistency has been reached between the simulation and experimental results. The temporal evolution of the kinetic energy, potential energy and the energy dissipations in the rod and ferrofluid are also calculated simultaneously. The results indicate that the damping effect in this case is due to the large viscous energy dissipation in the ferrofluid, which results from the out-of-phase oscillation of the secondary mass relative to the primary one. The increase of this energy dissipation can lead to further improvement of the damping behavior. The optimum damping condition can be found by simulations.

Defining the mass ratio $\mu=m_2/m_1$, the natural frequency of primary mass $\omega_n=(K_1/m_1)^{1/2}$, the damping ratios $\zeta_1=C_1/2m_1\omega_n$ and $\zeta_2=C_2/2m_2\omega_n$, the dimensionless form of the equations of motion can be written as:

$$\ddot{x}_1^* + 2\zeta_1\dot{x}_1^* + 2\mu\zeta_2(\dot{x}_1^* - \dot{x}_2^*) + x_1^* = 0, \quad (4)$$

$$\ddot{x}_2^* - 2\zeta_2(\dot{x}_1^* - \dot{x}_2^*) = 0, \quad (5)$$

where $x^*=x/x_{10}$ with x_{10} the initial displacement of the primary mass, and $t^*=t\cdot\omega_n$. The repulsive force has been neglected in the equations. It assumes that the inner length of the box is large enough so that the separation between the ends of the magnet and the plugs could never be smaller than the thickness of the coating ferrofluid. The experimental result in Fig.2(b) corresponds to this case.

In calculations, three different values are chosen for the parameter ζ_1. The first one is zero which corresponds to the standard Lanchester damper. The second one (4.28×10^{-3}) is the value calculated with the experimental parameters discussed in Sec.2, and the third one is taken to be ten times the second one. For each ζ_1, the optimal value of the damping constant ζ_{2opt} which gives the largest logarithmic decrement δ_{opt} is found for different mass ratio μ. The results for ζ_{2opt} and corresponding δ_{opt} are shown in Fig.3(a) and (b), respectively. It can be seen that for a fixed mass ratio μ, ζ_{2opt} decreases with the increase of the damping constant of the rod ζ_1, while δ_{opt} increases with it. But since C_1, and thus ζ_1, is relatively small in our experiment, the results obtained with the experimental parameters are very close to the ones with $\zeta_1=0$, i. e. the Lanchester damper. When the Lanchester damper is vibrating under a harmonic external force, analytical calculations had shown that the optimal damping constant required for most favorable operation is

$$\zeta_{opt} = \sqrt{\frac{1}{2(2+\mu)(1+\mu)}}.$$

This analytical result is well consistent with our numerical values for the free vibrating case ,cf. Fig. 3(a). It means that if ζ_1 of the damping system is small, this analytical result can be used to find the optimal damping constant ζ_{2opt} in both the free and forced vibration cases.

3.2 *Active control of vibration*

In the experimental setup of Sec. 2 where the mass ratio μ is 0.255, the damping constant ζ_2 is 0.76 which gives a logarithmic decrement of δ = 0.32. The simulation results suggest that δ can be further increased to 0.39 if the optimum value of ζ_{2opt} = 0.42 is used for this system. A direct way to do this is to decrease the viscous friction coefficient C_2 from 1.0 kg/s to 0.55 kg/s and fix all the other parameters. However, in experiments C_2 is determined by the viscosity and thickness of the ferrofluid attached on the magnet. The precise control of C_2 for a passive damper is quite difficult, because this can only be done by using proper sample of ferrofluid and controlling the amount of the sample on the magnet. Especially when the mass ratio is changed continuously due to different applications, the tuning of C_2 will be impractical.

An other solution would be to impose, by means of an appropriated device like an electromagnet located at the extremity of the tube, a force on the magnet which plays the role of a secondary mass. A proper choice of the amplitude and initial phase of this "internal" force can optimize the motion of the magnet in order to damp the motion of the primary mass. Then we have actually an active damper instead of a passive one.

To simplify the presentation, let us neglect the damping constant C_1 of the rod and suppose that the system is forced by an external harmonic force $F_{ext} = F_{0ext} sin\,\omega t$. The "internal" force which acts on the magnet is supposed to be sinusoïdaL $F_{mag} = F_{0mag}$ $sin(\omega t+\phi)$. Then the equation of motion of the two masses can be written as

$$m_1\ddot{x}_1 + C_2(\dot{x}_1 - \dot{x}_2) + K_1 x_1 - F_{0mag}\sin(\omega t + \phi) = F_{0ext}\sin\omega t,$$

$$m_2\ddot{x}_2 - C_2(\dot{x}_1 - \dot{x}_2) + F_{0mag}\sin(\omega t + \phi) = 0.$$

Assuming the solution to be $x_j = X_j exp(i\omega t)$, the steady-state solution for X_1 is obtained:

$$\frac{X_1}{\delta_{st}} = [\frac{g^2(1+\gamma_F\cos\phi)^2 + (2\zeta_2 - g\gamma_F\sin\phi)^2}{(g^3-g)^2 + 4\zeta_2{}^2(1-g^2-\mu g^2)^2}]^{1/2},$$

where $\delta_{st} = F_{0ext}/K_1$, $g = \omega/\omega_n$ and $\gamma_F = F_{0mag}/F_{0ext}$. When γ_F equals to zero, this solution will return to that of the Lanchester damper. It is easy to see that if the initial phase and relative magnitude of the magnetic force are given by:

$$\phi_{opt} = tg^{-1}(-2\zeta_2/g), \qquad \gamma_{Fopt} = -\frac{1}{\cos\phi_{opt}},$$

X_1 will be reduced to zero, i. e., there will be no vibration of the primary mass in the steady state. In practice, however, the frequency of the external force may be unknown or may keep on changing, thus the determination of ϕ_{opt} and γ_{Fopt} is not always easy; furthermore the amplitude of the internal force should become very large in order to keep a zero motion at low frequencies.

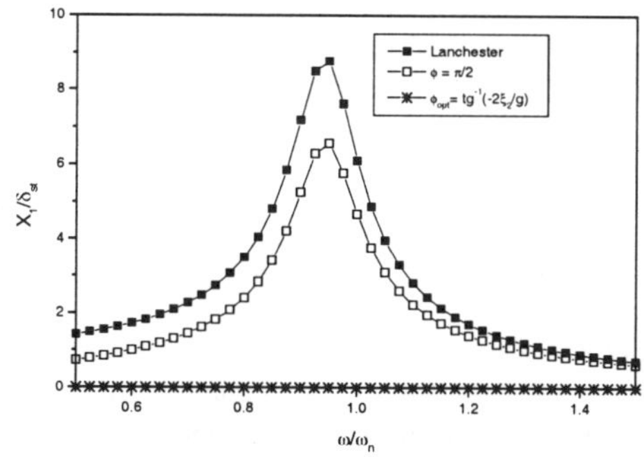

Fig.4 Simulation results of the optimal value of X_1/δ_{st} for the Lanchester damper and the two cases of applying additional magnetic interaction force to the secondary mass. The damping constant ζ_2 is taken to be the optimum value ζ_{2opt} of the Lanchester damper in all the three cases.

A more simple strategy is to fix the phase angle of the internal force at some given value and adjust the amplitude parameter to optimize the damping. For instance, if the phase ϕ is fixed to $\pi/2$, the condition required for getting a minimum X_1 is $\gamma_F = 2\zeta_2/g$. The corresponding normalized displacement is shown in Fig.4 and compared with the displacement obtained for a passive Lanchester damper. We see that the damping is well improved especially below the natural frequency. In practice the internal force on the permanent magnet can be more complicated than a pure sinusoidal force, but the results shown here indicate the possibility of getting an improvement of the damping effect with this device by introducing a magnetic interaction force on the secondary mass.

3.3 *Influence of the repulsive force*

Up to now, we have neglected the influence of the short-range repulsive interaction between the ferrofluid and the plugs located at the extremities of the tube. However, the length of the box L_1 is usually limited in practice. When the viscosity of the ferrofluid is low (but still strong enough to support the magnet), the secondary mass can move a relatively long distance in the box and touch the plugs frequently. In this case, the interaction between the ferrofluid and the plugs affects strongly the dynamics of the system. Simulations are carried out to investigate this effect by using Eq.(2-3) and the exponential form formula for F_{rep}. The initial displacement of the primary mass is chosen to be 1cm. L_l. is changing from 3cm to 7cm. The other experimental parameters are kept the same as before.

The results show that when L_l is equal to 6cm or larger, the repulsive force plays no role as expected. If L_l is much smaller – smaller than 3.5cm, the viscous energy dissipation is very low due to the limited relative displacement between the two masses. The damping behavior is similar to the blocked case. However, in the region between 3.5cm and 6cm, the repulsive force is found to be very helpful in improving the damping behavior of the system, especially for smaller ζ_2. For example, δ can be as high as 1.34 in the case of L_l=5cm and ζ_2=0.18. In fact, the motion of the primary mass is nearly stopped after only one cycle of vibration. The observation of the transient displacement in this case shows that there are two large impacts of the two masses during vibration due to their large relative velocity. During the shock, an important part of the kinetic energy of the primary mass is transferred to the secondary one. This energy is then quickly dissipated in the ferrofluid with the relative motion of the two masses. As a result, the potential energy, and thus the vibration amplitude of the primary mass, decreases quickly. Similar behavior happens for the cases of L_l=5cm with smaller ζ_2 and L_l=4cm with nearly all values of ζ_2 studied. It implies that by choosing L_l. and ζ_2 carefully, larger damping effects can also be achieved by taking advantage from the interaction between the ferrofluid and the plugs.

However, the influence of the repulsive force is very sensitive to experimental conditions. For example, if we increase the initial displacement of the box to 2cm and fix the other parameters, the region of L_l showing an improved damping effect is shifted between 5cm and 7cm. Thus the dimension of the free space between the magnet and the extremity of the box should be considered very cautiously in designing such a damper.

4 Conclusion

The use of ferrofluids in developing rod vibration damping systems has been studied. Based on the principle of anti-resonance, a prototype of one-dimensional ferrofluid damper was designed and experimentally studied. After measuring the repulsive force between the walls and a magnet surrounded by a ferrofluid,, a dynamic model is proposed which well reproduce the experimental measurements of the damping effect. Numerical simulations based on this model have been carried out to find the optimum damping conditions. It is also shown that this system can be modified to act as an active damper with still improved properties.

References

1. Magnetic Fluids and Applications Handbook, (Eds: B. M. Berkovski, V. G. (Bashtovoi). Begell House Inc.Publishers, NY, 1996.

2. V.G. Bashtovoi et al. Magnetic Fluid Shock Absorber. USSR Patent № 1692218, 1991,. Shock Absorber. USSR Patent № 1692217, 1991.

3. V.G. Bashtovoi, G.Bossis, D.Kabachnikov, M.Krakov,O.Volkova "Modelling of magnetic fluid support",J.M.M.M. (2002) to appear

PERFORMANCE EVALUATION OF ER SHOCK DAMPER SUBJECTED TO IMPULSE EXCITATION

H. J. SONG, S. B. CHOI, J. H. KIM
Smart Structures and Systems Laboratory, Department of Mechanical Engineering
Inha University, Incheon 402-751, Korea
E-mail: seungbok@inha.ac.kr

K. S. KIM
Impact Dynamics and Fatigue Fracture Laboratory, Department of Naval Architecture and Ocean Engineering, Inha University, Incheon 402-751, Korea
E-mail: ksukim@inha.ac.kr

This paper presents shock responses of a cylindrical shock damper featuring an electrorheological(ER) fluid. An appropriate size of the ER shock damper is designed and manufactured on the basis of the field-dependent Bingham model of the ER fluid. The damping force of the ER shock damper is evaluated with respect to the excitation frequency at various electric fields. Subsequently, the ER shock damper is incorporated with a 1 degree-of-freedom (DOF) dynamic system in order to investigate shock responses. A skyhook controller is formulated to attenuate acceleration level due to impulse excitation and empirically realized. Controlled acceleration magnitudes are evaluated and compared with uncontrolled ones at various excitation frequencies.

1 Introduction

Recently, numerous types of ER(electrorheological) or MR(magnetorheological) fluid dampers have been proposed in order to achieve vibration control for vehicles subjected to bump and random road conditions [1-6]. Most of dampers in this sense have high-performance controllability of vibration in the relatively low excitation frequency range(below 20Hz). However, when we deal with vibration control of dynamic systems subjected impulsive load or explosive load, the damper should respond in a rapid motion(up to 500Hz). Conventionally, in order to meet this requirement a rubber type mount or a passive oil damper is used as a primary mechanism to sustain high frequency transient vibration(displacement, acceleration and force) [7,8]. However, the conventional type mechanism has unsuitable performance deterioration in the low frequency range in which a relatively high vibration amplitude occurs. Consequently, a new type of damper mechanism to effectively achieve vibration/shock isolation in the wide frequency range needs to be devised.

On the contrary the works on the ER or MR dampers for vibration control, the research work on the shock isolation using ER or MR fluid damper is considerably rare. El Wahed et al. [9] investigated transient impulse responses of ER and MR fluids by constructing a test rig capable of providing a transient input time to the upper assembly by means of a falling mass impacting on to the striker plate. In this work, we propose a cylindrical type ER shock damper and its field-dependent damping force characteristics are experimentally investigated as a first step. The shock damper is then incorporated to the 1 DOF dynamic system and its shock response is evaluated by realizing a semi-active skyhook control scheme. Controlled acceleration magnitudes are compared with uncontrolled ones at various excitation frequencies.

2 ER Shock Damper

The schematic configuration of the ER shock damper proposed in this work is shown in Fig.1. Two pistons are fixed on the rod and coil spring is used to achieve reaction force after imposing impulsive load. In the absence of the electric field, the damping force is occurred due to the viscosity of the ER fluid. This is given by

$$F_R = \frac{12\eta L}{bh^3}(A_P - A_r)^2 \cdot \dot{x} \tag{1}$$

In the above, η is viscosity of the ER fluid, L is electrode length, b is electrode width, h is electrode gap, A_p is piston area, A_r is piston rod area and $\dot{x}$ is piston velocity. On the other hand, if a certain level of the electric field is supplied to the ER damper, additional damping force is occurred due to the yield stress of the ER fluid. This is given by

$$F_{ER} = 2\frac{L}{h}\alpha E^{\beta}(A_P - A_r)\,\mathrm{sgn}(\dot{x}) \tag{2}$$

where $E(kV/mm)$ is electric field, and α and β are intrinsic values of the ER fluid to be experimentally determined. In this study, a chemical starch-based ER fluid is employed, and the values of α and β are evaluated by 190.5 and 1.874, respectively. These values are obtained in the flow mode at 30°C.

Now, the total damping force of the proposed ER shock damper is obtained by

$$F_T = F_R + F_{ER} + k_c x \tag{3}$$

where k_c represents the stiffness constant of the coil spring. In general, many different sizes of ER damper can be devised for various applications. In this work, a

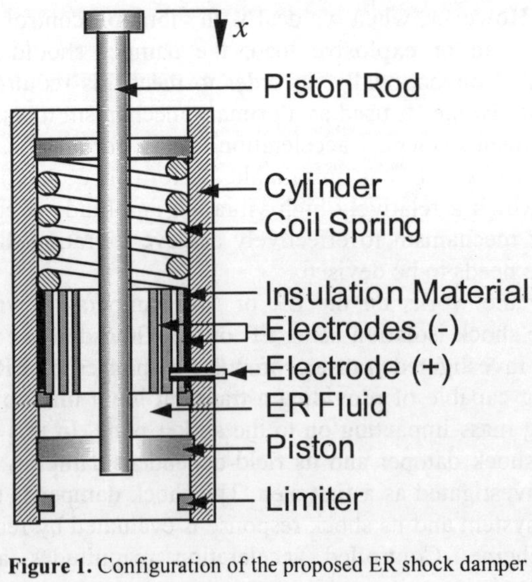

Figure 1. Configuration of the proposed ER shock damper

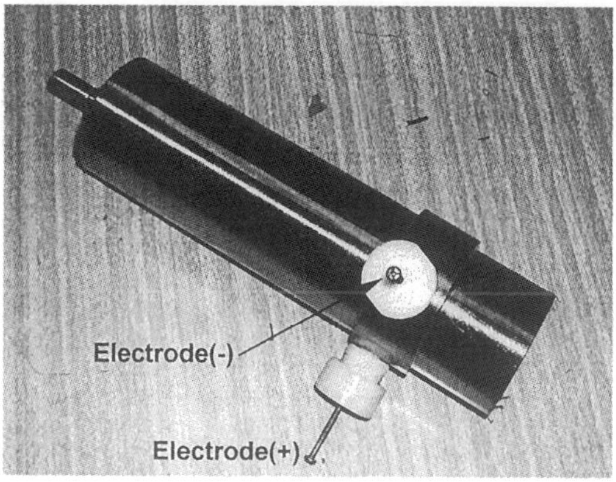

Figure 2. Photograph of the ER shock damper

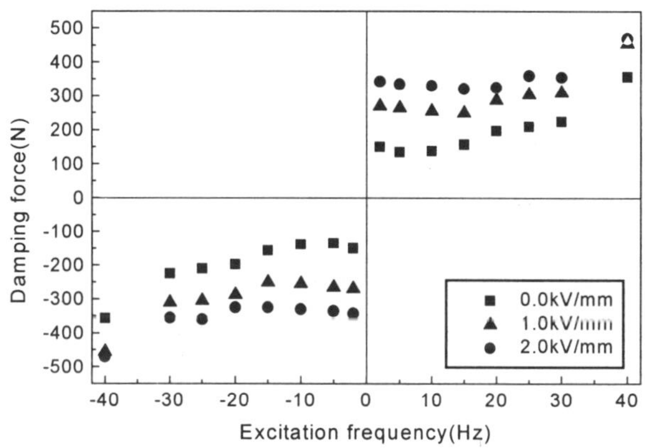

Figure 3. Damping force vs piston velocity

middle-sized damper which is applicable to impact isolation of home appliances such as air conditioner is designed. The principal design parameters are given as follows : h=0.001m, L=0.05m, A_p=63.4×10^{-4}m^2, A_r=4.52×10^{-4}m^2. The photograph of the ER shock damper manufactured in this study is shown in Fig.2. The piston rod and outer cylinder are made of the stainless steel and the MC(methyl cellulose) plastic is used for insulation material.

Figure 3 presents the measured damping force with represent to the piston velocity at various electric fields. This plot is obtained by calculating the maximum damping force at each velocity. The piston velocity is changed by increasing the excitation frequency up to 38Hz, while the excitation amplitude is maintained to be constant by ±1mm. It is clearly observed that the damping force increases as the electric field increases, as expected. The damping force at 10Hz is increased up to 200N by employing the electric field of 2kV/mm. Thus, we can achieve damping force control which is required to isolate the imposed shock or impulse excitation.

3 Shock Isolation System

In order to investigate shock isolation effectiveness of the proposed ER shock damper, 1

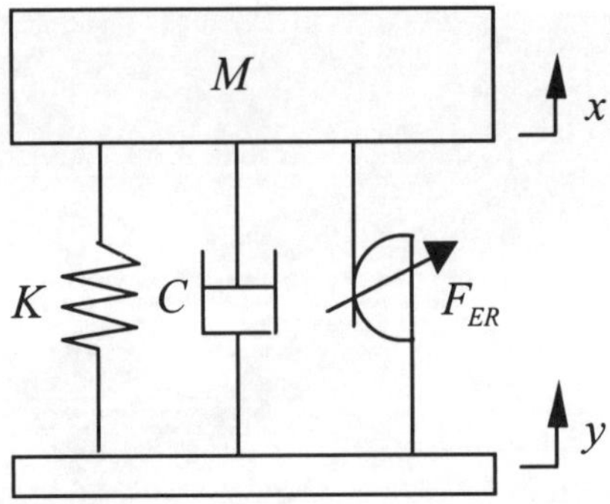

Figure 4. Mechanical model of the shock isolation system.

DOF system is constructed as shown in Fig.4. The dynamic system is simple and easy to implement in an experimental manner. The governing equation of motion of the system is given by

$$M\ddot{x} + C(\dot{x} - \dot{y}) + K(x - y) = -F_{ER} \tag{4}$$

where

$$C = R_e(A_p - A_r)^2 \,,\ K = k_c$$

The impulsive excitation is to be applied to the base plate and its mathematical expression is given by

$$y(t) = \begin{cases} \sin(\pi t/\tau)\,, & t < \tau \\ 0 \quad , & t > \tau \end{cases} \tag{5}$$

We see that above excitation is a half-sine function with the period τ .

The control performance is to isolate the dynamic motions of the sprung mass M from the input excitation. This can be qualified by the transmitted displacement and acceleration. To achieve this control target, we adopt a skyhook control scheme proposed by Karnopp et al. [10]. The input damping force is given by

$$F_{ER} = C_{sky}\dot{x} \tag{6}$$

where C_{sky} is control gain and it represents damping constant of the system. The damping force obtained from (6) is to be applied to the ER shock damper depending upon relative motion. Therefore, the following semi-active actuating condition is normally imposed.

$$F_{ER} = \begin{cases} F_{ER} & for \quad F_{ER}(\dot{x} - \dot{y}) > 0 \\ 0 & for \quad F_{ER}(\dot{x} - \dot{y}) \le 0 \end{cases} \tag{7}$$

The above condition physically implies that the activation of the control damping force only assures the increment of energy dissipation of the stable system. Once the control damping force is determined, the input electric field to be applied to the ER shock damper is obtained by

$$E = \left(\frac{h \cdot F_{ER}}{2\alpha L(A_p - A_r)} \right)^{1/\beta} \tag{8}$$

4 Experimental Results and Discussions

In order to evaluate the shock response in an empirical manner, an experimental set-up is established as shown in Fig.5. An electromagnetic shaker is used to generate impulse excitation given by Eq. (5), and its motion is controlled by PID (proportional-integral-derivative) controller associated displacement and velocity feedback signals. The sprung mass of 44kg is attached to the piston rod supported by the coil spring. The constant voltage or control voltage determined in the microprocessor is to be applied to the ER shock damper via high voltage(HV) amplifier.

Figure 6 presents time histories of the acceleration measured with the exciting period τ =0.05sec. It is clearly observed that the acceleration in the absence of the electric field is increased at the first and second peaks by imposing constant electric field of 2kV/mm. This is due to the increment of the total force of the ER shock damper. However, the acceleration magnitudes at two peaks are substantially reduced by activating the skyhook controller. This is because only dissipation energy is increased by activating the semi-active control input. Figure 7 presents acceleration levels of the first peak at various exciting frequencies(or period τ). We clearly see that the acceleration magnitude is increased as the electric field increases. The controlled response with the skyhook controller shows the minimum magnitude at each exciting frequency. Similar shock responses to the first peak are observed in the second peak as shown in Fig.8. From the results presented in this experimental investigation, we see that unwanted shock response can be effectively reduced by employing the ER shock damper associated with an appropriate control scheme.

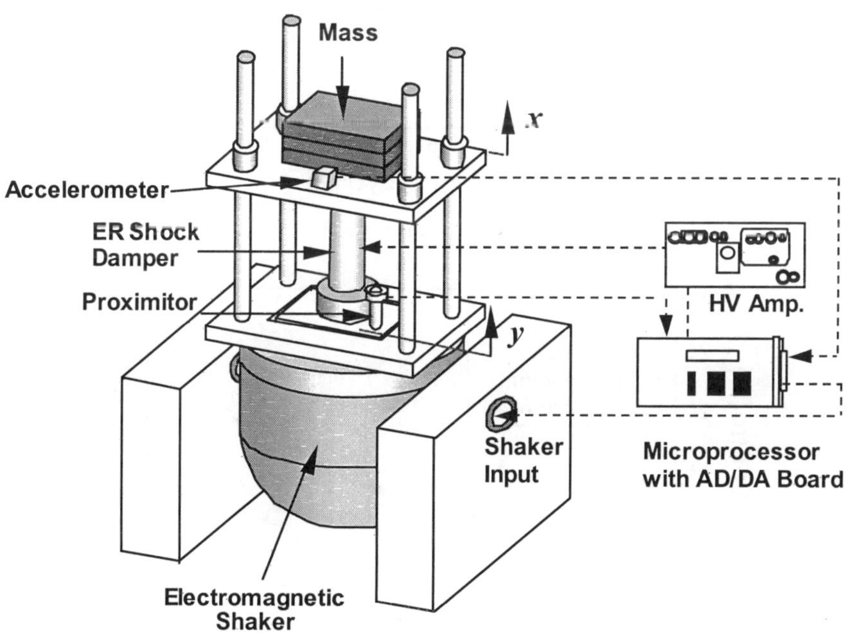

Figure 5. Experimental apparatus of the 1DOF shock isolation system

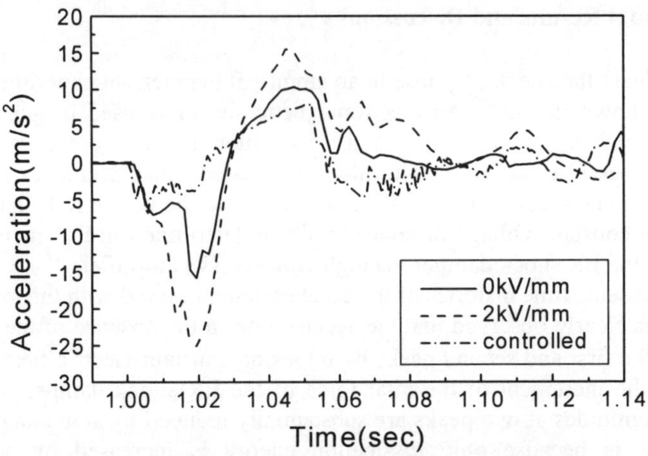

Figure 6. Acceleration with τ =0.05sec

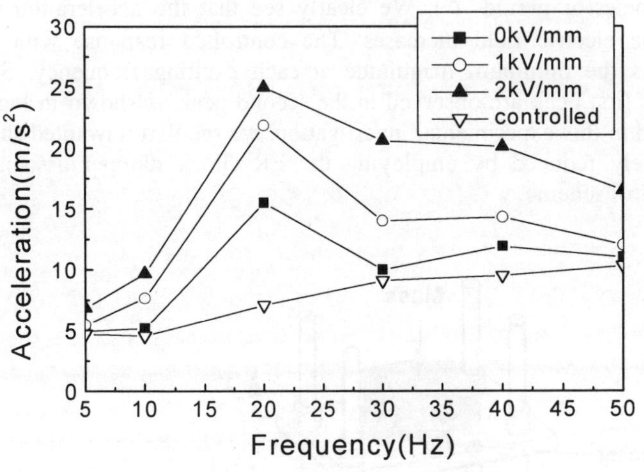

Figure 7. Acceleration at the first peak

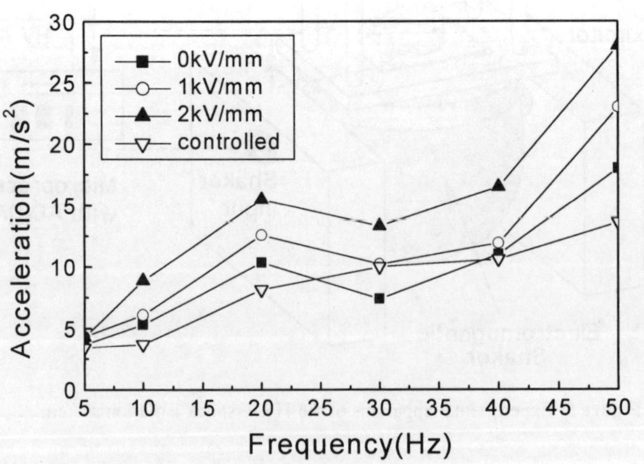

Figure 8. Acceleration at the second peak

5 Conclusion

A cylindrical ER shock damper was proposed and its shock isolation performance was experimentally investigated. After devising an appropriate size of the ER shock damper, the field-dependent damping forces were evaluated with respect to the piston velocity. The ER shock damper was then incorporated with 1 DOF system to analyze shock response. It has been demonstrated through implementation of the semi-active skyhook controller that unwanted shock responses such as acceleration of the sprung mass can be effectively reduced by activating the controller. The isolation performance of the proposed ER shock damper in the higher frequency range needs to be further explored.

Acknowledgements

This work was partially supported by the Korea Science and Engineering Foundation under Contract 1999-2-304-005-3. This financial support is gratefully acknowledged.

References

1. Choi, S. B., Choi, Y. T., Chang, E. G., Han, S. J. and Kim, C. S., Control Characteristics of a Continuously Variable ER Damper. *Mechatronics*, **8** (1998) pp. 143-161.
2. Choi, S. B., and Kim, W. K., Vibration Control of a Semi-Active Suspension Featuring Electrorheological Fluid Dampers. *Journal of Sound and Vibration*, **234** (2000) pp. 537-546.
3. Linder, J. E. and Wereley, N. M., Double Adjustable Shock Absorbers Using Electrorheological Fluid, *Journal of Intelligent Material Systems and Structures* **10** (1999) pp. 652-657.
4. Dyke, S. J., Spencer, B. F. Jr., Sain, M. K. and Carlson, J. D., Seismic Response Reduction Using Magnetorheological Damper. *Proceedings of IFAC World Congress* **1** (1996) pp. 145-150.
5. Kamath, G. M., Wereley, N. M. and Jolly, M. R., Characterization of Semi-Active Magnetorheological Fluid Lag Mode Dampers. *Proc. of the SPIE Conference on Smart Structures and Integrated Systems*, (1998) pp. 3329-2237.
6. Lee, H. S. and Choi, S. B., Control and Response Characteristics of a Magnetorheological Fluid Damper for a Passenger Vehicles. *Journal of Intelligent Material Systems and Structures* **11** (2000) pp. 80-87.
7. Rivin, E. I., Passive Engine Mounts Some Directions for Further Development. *SAE Technical Paper Series*, (1984), 940259.
8. Gillespie, T. D., Fundamentals of Vehicle Dynamics. *Society of Automotive Engineerins, Inc.*, (1992).
9. El Wahed, A. K., Sproston, J. L., and Schleyer, G. K., A Comparison between Electrorheological and Magnetorheological Fluids Subjected to Impulsive Loads. *Journal of Intelligent Material Systems and Structures* **10** (1999) pp. 695-700.
10. Karnopp, D. C., Crosby, M. J. and Harwood, R. A., Vibration Control Using Semi-Active Force Generators. *Journal of Engineering for Industry, ASME* **96** (1974) pp. 619-626.

VIBRATION CONTROL EVALUATION OF A COMMERCIAL VEHICLE FEATURING MR SEAT DAMPER

M. H. NAM, Y. M. HAN, S. S. HAN, H. G. LEE, S. B. CHOI AND C. C. CHEONG

Smart Structures and Systems Laboratory, Department of Mechanical Engineering
Inha University, Incheon 402-751, Korea
E-mail: seungbok@inha.ac.kr

This paper presents vibration control responses of a commercial vehicle installed with ER primary suspensions and a MR seat damper. After formulating the governing equation of motion of the full-vehicle system associated ER and MR dampers, the sky-hook controllers are designed by imposing semi-active actuating conditions. The control algorithms are then implemented via the hardware-in-the-loop simulation, and control responses of vertical displacement and acceleration at the driver's seat are evaluated under two road conditions; bump and random road.

1 Introduction

Numerous types of commercial vehicles, along with the extended driving hours and the exposure to severe working environments are used for various transport purpose such as industrial and agricultural ones. The increased number of these vehicles inevitably results in increase demand for improved ride comfort. The measured ride vibration at the driver's seat has a significant influence on the driver's fatigue and safety. It is observed that the ride vibration levels of commercial vehicles are 9-16 times higher than those of passenger cars, and most commercial vehicle drivers are exposed to the ride vibration for 10-20 hours a day [1]. Thus, the seat suspension to attenuate unwanted vibrations is normally employed for commercial vehicles. Recently, a very attractive and effective semi-active seat suspension system featuring ER (electro-rheological) or MR (magneto-rheological) fluids has been investigated for vibration isolation [2-8]. However, despite many investigations on the vibration control capability of each ER or MR damper, none deals with a full-vehicle suspension system installed with both primary and seat dampers utilizing ER or/and MR fluids.

Consequently, the main contribution of this work is to show vibration control performance of a commercial vehicle which is incorporated with ER and MR dampers for the primary and seat suspension, respectively. In order to achieve this goal, a cylindrical MR seat damper is devised and its damping force is experimentally evaluated with respect to the intensity of the input current (magnetic field). The seat damper is then integrated with the primary ER suspension to formulate a full-vehicle model, and its governing equation of motion is derived. Semiactive skyhook control algorithms for both primary and seat suspensions are designed to attenuate unwanted vibration at the driver's seat position, and the controllers are implemented via the hardware-in-the-loop-simulation (HILS). A commercial vehicle equipped with the devised MR seat damper and ER primary dampers is tested under both bump and random road conditions, and its control results are presented in time and frequency domains.

2 MR Seat Damper

When the magnetic field of H is applied to MR fluid, the yield stress (τ_y) can be expressed by

$$\tau_y(H) = \alpha_m H^{\beta_m} \tag{1}$$

where, α_m and β_m are intrinsic values of MR fluid to be experimentally determined. In this work, the product (MRF-132LA) from the Lord Corporation has been employed. The field-dependent yield stress given by Eq. (1) is to be incorporated with the damping force model of the proposed MR seat damper. Figure 1(a) presents the schematic configuration of the proposed MR seat damper. The piston head with magnetic cores is moving up and down, and the gas chamber is located outside as an accumulator of the MR fluid. In order to simplify the analysis of the MR seat damper, it is assumed that the MR fluid is incompressible and fluid inertia is assumed to be negligible. For laminar flow in the annular orifice, the flow resistance under the zero magnetic field is given by

$$R_e = \frac{12\eta_1 L_m}{b_1 h_m^3} \tag{2}$$

where, L_m is the orifice length, b_1 is the orifice width, h_m is the orifice gap, and η_1 is the viscosity of the MR fluid. On the other hand, the pressure drop due to the magnetic field is given by

$$\Delta P_{MR} = 2\frac{L_m}{h_m}\tau_y(H) = 2\frac{L_m}{h_m}\alpha_m H^{\beta_m} \tag{3}$$

where, $H = NI/2h_m$. Here, N is the number of coil turns, and I is the input current. Now, by neglecting the effect of the fluid inertia and frictional force, the damping force of the proposed MR damper is obtained by

$$F_d = \frac{A_{rm}^2}{C_g}x_p + (A_{pm} - A_{rm})^2 R_e \dot{x}_p + (A_{pm} - A_{rm})\Delta P_{MR}\,\mathrm{sgn}(\dot{x}_p) \tag{4}$$

where, C_g is the gas compliance, A_{rm} is the piston-rod area, A_{pm} is the piston area, and sgn(•) is the signum function. It is noted that the third term is controllable damping force by the intensity of magnetic field. The damping force given by Eq. (4) can be rewritten by

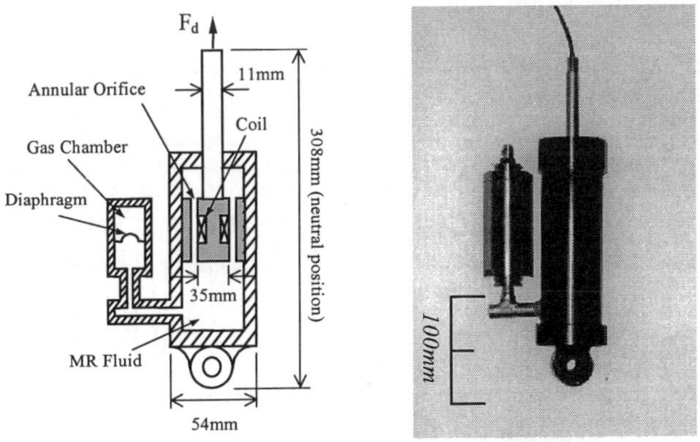

(a) schematic configuration (b) photograph

Figure 1. Schematic configuration and photograph of the MR seat damper.

$$F_d = k_e x_p + c_e \dot{x}_p + F_{MR} \tag{5}$$

where, k_e is the effective stiffness due to the gas pressure, and c_e is the effective damping coefficient due to the viscosity of the MR damper. The size and level of required damping force adopted in this work are chosen on the basis of a conventional passive oil seat damper for a commercial truck. The damping force given by Eq. (5) is evaluated with respect to the magnetic field in order to determine design parameters. The designed MR seat damper has the following specifications; orifice gap: 1.8mm, orifice length: 44mm, diameter of the piston head: 46mm, number of coil turn: 60, and diameter of the coil(copper wire): 0.75mm. Fig. 1(b) shows a photograph of the MR seat damper manufactured in this study.

Figure 2 shows the measured damping force characteristics of the proposed MR seat damper. Fig. 2(a) is obtained by increasing excitation frequency from 0.6Hz to 4.0Hz, where the excitation amplitude is maintained to be constant at 20mm($\pm$10mm). It is clearly observed that the damping force increases as the input current increases. This directly indicates that a certain desired damping force can be achieved by just controlling the input current regardless of the piston velocity. Fig. 2(b) presents time-histories of controlled damping force. The desired damping force to be tracked is set by

$$F_{desired} = C_{sky} \dot{x}_p \ [N] \tag{6}$$

where, C_{sky} stands for the gain of the skyhook controller and $\dot{x}_p$ indicates the piston velocity of the MR seat damper. The excitation displacement and frequency are chosen by 20mm and 1.4Hz, respectively. It is observed that controlled damping force fairly well tracks the desired damping force by supplying adequate control input current.

3 Controller Formulation

A full-vehicle model consisting of semi-active ER primary dampers and MR seat damper is shown in Figure 3. The excitation input from the road is transmitted through to the cabin floor. This then causes unwanted vibrations on the driver. For the simplification of the dynamic modeling, it is only considered the straight-going motion of vehicle. Accordingly, it is assumed that there exists only the vertical and the pitching motion of vehicle and the rolling motion is ignored in this study. Thus, we derive the equations of motion for the full-vehicle model installed with semi-active ER primary and MR seat suspensions [9]. Among many potential candidates for control algorithms, the skyhook

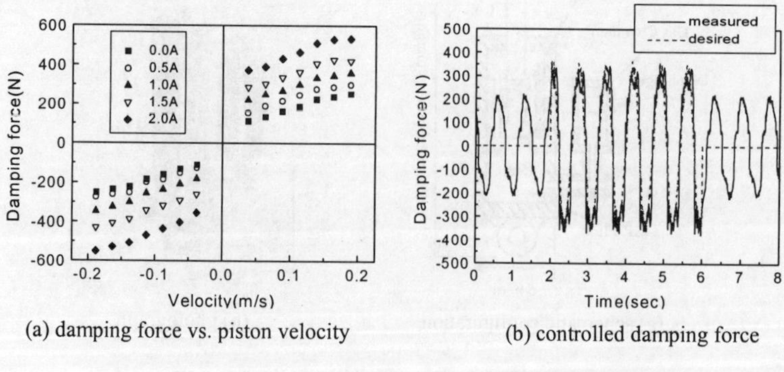

(a) damping force vs. piston velocity (b) controlled damping force

Figure 2. Damping force characteristics of the MR seat damper.

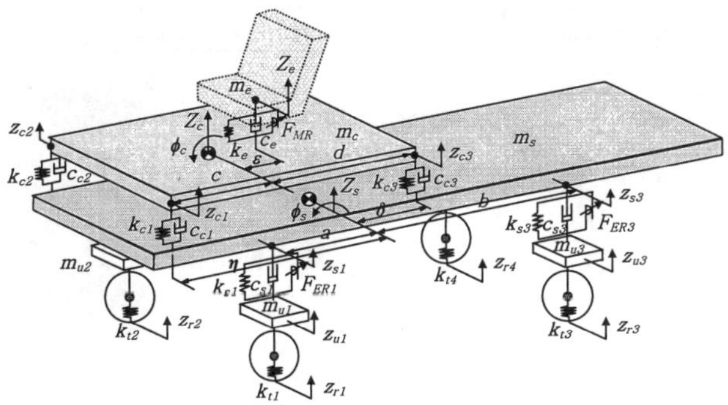

Figure 3. Mechanical model of the full-vehicle system featuring ER primary dampers and MR seat damper.

controller [10] is adopted in this work. It is well known that the logic of the skyhook is simple and easy to implement in practical field. The desired damping force for ER primary damper and MR seat damper is set by

$$u_i = C_{sky\_ER}(\dot{z}_s - a\dot{\phi}_s), \quad i = 1,2 \quad \text{(for front ER primary dampers)} \tag{7}$$

$$u_i = C_{sky\_ER}(\dot{z}_s + b\dot{\phi}_s), \quad i = 3,4 \quad \text{(for rear ER primary dampers)} \tag{8}$$

$$u_5 = C_{sky\_MR} \cdot \dot{z}_e \quad \text{(for MR seat damper)} \tag{9}$$

where, C_{sky_ER} and C_{sky_MR} are the gains of the skyhook controllers. The control gain physically indicates the damping coefficient. On the other hand, the damping forces of the primary and seat suspension systems need to be controlled depending upon the motion of suspension travel. Therefore, the following actuating conditions are normally imposed.

$$u_i = \begin{bmatrix} F_{ERi}, & for\ F_{ERi}(\dot{z}_s - a\dot{\phi}_s - \dot{z}_{ui}) > 0 \\ 0, & for\ F_{ERi}(\dot{z}_s - a\dot{\phi}_s - \dot{z}_{ui}) \le 0 \end{bmatrix} \quad (i = 1, 2) \tag{10}$$

$$u_i = \begin{bmatrix} F_{ERi}, & for\ F_{ERi}(\dot{z}_s + b\dot{\phi}_s - \dot{z}_{ui}) > 0 \\ 0, & for\ F_{ERi}(\dot{z}_s + b\dot{\phi}_s - \dot{z}_{ui}) \le 0 \end{bmatrix} \quad (i = 3, 4) \tag{11}$$

$$u_5 = \begin{bmatrix} F_{MR}, & for\ F_{MR} \cdot (\dot{z}_e - \dot{z}_c - \varepsilon\dot{\phi}_c) > 0 \\ 0, & for\ F_{MR} \cdot (\dot{z}_e - \dot{z}_c - \varepsilon\dot{\phi}_c) \le 0 \end{bmatrix} \tag{12}$$

The above actuating conditions physically imply that the activating of the controller only assures the increment of energy dissipation of the stable system. Once the control inputs are determined, the input electric field and input current to be applied to the ER primary damper and MR seat damper are, respectively, obtained as follows [9].

$$E_i = \left[u_i \cdot \frac{h_e}{2\alpha_e \cdot L_e(A_{pe} - A_{re})} \right]^{1/\beta_e}, \ (i = 1,2,3,4) \tag{13}$$

$$I = \frac{2h_m}{N} \left[u_5 \cdot \frac{h_m}{8\alpha_m \cdot L_m(A_{pm} - A_{rm})} \right]^{1/\beta_m} \tag{14}$$

4 Performance Evaluation via HILS

From a practical point of view, the ER primary dampers and MR seat damper should be

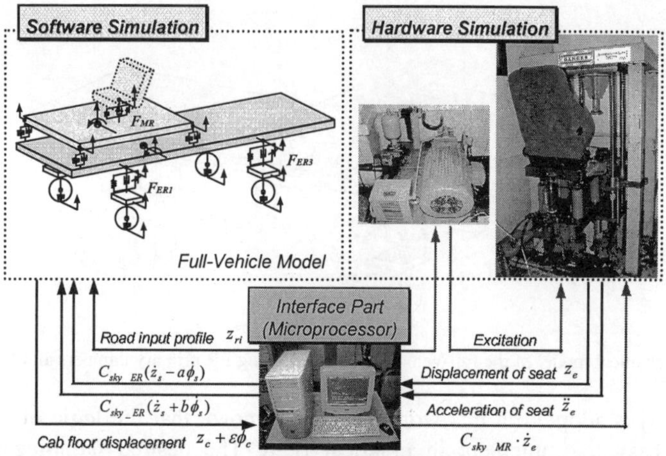

Figure 4. Schematic configuration of the HILS.

installed on a full-vehicle and it's effectiveness for the ride quality needs to be tested with respect to various road conditions. However, it is very expensive to completely build the entire full-vehicle system with hardware from the start. Therefore, we adopt a hardware-in-the-loop simulation (HILS) in which a full-vehicle system model consisting of the ER primary dampers and conventional cabin suspension is incorporated with actual hardware of the proposed MR seat damper. Figure 4 presents a schematic configuration of the HILS method. In the real-time simulation, the road and driving conditions are adopted for the conventional COE (cab over engine) type commercial vehicle model. The system parameters of the full-vehicle model adopted in this work are listed in the reference [9]. The full-vehicle model of a commercial vehicle in the computer system is to be excited from prescribed road profiles. The cabin floor displacement is calculated and then the displacement signal is converted to the hydraulic control unit for controlling the road input to the platform. When the vibration is transmitted to the MR seat suspension, the dynamic response signals, which are acquired from the LVDT (linear variable differential transformer) and accelerometer, are converted to digital signals via the A/D converter.

Figure 5 presents the bump responses of the full-vehicle system. As clearly observed from the results, both the vertical displacement and acceleration at the driver's seat has been significantly reduced by employing the proposed MR seat damper. The maximum displacement and acceleration with the passive seat damper are reduced by 45% and 30%, respectively. It is also seen from Figs. 5(c) and (d) that the electric field and input current are supplied to the ER primary damper and MR seat damper, respectively, according to the semi-active control conditions. Figure 6 shows the random road responses of the full-vehicle system. It is obvious that PSD of vertical displacement and acceleration at the driver's seat are considerably attenuated by activating the proposed MR seat damper. The maximum magnitude of both signals with the passive seat damper are reduced by 45% and 30%, respectively, at the resonant frequency by controlling the proposed MR seat damper.

5 Conclusion

In this study, a commercial vehicle equipped with semi-active ER primary dampers and MR seat damper was proposed, and its controllability for vibration isolation has been

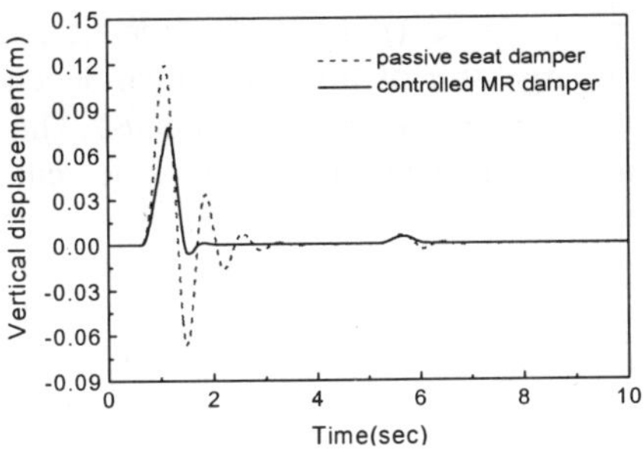

(a) vertical displacement at driver's seat

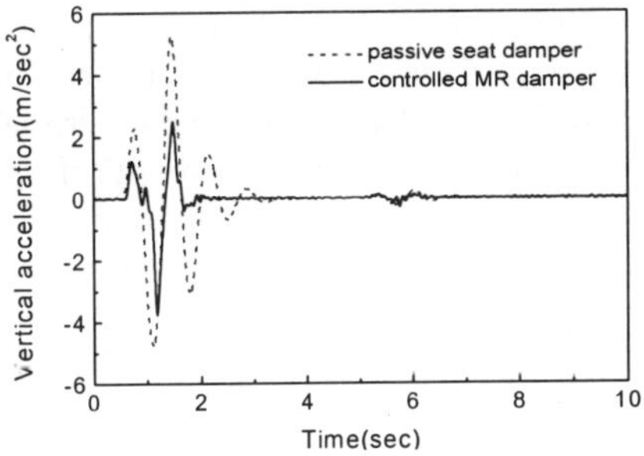

(b) vertical acceleration at driver's seat

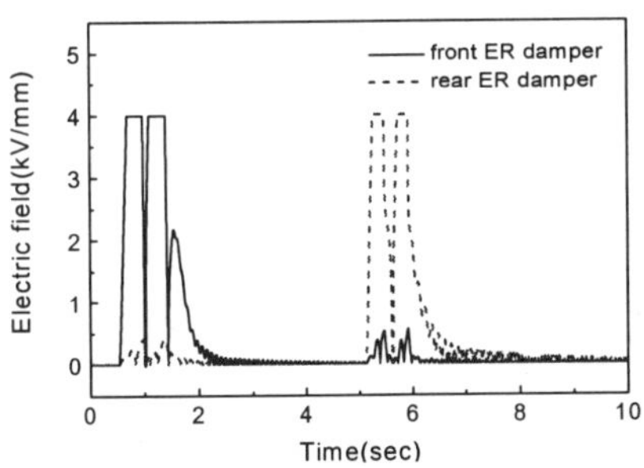

(c) control input for ER primary damper

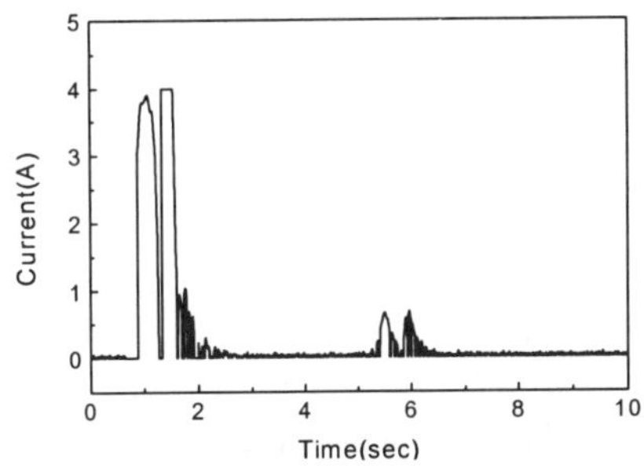

(d) control input for MR seat damper

Figure 5. Control responses of the full-vehicle system under bump excitation.

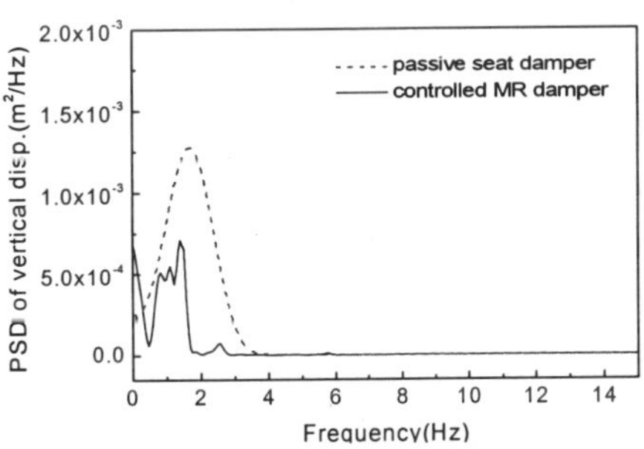

(a) PSD of vertical displacement at driver's seat

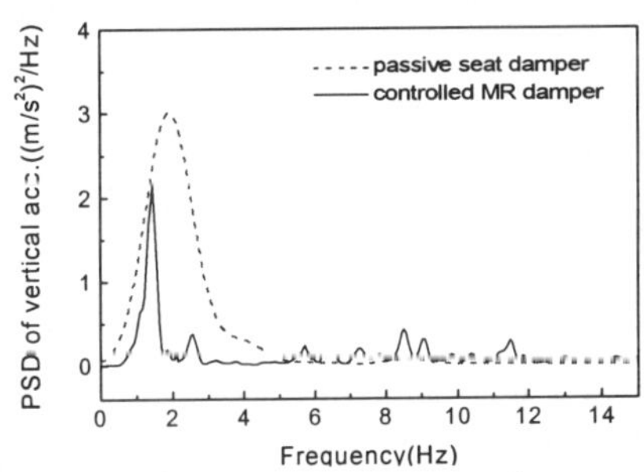

(b) PSD of vertical acceleration at driver's seat

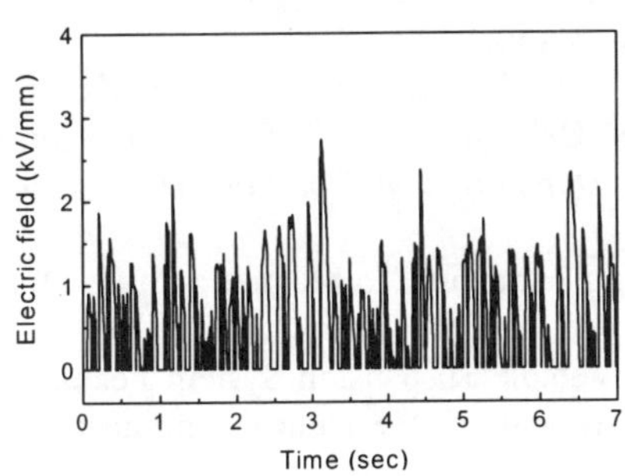

(c) control input for ER primary damper

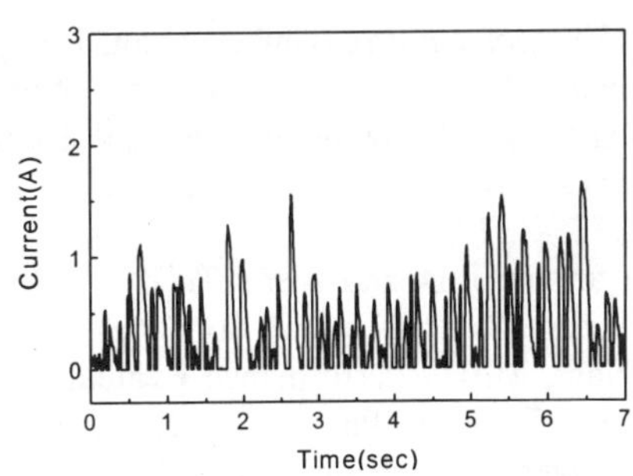

(d) control input for MR seat damper

Figure 6. Control responses of the full-vehicle system under random excitation.

demonstrated through the hardware-in-the-loop simulation (HILS). After manufacturing the MR seat damper, which is applicable to a large size of truck, the field-dependent damping force characteristics and controllability were experimentally evaluated. The MR seat damper was then integrated with the primary ER suspension in order to formulate a full-vehicle model, and its governing equation of motion was derived. The skyhook controller was designed for each damper and realized via HILS. It has bean demonstrated that both vertical displacement and acceleration are considerably reduced under the bump test by employing the proposed MR seat damper. In addition, the power spectrum densities of the acceleration at the driver's seat has been significantly reduced under the random test by activating the MR seat damper.

Acknowledgement

This work was partially supported by National Research Laboratory (NRL) Program. This financial support is gratefully acknowledged.

References

1. Queslati F. and Sankar S., Performance of a Fail-Safe Active Suspension with Limited State Feedback for Improved Ride Quality and Reduced Pavement Loading in Heavy Vehicles. *SAE Paper* **922474** (1992) pp. 796-804.
2. Wu, X. and Griffin, M. J., A Semi-Active Control Policy to Reduce the Occurrence and Severity of End-stop Impacts in a Suspension Seat with an Electrorheological Fluid Damper. *Journal of Vibration and Acoustics* **203** (1997) pp. 781-793.
3. Brooks, D. A., High Performance Damping Using Electro-Rheological Fluid. *Proceedings of International Conference on ERF, MRS and Their Applications* (Yamagata, Japan, 1997) pp. 250-251.
4. Choi, S. B., Choi, J. H., Nam, M. H., Cheong, C. C. and Lee, H. G., A Semi-Active Suspension Using ER Fluids for a Commercial Vehicle Seat. *Journal of Intelligent Material Systems and Structures* **9** (1998) pp. 601-606.
5. Dyke, S. J., Spencer, B. F. Jr., Sain, M. K. and Carlson, J. D., Seismic Response Reduction Using Magnetorheological Damper. *Proceedings of IFAC World Congress* **1** (1996) pp. 145-150.
6. Carlson, J. D. and Spencer, B. F. Jr., Magneto-rheological Fluid Dampers for Semi-active Seismic Control. *Proceedings of the 3rd Internal Conference on Motion and Vibration Control* (Chiba, Japan, 1996) pp. 35-40.
7. Pang, L., Kamath, G. M. and Wereley, N. M., Dynamic Characterization and Analysis of Magnetorheological Damper Behavior. *the 5th SPIE Symposition on Smart Materials and Structure, Passive Damping and Isolation* **3327** (1998) pp. 3325-3327.
8. Carlson, J. D. and Weiss, K. D., A Growing Attraction To Magnetic Fluids. *Machine Design* **8** (1994) pp. 61-66.
9. Nam, M. H., Performance Evaluation of Vehicle Suspension System Featuring MR Seat Damper. Ph. D. Dissertation, Department of Mechanical Engineering, Inha University, Incheon, Korea, 2001.
10. Karnopp, D., Crosby, M. J. and Harwood, R. A., Vibration Control Using Semi-Active Force Generators. *Journal of Engineering for Industry, ASME* **96** (1974) pp. 619-629.

MR INSERTS FOR SHOCK WAVE REDUCTION IN WARSHIP STRUCTURES

J. H. KIM, B. W. KANG, Y. K. KANG, S. B. CHOI

Smart Structures and Systems Laboratory, Department of Mechanical Engineering
Inha University, Incheon 402-751, Korea
E-mail: jaehwan@inha.ac.kr

K. S. KIM

Impact Dynamics and Fatigue Fracture Laboratory, Department of Naval Architecture and Ocean Engineering, Inha University, Incheon 402-751, Korea
E-mail: ksukim@inha.ac.kr

Underwater explosion that has high energy brings about the shock waves and the pulsating gas bubbles. In general, structural vibration from the shock waves is more serious than that of pulsating gas bubbles. The shock waves may damage the important fragile structures and equipment in the ship. This paper demonstrates the possibility of shock wave reduction in Magneto-Rheological (MR) inserts. To experimentally verify the MR insert, MR insert in an aluminum plate is made and two piezoelectric disks are used as wave transmitter and receiver. By measuring the transmission ratio between the transmitter and receiver, the shock wave attenuation possibility of MR insert in tested. Details of the experiments are addressed.

1 Introduction

Underwater explosion generates shock waves and pulsating gas bubbles such that they bring high energy into naval structures which causes serious structural damage or functional disorder of important equipments of warships [1,2]. It is much more dangerous when shock waves reach to a submerged structure for their impulsive loading with high frequencies. This is the main factor causing failure of survivability of warships [3]. To prevent damages caused by shock waves, somewhat heavy and over-designed built in mounting structures are used up to now. So, the performance of warship is decreased by the increased weight. By that reason, some attempts have been done for reducing the shock waves using smart structures. Kim et al.[4] have studied shock wave attenuation using ER inserts with numerical simulation and experimental comparison. Mace et al.[5] have studied the wave attenuation characteristics of ER and MR inserts. Generally, MR fluid exhibits similar physical properties with ER fluid except much higher viscosity, density and shear yield stress than ER fluid. Thus, MR inserts can possibly improve the wave attenuation performance. This study aims at demonstrating shock wave reduction of MR inserts experimentally. This will conclude the possibility of shock wave reduction on warship structures.

2 General Properties of MR Fluid

Since Rabinow discovered MR fluid in the late 1940's, of which yield stress and viscosity varies in the presence of magnetic field, various applications using MR fluids have been developed for example, shock absorbers, clutches and engine mounts, etc. Physical property changes of MR fluids are induced from the chain-like structures between paramagnetic MR particles in the low permeability solvent.

MR fluid with no fluid flow and shear stress under shear yield stress reveals viscoelastic behavior as given constitutive equation shown in Eq.(1) [6].

$$\tau = G^{*}\gamma, \qquad \tau \le \tau_y \tag{1}$$

where G^{*} is complex shear modulus in terms of storage modulus (G') and loss modulus(G'') as shown in Eq.(2). Storage modulus exhibits the ability of energy storing in the material while loss modulus does the ability of energy dissipation sometimes referred to as material damping property.

$$G^{*} = G' + jG'' \tag{2}$$

3 Physical dimensions

To investigate wave propagation in the warship structures, following experiments were performed. Piezoelectric transmitter and receiver are attached on an aluminum plate (1200mm×900mm×1mm) such that they are used to generate and detect the shock waves transmitting via aluminum plate, respectively (Fig.1(a)). Piezoelectric devices are made with PZT-4D and the diameter is 25mm and the thickness is 1mm. As the same manner, a MR insert is provided into the aluminum plate to compare the wave propagation characteristics (Fig.1(b)). MR insert of 90° circular arc, 1mm thickness on the half circle slit of 50mm width is made by sandwiching two aluminum plates of 0.3mm thick as shown in Fig.2. This hollow slit is arranged for reducing the effect of reflected waves on the plate boundaries and let the transmitted signal be detected directly while reflected or refracted waves are in detour. So, by catching the transmitted signal before the reflected signals at the plate boundaries are arrived, the plate can be assumed as infinite plate. MR fluid of MRF-132LD (Lord Corp., USA) is used and the density is 3,055kg/m^3.

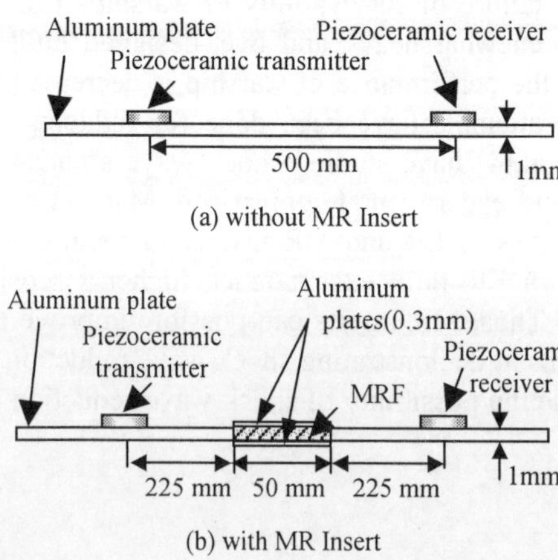

Figure 1. Cross-section of the plate without and with MR Insert

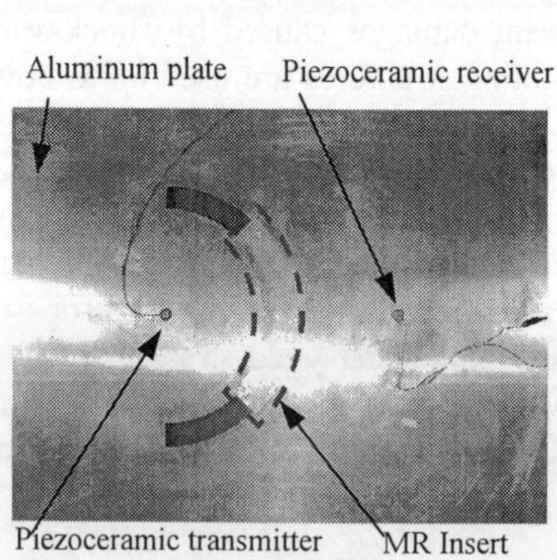

Figure 2. Photograph of the aluminum plate with MR Insert

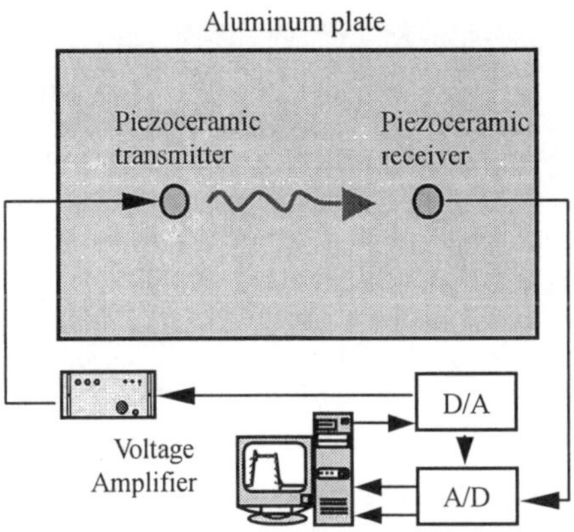

Figure 3. Schematic diagram of experimental setup for stress wave propagation test (without MR insert)

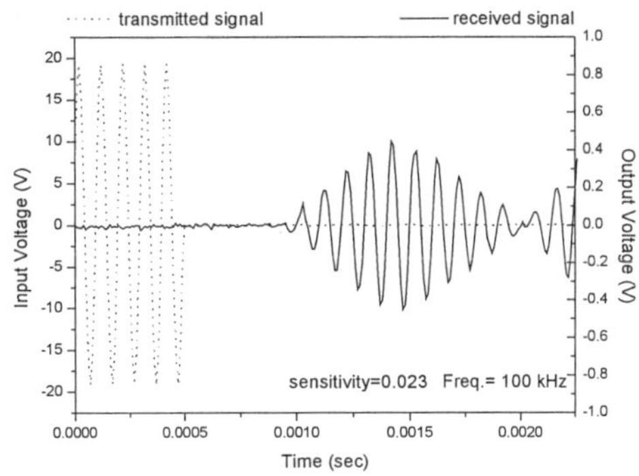

Figure 4. Transmitted and received signals in aluminum plate

4 Sensitivity of Piezoelectric transmitter/receiver

The sensitivity of aluminum plate without MR insert is tested as shown in Fig.3. Five sinusoidal pulse signals generated by on-board DSP are amplified to 20 volt and they are fed into piezoelectric transmitter. The transmitted signal is sensed by the piezoelectric receiver and the sensed signal is sent to the DSP board through an A/D converter. The sensitivity of piezoelectric sensor is defined as the ratio of output(receiver) voltage to input(transmitter) voltage. As shown in Fig.4, the sensitivity of piezoelectric sensor is 0.023 at 100kHz. The reason why the sensitivity is so low is due to the energy conversion of electrical energy to mechanical energy at piezoelectric transmitter and receiver. From the piezoelectric coupling factor, the ideal sensitivity can be calculated as Eq.(3) [7].

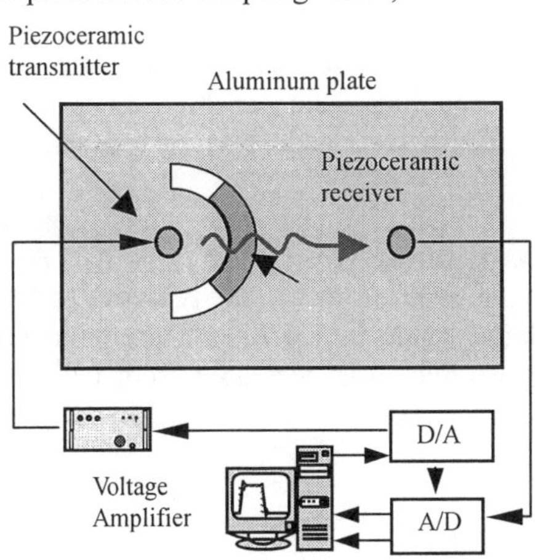

Figure 5. Schematic diagram of experimental setup for stress wave propagation test (with MR insert)

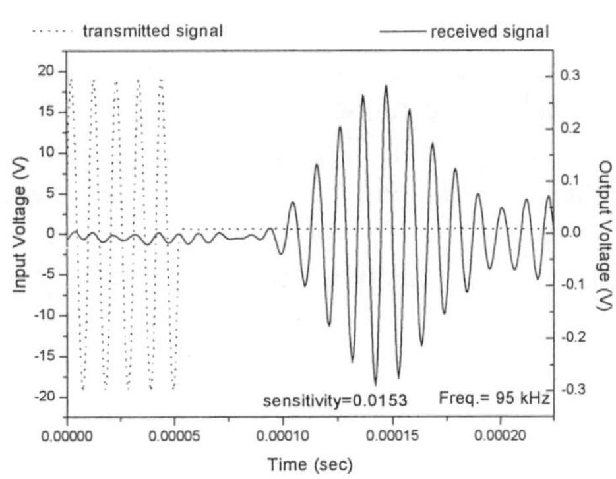

Figure 6. Transmitted and received signals of plate with MR insert

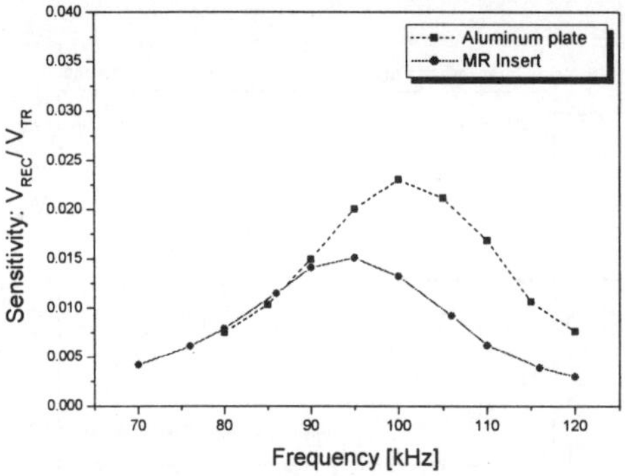

Figure 7. Sensitivity comparison of piezo electric receiver on the aluminum plate without and with MR Insert

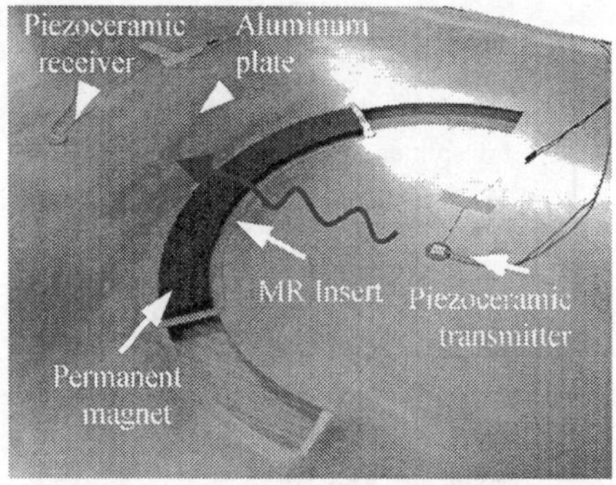

Figure 8. Photograph of the MR Insert with the magnetic field induced by permanent magnet (MFR=90°)

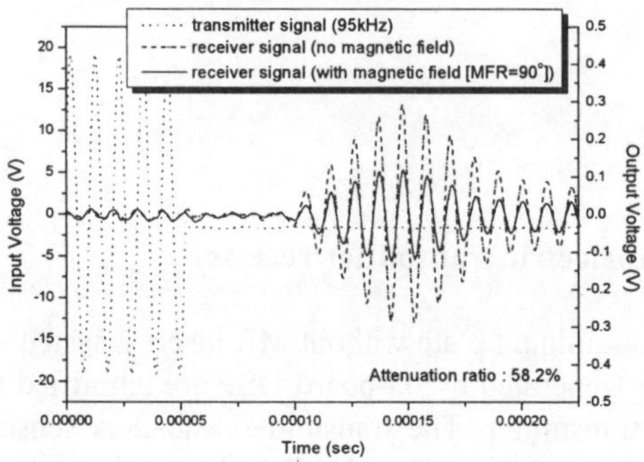

Figure 9. Shock wave attenuation under the applied permanent magnetic field (MFR=90°)

$$K^2 = \frac{mechanical\ energy\ stored}{electrical\ energy\ applied} \text{ or } K^2 = \frac{electrical\ energy\ stored}{mechanical\ energy\ applied} \tag{3}$$

Namely, the theoretical value of sensitivity in this case can be calculated as $(0.57)^2 \times (0.57)^2 = 0.1056$, where the piezoelectric coupling factor of PZT-4D is 0.57. However, the sensitivity is lowered to 0.023 due to the bonding layer between the piezoelectric ceramic disks and plate.

As shown in Fig.5, the sensitivity is also measured at aluminum plate with MR insert under no magnetic field. In this case, maximum sensitivity was found to be 0.0153 at 95kHz (Fig.6). To illustrate the sensitivity variation with respect to the presence of MR insert, each sensitivity of Figs.3 and 5 are plotted again in Fig.7. MR insert reduces rigidity of structure such that the maximum sensitivity was decreased from 0.023 to 0.0151 in lower frequency band.

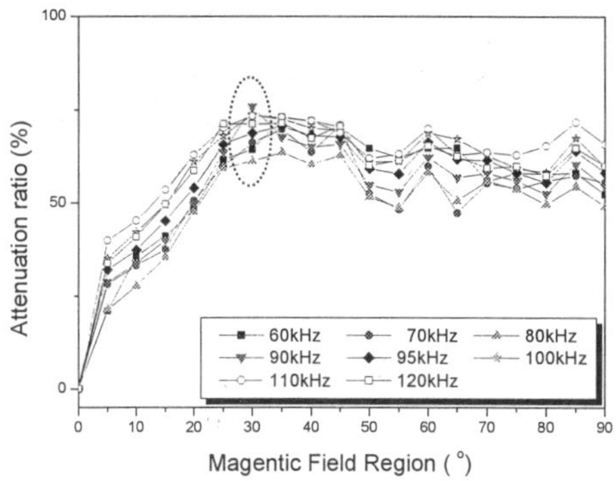

Figure 10. Attenuation ratio vs. magnetic field region with respect to the shock wave frequency

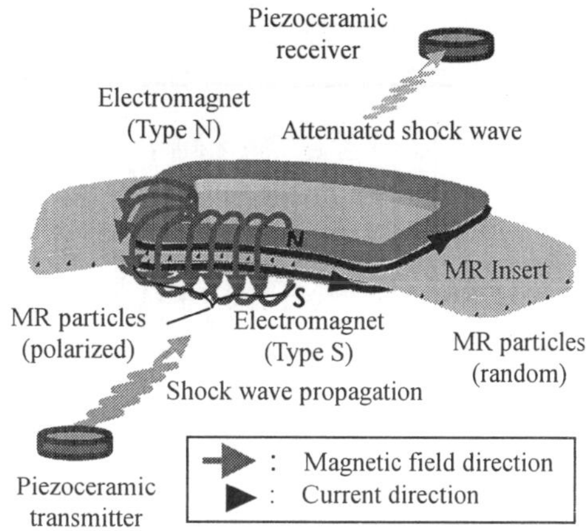

Figure 11. Schematic diagram of shock wave attenuation using MR Insert

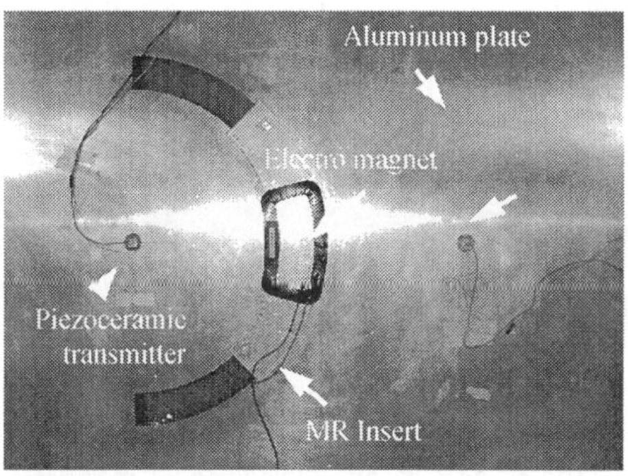

Figure 12. Photograph of the aluminum plate with electromagnet embedded in MR Insert

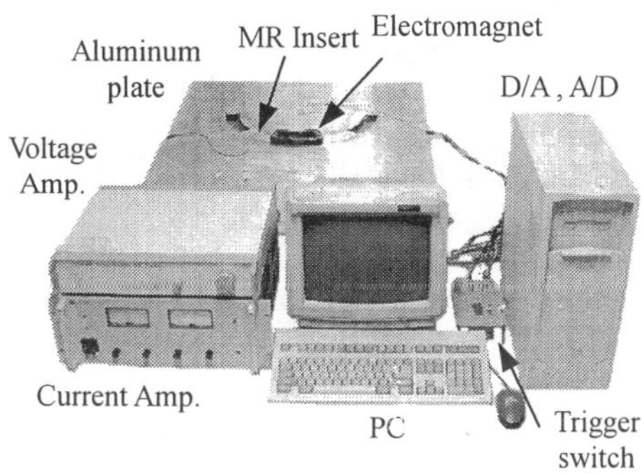

Figure 13. Photograph of shock wave attenuation test system

5 Performance Test for MR Insert

i) Shock Wave Reduction under permanent magnetic fields

Shock wave reduction characteristics of MR insert under permanent magnetic field was investigated using two rubber magnet patches of the same circular arc with MR insert and of 5mm thickness in each on both side of MR insert as shown in Fig.(8). Fig.(9) illustrates the reduction of shock wave propagation in the presence of permanent magnetic field. Directions of wave motion and chain-like structure of MR particles are perpendicular each other such that a remarkable shock wave reduction of 58.2% was achieved. It is important to find out the proper region of magnetic field to be applied. Thus, various types of permanent magnetic field region (MFR) were attempted, which is expressed as the angles of circular arc, from 90° to 5° with 5° decrement. The tested frequency band is changed from 60kHz to 120kHz.

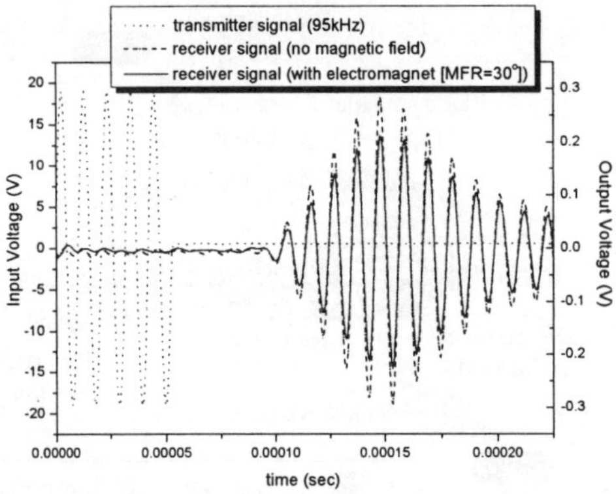

Figure 14. Shock wave attenuation under the applied field electromagnetic field (MFR=30°)

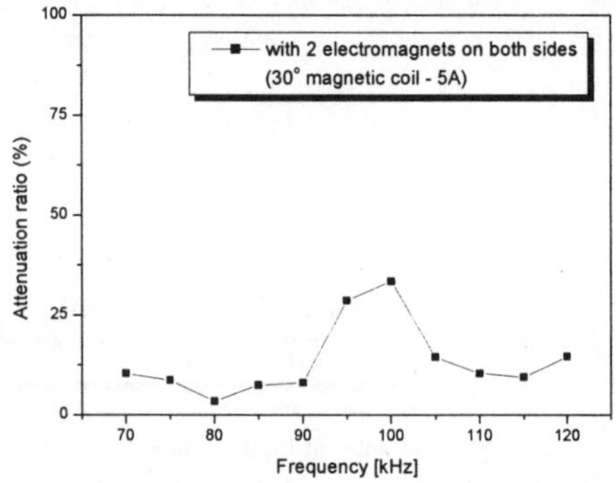

Figure 15. Attenuation ratio under the applied magnetic field using two magnetic coils on both sides (MFR=30°)

Shock wave reduction characteristics of MFR are shown in Fig. 10 with different MRF sector angles and wave frequencies. The performance of shock wave reduction is nearly saturation as indicated at dotted line in Fig.10. The performance is sluggish at a certain region of MFR -- 30° sector angle, and it shows 75.8% of the reduction at 90kHz and 68.8% at 95kHz.

ii) Shock Wave Reduction under Electromagnetic Fields

Magnetic field strength on MR insert should be able to control to reduce the shock waves transmitted through structures. Two electromagnets that are suitable for applying magnetic field to MFR of 30° are setup on both side of MR insert. Direction of magnetic flux induced by each electromagnet (type N and type S) were aligned such that a uniform magnetic field can be produced. Fig.11 shows the concept of shock wave reduction under controllable magnetic field. 0.8mm coated copper wire is wound 110 times for each electromagnet 5A current input is applied to the electromagnets connected in series.

By applying magnetic field with 30° MFR, 24.8% of shock wave reduction was achieved at 95kHz as illustrated in Fig.14. By attaching an additional electromagnet on each side of MR insert, the shock wave reduction of 28.7% at 95kHz and 33.5% at 100kHz were obtained, respectively. Fig.15 shows the shock wave reduction at different frequencies.. There is a stop band between 95kHz and 100kHz that exhibits large shock wave reduction. This means that shock waves propagating at a specific frequency band can be reduced by incorporating MR insert.

6 Conclusion

In this study, MR inserted smart structures were experimentally studied to reduce the shock waves propagating through structures. To test the wave transmission characteristics, piezoelectric transmitter and receiver were attached on an aluminum host plate and the sensitivity between the received signal and the transmitted signal was examined.

An MR insert was made in the host plate to test the shock wave reduction performance of the insert. Two ways of magnetic fields production were tried by using permanent magnets and electromagnetic coils. In the permanent magnet attempt, different areas of MFR were made to find the effective magnetic field imposing area for shock wave reduction.

As the result, the efficient region of MFR was found as 30° and 67-76% of shock wave reduction was achieved at 90-95kHz frequency band. When field controllable electromagnets are used, 29-34% of shock wave reduction was exhibited at 95-100 kHz frequency band. This demonstrates that the shock waves propagating through the structure can be reduced at a certain frequency band by applying magnetic fields on MR insert. Thus, it is possible to reduce shock waves in warship structures by using MR insert.

Acknowledgements

This work was supported by the Korea Science and Engineering Foundation under Contract 1999-2-304-005-3. This financial support is gratefully acknowledged.

References

1. Mehaute, B. L., and Wang, S., Water Waves Generated by Underwater Explosion, *World Scientific*, (1996).
2. Cole, R. H., Underwater Explosions, *Princeton University Press*, (1948).
3. Chung, J. H., Performance Evaluation for Shock-Proof of Navy Shipboard Equipment, *Journal of Korean Society for Noise and Vibration Engineering*, **8** (1998) pp. 21-28.
4. Kim, J. H., Kim, J. Y., Choi, S. B. and Kim, K. S., Smart Structures for Shock Wave Attenuation using ER Inserts, *SPIE's 8th Annual International Symposium on Smart Structures and Materials*, **4327** (2001).
5. Mace, B. R., Jones, R. W. and Harland, N. R., Wave Transmission thorough Structural Inserts, *Journal of the Acoustical Society of America*, **109** (2001) pp. 1417-1421.
6. Jolly, M. R., Bender, J. W. and Carlson, J. D., Properties and Applications of Commercial Magnetorheological Fluids, *Journal of Intelligent Material Systems and Structures*, **10** (1999) pp.5-13.
7. Piezoelectric Ceramics Data Book, *Morgan Matroc Limited, USA*, (1998).
8. Ferry, J. D., Viscoelastic properties of polymers, 2nd Edition, *John Wiley & Sons, INC., USA*, (1970).

VIBRATION CONTROL OF A WIRE CUT DISCHARGE MACHINE USING ER BRAKE ACTUATOR

K. S. KIM

Department of Mechanical Engineering, Chonan National Technical College, Chonan 330-717, Korea
E-mail: kskim@dragon.cntc.ac.kr

S. B. CHOI AND M. S. CHO

Smart Structures and Systems Laboratory, Department of Mechanical Engineering, Inha University, Inchon 402-751, Korea
E-mail: seungbok@inha.ac.kr

This paper presents vibration characteristics of a wire cut discharge machine in which an electrorheological brake actuator is used to control the wire tension. On the basis of the tension level required in the machine an appropriate size of the ER brake actuator is devised. The ER brake actuator is then incorporated with the machine and the field-dependent wire tension is experimentally evaluated. The straightness of the workpiece is also empirically investigated by changing the intensity of the electric field. In addition, the tension control performance of the discharge machine is simulated by utilizing a robust sliding mode controller.

1 Introduction

The wire cut discharge machine(WEDM) is widely used to machine various shapes of workpieces by discharging between the electrode wire and the workpiece. Since the wire has a diameter in the range of 0.05~0.3*mm*, it is possible to machine small, delicate and complicated specimens. For the high performance machining, the wire tension needs to be appropriately controlled during the machining. However, the occurrence of unwanted vibrations of the wire due to repulsive force and fluid injection is inevitable. This may cause the degradation of the surface roughness and straightness of the workpiece, and hence lead to the additional finishing work. Therefore, many researches have been undertaken to control the wire tension by reducing or eliminating the unwanted vibration[1-5]. One of most popular methods to control the tension is to utilize electromagnetic brake actuator. The control scheme of this method is the relatively simple, but it is weak to external disturbances such as repulsive force. Moreover, its response is relatively slow to take into account for discharging force which is occurred several hundreds times in a second. Consequently, an alternative actuator mechanism to accurately control the wire tension needs to be devised.

In general, the straightness of the workpiece is improved by increasing the wire tension. This, however, leads to the breaking of the wire. Therefore, an appropriate tension control actuator, which can judiciously adapt to varying external conditions with reasonable fast response characteristics, should be developed. One attractive method is to utilize an electrorheological(ER) brake actuator. As is well-known, the ER brake actuator has several advantages including design simplicity, fast response time and real-time controllability[6-9]. In this work, a cylindrical type ER brake actuator is devised and integrated with the WEDM. After evaluating the field-dependent wire tension, the straightness of the workpiece is empirically investigated by changing the intensity of the electric field. In addition, the wire tension control is undertaken by adopting a robust sliding mode controller.

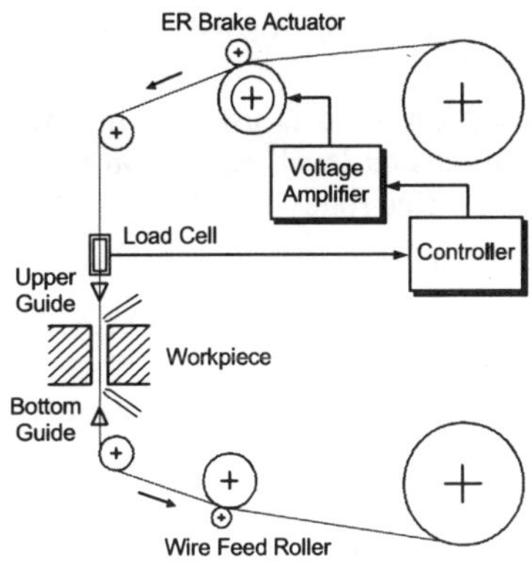

Figure 1. Tension control mechanism with the ER brake actuator

2 Model of ER Brake Actuator

A schematic configuration of the tension control mechanism in the WEDM is shown in Figure 1. A constant speed of the wire is generated by the wire feeding system and the speed(or equivalently tension) is to be controlled by the ER brake actuator. This is accomplished by applying an appropriate control electric field. The wire tension is measured by the load cell and is fed back to the microprocessor(controller). The actual tension is compared with the desired value, and an appropriate electric field to match two tension values is to be applied to the ER brake actuator. In this work, a cylindrical type of the ER brake is devised as shown in Figure 2. The inner cylinder electrode is fixed, while

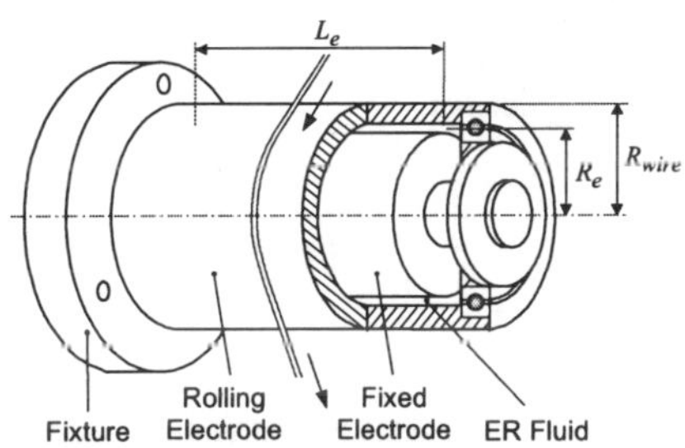

(a) schematic configuration

(b) photograph

Figure 2. The proposed ER brake actuator

the outer cylinder(rolling electrode) is rotated. Thus, a shear mode flow motion is occurred between two electrodes.

A chemical starch based ER fluid is used in this study. By considering Bingham property of the ER fluid and tension level required in the WEDM, the following design parameters are determined: electrode length (L_e) =100*mm*, radius of the inner cylinder(R_e)=28.5*mm*, radius of the wire roller(R_{wire})=41*mm*. It is noted that the outer cylinder is treated with silicone coating in order to provide a sufficient friction between the wire and the surface of the wire roller.

The braking tension of the proposed ER brake actuator mechanism is obtained by [9]

$$T_c(E)=\frac{2\pi R_e^2 L_e}{R_{wire}}\tau_y(E) \tag{1}$$

where $\tau_y(E)$ is the field(E)-dependent yield shear stress of the ER fluid which is given by $\tau_y(E)=\alpha \cdot E^{\beta}[Pa]$. The α and β are intrinsic values of the ER fluid which are to be experimentally determined. As for the chemical starch based ER fluid, α and β are evaluated as 389.5 and 1.48, respectively.

3 Performance of ER Brake Actuator

Figure 3 presents the wire tension of the WEDM. The conventional electromagnetic actuator is replaced by the proposed ER brake actuator. As clearly observed, the wire tension is increased as the electric field increases. For instance, the tension of 3.85*N* in the absence of the electric field is increased up to 16.2*N* by applying the electric field of

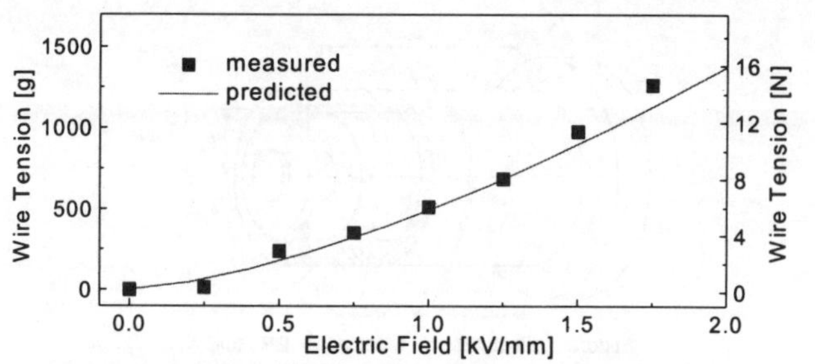

Figure 3. The field-dependent wire tension

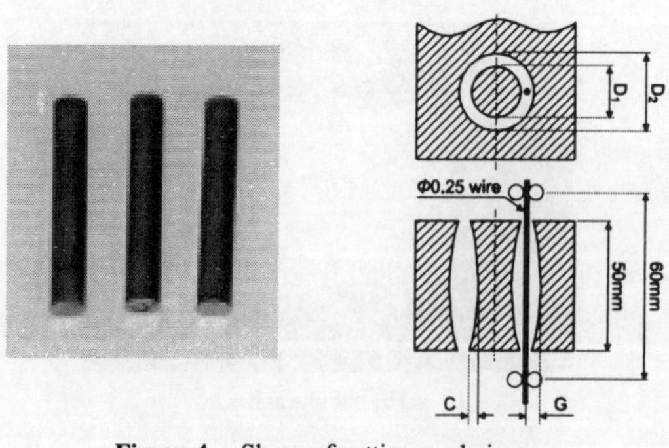

Figure 4. Shape of cutting workpieces

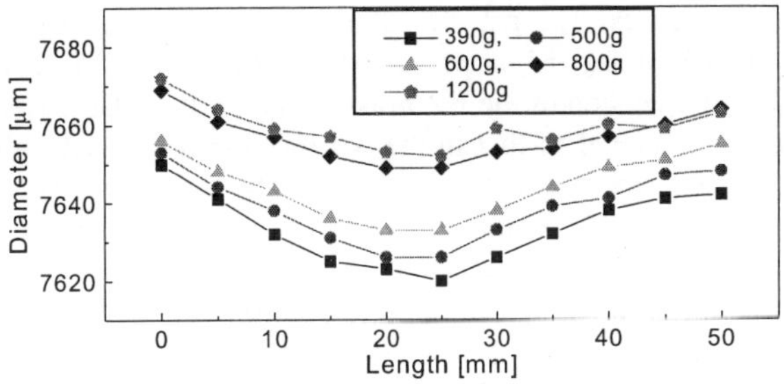

Figure 5. Workpiece contour for various tensions

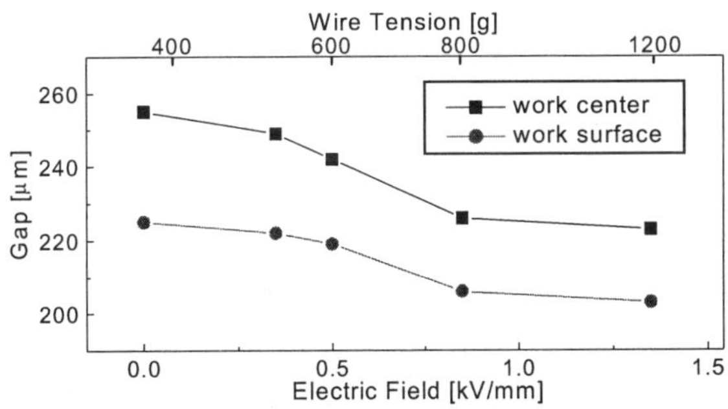

Figure 6. Discharging gap of the workpiece

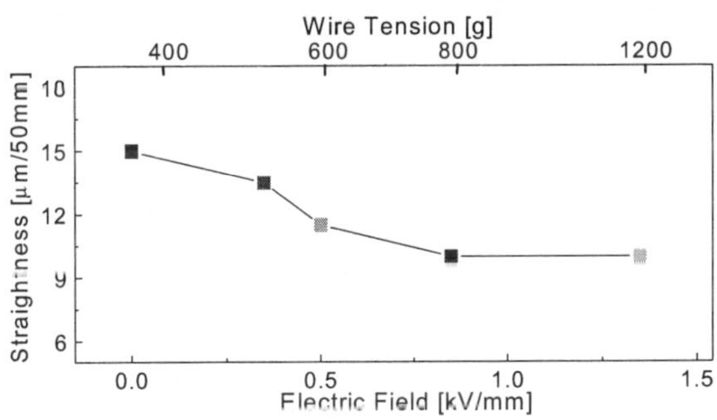

Figure 7. Straightness of the workpiece

1.75*kV/mm*. It is also observed that the predicted tension agrees well to the measured one. This advocates the validity of simple tension model given by Eq.(1). For the evaluation of the machining performance, a brass wire(diameter=0.25*mm*) is used as the wire electrode, and a carbon steel is adopted as the workpiece. The cutting speed of the machine is set by 0.5*mm/min*, while the traveling speed of the wire is fixed by 10*m/min*. The distance between rollers is set by 60*mm*. Figure 4 presents the photographs of the workpieces and the machined shapes. From the shape, we clearly see that the machined diameter(D_1) of the workpiece is decreased as the depth from the surface increases. The diameter becomes minimum at the middle point as shown in Figure 5. This is, of course, due to the deflection force of the wire. It is also found out that the diameter becomes larger as the electric field increases. This is due to that the deflection force becomes smaller as the tension due to the ER brake actuator increases.

Figure 6 presents the measured discharging gap(G in Fig.4) of the workpiece. The smaller gap is good for more accurate machining. We clearly see that the gap is decreased as the electric field increases. The gap of $225\mu m$ in the absence of the electric field is decreased to $203\mu m$ by applying the electric field of $1.35kV/mm$. This is, of course, due to the increment of the wire tension of the ER brake actuator. It is also observed that the gap in the work surface is smaller than that in the work center at the same electric field. The difference between two gaps is controllable band. The straightness(C in Fig.4) of the workpiece is also measured and presented in Figure 7. We clearly see that the straightness become smaller as the electric field increases. This implies that more accurate machining can be undertaken with higher electric field. The straightness of $15\mu m$ without the electric field is decreased to $10\mu m$ by applying the electric field of $0.85kV/mm$. From this result, we know that the straightness can be improved by increasing the wire tension. However, an optimal tension to achieve desired straightness and to reduce unwanted vibration should be determined by an appropriate control algorithm.

4 Control of Wire Tension

In order to control the wire tension of the proposed machine mechanism the following control model is considered [9].

$$\dot{\mathbf{x}}(t) = \begin{bmatrix} 0 & 1 & 0 \\ 0 & 0 & 1 \\ -\dfrac{\omega_1^2}{\tau} & -\dfrac{\omega_1^2\tau + 2\zeta_1\omega_1}{\tau} & -\dfrac{2\zeta_1\omega_1\tau + 1}{\tau} \end{bmatrix} \mathbf{x}(t) + \begin{bmatrix} 0 \\ 0 \\ \dfrac{C_t}{\tau} \end{bmatrix} T_c(E) \tag{2}$$

$$y(t) = \begin{bmatrix} \Phi_1(r_1) & 0 & 0 \end{bmatrix} \mathbf{x}(t)$$

In the above, ω_1 is the first mode natural frequency of the wire, ζ_1 the first mode damping ratio, τ the time constant of the ER brake actuator, and C_t the first mode shape of the wire. $\mathbf{x}(t)$ are state variables representing modal displacement and velocity of the flexible wire, and $T_c(E)$ is control input. In the control model, the output $\mathbf{y}(t)$ represents the tension of the wire. So, by controlling the state variables we can achieve desired tension of the electrode wire. Figure 8 presents the measured frequency response of the electrode wire. From this result we distill the system parameters such as the first mode natural frequency.

Now, in order to achieve desired tracking motion we define the tracking error as follows.

$$e_i = x_i - x_{id}, \quad i = 1,2,3 \tag{3}$$

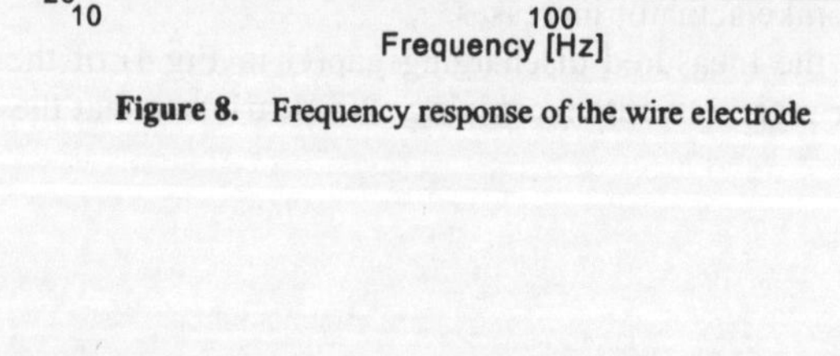

Figure 8. Frequency response of the wire electrode

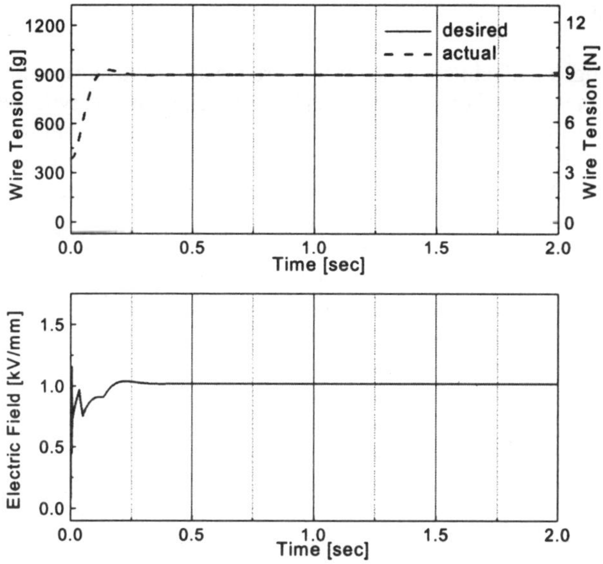

Figure 9. Tension control responses

where, x_{id} is the desired state corresponding to the x_i. Among many control candidates, a robust sliding mode controller[10] which is known to be robust external disturbances and uncertainties is adopted. As a first step, the following stable sliding surface is defined.

$$s = g_1 e_1 + g_2 e_2 + e_3, \quad g_i > 0 \tag{4}$$

Then, a sliding mode condition, which allows one to design a controller to make the tracking error be zero for arbitrary initial errors, is imposed as follows.

$$s \cdot \dot{s} < 0 \tag{5}$$

Now, on the basis of the control model and the sliding mode condition the following sliding mode controller is designed.

$$T_c(E) = -\frac{1}{C_t}\left[\tau(g_1 x_2 + g_2 x_3 - x_{3d}) - (2\zeta_1\omega_1\tau + 1)x_3 - (\omega_1^2\tau + 2\zeta_1\omega_1)x_2 - \omega_1^2 x_1\right] - K \cdot sign(s) \tag{6}$$

In the above, K is the discontinuous gain. It is noted that the sign function in Eq.(6) is normally replaced by saturation function in order to reduce unwanted chattering phenomenon of the controller[10]. Figure 9 presents regulating tension control response of the proposed system. It is clearly observed that the actual wire tension is well regulated on the desired set value of 900g by applying an appropriate control field. This tension controllability will enhance the machining performance of the WEDM in terms of surface roughness and straightness.

5 Conclusion

An electrorheological(ER) brake actuator was proposed for the tension control of the wire cut discharge machine. After devising a cylindrical shear-mode type ER brake, it was incorporated with the discharge machine. It has been demonstrated through experimental investigation that the wire tension is increased as the electric field increases. It was also shown that the discharging gap and straightness of the workpiece decrease as the electric field increases. In addition, tension control was demonstrated by adopting a robust sliding mode controller. The results presented in this work indicate that the machining performance of the wire cut discharge machine can be much improved by applying an appropriate control field associated with the proposed ER brake actuator.

References

1. Dauw, D. F., Sthioul, H., Delpretti, R. and Tricarico, C., Wire Analysis and Control for Precision EDM Cutting. *Annals of the CIRP*, **38** (1989) pp.191~194.
2. Magara, T., Yamada, H., Yatomi, T. and Kobayashi, K., Study on Improvement of Surface Straightness in Wire-EDM. *Journal of Japan Society of Precision Engineering,* **54(4)** (1993) pp.619~624.
3. Liao, Y. S., Chu, Y. Y., Yan, M. T., Study of Wire Brakinging Process and Monitoring of WEDM. *International Journal of Machine Tools & Manufacture*, **37(4)** (1997) pp.555~567.
4. Guo, Z. N., Lee, T. C., Yue, T. M., Lau, W. S., Study on The Machining Mechanism of WEDM with Ultrasonic Vibration of The Wire. *Journal of Materials Processing Technology*, **69** (1997) pp.212~221.
5. Gőkler, M. I., Ozanőzgü, A. M., Experimental Investigation of Effects of Cutting Parameters on Surface Roughness in The WEDM Process. *International Journal of Machine Tools & Manufacture*, **40** (2000) pp.1831~1848.
6. Stevens, N. G., Sproston, J. L. and Stanway, R., An Experimental Study of Electro-Rheological Torque Transmission. *ASME Journal of Dynamic Systems, Measurement and Control,* **110** (1988) pp.182~188.
7. Carlson, D., Duclos, T. F., ER Fluids Clutches and Brakes-Fluid Property and Mechanical Considerations. *Proceedings of the 2nd International Conference on ER Fluids*, (1990) pp.353~367.
8. Johnson, R., Bullough, W. A., Firoozian, R., Hosseini-Sianaki, A., Makin, J. and Xiao, S., Testing on a High Speed Electro-Rheological Clutch. *Proceedings of the 3rd International Conference on ER Fluids,* (1991) pp.424~441
9. Choi, S. B., Cheong, C. C., Kim, G. W., Feedback Control of Tension in a Moving Tape Using An ER Brake Actuator. *Mechatronics,* **7(1)** (1997) pp.53~66.
10. Choi, S. B. and Park, D. W., Moving Sliding Surfaces for Fast Tracking Control of Second-Order Dynamic Systems. *ASME Journal of Dynamic Systems, Measurement and Control,* **116** (1994) pp.154~158.

A SIMPLE ONE DIMENSIONAL ROBOT JOINT BASED ON THE ER LINEAR REVERSING MECHANISM

K. P. TAN, W. A. BULLOUGH, R. STANWAY, N. SIMS, A. R. JOHNSON AND R. C. TOZER.
Smmart Engineering Network, University, Sheffield S1 3JD UK.

The feasibility of using a well-known twin ER clutch linear reversing mechanism as a robotic actuator is demonstrated. High speed of response, displacement and positional accuracy can be obtained bi-directionally. A validated mathematical model of the apparatus provides a basis for the control strategy. Positional accuracy is enhanced by an ER brake which works in sequence with the clutch excitation switches. The robot arm, in rotational displacement is tested over a large number of cycles, speeds, displacements and inertial loads and the quality of control compared to that of competitive conventional servo motors.

1. INTRODUCTION

In providing a multi-functional manipulator [1] for a variety of tasks, DC servo driven robot arms experience positional inaccuracies due to the well known cogging torque phenomenon. Typical inaccuracies of 1.14° [2] and 0.4° [3] can arise. Lead-lag [4] and state space control techniques [5] can produce robot arm positioning errors of about 4°. To improve on this accuracy, the concept of an electro-rheological clutch based mechanism has been proposed.

Brookfield et al [6] employed an ER fluid clutch to raise a robot arm. By means of a Proportional-Integral-Derivative (P.I.D.) control strategy, the robotic positional error achieved was 0.75°. However, vibration was experienced during the spring-driven reversal. Choi et al [7] developed an ER clutch/brake mechanism to move a positioning table linearly. However, the reciprocating response of the table was degraded by the driving DC motor. To achieve flexible fast reciprocating linear motions, a twin ER clutch mechanism was investigated [8]. A mathematical model describing the output kinematics of this clutch mechanism was thus made available. This model was used in the design of a control strategy for positional control of the present robot arm: the resulting strategy is used to control an ER brake in oder to achieve accurate robotic position. The displacement response of the robot arm is examined as a trend study using different motor driving speeds and load inertias. Next, the robots repeatability is observed over a significant number of reciprocating tests. Finally, the speed of response and positional accuracy of the robot arm using both the ER mechanism and DC servomotors is compared. The clutch driven, belt connected mechanism is shown in Figures 1 and 2. Details are as in [8, 9] save that a K.S.M. DC power amplfier and switch units were employed to energise the ER devices alternately. The measurement instruments include a DC tachometer for measurement of clutch speed and an encoder fitted to the 61.25/1 gear reduction unit. The fluid used is Bayer AI3565 50% volume fraction ER fluid.

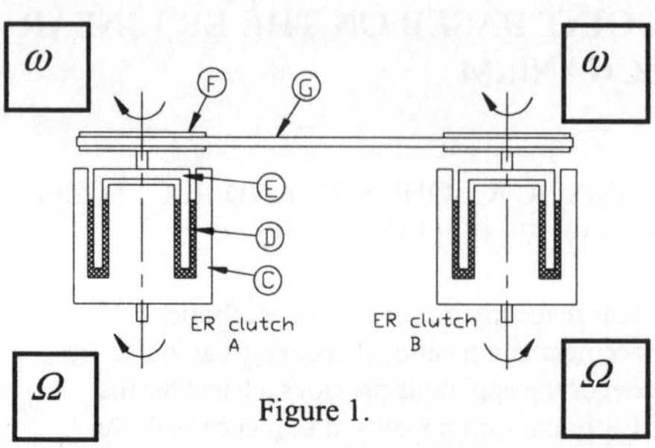

Figure 1.

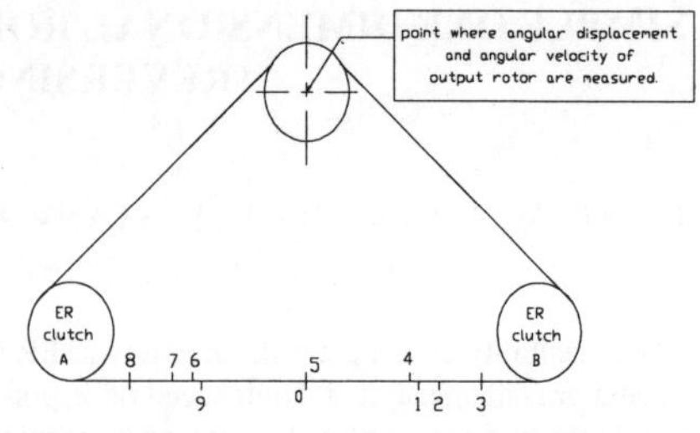

Figure 2.

Figures 1 and 2. Schematic view of the reciprocating ER clutch mechanism.

2. VALIDATION OF MEASURED ROBOTIC DISPLACEMENT RESPONSE USING A SIMULINK CONTROL PROGRAM.

The kinematic model is applied in a Simulink control program (Figure 3) in order to predict the robot performance and to aid implantation in the DSpace controller. This controller (Figure 4a) model consists of on/off relays, the kinematic model, gear ratio, robot arm inertia and an encoder. The simulated displacement response of the robot manipulator is shown in Figure 5.

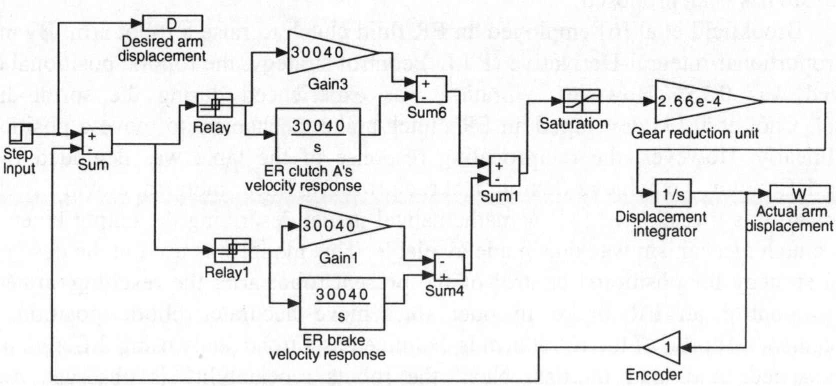

Figure 3. Simulink model to control the theoretical robotic displacement response.

To validate the predicted robotic model displacement, the experimental displacement of the robot arm was measured. This was done by using the control strategy and mechanism of the robot arm in Figures 4(a) and 4(b). This robot arm has a length of 250 mm and a weight of 0.33 kg. A parallel plate type ER brake has equivalent ER active and viscous active dianeter of 96 mm. By adding the inertia of the ER brake, gear train and robot arm with the ER output rotor units etc, the total output moment of inertia of the ER clutch/brake mechanism amounts to 1.9 x 10^{-4} kgm^2.

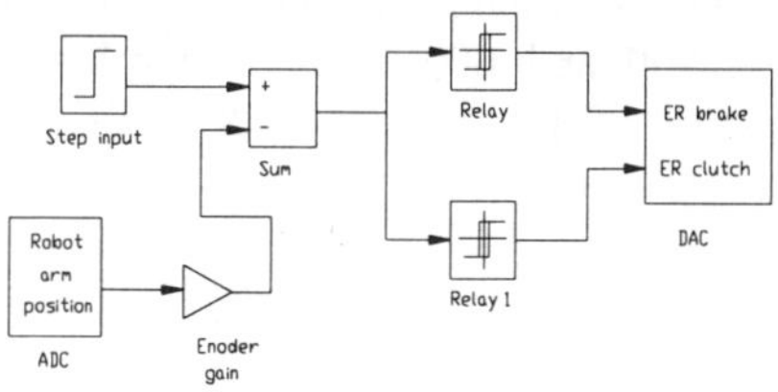

Figure 4(a).

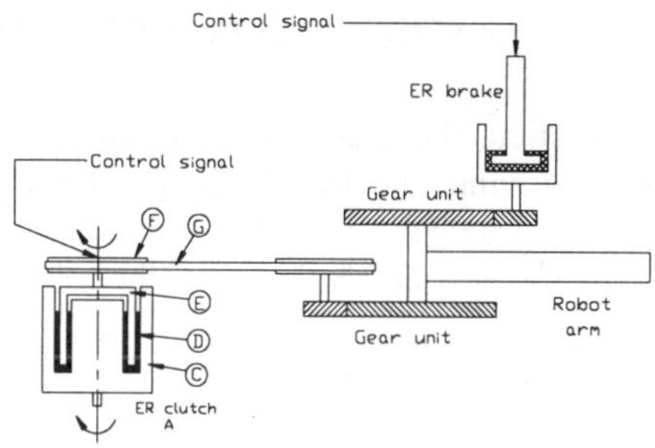

Figure 4(b).

Figures 4(a) and 4(b). Experimental setup and control strategy of the ER driven-robot arm.

A DSpace controller unit using the control strategy of Figure 4(a) controls the robot arm to achieve a set desired displacement. When this displacement is achieved, the encoder signals the on/off controller to switch off the driving ER clutch and to switch on the ER brake. Figure 5 illustrates validations of five measured robotic displacements with that of the model prediction, for 40° displacement using a driving speed of 400 rpm and 1 kV control signal.

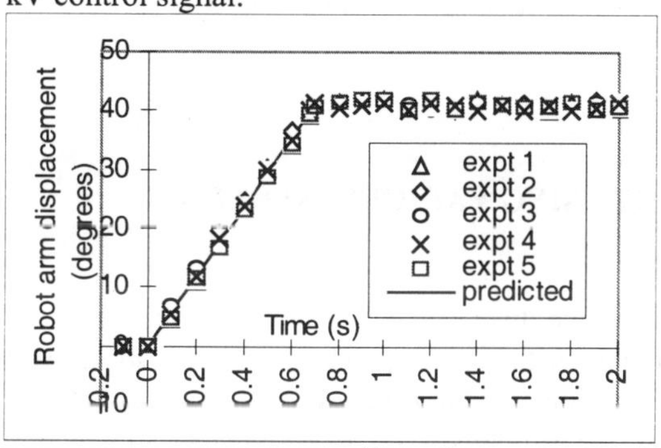

Figure 5.

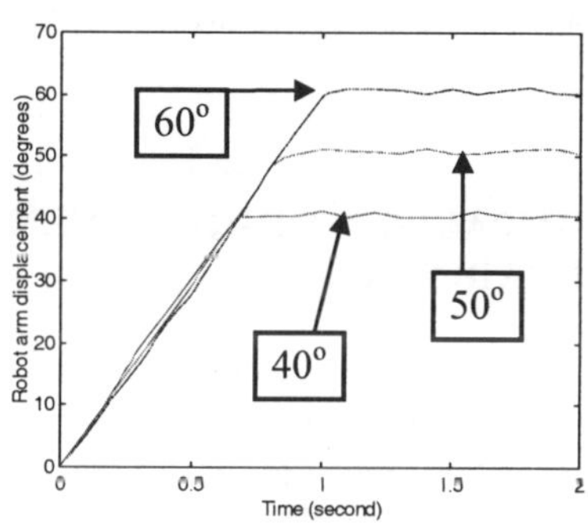

Figure 6.

Figure 5. Validation of the measured robotic displacement response. Figure 6. The behaviour of three different displacements at a particular driving speed .

3. ROBOT ARM DISPLACEMENTS FOR A SINGLE SPEED.

To achieve this demonstration, the twin ER clutches were driven at 600 rpm. Through the application of a control signal, a pulse of 1 kV voltage, the ER clutch actuated the robot manipulator to achieve three programmed displacement responses of 40°, 50° and 60°. On reaching these displacements, the ER brake is activated by the controller to hold the robot arm stationary. Figure 6 is used to illustrate the behaviour of the robot arm at three different displacements.

4. INFLUENCE OF MOTOR DRIVING SPEEDS.

In this section, the main aim is to study the displacement response of the robot arm at different driving speeds of the ER clutches (400, 500 and 600 rpm). The robot is controlled to achieve a common displacement response of 40°. Figure 7. Obviouly, the faster the rotating speed, the faster the robot arm reaches steady-state.

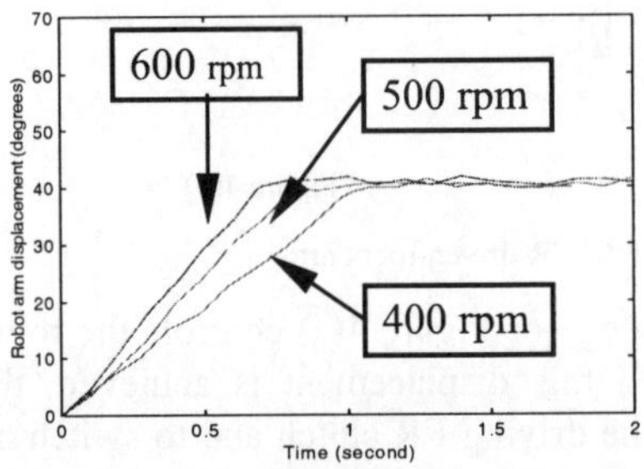

Figure 7.

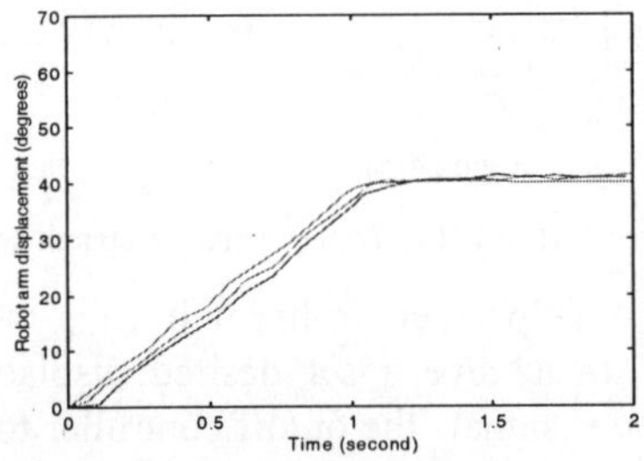

Figure 8.

Figure 7. Displacement response of the robot arm for three driving speeds. Figure 8. Effects of robotic displacement with respect to inertia of robot arm (———), robot arm with a 1 kg load attached (············) and robot arm with a 2 kg load attached (-·-·-·-·-·-).

5. COMPARISON OF DISPLACEMENT USING VARIOUS INERTIAL LOADINGS.

The objective of this section is to understand the displacement response of the robot arm when the manipulator is subjected to different inertial loadings. These loadings are the self inertia of the robot arm plus 1 kg and 2 kg loads. This trend study is conducted for a constant robotic displacement of 40°. Referring to Figure 8, the larger the inertial load, the longer the robot arm takes to reach the steady-state displacement.

6. DISCUSSION AND CONCLUSION.

The mechanism proposed and the method validated borrows from the work of others [6, 7]. Strategically, it offers more via true bi-directional flexible operation through fast voltage switching with respect to time, which produces a required change in angular displacement. Performance wise, the basic concept is improved since on reaching this position, the viscous stress tending to drive the belt backwards (from trailing rotor) assists the brake and nullifies the effects of any like stress imminent from the now switched off lead rotor. No doubt the improved/smoother action and speed of the operation owes much to this effectiveness of the DSpace controller, the quality of the output shaft encoder, the belt drive and the close fit between the modelled and experimental run-up and run-down performance [9]. This transient period through completed within a few milliseconds or so, is of crucial importance. However, it should be recognised that its duration depends on [10] the fluid temperature (if ERF rather than MRF is utilised) and future work should include the control of heat input to the fluid to suit the (flexible) operating cycle - this is probably best done by judicious lowering (50%) of the clutch input voltage for a selected time period between clutch lock up and brake application: in this sense structural hysteresis in the fluid becomes a benefit to the digital machine operation.

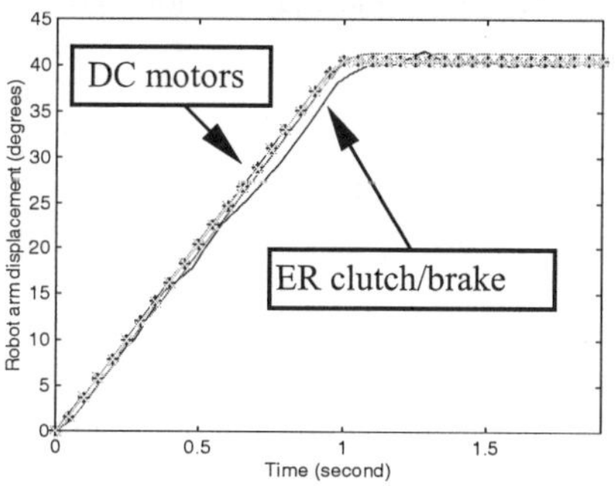

Figure 9.

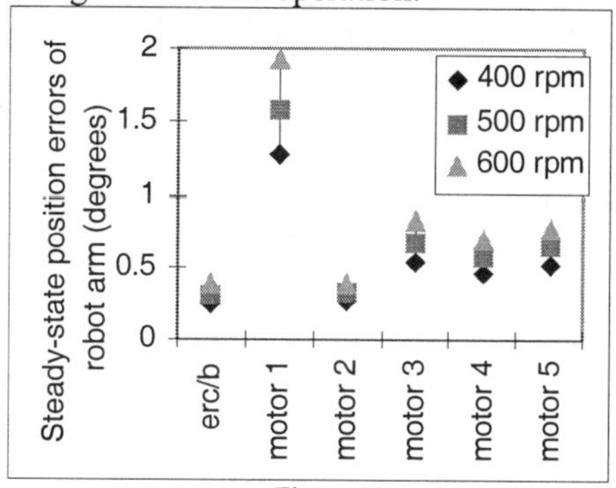

Figure 10.

Figure 9. Robotic displacement between ER mechanism and DC servomotors. Figure 10. Displacement accuracies between ER mechanism and DC servomotors. (——, ········, -·-·-·-·, XX and O O, Motors 1, 2 3 4 and 5 respectively).

Amongst the more important features of all machines are repeatability of operation and the ability to predict performance at the design stage. These are qualitatively illustrated in Figures 5, 9(a) and 9(b) respectively. Serviceability of the ER fluid proved satisfactory of the low voltages, speeds and frequency of operation experienced. The brake improves the tolerance on positional accuracy from typically 0.2° to 2.3° against the mechanism without a brake. Whilst probably slightly slower than a DC servomotor on the lighter duties on which the ER fluid device is best suited a better positional accuracy is expected - against a whole series of selected commercial available machines - see Figures 9 and 10. The device could probably be constructed on a common axis. Its speed of approach and acceleration can be changed by judicious choice of the gearing ratio.

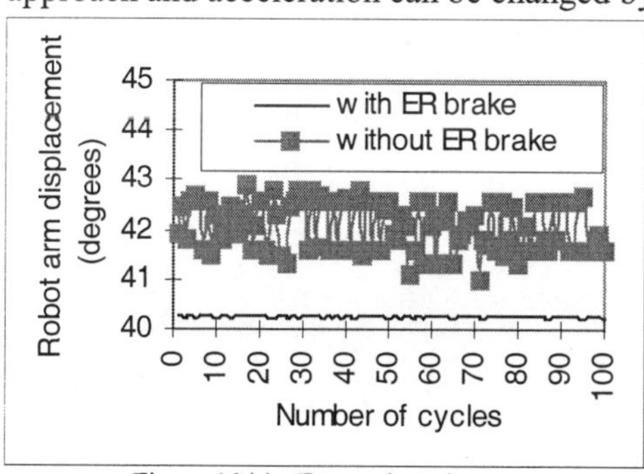

Figure 10(a). Forward motion

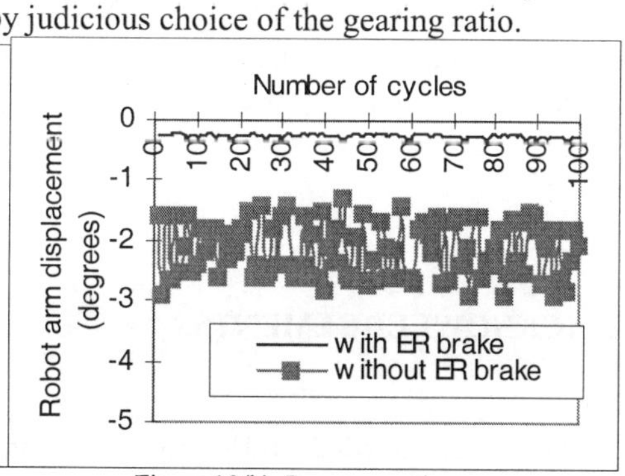

Figure 10(b). Reverse motion.

Figures 10(a) and 10(b). Forward and reverse motions of the robot arm at clutch driving speed of 600 rpm, fluid temperature of 30°C and input step voltage of 1 kV.

7. REFERENCE.

1. **Y. Chen.** 'Replacing a PID controller by a lead-lag compensator for a robot arm - a frequency reposne approach', IEEE Transactions of Robotics and Automation, Vol 5, no 2, April 1989, p174-182.

2. **T. J. Tarn, A. K. Bejczy, X Yun and Z Li.** 'Effect of motor dynamics on non-linear feedback robot arm control', IEEE Transactions of Robotics and Automation, Vol 7, no 1,April 1991, p114-121.

3. **M. Moallem and K. Khorasani.** 'An integral manifolf approach for tip-position tracking of flexible multi-link manipulator', IEEE Transactions of Robotics and Automation, Vol 13, no 6, December 1997, p823-837.

4. **J. H. Yang, F. L. Lian and L. C. Fu.** 'Nonlinear adaptive control flexible link manipulator', IEEE Transactions of Robotics and Automation, Vol 13, no 1, February 1997, p141-148.

5. **C. C. Cheah and D. W. Wang.** 'Learning Impedence control for robotic manipulator', IEEE Transactions of Robotics and Automation, Vol 14, no 3, June 1998, p452-465.

6. **D. J. Brookfield and Z. B. Dlodo.** 'Robot torque and position control using an Electro-rheological actuator', Proceedings of Institution of Mechanical Engineers, Vol 212, 1998, p 229-238.

7. **S. B. Choi, S. L. Kim, H. G. Lee and M. H. Nam.** 'Position control of a moving table using ER clutch and ER brake', Proceedings of 6th international conferences on electro-rheological fluids, magneto-rheological suspensions and their applications, Japan, July 1997, World Scentific Publications, p661-669.

8. **J. Makin.** 'Fast actuating controllers and rheological based devices,' Ph.D Draft Thesis, University of Sheffield, 2000.

9. **A. R. Johnson, J. Makin and W. A. Bullough.** 'Dynamic simulation and performance of an electro-rheological clutch based reciprocating mechanism', Smart Materials and Structures, Vol 8, 1999, p591-600, UK.

10. **J. Makin, W. A. Bullough, R. Firoozian, A.R. Johnson and A.H. Sianaki.** 'The Effect of Solid Fraction Concentration on the Time Domain Performance of an ER Fluid in the Shear Mode', Proceedings of the Fourth International Conference on Electro-Rheological Fluids, July 20-23, Feldkirch, World Scientific Publishing Company, 1994, p372-393.

8. ACKNOWLEDGEMENTS.

Thanks are due to Mr J. B. Hodgson for technical support.

II. MICROSTRUCTURES

STRUCTURES IN A MAGNETIC SUSPENSION SUBMITTED TO UNIDIRECTIONAL AND ROTATING FIELD

P. CARLETTO, G. BOSSIS

Laboratoire de Physique Matière Condensée, CNRS, Université deNice,06108 Nice Cedex 2, France

A. CEBERS

Institute of physics, Latvian Academy of Sciences, Salaspils, LV-2169, Latvia

Field induced structures are studied inside suspensions of magnetic colloidal particles of micronic size. We have characterized the average distance between aggregates in a thin cell with the magnetic field perpendicular to the plane and also in the presence of a rotating field with the plane of rotation perpendicular to the plane of the cell. The characteristic size of the mesostructure is predicted on the basis of a thermodynamic model. The theory well predicts the experimental results in the uniaxial case but not in the case of ae rotating field; in this last case, the surface tension which is needed to have a good fit is far too low compared to its expected order of magnitude. When the field is uniaxial and sinusoidal we have found a collective instability where all the aggregates are rotating simultaneously in a chaotic way.

1 Introduction

MagnetoRheological (MR) suspensions are composed of magnetic particles whose size, in the micronic range, is at least two orders of magnitude larger than in the better known magnetic suspensions, called ferrofluids. The dipolar force between two particles being proportional to the square of the radius of the particles, even in low fields, the magnetic force becomes larger then the thermodynamic force and we observe a phase separation with a pattern formation [1]. This pattern consists usually of a network of chains of particles or of cylindrical aggregates. The size of the aggregates and the period of the pattern are controlled by a balance between the long range demagnetization energy which tends to split the aggregates and the internal dipolar energy of the aggregates including the surface energy. Furthermore the entropic part must be taken into account in the minimization of the free energy of the suspension.

In magnetic liquids and mainly in ferrofluids, it is possible to successfully predict the change of period of striped pattern with the amplitude of the external magnetic field, if the effective surface tension at the boundary between the two phases is known [2][3].

More recently, well defined structures formed of cylindrical aggregates aligned on the field and arranged on an hexagonal pattern have been observed in MR suspension [4][5][6]. The transformation of these cylindrical structures into a striped pattern was observed in an oscillating shear flow and explained with the help of a model based on a minimization of the free energy [7]. A similar striped pattern has also been found in the presence of a rotating magnetic field [8][9]. In this paper, we present experimental results obtained with a MR suspension both in unidirectional and rotating magnetic field and the model we propose to interpret these results. In the last part, we present briefly new structures obtained in a slanted cell in the presence of a sinusoidal magnetic field and the collective rotation which can be observed in some frequency bands.

2 The model

The model we present here is a generalization of the one previously used to predict the structure in an unidirectional field[6],[7]. We shall recall the main steps in the case of the unidirectional field and we shall generalize to the case of a rotating field. Let us consider a two dimensional hexagonal array of N_a ellipsoidal aggregates with a semi-axis b and a length h=2l and separated by a distance d. The following notations are used : M_a is the magnetic moment of one aggregate and V_a its volume. The quantity $\varphi=N_aV_a/V$ is the part of volume which is occupied by the aggregates. It must not be confused with the internal volume fraction inside the aggregate $\Phi_a=\Phi_0/\varphi$ where Φ_0 is the average volume fraction of particles in the suspension. In our experiment $\Phi_0 = 0.04$.

In unidirectional field, the magnetic moment M_a of an ellipsoid aggregate is given by [10]:

$$M_a = \frac{\mu_a - 1}{4\pi}\left[\frac{H_0 - \sum_j T_{ij}^r M_a}{1+n_Z(\mu_a - 1)}\right]V_a = \alpha_a\left(H_0 - \sum_j T_{ij}^r M_a\right) \quad (1)$$

where $\mu_a=f(\Phi_a)$ is the permeability of the aggregate obtained by using the equation of Maxwell-Garnett. n_z is the demagnetization factor of an ellipsoid and for a hexagonal array $T_{ij}^r = \frac{1}{h^2}\left[\frac{2}{d_{ij}} - \frac{2}{\left(d_{ij}^2 + h^2\right)^{1/2}}\right]$ is the propagator of the field due to aggregate j on the aggregate i whose axes are separated by the distance d_{ij} [4]. The total magnetic energy per unit volume is then:

$$\frac{U_m}{V} = \frac{-1}{2V} N_a \vec{M}_a \vec{H}_0 = \frac{-\varphi H_o^2}{8\pi}\left[\frac{(\mu_a - 1)}{1+n_z(\mu_a - 1)}\right]\left[\frac{1}{1+n_r}\right] \quad (2)$$

with $n_r = \sum_j T_{ij}^r = \left[\frac{(\mu_a - 1)}{1+n_z(\mu_a - 1)}\right][\varphi]\left[\sqrt{\left(1+\frac{d^2}{h^2}\right)} - \frac{d}{h}\right]$ and $d = b\sqrt{\frac{2}{3\varphi}}$

Finally, we have also to take into account the surface tension of the aggregate σ which comes from the change of the local field on the particles belonging to the surface S_a of an aggregate. Indeed the magnetic field is weaker on the surface than inside the aggregate and the corresponding energy is [7]: $\frac{U_S}{V} = \frac{N_a \sigma S_a}{V}$ with $\sigma = \frac{M_a^2 a}{3\mu_o V_a^2}$ (3)

Finally, we have in unidirectional field a total magnetic energy : $\frac{U}{V} = \frac{U_m}{V} + \frac{U_S}{V}$

In rotating field the average interaction in the plane of rotation of the field is attractive and induces the formation of a layered pattern. We call e the thickness of the layer, and h its height ;the width L is very large compared to the other dimensions so we shall consider a two dimensional situation. Let us call $\vec{H}_0^n$ the component of the field normal to the plane of the cell and $\vec{H}_0^t$ the tangential component: $\vec{H}_0 = \vec{H}_0^n + \vec{H}_0^t$. These two components are sinusoidal with a phase difference of $\pi/2$. We know that our particles are superparamagnetic and that the magnetization does not present any hysteresis at the frequencies used (ν<100Hz), so we can treat each component independently. Due to this simple geometry we can reason at the level of one particle and write for the normal magnetic moment of a layer: $M_a^n = N_V m_V^n + N_S m_S^n$ with N_V, m_V^n, and N_S, m_S^n respectively the numbers of particles and the magnetic moments of a given particle in the volume of the aggregate and on its surface. The magnetic energy U_{layern} of a layer of paramagnetic particles in a rotating field is then:

$$U_{layer} = -\frac{1}{4}\left[M_a^n H_0^n + M_a^t H_0^t\right] \quad (4)$$

The magnetic moment of a particle inside the aggregate, will be the product of its polarisability ,α, times the local field: H_{Loc}

$$m_V^n = \alpha H_{V,loc} = \alpha\left[H_0^n - 4\pi\, n_d \frac{M_a^n}{V_a} + \frac{4\pi}{3}\frac{M_a^n}{V_a}\right] \quad \text{with} \quad \alpha = a^3\frac{(\mu_P - 1)}{(\mu_P + 2)} = a^3\beta \quad (5)$$

The second term in the right-hand side of (5) is the demagnetization field and the third one the Lorentz field inside the layer. For a particle on the surface, we divide the Lorentz field by two, so we have: $m_S^n = m_V^n - \frac{2\pi}{3}\alpha\frac{M_a^n}{V_a}$ (6)

Using Eqs. (5-6), we obtain the normal component of the magnetic moment of a layer per unit of volume :

$$M_n = \frac{M_a^n}{V_a} = \frac{3\Phi_a \beta/4\pi}{1 - \beta\Phi_a\left(1 - 3n_d - \frac{2a}{e}\right)} H_0^n \quad (7)$$

For the tangential component of the magnetic field, the method will be the same one with a significant simplification since in this case the demagnetizing factor is zero. We thus have result (7) with $N_d=0$ and H_0^t instead of H_0^n. Using equations (4-7), we can calculate total energy per unit of volume. In our case, $H_0^t = H_0^n$ (rotating field) and we obtain:

$$\frac{U}{V} = \frac{-3\beta\, {H_0^n}^2 \Phi_0 A}{16\pi} \quad \text{with } A = \frac{1}{1 - \beta\Phi_a\left(1 - 3n_d - \frac{2a}{e}\right)} + \frac{1}{1 - \beta\Phi_a\left(1 - \frac{2a}{e}\right)} \quad (8)$$

The value of the demagnetizing factor N_d for a stripe pattern with a period d was already calculated by A. Cebers[14].

$$n_d(\varphi, d^*) = 1 - \frac{d^*}{\pi^3 \varphi}\sum_{k=1}^{+\infty}\frac{\sin^2(\pi k \varphi)}{k^3}\left[1 - \exp\left(\frac{-2\pi k}{d^*}\right)\right] \quad \text{with } d^* = \frac{d}{h}$$

In both cases (unidirectional or rotating fields), the distance d which characterizes the mesostructure is obtained by a minimization of the free energy relatively to this distance.

Saying that the total pressure (magnetic pressure plus osmotic pressure) inside the aggregate is close to zero for well formed mesostructures , we get an other equation allowing to determine the second unknown which is the internal fraction, Φ_a , inside the aggregates. The magnetic and osmotic pressures are respectively:

$P_m = -\frac{\partial}{\partial V}(U_m)\Big|_{N,T}$ and $P_{osm} = \frac{kT}{v_p} 1.85 \frac{\Phi_a}{0.64 - \Phi_a}$ (v_p : volume of the particle) [11][12].

The system to solve is then : $\frac{\partial}{\partial d}\left(\frac{U}{V}\right)_{\varphi}=0$ and $P_m+P_{osm}=0$ (10)

3 Experimental results in a constant unidirectional field

The suspension is placed in a sealed cell between two plates making a small angle in order to have a variable thickness from one extremity of the cell to the other. This cell is placed at the center of Helmoltz coils and we look to the structure with the help of a microscope equipped with a video camera. The magnetic field is driven by computer which also record the pictures of the structure. The suspension we have used is made of polystyrene particles containing 63% by weight of magnetite. These particles suspended in water, and provided by Merck, are spherical but polydisperse with an average diameter 2a of 0.78 μm for all experiments presented in this paper. These particles behave as a super-paramagnetic material and their initial relative permeability was deduced from the measurement of the permeability of the suspension and Maxwell-Garnett theory. We have found $\mu_p=2.69$ for the permeability of the particles. The main quantity which controls the phase separation in the absence of shear is the ratio of magnetic energy of two aligned dipoles to thermal energy:

$$\lambda=\frac{2\,m_p^2}{4\pi\,\mu_0(2a)^3kT}=\frac{\pi\,\mu_0\,\beta^2a^3H_0^2}{kT} \quad (11)$$

which , for a = 0.39 μm, β = 0.36, and room temperature, gives unity for a field close to 4 Oersted. A typical view of a structure that we can see when the magnetic field is perpendicular to the plane of the cell is shown in Fig. 1. The black dots represent the projections of cylindrical aggregates in the plane of observation which is perpendicular to the field. The typical size of the aggregate is a few microns. The average size between the aggregates is obtained by image analysis ,using an average over about twenty images for each height of the cell. The change of the distance inter-aggregate with respect to the height of the suspension is shown in the Fig. 2 (solid squares). The two equations of the model are solved for different heights of the cell (i.e. different thickness of suspension) with a given value of λ=268 (H=76 Oersted), and a=0.39 μm. The theoretical result for d is shown in Fig. 2 (black circle) and is in very good agreement with the experiments. The theoretical as well experimental results can be fitted by a power law of exponent of 0.49.

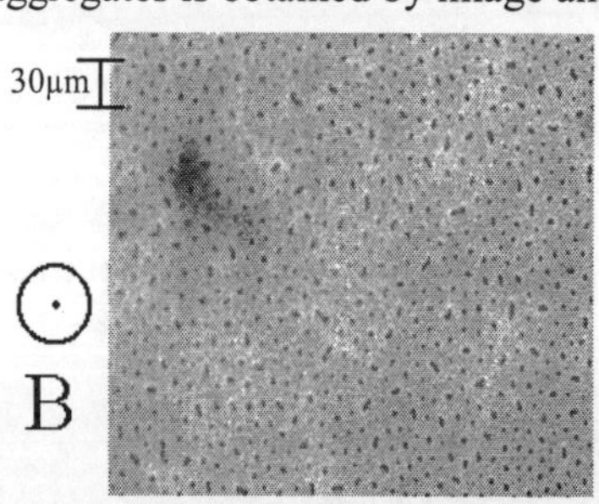

Figure 1 Top view of the suspension recorded by video camera. The field is perpendicular to the plane of the figure , H=76 Oersted.

It is worth stressing that this power law has nothing of universal since an other value of λ would give an other exponent. Now let us see what happen if instead of a constant field we apply a rotating field .

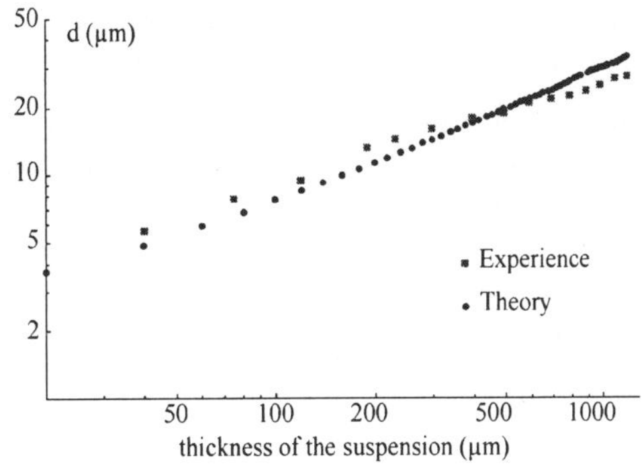

Figure 2 Comparison theory-experiment for the inter-aggregate distance versus the thickness of the cell.

4 Experimental results in a rotating field

A different situation happens in the case of rotating field if the frequency of rotation of the field is high enough for a particle does not move appreciably during the period. In this case the average energy is attractive in the plane of rotation of the field: $U=(m^2/2r^3)(3\cos^2\theta-1)$, where θ is the angle between the vector **r** joining two dipoles and the normal to the plane of the rotating field. This is just the opposite interaction of two parallel dipoles of strength $m/\sqrt{2}$ whose line of centers makes an angle θ with **r** [8]. This interaction is attractive for $\theta=\pi/2$ (dipoles in the plane of rotation) and repulsive for $\Theta=0$;so it induces a layered pattern with the normal to the layers parallel to the axis of rotation of the magnetic field [13]. The rotating field is obtained with two pairs of coils in Helmoltz configuration and supplied in quadrature of phase. In Fig. 4 we have plotted the period of the layered pattern versus the amplitude of the rotating field for different frequencies. The layered structure becomes visible for H=8 Oe; the period decreases significantly between 12Oe and 16Oe and then remains constant and equal to 8 μm for a cell height h=75μm. There is no dependence on frequency in the range we have used (between 1Hz and 35Hz).

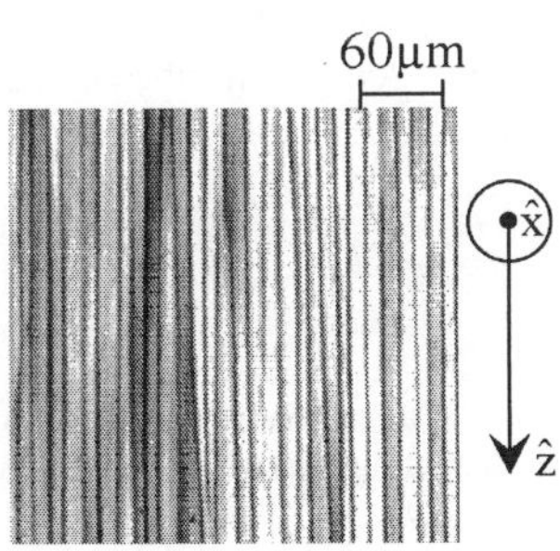

Figure 3 Top view of the suspension recorded by video camera . The field is in the plane (Ox,Oz) perpendicular to the plane of the figure , H=20 Oe and the height of the suspension is h=75μm.

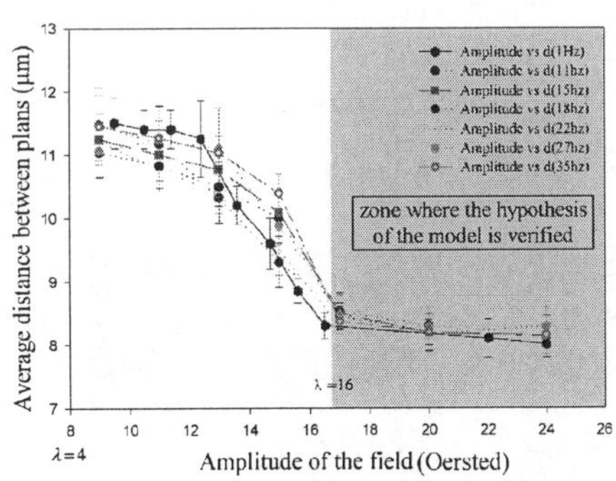

Figure 4 Average distance between plans versus the amplitude of the field for different frequencies of rotation of the field. The height of the suspension is 75 μm.

We can compare the experimental results d=f(h) with the theoretical predictions obtained by the model in the domain where the model is valid i.e. for $H_0 > 16\,\text{Oe}$ since above this field the phase separation is achieved and we can neglect the presence of particles between two stripes. Fig. 5 shows this comparison and we find a good agreement if we take the surface energy as an adjustable parameter. Actually in order to obtain this agreement we need to take a surface energy 100 times smaller than the one obtained from Eq.3. Some experiments realized recently show that the layers of particles are not homogeneous as can be seen in Fig.6 where we can see in the plane of rotation an important substructure which should be taken into account in the model both for the calculation of internal energy and surface energy.

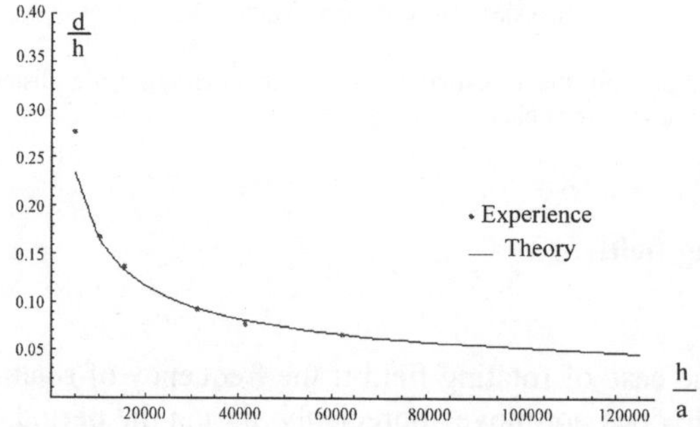

Figure 5 The change of average inter-aggregate distance normalized by the height of the cell h versus the thickness of the cell ,h, normalized by the radius of particles a.

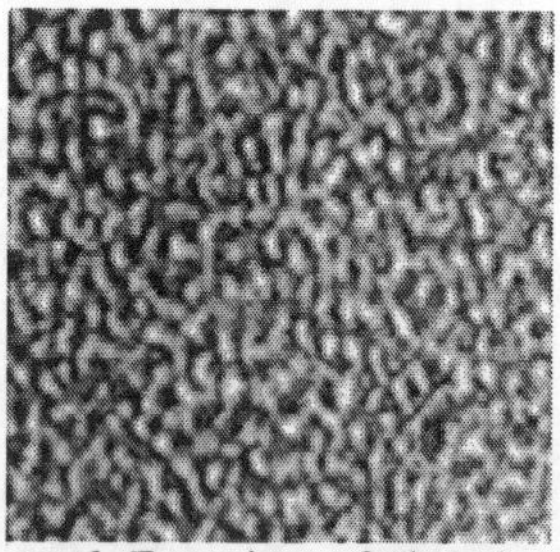

Figure 6 Top view of the plane of particles recorded by video camera . The field is in the plane of the figure. H=20 Oe ; the height of the suspension is h=50 μm

5 Structures in sinusoidal field

In a sinusoidal field, we do not expect a different behavior than in a constant field as long as the period of the field remains smaller than the desaggregation time which is a few seconds. This is the case if we use a flat cell, but when the suspension is placed between two glass plates which make a small angle α between each other , we observe a rotational instability in a given range of frequencies and positions inside the cell. The aggregates which were cylindrical at lower frequencies (Fig. 7 a) join together when we increase the frequency, cf Fig.7b, where they look like rods when seen from above .Then they undergo a chaotic rotation around an axis which is parallel to the one of the magnetic field.

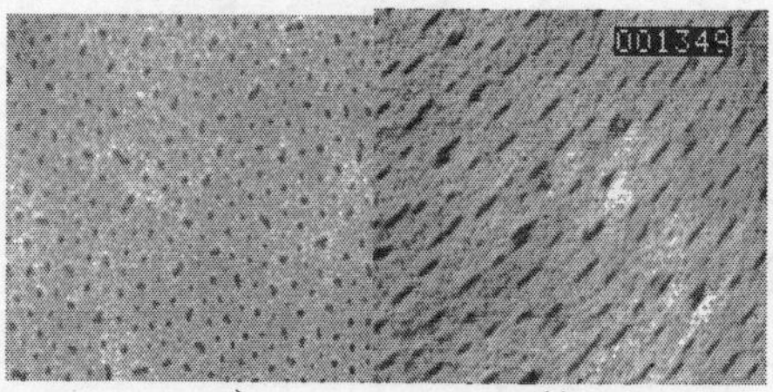

a) b)

Figure 7 a) Top view of the suspension recorded by video camera. The alternating field is perpendicular to the plane of the figure . The frequency and the amplitude of the field are 15 Hz and 60 Oe. The height of the cell is 800μm and the angle is 1.48 °.
b) Same view but the frequency of the field is now 25Hz.

We have measured the threshold field and the maximum field for which the instability appears and disappears versus the frequency of the field for different angles of the cell (Fig. 8). The spatial position of the instability in the cell depends on the amplitude of the field: increasing the field displaces the instability at higher heights inside the cell (Fig.9). We are not able to well explain these observations. A clue is that the angle inside the cell induces an alternative tangential field which can create these sheet like aggregates and their rotation if the magnetization of the aggregate has a characteristic relaxation time corresponding to the period of the field. At small heights the aggregates are smaller so they should have a shorter disaggregation time and then show the instability at higher frequencies. This is in qualitative agreement with our observation. A field threshold exists because the magnetic torque (generated by the delay of the magnetic moment of the aggregate with regard to the magnetic field) should overcome the hydrodynamic torque in order to initiate the rotation. A model based on this hypothesis will be worked out in order to see if it can explain our observations.

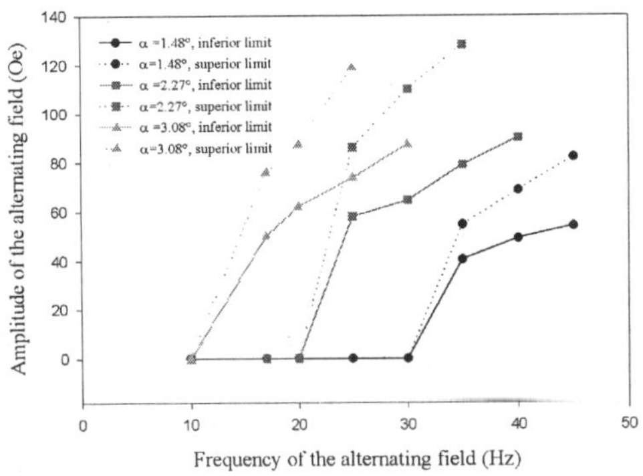

Figure 8 the variation of the threshold and maximum amplitudes for which appears and disappears the instability according to the angle of inclination and to the frequency of the field

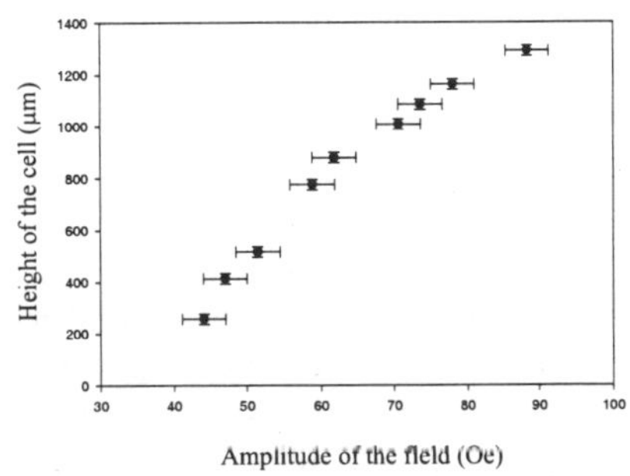

Figure 9 the spatial position in the cell of the instability versus the amplitude of the field. The angle of the cell is 1.48° and the frequency of the field is 35 Hz.

References

1. M. Seul, D. Andelman, *Science.* **267** (1995) 476.
2. A. Cebers, *Magnetohydrodynamics* **26** (1990) 309-313.
3. Yu. A. Buyewitch, Yu. A. Zubarev, *J. Phys. II.* **3** (1993) 1633.
4. E. Lemaire, G. Bossis, Y. Grasselli, *J. Phys. II.* **2** (1992) 359.
5. J. Liu and al., *Phys. Rev. E.* **74** (1995) 2828.
6. Y. Grasselli, G. Bossis, E. Lemaire , *J. Phys. II.* **4** (1994) 253.
7. S. Cutillas, G. Bossis, A. Cebers, *Phys. Rev. E.* **57** (1998) 804.
8. T. C. Halsey, R. A. Anderson, J. E. Martin, *Procee. Of 5th Intern. Conf. On E.R. fluids, M.R. suspensions and Associated Technology* (1995) 192.
9. G. Bossis, P. Carletto, S. Cutillas, O. Volkova, *Magnetohydrodynamics* **35** (1999).
10. L. D. Landau,M. E. Lifschitz, *Pergamon Press* (1984) 41.
11. W. B. Russel, D. A. Saville, W. R. Schowalter, *Cambridge Univ. Press* 1989.
12. S. Cutillas, G. Bossis, *EuroPhys. Lett.* **40** (1997) 465.
13. G. Bossis, A. Cebers, *J. M. M. M.* **82** (1999) 218.
14. A. Cebers, *J. M. M. M.* **149** (1995) 93-96.

LAMELLAR STRUCTURES INDUCED BY COUPLED ELECTROMAGNETIC AND SHEAR FIELDS: A DEFINING CHARACTER OF ER/MR ACTIVITY

S. HENLEY

Dept. of Materials Science and Engineering
The University of Michigan
Ann Arbor, Michigan 48109-2136(U.S.A

F. E. FILISKO

Dept. of Materials Science and Engineering
The University of Michigan
Ann Arbor, Michigan 48109-2136(U.S.A

In this article it will be demonstrated that columns of particles bridging the electrodes are not the characteristic feature of ER fluids under field. Instead a higher level of order, particle lamellae, must develop under shear and E field if a suspension is to be a legetimate ER fluid.

1. Introduction

The observation that particles in ER and MR suspensions aggregrate into chain or columnar structures which bridge the electrodes or poles under the influence of the electric or magnetic field is well recognized and has served as the basis for virtually all theoretical models involving the rheological properties of these materials under field. However these chains or columns have almost exclusively been observed in very dilute suspensions and under static conditions. There are however many observations and reports in the literature of a higher level of ordering of the particles into rows or lamellar structures when fluids of higher concentrations of particles are exposed to a electric or magnetic field and a shear field simultaneously[1-5]. This phenomenon was looked at in detail [4] and reported to apparently be a standard feature of ER fluids simultaneously under E and shear fields. In other words the particles in an ER fluid(and in fact in virtually any dielectric dispersion) under E field alone form chains or columns(depending upon particle concentration) of particles which bridge the electrodes. However under shear and E fields, the particles form into lamellae or walls, the length of which being in the shear direction and the height in the field direction. It was hypothesized that this phenomenon could be important in the ER(MR) phenomenon because the force needed to shear an object is proportional to the area of the shear plane which is the cross section of these lamellae parallel to the electrodes or the cross sectional area of a column. However the cross sectional area of a column can be orders of magnitude less than a lamellae. The formation of these lamellar structures was shown to be walls between parallel plates and rings(or more correctly cylinders) between parallel rotating discs for only a few materials, i.e. sulfonated polystyrene and zeolites both of which are ER active in the dry state. While columns form for materials in static conditions even for non ER active suspensions under E fields, it will be shown in this paper that for the systems studied, non ER active suspensions under shear and E field do not show particulate lamellae and in fact show no particulate aggregration at all. However the particles in ER active suspensions not only aggregrate into columns under static conditons but as well regiment into lamellar structures when exposed simultaneously to a shear field. The implication to be supported in this paper is that the distinguishing characteristic between suspensions which are ER active vs. those which are not is not the ability to form columns but the ability to form these lamellar structures under E field and shear.

2. Literature Survey

There have been few but substantial reports for the appearance of unique textures or higher ordered structures(i.e. beyond strings or columns) being formed in dispersions simultaneously exposed to an electric and a shear field. Bossis, et.al.(1994)[1] reported the occurrence of stripes of particles aligned in the flow direction for a dispersion exposed to an electric field and under oscillatory shear. They mention that these stripes, which occur in shear fields alone, may actually be sheets due to the imposed E field. In another study they show that under various amplitudes of oscillation and under an E field, these sheets may be perpendicular or parallel to the flow direction. More recently, Cutillas and Bossis(1998)[3] compared different structures observed in ER and MR fluids under field and oscillatory shear. This group refer in their papers as the phenomenon being a phase separation induced by the field which it is of course in the sense that the dispersed phase, normally uniformly dispersed, is caused to associate into unique structures by either/or E/M and shear fields. Another study involving steady shear and flow through a slit, concentric cylinders, and between parallel rotating discs, Tang,et.al.(1998)[5] report structures consisting of parallel lines of particles in the couette and concentric rings of particles between the parallel discs. A subsequent extensive study by Henley and Filisko(1999)[4] using parallel discs and concentric cylinders, clearly demonstrate the electric field and shear field induced phase separation(assembly) of the dispersed particles as well as the particulate structures resulting from the coupled E and shear fields.

The occurrence of particle aggregation and structuring within dispersions exposed to electric and shear fields has not only been observed experimentally but has been shown in computer simulations [6-9] . The simulations predict the occurrence of layered structures (of particles) in the idealized situation considering point dipole interactions, Brownian and shear forces but neglecting many-body interactions. These simulations are primarily aimed at relating the macroscopic rheological behavior to the dispersion microstructure.

3. Experimental

The paraffin wax is of a type commonly available for household use or candlemaking and available at craft or grocery stores . The barium titanate (average particle diameter of 7 μm), polysodium-4- styrenesulfonate (average particle diameter of 9 μm), and polystyrene (average particle diameter of 8 μm) were obtained from Aldrich Chemical Company. The paraffin oil was obtained from Fisher scientific and dried by being stored over molecular sieves.

A portion of each of the (as received) powders was placed in a 100 C vacuum oven over night. The wet barium titanate was prepared by placing a portion of it in an open container that was placed in a 90 C shallow water bath (this allowed the particles to absorb water onto their surfaces) for 2 hours. The ER fluids, for preparation of the solid discs, were made by adding 0.5 gm of each of the wet and dry powders individually to 10 gm molten paraffin wax at 75 C . Each of the suspensions while still molten was then placed under a vacuum for 10 minutes to remove trapped air bubbles which may have occurred during mixing or were trapped on the irregular particle surfaces. The samples were stored in sealed containers.

The disc samples were prepared on a Ferranti-Shirley rheometer fitted with parallel discs (3.0 cm. diameter and a 1 mm gap) using a Hipotronic high voltage D.C. power supply. Samples were prepared by first melting the solidified suspensions and then adding them to the device at 75 C. An E field of 2000 V/mm was applied for 2 minutes prior to beginning the rotation of 10 rpm. The rotation of the upper disc was applied for 2 minutes at which time the rotation was stopped and the voltage was reduced to 500 V/mm. The sample was allowed to cool to about 58 C, at a rate of 1 degree per minute, before it was removed. The melting temperature of the wax was 67-70 C. This procedure allowed a relatively short time to pass before the sample solidified. The voltage was maintained because the rings remained sharper, however it was reduced to 500 V/mm in order to minimize electrophoresis which may occur. The disc samples were removed by first lowering the lower disc of the parallel disc instrument and then gently detaching the sample from the upper disc using a razor blade. The disc invariably adhered to the upper disc during separation because the upper disc was cooler due to a lower thermal mass than the lower disc.

For the rheological measurements, samples were prepared as above except 0.5 gm each of the dry $BaTiO_3$, wet $BaTiO_3$, dry PSS and dry PS were placed in 10 gm dried paraffin oil. These samples were characterized at room temperature using parallel discs with the upper disc rotated at 10 rpm and varying the E field.

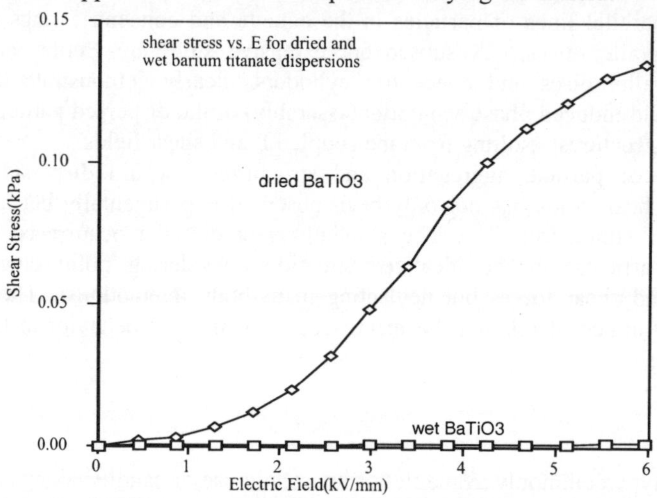

Figure 1: Shear stress vs. electric field for suspensions of dried and wet barium titanate in paraffin oil at 10 rpm.

4. Results

Figure 1 illustrates the relationship of shear stress versus the applied electric field strength at a angular velocity of 10 rpm for the paraffin oil suspensions of both the wet and dry barium titanate ($BaTiO_3$) suspensions. Initially both suspensions display relatively low and similar shear stresses at zero electric field. As the electric field is increased to 6 kV/mm, the dry $BaTiO_3$ suspension displays only a relatively small increase in shear stress over this range of the E field. However the wet $BaTiO_3$ suspension undergoes a significant increase in shear stress to about 0.14 kPa, a value over 100 times greater than for the dry suspension. This behavior is consistent with the performance of $BaTiO_3$ suspensions under E field and one reason we presented it here, i.e. a positive but very small increase in viscosity with E field for the dry material due to its high permittivity but a huge increase in viscosity due to the adsorbed water.

Figure 2A displays a solidified sample of the dispersion of dry $BaTiO_3$ powder in paraffin after being exposed to an E field of 2000 V/mm at 75C, in its molten state, for two minutes with no rotation or shear applied prior to cooling. A series of irregularly shaped black spots are observed that have an average diameter of around 0.61 mm. The white area surrounding the dark spots is the solidified paraffin wax virtually devoid of

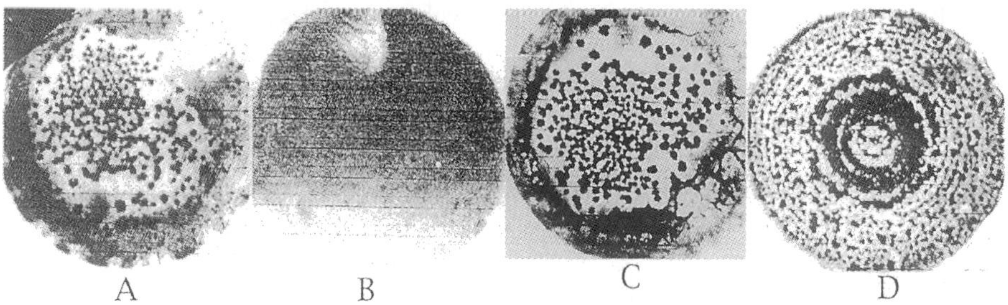

Figure 2: Sample discs prepared from wet and dried barium titanate/molten paraffin dispersions under various combinations of shear and E fields; A. dried barium titanate under E field alone; B. dried barium titanate under E and shear field; C. wet barium titanate under E field alone; D. wet barium titanate under E and shear fields.

powder. The dark spots are aggregated particles, presumeably the ends of the columns, which are aligned in the direction of the field that is normal to the picture.

Figure 2B displays the solidified wax disc of an identical sample as illustrated in figure 2A except that when its molten state at 75C, the sample was exposed to both an E field of 2000 V/mm and a shear field (clockwise) as detailed in the experimental section. In this case, there are no discernable structures or particle aggregrates, only a uniformly grey disc. The uniform dark color of the complete disc suggests that the individual particles under these conditions are uniformly dispersed rather than existing as aggregates as in figure 2A which was not sheared. Since the field was turned on prior to the beginning of rotation, a situation which produces columnar aggregates as shown in fig. 2A, then apparently the shearing, despite being relatively small, was involved in and adequate to cause the columnar aggregrates to redisperse and/or prevent the particles from aggregrating. This looks identical to a sample of the as mixed dispersion, that which has not been exposed to an E field and shear field, or that which has been sheared but not exposed to an E field.

Figure 2C displays the wax disc of a sample obtained in an identical manner to figure 2A except the molten wax suspension contained the wet barium titanate. Again, a series of black spots are seen. In this case, the spots have an average diameter of 0.83 mm. These spots are again the ends of columnar particulate aggregrates that are aligned with the field, which is normal to the picture. As in figure 2 these spots have no long-range order but are arranged in a random fashion.

Figure 2D illustrates the typical structures formed from a suspension containing the identical sample used to obtain figure 2C, i.e. wet $BaTiO_3$ in paraffin wax, except that when in the molten state at 75C, the sample was exposed to both an E field of 2 kV/mm and a shear field corresponding to an angular velocity of 10 rpm. Here the presence of dark spots or particle aggregrates are apparent on a white matrix which is solidified paraffin wax, apparently devoid of particles. This as contrasted with the sample illustrated in figure 2B, the particle aggregrates(dark regions) still persist upon shearing. Additionally the effect of shearing(with the E field on) is to result in the particle columns regimenting or aligning into equally spaced rings. In some cases the spots or column ends remain distinct while in others dashes of particles are observed oriented in the shear direction. Near the center continuous dark rings or aggregrates exist which are related to the relatively lower shear rate in this region.

Figure 3 illustrates the relationship of shear stress versus the applied electric field strength at a angular velocity of 10 rpm for the paraffin oil suspensions of sulfonated polystyrene and polystyrene particles, both types of particles dried as in the experimental section. Initially both suspensions display relatively low and similar shear stresses at zero electric field. As the electric field is increased to 6 kV/mm, the polystyrene suspension shows no increase in shear stress over this range of the E field. This is as expected because PS is not an intrinsically active material and has a low permittivity. However the sulfonated polystyrene suspension undergoes a significant increase in shear stress to about 0.6 kPa, an increase of over 400 times from the zero field value. Again, this behavior is as expected because sulfonated polystyrene is a well known intrinsically active ER particle[10].

Figure 4A displays a solidified dispersion of dry polystyrene (PS), a material with no reported ER activity associated with it, in paraffin wax. This sample was exposed to an E field of 2000 V/mm and no rotation or shear was applied. In this case, a series of irregularly shaped black spots are observed that have an average diameter of 0.63 mm. These spots are the ends of the columns of aggregrated particles that are aligned in the direction of the field, which is normal to the picture. Notice that even though this material shows no ER activity, the particles still aggregrate into columns under the influence of the electric field alone.

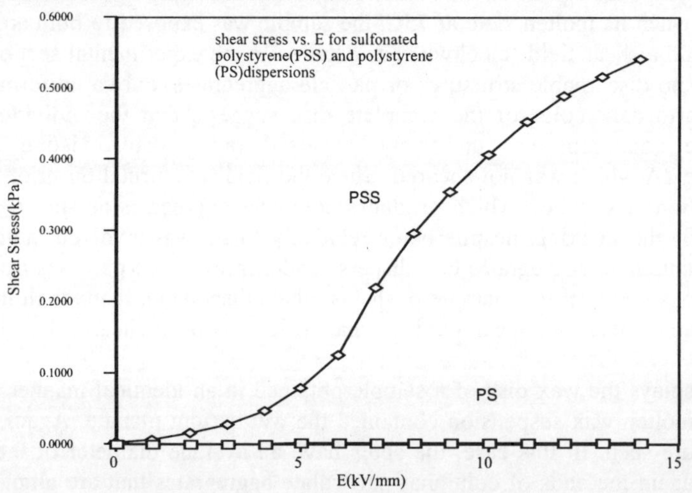

Figure 3: Shear stress vs. electric field for suspensions of polystyrene and sulfonated polystyrene in paraffin oil at 10 rpm.

Figure 4B displays the solidified wax disc of an identical sample as illustrated in figure 4A, a suspension of dry polystyrene in paraffin wax, except that when in its molten state at 75C, the sample was exposed to both an E field of 2000 V/mm and a shear field (clockwise) as detailed in the experimental section. In this case, there are no discernable structures or particle aggregrates, only a uniformly grey disc. The uniform dark color of the disc suggests that the individual particles under these conditions are uniformly dispersed rather than existing as aggregates as in figure 2A which was not sheared. This is identical to a similar sample of the as-mixeddispersion. Since the field was turned on prior to the beginning of rotation, a situation which produces columnar aggregates as shown in fig. 4A, then apparently the shearing, despite being relatively small, was involved in and adequate to cause the columnar aggregrates to redispersen and prevent their reaggregration.

Figure 4C displays the wax disc of a sample obtained in an identical manner to figure 4A except the powder used was a sulfonated polystyrene. Again, a series of black irregularly shaped spots are seen. In this case, the spots have an average diameter of 0.45 mm. These spots are apparently the ends of columnar aggregrates of particles that are aligned with the field, which is normal to the picture.

Figure 4D illustrates the typical structures formed from a suspension containing the identical sample used to obtain figure 2C, i.e. sulfonated polystyrene in paraffin wax, except that when in the molten state at 75C, the sample was exposed to both an E field of 2 kV/mm and a shear field corresponding to an angular velocity of 10 rpm. Here the presence of dark rings or particle aggregrates are apparent on a white matrix which is solidified paraffin wax, apparently devoid of particles. As contrasted with the sample illustrated in figure 4B,

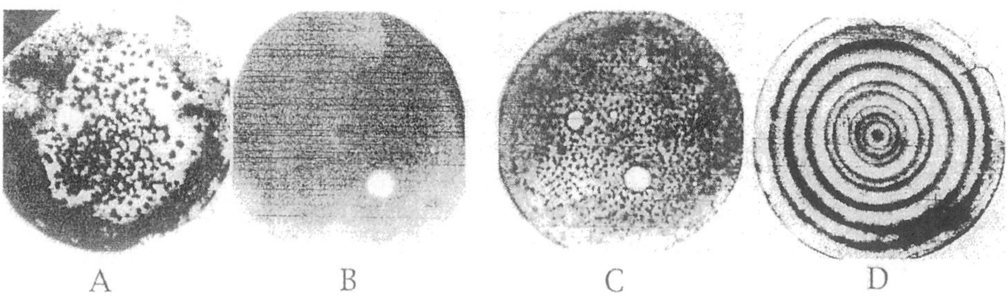

Figure 4. Sample discs prepared from dried polystyrene and sulfonated polystyrene/molten paraffin dispersions under various combinations of shear and E fields: A. dried polystyrene under E field alone; B. dried polystyrene under E and shear fields; C. sulfonated polystyrene under E field alone; D. sulfonated polystyrene under E and shear fields.

the particle aggregrates(dark regions) still persist upon shearing. Additionally the effect of shearing(with the E field on) results in the particle aggregrates going from columns to continuous lamellae. It is not known whether the columns essentially remain intact but migrate into the lamellae or whether they break up and reform as lamellae.

5. Discussion

The data presented is merely a small part of many hundreds of similar studies we have performed and observations made which allow us to generalize in making the statement that virtually all ER active suspensions under shear and E fields form these lamellar structures. The term ER active we are using here to mean suspensions which will increase their apparent viscosities significantly, from maybe 10 to 10000 times, more or less, when exposed to E fields on the order of 1-3 kV/mm. For instance the $BaTiO_3$ dry suspension does show a relative slight increase of viscosity with E field, however we would not consider that ER active. The wet material which is a well known ER active material, shows an increase in apparent viscosity of more than 100 times, much more than the dry material as illustrated in figure 1.

We have less observations on suspensions which are not ER active including dry Al_2O_3, dry bentonite, dry TiO_2, as well as dried $BaTio_3$ and dried polystyrene presented in this paper which show columns under E field and static conditions but under shear do not show lamellar particulate structures and none show significant, if any, ER activity. However we feel we have performed sufficient studies , as illustrated here for two situations, to make the following statement with a high degree of confidence. It is a distinguishing characteristic of ER active suspensions that under the combined effect of electric and shear fields , not parallel to each other, that the particles aggregrate or regiment into lamellar structures, the long dimension being in the shear direction. The height of the lamellae are arbitrarily chosen to be the distance between the electrodes and in the field direction.

Although the phenomenon presented here is unquestionable there is little known about it other than the mechanisms for interaction between the particles, i.e. induced dipoles in the particles and/or the current through the suspension, and even these are not totally resolved. Questions such as what determines the diameter of the columns, the width of the lamellae, the general periodic spacing of the lamellae and how they evolve with time of shearing are generally unknown and many are under current study. Further , how if at all these lamellar structures are important in the ER phenomenon is still to be determined although it has been hypothesized[4] that these lamellar structures slip at the electrode surface under shear as opposed to breaking between the electrodes. The types and characteristics of the particles and the matrix fluid, the concentration, the character of the electrodes and gap, the shear rate and time of shearing, the magnitude and character (AC, DC) of the electric field are all parameters which are apparently related in still to be determined ways to the column breadth and the character of the lamellae. Thus the reasons the diameter of the columns and the types of rings are different in the figures cannot at this time be explained, nor can generally be reproduced.

References

1. G. Bossis,. Y.Grasselli, E.Lemaire, J.Persillo, L.Petit," Phase separation and flow induced anisotropy in electrorheological fluids", Europhysics Lett.25(5), 335-340(1994).
2. S. Cutillas, and G.Bossis,"A comparison between flow-induced structures in electrorheological and magnetorheological fluids" Europhys.Letters 40 (4),pp 465- 470(1997).
3. S. Cutillas, G.Bossis, A.Cebers, E.LKemaire, A.Meumer,"Field and flow induced structures in relation to the rheology of ER and MR fluids", in ER Fluids, MR Suspensions, and their Applications, ed. M. Nakano and K.Koyama,World Scientific, Singapore(1998).
4. S. Henley, and F.E.Filisko,"Flow profiles of ER suspensions:An alternative model for ER activity", J.Rheol.43(5),1323-1336(1999).
5. X. Tang, W.H.Li, X.J.Wang, P.Q.Zhang,"Structure evolution of ER fluids under flow conditions", in ER Fluids, MR Suspensions, and their Applications, ed. M.Nakano and K.Koyama, 164-171, World Scientific, Singapore(1998).
6. M. Whittle.,"Computer simulation of an ER fluid", J. Non Newt. Fluid Mechanics 37, 233-263(1990).
7. J.R. Melrose,"Brownian dynamics simulation of dipole suspensions under shear: the phase diagram", Molecular Phys.76(3), 635-660(1992).
8. J.R. Melrose, and D.M.Heyes," Simulation of ER and particle mixture suspensions: agglomeration and layered structures", J. Chem. Phys.98(7), 5873-5886(1993).
9. J. Takimoto Computer Simulation of Model Electrorheological Fluids, Electrorheological Fluids Mechanisms, Properties, Structure, Technology, and Applications, 53-58 (1991)
10. U. Treasurer, Radzilowski,L.H. and F.E.Filisko," Polyelectrolytes as inclusions in electrorheologically active materials: Effect of chemical characteristics on ER activity", J.Rheol.35(6), 1051(1991).

CHAIN ROTATIONAL DYNAMICS IN MR SUSPENSIONS

SONIA MELLE[1,2], OSCAR G. CALDERÓN[3], MIGUEL A. RUBIO[1],
AND GERALD G. FULLER[2]

[1]*Dpto. Física Fundamental, UNED, Senda del Rey 9, Madrid 28040, Spain*
E-mail: smelle@fisfun.uned.es; mar@fisfun.uned.es

[2]*Dpt. Chemical Engineering, Stanford University, Stanford, CA 94305-5025, USA*
E-mail: smelle@chemeng.stanford.edu; ggf@chemeng.stanford.edu

[3]*Dpto. Óptica, UCM, Ciudad Universitaria s/n, Madrid 28040, Spain*
E-mail: oscargc@eucmax.sim.ucm.es

The dynamics of induced dipolar chains in magnetorhelogical suspensions subject to rotating magnetic fields has been experimentally studied combining scattering dichroism and video microscopy experiments. When a rotating field is imposed the chainlike aggregates rotate synchronously with the magnetic field. We found that the average size of the aggregates decreases with Mason number (ratio of viscous to magnetic forces) following a power law with exponent -0.5 being the hydrodynamic friction forces the cause of the chains break up. However the total number of aggregated particles shows two different behaviors depending on Mason number. For low Mason numbers, the total number of aggregated particles remains almost constant and above a critical Mason number, the rotation of the field prevents the particle aggregation process from taking place so the number of aggregated particles decreases with Mason number following a power law behavior with exponent -1. Athermal molecular dynamics simulations are also reported, showing good agreement with the experiments.

1 Introduction

Magnetic fluids exhibit interesting dynamical behavior when subjected to rotating magnetic fields as has been reported in previous experimental studies on magnetic holes [1,2] and magnetic droplets [3,4]. These systems show synchronous and non-synchronous regimes depending on the value of the driving frequency. J.C. Bacri et al. [3] found unexpected spiny, starfish shaped magnetic microdroplets when a high frequency rotating magnetic field was applied. At lower frequencies, O. Sandre et al. [4] found that the magnetic droplets break up to decrease their viscous drag and facilitate tracking the field rotation. Computer simulations of colloidal suspensions subjected to high frequency biaxial fields have been developed by J.E. Martin et al. [5-7]. They predicted the formation of 2D sheetlike structures aligned in the field plane.

In a previous work [8,9], we used scattering dichroism to study the orientation dynamics of low concentration magnetorheological (MR) suspensions ($\phi \sim 0.01$) when rotating magnetic fields of moderate frequencies (up to 10 Hz) were applied. In these suspensions, dichroism arises from the formation of optically anisotropic chains upon imposition of the magnetic field and gives information about the number of aggregated particles, N_a. In these experiments we simultaneously measured the dichroism ($\Delta n''$) and the orientation angle of the structures. We found that the induced chain-like aggregates rotate synchronously with the field but lag behind with a constant phase angle α. Furthermore, two different behaviors were found below or above a critical frequency.

Below this value, the dichroism remains almost constant but above the critical frequency, the viscous drag overcomes the magnetic force and reduces the dichroism following a power law with an exponent approximately -1.

In this work, scattering dichroism and video microscopy experiments have been combined to study the dynamics of induced dipolar chains in MR suspensions subject to rotating magnetic fields. Using scattering dichroism we studied the interplay between magnetic and viscous forces over the critical frequency separating these two regions. A dimensionless parameter that compares these two forces is the well-known Mason number (ratio of viscous to magnetic forces). This number has been defined with different proportionality factors in literature [10-12]. We have adopted the following definition:

$$\mathrm{Ma} \equiv \frac{12^2 \eta \omega}{\mu_0 M^2}, \tag{1}$$

where η is the solvent viscosity, ω is the field rotation frequency, M is the particle magnetization, and μ_0 is the vacuum magnetic permeability. We obtained a linear dependence of the critical frequency with the square of the magnetization and the inverse of the viscosity indicating that the Mason number governs the rotational dynamics. Consequently, we find a critical Mason number below which the dichroism remains almost constant and above which the viscous drag prevents the particle aggregation process from taking place.

We have also performed video microscopy experiments to visualize the dynamics of chains for a range of Mason numbers around the critical value (where both magnetic interaction and viscous drag are comparable). Direct observations show that chains rotate synchronously with the field adjusting their size to decrease the viscous drag. We find that the average size of the structures decreases with Ma following a power law with exponent close to -0.5 being the hydrodynamic forces the cause of the chains break up. However the total number of aggregated particles shows two different behaviors depending on Mason number in agreement with scattering dichroism experiments. Our experimental findings have been interpreted also with the help of athermal molecular dynamics simulations that show good agreement with the experimental results.

2 Materials and procedure

The experiments were performed using three suspensions of super-paramagnetic particles obtained diluting 10 % in weight suspensions supplied by *Estapor* (ref. M1-180/12, M1-070/60a and M1-070/60b). The particles content magnetite crystals Fe_3O_4 of small diameter (1-20 nm) dispersed in a monomer and polymerized using an emulsion process. The surface of the latex micro-spheres is composed of carboxylic acid (-COOH) groups with a surfactant coating layer of sodium dodecyl sulfate (SDS) to stabilize the dispersions. Due to their small average diameter and density (ρ_p ~ 1.3 g/ml) the sedimentation time (sedimentation velocity v_s ~ 0.02 μm/s) is long enough to neglect gravitational effects. The properties of the suspensions used in the experiments are detailed in Table 1.

Because the techniques we are using need optical transparency, the suspension was sandwiched between two circular quartz windows with diameter 6.5 mm, separated 100 μm along the vertical optical axis and held in place by a delrin attachment. The sample is surrounded by two orthogonal pairs of coils that generate rotating magnetic fields in the plane orthogonal to the optical axis. All experiments were performed at a temperature T = 282 ± 1 K.

Table 1. Properties of the magnetic suspension used in the experiments.

Experiment	Dichroism	Dichroism	Microscopy
Suspension name	d-M1-180/12	d-M1-70/60a	d-M1-70/60b
Particle mean diameter	1.01 μm	0.90 μm	1.24 μm
Particle magnetic content	13.0 %	62.0 %	61.0 %
Particle polymer nature	PS	PS	PS
Particle surface group content	20 μeq/g	145 μeq/g	117 μeq/g
Particle saturation magnetization	12 emu/g	52.7 emu/g	52 emu/g
Suspension volume fraction	0.016	0.01	0.0001
Suspension glycerol content	82.5 %	40-80 %	0 %

A full description of the scattering dichroism technique can be found in Refs. [8,13]. To perform video microscopy experiments we illuminated the sample from the top using a light source with flexible bundle. To amplify the images, we used a zoom system *Navitar* 12X with which a resolution of 0.34 - 4.2 μm for field of view around 170 - 2100 μm can be achieved. The images were recorded with a CCD video camera *Sanyo* VDC-3825 connected to a S-VHS VCR *Panasonic* AG 1975. We digitalized single frames on a computer for their subsequent analysis.

3 Experimental results

3.1 *Scattering Dichroism Results*

As we showed previously [8], two different behaviors for the dichroism and the phase lag are found below or above a critical frequency. We have studied the dependence of the critical frequency on the magnetization by applying magnetic fields with different amplitudes on suspension d-M1-180/12. The dependence of the critical frequency on the viscosity of the carrier fluid was analyzed applying a constant field on suspensions with different glycerol concentrations (d-M1-70/60a). We found that the critical frequency increases linearly with the square of the magnetization and decreases with the inverse of the viscosity, so the Mason number governs the dynamics. As expected from this result, we obtained a good collapse of the dichroism and the phase lag curves (measured at different magnetic fields and viscosities) with Mason number (see Fig. 1). The change in behavior of the dichroism and the phase lag occurs at a critical Mason number, $Ma_c \sim 1$, above which the viscous forces dominate and inhibit the aggregation process, being this one the mechanism responsible for the decrease of dichroism.

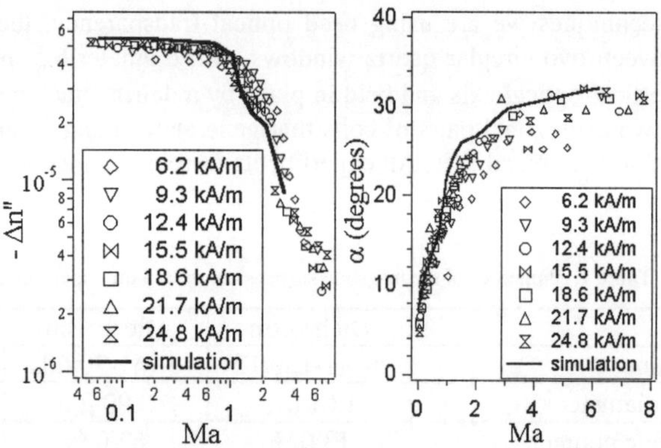

Figure 1. Collapse of dichroism (left) and phase lag (right) with Mason number for different magnetic field strengths. Suspension d-M1-180/12. Experimental results (markers), simulation results (solid line).

3.2 *Video microscopy Results*

In Fig. 2 video microscopy images show the behavior of the structures induced in the d-M1-070/60b suspension when we applied a magnetic field of H = 1.55 kA/m for two rotational frequencies corresponding to Mason numbers Ma = 1.2 10^{-3} (right) and Ma = 1.2 (left). These images were taken 300 sec after applying the magnetic field. Direct observations show that chains rotate synchronously with the field adjusting their size to decrease their viscous drag. The average size of the aggregates decreases with frequency following a power law behavior with an exponent close to -0.5, the same exponent predicted by the chain model developed by Martin et al. for ER fluids subject to steady shear [14]. For Ma<1, although the length of the chains decreases with frequency, the number of aggregated particles (dichroism) remains almost constant supporting the scattering dichroism experimental findings.

Once the critical Mason number is surpassed, the viscous forces start to dominate the magnetic ones and the aggregation process is inhibited. Furthermore, for Ma>1, the magnetic field is not strong enough for the structures to remain aligned in the field direction when rotating at these high frequencies and more isotropic structures are formed (see Ma = 1.2 in Fig. 2). These two mechanisms, the increase of the number of isolated particles and the formation of more isotropic structures, contribute to the decrease of the dichroism.

Fragmentation and aggregation processes are observed at low frequencies (Mason numbers below the critical value). As an example, the break up process of a chain rotating on a magnetic field with frequency f= 0.01 Hz is shown in Fig. 3. First the chain is bent (left picture), then it develops an S shape with the arms moving faster to follow the field (middle picture). Ultimately, the chain breaks up into smaller pieces (right picture). This S shape has been also reported for chains subject to shear in the direction perpendicular to the field [14], and also for magnetic droplets under rotating fields [4].

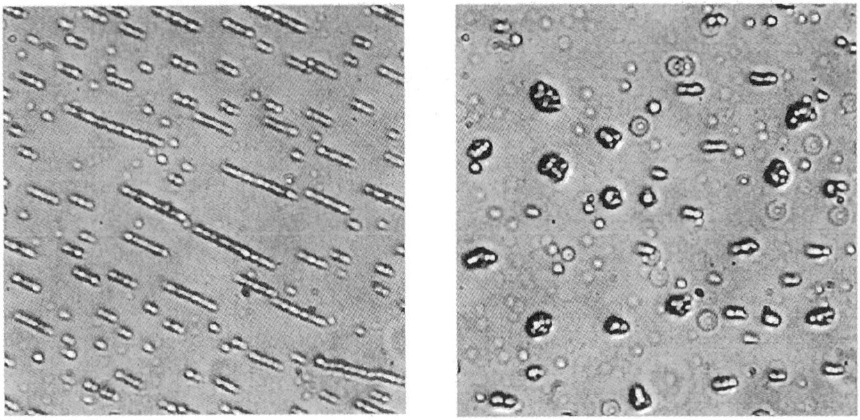

Figure 2. Video-microscopy images recorded after 300 s of applying a magnetic field with amplitude H=1.55 kA/m for two rotational frequencies which correspond to Ma=0.0012 (left) and Ma=1.2 (right). We used suspension d-M1-070/60b. Field of view: 53 x 53 μm.

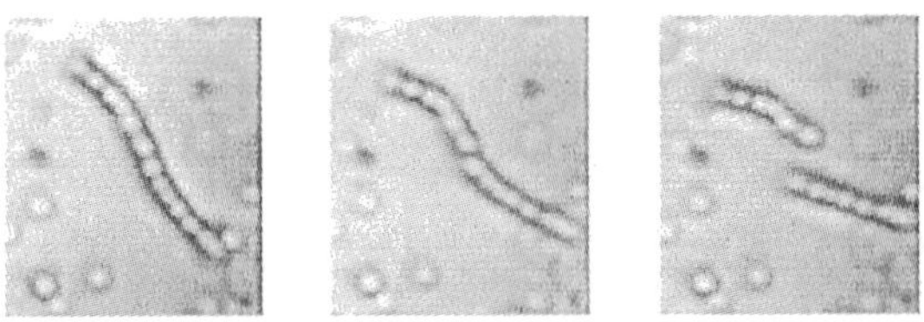

Figure 3. Detail of the S shape of a chain induced on the same suspensions when a magnetic field of amplitude H=1.55 kA/m rotates counter clockwise with f = 0.01 Hz. Field of View: 14.4 x 14.2 μm

4 Simulations

In order to interpret the experimental results we conducted athermal molecular dynamics simulations in 2D for a system of N=400 spherical particles of radius *a*. The particles are considered as "hard" spheres with induced dipolar interactions and Stokes friction against the solvent. In our experiments, the ratio between the magnetic and thermal energies, $\lambda = W_m / (k_B T)$, is much larger than one. Therefore Brownian motion should have a negligible effect on the evolution of the structures [15]. Similar models have been previously used to simulate MR fluids [16,17] as well as electrorheological fluids [18]. Under these assumptions, the equation of motion for the *i*th particle can be written as follow:

$$\gamma \frac{d\boldsymbol{r}_i}{dt} = \boldsymbol{F}_i^m + \boldsymbol{F}_i^{ev} \tag{2}$$

where $\gamma = 6\pi\eta a$ is the Stokes drag coefficient for a solid sphere, and $\boldsymbol{F}_i^m$ is the magnetic force in dipolar approximation. Note that the inertial term was neglected because the viscous drag term dominates. As we are using one-particle Stokes' hydrodynamics, an excluded-volume repulsive force, $\boldsymbol{F}_i^{ev}$, must be included to keep particles from overlapping. This force is calculated as in ref. [16]. Making the above equation dimensionless, a couple of scaling parameters immediately drop out. A spatial scale $r_s=2a$, and a time scale $t_s=12^2\ \eta/(\mu_0 M^2)$. This time scale leads to a dimensionless rotating frequency equal to the Mason number defined in Eq. (1). The same procedure has been reported in several works [17-19].

The simulations reveal that the magnetic dipolar interaction induces the formation of chainlike structures that follow the magnetic field rotating with the same frequency in agreement with the experimental findings. In Fig. 4 we plot the particles position in the XY plane for different frequencies (Mason numbers) at an arbitrary time. As demonstrated in this figure, the size of the structures becomes smaller as the rotating frequency increases. This indicates that hydrodynamic friction forces overcome the dipolar magnetic forces and therefore the chains break up to decrease their viscous drag.

According to the simulations, the average length of the chains decreases with frequency, scaling as the inverse square root of the Mason number (power law exponent equal to -0.45). This value is very close to the value obtained on the video microscopy experiments.

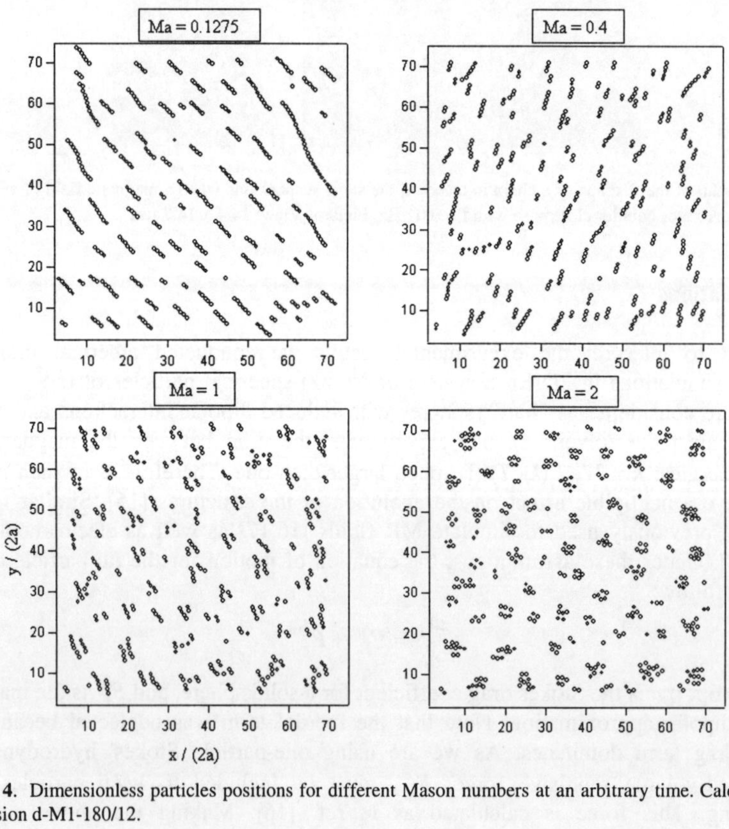

Figure 4. Dimensionless particles positions for different Mason numbers at an arbitrary time. Calculations for suspension d-M1-180/12.

The calculated dependence of the number of aggregated particles and phase lag on Mason number is shown in Fig. 1 (solid line). Two different regimes appear in agreement with the experimental dichroism. This shows that although the average length is decreasing monotonously with Ma, this diminution in length starts affecting the dichroism when the viscous forces dominates over the magnetic forces, i.e., for small average chain lengths (close to two particles).

5 Acknowledgements

This research was partially supported by MEC (Project no. BFM2000-0019). S. M. Was supported by a fellowship from MEC. We acknowledge fruitful discussions with J.E. Martin, P. Español, and I. Zuñiga.

References

1. Helgesen G., Pieranski P., and Skjeltorp A.T., *Phys. Rev. Lett.* **64** (1990) pp. 1425.
2. Helgesen G., Pieranski P., and Skjeltorp A.T., *Phys. Rev. A* **42** (1990) pp. 7271.
3. Bacri J.-C., Cebers A., and Perzynski R., *Phys. Rev. Lett.* **72** (1994) pp. 2705.
4. Sandre O., Browaeys J., Perzynski R., Bacri J.-C., Cabuil V., and Rosensweig R.E., *Phys. Rev. E* **59** (1999) pp. 1736.
5. Martin J.E., Anderson R.A., and Tigges C.P., *J. Chem. Phys.* **108** (1998) pp. 7887.
6. Martin J.E., Anderson R.A., and Tigges C.P., *J. Chem. Phys.* **110** (1999) pp. 4854.
7. Martin J.E., E. Venturini, J. Odinek, and R.A. Anderson, *Phys. Rev. E* **61** (2000) pp. 2818.
8. Melle S., Fuller G.G., and Rubio M.A., *Phys. Rev. E* **61** (2000) pp. 4111.
9. Melle S., Calderón O.G., Fuller G.G., and Rubio M.A., *J. Coll. Inter. Sci.* (submitted).
10. Mason number was first introduced in literature for ER fluids under steady shear. See Gast A.P., and Zukoski C.F., *Adv. Colloid Interface Sci* **30** (1989) 153.
11. Volkova O., Cutillas S., and Bossis G., *Phys. Rev. Lett.* **82** (1999) 233.
12. Martin J.E., *Phys. Rev. E* **63** (2000) pp. 011406.
13. Fuller G.G., *Optical rheometry of complex fluids*, Oxford University Press (1995).
14. Martin J.E., and Anderson R.A., *J. Chem. Phys.* **104** (1996) pp. 4814.
15. Martin J.E., Venturini E., Odinek J. and Anderson R.A., *Phys. Rev. E* **61** (2000) pp. 2818.
16. Mohebi M., Jamasbi N., and Liu J., *Phys. Rev. E* **54** (1996) pp. 5407.
17. Martin J.E., Anderson R.A., and Tigges C.P., *J. Chem. Phys.* **108** (1998) pp. 3765 and pp. 7887.
18. Klingenberg D.J., Swol F. van, and Zukovski C.F., *J. Chem. Phys.* **91** (1989) pp. 7888.
19. Larson R.G., *The structure and Rheology of Complex Fluids,* Oxford University Press, New York, 1999.

THE INTERACTION BETWEEN ELECTRORHOLOGICAL PARTICLES IN AC ELECTRIC FIELD STUDIED USING OPTICAL TWEEZERS

J. XU, X. DONG, Z.Y. CHEN, Y.G. JIANG, L.W. ZHOU

National Key Laboratory of Applied Surface Physics & Department of Physics
Fudan University, Shanghai 200433, China
E-mail: lwzhou@fudan.ac.cn

The dipole interaction between a pair of electrorheological (ER) particles under an external AC electric field was confirmed in a dynamic case using optical tweezers positioning and a system of high speed imaging and image processing. By measuring the interactions between two micron-sized ER particles with both central distance and structure forming time, the structure response time and the interaction strength were obtained for the particles under an AC electric field. The spacial resolution of the optical tweezers-high speed imaging-image processing system is 0.26μm. The sampling rate of the high-speed video recorder is up to 8000 frame/s with the corresponding time resolution being 0.125 ms.

1 Introduction

ER effect is usually described with the dipole interaction induced by the particle polarization in an electric field. The difference of dielectric constants of particles and solvent determines the response of ER fluids. The strength of ER fluids can be characterized by the interaction between two particles.

Tao [1] calculated the interaction between particles in the static situation by finite element approximation (FEA) method. It is found that, even when two particles is very close to each other, the result of FEA consists with that of dipole approximation when the ratio of dielectric constants between particle and liquid, $\kappa = \varepsilon_p/\varepsilon_f \leqslant 10$. When $\kappa = \varepsilon_p/\varepsilon_f \geqslant 1000$, the FEA result consists with conductance model.

Using 2D numerical simulation, Klingenberg *et al.* [2] drew a conclusion that the variation of structure does not depends on the electric field strength after certain time, but the time to form a structure is inversely proportional to the square of electric field, and they gave a time scale which is proportional to $\eta_f / \varepsilon_f \beta^2 E^2$, where η_f is the viscosity coefficient of the oil, and β, the dielectric mismatch of particles and oil. Lu *et al.* [3] studied the interactions between two $SrTiO_3$ particles and two $SrTiO_3$ coated metal particles in an electric field, the diameter of the particle is 6.3mm. They found that, when the two particles almost touched with each other, the interaction force is about seven times as much as that predicted by electric dipole approximation. Comparing the experimental results with the theoretical calculation, the interaction force predicted by dipole-interaction theory is much smaller than experimental result. If we use FEA method [1] instead of dipole-interaction theory, they meet well only when the distance between two particles is about 100μm, they have large difference when the distance is very small, this perhaps results from nonlinear effect of the oil's conductance. In case of large distance, they meet not too well. Perhaps from the many-body effect for the dipole-dipole interaction is a kind of long-range interaction while it is not accounted [1]. Atten [4] described the interaction between two semi-spheres with 7mm radius by using the conductance model, the sphere is coated with 0.1μm doped polyaniline layer. Under the same electric field, the interaction force grows with the increasing conductance of the layer. Under low field, $F \sim E^2$, while under high field, $F \sim E$. In case of sphere coated with an insulating layer, the interaction force from the experiment is larger than that predicted by theory, but they are in the same order.

Comparing the average force of Brownian motion, F_i^B, with that of dipole interaction, F_{ij}^{el}, the former can be neglected ($F_i^B / F_{ij}^{el} \sim O(kT/\varepsilon_0\varepsilon_f\sigma^3\beta^2E^2) \sim O(10^{-6})$, when T≈298K, $\varepsilon_f \approx 7.3$, $a \approx 10^{-5}$m, $\beta \approx 3.2$, $E \approx 10^6$V/m). By neglecting the Brownian motion and using similarity principle, we found that the ratio of the term of the viscosity to that of kinetic energy is proportional to the square of particle radius ($f = 6\pi\eta va$, $E_k = (1/2)mv^2 = (1/2)(4/3)\pi a^3\rho v^2$, $f / E_k \propto 1/a^2$).

For the particles whose radius a is the order of millimeter, the relaxation time, $\tau = 2a^2\rho/9\eta$, is about the order of 1s (supposing that specific weight $\rho = 5\times10^3$kg/m^3, and viscosity $\eta = 10^{-3}$Pa·s); while if radius is decreased to microns, the relaxation time is the order of 10^{-6}s. So only when particle radius is very small, say, as the order of microns, can the particle system be treated as a quasi-static state, and be used as a model system to study the motion of real ER fluids.

This work studies the interaction between ER particles with radius of microns in silicone oil. Optical tweezers can be used to control the position of micron particles with sub-micron resolution. The particles are released when the optical tweezers are turned off right before an electric field is applied, the trace of the particles is caught with a high speed CCD camera, and recorded in a hard disk, and so one can get the interaction between two particles from their time variation of position. Through the measurement of critical electric field we can also get the interaction force between particles.

2 Experiments and discussions

2.1 Setup of Optical tweezers

The optical tweezers are built in the same way described in Ref. 7. A 150mW, 532nm Nd:YAG laser (CrystaLaser LC GCL-150-S) is adopted. After expanded by a telescope, the parallel light is reflected at a right angle, and then splits into two beams with an equal power by a polarizing beam-split cube (PBSC). Each of the two beams turns at about a right angle by a mirror attached to a step motion device, goes through another PBSC, then enters the microscope (Olympus IX70) through an microscope objective of a type of Olympus ACH 100X/1.25 Oil before it hits a sample particle. A high speed CCD (RedLake PCI8000S with the highest sampling rate of 8000 frames/s) attached at the top of the microscope catches the images and sends the digitized data to a computer for recording and further processing. The distance between the two parallel electrodes of the sample cell is 1mm. By using dual trap optical tweezers, one can position two particles independently, and then analyze their tracks in an electric field. The position of each particle at different time is obtained, and the interaction of two particles can then be calculated and compared with various model interactions, such as the dipole interaction, in an electric field. In the experiment, the particle volume fraction of the sample is so small that when the interaction of two particles is measured, the interactions of the two particles with others can be neglected.

2.2 *Experiments with optical tweezers*

We studied the interactions between particles in ER fluids under both dynamic and static conditions. The sample we selected is glass beads/silicone oil system, the sample is made by putting glass beads (diameter of 2μm, relative dielectric constant of 4.5), that dried for 24 hours under 160℃, into silicone oil (motion viscosity of 10 mm^2/s, relative dielectric constant of 2.5). Two types of experiments have been performed. One is to find the critical electric field (see below) for different center distance, d, of particles trapped by optical tweezers; another is to determine the relation between the particle chain aggregation time and electric field.

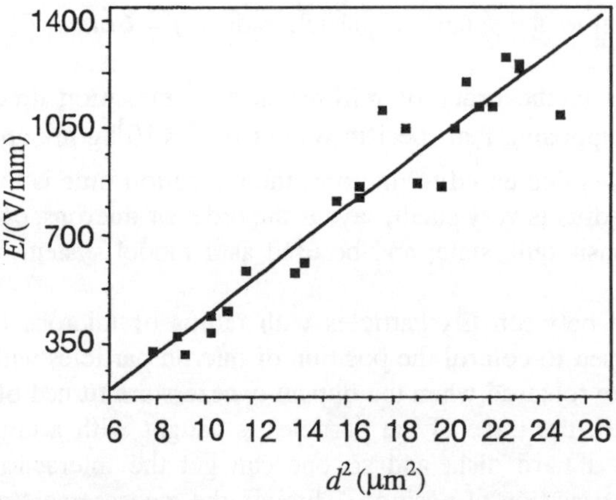

Figure 1. The relation between the critical electric field and the particle center distance when the power of the optical trap is constant

First, by trapping the particles in optical tweezers under increasing electric field, we found a relation between critical electric field and the center distance of two particles. In the experiments, the particles can be trapped by the dual optical tweezers, each of which has an optical power of 10.8mW, and placed with certain initial center distance *d* aligening in parallel to the electric field. The electric field is then slowly increased, so the interactive force between the two particles is also increased until one particle is escaped from optical tweezers, we get critical electric field which is the electric field value when one particle is escaping, we can treat critical electric field as a ruler of interaction force. Because the power of optical tweezers is kept invariant, we can get the relation between initial distance and critical electric field. The experimental result is shown in Fig. 1, where the black dots denote experiment data and solid line is a linear fit. One can find that E is proportional to d^2 when f is invariant, so the experiment result is consistent with the dipole modle.

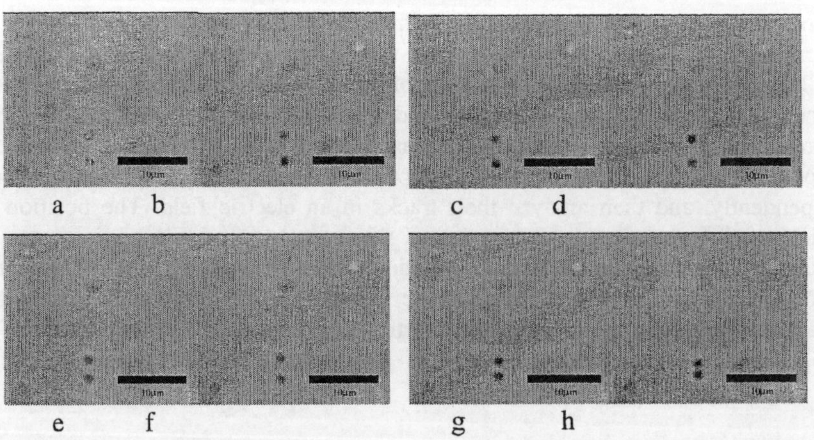

Figure 2. The moving process of two particles in an AC electric field

Figs. 2a to 2h are the motion of ER particles recorded by the high speed CCD camera under AC electric fields. Two ER particles (glass beads with diameter of 2μm) were placed with optical tweezers so that the connection line of the particles was parallel to the electric field, and the center distance between the two particles is about 5μm. The effective value of the AC electric field is 212-707V/mm, and the frequency, 1kHz (Fig. 2a); the optical tweezers were then released, the motion of ER particles under electric field was recorded (Figs. 2b to 2g); at last, the particles collided with each other (Figure. 2h).

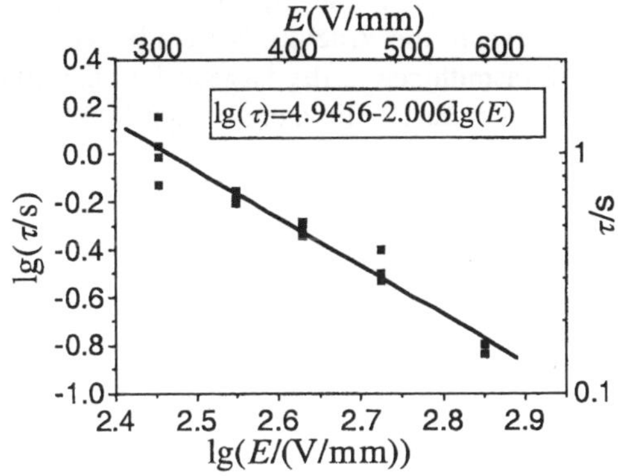

Figure 3. The relation between the aggregation time and the AC electric field (effective values)

The system can be treated as a quasi-static state, so the aggregation time should be inversely proportional to the square of electric field with the same center distance between particles (Fig. 3), 2.8*a* to 2.0*a*. The black dots are experimental data, and the solid line is a linear fit. The similar experiments were also done when particle center distance changes from 3.0*a* to 2.0*a*, from 2.5*a* to 2.0*a*, and from 2.3*a* to 2.0*a*. The same conclusion can also be drawn. This is consistent with the result of Klingenberg [2], who concludes that the structure formation time is inversely proportional to E^2. It is the first time to verify the dipole interaction of real ER particles with low dielectric constant in dynamic condition.

2.3 DWS experiments

We use a charge-coupled device (CCD) camera to study diffusive wave spectroscopy (DWS) [8]. This enables us to do DWS experiments in nonergodic ER fluids. For simple diffusion, average value of the square of position change, $<\Delta r^2>=6Dt$, where D is the particle diffusion coefficient, then the field autocorrelation function

$$g_1(t)=\sum_s P(s)\exp(-2k_0^2Dt\frac{s}{l^*}) \quad (1)$$

where k_0 is the scalar wavevector of the incident light. $g_1(t)$ can also be written as

$$g_1(t)=\sum_s P(s)\exp[(-2t/\tau)\frac{s}{l^*}] \quad (2)$$

where $\tau=(k_0^2D)^{-1}$ is the characteristic decay time reflecting the dynamics of the scattering particles, and it can be easily acquired from DWS experiments.

We observe the changes of transmittance using a sample cell with different sample thickness, L, as shown in Fig. 4, and the absorption length l_a is 4mm. In the experiment, L=2mm has been chosen. For sample thickness L such that $l_a>L\geqslant l^*$, the diffuse transmittance is given by

$$T\approx\frac{5l^*}{3L+4l^*} \quad (3)$$

where l^* is sensitive to S (q), the structure factor [9,10]. l^*=818μm calculated from Mie scattering theory [11], and Eqs. 1 and 2 are included in the deduction. Thus we gain the qualitative result: the structure is most likely to be the low energy lattice (BCT) when the saturate transmittance is the largest, i.e., the highest field is applied. Our experiment is performed in order to gain more insight into the interaction between particles.

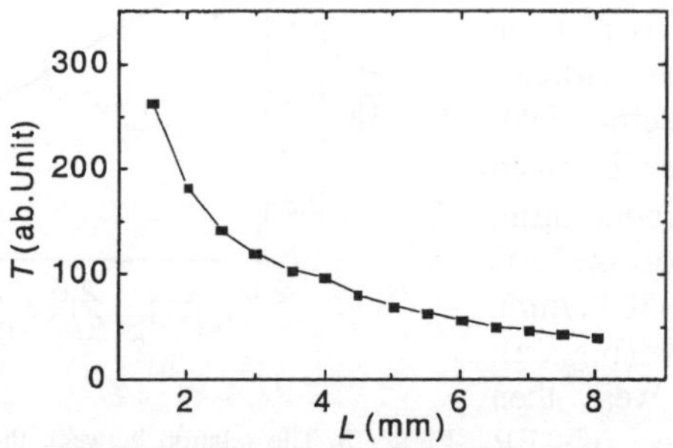

Figure 4. Transmittance vs. sample thickness

When the electric field is applied to the sample, the chaotic particles response at very short time scale and the system becomes nonergodic. From the CCD camera, the increase of the diffuse transmittance upon application of ac electric field is quite different. Fig. 5 is the time-dependent transmittance under different field. The curves increase gradually and then saturate at a special value. The higher the excited field, the steeper the slope is obtained, which reflects the faster response of particle structure formation. Also, the saturate values increase by two orders of magnitude as the voltage increases from 500V to 3000V.

Figs. 6 and 7 are, respectively, the time evolution of diffusion coefficient and characteristic time under different field from the real-time measurement of intensity autocorrelation function. [12] Both figures show the force response time is as short as several hundred milliseconds, identical with the result of Ref. 13.

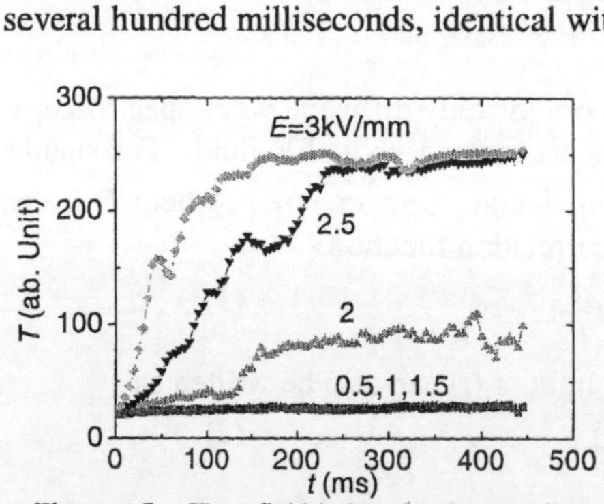

Figure 5. The field-induced change in transmittance measured at different field

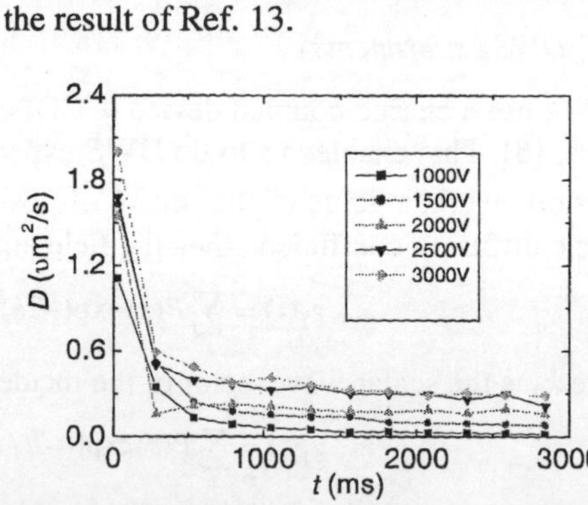

Figure 6. The time variation of diffusion coefficient under different field

Fig. 8 is derived from Fig. 6, where the averaged diffusion coefficient, D, is plotted upon the electric field E. The relation of D and E is fitted as lg$D$= -7.377+2.193lg E, that is, $D \propto E^{\alpha}$ (α =2.19), consistent with the prediction $D \propto E^2$. The transference of the net charges under electric field plays an important role in the contribution of particle interaction. The deviation may come from the dipole-dipole interaction [14].

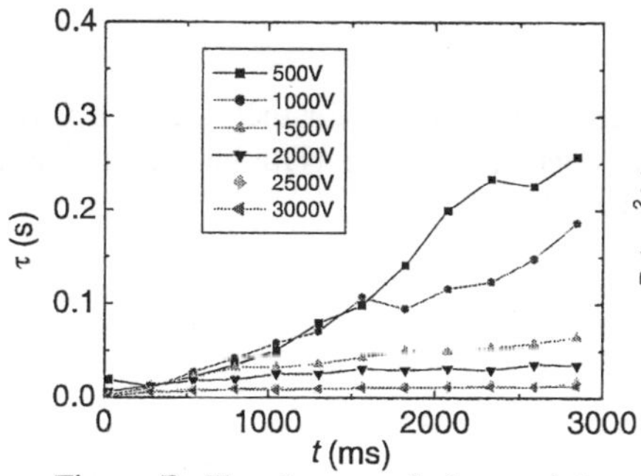

Figure 7. The changes of characteristic time at different field.

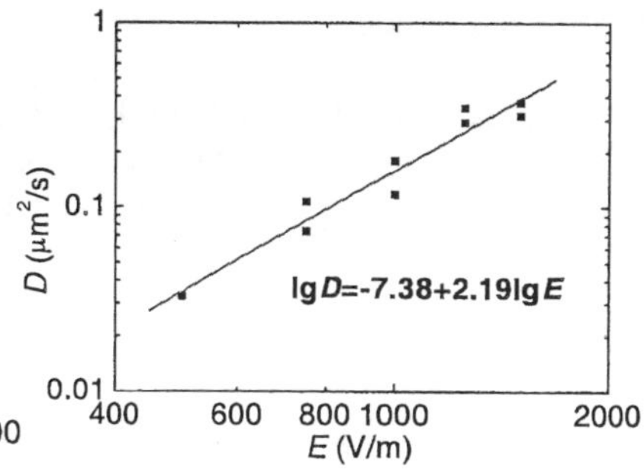

Figure 8. The average diffusion coefficient upon the electric fields

2.4 *Observation of dynamic chain formation*

A homemade electrorheometer was modified to observe the chain formation under several applied field while shear stress can be measured at the same time. The shear rate of he electrorheometer can be adjusted from 10^{-3} to 10^{3}. The sample used in this observation is the glass beads/silicone oil ER fluids. The motion pictures were caught with a regular CCD camera, and the pictures were then analyzed so that cluster size of glass beads can be determined as shown in Fig. 9. Generally speaking, larger cluster sizes have higher probability at higher electric field, while it was also observed that certain shear rate is more suitable for chain formation than both either higher or lower shear rates.

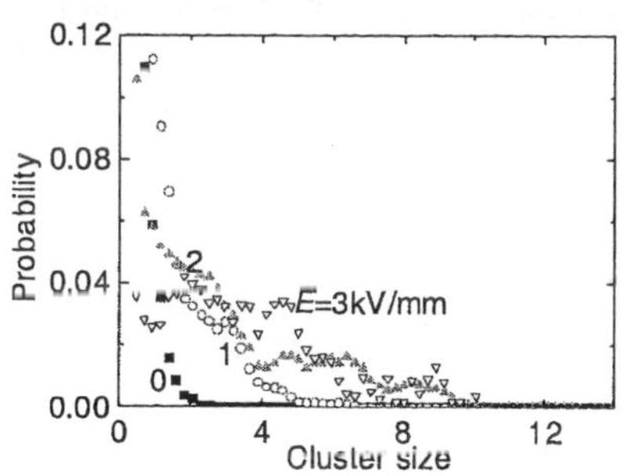

Figure 9. The probability of cluster size of glass beads when applied field E is changed while the shear rate, kept at 1.33 s^{-1}.

3 Conclusions

This work measures the interaction between particles in glass beads/silicone oil ER system by using optical tweezers and DWS methods; the relation between structural responding time and electric field, and that between interaction strength and electric field have been determined. The results show that when the ratio of particle/oil dielectric constant, κ, is small, the aggregation time of particles under different electric field is inversely proportional to E^2. The critical electric field is proportional to the distance between the centers of particles. Those are consistent with the dipole model. The purpose of this work is to study the relation between the polarization rate and other macroscopic properties of ER fluid, and give a proper description of structure stability in dynamic case. Further work will include to use above methods with samples of different conductivity, and to study directly the nonlinearity of electric field between particles.

4 Acknowledgement

The authors would like to thank L.F. Zhang and Q.Y. Fu for their work in DWS and dynamic structure formation. This work is supported by the National Natural Science Foundation of China under grant number of 19834020.

References

1. Martin J. E. and Odinek J., Structure and dynamics of electrorheological fluids, *Phys. Rev. E* **57** (1998) pp. 756-775. Tao R., Jiang Q. and Sim H. K., Finite-element analysis of electrostatic interactions in electrorheological fluids, *Phys. Rev. E*, **52**(1999) pp. 2727-2735.
2. Klingenberg D. J., Swol F. V. and Zukoski C. F., Dynamic simulation of electrorheological suspension, *J. Chem. Phys.* **91**(1989) pp. 7888-7895.
3. Lu K. Q., Wang J. M., Zhao X. P. and Wang Y. R., The interaction between two spheres in silicone oil under an electric field. In *Proceedings of 6th International Conference on Electro-Rheological Fluids, Magneto-Rheological Suspensions and their Applications*, ed. by M. Nakano M. and K. Koyama K. (World Scientific, Singapore, 1997) pp. 115-122.
4. Gonon P., Foulc J. N. and Atten P., A conduction model describing particle-particle interaction in the case of surface conducting particles. In *Proceedings of 7th International Conference on Electro-Rheological Fluids, Magneto-Rheological Suspensions and their Applications*, ed. by R. Tao R. (World Scientific, Singapore, 1999) pp. 161-170.
5. Cai W. Q., Optical trapping, optical levitation and optical manipulation of dielectric micro-particles, *Progress in Physics*, **17**(1997) pp. 64-82 (in Chinese).
6. Ashkin A., Forces of a single-beam gradient laser trap on a dielectric sphere in the ray optics, *Biophys. J.*, **61**(1992) pp. 569-582
7. Smith S. P., Bhalotra S. R. and Brody A. L., *et al.*, Inexpensive optical tweezers for undergraduate laboratories, *Am. J. Phys.* **67**(1999) pp. 26-35.
8. Zhang L. F., Dong X., Xu J., Jiang Y. G. and Zhou L. W., Short time dynamic structure formation of carbon-doped TiO_2 ER fluids with different conductivity using diffusing wave spectroscopy, in this proceeding.
9. Pine D. J., Weitz D. A., Chaikin P. M. and Herbolzheimer E., Diffusing wave spectroscopy, *Phys. Rev. Lett.* **60** (1988) pp. 1134-1137.
10. Saulnier P.M., Zinkin M. P. and Watson G. H., Scatterer correlation effects on photon transport in dense random media, *Phys. Rev. B.* **42** (1990) pp. 2621-2623.
11. Ishimaru A., *Wave propagation and scattering in random media*, (Academic Press, New York, 1978).
12. Zhang L. F., Dong X., XU J., Jiang Y. G., Zhou L. W. and ZHU Y. H., Short time dynamic structure formation of carbon-doped TiO_2 ER fluids with different conductivity using diffusing wave spectroscopy, in this proceedings.
13. Wen W. J., Zheng D. W. and Tu K. N., In situ time response measurement of the microspheres dispersed in electrorheological fluids, *Phys. Rev. E.* **57** (1998) pp. 4516-4519.
14. Ginder J. M., Diffuse optical probes of particle motion and structure formation in an electrorheological fluid, *Phys. Rev. E.* **47** (1993) pp. 3418-3429.

MAGNETIC INTERACTIONS OF CHAINS FORMED BY FERROMAGNETIC SPHERES

S.LACIS, E.ZAVICKIS

Physics Department, University of Latvia, Zellu str. 8, LV-1002 Riga, Latvia, E-mail: slacis@lanet.lv

G.BOSSIS

LPMC, UMR 6622, University of Nice, parc Valrose, 06108 Nice Cedex 2, E-mail: bossis@unice.fr

General properties of sphere-chain and chain-chain structures are analysed for homogeneously magnetized spheres. For field aligned structures the local maximum of the magnetic force is observed at $H_0/M_S \approx 0.5$. The magnetic saturation effects are accounted by the Frölich-Kennelly law. Outlines of the approximate methodology of multi-chain interaction are given together with the test results.

1 Introduction

The understanding of the character of the magnetic interaction between ferromagnetic particles leads to the understanding of the rheological properties of magnetorheological materials (fluids, elastomers). The cardinal properties for the magnetic interaction could be obtained calculating forces between ferromagnetic spheres placed in an external magnetic field. Accounting for the non-linear magnetization curve of a sphere material makes such calculations quite difficult. The finite element method (FEM) has already been used to calculate magnetic forces between ferromagnetic spheres taking into account magnetic saturation (for example, see references [1,2]). The advantage of the FEM is that an essentially exact solution is obtained that automatically includes two important effects: local-field corrections and all multipole terms [3,4]. The slightly different from pure FEM approach is realized in [5,6] by combining 3D FEM with the boundary element method (BEM). The advantage of this method is that it reduces total number of unknowns.

In this work we present new results concerning magnetic forces between particles in dipole approximation and also for 2D axisymmetric case calculated by FEM. 3D FEM-BEM formulation requires essentially more computer resources but enables to study chain-like structures placed in inclined field. In present paper 3D results are used only to test the results of an approximate approach. The approximate approach is based on the linear predictive correction of the chain mutual interaction. First of all the magnetic field problem is solved for every separate chain in 2D axisymmetric case. Obtained results are interpreted approximately for the set of chains, the nonlinear magnetization of materials is taken into account.

2 The model of the point-like dipoles

The very basics of sphere interactions inside chains and structures like chain-sphere or chain-chain could be obtained using the dipole particle model. It corresponds to the case of the high magnetic field value when spheres are completely magnetized with saturation magnetization and could be regarded as equivalent magnetic point dipoles situated in the centers of spheres. In such a case we can use the following formula

$$\vec{F} = \frac{3\mu_0}{4\pi R^5}\left[\left(\vec{M}_{D1}\vec{R}\right)\vec{M}_{D2} + \left(\vec{M}_{D2}\vec{R}\right)\vec{M}_{D1} + \left(\vec{M}_{D1}\vec{M}_{D2}\right)\vec{R} - 5\left(\vec{M}_{D1}\vec{R}\right)\left(\vec{M}_{D2}\vec{R}\right)\vec{R}/R^2\right] \quad (1)$$

to calculate the interaction force for two dipoles (this is the force acting on the second dipole). Here $\vec{M}_{Di}$ are magnetic dipole moments and $\vec{R}$ is the vector pointing from the first dipole to the second. In the case of homogeneously magnetized spheres $\vec{M}_{Di} = \vec{M}_i\, V_{SPH}$ (the dipole moment is equal to a magnetization vector multiplied by the volume of a sphere) and $\vec{R} = \vec{r}_1 - \vec{r}_2$ is the vector pointing from the center of the 1st sphere to the center of the 2nd one. The formula (1) gives the base value of the force between two spheres at contact magnetized with saturation magnetization and aligned with the field: $F_{B2D} = \frac{\pi\mu_0}{6} M_S^2 R^2$. In case of a chain of spheres the chain rupture force is formed by interaction of spheres in the one half-chain with spheres of the other half-chain. Here the reference force value is $F_{BCh} = f_C F_{B2D}$. The factor $f_C = \sum_{i=1}^{\infty} i^{-3} = \zeta(3) \approx 1.20206$ describes an increase of the magnetic force strength inside a chain in comparison with interaction of two spheres. Actually the interaction inside a chain is formed mainly by nearest neighbors: already 5 nearest spheres in both directions forms 99% of a total force.

The point-like dipole model describes well the main properties of the interaction between a single chain of finite size and a single sphere. In Fig.1 forces acting on one sphere at different positions are shown by vectors holding the same vector scale. One obvious property of such a system is that considerably strong forces will try to join a single sphere to the very tip of the chain. Another phenomena is that along the chain magnetic forces has the alternating attraction-repulsion character. Values of these forces are considerably smaller then attraction forces at the tip of the chain. So the conclusion could be drawn that a single sphere has preference to join some chain at the tip rather than at the side of chain. End effects are remarkable just for first 3-4 spheres at the tip of the chain. Closer to the middle part of the chain the periodic distribution of forces is to observe as it is shown in Fig.2.

The analysis of forces in Fig.2 gives the clue that tangential forces push a single sphere to regions where it could stick to the chain due to the attraction forces. It should be noted that an infinite chain essentially differs from a finite length chain. Every finite length chain has the dipole interaction nature at long distances. Outside the infinite chain the magnetic force acting on a single sphere (Fig.2) has the periodic character in the longitudinal direction. The magnetic force alternates along the chain but the magnitude of the force decays rapidly (at least exponentially) with the increase of the distance from the chain. For practical situations we could assume that mutual interaction is negligible if the distance between the single sphere and

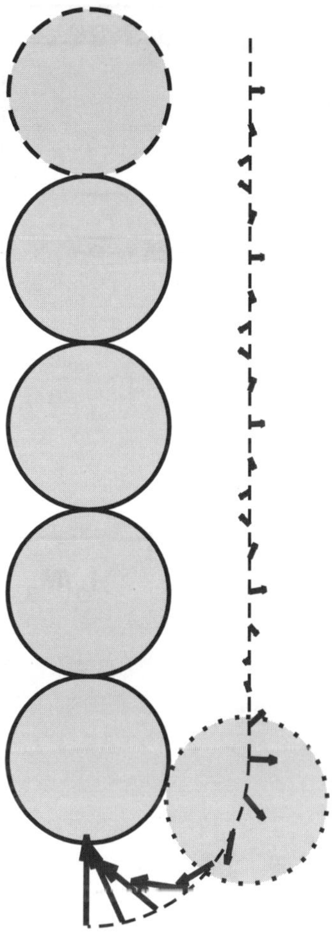

Figure 1. Force near a chain

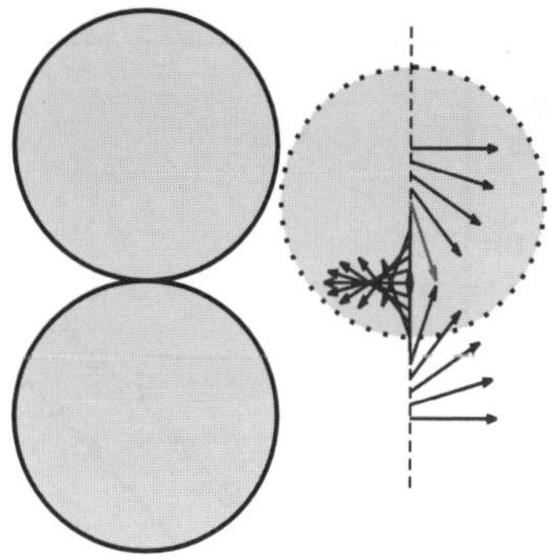

Figure 2. Periodic distribution of forces

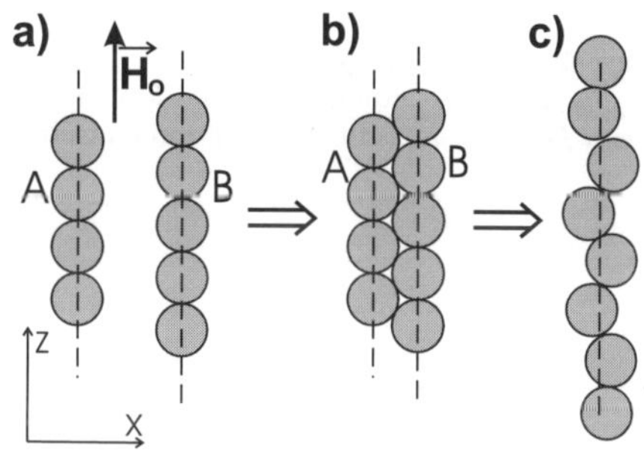

Figure 3. Two chain interaction

the chain is larger then 10 sphere radii. One more conclusion drawn form Fig.2 is the following: two chains will stick together according to Fig.3. After the sticking this new cluster of spheres could straighten under the action of an external magnetic field.

3 Models with nonlinear magnetization

The assumption of a homogeneous magnetization can explain the chain behavior at very high fields, but for the reasonable values of magnetic field intensity the nonlinear magnetization should be taken into account. It could be fulfilled by means of Frölich-Kennelly law [7]

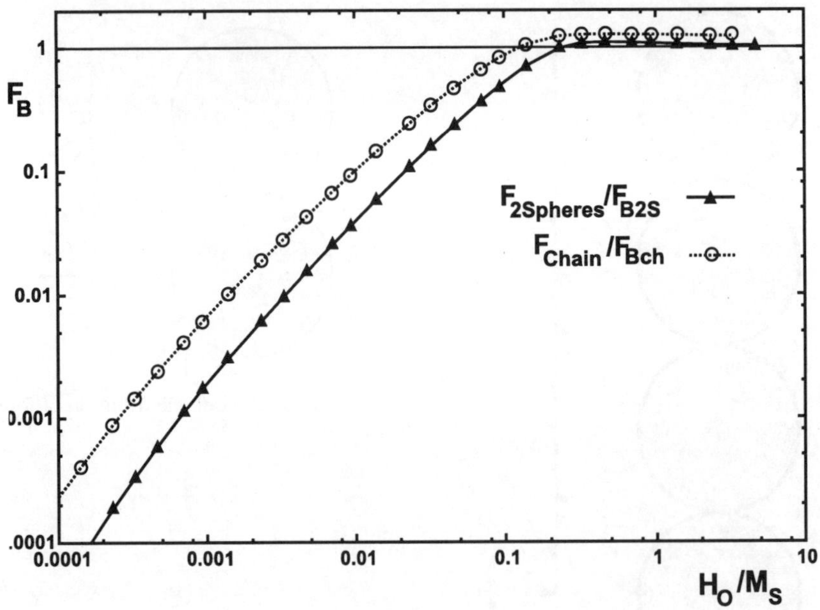

Figure 4. The rupture force:
$F_{2Spheres}/F_{B2S}$ - non-dimensional force between two spheres
F_{Chain}/F_{BCh} - non-dimensional force inside the chain of spheres

$$\chi(H) = \frac{M_S}{M_S/\chi_{init} + H}, \tag{2}$$

where χ is the magnetic susceptibility; χ_{init} is the value of the magnetic susceptibility at $H \to 0$ and M_S is the saturation magnetization of a material. Let us examine Fig.4 where the non-dimensional rupture force value is plotted versus the non-dimensional external magnetic field value for two cases: 1) two spheres ($F_{2Spheres}$) and 2) two spheres embedded in an infinite chain (F_{Chain}). In the both cases $\chi_{init} = 249$. This log-log plot shows that the magnetic forces do not possess power law character $F \propto H^\gamma$ for sufficiently large intervals of the magnetic field strength. It means that for engineering calculations of pure magnetic interaction of spheres more complex approximation is required (NB! the viscous interactions are not discussed). The remarkable feature for sphere interactions is the outcome that the non-dimensional force exceeds value 1 (see also reference [6]). Let us recall that the value 1 corresponds to the interaction force of saturated spheres. The maximal interaction force is reached at the non-dimensional field intensity $H_0/M_S \approx 0.5 \pm 0.05$ and values are $F_{2Spheres}/F_{B2S} = 1.09$ for two spheres and $F_{Chain}/F_{BCh} = 1.055$ for the chain of spheres. The further increase of the magnetic field intensity leads to the decrease of the non-dimensional force down to 1. This phenomenon is already described in [6] and the more detailed report is under preparation.

How to get a more detailed description of chain interaction including account for nonlinear magnetization law (e.g. Frölich-Kennelly law [7])? 2D axysimmetric calculations are limited to the configuration of one chain. One possibility is to find some approximation using 2D axysimmetric FEM results [2] obtained for single chain structures which are easy to calculate. In the case of multiple chains in 3D the additional distribution of magnetization is induced by neighbor chains. If the magnetization induced due to neighbors is small then it could be taken into account using methodology of linear perturbations. Magnetization vector $\overrightarrow{M}$ inside spheres of one chain could be regarded as

$$\overrightarrow{\widetilde{M}} = \overrightarrow{M} + \delta\overrightarrow{M}, \tag{3}$$

where $\overrightarrow{M}$ is the magnetization of a single chain placed in an external field and $\delta\overrightarrow{M}$ is the additional magnetization due to the additional magnetic field created by all other chains. For a single chain magnetization $\overrightarrow{M}$ could be easily calculated using 2D axisymmetric FEM code if the chain is aligned in the direction of the external magnetic field.

Inside the ferromagnetic material

$$\overrightarrow{M} + \delta\overrightarrow{M} = \chi\left(\left|\overrightarrow{H} + \delta\overrightarrow{H}\right|\right)\left(\overrightarrow{H} + \delta\overrightarrow{H}\right) \tag{4}$$

The average magnetic field intensity inside the i_{th} sphere created by the j_{th} sphere with the homogeneous magnetization $\left\langle \delta\overrightarrow{M_j} + \overrightarrow{M_j}\right\rangle$ (the term $\overrightarrow{M_j}$ is present only if both spheres are from different chains) is

$$\left\langle \delta\overrightarrow{H}_{ij}\right\rangle = \frac{3\left(\left\langle \delta\overrightarrow{M_j} + \overrightarrow{M_j}\right\rangle \cdot \overrightarrow{R_{ij}}\right)\overrightarrow{R_{ij}} - R_{ij}^2\left\langle \delta\overrightarrow{M_j} + \overrightarrow{M_j}\right\rangle}{4\pi R_{ij}^5} \cdot \frac{4\pi R_0^3}{3}, \tag{5}$$

where $\overrightarrow{R_{ij}} = \overrightarrow{r_i} - \overrightarrow{r_j}$ is the vector pointing from the center of the i_{th} sphere to the center of the j_{th} sphere ($i \neq j$). The brackets $\langle\ ,\ \rangle$ are used for the sphere averaged values. In the case of $i = j$

$$\left\langle \delta\overrightarrow{H}_{ii}\right\rangle = -\frac{\left\langle \delta\overrightarrow{M_i}\right\rangle}{3}. \tag{6}$$

The next approximation is performed with formula (4) under the assumption that $\delta\overrightarrow{H_i}$ is essentially smaller than $\overrightarrow{H_i}$:

$$\left\langle \delta\overrightarrow{M_i}\right\rangle = \left.\frac{\partial\chi}{\partial H}\right|_{H=\langle H_i\rangle} \frac{\left(\left\langle \overrightarrow{H_i}\right\rangle \cdot \left\langle \delta\overrightarrow{H_i}\right\rangle\right)}{\langle H_i\rangle}\left\langle \overrightarrow{H}_i\right\rangle + \chi\left(\langle H_i\rangle\right)\left\langle \delta\overrightarrow{H_i}\right\rangle, \tag{7}$$

here $\langle H\rangle = \left|\left\langle \overrightarrow{H}\right\rangle\right|$. Inserting (5) into the relation (7) provides the linear system of equations for unknowns $\left\langle \delta\overrightarrow{M_j}\right\rangle$. Once this system of equations is solved the

Table 1. Magnetic force between spheres at different field orientations

	Force (perp.)		Force (axial)	
φ	3D FEM+BEM	Lin.pert.	3D FEM+BEM	Lin.pert.
0	0.0000000	0.0000000	-0.0011848	-0.0012153
30	-0.0005130	-0.0004632	-0.0007405	-0.0007901
60	-0.0005130	-0.0004624	0.0001481	0.0000666
90	0.0000000	0.0000000	0.0005924	0.0004990

magnetic interaction forces could be calculated by means of the formula (1) (the magnetic dipole interaction).

Thus two simplifying assumptions are done: 1) values $\delta\overrightarrow{M}$ are constant for each sphere; 2) $\delta\overrightarrow{H}$ is created by the dipole moments of neighbor spheres and by magnetization $\delta\overrightarrow{M}$ homogeneously distributed throughout each sphere.

This method is tested for the case of two chains and promising results are obtained. The simplest test is carried out for two spheres at different magnetic field orientations (an angle φ is measured between the axis of two spheres and the magnetic field direction). Results are presented in the Table 1.

The distance between centers of spheres is $4R_0$ (R_0 is the radius of spheres), calculations are carried out for the external field $B_0 = 0.1T$, the magnetization curve parameters are $\chi_{init} = 249$ and $M_S = 1710000\,A/m$. The results obtained with linear perturbation method are marked as “Lin.pert.” and label “3D FEM+BEM” stays for the results obtained by 3D mixed finite-boundary element method [5,6]. Discrepancies depend on the magnetic field orientation and are usually smaller than 10% except $\varphi = 90$ where discrepancy is about 18%. The same behavior is observed for $B_0 = 1T$ and $B_0 = 0.01T$. The increase of the magnetic field intensity leads to more accurate approximate results because of a more homogeneous magnetization inside spheres caused by the magnetic saturation. The decrease of the distance between spheres causes a larger perturbation of a magnetic field inside spheres and thus it leads to the less accurate results. At the distance $3R_0$ between centers of spheres the discrepancy between results obtained by both methods increases up to 25%. Evidently this is the limit distance between the neighbor chains for obtaining reasonable results by the linear perturbation approach.

4 Conclusions

The qualitatively good description for the interaction of chain-like structures could be obtained on the basis of homogeneously magnetized spheres. This case could be magnetically described by point-like dipoles. Accounting for the nonlinear magnetization curve leads to intrinsic phenomena including the observation of the magnetic interaction force maximum at $H_0/M_S \approx 0.5$. The interaction between chains could be calculated approximately with reasonable accuracy if the distance between axes of chains is larger than 4 radii of a sphere.

Acknowledgments

This work was supported by Research Grant No. 01.0070 of the Latvian Council of Science

References

1. L. C. Davis, Appl. Phys. Lett. **60**, 319 (1992).
2. S. Lacis, O. Volkova, G. Bossis, in *Modelling of Material Processing - Proc. Int. Colloq.*, (University of Latvia, Riga, 1999), 222.
3. D. J. Klingenberg, F. van Swol, C. F. Zukosk, J. Chem. Phys. **94**, 6170 (1991).
4. H. J. H. Clercx, G. Bossis, J. Chem. Phys. **103**, 9426 (1995).
5. E. Zavickis, (In Latvian) Latv. J. Phys. Tech. Sci. **N 5**, 40 (2000).
6. E. Zavickis, S. Lacis, Latv. J. Phys. Tech. Sci. **N 4**, 3 (2001).
7. D. Jiles, *Introduction to Magnetism and Magnetic Materials* (Chapman & Hill, London, 1995).

PARTICLE DYNAMICS OF STRUCTURE FORMATION AND DISINTEGRATION IN A MODEL MAGNETORHEOLOGICAL FLUID

S. CUTILLAS AND J. LIU

Department of Physics and Astronomy, California State University
1250 Bellflower Boulevard, Long Beach, CA 90840, USA

Using diffusing-wave spectroscopy (DWS), we studied experimentally the particle dynamics for a density matched superparamagnetic polystyrene colloid in a refractive index nearly matched multiple-layered cell. Particle dynamics is probed during structure formation and disintegration, when a 0.2 Hz square-wave magnetic field was turned on for 4 s and off for 1 s. The correlation function shows that the particles move slower and more restricted when the magnetic field is on. Even during the *off* cycle of the magnetic field, the particles' motion is not free but still constrained with a less degree than that when the field is on. Thus, it takes more than 1 s for the induced structure to disintegrate. As the magnetic field strength increases, so does the degree of constrain for both the *on* and *off* cycle of the magnetic field and the differences between them. Modified telegrapher theory is found to be valid for our strongly absorbing and limited multiple scattering sample.

1. Introduction

Magnetorheological (MR) fluids have gained great attention in the last decade due to its application as a smart material: from shock absorber for cars to earthquake damage control, from damper for artificial knee to cancer therapy [1-2]. All of these applications are due to the novel property of MR fluids that switch from a liquid state to a solid state quickly and reversibly controlled by an applied magnetic field. This tunability in rheological property comes from the change of microstructures of the particles in the fluids. When the permeability difference exists between the particle and the carry fluid, an external magnetic field induces a magnetic dipole moment in each particle. The dipole-dipole interaction leads to the formation of anisotropic structures such as chains, columns or labyrinths while without magnetic field, the particles are in Brownian motion and are distributed randomly [3]. When the external field is switched off, Brownian forces destroy all previously formed structures and a liquid state is recovered.

Many studies have been reported on the rheological properties of MR fluids. Some are on the field-induced structure formation [4]. But very few are on the particle dynamics, due to the limitation of strong light scattering and absorption. Using single light scattering technique such as video-microscopy [5] and dynamic light scattering (DLS) [6-7], the study of particle dynamics is limited to very low particle volume fractions ($\phi \sim 10^{-4}$) thus to particles within single chains.

Taking advantage of multiple light scattering, diffusing-wave spectroscopy (DWS) was developed in the last decade to study particle dynamics in turbid complex fluids, such as suspensions, emulsions and foams where light scattering is strong and particle structure is isotropic [8-11]. In DWS, a laser light is directed to the fluids, a transmitted or reflected light is detected and sent to a digital correlator in a computer. From it an electric field auto correlation function, $g_1(\tau) = \langle E(0)E^*(\tau)\rangle/\langle|E|^2\rangle$, is calculated to measure the fluctuations of the scatted light as a function of the delay time τ. From $g_1(\tau)$ with proper theoretical model, the mean square displacement of particles is obtained. Due to multiple light scattering in DWS, the light transport inside a sample is not ballistic as it is in DLS, but a diffusive process [10]. Hence, the scattered light intensity is averaged over all possible scattering wave vectors. Thus DWS is capable of probing motions on a very short length (~ 10 Å) and time (~ 10 ns) scales in a turbid media.

Using DWS, Furst and Gast [12] studied particle dynamics of superparamagnetic emulsion droplets for ϕ = 0.5%, an order of magnitude higher than previous work. However, at this ϕ, the formed structures are still separated chains. They found that the particle displacements are characterized by "a constrained, sub-diffusive motion at long times that slows as the dipolar interactions are increased." The only experimental study reported on particle dynamics in a dense labyrinthine structure (ϕ = 30%) of MR fluids is presented in our earlier work [7]. We found a strongly trapped motion in the particle displacements that moves slower as the magnetic field increases.

Here we present another DWS study of a model MR fluid for particles in a column structure: 2.3% of superparamagnetic polystyrene colloid that are density matched by the carrier fluid in a multiple layered sample cell. This study is performed for both particle structure formation and disintegration when magnetic field was on and off in a square waveform of 0.2 Hz. During the *on* cycle of the magnetic field (4s), we found a constrained motion that slows as the magnetic field strength increases. During the *off* cycle of the magnetic field (1s), we found that the particle motion speeded up but were still constrained compared to before applying magnetic field indicating that the structure had not completely disintegrated. The higher the magnetic field, the more constrain are on the particle motion for both magnetic field on and off, reflecting the degree of tight binding of particles in a column due to magnetic interaction. The difference between the field being on and off widens as the applied magnetic field strength increases. Although brief, this work is the first to study the particle dynamics during structure disintegration.

2. Sample Preparation and Characterization

We used superparamagnetic polystyrene particles obtained commercially (Seradyn Inc., USA) that are re-dispersed in a buoyant carry fluid. As claimed by the manufacture, each particle contains 40wt% Fe_3O_4 grains and has an average diameter of $2a$ = 1.00 ± 0.05 μm and a particle density of 1.40 ± 0.01 g/cm^3. The commercial fluid is water based and each particle is coated with a layer of carboxyl chains (COOH). Before its use, all the small particles (~ 100 nm) are removed by magnetic separation. To avoid sedimentation and strong internal reflection from the sample holder, we made a new carrier fluid consisting of: 1.00 ± 0.01 cm^3 of de-ionized water, 6.00 ± 0.01 cm^3 of a saturated aqueous solution of sucrose (density: ρ_s = 1.330 ± 0.003 g/cm^3) and 4.00 ± 0.01 cm^3 of a saturated solution of potassium thiocyanate, KSCN (ρ_K = 1.436 ± 0.003 g/cm^3). The density of the carrier fluid thus prepared is 1.400 ± 0.003 g/cm^3 at 294 K and matches the density of the particles. When the particles are dispersed in this buoyant liquid, they are stable provided they were not in contact with other surfactant coatings such as Sodium Dodecyl Sulfate. After this, we measured the particle diameter using DLS and Contin algorithm analysis and found $2a$ = 0.76 ± 0.04 μm, which will be used in the analysis of the experiments. The viscosity of the buoyant liquid varied from 16 mPa.s at 14 ^{0}C to 9 mPa.s at 28 ^{0}C when measured with a Brookfield RV-III viscometer. At the temperature of 27.0 ^{0}C used in DWS experiments, the viscosity of the buoyant liquid is 9.5 ± 0.2 mPa.s, which is about 10 times higher than that of the water.

The requirement of multiple light scattering in DWS is easily satisfied in the absence of a magnetic field since the particles are in random Brownian motion and scatter light strongly. However, when an external magnetic field is applied, particles form chains along the field direction and chains aggregate to form anisotropic columns or 2-D labyrinth structures [3]. This widens the gap between aggregates and allows the light to penetrate with little scattering. In addition, the aggregates scatter light anisotropically, which complicates the theoretical interpretation.

In order to stay within the limit of multiple light scattering, we constructed a special sample cell of 8 layers as shown in Fig. 1. This cell consists of equally spaced layers of rectangular microscope cover glasses (22x30 mm^2 in area and (150 ± 5) μm in thickness). In between the glasses, spacers made of plastic sheets (100 μm) are placed to create sample holder. One side of the sample cell is sealed by epoxy glue and the other side is left open to fill in the sample. After filling all the empty layers with the suspension, the cell is completely sealed with epoxy and set aside to dry for 3 days before use.

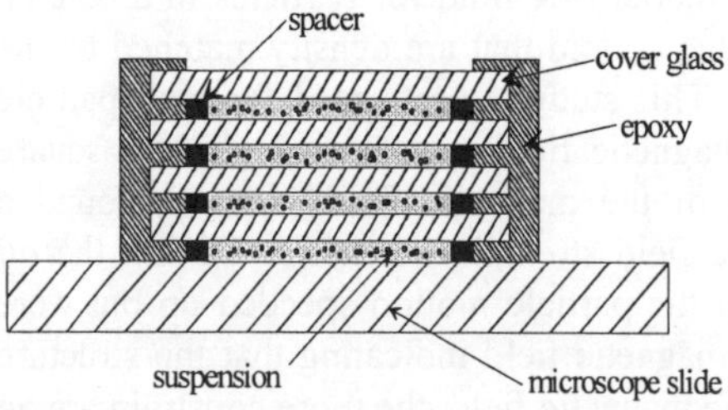

Fig. 1. Cross section of a multiple layered sample cell. Only 4 of the 8 layers are draw here.

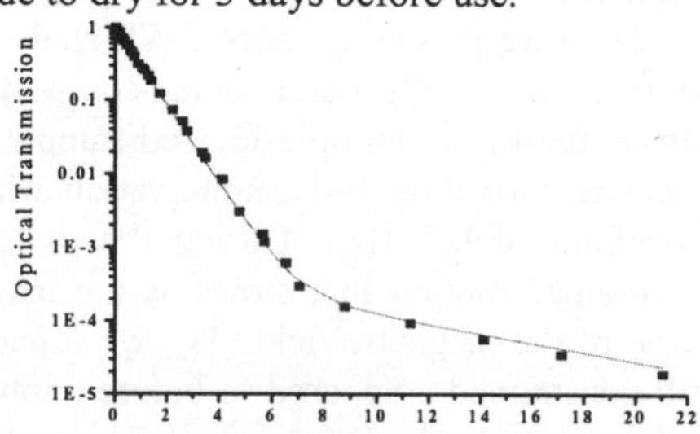

Fig. 2. Optical transmission vs. effective cell thickness at ϕ = 0.081%. Line is from fitting.

In this specially designed cell, when a magnetic field is applied normal to the surface of glasses, columns formed in different layers have no spatial correlations, as verified by video-microscopy. This insures isotropic and multiple light scattering when the incident light beam is parallel to the magnetic field direction. The buoyant liquid has a refractive index of 1.462 ± 0.01 as measured with an Abbe Refractometer at 300 K and at the wavelength of 632.8nm (using a filter). This is quite close to 1.52, the refractive index of the sample cell made of borosilicate glass. Thus we can neglect reflections between the glass cell surfaces and the suspension and consider the whole cell as an effective medium with a total cell thickness of L = 8x105 μm = 840 μm.

The magnetic hysteresis loop of the suspension is measured using an Alternating Gradient Magnetometer from Princeton Measurement. An aqueous suspension of volume of 4.23 mm^3 is prepared at ϕ = 1.9%. Measurement is taken at -13^0C. No remnant magnetization is found as expected from an iron oxide material. The relative magnetic susceptibility for particles at low field is found as χ_p = 1.29. This corresponds to the magnetite grain diameter of 12 nm based on Langevin Equation.

3. Experimental Set up and Results

3.1 Optical Transmission at zero magnetic fields

To understand the sample better, optical transmission is measured at the wavelength of 632.8 nm. A rectangular sample cell is used with the cell thickness (L) along the direction of incident light ranging from 80 μm to 4100 μm. Two particle volume fractions ϕ = 0.081% and 0.43%, are measured. Since the optical transmission depends on the product of ϕ and L, we converted the data obtained from ϕ = 0.43% to that of ϕ = 0.081% by effectively increasing L by a factor of 5.3, the ratio of the two volume fractions. The laser beam diameter was kept at 9 mm to ensure a factor of two or more greater than L required to be in the multiple scattering regime.

Figure 2 shows a semi-log plot of the optical transmission (T) of our suspension as a function of effective cell thickness L' at the equivalent particle volume fraction ϕ' = 0.081%. Two different exponential decays are observed corresponding to two different regimes: single light scattering and absorption at low L' and multiple scattering and absorption at high L'.

The first decay can be expressed as $T_1 = \exp(-L'/L_e)$ where L_e is the extinction length (or inverse of the turbidity τ) of the suspension. $L_e = 1/[N(C_s+C_a)]$, where C_s and C_a are respectively the scattering and absorption cross sections of particles and N is the number density of the scattering site (particle) [13]. From fitting the data of Fig. 2 we find that $L_e = 0.813 \pm 0.020$ mm. The second decay can be expressed as $T_2 = \exp(-L'/L_a)$ where L_a is the effective absorption length which accounts for both absorption and scattering. Since in this regime light scattered many times before exiting the cell, the actual absorption length, l_a, is much larger than the effective absorption length L_a. In the limit of a great number of scattering events (Boltzman limit): $l_a=3(L_a)^2/l^*$ [10] where l^* is the transport mean free path of scattered light that is the distance required for photons to be randomized. From Fig. 2, we found that $L_a=(6.25 \pm 0.07)$ mm. From fitting the DWS data (ϕ = 2.35 %, see section 3.2) and converting the values back to ϕ = 0.081%, we found that the scattering length l_s = 0.86 mm, l_a = 10.8 mm and l^*= 10.9 mm. Thus, in thin cells, scattering dominates absorption ($l_s < l_a$) while at thick cells multiple scattering and absorption are equally important ($l_a = l^*$).

Usually, with less absorptive particles, between these two exponential decays there is a region of $1/L$ decay that is a characteristic of the onset of multiple scattering [13]. The condition for strong multiple light scattering as required in DWS theory is $L >> l^*$. Unfortunately, strong absorption has limited us to $L \sim 1$ mm at ϕ = 2.35 %, in order to have a good signal to noise ratio. This gives $L \sim 3l^*$ which is in between the regime of DLS and DWS. The typical DWS theory does not apply. A different theory based on modified telegrapher equation (MTT) [14] will be used to account for the ballistic transport of photons. MTT applies for $2l^* < L$ and accounts also for strong absorption and anisotropic scattering from large particles. It is not clear if MTT works for scattering from large anisotropic structures such as columns formed when field is on.

3.2 DWS Experimental Set up and zero magnetic field

Figure 3 shows the DWS experimental set up. The sample cell is placed inside a water bath for precise temperature control (27 ± 0.1 ^{0}C), which is located at the center of a pair of magnetic coils. A waveform generator produces a square wave magnetic field output (H) and turns on or off the magnetic coils. A collimated beam of 8 mm in diameter from a He-Ne laser (λ_0=632.8nm) is directed vertically to the sample. The forward scattered light is collected 25 cm away from the sample directed at the center of the incident beam by a single mode optical fiber. The fiber is then divided equally into 2 different single mode optical fibers that are connected to two photo-detectors. We cross correlate the split signals in order to avoid spurious noise coming from the electronics. The detected photons are converted into electrical pulses. After amplification, the signal is further divided into two channels through a two-stage relay controlled by the TTL output of a waveform generator. Relay is used for measuring separately the particle dynamics during the *on* and *off* cycle of the magnetic field. One stage is normally open (no signal passes) when the field is off and becomes closed once the field is on (TTL level is higher than 3 V). The other stage is normally closed and opens when the field is on. After the relay, one signal from each detector enters a digital correlator board located inside a personal computer for cross correlation. Thus, when H = 0 G, data is processed by correlator B, whereas when H ≠ 0 G, it goes to correlator A. Using a white light as the source, a CCD camera with a objective lens allows direct view of the particle structure.

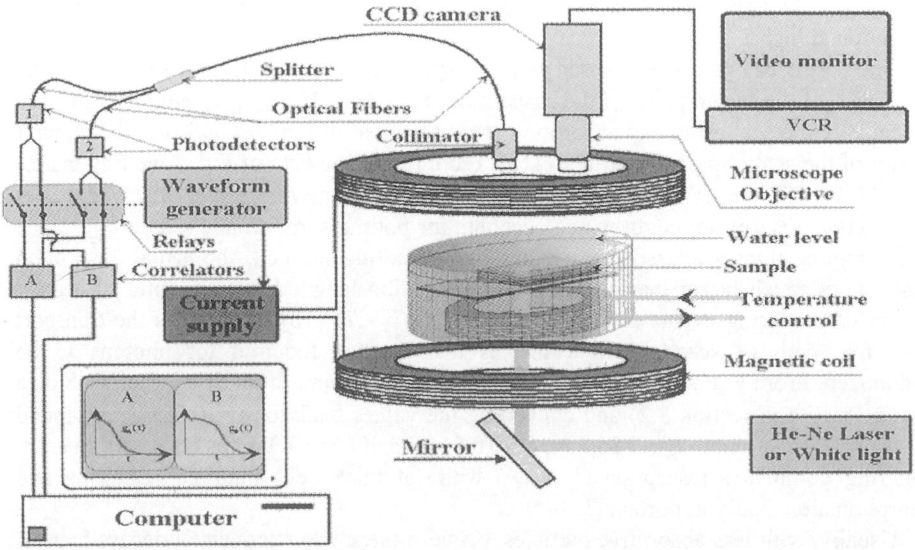

Fig. 3. Experimental Setup for DWS measurement for fields both *on* and *off* and for structure visualization.

For comparison, two sample cells are used in the DWS experiments: a single cell (L=1.04 mm) and an 8-layers cell with 105 μm each (L = 0.84 mm). Particle volume fraction for both cases is 2.35%, which corresponds to L' = 30.2 mm for single cell and 24.4 mm for multiple cell at ϕ' = 0.081%. Based on Fig. 2 they are in the multiple scattering regime. Figs. 4 and 5 show the result of normalized electric field-field auto correlation function, $g_1(\tau)$, as a function of correlating time between two detected light signals, where we included fitting curves based on both DWS and MTT theories [14].

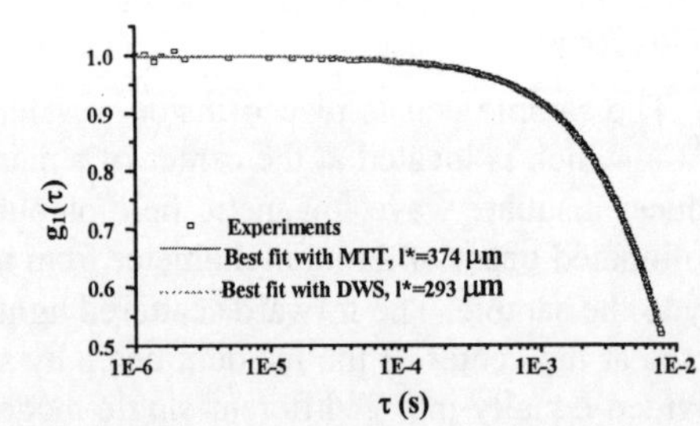

Fig. 4. Field-field correlation function at H=0 G in a single sample cell L = 1.05 mm.

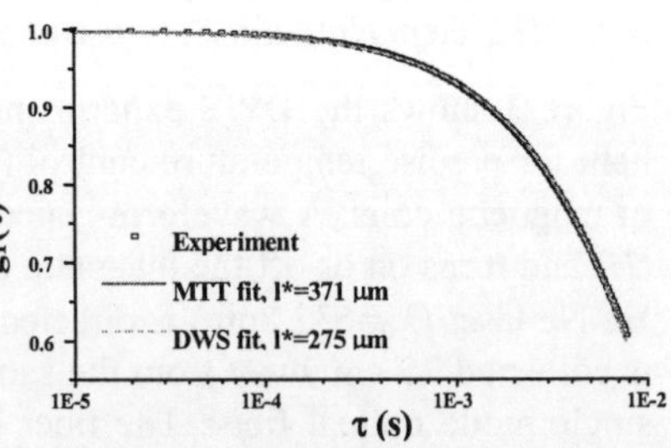

Fig. 5. Field-field correlation function at H=0 G in a 8 layered sample cell, total L = 0.84 mm.

Two fitting parameters, A and l^* are used to fit $g_1(\tau)$. A is a normalization parameter and equals to 1 in an ideal case. We found 0.99<A<1 for both cases. There is additional fitting parameter in MTT: l_s, the scattering length. By fitting the data in Fig. 4, we found that l^* = 293 μm using DWS and 374 μm using MTT. Similar results are found for fitting the data of Fig. 5: l^* = 275 μm for DWS and 371 μm for MTT. To verify that MTT is valid for our case, we calculated l^* based on the form factor from Mie scattering [15] for the data of Fig. 4. We found that l^*= 366 μm using the l_s and l_a value estimated from the l^* fitted by the DWS theory above and 373 μm for MTT (l_s = 30 μm). It is clear that MTT works in our case. Since L/l^*~3, DWS theory does not apply to our experiments as expected and verified here. The Mie calculation for l^* is first checked for validity by analyzing the published data and good agreement is obtained.

Based on MTT, both samples give a consistent l^* to within 1% uncertainty. We can neglect effects coming from multiple internal reflections between cover glasses and the liquid in our multiple layered sample cells. To check the effect of using relays on DWS experiments, we measured standard polystyrene particles ($2a$ = 298 ±3 nm). Two TTL signal frequencies were tested: 0.05 and 0.5 Hz. After adjusting the baseline according to the frequency used, all the data with and without relays were superimposed over one another quite well at short time up to $\tau \sim 0.01$ s. The same l^* was obtained in all cases. At very long time $\tau > 0.01$ s, a slight increase is found in both data using relays. This is due to conversion of the directly measured intensity autocorrelation function $g_2(\tau)$ to $g_1(\tau)$ using the relation: $g_1(\tau) = (|g_2(\tau)-1|)^{1/2}$. The *on* and *off* TTL signal cuts pulse streams which leads to reduced counts at some channels and zero at others and affects the base line. Even after adjusting the base line according to the frequency used, $g_2(\tau)$is still less than 1 at higher τ . Thus absolute value is used in the Siegert relation [9]. This leads to the increase after the regular decay. This effect is amplified drastically after the field is on as seen in Fig. 6. However, alternating and interrupting the data collection seems not affecting the computation of the correlation function (l^*) at short times.

3.3 Appling a Square-Wave Magnetic Field

A square wave magnetic field H (0, 15.3 G) is applied to the multiple layered sample used for obtaining the data for Fig. 5. The frequency of the magnetic field is 0.2 Hz with a duty cycle of 80%, that is, the field was *on* for 4 s and *off* for 1 s for each 5 s period. A series of correlation functions were recorded and averaged for 10 min during both the *on* and *off* cycle of the magnetic field using relays. Fig. 6 shows the results for different measuring times after the field is applied to check the effect of kinetics. The zero-field data of Fig. 5 is also included for comparison.

Except the long time tail, good overlapping are seen in all the correlation functions. Thus, applying the magnetic field and varying the measuring time does not affect $g_1(\tau)$. If a constant magnetic field of H = 15.3 G is applied here, it would be strong enough to change the structure and affect the dynamics significantly since the magnetic dipolar energy is 14 times larger than the thermal energy (coupling constant λ= 14 [4]). When a square wave of magnetic field is applied, the field is not strong enough to maintain a structure even after 60 minutes of application of the *on-off* field.

Next, we increased the maximum field strength from 0 to 118G in 9 steps. After the preset value of the field strength is reached, we waited 50 min before taking data to allow a steady structure to form. Each correlation function is averaged for 10 minutes. For clarity, the artefacts are removed at very long time and only 4 of the 9 data are shown in Fig. 7. Beginning at 29.5 G (λ= 53), magnetic field effect is quite obvious upon $g_1(\tau)$: as the magnetic field increases, the decay in $g_1(\tau)$ not only slows down with time but also shows an increasing split between the *on* and *off* cycles of the magnetic field. The characteristic decay time τ_c in $g_1(\tau)$ varies from ~ 0.01 s at H = 0 to ~ 1 s at H = 118 G (*on* cycle, λ= 767). The particle motion is constrained and slows as the dipolar interaction increases. Even during the 1s that magnetic field was off, $\tau_c \sim 0.1$ s at $H = 118$ G (*off* cycle), an order of magnitude different from the two extreme cases. The particles are still constrained although with less degree compared to that when the field is on. The stronger the magnetic field, the more constrained the particles' motion is. It reflects the degree of particle binding in a column during the *on* cycle. Thus 1 s is not enough to completely disintegrate the structure but definitely gives more space and speed for the particle to move as compared to when the field is on.

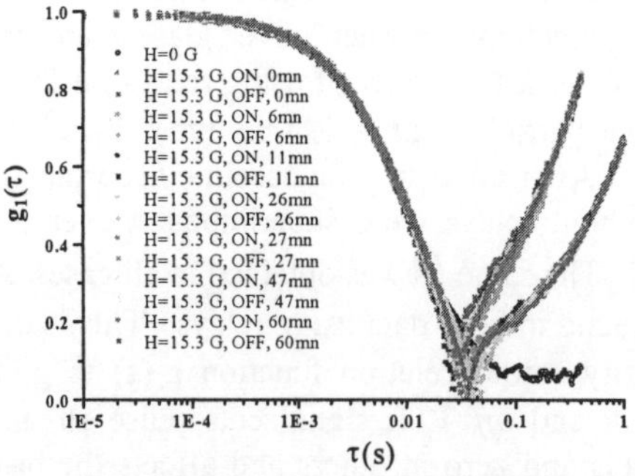

Fig. 6. Effect of different measuring time after a square-wave magnetic field is applied.

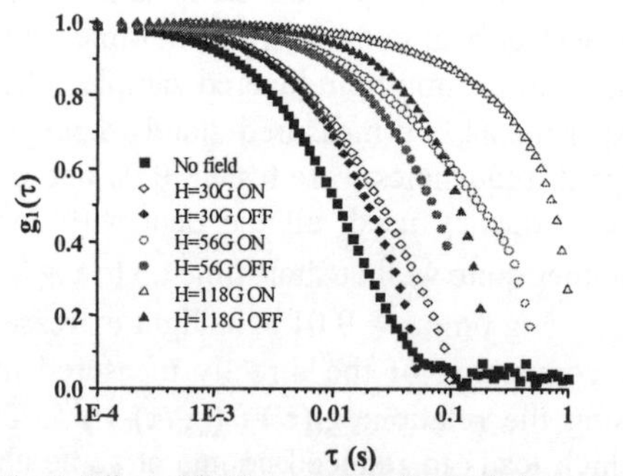

Fig. 7. Effect of field strength during both the *on* and *off* cycle of the square-wave magnetic field.

In summary, DWS method with MTT theory can be used to study particle dynamics in a strongly absorbing medium. Using a specially designed sample cell and relays, particle dynamics can be probed during both the structure formation and disintegration period.

Acknowledgement

We gratefully acknowledge the financial support of NASA (NAG3-1830), Research Corporation (CC4801), and California State University Long Beach as well as technical help in preparation of the manuscript from Sadayoshi Okumoto.

References

1. See Lord Corp.'s web page www.mrfluid.com.
2. J. Liu, G. Flores and R. Sheng, *J. Magnetism and Mag. Mat.* 225 (2001) 209-217.
3. G.A. Flores, J. Liu, M. Mohibe and N. Jamasbi, *Phys. Rev. E.* 59 751-762 (1999).
4. J. Liu et al, *Phys. Rev. Lett.*, 74 **(14)**, 2828 (1995).
5. A. S. Silva et al, *Phys. Rev.* E **54** (5), 5502 (1996).
6. M. Hagenbüchle and J. Liu, *Applied Optics*, **36**, 7664 (1997).
7. S. Cutillas and J. Liu, *Int. J. Mod. Phys.* **B 15** 803-810 (2001)
8. G. Maret and P.E. Wolf, *Z. Phys. B* **65**, 409 (1987).
9. D.J. Pine, D.A. Weitz, P.M. Chaikin and E. Herbolzheimer, *Phys. Rev. Lett.* **60**, 1134 (1988).
10. D.J. Pine, D.A. Weitz, J.X. Zhu and E. Herbolzheimer, *J. Phys. France* **51**, 2101 (1990).
11. P.D. Kaplan, M. Kao, A.G. Yodh and D.J. Pine, *Applied Optics* **32**, 3828 (1993).
12. E. M. Furst and A. P. Gast, *Phys. Rev.* E **58** (3), 3372 (1998).
13. A.Z. Genack, *Scattering and Localization of Classical Waves in Random Media*, P. Sheng ed. (World Sci., Singapore, 1990).
14. P.A. Lemieux, M.U. Vera and D.J.Durian, *Phys. Rev.* E **57** (4), 4498 (1998).
15. H.C. Van de Hultz, *Light Scattering by Small Molecules* (Dover, New York, 1957).

SHORT TIME DYNAMIC STRUCTURE FORMATION OF CARBON-DOPED TiO_2 ER FLUIDS WITH DIFFERENT CONDUCTIVITY USING DIFFUSING WAVE SPECTROSCOPY

L.F. ZHANG, X. DONG, J. XU, Y.G. JIANG, L.W. ZHOU

The State Key Laboratory of Applied Surface Physics & Physics Department
Fudan University, Shanghai 200433, China
E-mail: lwzhou@fudan.ac.cn

Y. H. ZHU

East China University of Science and Technology, Shanghai,200237, China

A new method of doing DWS (diffusing wave spectroscopy) measurement has been developed by using high speed CCD to investigate the structure and the dynamics of nonergodic and concentrated system such as electrorheological (ER) fluids under high electric field. The DWS results from the CCD method and the traditional method using a single-mode optical fiber agree well for ergodic system. The structure response time is obtained by measuring the transmittance of the sample under different electric field. The force response time of about 100ms is obtained from the time-dependent diffusion coefficient. In this paper, by comparing the time variation of transmittance and diffusion coefficient obtained under different conditions, a criterion of fast and strong responses of practical ER fluids to electric field is provided.

1 Introduction

Electrorheological (ER) fluids are typically comprised of suspending particles in a liquid whose dielectric constant or conductivity is mismatched with that of particles in order to create dipolar interactions under high electric field [1]. Many studies have shown that the rheological response is greatly related to the field-induced structure formation [2]. When high electric fields are applied to ER fluids, the particles rapidly aggregate into fibrous columns aligned with the direction of electric field. These structures span the electrode gap in quiescent suspensions (above certain particle concentration). Following the increment of the electric field or the concentration increasing, these columns become thicker, and individual columns are not discernable at sufficiently large concentrations. It has been appreciated that field-induced particle polarization can result not only from a dielectric mismatch between particle and suspending fluid, but also their conductivity mismatch. Different particle conductivity leads to different polarization rate, which may be related with different structure stability, and then with different dynamic shear stress. It is our purpose to verify the relation of structure stability with the particle conductivity.

In this paper, a new method of doing DWS experiment using high speed CCD has been proposed. The influences of electric field, shear rate and particle conductivity of the material on the particle structure formation are studied especially at short time scales from microseconds to milliseconds, as well as at long time scales from seconds to minutes, after the electric field is turned on. The long time scaled particle interactions are cross-checked with a fast video camera (RedLake) with highest frame rate of 8000 frames/s. A good agreement is found between our experimental results and the theoretical analysis. [2]

2 DWS Concepts

Diffusing-wave spectroscopy (DWS) is an extension of conventional dynamic light scattering (DLS) to the multiple scattering regime. It provides information about the structural and dynamic properties of the sample by detecting the intensity of a single speckle spot of the scattering light and its temporal fluctuations, which are characterized by their temporal autocorrelation function. Since DWS is very sensitive to the particle motion of the order of a few angstroms, the method has been widely used in concentrated system such as ER fluids [3], MR fluids [4], sand dunes [5] and colloidal suspensions [6].

Here we give a brief overview of the DWS theory developed by Weitz and Pine [7] and the extensive application in nonergodic systems proposed [8].

The whole theory is based on two approximations: one is that the light propagation is described as diffusion through the scattering media. The other is that every individual scattering event is taken as the contribution of an average scattering event. We consider the autocorrelation function of the electric field of light. Take, for example, a particular sequence of scattering events occurring at points $\boldsymbol{r}_1, \boldsymbol{r}_{2,\ldots,}\boldsymbol{r}_{\rm n}$, for which the scattering wave vectors are $\boldsymbol{q}_1, \boldsymbol{q}_{2,} \ldots , \boldsymbol{q}_{\rm n}$. The field E(t) at the detector carries a phase shift

$$E(t) \sim \sum_{p} \exp[\mathrm{i}\sum_{j} \boldsymbol{q}_{\rm j} \cdot \boldsymbol{r}_{\rm j}(t)] \qquad (1)$$

here the first sum is over all possible scattering paths and the second is over all scattering events through a single path. For noninteracting or weakly interacting particles we find the decay of the normalized autocorrelation function

$$g_{(1)}(t) \equiv \frac{\left\langle E(t)E^*(0)\right\rangle_{\rm T}}{\left\langle |E|^2\right\rangle_{\rm T}} = \left\langle \sum_{\rm p} \exp\left\{\mathrm{i}\sum_{\rm j} \boldsymbol{q}_{\rm j} \cdot \left[\boldsymbol{r}_{\rm j}(t) - \boldsymbol{r}_{\rm j}(0)\right]\right\}\right\rangle \qquad (2)$$
$$= \sum_{\rm p} e^{-N<q^2><\Delta r^2>/6}$$

here <...>$_{\rm T}$ denotes *time* averaging which can be different from the *ensemble* averaging <...>$_{\rm E}$. In the second equality, we have made the assumption that $\Delta\boldsymbol{r}$ is a Guaussion random variable. For isotropic scattering, $<q^2>=2k_0^2$, where k_0 is the scalar wavevector of the incident light. N is the number of scattering events of a single path s. It can be estimated as N=$s/l*$ by drawing into an important parameter l', the transport mean free path. It is defined as the length scale over which the direction of light propagation is randomized. For large N, $g_{(1)}$ (t) depends on the path length s and not on any other path-dependent properties. So we replace the sum over paths with the fraction of scattered intensity in path of length s P (s). It must be determined for experimental configuration. Thus, we get

$$g_{(1)}(t) = \sum_{s} P(s)\exp(-\frac{1}{3}k_0^2 < \Delta r^2(t) > \frac{s}{l^*}) \qquad (3)$$

$g_{(2)}(t)$ is the intensity autocorrelation function, which is given by Siegert relation[6],

$$g_{(2)}(t) \equiv \left|g_{(1)}(t)\right|^2 \qquad (4)$$

In traditional ways, a small speckle of the scattered light is transmitted by a single-mode fiber (SMF), detected and amplified by a photomultiplier, and then calculated in a correlator. Numerical results can be obtained if the sample system is ergodic, for example, an ER fluid under zero electric field. Indeed, here is a theoretical assumption that ensemble average is equal to time average, while time average can be easily obtained in experiments. However, for nonergodic systems such as ER fluids under high electric field, the equality no longer exists. To overcome this problem, Ref. 8 has developed a double-layer technique and established a "multiplication rule" to interpret the experimental data. This method is not suitable to study the fast response of ER fluid shortly after electric field is turned on since its average time is usually 3 or 4 orders of magnitude of the decay time. In section 3, we will introduce a different method using a CCD camera [9] to deal with the fast response of nonergodicity in ER fluids.

3 Experimental setup

One sample used in this study consists of micron particles (radius d=0.5μm) of glass spheres suspended in silicon oil with a volume fraction Φ=2%. The refractive indexes of the particles and the base fluid are close (n_{glass}=1.45; n_{oil}=1.4). The particles have been dehydrated at 160℃ for 24 h. We prepare two sample cells labeled A and B. Both are rectangular cuvette with a pair of parallel, transparent conducting glass electrodes. The separation between the electrodes of cell A is 2mm, while the electrodes of cell B are sealed with a latex rubber membrane to allow the separation to be changed from 1.5mm to 8mm.The transmission geometry has some requirements for the sample cells: the thickness L satisfies $L/l^* \geqslant 5$, and the lateral extent is much greater than L.

Fig. 1 shows the typical arrangement for a traditional DWS experiment. We use a 150 mW YAG laser (CrystaLaser GCL-150-S, wavelength λ_0=532nm) with a coherent length more than 100m. The laser light is expanded to a plane wave with 6mm in width by a lens pair, and then illuminates the front electrode along the field direction. A single-mode optical fiber is fixed behind the rear electrode to detect a small portion of the light that is diffusely transmitted though the sample. The intensity fluctuations are measured by a single photon detector (ALV SO-SIPD). We use a commercial correlator (ALV-5000E) to calculate the autocorrelation function $g_{(2)}(t)$.

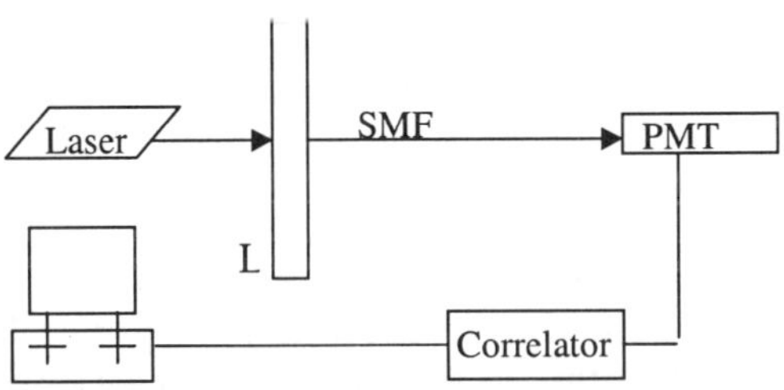

Figure 1. The arrangement of traditional DWS experiment for transmission geometry

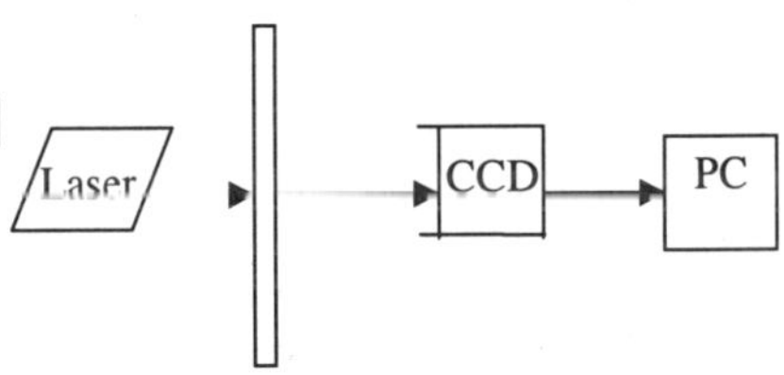

Figure 2. CCD based DWS experiment for transmission geometry

In our method, as shown in Fig. 2, a charge-coupled device (CCD) camera with a highest frame rate up to 8000 frames/s is used to catch the temporal evolution of many speckle spots simultaneously. The captured images are analyzed first by ensemble averaging over 130X30 pixels in two different frames taken at two different time t_i' and t_i'+t. A time average is then performed over 128 video frames, which takes 16 ms at frame rate of 8000 frames/s. Structure formation gives the evolution of anisotropic light transport after the field is applied and the chainlike structure begins growing. The method has the advantage that one can perform correctly both time and ensemble averages (the latter is the average over a large number of speckle spots), which enable us to do DWS experiments in nonergodic ER fluids. We also develop a set of software to get the time-resolved autocorrelation function.

When the electric field is zero, the particles disperse in ER fluid randomly, and the sample can be considered as an ergodic system. In this case, the DWS measurement using optical fiber and that using CCD have been compared with each other. Fig. 3 shows the comparison of autocorrelation function gained by these two methods. The decay time acquired using single-mode optical fiber and ALV correlator is about 22.0ms, while that using CCD, about 21.7ms, leading to a very good agreement with the warp less than 2%. This ensures that the CCD-based DWS measurement is reliable.

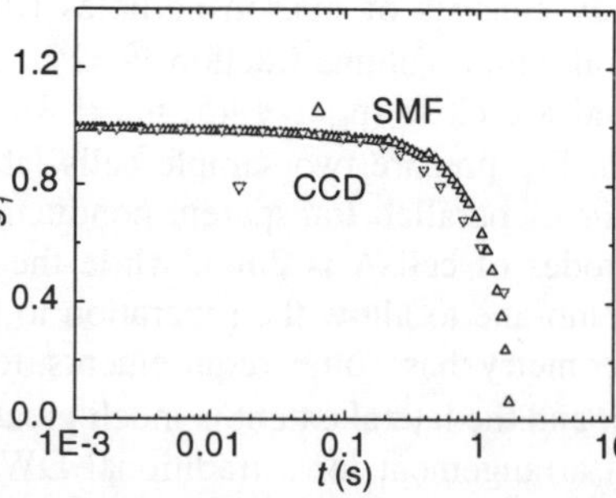

Figure 3. Intensity autocorrelation function $g_1(t)$ obtained by single mode optical fiber and CCD method. The decay time of $g_2(t)$ from single mode fiber is 22.0ms, while that from CCD DWS is 21.7 ms.

4 Results and Discussion

4.1 *Static DWS measurement*

Samples of ER fluids consisting of carbon doped TiO_2 particles and silicone oil were used in DWS study. Carbon doped TiO_2 is used because the TiO_2 has quite large dielectric constant (80-100), and carbon can be homogenously inserted into the pores of TiO_2 via organic cabonization at proper temperature (400℃ as we used). ER fluids consist of silicone oil and carbon-doped TiO_2 particles with different particle conductivity made with TiO_2 powders that have different composition of DDA. The shear stresses (at electric field of 1kV/mm) of the ER fluids at volume fraction of 10% were measured, and the particle conductivity (at frequency of 2kHz using HP-4284A LCR-meter) of a part of the ER fluids were also measured. DWS measurement of transmission geometry was performed with the TiO_2 ER fluids at volume fraction of 0.33%.

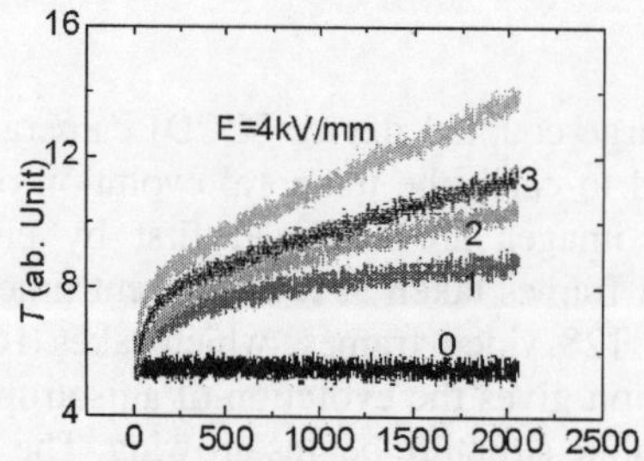

Figure 4.Time variation of transmittance at different peak to peak values of AC field

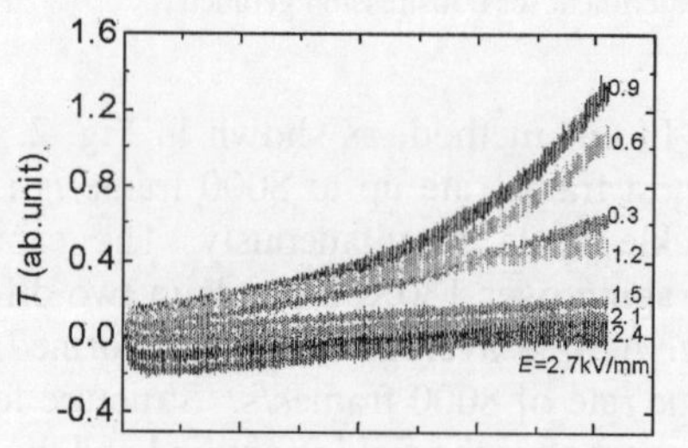

Figure 5. Time variation of transmittance at different DC field

Fig. 4 gives the time variation of light transmittence of the fluids for every 0.125ms after AC electric field is turned on which reflects the particle structure change of the system. The transmittence increases rapidly in about 100 millisecond, and then slows down. This reflects that the structure forms in two steps.

Fig. 5 displays the time variation of transmittance at different DC field of the sample with carbon content 2%. From the figure we found that, in the region of low electric field, the transmittance increases monotonically; in the high electric field, it increases after a period of decreasing. The increasing reflects the structure formation, and the period of decreasing is the result of the particle aggregation on the electrode surface. This phenomenon was observed only in the DC condition, but not in the AC condition, so we can draw the conclusion that the phenomenon results from the existence of net charges, the effect of the net charges in DC condition is much more important than that in AC condition.

Fig. 6 shows the time variation of diffusion coefficient using samples of different conductivity under a DC field of 600V/mm. One can find that there are more complicated behaviors than those in AC condition, while keeping the position order of the curves about same. Each curve in the figure has a valley and a peak, the peak is the result of the dipole-dipole interaction as same as in the AC condition. As shown in Fig. 8, as electric field is increasing, the peaks of the curves vanish step by step, and just the valleys survive. This agrees with the fact that the effect of the dipole moments is much smaller than that of net charges, so one can say that valleys result from the net charges.

Fig. 7 gives a time variation of diffusion coefficient, D, under AC electric field of 3000 V/mm peak-to-peak value using samples of different conductivity due to different carbon content. It displays that diffusion coefficient is higher for lower carbon content. Each of the three curves shows that D increases rapidly first after field is on, and followed by a sharp change in slope at about the same time when curve 3 in Fig. 4 is turning. So there is a peak in the curve. After the turning point in Fig. 7, the change of $D(t)$ for lower carbon content is faster than those have higher carbon content. D represents the kinetic properties of particles, when the structure formation is completed, D should be smaller than before. As the carbon content increases, the conductance increases, the formation time decreases, and the number of "free" particles (those that still exist in the system but are not strongly bound by the structure) decreases. So one can conclude that the peak is the result of the dipole-dipole interaction.

Fig. 8 shows time variations of diffusion coefficient, $D(t)$, of sample 1 under different electric field. There are valley for some of $D(t)$ curves, and this is quite different with the case for glass beads/silicone oil fluids. We found that the valley depths of $D(t)$ increase when electric field increases. This shows that the particle interaction increases with field as the diffusion coefficient reflecting the interaction. The $D(t)$ valley positions shift to shorter time when the field increases because response time gets shorter at higher field. Diffusion coefficient D reflects the particle velocity. The descending region of $D(t)$ shown in the figure implies that the particle interaction rises with time after electric field is applied, and reaches the maximum when $D(t)$ reaches its valley at certain time (valley time). The higher the field is, the stronger the particle interaction is, and the sooner the $D(t)$ reaches its valley position.

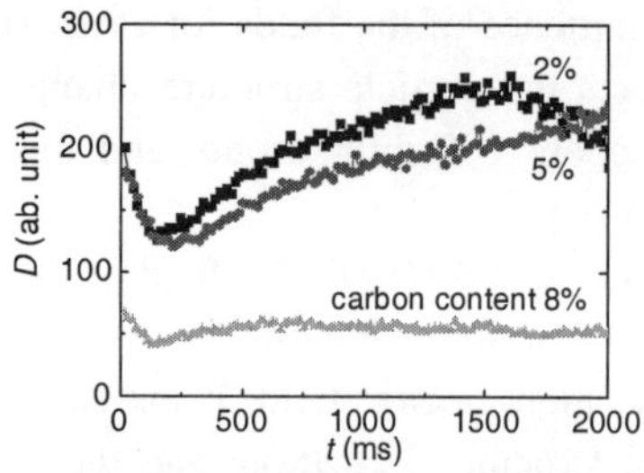

Figure 6. Time variation of diffusion coefficient using samples of different conductivity. (The DC field is 600V/mm)

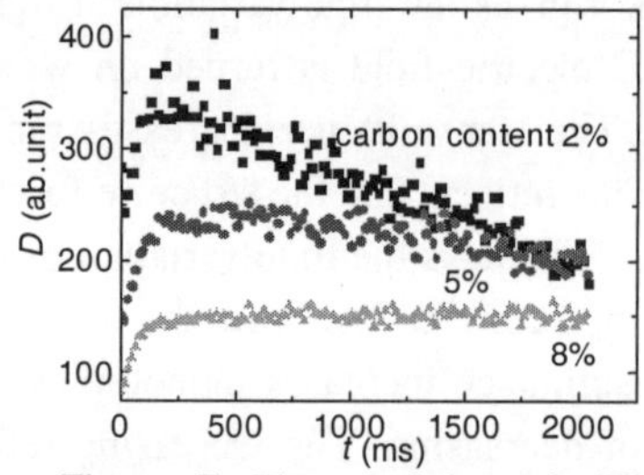

Figure 7. Time variation of diffusion coefficient using samples of different conductivity. (The p-p AC field is 3kV/mm)

4.2 *Dynamic DWS experiments*

We use a pump (Gilson, Minplus 3) to make the ER fluids flow in a cell at controllable shear rates. The laser illuminates on the flowing sample in the direction orthogonal to that of the shear, with parallel to that of electric field. For different shear rate, we collect the scattered light using CCD camera when AC voltage is applied. The behaviors of the ER fluid under shear make vital different with that of the quiescent fluid.

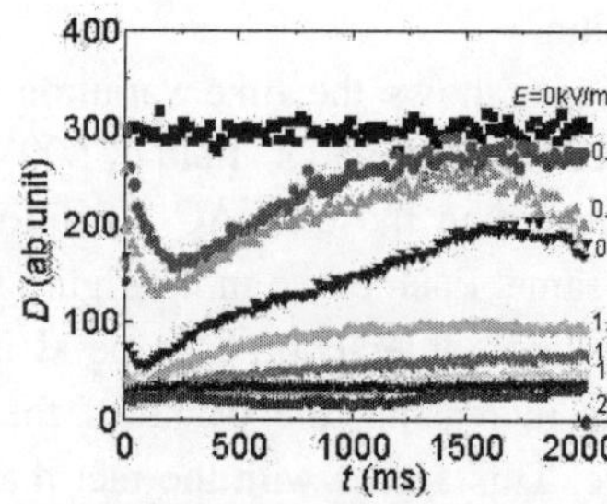

Figure 8. Time variation of diffusion coefficient at different DC field using ER fluid of carbon content of 2%

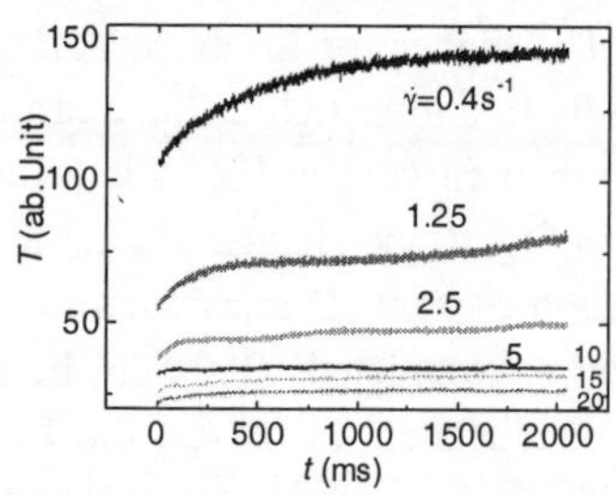

Figure 9. Time variation transmittance at different shear rate (at peak-to-peak AC electric field of 3kV/mm, carbon content of 5%).

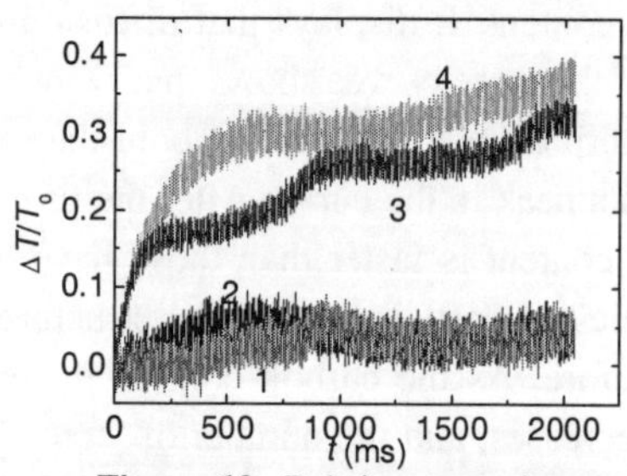

Figure 10. Relative transmittance vs. time. (sample 1: carbon content 2%, shear rate $2.5s^{-1}$; sample 2: 2%,1.25; sample 3:5%, 2.5; sample 4: 5%, 1.25)

Fig. 9 displays a time variation transmittance at different shear rate (at peak-to-peak AC electric field of 3kV/mm, carbon content of 5%). From the curve we can find that, with the increment of the flow rate, the transmittance decreases, in a fixed flow rate, the transmittance increases with time.

Fig. 10 displays the relative transmittance versus time. (sample 1: carbon content 2%, shear rate of $2.5s^{-1}$; sample 2: 2%,1.25 s^{-1}; sample 3:5%, 2.5 s^{-1}; sample 4: 5%, 1.25 s^{-1}). The higher the carbon content the samples have, the higher the transmittance is, and the less the time to get the saturated value of transmittance would be.

5 Conclusions

This work has developed a new method of CCD-based DWS measurement that is suitable for study of nonergodic systems such as ER or MR fluids under high fields. The method can be used to study short time variations of structure formation of the systems in both static and dynamic cases. Using CCD-based DWS method, we have studied the short time structure formation of carbon-doped TiO_2 ER fluids in both static and dynamic cases. For this ER fluids, whose carbon content is within 2 to 8 percent, there are at least two factors that affect ER process – one is from the net charges, and the other is from the dipole moments of the solid particles. For the sample we studied, in low DC field, both net charges and dipole moments are important; in high DC field, net charges dominate; and in AC field, only dipole moments are important. For the samples we studied, the samples with higher carbon contents, therefore, higher conductivity, form structure faster under the same shear rate and the same field, and also have more stable structure reflected from their higher transmittence. Further work should involve the DWS measurement with samples of much wide span of conductivity and shear rate to give a general idea of structure stability.

6 Acknowledgement

The authors would like to thank F. Lu for his cooperation and fruitful discussion. This work is supported by the National Natural Science Foundation of China under grant number 19834020.

References

1. Martin J. E. and Odinek J., Structure and dynamics of electrorheological fluids, *Phys. Rev. E* **57** (1998) pp. 756-775.
2. Klingenberg D. J., Zukoski C. F. and Hill J. C., Kinetics of structure formation in electrorheological suspensions, *J. Appl. Phys.* **73**(1993) pp. 4644-4648.
3. Ginder J. M., Diffuse optical probes of particle motion and structure formation in an electrorheological fluid, *Phys. Rev. E* **47** (1993) pp. 3418-3421.
4. Furst E. M. and Gast A. P., Particle dynamics in magnetorheological suspensions using diffusing-wave spectroscopy, *Phys. Rev. E* **58** (1998) pp. 3372-3376.
5. Menon N. and Durian D. J., Diffusing-Wave Spectroscopy of Dynamics in a Three-Dimensional Granular Flow, *Science* **275** (1997) pp. 1920-1922.
6. Romer S., Scheffold F. and Schurtenberger P., Sol-Gel Transition of Concentrated Colloidal Suspensions, *Phys. Rev. Lett.* **85** (2000) pp. 4980-4983.
7. Weitz D.A. and Pine D. J., Diffusing-wave spectroscopy. In *Dynamic Light Scattering, The method and some applications*, ed. by W. Brown W. (Clarendon, Oxford, 1993) pp. 652-719.
8. Romer S., Scheffold F. and Schurtenberger P., Sol-Gel Transition of Concentrated Colloidal Suspensions, *Phys. Rev. Lett.* **85** (2000) pp. 4980-4983.
9. Kirsch S., Frenz V., Schaertl W., Bartsch E. and Sillescu H., Multispeckle autocorrelation spectroscopy and its application to the investigation of ultraslow dynamical processes, *J. Chem. Phys.* **104** (1996) pp. 1758-1761.

MONODISPERSE COLLOIDAL SUSPENSIONS OF SILICA AND PMMA SPHERES AS MODEL ELECTRORHEOLOGICAL FLUIDS: A REAL-SPACE STUDY OF STRUCTURE FORMATION

ANAND YETHIRAJ AND ALFONS VAN BLAADEREN

Soft Condensed Matter, Utrecht University, and FOM Institute for Atomic and Molecular Physics, Amsterdam, The Netherlands. Email: yethiraj@amolf.nl, A.vanBlaaderen@phys.uu.nl

Colloidal particle coordinates in three dimensions can be obtained in 3D samples with a combination of the increased resolution and optical sectioning capabilities of confocal microscopy and fluorescently labeled model core-shell silica colloids. In this work we show how this capability can be used to analyze structure formation in electrorheological fluids on a quantitative basis. We find body-centered-tetragonal (BCT) crystals for colloidal particles in an electric field. Metastable sheet like structures were identified as an intermediate phase prior to BCT crystal formation. Due to finite-size effects induced by the electrode surface the sheets are not randomly oriented, but grow preferentially with a 60° tilt with respect to the electric field. Preliminary measurements indicate that flow-aligned sheets form under shear. Finally, we show that in the case that the ionic strength is very low, electric-field-induced dipolar interactions can be present in addition to long-range repulsions between the colloids leading to interesting metastable and equilibrium structures with possibilities for applications in photonic bandgap crystals as well as in model ER studies.

1 Introduction

Electrorheological (ER) [1] and magnetorheological (MR) fluids have been studied since the 1950's with a view to developing wide-ranging applications that relate to the reversible millisecond field-responsive control of the apparent fluid viscosity. Commercial ER and MR fluids are "messy" systems of odd-shaped and polydisperse sub-millimeter particles in a fluid suspension. The detailed physical mechanisms and the role of variables such as size and shape polydispersity, the role of a mobile surface-charge layer in the ER case, and the role of Brownian motion, are only now beginning to be understood (see [2] for a recent review). Moreover, while rheological measurements have yielded a lot of information, there is yet only limited information about the kinds of structures formed [3-5] and their dynamics. In this paper, we describe a technique and two model systems for studying the kinetics of the ER response.

2 Methods

Two model ER fluids are used in our studies. The first is a suspension of monodisperse 1.4 μm silica spheres, where the spheres have a fluorescent-labeled core surrounded by a non-fluorescent shell [6]. The solvent is a mixture of water and glycerol, with 16% water by weight, in order to match the refractive index of the silica spheres, enabling imaging in the single-scattering limit. The inter-particle interactions here are known to be hard-sphere like in the absence of the field. The second system is a suspension of 2 μm polymethyl methacrylate (PMMA) fluorescent spheres [7], sterically stabilized with a poly-hydroxysteric acid (PHSA) surface coating, in an organic solvent (a mixture of cyclo-heptyl bromide and cis-decalin, where simultaneous matching of refractive-index and density to that of PMMA is possible) where the inter-particle interactions are soft due to the combination of charge on the sphere surface and a large Debye length in the solvent.

Images are obtained via fluorescent confocal laser scanning microscopy, three-dimensional structure information (see Ref. [8] for details) of a sample volume of 100 μm x 100 μm x 50 μm in several seconds, and two-dimensional structure information in a sample area of 100 μm x 100 μm on the timescale of a second.

The field is varied up to E_{peak} =1 V /μm (E_{rms} = 0.707*E_{peak}) in the silica system and 2 V/μm in the PMMA system, while the frequency is maintained at 1 MHz to be well above the low-frequency electrohydrodynamic effects associated with the double layer. The applied field is monitored with an oscilloscope. The electrode gap is maintained at about 100 μm. While the silica spheres and the solvent are index-matched at light frequencies, the dielectric constant mismatch at 1MHz is still essentially the same as at zero frequency and enough to explain the strong ER effect. In the PMMA suspension, the zero-frequency dielectric mismatch is an order of magnitude lower, and the mechanism for the relatively strong ER response reported here is as yet unknown.

3 Results

3.1 The ER Effect in The Hard-Sphere System: Strings, Two-Strings, Sheets and BCT Crystallites.

In the silica/water-glycerol system, we observe the almost instantaneous formation of strings (timescale ~ milliseconds) and the subsequent slower formation of sheets and BCT crystals (with the shortest interparticle spacing in the c direction and a:c≡ b:c=.√3/√2; first predicted as a minimum energy structure in Ref. 9). This has been reported in an earlier work [5]. Time scales associated with the aggregation of columns is expected [10] to be on the order of $\tau \sim (\eta/k_BT)\, r_\perp^3 \sim 15$ sec. for single-chain columns of 1.4 μm diameter spheres, given the 230 mPa.s viscosity of the index-matched water-glycerol mixture. The sedimentation rate of silica spheres in the viscous solvent mixture is on the order of μm/minute. This is consistent with the calculated sedimentation velocity
$v = mg/\gamma_0 = (0.1\text{x } 10^{-12})/(5 \text{ x } 10^{-6}) \sim 0.02$ μm/sec, where $\gamma_0 = 6\pi\eta R$ is the viscous drag coefficient.

Unless specifically stated, all images shown in this work are XY slices in the mid-

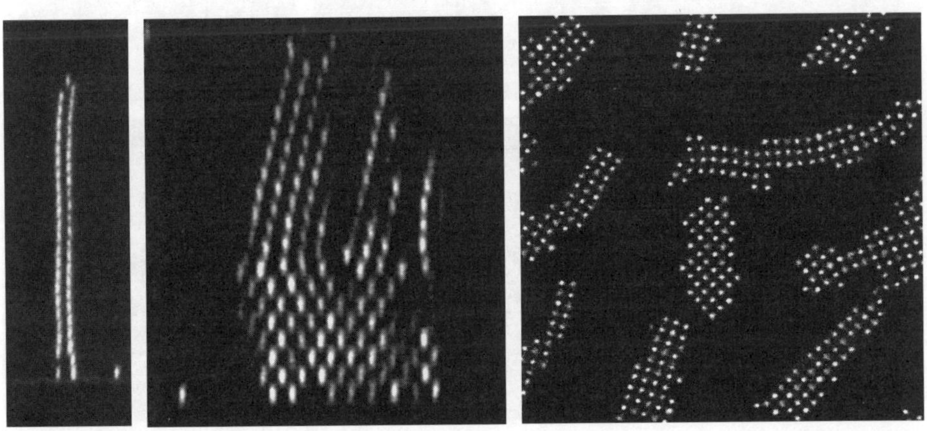

Figure 1(a) x-z of two-string (b) x-z of sheets in tilt transition (c) x-y of BCT crystallites. Particle volume fraction φ≈ 15% ; E_{peak} =1 V /μm.

plane of the sample, perpendicular to the applied electric field (along Z). Thus, a string appears as a single particle, a sheet looks like a chain of particles, and the BCT structure looks like two interpenetrating square lattices of more-in-focus and less-in-focus spheres. The formation of sheets is particularly interesting. First a string attracts a neighbor to form a two-string. Two-strings are typically not vertical but slightly bent near the top electrode (see Figure 1a). The two-strings further coarsen to form sheets. As reported in Ref. 5, the sheets are composed of 60°-tilted strings. This effect arises from the fact that while the bulk dipolar interaction is insensitive to the angle of the hexagonal sheet with respect to the field direction, the tilted sheets are energetically favorable due to boundary effects. In fact, we see a transition from vertical strings, to sheets composed of tilted strings and finally to a BCT structure (Fig. 1c) composed of vertical strings. Fig. 1(b) captures this tilt transition.

3.2 Coarsening of Structures: The Time-Evolution in the Formation of the BCT crystallites

The time evolution of the structures is shown in two-dimensional snapshots in Fig. 2, all taken in a plane perpendicular to the field direction. The time sequence shows snapshots taken at t=0 s, 1s, 100s, 400s, 2400s, 36000s, respectively. Sheet formation has already begun in Fig. 2(b), while a few small BCT crystallites can be in seen in Fig 2(c) in the form of right-angle kinks in the meandering sheet structure. Further structure formation is extremely slow, and the structure at 36000 s (10 hours) is not significantly different from that at 180000 s (~3 days). In fact, because there is no further significant time-evolution between 10 hours and 3 days, the data therein may be used to study fluctuations in the final structure, which is a network of three-dimensional BCT crystallites. A pictorial representation of this is provided in Fig. 3. Fig 3(a) is a snapshot of the final structure,

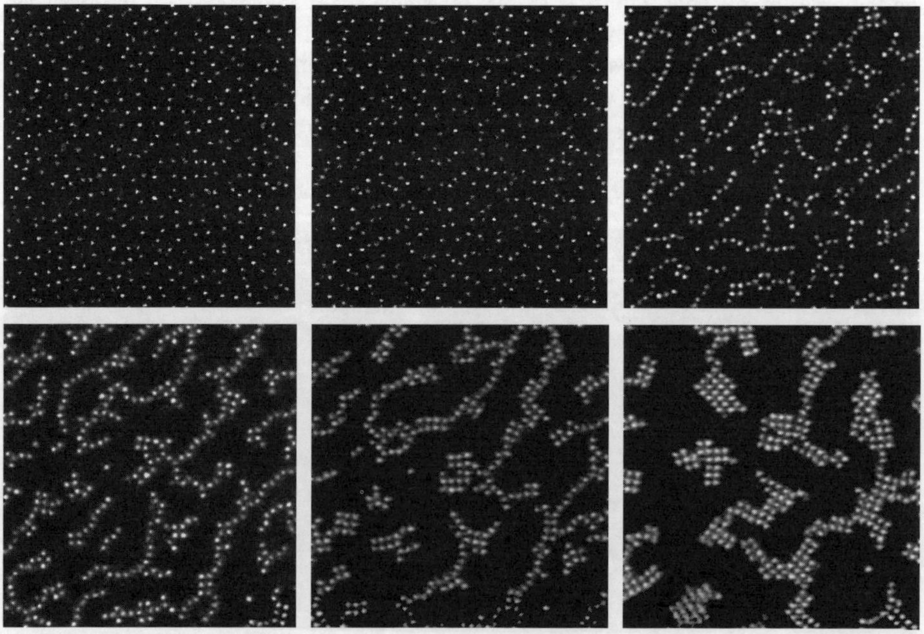

Figure 2 Coarsening of structures beginning from (a) t=0s, (b) t=10 s, (c) t=100s, (d) t=400s, (e) t=2400, f) t=36000s. The coarsening diplays an evolution from strings to sheets to a network of BCT (body-centred cubic) crystallites. $\varphi \approx 15\%$, $E_{peak} = 1 V/\mu m$.

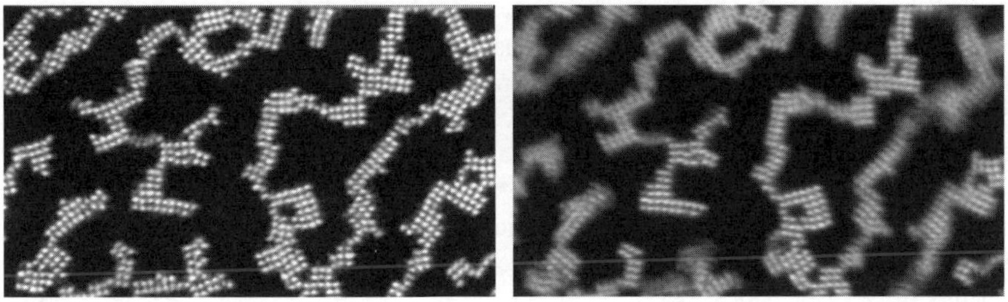

Figure 3 (a) Snapshot of the end state of coarsening observed over three days. (b) An average of the time series of 90 frames once coarsening is too slow to observe exhibits that the large crystallites fluctuate very little, while the single sheets fluctuate the most. $\varphi \approx 15\%$,; $E_{peak} = 1$ V /μm.

while Fig. 3(b) is an average from a time series of 87 frames. If a string (seen in the xy image as a sphere) moves little during these frames its average image is not much different from that in a snapshot. However, if its position fluctuates the corresponding region in the average image is blurred. Comparing Fig. 3(a) and 3(b) we notice that the thick columns fluctuate the least (are sharpest defined in the average image) while the single sheets are the most blurred. In order for a structure to evolve, it must sample the space of possible structures, and thus it will take a long time (weeks!) for the small 5-particle-wide BCT crystallites to coarsen further.

3.3 Tuning interactions: ER Effect in the Soft Sphere System

Suspensions of PMMA spheres in the refractive-index and density-matched solvent-mixture of cyclo-heptyl bromide and decalin exhibit a surprisingly strong ER response. At 1 V/μm, the system forms strings that completely span the electrodes (Fig. 4(a)). In contrast with the "hard-sphere" system reported above, and in Ref. 5, these strings in this

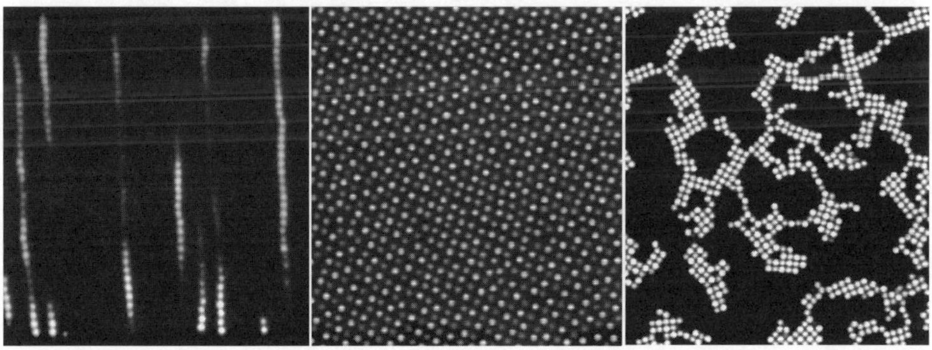

Figure 4 (a) String formation; $\varphi \approx 10\%$, $E_{peak} = 1$V/μm (XZ snapshot) (b) A space-filling tetragonal crystal; $\varphi \approx 25\%$, $E_{peak} = 1$V/μm. (c) Network of BCT crystallites; $\varphi \approx 25\%$, $E_{peak} = 2$ V/μm.

field regime are stable, and do not in time agglomerate to form sheets. As the concentration is increased the strings that form do eventually order in a quasi-two-dimensional fashion (the third dimension already being ordered by the formation of strings) to form space-filling crystals, such as the tetragonal crystal shown in Fig 4(b). This is different from the BCT crystals reported earlier in that these crystals do not have the ratio a:c≡ b:c=.√3/√2. *Instead this ratio is determined purely by the constraint of filling space.* The likely reason for this is that charge repulsion of the spheres at low fields is still strong compared to the dipolar interaction. As the field is increased, the dipolar interaction first wins along the field direction, but due to its anisotropic nature, there is a range of field values where the soft repulsion dominates in the plane perpendicular to the field. At a higher field of ~ 2 V/μm, these space-filling structures eventually give way to the expected BCT crystallites characteristic of the dipole-interaction dominated regime (Fig. 4(c)). The BCT crystallites seen here are very similar to the ones observed in the aqueous silica suspensions.

3.4 The Effect of Shear

Next, we explore the effect of shear. Preliminary experiments reveal a qualitative feature of shear: the lining up of sheets along the shear field. There also appears to be a characteristic spacing between sheets that appears at the onset of shear. This is seen in both model systems (Fig. 5). The spacing in both cases is set simply by the particle volume fraction, or when seen in a two-dimensional snapshot (the third dimension being irrelevant because the sheets span the entire region between the electrodes), the particle area fraction which in turn determines the number of particles in focus un the field of view. The applied shear aligns the sheets along the shear direction. The total number of sheets (and hence the spacing) is simply set by the total number of particles in the field of view.

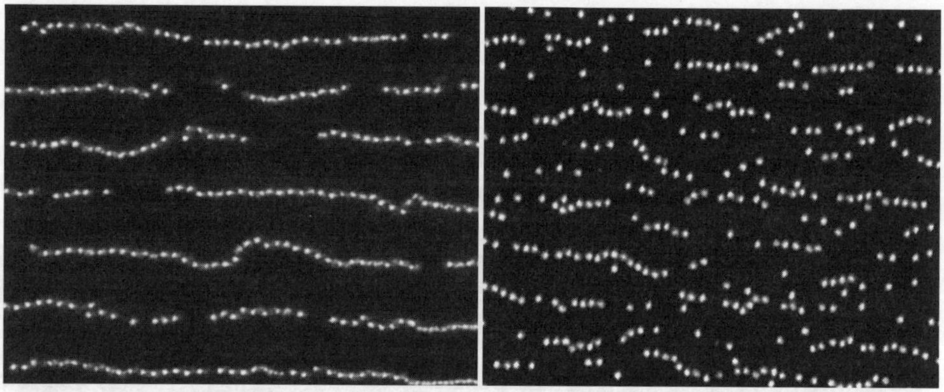

Figure 5 Sheets after shear (a) aqueous silica system (b) non-aqueous PMMA system. E_{peak} = 1V/μm.

4 Discussion and Conclusions

We have explored two systems in this study, both versatile systems for studying ER mechanisms. The first is, in zero field, a suspension of hard spheres. This system behaves much like commercial ER fluids, except that the appreciable size polydispersity in the latter likely prevents the formation of the three-dimensional BCT crystal. Here we observe the transition from strings to two-strings, sheets (with strings tilted 60°), double-sheets, and finally crystallites of the BCT structure, which coarsen extremely slowly in time. The second system, which in zero field is a charged suspension with soft, repulsive interactions, exhibits a novel intermediate-field regime where the strings remain individual strings and do not agglomerate into sheets, as well as new crystal structures that are of interest too in engineering photonic band gap crystals [11], and in the more pedagogical study of a martensitic crystal-to-crystal transition. At high fields, this system also exhibits the string-to-sheet-to-BCT coarsening, establishing qualitatively similar structure formation.

Acknowledgements

We acknowledge Hans Wisman (electronics), Jacob Hoogenboom and Gilles Bosma (silica and PMMA particle synthesis respectively). This work was supported by the Stichting voor Fundamenteel Onderzoek der Materie (FOM), which is part of the Nederlandse Organisatie voor Wetenschappelijk Onderzoek (NWO).

References

1. Winslow W. M., *J. Appl. Phys.* **20**, 1138 (1949).
2. Parthasarathy M and Klingenberg D, *Mater. Sci. Eng.* R. **17**, 57 (1996).
3. Martin J. E., Anderson R. A., and Tigges C. P., *J. Chem. Phys.* **108**, 3765 (1998).
4. Martin J. E., Anderson R. A., and Tigges C. P., *J. Chem. Phys.* **110**, 4854 (1999).
5. Dassanayake U., Fraden S., and van Blaaderen A., *J. Chem. Phys.* **112**, 3851 (2000).
6. van Blaaderen A. and Vrij A., *Langmuir* **8**, 2921 (1992).
7. Bosma G., Pathmamanoharan C., de Hoog E. H. A., Kegel W. K., van Blaaderen A., Lekkerkerker H. N. W., *Langmuir*, accepted for publication; de Hoog E. H. A., Kegel W. K., van Blaaderen A, Lekkerkerker H. N. W., *Phys. Rev. E.* **64,** 1407 (2001).
8. van Blaaderen A.and Wiltzius P., *Science* **270**, 1177 (1995).
9. Tao R. and Sun J. M., *Phys. Rev. Lett.* **68**, 2555 (1992).
10. Halsey T. C. and Toor W., *Phys. Rev. Lett.* **65**, 2820 (1990).
11. van Blaaderen A., Velikov K. P., Hoogenboom J. P., Vossen D. L. J., Yethiraj A., Dullens R., van Dillen T., and Polman A., In: C.M. Soukoulis (ed.), "Photonic Crystals and Light Localization in the 21st Century ", pp.239-251 (Kluwer Academic, Dordrecht, 2001).

POLARIZATION OF NON-EQUILIBRIUM DOUBLE LAYER AND AGGLOMERATION OF POLYELECTROLYTE BALLS

A.CEBERS
Institute of Physics, Salaspils, LV-2169, Latvia
E-mail: aceb@sal.lv

I.RUBINSTEIN
Blaustein Institute for Desert Research, Ben-Gurion University of the Negev, Sede Boqer Campus, 84990, Israel
E-mail:robinst@bgumail.bgu.ac.il

Non-equilibrium polarizability of a polyelectrolyte ball is calculated in the framework of a quasi-onedimensional model, taking into account the concentration polarization phenomenon. This non-equilibrium polarizability is shown to be proportional to the volume of a polyball and to have a negative sign. Based on these results, the available experimental data on the aggregation phenomena in DNA solutions are considered in terms of the electric field induced phase separation. Numerical estimates for the critical electric field strength necessary for the aggregation to take place and the frequency range where it is observed are in a reasonable agreement with experiment. In the framework of this model, the internal circulations observed in the DNA aggregates are interpreted as being due to the spontaneous rotation of polyelectrolyte balls with a negative polarizability.

1 Introduction

Studies of various magnetizable and polarizable particle systems have shown the possibility of specific field induced phase separation [1], and the importance of the flows caused by the particle rotation due to the finite polarization relaxation time [2]. On the other hand, quite intriguing phenomena induced by the applied electric fields have been reported for polyelectrolyte, e.g., DNA, solutions [3,4]. In these studies, formation of aggregates of rotating particles with the internal liquid circulation has been observed. In this presentation, based on our understanding of the nonequilibrium electric double layer, developing at a polyelectrolyte particle in the course of concentration polarization of the surrounding electrolyte solution by an external electric field [5], an attempt is made to interpret these phenomena as the electric field induced phase separation. The possible link of the formation of aggregates with the internal circulation due to the finite relaxation time of the nonequilibrium double layer is also discussed.

2 Concentration polarization and nonequilibrium double layer

We propose the following model of polarization of a polyelectrolyte ball by an external electric field. Let the polyelectrolyte ball be negatively charged and, thus, impermeable for anions. On the other hand, the cations driven by the external electric field, may penetrate the ball , which, thus, acts for them as an ideal cation-exchange membrane under electric current. The solution becomes depleted of the electrolyte at the current entrance side of the ball and enriched at the exit from it a phenomenon known as concentration polarization in electrodialysis [5]. For an electric field strong enough, when the interface electrolyte concentration approaches zero, a nonequilibrium electric double layer developes at the depletion side of the particle [6]. This double layer is typified by the formation of an extended space charge region, reaching into the solution for a distance of up to a few hundred angstroms, on top of the usual quasiequilibrium diffuse space charge layer, adjacent to the interface and typically extending to up to a few tens of angstroms only [5-8]. This extended space charge causes the so called electrokinetic effects of the second kind [9], and the related instabilities of the diffusion layer at the ion exchange membrane [5,10]. Our basic idea here is that, due to its large size, compared to that of its quasiequilibrium counterpart, the extended nonequilibrium space charge at the depleted side of the ball will dominate its dipole moment induced by the external electric field.

To estimate this dipole moment, we use the following approximate model (see [5] for the details of the extended space charge calculations). Let the halfspace $y < 0$ be filled by a polyelectrolyte. Let the potential of the polyelectrolyte be $-V$. Equations for steady state electrodiffusion of univalent ions are

$$D\Big(-n_{+,y}-\frac{e\phi_{,y}}{k_BT}n_+\Big)_{,y}=0 \tag{1}$$

$$D\Big(-n_{-,y}+\frac{e\phi_{,y}}{k_BT}n_-\Big)_{,y}=0 \tag{2}$$

$$dE_{,y}=4\pi e(n_+-n_-) \tag{3}$$

$$E_{,y}=-\phi_{,y}$$

Here, d is the dielectric permeability of the liquid, and, for simplicity, equal diffusivities D have been assumed for cat- and anions.

Let us use, for the characteristic length scale L, the thickness of the diffusion layer at the particle (this will be later identified with the radius of the polyelectrolyte ball). Thus, the equations are put into a dimensionless form by introducing the following scales: potential - k_BT/e; concentration - n_0; length- L; electric field strength - $4\pi en_0r_d/d$ with $r_d=\sqrt{\frac{dk_BT}{4\pi e^2n_0}}$ - Debye screening length; electric current density - en_0D/L. As a result, we arrive at the following set of the dimensionless equations

$$(-n_{+,y}-\phi_{,y}n_+)_{,y}=0 \tag{4}$$

$$(-n_{-,y}+\phi_{,y}n_-)_{,y}=0 \tag{5}$$

$$\varepsilon E_{,y}=n_+-n_- \tag{6}$$

$$E = -\varepsilon\phi_{,y} \tag{7}$$

Here the dimensionless Debye length $\varepsilon = \frac{r_d}{L}$ is the small parameter.

Since the polyelectrolyte is negatively charged, the flux of the anions is assumed zero

$$-n_{-,y} + \phi_{,y}n_{-} = 0 \tag{8}$$

Integration of (4) gives

$$-n_{+,y} - \phi_{,y}n_{+} = j \tag{9a}$$

which yields, using (6)-(8),

$$n_{+,y} = -\frac{1}{2}j + \frac{1}{2}EE_{,y} + \varepsilon\frac{1}{2}E_{,yy} \tag{9b}$$

Integrating (9b), we obtain

$$n_{+} = -\frac{1}{2}j(y - y_0) + \frac{1}{4}E^2 + \frac{1}{2}\varepsilon E_{,y} \tag{10}$$

Substitution of (10) into (9a) yields the following equation for the electric field strength

$$\varepsilon^2 E_{,yy} = -\varepsilon j - Ej(y - y_0) + \frac{1}{2}E^3 \tag{11}$$

Here y_0 is to the leading order the coordinate of the outer edge of the extended space charge zone. Seeking a solution of (11) for small ε, we point out that, whenever $y_0 - y \gg \varepsilon^{2/3}$, the second and third terms in the equation (11) are balanced, which yields for the scaled electric field strength

$$E = -\sqrt{2|j|(y_0 - y)} \tag{12}$$

Since the concentration of cations in the extended space charge zone is small, practically all the potential falls on this region and, thus, y_0 may be found from the condition

$$y_0^{2/3} = \frac{3\varepsilon V}{2\sqrt{2|j|}}$$

The dipole moment per unit surface of the interface on the depletion side of the particle thus is

$$p = L^2 \int_0^{y_0} \rho y \, dy$$

where the dimensional charge density ρ in the extended space charge region is $en_0\varepsilon E_y$. By integration, we obtain

$$p = en_0\varepsilon^2 L^2 V$$

Finally, the total dipole moment induced on the depletion side of the particle is obtained by multiplying this value by the projected area of the sphere $\pi L^2 \quad (a = L)$ which yields

$$P = \pi e n_0 r_d^2 a^2 V$$

Relating the potential drop V to the electric field E_0 near the particle as $V = -\frac{eE_0 a}{k_B T}$ and substituting the expression for the Debye screening length, we finally obtain

$$P = -\frac{1}{4} d a^3 E_0$$

We notice that the induced dipole moment is proportional to the volume of the ball and opposite to the electric field strength.

3 Phase diagram

The calculated polarizability of a polyball, combined with the mean field model of phase separation due to dipolar interactions [1], allows us to predict the critical value of the electric field above which a system of polyballs is unstable. In terms of this mean field model, the polarizability of a suspension of polyballs is expressed as

$$\kappa = \frac{n\kappa_0}{1 - \frac{4\pi n \kappa_0}{3d}} \tag{13}$$

Here, $\kappa_0 = -\frac{1}{4} d a^3$ - polarizability of a single particle, n - number of balls per unit volume.

General expressions for the free energy per unit volume and the chemical potential in an electric field read [11]

$$f = f_0(n, T) - \frac{1}{2}\kappa E^2 \tag{14}$$

$$\psi = \psi_0(n, T) - \frac{1}{2}\frac{\partial \kappa}{\partial n} E^2 \tag{15}$$

Accounting for the excluded volume effect for the field independent parts of free energy and chemical potential (14), (15) by means of the Van der Waals approach yields ($v_0 = 4v = \frac{16\pi a^3}{3}$)

$$f_0 = k_B T \ln\left(\frac{n v_0}{1 - n v_0}\right) \tag{16}$$

and

$$\psi(n, T) = k_B T \ln\left(\frac{n v_0}{1 - n v_0}\right) + \frac{k_B T}{1 - n v_0} \tag{17}$$

The relations (14) - (17) yield for the osmotic pressure of the suspension

$$p = \frac{k_B T n}{1 - n v_0} + \frac{1}{2}\left(\kappa - n\frac{\partial \kappa}{\partial n}\right) E^2 \tag{18}$$

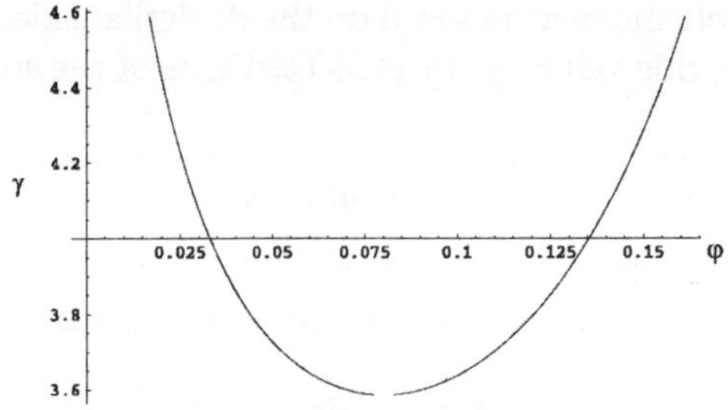

Figure 1. Phase diagram of the suspension of polyballs

The first term in the expression (18) corresponds to the osmotic pressure of an ideal mixture, the second term comes from electrostriction of the suspension due to interaction between the particles. Accounting for the particles interaction in the frame of the mean field model amounts to substitution of (13) into (15), (17), (18), which yields

$$p = \frac{k_B T}{v}\Big(\frac{\varphi}{1-4\varphi} - \gamma\frac{4\varphi^2}{(1+\varphi/4)^2}\Big) \tag{19}$$

$$\psi = k_B T\left(\ln\left(\frac{4\varphi}{1-4\varphi}\right) + \frac{1}{1-4\varphi} + 16\gamma\frac{1}{(1+\varphi/4)^2}\right) \tag{20}$$

Here $\gamma = \frac{\kappa_0^2 E^2}{d(2a)^3 k_B T}$ is the dipolar interaction parameter and $\varphi = n\frac{4\pi}{3}a^3$ is the volume fraction of polyballs. With the equations of state (19) and (20), a homogeneous state of the system turns out to be unstable with respect to phase separation, that is the system splits into spatial regions of two types, with polyball volume fractions φ_1 and φ_2, respectively. These latter values may be found from the phase equilibrium conditions

$$p(\varphi_1) = p(\varphi_2), \quad \psi(\varphi_1) = \psi(\varphi_2) \tag{21}$$

Some remarks are due concerning the applicability of the equilibrium conditions to the phases containing polyballs with nonequilibrium double layers. Justification stems from the following reasoning: the second condition follows from the absence of net transfer of particles between the phases; the first condition comes from the mechanical equilibrium. Indeed, it is known from electrodynamics that the force on a polarized medium due to inhomogeneity of the effective local field may be expressed as follows (a constant external field is assumed) [11]:

$$\vec{f}_e = (\vec{P}\nabla)\vec{E}_e = \frac{1}{2}n\frac{\partial^2\kappa}{\partial n^2}\nabla n E^2$$

The latter expression may be rewritten as

$$\vec{f}_e = \frac{1}{2}\nabla\left(n\frac{\partial\kappa}{\partial n}\right)E^2 - \frac{1}{2}\nabla\kappa E^2$$

which yields the following force balance equation

$$-\nabla p + \vec{f}_e + \vec{f}_s = 0 \tag{22}$$

Here $\vec{f}_s$ is the contribution to the force due to the nonuniformity of concentration. When this contribution has a potential, integration of (22) yields for mechanical equilibrium, assuming vanishing surface forces in the homogeneous phases,

$$p_0 + \frac{1}{2}\left(\kappa - n\frac{\partial \kappa}{\partial n}\right)E^2 = Const$$

that is, the first of the equations (21). Phase diagram calculated from the equations (21), taking into account the relations (19), (20), is presented in Fig.1. The critical point of the phase transition may be found analytically, yielding following critical values for polyboll volume fraction and the dipolar interaction parameter

$$\varphi_c = 2/25\,,\ \gamma_c = \frac{(1+\varphi_c/4)^3}{8\varphi_c(1-4\varphi_c)^2}$$

4 Discussion

According to the results of the two previous sections, the suspension of polyballs becomes unstable with respect to phase separation in an electric field above some critical value.

Below, are some estimates and suggestions concerning the applicability of our model to interpretation of experimental data on phase separation in DNA solutions. Refs. [3], [4] contain experimental data on the critical field for phase separation of the λ bacteriophage DNA (48.5 kb pairs) with concentration 10 $\frac{\mu g}{ml}$. Using the above relation for the dipolar interaction parameter γ and the calculated value of particle polarizability, and assuming that the aforementioned concentration corresponds to a volume fraction of polyballs close to the critical value φ_c (the radius of a polyball estimted at 0.45 μm, in a good accord with the gyration radius of λ DNA as determined by the light scattering [12]) we obtain for the critical value of the electrical field 470 V/cm a value sufficiently close to that of about 200 V/cm given in [4].

Another aspect of application of the present model to DNA suspensions concerns the observed rotation of individual particles in the agglomerates, which results in an internal circulations in them. Since, as shown above, the susceptibility of a single particle is negative, its rotation has a natural explanation in terms of the electrorotation model of Ref [2]. According to this model the polarization relaxation equation reads

$$\frac{\partial \vec{P}}{\partial t} = [\vec{\Omega} \times \vec{P}] - \frac{1}{\tau}(\vec{P} - \kappa_0 \vec{E})$$

Here Ω is the angular velocity of a particle, and τ, the characteristic time of the polarization relaxation, is estimated as $\tau = \frac{a^2}{D}$ (D - ionic diffusivity). This equation, with $\vec{\Omega}$ related to $\vec{P}$ and $\vec{E}$ through the torque balance on the particle

$$\alpha\vec{\Omega} = [\vec{P} \times \vec{E}]$$

allows us to estimate the critical value of the electric field above which a spontaneous rotation of particles sets on:

$$E_{rc}^2 = -\frac{\alpha}{\tau \kappa_0}$$

Here $\alpha = 8\pi\eta a^3$ is the rotational friction coefficient of a spherical particle. We point out that the frequency related to the relaxation time τ ($\nu \cong 1/\tau \approx 10^2$ Hz for $a = 0.5\,\mu$m, $D = 10^{-6}\,\mathrm{cm}^2/\mathrm{sec}$) compares well by the order of magnitude with the experimental value ≈ 200 Hz at which the increase of the critical electric field strength for phase separation is observed [4]. For the same values of physical parameters as in the above estimate of the critical field strength for phase separation, we arrive at the threshold value of the electric field strength 740 V/cm, above which spontaneous rotations of particles should occur. This value is too within the range of the electric field strength at which the formation of aggregates with internal circulations is observed [3].

5 Conclusions

Numerical estimates based on the proposed model of polarization of polyballs by the external electric field are compatible with the existing experimental data on the behaviour of DNA suspensions. If correct, this model suggests some intriguing questions concerning the phase behaviour of a system of rotating polyballs and formation of their agglomerates with internal circulation.

References

1. A. Cebers, Magnitnaya Gidrodinamika **18**, 42 (1982)
2. A. Cebers, E. Lemaire, L. Lobry, Magnitnaya Gidrodinamika **36**, 347 (2000)
3. L. Mitnik, C. Heller, J. Prost, J. L. Viovy, Science **267**, 219 (1995)
4. S. Magnusdottir, H. Isambert, Ch. Heller, J. -L. Viovy, Biopolymers **49**, 385 (1999)
5. I. Rubinstein, B. Zaltzman,in: *Mathematical Models and Methods in Applied Sciences*11263 2001
6. I. Rubinstein and L. Shtilman, J.Chem. Soc., Faraday Trans. **2**, 75 (1979)
7. V. V. Nikonenko, V. I. Zabolotsky, and N. P. Gnusin, Elektrokhimia **25**, 301 (1989)
8. A. V. Listovnichy, Elektrokhimia **25**, 1651 (1989)
9. S. S. Dukhin, Adv. Colloid Interface Sci. **35**, 173 (1991)
10. I. Rubinstein and B. Zaltzman, Phys.Rev.E **62**, 2238 (2000)
11. L. D. Landau, E. M. Lifshitz and L. P.Pitaevski, *Electrodynamics of continous media*(Pergamon Press, Oxford, 1984) p.47
12. R. Verma, J. C. Crocker, T. C. Lubensky, A. G. Yodh, Phys.Rev.Lett. **81**, 4004 (1998)

ELECTRIC-FIELD-INDUCED LAMELLAR STRUCTURES IN MAGNETIC FLUIDS– A 2D DIFFUSION MODEL

BRETT RILEY, ANIKET BHATTACHARYA, MICHAEL JOHNSON, XIAODONG DUAN, and WEILI LUO
Department of Physics, University of Central Florida,
Orlando, Florida, 32816, USA

A lamellar pattern can form in a thin magnetic fluid layer when the applied electric field is above a critical value. A 2D simulation is performed to study the field-induced phase separation and the pattern by using the mass continuity equation. The simulation produces the similar structure in field but does not match the experimental growth law.

Most studies of field-induced structures in magnetic fluids have investigated the role of magnetic fields. Much less is known about the effects of electric fields applied to magnetic fluids[1]. Recent experiments have shown that a DC electric field can induce phase separation and lamellar structures in magnetic fluids[2]. In order to understand the physical mechanism for this behavior, we have carried out computer simulations on a lattice gas model, which is based on a two dimensional diffusion equation.

We begin our study from the experimental observations that the lamellar structure is observed in very thin cells with an electric field imposed perpendicular to the ferrofluid layer, and is associated with the migration of magnetic particles within the fluid[1]. In the presence of a uniform electric field E, at temperature T, the free energy density of the system can be written as

$$F=F_{bulk} + F_{inhomo}, \text{ with } F_{bulk}= U_E - TS \text{ and } F_{inhomo} = k\ \nabla^2 C. \qquad (1)$$

Here U_E is the electrostatic energy contribution due to the electric field, and is given by:

$$U_E = -\ 1/2\ \ \varepsilon(C)\ E^2, \qquad (2)$$

where the dielectric constant $\varepsilon(C)$ depends on the local particle concentration, $C(\boldsymbol{r})$ in a nonlinear fashion. (The negative sign in U_E is a consequence of choosing the electric field as a thermodynamic variable.) For our choice of $\varepsilon(C)$, we use the expression

$$\varepsilon(C) = exp\ [\ C\ ln\ \varepsilon_p + (1 - C)\ ln\ \varepsilon_f\ ] \qquad (3)$$

which has been used previously to study phase separation process under the influence of an electric field[2].

For the entropy, S, we use the entropy of mixing from a lattice-gas model for a binary system.

$$S = (k_B / v)T [C \ln C + (1 - C) \ln (1 - C)] \quad (4)$$

Here v is the average correlation volume. (For a non-interacting system, v is the volume of a single particle). The inhomogeneous term F_{inhomo} represents the energy of interfaces which results from phase separation.

The time-dependent Ginzburg-Landau equation for the concentration, C(**r**, t), can be written as

$$dC/dt = M \nabla^2 \mu, \quad (5)$$

where M is the mobility and the local chemical potential $\mu(\mathbf{r})=\delta F/\delta c(\mathbf{r})$ is obtained from the free energy functional above, which results in the following equation:

$$dC/dt = \nabla^2(-R \exp(\ln(\varepsilon_p/\varepsilon_f)\, C)\, E^2 + \ln (C / (1-C)) - \nabla^2 C) \quad (6)$$

Both length and time are scaled by λ and τ which are defined as:

$$\tau = v^2 K / M (k_B T)^2 \quad (7)$$

$$\lambda^2 = K\, v / k_B T, \quad (8)$$

and the control parameter R is given by:

$$R = (\varepsilon_f \ln(\varepsilon_p/\varepsilon_f) E^2\, v / 2\, k_b t) \quad (9)$$

Temperature is held constant at T=298 K. The two main variables used are voltage and time; voltage is used in place of electric field, assuming a separation between plates of d ≈ 10 μm. For the dielectric constants we use $\varepsilon_p = 20$, the value for magnetite particles; and $\varepsilon_f = 2$, the approximate value for kerosene at room temperature. For the average correlation volume, we choose it with a radius of 10 times the average particle radius as found from other calculations [3]. The average concentration is set to 7% to match experiment. The surface energy constant, K, is set to one as it is included with the scaling factors of the length and time. As is apparent from the control parameter, the value of K does not affect the critical values for phase separation. However, for direct comparison with experiment, the value of K is needed to convert time and distance from scaled to real units.

The initial configuration at t =0 is drawn from a random fluctuation of 1% around the mean concentration. We observe phase separation beyond a critical value of the electric field, E_c, consistent with experimental result[4, 5]. Below the critical field, the system always returns to the same original local concentrations, even if there is a momentary surge away from the average. At equal to or above the critical value, the deviation of local concentration from the average grows with time.

Figure 1 and figure 2 show the time development of the lamellar pattern. The dark/light regions represent concentrations greater/less than the average value. The amplitude is defined as the difference between the averages of the high and the low regions. Below E_c, no phase separation occurs and the amplitude difference between the two regions is not visible. For an average concentration of 7%, the critical voltage in the simulation is 10 volts at the room temperature. At this value, lamellar structures form at time $t \sim \tau$, indicating that the labyrinthine pattern observed in experiment can be explained by a diffusion process.

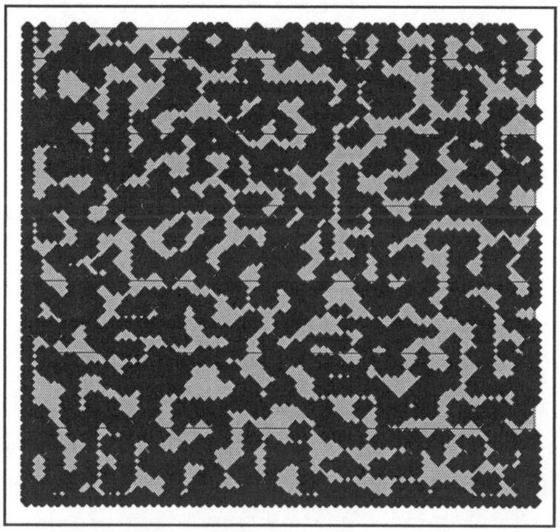

Fig. 1 Structure formation at $t=0.25\tau$, c=7%, V=10 V, T=298 K

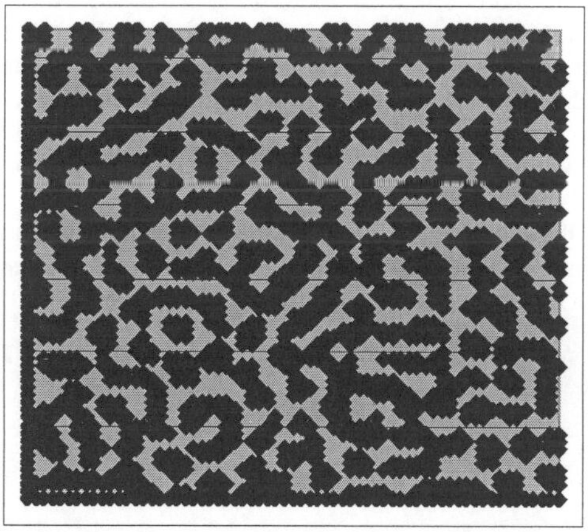

Fig. 2 Structure formation at $t=1.5\tau$, c=7%, V=10 V, T=298 K

The width of the stripes increases at short time until $t \sim \tau$, then decreases slightly. Experimentally, the stripe width grows with time, following a power law with an exponent of 1/3 [6]. This may indicate that electric field is not the only driving force for the phase separation.

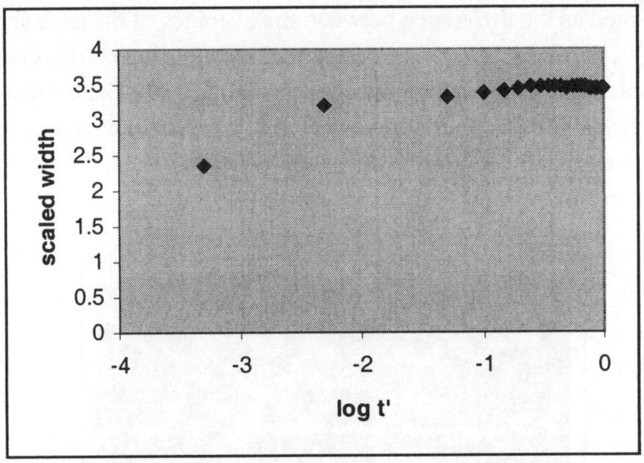

Fig. 3 Stripe width vs. time; c=7%, V=10 V, T=298 K

Future work with the simulation will include studies of varying voltage and temperature, and a direct comparison of the simulation with experiment. This work is supported by US NSF DMR under grant #0072901.

References

1. E. Lemaire, Y. Grasselli, and G. Bossis, *J. Phys. II France* **2**, 359-369 (1992); S. Cutillas, G. Bossis, E. Lemaire, A. Meunier, and A. Cebers, *Inter. J. Mod. Phys.* B **13**, 1791 (1999).
2. Xiaodong Duan and Weili Luo, "Evidence of second order phase transition of ferrofluid in external electric field," *International Journal of Modern Physics B* **15**, 837-841(2001).
3. K. Lichteneker, *Phys. Z* 25, 225 (1924). C. Brosseau, *J. Appl. Phys.* **75**, 672 (1994).
4. Xiaodong Duan and Weili Luo, "*Field-Induced Universal Labyrinthine Patterns In Nanocolloids*," Cond-mat/0105484, 2001.
5. The critical value of the electric field $E_c \sim C$, in the dilute regime and can be inferred by studying the spinodal of this model; the result will be published in a separate paper.
6. Xiaodong Duan, Ph.D. Dissertation, University of Central Florida, 2001.

HEXAGON-STRIPE TRANSITION AT THE MAGNETIC FIELD INDUCED PHASE TRANSFORMATIONS OF THE MAGNETORHEOLOGICAL SUSPENSIONS

A.CEBERS

Institute of Physics, Salaspils-1,LV-2169,Latvia

E-mail: aceb@sal.lv

The phase diagram of the magnetorheological suspension allowing for the modulated phases in the Hele-Shaw cell under the action of the normal field is calculated. The phase boundaries between the stripe, the hexagonal and the unmodulated phases in dependence on the layer thickness and the magnetic field strength are found. The existence of the transitions between the stripe and the hexagonal phases at the corresponding variation of the physical parameters is illustrated by the numerical simulation of the concentration dynamics in the Hele-Shaw cell. It is remarked that those transitions in the case of the magnetorheological suspensions can be caused by the compression or the expansion of the layer. Among the features noticed at the numerical simulation of the concentration dynamics in the Hele-Shaw cell are: the stripe patterns formed from the preexisting hexagonal structures are more ordered than arising from the initial randomly perturbed state; at the slightly perturbed boundary between the concentrated and diluted phases the hexagonal and the inverted hexagonal phases are formed and others.

1 Introduction

Patterns formed by the magnetic soft media under the action of the magnetic field are causing the big interest (for the recent review see the special issue on the labyrinthine pattern formation [1]). The special class of the labyrinthine patterns is formed by the magnetic liquids in the Hele-Shaw cells under the action of the normal field [2]. In spite of the dipolar repulsion forces arising in that case the break of the magnetic fluid stripes into the separate droplets never is observed in the monophasic limit. The explanation of this fact comes from the numerical simulation of the dynamics of the overextension deformations which shows that their development is stabilized at the nonlinear stage due to the vertex splitting instability [3]. Nevertheless there are some indications that the transformation of the labyrinthine patterns to the hexagonal ones exists. For example in [4] the transformation of the labyrinthine pattern of the concentrated phase of the magnetic colloid into the hexagonal array of the droplets has been observed. Inverse process of the transformation of the hexagonal pattern into the stripe or the labyrinthine patterns is mentioned more rarely. There is of course the opportunity to transform the hexagonal array of the magnetic particle agglomerates into the stripe patterns by the shear flow [5] or the rotating magnetic fields [6]. Nevertheless the process of the hexagonal array transformation into the stripe patterns as indicates the experimental data for 2D system of the quantum dots - silver nanocrystals on the air-water interface can arise also spontaneously at the compression of the system [7]. These processes of the structure transformation of the particle arrays have the interest from the point of view of their potential application in the production of the nanowires, 2D photonic crystals and the other devices for the nanotechnology.

2 Phase diagram

Here the process of the structure transformation of the magnetorheological suspension in the flat layer is considered from the point of view of its phase diagram in the Hele-Shaw cell. Considering the particle interaction in the framework of the mean field model [8] it predicts that the phase separation takes place if the magnetodipolar interaction parameter $\gamma = \frac{\kappa^2 H^2}{d^3 k_B T}$ is greater as critical value. The value of the concentration at the critical point then is about 0.1 and depends slightly by which model the excluded volume effects are accounted for - Carnahan-Starling, van der Waals etc., but this does not make any qualitative difference. We should remark here that opposite to the case of the magnetic colloids the critical concentration does not depend on the magnetic field strength. That means that the important parameter - the difference of the concentration and its value at the critical point in the case of the magnetorheological suspensions can be changed only by the compression of the system but not by the change of the magnetic field. To calculate the phase diagram of the system we use the expansion near the critical point. Since at fixed size of the particles and the temperature there is the critical value of the magnetic field strength determined by the critical value of the magnetodipolar interaction parameter than near this critical value of the magnetic field strength the thermodynamic potential of the suspension can be written as follows

$$F = -\frac{1}{2}\int \alpha\delta n^2 dS + \frac{1}{4}\int \gamma\delta n^4 dS + \frac{1}{2}\int \beta(\nabla\delta n)^2 dS +$$

$$\left(\frac{\partial M}{\partial n}\right)^2 \int dS \int dS' \delta n(\vec{\rho})\delta n(\vec{\rho}')\left(\frac{1}{\sqrt{(\vec{\rho}-\vec{\rho}')^2}} - \frac{1}{\sqrt{(\vec{\rho}-\vec{\rho}')^2 + (h/l)^2}}\right) \quad (1)$$

where ($\alpha = \frac{\partial^2 M}{\partial n^2}(H - H_c), \gamma = \frac{1}{6}\frac{\partial^3 \phi}{\partial n^3}$), β - the parameter describing the contribution of the particle concentration nonhomogenity into the thermodynamic potential. δn is the deviation of the concentration from the critical one, but h is the thickness of the Hele-Shaw cell. The last term in the relation (1) describes the long-range magnetic forces arising in the Hele-Shaw cell. Since the thermodynamic potential (1) allows the formation of the modulated phases [9] then the phase diagram of the system in dependence on the magnetic field strength and the layer thickness can be calculated by introducing the concentration modulations in the following form:
a) for the stripe phase

$$\phi = \phi_0 + \phi_q \cos qx \quad (2)$$

b) for the hexagonal phase

$$\phi = \phi_0 + \sum_{j=1}^{3} A_j e^{i\vec{k}_j\vec{\rho}} + c.c. \quad (3)$$

where the wave vectors of the hexagonal pattern satisfy the relation $\sum_{j=1}^{3} \vec{k}_j = 0$ ($A_1 = A_2 = A_3$). Substituting the expressions (2) and (3) in (1) in the case when the modulated structure period is much less than the thickness of the cell the following expressions for the free energy surface densities are obtained:

1. Stripe phase

$$f_S = -\frac{1}{2}\phi_0^2 + \frac{1}{4}\phi_0^4 + \frac{1}{3\sqrt{3}}\frac{Bm}{Bm_c}\frac{h}{l}\phi_0^2 +$$
$$+\frac{1}{4}\phi_q^2\left(\left(\frac{Bm}{Bm_c}\right)^{2/3} - 1 + 3\phi_0^2\right) + \frac{3}{32}\phi_q^4 \quad (4)$$

2. Hexagonal phase

$$f_H = -\frac{1}{2}\phi_0^2 + \frac{1}{4}\phi_0^4 + \frac{1}{3\sqrt{3}}\frac{h}{l}\frac{Bm}{Bm_c}\phi_0^2 +$$
$$3A^2\left(\left(\frac{Bm}{Bm_c}\right)^{2/3} - 1 + 3\phi_0^2\right) + 12\phi_0 A^3 + \frac{45}{2}A^4 \quad (5)$$

3. Unmodulated phase

$$f_L = -\frac{1}{2}\phi_0^2 + \frac{1}{4}\phi_0^4 + \frac{1}{3\sqrt{3}}\frac{h}{l}\frac{Bm}{Bm_c}\phi_0^2 \quad (6)$$

The relations (2-6) are written in dimensionless form. For that the following scales are introduced - the length - $l = \sqrt{\beta/\alpha}$; the concentration $\sqrt{\alpha/\gamma}$; the volume density of the free energy: α^2/γ^2. Magnetic Bond number Bm appearing in the relations above is defined as $Bm = \left(\frac{\partial M}{\partial n}\right)^2 \frac{h}{l\alpha}$. The critical magnetic Bond number Bm_c is defined by the relation

$$Bm_c = \frac{(h/l)^2}{6\sqrt{3}\pi} \quad (7)$$

The wave number of the arising structure can be found from the relation

$$q = \left(\frac{2\pi Bm}{(h/l)^2}\right)^{1/3}$$

From the relations (4-6) the chemical potential and the osmotic pressure of the phases are calculated according to the expressions ($i = (S, H, L)$)

$$\mu_i = \frac{\partial f_i}{\partial \phi_i}; p_i = \mu_i \phi_i - f_i$$

where the surface densities of the free energy of the modulated phases are found after minimizing the expressions (4) and (5) with the respect to the order parameters of the phases. Resulting phase diagram can be calculated by the double tangent construction:

$$\frac{\partial f_i}{\partial \phi_i} = \frac{\partial f_j}{\partial \phi_j} = \frac{f_i - f_j}{\phi_i - \phi_j} \quad (8)$$

The resulting phase diagram is quite complicated and we can mention here only some of its properties. The phase separation takes place at $Bm > Bm_c$. This corresponds to the power law for the critical field strength in dependence on the

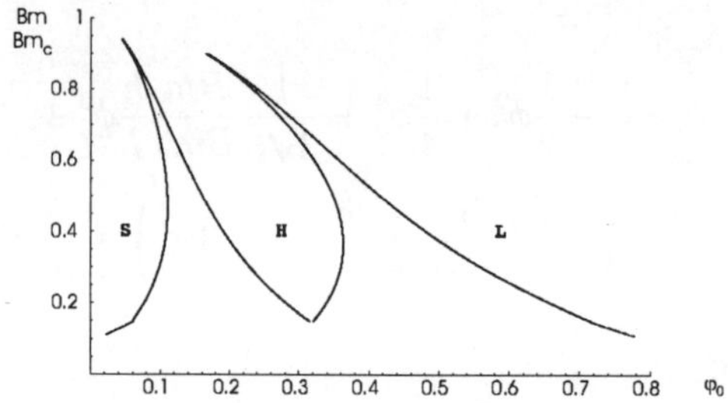

Figure 1. Phase diagram of the magnetic colloid in the Hele-Shaw cell. $\frac{h}{l} = 10$.

layer thickness found for the phase separation in the Hele-Shaw cell in [9]. If only the stripe and the unmodulated phases are considered then there is tricritical point for the phase transition between the stripe and the unmodulated phase defined by the relation [10]

$$3\left(\frac{Bm_*}{Bm_c}\right)^{\frac{2}{3}} = 4 - \frac{2}{3\sqrt{3}}\frac{Bm_*}{Bm_c}\frac{h}{l} \tag{9}$$

At the magnetic Bond number value $Bm > Bm_*$ the transition from the unmodulated to the stripe phase is the II order transition, for $Bm < Bm_*$ it is the I order transition. The modulated phases exist only if the magnetic Bond number is larger than some critical value. That probably is connected with the limited region when the free energy expansion near the critical point is valid. In principle according to the suggested approach the phase diagram could be calculated without using the free energy expansion near the critical point.

The resulting phase diagram calculated here is shown in Fig.1. By the letters S, H, L the regions where the stripe, the hexagonal and the unmodulated phases are stable are denoted. The regions of the stability of these phases depend on the thickness of the cell. Since as was remarked above the critical concentration of the phase separation of the magnetorheological suspension does not depend on the magnetic field strength then the natural way to cause the corresponding phase transitions is to change ϕ_0 by compression of the system as it is realized for the system of the polarized nanoparticles in the monolayer [7]. The phase transitions occuring according to our model and the calculated phase diagram are illustrated by some numerical simulation results in the next part of the work.

3 Numerical simulation results

The theoretical model describing the magnetic particle concentration evolution in the framework of the Cahn-Hilliard model with account for the long-range magnetic interaction forces has been proposed in [9,11] and in the accepted above dimensionless form reads

$$\frac{\partial \phi}{\partial t} + \Delta\left(\phi - \phi^3 + \Delta\phi - \frac{2Bm}{(h/l)^2}\int J(\vec{\rho} - \vec{\rho}', \frac{h}{l})\phi(\vec{\rho}')dS'\right) = 0 \tag{10}$$

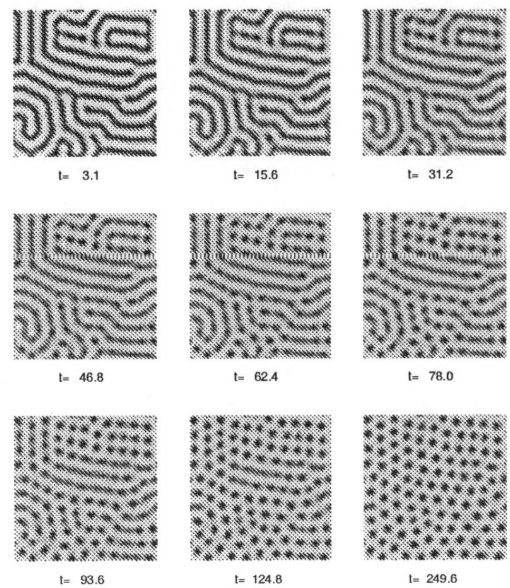

Figure 2. Transformation of the stripe structure to the hexagonal pattern ($\frac{Bm}{Bm_c} = 0.78, \frac{h}{l} = 10, \phi_0 = 0.2$)

where $J(\vec{\rho}, h) = \dfrac{1}{\sqrt{\vec{\rho}^2}} \quad \dfrac{1}{\sqrt{\vec{\rho}^2 + h^2}}$. The numerical simulation of the concentration dynamics according to the equation 10 is carried out by the pseudospectral technique. Mesh 128×128 usually is used, the step size is 1 in the used dimensionless coordinates. At the values of ϕ_0 corresponding to the equal volumes occupied by the phases according to the phase diagram shown in Fig.1 the stripe phase develops. Since there is not the long-range order in the orientations of the stripes the resulting pattern is labyrinthine-like. If for the such pattern the state of the system is brought to the region where the hexagonal phase is stable the transition from the stripe phase to the hexagonal one takes place as it is shown in Fig.2. The reverse process as observed for the polarized system of the quantum dots [7] also is possible and is shown in Fig.3. At the process shown in Fig.3 expansion of the system leading to the diminishing of the ϕ_0 value was carried out and as result the pattern from the hexagonal has transformed to the stripe one. It is interesting to note that the stripes formed from the hexagonal pattern in the centre of the sample in Fig.3 are well ordered. It could be remarked that the similar trend has been seen at the numerical simulations of the phase behaviour of the similar systems - block copolymers [12].

The phase diagram shown in Fig.1 allows to choose the values of the physical parameters for which the system finds itself in the intermediate state. The formation of the such state is shown in Fig.4 where from initially randomly perturbed state the pattern containing the mixture of the bubbles and the dumbbells forms. The formation of the such mixture for the chosen set of the parameters is in the good agreement with the phase diagram in Fig.1 since the point corresponding to

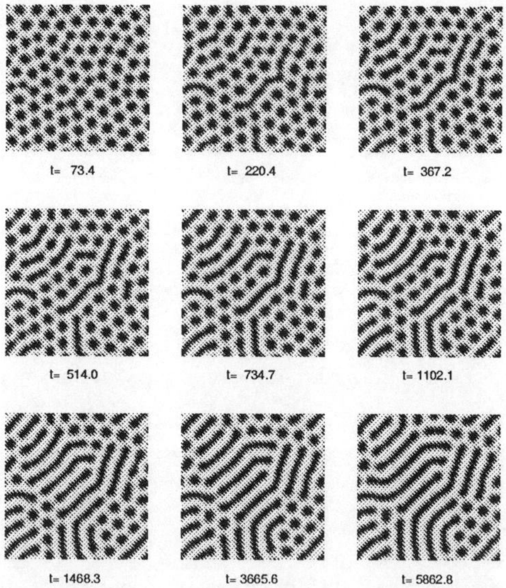

Figure 3. Transformation of the hexagonal structure to the stripe structure at the decrease of the mean concentration. $\frac{Bm}{Bm_c} = 0.78, \frac{h}{l} = 5, \phi_0 = 0.05$. Initial pattern - the hexagonal structure formed at $\frac{Bm}{Bm_c} = 0.78, \frac{h}{l} = 5$ and $\phi_0 = 0.2$.

Figure 4. Mixed intermediate state of the magnetic colloid ($\frac{Bm}{Bm_c} = 0.3, \frac{h}{l} = 10, \phi_0 = 0.075$).

those values of the physical parameters corresponds to the region of the equilibrium between the stripe and the hexagonal phases.

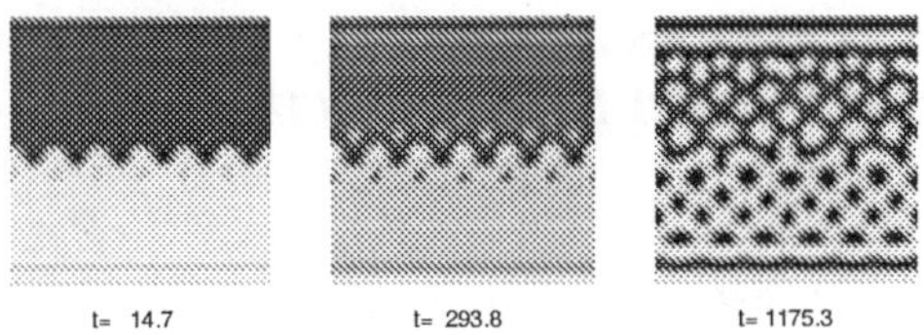

Figure 5. Foam formation at the interface between diluted and concentrated phases ($\frac{Bm}{Bm_c} = 0.3, \frac{h}{l} = 5$).

Finally it should be remarked that the hexagonal and the inverted hexagonal phases are formed on the slightly perturbed boundary between the concentrated and the diluted phases as it is shown in Fig.5.

4 Conclusions

The thermodynamical model of the magnetorheological suspension proposed allows the existence in definite range of the physical parameters of the stripe, the hexagonal and the unmodulated phases. In the case of the magnetorheological suspensions the transitions between them can be caused by the compression or the expansion of the layer with the magnetorheological suspension. The existence of the such transitions is confirmed by the numerical simulation of the concentration dynamics of the suspension with account for the long-range magnetic forces. Several interesting features of the phase transformation dynamics were remarked in the numerical simulations: the stripe patterns formed at the change of the physical parameters from the initial hexagonal pattern are more ordered as formed from the random initial conditions; the hexagonal and the inverted hexagonal phases are formed from the slightly perturbed boundary between the concentrated and the diluted phases.

References

1. Special issue on the labyrinthine pattern formation in the magnetic liquids Magnitnaya Gidrodinamika **35**, 305-390 (1999)
2. A.Cebers, M.M.Maiorov, Magnitnaya Gidrodinamika **16**, 15 (1980)
3. A.Cebers, I.Drikis,Magnitnaya Gidrodinamika **32**, 11 (1996)
4. J.C.Bacri, R.Perzynski, D.Salin, Recherche **18**, 1150 (1987)
5. S.Cutillas, G.Bossis, A.Cebers, Phys. Rev. E **57**, 804 (1998)
6. P.Carletto et al., in: *The Eight International Conference on Electrorheological (ER) Fluids and Magnetorheological (MR) Suspensions*,(France, Nice,2001)
7. R.P.Sear, S.W.Chung, G.Markovich, W.M.Gelbart, J.R.Heath, Phys. Rev. E **59**, R6255 (1999)
8. A.Cebers, Magnitnaya Gidrodinamika **18**, 42 (1982)
9. A.Cebers, Magnitnaya Gidrodinamika **22**, 132 (1986)
10. A.Cebers, Magnitnaya Gidrodinamika **35**, 344 (1999)
11. A.Cebers, Phys. Rev. E **61**, 700 (2000)
12. M.Nonomura, T.Ohta, Journ. Phys. Soc. Japan **70**, 927 (2001)

EQUILIBRIUM PATTERNS OF CHAIN CLUSTERS COMPOSED OF FERROMAGNETIC PARTICLES IN AN EXTERNAL MAGNETIC FIELD

TOMOFUMI UKAI, TORU MAEKAWA

Bio-Nano Electronics Research Centre, Toyo University
2100, Kujirai, Kawagoe, Saitama 350-8585, Japan

HISAO MORIMOTO

Department of Mechanical Engineering, The University of Tokyo (JSPS Research Fellow)
7-3-1 Hongo, Bunkyo-ku, Tokyo, 113-8656, Japan

We carry out Monte Carlo simulations of a ferromagnetic colloidal system, which is subjected to an external magnetic field, to investigate the structures formed by chain clusters. The control parameters are the ratio of the dipole moment energy to thermal energy, λ, and the ratio of the interactive energy between the dipole and the external magnetic field to thermal energy, ξ. We investigate the effect of the system height on the pattern formations for $\lambda = 18$ and $\xi = 30, \infty$. Note that the system becomes paramagnetic when $\xi = \infty$. We find that as the system height increases, chains coagulate to form fat clusters and spatially ordered structures are created when $\xi = 30$, whereas chains form thin meandering walls when $\xi = \infty$.

1. Introduction

It is known that meandering chain clusters are formed in ferromagnetic colloidal systems when the dipole-dipole interactions are strong and the chain clusters become rather straight in a strong magnetic field [1,2]. Various secondary structures are created in the ferromagnetic systems through chain-chain interactions when the number densities of particles are high [3,4]. Although experimental, theoretical and numerical studies on the secondary structures have been done [5-7], the conditions for the formations of different structures have not yet been understood clearly.

In this paper, we focus on the patterns formed by chain clusters and investigate the effect of the system height on the secondary structures. In section 2, we explain a numerical modelling and develop a Monte Carlo calculation method. In section 3, we investigate the secondary structures and discuss the effect of the control parameters on the pattern formations. In the final section, we summarize the result of the analysis.

2. Numerical modelling

Ferromagnetic particles are shown in Fig.1. Each particle is composed of a magnetic dipole and has degrees of freedom of translational and rotational motions. The particles are dispersed in a solvent which is sandwiched between two parallel plates (see Fig.2). The interactive potentials are the dipole-dipole potential, u_d, the repulsive potential between particles, u_c, the repulsive potential between a particle and the wall, u_w, and the interactive energy between a dipole and the external magnetic field, u_H:

$$u_d = \frac{1}{4\pi\mu_0}\left\{\frac{\boldsymbol{m}_i \cdot \boldsymbol{m}_j}{r_{ij}^3} - \frac{3}{r_{ij}^5}\left(\boldsymbol{m}_i \cdot \boldsymbol{r}_{ij}\right)\left(\boldsymbol{m}_j \cdot \boldsymbol{r}_{ij}\right)\right\}, \tag{1}$$

$$u_c = \frac{|\boldsymbol{m}|^2}{4\pi\mu_0 d^3} exp\left\{-40\left(\frac{r_{ij}}{d} - 1\right)\right\}, \tag{2}$$

$$u_w = \frac{|\boldsymbol{m}|^2}{4\pi\mu_0 d^3}\left[exp\left\{-40\left(\frac{2z}{d} - 1\right)\right\} - exp\left\{-40\left(\frac{2(L_z - z)}{d} - 1\right)\right\}\right], \tag{3}$$

$$u_H = -\boldsymbol{m}_i \cdot \boldsymbol{H}, \tag{4}$$

where μ_0, $\boldsymbol{m}_i$, $\boldsymbol{r}_{ij}$, d, L_z, z and $\boldsymbol{H}$ are, respectively, the permeability, the magnetic dipole moment of particle i, the displacement vector from particle j to particle i, the diameter of

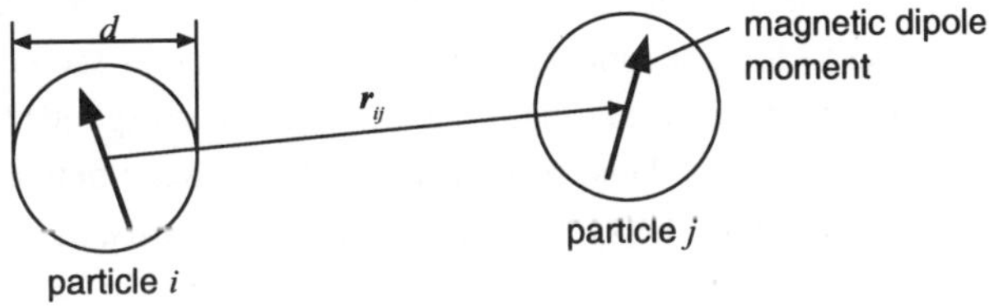

Fig. 1. Ferromagnetic particles.

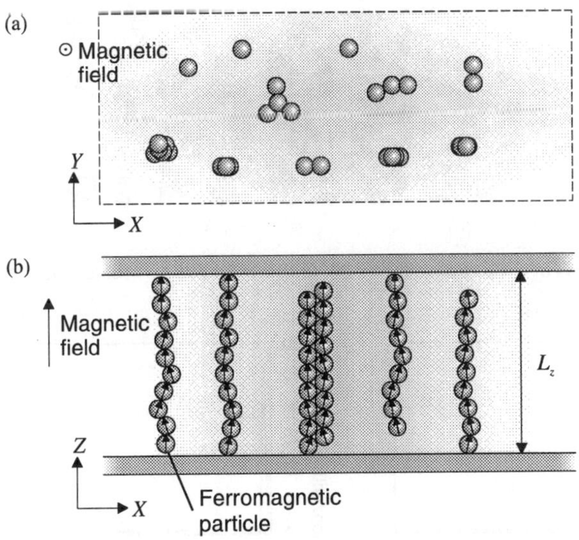

Fig. 2. Calculation system. A ferromagnetic colloidal system is sandwiched between two parallel plates. A uniform dc magnetic field is applied in the z-direction. Periodic boundary conditions are employed in both the x- and y-directions. (a) Top view; (b) side view.

a particle, the depth of the cell, the z–coordinate and the external magnetic field. The control parameters are the ratio of the dipole moment energy to thermal energy, λ, and the ratio of the interactive energy between the dipole and the external magnetic field to thermal energy, ξ:

$$\lambda \equiv \frac{|\boldsymbol{m}|^2}{4\pi\mu_0 d^3 kT}, \quad \xi \equiv \frac{|\boldsymbol{m}||\boldsymbol{H}|}{kT}, \tag{5}$$

where k and T are the Boltzmann constant and temperature.

Since the interactions both between particles and between chain clusters are strong and therefore the calculation time based on the Metropolis scheme becomes long, we employed Biased Monte Carlo and Cluster-Moving Monte Carlo methods [8]. We fixed the control parameters at $\lambda = 18$ and $\xi = 30, \infty$, the volume fraction of particles at $\rho = 0.1$ and the horizontal lengths of the calculation cell at $L_x/d = L_y/d = \{49\pi/(6\rho)\}^{1/2}$ (= 16.0). We changed the height of the cell; $L_z/d = 5$, 10, 20 and 40.

3. Result and discussion

Snapshots of the top views of chain clusters are shown in Fig.3 for $\xi = 30$. In this case, chain clusters are not straight but slightly winding. As the system height increases, chain clusters coagulate to form secondary clusters and the size of each cluster becomes larger. The structure factors are shown in Fig.4. The first peak corresponds to the distance between the secondary clusters and the second peak to the distance between particles. The first peak becomes higher with the system height. In other words, as we mentioned, spatially ordered secondary structures are created as the system height increases.

Snapshots of chain clusters are shown in Fig.5 for $\xi = \infty$. In this case (paramagnetic

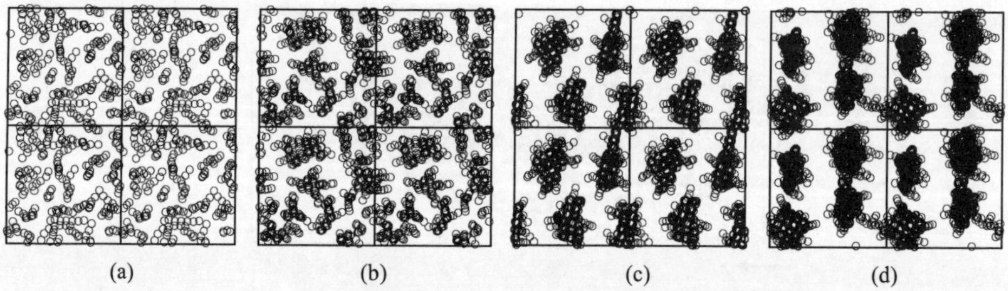

Fig. 3. Snapshots of top views of chain clusters for $\xi = 30$. (a) $L_z/d = 5$; (b) $L_z/d = 10$; (c) $L_z/d = 20$; (d) $L_z/d = 40$.

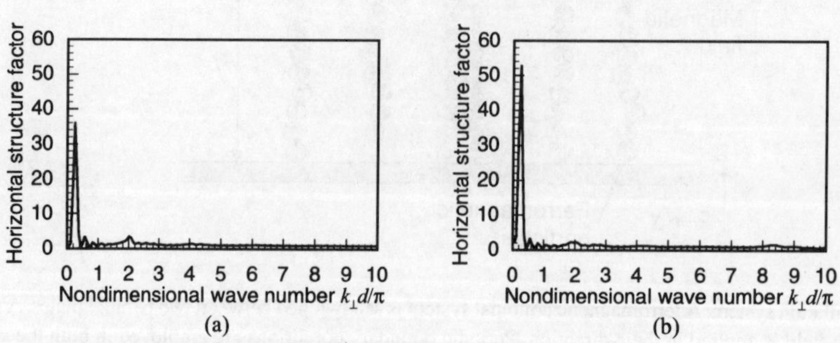

Fig. 4. Structure factors. (a) $L_z/d = 20$; (b) $L_z/d = 40$.

case), chain clusters are almost straight in the direction of the external magnetic field. However, unlike the ferromagnetic case, chain clusters form walls and the walls become bent as the system height increases. When the system is high, those walls coagulate to form thicker walls. To understand the basic mechanism of the secondary structure formations in the paramagnetic system, we calculated the interactive potentials between two chain clusters, which are composed of paramagnetic particles and are lined in the direction of the external magnetic field (see Fig.6). When the particles' levels are the same between parallel straight clusters (see Fig.6 (a)), the potential between two clusters are repulsive (see Fig.7 (a)), whereas when the particles' locations are staggered (Fig.6 (b)), the potential between chain clusters are attractive (Fig.7 (b)). When two staggered chain clusters stick

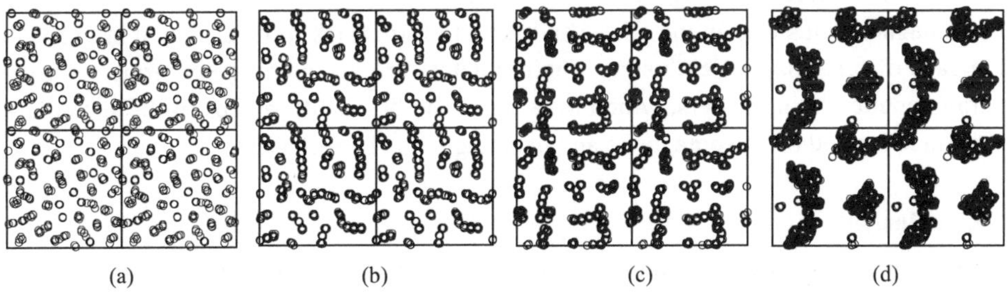

Fig. 5. Snapshots of top views of chain clusters for $\xi = \infty$. (a) $L_z/d = 5$; (b) $L_z/d = 10$; (c) $L_z/d = 20$; (d) $L_z/d = 40$.

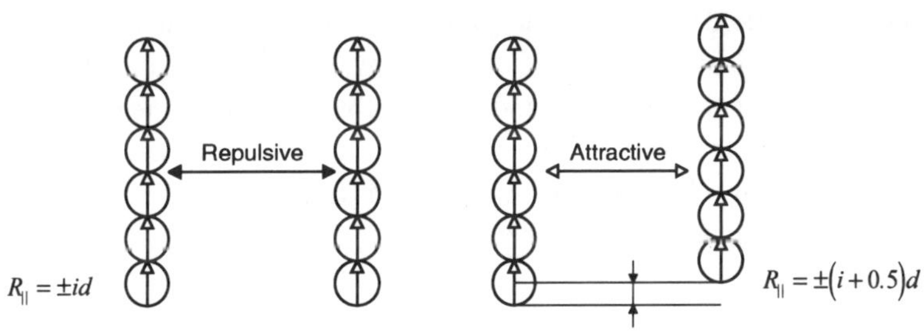

Fig. 6. Locations of chain clusters. (i = 0, 1, 2, ···). (a) Particles are leveled. (b) Particles are staggered.

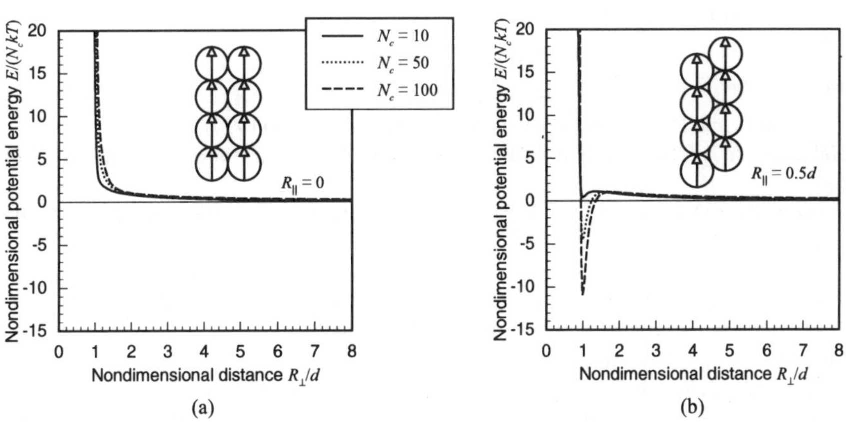

Fig. 7. Potential between chain clusters. N_c is the number of particles in a chain clusters. (a) Particles are leveled. (b) Particles are staggered.

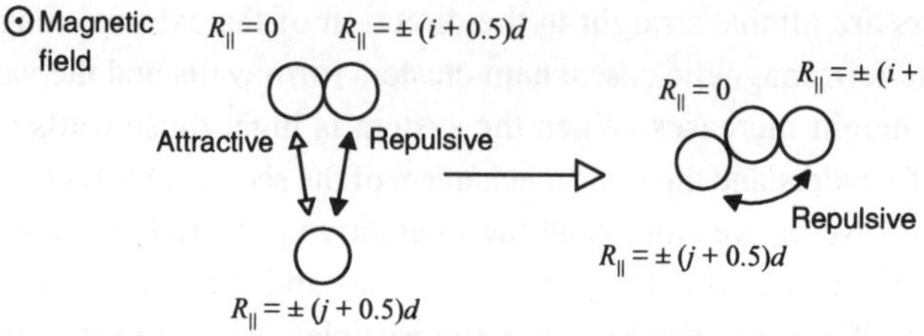

Fig. 8. Schematic of the formation of a wall structure. Top views of three chains. ($i, j = 0, 1, 2, \cdots$).

together, a third chain cluster cannot be placed on them simultaneously due to the repulsive potential. Therefore, paramagnetic chain clusters tend to form wall structures (see Fig.8). Ferromagnetic clusters, on the other hand, tend to form thick rods, as shown in Fig.3. In the case of ferromagnetic clusters, chain clusters are not completely straight unlike paramagnetic chain clusters, which may cause the basic differences in the secondary structures formed by ferromagnetic and paramagnetic chain clusters.

4. Conclusions

We studied the secondary structures formed by ferromagnetic chain clusters which are subjected to dc magnetic fields and obtained the following results: (1) In the case of $\xi = 30$, where ξ is the ratio of the interactive energy between the dipole moment and the external magnetic field to thermal energy, chain clusters form fat clusters as the system height increases. (2) In the case of $\xi = \infty$, thin walls are formed as the system height increases. The wall length increases and walls become bent with the system height.

We are going to investigate the coagulation processes in more detail in order to understand the basic mechanism which causes the differences in pattern formations between ferromagnetic and paramagnetic systems.

Acknowledgements

This study has been supported by the High-Tech Research Centre Fund organized by the Ministry of Education, Science, Culture and Sports, Japan since 1996 and by the Toyo University Special Fund since 1999.

References

1. P.G. de Gennes and P.A. Pincus, *Phys. Kondens. Mater.*, **11**, 189 (1970).
2. G. Helgesen, A.T. Skjeltorp, P.M. Mors, R. Botet and R. Jullien, *Phys. Rev. Lett.*, **61**, 1736 (1988).
3. H. Wang, Y. Zhu, C. Boyd, W. Luo, A. Cebers and R.E. Rosenssweig, *Phys. Rev. Lett.* **72**, 1929 (1994).
4. G.A. Flores, J. Liu, H. Mohebi, and N, Jamasbi, *Proc. of the 6th Int. Conf. on ERF, MRS and their Applications,* ed. M. Nakano and K. Koyama (World Scientific, Singapore, 1997), p. 611.
5. M. Gross and S. Kiskamp, *Phys. Rev. Lett.*, **79**, 2566 (1997).
6. R. Tao and J.M. Sun, *Phys. Rev. Lett.*, **67**, 398 (1991).
7. M. Mohebi, N. Jamasbi and J.M. Sun, *Phys. Rev. Lett.*, **54**, 5407 (1996).
8. D. Frenkel and B. Smit, *Understanding Molecular Simulation: From Algorithms to Applications,* (Academic Press, San Diego, 1996).

EQUILIBRIUM AND DYNAMIC ANALYSES OF MAGNETIC CHARACTERISTICS OF A TRIANGULAR FERROMAGNETIC LATTICE SYSTEM

TOMOFUMI UKAI, TORU MAEKAWA

Bio-Nano Electronics Research Centre, Toyo University
2100, Kujirai, Kawagoe, Saitama 350-8585, Japan

HISAO MORIMOTO

Department of Mechanical Engineering, The University of Tokyo (JSPS Research Fellow)
7-3-1 Hongo, Bunkyo-ku, Tokyo, 113-8656, Japan

We investigate equilibrium and dynamic magnetic characteristics of a triangular ferromagnetic lattice system by the Monte Carlo and Brownian dynamics methods. We find that dipoles form hexagonal vortex structures under certain conditions. We investigate the effect of the control parameters on the specific heat and linear and nonlinear magnetic susceptibilities. We also investigate the effect of an external dc magnetic field on the magnetization. We find that when the dipole-dipole interaction is strong, an hysteresis appears in the magnetization-magnetic field relation, which is similar to the magnetic characteristics of spin glass.

1. Introduction

Ferromagnetic colloidal systems generate different phases depending on the control parameters such as the system temperature, the number density of colloidal particles and external magnetic fields [1-5]. Since the interparticle potential is directional based on the dipole-dipole potential, several different structures are formed in the system and, as a result, the physical properties of the colloidal system may be changed dramatically by changing the control parameters [5,6]. Previously, we developed a Brownian dynamics calculation method and investigated the magnetic and cluster structures of a ferromagnetic colloidal system [7-9], according to which magnetic domains were formed in the case of a close-packed colloidal system. However, it is still not clear whether the magnetic structures obtained by the Brownian dynamic simulation are those in the equilibrium state or not since frustrations may be created in the case of high number density systems. In this paper, we study the magnetic characteristics of a triangular ferromagnetic lattice system, which is a model of a close-packed ferromagnetic colloidal system, and investigate the equilibrium and dynamic magnetic structures by the Monte Carlo and Brownian dynamics methods.

2. Numerical modelling and calculation method

In this section, we develop a phenomenological model of a close-packed ferromagnetic

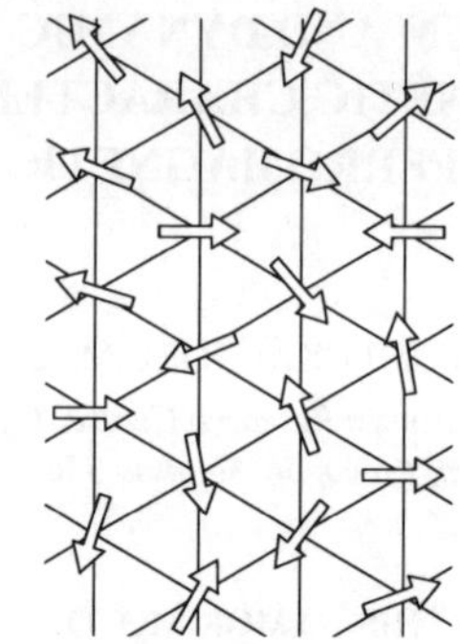

Fig. 1. Triangular ferromagnetic lattice system.

colliodal system, that is, a triangular ferromagnetic lattice system. Dipoles, which can rotate continuously in the azimuthal direction, are placed at triangular lattice points (see Fig.1). The interactive potentials and the Monte Carlo and Brownian dynamics methods are described in the following.

2.2. Interactive potentials

The interaction between particles is expressed by the dipole-dipole potential, u_d:

$$u_d = \frac{1}{4\pi\mu_0}\left\{\frac{\boldsymbol{m}_i \cdot \boldsymbol{m}_j}{r_{ij}^3} - \frac{3}{r_{ij}^5}\left(\boldsymbol{m}_i \cdot \boldsymbol{r}_{ij}\right)\left(\boldsymbol{m}_j \cdot \boldsymbol{r}_{ij}\right)\right\}, \tag{1}$$

where μ_0, $\boldsymbol{m}_i$ and $\boldsymbol{r}_{ij}$ are the permeability, the dipole moment at site i and the displacement vector from site j to site i. The interaction between the dipole moment and the external magnetic field is expressed as follows:

$$u_H = -\boldsymbol{m}_i \cdot \boldsymbol{H}, \tag{2}$$

where $\boldsymbol{H}$ is the external magnetic field. The control parameters are the ratio of the dipole-dipole interactive energy to thermal energy, λ, and the ratio of the interactive energy between the dipole and the magnetic field to thermal enery, ξ:

$$\lambda \equiv \frac{|\boldsymbol{m}|^2}{4\pi\mu_0 d^3 kT}, \quad \xi \equiv \frac{|\boldsymbol{m}||\boldsymbol{H}|}{kT}, \tag{3}$$

where k, T and d are, respectively, the Boltzmann constant, temperature and the diameter of a particle, that is, the length of the equilateral triangular side.

2.3. Monte Carlo method

The direction of dipole i is determined by the following probability density:

$$P(\varphi) = \frac{exp(\boldsymbol{n}_i \cdot \boldsymbol{h})}{Z}, \tag{4}$$

where φ is the angle between $\boldsymbol{n}_i$ and $\boldsymbol{h}$, $\boldsymbol{n}_i$ is the unit vector in the direction of dipole i, $\boldsymbol{h}$ is the magnetic field produced at lattice i by the other dipoles and the external magnetic field and Z is defined below:

$$\boldsymbol{n}_i = \frac{\boldsymbol{m}_i}{|\boldsymbol{m}_i|}, \quad h = -\lambda \sum_{j \neq i} \left\{ \frac{\boldsymbol{n}_j}{R_{ij}^3} - \frac{3}{R_{ij}^5} \left(\boldsymbol{n}_j \cdot \boldsymbol{R}_{ij} \right) \boldsymbol{R}_{ij} \right\} + \xi h_{ext}, \tag{5}$$

$$Z = \int_{-\pi}^{\pi} exp\left(\boldsymbol{n}_i \cdot \boldsymbol{h} \right) d\varphi. \tag{6}$$

Here, $\boldsymbol{R}_{ij}$ is the nondimensional displacement vector and $\boldsymbol{h}_{ext}$ the normalized external magnetic field:

$$\boldsymbol{R}_{ij} = \boldsymbol{r}_{ij} / d, \quad \boldsymbol{h}_{ext} = \boldsymbol{H} / |\boldsymbol{H}|. \tag{7}$$

The calculation procedure is as follows: (1) The directions of dipoles at lattice points are randomly distributed initially. (2) Dipole i is chosen at random. (3) The angle of dipole i, φ, is determined by the probability density, $P(\varphi)$. We repeat processes (1) to (3) until the system reaches the equilibrium state.

2.4. Brownian dynamics method

Dynamics of each dipole is calculated by the Langevin equation:

$$\frac{\partial \boldsymbol{\omega}}{\partial t} = -\frac{\zeta_r}{I} \boldsymbol{\omega} + \frac{\boldsymbol{T}}{I} + \boldsymbol{\tau}, \tag{8}$$

where $\boldsymbol{\omega}$, t, ζ_r, I, $\boldsymbol{T}$ and $\boldsymbol{\tau}$ are, respectively, the angular velocity of a dipole, time, the friction coefficient, the moment of inertia of a particle, the torque acting on a dipole and the random torque. We integrate Eq.(8) by Ermak's procedure [10].

3. Result and discussion

We investigate the equilibrium structures by the Monte Carlo method and the dynamic structures by the Brownian dynamics method. The system size nondimensionalized by the side of the equilateral triangle, L_y ($= 2L_x/3^{1/2}$), and the control parameter, λ, are changed as follows: $L_y = 20$, 100, 200, 400 and $\lambda = 0.1 \sim 10$.

3.1. Equilibrium structures in the absence of an external magnetic field

Snapshots of dipoles are shown in Fig.2 for $L_y = 20$. When the dipole-dipole interaction is weak ($\lambda = 0.1$), the directions of dipoles are randomly distributed, whereas hexagonal patterns are formed when the interaction is strong ($\lambda = 10$). We carried out calculations changing the system size, but the basic features were the same as those mentioned above. The size of the hexagonal cells did not change with an increase in the system size. In order to understand the phase transition obtained in the triangular lattice system in more detail, we examined the magnetic susceptibility and the specific heat. The nondimensional magnetic susceptibility, χ_1, and the specific heat, c, are calculated by the following equations:

$$\chi_1 = N\left\{ \left\langle \boldsymbol{\mu}^2 \right\rangle - \left\langle \boldsymbol{\mu} \right\rangle^2 \right\}, \tag{9}$$

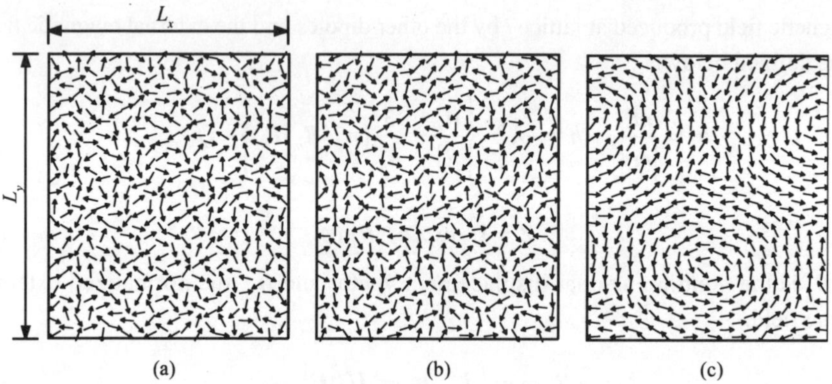

Fig. 2. Dipole distributions. (a) $\lambda = 0.1$; (b) $\lambda = 1.0$; (c) $\lambda = 10$.

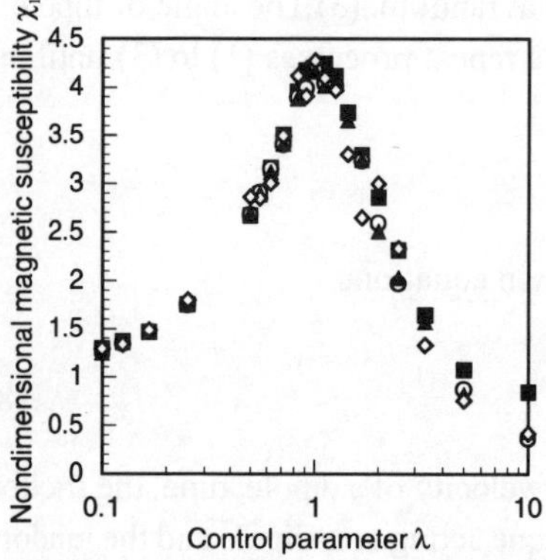

Fig. 3. Dependence of linear magnetic susceptibility on control parameter λ. ■; $L_y = 20$. O; $L_y = 100$. ▲; $L_y = 200$. ◇; $L_y = 400$.

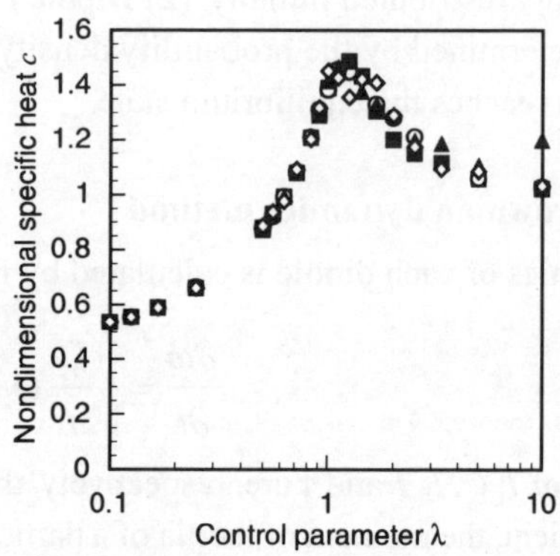

Fig. 4. Dependence of specific heat on control parameter λ. ■; $L_y = 20$. O; $L_y = 100$. ▲; $L_y = 200$. ◇; $L_y = 400$.

$$c = N\left\{\left\langle \Phi^2 \right\rangle - \left\langle \Phi \right\rangle^2\right\} + \frac{1}{2}, \tag{10}$$

where N, $\boldsymbol{\mu}$ and Φ are, respectively, the number of particles or lattice points, the nondimensional magnetization and the nondimensional potential energy:

$$\boldsymbol{\mu} \equiv \frac{\sum_{i=1}^{N} \boldsymbol{m}_i}{N|\boldsymbol{m}|}, \quad \Phi \equiv \frac{\phi}{NkT}. \tag{11}$$

ϕ is the potential energy of the system. Figs.3 and 4 show the dependence of the magnetic susceptibility and the specific heat on the control parameter λ. The susceptibility and specific heat become maximum at $\lambda \approx 1$ and there is no size effect unlike second-order phase transitions. We also examined the phase transition defining the order parameter, $(\Delta k)^2$:

$$k_w(\Delta k)^2 \equiv \frac{\int \boldsymbol{k}^2 |\Psi|^2 \, dk}{\int |\Psi|^2 \, dk} - \left\{\frac{\int \boldsymbol{k} |\Psi|^2 \, dk}{\int |\Psi|^2 \, dk}\right\}^2, \tag{12}$$

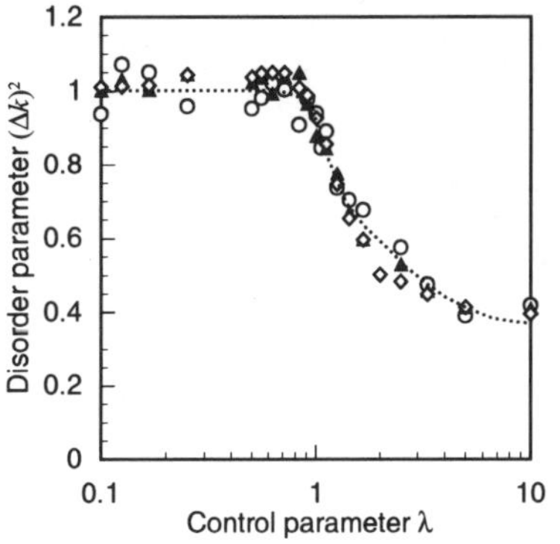

Fig. 5. Dependence of order parameter $(\Delta k)^2$ on control parameter λ.

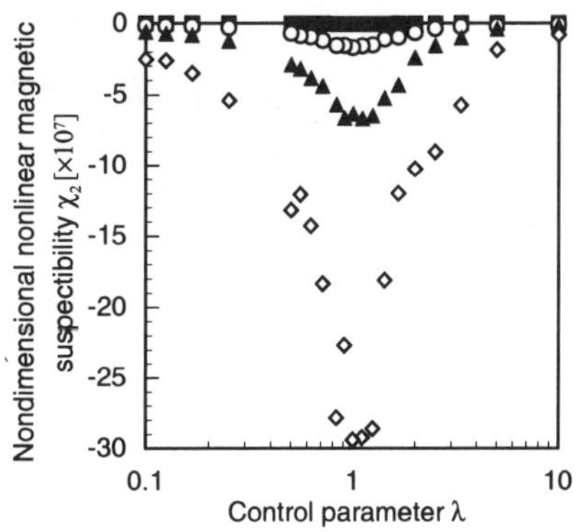

Fig. 6. Dependence of nonlinear magnetic susceptibility on control parameter λ.

where Ψ is defined as follows:

$$\Psi = \int \theta e^{-i\boldsymbol{kr}} d\boldsymbol{r}, \quad \theta \equiv e^{i\varphi}. \tag{13}$$

k_w is the value of the right-hand side of Eq.(12) calculated for a constant Ψ, which corresponds to white noise. Therefore, $(\Delta k)^2$ is normalized so that $(\Delta k)^2 = 1$ when the dipoles are randomly distributed. When some structure is formed in the system, $(\Delta k)^2$ decreases. The dependence of the order parameter, $(\Delta k)^2$, on the control parameter is shown in Fig.5. When $\lambda < 1$, the dipoles are randomly distributed, whereas ordered structures are created when λ is large.

Finally, we investigated the nonlinear susceptibility, χ_2:

$$\chi_2 = 96N^3 \left\{ \left\langle \boldsymbol{\mu}^4 \right\rangle - 3\left\langle \boldsymbol{\mu}^2 \right\rangle^2 \right\}. \tag{14}$$

Fig.6 shows the dependence of the nonlinear susceptibility on the control parameter. The nonlinear susceptibility becomes minimum at $\lambda \approx 1$ and decreases as the system size increases. The nonlinear susceptibility shows negative divergence at $\lambda \approx 1$ and the absolute values of minima of χ_2 increase in proportion to the square of the system size; $|\chi_2| \sim L^2$.

3.2. Magnetization curve

We calculated magnetization curves by both Monte Carlo and Brownian dynamics methods. A magnetization curve obtained by the Brownian dynamics method is shown in Fig.7 for $\lambda = 10$ where the intensity of the external magnetic field is increased from $\xi = 0$ to 2 and then decreased from $\xi = 2$ to 0. The solid curve corresponds to the one when there is no interactions between dipoles. Magnetization is increased from zero with an increase of the intensity of the magnetic field, following the equilibrium magnetization curve which was obtained by the Monte Carlo method. However, magnetization jumps to a higher value discontinuously at $\xi = 2$ and is decreased with an decrease of ξ, taking a different route. Magnetization jumps to zero at $\xi = 0$. The above hysteretic feature is very similar to that of spin glass [10].

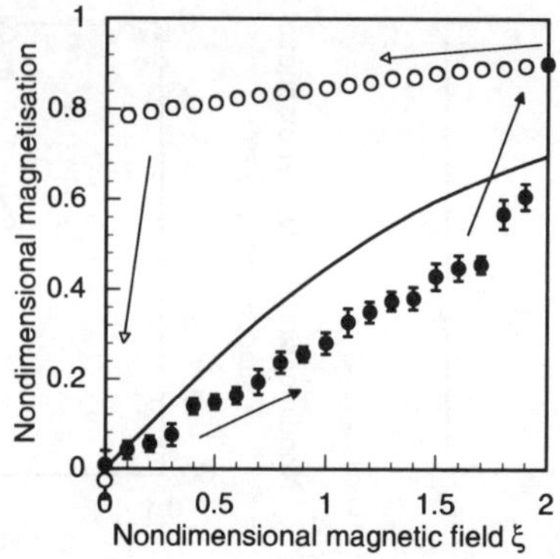

Fig. 7. Magnetization curve obtained by the Brownian dynamics method.

4. Conclusions

We analysed the magnetic characteristics of a triangular ferromagnetic lattice system by both biased Monte Carlo and Brownian dynamics methods and obtained the following results: (1) A phase transition occurs at $\lambda \approx 1$ where hexagonal patterns are formed in dipole distributions. The hexagonal cell size is independent of the system size. (2) The linear magnetic susceptibility and the specific heat become maximum at the phase transition point and the order parameter starts decreasing when the control parameter exceeds the critical value. There is no size effect on the susceptibility, the specific heat and the order parameter. (3) The nonlinear magnetic susceptibility shows negative divergence at the phase transition point. (4) An hysteresis appears in the magnetization curve, which is similar to that observed in spin glass, when the interaction between dipoles is strong.

Acknowledgements

This study has been supported by the High-Tech Research Centre Fund organized by the Ministry of Education, Science, Culture and Sports, Japan since 1996 and by the Toyo University Special Fund since 1999.

References

1. D. Levesque and J.J. Weis, *Phys. Rev. E*, **49**, 5131 (1994).
2. B. Groh and S. Dietrich, *Phys. Rev. E*, **50**, 3814 (1994).
3. D. Wei and G.N. Patey, *Phys. Rev. Lett.*, **68**, 2043 (1992).
4. A.G.-Villegas, S.C. McGrother and G. Jackson, *Chem. Phys. Lett*, **269**, 441 (1997).
5. M.J. Stevens and G.S. Grest, *Phys. Rev. E*, **51**, 5962 (1995).
6. F. Gazeau, B.M. Heegaard, J.-C. Bacri, A. Cebers and R. Perzynski, *Europhys. Lett.* **35**, 609 (1996).
7. H. Morimoto and T. Maekawa, *J. Phys. A*, **33**, 247 (2000).
8. T. Ukai, H. Morimoto and T. Maekawa, *Int. J. Mod. Phys. B*, **15**, 817 (2001).
9. H. Morimoto and T. Maekawa, *Int. J. Mod. Phys. B*, **15**, 823 (2001).
10. D.L. Ermak and H. Buckholz, *J. Comput. Phys.*, **35**, 169 (1980).
11. K. Binder, *Rev. Mod. Phys.*, **58**, 801 (1986).

FIELD INDUCED PHASE TRANSITION FOR SUSPENSION OF MONOSIZED SPHERES

S. MEN, A. MEUNIER, C. MÉTAYER AND G. BOSSIS

Laboratory of Condensed Matter Physics, University of Nice-Sophia Antipolis, Parc Valrose, 06108 Nice Cedex 2, France

E-mail: georges.bossis@unice.fr, shqmen@yahoo.com

W have developed a new Ewald summation for a three-dimensional dipolar system with two-dimensional periodicity in a uniaxial field and a rotating field in a horizontal plane. Under a constant pressure and temperature, Monte Carlo simulation has been carried out; phase transitions are found and chainlike structure for a uniaxial field and monolayer or multilayer for rotating field are obtained, which are well consistent with experiments.

1 Introduction

Ewald summation [1] is the most widely used method in computer simulation of long-ranged interactions in a Coulomb system consisting of charged particles [2-12] or in a dipolar system comprising dipoles only [13,14]. Among the Ewald sum techniques, periodic boundary conditions (PBC) are applied to avoid finite-size effect of the unit cell, which is generally a cubic box with base L in a conventional three-dimensional (3D) Ewald summation (EW3D) [2-5,13-15], and a slab with infinite size in x-y plane but finite length in z-the third direction in the type of two-dimensional (2D) Ewald summations (EW2D) [4,6-12].

For the interaction of a system with slab geometry, there have already been a great number of methods to deal with it [4,6-12]. Most authors [6-8] employed the regular EW2D to treat the problem. It is the most accurate but very time consuming. Others [4,9] developed the EW3D approaches to simulate approximately the 2D periodic systems. Sphor [4] computed the electrostatic potentials of a film consisting of water molecules in 2D PBC and 3D PBC, respectively, he found that the potentials for EW3D approaches the results of EW2D gradually with the increase of size L_z along the z direction, it almost converges if L_z is 5 times the lateral box size $L_x=L_y$. Yeh [9] introduced a shape-dependent correction term in an EW3D method to calculate the electrostatic forces of a 2D periodic system, which is similar to the surface term in the traditional EW3D expression just with a different proportional coefficient and including only the z component of the total dipole moment. In comparison with the results from standard EW2D, the accuracy and time consumed are both greatly improved. However, the formula of the interaction derived by the authors above are only for a Coulomb system or for a dipolar system but the expressions are obtained on the basis of the formula for the system consisting of charged particles [10-12].

In this paper, we present a new Ewald summation method for a slab dipolar system in 2D periodic boundary conditions. The interaction potential is derived directly from the interaction between point dipoles like what have been done in refs. [13-15] for 3D PBC. The accuracy is ensured by comparing with the results from the summation of the dipole-dipole interaction in whole real space, and the speed of the calculation is an order of magnitude faster than the direct sum in the real space. The paper is organized as follows:

in Section 2, we derive the expression of the interaction of a dipolar system in 2D PBC, and in Section 3, analysis of accuracy and simulation results are given. At last, in Section 4, we will draw some conclusions.

2 Methods

For a dipolar system in a unit cell consisting of N particles in a uniaxial field along z direction, assuming that the directions of all particles align along z axis-the direction of the external field, the dipole-dipole interaction energy between two particles with point dipole moment $\boldsymbol{m}$ at the center of the particles $\boldsymbol{r}_i(x_i, y_i, z_i)$, $\boldsymbol{r}_j(x_j, y_j, z_j)$ can be written as

$$u_{ij} = 8\gamma\left(1-3\cos^2\theta_0\right)/r_{ij}^{*3} \, , \tag{1}$$

where $\gamma = \dfrac{\mu_0 m^2}{4\pi d^3 kT}$, d denotes the diameter of the particles, k Boltzmann's constant and T the temperature of the system; θ_0 represents the angle between $\boldsymbol{r}_{ij}=\boldsymbol{r}_i-\boldsymbol{r}_j$ and z axis and $\cos\theta_0 = z_{ij}/r_{ij}$; Definitely, $z_{ij}=z_i-z_j$, and $r_{ij}^* = \dfrac{r_{ij}}{d/2}$ is the unitless center-to-center distance between particles i and j.

Whereas, in a field rotating in the Oxy plane, the time-averaged dipole-dipole interaction energy of these two particles becomes:

$$\bar{u}_{ij}^* = -4\gamma\left(1-3\cos^2\theta_0\right)/r_{ij}^{*3} \tag{2}$$

Taking account of 2-dimensional periodic boundary conditions, i.e., the system comprises infinite replicas of the unit cell along x and y directions, the total interaction energy of the N particles in the simulation box with the whole infinite system can be expressed as

$$U = \sum_{\vec{l}}{}' \sum_{i<j} \bar{u}_{ij}^* = \frac{1}{2}\sum_{\vec{l}}{}' \sum_{i=1}^{N}\sum_{j=1}^{N} \bar{u}_{ij}^* = C\sum_{\vec{l}}{}' \sum_{i=1}^{N}\sum_{j=1}^{N}\left(1-3\cos^2\theta_0\right)/r_{ijl}^{*3} \tag{3}$$

or be written

$$U = C\sum_{\vec{l}}{}' \sum_{i=1}^{N}\sum_{j=1}^{N}\left(\frac{1}{\left|\vec{l}+\vec{r}_{ij}\right|^3} - \frac{3z_{ij}^2}{\left|\vec{l}+\vec{r}_{ij}\right|^5}\right) \tag{4}$$

where C=4γ in a uniaxial field and equals –2γ in a rotating field, $\vec{l} = l_x\vec{L}_x + l_y\vec{L}_y$, l_x, l_y are integers and $\vec{L}_x, \vec{L}_y$ indicate the vector of the unit cell. The primed summation indicates the omission of the term $i=j$ when $l=0$.

As the standard treatment of the regular Ewald summations, by a certain transformation, Eq. (4) can be written as a sum of four terms:

$$U = U^{(r)} + U_{k\neq 0}^{(k)} + U_{k=0}^{(k)} + U^{(s)} \, ,$$

where $U^{(r)}$ is the contribution from the real space, $U_{k\neq 0}^{(k)}$ and $U_{k=0}^{(k)}$ are the contributions from the reciprocal space with reciprocal vector $\boldsymbol{k}\neq 0$ and $\boldsymbol{k}=0$, respectively, and $U^{(s)}$ is the self-correction term. In a uniaxial field or in a field rotating in x-y plane, they can be written as follows:

$$U^{(r)} = C\sum_{\vec{l}}{}'\sum_{i=1}^{N}\sum_{j=1}^{N}\left[\frac{2\eta}{\sqrt{\pi}}\left(1-\eta^2 z_{ij}^2-\frac{3z_{ij}^2}{\left|\vec{l}+\vec{r}_{ij}\right|^2}\right)\frac{e^{-\eta^2\left|\vec{l}+\vec{r}_{ij}\right|^2}}{\left|\vec{l}+\vec{r}_{ij}\right|^2}+\left(1-\frac{3z_{ij}^2}{\left|\vec{l}+\vec{r}_{ij}\right|^2}\right)\frac{erfc\left(\eta\left|\vec{l}+\vec{r}_{ij}\right|\right)}{\left|\vec{l}+\vec{r}_{ij}\right|^3}\right],$$

$$U_{k\neq 0}^{(k)} = C\frac{2\pi}{A}\sum_{\vec{k}\neq 0}\sum_{i=1}^{N}\sum_{j=1}^{N} e^{i2\pi\vec{k}\cdot\vec{\rho}_{ij}}\left\{\frac{2\eta}{\sqrt{\pi}}e^{-\pi^2k^2/\eta^2}e^{-\eta^2 z_{ij}^2}\right.$$

$$\left.-\pi|k|\left[e^{2\pi|k|z_{ij}}erfc\left(\frac{\pi|k|}{\eta}+\eta z_{ij}\right)+e^{-2\pi|k|z_{ij}}erfc\left(\frac{\pi|k|}{\eta}-\eta z_{ij}\right)\right]\right\},$$

$$U_{k=0}^{(k)} = C\frac{2\pi}{A}\frac{2\eta}{\sqrt{\pi}}\sum_{i=1}^{N}\sum_{j=1}^{N}e^{-\eta^2 z_{ij}^2},$$

$$U^{(s)} = -CN\frac{4}{3\sqrt{\pi}}\eta^3; \qquad (5)$$

where $A=\left|\vec{L}_x\times\vec{L}_y\right|$ denotes the area of the box in x-y plane, η is the so-called splitting parameter or convergent parameter, and $erfc(x)=\frac{2}{\sqrt{\pi}}\int_x^{\infty}\exp(-t^2)dt$ is the complementary error function, $\boldsymbol{\rho}_{ij}$ means the projection of $\boldsymbol{r}_{ij}$ in x-y plane.

Next step we will check the validity of the expressions above by comparing the interaction energy of a dipolar system obtained by Eq. (5) with that given by Eq.3. After that the Monte Carlo simulation of phase transitions of monodispersed magnetorheological suspensions is carried out.

3 Numerical accuracy and simulation results

3.1 *Numerical accuracy*

The system employed in the estimate of the numerical accuracy is 320 particles in a box with constant L_z=10 (unitless length, the unit is *a*-the radius of one particle) and changeable L_x=L_y according to the volume fraction ϕ, the dipole moment m is a constant. Under these circumstances, the interaction energies with different volume fraction are computed according to Eq. 3 and Eq. 5, and the choice of the value for parameter η is tested. The comparison of the results (for γ=1) are listed in Table 1. In this table, l, k implies l_i (i=x, y, z) is set to $-l\leq l_i\leq l$ and $-k\leq k_i\leq k$, respectively, the reason why the unique value k=1 is given is that this value is large enough to ensure the accuracy of the total interaction. Let L_0, V_0 and P_0 be the real size, volume and pressure of the unit cell, in the following calculation, the reduced units are used : reduced length : $L=L_0/a$; reduced pressure : $P=P_0V_0/NkT$, γ reduced dipolar energy or dipole moment.

From Table 1 we can see that the total interaction energy by Ewald summation in a 2D periodic slab saturated gradually with the increase of the convergent parameter η, and the saturated Ewald sum is quantitatively consistent with the direct sum ($l\geq 25$ to meet the accuracy 10^{-5}), the relative error is 0.8% and 1.3% corresponding to ϕ=0.05 and 0.20, respectively. Besides, for a smaller value of η, a larger number of cells will be needed, on the contrary, l=1 is enough when η=3.0 ; in this case the speed of calculation is faster than when a smaller η is applied. Therefore, η=3.0 and l=k=1 will be used in the whole simulation.

Table 1. Comparison of the interaction calculated by Eq.3 and Eq. 5

Volume fraction ϕ	Splitting parameter η	cells l	Reciprocal k	Total interaction energy	
				Ewald sum	Direct sum
0.05	2.0	1	1	-70.151	-70.0884
	2.0	1	2	-70.151	
	2.0	2	1	-70.1526	
	2.0	2	2	-70.1526	
	3.0	1	1	-70.1523	
	3.0	1	3	-70.1527	
	4.0	1	1	-70.1457	
	4.0	1	3	-70.1526	
0.20	2.0	1	1	-257.5729	-257.1074
	2.0	1	1	-257.5728	
	2.0	2	1	-257.5826	
	3.0	1	1	-257.5838	
	3.0	1	3	-257.5826	
	4.0	1	1	-257.5621	
	4.0	1	2	-257.5843	
	4.0	1	3	-257.5829	

3.2 *Simulation results*

By applying Eq 5, Monte Carlo simulation of phase transitions of magnetorhological suspension with 300 particles at a constant pressure and temperature under a uniaxial field and a horizontally rotating field is carried out, respectively. As mentioned above, the splitting parameter η=3 and the number of cells in real space and reciprocal space are both unity.

Fig. 1 shows the simulation result of the system in the uniaxial field along z axis. The size of the box along z axis is kept constant, i.e. L_z=10, the reduced pressure is 0.004, and the original volume fraction is 0.05. The simulation starts from a simple cubic lattice, after the equilibrium process, all the particles distribute homogeneously in the box. When we increase γ from 1 to 16, the particles tend to align along the field direction and form individual chains of particles. In all these simulations we do not take into account the channge of dipole due to the presence of the local field because it is a second order

perturbation and not needed if we want to get some reference system in order to well understand the mechanism of phase separation in a magnetic field.
Chain-like structure is formed by the particles in the uniaxial field as shown in Fig. 1, with a final volume fraction close to the initial one. There is no double chains or columns as in

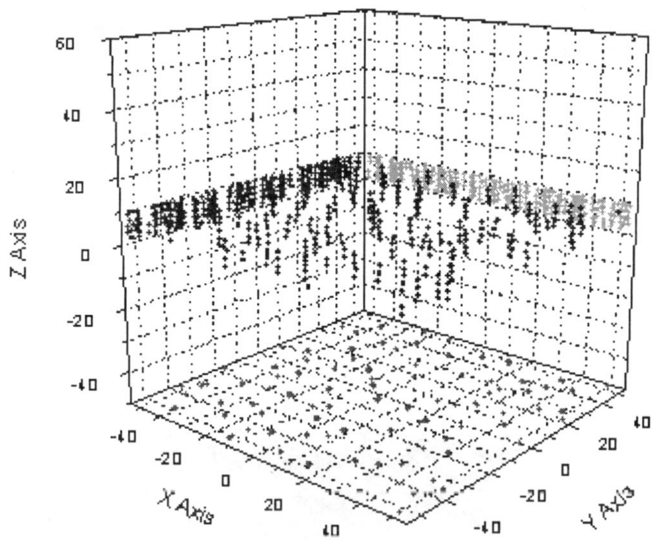

Fig. 1 chain-like structure in a uniaxial field along z axis

ER fluids because of the demagnetizing field of each chain which makes the chains to repel each other. The density profile along the z direction is depleted close to the two walls of the slab whereas it is the contrary for ER fluids[16].

If the external field is a time-dependent field rotating in the horizontal x-y plane, the structure formed by the particles is completely different from that found in the uniaxial field as shown in Fig. 2. Attractive interaction in the plane of rotation of the field [17-19] is observed, i.e., with the formation of a layered structure. Indeed if we take the derivative of Eq. 2, we obtained $f_{ij} = 24\gamma / z_{ij}^4 > 0$ and $f_\rho = -12\gamma / \rho^4 < 0$, which means that the interaction force between two particles along z direction is repulsive while the interaction is attractive if the particles are in the same x-y plane. Moreover, the detailed layer structure is related to the length of L_z. In the case of a smaller L_z, the configuration of the system is a monolayer, but the number of layers increases with the elongation of L_z as can be seen in Fig. 3D. Theoretical calculation shows that the net force between one particle outside a layer and a perfect monolayer arranged in a 2D hexagonal crystal falls very quickly to zero. However, there are some fluctuations in the z position of the particles inside one layer which could generate an attractive net force and explain that we have aggregation and formation of thick layers. Further work is being done to clarify the relationship between L_z and the number of layers.

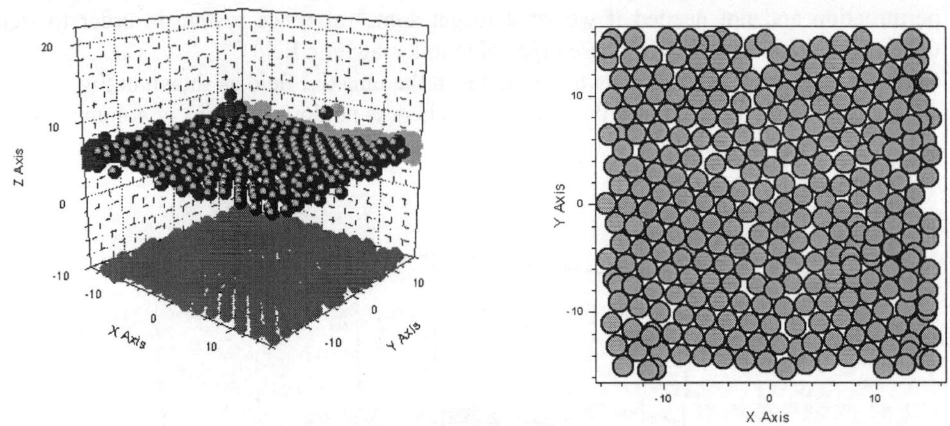

Fig. 2 3D configuration and its projection in x-y plane of a dipolar system with L_z=10 in a rotating field

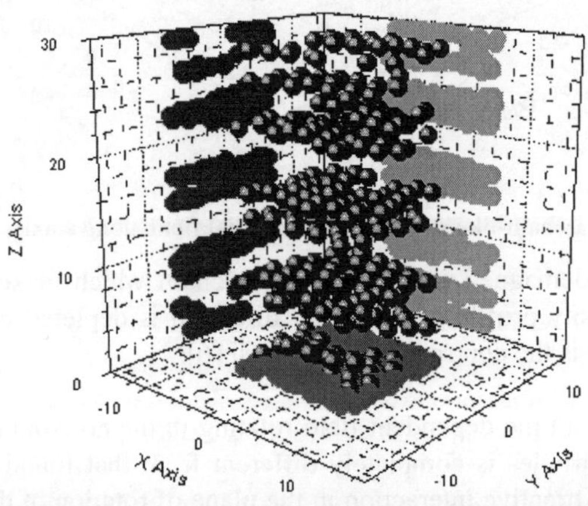

Fig. 3: 3D configuration of the dipolar system with elongated size L_z=30 in the rotating field

4 Conclusions

A new Ewald summation technique has been developed to calculate the interaction energy of a pure dipolar system without any charge in two-dimensional periodic boundary conditions with a finite length in the third direction. The accuracy of the method is examined by comparing this energy with that computed by a direct sum over replicas of the unit cell in real space. It can be found that they agree with each other quantitatively. Applying the efficient splitting parameter and the right number of replicas in real and reciprocal space, Monte Carlo simulations for the dipolar system in a uniaxial field and in a rotating field were carried out for MR like systems. The chain-like structure as well as the profile density are different from those obtained in ER fluids. In a rotating field the structure depends on the thickness of the slab: for thin slabs, we obtain monolayer of particles whereas for thick slabs we have the formation of thick layers of particles.

5 Acknowledgements

This study was funded by the program "materials" of the CNRS and one of us, S.Men, thanks the Ministry of Education for a post-doctoral fellowship. We also thank the computer Center IDRIS for free computing time.

References

1. P. P. Ewald, *Ann. Phys.* **64**, 253 (1921).
2. S. W. de Leeuw, J. W. Perram and E. R. Smith, *Proc. R. Soc. London, Ser. A* **373**, 27 (1980); **388**, 177 (1983).
3. J. Kolafa and J. W. Parram, *Molecular Simulation* **9**, 351 (1992).
4. E. Spohr, *J. Chem. Phys.* **68**, 385 (1998).
5. G. Hummer, N. Grϕnbech-Jensen, and M. Neumann, *J. Chem. Phys.* **109**, 2791 (1998).
6. D. E. Parry, *Scif. Sci.* **49**, 433 (1975); **54**, 195 (1976).
7. D. M. Heyes, M. Barber, and J. H. R. Claeke, *J. Chem. Soc., Faraday Trans. II* **73**, 1485 (1977).
8. S. W. de Leeuw and J. W. Perram, *Mol. Phys.* **37**, 1313 (1979).
9. I. Yeh, M. L. Berkowitz, *J. Chem. Phys.* **111**, 3155 (1999).
10. J. Hautman, M. L. Klein, *Mol. Phys.* **75**, 379 (1992).
11. S. Y. Liem, H. H. R. Clarke, *Mol. Phys.* **92**, 19 (1997).
12. M. Kawata, U. Nagashima, *Chem. Phys. Lett.* **340**, 157 (2001).
13. J. Perram, H. G. Petersen, and S. W. de Leeuw, *Mol. Phys.* **65**, 875 (1988).
14. M. P. Allen and D. J. Tildesley, *Computer Simulation of Liquids, Oxford Science Publications, 1 ed.* (Clarendon Press, Oxford, 1987).
15. Z. W. Wang, C. Holm, *J. Chem. Phys.* (in press).
16. S.Men, C.Métayer, A.Meunier,G.Bossis to be submitted
17. T. C. Halsey, R. A. Anderson, and J. E. Martin, *Int. J. Mod. Phys. B* **10**, 3019 (1996).
18. J. E. Martin, R. A. Anderson, and C. P. Tiggers, *J. Chem. Phys.* **108**, 7887 (1998); 110, 4854 (1999).
19. O. Volkova, S. Cutillas, P. Carletto, G. Bossis, A. Cebers, and A. Meunier, *J. Magn. Magn. Mater.* **201**, 66 (1999).

RELATIONSHIP BETWEEN CHAIN STRUCTURES AND VISCOSITY IN THE MAGNETO-RHEOLOGICAL SUSPENSIONS STABLE DISPERSING DIFFERENT CONCENTRATION OF IRON PARTICLES WITH SMECTITE

A. SHIBAYAMA, T. MIYAZAKI, T. OTOMO AND T FUJITA

Faculty of Engineering and Resource Science, Akita University, 1-1 Tegata-Gakuen cho, Akita, 010-8502, Japan,
E-mail: sibayama@ipc.akita-u.ac.jp

K. SHIMADA

Department of Machine Intelligence and System Engineering, Akita Prefectural University, 84-4, Aza-Sanuki, Tsuchiya, Honjyo, 015-0055, Japan,
E-mail: shimada@akita-pu.ac.jp

Different concentration of spherical iron particles (7 to 8 μ m) are dispersed in silicone oil to increase the stability with smectite. The concentration of iron particles is maintained between 2 to 40 volume %. The solenoid coil surrounding the cylinders applies the magnetic field in the longitudinal direction of cylinder, when the cylindrical viscometer is employed. As the magnetic field increases, the diameter of chain structure also increases. The shear stress versus shear rate is proportional relation since the viscosity enlarges as the magnetic field increases at small concentration of iron in MR fluid. Additionally, the more increase of the magnetic field strength caused the decrease of the shear stresses at large concentration of iron in MR fluid, while the share rate is increased for the open system of the cylinders. It is assumed that wider clusters or wider chain structures are partly produced in the cylinder under the certain magnetic field. Therefore, the shear stress is not uniformly increased. On the other hand, when the cone and plate viscometer is used, the magnetic field direction is perpendicular to the flow direction. The shear stress versus share rate behaves like a Bingham fluid type under the effect of the magnetic field. As increasing the iron particles volume %, the chain structures cannot increase and make other ring type structures. When the magnetic field is measured by hall probe, the magnetic field decreased at some amount of volume percent of iron in suspension because of shield effect.

1 Introduction

Magnetorheological fluid (MR fluid) is one of the well-known functional fluids. It is a suspension, which is created by dispersing several μ m size particles of iron at some volume percentage in liquid. It is generally known that the yield stress of MR fluid is quite high, (10 – 100kPa) [1,2,3,4]. It is difficult to stably disperse the magnetic particles in liquid of MR at low volume fraction since that the μ m size particles sedimentate due to gravitating force. Although a large number of studies have been made on the behavior of magnetic particles and its concentration in MR fluid. However, the influence of magnetic field direction on characteristics of MR fluids is less reported [5,6].

In this study, in order to prepare the MR fluid, different concentration of spherical iron particles (7 to 8 μ m) are dispersed in silicone oil (used as solvent) mixed with smectite (used to increase the stability). The concentration of iron particles is maintained between 2 to 40 volume %. When the cylindrical viscometer is employed to investigate the relationship between shear stress and shear rate, the solenoid coil surrounding the cylinders applied the magnetic field in the longitudinal direction of cylinders. On the other hand, when the cone and plate viscometer is used to measure the viscosity, the magnetic field direction is perpendicular to the flow direction.

In this paper, the relationship between the chain structures and viscosity (shear stress/shear rate) in the MF fluid stable dispersing various concentrations of iron particles based on the silicone oil and smectite is investigated and reported.

2 Experimental Methods

2.1 Preparation of MR fluid

Some different concentrations of spherical iron particles are prepared by the dispersing in silicone oil stabled with smectite as shown in following procedure (Fig. 1). At first, smectite modified by surfactant (specific gravity: 1.64×10^3kg/m^3) and methyl phenyl silicone oil (specific gravity: 0.995×10^3kg/m^3) is mixed and stirred while heating at 383K. Next, the iron particles of about 7–8 μm in size are added in the methyl phenyl silicone oil and smectite solution. Consequently, the MR fluid is produced as solvent of silicone oil mixed with smectite of constant concentration of 3-volume % in solvent. In this procedure, the concentration of iron particles is maintained at between 2 to 40 volume %.

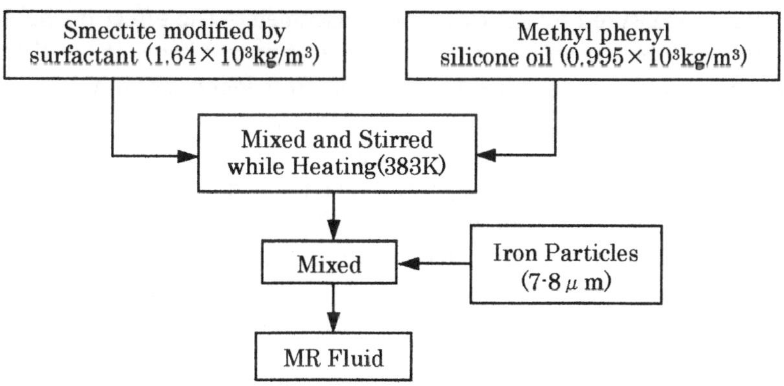

Figure 1. The preparation flowsheet for the MR fluid.

2.2 Measurement of viscosity of MR fluid by using cylindrical and cone-plate viscometer

The viscosity of the MR fluid is measured by the cylindrical and cone-plate rotational viscometer in the magnetic field while the solenoid coil surrounded the viscometer. Then, the magnetic field has two directions. One is parallel to the fluid flow, and the other one is perpendicular to the fluid flow. The schematic illustrations are shown in Figure 2 and 3. The magnetic fields were applied at the range of 0 to 0.08 or 0.1 (T). Following that, the relationship between shear rate and shear stress of the fluid is observed.

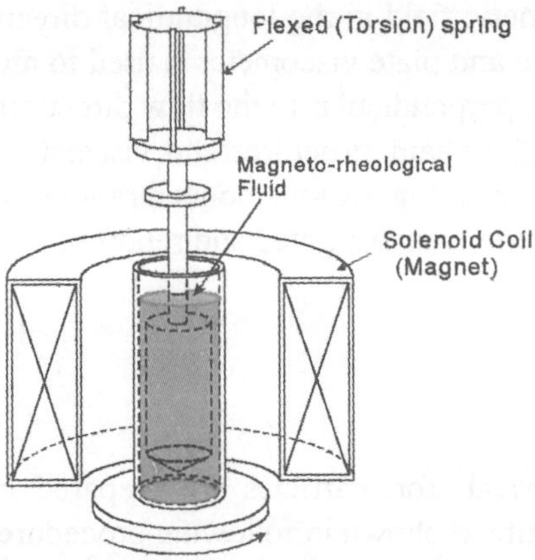

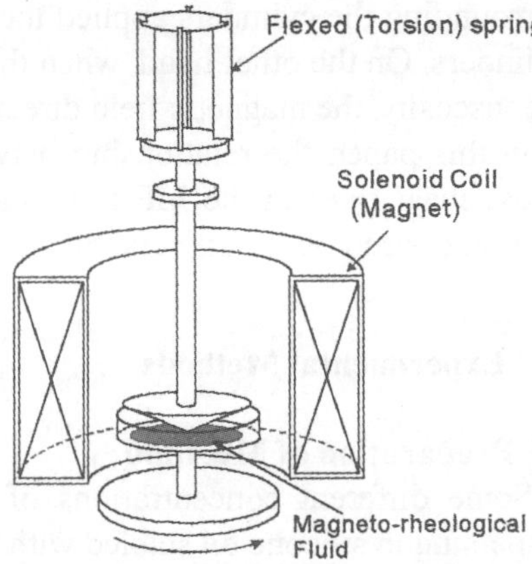

Figure 2. Schematic view of a cylindrical viscometer. **Figure 3**. Schematic view of a cone and plate viscometer.

3 Results and Discussions

3.1 Magnetic field direction: parallel to the fluid flow

At first, the shear stress versus shear rate of MR fluid at different concentration of iron particles is measured by the cylindrical viscometer. The effect of the different concentration of spherical iron particles on the relationship between the shear stress and shear rate is depicted in Figure 4. The shear stress increases with the increase of the amount of the iron volume percentage and the viscosity (shear stress/shear rate) raise with the magnetic field. When iron volume percentage in MR fluid is very small, the increase of shear stress is almost proportional to the shear rate for the used range of the applied magnetic field, therefore, the MR fluid behaves like a Newtonian fluid. However, while the iron volume percentage included in MR fluid is greater than 32.2%, the relationship between shear rate and shear stress is not proportional. In the open system of cylinders, the shear stress decreases at more than certain amount of magnetic field. If volume percentage of the iron particles contained in MR fluid is high, the viscosity increases greatly even if the small magnetic field is supplied. It is considered that at a large amount and high density of iron particles, the fluid behaves like an apparent solid. Moreover, it is proposed that the diameter of chain structure can be increased with the magnetic field. Figure 4 (e) and (f) show that the variation of the magnetic field strength caused the decrease of the shear stresses, while the share rate is increased. It is assumed that wider clusters or wider chain structures (comparing to the case when the iron volume percentage is smaller) are partly produced in the cylinder under the certain magnetic field. As a result, the fluid will not be able to contact the cylindrical inner wall due to the wider cluster formation. Therefore, the shear stress is not uniformly increased and decreased for the open system of the cylinders. One indication of this phenomenon is depicted in Figure 5. It shows the formation of the chain structure at concentration of 2.0 vol%, under the magnetic field of 0 T or 0.08 T applied vertically to MR. However, this phenomenon can not be clearly observed for the MR of high volume % due to larger number of iron

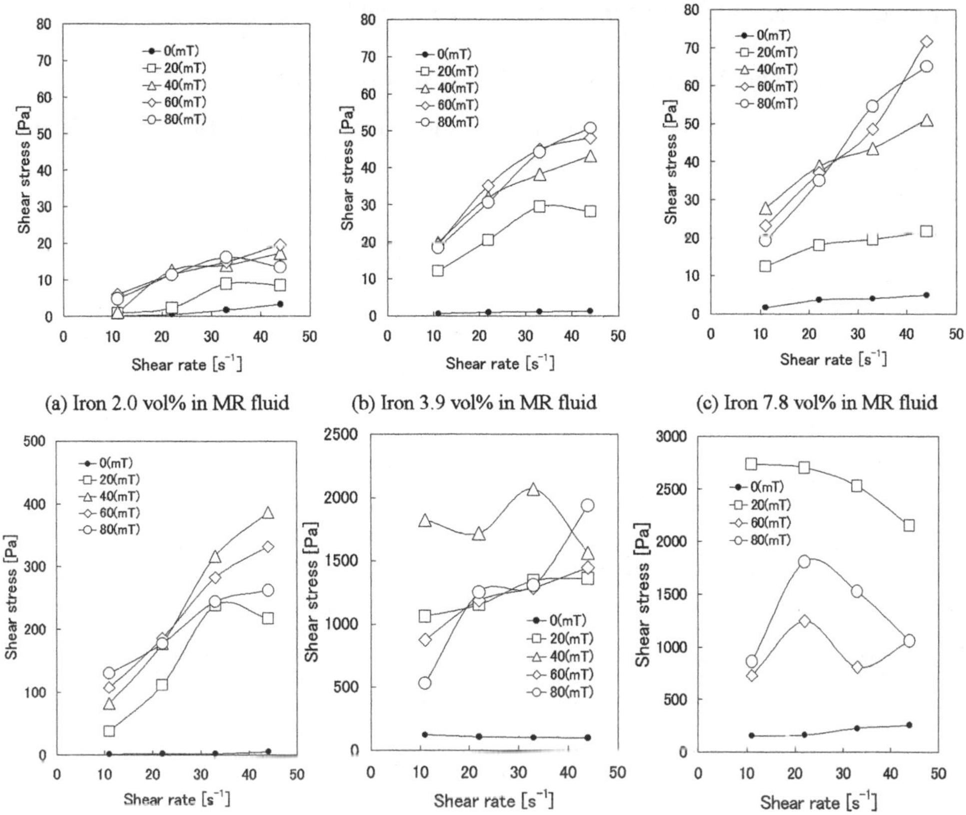

(a) Iron 2.0 vol% in MR fluid (b) Iron 3.9 vol% in MR fluid (c) Iron 7.8 vol% in MR fluid

(d) Iron 13.7 vol% in MR fluid (e) Iron 32.2 vol% in MR fluid (f) Iron 40.1 vol% in MR fluid

Figure 4. Shear stress versus shear rate of MR fluid by measuring cylindrical viscometer.
<u>Magnetic field direction : parallel to the fluid flow</u>
(different iron volume percentages in MR fluid).

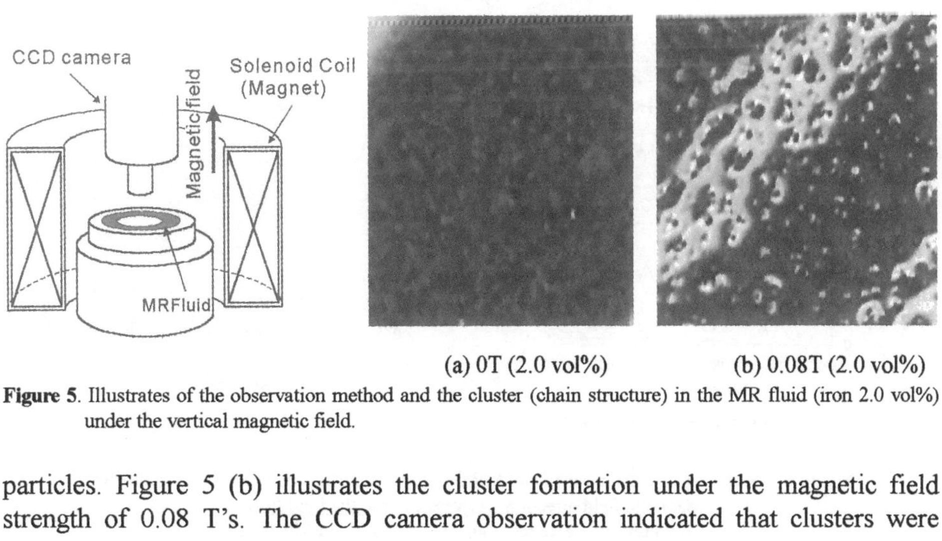

(a) 0T (2.0 vol%) (b) 0.08T (2.0 vol%)

Figure 5. Illustrates of the observation method and the cluster (chain structure) in the MR fluid (iron 2.0 vol%) under the vertical magnetic field.

particles. Figure 5 (b) illustrates the cluster formation under the magnetic field strength of 0.08 T's. The CCD camera observation indicated that clusters were arranged parallel to the direction of the magnetic field.

3.2 *Magnetic field direction: perpendicular to the fluid flow*

Next, the relationship between shear rate and shear stress of MR fluid under the magnetic field, which its direction is perpendicular to the fluid flow, is investigated by the cone and plate viscometer and the results are plotted in Figure 6. In case of the Fig. 6 (a) – (d), it can be seen that amount of iron volume percentage in MR fluid is small, the fluid behaved like Bingham fluid for each applied magnetic field.

When in the fluid of 13.7 volume % (Fig. 6 (d)) the shear stress reached the highest value compare to others iron concentrations. In addition, the yield stress is saturated at some amount of iron volume concentration in suspension. As increasing the iron particles of volume concentration in MR fluid, the chain structures cannot increase and make other ring type structures. In order to explain this phenomena, the following model in introduced. That is to say that the shield effect is occurred when the iron particles are in a large amount of volume percent in fluid. When the magnetic flux density is measured by hall probe as shown in Fig.7, the magnetic flux density decreased at some volume percent of iron in suspension because of shield effect.

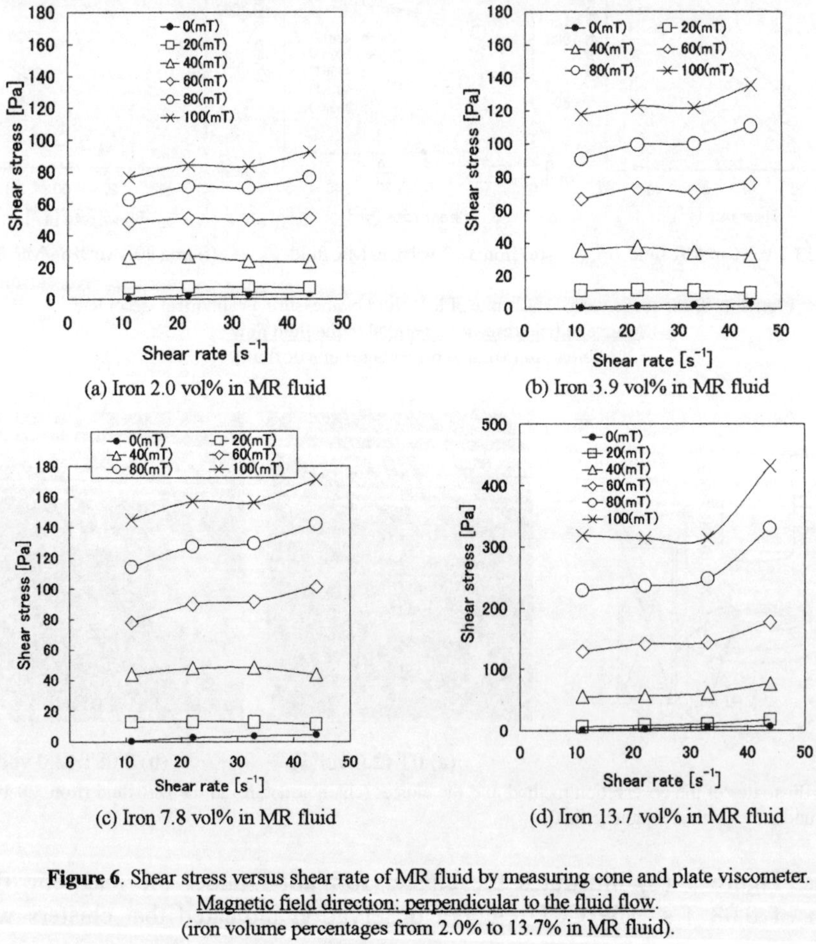

(a) Iron 2.0 vol% in MR fluid

(b) Iron 3.9 vol% in MR fluid

(c) Iron 7.8 vol% in MR fluid

(d) Iron 13.7 vol% in MR fluid

Figure 6. Shear stress versus shear rate of MR fluid by measuring cone and plate viscometer. <u>Magnetic field direction: perpendicular to the fluid flow.</u> (iron volume percentages from 2.0% to 13.7% in MR fluid).

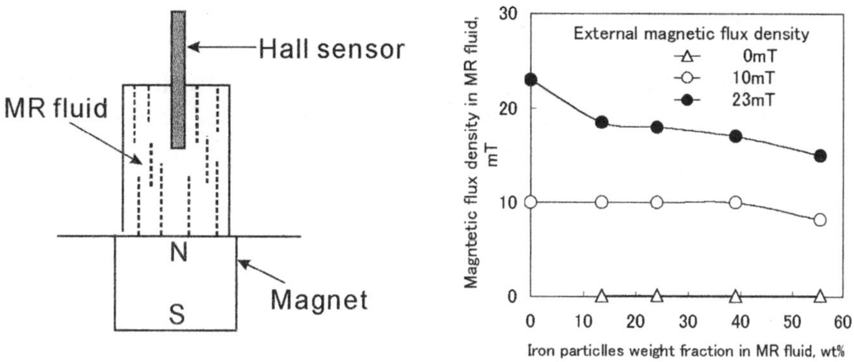

Figure 7. Magnetic flux density in MR fluid that containing different iron weight fraction measured by Hall sensor.

Finally, Figure 8 shows the photographs of iron particles and clusters in the MR fluid under non-magnetic (0T) and magnetic field (0.01T), while the concentration of iron particles was 3.9 % and 13.7 %, respectively. In these photos illustrates the chain structures created by the iron particles. It is apparent that the clusters are clearly formed while the concentration of MR is maintained at 3.9 vol%. However, when the concentration of MR increased at 13.7 vol%, the cluster formation is unclear due to the coagulation or wider cluster formation as the volume percentage increased.

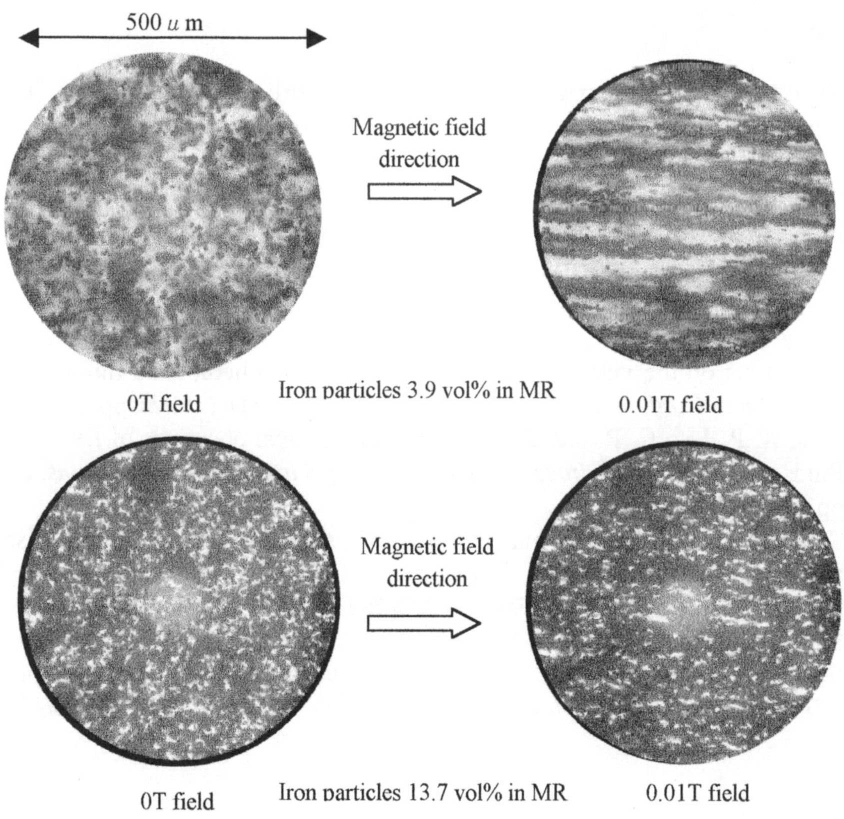

Figure 8. Photographs of iron particles and cluster in the MR fluid with non-magnetic (0T) and magnetic (0.01T) field.

4 Conclusion

In order to investigate the relationship between the chain structures and viscosity in the magneto-rheological suspension, the MR fluids containing different concentration of spherical iron particles dispersed in methyl phenyl silicone oil mixed with smectite to increase the stability are prepared. When the cylindrical viscometer, which its the magnetic field direction is parallel to the fluid flow, is employed, the relationship between the shear rate and shear stress is almost proportional and the viscosity raised as the magnetic field increases. Additionally, the more increase of the magnetic field strength caused the decrease of the shear stresses at large concentration of iron in MR fluid, while the share rate is increased. It is assumed that wider clusters or wider chain structures are partly produced in the cylinder under the certain magnetic field. Therefore, the shear stress is not uniformly increased. On the other hand, when the cone and plate viscometer, which its the magnetic field direction is perpendicular to the fluid flow, shear stress versus shear rate behaves like a Bingham fluid type under the magnetic field. Moreover, the yield stress is saturated at some amount of iron volume concentration of suspension. Finally, it is considered that the chain structures cannot increase and make other ring type structures as the volume percentage of the iron particles increase.

References

1. Fujita T., Yoshimura K., Seki Y., Dodbiba G., Miyazaki T. and Numakura S., Characterization of Magnetorheological suspension for seal, *Jour. Of Intelligent Material Systems and Structures,* **10** (2000), pp.770-774.
2. Fujita T., Ito R., Kuzuno E. and Kawada K., Preparation and properties of silicone oil based magnetic fluid, *Jour. of Japan Society of Powder and Powder Metallurgy,* ***43*** (1996), pp.757-760.
3. Fujita T., Jeyadevan B., Kuzuno E. and Nakagawa N., Formation of clusters and their effect on the viscosity of magnetic and electrorheological fluids, *Proc. Of the 6th Inter. Conf. on ERF, MRS and their Applications,* (1997), pp. 578-585.
4. Zhao X. P., Luo C. R. and Zhang Z. D., The optical character of ER and EMR fluids, *Proc. Of the 6th Inter. Conf. on ERF, MRS and their Applications,* (1997), pp. 827-833.
5. Fujita T., Jeyadevan B., Yamaguchi K. and Nishiyama H., Preparation, Viscosity and damping of functional fluids that respond to both magnetic and electric field, *Powder Technology*, **101** (1999), pp.279-287.
6. Shimada K., Nishida H. and Fujtia T., Differences in steady characteristics and response time of ERF on rotational flow between rotating disk and concentric cylinder, *Proc. Of the 7th Inter. Conf. on ERF and MRS,* (1999a), pp. 556-563.

III. MATERIAL PROPERTIES

EFFECT OF RARE EARTH SUBSTITUTION ON ELECTRORHEOLOGICAL PROPERTIES OF TIO_2

X. P. ZHAO, J. B. YIN, L. Q. XIANG, Q. ZHAO

Institute of Electrorheological Technology, Department of Applied Physics, Northwestern Polytechnical University, 710072 Xi'an, China

Email: xpzhao@nwpu.edu.cn

This paper describes a new class of water-free electrorheological (ER) fluids based on nonaqueous doped TiO_2 with rare earth (RE) in silicone oil. The thermal character and crystal structure of these materials are investigated with DSC, TG and XRD. The doped TiO_2 crystals possess anatase phase and their lattice spacing varies significantly with the content of rare earth. The rheological measurements show that the doped TiO_2 ER fluid exhibits an obviously higher shear stress than that of pure TiO_2 ER fluid under dc electric field. Especially, substitution with 10mol% cerium or 8mol% lanthanum for Ti can obtain a relatively high shear stress. On the basis of dielectric and conduction measurements, we preliminarily discuss the influence of the doping of rare earth on ER effects of TiO_2.

1 Introduction

Electrorheological (ER) fluids consist of a suspension of dielectric particles in insulating liquid of low dielectric constant [1]. Its viscosity increases dramatically in the presence of an applied electric field. This property makes ER fluids have potential for electrically controllable shock absorbers, clutches, valves, damping devices and so on [2]. However, there exist a number of limitation in present ER fluids including low yield stress, poor temperature dependence and sedimentation [3]. Much effort has been made to improve the properties of ER fluids. Except water-containing ER materials, the water-free ER materials including alumino-silicate, conjugated semiconducting organic compounds, micro-structured composite particles, even nano-composites (whose ER mechanism is attributed to a interface polarization) have been investigated in ER system in past years [4~8]. Continuous attention has been paid to inorganic crystalline such as TiO_2, $BaTiO_3$ etc. as potential ER materials due to their high dielectric constant [9,10]. Unfortunately, the ER activity of these kinds of materials is fairly weak in dry state [9,11]. With the help of water or surfactant, the ER activity can be improved [3,10,12]. But this extrinsic effect also results in the poor temperature dependence. In the paper, we will present a new ER system composed of doped TiO_2 with rare earth (RE) dispersed in silicone oil. The choice of substitution for Ti with large radius rare earth ions is aimed at modifying the intrinsically dielectric and polarization properties of TiO_2, so as to improve its ER activity. The basic rheological measurement shows that the shear stress of doped TiO_2 ER fluid is obviously higher than that of pure TiO_2. The maximum value appears when the content of cerium dopant is closed to 10mol% and lanthanum dopant is closed to 8mol%. Based on investigation about dielectric and conduction of doped TiO_2 ER fluids, we discuss the effect of rare earth substitution on ER properties of TiO_2.

2 Experiment

2.1 Materials Preparation

The particles of pure and doped TiO_2 with different rare earth content were synthesized by the means of sol-gel. Tetrabutyl titanate, rare earth nitrate (or RE chloride, RE is La, Ce, etc light rare earth elements) and dilute water were solved into anhydrous alcohol.

The H_2O/Ti molar ratio was 1.5 and Ti concentration was 0.8~1.0 M. The resulting clear yellow solution, whose pH value was controlled in range of 3~4 with HNO_3, was aged at room temperature for 12~48h to form transparent gel. The gels were then dried at 80°C and subjected to the following stepwise calcination programs in air in a furnace: 2h at 200°C, 2h at 400°C, and 2h at 550°C.

DSC-TG technique was used to study the thermal process of xerogel and the thermal character of particles. The crystalline pure and doped TiO_2 have been produced and the result is in good agreement with X-ray diffraction patterns. XRD (D/MAX- γ A, Japan) patterns of particles obtained after calcination show no organic compounds in products. The pure and doped TiO_2 particles exhibit anatase phase, and no other phases are found when Ce/Ti is smaller than 10mol% and La/Ti is smaller than 8mol%. Fig 1 gives the XRD pattern for typical doped TiO_2 with 6 mol% La and 8 mol% Ce . The value of lattice spacing (d) increases significantly with the content of rare earth and becomes saturated when Ce/Ti is beyond 10mol% and La/Ti is beyond 8mol%, which indicates that more rare earth fails to substitute for Ti.

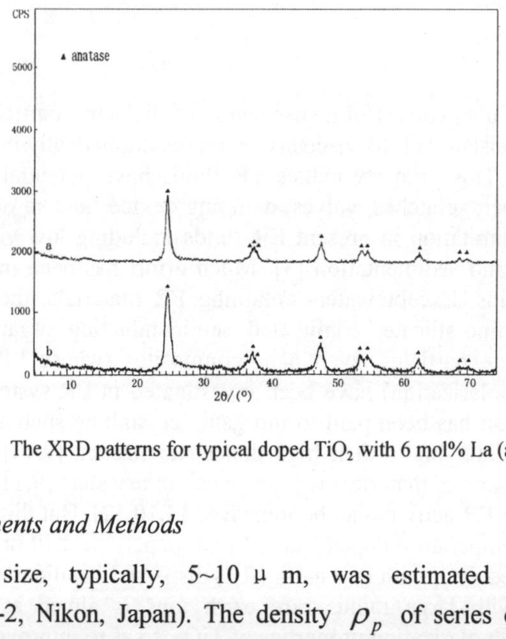

Figure.1 The XRD patterns for typical doped TiO_2 with 6 mol% La (a) and 8 mol% Ce (b)

2.2 *Measurements and Methods*

The particles size, typically, 5~10 μ m, was estimated by optical microscopy (ALPHAPHOT-2, Nikon, Japan). The density ρ_p of series of pure and doped TiO_2 powder was in the range of 3.68~3.80g/cm^3. The powders were dehydrated for 8h in vacuum at 150°C. A Fourier Transform Infrared Spectroscopy Spectra (EQUIOX55, Germany) was used to investigate the moisture adsorption of dried particles. In IR spectra we can note that the H_2O-banding adsorption band at about 1640 cm^{-1} is not seen, which indicating the water-free particles have been obtained after the treatment above. Then the particles were mixed quickly with dimethy-silicone oil (also dehydrated for 4h in vacuum at 150°C) in vacuum. The dielectric constant ε_f and conductivity σ_f of silicone oil were 2.70 and 10^{-12}S/m, respectively, the density ρ_f 1.003g/cm^3, and appearing viscosity η about 500mPa • s at 25°C. No additives were added into these ER fluids.

Rheological properties were measured with a coaxial cylinder rotational viscometer (NXS-11, China) modified for electrorheological measurements. The inner bob and outer cup were applied as two electrodes the gap between bob and cup was 2.0mm. The current density through ER fluids was recorded by a micro-ammeter.

Due to the difficulty of measuring dielectric properties of particles, a LCR meter (WK-4225) measured dielectric properties of the ER fluids. Taking care of the dielectric properties of ER suspensions depended on the structure induced by the applied electric field [13], we keep the suspensions in randomly dispersed system in order to make a relatively reasonable comparison between pure and doped TiO_2. The dielectric constant was derived from the measured capacitance C according to the relation, $\varepsilon = Cd/(\varepsilon_0 S)$, where ε_0 is the dielectric constant of vacuum, *ie* 8.85 $\times 10^{-12}$ F • cm^{-1}, d the thickness of gap between electrodes and s the contact areas of electrodes. The dielectric loss tangent (tan δ) was measured directly from LCR meter .

3 Results and discussion

3.1 Effect of rare earth substitution on ER properties of TiO_2

The measurement results of rheological properties reveal that the shear stress of pure TiO_2/silicone oil ER fluid is very low evenly under high dc electric field. Doping of rare earth is found to obviously improve the ER effect of TiO_2. The typical results of the shear stress of La-doped TiO_2 ER fluid as a function of field strength are given in Fig.2. At the same time, we find that the lowest electric field strength E_c inducing ER effects for doped TiO_2 ER fluid is obviously smaller than that for pure TiO_2. It indicates the ER effect is more easily induced by the applied electric field for doped TiO_2.

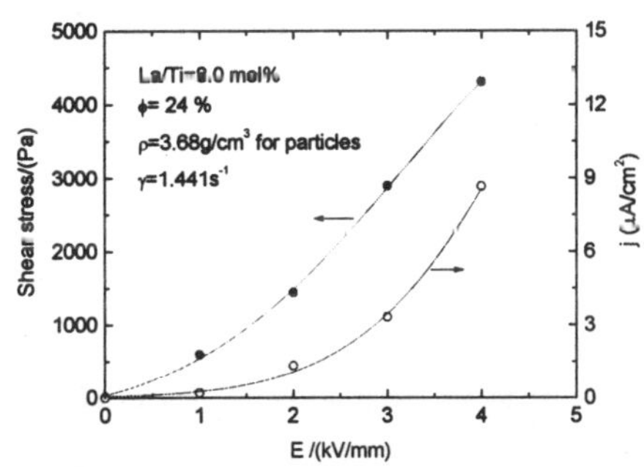

Figure.2 The shear stress of doped TiO_2 ER fluid as a function of dc electric field strength at 25°C

Especially, we find that substitution for Ti with 10mol% Ce or 8mol% La can obtain a relatively high shear stress. Fig.3 shows the dependence of shear stress of doped TiO_2/silicone oil ER fluid on the content of rare earth. Here, the volume fraction ϕ of ER fluids is defined 18% which remains ER fluids in low viscosity at zero fields in order to reasonable comparison. Worthy of note that other phases of rare earth oxide appear in XRD pattern when RE/Ti molar ratio exceeds the same limit concentration, which reveal further substitution for Ti fails when Ce/Ti or La/Ti exceeds the saturation value, 10mol%

or 8mol% respectively. It follows that TiO_2 crystal may be coated by rare earth oxide, which results in preventing the modification. Fig.4 shows the relationship between shear stress and shear rate for typical Ce-doped TiO_2/silicone oil ER fluid at ϕ 27% and dc electric field. In the range of the shear rate measured, a remarkable ER effect is seen. The yield stress is about 4.8kPa at 3kV/mm and 7.4kPa at 4kV/mm, which is 10 times larger than that of pure TiO_2 ER fluid.

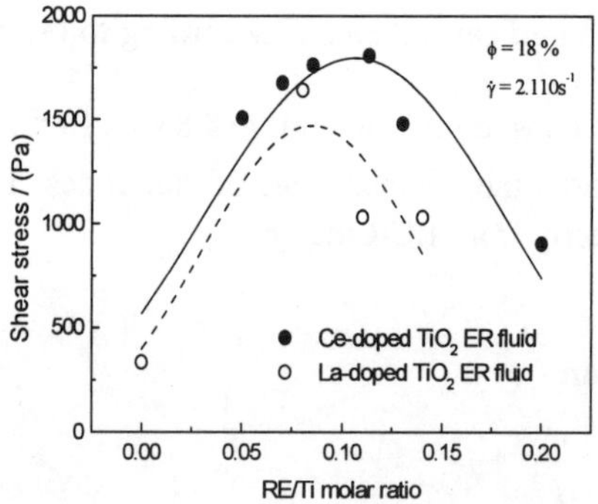

Figure.3 The dependence of shear stress of doped TiO_2/silicone oil ER fluids on the content of rare earth at 20°C and 3kV/mm dc field

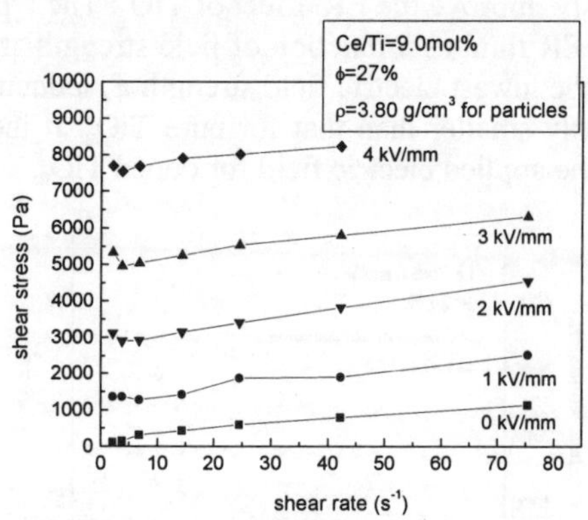

Figure. 4 The shear stress of a typical Ce-doped TiO_2 ER fluid as a function of shear rate at 25°C

In addition, a striking difference of temperature effect of shear stresses between pure and RE-doped TiO_2 ER fluids is also observed. The operational temperature range of ER fluids can be amplified through doping with rare earth. The maximum shear stress of doped TiO_2 ER fluid appears around 80°C, and the shear stress are still higher after 100°C than at room temperature. But the shear stress of pure TiO_2 ER fluid declines when temperature exceeds 40°C. More detailed investigation about the temperature effect of doped TiO_2 ER fluid has been reported in Ref. [14].

3.2 Dielectric and conduction properties

Dielectric constant of pure TiO_2 ER fluid is almost independent of frequency. The dielectric constants of doped TiO_2 ER fluids are much larger than that of pure TiO_2 ER fluid at 100 Hz, decrease with frequency and become same at high frequency. It respects the conduction influence at low frequency. More worthies to note the dielectric loss of ER fluids are obviously enhanced through doping rare earth into TiO_2. Fig.5 gives the dependence of dielectric loss tangent ($\tan\delta$) of ER fluid on cerium content. The dielectric loss tangent reaches the maximum when Ce/Ti is about 10 mol%. The striking difference in temperature dependence of dielectric properties between pure and doped TiO_2 ER fluids is also found in Fig.6. The dielectric constant of doped TiO_2 ER fluid increases with temperature and becomes saturate after 85°C. The dielectric loss tangent appears the peak about 70°C, indicating the remarkable dielectric relaxation over the measured temperature range. On the contrary, the dielectric constant and loss tangent of pure TiO_2 ER fluid are almost independent of temperature. Because the dielectric constant (~2.7)and loss ($\tan\delta$~0.002) of silicone oil remain relative independent of frequency and temperature in 0~100°C, it is believed that the dielectric properties of ER fluids are mainly influenced by the dielectric properties of dispersal particles in silicone oil according to the mixture model [15,16]. Therefore, the results of dielectric measurements represent that doping with rare earth causes changes in polarization properties of TiO_2. Furthermore, the increase of dielectric loss and the occurrence of dielectric relaxation at high temperature in doped TiO_2 ER fluid somehow mean a large proportion of the defect dipolar orientional polarization and the interfacial polarization. To further understand, we also measure the dc conductivity and ac conductivity of ER fluids and their temperature effects [14]. Excitingly, we find doping not only increase the conductivity but also cause a striking difference in temperature dependence. The dc conductivity and ac conductivity of doped TiO_2 ER fluid increase with temperature. In contrast, the dc conductivity and ac conductivity of pure TiO_2 ER fluid remain independent of temperature. This infers the thermally activated charge carriers in doped TiO_2 are more than in pure TiO_2. The thermally activated charge carriers may help to enhance Maxwell-Wagner interface polarization that is revealed with a dielectric relaxation in temperature range investigated.

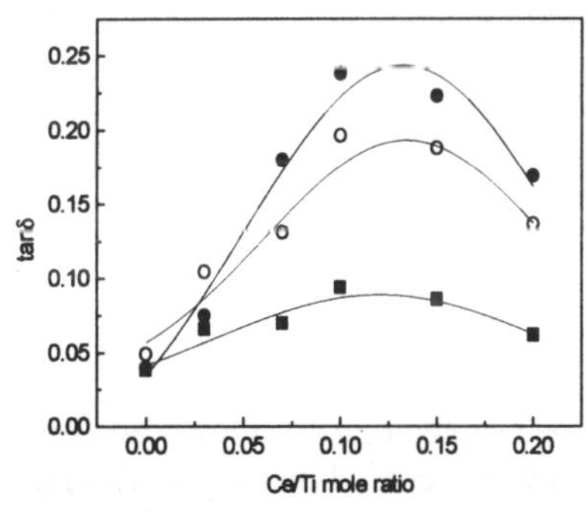

Figure. 5 The curve dielectric loss tangent of properties ER fluids (18% volume fraction) with different Ce/Ti mole ratio vs. frequency at 25°C (● for 100Hz, ○ 1kHz and ■ 10kHz)

According to the widely accepted studies [18~21], particles polarization is considered to be responsible for the interaction force that lead to the rheological change of ER fluid. The parameter in connection with articles polarization, such as large dielectric constant and suitable dielectric loss (>0.1 at 1kHz) or conductivity (~10^{-7} S/m) dominate ER effect. Comparing rheological with dielectric or conduction properties, we can clearly find doping with rare earth remarkably enhances the dielectric loss or conductivity of TiO_2, which result in increasing ER effect. Especially, the increase of dielectric constant, dielectric loss or conductivity with temperature in doped TiO_2 ER fluid can induce the further improvement of ER effect. It's cleared that weak ER activity of pure TiO_2 ER fluid is attributed to its low dielectric loss or conductivity. Moreover, the dielectric properties of pure TiO_2 is independent of temperature, which can not reinforce ER activity as temperature increase. Thus the shear stress rapidly decrease due to thermal convection.

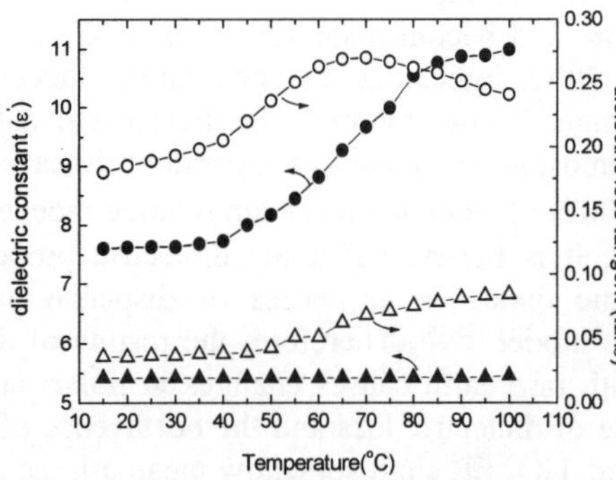

Figure. 6 Temperature dependence of dielectric of pure TiO_2 ER fluids (▲ and △) and typical Ce-doped TiO_2 ER fluid (● and ○) at 1kHz (Ce/Ti =9 mol %, 18% volume fraction)

Practically, the dielectric properties are strongly influenced by the presence of crystal state such as lattice distortion, defects and impurity [16,16]. It is found in the XRD patterns that substitution for Ti^{4+} by rare earth ions having large ionic radius (much larger than that of Ti^{4+}) cause expanding lattice distortion or defects in TiO_2. Thus, the potential factor influencing polarization or charge carriers state may be responsible to modification of dielectric and conduction properties of TiO_2 [17]. Unfortunately, the appearance of other phase shows that the further substitution of rare earth ions for Ti^{4+} fails when Ce/Ti mole ratio exceeds the saturation value, 10 mol%. It follows that TiO_2 crystal may be coated by cerium oxide, which make dielectric constant, loss and conductivity decrease so that ER activity becomes weak. But the work of understanding further mechanism about this new kind of ER material should be done.

4 Conclusion

A new class of water-free electrorheological fluids based on water-free doped TiO_2 with rare earth (RE) in silicone oil was prepared by means of Sol-gel technique. It is found that doping with rare earth can enhance the ER effect of TiO_2 and amplify the operational temperature range. More interesting is the impotent influence of content of rare earth substitution for Ti on ER effect. Substitution for Ti with 10 mol% Ce or 8 mol% La obtained a relatively high shear stress. We attributed the improvement of ER effect of doped TiO_2 to the modification of dielectric and conduction properties, particularly increase of dielectric loss or conductivity, by doping with rare earth. The striking difference in temperature dependence of dielectric and conduction properties can explain the different temperature effect between pure and doped TiO_2 ER fluids.

5 Acknowledgments

The authors are grateful to the National Natural Science Foundation of China (No.59832090) and the National Natural Science Foundation of China for distinguished Young Scholars (No.50025207) for supporting the research.

References

1. Halsey T. C., *Science*, 258, (1992):761.
2. Couter S. P., Weiss K. D., and Carlson J. D., *J. of Intelligent Materials System and Structure*, 4(2), (1993)248.
3. Block H., and Rattay P., *Progress in Electrorheology*, ed. by Havelka K. O. and Filisko F. E., (Plenum Press, New York, 1995) pp19.
4. Filisko F. E., and Radzilowski L. H., *J. Rheol.* 34, (1990):539.
5. Zhang Y. L., Ma Y., Lan Y. C. and Lu K. Q., *Appl. Phys. Lett.* 75, (1998):1326.
6. Block H., Kelly J. P., Qin A., and Wastson T., *Langmuir*, 6, (1990):6.
7. Tam W. Y., Yi G. H., Wen W., Ma H., Loy M. M. T., and Sheng P., *Phys. Rev. Lett.*, 78, (1997):2987.
8. Kim J. W., Kim S. G., Choi H. J., et al, *Proc. of the 7th Int. Conf. on ER fluid and MR suspensions*, ed. by Tao R., (World Scientific Press, Singapore, 2000) pp111.
9. Filisko F. E., *Progress in electrorheology*, ed. by Havelka K. O. and Filisko F. E., (Plenum Press, New York, 1995) pp3.
10. Rankin P. J. and Klingenberg D. J., *J. Rheol.*, 42(3), (1998):639.
11. Garino T, *ER fluids-mechanisms, properties, structure, technology and application,* ed. by Tao R. (World Scientific Press, Singapore, 1992) pp16.
12. Ostub Y., *J. of Colloids and Surfaces*, 58(1&2), (1991):73.
13. Blackwood K. M., Block H., Rattray P., et al, *Proc. Of the 4th Int. Conf on ER ER fluids-mechanisms, properties, structure, technology and application,* ed by Tao R. and Gay G. D., (World Scientific Press, Singapore, 1992)pp1
14. Yin J. B. and Zhao X. P., *J. of Phys D: Appl. Phys.*, 34(7), (2001)(in press)
15. Conrad H., Spreeher A. F., Choi Y., and Chen Y., *J. Rheol.*, 35(7) 1991:1393
16. Guan Z. Z., Zhang Z. T., and Jiao J. S., *Physics Properties of Inorganic Materials* (Beijing: Tsinghua University Press,1992)
17. Lee S. J., Kang K. Y, , and Han H. K., *Appl. Phys. Lett.,* 12,(1998):1784
18. Block H., and Kelly J. P., *J. of Phys D: Appl. Phys.*, 21,(1998):1663
19. Wu C. W. and Conrad H., *Phys. Rev. E*, (56),(1997):5789.
20. Whitte M., Bullough W. A. Peel, and D. J. and Froozian R., *Phys. Rev. E*, 49(6), (1994):5249.
21. Hao T., *Appl. Phys. Lett.*, 70,(1997):1956

ELECTRORHEOLOGICAL EFFECT IN SUSPENSIONS OF $AgI/Ag_2O/V_2O_5/P_2O_5$ GLASSES

MARCIN OSUCHOWSKI AND JANUSZ PŁOCHARSKI

Department of Chemistry, Warsaw University of Technology, ul. Noakowskiego 3, 00-664 Warszawa, Poland

The ER effect results from bulk and surface electric polarization processes in solid grains of ER suspensions but detailed mechanisms are not very clear. The aim of the present study was to find correlations between the character of bulk charge transport processes (electronic and/or ionic) in particles of a dispersion and parameters of the ER effect. As the dispersed phase we used glasses comprising oxides of silver, vanadium and phosphorus with addition of silver iodide. Bulk electric properties of this material could be modified without changing other properties influencing the ER effect like porosity, shape of grains, hardness, affinity to a liquid matrix etc. Variations of concentration of the components result in changes of ionic and electronic conductivity whilst other properties remain constant. The suspensions of the powdered glasses showed relatively high ER effect. The dynamic yield stress figured from 35 to 160 Pa at 2.0 kV/mm for 12 % concentration by volume. The highest values were observed for ionically conducting samples. The values of relaxation frequencies (f_R) based on bulk properties of the glass samples were calculated and correlated with the yield stress whose maximum was obtained for samples of f_R close to 100 kHz. High ER effect was observed also for samples of f_R in the MHz range but in this case different polarization mechanism was postulated. The influence of polarization mechanisms on rheological behavior of the prepared fluids was discussed.

1 Introduction

It is believed that the ER effect results from bulk and surface electric polarization processes in solid grains of ER suspensions. The detailed relationships are, however, not very clear and significantly more research is required to establish a firm basis for a successful development of technical applications. The phenomenon seems to be more related to transport of electrons and/or ions within particular regions of an ER fluid than to a high mismatch of electrical permittivity. The apparent lack of the effect in dry suspensions of ferroelectric materials [1] supports this idea. In case of water activated systems the polarization processes are confined to the water based phase absorbed or adsorbed on grains of highly polar nature. The water free systems also contain polar molecular units or structures. The examples are polyacenequinones [2] or aluminosilicates [3]. The water free ER fluids utilizing polar polyaniline [4] (or its derivatives) are also of this kind. Some exception are suspensions of doped non-polar poly(*p*-phenylene) in which case ionic contribution to overall conductivity is required to get the ER effect [5,6].

One of the characteristic features of the ER research is that a great variety of substances was investigated but comparative studies were scarce. This situation is not helpful for generalizations that is the prerequisite for the practical implementation of the ER technology. The particular aim of the present study was to find correlation between mechanisms and magnitude of bulk charge transport processes (electronic and/or ionic) in the grains and parameters of the ER effect in their suspensions. To realize this task we had to find a material whose bulk electric properties could be modified without changing other properties influencing the ER effect like porosity, shape of grains, hardness, affinity to a liquid matrix etc. Our attention has been drawn by the glassy system comprising oxides of silver, vanadium and phosphorus with addition of silver iodide [7]. Variations of concentration of the components result in significant changes of ionic and electronic conductivity whilst other properties remain almost constant. The analysis of obtained results should be helpful for understanding the detailed mechanism of the ER phenomenon.

2 Experimental

Preparation of bulk glasses:

Weighted amounts of AgI, $AgNO_3$, V_2O_5 and $NH_4H_2PO_4$ were ground together in a mortar and placed in a porcelain crucible that was then put into a furnace heated to 800°C. After one hour anneal in air atmosphere the molten content was quickly poured out onto a thick steel slab and additionally covered with a stainless steel plate. The rapid cooling of the molten material was required to prevent crystallisation. About 1 mm thick flat pieces of glassy material were obtained. In all cases their structure was checked by X-ray diffraction not showing any presence of crystalline phases. The composition of initial mixtures of components is given Table 1. The final composition of the glasses can be slightly different due to the volatility of some ingredients.

The rheological properties of a dispersion depend on volume concentration of solids and all prepared glasses had different values of specific weight. Therefore, density of all glasses was measured by means of the pycnometer method.

In order to estimate the electric properties of the prepared glasses gold electrodes were first vacuum evaporated onto the glass platelets and then the impedance spectroscopy system comprising the frequency response analyzer Solartron 1255 and EG&G PARC 273A potentiostat was applied. The complex impedance of all samples was measured within the frequency range of 1 mHz to 100 kHz at ambient temperature.

Preparation of ER fluids:

The prepared glass ingots were broken into smaller pieces and then ground to obtain powders. The granulometric parameters of the powders were measured in dilute suspensions in silicone oil by means of an optical microscope Nikon Alphaphoto2 equipped with a digital camera and a computer program for picture analysis. The results were very similar for all prepared samples. Most of grains (>90%, averaged by number) were below 30 μm with maximum of the distribution at about 12 μm. In all samples we found also few grains as big as 70 μm.

The silicone oil (polydimethylsiloxane) was dried by prolonged heating in high vacuum. All components were stored in a desiccator and their contact with laboratory air was as short as possible.

The ER fluids were prepared by mixing in a mortar weighted amounts of silicone oil of 0.1 Pas viscosity and a powdered glass sample. Concentration of all dispersions was 12 % by volume.

Measurements of ER effect:

ER properties of the fluids were studied by means of a PC controlled Bohlin Visco 88 viscometer with home modified measuring system operating in a two concentric cylinders geometry capable to apply a *dc* electric field up to 2.5 kV/mm. The data were taken within the shear rates of 12 – 1100 s^{-1}. The values of dynamic yield stress were calculated by extrapolation. All experiments were carried out at ambient temperature of 20-23 °C.

3 Results and Discussion

The interpretation of the impedance spectra was based on the work of Wasiucionek *et al.* [7]. The properties of the glass platelets with gold electrodes could be approximated by the general equivalent circuit shown in Fig.1. The physical sense of the individual elements of the circuit are as follows:

Re – resistance resulting from electronic contribution to conductivity;
Ri – resistance resulting from ionic contribution to conductivity;
Rc – contact resistance at gold/glass interface;
Qg – constant phase element related to geometric capacity of sample;
Qdl – constant phase element related to double layer capacity at gold/glass interface;
Qc – constant phase element related to contact capacity at gold/glass interface.

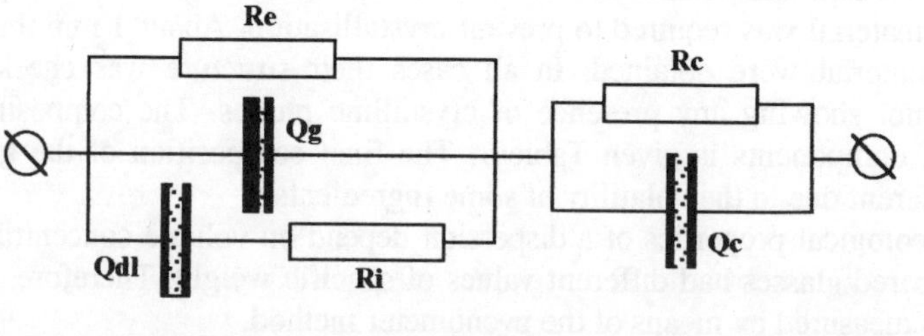

Figure 1. General equivalent circuit applied for interpretation of the impedance measurements.

The particular impedance spectra of the glass samples did not contained all the elements of Fig.1. Three groups of the spectra could be distinguished. The samples of pure electronic conductivity gave spectra of the type shown in Fig.2. In this case **Ri** = ∞. The typical spectrum of samples of pure ionic conductivity is presented in Fig.3. In this case **Rc** = 0 and **Re** = ∞. The samples of mixed conductivity gave spectra with two time constants of the type presented in Fig.4. In this case **Rc** = 0. In order to obtain the individual values of the equivalent circuit elements the program Equivalent Circuit v.4.55 by B.A.Boukamp was used. Then the bulk electrical parameters of the glass samples were estimated. Depending on the composition the total conductivity varied from 10^{-7} to 10^{-5} S/cm resulting from electrons, ions or both of them as the charge carriers. The basic electric parameters of the studied glasses are given in Table 1. In the studied system the type of conductivity was mostly determined by silver iodide whose increasing concentration changed conductivity from electronic to ionic.

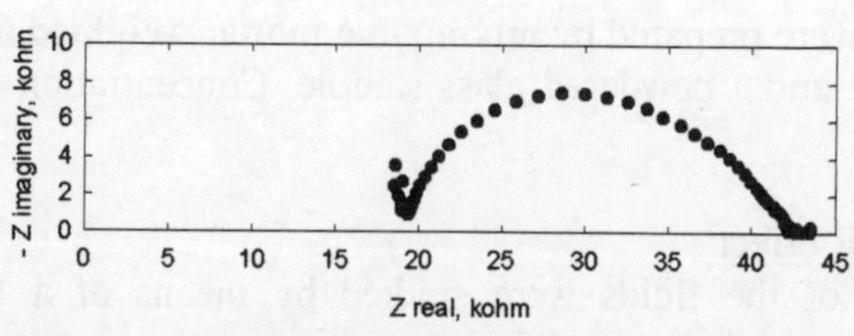

Figure 2. Impedance spectrum of the glass 3/15/67/15. Pure electronic conductivity.

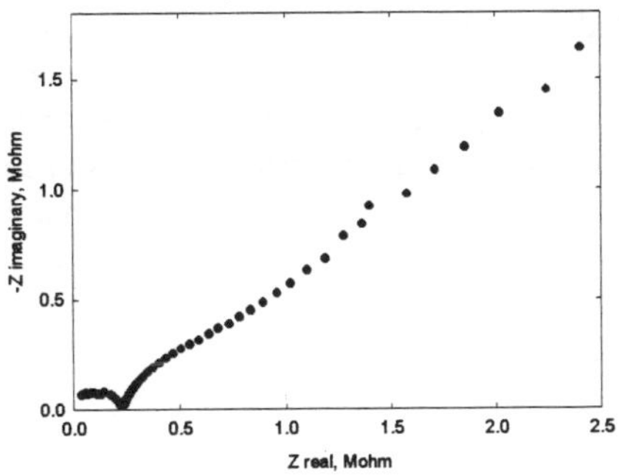

Figure 3. Impedance spectrum of the glass 30/21/18/31. Pure ionic conductivity.

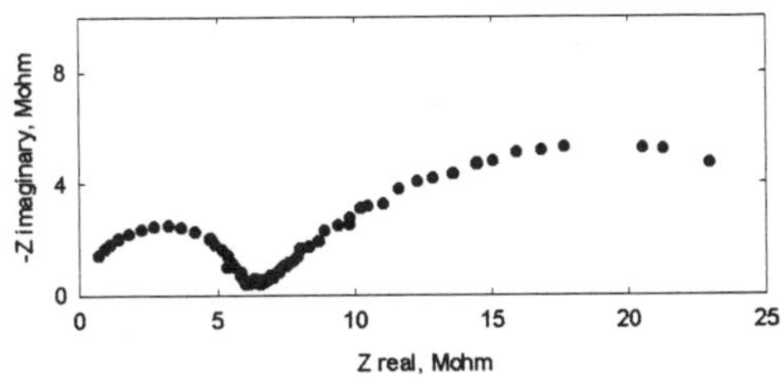

Figure 4. Impedance spectrum of the glass 5/10/50/35. Mixed electronic and ionic conductivity.

Table 1. Properties of the studied glasses.

Nominal composition (molar %) $AgI/Ag_2O/V_2O_5/P_2O_5$	specific weight [g/cm^3]	total conductivity [S/cm]	type of conductivity	relative permittivity	dynamic yield stress at 2 kV/mm [Pa]	exponent "n" in field dependence
0/15/70/15	3.4	3.1e-6	electronic	109	64	2.5
3/15/67/15	3.6	6.1e-6	electronic	27	134	2.5
5/10/50/35	3.9	1.4e-7	mixed	215	35	2.4
7/24/45/24	4.2	9.0e-8	mixed	77	63	3.1
8/23/44/25	4.0	3.1e-7	mixed	77	100	2.6
9/22/44/25	3.7	3.8e-7	mixed	140	67	1.9
10/23/45/22	4.1	1.0e-7	mixed	127	98	2.7
20/15/15/50	4.0	3.7e-6	ionic	465	151	2.1
30/21/18/31	4.5	8.1e-7	ionic	229	164	2.7
35/15/15/35	4.3	6.5e-6	ionic	470	161	2.0

The rheological behavior of the prepared suspensions is presented in Figs.5-8. The analysis of the presented curves show differences in dynamic yield stress. In general they can be attributed to differences in conductivity of the bulk glass. In most cases higher conductivity resulted in higher yield stress. The most impressive difference is, however, the shape of the obtained curves. The glasses with low concentration of AgI ("electronic") gave straight lines of Bingham type and the glasses with high content of AgI ("ionic") exhibited some bumps with maxima depending on field strength. The samples with mixed conductivity showed intermediate behavior. We believe that the observed differences are rather due to the material properties of the glass samples and not to particular grain size distribution of the dispersed powders because according to our measurements all powders had very similar grain shapes, sizes and their distribution.

We have also estimated the field dependence of the measured yield stress. We assumed that the relation took the form $\tau_0 = \text{const}\, E^n$ and calculated the value of the exponent for the field strength range 0–2 kV/mm. The obtained data are included in Table 1. For most cases the yield stress grew faster than E^2.

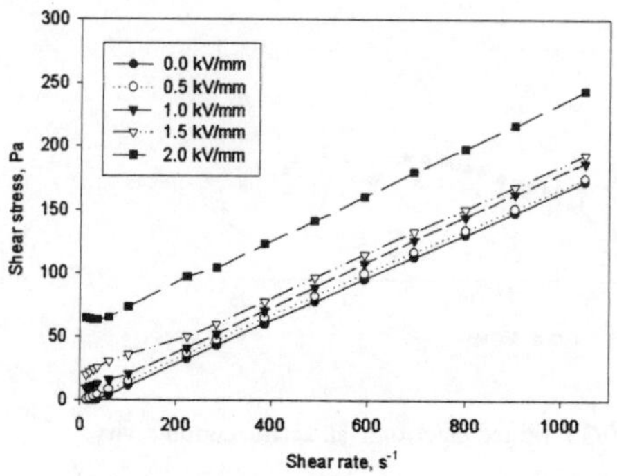

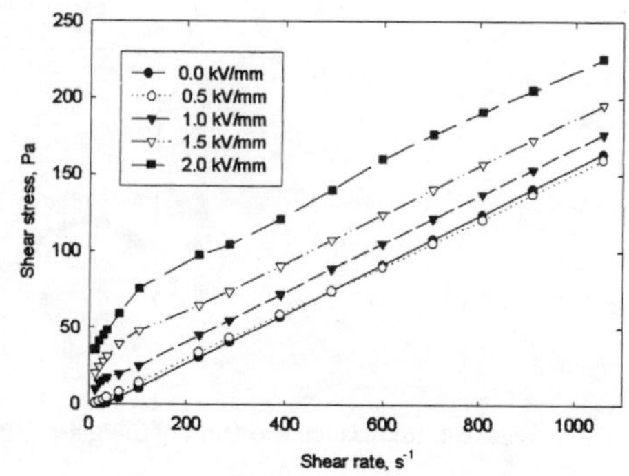

Figures 5 and 6. Flow curves for 12 % suspension of 0/15/70/15 and 5/10/50/35 glasses, respectively.

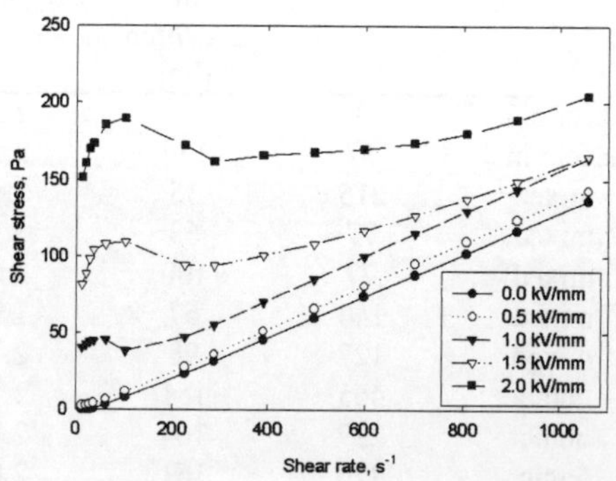

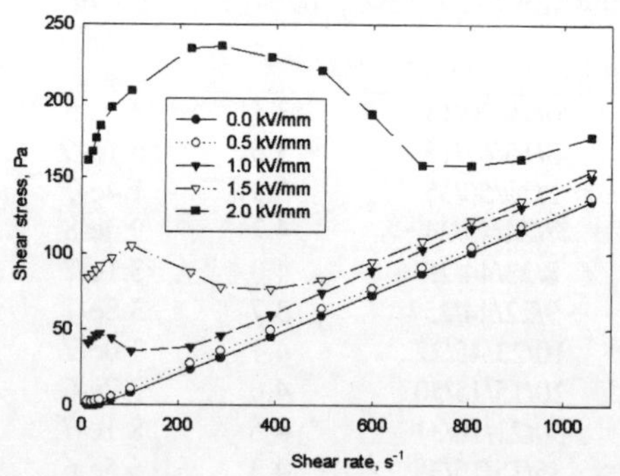

Figures 7 and 8. Flow curves for 12 % suspension of 20/15/15/50 and 35/15/15/35 glasses, respectively.

The grains of glasses studied in this work were not porous so the bulk polarization of a single particle must be due to the electric properties of the powdered glass. As showed earlier there were different charge carriers whose movement within a grain was the cause of bulk polarization. There is also a question of a grain surface. The glasses were synthesized at very high temperature, quickly powdered and stored in a dry container so water could not be present there in significant amounts. The same applies to the silicone oil. Therefore, it is very improbable that some kind of water based film was created on the grains. However, polarizability can be an intrinsic feature of the glass surface.

The structure of the studied glasses consists of tetrahedral molecular units with phosphorus or vanadium atoms inside a pyramid and oxygen atoms in corners. Each oxygen atom belongs to two such units. Breaking the glass sample into small pieces results in creation of a new surface with uncompensated oxygen bonds that immediately react with polar species from the closest vicinity and a monomolecular polar layer surrounding a glass grain is created. The polarization processes related to this layer must be considered when the ER effect is studied. In order to check whether this layer alone can be a source of the effect we ground a solid piece of quartz glass and dispersed the obtained powder in a dried silicone oil. The ER effect was small but easily measurable. The value of the dynamic yield stress was about 18 Pa at 2 kV/mm for 12 % suspension. When the quartz glass powder was additionally treated with trimethychlorosilane in order to reduce the polarity of the surface layer all the ER effect vanished. This simple experiment clearly revealed the importance of this layer.

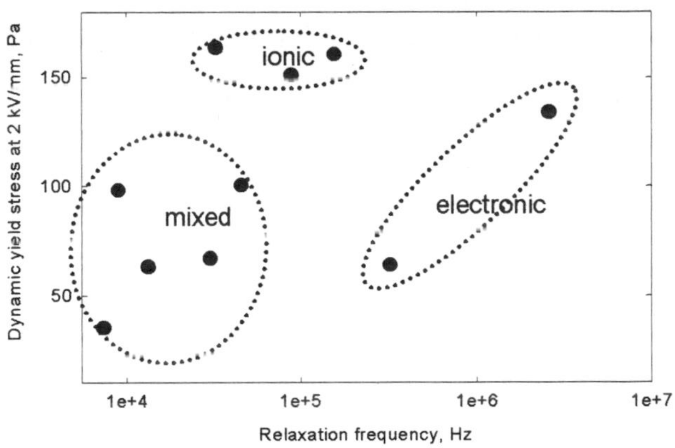

Figure 9. The relation between dynamic yield stress and bulk relaxation frequency for all studied samples.

One of the concepts used to explain the ER phenomenon relies on the assumption that the polarization processes responsible for the effect must happen in a suitable time domain. Too slow or too fast polarization does not lead to increase of viscosity of a dispersion under electric field [8]. The rate of a polarization process is usually estimated by relaxation frequency defined as $f_R = \sigma/\varepsilon\varepsilon_0$. The difficult question is, however, which particular process creates most of the observed yield stress and which values of conductivity and permittivity should be considered.

In Fig.9 we have plotted the values of the dynamic yield stress versus the relaxation frequency calculated for the bulk electric properties of the glasses given in Table 1. Three clearly distinguishable sets of data can be seen. The highest yield stresses are obtained at medium values of f_R (about 100 kHz). These data belong to the "ionic" samples. We believe that in this case the whole ER effect can be attributed to bulk polarization. Medium values of yield stress combined with very high f_R (in MHz range) are related to "electronic" samples. In this case the values of f_R are so high that it is rather not probable that the bulk polarization is responsible for the effect. We believe that the polar surface of the samples plays the main role here. The problem of "mixed" samples is more complicated because of obvious reasons and more work is required to clarify the situation.

There is also the question of the "bumps" observed on some flow curves. If they are not caused by some peculiar distribution of grain sizes the nature of the mentioned above polarization processes must be considered. At the moment it is rather a case of speculation but we can assume that such increase of shear stress is caused by some interplay between the time domain related to the particular range of shear rates and the rate of polarization processes within the dispersed grains. Because of high homogeneity of the material the distribution of electric properties is narrow and the shear stress enhancement is observed only for some range of shear rates. Definitely more work is required.

4 Conclusions

1. We have observed for the first time an ER effect in suspensions of powdered glasses of the $AgI/Ag_2O/V_2O_5/P_2O_5$ system whose type and value of conductivity depends on particular shares of the components.
2. The highest values of yield stress was observed for samples of ionic type conductivity of about 10^{-6} S/cm.
3. The nature of polarization processes influence the rheological properties of the ER suspensions under electric field. This applies not only for the dynamic yield stress value but for the whole flow curve.

5 Acknowledgements

The authors thank the Warsaw University of Technology in general and Department of Chemistry in particular who financially and organizationally supported the work described.

References

1. Filisko F.E. in "Progress in Electrorheology" eds. Havelka K.O. and Filisko F.E, Plenum Press, New York 1995, p 3.
2. Block H. , Kelly J.P., *J.Phys.D Appl.Phys.*, **21** (1988) 1661.
3. Filisko F.E, Radzilowski L.H., *J.Rheol.* 34 (1990) 539.
4. Gow C., Zukoski C.F., *J.Colloid Interface Sci.* **136** (1990) 175.
5. Płocharski J., Drabik H., Wyciślik H., CiachT., *Synthetic Metals*, **88** (1997) 139.
6. Płocharski J., Różański M., Wyciślik H., *Synthetic Metals*, **102** (1999) 1354.
7. Wasiucionek M., Garbarczyk J., Wnetrzewski B., Machowski P., Jakubowski W., *Solid State Ionics*, **92** (1996) 155.
8. Ikazaki F. Kawai A, Kawakami T., Edamura K., Sakurai, K. Anzai H., Asako Y., *J.Phys.D Appl.Phys.* **31** (1998) 336.

CHAIN BEHAVIOR IN MODEL HOMOGENEOUS ER FLUIDS DEPENDING ON TEMPERATURE

HIROKAZU OKAMURA, KAORI SUZUKI, TAKESHI MORI, KEIJI MINAGAWA, SEIZO MASUDA
Department of Chemical Science and Technology, The University of Tokushima, Tokushima 770-8506, Japan

MASAMI TANAKA
Faculty of Pharmaceutical Science, Tokushima Bunri University, Tokushima 770-8514, Japan

Water-soluble urethane-modified polyethers were prepared by addition of poly(ethylene oxide)-*co*-poly(propylene oxide) and aromatic isocyanate compounds. These polymers were found to dissolve in water at lower temperature and separate from solution upon heating. The temperature showing this unusual solubility change is called lower critical solution temperature (LCST). These chemical structures of thermo-responsive polymers were similar to those of urethane-modified ER active polymers containing poly(tetramethylene oxide) and aromatic urethane moiety. The thermo-responsive and ER polymers may have various intra- and intermolecular interactions through the urethane moiety. It is considered that both thermo-responsivility and ER effect are dependent on the conformational stability of the polymers under different conditions possibly related to these stimuli-responsivility through the molecular interactions. In order to clarify molecular motion of thermo-responsive polymers near the LCST, ^{1}H-NMR spin-lattice relaxation time (T_1) was measured in D_2O. The result indicated that hydrophobic interaction of terminal urethane moiety would strongly affect the LCST behavior.

1. Introduction

The electrorheological/magnetorheological (ER/MR) effects of homogeneous polymer systems,[1-3] e.g. polymer liquids and solutions, could be related to the conformational stability of the polymers under different conditions. We have previously reported the analysis of molecular motions of some urethane-modified poly(propylene glycol) (PPG) and poly(tetramethylene glycol) (PTMG) prepared as model ER fluids.[4-6] Either positive or negative ER effect was observed for these modified polyethers, depending on the chemical structure of the terminal urethane moiety which affects the molecular-chain behaviors. It is thus interesting to analyze the molecular-chain behaviors, such as chain motions and intra- and intermolecular interactions, of the polymers applicable to homogeneous ER/MR fluids. For last several decades, thermo-responsive polymers, such as polymethacrylate or methyl cellulose have gathered attention of many workers.[7,8] In particular, it is well known that poly(*N*-isopropylacrylamide) (PNIPAM) aggregates and precipitates in an aqueous solution upon heating above about 32°C because of hydrophobic interaction.[9-12] The solubility change is mainly due to some conformational change of the polymers governed by the affinity of the polymer chain and the solvent.

Therefore, it is important to study on solubility of thermo-responsive polymer in consideration of potential applicability of these polymers to novel ER/MR and other field-responsive materials.

In this work, a number of model compounds for homogeneous ER/MR fluids were prepared with some water-soluble polymers such as poly(ethylene oxide-*co*-propylene oxide) (PEG-*co*-PPG). The chemical structures are very similar to those of the urethane-modified ER active polymers. Our attention was focused on the temperature dependence because most of the aqueous solution properties of these polymers were found to be sensitive to temperature. These polymers were soluble in water at low temperature, while they showed a decrease of solubility at high temperature. The temperature at which a phase transition of these polymer solutions takes place is called the lower critical solution temperature (LCST). Here the LCST behavior of some PEG-*co*-PPG type polymers in water are mainly discussed from molecular interaction by measuring the turbidity as a function of temperature. Moreover, the application of the NMR analysis of polymer chain motions is discussed, which can be very useful for phase-transition observation of these polymers. The results are compared with those of ER active polyethers reported previously.[4-6]

2. Experimental

2.1. *Preparation of urethane-modified polyethers*

PEG-*co*-PPG's (weight-average molecular weight (Mw): 12000 and 2500) were purchased from Aldrich. These polymers were linear random copolymer and soluble in water. PEG-*co*-PPG's with Mw=12000 and 2500 were abbreviated as EP12 and EP2.5, respectively. For the purpose of increasing the interaction among the polymer chain, 4,4'-diphenylmethane diisocyanate (MDI), phenyl isocyanate (PhI) or *p*-chlorophenyl isocyanate (*p*-CPhI) was added to the terminals of PEG-*co*-PPG through a urethane-bond. The urethane-modified copolymers were prepared by 2:1 addition of the isocyanate compounds and PEG-*co*-PPG in bulk. In the case of the 2:1 addition of MDI and PEG-*co*-PPG, further 2 equivalent PEG-*co*-PPG was added to the reaction mixture after the formation of 'isocyanate-prepolymer'. The resulting copolymers were oily liquids and used without further purification. The preparation scheme are shown in Figure 1 and these copolymers are abbreviated as follows;

UEP 1: EP2.5 and R= phenyl group
UEP 2: EP2.5 and R= *p*-chlorophenyl group
UEP 3: EP12 and R= phenyl group
UEP 4: EP2.5 and R= 4,4'-diphenylmethane- (PEO)-*co*-(PPO).

2.2 *Transmittance measurements*

The LCST's of samples containing 0.5 or 0.1 wt-% polymer in water were determined from the turbidity of polymer aqueous solutions as a function of temperature. The turbidity of the polymer solutions was monitored by UV spectrometer (Shimadzu) at wave length 500 nm with a thermal controlled circulating water bath using a jacket cell.

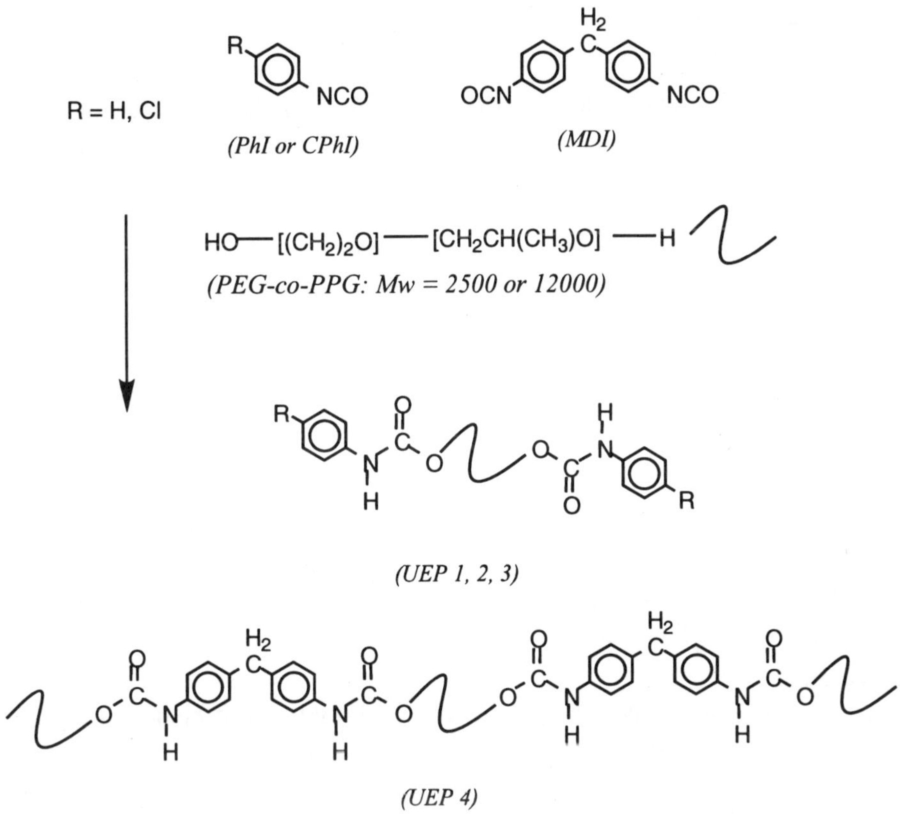

Figure 1. Preparation scheme for urethane-modified polyethers.

The heating rate was regulated around 1°C/min. In this study, the LCST was defined as the temperature getting to 50% of the transmittance of tungsten beam passing through the solution.

2.3. *Rheological measurements*

The shear viscosity was measured for urethane-modified polymers at 25°C with a rheometer (ARES Rheometrics). The measurements were carried out on a strain-controlled rheometer using a rotating parallel plates geometry. The applied shear rate values were varied in the range of 0.1-100s^{-1}.

2.3. *NMR relaxation measurements*

^1H NMR spin-lattice relaxation time (T_1) of the 0.5 wt-% polymer solution in deuterium oxide (D_2O) was examined as a function of temperature by use of a Varian unity 600 spectrometer (600MHz). For measuring T_1, the inversion recovery method (180°- t - 90° - FID) was used with delay time (t), varying at least 5T_1s. The repetition time was 20s. The T_1 values were determined for each proton of EP12 and UEP 1.

3. Results and Discussion

3.1. *Shear viscosity*

Figure 2 shows the relationships of shear viscosity and shear rate for all samples measured at 25°C. These samples were found to show a Newtonian-behavior over the shear rates examined here. The shear viscosity values of urethane-modified EP2.5 (UEP 1, 2, 4) were higher than that of EP2.5. Also, the shear viscosity values of EP12 and UEP 3 were very similar to each other. The comparison of EP2.5 and UEP 1, 2, 4 indicates clearly the relations of interaction of modified compounds because of sample having the same polyether chain (EP2.5) and the different urethane moiety (none, PhI, CPhI, MDI). The result indicates that the presence of intra- and intermolecular interactions are emphasized by urethane modification. The values of the shear viscosity of urethane moieties or strength of molecular interaction were following order:

MDI > CPhI > PhI > none (UEP 4 > UEP 2 > UEP 1 > EP2.5)

The urethane bond was found to be efficient for molecular interaction through hydrogen bonds. In the comparison of EP12 and UEP 3, the enhancement of shear viscosity with urethane moiety was relatively small, probably because of the low ratio of urethane-moiety within the polymer chain.

3.2. *Turbidity*

The LCST values of sample solutions are shown in Table 1 and plotted in Figure 3. Some of the LCST's were measured by eye because of thermal limit of the controlled water bath. It was found that the LCST's of UEP 1, 2, 3, 4 having functional groups such as PhI, *p*-CPhI, MDI were lower than that of EP2.5 or EP12 only. This result indicates that intra- and intermolecular interactions of polymer chains are emphasized by

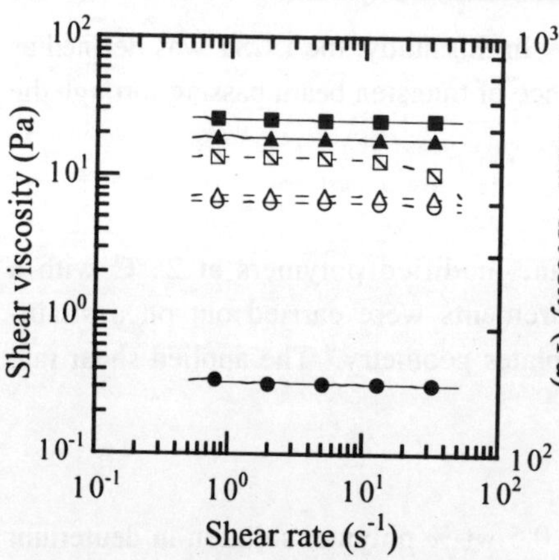

Figure 2. Dependence of shear viscosity on shear rate for PEG-co-PPG and urethane-modified polymers.

Y_1 ordinate: 10^{-1}-10^{2} (Pa) Y_2 ordinate: 10^{2}-10^{3} (Pa)
EP2.5, UEP 1, UEP 2, EP12, UEP 3, UEP 4

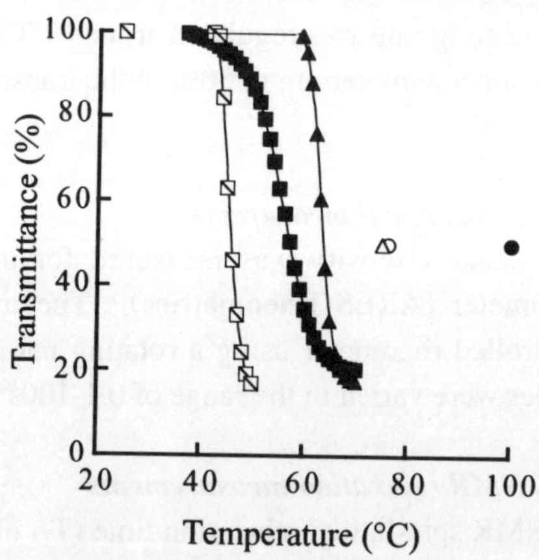

Figure 3. Turbisity against temperature aqueous solutions of urethane-modified polymers.

EP2.5, UEP 1, UEP 2, EP12, UEP 3, UEP 4

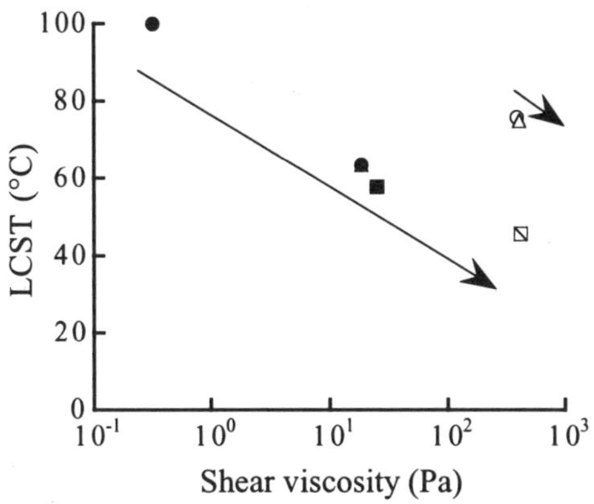

Figure 4. Relationships of LCST and shear viscosity for EP2.5, EP12 and UEP 1, 2, 3, 4: Shear rate = $50s^{-1}$.

●	EP2.5	○	EP12
▲	UEP 1	△	UEP 3
■	UEP 2	⧅	UEP 4

introduction of the aromatic hydrophobic groups. Further, the influence on the LCST of EP2.5 and UEP 2, 3, 4 having different terminal groups were larger when their interaction was increased by a presence of urethane moiety. The magnitude of the effect of urethane moieties on the LCST was following order:

MDI > *p*-CPhI > PhI > none (UEP 4 > UEP 2 > UEP 1 > EP2.5)

In comparison of PEG-*co*-PPG with different Mw, the LCST of EP2.5 was higher than that of EP12, while those of the UEP 1 and 2 having the same PhI urethane moiety showed the tendency to the contrary. It is considered that the difference of interaction were caused from difference of the relative quantity of the urethane moiety in the high Mw polymer and the low Mw polymer. The dependence of the LCST on the concentration of EP2.5 aqueous solution was examined in the presence of salts and plotted in Figure 4. The LCST's of higher and lower concentration samples were observed at lower and higher temperature regions, respectively. The turbidity curves at low polymer concentration (0.5 wt-%) was relatively broad, while that of high concentration sample (1 wt-%) was sharp. It is indicated that the phase-transition of the EP2.5 depends on the distance among polymer chains. Summarizing these results, it is demonstrated that the LCST is strongly influenced by the magnitude of the interaction among the terminal groups in solvent. There is a certain resemblance between thermo-responsive PEG-*co*-PPG type polymer and PTMG type ER active polymers. Thus it is expected that the stimuli responsibility can be improved with the introduction of functional groups possessing hydrophobic, hydrogen bond and van der Waals interactions. In our previously paper, the shear viscosity of two urethane-modified PTMGs (Mw = 650) with PhI and *p*-CPhI was measured at various shear rate under an electric field.[6] In no electric field, PhI and PTMG moiety showed Newtonian behavior while *p*-CPhI and PTMG moiety showed Non-Newtonian behavior such as shear-

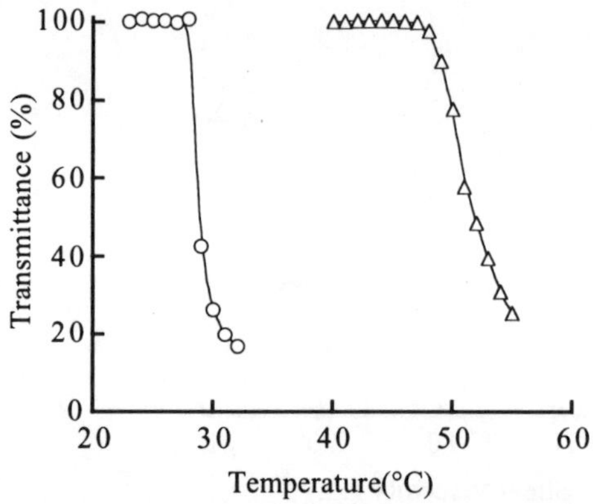

Figure 5. Turbidity curves as a function of tempearture of EP2.5 aqueous solution at 0.5 wt-% (circle) 0.1 wt-% (triangle): heating rate = 1°C/min.

thinning process. On the other hand, in the presence of an electric field, the shear viscosity of PhI-modified PTMG decreased and that of *p*-CPhI-modified PTMG increased, respectively, at low shear rate region. The shear viscosity of PTMG without urethane moiety did not show any ER effect. The influence on the ER effect of PTMG with different functional groups was larger when the molecular interaction of terminal increased. The magnitude of the influence of urethane moieties on the ER effect was following order:

p-CPhI > PhI > none (CPhPTCPh > PhPTPh > PTMG)

It was demonstrated that the ER effect is strongly influenced by the magnitude of interaction among the functional groups. In all of the ER active polymer samples, it is found that the ER effects increased with increasing electric field strength. This result is resemble to dependence of temperature for thermo-responsive polymers. In the case of ER active polymer, urethane bonds were used as molecular interaction with/without an electric fields although interaction would be emphasized with electric fields. Benzene ring would act as hydrophobic interaction and, chlorine would act as interaction with the polarity emphasized by an electric field.

3.3. *NMR relaxation time*

The NMR spin-lattice relaxation time (T_1) was measured for 0.5wt-% EP12 and UEP 1 in deuterium oxide at temperature range from 60°C to 90°C. Here large and small T_1 values correspond to the high and low activity of molecular motions, respectively. As the temperature increased, initially the T_1 values of all the protons increased, and then decreased around 70°C for EP12 and around 55°C for UEP 1, respectively. By further heating, the T_1 values of all the protons began to increase again. The initial increase of T_1 is induced from the activated molecular motion upon heating, and the decrease of T_1 around 70°C or 55°C corresponds to the LCST's. From the result of turbidity measurement, it is found that most of the polymers in water keep precipitating above these critical temperatures. The increase of T_1 at the temperatures higher than LCST is

Table 1. Shear viscosity, LCST and ratio of urethane bond of EP2.5, EP12 and UEP 1, 2, 3, 4.

Sample	η (Pa)	LCST (°C)	Ratio of urethane bond[a]
EP2.5	0.3	100	0
UEP 1	18.2	63.5	4.8
UEP 2	25.0	57.8	4.8
EP12	384.8	76.0	0
UEP 3	401.3	75.0	1
UEP 4	416.5	45.6	3.2

[a] Urethane bond ratio in one molecule of each compounds to the number for that of UEP 3 as a standard.

thus interpreted as the result of the remaining dissolved polymer in water. The LCST value shown by NMR measurement was lower than that of the observation of the turbidity by the eye. The observation of phase transition such as the turbidity reflected visual macroscopic behavior, e.g. the aggregation of polymers, while that of the NMR relaxation reflected microscopic individual polymer chain motion. In the ER active PTMG type polymer in deuterated acetone, the molecular motions were examined for each segments, e.g. terminal urethane moieties and flexible polyether main chain. It is found that T_1 of the hydrophobic terminals were longer than that of main chains. And it is also found that T_1 of both terminals and main chains were larger with increasing interaction of terminal groups.

4. Conclusion

This work has shown that the use of urethane modified PEG-*co*-PPG allows the reversible phase transition of polymer aqueous solutions. The LCST's could be controlled by varying the molecular structure of the copolymers. This versatility should allow the LCST and the ER effects to be finely tuned to suit specific applications.

References

1. A. Inoue and S. Maniwa, *J. Appl. Polym. Sci.* 55,113 (1994).
2. T. Uemura, K. Minagawa, and K. Koyama, *Chem. Lett.* 1995 (1994).
3. T. G. Lazareva, O. V. Yakovleva, S. A. Prodan, I. A. Il'yushenko, J. S. Gayduk, Int. J. Mod. Phys. B, 13, 1956 (1999).
4. K. Minagawa, H. Okamura, S. Masuda, M. Tanaka, N. Gohko, Y. Masubuchi, J. Takimoto, K. Koyama, T. Uemura, Int. J. Mod. Phys. B, 13, 1998 (1999).
5. K. Minagawa, H. Okamura, S. Masuda, M. Tanaka, Int. J. Mod. Phys. B, 13, 1975 (1999).
6. K. Minagawa, H. Okamura, S. Masuda, M. Tanaka, J. Int. Mater. Sys. Struct., 9, 626 (1998).
7. J. Eliassaf and A. Silberberg, Polymer, 3, 555 (1962).
8. N. Sarkar, J. Appl. Polym. Sci., 24, 1073 (1979).
9. M. Heskins and J. E. Guillet, J. Macromol. Sci. - Chem., A2(8), 1441 (1968).
10. S. Fujishige, K. Kubota and I. Ando, J. Phys. Chem., 93, 3311 (1989).
11. K. Kubota, S. Fujishige and I. Ando, J. Phys. Chem., 94, 3311 (1990).
12. S. Lin, K. Chen and R. Liang, Polymer, 40, 2619 (1999).

BEHAVIOR OF MAGNETORHEOLOGICAL FLUIDS UTILIZING NANOPOWDER IRON

NICHOLAS ROSENFELD AND NORMAN M. WERELEY
Smart Structures Laboratory, Dept. of Aerospace Engineering
University of Maryland, College Park MD 20742 USA Email: wereley@eng.umd.edu

RADHAKUMAR RADAKRISHNAN AND TIRULAI S. SUDARSHAN
Materials Modification Inc., 2721-D Merrilee Drive
Fairfax VA 22031 USA Email: radha@matmod.com

Iron nanopowders for use in magnetorheological (MR) fluids were synthesized using a Microwave Plasma Synthesis technique developed at Materials Modification Inc (Fairfax VA). Transmission electron microscopy and surface area analysis measured iron particle size at 15-25 nm. The nanopowders were mixed into hydraulic oil to create nano-scale MR fluid. A micron-scale fluid was created using 45 μm iron particles as well as a hybrid fluid using a 50/50 mix of micro- and nanoparticles. All three fluids had a solids loading of 60% (w/w or weight by weight fraction). The fluids were tested in a flow mode rheometer fabricated from a modified damper using a sinusoidal input dynamometer over a speed range of 12.7 to 177.8 mm/s (0.5 to 6 in/s) and an input current range of 0 to 2 A. The yield stress and plastic viscosity of the MR fluid were characterized using a Bingham plastic model.

1 Motivation

Magnetorheological (MR) fluids can be implemented in a variety of smart actuation systems, including optics polishing, fluid clutches, and aerospace, automotive, and civil damping applications. The overwhelming majority of existing MR fluids are composed of micron-scale (1-7 μm) particles of iron or iron compounds suspended in a carrier fluid. The size of the particles makes these fluids susceptible to settling when not in use. Once settled, remnant magnetic attraction between particles makes re-dispersion difficult. [1] The large particles also could lead to unwanted abrasion to components in contact with the fluid. Reducing the size of the magnetic particles in the fluid could eliminate these undesirable effects. Several methods for preparing nanopowders for use in MR fluids have been suggested, including chemical precipitation of ferrite powders, [2] hydrolysis of metal alkoxides into yttrium iron garnet particles, [3] and vaporization of steel in plasma. [4]

The synthesized powders were characterized for composition, particle size, shape and surface area. Compositional analysis was carried out by X-Ray Diffraction analysis (XRD). Particle size distribution and surface area were measured using a Horiba LA-910 laser light scattering particle size analyzer and a BET Coulter SA 3100 surface area analyzer, respectively. Transmission electron microscopy (JEOL Model 2010, Supertwin TEM, with point-to-point resolution of 0.23 nm, line resolution of 0.14 nm) was used to characterize iron nanopowders for their shape.

2 Fluid Production

Iron nanopowders were synthesized using the Microwave Plasma Synthesis (MPS) technique. This technique utilizes microwave energy to generate a plasma by ionization, dissociation and recombination of gas molecules. The high temperature vaporizes the precursors, which promotes chemical reactions at the molecular level in the presence of microwaves. These vapors are then rapidly quenched to form nanopowders. Iron nanopowders were synthesized from iron carbonyl precursors according to:

$$Fe(CO)_5 \rightarrow Fe(s) + 5CO \uparrow (g) \tag{1}$$

Vapors of iron pentacarbonyl were fed directly into the plasma zone, wherein the pentacarbonyl dissociated into iron and carbon monoxide. Argon gas was used to carry the precursor into the plasma chamber, where rapid heating of the powders resulted in its dissociation.

Fig. 1 is an X-Ray Diffraction (XRD) pattern of iron synthesized using the microwave plasma synthesis method. Only peaks corresponding to iron were present in the XRD pattern. The mean particle size of the iron powders measured by laser light scattering (Horiba LA-910) technique was of the order of several microns. The measuring cell in a Horiba LA-910 does not have the ultrasonication capability in its measuring cell and, hence, the data obtained is a measure of the agglomerate particle size. A dispersant was used to reduce the agglomeration but yielded only limited results. Reliable information on particle shape was therefore obtained using Transmission Electron Microscopy (TEM). A TEM micrograph of the iron powders is shown in Figure 2. The particle size of the iron powders in these micrographs is between 15 and 20 nm.

The surface area of the powders measured using BET was 42 m^2/g. If all the particles are assumed to be spherical, then the particle size measured from this surface area is 26 nm. This agrees well with the data obtained from TEM measurements.

To prepare stable MR fluids, hydraulic oil was chosen as a carrier fluid. The Mobil DTE 20 series is used extensively in high-pressure systems, in industrial, marine and mobile service, especially in servo valves and robotics because of their excellent anti-wear properties, multimetal compatibility and corrosion resistance.

Based on discussions with Central Soya Inc., Fort Wayne, IN, lecithin was used as a dispersing agent for preparing MR fluids. Lecithin (2 wt% - optimized) was mixed in hydraulic oil using a high-speed emulsifier at speeds close to 11,000 rpm. Iron nanopowders obtained from the microwave plasma synthesis technique were added to the oil and the mixing continued. The mixing speed was kept constant at 11,000 rpm for all experiments along with a constant mixing time of 30 minutes.The solids loading in the MR fluid was 60% (w/w).

3 Experimental Setup

The following experimental results were obtained using a modified Lord Motion Master MR flow-mode damper. A diagram of the original damper is shown in Fig. 3. The piston head contains an annular orifice valve concentric to the damper shaft. The MR valve consists of a bobbin with an electromagnetic coil concentric to a tubular flux return. When current is applied to the coil, the resulting magnetic field across the annular gap activates the yield stress in the MR fluid. The pressure

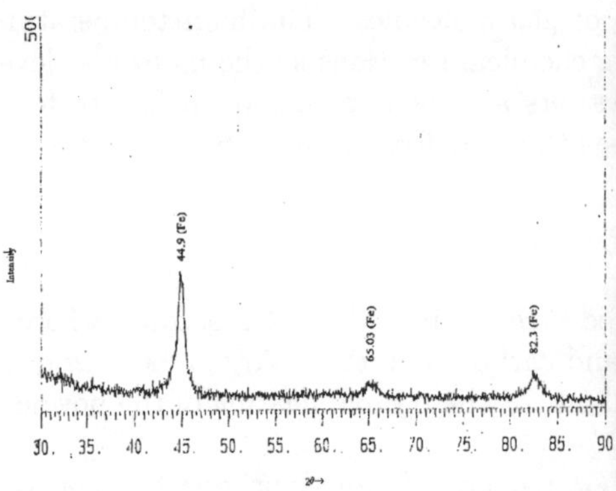

Figure 1. XRD of iron nanopowders synthesized using a microwave plasma

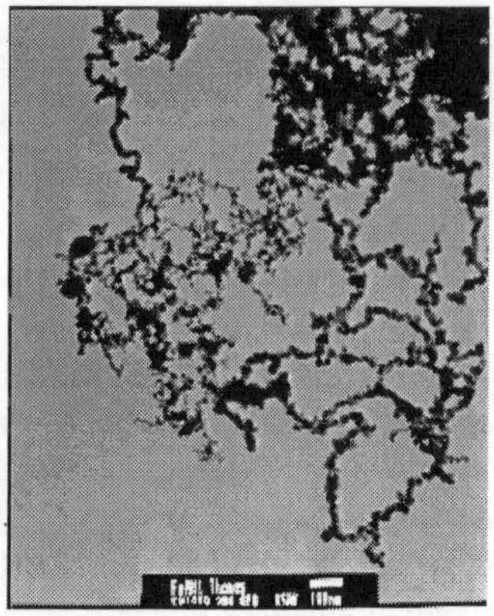

Figure 2. Transmission electron micrograph indicate spherical particles of average diameter 26 nm.

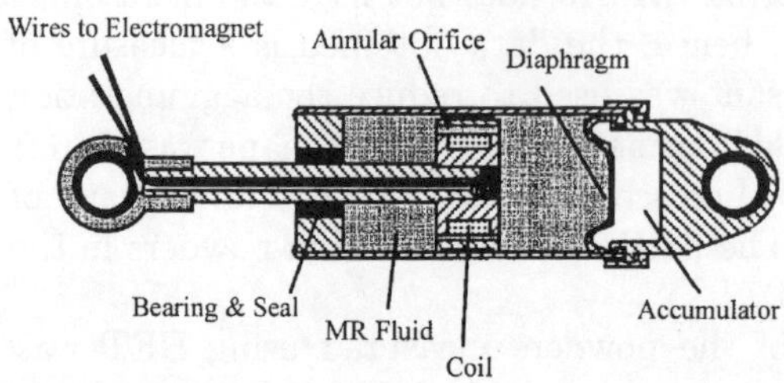

Figure 3. Schematic of stock truck seat damper from Lord Corporation.

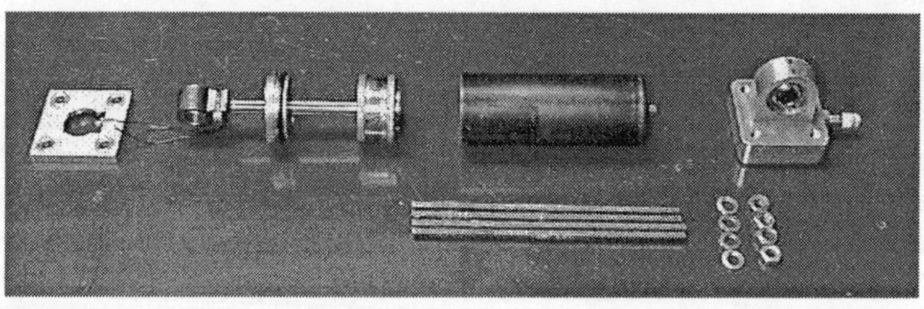

Figure 4. Flow mode rheometer fabricated from a modified Lord Corporation truck seat damper.

drop through the MR valve creates the velocity-dependent damping force.

The outer casing of the damper was modified to allow the insertion of different MR fluids. An exploded view of the damper is shown in Fig. 4. A custom endcap with a valve connecting to a nitrogen accumulator was fashioned to permit the regulation of its pneumatic pressure. The damper had a piston area of 360.5 mm^2 and a shaft area of 22.6 mm^2. The damper had an annulus of thickness of 0.432 mm and an average diameter of 28.9 mm.

The damper was tested on a Roehrig Engineering 5 HP dynamometer with a measurement stroke of 12.7 mm (0.5 in). A sinusoidal input was used to approx-

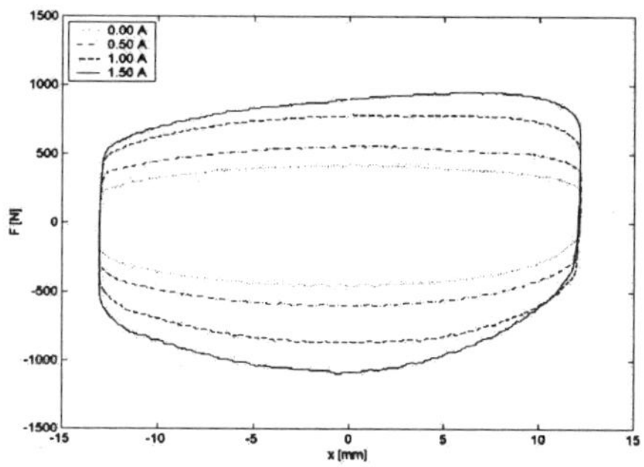

Figure 5. Typical force vs. displacement under sinusoidal loading for flow mode rheometer.

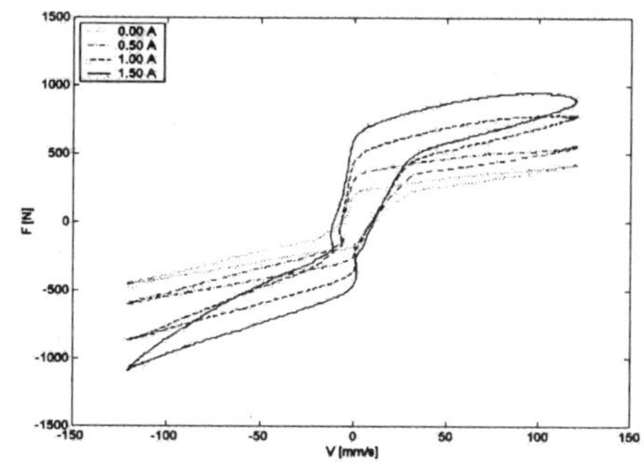

Figure 6. Typical force vs. velocity under sinusoidal loading for flow mode rheometer.

imate a constant-velocity flow when the piston was at its maximum shaft speed at the half-stroke position. Each fluid was loaded into the damper and the pneumatic reservoir pressure was set to 1.724 MPa (250 psi) with nitrogen gas. The damper was loaded into the dynamometer, where it was subjected to a sine-wave input. The dynamometer simultaneously measured the shaft position, shaft speed and damping force, which were collated into force vs. displacement (Fig. 5) and force vs. velocity charts (Fig. 6). Since the strength of the magnetic field across the annulus is a direct function of applied current, current was used as an independent input analogous to magnetic field strength. Each of the micron-scale, hybrid, and nano-scale fluids were tested with seven velocity inputs, with maximum speeds of 12.7 mm/s (0.5 in/s) and 25.4 to 177.8 mm/s (1 to 6 in/s) in 25.4 mm/s (1 in/s) increments. Each of these speed inputs were tested with input currents ranging from 0 A to 2 A at increments of 0.25, resulting in 63 different test conditions for each fluid.

4 Parameter Identification

To characterize the MR fluid properties, it was necessary to determine the yield stress, τ_y, and plastic viscosity, μ, over a range of steady-state operating conditions. These properties may be functions of the applied magnetic field strength, or the analogous current. These parameters can be calculated for Bingham plastic fluids by modeling the force response of the damper in terms of yield force, F_y, and post-yield damping, C_{po}, and using these properties to determine yield stress and plastic viscosity, respectively [5]. Assuming parallel plate geometry, the force response is:

$$f(t) = \begin{cases} C_{po}v - F_y & v < 0 \\ -F_y < f(t) < F_y & v = 0 \\ C_{po}v + F_y & v > 0 \end{cases} \tag{2}$$

The yield force, F_y is a function of yield stress, τ_y and damper geometry, while the postyield damping, C_{po}, is a function of plastic viscosity, μ, and damper geom-

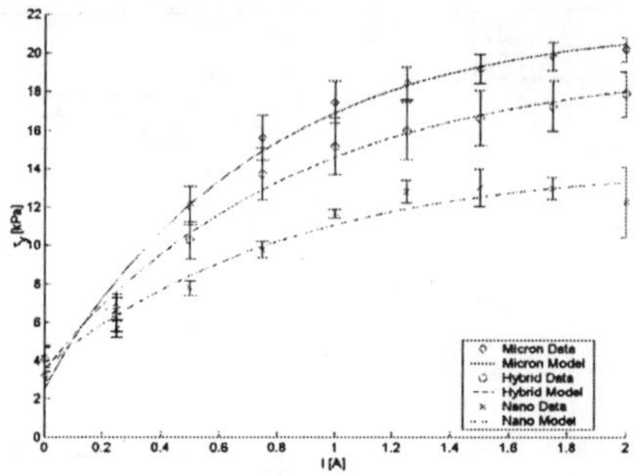

Figure 7. Comparison of yield stresses for the three magnetorheological fluids: 30 micron average particle size, 26 nm average particle size, and hybrid.

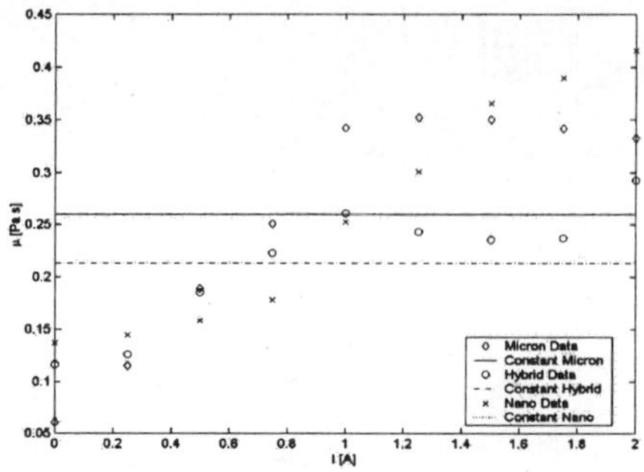

Figure 8. Plastic viscosity is plotted versus applied field. Average values of plastic viscosity are shown as lines for each material: micron particles, nanometer particles, and hybrid material.

etry, such that:

$$\tau_y = \frac{d}{2AL} F_y \tag{3}$$

$$\mu = \frac{A_d d^2}{12A^2 L} C_{po} \tag{4}$$

Note that the area A represents the area upon which the higher fluid pressure acts. As the piston moves inward, A is the piston head area; as the piston moves outward, A is the piston head area less the shaft area. For the given experimental damper, the force response was written as a function of velocity, yield stress and plastic viscosity:

$$f(t) = f(t, v, \tau_y, \mu) \tag{5}$$

The model parameters, τ_y and μ, are those that minimized the mean squared error between the measured and predicted force. A least mean squared error minimization routine was used to compute the yield stress and plastic viscosity as a function of applied current. Fig. 7 is a plot of the mean yield stresses for each fluid over the experimental velocity range. Error bars on this plot represent one standard deviation of yield stress over the velocity range. A representative plot of plastic viscosity at a steady-state speed has been plotted in Fig. 8.

5 Results

The experimental yield stress results showed strong correlation with both fluid composition and input current. All fluid and velocity combinations show that as input current increased from zero, the measured yield stress initially increased rapidly but saturated at high currents. For analysis, this relationship was approximated as an exponential function of current incorporating zero-field yield stress, $\tau_{y,0}$, field dependent yield stress, $\tau_{y,\infty}$, and growth parameter, α:

$$\tau_y = \tau_{y,0} + \tau_{y,\infty} \left[1 - \exp\left(-\alpha I\right)\right] \tag{6}$$

Note that at zero current, $\tau_y = \tau_{y,0}$, and as current increases approaches the saturation yield stress asymptotically. A least mean squared error minimization routine was again utilized to fit this model to the experimental yield stress vs. current data for each individual data set. The curves on Fig. 7 are the mean curves for each fluid over the experimental velocity range. Additionally, the calculated field dependent or saturation yield stress, plastic viscosity, zero-field yield stresses, and growth parameters have been plotted on Fig. 9, Fig. 10, Fig. 11, and Fig. 12, respectively. The magnitudes of the zero-field yield stresses range between 2.1 and 4.3 kPa while the saturation yield stresses range between 12 to 23 kPa.

At zero current or zero magnetic field conditions, the micron-scale fluids yield stress is lower overall than either the hybrid- or nano-scale fluids over the entire range of velocities. Taking the mean yield stress over the range of experimental velocities, the micron-scale fluid has a zero-field yield stress of 2.56 kPa, more than 25% less than the hybrid fluid average of 3.48 kPa and the nano-scale average of 3.60 kPa. This trend is most likely due to the greater internal friction forces in the hybrid- and nano-scale fluids caused by smaller particle size, and therefore greater surface area, per solid content.

MR effects quickly dominated the zero-field trends as current was applied to the magnetic coil. At all but the lowest applied currents, the micron-scale fluid exhibited a greater yield stress than the hybrid- and nano-scale fluids. These trends were most apparent when the fluid was at its saturation yield stress. Again taking the average over all experimental velocities, the micron-scale mean saturation yield stress of 21.7 kPa was 11% greater than the hybrid mean of 19.5 kPa and 45% greater than the nano-scale mean of 14.8 kPa. This trend agrees with previous data showing that yield stress decreases with decreasing particle size [6]. The growth parameter is a measure of how fast the yield stress tends to reach its saturation level; a larger growth parameter indicates a faster rise to saturation as a functin of field. The experimental results do not show a consistent trend between fluid composition and growth parameter. However, the hybrid fluid on average has a smaller growth parameter than the nano-scale fluid, while the micron-scale fluid has the largest growth parameter.

The plastic viscosities over the range of experimental data were seen to be very slight functions of applied magnetic field for a given velocity. This functionality was slight enough that plastic viscosity was simplified to the mean constant value of the experimental data at each velocity. At every experimental velocity, the constant plastic viscosities for all fluids were close, typically within 0.05 Pa s of each other. These similar results were likely due to the equal particle solids loading in each fluid. In Fig. 8, the plastic viscosities of the fluids are seen to decrease with increasing velocity, indicating slight shear thinning in the fluids.

6 Conclusions

Magnetorheological fluids were proposed based on nanometer-sized powders, micron-sized powders, and mixtures of the two. Nano- and hybrid scale MR fluids have a higher zero-field yield stress than micron- scale fluids. When a field is applied to the fluid, the micron-scale fluid exhibits the highest yield stress, and the

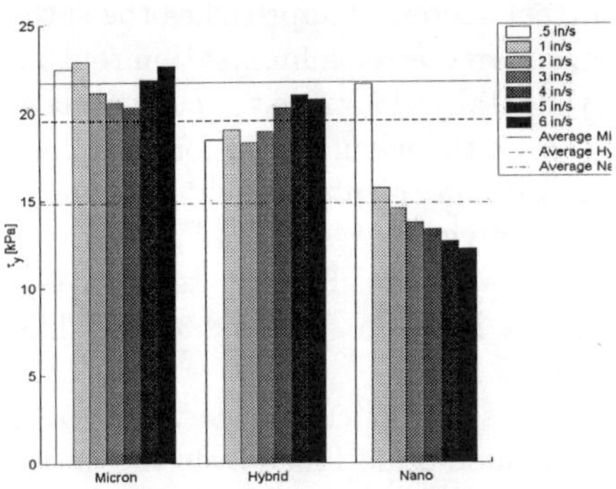

Figure 9. Field dependent yield stress.

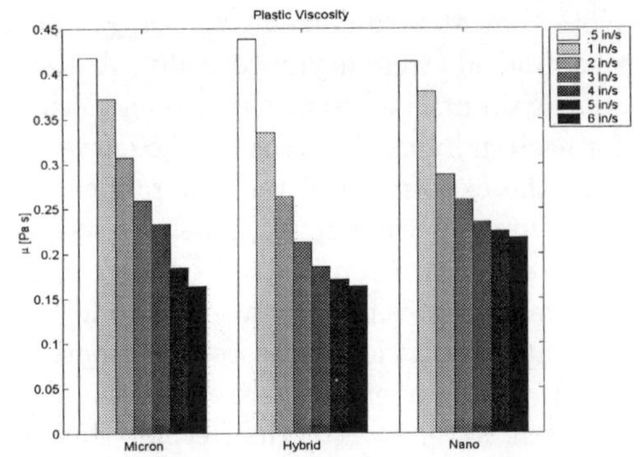

Figure 10. Plastic viscosity

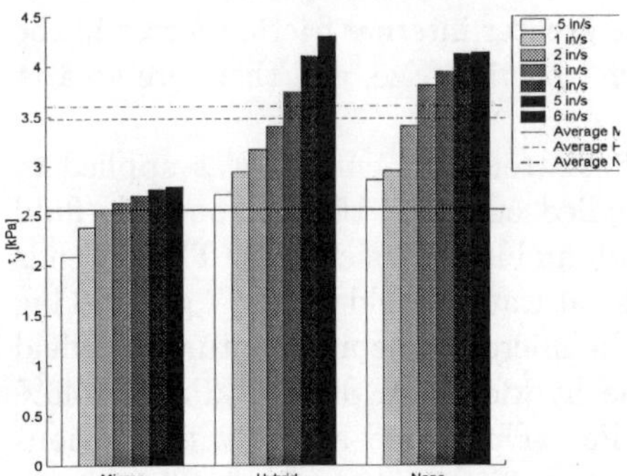

Figure 11. Off-state yield stress

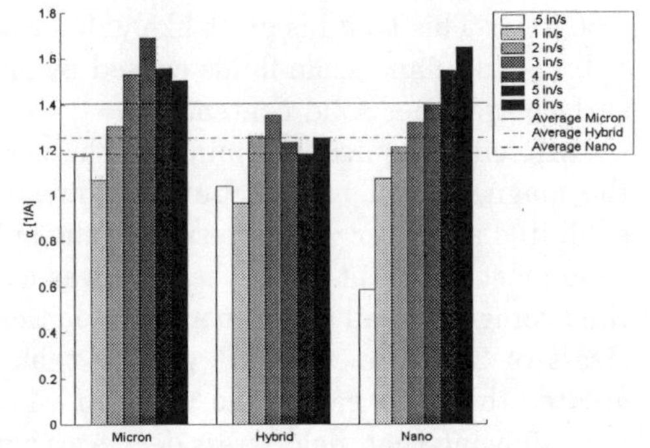

Figure 12. Growth Parameter

nano-scale fluid exhibits the lowest yield stress. At maximum field, the micron-scale fluid exhibited a peak field dependent change in yield stress of around 17 kPa, whereas the nanofluid exhibited peak field dependent change in yield stress of approximately 7 KPa. Plastic viscosity is similar for all three fluids, and all fluids showed signs of shear thinning.

References

1. P. P. Phule: MRS Bulletin, August 1998, 23-25.
2. Cl. Kormann, H. M. Laun and H. J. Ritcher: 5th International Conference on ER Fluids, MR Suspensions and Associated Technology, (1996), 362-67.
3. S. Taketomi, Y. Ozaki, K.Kawasaki, S. Yuasa and H. Miyajima: Journal of Magnetism and Magnetic Materials, vol 122 (1993), 6-9.
4. I. Bica: Materials Science and Engineering, vol B68 (1999), 5-9
5. J. Lindler and N. M. Wereley (2000). J. Intell. Systems and Structures, 10(5):363-76.
6. E Lemaire, A. Meunier, G. Bossis, J. Liu, D. Felt, P. Bashtovoi and N. Matoussevitch: Journal of Rheology, vol 39, no 5 (1995), 1011-20.

PROTEIN- F_e MR FLUID AND ITS PROPERTIES

HU LIN, CAI SHAOHONG AND ZHANG CHAOPING

Department of Physics, Department of Chemistry, Guizhou University, Guiyang. 550025, China
E-mail: lhu@tmail.gzu.edu.cn

The MRF consisted of protein-Fe particles prepared by micro emulsion technique has been studied. The process of protein-Fe preparation and its characterization with XRD, SEM and IR analysis are reported. The magnetorheological and temperature effects, as well as sedimentation stability of protein-Fe MRF have been investigated. The results show that the protein-Fe particles are composite of crystalline α-Fe coated with non-crystalline protein. The protein-Fe MR fluid has strong magnetorhological effect, while its sedimentation stability is much improved comparing to that of carbonyl iron MRF.

1 Introduction

Magnetorheological fluids (MRF) and electrorheological fluids (ERF) are rapidly developing intelligent materials in recent years. Under the effect of external magnetic field or electric field, their rheological behavior changes dramatically. Thus it is a much better way of controlling dynamic properties than the present technology [1]. The yield stress of MRF is about ten times greater than that of ERF, especially in the need of greater yield stress [2-4]. Lord Company [2] made a MRF with different proportions of iron, cobalt, nickel, of which the static yield stress is 20 to 80 kPa at volume fraction Φ=25% and H= 7000 Oe. The yield stress of MRF made by Shulman et al.[5] with iron and γ-Fe_2O_3 was four times greater than that of using iron only[6]. It is generally agreed that the key issue to promote the MRF technology is to solve the sedimentation problem of MRF to fit practical needs. The following study shows that the protein-Fe MRF has an improved sedimentation property.

2 Preparation and characterization of protein-Fe composite particle MRF

2.1 Preparation of protein-Fe composite particles The basic model makes the following assumptions:

Materials: SDS, isopentanol , n-heptane, TBP, $NaBH_4$ with a weight percentage 79.60%, gelatin and $FeCl_2 \cdot 4H_2O$. The water used in the sample preparation is double distilled water.

Methods: First, a certain quantity of SDS was put in a 300ml flask, in which a mixture of TBP, isopentanol and $FeCl_2 \cdot 4H_2O$ was originally prepared. Then, 30ml of gelatin, which was heated and dissolved at 50℃, was added slowly with stirring and keeping at 50 ℃. Afterwards, stirring for more than one hour was needed to form solution A. A certain quantity of $NaBH_4$ was added in a flask followed by 9ml of isopentanol , 60ml of n-heptane and several grams of carbonyl iron particles (<2.5μm), which have been drenched in non-polarized solvent. Stirring for an hour is then followed to form fluid B. Finally A is mixed into B together with stirring while the mixture was kept in water bath under a flow of nitrogen for one hour. Acetone and hot water is used to scour off the remainder for

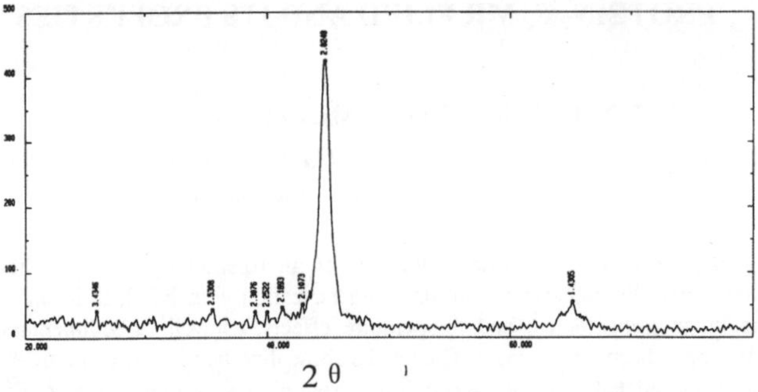

Figure 1. The XRD spectrum of iron/protein composite particle.

many times, then iron/protein powder was formed after 80℃ drying in vacuum for certain time. Then the protein-iron MRF can be made by mixing iron/protein powder with silicone oil.

2.2 *Structure and Magnetic Property of Iron/protein Composite Particles*

X-ray diffraction (XRD) measurement of protein- iron particles indicates that the particles contain mainly crystalline α–Fe as shown in Fig. 1. While there are some features from the scattering background showing protein being absorbed on the surface of Fe particles. Surface observations show that iron/protein composite particle is hard to be oxidized by water wash because of the coating of protein, as seeing little Fe_2O_3 contained in XRD spectrum. Tested by a high resolution SEM , the average diameter of particles is 1 to 1.7μm, in sphere shape, as Fig. 2 shows. In SEM photo, the dark-colored spheres are iron particles, and the light-colored fibers, macromolecule double-helix protein. From magnetic measurement under the magnetic field up to 8000 Oe, the saturated magnetization of the sample is $92.4 Am^2/kg$.

2.3 *IR test of protein-Fe composite particle*

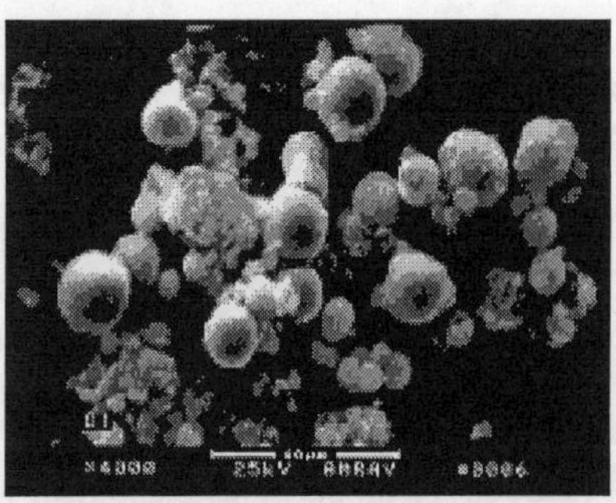

Figure 2. The SEM photo of iron/protein composite particle

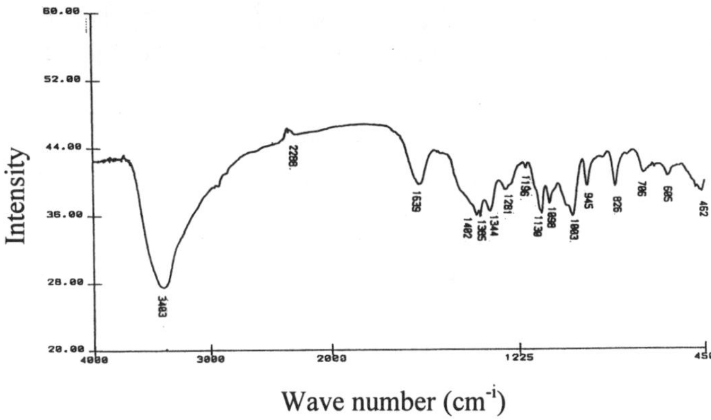

Figure 3. The IR spectrum of composite particle.

Known from the structure, protein molecule has three possible locations to coalesce with Fe^+ ions, namely the O of carboxyl in the carbon part, the N of amino in the nitrogen part and the N of acyl. The test result of the sample by IR spectrum shows in Fig 3. The absorption peak of iron/protein particle in the range of 1700-1100cm^{-1} indicates that the O and N on the amino aid connected with acyl have a certain bond effect with iron. 1639 cm^{-1} intensive band corresponds to the flex vibration of amide I of protein; The band at 1402 cm^{-1} corresponds to the flex vibration of symmetrical COO^- or C-N; The absorptions at 1385cm^{-1} and 1344cm^{-1} correspond to the symmetrical deformation of N-H or flex of C-N on the acyl-amide band IV of protein; 1281cm^{-1} corresponds to C-N-C or C-O flex vibration on the acyl-amide band V. All above shows that iron has a direct bond effect with N on amide and O on carboxyl in the protein molecule. 3403cm^{-1} band is due to the flex vibration of N-H bond; 1196cm^{-1} and 1130cm^{-1} bands are due to the flex vibration of CO or C-N-C in protein; 945cm^{-1} band is due to the flux of υ_{cH} or CH_2; 826cm^{-1} and 706 cm^{-1} bands are due to slight oscillation and deformation υ_{COO-}; 605 cm^{-1} band is due to CIS-CH vibration or CH, CH_2 oscillation; 462 cm^{-1} band is due to the deformation of $\upsilon_{cc\text{-}N}$. The results of IR spectrum indicate that protein molecule has bond effect with iron.

3 Properties of protein-Fe MRF

The rheological experiment of protein-iron MRF is performed on a home-made static yield stress testing device. Samples are placed in a copper sample chamber with the volume of 100mm x 80mm x 20mm. During the test, the sample chamber is located between the two magnetic poles of electric magnet, which can produce a horizontal magnetic field from zero to over 10,000 Oe when the distance of the two poles is about 20 mm. Remnant magnetism is delimited with a demagnetization process. Fig. 4 shows that the yield stress of protein-iron MRF is similar to that of carbonyl iron MRF under the same magnetic field. When the volume fraction of iron in MRF is 9% and the magnetic field is 5000Oe, the static yield stress is 4.5 kPa. It shows that when the encysted layer is not very thick, the effect of magnetic field is much greater than the spatial damping of protein to particles. The magnetized particles line alone the direction of magnetic field and the yield stress of MRF system becomes higher. The shear stress is sensitive to the volume fraction (Φ) of magnetic particle. The larger the ϕ, the greater the yield stress.[7] For example, when ϕ=40% and H = 4000 Oe, the static yield stress τ=60 kPa in carbonyl iron MRF[7].

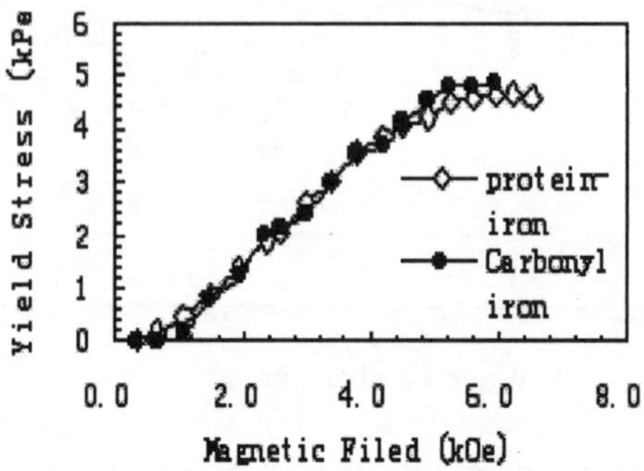

Figure 4. Static yield stress of MRF with protein-iron and Carbonyl iron.

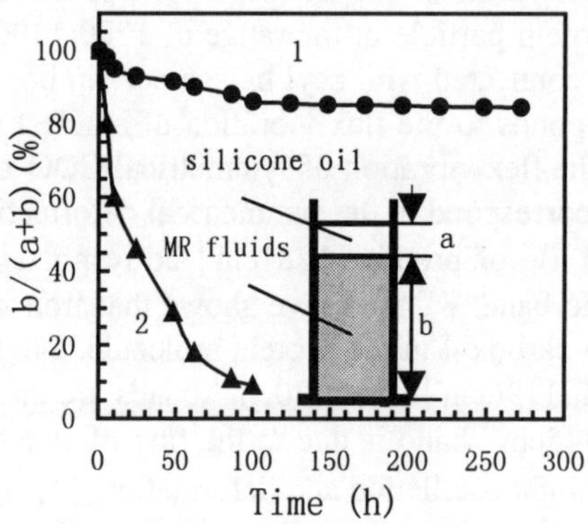

Figure 5. The IR spectrum of composite particle.

Since the sedimentation was obvious problem, we had added various polymer surface-active reagents to carbonyl iron MRF to improve the sedimentation stability in our previous study[8]. In present iron/protein MRF the clad disposal is taken on the iron surface using natural protein polymer. The protein polymer is absorbed on the surface of iron particle and forms a protecting layer to prevent particles approaching. Known from our experiment, the percentage of iron after clad disposal is 72.07% and percentage of protein is 27.93% with a total density to be 4.58g/cm^3, which is much smaller than that of carbonyl iron (6.81g/cm^3). Meanwhile, the clad disposal of protein increases the surface space of solid particles, thus the sedimentation rate of solid particles decreases a lot. When the density of solid particles becomes smaller, the density disparity of solid to fluid in dispersing system reduces. This conforms the theoretical prediction of the decrease for sedimentation rate[9]. The measurement shows that the sedimentation rate of protein-iron we made is much smaller than that of classical carbonyl iron, of which the comparison can be seen in Fig. 5. The classical carbonyl iron particles settle in to 30% after 50 hours while the particles in protein-Fe/silicone oil MRF only settle to 85% in 10 days and even in five mouths. The temperature dependence of iron/protein MRF is shown in Fig 6. It can be seen that static yield stress vs. temperature in the range of –20℃ to 100℃ is almost a constant. Fig. 7 plots the variation of the shear stress with shear rate for the sample.

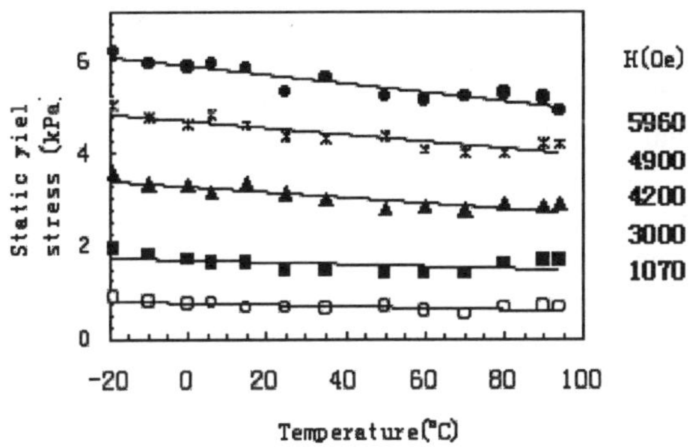

Figure 6.

Figure 6. Variation of static yield stress for protein-iron MRF with temperature (ϕ=9%)

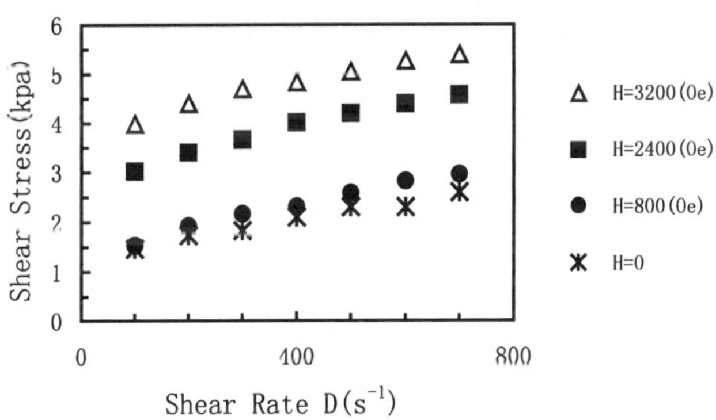

Figure 7. Variation of the shear stress with shear rate for the sample with ϕ=12%.

4 Summary

The protein-Fe particles have been prepared with micro-emulsion technique. Confirmed with XRD, SEM and IR, the bond reaction has been taken place between protein molecules and iron. The protein coated iron particles with low density, which are hard to be oxidized, can much improve the anti-sedimentation property of MRF suspension, while the shear stress is as same as carbonyl iron. In the range of –20℃ to 100℃, the static yield stress of protein-Fe MRF almost keeps as a constant. Therefore, protein-Fe is a good MRF material with comprehensive properties. Further analysis and study found that this material shows non-equilibrium dissipation property in the magnetorheological process, and critical phenomena of generalized phase transition appeared at the transition point[10-11].

5 Acknowledgements

The work is supported by both National Natural Science Foundation of China with grant number of 19872010 and Science & Technology Foundation of Guizhou Province, China (No. 983078). One the author (L.H.) would also thank the Institute of Physics of the Chinese Academy of Sciences, Physics Department, Fudan University and the Institute of Geochemistry of the Chinese Academy of Sciences.

References

1. K.Q. Lu, and X.W. Zhou, *Progress of Liquid State Physics* (Wuhan University Publishing House, Wuhan, 1997), in Chinese.
2. A.J. Margida, K.D. Weiss, and J.D. Carlson, *Inter. J Mod. Phys. B*, **10**, 544 (1996).
3. T.C. Jordan, Ph.D. Dissertation, Univ. of Connecticut, 1989.
4. Z.P. Shulman, *et al.*, *Magnetohydrodynamics*, **20**, 233 (1984).
5. J.B. Hayter and R. Pynn, *Phys. Rev. Lett*, **49**, 1103 (1983).
6. W.L. Kordonskii, and S.A. Demchyk, *Int. J. of Mod. Phys. B*, **10**, 613 (1996).
7. L. Hu, *et al*, *Progress in Fluidity Physics*, 1997，P.276-282.
8. L.Hu, *et al.*, Ma*chine Science and Technique Article Assembly*, **17**， 136 (1998), in Chinese
9. N.C. Laplov, *Chemistry Experiment* (Gaodeng Education Publishing House, Beijing, 1992), in Chinese.
10. Shaohong Cai, *et al.*, *Progress in Physics*, **19**, 270 (1999), in Chinese.
11. Shaohong Cai, L. Hu, and L. J. Dai, *Chinese Physics*, 9, 450 (2000).

ELECTRO-MAGNETORHEOLOGICAL FLUIDS DISPERSING ZEOLITE PARTICLES CONTAINING IRON

A. SHIBAYAMA, T. MIYAZAKI, K. YAMAGUCHI, K MURAKAMI AND T. FUJITA

Faculty of Engineering and Resource Science, Akita University, 1-1 Tegata-Gakuen cho, Akita, 010-8502, Japan,

E-mail: sibayama@ipc.akita-u.ac.jp

Some functional fluids that respond to both magnetic and electric fields have been prepared and their characteristics are described. In this study, an electro-magnetorheological fluid (EMRF) dispersing zeolite particles containing metallic iron by reducing precipitated magnetite has been investigated. When the viscosity is measured by cone plate viscometer and cylindrical viscometer, electric and magnetic fields are applied both between cone and plate or two cylinders. In case of cone plate, the shear stress at constant shear rate increased with the increase of both magnetic field and electric field. On the other hand when the viscosity is measured by cylindrical viscometer, the shear stress at constant shear rate increased with the increase of electric field, however, the increase rate of shear stress by magnetic field is very small. In this case the magnetic field direction is perpendicular to electric field. The EMRF has typical characteristics to respond with magnetic and electric field. The shear stress of EMRF in electric field is stronger than that of magnetic field. Additionally, the inflection and peak point in the shear rate-shear stress curve are appeared and the behaviors of the clusters in the electric field are observed. The experimental results suggested that the fluid viscosity (shear stress/shear rate) is affected by the arrangement of clusters parallel or perpendicular to the direction of the EMRF flow.

1 Introduction

In recent years the development of intelligent materials, especially, the magnetic fluids, electrorheological fluid and magneto-rheological suspension representative of functional fluids, have been required in various industrial fields [1,2]. The functional fluids are practically used in many fields, such as damper, sealing, actuator, crutch, etc. The important point of preparing the new useful functional fluids is to develop the dispersing particles [3,4,5,6]. Moreover in order to produce the electro-magnetorheological fluids, the particles of responsive to electric field and magnetic field should be developed. In this study, a new electro-magnetorheological fluids (EMRF) dispersing zeolite particles containing metallic iron has been prepared and investigated about shear stress and shear rate in the field of electric and magnetic.

On the other hand, zeolite is known as one of aluminosilicates and actually used as absorptive mineral for some industrial areas. In addition, although a lot of natural zeolite is produced in Japan, therefore it is expected to be developed a new effective technique. Consequently, in this study the zeolite was used to dispersant particles of electro-magnetorheological fluid [4,5].

At first, the EMRF prepared by zeolite containing magnetite is reduced in hydrogen atmosphere. Next, the characteristic of the new fluid is clarified by the examination of viscosity for rheological fluids using the cone-plate and cylindrical viscometer in the electric field and magnetic field.

2 Methods

2-1. Preparation of electro-magnetorheological fluids (EMRF)

The preparation of EMRF dispersing zeolite particles containing metallic iron by reducing precipitated magnetite is done by the following procedure (Fig. 1):

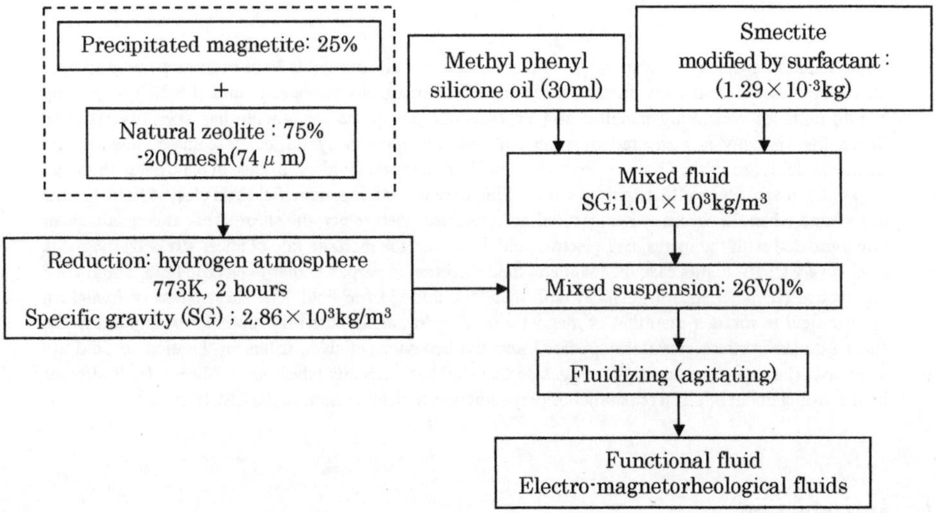

Figure 1. Flowsheet of the preparation of electro-magnetorheological fluids using natural zeolite.

At first, the magnetite is precipitated in the natural zeolite of under 30 μ m suspension. The prepared plus charged magnetite can be coagulated with minus charged natural zeolite in acidic region. The mixing ratio of precipitated magnetite and natural zeolite is 25% and 75%, respectively. Next, the dried zeolite containing magnetite is reduced in hydrogen atmosphere at 773K. After reduction, the well-sintered particles (up to 10 μ m in size) are produced. Those zeolite particles containing iron of approximately 50-weight % (or 26 vol%) are dispersed in methyl phenyl silicone oil (30ml, specific gravity: 0.995×10^3kg/m^3) by mixing with smectite modified by surfactant (1.29×10^{-3}kg, specific gravity: 1.64×10^3kg/m^3) to obtain the stable suspension. Finally, the new EMRF with methyl phenyl silicone oil and smectite based on the zeolite and iron particles was fluidized by automated mortar.

2-2. Measurement of viscosity of EMRF by using cone-plate and cylindrical viscometer

The viscosity of the EMRF has been measured by the cone-plate and cylindrical type rotational viscometer in the solenoid coil and electrode as shown in Figure 2 and Figure 3 [3,6]. The electric and magnetic fields were applied by range of 0 to 1 (kV/mm) from DC high voltage supply unit, and 0 to 0.08 (T) from solenoid coil respectively. In this condition, the relationships between shear rate and shear of the fluid stress has been examined.

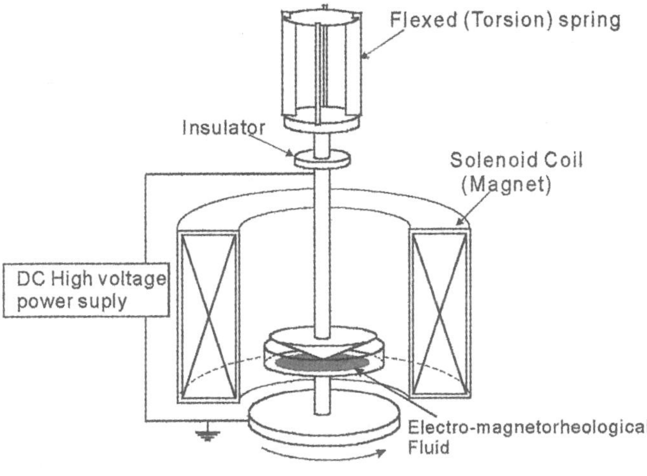

Figure 2. Schematic illustration of cone and plate viscometer.

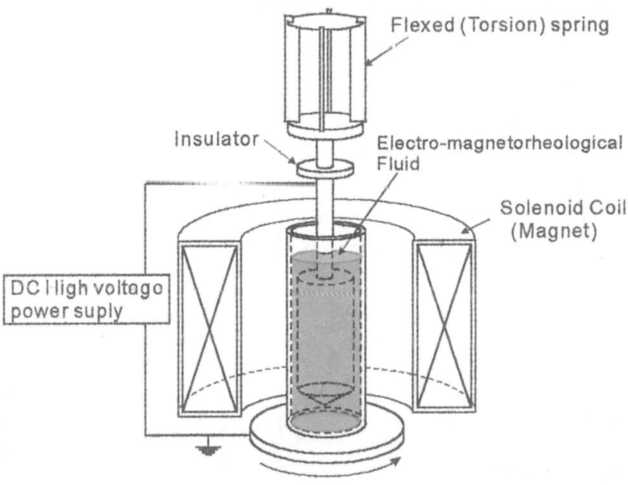

Figure 3. Schematic illustration of cylindrical viscometer.

3 Results and Discussions

3.1 The results of cone and plate viscometer on the electric or magnetic field

The characteristics of the EMRF with zeolite containing iron have measured by cone and plate viscometer on the independence of electric or magnetic fields. Figure 4 shows the relationship between the shear rate and shear stress. The stronger magnetic and electric fields are, the more shear stress increases. The new EMRF is reactive to the magnetic and electric fields as the shear stress increases with the shear rate in each applied field. In other words, the shear stress of electric field is stronger than that of magnetic field. Hence, EMRF responds more effectively with electric filed.

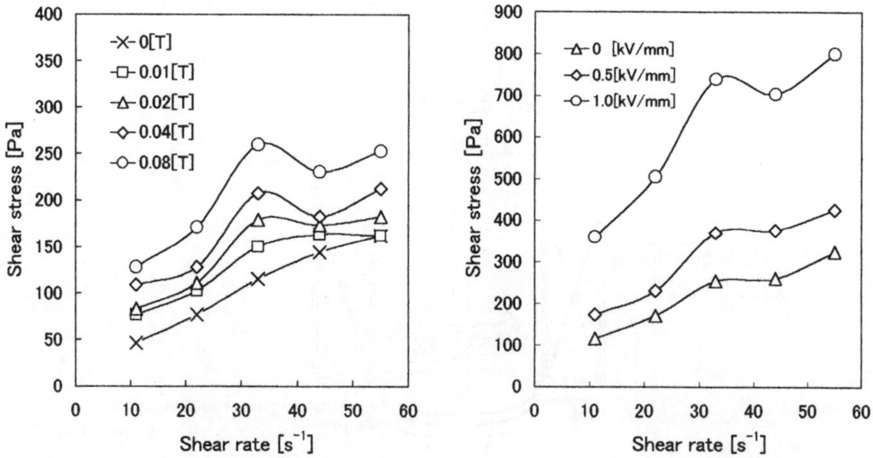

(a) Magnetic field (b) Electric field

Figure 4. Viscosities of EMRF dispersing zeolite particles containing iron and smectite in silicone oil using cone and plate viscometer.

3.2 The results of cylindrical viscometer on the electric or magnetic field

Next, the performance of the EMRF is examined by the cylindrical viscometer with electric or magnetic field. The shear stress—shear rate relationship under various applied electric or magnetic fields strengths for EMRF is shown in Figure 5. It can be seen that the shear stress is proportionally to both magnetic and electric fields.

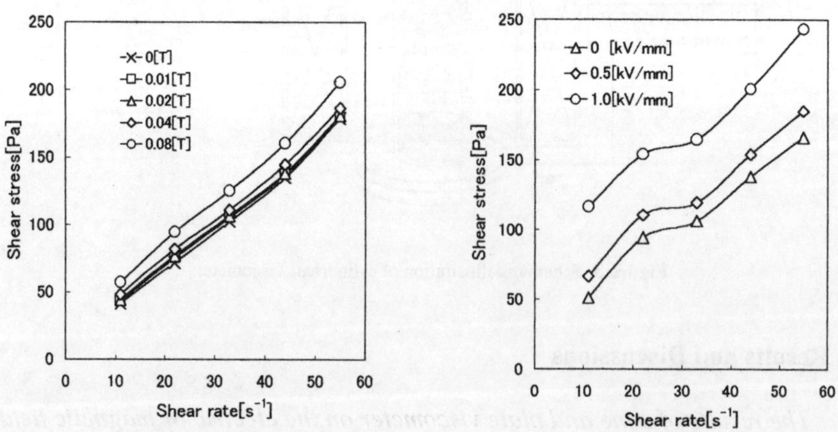

(a) Magnetic field (b) Electric field

Figure 5. Viscosities of EMRF dispersing zeolite particles containing iron and smectite in silicone oil using cylindrical viscometer.

Moreover, it is clearly apparent that the shear stress obtained by using cone and plate viscometer (Fig. 4) is stronger than the one measured by cylindrical viscometer. In case of applying magnetic field (Fig. 5(a)), the each shear stress of different magnetic strength exhibited almost the same increasing tendency. It is considered that the phenomenon is due to the direction of fluid flow in cylinder, which is perpendicular to direction of the magnetic field.

3.3 The characteristics of EMRF by cone-plate and cylindrical viscometer with both applied electric and magnetic field

It is important to investigate the responsibility of EMRF with both applied magnetic and electric field. Therefore the shear stress-shear rate relationship by using cone-plate and cylindrical viscometer under various magnetic and electric fields strength for the new EMRF with methyl phenyl silicone oil and smectite based on the zeolite and iron particles is measured (Fig. 6). In the absence of electric field (0 kV/mm), EMRF behaved like Newtonian fluids. However, EMRF changed to a Bingham fluid when it is exposed to an electric field. It is observed that the shear stress measured by cone-plate viscometer (Fig 6 (a)), is higher than that one measured by using cylindrical viscometer for both applied fields. It is considered that the direction of fluid flow is perpendicular to both directions of the electric field and magnetic fields. The shear stress at constant shear rate increased with the increase of both magnetic field and electric field. On the other hand, when the viscosity (shear stress/shear rate) is measured by cylindrical viscometer, while fluid is filled between two cylinders, the shear stress increased at constant shear rate with the increase of electric field. However, the increase rate of shear stress by magnetic field is very small. In this case the magnetic field direction is perpendicular to electric field.

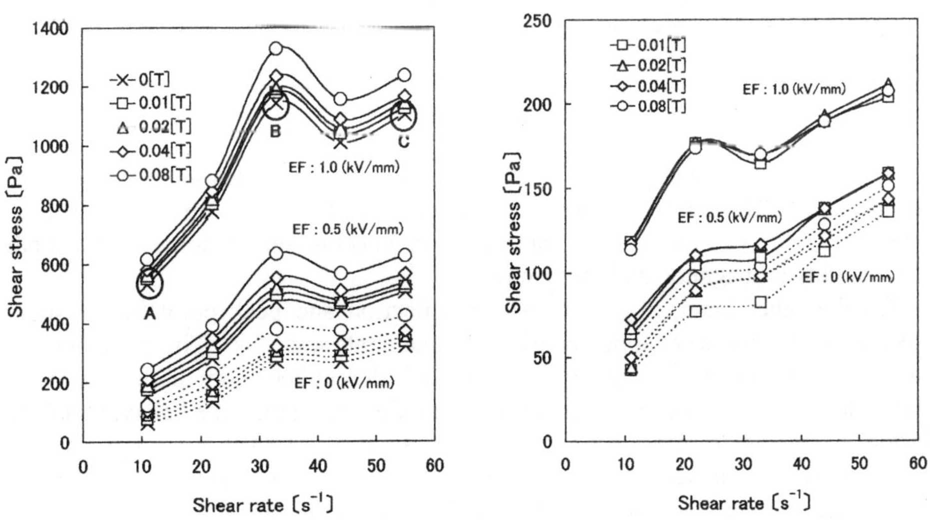

(a) In case of cone and plate viscometer

(b) In case of cylindrical viscometer

Figure 6. Characteristics of EMRF dispersing zeolite particles containing iron and smectite in silicone oil on the applied both magnetic and electric fields. [Electric Field (EF) :0, 0.5, 1.0 (kV/mm)]

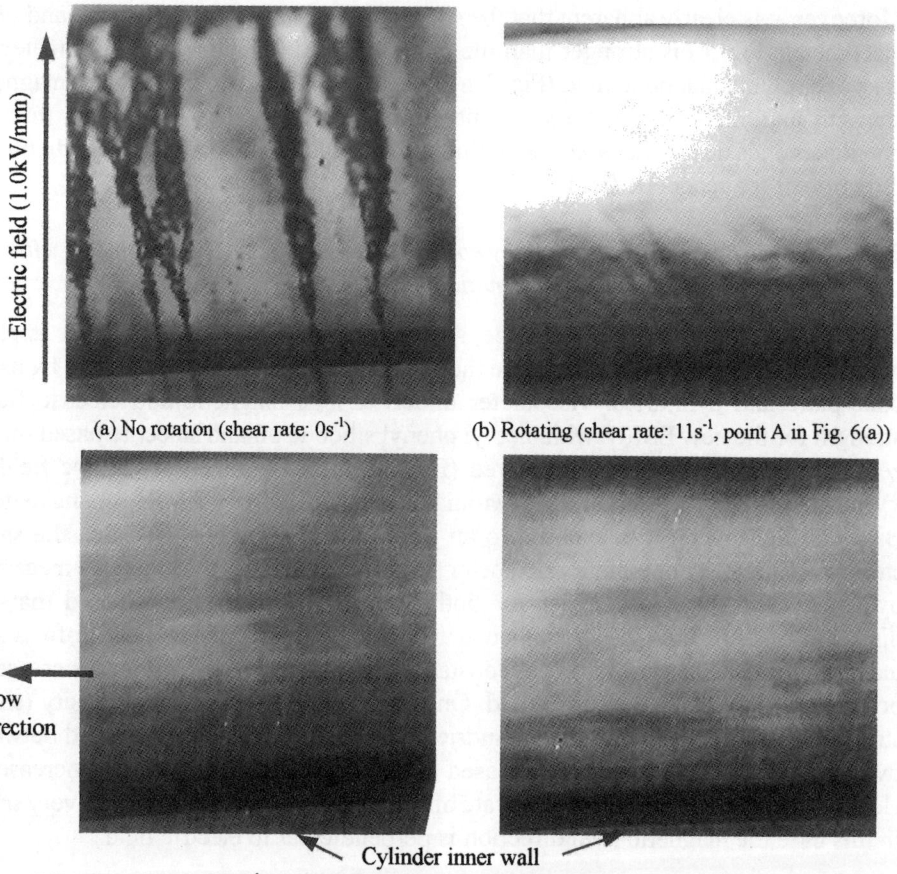

(a) No rotation (shear rate: $0s^{-1}$) (b) Rotating (shear rate: $11s^{-1}$, point A in Fig. 6(a))

(c) Rotating (shear rate: $33s^{-1}$, point B in Fig. 6(a)) (d) Rotating (shear rate: $55s^{-1}$, point C in Fig. 6(a))

Figure 7. Photographs of the cluster in EMRF by using cone and plate viscometer with electric field (1.0kV/mm). (Photo (b), (c) and(d) indicate the A, B and C in Fig.6.)

That is to say, when the both of magnetic and electric fields were applied, the viscosity of EMRF changes effectively. In addition, the EMRF, which is developed by using the zeolite containing iron particles with smectite and silicone oil, has both responsibility to magnetic and electric fields.

On the other hand, in Fig. 6 (a) the inflection and peak point in the shear rate-shear stress curve are appeared when the share rate was 33 s^{-1}. This plot suggests that the above phenomenon occurred due to the different behaviors of clusters in the EMRF flow under the electric or magnetic field. Following that, the behaviors of the clusters in the electric field are observed by using the CCD camera (Fig. 7). The observations were carried out at the points A (shear rate: 11 s^{-1}), B (33 s^{-1}) and C (55 s^{-1}) which are shown in Figure 6 (a). Figure 7 (a) shows the clusters formed under the electric field of 1.0 kV/mm, while the plate of the cone and plate viscometer did not rotate. Figure 7 (b) shows that the some clusters were retaining in perpendicular to the direction of the fluid flow. Figure 7 (c) illustrates the situation when the clusters are uniformly dispersed. However some clusters are retained along the wall. It is assumed that the shear stress at point B is stronger than the other ones due to increasing of the fluid viscosity (shear stress/shear rate) as the clusters are arranged in both parallel and perpendicular direction of fluid flow. Next, the shear stress decreased when the shear rate was $44s^{-1}$. Nevertheless, the shear stress increased with the shear rate forming the uniform clusters parallel to the flow direction (Fig. 7 (d)). Hence, the observations of the clusters in EMRF, under the electric field, demonstrated the existence of the peak point between the shear rate and shear stress.

4 Conclusions

The new electro-magnetorheological fluid (EMRF) is prepared by using the zeolite one of aluminosilicates containing iron particles with silicone oil and smectite. Especially, the dried zeolite containing magnetite is reduced in hydrogen atmosphere at 773K. The EMRF had typical characteristics to respond with magnetic and electric fields. The shear stress of EMRF in electric field is stronger than that of magnetic field. Additionally, the inflection and peak point in the shear rate-shear stress curve are appeared and the behaviors of the clusters in the electric field are observed. The experimental results suggested that the fluid viscosity (shear stress/shear rate) was affected by the arrangement of clusters parallel or perpendicular to the direction of the EMRF flow.

References

1. Fujita T., Yoshimura K., Seki Y., Dodbiba G., Miyazaki T. and Numakura S., Characterization of Magnetorheological suspension for seal, *Jour. Of Intelligent Material Systems and Structures*, **10** (2000), pp.770-774.
2. Akagami T., Asari K., Jeyadevan B. and Fujita T., Characterization of particle motion for polishing and texturing undr AC field by using particle dispersion type ER fluid, *Jour. Of Intelligent Material Systems and Structures*, **9** (1998), pp.672-675.
3. Fujita T., Jeyadevan B., Kuzuno E. and Nakagawa N., Formation of clusters and their effect on the viscosity of magnetic and electrorheological fluids, *Proc. Of the 6th Inter. Conf. on ERF, MRS and their Applications*, (1997), pp. 578-585.
4. Böse H., Investigations on zeolite-based ER fluids supported by experimental design, *Proc. Of the 6th Inter. Conf. on ERF, MRS and their Applications*, (1997), pp. 240-247.
5. Böse H. and Trendler A., Comparison of rheological and electric properties of ER fluids based on different materials, *Proc. Of the 7th Inter. Conf. on ERF and MRS*, (1999), pp. 80-87.
6. Fujita T., Jeyadevan B., Yamaguchi K. and Nishiyama H., Preparation, Viscosity and damping of functional fluids that respond to both magnetic and electric field, *Powder Technology*, **101** (1999), pp.279-287.

MAGNETOSTRICTIVE PHENOMENA IN MAGNETORHEOLOGICAL ELASTOMERS

J. M. GINDER, S. M. CLARK, W. F. SCHLOTTER, AND M. E. NICHOLS

Ford Research Laboratory, Ford Motor Company, P. O. Box 2053, SRL MD 3028, Dearborn, MI 48121-2053, U. S. A.

A host of fascinating and useful magnetic phenomena are found in composites containing magnetizable particles in viscoelastic solids. Embedding magnetically soft iron particles in natural rubber produces a class of magnetostrictive composites sometimes termed magnetorheological (MR) elastomers. We have previously shown that these materials can exhibit viscoelastic moduli that increase substantially in an applied magnetic field. In this paper, we incorporate MR elastomers in a simple resonant structure called a tuned vibration absorber to measure the complex dynamic shear moduli of these materials at high frequencies. We find that the field-induced modulus increase in MR elastomers is substantial even at kilohertz mechanical frequencies. As in previous measurements at low frequencies, the moduli are generally found to decrease with strain amplitude. We also report preliminary measurements of the relatively large elongation of these materials in applied magnetic fields.

1 Introduction

Composites comprising magnetically 'hard' or 'soft' particles embedded in polymeric media find use in a remarkably wide number of applications, including media for magnetic data storage, magnetic position sensors, flexible magnets, touch-screen displays, and electromagnetic shielding. By virtue of the processability afforded by their polymeric hosts, they can be formed into objects with myriad shapes and sizes. If processed in a magnetic field, these materials can possess anisotropic mechanical, electrical transport, and magnetic properties due to the chain-like particle structures that result from the magnetic dipole interactions between particles.

The magnetic interactions between particles in these composites depend on the orientation of each particle and their spatial relationship, coupling the magnetic and strain fields in these materials and giving rise to a number of interesting magnetomechanical phenomena that are often lumped under the term magnetostriction.[1] Magnetic materials can exhibit length and volume changes in homogeneous magnetic fields, the effect most traditionally associated with magnetostriction. In addition, deforming a magnetic material can modify the magnetization within it. Finally, their viscoelastic moduli can change with the applied magnetic field, the so-called 'delta-E effect.'

This magnetoviscoelastic effect has been observed in composites containing magnetically soft particles dispersed in silicone elastomers,[2,3,4] poly(vinyl alcohol),[5] and natural rubber,[6] materials often termed magnetorheological (MR) elastomers. The present authors have focused on a class of these materials involving carbonyl iron particles dispersed in natural rubber. We previously reported measurements of the dynamic shear modulus at mechanical frequencies below 20 Hz, finding that applying a magnetic field can substantially increase the shear modulus.[6] We have also constructed proof-of-concept components such as automotive mounts using these elastomers, showing that applied currents could tune their stiffness, and that the apparent stiffness increase responded within milliseconds of the application of the magnetic field.[7] Magnetostriction in these materials has been modeled using thermodynamic approaches for composites containing both chain-aligned[8] and randomly dispersed[9] particles.

In this paper, we use a simple spring-mass system, often termed a tuned vibration absorber (TVA), to extend our measurements of the dynamic shear modulus of MR elastomers to mechanical frequencies beyond one kilohertz. We find that the MR effect is substantial at these frequencies, suggesting that it may be utilized to construct devices for noise and vibration control. We also report initial measurements on the large, positive magnetostrictive length change induced by uniform applied fields in these materials.

2 Experimental Methods

2.1 Sample preparation

MR elastomer materials were prepared by mixing carbonyl iron particles of ~ 3 μm nominal diameter (ISP S-3700), natural rubber or cis-poly(isoprene), and additives including crosslinking agents, antioxidants, and mixing aids in appropriate concentrations on a two-roll mill. In all the measurements described in this paper, the iron volume fraction of the elastomers was roughly 0.27. A mold, Figure 1, was constructed to enable cross-linking of 60 mm disc-shaped samples in a uniform magnetic field to permanently lock the particles into chainlike structures. The magnetic field was generated using a ~ 350 turn, 20 gauge insulated copper coil housed in the mold and powered by a constant-current power supply. A ten-turn sense coil was wound around the perimeter of the sample volume; integrating the voltage induced on the sense coil using an integrating fluxmeter (Walker MF-3D) yielded the average flux density in the sample. Samples were placed in the pre-heated mold; the mold was held at 150 °C between the platens of a hydraulic press for 15 minutes to allow cross-linking to proceed. With a sample thickness of 2 mm, flux densities as high as 0.8 T were achievable using this mold.

2.2 Dynamic modulus

To complement earlier low-frequency measurements of the dynamic shear modulus of these materials, and to explore their use in components for vibration abatement, a tuned

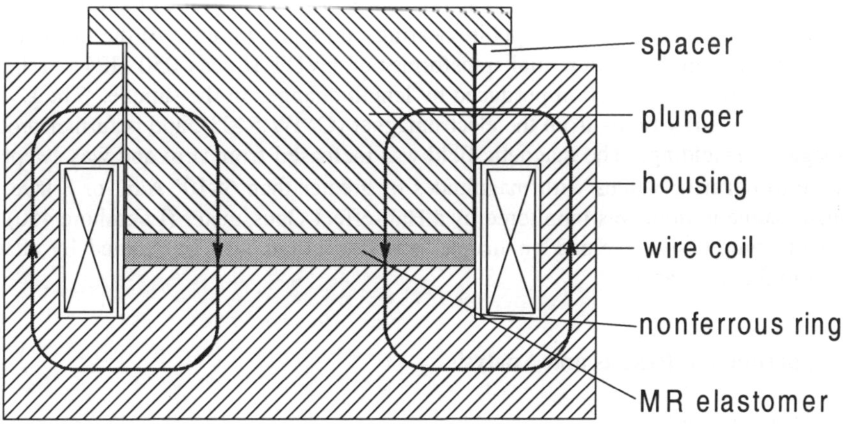

Figure 1: Schematic cross-section of the mold used to cure MR elastomer samples in strong, uniform magnetic fields. The loops illustrate the direction of magnetic flux flow in the mold.

vibration absorber (TVA) was constructed using them.[10] At their simplest, TVAs consist of a reaction mass coupled to a moving base through one or more spring elements. In our TVA, the reaction mass was a block of low-carbon steel bonded to two slabs of MR elastomer using cyanoacrylate adhesive. The elastomer slabs were bonded to a rigid 'U'-shaped base also constructed of low-carbon steel. A ~ 600 turn wire coil was placed around one arm of the U in order to generate a magnetic field in the elastomer. A 10 turn sense coil was wound around the other arm to provide a measure of the flux density induced in the magnetic circuit. The base of this assembly was driven sinusoidally using an electrodynamic shaker (Labworks pa-141); the magnitudes of the base acceleration $|a_b|$ and the resulting acceleration $|a_m|$ of the reaction mass were measured using accelerometers (PCB 352-C65) and associated amplifiers. The mechanical frequency was increased in steps at several different acceleration amplitudes and flux density levels; at each step, the transmissibility $T \equiv |a_m| / |a_b|$ and phase was determined.

As will be demonstrated, the transmissibility and phase information can be used to extract the dynamic stiffness of the elastomer, and thus its shear modulus. To verify the validity of this approach, the reaction mass was subsequently fixed through a load cell (PCB 208-C01). The base was driven mechanically as before, and the reaction force generated by the elastomer was recorded as a function of frequency and applied flux density. The elastomer stiffness inferred in this manner corresponded closely to those inferred from the TVA measurements.

2.3 *Magnetostriction*

The difficulty involved in measuring the magnetostriction of highly deformable materials such as MR elastomers is not insignificant. For example, attaching strain gauges to such samples drastically alters their compliance and thus their field-induced length changes. Very sensitive length changes can be measured capacitively,[11] but making robust contact between a moveable capacitor plate and the sample is difficult, particularly when the sample surface is not perfectly flat. While optical techniques can also be used to measure length changes, the magnet used to generate the magnetostriction can block optical access. We constructed a dilatometer to enable measurement of the field-induced length changes in cylindrical MR elastomer samples induced by the uniform magnetic field of an electromagnet. In our fixture, the sample contacts the polished lower surface of a lever arm equipped with low-friction sapphire bearings at its pivot point. At the free end of the lever, a linear variable differential transformer or LVDT (Lucas 050 HR) is affixed in order to monitor the deflection produced by length changes in the sample. The LVDT is subject only to stray fields from the electromagnet (GMW 5403), and is further protected by magnetic shielding. The magnetic field was increased from zero and was swept from positive to negative values with magnitudes of greater than 6×10^5 A/m, or 8 kOe. The applied magnetic field was monitored by a Hall-effect Gaussmeter (Lakeshore 450). The effect of compressive preload on magnetostriction could also be studied by attaching masses to the lever arm.

3 Experimental Results and Analysis

3.1 *Dynamic modulus*

Figure 2 shows the frequency-dependent transmissibility of the MR elastomer TVA at a peak base acceleration of 0.2 g and at average flux densities ranging from 0 to 0.56 Tesla applied parallel to the uniaxial direction of the elastomer. This mass-spring system is clearly tunable by a magnetic field, with a resonance that shifts upward in frequency by over 20% at 0.56 Tesla. The phase also reflects the tunability of the resonance. The transmissibility is found to decrease as the amplitude of the base acceleration increases. The dynamic spring rate can be extracted from the measured transmissibility and phase using the equation of motion for this system,

$$M\frac{d^2 z_m}{dt^2} + K^*(z_m - z_b) = 0,$$

where t is time, M is the mass of the reaction mass, $K^* = Ke^{i\delta}$ is the complex spring

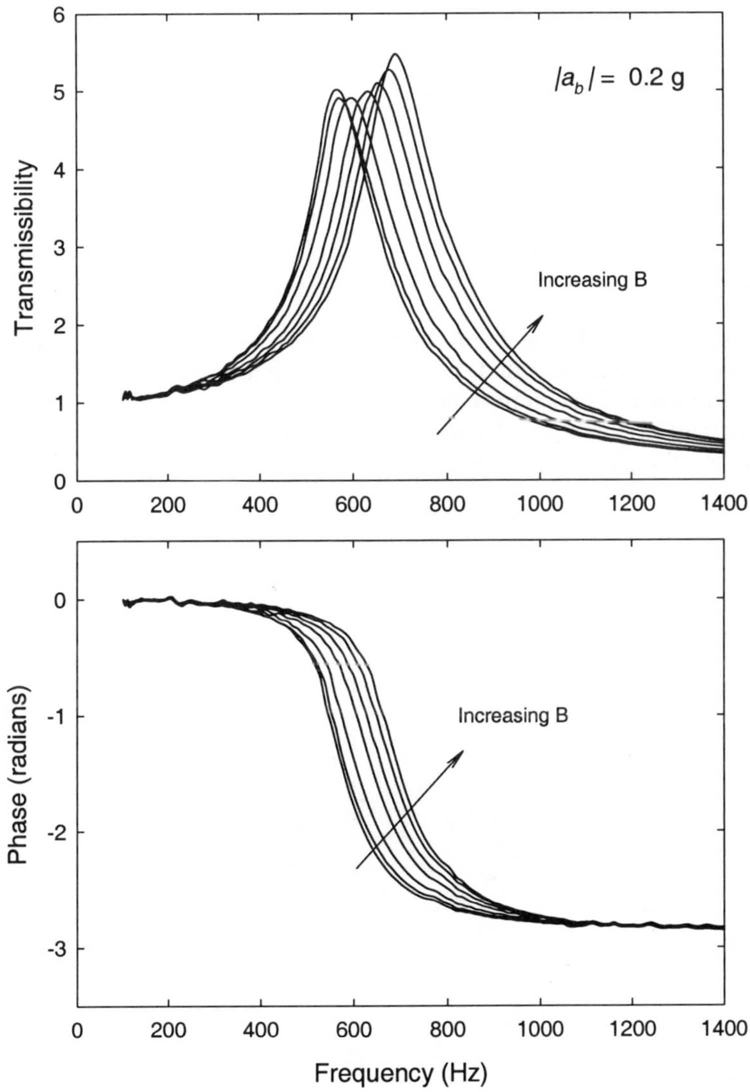

Figure 2: MR elastomer TVA transmissibility (top) and phase (bottom) versus frequency for a fixed base acceleration of 0.2 g. The resonance shifts to higher frequencies as the average flux density in the sample is increased from 0 to 0.56 Tesla.

constant of the elastomer slabs, and δ is the loss angle. Solving for the spring rate,

$$K^* = M \frac{dz_m^2}{dt^2}(z_b - z_m)^{-1}.$$

Assuming sinusoidal motion at an angular frequency ω, the magnitude of the dynamic stiffness K is

$$K = \frac{M\omega^2 T}{\sqrt{T^2 - 2T\cos\varphi + 1}},$$

where T is the transmissibility and ϕ is the phase difference between the motion of the reaction mass and the base. The elastomer loss tangent can also be calculated:

$$\tan\delta = \frac{\sin\varphi}{\cos\varphi - T}.$$

The peak shear strain on the elastomer slabs is given by

$$|\gamma| = \sqrt{T^2 - 2T\cos\varphi + 1}\,|a_b|/(h\omega^2),$$

where h is the elastomer thickness. Ignoring edge effects, the shear modulus magnitude is obtained from the spring rate by $G = Kh/A$, where A is the area of the elastomer bonded to the reaction mass.

Applying the approach outlined above to the measured transmissibility, the MR elastomer shear modulus magnitude at a mechanical frequency of 1 kHz was obtained, Figure 3. The increase in shear modulus is substantial – roughly 2 MPa. It initially varies roughly quadratically with flux density, then begins to saturate at higher flux densities. The shear modulus generally decreases with increasing strain, a phenomenon widely observed in these materials.[3,4,6]

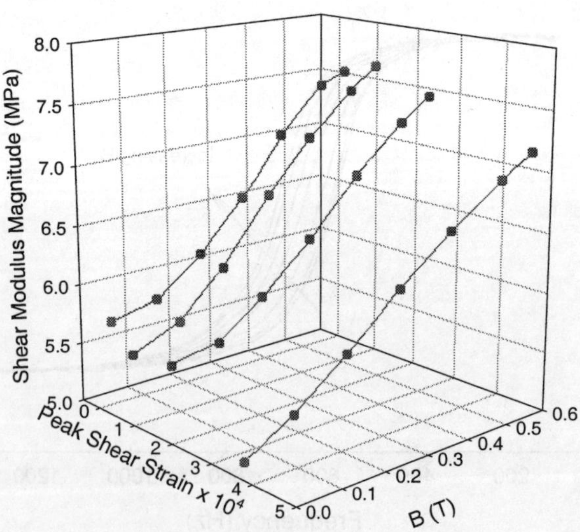

Figure3: The variation of the 1 kHz shear modulus magnitude with strain and flux density determined by measuring the transmissibility and phase of the MR elastomer TVA at four different accelerations.

Magnetostriction

Figure 4 shows the measured fractional length change induced by an applied magnetic field in a MR elastomer cylinder of length 11.0 mm and diameter 9.5 mm. The external field was applied parallel to the cylinder axis, which coincided with the axis of uniaxial symmetry in the material. The magnetostriction data were acquired with a 100 kPa compressive preload applied to the sample. The sample extends substantially at low fields; saturation effects are apparent at higher fields, culminating in an extension of about 3000 parts per million (ppm) at the highest applied fields. By comparison, magnetostrictive Terfenol-D typically exhibits extensions of 2000 ppm or less, though it does have the ability to generate higher loads since it is a much stiffer material. Significant hysteresis in the extension of the MR elastomers is observed; it is likely due to mechanical hysteresis of the host elastomer rather than magnetic effects, since the magnetic coercivity of iron-based MR elastomers is quite small – less than 10 Oe.[6] While the magnitude and hysteresis of the elongation observed here is reminiscent of that obtained in elastomers comprising 0.1-0.2 mm silicon steel particles in silicone elastomers,[12] the latter were randomly dispersed composites exposed to very high fields (~ 80 kOe). In both natural rubber and silicone-based elastomers, the magnitude of the magnetostriction is much larger than that obtained in bulk iron, suggesting that the effect is due to interparticle or shape effects. One might expect field-induced compression or shrinkage in oriented materials like the MR elastomers studied here, as particles in chains should experience attractive forces when fields are applied along the chain axis. Experiments are underway to explore the magnetostriction of samples with different aspect ratios, field directions, and degrees of particle alignment.

4 Summary

We have experimentally explored two aspects of magnetostriction in MR elastomers based on natural rubber: the field-induced increase in dynamic modulus, or delta-E

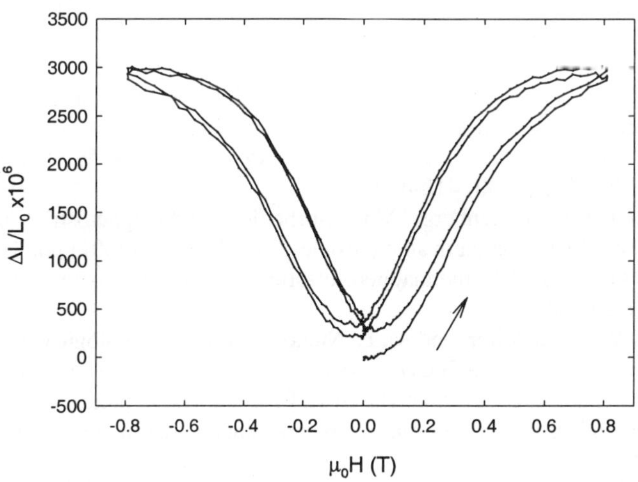

Figure 4: Dependence of the fractional length change in a cylindrical MR elastomer sample on external applied field. The sample's uniaxial axis was oriented parallel to the field, and it was compressively preloaded by 100 kPa.

effect, and the length change traditionally associated with these phenomena. An MR-based tuned vibration absorber was shown to be tunable with magnetic field. The corresponding dynamic modulus increase in these materials was found to be quite substantial at the audio frequencies tested. The field-dependent length change of an MR elastomers cylinder was measured using a lever-arm-based dilatometer. This magnetostriction was found to be positive, relatively large, and somewhat hysteretic.

5 Acknowledgements

The authors thank Craig Davis, Larry Elie, and Will Hockett of Ford Motor Company for their support of and contributions to this work.

References

1. See e.g. B. D. Cullity, *Introduction to Magnetic Materials*, (Addison-Wesley, Reading, MA, 1972).
2. T. Shiga, A. Okada, and T. Kurauchi, “Magnetroviscoelastic Behavior of Composite Gels,” *J. Appl. Polym. Sci.* **58** (1995) pp. 787-792.
3. M. R. Jolly, J. D. Carlson, B. C. Munoz, and T. A. Bullions, “The Magnetoviscoelastic Response of Elastomer Composites Consisting of Ferrous Particles Embedded in a Polymer Matrix,” *J. Int. Mat. Sys. Struct.* **7** (1996) pp. 613-622.
4. G. Bossis, C. Abbo, S. Cutillas, S. Lacis, and C. Métayer, "Electroactive and Electrostructured Elastomers," in *Proceedings of the 7th International Conference on ER Fluids and MR Suspensions*, ed. by R. Tao (World Scientific, Singapore, 2000), pp. 18-27.
5. T. Mitsumata, K. Ikeda, J. P. Gong, Y. Osada, D. Szabó, and M. Zrínyi, "Magnetism and compressive modulus of magnetic fluid containing gels," *J. Appl. Phys.* **85** (1999) pp. 8451-8455.
6. J. M. Ginder, M. E. Nichols, L. D. Elie, and J. L. Tardiff, "Magnetorheological Elastomers: Properties and Applications," in *Smart Structures and Materials 1999: Smart Materials Technologies*, ed. By M. Wuttig, (1999) pp. 131-138.
7. J. M. Ginder, M. E. Nichols, L. D. Elie, and S. M. Clark, "Controllable-Stiffness Components Based on Magnetorheological Elastomers," in *Smart Structures and Materials 2000: Smart Structures and Integrated Systems*, ed. by N. M. Wereley, Proceedings of SPIE **3928** (2000) 418-425.
8. Y. M. Shkel and D. J. Klingenberg, "Magnetorheology and magnetostriction of isolated chains of nonlinear magnetizable particles," *J. Rheol.* **45** (2001) pp. 351-368.
9. L. Borcea and O. Bruno, "On the magneto-elastic properties of elastomer-ferromagnet composites," to be published.
10. J. M. Ginder, W. F. Schlotter, and M. E. Nichols, "Magnetorheological Elastomers in Tunable Vibration Absorbers," in *Smart Structures and Materials 2001: Smart Structures and Integrated Systems*, ed. By L. P. Davis, SPIE Proceedings, in press.
11. See e.g., G. Brändli and R. Griessen, "Two capacitance dilatometers," *Cryogenics* **25** (1973) pp. 299-302.
12. S. Bednarek, "The giant magnetostriction in ferromagnetic composites within an elastomer matrix," *Appl. Phys. A* **68** (1999) pp. 63-67.

COMPRESSIVE MODULUS OF FERRITE CONTAINING POLYMER GELS

TETSU MITSUMATA[1], KENTA FURUKAWA[1], ETIENNE JULIAC[2], KENJI IWAKURA[1], AND KIYOHITO KOYAMA[1, 2]

[1]*Department of Polymer Science and Engineering, Faculty of Engineering, Yamagata University, 4-3-16 Jonan, Yonezawa 992-8510, Japan*

[2]*Venture Business Laboratory, Graduate School of Engineering, Yamagata University, 4-3-16 Jonan, Yonezawa 992-8510, Japan*

The mechanical properties of magnetic gel have been investigated. Magnetic gels, which consist of finely dispersed powder of barium ferrite ($BaFe_{12}O_{19}$) and poly vinyl alcohol (PVA), have been synthesized. The diameter of barium ferrite is less than 45 μm. The magnetic gels varying with ferrite concentration, crosslinking densities were prepared by mixing 10 wt.% PVA aqueous solution and barium ferrite using glutaraldehyde as a crosslinking agent in the presence of HCl. The diameter of barium ferrite is large enough to have a permanent magnetic moment. We applied a 10 kOe magnetic field in order to saturate the magnetic moment of barium ferrite. After magnetization, the compressive modulus was estimated with an ultrasonic method in order to find the influence of magnetization. Ultrasonic measurements were carried out using burst waves at 10 MHz and 295.5 K. The modulus of magnetized gel was found to depend on the concentration of magnetic substance, the crosslinking density, and the degree of swelling. It was clear that the modulus of magnetized gel was higher than the gel without magnetization for all samples. The change in modulus to the initial modulus $\Delta M'/M'_0$ for 10 wt.% and 15 wt.% of ferrite concentration was about 0.28% and 0.4% in a lower density region, respectively. Moreover, the change in modulus $\Delta M'/M'_0$ was constant in a lower density region however it strongly depends on the density in a higher density region. When the stress direction is perpendicular to the magnetization, the change in modulus increased. On the contrary, the change in modulus decreased when the stress direction is parallel to the magnetization. As increasing the density, the distance between magnetic substances become short and therefore the magnetic interaction is more significant in a higher density region.

1 Introduction

The stimuli-responsive properties of polymer gels such as electric and magnetic field, temperature, pH, salt, solvent concentration, visible light, have been widely investigated in the past decade [1-4]. The ferro fluid containing gel called ferrogel is one of the stimuli responsive gels, and it undergoes an abrupt elongation and contraction behavior under non-uniform magnetic field [5, 6]. These characteristic phenomena that can be seen in ferrogel, come from the fact that the finely distributed ferromagnetic particles are attached to the flexible network by an adhesive force, and so it occurs a direct coupling between magnetic and mechanical properties. According to this, we can expect that the modulus of the gel containing the magnetic substance is strongly correlated with both the strength and the direction of magnetization. One of authors has been reported about the magnetism and compressive modulus of magnetic fluid containing gels [7]. The results show that the compressive modulus with field of 4 kOe was much higher than that without field. The mean change in modulus increased with increasing magnetic field and it saturated above 2 kOe.

However, the relationship between the crosslinking density of gel or the concentration of magnetic fluid and the induced modulus by magnetic field has not been well understood. In this study, we have focused on the relation between the magnetization and the change in modulus using ferrite containing polymer gels called magnetic gel. Magnetic gels, which consist of finely dispersed powder of barium ferrite and poly vinyl alcohol (PVA), have been synthesized. The diameter of barium ferrite is much larger than in the case of ferro fluid, so it enables the gel to have a permanent magnetic moment. We applied a 10 kOe magnetic field in order to saturate the magnetic moment of barium ferrite. After magnetization, the compressive modulus was estimated with an ultrasonic method in order to find the influence of magnetization. Ultrasonic measurements were carried out using burst waves at 10 MHz and 295.5 K. The modulus of magnetized gel was found to depend on the concentration of magnetic substance, the crosslinking density, and the degree of swelling. We are discussing on the influence of magnetization direction on the compressional wave velocity and attenuation.

2 Experimental

2.1 Magnetic Gel Preparation

The magnetic gel was prepared by mixing 10wt.% PVA aqueous solution and magnetic particle of barium ferrite $BaFe_{12}O_{19}$ (Sigma-Aldrich Co.). The diameter of a magnetic particle was less than 45 μm. The weight ratio of ferrite to that of PVA c_f was kept as 5 wt.%, 10 wt.%, and 15 wt.%. A Glutaraldehyde was used as a crosslinking agent in the presence of HCl. The concentration of HCl was constant to be 15 wt.% to the PVA aqueous solution. The crosslinking density ν was 0.5 mol%, 1 mol%, 1.5 mol%, and 2 mol%, and the crosslinking reaction was carried out at the room temperature. We obtained PVA gels with various crosslinking densities in a similar manner with magnetic gels without magnetic particles. The shape of samples for the ultrasonic measurement is a disk with a diameter of 9~15 mm and a thickness of 4~5 mm. The magnetic gel was immersed in purified water during few days until the gel had reached an equilibrium swelling.

2.2 Measurement of Longitudinal Modulus

We have measured the longitudinal modulus of magnetic gel using an ultrasonic pulse-echo method. We used a transducer at 10 MHz (Panametrics® V112 SM – Ø 3 mm) working as longitudinal waves both emitter and receiver with a –6 dB bandwidth equal to 9.7 MHz. The signal generation system was provided by Panametrics® company (Model 5077 PR) and was used as a pulse emitter-receiver; it sent initially a 300 V rectangular pulse to the transducer and did the treatment of the response before sending it to the oscilloscope. The digital oscilloscope used was a Tektronix® TDS 744A (500 MHz, 2 GS/s) which permits an acquisition of a maximum of 500000 points on one waveform.

3 Results and Discussion

Figure 1 shows the schematic illustrations, which represent the combinations between the directions of the magnetization m and the strain γ due to the ultrasonic measurement. As shown in the figure, we prepared both gels without magnetization and gels with different magnetization directions. The magnetization of gels is due to the remanent magnetization of barium ferrite dispersed in the polymer gel. Since the size of the barium ferrite is larger than that of the polymer network, the ferrite particle is fixed by the network and it can neither move nor round in the medium.

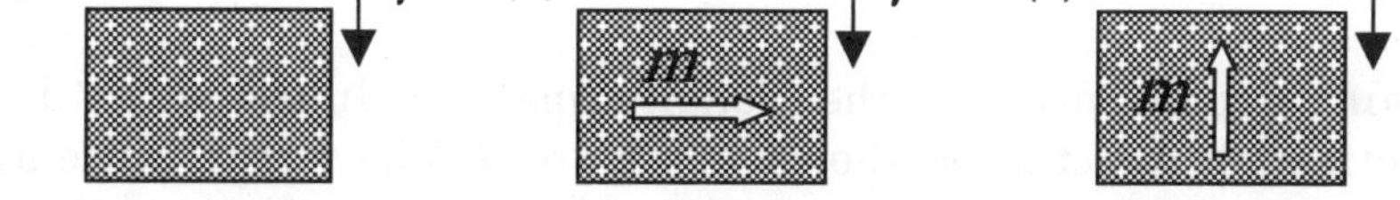

Figure 1. Magnetic gels used in the present study. (a): No field was applied to the magnetic gel. (b)(C): Magnetization direction is perpendicular and parallel to the deformation, respectively.

Figure 2 shows the strain dependence of the sound velocity for magnetic gels with and without magnetization. The ferrite concentration c_f and the crosslinking density ν are 10 wt.% and 0.5 mol% respectively. Closed circles represent the sound velocity for magnetic gel having no magnetization. Open circles stand for the sound velocity with magnetization. The measurements, with and without magnetization, were carried out using the same sample. It is difficult to observe the influence of magnetization using different samples, because the measurement technique is very sensitive to the modulus variation of gel. As we measure the equilibrium sound velocity after relaxation, the sound velocity is almost independent of the strain in the region of γ<10%. Each thickness step being less than 10 μm, the relaxation is finished within few seconds. It is evident that the sound velocity of the magnetic gel with magnetization is higher than without magnetization. This indicates that the longitudinal modulus of magnetized gel is much higher than gel modulus without magnetization.

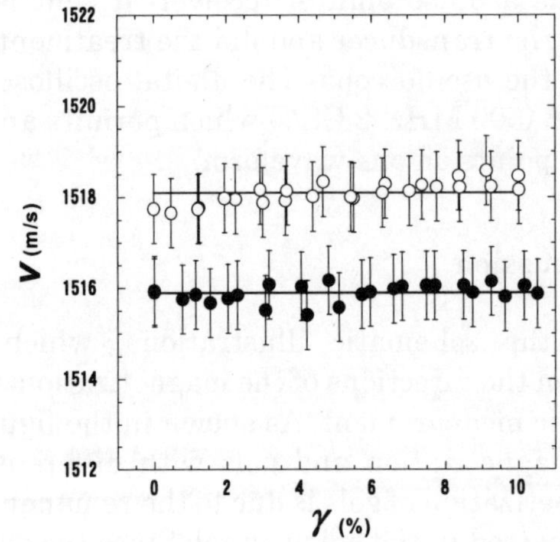

Figure 2. Strain dependence of the sound velocity for magnetic gels (c_f=10 wt.%, ν=0.5 mol%) with (○) and without (●) magnetization. The magnetization direction is perpendicular to the deformation direction.

Strain dependence of the longitudinal modulus with different magnetization direction is shown in Figure 3. The ferrite concentration and the crosslinking density were 10 wt.% and 0.5 mol% respectively. The symbols in the figure are similar to that of Figure 2. According to the elastic theory, the longitudinal modulus M' can be explained by the following simple formula,

$$M' = \rho v^2 .$$

Here, M' is the longitudinal modulus, ρ and v stand for the density and the velocity of the magnetic gel, respectively. One can see that the longitudinal modulus of magnetized gel is clearly larger than that of the gel without magnetization, independently of the magnetization direction. The change in modulus under "perpendicular strain" (Figure 3(a)) is equal to the one under the "parallel strain" (Figure 3(b)). Therefore, the longitudinal modulus of magnetic gel with c_f=10 wt.%, ν=0.5 mol% does not depend on the magnetization direction.

Strain dependence of the longitudinal modulus with different magnetization direction is shown in Figure 4. The ferrite concentration and the crosslinking density are 10 wt.% and 2.0 mol% respectively. The

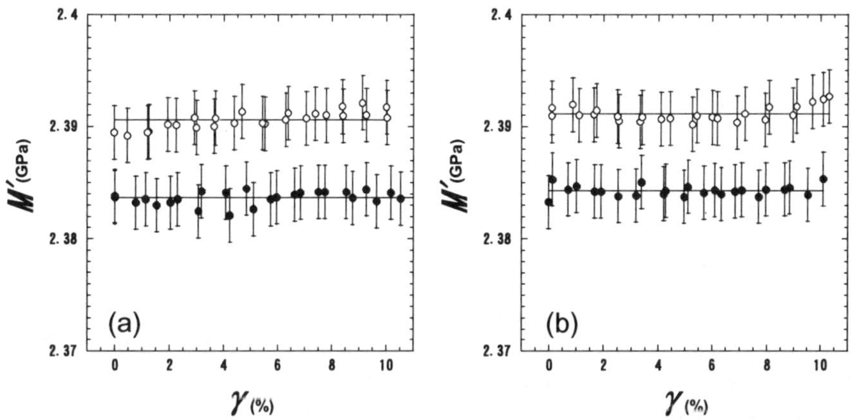

Figure 3. Strain dependence of the longitudinal modulus for magnetic gels (c_f=10 wt.%, ν=0.5 mol%) with (○) and without (●) magnetization. The magnetization direction is (a) perpendicular and (b) parallel to the deformation direction.

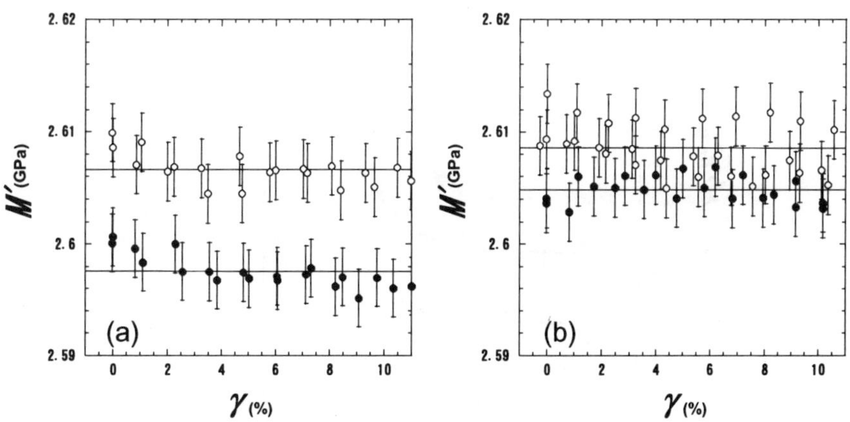

Figure 4. Strain dependence of the sound velocity for magnetic gels (c_f=10 wt.%, ν=2.0 mol%) with (○) and without (●) magnetization. The magnetization direction is (a) perpendicular and (b) parallel to the deformation direction.

symbols in the figure are similar to that of Figure 2 and 3. When the magnetization direction is perpendicular to the deformation direction, the mean change in modulus is about 9 MPa corresponding to 0.35% of initial modulus in a lower density region. On the other hand, the modulus shows small change (3.8 MPa corresponding to 0.15%) when the magnetization direction is parallel to the deformation direction. This suggests that the change in modulus depends on the magnetization direction since the concentration of magnetic substance and the crosslinking density are completely the same.

Figure 5 shows the density dependence of the change in modulus for magnetic gels with different magnetization directions. The change in modulus $\Delta M'/M'_0$ can be defined by the following equation:

$$\frac{\Delta M'}{M'_0} = \frac{M' - M'_0}{M'_0},$$

where $\Delta M'$ is the change in longitudinal modulus, M'_0 represents the modulus without magnetization. As mentioned above, it is clear that the modulus of magnetized gel is higher than that of the gel without magnetization for both parallel and perpendicular cases. The change in modulus to the initial modulus for c_f =10 wt.% and 15 wt.% is about 0.28% and 0.4% in a lower density region, respectively. Moreover, the change in modulus is constant in a lower density region however it strongly depends on the density in a higher

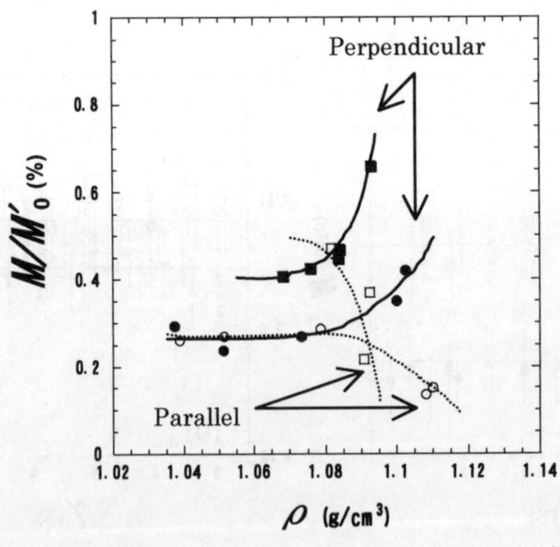

Figure 5. Density dependence of the change in longitudinal modulus for ferrite containing gels. Solid and dotted lines show the perpendicular and parallel situations, respectively. Circle and square symbols represent 10 wt.% and 15 wt.% of ferrite concentration, respectively.

density region. When the stress direction is perpendicular to the magnetization, the change in modulus increased. On the contrary, the change in modulus decreases when the stress direction is parallel to the magnetization. As increasing the density, the distance between magnetic substances becomes short and, therefore, the magnetic interactions are more significant in a higher density region.

Acknowledgements

This research is supported by both the Charitable Trusts Ono Research Fund for Acoustics and a Grant-in-Aid for Encouragement of Young Scientist from Japan Society for the Promotion of Science.

References

1. Y. Osada, H. Okuzaki, H. Hori, *Nature* **355** (1992) 242.
2. Y. Osada, A. Matsuda, *Nature* **376**, (1995) 6537.
3. T. Mitsumata, K. Ikeda, J. P. Gong, Y. Osada, *App. Phys. Lett.* **73**, (1998) pp. 2366-2368.
4. T. Mitsumata, J. P. Gong, and Y. Osada, *Polym. Adv. Technol.*, **12** (2001) pp.136-150.
5. M. Zrinyi, L. Barsi, A. Buki, *Polym. Gels and Networks* **5** (1997) 415.
6. D. Szabo, G. Szeghy, and M. Zrinyi, *Macromolecules* **31** (1998) pp. 6541-6548
7. T. Mitsumata, K. Ikeda, J. P. Gong, Y. Osada, D. Szabo, M. Zrinyi, *J. App. Phys.* **85** (1999) pp. 8451-8456.

THE DYNAMICS OF MAGNETORHEOLOGICAL ELASTOMERS STUDIED BY SYNCHROTRON RADIATION SPECKLE ANALYSIS

W. F. SCHLOTTER, C. CIONCA, S. S. PARUCHURI, J. B. CUNNINGHAM, E. DUFRESNE, S.B. DIERKER, D. ARMS, AND R. CLARKE
University of Michigan, Ann Arbor, MI 48109-1120 USA
E-mail: wschlott@umich.edu

J. M. GINDER AND M. E. NICHOLS
Ford Motor Company, Research Laboratory, Dearborn, MI 48121-2053 USA

We introduce a new technique for probing the microscopic relaxation of magneto-viscoelastic materials consisting of magnetic particles embedded in a natural rubber matrix. Transversely coherent x-rays from a high brilliance synchrotron source are scattered by the magnetic particles, forming a speckle pattern at low scattering angles. The time dependence of this pattern is recorded with a CCD area detector while the sample is cyclically perturbed by a reversal of the magnetic field direction. The corresponding time-resolved scattering pattern probes both the dynamics of the particles and the relaxation of the matrix in which they are embedded. X-ray photon correlation spectroscopy (XPCS) reveals characteristic time scales for this relaxation by applying the intensity auto-correlation function to the time dependent speckle pattern. For low angle scattering, the wave vector dependence of the relaxation rate exhibits power law length scaling.

1 Introduction

Magnetic particles embedded in a polymer matrix enable myriad applications such as data storage media, heat activated printing toner and flexible permanent magnets. When these particles are embedded in a cross-linked elastomer the composite material displays an increase in dynamic stiffness when introduced to an applied magnetic field.[1] Such composites are known as magnetorheological (MR) elastomers or magneto-viscoelastic materials and are the solid state analog of MR fluids.[2] During cross-linking or curing, each particle in an MR elastomer is locked in position until magnetic or mechanical perturbations induce particle configurations of lower energy. Following such a disturbance each particle will approach its original energy state as the elastomer relaxes. Weakly interacting embedded particles will yield to the path dictated by the matrix material. This suggests a method for measuring the relaxation dynamics of any matrix on a microscopic level.[3]

Photon correlation spectroscopy (PCS) has previously been employed to measure particle dynamics in ER and MR fluids.[4,5] The wavelength of visible light limits such studies to minimum length scales of a few hundred nanometers. Moreover, the optical properties of ferromagnetic particulates and the natural rubber matrix make MR elastomers opaque to visible light; however, high intensity x-ray synchrotron radiation can penetrate bulk samples thereby providing a microscopic probe of the composite structure and its dynamics.

X-ray photon correlation spectroscopy (XPCS) was recently employed to measure the diffusive behavior of gold colloids and binary fluids.[6] When coherent x-ray wave fronts are scattered at the particle-host interface, they produce a speckle interference pattern.[7] The time dependence of this intensity pattern can be correlated to explore the characteristic times associated with the motion of scatterers in the sample. We have extended the use of this technique to MR elastomer samples at the Advanced Photon Source, Argonne National Laboratory. This third generation x-ray synchrotron source accelerates relativistic electrons through arrays of permanent magnets known as undulators, thus producing extremely intense quasi-monochromatic x-ray radiation. Excellent transverse and longitudinal coherence of the beam enables interfering wave fronts at the scattering interface to produce a speckle interference pattern.

2 Experimental Method

2.1 MR Elastomer Samples

Magnetite particles (Magnox Pulaski, TMB-X-1539) with an average size of ~200nm were embedded in a natural rubber, cis-polyisoprene, matrix with a $\phi \approx 0.25$ particulate volume fraction. Samples with an average thickness of 100 μm were mounted on a glass slide and cured at 140°C in the absence of an applied magnetic field.

2.2 Small Angle Scattering Setup

Monochromatic x-rays were used to probe the sample as shown in the configuration in Figure 1. The transverse coherence of the beam is ~10 μm in the vertical direction and ~4 μm in the horizontal. Coherence (S_1) and guard slits (S_2) are used to select only the transversely coherent section of the beam. Both of these square apertures accept a 100 μm^2 beam cross section. For this geometry and x-ray wavelength, the size of a single speckle spot was calculated to be 8 μm. The coherent beam enters a hole in the first pole piece of the dipole electromagnet, GMW-3470, and travels

18 cm to the sample (MR_1). The sample was rigidly mounted in a 6 mm gap between the pole pieces and the direction of the applied magnetic field was reversed periodically. Tapered axial holes were bored in each pole piece, allowing the beam to travel parallel to the field lines in the sample.

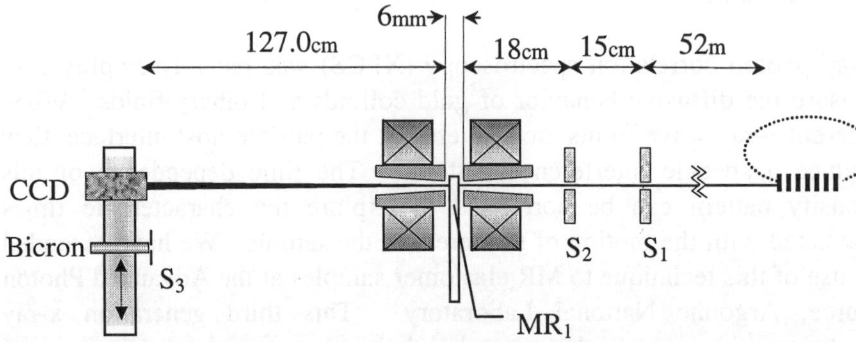

Figure 1: Schematic arrangement for small-angle XPCS measurements. X-rays from the undulator source at right are scattered from the MR elastomer sample in transmission geometry. Slits are used to isolate the coherent section of the beam for scattering. Time dependent scattering is recorded with a CCD area detector or a Bicron photo multiplier point detector mounted on a translating stage.

2.3 CCD detector

Time dependent speckle patterns were measured with a CCD chip positioned to capture directly the small-angle scattering. The sample-detector distance shown in Fig. 1 is chosen so that the speckle size is a good match to the CCD pixel resolution. A Bicron photomultiplier detector with 64 μm^2 slits (S_3) was used for calibration, but the CCD was selected for data acquisition due to the following attributes:

- Efficient parallel data acquisition
- Ability to access very small scattering angles, corresponding to small scattering wave vectors, $0.001\ Å^{-1} < q < 0.01\ Å^{-1}$
- Enhanced statistics enabled by averaging over annular regions of fixed q

A Texas Instruments TC253 low noise, high sensitivity CCD with 7.4 μm pixel width in a 658 x 494 array was controlled by a 12-bit QMAX 650 camera.[8] Digital images were transmitted to a microcomputer and recorded as a movie with KSA 400 image acquisition software.[9] 100-frame movies were recorded at 0.42 Hz with a 1.8 second exposure duration.

3 Experimental Results

We first consider the static small-angle intensity profile, which is generated by averaging the intensity of each pixel in every frame of a scattering movie with respect to time. The resulting two dimensional matrix can be represented as the time averaged image shown in Fig. 2. The main beam is observed in the middle of the rectangular shadow of the beam stop and shows weak Fraunhofer diffraction features characteristic of a square aperture. This is a signature of the coherent light necessary for speckle interference. A profile of radial intensity with respect to wave vector was extracted from the data of Fig. 2 and fit to a Guinier curve.[10] This predicts the particle size to be 300±100nm, consistent with SEM micrographs and product literature.

Figure 2: Integrated scattering intensity pattern. The image is 200 pixels on a side and subtends 1.1 mrad. Note the weak Fraunhofer diffraction feature for a square aperture. The horizontal area of low intensity dividing the circular pattern is the shadow of the beam stop. By summing the intensity of each pixel at a constant angle the average intensity at a specific wave vector was determined. These plots were then used to determine the average particle size in the Guinier approximation

4 Analysis

4.1 Intensity correlation

Temporal fluctuations in intensity were recorded for each pixel and characterized with the intensity auto-correlation function,[11]

$$g_2(\Delta t) = \frac{< I(t+\Delta t) I(t) >}{< I(t) >^2},$$

where *I(t)* is the intensity at a time *t* and all averages are over time. Correlation results for each pixel were averaged with others of similar wavevector magnitude, q where $q=(4\pi/\lambda)\sin\theta$. The time dependence of these data reflects the scatter motion at specific wave vectors, Fig. 3.

Fitting these data to exponential decay curves, also plotted in Fig. 3, quantifies the characteristic time constants. For weakly interacting particles this relaxation is characteristic of the matrix material.

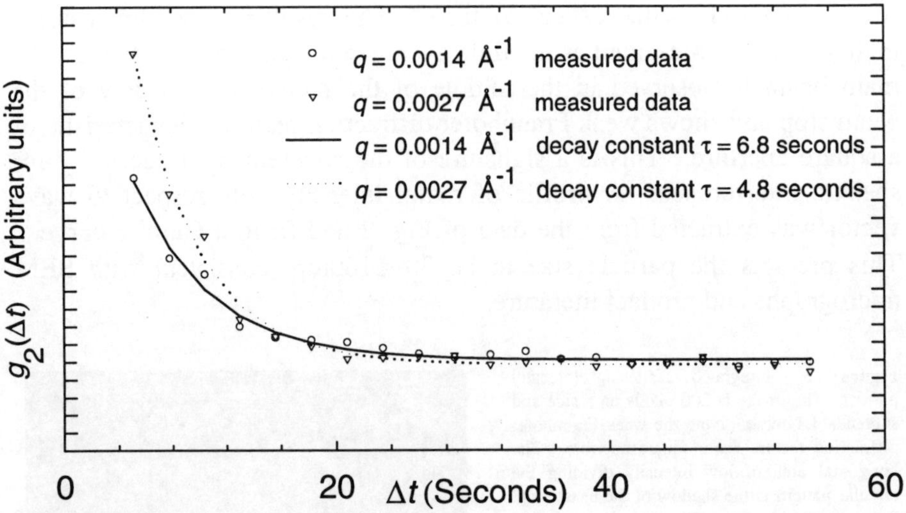

Figure 3: Plot of the azimuthally averaged auto-correlation results for q = 0.0014 $Å^{-1}$ and q = 0.0027 $Å^{-1}$ with corresponding exponential relaxation curves.

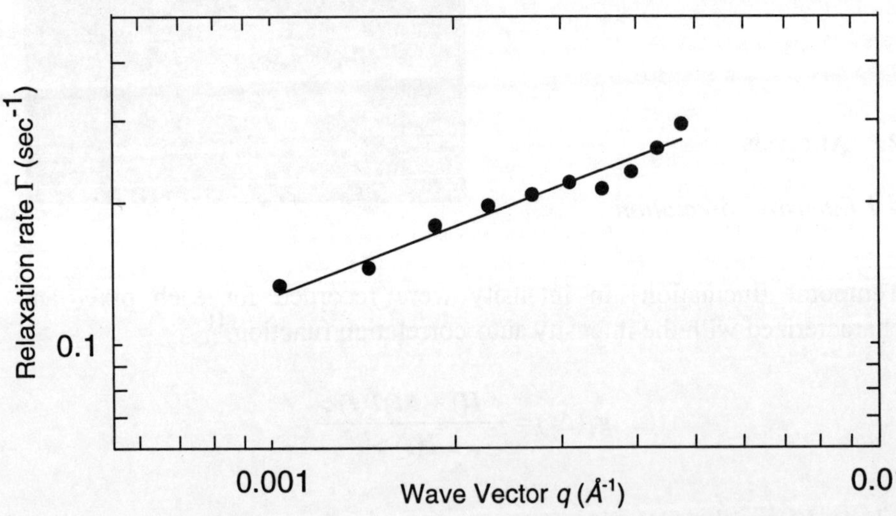

Figure 4: Scaling of the relaxation rate with scattering wave vector. The solid line is a fit to $q^{0.49}$

4.2 Wave vector dependence of time decay

A characteristic relaxation rate, Γ, for the material is found by taking the inverse of the time constant. The relaxation rate is found to vary as a power law in q, the magnitude of the scattering wave vector, as shown in Fig. 4.

5 Discussion

Qualitatively similar power law dependence is observed in other scattering systems. For example, colloidal media undergoing Brownian motion show a q^2 relaxation rate dependence.[12] While thermal excitations may contribute to the dynamics of our elastomer samples, the changing magnetic field provides the dominant source of excitation for the particles. In contrast to the behavior of diffusive systems, where the relaxation rate scales like q^2, we observed ~ $q^{0.5}$. Whether this represents true power law scaling, or a crossover to q independent behavior at small q remains to be checked with more detailed studies.

From the data, the relaxation rate for a macroscopic wave vector $q = 0$ can be approximated to be around 0.02 Hz. This limiting rate is consistent with the effects of macroscopic mechanical perturbation of the material, which relax on time scales on the order of one minute. This quasi-static behavior is expected in the viscoelastic terminal region of cross-linked polymers.[13] In this region viscoelastic losses dominate the elastic restoration force resulting in an overdamped system.

6 Conclusions

Clearly, these results show that XPCS is a practical technique for measuring the dynamic behavior of particles embedded in viscoelastic materials. The use of high intensity x-ray synchrotron radiation and CCD area detectors allows for real-time measurements on visibly opaque media. Intensity auto-correlation functions were calculated from time dependent speckle patterns, which were recorded as movies. Exponential time decays were fit to these correlations at specific wave vectors. The inverse of the decay time, the relaxation rate Γ, has a power law relationship to wave vector. Unlike the quadratic q dependence characteristic of diffusive media, sub-linear behavior is observed in this case. We are in the process of formulating a quantitative model for this observed behavior. One preliminary conclusion is that algebraic scaling of the kind we observed is indicative of dynamics governed by affine processes. (i.e. self similar structure which relaxes without selecting a particular length scale) Through a better understanding of particle dynamics in weakly interacting systems this technique serves as a novel probe of the viscoelastic properties of MR materials.

7 Acknowledgment

The authors would like to thank Larry Elie and Teamour Nurushev. Use of the Advanced Photon Source was supported by the U.S. Department of Energy, Office of Science, Office of Basic Energy Sciences, under Contract No. W-31-109-ENG-38. The experiments were conducted at MHATT-CAT Sector 7, which is supported by Department of Energy Grant DE-FG02-99ER45743.

8 References

[1] J.M. Ginder, M. E. Nichols, L. D. Elie, and J. L. Tardiff, "Magnetorheological Elastomers: Properties and Applications", *Smart Structures and Materials 1999: Smart Material Technologies*, ed. by M. Wuttig, (1999) pp. 131-138.

[2] J.M. Ginder, "Rheology Controlled by Magnetic Fields", *Encyclopedia of Applied Physics*, VCH Publishers, Inc., **16,** (1996) pp. 487-503.

[3] P.D. Kaplan, V. Trappe, and D.A. Weitz, "Light-scattering microscope," Applied Optics, **38**, (1999) pp. 4151-57.

[4] M. Hagenbüchle, P. Sheaffer, Y. Zhu and J. Liu, "Static and Dynamic Light Scattering of Dilute MR Emulsions," *Proceedings of the 5th International Conference on ERMR*, ed. by W. Bullough, World Scientific, (1996) pp. 236-239.

[5] J.M. Ginder, "Diffuse optical probes of particle motion and structure formation in an electrorheological fluid", Phys. Rev. E, **47**, (1993) pp. 3418-3447.

[6] S.B. Dierker, R. Pindak, R.M. Fleming, I.K. Robinson, L. Berman, "X-Ray Photon Correlation Spectroscopy Study of Brownian Motion of Gold Colloids in Glycerol", Phys. Rev. Lett. **75**, (1995) pp. 449-452.

[7] M. Sutton, S.G.J. Mochrie, T. Greytak, S.E. Magler, L.E. Berman, G.A. Held, G.B. Stephenson, "Observation of speckle by diffraction with coherent X-rays", Nature **352**, (1991) pp. 608-610.

[8] CCD Direct, Ann Arbor, MI.

[9] k-Space Associates Inc. Ann Arbor, MI.

[10] O. Glatter, O. Kratky, *Small Angle X-ray Scattering,* Academic Press, (1982) p. 155.

[11] see B. Chu, *Laser Light Scattering*, Academic Press, (1994) p. 94.

[12] T. Nurushev, *Studies of Static and Dynamic Critical Behavior of Simple Binary Fluids and Polymer Mixtures using XPCS*, PhD Thesis, University of Michigan, (2000).

[13] J.D. Ferry, *Viscoelastic Properties of Polymers*, Wiley, (1970).

Electroviscoelastic Properties in ER Gel of Disperse System under DC Electric Field

R. HANAOKA, S. TAKATA, H. FUJITA, T. FUKAMI
Division of Electrical Engineering, Kanazawa Institute of Technology, Nonoichi, Ishikawa 921-8501, Japan

K. SAKURAI
R & D Department, Fujikura-kasei Co., Ltd. 13-1, Sakurada 5, Washimiya, Kitakatsushika, Saitama 340-0203, Japan

K. ISOBE AND T. HINO
Silicones Division, Nippon Unicar Co., Ltd. 6-1, Ohte 2, Chiyoda, Tokyo 100-0004, Japan

ABSTRACT

The silicone oil-based electrorheological (ER) gel containing the non-aqueous fine particles was newly created in the present study. After these particles were dispersed at 30wt.% in the dimethylsilicone oil, the ER gel was produced by the hydrosilylation reaction in the mixture of the modified silicone oil. This reaction could considerably be promoted by heating at 90°C.

The behavior of particles in the ER gel was observed by a microscopical method. When an electric field was applied to the ER gel, the gap between the electrodes was bridged by the chains of particles arranged in the direction of the electric field. The dynamic properties of the ER gel were also examined under the applied dc electric field up to 2kV/mm using the oscillating rheometer with the low frequencies of 1Hz or less. Consequently, it is shown that the electroviscoelastic effect of the gel can be controlled by the electric field strength.

1 Introduction

In electrorheological (ER) fluids, the ER effect can be characterized as a rapid and reversible increase in the apparent viscosity upon the applied electric field. It is known that this effect arises from the strong cohesion between particles as well as particles and electrodes.[1, 2] The ER fluids are expected to be instrumental in the development of numerous devices because of the usefulness of electrical responses.[3] Many investigations on this subject have been devoted to development of ER particles with a high performance for useful applications of the ER response.[4, 5] However, since the coagulation of particles gives rise to an instability of the ER response and depresses the reliability of the system, the poor dispersion stability of particles often limits the practical development to the fluids usefulness. This key matter can be settled by the development of the chemically cross-linked gel containing ER particles (this will be referred to as ER gel). The particles in ER gel are preserved in just the state they were dispersed in place by the nature of the gel matrix, but the viscoelastic properties of the ER gel will be changed by the interaction between particles induced by the electric field.

The purpose of the present study is to examine the electroviscoelastic effect for our created ER gel. There are only a few reports on the viscoelastic behaviors under electric fields for the silicone gel containing fine particles.[6, 7] In this paper, we report experimental results regarding the dynamic viscoelasticity of the ER gel under the dc electric field by a shearing test using the oscillating rheometer with the frequencies up to 1Hz.

2. Experimental setup and procedure

2.1 Sample ER gel

The dispersing particles were the non-aqueous spherical solid particles designed for ER fluids and produced by Fujikura-kasei Co., Ltd.[8] These particles consisted of a double structure in which an organic core sphere was coated with an inorganic layer shell of approximately 1μm in thickness. The organic-core was the cross-linked copolymer sphere of butyl acrylate and 1,3-butanediol dimethaclylate, and an average diameter was approximately 10μm. The inorganic-shell was composed of titanium oxide and tin oxide. The surface of these composite particles was also complexed with organic dyes.

After these ER particles were well dispersed at 30wt.% in the dimethylsilicone oil with the viscosity of 97mPa•s at 25°C, the ER gel was produced by the hydrosilylation reaction with Si-H bonds in the mixture of the hydrogensilicone and unsaturated group compounds. The hydrosilylation is a very convenient method for the synthesis of various organosilicon compounds for the study of silicon chemistry because of an advantage of a large dependence on temperature in the reaction velocity.[9] In the present study, this reaction for the gelation could considerably be promoted by a slight addition of platinum catalyst and heating at 90°C for 6 hours, and the particle sedimentation during this curing process could be negligible by a prompt reaction. The typical properties of the ER gel specimen are shown in Table 1.

Table 1 Typical properties of ER gel specimen.

Test samples	Density (g/cm^3)	Relative permittivity (f=100Hz)	Conductivity ($\mu S/cm$)
solid particle	1.17	22.5	1.62×10^{-3}
dimethylsilicone	0.97	7.3	1.40×10^{-9}
hydrogensilicone	0.96	6.4	1.47×10^{-9}
unsaturated group compound	1.16	6.2	6.26×10^{-9}
ER gel	0.99	9.3	1.18×10^{-3}

2.2 Viscoelastic test and apparatus

The viscoelasticity of the ER gel can be affected by the applied electric field because of the cohesive effect of the dispersing particles. First, the shear modulus G was measured to examine the elastic properties of the sample gel with and without an electric field, on the basis of a customary method for a Hookean elastic solid. The G was determined from the shear stress divided by the corresponding elastic shear strain. A Poisson's ratio p was also determined from the measurement of deformations for the corresponding tensile (or compressive) stress, for the gel free from electric field. Second, the dynamic viscoelastic properties (the storage and the loss moduli G' and G'' and the loss tangent $tan\delta$) were examined under the dc electric fields up to 2kV/mm using the parallel-plate type rheometer with a sinusoidal oscillation in a range of 0 to 1Hz in frequency f.

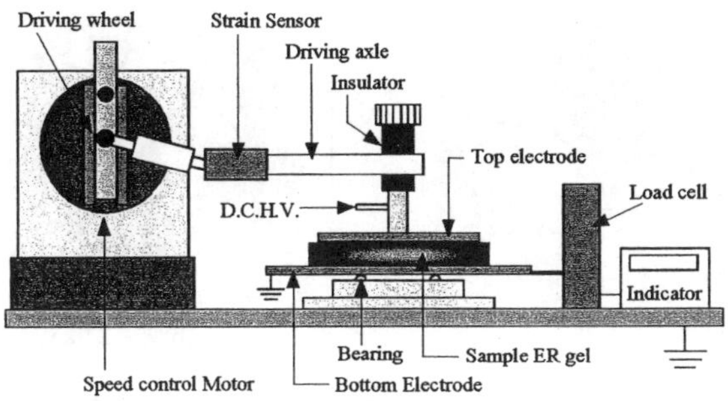

Fig. 1 Schematic diagram of apparatus used for measurement of dynamic properties.

Figure 1 shows a schematic diagram of the apparatus used for the measurement of the dynamic properties. A horizontal motion of the shearing test electrode (top electrode: 50mm x 70mm) with the sinusoidal oscillation (f=0 to 1Hz) was obtained from a rotary motion of the speed control motor through a driving axle and a driving wheel. Then, the shear strain γ was given sinusoidally to the sample ER gel between the parallel-plate electrodes (top and bottom electrodes with the gap spacing d=2mm) by the forcible oscillatory stress input as shown in Eq. (1).

$$\gamma = \gamma_0 exp(i \omega t) \tag{1}$$

where γ_0 denotes the amplitude of shear strain, i the imaginary unit, ω (=$2\pi f$) the angular frequency of strain and t the time. The γ_0 was controlled in a range of 0 to 0.625 by adjusting the radius of the driving wheel. The top electrode and the driving axle were insulated electrically, and a dc high voltage was applied to the top electrode. The bottom electrode was grounded, and it could be slid along the horizontal direction without a hitch. The stress τ acting on the bottom electrode was also measured using a 3-kgf load cell which was coupled directly on the bottom electrode. It is assumed that this stress output τ is equivalent in magnitude to the oscillatory stress input applied to the top electrode. All experiments were done at room temperature.

3. Results and Discussion

3.1 Behavior of dispersing particles in ER gel

The behavior of particles in the sample gel was observed using a microscope under the applied electric field. The microscopical results have revealed that the particles line up in the direction of the electric field and the gap between the electrodes is bridged by the chains of particles. A typical example of the appearance of the particles in the gel with and without an electric field is given in Fig. 2. The dispersing particles existed in a random arrangement at zero electric field as if they were caged in the gel matrix (Fig. 2 (a)). The moment an electric field was applied to the gel, there occurred a rearrangement of the particles in the direction of the field, which showed a chain-like structure (Fig. 2 (b)). The particles, however, were never linked in a beeline across two parallel electrodes.

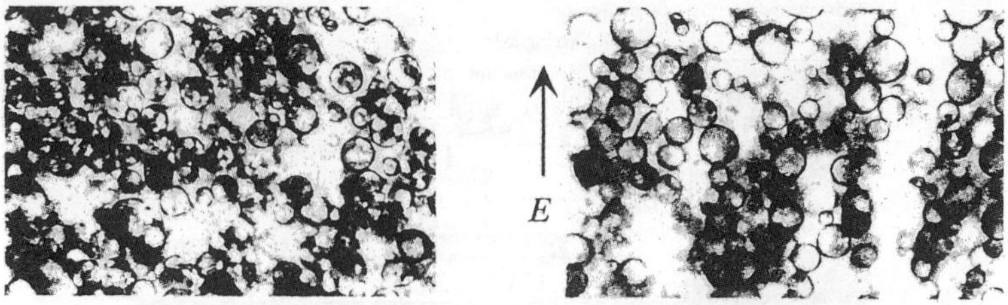

(a) $E = 0$ (b) $E = 2.0$kV/mm

Fig. 2 Typical example of appearance of particles in ER gel.

It seems that this is attributable to the curing process of the silicone gel matrix. The hydrosilylation reaction for the gelation in the present study appeared to form the connective tissue with the cell walls which wraps up the several particles. By applying a dc electric field to the gel, the particles in each cell can be aligned with the electric field as the chainlet because of a low density network structure in the cell including a liquid phase. This chainlet is connected in turn to that in adjacent cell so that the chain which bridges across two parallel electrodes takes the tortuous shape spatially.

3.2 Effects of electric field on viscoelastic properties in ER gel

A Poisson's ratio p of the sample gel in zero electric field was approximately 0.46. This means that there is hardly any density change for a simple tensile (or compressive) stress. The shear modulus G at zero electric field also revealed that the sample gel was very soft material, but the G values increased greatly by the applied electric fields E as shown in Table 2. This suggests that the cohesive behavior of the particles in the sample gel gives rise to the ER effect.

Table 2 Shear modulus G of ER gel.

electric field E (kV/mm)	0.0	0.5	1.0	1.5	1.75	2.0
shear modulus G (kPa)	0.15	0.54	1.2	1.9	2.4	3.1

When the sinusoidal shear stress is given to the gel, the internal time-scale of the gel changes by the oscillatory shear strain varying with the time, and the gel will show the viscoelastic behavior depending on the time scale of the deformation process. Since the viscoelasticity of the gel has the properties between a Hookean elastic solid and a Newtonian viscous fluid, there exists the phase difference (loss angle δ with $0<\delta<90°$) between the shear strain and the shear stress. The shear stress output τ will have a similar form as an applied oscillatory strain (Eq. (1)), but its phase will be in advance of the strain by an angle δ. The τ then is given by Eq. (2) .

$$\tau = \tau_o \exp \{i(\omega t+\delta)\} \tag{2}$$

where τ_o denotes the amplitude of shear stress. The loss angle δ can be determined from the phase difference between the sinusoidal oscillation at top electrode and the stress output τ acting on the bottom electrode.

In characterizing electroviscoelastic response of the ER gel, a convenient method is to examine the effect of the electric field on a complex shear modulus G^* ($=G' + i\,G''$) and $tan\delta$.[10] If we choose to evaluate with the simple Kelvin model, G^* and $tan\delta$ are represented as follows, respectively.

$$\begin{aligned} G^* &= G' + i\,G'' \\ &= G\,(1+\omega^2\lambda^2)^{1/2}(cos\delta + i\,sin\delta) \\ &= G\,(1+ i\,tan\delta) \end{aligned} \tag{3}$$

$$tan\delta = G''/G' = \omega\lambda \tag{4}$$

where λ denotes the retardation time on the deformation.

Figure 3 shows the E dependence of the relationship between τ_o and f. The τ_o increased greatly with increasing E, but they were almost a constant independently of f under a fixed E, except for a slight change of τ_o at $E \leq 0.5$kV/mm and $f \leq 0.25$Hz. The yield stresses τ_y are also indicated in Fig. 3, where τ_y is defined as τ_o which is obtained when the each curve is extrapolated to $f=0$. In the case of $E=0$, the relationship between τ_o and f is almost the same as that of a non-Newtonian curve which τ_y is approximately zero. When the electric fields were applied to the gel, especially in $E \geq 1$kV/mm, τ_y-values become very large. It is evident that the ER effect causes by the cohesive behavior of the dispersing particles in the gel upon the applied electric field. Figure 4 shows the relationship between $tan\delta$ and E. The $tan\delta$ were roughly a constant over the electric field up to 2kV/mm for $f=0.03$Hz, but they decreased with increasing E for $f \geq 0.25$Hz. They also increased with increasing f under a fixed E. These results denote that an elastic

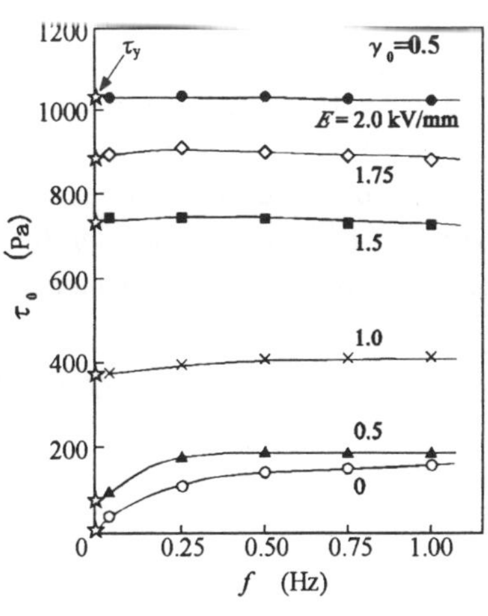

Fig. 3 E dependence of relationship between τ_o and f.

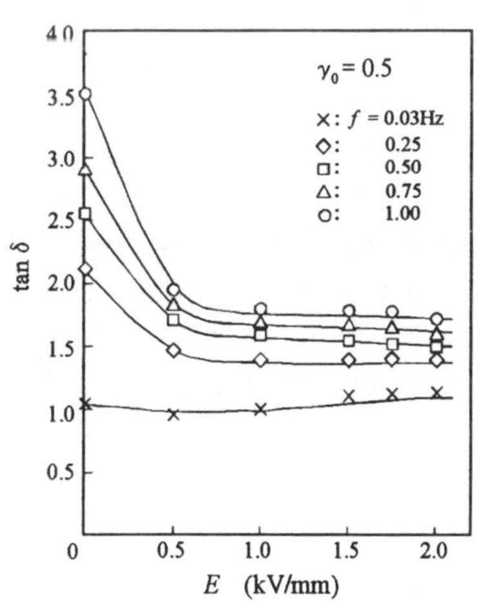

Fig, 4 Relationship between $tan\delta$ and E.

element in the sample gel is reinforced by increasing E (or decreasing f), since the $tan\delta$ represents the rate which the elasticity of viscoelastic materials is lost by the internal friction. However, all $tan\delta$ in Fig. 4 were always $tan\delta \geq 1$. This means that the sample gel is essentially the material with a large viscous element. The relationship between $tan\delta$ and E will be reflected in the properties of a complex shear modulus G^*. The relationship between G^* (G' and G'') and E is shown in Fig. 5. The G' and G'' increased with increasing E, and G'' increased with increasing f. The G'' also were always larger than G'. In the Kelvin model, the material shows the viscoelastic property when the retardation time λ is roughly the same as the time scale of the experiment T. The sample gel reveals the viscoelastic behavior, because $T \leq 6\lambda$ is satisfied with the scopes of E and f used in this study.

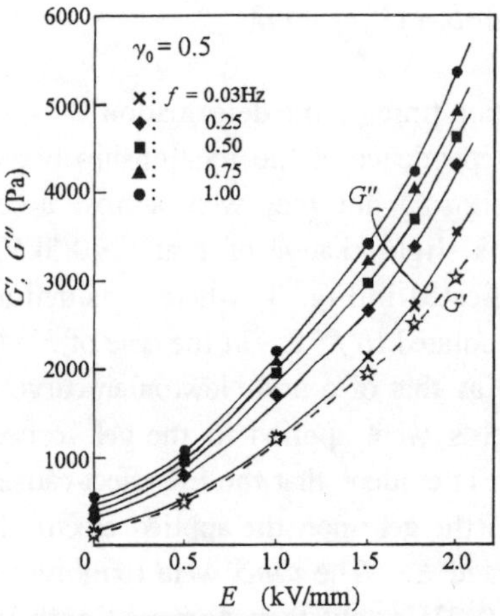

Fig. 5 Relationship between G^* (G' and G'') and E.

The effect of γ_o on $tan\delta$ and G^* was also measured as a function of E. Though the results showed the same tendency as Fig. 4 and Fig. 5, the $tan\delta$ and G'' increased by an increase of γ_o. For instance, in E=1.5kV/mm and f=0.5Hz, when γ_o varied 0.25 to 0.625, $tan\delta$ and G'' changed approximately 1.2 to 1.7 and 2.3 to 3.2kPa, respectively.

On the other hand, when an oscillatory strain is applied to the sample gel, the mechanical energy will be changed to the thermal energy which depends on $tan\delta$ (or G'') by the existence of the viscous element. In addition to this thermal dissipation, the electrical energy upon the applied electric fields will also be dissipated thermally. The total energy dissipation W per one cycle in the oscillatory strain is then given by Eq. (5).

$$W = \int_0^{2\pi/\omega} \{\tau\,(d\gamma/dt) + E\cdot J\}dt$$

$$= \pi\,\gamma_o^2 G' tan\delta + E\cdot J/f \tag{5}$$

where J denotes the current density. On the right-hand side in Eq. (5), the first term represents the thermal dissipation due to the mechanical energy, W_V, and the second term the one due to the electrical energy, W_P. The J-values in the sample gel showed a non-linear variation in J=0 to 2.2μA/cm^2 for E=0 to 2kV/mm. Consequently, the W_P were larger than W_V. For instance, in E=1.5kV/mm, γ_o=0.5 and f=1.0Hz, W_V and W_P were approximately 2.6x10^{-3}J/cm^3 and 1.4x10^{-2}J/cm^3, respectively.

4. Conclusion

The silicone gel containing the non-aqueous ER particles was newly created in this study. The problem of the poor dispersion stability of particles in the ER fluids can be neglected by the gelation of the ER fluid. The behavior of particles in the ER gel was observed microscopically, and the viscoelastic properties of the gel were studied under the application of dc electric fields by an oscillatory shearing test in the low frequency up to 1 Hz. It was evident that the viscoelasticity of the ER gel was controlled by the electric field. In this work, we conclude that the ER gel will be helpful in the development of vibration-proof devices for a small-amplitude oscillatory shear at a low frequency.

Acknowledgments

The authors thank the Ministry of Education of Japan for support in a Grant-in-Aid of scientific research.

References

1. Gast A. B. and Zukoski C.F., Electrorheological Fluids as Colloidal Suspensions, *Adv. Colloid Interface Sci.* **30,** 153-202 (1989)
2. Hanaoka R, Hotta K, Anzai H, Sakurai K and Kuroda S, Internal structure and ER properties in Disperse System under dc Electric Field, *Elect. Eng. in Japan,* **132**, No.5, 9-18 (2000).
3. Havelka K. O, in *Novel Materials for Electrorheological Fluids—Progress in Electrorheology*, ed. Havelka K. O. *et al.* (Plenum Press, New York, 1995), pp. 43-54.
4. Block H and Kelly J. P, Electrorheology, J. Phys. D: Appl. Phys., **21,** 1661-1677 (1988).
5. Xu Y.Z and Liang R. F, Electrorheological Properties of Semiconducting Polymer-based Suspensions, J. Rheol. **35**, No. 7, 1335-1373 (1991).
6. Shiga T, Ohta T, Hirose Y, Okada A and Kurauchi T, Electroviscoelastic Effect of Polymeric Composites Consisting of Polyelectrolyte Particles and Polymer Gel, J. Mater. Sci. **28,** 1293-1299 (1993).
7. Shiga T, Okada A and Kurauchi T, Electroviscoelastic Effect of Polymer Blends Consisting of Silicone Elastomer and Semiconducting Polymer Particles, Macromolecules, **26**, 6958-6963 (1993).
8. Hanaoka R, Takata S, Koike M, Sakurai K and Anzai H, Temperature Dependence of Electrical Response of Aqueous and Non-Aqueous ER Fluids,—*Proc. of 2nd Conference on Electrostatics*, (Montpellier, France 2000), pp. 161-165.
9. Ojima I, in *The Chemistry of Organic Silicone Compounds, Part 2*, ed. Patai S *et al.* (John Wiley & Sons, NewYork 1989) pp. 1479-1483.
10. Barnes H. A, Hutton J. F and Walters K, *An Introduction to Rheology*, (Elsevier, New York 1998).

SOFT GELS WITH ORDERED IRON PARTICLES: FABRICATION AND ELECTRORHEOLOGICAL RESPONSE

YUXIAN AN AND BO LIU

Polymer Program, Institute of Materials Science, University of Connecticut, Storrs, CT USA

MONTGOMERY T. SHAW

Polymer Program, Institute of Materials Science, and Department of Chemical Engineering, University of Connecticut, Storrs, CT USA
E-mail: shawmt@uconnvm.uconn.edu

For a number of years we and others have investigated electrorheological (ER) gel systems comprising low-modulus elastomers mixed with various particles. Unlike ER fluids, the particles in gels do not change relative positions, thus creating model systems that can be used to check equilibrium theories of ER response. To make such materials, a magnetic field was used to align iron particles during the cure of a liquid silicone prepolymer in a sheet mold. Particle chains were made with as little as 0.5% particles, and aligned at angles from 0 to 45° to the normal. The gap between the particles was controlled by swelling the crosslinked material with additional prepolymer. The dependence of the ER effect on particle structuring in these materials was investigated experimentally using rheometry and dielectrometry. Also investigated were the properties of these materials as transducers and actuators.

1 Introduction

Conventional electrorheological (ER) fluids are concentrated suspensions of micrometer-sized, polarizable particles, such as silica, barium titanate, zeolites, etc., dispersed in nonconductive liquids, such as mineral and silicone oils [2,5,6,8,14,18]. An externally applied electric field induces dipoles on the suspended particles and the dipolar interactions force the particles into chains. The increase in the viscosity as well as the increase in the stiffness (modulus) of these suspensions is due to the strong particle-particle interactions within these chains on application of the field. In ER fluids, sedimentation and electrophoresis of the dispersed particles can lead to inefficient particle arrangement and reduced effectiveness. ER gels, i.e., soft solids containing particles pre-organized in the solid matrix, can eliminate these problems [3,7,11,13,15,16,17]. As magnetic fields have qualitatively the same effect on magnetically susceptible particles, it was reasoned that strong fields applied uniformly and suddenly to a suspension could result in very efficient patterns that would also be responsive to electric fields if particle-particle contacts could be eliminated [6]. This was done in the present investigation by swelling the gel with additional prepolymer.

In this paper, we were specifically interested in the influence of particle content and chain angle on the ER properties of the gels, and the opportunities for using these materials as transducers and actuators. Of particular interest was the possibility of using angled particle chains to create a shear actuator.

2 Experimental

Silicone gels were prepared from a polydimethyl siloxane prepolymer with vinyl groups (Bayer Baysilone®U basic H6, 6000 mPa s, density = 0.98 g/cm^3) and a vinyl-terminated polydimethylsiloxane prepolymer (Baysilone®U1, 1000 mPa s, density = 0.98 g/cm^3). Various ratios of these were used to control the crosslink density of the gel. The crosslinking agent was poly (dimethyl siloxane-co-hydromethylsiloxane) (Baysilone®U Cross-linking Agent 430, 40 mPa s, density = 0.96 g/cm^3). A Pt catalyst (Baysilone®U Catalyst Pt/L, 2% Pt) was used to promote the reaction of the Si-H group on the crosslinking agent with the vinyl groups of the prepolymers. Xylene was added to lower the viscosity of the mixture and to ease the swelling of the gel after cure. The mixture of silicone prepolymer and iron particles, confined in a nylon mold (Figure 1, left), was placed between the poles of a large electromagnet. The mixture was subjected to 1.6-1.8 T for 10 min to align the particles and to 0.9-1.0 T for another 120 min to prevent sedimentation of the iron particles during the cure. The total curing time was 24 h. Upon application of the magnetic field, the iron particles aligned in the silicone prepolymer in less than a second. Various angles of orientation of particle chains relative to the field direction were achieved by tilting the mold in the magnetic field during cure. After removing the gel from the mold the sample was immersed in prepolymer (U1) for up to 3 days to swell the network and separate the iron particles within the chain. The xylene, meanwhile, escaped.

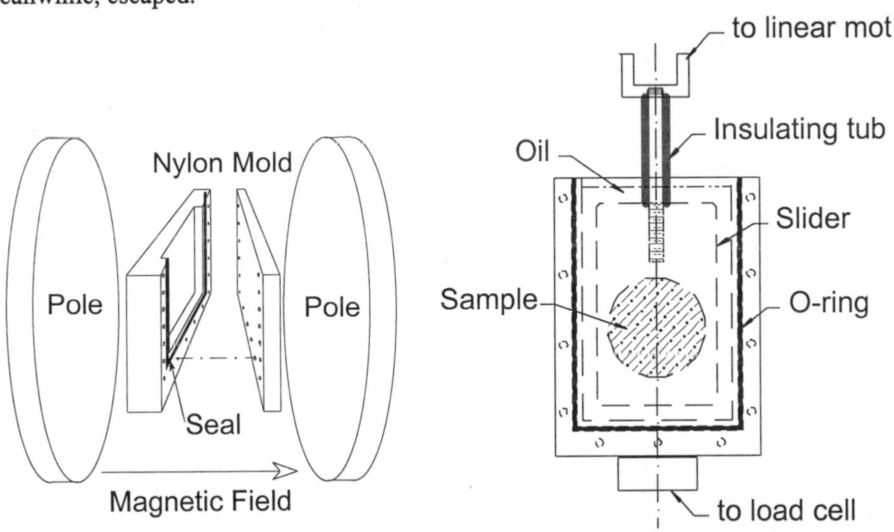

Figure 1. Left: Pictorial of mold used to form gels while holding the reacting mixture in the magnetic field. Right: Shear sandwich for ER and actuator measurements.

Rheological characterisation was conducted with a modified Rheometrics System IV at ambient temperature. A parallel-plate fixture was used for measuring the torque and normal force because it can easily accommodate the restrictions imposed by the elastic samples. The plates were sandblasted to avoid slippage of sample during the measurement. Using a Hitachi power supply and a Trek amplifier, a DC voltage of up to 5 kV was applied between the plates. The resulting current density was recorded during the ER measurement. Additional details of the ER characterization have been published previously [9,11].

The response of the gel as a force transducer was characterized by subjecting the sample to a variable normal load using the same setup. Current was measured at a fixed potential of 60 V. To estimate the actuation characteristics of the gels, specimens were compressed slightly between the parallel plates of the rheometer and the resulting normal force was recorded as a function of applied voltage. For the specimens featuring angled strings of particles, a shear sandwich fixture (Figure 1, right) was used at fixed gap. The developed shear forces were measured as a function of applied voltage.

3 Results and Discussion

The structure of the ER gels is illustrated in Figure 2, which shows both a side view (left) and an end view (right) of the particle chains.

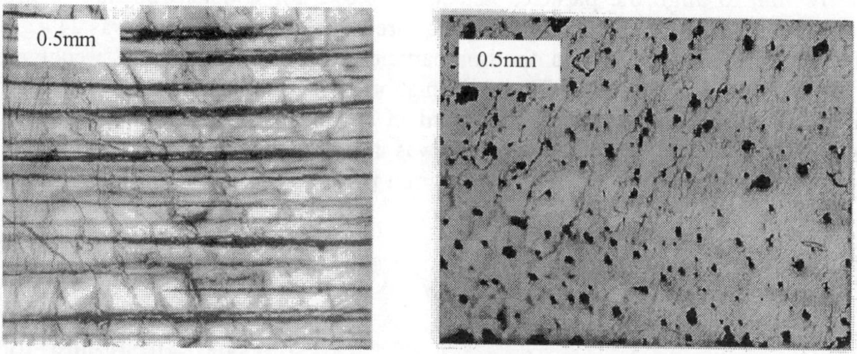

Figure 2. Side view (left) and end view (right) of particle chains in silicone-iron ER gels.

As a result of large magnetic polarizability and high saturation magnetization ($\mu_0 M_s = 2.1$ Tesla), the iron particles aligned to form nearly perfect chains along the direction of magnetic field. The space between the chains or columns was uniform throughout the entire sample as illustrated by the end view of Figure 2. The photomicrograph on the left of Figure 3 shows particle chains at an angle relative to the direction of the field. The relative transparency of the samples with different tilting is shown in Figure 3, right. It is seen that with the increase of the tilting angle, the sample becomes noticeably darker. This effect might be exploited in an electrically actuated optical filter.

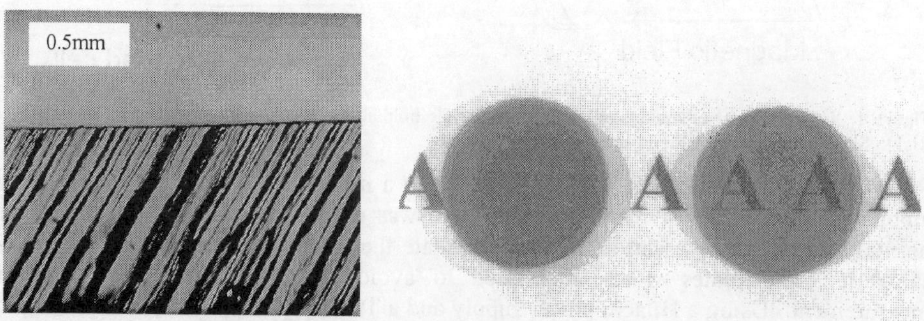

Figure 3. Left: Cross section of gel featuring particle chains permanently aligned at a 30° angle. Right: disk on the left features particle chains aligned at an angle of 30°, whereas that on the right has chains parallel to the axis of the disk. In all gels shown the particle content is 0.5 vol.%.

The left panel of Figure 4 shows that swelling of the cured gels lowered conductivity markedly. In the case shown, the conductivity along the direction of particle chains fell by several orders of magnitude on swelling a few percent. The solid line is based on a simple one-dimensional model (Figure 5) wherein the particles and the polymer matrix in the chains are replaced by series resistors. This combination is then hooked in parallel with a resistor representing the matrix in between the chains. Although the matrix material between the chains makes a negligible contribution to the conductivity, it acts to pry the particles apart during swelling, allowing the gaps to fill with the diluent. (Microscopic evidence of this process was presented in a previous publication [11].) By this means, the conductivity of particle becomes very sensitive the amount swelling. The somewhat more gradual change suggested by the data could be due to the distribution of particle spacing in the chains.

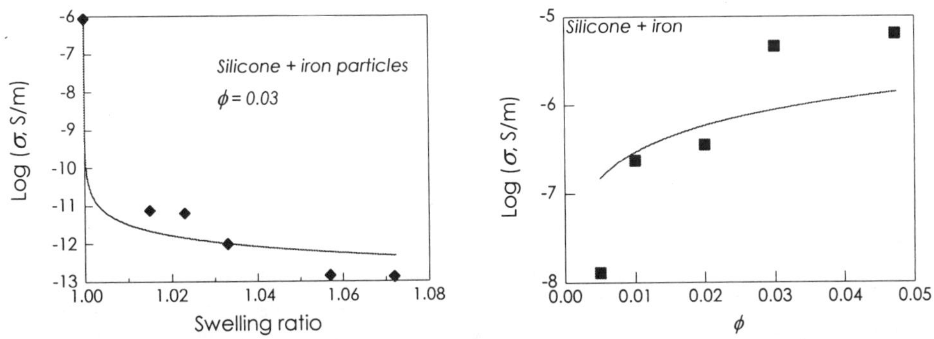

Figure 4. Left: Conductivity of gel with 3% aligned iron particles. Right: Influence of iron particle concentration on conductivity of silicone-iron ER gels. The curves are based on the parallel-series model illustrated in Figure 5.

The conductivity of the gels can also be decreased by decreasing the volume fraction of particles. As the volume fraction becomes very low, not only will the number of chains decrease, but also the average spacing between the particles will necessarily increase. Eventually, many of the particle chains may not be complete, and the conductivity will drop drastically. This effect can be seen in right panel of Figure 4, where the curve has been calculated using the model (Figure 5), which assumes that only the number of columns changes as the volume fraction changes.

Gels can be compressed to increase their conductivity and thus act as transducers. Figure 6 describes the result of this experiment using a gel with 3% aligned iron particles. The model in Figure 5 can also be used to interpret the response of the gels when subjected to deformation. The model predicts a sharp increase with strain as the particles in the chains are forced to contact; however, the reality is a lower response. Undoubtedly, the columns buckle locally, preventing metal-to-metal contact.

The ER response of a gel similar to the ones described is shown in Figure 7. The curve fitted to the data is based on the point-dipole approximation, and must obviously fail at higher fields [1,10]. The parameter E_0 is the key for performance; the lower the better. The value of 2.5 kV found for this specimen means that the modulus will double at 2.5 kV. Values as low as 1.3 kV have been reported on similar gels [11].

ER gels have been proposed by several groups as materials that might be appropriate for application in actuators [3,12,17]. The present ER gels with permanent strings of particles are close in structure to those envisioned by Martin and Anderson [12] in a recent analysis of the electrostriction. Although they presented specific results for much

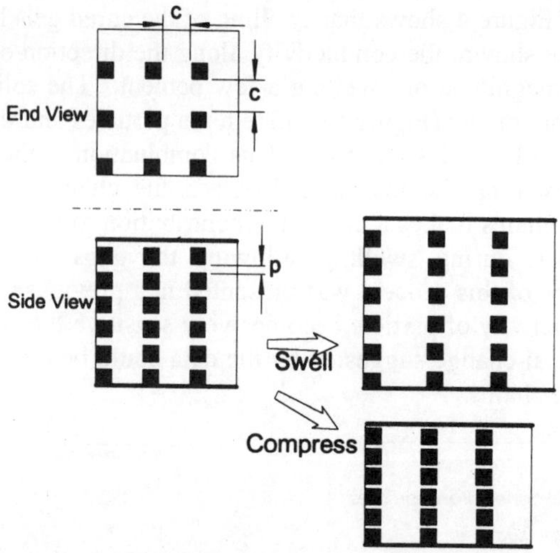

Figure 5. Model of gels with columns of conducting particles. The value of p may be zero in the unswollen sample.

higher particle concentrations, the stresses appeared to scale nearly linearly with concentration, allowing an estimate of the stresses expected for a uniaxial arrangement of 1% particles. This result is depicted in Figure 8 as the horizontal line. The observations agree with the calculation at low stresses, but are higher at higher stresses, suggesting other sources of apparent stress. As the experimental situation in the rheometer is not clearly defined due to axial compliance, these deviations could be spurious and need further investigation. In addition, there is a continuing concern in any experiments with ER gels on the integrity of the connections at the surface of the gel as the applied voltage is changed.

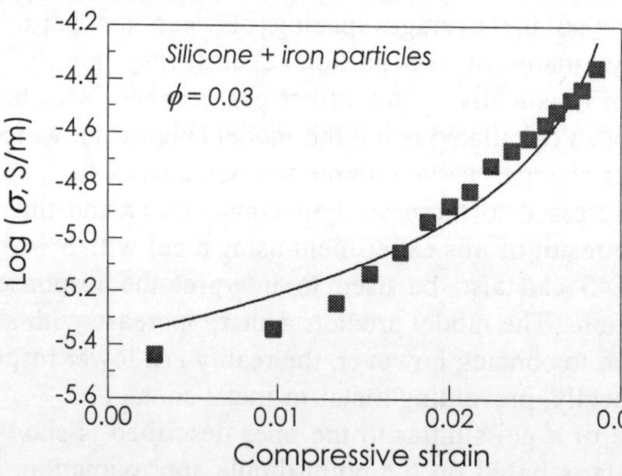

Figure 6. Conductivity of silicone/iron ER gel (3 vol.% iron, no swelling) when compressed. The curve is calculated from the parallel-series model shown in Figure 5.

Unlike the parallel-plate fixtures, the shear sandwich cell (Figure 1, right) is not sensitive to compliance due to electrostatic attraction between the plates; however, it requires a sample with tilted chains. The results for a sample with chains tilted at 45 ° are shown in Figure 8, right. Interestingly, this geometry reproduces roughly the situation in an solidified ER fluid that is sheared one strain unit, as the gel matrix contributes nothing to

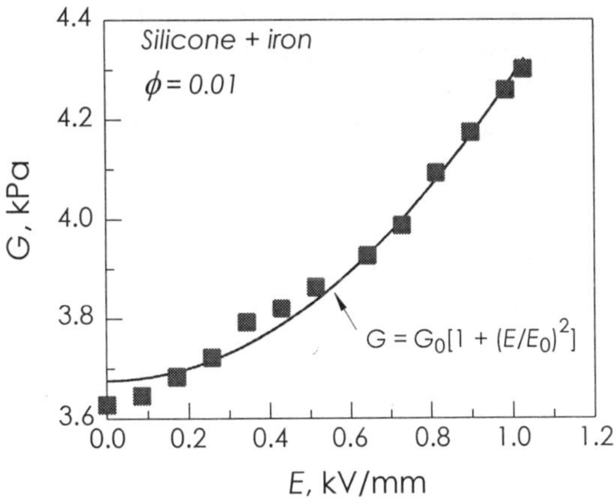

Figure 7. ER response of silicone-iron ER gel, swollen 10 vol. %.. The value of E_0 is about 2.5 kV/mm.

the stress level. While the results are nearly linear, corresponding to a conductivity limit, there is a slight suggestion of an upward curvature. By examining the influence of tilt angle, the entire stress-strain curve for an equivalent ER fluid could be simulated.

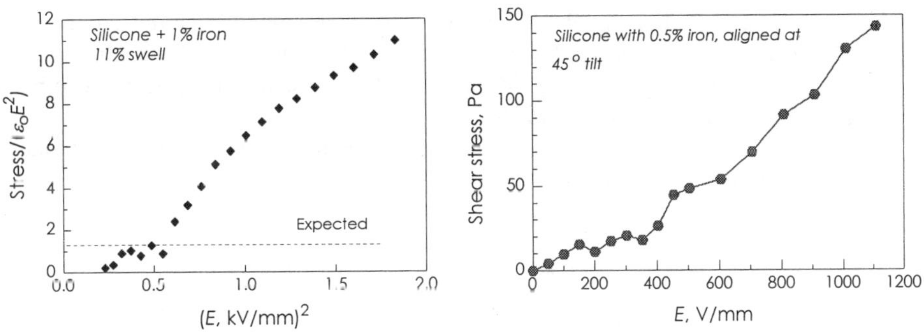

Figure 8. Left: Normal stress produced by ER gel with 1% iron particles. The expected response is based on the calculations of Martin and Anderson [12]. Right: shear actuation stress produced by a gel with chains tilted at 45 ° to the normal. The shear cell of Figure 1 (right) was used

4 Acknowledgements

The authors thank Prof. William Hines of the University of Connecticut for advice regarding the operation of the electromagnet in the IMS magnet facility.

References

1. Atten P., Gonon P., Foule J.-N. and Boissy C., Estimate of the electric stress in the liquid lying between particles of an electrorheological fluid. *Proc. 1999 Conf. Electr. Insul. Dielect. Phenomena*, (1999) pp. 333-336.
2. Block H. and Kelly J. P., Electro-rheology. *J. Phys. D:Appl. Phys.* **21** (1988) pp. 1661-1677.

3. Bohon K. and Krause S., An electrorheological fluid and siloxane gel based electromechanical actuator: working toward an artificial muscle. *J. Polym. Sci. Part B: Polym. Phys.* **36** (1998) pp. 1091-1094.
4. Ceccio S. L. and Wineman A. S., Influence of orientation of electric field on shear flow of electrorheological fluids. *J. Rheol.* **38** (1994) pp. 453-463.
5. Conrad H. *MRS Bulletin* **23**(8) (1998) pp. 35-42.
6. Davis L. C., The metal-particle/insulating oil system: an ideal electrorheological fluid. *J. Appl. Phys.* **73** (1993) pp. 680-683.
7. Fisher M. D., Eckstein M. H. and Schmidt R. G., Electrorheological Gels and a Method for the Preparation Thereof. US Patent 5,412,006 (1995).
8. Klingenberg D. J., Particle Polarization and Nonlinear Effects in Electrorheological Suspensions. *MRS Bulletin* **23**(8) (1998) pp. 30-34.
9. Liu B., Electorheology of Filled Silicone Elastomers. Ph.D. Thesis, University of Connecticut, Storrs, CT USA (1999).
10. Liu B., Boggs S. A. and Shaw M. T., Numerical simulation of electrorheological properties of anisotropically filled elastomers. *IEEE Trans. D&EI*, **8** (2001) pp. 173-181.
11. Liu B., Somani R. H. and Shaw M. T., Electrorheological behavior of gels with ordered particles. *Proc. XIIIth International Congress on Rheology* **4** (2000) pp. 187-189.
12. Martin, J. E. and Anderson, R. A., Electrostriction in field-structured composites: Basis for a fast artificial muscle? *J. Chem. Phys.* 111 (1999) pp. 4273-4280.
13. Nichols M. E., Elie L. D., Ginder J. M. and Pett R. A., Electrorheological Elastomers. European Patent EP0706189A1 (1996).
14. Parthasarathy M. and Klingenberg D. J., Electrorheology: mechanisms and models. *Mat. Sci. Eng. Res.* **R17**(2) (1996) pp. 57-102.
15. Shiga T., Deformation and viscoelastic behavior of polymer gels in electric fields. *Adv. Polym. Sci.* **134** (1997) pp. 133-163.
16. Shiga T., Ohta T., Hirose Y., Okada A. and Kurauchi T., Electroviscoelastic Effect of Polymeric Composites Consisting of Polyelectrolyte Particles and Polymer Gel. *Mater. Sci.* **28** (1993) pp. 1293-1299.
17. Shiga T., Okada A. and Kurauchi T., Magnetoviscoelastic behavior of composite gels. *J. Appl. Polym. Sci.* **58** (1995) pp. 787-792.
18. Winslow W. M., Induced fibration of suspensions. *J. Appl. Phys.* **20** (1949) pp. 1137-1140.

FIELD DEPENDENCE OF VISCOELASTIC PROPERTIES OF MR ELASTOMERS

BELLAN.C, BOSSIS G.

LPMC (UMR6622) Université de Nice, Parc Valrose, 06108 Nice cedex2 (France)

A magnetoactive elastomer made of micronic carbonyl iron particles, structured in elongated clusters and embedded in a silicon elastomer matrix is studied under traction both in static and dynamic modes. The application of a magnetic field of 120kA/m induces a change in elastic moduli of about 0.6MPa at strains of 4 to 5% . Still higher changes(4MPa) are observed in dynamic storage modulus at low strains (10^{-4} to 10^{-3}). The shape of the stress-strain curves are explained by taking into account the existence of a fiber like structure .

1 Introduction

Magnetoactive elastomers are composite materials made of magnetic particles embedded in an elastomeric matrice. Mechanical properties of this material are adaptable through the application of a magnetic field since magnetic interactions between the particles will increase the strength of the material. This possibility has been demonstrated several years ago by different teams[1-4] and some applications for the active control of vibrations are being developped. We can also envisage the use of electro-active elastomers but the need of high electric field imposes the use of sheets of thicknesses smaller than one millimeter which reduces their range of application for the control of vibrations.
Most of the investigations on these elastomers are dedicated to the shear modulus G' and its change with the magnetic field[5,6]. In this paper we shall study the behavior of magneto-active elastomers in the traction-compression mode.

2 Materials

We have used carbonyl iron particles ,grade HQ, from BASF embedded in a silicone elastomer. The saturation magnetisation of the particles has been measured (M_s=1710kA/m), these particles are well spherical and quite monodisperse with an average diameter of 2µm. The matrice is made of a blend of silicone RTV(room temperature vulcanising) 581 (Rhone Poulenc) and 45% in volume of silicone oil 47V20. the viscosity of the blend at room temperature is 1.2Pa.s and the elastic modulus of the cured blend is 0.2MPa instead of 1MPa for the RTV581 cured without silicone oil.. The iron particles are introduce inside the blend in the presence of a surfactant molecule (octadecylméthoxysilane) in order to avoid agregate formation. After adding the catalyst the fluid is poured in a cylindrical mould and placed under vacuum a few minutes in order to remove air bubbles. Then the mould is placed between the poles of an electromagnet and continuously rotates during the curing process in order to prevent sedimentation. Two kinds of samples are prepared; the first one ,called homogeneous is cured without magnetic field; wheras the second kind ,called "structured" is cured in the presence of a magnetic field. In order to avoid a branched structure due to strong aggregation, we raise the field progressively (from 0 to 200kA/m in about 10mn)[7]. The field is applied along the revolution axis of the cylinder, which will be the axis of traction. The views in Fig. 1 are obtained by transmission electronic microscopy and confocal microscopy on thin slides (10microns) of the composite.

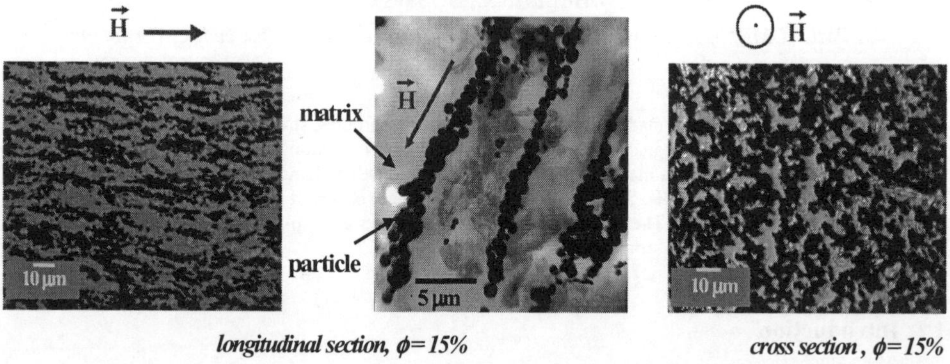

Fig. 1. Microstructure for a volume fraction of 15% cured under magnetic field.

The volume fraction of iron particles is 15% and we can clearly see the columnar structure with a thickness of a few particles in the central picture. On the left, with a smaller magnification, it is also clear that these columns have a broad size distribution and do not percolate through the sample at this volume fraction.

3 Elongation tests

We use cylindrical samples of diameter d=7mm and length L=50mm. At each extremity the elastomer has polymerized on an aluminium cylinder which is attached to the traction device.

We have done the measurements of force versus elongational strain $\Delta L/L$ both for structured and unstructured samples with vomume fractions of magnetic particles between 5% and 25%.

3.1 Comparison homogeneous-structured

In fig.2 we have plotted the stress versus the strain for one sample without particles(lower curve) and for two samples containing each 15% volume fraction of particles, but one has been cured in the absence of magnetic field whereas the other was cured in the presence of the field. On the other hand for each of these two filled elastomers we have realized the traction experiment without applied field and in the presence of a field (H=123kA/m)

First let us consider the behavior of the homogeneous sample (the two curves with squares): there is almost no change when we apply the magnetic field during the traction test. On the contrary if we look at the behavior of the structured sample(upper curves) we see that there is a large change of stress when the traction is done in the presence of the field. We have divided this increase of stress $\Delta\sigma$ in two parts: one $\Delta\sigma_i$ is an initial stress at very low strain(10^{-4}) and the remaining part $\Delta\sigma_{st}$ which depends on the strain.:

$$\Delta\sigma=\sigma(\varepsilon,H)-\sigma(\varepsilon,H=0)=\Delta\sigma_i+\Delta\sigma_{str} \quad (1)$$

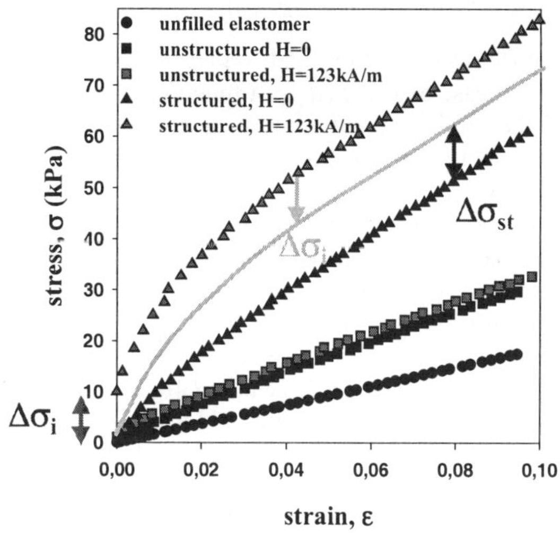

Fig. 2. Traction test, ϕ=15%

It is also interesting to note that, in the absence of field, the Young modulus is much higher for the structured material than for the homogeneous one. This is a classical effect due the presence of fibers in the matrix (in the simplest model we could say that the Young modulus is given by $E=E_m(1-\Phi)+\Phi E_f$,where E_f is the modulus of the fibers which is much higher than the one of the matrix). Actually the fibers are here chains of particles whose stiffness will be very sensitive to the amount of polymer in the gap between the particles which can locally be strained at very high values and so give a higher total stress. Of course this stiffness is reinforced by the application of the magnetic field due to the magnetic attractive force between the particles.

3.2 Influence of the volume fraction

In fig.3 we have plotted the change of stress,$\Delta\sigma$, for an applied filed of 123kA/m as a function of the strain for three volume fractions: 5, 15 and 25%. The maximum of the curve is respectively at 13, 20, 28kPa. In a chain model we should expect a linear increase with the volume fraction, whereas we see that it is slower especially between 5 and 15% .The reason for that is not obvious, an hypothesis would be that the chain formation before curing is hindered at high volume fraction- the structure is more interconnected with lateral branches which do not participate to the magnetic attractive force between particles.

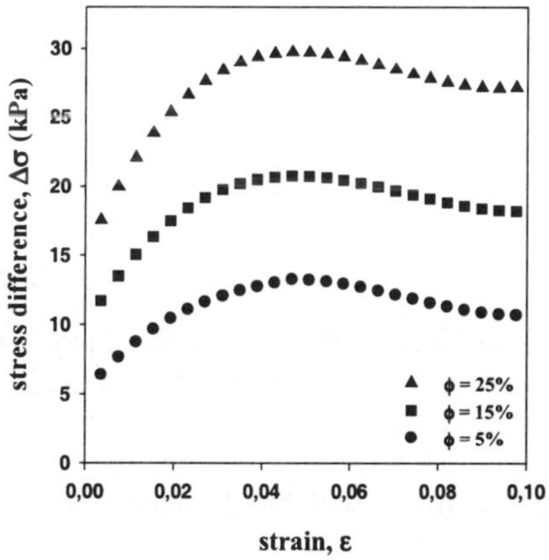

Fig. 3. Effect of the volume fraction; H=123kA/m

3.3 Influence of field intensity

The curves plotted in Fig.4 represent the increase of stress $\Delta\sigma$ as a function of the strain for different magnetic fields and a volume fraction of 25%. We note that the initial stress $\Delta\sigma_i$ increases with the magnetic field and also that the presence of a maximum which shifts towards higher strain when we increase the field. In order to interpret these results we have used a model proposed by Cox (1952)[8] and based on the calculation of local stresses around a cylindrical fiber.

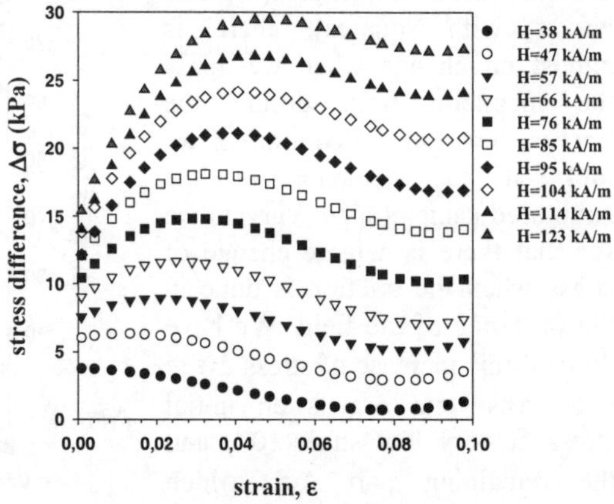

Fig. 4.Increase of stress for different fields

3.4 model

We consider the stress on a solid fiber of length L. As long as the matrix is attached on the fiber, the average strain- and so the average stress in the matrix around the fiber- grows linearly with the length of the fiber. Integrating this stress along the fiber shows that the mechanical stress at the center of the fiber will grow as $\sigma_f^L \propto L^2\varepsilon$. On the other hand the magnetic stress, σ_H, prevents the « fiber » to break as long as the mechanical stress at the center is lower than the magnetic stress. The rupture will occur when the mechanical stress will be larger than the magnetic one (Fig. 5b).

$$\sigma_f^L \propto L^2\varepsilon$$

$$\sigma_H \propto H^2$$

$$\varepsilon_{max}^L \propto \sigma_H / L^2$$

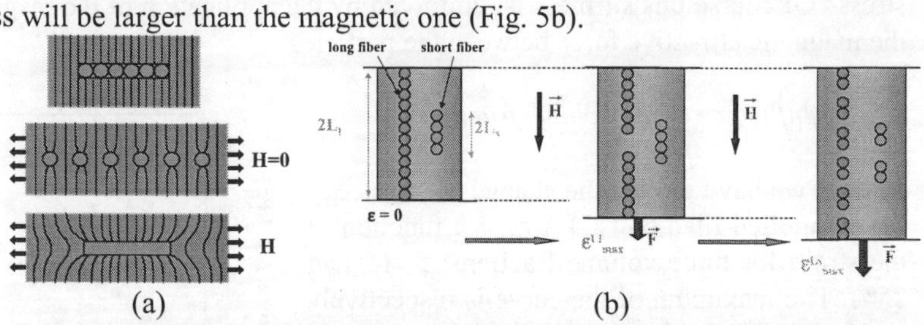

Fig. 5. model based on the rupture of fibers.

According to this model the initial stress $\Delta\sigma_i$ is related to the existence of long chains which will break at such a low strain that it looks like if we had a stress at zero strain. Increasing the strain these fibers will break again and again keeping the average stress at a constant level. Now if we suppose that we have also from the beginning a part of short fibers, these last ones will give rise to an increase of stress till the strain ε_{max}^{Ls} where they begin to break. Such a model with two typical (long and short) lengths of chains explain the

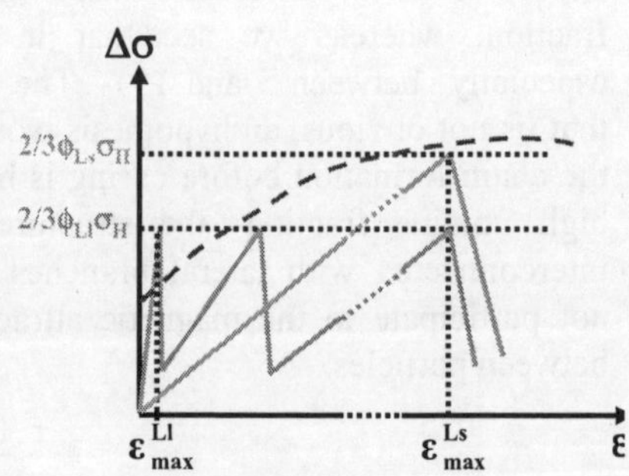

Fig. 6 Predicted behavior

initial stress and the presence of a maximum.

The magnetic force, F_m, between two spheres inside a chain can be calculated by finite elements and the corresponding magnetic stress is:

$$\sigma_H = N_c \frac{F_m}{S} = \frac{3}{2}\Phi \frac{F_m}{\pi a^2}$$

Where N_c/S is the number of chains per unit cross section. Using the model described above it is simple to calculate the initial stress $\Delta\sigma_i$; we obtain $\Delta\sigma_i = 2/3\ \Phi_L \sigma_H$ where Φ_L is the volume fraction of long chains. In practice this is a parameter that we can obtain from a fit between the experimental values of $\Delta\sigma_i$ and the theoretical ones.

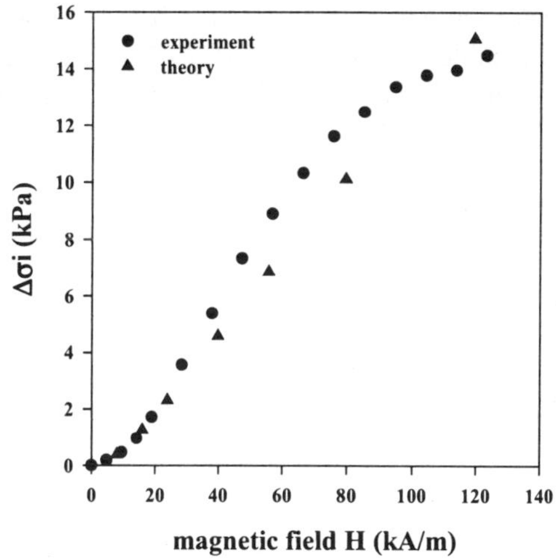

Fig.7 Comparison theory-experiment for the initial stress dependence on imposed field.

The result of this fit, shown in Fig.7, gives Φ_L=8% which seems a reasonable value when we look at the microstructures shown in Fig.1. Furthermore we find at low fields ($0<H<18$kA/m) a behavior in $H^{1.8}$ which is close to the expected power 2, whereas it becomes linear at intermediate field and then begins to saturate, a behavior which is observed for the yield stress of MR suspensions. Note also that this calculation of the magnetic stress has been done for two spheres at contact; when the short fibers have begun to break, the number of gaps between particles and the correlated decrease of magnetic interparticle force dominates and lead to a decrease of the total stress with the strain.

4 Dynamic tests

We have measured the storage and loss modulus E' and E'' of our magnetic elastomer for different frequencies and strains with and without magnetic field. We shall only show here the results obtained for a given frequency as a function of the strain. The sample was cylindrical with a higher diameter (d=20mm) and a smaller length (L=26mm) in order to prevent buckling. We have done these measurements for different samples containing a volume fraction between 5 and 30% .

In Figs 8. and 9 we show typical curves for storage and loss moduli obtained with a sample containing 30% in volume of carbonyl iron and a frequency of 5Hz. The lower curve corresponds to the unfilled elastomer. Then we have two flat superposed curves which correspond to the homogeneous sample with and without field. These curves do not show a dependence on strain neither on field. On the contrary for the structured elastomer we have a very large value of E' at low strain and also an increase of 4MPa for H=42kA/m when applying the magnetic field. The storage modulus then decreases with the strain, a behavior, known as Payne effect, and related to the opening of the network structure of aggregates when the strain increases. A similar behavior is observed for the loss modulus. A plot of tgδ=E''/E' which is not shown here would give a quite marked peak for tgδ at a strain of about 10^{-3} with a value of 0.16 without field and 0.22 with a field of 42kA/m. We see that despite the quite low field we apply the effect of the field is much higher than in the quasi static test. Actually it is because we are working at much lower strains in a domain where the fibers are still not broken; the total change in magnetic stress is close to the one observed in static case for a strain of 10^{-2}.

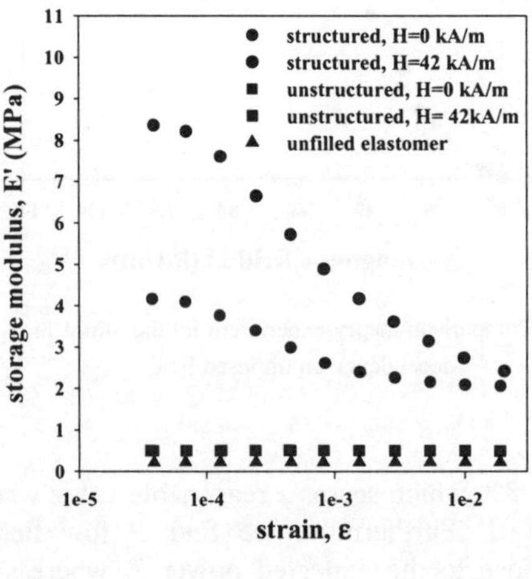

Fig. 8. Storage modulus versus strain, ν=5Hz

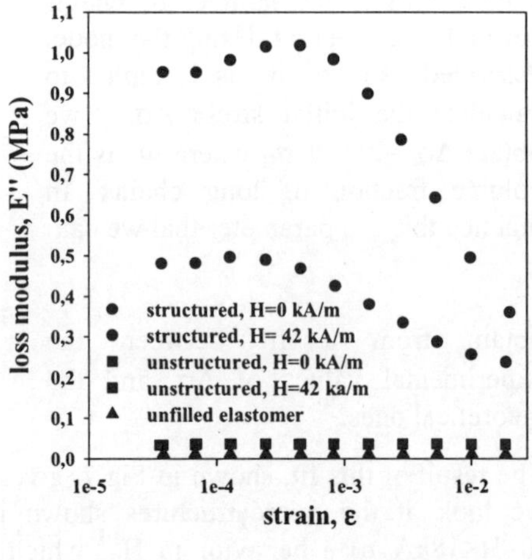

Fig 9 Loss modulus versus strain;ν=5Hz

5 Conclusion

In this paper we have shown that in tensile tests the application of a magnetic field H=123kA/M can give an increase of stress of about 30kPa at a strain of 5% and for a volume fraction of 25%. The behaviour of this magnetic stress as a function of the strain can be understood by considering the existence of two populations of chains of spheres : long ones which explain the initial stress and short ones which explain the increase of magnetic stress till a strain of a few percent; the decrease at higher strain coming from the progressive separation of the particles. The dynamic test has shown the same kind of behaviour but with a larger difference on the storage and loss modulus between the samples which have been polymerized in the presence of the field and the unstructured ones. This larger difference is due to the small strains explored in dynamic tests. A very large non linear behaviour is present in the structured sample and also a large amplification of the dissipation.

The range of modulation of elastic properties by the application of a magnetic field is already large enough to make these materials interesting for numerous applications in semi-active or active dampers. They can furthermore be improved by a better control of structuration before polymerization.

Acknowledgments

This study was granted by the company Technofirst and by the Conseil Général Provence –Côtes d'Azur.

References

1 Japanese Patents : M Shiga, M.Hirose, K. Okada; Tokkai Hei 04266970 22Sept 1992: A.Shiga, Y.Fujimoto, M.Hirose,T okai Hei5-25315, 2 Feb 1993

2 G.Bossis, E.Lemaire,. SFP Colloquium, 4iéme journées de la matière condensée-Rennes 31 August-2 September 1994

3 T. SHIGA, A.OKADA ET T. KURAUCHI, « Magnetoviscoelastic Behavior of Composite Gels », *J. appl. Polym. Sci*, **vol. 58**, p. 787-792 (1995)

4 M.R .JOLLY, J. D. CARLSON, B.C. NUÑOZ ET T.A. BULLIONS, « The Magnetoviscoelastic Response of Elastomer Composites Consisting of Ferrous Particles Embedded in a Polymer Matrix », *J. Int. Mat. Sys. Struct*, **vol. 7**, p. 613-622 (1996)

5 PL.C.Davis "model of magnetorheological elastomers";J.of Applied Physics, vol 85,p-3348-3351 (1999)

6 M.Ginder,M.E.Nichols,L.D.Elie and J.L.Tardiff," « Magnetorheological Elastomers : Properties and Applications », *SPIE Conference on Smart Material Technologies*, **vol. 3675**, p. 131-138 (1999)

7 G. Bossis, C. Abbo, S. Cutillas, S. Lacis, C Métayer "Electroactive and electrostructured elastomers" Int.J.Mod Phys **B, 15,** 564- (2001)

8 D. HULL ET TW. CLYNE, An Introduction to Composite Materials, Cambridge University Press (1996)

9 M.J. WANG, « Effect of Polymer-Filler and Filler-Filler Interactions on Dynamic Properties of Filled Vulcanisates », *Rubber Chem. Technol.*, **vol. 71**, p. 520-589 (1998)

ISOTROPIC MAGNETORHEOLOGICAL NITRILE RUBBERS WITH LARGE IRON PARTICLES

MATTIAS LOKANDER AND BENGT STENBERG

Department of Polymer Technology, KTH, SE-100 44 Stockholm, SWEDEN.
E-mail: lokander@polymer.kth.se

Magnetorheological rubber is the solid analog of magnetorheological suspensions. The incorporation of magnetically polarizable particles in a rubber matrix results in a material for which the shear modulus can be increased by an applied magnetic field. The main advantage of magnetorheological rubbers in comparison with magnetorheological suspensions is that since the matrix is crosslinked the particles do not sediment. However, magnetorheological work only in the pre-yield region while suspensions typically work in the post-yield region, so the two groups of materials are more complementary than competitive.

We have produced and analyzed magnetorheological rubbers, made from nitrile rubbers with different acrylonitrile content and large (over 50 μm) iron particles. The acrylonitrile content in the material influences material properties as the modulus and the damping. The materials were conventionally sulfur vulcanized. The increase in the dynamic shear modulus caused by a magnetic field of 0.2T was measured. Even though the particles have not been aligned within the material by an applied magnetic field during the curing, the increase in modulus is substantial.

1 Introduction

Magnetorheological rubber is the solid analog of magnetorheological suspensions. The incorporation of magnetically polarizable particles in a rubber matrix results in a material for which the shear modulus can be increased by an applied magnetic field. The main advantage of magnetorheological rubbers in comparison with magnetorheological suspensions is that since the matrix is crosslinked the particles do not sediment. However, magnetorheological work only in the pre-yield region while suspensions typically work in the post-yield region, so the two groups of materials are more complementary than competitive.

In previous work on MR rubbers, the matrix material has in most cases been a very soft silicone rubber or gel.[1, 2] When the matrix material has been a harder construction rubber as natural rubber the iron particles have been carbonyl iron, as in MR-suspensions.[3] Before the curing the particles have been aligned within the matrix by the application of a magnetic field.[1 - 3]

2 Experimental

The nitrile rubbers were Perbunan 1845 (18% ACN), Perbunan 3445 (34% ACN), Krynac 4045 (45% ACN) and Krynac 5075 (50% ACN). DMTA-curves for these materials are showed in Figure 1. Further, a mix of Polybutadiene (Buna CB55) and natural rubber was used as a material with 0% ACN. All materials except the natural rubber are made by Bayer.

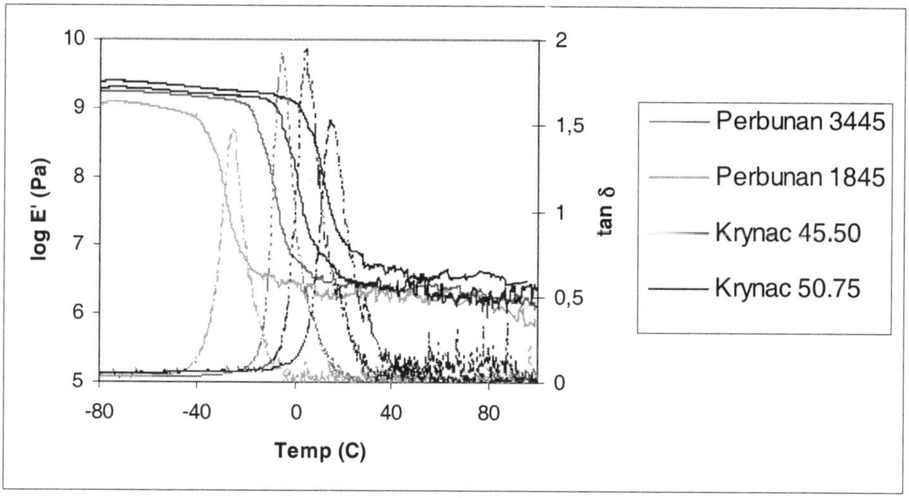

Figure 1. Dynamic-mechanical properties of the pure nitrile rubbers.

The iron particles were irregularly shaped pure iron particles from Höganäs (ASC300, size <60 μm) and spherical carbonyl iron (CI, size 1-10 μm) from BASF.

The shear modulus with and without an applied magnetic field have been measured using an Instron 8032 equipped with an electromagnet capable of applying a magnetic field of 170 kA/m over the samples. The total magnetic induction over samples containing 28% by volume of iron particles was 0.24 T. The frequency during the measurements swept from 1 to 21 Hz, the amplitude varied with frequency. It was 1% at 1Hz and smaller at the higher frequencies.

3 Results and Discussion

The increase in modulus for materials of Perbunan 3445 with varying content of ASC300 and carbonyl iron respectively is shown in figure 2. The increase in modulus is significantly larger for the materials with ASC300 than for the materials with carbonyl iron. The increase for the materials with ASC300 is also larger than expected, considering the low magnetic field and that the particles are not aligned within the material.

The optimal increase in modulus for the material with ASC300 is obtained at an iron content just below 30% by volume. This is similiar to a prediction for materials with aligned carbonyl iron.[4] And the mechanical properties deteriorate rapidly with increasing iron content above 30% by volume (Figure 3). Therefore 28% by volume was considered as a suitable iron content for further experiments.

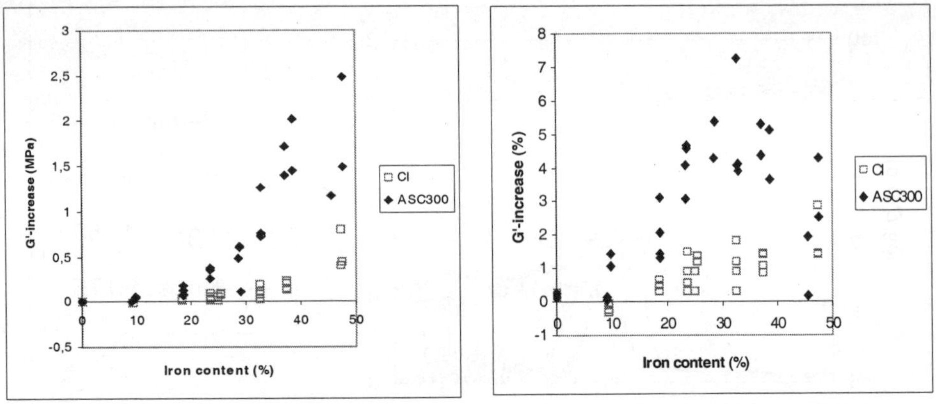

Figure 2. The magnetorheological effect of Perbunan 3445 with ASC300 and carbonyl iron measured at 7 Hz. To the left the absolute increase, in MPa, is shown and to the right you see the relative increase, in %.

The matrix material does not influence the magneto-rheological effect when the particles from Höganäs is used. When carbonyl iron is used a larger increase in modulus is obtained for the hardest matrix materials. (Figure 4)

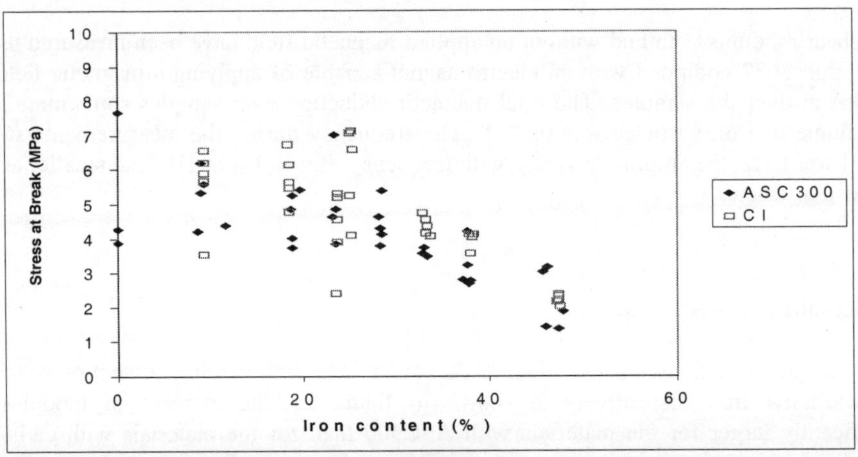

Figure 3. Stress at Break for Perbunan 3445 with ASC300 and carbonyl iron.

4 Conclusions

A larger increase in modulus when a magnetic field is applied (MR effect) is obtained when large particles are used instead of carbonyl iron in materials where the particles are not aligned within the material prior to the curing.

The maximum MR effect is obtained at an iron particle concentration of about 30 % by volume.

The matrix material does not influence the MR effect for isotropic MR-rubber with large iron particles. But when carbonyl iron is used a larger increase in modulus is obtained when harder matrix materials are used.

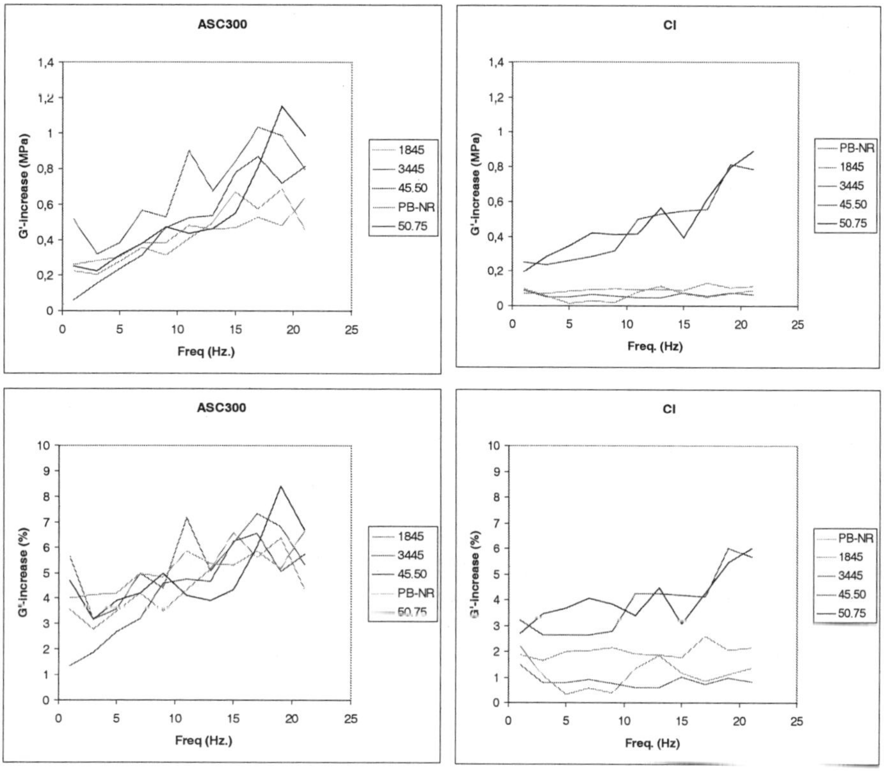

Figure 4. Magnetorheological effect of different matrix materials with 28% by volume of ASC300 and carbonyl iron.

5 Acknowledgements

The European Commission is greatfully acknowledged for financial support (Contract no: G5RD-CT-1999-00125). Kent Lindgren at the Marcus Wallenberg Laboratory for Sound and Vibration Research is thanked for help with the MR-measurements.

6 References

1. Shiga T., O.A., Kurauchi, T.,, *Magnetroviscoelastic Behaviour of Composite Gels.* J. Appl. Polym. Sci., 1995. **58**: p. 787-792.

2. Jolly M. R., C.J.D., Muñoz B. C., Bullions T. A.,, *The Magnetoviscoelastic Response of Elastomer Composites Consisting of Ferrous Particles Embedded in Polymer Matrix.* J. Intel. Mater. Syst. Struct., 1996. **7**(11): p. 613-622.

3. Ginder J. M., N.M.E., Elie L.D., Tardiff J. L.,, *Magnetorheological Elastomers: Properties and Applications.* SPIE, 1999. **3675**: p. 131-138.

4. Davis L.C., *Model of magnetorheological elastomers.* J. Appl. Phys., 1999. **85**(6): p. 3348-3351.

A NEW ORGANIC/INORGANIC HYBRID WITH HIGH ELECTRORHEOLOGICAL ACTIVITY

X.P. ZHAO X. DUAN

Institute of Electrorheological Technology, Department of Applied Physics, Northwestern Polytechnical University, Xi'an 710072, P.R.China
E-mail: xpzhao@nwpu.edu.cn

In-situ sol-gel method to prepare colloidal hybrids of surfactant modified polysucchride and titanium oxide has been presented, and experiments indicated these highly ER active particles exhibited a remarkable ER effect. The static shear stress can be up to 37 k Pa (shear rate 5 S^{-1}) under DC field of 4 kV/mm at room temperature, well above that of simple blends of starch and TiO_2. In the meanwhile, temperature dependence and sedimentation stability were also greatly improved. Based on recent experimental facts, we find that dielectric properties and surface (interface) activity are two necessary conditions fulfilling the requirement of high ER activity. Adequate grinding of particles with oil can effectively enhance the shear stress, which may be owed to the decline of the activation energy needed for restructuring. It has provided us a new horizon for preparation of excellent ER materials and further studies should be continued to make.

1 Introduction

Electrorheological fluids (ERF) have been widely recognized as a promising kind of soft matter with a variety of potential applications [1,2]. So far, a series of research articles have represented fruitful achievements in the realm of electrorheology, both theoretically and experimentally [3-8]. However, in current ER mechanism dielectric properties seem to be overemphasized, at the same time, the influence of particle surface on ERF restructuring comparatively overlooked. As a result, it often leads to a big deviation of tentative predictions from experimental facts [9,10]. Several available ER materials such as polyaniline [11] and porous zeolite [12] are hard to be applied in practical occasions due to their relatively low shear stress and other restricts. So it remains a demanding task to design and prepare new types of high-performance ERF suitable for engineering.

To solve this "bottle neck" of material research, Filisko suggested that special attention should be paid to the property of particle surface or particle/oil interface. The key role of polar functional groups in particle surface has attracted much interest [13]. Current experiments revealed that sufficient active polar groups like hydroxyl group (–OH) on the surface of particles would enhance interfacial polarization, both in strength and in stability

[14]. In this study, colloidal hybrids of modified polysaccharide and titanium oxide were prepared by in-situ sol-gel method, and a remarkable ER effect was gained — the static yield stress can be up to 37 k Pa. In the meanwhile, temperature dependence and sedimentation stability were optimized simultaneously. Based on experimental facts above, we propose that suitable dielectric properties and active particle surface are two necessary prerequisite of high ER activity.

2 Experiments

In situ sol-gel reaction has been employed to synthesize hybrids of surfactants modified carboxylmethylstarch and colloidal titanium oxide. All chemical reagents used in the experiment were of analytical grade. Carboxylmethylstarch (CMS) was firstly synthesized with soluble starch, sodium hydroxide (NaOH) and chloroacetic acid ($CH_2ClCOOH$) in a two-step process described in Refs. 15. 10 m L tetrabutyl titanate ($Ti(OBu)_4$), as inorganic precursor, was dissolved in 30 m L alcohol to form a clear sol. Another homogeneous solution containing 2 g CMS and a proper amount of polar additives (including polyethylene glycol, citric, amide, surfactants, *etc*) was poured rapidly into it with stirring for a fixed duration at 20 ℃. After aging in air, wet gels were then dried in a vacuum oven for 4 h at 70 ℃ to obtain dry colloidal particles. Fourier transform infrared spectroscopy (FT-IR, EQUINOX55) was used to identify the vibrant features of the functional groups in hybrids, indicating there is group interaction between starch-COOH, surfactant-OH, surfactant-NH_4^+ and Ti-OH in the components. The bulk density of hybrids is 2.2 g/cm^3, and the size and shape are strongly dependent on pH value. Finally these hybrid particles were ground into powder about 1 μ m and mixed with dimethyl silicone oil in a volume ratio of 39 %.

3 Results and Discussions

Dielectric measurements were carried out at room temperature through the frequency range 100 Hz to 10^4 Hz using LCR-4225 automatic meter. The concerned experimental data were summarized following exponential functions given below:

$$\varepsilon' = A_1 \times \exp(-\omega/c_1)$$
$$\sigma = A_2 \times \exp(\omega/c_2)$$

Where A_1, A_2, c_1, c_2 are some constants. Thus dielectric relaxation time (defined as ε_s'/σ) can also be described as an exponential function of the frequency. T.Hao owed ER effect to dielectric loss in the case of AC field [16]. While in a series of related experiments [17] we concluded that under static or low frequency alternating electric field (when $\omega \to 0$), a moderately large mismatch of σ/ε' promises contribution to interfacial polarization and induced shear stress. Figure 1 gives the typical dielectric spectra of modified CMS/TiO hybrids, the conductivity of which is 0.6×10^{-7} S/m ($\omega = 100$ Hz, $T = 20$ ^oC), in good accordance with H.Block's analysis that dispersed particles with a conductivity of around 10^{-7} S/m were usually accompanied by the largest ER effect [11].

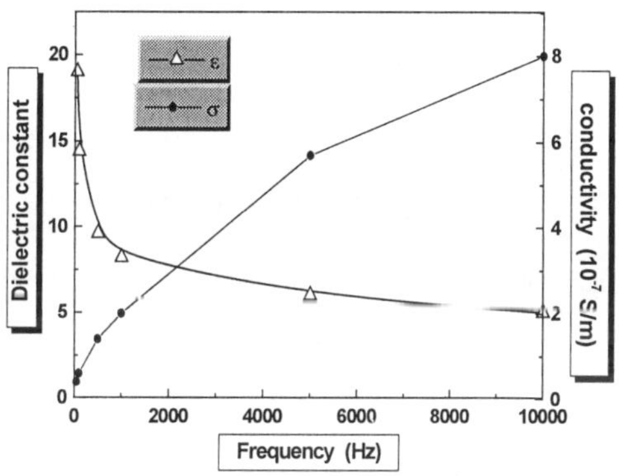

Figure 1. The dielectric spectra of modified CMS/TiO hybrids (T=15 ℃)

The mechanic and rheological behaviors under steady and dynamic shear flow were investigated using rotary rheometer (NXS-11A, the gap between the outer cup and inner bob is 2 mm) and high-voltage DC power source (GYW-010). The rheometer was modified to facilitate the supply of electric field and the measurement of temperature dependence of ER response. Figure 2 shows the yield stress, leaking current density of modified CMS/TiO ERF *vs*. DC field. It can be seen that its mechanic strength has been improved remarkably compared with simple blends of starch and titanium dioxide. The static yield stress has been up to 37 k Pa (shear rate 5 S^{-1}) under 4 kV/mm DC at 20 ℃, with leaking current density limited to no more than 40 μ A/cm^2. In the experiments we discover that adequate grinding plays a vital role in the improvement of ER activity. By grinding of particles and oil together, interfacial affinity will be greatly improved so that the activation energy required for phase transition maybe effectively reduced. Furthermore, many carboxyl and hydroxyl in the hybrids also promote surface activity because they produce synergetic effect (form hydrogen bond) or could react with each other to be polymerized.

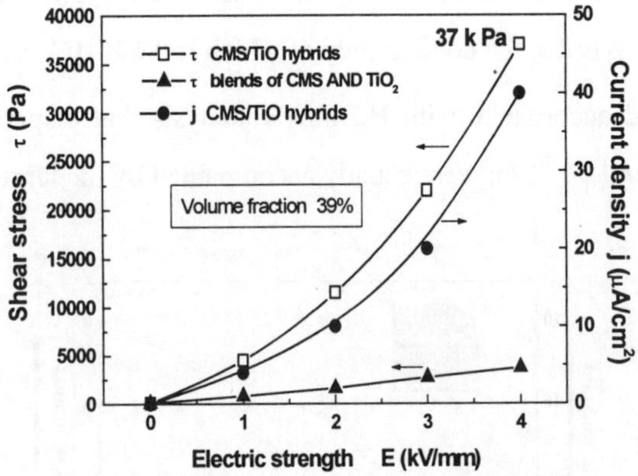

Figure 2. The shear strength, leaking current density of modified CMS/TiO hybrid ERF *vs.* Electric intensity (DC, $\gamma=5\ S^{-1}$, T=15 ℃)

Energetically, the idea of yield stress directly related to dielectric mismatch sounds reasonable, and has been supported by some experiments [18,19], but practically attempts to secure high ER effect due to this principle haven't proved very successful. For example, pure TiO_2 and $BaTiO_3$ *etc.* after adsorption of water still exhibit little ER effect [13], granted their conductivity increases by several orders of magnitude. Here particle surface activity (or particle/oil interface) is a subtle factor deserving serious consideration. Much likely, the phase transition of ERF [3,5] has similar activation energy responsible for the ease of restructuring or not. The combination of surfactants with polar liquids (water, amide, and so on) favors the decline of activation energy because they can readily link particles as a whole due to the theory of minimum free energy.

The dependence of shear stress on shear rate and operating temperature, shown in figure 3 and 4, can be approximately expressed as follows:

$$\tau(Pa)=\tau_y(E)+\eta_a\gamma=\frac{(a-bE)\times E^n}{(T-T_{\max})^2+c}\times10^6+\frac{d}{T^2}\times\gamma$$

Where a, b, c, d, n are specific constants, and T_{max} corresponds to the temperature at which the maximum shear stress appears. When a=50, b=2.5, c=7500, d=4, n=1.6, T_{max}=75, the above equation tallies well with the experiment in the scope of our test. In figure 3 hybrid ERF behaves as quasi-Bingham fluid. Its shear stress increases with the shear rate in the range of 100 S^{-1} and has a minimum value close to the yield point. In figure 4 the temperature dependence has been obviously optimized. Its maximum stress appears at around 70 ℃, and remains a considerable value even at 90 ℃. In addition, the presence of suspension stabilizers in our hybrids endows their suspensions with excellent sedimentation stability. The suspension particles don't deposit during three months, seen in figure 5. We also observed that after half a year, the sedimentation ratio is still above 80 %.

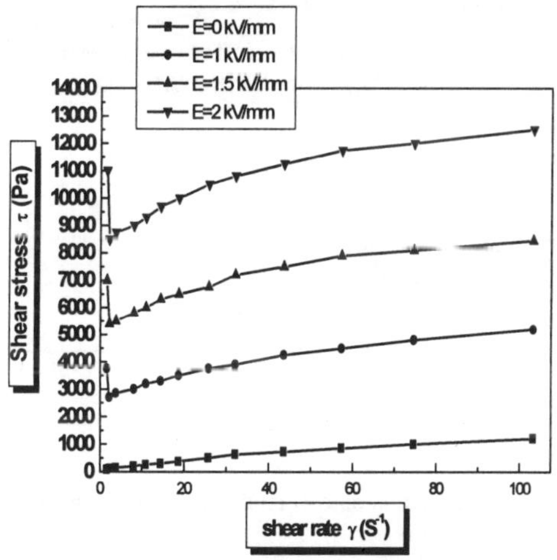

Figure 3. The curve of shear stress of modified CMS/TiO hybrid ERF vs. shear rate (DC, T=15 ℃)

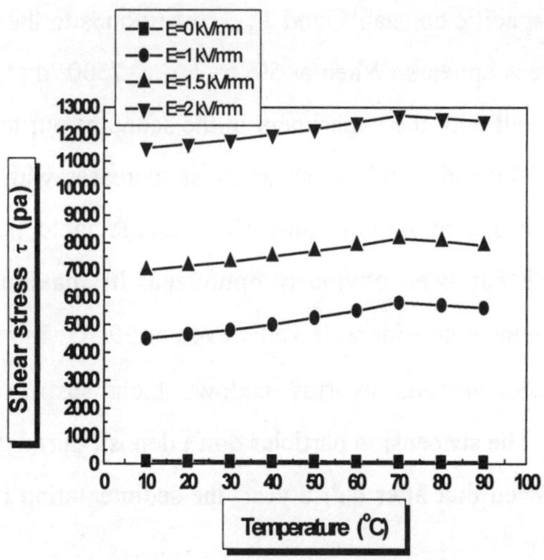

Figure 4. The curves of the dependence of shear stress on temperature

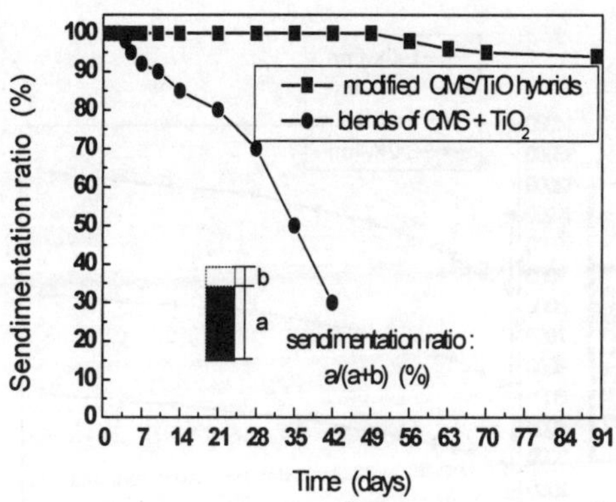

Figure 5. The sedimentation ratio of modified CMS/TiO hybrid ERF vs. time

4 Conclusions

In summary, we have developed a new kind of highly ER active particles and laid foundation for an optimal criterion empirically satisfied to acquire ideal ER behaviors, *i.e.* (i) long chainlike molecular structure; (ii) a large mismatch of σ/ε' between particles and oil; (iii) high surface activity of particles. Experiments have showed that our hybrid

ERF has rather a large shear stress, a wide operating temperature and is free from abrasion and particle sedimentation that are encountered in conventional ERF. Furthermore, activation energy was put forward to explain the big discrepancy of different materials in ER activity, when their dielectric properties are not far off. Maybe it provided us a good approach to the puzzling problem concerning particle surface activity, although strict studies should be further taken up. We hope it would be beneficial to understand ER mechanism correctly, and resolve traditional difficulty of material research soon.

5 Acknowledgements

This research was supported by the Chinese National Science Foundation under Grant No 59832090, the National Natural Science Foundation of China for Distinguished Young Scholars (50025207), and the Doctorate Foundation of Northwestern Polytechnical University.

References

1. Halsey T C, *Science*, **258** (1992) pp.761.
2. Tada H, Saito Y, Hirata M, Hyodo M, Kawahara H, *J. Appl. Phys.*, **73** (1993) pp. 489.
3. Chen T J, Zitter R N, Tao R, *Phys. Rev. Lett.*, **68** (1992) pp. 2555.
4. Zhou L, Wen W, Sheng P, *Phys. Rev. Lett.*, **81** (1998) pp. 1509.
5. Whittle M, Bullough W A, *Nature*, **358** (1992) pp. 373.
6. Dodd S J, Dissado L A, Champion J V, Alison J M, *Phys. Rev.* **B**, **52** (1995) pp. 16985.
7. Davis L C, *J. Appl. Phys.*, **72** (1992) pp. 1334.
8. Tian Hao, *Appl. Phys. Lett.*, **70** (1997) pp. 1956.
9. Trau M, Sankaran S, Saville D A, Aksay I A, *nature*, **374** (1995) pp. 437.
10. Wu W C, Conrad H, *J. Appl. Phys.*, **81** (1997) pp. 383.
11. Block H, Kelly J P, Qin A and Waston T, *Langmuir,* **6** (1990) pp. 6.
12. Conrod H, Sprecher A F, Choi Y, *J. Rheol.* **35** (1991) pp. 1393.
13. Hao T, Xu Y Z, *Appl. Phys. Lett.*, **69** (1996) pp. 2668.
14. Zhang Y L, Ma Y, Lan Y C, Lu K Q, Liu W, *Appl. Phys. Lett.*, **73** (1998) pp. 1326.
15. Zhao X P, Duan X, Yin J B. Chinese Patent 00113763.8 (2000)
16. Tian Hao, *Appl. Phys. Lett.*, **70** (1997) pp. 1956.
17. Zhao X P, Yin JB, Xiang L Q, Zhao Q, *Journal of Material Research* (in Chinese), **14** (2000) pp. 653.
18. Tang X, Conrad H, *J. Appl. Phys.*, **80** (1996) pp. 5240.
19. Davis L C, *Appl. Phys. Lett.,* **60** (1992) pp. 319.

INFLUENCE OF SURFACTANTS ON PROPERTIES OF ELECTRORHEOLOGICAL FLUIDS CONTAINING POLYANILINE

MAŁGORZATA BOCIŃSKA, HENRYK WYCIŚLIK, MARCIN OSUCHOWSKI AND JANUSZ PŁOCHARSKI

Department of Chemistry, Warsaw University of Technology, ul. Noakowskiego 3, 00-664 Warszawa, Poland

Sedimentation which is a natural process in most of ER fluids can be reduced by addition of surfactants that influence also other properties of the fluids. To study both the ER effect and the rate of sedimentation was the aim of the investigations. The ER fluids comprised powdered polyaniline and silicone oil to which surfactants of different polarity were added. The rate of sedimentation was measured by a sedimentation balance. The flow curves were recorded under electric field up to 2.5 kV/mm. Current density was also measured as a function of shear rate. It was found that the activity of a surfactant depends strongly on its polarity. The lipophylic surfactants stabilized the suspension very well but about 30% decrease of the dynamic yield stress was observed. The current density was reduced as well by almost one order of magnitude. The hydrophylic surfactants hardly stabilized the suspension but increase of yield stress was observed that was not followed by increase of current density. The role of different types of non-ionic surfactants was discussed.

1 Introduction

Almost all electrorheological (ER) fluids comprise suspensions of particles in liquid matrices. The intrinsic feature of such systems is their aptitude for sedimentation being one of the limits of their technical applicability. A density matching which seems to be a natural method to solve the problem cannot be used for many systems and introduction of surfactants must be considered. These substances obviously do influence all basic properties of ER fluids but systematic studies of the role of surfactants are rather scarcely presented in the literature. The general prerequisite resulting from the requirement of low conductivity of an ER system is application of non-ionic surfactants. Other requirements justifying a particular choice of a surfactant are considerably less clear and the aim of our study was to address relations between the chemical nature of a surfactant and sedimentation rate as well as ER effect.

The general role of surfactants in ER fluids was studied by Kim and Klingenberg [1]. They observed increase of the yield stress after addition of a few percents of a detergent followed by the decrease at higher concentrations. In all cases surfactants enlarged the ER effect as compared to the initial suspensions. The influence of detergents was explained on the basis of surfactant-enhanced interfacial polarization for the lower concentrations and field-induced phase separation for the higher concentrations.

The particular problem of surfactants in polyaniline based ER dispersions was studied by Lee *et al.* [2]. The authors applied two methods to estimate the sedimentation rate and correlated the results with adsorption isotherms of the detergents as well as ER activity of the fluids. The measurements of sedimentation were based on changes of optical transmittance of an ER fluid in a spectrophotometer cell and on observations of clear layers of oil often formed at the top of a glass cylinder filled with the tested suspension. A flat maximum of the dynamic yield value resulting from surfactant addition was observed for concentrations of a few percents. The most effective stabilization was observed for the surfactant molecule with a long and bulky lipophilic tail giving rise to steric hindrance preventing aggregation of the polyaniline particles.

Many water free ER systems reported recently contain dispersions of conjugated polymers to mention polyphenylene, polypyrrole, polyaniline and polythiophene as a few examples and we used such a polymer in our investigations. In the present work we have chosen protonated polyaniline in silicone oil which gives mixtures of stable properties and relatively high yield stress [2-4]. In order to measure the sedimentation rate we first applied the method with glass cylinders described by Lee *et al.* [2], however, it worked only for polyemeraldine base powder and not for the protonated polymer. In the latter case we observed an opaque deposit settling on the walls of the vessel making this measurement impossible as well as other optical methods. Therefore, we had to elaborate a more direct method to measure the settling rate. Our experiments were supplemented by measurements of rheological properties and current density.

2 Experimental

Polyaniline was synthesized by chemical oxidation of aniline with $(NH_4)_2S_2O_8$ in 1.5 M HCl at 0-5 °C. The obtained polyemeraldine salt was deprotonated with 3% ammonia and thoroughly washed with water. Then a suspension of this polymer in water was prepared to which drops of 1M H_2SO_4 were added until the pH value was 5.7. The mixture was stirred for a few days during which the pH value was kept constant. The protonation level resulting from this acidity was believed to give the highest ER effect [4]. The protonated polyaniline was filtered out, washed with methanol and dried at ambient temperature in high vacuum (about 10^{-5} Tr) for 20 hours. The size of grains of the powder was within 2 and 5 μm (90%, averaged by number) with maximum of distribution at 2.7 μm.

The silicone oil (polydimethylsiloxane) of 0.1 Pas viscosity was dried under dynamic vacuum of about 10^{-5} Tr at 120 °C for a few hours with continuous agitation. All components were stored in a desiccator. The ER suspensions were prepared by homogenisation of weighted amounts of the protonated polyaniline powder and the silicone oil in a mortar with a pestle. All prepared suspensions contained 10% w/w of polyaniline in the oil.

Surfactants were added to the oil before the powder was introduced. The list of the used detergents is presented in Table 1.

Table 1. Surfactants used in the study.

Name and manufacturer	Composition	Properties
LUTENSOL AP9 BASF Ludwigshafen, Germany	$C_9H_{19}(C_6H_4)(OCH_2CH_2)_9OH$ adduct of nonylphenol and ethylene oxide	viscous liquid well soluble in water; high polarity
PLURONIC 10100 BASF Ludwigshafen, Germany	$HO(CH_2CH_2O)_{10}(CH(CH_3)CH_2O)$- $-(CH_2CH_2O)_{10}H$ ethylene and propylene oxide copolymer	viscous liquid insoluble in water; low polarity
ROKWIN 80K Z.Ch."Rokita" Brzeg Dolny, Poland	mixture of esters of sorbitol and stearic and oleic acids	viscous liquid poorly soluble in water; low polarity

Because of the reasons mentioned we decided to elaborate a direct method of measurement of sedimentation rate. We designed and built a miniature sedimentation balance shown in Fig.1. The test consisted in weighting the powder settled in a small container immersed in the ER fluid. Comparison the results of the glass cylinder method and the sedimentation balance for the undoped polymer gave very similar results that encouraged us to use the latter method for the protonated polyaniline.

ER properties of the fluids were studied by means of a PC controlled Bohlin Visco 88 viscometer with home modified measuring system operating in a two concentric cylinders geometry capable to apply an electric field up to 5 kV/mm.

All experiments were carried out at ambient temperature of 20 – 23 °C.

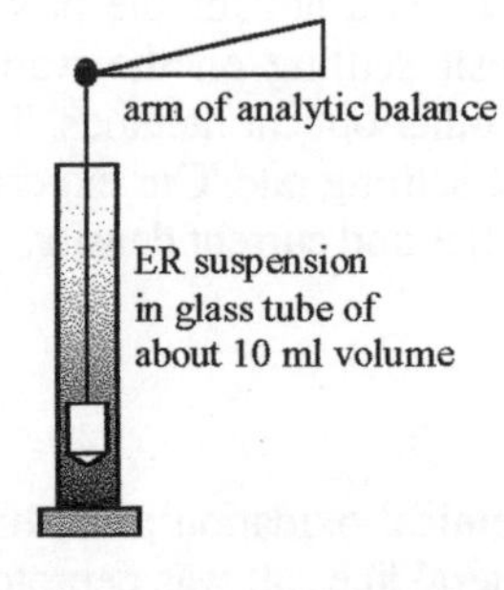

Figure 1. Scheme of sedimentation balance.

3 Results and Discussion

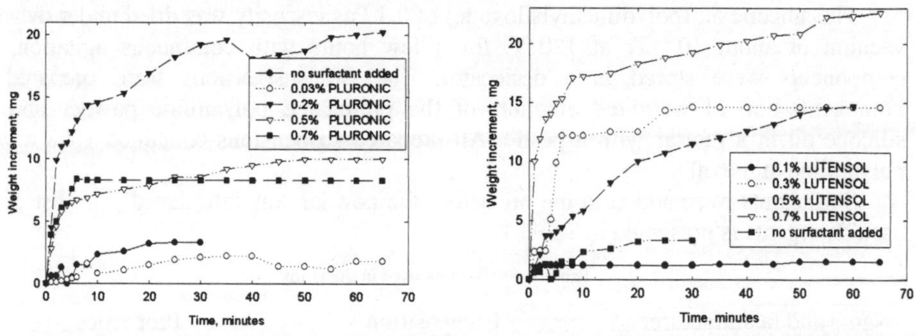

Figure 2. Weight increment measured by sedimentation balance for various concentrations of (a) PLURONIC and (b) LUTENSOL added to basic oil (concentrations given in %w/w in respect to the oil).

Figs. 2a,b and 3a show the sedimentation rates for the three surfactants. The collected results are given in Fig.3b. The different chemical composition of the applied surfactants results in their different stabilization activity. Poor properties were observed for PLURONIC (Fig.2a). Its linear molecule comprises propylene and ethylene glycol units and has no distinguished hydrophilic and lipophilic ends. Therefore, it is probable that the chains are adsorbed flat on a polyaniline grain surface and only very limited steric hindrance can be expected. Some improvement of the dispersion stability is observed only at very low concentrations (about 0.03 %w/w in oil) whilst the higher amount of the surfactant considerably accelerate the settlement process. The LUTENSOL modified samples behaved in a similar way (Fig.2b) that in this case was rather due to the high hydrophilic properties of this compound. The third series of samples modified with

ROKWIN revealed totally different features. All concentrations of the surfactant exceeding 0.3 %w/w stabilized effectively the dispersions. Additional experiments proved that no change in ER behavior of a 17% dispersion was observed after 20 hours of storage in quiescent conditions and some significant difference appeared only after 72 hours.

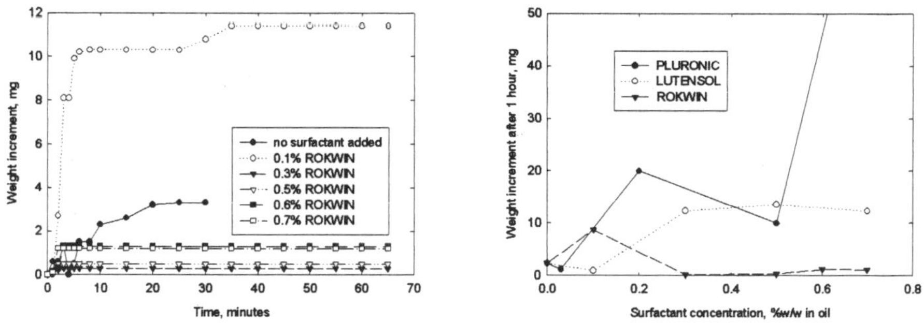

Figure 3. (a) Weight increment measured by sedimentation balance for various concentrations of ROKWIN added to basic oil (concentrations given in %w/w in respect to the oil). (b) Change of sedimentation rate resulting from surfactant additive measured as weight increment after 1 hour test.

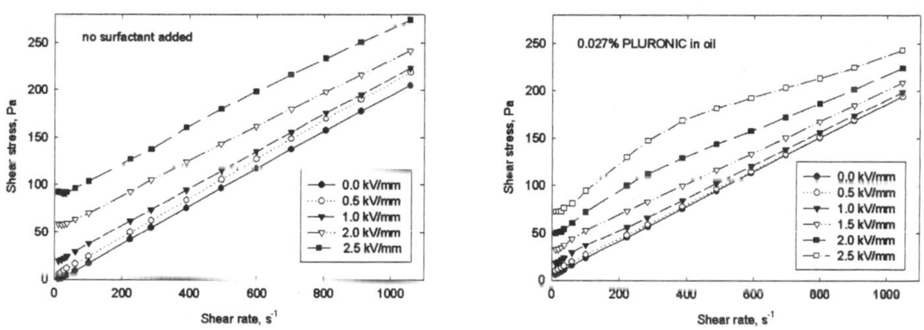

Figure 4. Flow curves of polyaniline dispersions under electric field: (a) basic fluid with no surfactant added; (b) PLURONIC added to basic oil in concentration of 0.027 %w/w in respect to the oil.

In the next step rheological properties of the prepared fluids were studied. Fig.4a presents the results obtained for the basic fluid. A set of parallel straight lines resembling the Bingham model was obtained and the dynamic yield stress values estimated by extrapolation of the recorded curves grew almost perfectly with field strength squared. Figs.4b and 5a,b show analogous curves for the dispersions containing surfactants at stabilizing concentrations. Their shape is different with apparent shear thinning at higher shear rates. The absolute values of dynamic yield stress are lower and they increase slower than E^2. The collected data are listed in Table 2. Kim and Klingenberg [1] described a strong increase of the magnitude of the ER effect due to surfactants but in the case of the system they studied the whole ER activity could be attributed to the polar species absorbed and adsorbed on the dispersed grains of alumina. In the case of polyaniline the ER activity seems to be intrinsic feature of the polymer particles. Lee *et al.* [2] who used polyaniline did report some increase, but not that high as well as a flat maximum of the ER effect for a few percents of detergent added. Our investigations univocally indicated the presence of the maximum only for PLURONIC at 0.6%w/w but this concentration did not stabilize the dispersion.

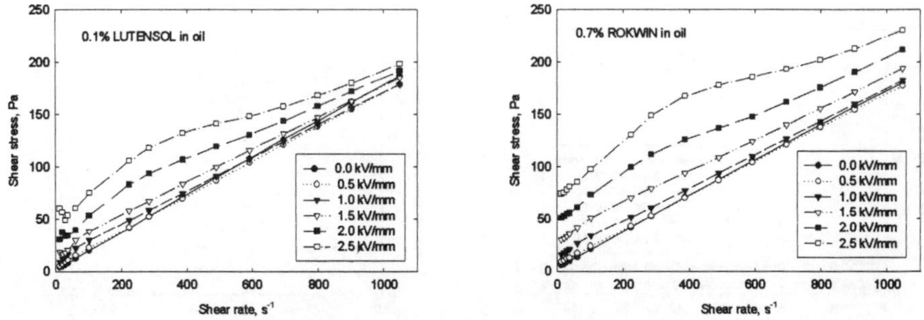

Figure 5. Flow curves of polyaniline dispersions under electric field: LUTENSOL (a) and ROKWIN (b) added to basic oil in concentrations of 0.1 and 0.7 %w/w, respectively.

Introduction of a surfactant changes also the current flowing through an ER fluid under electric field. Fig.6a presents current densities as a function of shear rate and field strength for the basic fluid and Fig.6b shows the changes caused by the surfactants. In all cases a considerable decrease of current density was observed in some contradiction to the results of Lee *et al.* [2] who, however, used higher concentrations of surfactants.

Table 2. Dynamic yield stress of ER fluids containing surfactants.

Concentration of surfactant [%w/w]	Dynamic yield stress [Pa]			
	1.0 kV/mm	1.5 kV/mm	2.0 kV/mm	2.5 kV/mm
	PLURONIC PE 10100			
no surfactant	19.3	--	58.2	92.3
0.2	17.9	32.0	50.6	72.3
0.5	21.3	37.8	63.0	90.1
0.6	30.0	51.2	83.6	124.3
0.7	21.6	40.1	65.2	91.3
	LUTENSOL AP9			
0.1	8.9	18.4	30.4	60.3
0.3	16.3	28.7	45.5	66.0
0.5	15.3	25.0	43.5	64.6
	ROKWIN 80K			
0.1	14.4	26.4	43.1	66.2
0.3	15.2	26.9	44.8	63.2
0.5	14.8	27.9	46.3	70.8
0.6	15.2	28.0	49.2	70.6
0.7	15.9	30.0	50.8	74.2

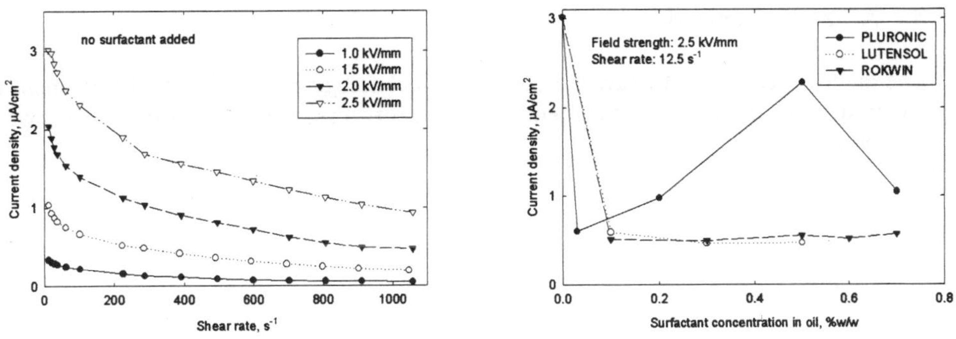

Figure 6. Density of current flowing through polyaniline dispersions under electric field: (a) basic fluid with no surfactant added; (b) dependence on type of surfactant and its concentration.

Polyaniline is a conjugated polymer that after protonation gains electrical conductivity of values from a semiconductor region. Therefore, it seems that at least a part of the ER effect successfully observed in suspensions of this polymer is due to electronic polarization processes within the bulk of polymeric particles. If so, there must be some optimum conductivity that we attempted to estimate in our previous work [4] receiving the value of about 10^{-8} S/cm being the result of uniform protonation in a solution of sulfuric acid of pH=5.7. Introduction a surfactant modifies the character of the polarization processes responsible for the ER effect as is suggested by the presented change of the shape of the flow curves. A surfactant rich layer is probably created around a grain and the polarizability of this cover depending on the chemical nature of the surfactant takes the leading role in the ER effect. From that moment the interfacial polarization model prevails which, however, seems to be ER ineffective at low concentrations of a surfactant. In the non-modified basic fluid under field a large part of total current is transported probably by electronic carriers typical of a conjugated polymer. The surfactant layer at a grain surface forms a barrier for the electrons and the current must flow mostly due to ionic carriers what explains the reported current reduction.

4 Conclusions

1. We have built and tested a miniature sedimentation balance suitable for measuring the sedimentation rate of an ER fluid. We believe that due to the direct character of the method the obtained results are less prone to various errors than for example indirect optical measurements.

2. Good stabilization of the polyaniline based ER dispersions was achieved only with highly hydrophobic surfactants of low HLB number.

3. Introduction of a surfactant at concentrations below 1% significantly reduced the current drawn by our ER system.

4. According to our experiments it was not possible to obtain a stable dispersion and simultaneously profit from the enhancement of ER effect caused by a surfactant additive.

5 Acknowledgements

The authors thank the Warsaw University of Technology in general and Department of Chemistry in particular who financially and organizationally supported the work described.

References

1. Kim Y. D. and Klingenberg D. J. *J. Colloid Interf. Sci.* **183** (1996) 568-578.
2. Lee H-J., Chin B. D., Yang S-M. and Park O.O. *J. Colloid Interf. Sci.* **206** (1998) 424.
3. Gow C. J. and Zukoski IV C. F. *J. Colloid Interf. Sci.* **136** (1990) 175.
4. Gozdalik A., Wyciślik H. and Płocharski J. *Synthetic Metals* **109** (2000) 147.

REVERSIBILITY OF THE ER EFFECT IN IMMISCIBLE LIQUID BLENDS

SHUHEI YAMAMOTO, YUICHI MASUBUCHI, JUN-ICHI TAKIMOTO AND KIYOHITO KOYAMA

Department of Materials Science and Engineering, Faculty of Engineering, Yamagata University
4-3-16 Jonan, Yonezawa 992-8510, Japan

. ABSTRACT : Blends of immscible liquids with different dielectric constants and viscosities were known to show the ER effect due to the change of the domain structure by the electric field. One of major problems of this type of ER fluids is that the stress dose not drops to the original value even long after the removal of the field. In this paper, we report on our attempts to improve the reversibility of the ER effect in immiscible blends.

1 Introduction

Electrorheological (ER) effect is defined as a reversible change of the rheological properties of a fluid by an application of electric field, and fluids which show the ER effect are called ER fluids [1]. Recently, it has been found that blends of two liquids with different viscosities and different electric properties (dielectric constants or conductivities) show the ER effects [2]. The origin of this ER effect is the change of the domain structure by the electric field. One of the problems, however, of this type of ER fluids is that the recovery of the viscosity to its zero-field value after removal of the field is very slow, especially under low shear rate, because the domain structure does not easily return to the original low-viscosity structure.

In this paper, we report that the reversibility of the ER effect in immiscible blends can be improved by using ternary blends. By adding the third component liquid, we can modify the miscibility and the domain structure of the blends, and the reversibility under low shear rate can be improved at appropriate composition.

2 Experiments

2.1 *Materials*

The first component of our ternary blend was urethane-modified polypropylene glycol (UPPG) with viscosity of about 100 Pa s. The molecular structure of UPPG is shown in Fig. 1. The other components were two poly(dimethylsiloxane) (PDMS) with different viscosities: 1000 cSt (~ 1 Pa s) and 10 cSt (~ 0.01 Pa s). The weight fraction of UPPG, ϕ_1, was fixed at 67 wt%, and the weight fractions of 1000 cSt and 10 cSt PDMS (ϕ_2 and ϕ_3)

were varied under the condition $\phi_2 + \phi_3 = 33$ wt%. These three fluids were mixed by a mechanical stirrer with rotation speed of 360 rpm at room temperature.

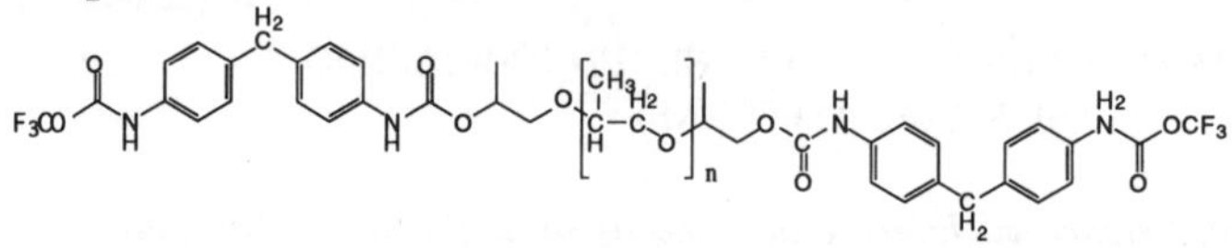

n ~ 50   μ ~100Pa s

Fig.1 Chemical structure of urethane-modified polypropylene glycol (UPPG).

2.2 *Rheological measurements*

Rheological measurements were carried out under room temperature by using a rotating rheometer (MR-300 V.2, Rheology Co., Ltd.). The geometry was parallel disks, and the gap distance was 1 mm. The shear rate was varied in the range of 0.5 ~ 32 s^{-1}, and an electric field of 0 ~ 2 kV/mm was applied.

3 Results and Discussion

3.1 *ER effect and its decay time*

Figure 2 shows the flow curves of the UPPG/PDMS-1000cSt/PDMS-10cSt ternary blends with three different compositions: $\phi_3 = 0$, 13 and 26 wt%. All of the blends show the ER effect of comparable magnitude. Under no electric field, the viscosities of the blends are much lower than that of UPPG (μ_{UPPG} ~ 100 Pa s), and they decrease as the fraction ϕ_3 of the lower viscosity PDMS increases. Under the DC electric field of 1 kV/mm or higher, on the other hand, the viscosities of the blends are about 50 to 100 Pa s, i.e., of the order of μ_{UPPG}.

Figure 3 shows, for the same three blends as Fig.2, the time dependence of the shear viscosity after cutting off the electric field of 2 kV/mm. For all the blends, the decay of the viscosity becomes faster as the shear rate increases. By comparing among the three blends, we can see that the decay is fastest for the blend $\phi_3 = 13$ wt% (Fig. 3(b)), and slowest for $\phi_3 = 26$ wt% (Fig.3(c)).

In order to compare the decay speed more quantitatively, we define the recovery time $T_{1/2}$ as the time for the viscosity to decay from μ_E to $(\mu_E + \mu_0)/2$, where μ_E and μ_0 are the viscosity under the field and that long after the removal of the field, respectively. If the viscosity does not return to its original value within an experimental time, we define $T_{1/2}$=inf. Figure 4 shows $T_{1/2}$ as a function of the fraction ϕ_3 of 10cSt PDMS. It can be clearly seen that the recovery time $T_{1/2}$ is shortest at around ϕ_3 ~ 10 wt%.

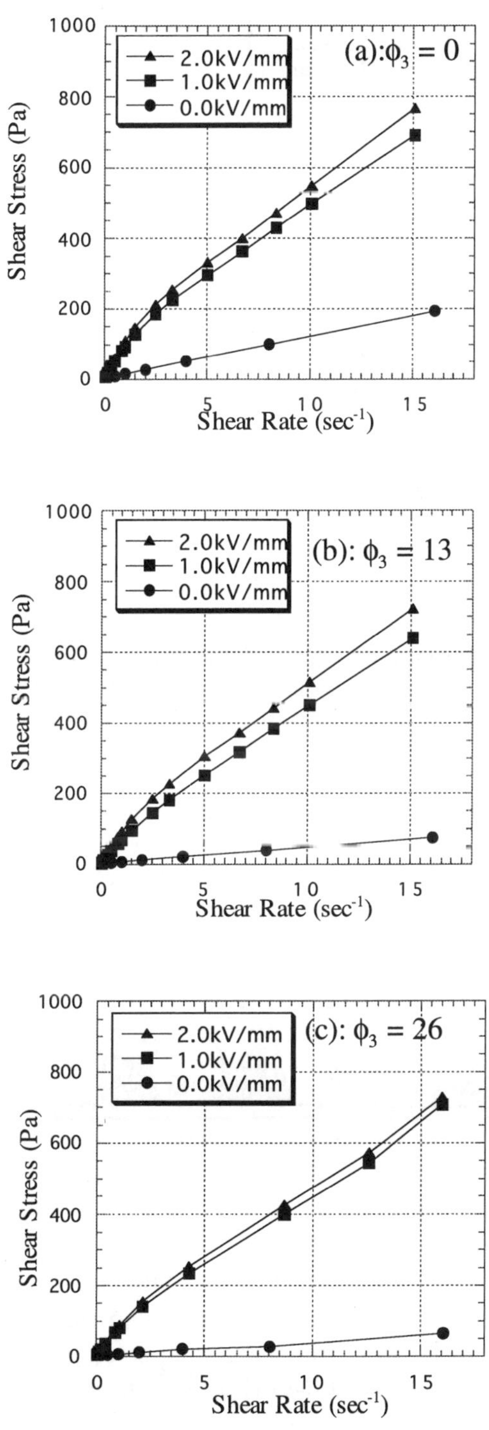

Fig.2 Shear rate dependence of shear stress under various fields. (a):ϕ_3 = 0, (b): ϕ_3 = 13, (c): ϕ_3= 26

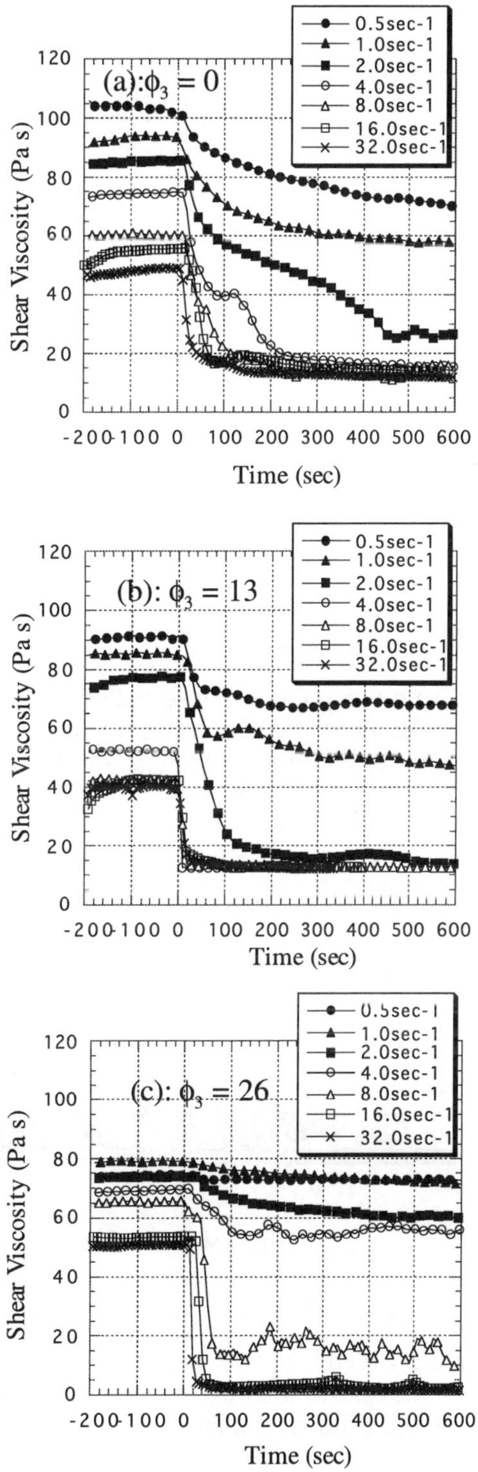

Fig.3 Decay of shear viscosity after switching off the electric field. (a):ϕ_3 = 0, (b): ϕ_3 = 13, (c): ϕ_3= 26

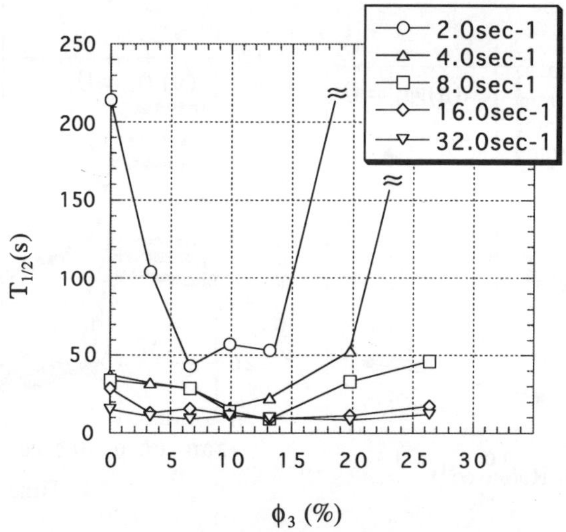

Fig.4 Recovery time $T_{1/2}$ vs. weight fraction of 10cSt PDMS ϕ_3.

3.2 Domain structure observation

From our previous observation of the domain structure of the binary blend under shear and electric field [3], we can summaries the mechanism of the ER effect in the binary blend as shown in Fig.5. Under no electric field, UPPG droplets are dispersed in a PDMS (1000cSt) matrix. If an electric field is applied, UPPG, which has higher conductivity and dielectric constant than PDMS, connects the electrodes to form column-like structure, which is deformed, broken, and reformed under shear flow. If the shear rate is low, UPPG continues to connect the electrodes even after the field is turned off, and the viscosity of the blend does not return to its original value.

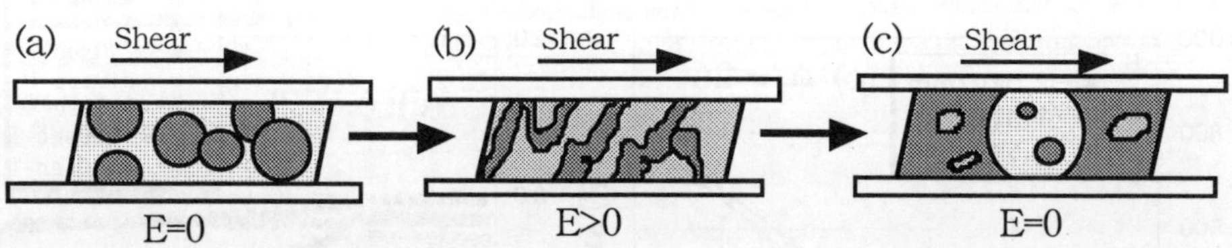

Fig.5 Mechanism of the decay in the binary blend (ϕ_3 =0 wt%).

Figure 6 shows the optical micrographs of the domain structure the binary blend (ϕ_3 =0) and the ternary blend with ϕ_3 = 13 wt% under no field. In the case of the binary blend (Fig.6 (a)), UPPG droplets are dispersed in a PDMS (1000cSt) matrix. In the case of the ternary blend (Fig. 6(b)), on the other hand, the domain structure becomes rather complicated. Although PDMS (1000cSt) and PDMS (10cSt) are miscible, it seems that they segregate with each other under the existence of UPPG. In Fig. 6 (b), the left part mainly consists of UPPG and PDMS (1000cSt), while the right part mainly consists of PDMS (10cSt). This was confirmed by taking each part by a syringe and examining them by a microscope. In the left part, local fraction of UPPG is so high that UPPG becomes the matrix and PDMS becomes the droplets.

From this observation, we can conjecture the origin of the short recovery time $T_{1/2}$ in the ternary blend (ϕ_3 = 13 wt%) as shown in Fig. 7. As observed in Fig. 6 (b), the lowest viscosity component, PDMS (10cSt), segregates from the other components in the ternary blends. Under shear flow, we expect that this low-viscosity component forms a thin layer as shown in Fig. 7 (a). Under the electric field, the domain structure would be almost the same as in the binary blend. After turning off the field, however, the layer of low viscosity component would quickly form again, resulting in the fast decay of the viscosity.

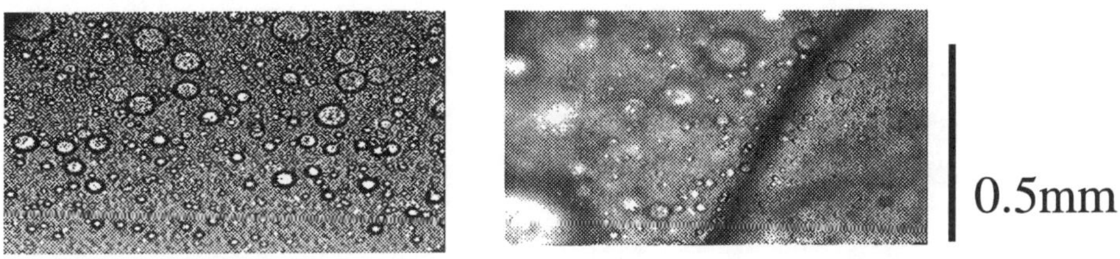

(a) (b)

Fig.6 Domain structures of (a): UPPG / PDMS (1000cSt)., and (b): UPPG / PDMS (1000cSt) / PDMS (10cSt)

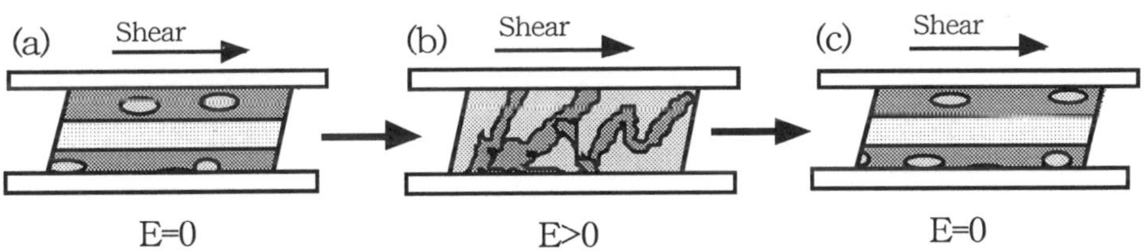

Fig.7 Mechanism of the fast decay in the ternary blend (ϕ_3 ~ 13 wt%).

3.3 Conclusion

Ternary blends have ER effect comparable to binary blends, and the recovery time of ternary blend (ϕ_3 ~ 13 wt%) is shorter than that of the binary blend.

Reference

1. W. M. Winslow, *J. Appl. Phys*. **205**, (1949) 1137.
2. K. Tajiri, M. Sugaya, H. Orihara, T. Nagaya, Y. Ishibashi, M. Doi and A. Inoue, *J. Rheol.*, **41**(2), (1997) 335.
3. H. Kimura, K. Aikawa, Y. Masubuchi, J. Takimoto, K. Koyama, and T. Uemura, *J. Non-Newtonian Fluid Mech.*, **76**, (1998) 199.

ELECTRIC-FIELD RESPONSE OF THE STRUCTURE AND RHEOLOGICAL PROPERTY OF SILICONE/POLYETHER BLENDS

KEIJI MINAGAWA, KOICHI SAITOH, HIROKAZU OKAMURA, TAKESHI MORI, SEIZO MASUDA
Department of Chemical Science and Technology, The University of Tokushima, Tokushima 770-8506, Japan

MASAMI TANAKA
Faculty of Pharmaceutical Science, Tokushima Bunri University, Tokushima 770-8514, Japan

The ER effects were studied for various immiscible blends composed of urethane-modified polyethers and silicones. Some polyethers and silicones were prepared from polypropylene glycol (PPG) and modified poly(dimethyl siloxane) having terminal hydroxyl groups, respectively. The influence of blend composition on the ER effects were examined for some binary and ternary blends based on these materials. The ER effects of these blends varied with the modification of polyethers and silicones, and also with the blend composition. In particular, the response of viscosity on applying or removing an electric field was found to vary with the blend composition, which should reflect some time-dependent change in the phase structure during the application and removal of the electric field. The variety of field response are discussed on the basis of the miscibility of the blend components.

1. Introduction

The electrorheological (ER) fluids composed of two immiscible fluids are interesting materials showing some different feature from usual ER suspensions.[1,2] The ER effect of the immiscible ER fluids appears as an increase (or a decrease in some cases) of apparent viscosity without obvious yield stress. This property can be useful in some kinds of applications which do not require the Bingham type ER response characteristic for suspensions. The regulation of ER effects of immiscible blends are, however, rather difficult because an irreversible change in the phase structure can occur after applying/removing an electric field and/or shear deformation. It is necessary to control the dispersibility of the blend to obtain stable ER materials.

We have been studying ER effects of some polyether liquids modified with urethane groups,[3-5] and have found that these materials are applicable as homogeneous ER fluids and also as a component of immiscible ER fluids.[6,7] In this paper, we report the ER effects of some immiscible polyether/silicone blends, focusing on the influence of modified silicone as an additive for the blends.

2. Materials and Methods

2. 1. *Preparation of urethane derivatives of polyether and silicone*

Urethane-modified polyethers and silicones were prepared by addition of them with 4,4'-diphenylmethane diisocyanate (MDI). The preparation schemes are shown in Fig. 1. For modification of polyether, propylene glycol (PPG; Mw = 3000) and MDI were added at the molar ratio of PPG/MDI = 1/2. The resulting urethane prepolymer having isocyanate groups at its terminal was further reacted with ethyl lactate to form terminal modification with the polar group. Similarly dimethyl silicone with a reactive terminal hydroxyl group (Shin-Etsu Chemical, X-22-170B) was added with MDI at the molar ratio of DMS/MDI = 2/1; in this case, no further modification was necessary because of the absence of unreacted isocyanate groups after this reaction. The resulting compounds have urethane bonds between the main-chain and aromatic groups.

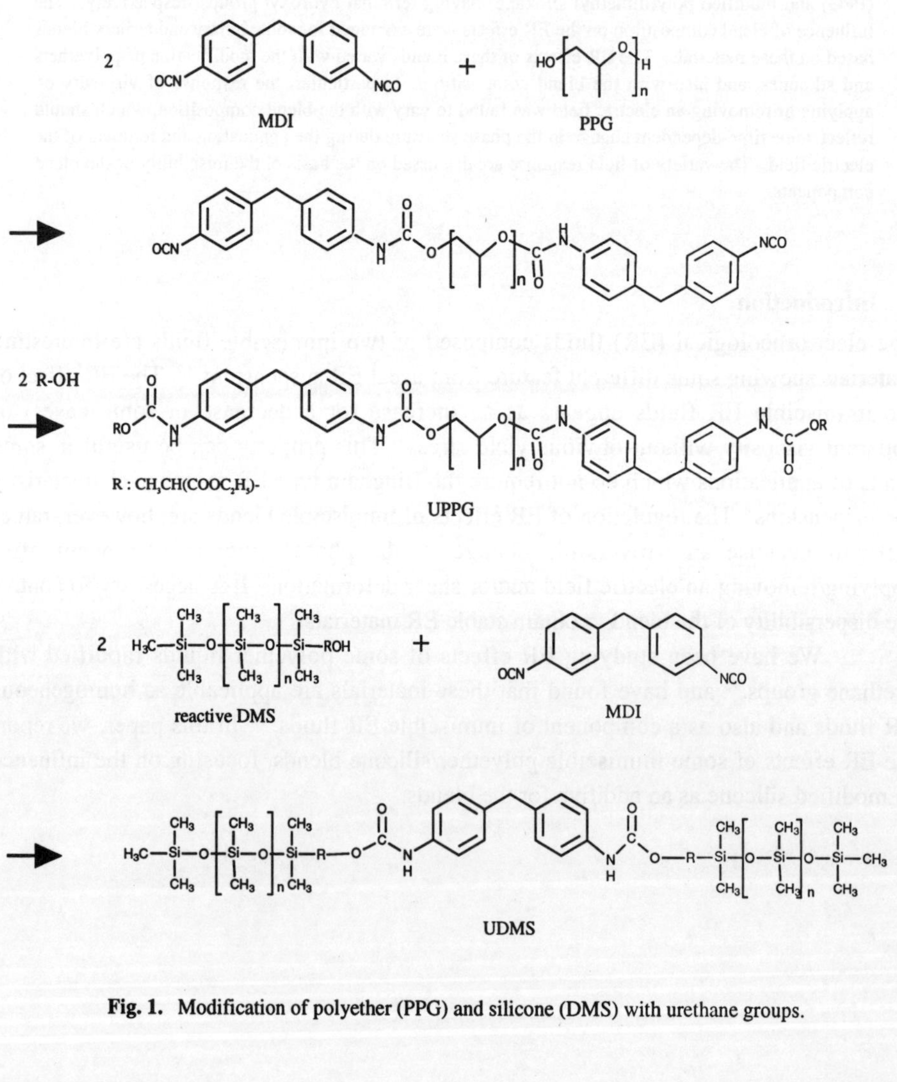

Fig. 1. Modification of polyether (PPG) and silicone (DMS) with urethane groups.

2. 2. *Preparation of ternary immiscible blends*

Ternary immiscible blend was prepared by mixing the modified polyether (UPPG), modified silicone (UDMS), and dimethyl silicone (DMS; Shin-Etsu Chemical, KF-96-1000cSt), at the weight ratio of UPPG/DMS/UDMS = 10/9/1. For comparison, binary blend of UPPG and DMS was also prepared (UPPG/DMS ratio = 10/10).

2. 3. *Measurements*

The ER measurements were carried out by use of a rotating parallel-disc rheometer (Rheometric ARES) under a DC electric field. The gap between the parallel discs was 0.5 mm. The shear viscosity was measured at various shear rates ranging between 5 and 100 s^{-1}.

3. Results and Discussion

3. 1. *Shear viscosity*

The viscosity under various field strength ranging from 0 to 2 kV/mm was plotted against shear rate in Figures 2 and 3, which represent the results of UPPG/DMS(10/10) and UPPG/DMS/UDMS(10/9/1), respectively. In the both fluids, the viscosity increased in the presence of an electric field at low shear rate region (positive ER effect). The increase of viscosity for the binary UPPG/DMS fluid (Fig. 2) was slightly larger than that of the ternary UPPG/DMS/UDMS fluid (Fig. 3), particularly at rather high shear rate above 50 s^{-1}. In the UPPG/DMS(10/10) immiscible blend, the UPPG phase forms droplets in the DMS matrix, and the droplets can fuse with one another to form bridging structure under an electric field. The decrease of ER effect at higher shear rate obviously seen in the UPPG/DMS/UDMS ternary fluid indicates that the bridges formed under the electric field can be easily destroyed by shear flow, comparing to the binary fluid containing no UDMS. This behavior is interpreted by assuming that the UDMS added to the UPPG/DMS blend acts as a surfactant and stabilizes the dispersion of UPPG droplet in DMS matrix.

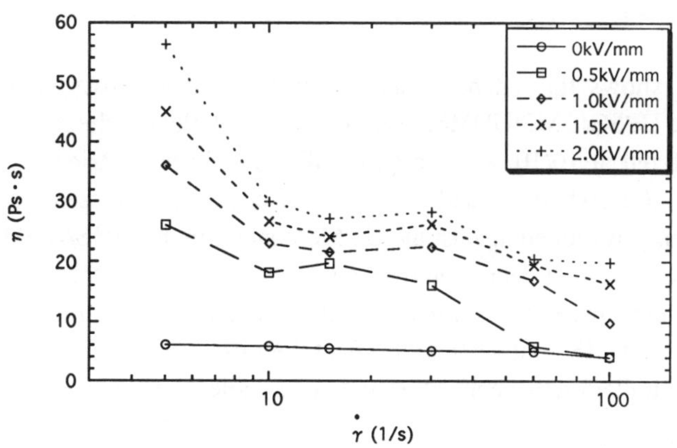

Fig. 2. Shear viscosity of UPPG/DMS(10/10) blend under various electric fields.

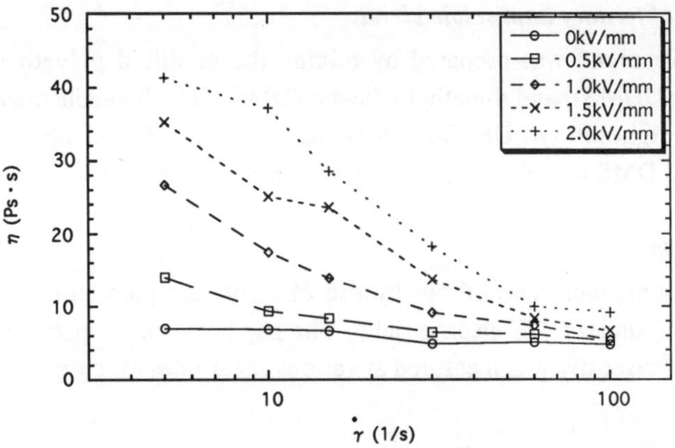

Fig. 3. Shear viscosity of UPPG/DMS/UDMS(10/9/1) blend under various electric fields.

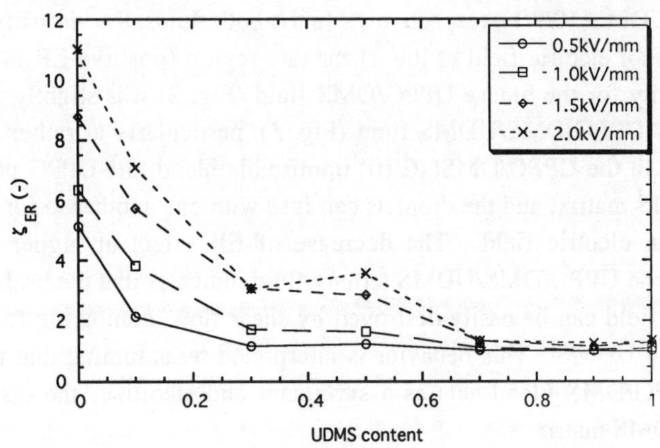

Fig. 4. Composition dependence of the ER effect for UPPG/DMS/UDMS ternary blends.

Figure 4 shows the dependence of the induced viscosity change on the composition of the UPPG/DMS/UDMS. Here the ratio of UPPG to the mixture of DMS and UDMS was fixed at 10/10, and the DMS/UDMS ratio was varied. The relative viscosity $\zeta_{ER} = \eta_E / \eta_0$, where η_E and η_0 are the viscosity in the presence and absence of the electric field, was plotted against the UDMS content in the DMS/UDMS mixture. It is found that the ER effect decreased with an increase of the UDMS content. Since DMS and UDMS used here have similar viscosity under no electric field, the reduced ER effect at higher UDMS content should reflect the miscibility of UPPG and UDMS, both of which has similar urethane moieties within the molecular chain.

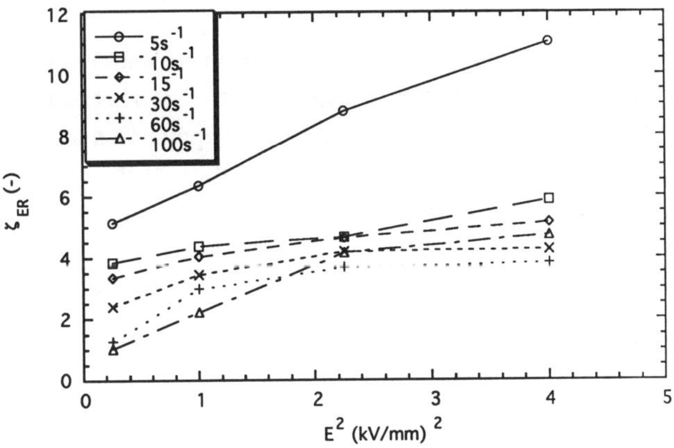

Fig. 5. Dependence of the induced viscosity change on the strength of the square of applied electric field for UPPG/DMS(10/10) blend.

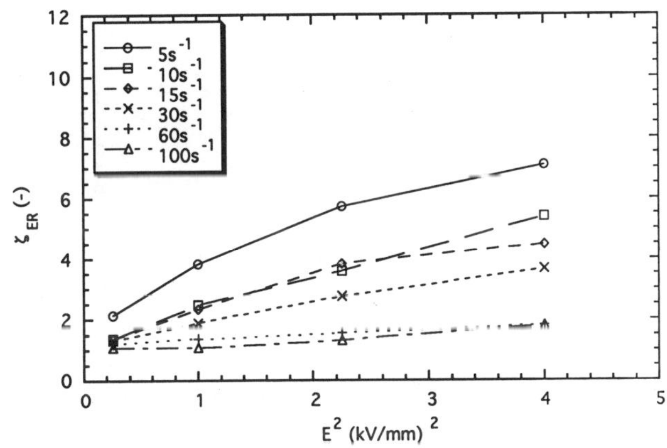

Fig. 6. Dependence of the induced viscosity change on the strength of the square of applied electric field for UPPG/DMS/UDMS(10/9/1) blend.

Figures 5 and 6 show the dependence of the relative viscosity ζ_{ER} on the strength of the square of applied electric field for binary UPPG/DMS(10/10) blend and UPPG/DMS/UDMS(10/9/1) blend, respectively. The ER effect, shown as the relative increase of viscosity from the no electric field, increased with the increase of the field strength. The dependence on the square of the field strength deviated from linearity for the both fluids, which could be due to the complicated nature, i.e. deformation, aggregation, fusion, collapse, etc., of UPPG droplets in DMS matrix under the electric and shear flow.

3. 2. *Response time*

Figures 7 and 8 show the time change in the viscosity of binary and ternary immiscible fluids, respectively, during the successive measurements. The shear rate was 10 s^{-1}. From these figures, the response of shear viscosity to the application and removing of the electric field can be compared clearly. Upon application of the electric field, the viscosity of the binary blend increased with time, and reached a steady value at around 20 seconds after the application of the field. On the other hand, in the case of the ternary blend, more time was necessary for reaching a constant viscosity under the field. The initial stage of the response to the electric field was similar to that of the binary blend, i.e., relatively rapid increase of viscosity within a few seconds was observed. In the following measurement, the viscosity still increased gradually, and reached a constant value after 200 seconds from the field application.

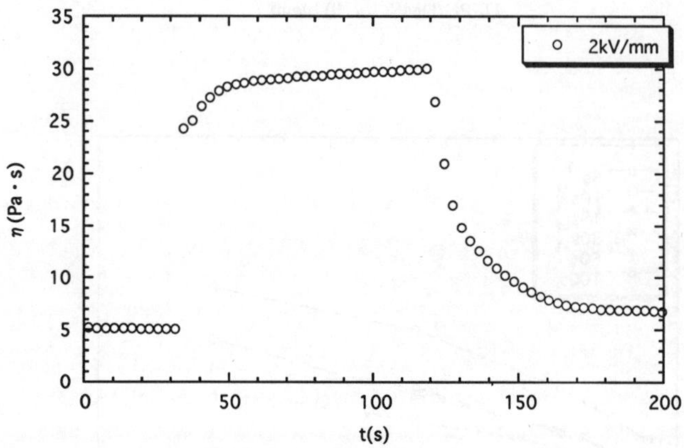

Fig. 7. Response of shear viscosity to a 2 kV/mm electric field for UPPG/DMS(10/10) blend.

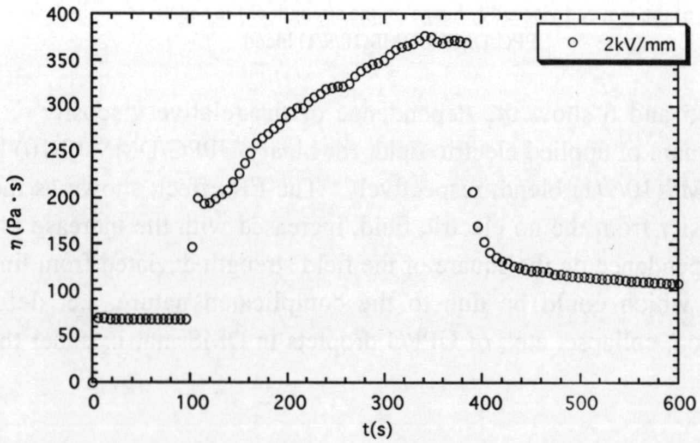

Fig. 8. Response of shear viscosity to a 2 kV/mm electric field for UPPG/DMS/UDMS(10/9/1) blend.

This result indicates that the structure of the UPPG/DMS blend is influenced by the UDMS, which has both urethane groups and siloxane chains within the molecule. The affinity among the urethane moieties in the polyether and silicone components could play important role for the stabilization of the dispersion.

In conclusion, urethane-modified silicone acts as a surfactant of the liquid/liquid immiscible ER blend composed of urethane-modified polyether and silicone oil. Though the ER effect was decreased by the addition of UDMS, the stability and redispersibility of the polyether droplet may be improved because of the similar urethane structure of the modified polyether and silicone.

References

1. K. Tajiri, M. Sugaya, H. Orihara, T. Nagaya, Y. Ishibashi, M. Doi, and A. Inoue, *J. Rheol.* **41**(2), 335 (1997).
2. H. Kimura, K. Aikawa, Y. Masubuchi, J. Takimoto, K. Koyama, and K. Minagawa, *Rheol. Acta* **37**, 54 (1998).
3. T. Uemura, K. Minagawa, and K. Koyama, *Chem. Lett.* 1995 (1994).
4. T. Uemura, K. Minagawa, J. Takimoto, and K. Koyama, *J. Chem. Soc. Faraday Trans.* **91**(6), 1051 (1995).
5. K. Minagawa, H. Okamura, S. Masuda, M. Tanaka, N. Gohko, Y. Masubuchi, J. Takimoto, K. Koyama, and T. Uemura, *Proc. Int. Conf. on ERF, MRS and their Applications,* eds. M. Nakano and K. Koyama (World Scientific, Singapore, 1998), p. 416.
6. H. Kimura, K. Aikawa, Y. Masubuchi, J. Takimoto, K. Koyama, and T. Uemura, *J. Non-Newt. Fluid Mech.* **76**, 199 (1998).
7. K. Minagawa, S. Masuda, N. Gohko, K. Koyama, and M. Tanaka, in *Proc. 7th Int. Conf. on Electrorheological Fluids, Magnetorheological Suspensions,* ed. R. Tao (World Scientific, Singapore, 2000), p. 95.

NOVEL ELECTRORHEOLOGICAL MATERIALS WITH TERNARY BLENDS CONTAINING POLYETHERS AND URETHANES

KEIJI MINAGAWA, TETSUO KANNO, HIROKAZU OKAMURA, TAKESHI MORI, SEIZO MASUDA
Department of Chemical Science and Technology, The University of Tokushima, Tokushima 770-8506, Japan

HIDENAO TAKAGI, KIYOHITO KOYAMA
Department of Polymer Science and Engineering, Yamagata University, Yonezawa 992-8510, Japan

MASAMI TANAKA
Faculty of Pharmaceutical Science, Tokushima Bunri University, Tokushima 770-8514, Japan

Ternary ER blends composed of two different liquids and solid dispersoid were prepared by blending silicone oil, viscous polyether liquid, and urethane powder at various compositions. The polyether liquid used here was polytetramethylene glycol modified with phenyl isocyanate, and the powdér was urethane derivative of 4,4'-diphenylmethane diisocyanate. Under certain conditions, some of the ternary blends showed larger increase of viscosity under an electric field than those of binary mixtures prepared from each component of the ternary one. In this case, the viscous polyether and urethane powder act cooperatively by forming some particular structures within the blends. Microscopic observation revealed that each urethane particle was covered with polyether liquid in the absence of electric field, and that the binary droplets aggregated to form viscous bridges between the electrodes keeping powders within them.

1. Introduction

Typical electrorheological (ER) materials are inhomogeneous fluids such as suspensions containing solid particles and immiscible liquid/liquid (or liquid crystal) blends, and their apparent viscosity usually increases with the application of an electric field.[1] The increase in the apparent viscosity is caused by certain structural changes of the dispersoid, e.g. formation of clusters bridging the electrodes. Besides these suspensions and immiscible blends, terminal-modified polyether liquids with urethane groups can also be used as homogeneous ER fluids.[2-5] The polyether-based homogeneous ER fluids show an increase or a decrease of apparent viscosity in an electric field depending on the molecular factors such as terminal groups, main-chain structure, chain length, and so on. In particular, the decrease of viscosity, or negative ER effect, is an interesting property of some of these polyether liquids. It has been found that the ER effect of the polyethers can be changed further with some additives such as organic compounds, and that liquid/liquid immiscible ER fluids can also be prepared by mixing with silicone oil. Combination of these materials may give new multi-phase ER fluids, which could be useful for improvement of conventional ER materials.[6] In this paper, we report modification of immiscible polyether/silicone blends, by adding urethane powders as a solid phase in one of the liquid phases.

2. Materials and Methods

2. 1. *Preparation of polyether liquids and urethane powders*

Commercially available polyether and isocyanates were used without further purification. The urethane-modified polyether was prepared by addition of polytetramethylene glycol (MW=650) with phenyl isocyanate (PhI), in a similar manner reported previously.[4] The resulting polyether has urethane bonds between the main-chain and terminal aromatic groups. Urethane powder was similarly prepared from 4,4'-diphenylmethane diisocyanate (MDI) and ethylene cyanohydrin by the addition reaction. The preparation schemes of the samples are shown in Fig. 1.

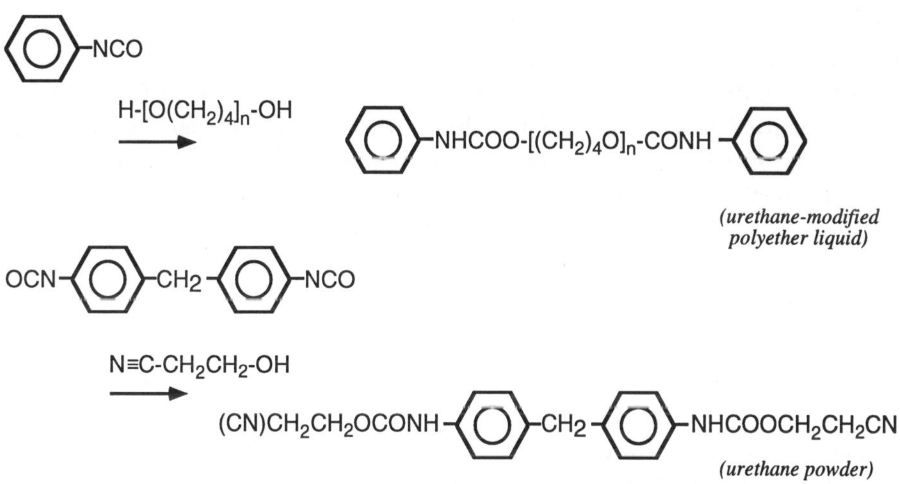

Fig. 1. Preparation scheme for urethane compounds.

2. 2. *Viscosity and electric properties of components*

The viscosity and electric properties of each component were measured prior to preparation of the blend samples. The viscosity was measured with a rotating parallel-disc rheometer (Rheometric ARES) at the constant shear rate of $1.0 s^{-1}$ at 25°C. The dielectric constant was measured with an LCR meter (HP4284A), and the conductivity was evaluated from the dielectric constant by using the following equation;

$$\chi = 2\pi f \varepsilon'' \varepsilon_0 \tag{1}$$

where f is the electric frequency, and ε_0 is the dielectric constant in vacuum, 8.854185

$\times 10^{12}$F/m. Table 1 shows the values determined.

Table 1. Viscosity, dielectric constant and conductivity of each component of the ternary ER fluid

	η (poise)	ε	χ ($10^9/\Omega$ m)
Polyether	8.5×10^2	13.2	10.3
Urethane powder	-	10.0	3.7
DMS	9.0	2.8	1.8×10^{-2}

2. 3. *Preparation of blends*

Binary and ternary blends were prepared by mixing the modified polyethers, urethane powders, and dimethyl silicone (DMS; 500cSt). The total content of the dispersoids, i.e. urethane powder and polyether liquid, were fixed at 10, 20, and 30wt%, and the ratio of these dispersoids was varied.

2. 4. *Measurements and observation*

The ER measurements were carried out by use of a rotating parallel-disc rheometer (Rheometric ARES) in the absence and presence of 2kV/mm DC electric field. The gap between the parallel discs was 0.5 mm. Shear viscosity was measured at various shear rates ranging between 0.05 and 200 s^{-1}. Observation of the structure was carried out by using an optical microscope equipped with a rotating disc and a ring outside the disc, both of which act as electrodes.[7]

3. Results and Discussion

3. 1. *Shear viscosity*

The apparent viscosity of the binary and ternary blends increased in response to an electric field, depending on the composition. The positive ER effect was relatively large when the blends contain 20wt% and 30wt% of dispersoid. On the other hand, in the case of blends with 10wt% dispersoid, a characteristic dependence of the ER effect on the composition was observed. Here we focus on the behavior of the 'synergistic' effect of the urethane powder and the polyether liquid.

Figure 2 shows an electrically induced viscosity change for a binary blend composed of the urethane powder and 500cSt silicone oil. The concentration of the urethane powder was 10wt%. In Fig. 3, similar plots are shown for a ternary blend containing urethane powder (7wt%) and polyether (3wt%) in 500cSt silicone oil. It is found from these figures that the viscosity increased at low shear rate, and that the increase (positive ER effect) of the ternary blend is larger than that of the binary one under the similar conditions. This result indicates the existence of some synergistic effect of the two different dispersoids, due to the change in their combined structure under an electric field and shear flow.

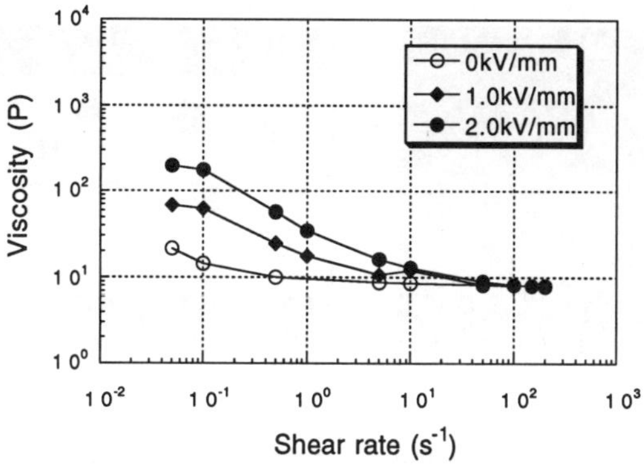

Fig. 2. Apparent viscosity change of a binary suspension composed of urethane powder and silicone oil in the presence of an electric field.

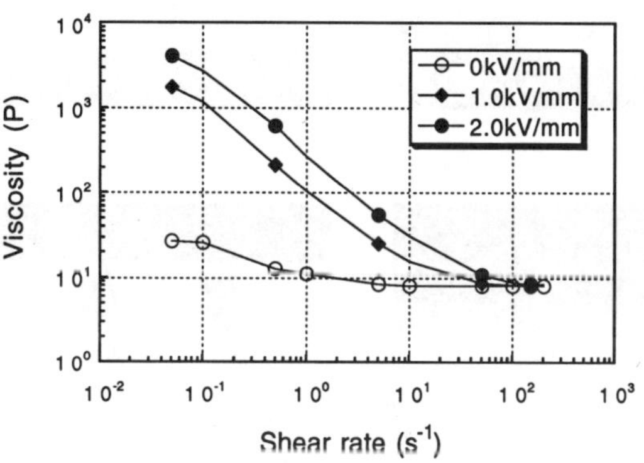

Fig. 3. Apparent viscosity change of a ternary suspension composed of urethane powder, modified polyether and silicone oil in the presence of an electric field.

The ER effects of ER fluids containing particles and liquid drops were reported by Chin and Park,[6] describing that the structure formed by the particles and drops depend on the conductivity of the liquid drops and the continuous phase. When the drops have higher conductivity than that of continuous phase, the drops form bridging structure between the electrodes, while the continuous phase is more conductive than the drops, deposition of drops occurs at one of the electrodes. The particles make bridging structure in the both cases, as they have higher conductivity than the drops and the continuous phase. In our ternary blends containing urethane powders and polyether drops, the conductivity of the liquid drop is higher than those of the particles (powders) and the continuous phase (silicone oil). It is important to clarify the structure under an electric field, which affects the increase of viscosity. In the next section we show the result of direct observation of these blends under an electric field and shear flow.

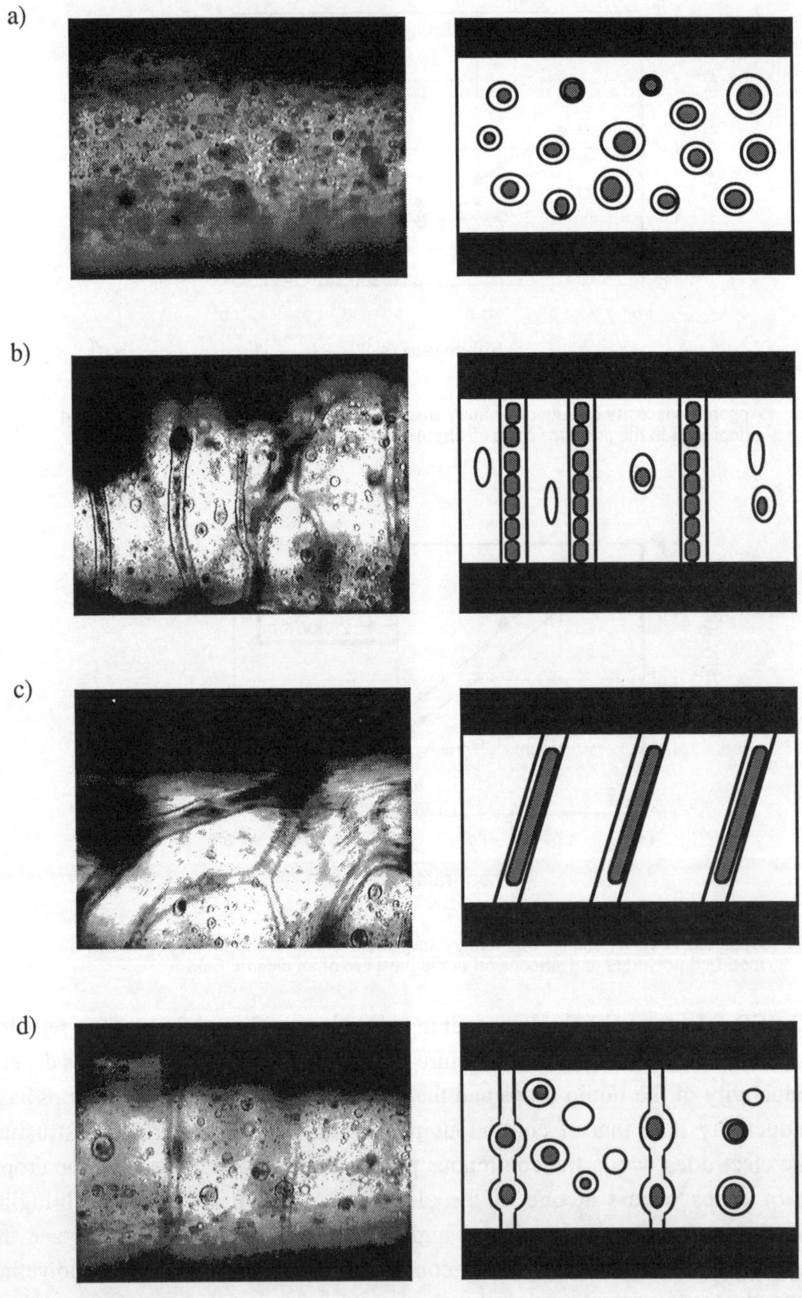

Fig. 4. Photographs and schematic illustrations showing the changes of the structure of a ternary ERF. (a) E = 0 kV/mm, $\bar{\gamma} = 0$ s^{-1}; (b) E = 2 kV/mm, $\bar{\gamma} = 0$ s^{-1}; (c) E = 2 kV/mm, $\bar{\gamma} = 5$ s^{-1}; (d) E = 0 kV/mm, $\bar{\gamma} = 0$ s^{-1}.

3. 2. *Observation of the structure*

The structure of the ternary blend was observed with an optical microscope in the absence and presence of an electric field and shear flow. The photographs and schematic illustrations showing the changes of the structure of these suspensions are shown in Fig. 4. In no electric field and no shear flow, the urethane powders dispersed in viscous polyether liquid to form droplets containing particles (Fig. 4a). Under an electric field, the droplets aggregates and form many bridges between the two electrodes (Fig. 4b). The blend was then subjected to a shear flow, resulting in a repeated deformation and recombination of the liquid bridges, where the particles were still kept within the bridges (Fig. 4c). After the application of the electric field and shear flow, the drops and particles were redispersed in the medium forming a similar structure to the initial state (Fig. 4d). These observations show that the urethane powders are kept in the polyether phase in the both cases that the polyether form droplets or bridged structures. The stable structure would be caused by the affinity of the powders and the liquid drops, both of which have urethane bonds contributing the large interactions between particles and also between particle and drops. The synergistic effect on the viscosity change is interpreted as the reinforcement of liquid bridges by the powders.

In conclusion, the introduction of solid core to the liquid/liquid immiscible blend was found to be effective for the increase of viscosity under the electric field. Although the ternary fluid presented here show relatively small ER effect, the utilization of particles to various liquid/liquid immiscible ER fluids may be useful for enhancement of the strength of the brides, causing improved ER effect.

References

1. *For example, Proc. 7th Int. Conf. on Electrorheological Fluids and Magnetorheological Suspensions,* ed. R. Tao (World Scientific, Singapore, 2000); *Proc. 6th Int. Conf. on Electrorheological Fluids, Magnetorheological Suspensions and their Applications,* eds. M. Nakano and K. Koyama (World Scientific, Singapore, 1998).
2. T. Uemura, K. Minagawa, and K. Koyama, *Chem. Lett.* 1995 (1994).
3. T. Uemura, K. Minagawa, J. Takimoto, and K. Koyama, *J. Chem. Soc. Faraday Trans.* **91**(6), 1051 (1995).
4. K. Minagawa, H. Okamura, S. Masuda, M. Tanaka, N. Gohko, Y. Masubuchi, J. Takimoto, K. Koyama, and T. Uemura, in *Proc. 6th Int. Conf. on Electrorheological Fluids, Magnetorheological Suspensions and their Applications,* eds. M. Nakano and K. Koyama (World Scientific, Singapore, 1998), p. 416.
5. K. Minagawa, H. Okamura, S. Masuda, and M. Tanaka, in *Proc. 6th Int. Conf. on Electrorheological Fluids, Magnetorheological Suspensions and their Applications,* eds. M. Nakano and K. Koyama (World Scientific, Singapore, 1998), p. 393.
6. B. D. Chin and O. O. Park, *J. Rheol.,* **44**(2), 397 (2000).
7. H. Kimura, K. Aikawa, Y. Masubuchi, J. Takimoto, and K. Koyama, in *Proc. 6th Int. Conf. on Electrorheological Fluids, Magnetorheological Suspensions and their Applications,* eds. M. Nakano and K. Koyama (World Scientific, Singapore, 1998), p. 436.

ER RESPONSE OF SILICONE GEL CONTAINING DIELECTRIC PARTICLES

Jun Koide, Tetsu Mitsumata, Yuichi Masubuchi, Jun-ichi Takimoto, and Kiyohito Koyama

Department of Material Science and Engineering, Yamagata University, 4-3-16 Jounan
E-mail: koide@ckpss.yz.yamagata-u.ac.jp

In order to investigate relationship electro-rheological (ER) response and stiffness of matrix gel (ex. silicone gel) we synthesized two types of silicone gel containing dielectric particles as a sample and measured modulus of these samples. We suggested ER response does not depend on modulus of matrix gel but structures in matrix.

Introduction

Electro-rheological(ER) fluids were classified into three types; dispersion systems, homogeneous systems, and immiscible blend systems. Previous studies of ER fluid suggest that dispersion system show largest ER effect although it has the problems such as sedimentation and aggregation of dispersed particles with time. A new concept to overcome this problem is ER gel composed of dispersed particles and gel instead of polydimethylsiloxane (PDMS). T. Shiga et al. reported rheological properties of ER gel, and found that ER effect is about 40 times smaller than that of normal dispersed systems. [1,2] They investigated concentration dependence of dispersed particle on rheological properties, and ER effect increased up to dispersed particle concentration of 48wt%, however, the smallness of ER effect should not be negligible. Moreover, they have not investigated the effect of stiffness of matrix component on rheological properties. The small number of reports on ER gel makes it difficult to understand the nature and therefore the enhancement of ER effect. The purpose of this study is to investigate the effect of stiffness of matrix gel on rheological properties under electric field. The main story and its procedure are as follows: At first, we synthesized a ER gel, however, it was impossible to measure rheological properties, instead, Young's modulus after cutting electric field was measured. Next, we synthesized ER gel having different stiffness of matrix, which made it possible to measure by using commercial shear rheometer.

Experimental

At first, we synthesized a ER gel composed of PDMS/TEOS/suspension ER fluid. That is, PDMS is simple silicone oil which has 5Pa.s at room temperature, and TEOS (tetraethoxysilane; Kanto Chemical Co., Ltd.) is a crosslinker, and suspension ER fluid is composed of carbon black particle (particle size: 10~20 m; Bridgestrone Co., Ltd.) and PDMS (about 1Pa.s at room temperature). These three components were synthesized through process of hydrolysis and condensation at 50℃. Young's modulus is calculated from the equation as follows;

$$K = \rho \,.\, V^2 \qquad (1)$$

Where ρ is density, K is Young's modulus, and V is sound speed. ρ was measured by using pycnometer, and sound speed was measured by ultrasound .Figure 1 shows a equipment to measure V consists of brass and transducer(MLB 35ELKB-AEROTECH Co., Ltd.), and the sample after cutting electric field off was between them.

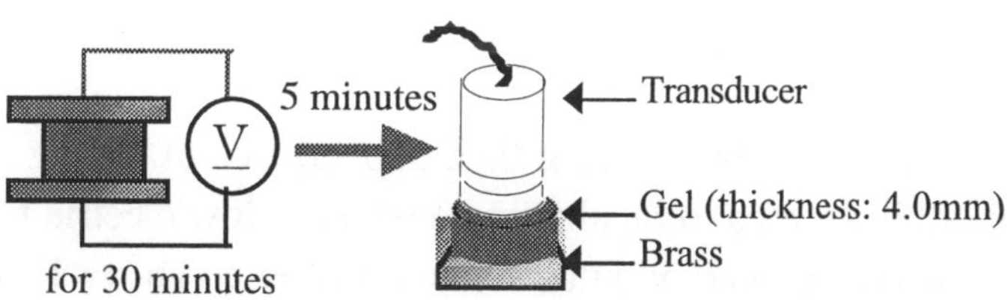

Fig.1 Schematic of measurement by ultrasound

For the second experiment, we synthesized other ER gels by using two kinds of polysiloxane and $BaTiO_3$ particles (size: 1~1.4 m, Kanto Chemical Co., Ltd.) . The chemical structures of two kinds of polysiloxane are partially different as shown in Figure 2. and composition of two samples are shown in Table 1. These three samples were mixed for 30min, and then these were made thermoset at 140℃ for 5 min. The resulting samples had the thickness of 1.3mm and the weight of 2.1g. Time dependence of small strain oscillatory in shear was carried out by using a commercialized rotational shear rheometer(Rheology Co., Ltd.). The geometry was parallel disks whose gap has 1.2mm. Strain amplitude was 0.01, angular frequency was 10Hz, and the electric filed was DC 2.5kV/mm were selected as a measurement condition.

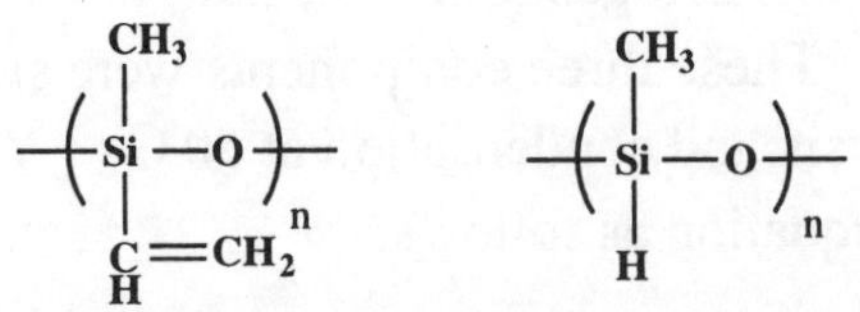

Methylvinylpolysiloxane Dimetylhydrogensiloxan

Fig.2 Chemical Structures of Polysiloxane

Table.1 Compositions of silicone/$BaTiO_3$ gel (wt%)

	A	B	$BaTiO_3$
Sample1	72.0	0.80	20.0
Sample2	72.7	0.73	20.0

Results and Discussion

Figure 4 shows electric filed dependence of Young modulus. Young modulus seems to be almost constant below electric field of 1.5kV/mm, however, Young modulus steeply increased above 1.5kV/mm up to 3.0kV/mm. The rate of increase of Young modulus was 82.5%. It will be speculated that structure of silicone gel is heterogeneous, which may allow us to interpret that the dispersed particles moves to the spatially nonconfined zones. That is, homogenization of system will leads to increase of density as a result, increase of Young modulus since it is proportional to density according to equation (1). Next, we will discuss another experimental results.

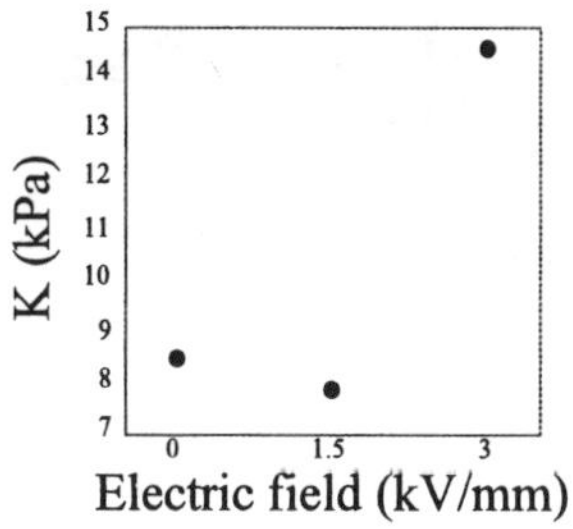

Fig.3 Electric field dependence of Young's Modulus

Figure 4 shows time dependence of storage modulus, G', for two different stiffness of samples. Before imposing strain below 500s, G' of each sample is constant, and the value of G' for sample-2 is apparently large, that is, stiffness of sample-2 is larger than that of sample-1. After 500s, G' of each sample increased with time and takes overshoot and then seems to approach to constant value. For both samples, reduced increase fraction is defined as follows;

$$\xi_{ER}=(G'_1-G'_0)/G' \quad (2)$$

Where G'_0 is storage modulus before imposing strain, and G'_1 is storage modulus after imposing strain at arbitrary time. Reduced increase of two samples was about 15~20wt%, and was quite similar each other. However, increase fraction of G', that is, $G'_1-G'_0$, of sample-2 was larger than that of sample-1. We expected that lower stiffness of matrix would lead to large increase fraction, however, Figure 4 showed opposite result. This result might suggest that mobility of dispersed particles in matrix is not important factor. The specific mechanism may exist, thus, this point needed to be studied further by synthesizing a series of ER gel which has different stiffness of matrix.

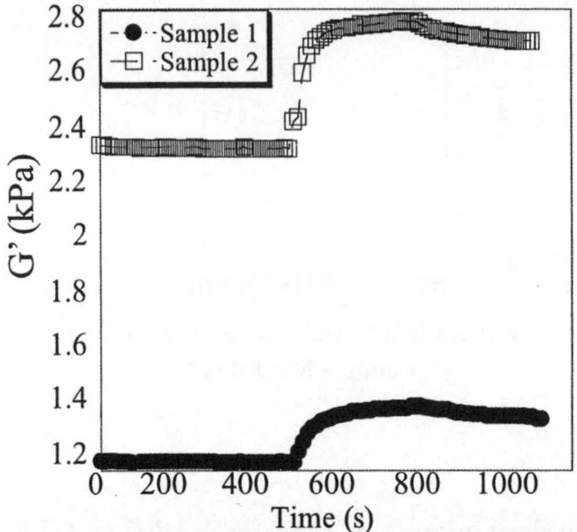

Fig.5 ER response of Silicone/ $BaTiO_3$ gel

Conclusion

The present study provided Stiffness of matrix does not depend on ξ_{ER}. In short, the density of crosslinking point of matrix does not depend ξ_{ER}. In order to improve ξ_{ER} we need to synthesize new gel having heterogeneous structures.

References

1. T. Shiga, T.Ohta , A. Okada, and T. Kurauchi. *J.Mater. Sci* **28** (1993)pp.1293-1299
2. T. Shiga, A. Okada, and T. Kurauchi. *Macromolelcules*. **26** (1993)pp.6958-6963

SYNTHESIS AND ELECTRORHEOLOGY OF PHOSPHATE CELLULOSE SUSPENSIONS

J. W. KIM, Y. H. CHO AND H. J. CHOI

Department of Polymer Science and Engineering., Inha University, Inchon, 402-751, Korea
E-mail: hjchoi@inha.ac.kr

M. S. JHON

Department of Chemical Engineering, Carnegie Mellon University, Pittsburgh, PA 15213-3890, USA

As a candidate for the development in electrorheological (ER) materials, phosphate cellulose (PC) particles were synthesized from the phosphate–ester reaction between cellulose particles and ammonium phosphate solution prepared from six different molar concentrations of phosphoric acid. The synthesized particles were then dispersed in silicone oil to prepare the PC- based anhydrous ER fluid. The effects of electric field strength and particle concentration on the ER performance were investigated with both steady shear and dynamic oscillatory shear modes. The ER system with PC particles formed with a 2 M phosphoric acid solution exhibited the highest yield stress. A difference in the flow behavior of the PC-based ER fluids was also analyzed via the dielectric spectra using an impedance analyzer. We adapted the PC-based ER fluids to the the semi-active ER damper and then examined their damping characteristics.

1 Introduction

Electrorheological (ER) fluids are ones whose rheological properties depend strongly and reversibly on an applied electric field [1-3]. In general, ER fluids solidify in milliseconds under the applied electric field, and they fluidize under applied deformation that destroys the particle chain structure. Because of their fast dynamic structural response and their tunable viscosity, ER fluids are used as a smart material for active devices that transform electrical energy to mechanical energy, such as actuators, torque transducers, and dampers [4].

Most ER particles, especially wet-base systems, require the addition of small amount of water or other polar additives, such as surfactant, to produce the desirable ER performance. In wet-base systems with hydrophilic particles, including mesoporous molecular sieve [5] and cellulose [6], the particle chain structure develops by the migration of ions in the absorbed water. However, hydrous ER fluids possess several drawbacks in practical engineering applications due to the presence of water. Various anhydrous ER fluids have therefore been developed to overcome these disadvantages. Examples include zeolite [7,8], annealed diphenyldiacetylene [9], and several semiconducting polymers, such as polyaniline [10-12], copolyaniline [13-15], chitosan [16], poly(acene quinone) radicals [17,18], polyphenylenediamine [19], poly(p-phenylene) [20,21] and polymer-clay nanocomposites [22-24].

These materials possess branched polar groups such as amine ($-NH_2$), hydroxy (-OH), and amino-cyan (-NHCN). The polar groups may affect the ER response by playing the role of the electronic donor/accepter in the presence of an electric field. The chemical structure of the organic materials, is therefore an important factor in the determination of ER effect. The difference between dry-base and wet-base systems is due to the type of carrier species involved in the particle polarization; the particle chain structure formation is induced by the migration of ions in the absorbed water for wet-base ER fluids, while the electrons move inside the molecules of the particles for dry-base ER fluids.

In this paper, we investigated phosphate cellulose (PC) particles as a candidate for high performance dry-base ER systems. We investigated the effect of particle concentration and electric field strength via both steady and dynamic shear tests. Ahn *et al.* [25] recently examined the ER behavior of PC suspensions and found that current density and conductivity are controlled by the phosphoric acid concentration in phosphate-ester reactions with cellulose. Note that microcrystalline cellulose has been used as a wet-base ER material [6].

The effect of the phosphate-ester reaction on ER response was examined under several applied electric field strengths. From measurements in both controlled shear rate and shear stress modes, we found that an ER system with PC particles formed in a 2 M phosphoric acid solution exhibits the highest yield stress (or ER performance).

2 Experimental

PC particles were synthesized by a phosphate-ester reaction. Six different H_3PO_4 molar concentrations (0.5, 1.0, 1.5, 2.0, 2.5, and 3.0 M) of phosphoric acid (H_3PO_4) (orthophosphoric acid, 85 % assay) and 4 M of urea ($(NH_2)_2CO$) solution were prepared by stirring for one hour at 25 °C to make a homogeneous solution. Ammonium phosphate was generated from the reaction of phosphoric acid and urea. 40 g of cellulose particles with average particle size of 20 μm (Sigmacell, Type 20, Sigma Chemical Co., USA) were mixed with the 400 m*l* ammonium phosphate solution and maintained 48 hours for the phosphate ester reaction of cellulose as shown in Fig. 1. Note that all measurement were performed at 25 °C unless specified.

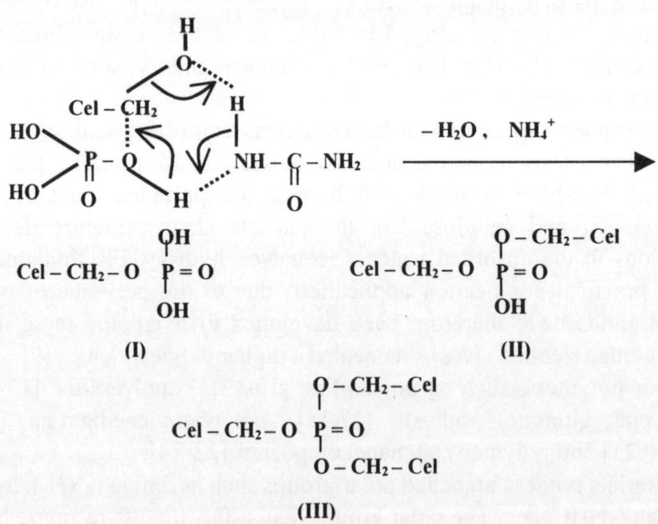

Figure 1. Esterification reaction steps of PC. Non-crosslinked monosubstituted (I), crosslinked disubstituted (II), and crosslinked trisubstituted (III) esters with cellulose.

A complex stabilized by a hydrogen bond between H_3PO_4, $(NH_2)_2CO$ and cellulose is assumed to be a transition state in this cellulose phosphorylation. However, the esterification of cellulose with an phosphoric acid-urea mixture involves a complex chemical process, such that H_3PO_4 can form. After completion of reaction, PC particles were washed several times with distilled water to remove unreacted $(NH_2)_2CO$ and H_3PO_4 that remained on the surface, and they were then filtered using an aspirator. After a drying procedure in a vacuum oven at 25 °C for 72 hours for complete removal of water, the PC particles were ground using a mixer and passed through a 38-μm sieve shaker to control the particle size and its distribution.

Conductivities of the PC particles treated with six different phosphoric acid solution molar concentrations were measured to be 1.5×10^{-10}, 2.7×10^{-10}, 6.9×10^{-10}, 8.7×10^{-10}, 1.4×10^{-9}, and 3.5×10^{-9} S/cm, by the 2-probe method, using a picoammeter (Keithley 487, USA) [26]. Conductivity obviously increased with phosphoric acid concentration. The densities of the PC were also measured, using a pycnometer, to be from 1.24 to 1.34 g/m*l*. They increased linearly with phosphoric acid concentration in PC, since a decrease of phosphate particle mass was generated with the increase of phosphoric acid concentration.

ER fluids were prepared by dispersing the PC particles in silicone oil (density of 0.96 g/m*l* and kinematic viscosity of 50 cS). A rotational rheometer (Physica MC 120,Stuttgart, Germany) with a Couette-type geometry, a high-voltage generator, and an oil bath for temperature control were used for rheological property measurements [27]. We also measured the dielectric spectra of the ER fluids using an impedance analyzer (Hewlett Packard 4284A, USA) with a liquid test fixture in the frequency range of 20 ~ 10^6 Hz.

Furthermore, a semi-active, cylindrical flow-mode type ER damper filled with the PC-based ER fluid was investigated under various electric field strengths. The ER damper is divided into the upper and lower chambers by the piston, and is filled completely with the ER fluid flowing through the duct between inner and outer cylinders from one chamber to the other. On the other hand, a gas chamber positioned at the lower parts acts as an accumulator of the ER fluid induced by the piston motion.

3 Results and Discussion

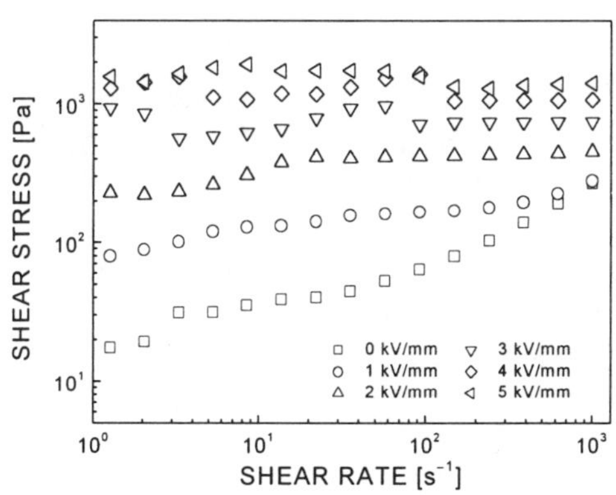

Figure 2 Flow curves for PC-based ER fluid with six different electric fields strengths (2 M of phosphoric acid).

Figure 2 shows the shear stress *vs.* shear rate curves of PC based ER fluid at six different electric field strengths. Dynamic yield stress, which is obtained by extrapolation from the constant stress region, increased with electric field strength.

The shear stress behavior of the ER fluid is generally divided into three regions. In the low shear rate region, the shear stress remained constant or decreased slightly with shear rate. In the intermediate shear rate region, the shear stress decreased with shear rate before chain breaks. Finally, in the high shear rate region, the shear stress increased with shear rate. This behavior is normally observed for the case of high performance ER fluids. However, for the case of PC-based ER fluid, the chain did not break at the high shear rate region.

To explore the difference in flow properties of ER fluids as shown in Fig. 2, we examined the polarization properties. We obtained the dielectric spectra of these ER fluids in the frequency range of 10 to 10^6 Hz using an impedance analyzer. Figures 3(a) and (b) show the permittivity (ε') and loss factor (ε'') as a function of frequency, respectively.

The dielectric relaxation spectra given in Fig. 3, can be fit to the following three parameter model [28]:

$$\varepsilon^* = \varepsilon' - i\varepsilon'' = \varepsilon_\infty + \frac{\varepsilon_0 - \varepsilon_\infty}{1 + (i\omega\tau)^{1-\alpha}} \qquad (0 \le \alpha < 1). \qquad (1)$$

Here, ε_0 and ε_∞ are limiting values of ε at $\omega \to 0$ & ∞, respectively. τ is the relaxation time for the interfacial polarization, and the exponent α determines the broadness of the relaxation spectra. Equation (1) reduces to a well-known single relaxation model (Debye dielectrics) for $\alpha = 0$ [29].

From Eq. (1), the relaxation times are calculated as 2.2, 2.0, and 2.3×10^{-3} s for 1.5 M, 2.0 M, and 2.5 M of phosphoric acid, respectively. From Fig. 3, we observed that 2.0 M phosphoric acid PC-based ER fluid gives the highest performance. From the relaxation time estimated from dielectric spectra contributes to polarization explains the flow behavior shown in Fig. 2. Samples above 1.5 M phosphoric acid have very short relaxation times (approximately 10^{-3} s). The shear stress of 2.0 M phosphated cellulose ER fluid was sustained at the high shear rate region (>10 s^{-1}).

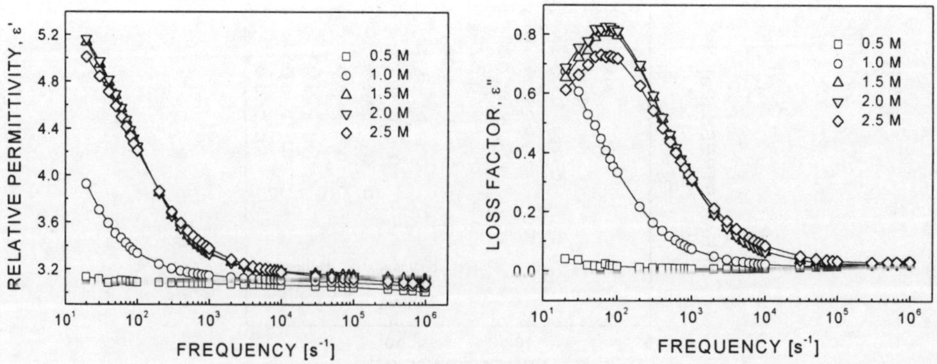

Figure 3. Dielectric spectra of the ER fluids; (a) permittivity and (b) loss factor *vs.* frequency.

Figure 4(a) presents the damping forces with time at the operating temperature of 25 °C. As the electric field increases, the damping force of the ER damper, filled with PC based ER fluid, increases. Thus, the force required to move the piston increased due to the increment of the yield stress. We found that the damping force is 0.2 kN at zero applied electric field strength. On the other hand, damping force is approximately 1.1 kN at 5kV/mm (the increase is approximately five times). This result implies that a large range of the damping force of the ER damper can be continuously controlled by tuning the electric field strength.

Figure 4(b) illustrates the electric field strength dependence on damping force with the piston velocity. The piston velocity is changed with the excitation frequency (ranges from 0.4 to 3.0 Hz), while the excitation amplitude is maintained. Damping force increases with as electric field strength and piston velocity (except 5kV/mm).

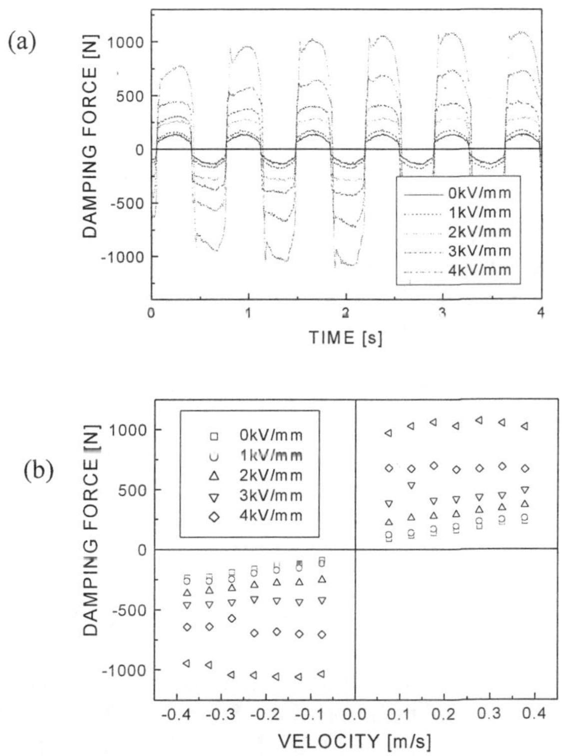

Figure 4. The damping characteristics of PC-based ER fluids. (a) damping force *vs.* time and (b) damping force *vs.* velocity.

4 Conclusion

We examined ER behavior for PC-based ER fluids with six different electric field strengths. Understanding the characteristic of flow behavior for these ER fluids assists in sustaining ER properties at the high deformation condition. From the dielectric spectra, we obtained the relaxation times and then adapted them to the polarization process. The dielectric spectra and the conductivity of the particles are used in the examination of the damping characteristics for the ER-based device.

5 Acknowledgment

This study was supported by research grants from the Korea Science and Engineering Foundation (KOSEF) through the Applied Rheology Center (ARC), an official KOSEF-created engineering research center (ERC) at Korea University, Seoul, Korea.

References

1. Goodwin J. W., Markham G. M. and Vinent B., Studies on Model Electrorheological Fluids, *J. Phys. Chem. B* **101** (1997) pp. 1961-1967.
2. Tao R., Zhang J., Shiroyanagi Y., Tang X. and Zhang X., Electrorheological Fluids under Shear, *Int. J. Mod. Phys. B* **5** (2001) pp.918-929.
3. Bullough W. A., Atkin R. J., Urang S., Kum T, G., Musch C. and Rober T., Two Dimensional Flow of an ESF: Experiment, CFD, Bingham Plastic Analysis, *Int. J. Mod. Phys. B* **15** (2001) pp.745-757.
4. Bossis G., Abbo C., Cutillas S., Lacis S. and Métayer C., Electroactive and Electro structured Elastomers, *Int. J. Mod. Phys. B* **15** (2001) pp.564-573.
5. Choi H. J., Cho M. S., Kang K. K., and Ahn W. S., Electrorheological Characteristics of Mesoporous Molecular Sieve (MCM41) Suspension, *Micropor. Mesopor. Mater.* **39** (2000) pp.19-24.
6. Ikazaki F, Kawai A., Uchida K., Kawakami T. Sakurai K., Anzai H., Asako Y. *J. Phys. D: Appl. Phys.* **31** (1998) p.336.
7. Cho M. S., Choi H. J., Chin I. J. and Ahn W. S., Electrorheological Characterization of Zeolite Suspensions, *Micropor. Mesopor. Mater.* **32** (1999) pp.233-239.
8. Böse H., Investigations on Zeolite-Based ER Fluids Supported by Experimental Design, *Int. J. Mod. Phys. B* **13** (1999) pp.1878-1885.
9. Kojima Y., Matsuoka T. and Takahashi M. Electrorheological Properties of Suspension of Diphenyldiacetylene Annealed Under Elevated Pressure, *J. Appl. Polym. Sci.* **53** (1994) p.1393-1396.
10. Quadrat O., Stejskal J., Kratochvíl P., Klason C., McQueen D., Kubát J. and Sáha P., Electrical Properties of Polyaniline Suspensions, *Synth. Met.* **97** (1998) pp.37-42.
11. Lee J. H., Cho M. S., Choi H. J., Jhon M. S. , Effect of Polymerization Temperature on Polyailine Based Electrorheological Suspensions, *Colloid. Polym. Sci.* **277** (1999) pp.73-76.
12. Choi H. J., Lee Y. H., Kim C. A. and Jhon M. S., Microencapsulated Polyaniline Particle for Electrorheological Material, *J. Mater. Sci. Lett.* **19** (2000), pp.533-535.
13. Cho M. S., Choi H. J. and To K., Effect of Ionic Pendant Groups on a Polyaniline-based Electrorheological Fluid , *Macromol. Rapid Commun.* **19** (1998) pp. 271-273.
14. Choi H. J., Kim J. W., To K., Synthesis and Electrorheological Behavior of Semiconducting Poly(aniline-co-o-ethoxyaniline), *Synth. Met.* **101** (1999) pp.697-698.

15. Choi H. J., Cho M. S., Kim J. W., Webber R. M. and Jhon M. S, Effect of Ionic and Nonionic Substituents on the Electrorheological Characteristics of Polyaniline Derivatives, *Int. J. Mod. Phys. B* **15** (2001) pp.988-995.
16. Choi U. S., Ahn B. G. and Kwon O. K., Electrorheological Behavior of Chitosan Derivatives Suspensions, *Int. J. Mod. Phys. B* **15** (2001) pp.1025-1032.
17. Block H., Kelly J. P., Qin A. and Watson T., Materials and Mechanisms in Electrorheology, *Langmuir* **6** (1990) pp. 6-15.
18. Choi H.J., Cho M.S, Jhon M.S., Hysteresis Behaviors of Poly(naphthalene quinone) Radical Electrorheological Fluid, *Int. J. Mod. Phys. B* **13** (1999) pp.1901-1907.
19. Trlica J., Sáha P., Quadrat O. and Stejskal J., Electrorheological Activity of Polyphenylene diamine Suspensions in Silicone Oil, *Physica A* **283** (2000) pp.337-348.
20. Sim I. S., Kim J. W., Choi H. J., Kim C. A. and Jhon M.S., Preparation and Electrorheological Characteristics of Poly(p-phenylene)-Based Suspensions, *Chem. Mater.* **13** (2001) pp.1243–1247.
21. Choi H. J., Cho M. S., Kim J.W., Kim C. A. and Jhon M. S., A Yield Stress Scaling Function for Electrorheological fluids, *Appl. Phys. Lett.* **78** (2001) pp.3806-3808.
22. Kim J. W., Kim S. G., Choi H. J. and Jhon M. S., Synthesis and Electrorheological Properties of Polyaniline-Na^+-Monmorillonite Suspensions, *Macromol. Rapid Commun.* **20** (1999) pp.450-452.
23. Kim J. W., Noh M. H., Choi H. J., Lee D.C. and Jhon M. S., Synthesis and Electrorheological Characteristics of SAN-clay composite Suspensions, *Polymer* **41** (2000) pp.1229-1231.
24. Choi H. J., Kim J. W., Joo J. and Kim B. H., Synthesis and Electrorheology of Emulsion Intercalated PANI-clay Nanocomposites, *Synth. Met.* **121** (2001) pp.1325-1326.
25. Ahn B. G., Choi U. S., Kim C. H. and Kwon O. K., The Effect of Activation Time of Phosphoric Ester Cellulose Particles on the Electrorheological Properties of Anhydrous ER Fluids, *Int. J. Mod. Phys. B* **15** (2001) pp. 1009-1016.
26. Lee J. M. and Lim K. H., Electrochemical Synthesis of Conducting Polythiophene in an Ultrasonic Field, *J. Ind. Eng. Chem.* **6** (2000) pp. 157-162.
27. Kim S. G., Kim J. W., Choi H. J., Suh M. S., Shin M. J. and Jhon M. S., Syctlıesis and Electrorheological Characterization of Emulsion-polymerized DBSA-doped Polyaniline Based Suspensions, *Colloid Polym. Sci.* **278** (2000) pp. 894-898.
28. Choi H. J., Cho M. S. and To K., Electrorheological and Dielectric Characterizations of Semiconductive Polyaniline-Silicone Oil Suspensions, *Physica A* **254** (1998) pp. 272-279.
29. Wen W., Ma H., Tam W. Y. and Sheng P., Frequency-induced Structure Variation in Electrorheological Fluids, *Appl. Phys. Lett.* **77** (2000) pp. 3821-3823.

EFFECT OF SURFACE MODIFICATION OF COLLOIDAL SILICA ON ELECTRORHEOLOGICAL PROPERTIES

C. GEHIN, J. PERSELLO

Laboratoire de Chimie des Matériaux et Interfaces, Université de Franche Comté, UFR Sciences et Techniques, 16, route de Gray, 25030 Besançon cedex, France.

Electrorheological (ER) effect of surface modified silica particles dispersed in silicone oil was experimentally studied. Monodispersed silica particles were prepared by hydrolyzing tetraethylorthosilicate in an ethanol solution. Rheometry measurements of shear stress and viscosity under shear rates of the suspensions were carried out with a rotational double concentric rheometer. Electric field is applied transverse to the direction of the shear. Flow curves reveals that there is no field effect on dispersion of dried particles. When water is adsorbed on the silica particles surface, the ER effect exists, but beyond a rate of 5 wt% of water, the capillary forces produce the aggregation of the particles. This agrees with the theory where it has been shown that the ER behaviour is proportional to the dielectric constant of the particles. Moreover hydrated lithium cations adsorbed on the silica surface involve a reversible ER effect with short relaxation times.

1. Introduction

Electrorheological fluids (ERF) have drawn a great deal of attention due both to their unique properties and to their potential technological applications.
The design of ERF must combine a number of factors that affect the properties of the fluids.[1] This includes the absolute magnitude of the electrorheological (ER) effect, the time scale of structure formation, and the stability of the fluids.
Large research effort is directed toward the understanding of ERF macroscopic behavior. In particular ERF exhibit a wide range of mechanical and optical behavior in a multiplicity of practical applications.[2] Consequently, a microscopically based understanding of these fluids is desirable to further the development of such materials for specific uses. For example, the development of an ERF for optical applications could benefit substantially from understanding the flow behavior changes, associated to the structure transitions, resulting from variations in the surface chemical properties of the particles. The majority of those fluids displaying large reversible structural changes are composed of silica particles suspended in silicone oil.
An induced polarization results at the interface between two materials that differ in their dielectric constants. In the presence of an electric field the induced dipole moments cause the formation of strings aligned in the direction of the field thus increasing viscosity and induce optical anisotropy in the system. The magnitude of the distortion depends on the relative dielectric constant of the particles. In the case of silica particles, there is a significant requirement that the particles must be covered with substantial amounts of water. Since water has a high dielectric constant, the

presence of small amounts of water on a particle will significantly enhance the electric polarizability, producing stronger inter-particle interactions and increasing to increased viscosity.
The adsorbed water plays an essential role in the silica in silicone oil ERF, but critical limitations associated with water arise in designing devices. When silica particles are dispersed in non-polar solvent, for particles close to contact, water in the contact region forms bridges that connect the particles through surface tension forces. Capillary forces induce non-reversible particles aggregation, which affects the strength and the time development of the structures.
Motivated by experimentally based intuitive guidelines that arose from studies of colloidal silica suspensions, a new concept has been developed. The basic idea is to structure the water molecules at the surface of silica by adsorption of lithium cations.
This paper reports the results of investigation of the relation between dielectric and ER properties of suspensions made of spherical nano-structured colloidal silica dispersed in a polydimethylsiloxane oil.
The attention is focused on the relation between surface conditions of the particles and ER behavior of suspensions. The effects of water and lithium ions are examined for silica suspension in a silicon oil under steady shear conditions, over a wide range of water, lithium and particle concentrations. The fluids are studied by rheological measurements under constant electric field.
Our goal is to provide the critical concentrations of water and density of adsorbed lithium, which produces reversible electric-field induced structures.
In the following, the effect of water content on the nonreversible aggregation of the silica particles will be investigated by light scattering measurements. We will detail the effect of water and adsorbed lithium ions on the rheological behavior of the silica suspensions when an electric field is applied.

2. Experimental

2.1. Materials

The ERF are suspensions of silica particles dispersed in silicone oil. The silicon oil (DC 200, from Fluka) has a Newtonian viscosity of 10mPa.s and a density of 0.937, was used as received. This oil is consisting in a polymer: the poly(dimethylsiloxane). The particles are prepared according to the method of Stöber[3] and Matsoukas[4] by hydrolyzing tetraethylorthosilicate (TEOS) in a mixture of water, ammonia and ethanol. The TEOS is regularly added with a flow rate of 1mL/s. The initial mixture is clear and homogenous. As the condensation reaction is running, the mixture becomes turbid. Shaking is continued at 25°C for 24 hours to give an ethanolic suspension of colloidal silica. After, the silica is vacuum-dried at 60°C under pressure of about 110 Pa for 6 hours. We obtained spherical, monodispersed and nanometric particles.
In order to adsorb some lithium cations on the silica, the dried silica is dispersed in aqueous solutions containing different amounts of lithium hydroxide. The suspension is shaked over 7 days. Then, the suspension is centrifuged, the silica is recovered, dispersed again in ultra pure water and centrifuged in order to eliminate the non-adsorbed lithium ions. This cycle of manipulations continues until the conductivity of the supernatant becomes closed to the conductivity of the water. Some particles are dried at different temperatures (80 or 600°C). All particles are transferred in the silicone oil using a centrifugal ball mills and diluted at different concentrations. The

resulting ERF are listed in table 1 which shows the particles radius obtained by SEM, the number of lithium ions per nm^2 evaluated from potentiometric measurements for different silicas and the drying.

ER fluid	S24	S26	S28	S29	S30	S32
Radius (nm)	162	162	150	300	206	150
$Li^+(n_b/nm^2)$	0	0	0,07	0,17	0,54	0
Drying (°C)	80	600	60	60	60	60

Table 1. Presentation of the different characteristics of the studied fluids

2.2. Techniques

In every measurement discussed in this paper, the electric field is applied transverse to the direction of the shear. A rotational double concentric cylinder rheometer is used to examine the ER properties of the suspensions. The gap between the inner and the outer cylinders is of 2 mm. A high voltage is applied between the concentric cylinders producing an electric field in the gap. Each cylinder is insulated from the rest of the rheometer. Steady shear rate rheological measurements are performed in the following manner: 100-4000V was used at a frequency of 1000 Hz. The dispersions are so dielectric that the current is of order of μA.

3. Theoretical considerations

The ER effect of a fluid is the result of the electric polarization forces.[5,6,7] The studied ERF are composed of uncharged, spherical, dielectric particles of radius a and dielectric constant ε_p suspended in a non-conducting dielectric fluid of dielectric constant ε_m. The particles in an ER suspension are polarized by the external electric field and there is interaction between the resulting dipoles.

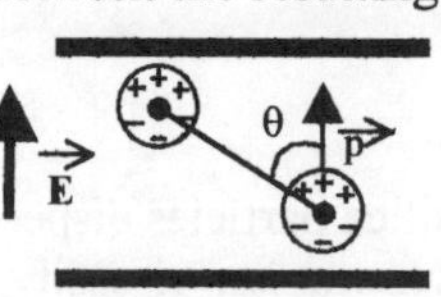

Fig. 1. Effect of applied electric field on the particles

The expression of the resulting dipole $\vec{p}$ is: $\vec{p}=\varepsilon_0.\varepsilon_m.\,\beta.a^3.\vec{E}$ (1)

where ε_0 is the permittivity of free space and β the particle dipole coefficient given as:

$$\beta=(\varepsilon_p-\varepsilon_m)/(\varepsilon_p+2\varepsilon_m) \quad (2)$$

This coefficient is a measure of the polarizability of the particle in an external electric field. The dipole-dipole interaction energy is:

$$U(r,\theta)= p^2((1-3\cos^2\theta)/r^3) \quad (3)$$

where r is the distance between two particles and θ is the angle between the directions of the dipole moment and the electric field.

The interaction energy is thus directly proportional to the coefficient β. If $\varepsilon_p=\varepsilon_m$, β=0, then there is no dipolar interaction between the particles. If $\varepsilon_p\neq\varepsilon_m$, the different abilities of both materials to polarize produce a charge separation on the particle surface. In our case: silica particles in silicon oil, $\varepsilon_m<\varepsilon_p$.

$$\delta U/\delta\theta = (3p^2/r^3).\sin 2\theta \qquad (4)$$

At a given r, $\delta U/\delta\theta$ is equal to 0 at 2 angles: $\theta=0$ rad and $\theta=\pi/2$ rad where the particles are in equilibrium positions. The function being bijective, there is a critical angle θ_c below which the dipolar energy of interaction becomes attractive. Thus, the interaction aligns both particles in the direction of the electric field. Beyond θ_c, the energy of interaction will be repulsive. In this case, the two particles are aligned perpendicularly to the electric field.
Then, the presence of many particles involves the formation of chains of particles parallel to the direction of the electric field, regularly spaced[8] and the intensity of the potential interaction energy is proportional to the difference ε_p-ε_m. One way to improve the ER effect is then to increase the difference of the dielectric constants
The degradation of these structures by shear stress gives rise to the observed enhancement of suspension shear rate.

4. Results and discussion

3.1 *Research of the critical shear rate*

τ_E/τ_0 is the ratio of the shear stress in an applied electric field of 1kV/mm to the zero-field shear stress.

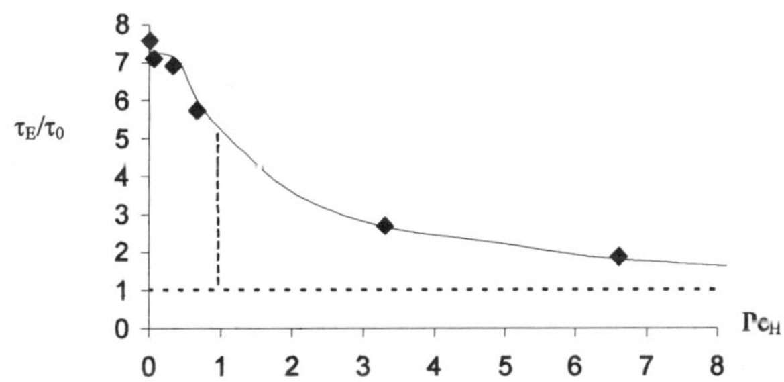

Fig.3. Relative shear stress versus hydrodynamic Peclet number for the suspension S32 at Φ=18.24 vol%

Its dependence with the hydrodynamic Peclet number for the suspension of silica particles (S32) at a volumic concentration of 18.24% can be seen on the figure 3. The hydrodynamic Peclet number (Pe_H) compares the convection motion due to the shear rate to the diffusion of particles due to the Brownian motion (Equation (5) where η is the oil viscosity, T the temperature and k the Boltzman constant).

$$Pe_H = \frac{6\pi\eta a^3\dot{\gamma}}{kT} \qquad (5)$$

At $Pe_H<1$, where the particles motion is mainly depending of the Brownian motion, the ER effect is large and almost independent of the shear rate. At $Pe_H>1$, the ER effect is strongly dependent on the shear rate. The ratio τ_E/τ_0 tends to reach the limit value of 1: the ER effect becomes negligible. If Pe_H is equal to 1, there exists a critical shear rate $\dot{\gamma}_c = 6.80 s^{-1}$. Below $\dot{\gamma}_c$, ER effect is significant. In order to observe a large ER effect, the shear rate applied should not exceed $\dot{\gamma}_c$.

3.2. *Effect of adsorbed water on electrorheology*

Figure 4. shows the effect of the adsorbed water on the silica particles on flow curves.

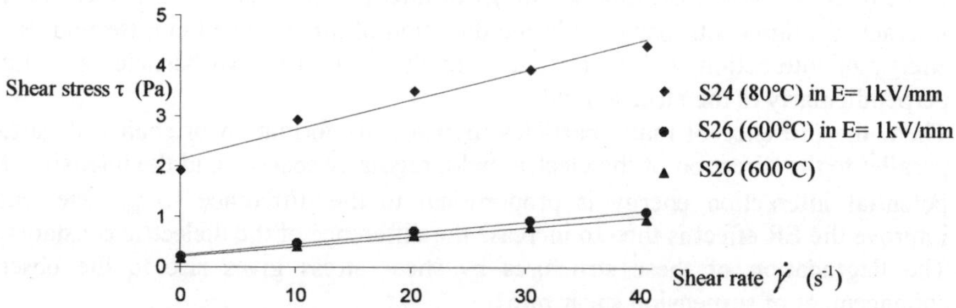

Fig.4. Flow curves of 18.24 vol% of silica in suspension

Little ER effect is produced by suspension containing samples heat-treated at 600°C (S26). The same particles dried at 80°C (S24) reveal the effect of the adsorbed water on the electrorheology: applying an electric field involves the increase of the shear stress at the same share rate.

Adsorbed water on the silica surface contributes to the ER effect.[9, 10] But with heat treatment at 600°C, adsorbed water on the silica surface is removed and then the ER effect is negligible.

This can be explained by the fact that adsorbed water contributes to the increase of the particle dielectric constant by the high value of the water dielectric constant. This leads to the increase of the interaction energy between particles and thus of the ER effect.

Figure 5 shows the dependence of the transmission on the rate of adsorbed water. We can see the decrease of the transmission of the suspension with the increase of the rate of adsorbed water. Below 5wt% of adsorbed water, the transmission is constant and equal to 100%. But, beyond 5wt% water involves the decrease of the transmission. This can be explained by the fact that the presence of a lot of adsorbed water (more than 5wt%) involves the irreversible aggregation of the particles caused by capillary bonding.[11, 12]

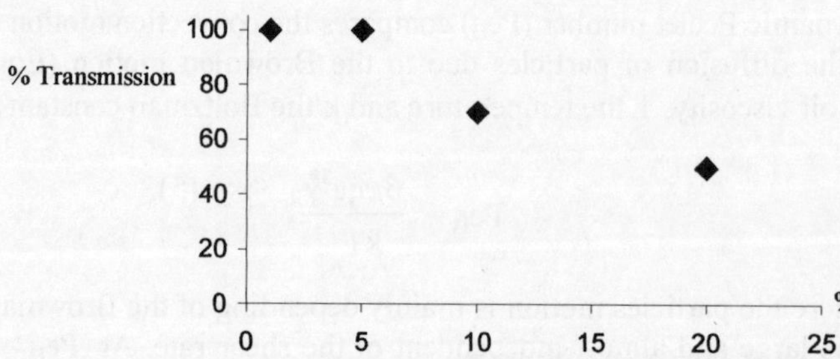

Fig.5. % Transmission versus % water adsorbed on silica at ϕ= 0.01%, e= 1 mm

But even if the presence of water on silica particles surface induces an increase of the ER effect due to the increase of dielectric constant, too much water causes an irreversible aggregation of the particles and a destruction of the ERF. By this way, the increase of the ER effect is limited.

3.3. *Effect of adsorbed lithium cations on electrorheology*

Adsorbing hydrated lithium cations on the silica surface is another way to increase the ER effect: figure 6 shows the effect of the number of adsorbed hydrated lithium on the silica particles (S32) on the ratio τ_E/τ_0. The shear rate $\dot{\gamma}$ is equal to $0.045 s^{-1}$. The volume fraction of silica in the suspension is 18.24%.

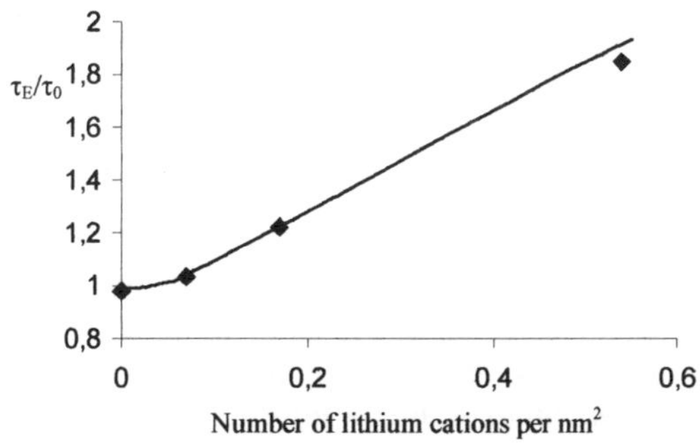

Fig. 6. Relative shear stress versus number of lithium cations adsorbed on the silica surface of the suspension S32 at Φ=18.24 vol%

The increase of adsorbed lithium cations on the silica particles enhances the ER effect. Lithium cations are adsorbed on ionized silanol groups of the silica surface. The hydratation number of lithium cation is 5-6.[13] Thus the adsorption of lithium cations on the silica surface increases the dielectric constant of the silica because of the presence of water bound to them. This explains why the ER effect increases with the rate of hydrated lithium cations adsorbed on the silica surface particles.

During an relaxation experiment, different electric field intensities are applied between the electrodes: 0 to 3.5 kV step by step and then 3.5 to 0 kV. On the figure 7 can be seen that the ratio τ_E/τ_0 increases step by step when the electric field increases. Its values are noted on the curve (kV).

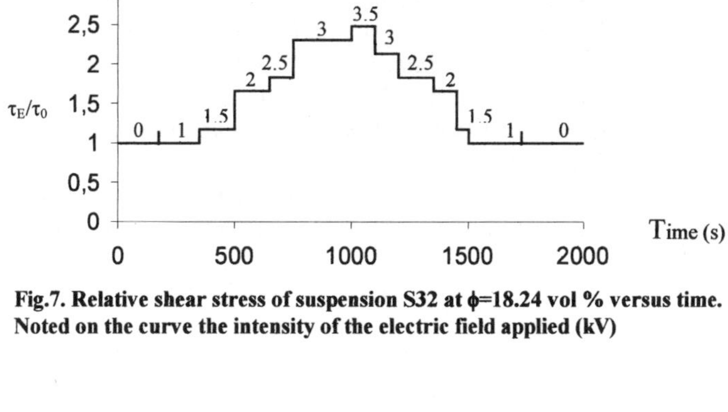

Fig.7. Relative shear stress of suspension S32 at ϕ=18.24 vol % versus time. Noted on the curve the intensity of the electric field applied (kV)

When the field is constant, the ratio is constant, and when the electric field decreases, it decreases as well. The relaxation times are very short: inferior to 1/10 second. This reveals that an ER system with hydrated cations adsorbed on the particles surface is totally reversible with very short relaxation times. The silica ionic electric charge increased by the charges of lithium cations prevents the particles from aggregating when there is no electric field applied in making an electrostatic energy barrier.

5. Conclusion

In this paper, the effect of surface modification of colloidal silica on ER properties was investigated. These ERF were constituted by spherical, nanometric and monodispersed silica particles dispersed in silicon oil. A theoretical approach shows that the ER effect is proportional to the dielectric constant of the particles. Studying the effect of adsorbed water on the silica surface reveals that the water permits to increase the ER effect by modifying the dielectric constant of the silica. But beyond 5 wt% of water adsorbed on the silica , the water induces irreversible particles aggregation and then the destruction of the ERF. This phenomenon is thus limited. Adsorbed hydrated lithium cations permit to enhance the ER effect. The reversibility of the phenomenon is total and the relaxation times are very short.
These properties could be useful where mechanical applications of ER fluids need an increase and a reversibility of the ER effect.

References

[1] M.Parathasarathy, D.J.Klingenberg, *Mater. Sci. Eng.*, **R17**, p.57, (1996)
[2] T.C. Jordan, M.T. Shaw, *IEEE Trans. Elect. Insul.*, **24**, 5, p.849, (1989)
[3] W.Stöber, A.Fink, E.Bohn, *J. Colloid Interface Sci.* **26**, p.62, (1968)
[4] T.Matsoukas, E.Gulari, *J. Colloid Interface Sci.* **124**, 1, p.252, (1988)
[5] T.Hasley, J.Martin, *Pour la Science,* **194,** p.60, (1993)
[6] A.P.Gast, C.F.Zukoski, *Adv. Coll. Int. Sci.*, **30**, p.153, (1989)
[7] L.C.Davis, *Appl. Phys. Lett.*, **60**, 3, p.319, (1992)
[8] H.See, M.Doi, *J. Phys. Soc. Japan*, **60**, p.2778, (1991)
[9] A.Kawai, K.Uchida, K.Kamiya, F.Ikazaki, *Adv. Powder Technol.*, **5**, 2, p.129, (1994)
[10] V.Pavlinek, O.Quadrat, B.Porsch, P.Saha, *Colloids and surfaces A*, **155**, p.241, (1999)
[11] H.See, H.Tamura, M.Doi, *J. Phys. D :Appl. Phys.*, **26**, p.746, (1993)
[12] Y.Otsubo, M. Sekine, S. Katayama, *J. Colloid Interface Sci.,* **150**, 2, p.324, (1992)
[13] J.Israelachvili, *Intermolecular and surface forces* (Academic Press, (1991))

ELECTRORHEOLOGY OF AMINE DISPERSANT COATED CHITOSAN PHOSPHATE SUSPENSION

UNG-SU CHOI

Tribology center, Korea Institute of Science and Technology, PO Box 131, Cheongryang, Seoul, Korea
E-mail: uschoi@kist.re.kr

YOUNG-GUN KO

Tribology center, Korea Institute of Science and Technology, PO Box 131, Cheongryang, Seoul, Korea
E-mail: y-gun@kist.re.kr

The electrorheological (ER) behavior of amine coated chitosan phosphate suspension in silicone oil was investigated. Amine coated chitosan phosphate suspension showed a typical ER response (Bingham flow behavior) upon application of an electric field due to the polarizability of phosphoryl polar group. The shear stress for the suspension exhibited a linear dependence on an electric field power of 1.94. The values of structure factor, A_s obtained 2 for amine coated chitosan phosphate suspension and it may be due to the formation of multiple chains upon application of the electric field. Throughtout the experimental results, The suspension showed good dispersion stability and found to be an anhydrous ER fluid.

1. Introduction

ER fluids are nonaqueous suspensions composed of electrically polarizable particles dispersed in a dielectric fluid and especially, the disperse phase plays an important role in the ER phenomenon.[1,2,3] The organic polymers, polyaniline[4,5] and polyurethane[6] as the disperse phases have been widely used for the formulation of anhydrous ER fluids. Because they have the polar groups such as amino(-NH_2) and amino-cyan(-NHCN) radicals, respectively, suspensions of these particles provide the ER effect upon application of the electric field. The chemical structure of the organic polymers is therefore important in the ER effect.

Chitosan as the base material is a natural organic biocompatible polymer from chitin by N-deacetylation and composed of poly D-glucosamine. It has been widely used for applications to the fields of biochemisry, pharmacology, emzymology, microbiology, agricurture and environment.[7] As the new anhydrous ER suspension, chitosan suspension provided the ER behavior upon application of an electric field due to the polarizability of the branched amino group.[8] And also a chitosan derivative suspension composed of the anhydrous disperse phase, chitosan phosphate synthesized by phosphorylation reactions between chitosan and phosphorous acid introduced and its ER behavior investigated. Chitosan phosphate suspension showed execllent ER performance under the electric field due to the phosphoryl polar group.[9]

However, it has certain problem, such as dispersion stability inspite of its high ER performance and nonadhesion to the electric cell. To solve this problem, amine dispersant coated chitosan phosphate as the disperse phase has been prepared and the electrical and rheological properties of the suspension investigated. The amine dispersant coated chitosan phosphate suspension provides the ER effect under the electric field and also stable dispersion stability.

This study is to describe the ER behavior of amine dispersant chitosan phosphate suspension and to investigate the possibility of a newly anhydrous ER fluid.

2. Experimental

2.1. *Materials*

The base fluid was silicone oil provided by Dow Corning with a specific gravity of 0.97, a kinematic viscosity of 50 cst at 40 °C and a dielectric constant of 2.61 at 25 °C. Chitosan used as the raw material was a commercial powder provided by Jaekwang Co.(Korea) and contained nitrogen content of 4.8 wt%. Chitosan phosphate as the disperse phase was synthesized by phosphorylation reactions between chitosan and phosphorous acid(2 mole soln') and contained phosphorous content of 9.5 wt%. After the reaction, chitosan phosphate was dispersed in aceton solution contained 3% dioctyl amine for 1hour. And then amine coated chitosan phosphate was dryed in oven. Its particle sizes was in 5μm average diameter. Prior to mixing in silicone oil, amine coated chitosan phosphate particle was dried for 5h at 150 °C and the silicone oil for 3h at 130 °C to remove moisture in vacuum oven. Amine coated chitosan phosphate suspensions were then prepared at volume fractions of 0.1 to 0.3. After vigorous mixing, the suspensions were stored in a dessicator to maintain the dry state.

2.2. *Tests*

The dc current density J and the conductivity σ of the silicone oil and of amine coated chitosan phosphate suspension were determined at room temperature by measuring the current passing through the fluid upon application of the electric field E_0 and dividing the current by the area of the electrodes in contact with the fluid. The current was determined from the voltages drop across a 1MΩ resistor in series with the metal cell containing the oil using a voltmeter with a sensitivity of 0.01mV. This method gave a current measuring sensitivity of 0.01nA. The dc conductivity was taken to be $\sigma = J / E_0$.

The rheological properties of the suspension were investigated in a dc field using the Physica Couette-type rheometer with a 1mm gap between the bob and cup. The resistance to shear produced by the suspensions was measured as a torque on the drive shaft and then converted to shear stress and viscosity. The shear stress for the suspensions was measured under shear rates of 1 to 300 s^{-1}, electric fields of 0 to 3.0 kV/mm and volume fractions of 0.1 to 0.3, respectively.

The dispersion stability of the suspension was monitored on time for sedimentation of the disperse phase within glass tube at room temperature and represented as the sedimentation degree:

$$\text{Sedimentation degree (\%)} = a / (a + b) \times 100 \qquad (1)$$

where a is the oil layer separated from the suspension and b the disperse layer of the suspension.

3. Results

The electrical properties of ER fluids are important for predicting the power requirements for the design of an ER device and also identifying the ER mechanism. The conductivities of the chitosan phosphate and amine coated chitosan phosphate suspensions for a volume fraction ($\phi = 0.3$) vs electric field are given in Fig.1. The results in Fig.1 indicate that the conductivity of amine coated chitosan phosphate suspension is about 1 order of magnitude lower than that of chitosan phosphate suspension and about 3 orders higher than the silicone oil. The decrease in conductivity of amine coated chitosan phosphate suspension is considered to result from the lower ploarizability caused by the insulation effect of dioctyl amine coated chitosan phoaphate particles.

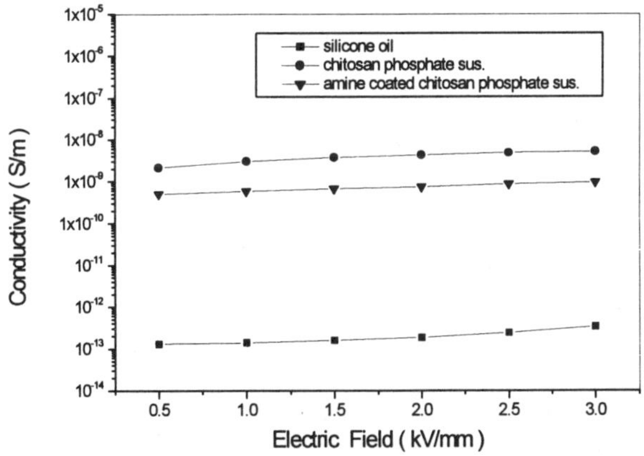

Fig. 1. Effect of electric field on conductivity for silicone oil and chitosan phosphate suspensions.

To investigate the effect of amine coated chitosan phosphate suspension on the rheological properties, studies were carried out by varying shear rates, electric fields and volume fractions. The effect of the shear rate on the shear stress for amine coated chitosan phosphate is illustrated in Fig. 2. As seen in Fig. 2, the rheological behavior of the amine

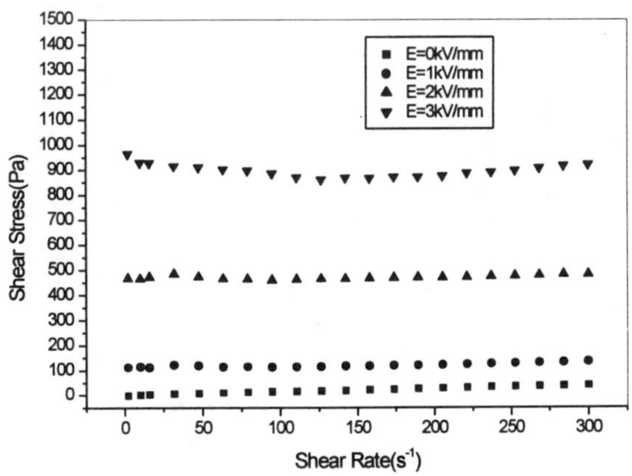

Fig. 2.Effect of shear rate on shear stress for amine coated chitosan phosphate suspension (ϕ=0.3).

coated chitosan phosphate suspension behaves as a Newtonian fluid without the electric field, but upon application of the electric field, it exhibits a yield stress τ_E, that is, this suspension approximates a Bingham-type behavior, which is described by the equation

$$\tau = \tau_E(E, \dot{\gamma}) + \eta\dot{\gamma} \tag{2}$$

And the resulting amine coated chitosan phosphate suspension shows excellent ER response upon application of the electric field due to phosphoryl polar group. Figure 3 shows a log-log plot of the shear stress versus the square of the electric field for the amine coated chitosan phosphate suspension. The results in Fig.3 indicate that the shear stress was proportional to an electric field power of 1.94, that is, $\tau \propto E^{1.94}$.

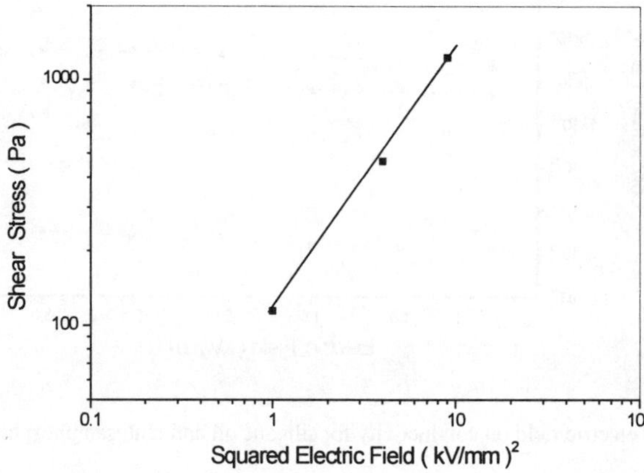

Fig. 3. Shear stress vs squared electric field for amine coated chitosan phosphate suspension (ϕ= 0.3, $\dot{\gamma} = 10s^{-1}$).

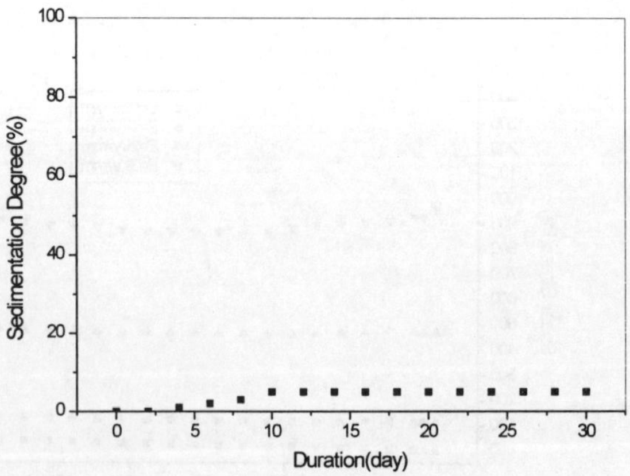

Fig. 4. Effect of duration on sedimentation degree for amine coated chitosan phosphate suspension (ϕ=0.3).

The sedimentation degree of amine coated chitosan phosphate suspension was investigated for dispersion stability of the suspension and the results are illustrated in Fig.4. The results in Fig.4 show that sedimentation degree slowly increases until 5% on time and then becomes constant over 10 days.

4. Discussion

To explain the ER behavior of amine coated chitosan phosphate suspension, we will examine the results obtained with the assumption that the base fluid and particles behave as ideal dielectric materials, and particles are aligned in chains or columns between electrodes. With these assumptions, the theological analysis of Conrad etal.[10] gives for the polarization component of the yield stress

$$\tau_E = 44.1\, A_s\, \phi\, \varepsilon_0\, K_f\, (\beta E)^2 \,|\, \{ \exp [\, (14.84 - 6.165(R/a))\, \beta^2 \,] \} \times [1/(R/a)^8 - 4/(R/a)^{10}\,]^{1/2} \,|_{max} \quad (3)$$

where A_s is taken to be a structure factor pertaining to the alignment of the particled. It is equal to one for perfectly aligned single –row chains and may a value of the orders of 10 for multiple chains or columns. K_f is the dielectric constant, β the relative polarizability (β=1) and R/a the ratio of the separation of the particles center to their radius (R/a ≅ 2.05). The structure factor, A_s is obtained from the ratio value of measured to calculated shear stress using Eq.(3), that is, $A_s = \tau_{meas.} / \tau_{calc.}$. We obtain $A_s = 2$ for amine coated chitosan phosphate suspension under the test conditions at a shear rate of 10 s^{-1}, the electric fields of 1 to 3 kV/mm and a volume fraction of 0.3. It may be due to the formation of multiple chains aligned between electrodes for the suspension.[10,11]

5. Summary

This study was conducted to investigate the electrorheological behavior of amine coated chitosan phosphate suspension and the following results are found;

(1) Amine coated chitosan phosphate suspension in silicone oil showed the ER response upon the application of the electric field due to the polarizability of phosphoryl polar group.
(2) The shear stress of amine coated chitosan phosphate suspension increased linearly with 1.94 power of the electric field.
(3) Amine coated chitosan phosphate suspension showed good dispersion stability.
(4) The values of structure factor, A_s were 2 for amine coated chitosan phosphate suspension and it may be due to the formation of multiple chains upon application of the electric field.

Acknowledgements

This research was sponsored by NRL project of the Korean Ministry of Science and Technology.

References

1. W. M. Winslow, *J.Appl.Phys.* **20**, 1137 (1949).
2. H. Block and J. P. Kelly, *J. Phys. D:Appl. Phys.* **21**, 1661 (1988).
3. D. J. Klingberg and C. F. Zukoski, *Langmuir* **6**, 15 (1990).
4. H. Block and J. P. Kelly, *Langmuir* **6**, 6 (1990).
5. C. Dearmih, S. P. Arme, J. Winter and F. A. Uribe, *Macromolecules* **22**, 649 (1976).
6. R. Bloodworth and E. Wendt, in *Progress in Electrorheology*, ed. K. O. Havelka and F. E. Filisko (Plenum Press, New York, 1995), p. 185.
7. C. J. Brine, "Chitin: Accomplishment and Perspectives," in *Advances in chitin and Chitosan*, ed. C. J. Brine, P. A. Sandford and J. P. Zikakis (Elsevier Applied Science, New York and London, 1992), p. 1.
8. U. S. Choi, *Colloids and Syrfaces A: Phys. and Eng. Aspects* **157**, 193 (1999).
9. H. S. Jee, Y. G. Ko, S. S. Lee and U. S. Choi, *J. Korean Ind. Eng. Chem.* **11**, 605 (2000).
10. H. Conrad and Y. Chen, in *Progress in Electrorheology*, ed. K. O. Havelka and F. E. Filisko (Plenum Press, New York, 1995), p. 55.
11. H. Conrad, Y. Chen and A. Sprecher, *J. Modn. Phys. B* **16**, 2575 (1992).

ELECTRORHEOLOICAL BEHAVIOR OF CHITOSAN DICARBOXYLATE SUSPENSIONS

UNG-SU CHOI

Tribology center, Korea Institute of Science and Technology,
PO Box 131, Cheongryang, Seoul, Korea
E-mail: uschoi@kist.re.kr

YOUNG-GUN KO

Tribology center, Korea Institute of Science and Technology,
PO Box 131, Cheongryang, Seoul, Korea
E-mail: y-gun@kist.re.kr

The electrorheological (ER) behavior of chitosan dicarboxylate suspensions in silicone oil was investigated by varying the electric fields, volume fractions of particles, and shear rates, respectively. The chitosan dicarboxylate susepnsions showed a typical ER response caused by the polarizability of an amide polar group and shear yield stress due to the formation of multiple chains upon application of an electric field. Of these, chitosan malonicate suspension represented slightly higher rheological performance than any other suspensions due to dependent upon the carbon chain length. The shear stress for the suspension exhibited a linear dependence on an electric field power of 1.88. On the basis of the results, the newly synthesized chitosan dicarboxylate suspensions were found to be an anhydrous ER fluid.

1. Introduction

Electrorheological (ER) fluids consist of highly polarizable particles in an insulating fluid and their ER mechanism is characterized by the formation of particles upon application of an electric field.[1,2,3]

Since Winslow's discovery of the ER effect in 1947,[1] hydrous ER fluids composed of cellulose[4] and corn starch[5] as the organic disperse phases and silica gel[6] and zeolite[7] as the inorganic phases have been widely used and studied for a long time. Their ER performance is dependent upon the activation of a low molecular solvent and mostly frequently water under an electric field, plus they have lots of problems, including dispersion stability, durability, corrosion, abrasion, and a limited temperature in actual use. Recently, anhydrous ER fluids, which do not contain water or polar solvent in the disperse, have been introduced composed of synthetic polymer,[8] polyaniline,[9] and polyurethane[10] as the organic disperse phases. However, they also have certain problems, such as dispersion stability and adhesion to the cell inspite of their high ER performance.

The current study attempted to solve the basic problems of conventional ER fluids and introduced a chitosan derivative based on a natural biocompatible polymer, as the organic disperse phase. Chitosan dicarboxylates as the disperse phase were synthesized by chemical reactions between chitosan and dicarboxyl acids, and the electrical and rheological properties of the synthesized chitosan dicarboxylates in silicone oil were investigated. The chitosan dicarboxylate suspensions provided an ER performance under an electric field due to the polarizability of the branched amide group.

This study describes the ER behavior of the synthesized chitosan dicarboxylate suspensions and estabilishes the ER mechanism to investigate the possibility of an anhydrous ER fluid.

2. Experimental

2. 1. *Materials*

The base fluid used was silicone oil (Dow Corning Co.) with a specific gravity of 0.97, kinematic viscosity of 50cst at 40°C, and dielectric constant of 2.61 at 25°C. The chitosan as a raw material was a commercial powder (Jaekwang Co., Korea) with a nitrogen content of 4.8wt% and molecular weight of 100,000. The dicarboxyl acids are malonic acid, adipic acid and sebacic acid, respectively and these were provided by Sigma Aldrich. The chitosan dicarboxylates were synthesized by an amide reaction between chitosan and dicarboxyl acids of 0.5 mol ratio under the catalysis of TPP (Triphenyl phosphine) and DEAD (Diethyl azodicarboxylate). Their chemical reaction mechanism and chemical structure is shown in Scheme 1. The synthesized particle size was on average 25μm in diameter. Prior to mixing in silicone oil, the chitosan dicarboxylate particles were dried for 5h at 150°C and the silicone oil for 3h at 130°C to remove any moisture in the vacuum oven. The chitosan dicarboxylate suspensions were then prepared at volume fraction of 0.1 to 0.3.

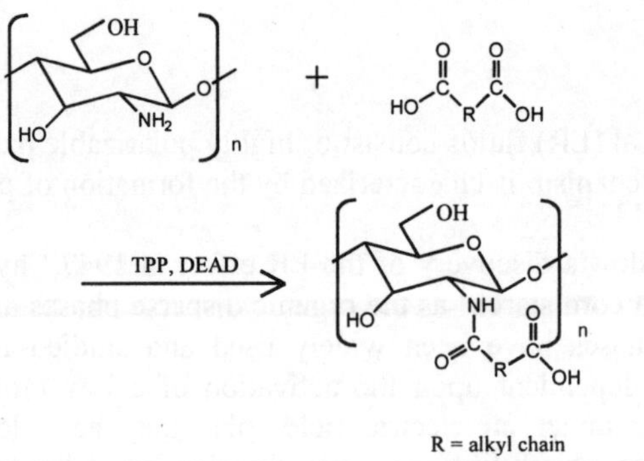

Scheme 1. Chemical reaction mechanism and chemical structure of chitosan dicarboxylate.

2. 2. *Tests*

The dc current density of the chitosan dicarboxylate suspensions was determined at room temperature by measuring the current passing through the fluid upon application of an electric field E_o and then dividing the current by the area of the electrodes in contact with the fluid. The current was determined from the voltage drop across a 1MΩ resistor in series with the metal cell containing the oil using a voltmeter with a sensitivity of 0.01mV. DC conductivity was taken to be ($\sigma = J/E_0$).

The rheological properties of the suspension were investigated in a dc field using a Physica Couette-type rheometer with a 1mm gap between the bob and the cup. The resistance to shear produced by the suspensions was measured as the torque on the drive shaft and then converted to shear stress and viscosity. The shear stress for the suspensions was measured under shear rate of 1 to 300s^{-1}, electric fields of 0 to 3kV/mm, and volume fractions of 0.1 to 0.3, respectively.

3. Results

The electrical properties of ER fluids are important for predicting the power requirements for the design of an ER device and also to identify the ER mechanism. The current densities of chitosan malonicate, chitosan adipicate and chitosan sebacicate suspensions for the volume fraction (ϕ=0.3) with 0.5mol ratio of dicarboxyl acids under the electric field are given in Fig. 1. As seen in Fig. 1, the current density of the chitosan dicarboxylate suspensions increased with an increasing the electric field and showed slightly decreasing trend with increasing the number of carbon atoms. This appeared to result from the polarizability of the polar group, an amide radical, and the less chemically active of the long carbon chains than short carbon chains.

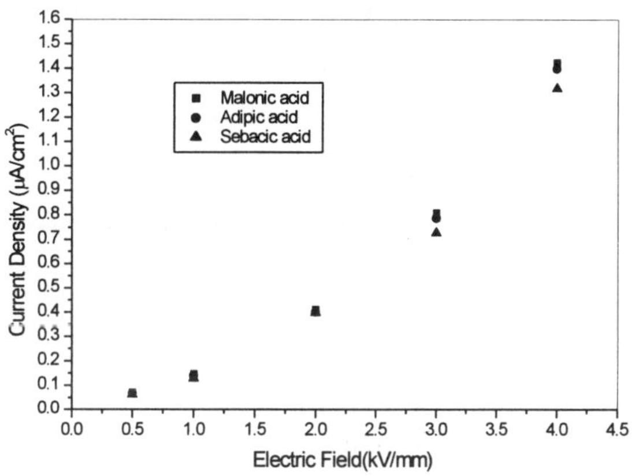

Fig.1. Effect of electric field on current density withchitosan dicarboxylate suspensions (ϕ=0.3).

To determine the effect of carbon chain length of chitosan dicarboxylate suspensions on the rheological properties, studies were conducted by varying the alkyl group between one and eight carbon atoms, namely, malonic, adipic and sebacic and the results were obtained by varying the shear rates, electric fields, and volume fractions. The effect of the shear rate on the shear stress for the chitosan dicarboxylate suspensions is illustrated in Fig. 2. The suspensions exhibited a Bingham flow behavior upon application of the electric field (E= 3kV/mm) and the shear stress of the suspension showed a similar trend. But, chitosan malonicate suspension represented slightly higher reological performance than any other suspensions. The result being that the reological performance is dependent upon the carbon chain length. Fig. 3 shows a log-log plot of the shear stress versus the square of the electric field for the chitosan dicarboxylate suspensions. The results in Fig. 3 indicate that the shear stress was proportional to an electric field power of 1.88, that is, $\tau \propto E^{1.88}$. This follows from the fact that the interaction force for the dipole in an electric field is proportional to the electric field intensity.

4. Discussion

Polarization models based on a point-dipole approximation, with a focus on the mismatch between the real components of the dielectric permittivities of the particles and

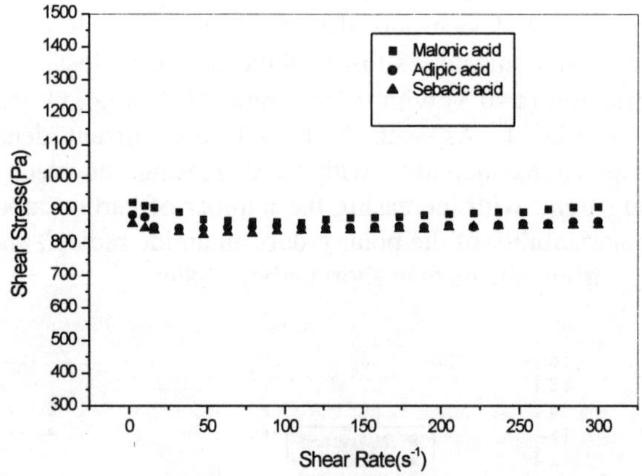

Fig.2. Effect of shear rate on shear stress for chitosan dicarboxylate suspensions (ϕ=0.3, E= 3kV/mm).

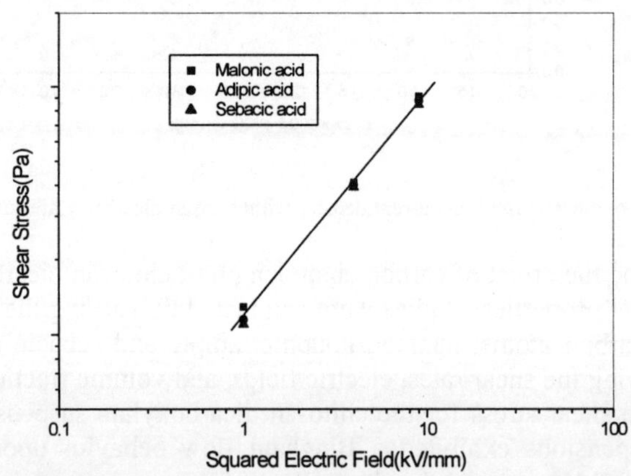

Fig.3. Effect of squared electric field on shear stress for chitosan dicarboxylate suspensions ($\dot{\gamma} = 10\ s^{-1}$, $\phi = 0.3$).

the based fluid, have been proposed to explain the behavior and can be described by the equation[3,11]

$$\tau \propto \phi K_f E^2 \beta^2 \quad (1)$$

where ϕ is the volume fraction of particles, K_f the dielectric permittivity of the base fluid, E the electric field, and β the relative polarizability at dc or low frequency ac fields given by [3]

$$\beta = (\sigma_p - \sigma_f) / (\sigma_p + 2\sigma_f) \tag{2}$$

where σ_p is the conductivity of the particles and σ_f the conductivity of the base fluid.

The general flow behavior of ER fluids upon application of an electric field can be described using the Bingham plastic equation and the polarization model given in Eq. (1). To explain the ER mechanism of the chitosan dicarboxylate suspensions under an electric field, the results obtained with the suspension were examined based on the assumption that the base fluid and particles behaved as ideal dielectric materials, and the particles were aligned in chains or columns between electrodes. Using these assumptions, the theological analysis of Conrad et al.[3] for the polarization component of the shear yield stress gives

$$\tau_E = 44.1\, A_s\, \phi\, \varepsilon_o K_f(\beta E)^2 \big| \{ \exp [(14.84 - 6.165(R/a))\beta^2] \} \times 1/ (R/a)^8 (1- 4/ (R/a)^{10})^{1/2} \big|_{max} \tag{3}$$

where A_s is taken to be the structure factor pertaining to the alignment of the particles. This is equal to one for perfectly aligned single-row chains and can have a value of the order of ~10 for multi-chains or columns. K_f is the dielectric constant, β the relative polarizability ($\cong$ 1), and R/a the ratio of the separation of the particle center to the radius ($\geq$ 2.05). The structure factor, A_s is obtained from the ratio value of the measured to the calculated shear stress using Eq. (3), that is, $A_s = \tau_{meas.} / \tau_{calc.}$ A_s= 2 was obtained for all the test conditions at a shear stress of 10 s^{-1}, electric fields of 1 to 3 kV/mm, and a volume fraction of 0.3, which may have been due to the formation of multi-chains aligned between the electrodes.[12]

5. Conclusions

This study was conducted to deduce the ER behavior of a chitosan dicarboxylate suspensions, estabilished the ER mechanism, and investigated their potential as an ER fluid. The following is a summary of the results:

(1) Chitosan dicarboxylate suspensions in silicone oil showed an ER response upon the application of an electric field and the suspension exhibited a Bingham flow behavior. This was considered to result from the polarizability of the branched polar group, an amide radical.

(2) Chitosan malonicate suspension showed slightly higher rheological performance than any other suspensions due to dependent upon carbon chain length.

(3) The shear stress of the chitosan dicarboxylate suspensions increased linearly with an electric field power of 1.88.

(4) The value of the structure factor, A_s was 2 and this may have been due to the formation of multi-chains aligned between the electrodes upon the application of an electric field.

Acknowledgements

This research was sponsored by NRL project of the Korean Ministry of Science and

Technology.

References

1. W. M. Winslow, *J. Appl. Phys.* **20**, 1137 (1949).
2. H. Block and J. P. Kelly, *J. Phys. D: Appl. Phys.* **21**, 166 (1988).
3. H. Conrad and Y. Chen, in *Progress in Electrorheology*, ed. K. O. Havelka and F.E. Filisko (Plenum Press, New York, 1995), p. 55.
4. H. Uejima, *Jpn. J. Appl. Phys.* **11**, 319 (1972).
5. Y. Li, Y. Chen and H. Conrad, *ASME* **235**, 29 (1995).
6. X. Tang, C. Wu and H. Conrad, *J. Rheol.* **39**, 1059 (1995).
7. F. Filisko, in *Progress in Electrorheology*, ed. R. Tao (World Scientific Co., 1991), p. 116.
8. T. Kawakami, R. Aizawa, M. Konishi and Y. Asako, in *Progress in Electrorheology*, ed. M. Nakano and K. Koyama (World Scientific Co., 1997), p. 58.
9. H. Block and J. P. Kelly, *Langmuir* **6**, 6 (1990).
10. M. Leclerc, J. Guay and L.H. Dao, *Macromolecules* **22**, 649 (1976).
11. D. J. Klingberg and C. F. Zukoski, *Langmuir* **6**, 15 (1990).
12. H. Conrad, Y. Chen and A. Sprecher, *J. Modn. Phys. B* **16**, 2575 (1992).

ELECTRORHEOLOGY OF POLYANILINE-COATED POLY(METHYL METHACRYLATE) MICROSPHERE SUSPENSIONS IN SILICONE OIL

HYOUNG J. CHOI, YUN H. CHO, MIN S. CHO

Department of Polymer Science and Engineering, Inha University, Inchon, 402-751, Korea.
E-mail: hjchoi@inha.ac.kr

MYUNG S. JHON

Department of Chemical Engineering, Carnegie Mellon University, Pittsburgh, PA 15213, U.S.A.

Electrorheological (ER) suspensions containing spherical-monodispersed polymer particles in silicone oil were investigated. The core was poly(methylmethacrylate) (PMMA), obtained by dispersion polymerization method, and the shell, coated on monodisperse PMMA particles, was polyaniline(PA) prepared by in-situ polymerization of aniline in aqueous acidic medium. Both steady shear and dynamic tests were conducted under various electric field strength and particle concentrations in order to study flow and viscoelastic properties of the PAPMMA based ER fluids. Dielectric spectra of the ER fluids obtained were found to give a clue both to analyze their electrical polarization properties and to interpret flow behavior of the PAPMMA ER fluids.

1 Introduction

ER fluids are suspensions of polarizable particles in a non-conducting oil, which exhibit a dramatic change in rheological and electrical properties in the presence of an electric field, due to formation of a chainlike or columnar structure oriented along the electric field direction over the gap between the electrodes [1,2,3]. Clearly, the ER performance of an ER fluid crucially depends on the electrical properties of the suspending particles [4]. Various anhydrous ER fluids have been developed. These new ER materials, which possess advantages of a wide working temperature range, reduced device abrasion, and a relatively low current density, have intrinsic charge carriers in either the bulk particles or their surfaces that can move locally under an applied electric field. Examples include dehydrated zeolite [5,6], inorganic materials [7], carbonaceous material [8,9] and various semiconducting polymers, such as polyaniline [10,11], phosphate cellulose [12,13], poly(p-phenylene) [14], chitosan [15] and poly(acene quinone) radicals [16]. However, the most common particles in ER fluids are polydispersed and irregularly shaped. Therefore, there were some attempts to use ideal, well-characterized particles, such as polyaniline-coated silica [17], PMMA with double layered shell [18] and polyaniline/clay nanocomposite [19,20].

In this paper, electrorheological and polarization properties of suspensions of polyaniline (PA) coated PMMA particles (PAPMMA) in silicone oil were investigated. The PAPMMA particles were made by in-situ polymerization of aniline in presence of monodispersed PMMA microsphere. For the PAPMMA based ER fluids, the polarization under an applied electric field, therefore, was observed to occur on the semiconductive PA surface only.

2 Experimentals

2.1 Preparation of PMMA Particles

At first, PMMA particles were synthesized by dispersion polymerization that is one of most useful methods to make monodispersed polymer particles. The purified methyl methacrylate monomer and a radical initiaor (azoisobutyronitrile) were dissolved in methanol containing poly(vinyl pyrrolidone) as a stabilizer at room temperature. After removing oxygen by nitrogen purging in double jacket reactor, the reaction mixture was heated up to 55°C and kept at the same temperature for 24 hours. The monodispersed particles of 2 μ m obtained were sufficiently washed by methanol and dispersed in water prior to the polyaniline polymerization.

2.2 Preparation of Polyaniline Shell on PMMA Surface

The PMMA particles were dispersed in water containing sodium dodecylsulfate (SDS) and kept under mild stirring to allow SDS contacting and absorbing to the PMMA surface. Aniline was then added to this suspension by 4, 20 and 30 parts based on PMMA 100. Small amount of hydroquinone was also added to the reaction system in order to minimize polymerization of aniline in the water phase causing polyaniline particles [21]. Finally, it was acidified to pH 0.7 using HCl and then ammonium peroxysulfate initiator was put in the reaction system as a solution in aqueous HCl of pH 0.7 as well. The polymerization was continued for 24 hours at 25°C. The PAPMMA products were washed and their conductivities were controlled by same procedure of polyaniline [11] to the range from 10^{-9} to 10^{-10} S/cm, which is in the favorable conductivity range for particle materials of ER fluids. After the conductivity adjustment, products were dried under vacuum at room temperature. ER fluids were prepared by dispersing the dried PAPMMAs in silicone oil by 10 vol%.

2.3 Particle Characterization

From scanning electron microscope (SEM) (Fig. 1), the PAPMMA was found to have very rough surface and there were few aggregates of PMMA particles. Conductivities and dielectric constants of each PA-PMMA were measured using pressed disks of PAPMMAs. The amount of polyaniline on PMMA surface was calculated from thermal gravity analysis (TGA) results at 400°C, where PMMA is almost decomposed while polyaniline only shows about 5% of weight loss.

2.4 Electrorheological Characterization

Electrorheological properties of ER fluids using each PAPMMAs and PA were measured by a rotational rheometer (MC120, Physica) equipped a high voltage generator (DC) and concentric cylinder type measuring unit [22]. Polarization processes of each ER fluid were also determined using an impedance analyzer (HP) equipped with a liquid test fixture from 20Hz to 1MHz of electrical frequency. The input voltage is about 1 [V] so that there is no structuring in the ER fluids during measurement.

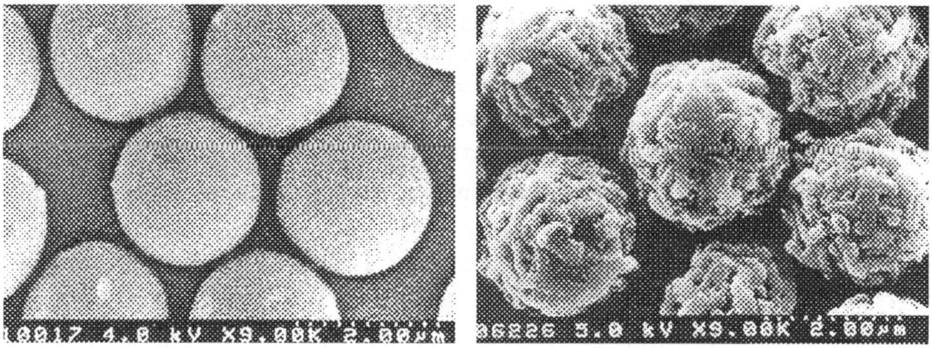

Figure 1. SEM photos of PMMA cores and PA-PMMA particles.

3 Results and Discussions

Flow curves of the ER fluids of PAPMMA20 (PA thickness; 52nm, conductivity; 8×10^{-9} S/cm) and PA (conductivity; 1.0×10^{-11} S/cm) dispersed in silicone oil by 10 vol%, were measured at various electric field and shown in the following Fig. 2 (a).

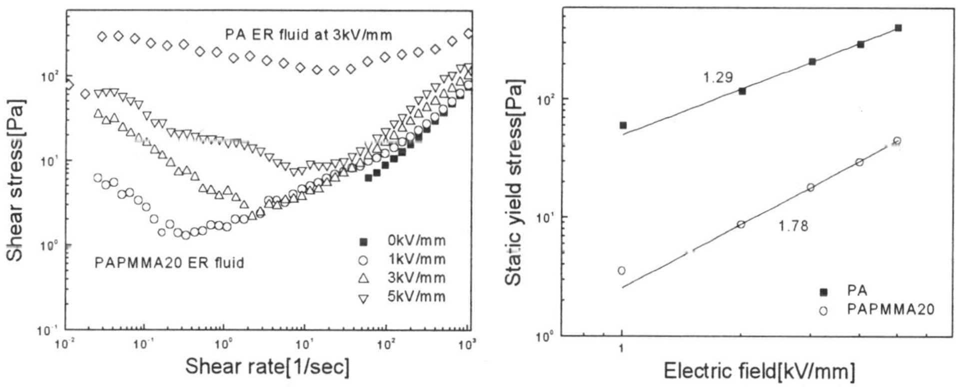

(a) (b)

Figure 2. Shear stress as a function of shear rate for PAPMMA20 ER fluid and PA ER fluid with various electric fields at 10vol% (a) and static yield stress as a function of shear rate for PAPMMA20 ER fluid and PA ER fluid with various electric fields (b).

Both PAPMMA4 (PA thickness; 11nm, conductivity; 8.6×10^{-9} S/cm) and PAPMMA30 (PA thickness; 83nm, conductivity; 3.2×10^{-10} S/cm) ER fluids also show flow behavior similar that of the with PAPMMA20 under an electric field. In addition, it can be noted that PA ER fluid shows different flow properties and its stress enhancement is bigger than that of PAPMMA series. For the both cases, shear stress increases as the field strength increases in the whole shear rate range and the enhanced shear stress is reduced as the shear rate increases up to a certain shear rate (a critical shear rate), which is a well-known trend of ER fluids. It is due to the competition between electrostatic and hydrodynamic interaction between the particles. Above the critical shear rate, the hydrodynamic interaction becomes dominant, and then shear stress monotonically increases with shear rate like Newtonian fluids. In addition, from the study of effect of particle concentration on the ER property at 3kV/mm, shear stresses were observed to increase with the particle concentration over the entire shear rate range as general ER fluids, and the critical shear rate was not severely changed by the particle concentration.

Fig. 2 (b) shows a plot of the static yield stress (τ_s) as a function of the electric field strength for the PAPMMA20 and PA ER fluids. The τ_s is the minimum shear stress required to flow ER fluids under an electric field and it is measured by the controlled stress mode, which shear rate is measured with sweeping shear stress. The correlation of τ_s with the electric field strength (E) at a given particle concentration is expressed as τ_s =cE^n [23]. Not only the ER properties of PAPMMA were much smaller than that of the PA based ER fluid at the same particle concentrations, but also the flow behavior was also different from the PA system. Recently, Gonon et al. [24] developed a simple, analytical conduction model for the case of ER fluids based on surface conducting particles; insulating spheres coated with a thin conducting polyaniline film. Although the electrical characteristics of the PAPMMA particles are more suitable than the PA particles for ER performance, τ_s of PA ER fluid are higher than that of PAPMMA ER fluid. It means that the electrostatic attraction between the PA particles is stronger than the PAPMMA particles. For confirming the reason, we measured dielectric spectra for each ER fluid in following session.

On the other hand, the exponent ns in the relationship between electric field and τ_s are about 1.78 for PAPMMA series and 1.29 for PA system as shown in Fig. 3. It is supposed to be related to the particle shape and size distribution, so that each particle fibril of the PAPMMA system is more compact than that of the PA system because its particle is spherical and monodisperse. PAPMMA ER fluids, therefore, can be considered to be more ideal and controllable than the PA ER fluid.

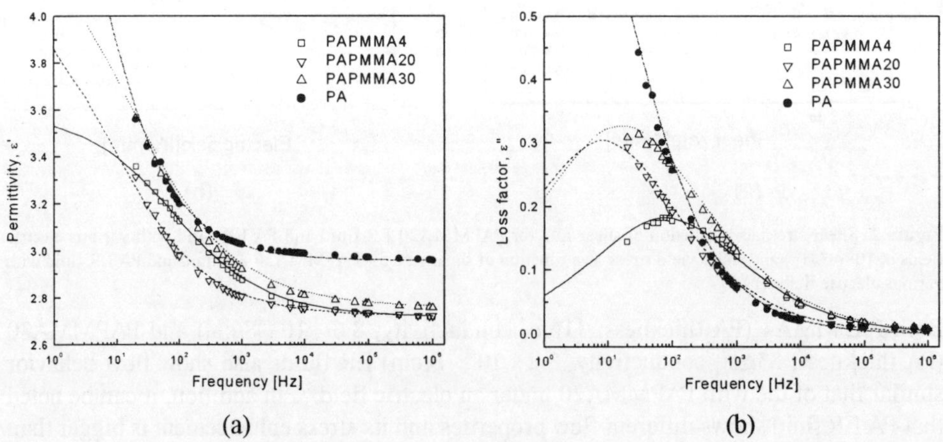

Figure 3. Permittivity (a) and loss factor (b) as a function of the electrical frequency.

Figs. 3(a) and (b) show permittivity (ε') and loss factor (ε'') respectively as a function of the electrical frequency for PAPMMA series and PA ER fluids. It is a typical result from the interfacial polarization of suspensions such as ER fluids consisting of a polarizable-dispersed phase and insulating medium. The dielectric relaxation of the ER fluids from the shape of ε' and ε'' as given in Fig. 3, indicating the characteristics of interfacial polarization could be deduced.

$$\varepsilon^* = \varepsilon' - i\varepsilon'' = \varepsilon_\infty + \frac{\varepsilon_0 - \varepsilon_\infty}{1+(i\omega\lambda)^{1-\alpha}} \qquad (0 \le \alpha \le 1) \qquad (1)$$

where the exponent $(1-\alpha)$ determines the broadness of the relaxation time distribution. In the $\alpha = 0$ case, this equation represents a single relaxation time (Debye dielectrics). The interfacial polarization induced by external electric field (AC) is delayed as the frequency (ω) increases and becomes absent at an ultimate frequency, which is expressed by the decrease of ε' from ε_0 (ε' as $\omega \rightarrow 0$) to ε_∞ (ε' as $\omega \rightarrow \infty$) or the peak of ε'' in dielectric spectrum. The relaxation time (λ) of the interfacial polarization can be obtained from the relation $\lambda = 1/2f_{max}$, where f_{max} is the frequency when ε'' is maximum [25]. In general, λ of ER fluid is known to be related with yield stress and stress enhancement under the applied electric field, as the shorter λ within adequate range, the higher stress enhancement is achieved [26,27]. However, PA ER fluid has stronger ER effect, although it has the longer λ than that of PAPMMA ER fluid. The reason of this phenomenon is an interfacial polarizability. The fact that PA suspension has best ER properties in spite of its unsuitable electrical properties for ER fluid is the superior interfacial polarizability to PAPMMA systems. The electrical properties of PAPMMAs mainly stem from the PA layer on the insulating PMMA surface. Under an electric field, PA-PMMA particles polarize with movement of the electrons in the PA layer and the imaginary counter charges [28] are induced at the interface between medium oil and PA layer as well as between PA and PMMA. The total polarization in whole particle must be reduced by the latter interfacial polarization and then, the electrostatic interaction within PAPMMA particles becomes weaker than that within PA particles. It is also the reason of the difference in flow behaviors and ER properties between PA and PA-PMMA ER fluids.

4 Conclusion

Monodispersed and spherical PMMA particles coated by PA were prepared and used to particle materials of ER fluids. PAPMMA particles had same particle size distribution as PMMA core and their surfaces were very rough due to PA layer. Static yield stresses (τ_s) of PAPMMA suspensions were smaller than that of PA ER fluid at same particle concentration and the flow behavior of PAPMMA system was also different from PA system. Although electrical characteristics of PAPMMA particles are better than PA particles for ER fluids within general fact, the electrostatic attraction between PA particles should be stronger than PAPMMA particles considering the interfacial polarizability of each suspension obtained from dielectric spectrum. The lower interfacial polarizability of PAPMMA ER fluids is supposed to be due to the insulating PMMA core. The exponent ns in the relationship between electric field and ns are about 1.78 - 1.8 for PAPMMA series and 1.29 for PA system. It must be resulted from the particle shape and size distribution of PAPMMA closer to ideal situation than PA.

5 Acknowledgements

This study was supported by research grants from the Korea Science and Engineering Foundation (KOSEF) through the Applied Rheology Center (ARC), an official KOSEF-created engineering research center (ERC) at Korea University, Seoul, Korea.

References

1. Tao R., Zhang J., Shiroyanagi Y., Tang X. and Zhang X., Electrorheological Fluids under Shear, *Int. J. Mod. Phys. B* **15** (2001) pp. 918-928.
2. Bossis G., Abbo C., Cutillas S., Lacis S. and Métayer C., Electroactive and Electrostructured Elastomers, *Int. J. Mod. Phys. B* **15** (2001) pp. 564-573.
3. Choi H. J., Cho M. S., Kim J. W., Webber R. M. and Jhon M. S., Effects of Ionic and Nonionic Substituents on the Electrorheological Characteristics of Polyaniline Derivatives, *Int. J. Mod. Phys. B* **15** (2001) pp. 988-995.
4. Gonon P., Foulc J. N. and Atten P., A Conduction Model Describing Particle-Particle Interaction in the Case of Surface Conducting Particles, *Int. J. Mod. Phys. B* **15** (2001) pp. 704-713.
5. Cho M. S., Choi H. J., Chin I. J. and Ahn W. S., Electrorheological Characterization of Zeolite Suspensions, *Micropor. Mesopor. Mater.* **32** (1999) pp. 233-239.
6. Qiu Z. Y., Liu L. W., Wang Z. W. and Zhou L. W., Rheological and Electrical Properties of NaY Zeolite Electrorheological Fluid, *Int. J. Mod. Phy. B* **15** (2001) pp. 610-617.
7. Zhang Y., Zhang S. and Lu K., The ER Behavior of Surface Modified Complex Strontium Titanate Suspensions, *Int. J. Mod. Phy. B* **15** (2001) pp. 596-598.
8. Choi H. J., Kim J. M., Yoon S. H., Fujiura R., Komatsu M. and Jhon M. S., Synthesis and Electrorheological Characterization of Carbonaceous Particle Suspensions, *J. Mater. Sci. Lett.* **18** (1999) pp. 1445-1447.
9. Kim J. W., Choi H. J., Yoon S. H. and Jhon M. S., Electrorheological Behavior of Carbonaceous Particle-Based Suspensions, *Int. J. Mod. Phys. B* **15** (2001) pp. 634-640.
10. Kim S. G., Kim J. W., Choi H. J., Suh M. S., Shin M. J. and Jhon M. S., Synthesis and Electrorheological Characterization of Emulsion polymerized DBSA-doped Polyaniline Based Suspensions, *Colloid Polym. Sci.* **278** (2000) pp. 894-898.
11. Choi H. J., Lee J. H., Cho M. S. and M. S. Jhon, Electrorheological Characterization of Semi-conducting Polyaniline Suspensions, *Polym. Eng. Sci.* **39** (1999) pp. 493-499.
12. Kim S. G., Choi H. J. and Jhon M. S., Preparation and Characterization of Phosphate Cellulose Based Electrorheological Fluids, *Macromol. Chem. Phys.* **202** (2001) pp. 521-526.
13. Kim S. G., Kim J. M., Jang W. H., Choi H. J. and Jhon M. S., Electrorheological Characteristics of Phosphate Cellulose-Based Suspensions, *Polymer* **42** (2001) pp. 5005-5012.
14. Sim I. S., Kim J. W., Choi H. J., Kim C. A. and Jhon M. S., Preparation and Electrorheological Characteristics of Poly(p-phenylene)-Based Suspensions, *Chem. Mater.* **13** (2001) pp. 1243-1247.

15. Choi U. S., Ahn B. G. and Kwon O. K., Electrorheological Behavior of Chitosan Derivative Suspensions, *Int. J. Mod. Phys. B* **15** (2001) pp. 1025-1032.
16. Choi H. J., Cho M. S. and Jhon M. S., Hysteresis Behaviors of Poly(naphthalene quinone) radical Electrorheological Fluid, *Int. J. Mod. Phys. B* **13** (1999) pp. 1901-1907.
17. Trlica J., Sáha P., Quadrat O. and Stejskal J., Electrorheology of Polyaniline-coated Silica Particles in Silicone Oil, *J. Phys. D: Appl. Phys.* **33** (2000) pp. 1773-1780.
18. Saito T., Anzai H., Kuroda S. and Osawa Z., Electrorheological Particles Composed of Polymer Core with Controlled Diameter and Electro-Conductive/ Nonconductive Double Layers Shell, *Int. J. Mod. Phys. B* **13** (1999) pp. 1698-1696.
19. Kim J. W., Kim S. G., Choi H. J. and Jhon M. S., Synthesis and Electrorheological Properties of Polyaniline-Na^+-Montmorillonite Suspensions, *Macromol. Rapid Commun.* **20** (1999) pp. 450-452.
20. Choi H. J., Kim J. W., Joo J. and Kim B. H., Synthesis and Electrorheology of Emulsion Intercalated PANI-caly Nanocomposites, *Synth. Met.* **121** (2001) pp. 1325-1326.
21. Choi H. J., Kim T. W., Cho M. S., Kim S. G. and Jhon M. S., Electrorheological Characterization of Polyaniline Dispersions, *Eur. Polym. J.* **33** (1997) pp. 699-703.
22. Choi H. J., Cho M. S., Kang K. K. and Ahn W. S., Electrorheological Characteristics of Mesoporous Molecular Sieve (MCM-41) Suspension, *Micropor. Mesopor. Mater.* **39**(2000) pp. 19-24.
23. Choi H. J., Cho M. S., Kim J. W., Kim C. A. and Jhon M. S., A Yield Stress Scaling Funtion for Electrrheological Fluids, *Appl. Phys. Lett.* **78**(2001) pp. 3806-3808.
24. Gonon P., Foulc J. W., Atten P. and Boissy C., Particle-particle Interactions in Electrorheological Fluids Based on Surface Conducting Particles, *J. Appl. Phys.* **86** (1999) pp. 7160-7169.
25. Choi H. J., Cho M. S. and To K., Electrorheological and Dielectric Characteristics of Semiconductive Polyaniline-silicone Oil Suspensions, *Physica A* **254** (1998) pp. 272-279.
26. Block H., Kelly J. P., Qin A. and Watson T., Materials and Mechanisms in Electrorheology, *Langmir* **6**(1990) pp. 6-14
27. Lee J. H., Cho M. S., Choi H. J. and M. S. Jhon, Effect of Polymerization Temperature on Polyailine Based Electrorheological Suspensions, *Colloid Polym. Sci.* **277**(1999) pp. 73-76.
28. Boissy C., Atten P. and Foulc J-N., On a Negative Electrorheological Effect, *J. Electrostat.* **35**(1995) pp. 13-20.

SYNTHESIS AND ELECTRORHEOLOGY OF MESOPOROUS PARTICLE SUSPENSIONS

HYOUNG J. CHOI AND MIN S. CHO

Department of Polymer Science and Engineering, Inha University, Inchon, 402-751, Korea.
E-mail: hjchoi@inha.ac.kr

WHA-SEUNG AHN

Department of Chemical Engineering, Inha University, Inchon, 402-751, Korea

Electrorheological (ER) properties of a fluid composed of mesoporous MCM-41 particles suspended in silicone oil under a high electric field were investigated. Typical properties of ER fluids such as viscosity enhancement and viscoelasticity under applied electric fields in MCM-41/silicone oil suspensions were observed. In the dynamic and static yield stress measurements, a strain hardening effect was also observed and subsequently confirmed by elasticity measurement as a function of time under a linear viscoelastic condition.

1 INTRODUCTION

Electrorheological (ER) fluids are colloidal dispersions whose rheological properties can be changed by an imposed electric field [1]. Commonly they are composed of a suspension of micron-sized particles in a non-conducting fluid, and single phase ER fluids such as solutions of liquid crystalline polymer were reported [2]. The variation of rheological properties of ER fluids under an electric field is caused by a characteristic fibrillation; the strings of the particles attracting each other are oriented along the direction of the electric field applied. In the absence of an electric field, the fluid exhibits either a Newtonian or non-Newtonian fluid-like behavior, depending on the particle concentration.

ER suspensions can be classified either an intrinsic polarizable (anhydrous) or extrinsic polarizable (hydrous) system. The former usually contains semiconducting polymer particles such as polyanilines and its derivatives [3-6], poly(acene quinone) radicals [7], poly(p-phenylene) [8] and inorganic particles including PZT [9] and carbonaceous particle [10] which are polarized in an electric field due to electrons within the polymer molecule, whilst those materials that can be polarized in the electric field with additives such as water and alcohol, are known as hydrous systems. It is generally believed that the anhydrous ER systems are better than the hydrous counterparts in practical applications due to higher thermal and dispersion stabilities.

Commercial zeolites have also been tested as a suspension material for ER fluids [11,12]. Recently, Cho et al. [12] studied rheological and dielectric properties of the ER fluids prepared from commercial zeolites 3A, 5A and 13X suspended in silicone oil and found that the ER fluid using 13X showed the best ER performance among the zeolite based ER fluids investigated. It can be noted that porous solid materials including zeolite have been used technically as adsorbents, catalysts and catalyst supports owing to their high surface areas. However, since their applications are limited by the relatively small pore openings, pore enlargement was one of the main aspects in zeolite research [13], and a myriad of synthetic efforts to enlarge the pore size led to the preparation of ordered mesoporous materials with a wide range of pore sizes up to 30nm [14].

Therefore, a recent discovery of a new family of crystalline mesoporous molecular sieve materials M41S by researchers at Mobil Oil Corp. has attracted considerable interest. They are either a hexagonal phase, a cubic phase, or a unstable lamellar phase which are known as MCM(Mobil Composition of Matter)-41, MCM-48 and MCM-50, respectively [15]. Since mesoporous materials possess high surface areas and the size of pores which can be tuned within 20-100Å in very narrow distribution by varying the preparation conditions, a large number of potential applications of these materials are expected to emerge in the area of catalysis, separation, adsorption and advanced materials [16]. A potential application of MCM-41 to ER fluids is also suggested by Choi et al. [17].

In this study, the synthesized MCM-41 was dispersed in silicone oil with 5 ~ 20 wt% particle concentration and these suspensions were evaluated as ER fluids. We also estimated the potential of mesoporous materirals as a dispersed phase of ER suspensions.

2 EXPERIMENTAL

MCM-41 was synthesized by the literature procedure of Ryoo et al. [18] using an aqueous solution of hexadecyltrimethylammonium chloride (HTACl, 25wt%, Aldrich), a concentrated ammonia solution (28wt% NH_3), and sodium silicate. HTA-silicate gel with a molar composition of $4SiO_2$:1HTACl:$1Na_2O$:$0.15(NH_4)_2O$: $200H_2O$ was placed into a polypropylene bottle and heated to 370 K in an oven for 3 days. $EDTANa_4$ and acetic acid treatments were performed during the synthesis to enhance crystallinity and hydrothermal stability. The precipitated product was filtered, washed with distilled water and dried for 12 h in an oven kept at 370 K, followed by calcination in O_2 at 823 K for 10 h.

XRD pattern of the calcined MCM-41 sample was obtained with a powder X-ray diffractometer (Philips, PW-1700) using a CuK_α radiation (40kV, 25mA). Nitrogen adsorption isotherm and specific area were determined using Micromeritics ASAP 2000 automatic analyzer.

A series of fluid mixture were prepared by dispersing the MCM-41 particles in anhydrous silicone oil with 5, 10, and 20 wt% of particle concentrations. ER properties of the fluid were then measured at 25±0.1°C using a rotation type rheometer (MC120, Physica) equipped with a high voltage generator. Measuring unit was a concentric cylinder having 1.06 mm gap between a set of bob and cup, and static high electric field perpendicular to the flow direction was applied to the measuring unit through the cup; the inner wall of the cup is the positive electrode, the bob being grounded.

Initially, an electric field was applied to the ER fluid for 3 minutes before the measurement in order to establish the equilibrium internal structure of particles in the ER fluid. Flow curves for each ER fluid were obtained in a controlled shear rate. Static yield stresses were measured in controlled shear stress experiments, and these corresponded to the smallest stress to fluidize an ER fluid under the applied electric field.

3 RESULTS and DISCUSSION

Figure 1 shows the XRD pattern and TEM photo of the calcined MCM-41 prepared. Regular hexagonal channels of MCM-41 structure were clearly implicated by the strong (100) peak followed by three (110) (200) and (211) peaks in the 2θ ranges of 2 – 7° in the diffractogram (Fig. 1(a)). BET surface area of the sample was ca 950 m^2/g with average

BJH pore diameter of 31 Å. TEM image of Fig. 1(b) shows uniform hexagonal shaped pore structure of MCM-41 used in the ER experiments. Particle morphology of MCM-41 materials is reported variable and surface textures exhibit a range from smooth through pitted [19]. According to our SEM analysis (not shown), MCM-41 crystals prepared were mostly made up of particles, 80 – 120 nm in size, and existed as agglomerated groups [17].

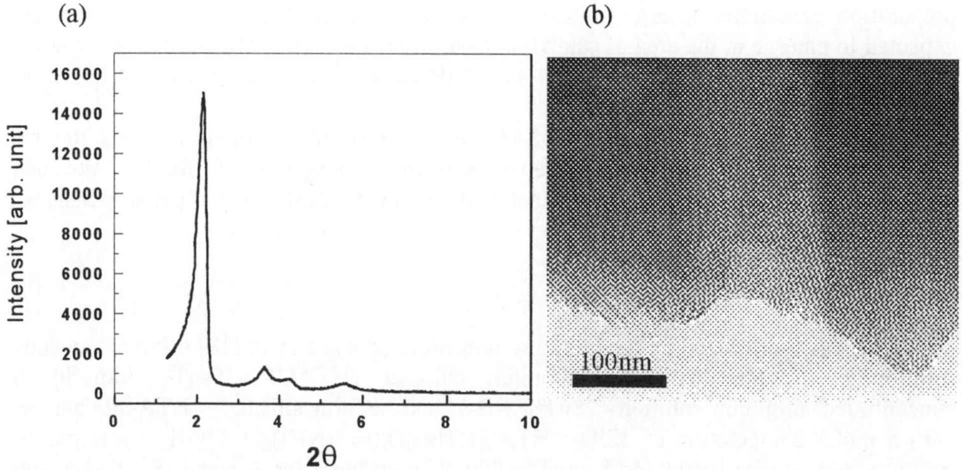

Figure 1. XRD pattern (a) and TEM image (b) for MCM-41 used herein.

Figures 2(a) and (b) show the flow behavior of shear stresses and shear viscosities as a function of shear rate under various applied electric fields for the MCM-41 ER system with 10% particles (w/w) in silicone oil. Typical ER trends such as stress increase with electric field and shear stress decrease or plateau at low shear rate region [20,21,22] are observed as shown in Figs. 2. MCM-41 is hydrophilic but has no polarizable species such as electrons and ions, so that the MCM-41 ER fluid is believed to be activated by absorbed moisture. The moisture content in MCM-41 was ca. 6% (w/w) from TGA thermogram.

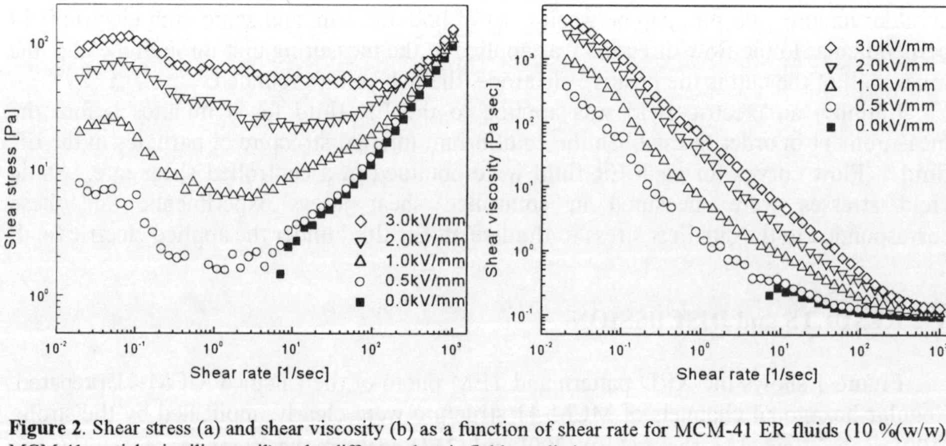

Figure 2. Shear stress (a) and shear viscosity (b) as a function of shear rate for MCM-41 ER fluids (10 %(w/w) MCM-41 particles in silicone oil) under different electric fields applied.

Shear stress data of MCM-41 ER fluids having three different particle concentrations (5, 10, and 20 wt%) are represented in Fig. 3(a). A sample with 5wt% MCM-41 showed little ER effect as is obvious from its Newtonian flow behavior; the slope of stress versus shear rate in log-scaled plot was almost 1.0. The rest demonstrate similar flow behaviors of ER fluids with the shear rate being the same as the results given in Fig. 2. The "critical particle concentration" above which such a large ER property is observed is noteworthy. More detailed measurement on this observation will be attempted in future communication.

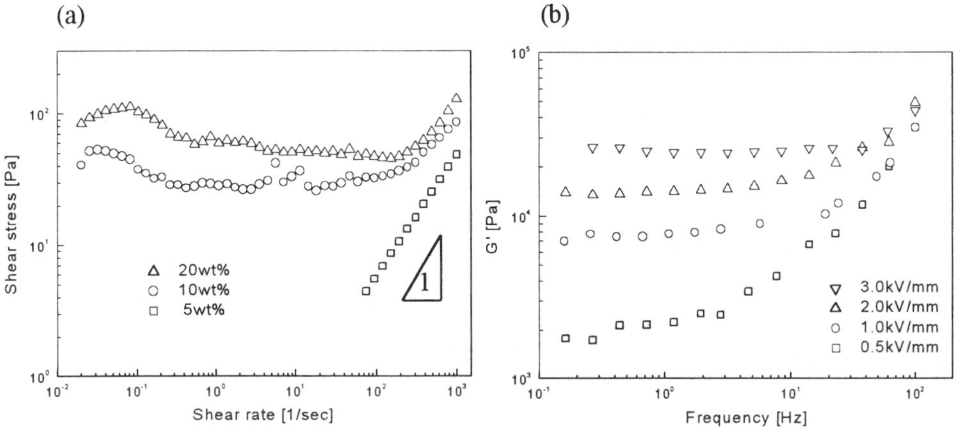

Figure 3. (a) Shear stress as a function of shear rate for MCM-41 ER fluids with various particle concentrations (5, 10 and 20 % (w/w)) at 3kV/mm. (b) G' as a function of deformational frequency with strain of 0.0015 for MCM-41 10 %(w/w) ER fluid.

Dynamic tests employing an oscillatory shear were also conducted and they were used to study the viscoelastic properties of the solidified ER fluid under an applied electric field. Linear viscoelastic strain is obtained as 0.0015 from the results of the strain amplitude sweep, which is a measurement of stress as a function of sinusoidal strains at a constant frequency. The storage modulus (G'), the in-phase stress component with the strain, is observed to be larger than the loss modulus (G"), i.e., the out-of-phase stress component, and these values were independent of the strain in the linear viscoelastic region. Furthermore, as we increased the applied strain, G" becomes bigger than G', and these moduli sharply decreased. This phenomenon can be explained in terms of the elasticity of ER fluid, which is generated by particle chain structures in an imposed electric field [23,24]. When the fibril structures of the suspended zeolite particles sustain the applied strain, the elasticity is dominant in the linear viscoelastic region. However, as the strain is increased, the deformation starts to distort the structure, and the structure breaks down beyond a certain degree of deformation, and finally the elasticity of ER fluid disappears abruptly [10]. Martin et al. [25], reported that they could not observe any linear viscoelastic region and, therefore, claimed that the energy of ER fluid was stored in G". Based upon our experimental result, however, it seems their strain amplitude sweep starting from 0.01 too large to find the linear viscoelastic region.

Figures 3(b) shows plot of G' as a function of frequency with small strain in linear viscoelastic region. It was shown that G's either remained constant or increased slightly as deformation frequency was increased up to 100 Hz. Since the relaxation time for deformation was too long, it is expected that the internal chain structures of ER fluids are not destroyed by deformation under the given conditions. The increase in G' with the applied electric field indicate that the ER fluid becomes more elastic with the electric fields under linear viscoelastic conditions.

There are two kinds of yield stress for ER fluids, dynamic yield stress (τ_d) and static yield stress (τ_s), depending on the measurement methods [26]. τ_d is an extrapolated value from shear stress vs. shear rate curve, especially the plateau region of stress in log-scale plot. τ_s is closer to the definition of yield stress than τ_d because it is determined as a stress to flow a solidified ER fluid with increasing stress by using controlled shear stress (CSS) type rheometers or specially designed equipments. In Figure 4, τ_d and τ_s for MCM-41 ER fluids are plotted at different applied electric fields. τ_s was measured by a fast sweep method of stress increasing from 0 to $2\tau_d$ [Pa] within 5 seconds, in which the τ_d is obtained from flow curve in same electric field. Reproducible data could be obtained from this method, and furthermore, we were able to make ER fluids yield at similar condition for each electric field by varying the stress sweep range up to $2\ \tau_d$. In Fig. 4, τ_d's are higher than the corresponding τ_s's in the whole electric fields, and it manifests a reinforcing effect of the solidified ER fluid by a small strain as shear rate starts applying in initial stage, so called a "strain hardening effect". The sufficiently small strain in this stage deforms the particle chain structures in the ER fluid under electric field, and it is expected to rearrange and stabilize the chain structures until breakage of them with higher strain followed.

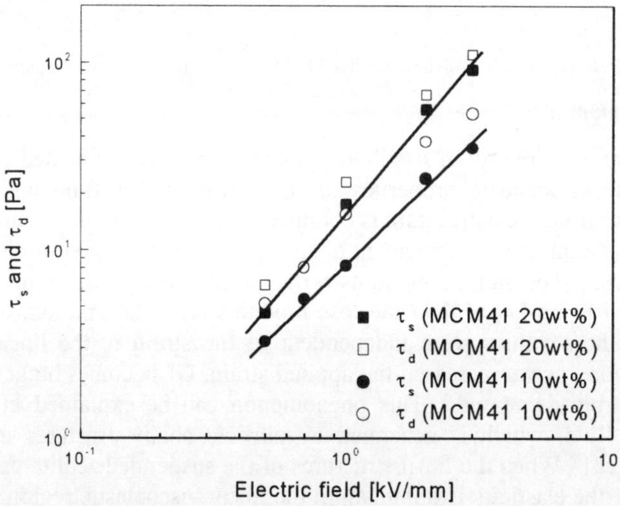

Figure 4. τ_d (open symbols) and τ_s (solid symbols) of each MCM-41 ER fluid with various electric fields. The solid lines represent the relationship between τ_s and electric field such as $\tau_s \sim E^{1.67}$ for MCM-41 20wt% and $\tau_s \sim E^{1.35}$ for MCM-41 10wt%.

4 Conclusions

MCM-41 mesoporous molecular sieve was synthesized and its ER performance was investigated. The synthesized MCM-41 had well-defined channel structure according to XRD and TEM results. Its suspension in silicone oil showed typical ER properties and moisture absorbed in hydrophilic MCM-41 is the polarization species under a high electric field. The static yield stresses measured in CSS mode of rheometer, and these were related to applied electric fields by $\tau_s \sim E^{1.67}$ for MCM-41 20wt% ER fluid and $\tau_s \sim E^{1.35}$ for MCM-41 10wt% ER fluid. In addition, τ_s's were lower than τ_d's under each condition for both ER fluids and it is a consequence of the strain hardening effect under the small strain range. The linear viscoelastic properties of MCM-41 ER fluid under electric fields were also determined by G' and G" measurement at small strain. The elasticity of solidified ER fluids increased with applied electric fields under the linear viscoelastic condition.

5 Acknowledgement

This study was supported by research grants from the Korea Science and Engineering Foundation (KOSEF) through the Applied Rheology Center (ARC), an official KOSEF-created engineering research center (ERC) at Korea University, Seoul, Korea.

References

1. Cutillas S. and Bossis G., A Comparison between Flow-induced Structures in Electrorheological and Magnetorheological Fluids, *Europhys. Lett.*, **40** (1997) pp. 465-470.
2. Yamada H., Taniguchi Y., and Inoue A., Behavior of ER fluids Based on Three Variants of a Liquid Crystalline Polysiloxane, *Int. J. Mod. Phys. B*, **15** (2001) pp. 947-964.
3. Choi H. J., Kim T. W., Cho M. S., Kim S. G. and Jhon M. S., Electrorheological Characterization of Polyaniline Dispersion, *Eur. Polym. J.* **33** (1997) pp. 699-703 .
4. Choi H. J., Cho M. S., and To, K., Electrorheological and Dielectric Characteristics of Semiconductive Polyaniline-silicone Oil Suspensions, *Physica A* **254** (1998) pp. 272-279.
5. Choi H. J., Kim J. W., Suh M. S., Shin M. J. and To, K., Synthesis and Viscoelastic Behaviors of Poly (aniline-co-o-ethoxyaniline) Particles Suspended Electrorheological Fluid, *Int. J. Mod. Phys. B* **15** (2001) pp. 649-656.
6. Choi H. J., Cho M. S., Kim J. W., Webber R,. M. and Jhon M. S., Effect of Ionic and Nonionic Substituents on the Electrorheological Characteristics of Polyaniline Derivatives, *Int. J. Mod. Phys. B* **15** (2001) pp. 988-995.
7. Choi H. J., Cho M. S. and Jhon M. S., Hysteresis Behaviors of Poly(naphthalene quinone) radical Electrorheological Fluid, *Int. J. Mod. Phys.B* **13** (1999) pp. 1901-1907.
8. Sim I. S., Kim J. W., Choi H. J., Kim C. A. and Jhon M. S., Preparation and Electrorheological Characteristics of Poly(p-phenylene)-Based Suspensions, *Chem. Mater.*, **13** (2001) pp. 1243-1247.
9. Wen W., Wang N., Tam W. Y. and Sheng P., Magnetic Materials-Based Electrorheological Fluids, *Appl. Phys. Lett.* **71** (1997) pp. 2529-2531.

10. Kim J. W., Choi H. J., Yoon S. H. and Jhon M. S., Electrorheological Behavior of Carbonaceous Particle-Based Suspensions, *Int. J. Mod. Phys. B* **15** (2001) pp. 634-640.
11. Klingenberg D.J., Pakdel P., Kim Y. D., Belongia B. M. and Kim, S., Protein-Enhanced Electrorheological Fluids, *Ind. Eng. Chem. Res.* **34** (1995) pp.3303-3306.
12. Cho M. S., Choi H. J., Chin I. J. and Ahn W. S., Electrorheological characterization of zeolite suspensions, *Micropor. Mesopor. Mater.* **32** (1999) pp. 233-239.
13. Kruk M., Jaroniec M. and Sayari A., New Insights into Pore-size Expansion of Mesoporous Silicates using Long-chain Amines, *Micropor. Mesopor. Mater.* **35/36** (2000) pp. 545-553.
14. Danumah Ch., Zaidi S. M. J., Xu G., Voyer N., Giasson S. and Kaliaguine S., Siliceous Mesoporous Molecular Sieves Derived from Crown Ether Surfactants, *Micropor. Mesopor. Mater.* **37** (2000) pp. 21-32.
15. Ciesla U. and Schűth F., Ordered Mesoporous Materials, *Micropor. Mesopor. Mater.* **27** (1999) pp. 131-149.
16. Zhang W., Fr ö ba M., Wang J., Tanev P. T., Wong J. and Pinnavaia T. J., Mesoporous Titanosilicate Molecular Sieves Prepared at Ambient Temperature by Electrostatic and Neutral Assembly Pathway: A comparison of Physical Properties and Catalystic Activity for Peroxide Oxidations, *J. Am. Chem. Soc.* **118** (1996) pp. 9164-9171.
17. Choi H. J., Cho M. S., Kang K. K. and Ahn W. S., Electrorheological Properties of a Suspension of a Mesoporous Molecular Sieve (MCM-41), *Micropor. Mesopor. Mater.* **39** (2000) pp.19-24.
18. Ryoo, R., Ko, C. H., and Howe, R. F., Imaging the Distribution of Framework Aluminum in Mesoporous Molecular Sieve MCM-41, *Chem. Mater.* **9** (1997) pp. 1607-1613.
19. Huo, Q., Margolese, D. I., Ciesla, U., Feng, P., Gier, T. E., Sieger, Firouzi, A., Chmelka, B. F., Schüth, F., and Stucky, G. D., *Chem. Mater.* **6** (1994) pp. 2317.
20. Kim S. G., Kim J. W., Choi H. J., Suh M. S., Shin M. J. and Jhon M. S., Synthesis and Electrorheological Characterization of Emulsion-Polymerized DBSA-doped Polyaniline Based Suspensions, *Colloid Polym. Sci.* **278** (2000) pp. 894-898.
21. Kim S. G., Choi H. J. and Jhon M. S., Preparation and Characterization of Phosphate Cellulose based Electrorheological Fluids, *Macromol. Chem. Phys.* **202** (2001) pp. 521-526.
22. Lee Y. H., Kim C. A., Choi H. J. and Jhon M. S., Synthesis and Electrorheological Characteristics of Microencapsulated Polyaniline Particles with Melamine-Formaldehyde Resins, *Polymer* **42** (2001) pp. 8277-8283.
23. Cho M. S., Choi Y. J., Choi H. J., Kim S. G. and Jhon M. S., Viscoelasticity of an Electrorheological Fluid using a Vertical Oscillation Rheometer, *J. Molecular Liq.* **75** (1998) pp. 13-24.
24. Kim S. G., Kim J. W., Cho M. S., Choi H. J. and Jhon M. S., Viscoelastic Characterization of Semiconducting DBSA-doped Polyaniline Electrorheological Suspensions, *J. Appl. Polym. Sci.* **79** (2001) pp. 108-114.
25. Martin J. E., Adolf D. and Halsey T. C., Electrorheology of a Model Colloidal Fluid, *J. Colloid Interface Sci.* **167** (1994) pp.437-452.
26. Choi H. J., Cho M. S., Kim J. W., Kim C. A. and Jhon M. S., A Yield Stress Scaling Function for Electrorheological Fluids, *Appl. Phys. Lett.* **78** (2001) pp. 3806-3808.

ELECTRORHEOLOGICAL BEHAVIORS OF POLYANILINE-MONTMORILLONITE CLAY NANOCOMPOSITE

JUN LU XIAOPENG ZHAO

Institute of Electrorheological Technology, Department of Applied Physics 141#, Northwestern Polytechnical University, Xi'an, 710072 P. R. China

Email: xpzhao@nwpu.edu.cn

Polyaniline-montmorillonite nanocomposite (PANI-MMT) particles were synthesized by an emulsion intercalation method and characterized by IR, XRD and TEM spectrometry. TEM showed that the particle's size of MMT-PANI particles was about 100 nm. The dielectric constant of PANI-MMT nanocomposite was increased 2.4 times than that of MMT and 7 times than PANI, the conductivity of PANI-MMT particles was increased 10 times than that of MMT. Meanwhile, the dielectric loss tangent was also increased about 1.36 times than that of PANI. The electrorheological behaviors of the suspensions of PANI-MMT nanocomposite particles in silicone oil with a 30% weight fraction were investigated under DC electric fields. In 3 kV/mm DC field at room temperature, the yield stress was 8.26 kPa (shear rate 5 s^{-1}). In 4 kV/mm DC field, the shear strength was 8.30 kPa (γ=103.1 s^{-1}, T=20 ℃), and much higher than that of pure polyaniline (PANI), montmorillonite (MMT) and mixture of polyaniline with clay (MMT+PANI). The sedimentation experiment showed that the PANI-MMT nanocomposite particles did not deposit during about two months. The relevant influential factors between shear stress and electric fields, between shear stress and shear rate, between shear stress and temperature was also discussed preliminarily. The results showed that the MMT-PANI ER fluid displays a notable ER effect under DC electric field.

1 Introduction

Electrorheological (ER) fluids typically consist of electrically polarizable particles dispersed in low-dielectric oils. Application of an electric field induces polarization of the suspended particles, which, upon interaction, form in an increase in the apparent viscosity. Because of their controllable viscosity and short response time, ER fluids are regarded as a smart material for active devices, which transform electric energy to mechanical energy. When an electric fluid is applied, the fluid exhibits an enhanced shear stress, and in particular develops a yield stress at zero shear rate. A fluid that has this characteristic is called a Bingham fluid. Thus, ER fluid behaves like a Bingham fluid under an applied field, whereas it exhibits the characteristics of a Newtonian fluid when the electric field is turned off. During the past decade, ER fluids have gained much attention due to their promising applications in mechanical systems such as brakes, clutches and shock absorbers [1]. Among various ER materials, polymeric particles have attracted great interest because of thermal stability at high temperature, lower density and easy preparation. Recently, semiconducting polymers including poly (acene quinone) radicals [2], polyaniline (PANI) [3], copolypyrrole [4] and copolyanilines [5] have been adopted as anhydrous ER fluids. In particular, PANI is considered as one of the most promising ER fluids [6]. Unfortunately, some available ERF materials, part with relatively low shear stress and narrow operating temperature, part with poor suspension stability, are not satisfactory, much less suitable for engineering.

In recent years, clay minerals have attracted great interest from researchers in the field of organic-inorganic nanocomposites because of their small particle size and intercalation properties, which frequently exhibit unexpected hybrid properties synergistically derived from two components, especially in the application of reinforcement materials with polymers [7,8,9]. Kuramoto [10,11] *et al.* prepared polyaniline-coated silica particles and study their ER behavior. The most widely used method to prepare the organic-inorganic nanocomposite was emulsion intercalative polymerization or a melt intercalation of polymers into layered clays. Montmorillonite (MMT) clay, which is an important class of lamellar inorganic compounds whose interlayer spacing can be modified by cation exchange, has been widely used for this purpose. Montmorillonite contains negatively charged layered silicates bonded by ionic bonding either metal cations such as Na^+ or Ca^{2+}. When the clay is exchanged with Na^+, it possesses a large swelling rate water, and its inner layer spacing becomes large enough to be penetrated by a monomer. The emulsion system consisting of an aqueous medium can contribute to maximization of the affinity between the hydrophilic host and hydrophobic guest by the action of the emulsifier. Furthermore, the emulsion polymerization has the advantage of being able to simultaneously attain both high molecular weight and high reaction rates. According to this concept, some polymer/ montmorillonite material had been prepared and studied their properties [12]. Wu *et al.* introduced a sample emulsion intercalative polymerization technique using aniline and Na^+-MMT to synthesize a polyaniline–MMT hybrid nanocomposite [13]. They claimed that in the nanocomposite polyaniline exists in single chains with extended-chain conformation owing to the confined environment in the nanometer size gallery. Choi *et al.* synthesized styrene-acrylonitrile (SAN)-clay nanocomposite particles and studied their ER properties [14]. Dielectric and polarization properties are closely related to molecular or structure of materials. Meanwhile, the size of particles also play a important role for particle's polarization under the electric field. According to this idea, in our previously work, we found that doping of cerium ion into anatase TiO_2 can improve the electrorheological effects of TiO_2 and broaden operational temperature range [15]. That was because dieletric loss and dielectric constant at low frequency of TiO_2 was improved by doping of cerium.

In order to continue our research project, according to the physics and chemistry design, we prepared and characterized the high dielectric constant and conductivity polyaniline-montmorillonite clay nanocomposite (PANI-MMT) using an emulsion intercalation polymerization, and investigated the ER effect of a suspension of PANI-MMT dispersed in silicone oil as well as the dielectric properties of the PANI-MMT particle.

2 Experimental

IR spectra were obtained by use of neat capillary cells on a Bruker Equinox-55 instrument. X-ray diffraction study of the samples was carried out with a Rigaku D/MAX/2400 X-ray diffractometer (45kV, 80mA) with a copper target. The sample for transmission electron microscopy (TEM) observation was obtained on a H-800 at an accelerated voltage of 100 kV. The rheological properties of the ER fluids were examined via a system consisting of a rotary rheometer (NXS-11A, the gap between the outer cup and inner bob is 2 mm) and a high-voltage DC power source (GYW-010). The static yield stress was measured with a parallel-plate force transducer.

After Swelling 10 g Na^+-montmorillonite (cationic exchange capacity: CEC=80 meq/100g) in 500 mL deionized water and stirring for 2 h at 80℃, 2 mL aniline and 0.04 mmol HCl were added and continue stirring for 6 h at 80℃. The suspension was filtered and the collected solid washed with a mount of deionized water until no aniline can been found in aqueous solution, then dispersing to 500 mL deionized water. The pH of aqueous solution was adjusted to about 2 by the addition of HCl solution, 1.6 g (0.007 mol) $(NH_4)_2S_2O_8$ was added and stirred for 12 h at room temperature. The solid was filtered, washed with deionized water, EtOH, toluene and then immersed in NH_3 aueous solution (pH=10). The suspension was filtered and dried in vacuum. ER fluids were prepared by dispersing the PANI-MMT particles in a nonconductive dielectric silicone oil (ε_f = 2.60~2.80, $\sigma = 10^{-12} \sim 10^{-13}$ S/m, ρ = 0.9~1.0 g/cm^3, $\eta \approx$ 500 mPa • s (25℃)) with a 30% weight fraction, which was dried in a vacuum oven before use.

3 Results and discussion

The X-ray diffraction curves of MMT and PANI-MMT are shown in figure1. The evidence of insertion depends on the change of clay interlayer spacing. According to the Bragg's diffraction formula: $\lambda = 2d \sin \theta$, we can estimated the variation of the interlayer spacing. A strong peak appears at $2\theta = 5.8°$ representing the diffraction from the (001) crystal surface of the PANI-MMT, the corresponding d-spacing for this peak was 1.52 nm. However, the d-spacing for the peak of free MMT (001 crystal surface) was 0.96 nm. From the results of X-ray diffraction experiments, we observed the insertion of PANI between the interlayer of the MMT, whose inner layer distance consequently becomes larger (1.52 nm) than in montmorillonite (0.96 nm).

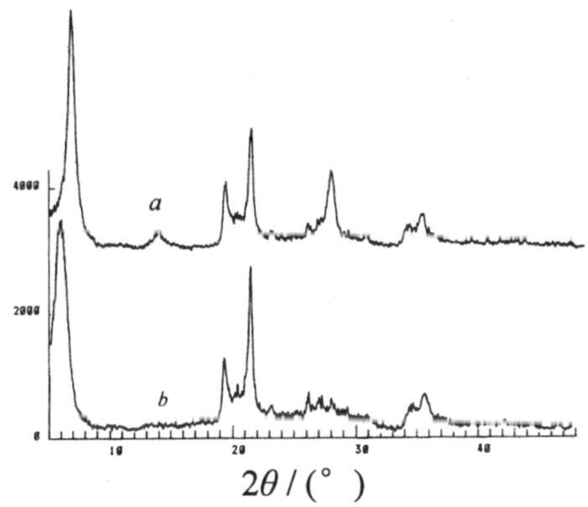

Figure 1. XRD patterns of different materials (a) MMT (b) PANI-MMT

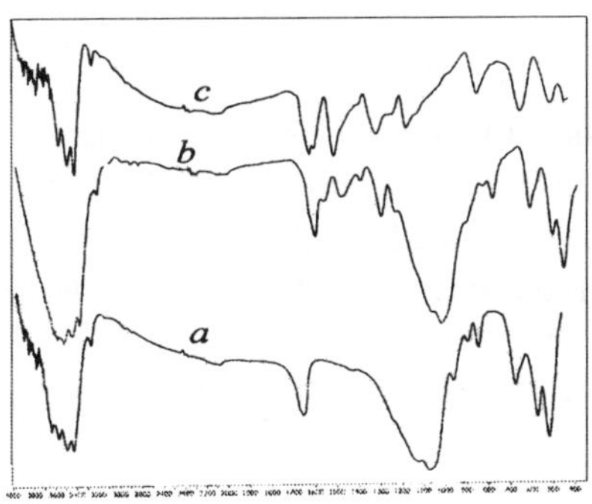

Figure 2. IR spectra (a) MMT; (b) PANI-MMT; (c) PANI

Figure 2 shows the FT-IR spectra of the PANI particle, MMT clay particle and PANI-MMT nanocomposite, respectively. For the PANI-MMT particles, the peak at wave number 1494.4 cm^{-1}, 1615.8 cm^{-1} and 1310.2 cm^{-1} are associated with aromatic amine N-B-N, N=Q=N and C-N stretching vibrations (B is the structure of benzenoid and Q is quinonoid), It shows that polyaniline is on the form of emerldine salt in PANI-MMT. TEM spectrometry provides a evidence that we obtained is really a nanocomposite with polyaniline inserted between the clay layers and the diameter of MMT-PANI particles is about 100 nm. The content of polyaniline in the PANI-MMT nanocomposites was 6.9wt% with the measurement of quantitative ultraviolet spectra. This measurement was slightly less than that of calculation (7.36wt%) obtained from the cationic exchange apacity of MMT. Apparently, we obtained is really a nanocomposite with polyaniline inserted between the clay layers of montmorillonite, and not the surface coating of the clay by polyaniline.

Dielectric properties measurements were carried out at room temperature through the frequency range 100 Hz to 10^4 Hz. Three basic electrical parameters (dielectric constant ε_p, conductivity σ_p and dielectric loss tangent D) of the PANI-MMT nanocomposite particles were measured directly or deduced from capacitance. The dielectric constant of PANI-MMT nanocomposite is 146.8 and conductivity 23.2×10^{-7} S/m (f=100 Hz, T=20 ℃). ε_p has increased 2.4 times than MMT and 7 times than PANI, σ_p has also increased about 10 times than PANI. Meanwhile, the dielectric loss tangent also increased about 1.36 times than PANI. Because polyaniline is in extended chain conformation between the interlayer of the MMT, It has well electric conduction in the direction of parallel layer. Wnek [16] think that the electric mechanism of polyaniline is radical of semi quinonoid cation. Because of reductive and oxidation in the chain of polyaniline, redox reaction can be taken please between the chain of polyaniline in the exist acid.

Block [17] proposed that suspension particles with a conductivity of around 10^{-7} S/m were usually accompanied by the largest ER effect, because in that case the strength of interfacial polarization can reach a maximum. For the PNNI-MMT ER fluid, $\Gamma(\sigma_p/\sigma_f)$ is 1×10^7, Γ_ε $(\varepsilon_p/\varepsilon_f)$ is 68.5. As we know, when $\Gamma >> \Gamma_\varepsilon$, shear stress ($\tau$) is given proportional to $\varepsilon_f\beta^2E^2$, *i.e.*

$$\tau \propto \varepsilon_f\beta^2E^2$$

where ε_f denotes the dielectric constant of continuous medium, E is the strength of electric field, conductivity mismatch coefficient β is defined as:

$$\beta = \frac{\sigma_p - \sigma_f}{\sigma_p - \sigma_f} = \frac{\Gamma - 1}{\Gamma + 2}; \qquad \Gamma = \frac{\sigma_p}{\sigma_f}$$

σ_p and σ_f are the conductivity of particles and silicone oil, respectively. Therefore, a moderately large value of Γ_ε is preferred. The dielectric constant of PANI-MMT nanocomposite is 146.8 and conductivity 23.2×10^{-7} S/m (f=100 Hz, T=20 ℃), the dielectric constant of MMT and PANI are 61.5 and 21.4, the conductivity are 7.89×10^{-7} S/m and 2.71×10^{-7} S/m (f=100 Hz, T=20 ℃), respectively. It can be seen that dielectric constant and conductivity of PANI-MMT increased much more than that of MMT or PANI.

As for evaluation of ER effect, the yield stress and corresponding leaking current density under a fixed shear rate is widely accepted as the most common standard. Figure 3 shows the yield stress of PANI, MMT, PANI+MMT and PANI-MMT ER fluids vary with DC field and, leaking current density of PANI-MMT ER fluids (30%(w/w)). It can be seen from the curves that the PANI-MMT ER fluid displays a notable ER effect under DC electric field. The yield stress is 8.26 kPa (shear rate 5 s^{-1}) in 3 kV/mm DC field at room temperature, much higher than that of pure polyaniline (PANI), montmorillonite (MMT), mixture of polyaniline with clay (PANI+MMT). The leaking current density of PANI-MMT ER fluids limited to no more than 30 μ A/cm^2. The static yield stress of PANI+MMT is 2.66 kPa (shear rate 5 s^{-1} , 3 kV/mm (DC), 20 ℃), only one third of PANI-MMT ER fluid.

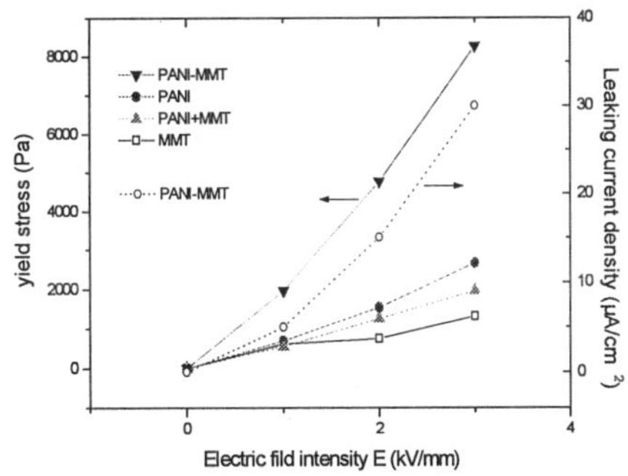

Figure 3. The yield stress *vs*. electric field intensity curve (5 s^{-1}, 20 ℃)

Figure 4 shows the relation between dynamic shear stress and shear rate of the 30%(w/w) PANI-MMT ER fluid in effect of applied electric field. In the absence of electric field, the PANI-MMT ERF behaves as Newtonian fluid; once subjected to electric field, it works in a fashion of Bingham fluid, namely. The shear strength increases with increasing electric field at fixed shear rate. In 4 kV/mm (DC) field, the shear strength is 8.3 kPa (γ=101.3 s^{-1}, T=20 ℃), and much higher than that of Kim's polyaniline-montmorillonite clay nanocomposite [18].

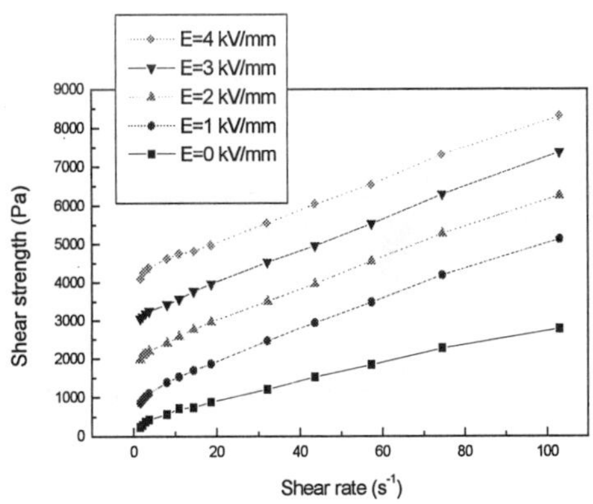

Figure 4. Shear stress of PANI-MMT ERF as a function of shear rate ERF at the different electric field

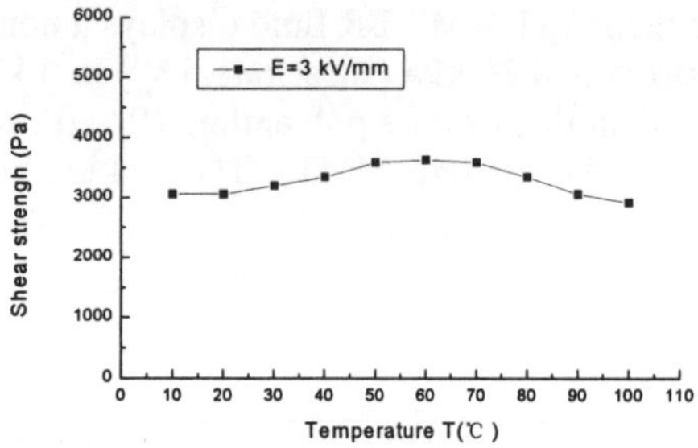

Figure 5. The curve of shear stress of PANI-MMT nanocomposite ERF vs. Temperature

Figure 5 shows the variation of shear stress of PANI-MMT ER fluid with the increase of the temperature under 3 kV/mm (DC) and shear rates of 1.61 s^{-1}. It can be seen from the fig.7 that τ increases gradually with the increase of the temperature, and it increased to the max when the temperature arrived about 60℃. After this temperature point, however, the τ decreased gradually with the increase of the temperature. In addition, the leaking current density increased abruptly with the increase of the temperature. Generally, the temperature has two effects on the ER fluid: one is the effect of the temperature on the polarization intensity of particle. As a consequence, the increase of the temperature results in the decrease of activation energy polarized by electric filed and the increase of the polarization ability of the particle. On the other hand, the temperature also has the effect on Brown motion, i.e. Brown motion is not use for the chain formation of the particle. As a result, the effect of ER decreases while the leaking current density increases. To summarize, the effect of the temperature on the ER fluid depends on the integrative effects of the aforementioned two factors.

The PANI-MMT ER fluid shows their suspensions with excellent sedimentation stability, for MMT itself is a good suspension stabilizer due to lamellar structure. As the intercalation fraction of PANI, its suspension ability also increases. The particles of PANI-MMT don't deposit during about two months (Figure 6).

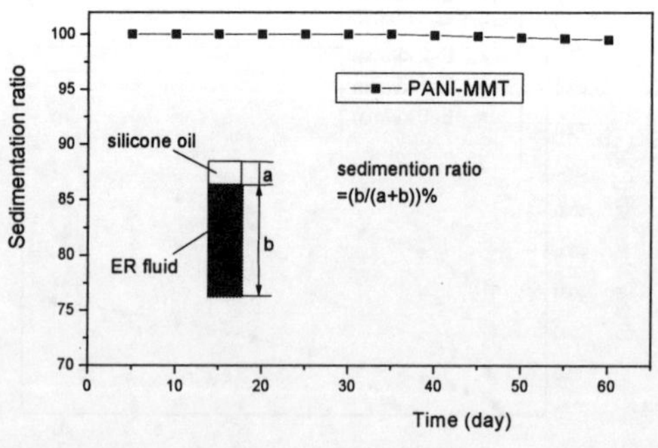

Figure 6. The sedimentation ratio of PANI-MMT nanocomposite ERF vs. time

In summary, the high conductivity polyaniline-montmorillonite clay nanocomposite (PANI-MMT) with extended chain conformation was synthesized using emulsion polymerization and the ER properties of the PANI-MMT/silicone oil suspension were then investigated. Furthermore, this ER fluid exhibited a notable ER effect under DC electric field. In 4 kV/mm (DC) field, the shear strength is 8.3 kPa (γ=101.3 s^{-1}, T=20 ℃), leaking current density limited to no more than 30 μ A/cm^2. Meanwhile, the antisedimentation of PANI-MMT ER fluids is much excellent and not deposit during about two months. PANI-MMT particles that can show an electrorheological effect without water has a preferable property in an electrorheological fluid.

4 Acknowledgment

This work was supported by the national natural science foundation of China under grant No. 59832090 and the national natural science foundation of China for distinguished young scholar under grant No. 50025207.

Reference

1. S. P. Coulter, K. D. Weiss and J. D. Calson, *J. Intelligent Material System and Structure*, **4**, 248 (1993).
2. H. J. Choi, M. S. Cho and M. S. Jhon, *Polym., Adv. Tech.*, **8**, 697 (1997).
3. H. J. Choi, M. S. Cho and K. To, *Physica A*, **254**, 272 (1998).
4. J. W. Goodwin, G. M. Markham and B. Vinent, *J. Phys. Chem. B*, **101**, 1961 (1997).
5. M. S. Cho, H. J. Choi and K. To, *Macromol. Rapid Commun.*, **19**, 271 (1998).
6. H. J. Choi, T. W. Kim, M. S. Cho, et.al, *Eur. Polym. J.*, **33**, 699 (1997).
7. M. Kawasumi, N. Hasegawa, M. Koto, et.al, *Macromolecules*, **30**, 6333 (1997).
8. R. A. Vaia, B. B. Sauer, et.al, *J. Polym. Sci: Part B: Polym. Phys.*, **35**, 59 (1997).
9. H. L. Tyan, K. H. Wei, et.al, *J. Polym. Sci: Part B: Polym. Phys.*, **38**, 2873 (2000).
10. N. Kuramoto, M. Yamazaki, K. Nagai, et.al, *Thin Solid Films*, **239**, 169 (1994).
11. K. Yatsuzuka, K. Miura, et.al, *IEEE Trans. Ind. Applicat.*, **31**, 457 (1995).
12. (a) M. S. Wang, T. J. Pinnavaia, *Chem. Mater.*, **6**, 468 (1994). (b) N. Ogata, S. Kawakage, T. Ogihara, *Polymer*, **38**, 5115 (1997). (c) A. Usuki, Y. Kojoma, M. Kawasumi, A. Okada, *Polym. Prepr.*, **28**, 447 (1987).
13. Q. J. Wu, Z. J. Xue, Z. N. Qi and F. S. Wang, *Acta Polymerica Sinica*, 599 (1999).
14. H. J. Choi, J. W. Kim, M. H. Noh, et.al, *J. Mat. Sci. Lett,*. **18**, 1505 (1999).
15. J. B. Yin, and X. P. Zhao, *Chin. Phys. Lett.* **18**, 1144 (2001).
16. Wnek, G. E.; Chien, J. C. W.; Gooding. *Synth. Met.*, **15**, 213 (1986).
17. Block H, Kelly J P, Qin A and Waston T, *Langmuir,* **6,** 6 (1990).
18. J. W. Kim, S.G. Kim, H. J. Choi, et.al, "in proceeding of the 7th International conference on Electro-rheological fluids and magneto-rheological suspensions", ed. R. Tao, World Scientific, Singapore, p111 (2000).

IV. PHYSICAL MECHANISMS

Electric-Field Induced Formation of Superconducting Granular Balls

R. Tao , X. Xu, and Y. C. Lan

Department of Physics, Temple University, Philadelphia, PA 19122, USA

When a strong electric field is applied to a suspension of micron-sized high T_c superconducting particles in liquid nitrogen, the particles quickly aggregate together to form millimeter-size balls. The balls are sturdy, surviving constant heavy collisions with the electrodes, while they hold over 10^6 particles each. The phenomenon is a result of interaction between Cooper pairs and the strong electric field. The strong electric field induces surface charges on the particle surface. When the applied electric field is strong enough, Cooper pairs near the surface are depleted, leading to a positive surface energy. The minimization of this surface energy leads to the aggregation of particles to form balls.

1. INTRODUCTION

Superconductors strongly repel magnetic field. This phenomenon, as known as the Meissner effect, is the scientific basis for the maglevel superconducting trains, moving at a speed of 500 km/h. According to electromagnetic theory, electric field and magnetic field are deeply related. The interaction between superconductors and electric field is thus also of great interest and importance.

The London brothers first suggested that superconductors and normal conductors could respond to static electric field differently; static electric field could penetrate into superconductors as far as the magnetic penetration depth, much deeper than that into a normal conductor. However, they later gave up the idea after the experiment failed to detect the prediction [1]. Later works by Landau [2] and Pippard et al [3] showed the electric field penetration effect indirectly [4]. Since the discovery of high temperature superconductor (HTSC) [5], there have been extensive studies about the electric field effect in the high-T_c cuprates [6] and measurement of electric-field penetration into the superconducting surface of YBaCuO [7,8]. These works show the importance and wide spectrum of the topic.

The recent discovery of formation of HTSC granular balls in a strong electric field further illustrates the rich physics in this area [9] The new phenomenon of HTSC ball formation confirms that superconductors are different from normal conductors in responding to a dc electric field or ac field of low frequency.

2. EXPERIMENT

The particles used in our experiment were $YBa_2Cu_3O_{7-x}$ (99.99% purity), $NdBa_2Cu_3O_x$ (99.9% purity), $Bi_2Sr_2CaCu_2O_{8+x}$ (99.9% purity), and $YbBa_2Cu_3O_x$ (99.9% purity) with T_c around *92*K, *94*K, *84*K, and *87*K, respectively. As shown in Fig.1, the particle size is of micrometer. They were prepared either by a solid state method [10] or the Sandia Chemical Preparation [11].

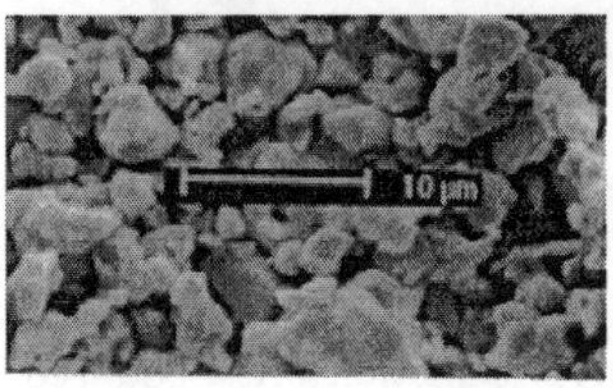

Fig. 1 Scanning electronic micrograph of $NdBa_2Cu_3O_x$ particles.

The experimental setup is sketched in Fig.2. On a horizontal microscope Teflon slide we mounted two parallel brass electrodes which were 10 mm wide and 4 mm apart. Teflon spacers were used between the electrodes to form a cell in a region of uniform electric field. This cell was originally used for cryogenic electrorheological experiments [12]. Only a small amount of slurry of liquid nitrogen and HTSC particles was needed to fill the cell, having 4 mm x 5 mm horizontally and 5 mm in vertical depth. The particle volume fraction is about 10%. They were well dispersed in liquid nitrogen before the electric field was applied (Fig.3). The whole cell was submerged in liquid nitrogen. The top surface of the cell was open to allow liquid nitrogen to flow in so that a constant temperature was maintained at 77K. The experiment was recorded by a high-speed video camera, which could take up to 1000 frames per second and enabled us to examine the dynamic process in detail.

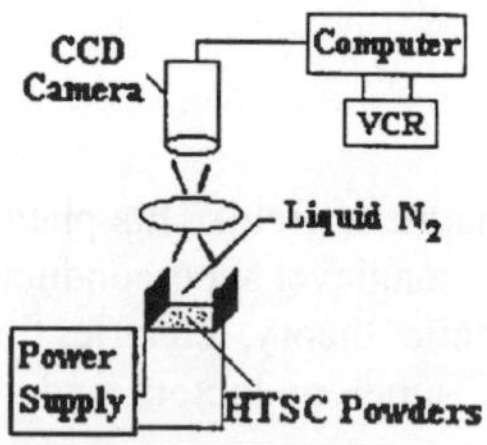

Fig. 2 The experimental setup.

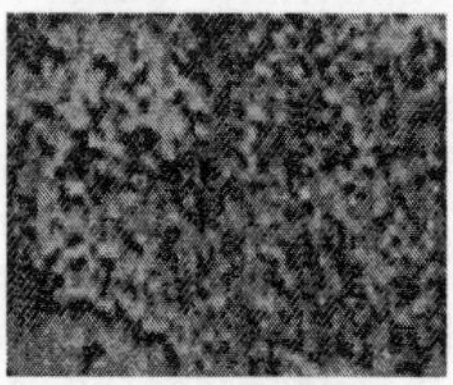

Fig.3. The particles were well dispersed in liquid nitrogen before the electric field was applied.

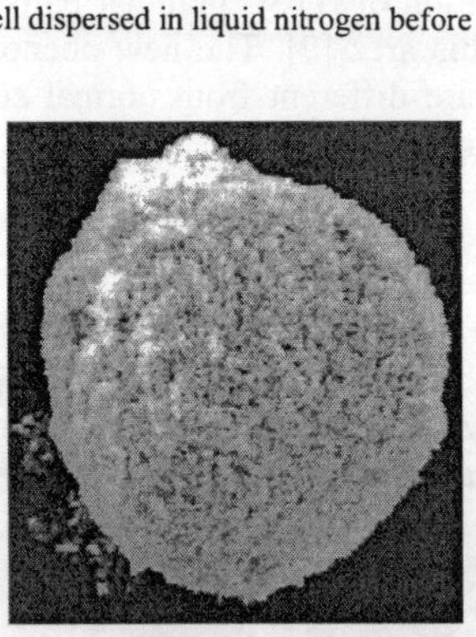

Fig. 4 SEM picture of a ball of $Bi_2Sr_2CaCu2O_{8+x}$ particles. The diameter is 0.25 mm and particles are closely packed. The ball was trapped in an ac field first. The granular friction sustained the ball, but the ball was very fragile after it was removed from liquid nitrogen.

When a dc electric field is turned on, HTSC particles first bounced between the two electrodes as a cloud. During the collision, they aggregate into one big ball in milliseconds. The ball then bounces between the two electrodes as one metallic particle. A typical ball is shown in Fig.4 which has a diameter about 0.25 mm. As the particle's average size is only 1-2μm, the ball consists of several million particles. Figure 5 shows the ball bouncing between the two electrodes and layers of particles formed on each electrode. Recently we have also found the ball formation in a strong ac field of low frequency. The phenomenon is quite similar, except that the HTSC particles first form chains in an ac field. As the electric field further increases to pass a critical value, the chains break and the particles assemble together to form balls.

The ball's speed was estimated about 0.2 m/s from the video recording at 0.8kV/mm. As soon as the ball touched the solid-brass electrode, it quickly discharged and acquired opposite charge, then bounced back to the other electrode (Fig.6). The whole process for the ball to strike at and bounce back from one electrode took only a couple of milliseconds. Therefore, the average acceleration in this collision was at least 10 times stronger than the gravity ($9.8m/s^2$), equivalent to the impact force on an automobile when it hits a soft bump at a speed of 15 miles per hour. Under such an enormous impact force, several million particles are still able to hold together, indicating that the force holding these particles is strong. However, the ball was fragile after it was removed from liquid nitrogen, although the granular friction may sustain the ball. If the particle concentration was high, two or more balls could be formed under a strong electric field. During the whole process, the electric current density was very low, about several $\mu A/cm^2$.

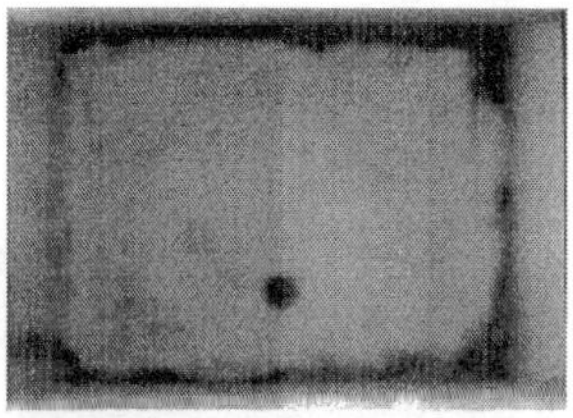

Fig.5. The ball is bouncing between the two electrodes. Strong layers are also formed on the electrodes.

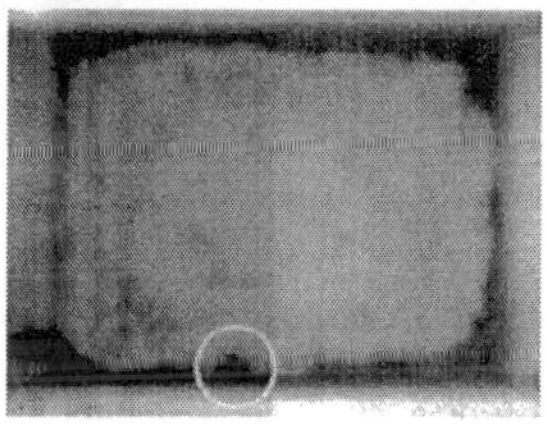

Fig.6 The ball is colliding with an electrode. At 0.8kV/mm, the ball's speed is about 0.2 m/s.

In order to verify that the ball formation is a result of superconductivity, we conducted several experiments. First, we placed the same cell with the same HTSC particles in silicon oil. At room temperature, under the same field, these particles only form chains (Fig.7). This is easy to understand. At room temperature, these particles are ceramics. The ceramic-oil suspensions are electrorheological fluids [12]. Under a strong electric field, these dielectric particles are polarized and form chains along the field direction. Also our test confirmed that normal metallic powders, such as copper, iron, and aluminum, do not form balls in liquid nitrogen under an electric field. Instead, they bounce between the two electrodes to transport charges and move separately in the same way as they do at room temperature.

Fig.7. At room temperature, these granular particles form chains in silicon oil and do not form balls

We also used liquid argon to raise the temperature from 77K through the T_c of BiSrCaCuO. The boiling point of argon is *87.3*K, just above the T_c of BiSrCaCuO. When the temperature was below the T_c, a big ball was formed and bounced between the two electrodes. We then raised the temperature to pass T_c and used the high-speed camera to examine the situation. We saw two frames in sequence. The first frame showed that the big ball was going to strike an electrode (Fig.8a). The next frame showed that in less than one millisecond, the ball disappeared: it broke into pieces in the collision (fig.8b). This concludes that once the particles are no longer in the superconducting state, the strong force holding these particles disappears.

(a) (b)

Fig.8 (a) A ball is going to strike an electrode as the temperature of liquid argon rose to pass T_c . (b) The ball disappeared at the collision, broken into pieces.

3. Theoretical Explanation

As shown in Figs.4 and 5, the shape of these balls is spherical, quite different from structures formed under electric polarization (Fig.7). The dipolar polarization thus cannot be the mechanism for the ball formation. The ball's round shape must be from the minimization of a positive surface energy. In the subsequent section, we will mainly discuss how a strong dc field induces a positive surface energy on superconducting particles, while the physics principle applies to ac field of low frequency as well.

When an electric field is applied, it induces surface charge distributions on the particle surface. In the dc field, the particles collide with the electrodes. In doing so, they pick up net charges. The acquired net charge is $q = \pm \gamma E a_t^2 \kappa_L$, where a_t is the ball radius when it hits the electrodes (Fig.6), $\kappa_L = 1.44$ is the dielectric constant of liquid nitrogen, E is the applied electric field, and γ is a constant measuring the deformation during the collision. If the ball is rigid when touching the electrode, it picks up a charge close to $0.75\, E a_t^2 \kappa_L$ as it leaves the electrode. If the ball is soft enough to deform into a hemisphere with its flat surface in full contact with the electrode, it picks up a charge close to $3(4)^{1/3} E a_t^2 \kappa_L/4 = 1.19\, E a_t^2 \kappa_L$ as it leaves the electrode. On the other hand, from the ball's bouncing speed between the two electrodes, we can estimate the actual net charge on the ball, hence find $\gamma \approx 1$. Therefore, the true deformation is just between the above two cases.

The polarization induces additional surface charge distributions on the ball surface, which change the local field near the surface. We denote the local field on the surface as ηE, where constant η varies with position. At one pole η reaches the maximum $3+\gamma$. At the equator η has the minimum value γ. The electric field also penetrates into the inside of superconductor. Let us assume that the screening length of electric field penetrating into the superconductor is l_s. The electric field inside the superconductor can be estimated as

$$E(x) \sim \eta E \exp(-x/l_s) \qquad (1)$$

where x is the depth from the surface into the superconductor. The Cooper pairs are deleted near the surface if the following condition is satisfied,

$$e\int_0^{\xi} E(x)dx \geq \Delta(T) \qquad (2)$$

where ξ is the superconductor's coherent length and $\Delta(T)$is the binding energy of Cooper pairs at temperature T, $\Delta(T)=2\varepsilon_f - \varepsilon$. Assuming $\xi >> l_s$, we then have the critical electric field

$$E_c = \Delta(T)/(\eta e l_s)\,. \qquad (3)$$

This implies that when $E > E_c$, the region near the surface is normal. In comparison with the normal region, the energy density in the superconducting region is lowered by - $\Delta(T)\, n_s(T)$, where $n_s(T)$ is the Cooper density in absence of electric field. In another words, when $E > E_c$ the normal surface region has a positive surface energy. The minimization of this positive surface energy leads to the ball formation. While we will estimate this surface energy later, it is first worthwhile to note that the above argument implies a critical electric field for the ball formation. When the electric field is below E_c, this effect disappears because the electric field is not strong enough to deplete Cooper pairs. The experiment has confirmed this conclusion. The critical fields for $NdBa_2Cu_3O_x$, $Bi_2Sr_2CaCu_2O_{8+x}$, and $YbBa_2Cu_3O_x$ are 0.76 kV/mm, 0.65 kV/mm and 0.45 kV/mm respectively.

Now let us estimate the critical field from Eq.(3). We need to know l_s first. With the two-fluid model for superconductors, the screening length l_s is affected by both normal electrons and Cooper pairs. The Thomas-Fermi screening length l_{TF} characterizes the screening effect of normal electrons, $l_{TF} = (\pi/3)^{1/6}[a_0/(4n_n^{1/3})]^{1/2}$, where the Bohr radius $a_0 = \hbar^2/(me^2)$ and the normal electron density $n_n = n_n(T)$. If we take London's assumption [1] that the electric-field penetration depth for Cooper pairs is the same as the magnetic penetration depth,λ, then $1/l_s = 1/l_{TF} + 1/\lambda$. When T is below the T_c, normal electron density $n_n(T)$ is reduced. Thus as T lowers, l_s increase with l_{TF}, but is still shorter than λ. The experiment on YbaCuO has found that l_s depends on temperature and can be as high as 100 Å [8]. If, for example, we take l_s close to 16 Å for $Bi_2Sr_2CaCu_2O_{8+x}$, at $T=77K$, then the critical electric field estimated from Eq.(3) is very close to our experimental result.

Now let us to estimate the surface energy. Because $\xi > l_s$, there is a sharp boundary the normal region and the superconducting region. Therefore, the concept of surface charge is still correct. As mentioned before, in comparison with the normal region, the superconducting region has the energy density lowered by - $\Delta(T)\ n_s(T)$. When $E > E_c$, the surface region of thickness ~ ξ is normal and has the surface energy per unit area, i.e., the surface tension, is estimated by [13]

$$\sigma = \Delta(T) n_s(T) \xi . \tag{4}$$

We now show how to determine the surface tension from the experiment. Once a ball leaves an electrode, its charge is fixed until it hits the other electrode. If a ball has a radius a and charge q in the middle between the two electrodes, the total energy is given by

$$U = 4\pi\sigma a^2 + q^2/(2\kappa_L a) - E_0{}^2 a^3 \kappa_L/2. \tag{5}$$

The second term is the Coulomb energy, while the third term is the polarization energy. A ball made of granular particles has some freedom to fluctuate its size. The radius of a stable ball inside the electric field (Fig.5) is determined by $\delta U/\delta a = 0$, then

$$\sigma = q^2/(16\pi\kappa_L a^3) + 3E_0{}^2 a \kappa_L/(16\pi). \tag{6}$$

Substitute q here. While Eq.(4) is not applicable to the situation when the ball collides with the electrodes, a_t and a should be almost the same. We then have

$$E_0{}^2 a = 16\pi\sigma/[(\gamma^2+3)\kappa_L]. \tag{7}$$

As shown in Fig.9, this relationship holds very well in our experiments. For $Bi_2Sr_2CaCu_2O_{8+x}$, $E_0^2 a$=0.12 kV^2/mm = 13.33 dyne/cm and its surface tension σ =1.52 dyne/cm. This magnitude is close to the value estimated from Eq.(4). For $NdBa_2Cu_3O_x$, $E_0^2 a$ = 17.3 dyne/cm and its surface tension is 1.99 dyne/cm. When E_0 is below the critical field, σ = 0 and hence a = 0 from Eq.(6).

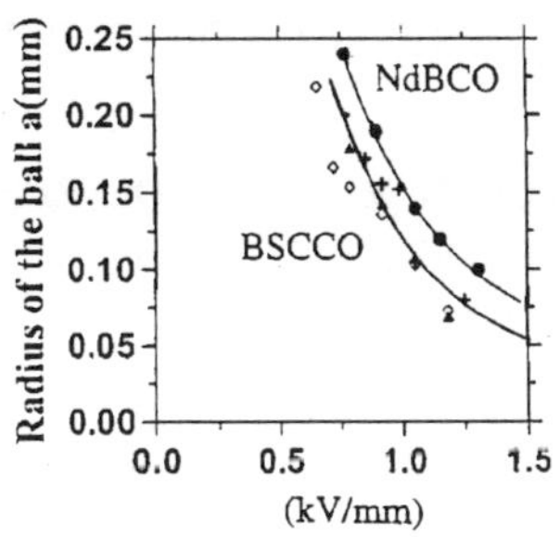

Fig.9 The sample BiSrCaCuO holds E_0^2a=0.12 kV^2/mm, while NdBaCuO has E_0^2a=0.156 kV^2/mm.

If we apply the parameters for $Bi_2Sr_2CaCu_2O_{8+x}$ and $NdBa_2Cu_3O_x$ into Eq.(4), we have found that the surface energy density for both these two HTSC materials close to the experimental results.

Our theoretical explanation also implies that the particle surface must be conductive to redistribute the charges to minimize the normal surface area in the ball formation. To verify this, we let NdBaCuO particles absorb some moisture. In liquid nitrogen, the moisture becomes a thin ice surface layer. Although they still show the strong Meissner effect, the NdBaCuO particles with an insulate layer do not form balls in a strong electric field, instead, form chains. Similarly, the surface of YBaCuO particles can easily deteriorate; once their surface becomes insulate, they stop the ball formation, too.

This research is supported by a grant from NSF DMR-0075780.

References

1. F. and H. London, Proc. Roy. Soc. A, **152** 24 (1935); H. London, Proc. Roy. Soc. A **155**, 102 (1936).
2. I. L. Landau, Sov. Phys. JETP **11**, 295 (1970).
3. A. B. Pippard, J. G. Shepherd, and D. A. Tindall, Proc. R. Soc. London,Ser.A **324**, 17 (1971).
4. S. N. Artemenko and A. F. Volkov, Sov. Phys. Usp. **22** (5), 295 (1979).
5. J. G. Bednorz and K. A. Müller, Z. Phys. B **64**, 189 (1986).
6. M. K. Wu et al, Phys Rev. Lett. **58**, 908 (1987).
7. For example, see, T. Frey et al, Phys. Rev. B **51**, 3257 (1995); S. Sakai, Phys. Rev. B **47**, 9042 (1993); T. Yang et al, Physica C f **341-348**, 291 (2000).
8. W. G. Jenks and L. R. Testardi, Phys. Rev. B **48**, 12993 (1993).
9. R. Tao, X. Zhang, X. Tang, and P. W. Anderson, Phys. Rev. Lett. **83**, 5575 (1999).
10. U. Balachandran, et al, Materials Lett., **8**, 454 (1989).
11. B. C. Bunker et al, "High Temperature Superconducting Materials: Preparation, Properties and Process," (Ed. by W. E. Hatfield and J. H. Miller, Jr.; Marcel Dekker, Inc., NY, 1988).
12. T.J. Chen, X. Zhang, and R. Tao, Intern. J. of Mod. Phys. B, **13**, 1697 (1999)
13. P. G. de Gennes, *Superconductivity of Metals and Alloys,* (Addision-Wesley, 1989) p46.

GRANULAR FLOW IN THE PRESENCE OF AN ELECTRIC FIELD

KUNQUAN LU, WEI CHEN, MEIYING HOU, ZEHUI JIAN

Institute of Physics, Chinese Academy of Sciences, Beijing 100080, China.
E-mail: lukq@aphy.iphy.ac.cn

The granular flow in a vertical pipe in the presence of electric field E is studied. Depending upon its initial state and the applied field voltage the controlled flow rate remains in two phases, dilute flow or dense flow. For dilute flow, the electric field has no effect on the flow rate until V reaches a critical value V_1. At $V = V_1$, the flow rate drops abruptly and a transition of the particulate from dilute to dense flow occurs. For dense flow, the flow rate decreases monotonically with increasing V. A two-dimensional computer simulation has been done and the results agree qualitatively well with the experimental measurements.

1 Introduction

Granular matter, which can exhibit solidlike and liquidlike behavior, has attracted much attention in recent years not only because of its ubiquity in nature and application in technology, but also because of the poor understanding in its basic mechanism [1-4]. The study of granular flow in a vertical pipe is one of the most challenging projects in the field, and many interesting phenomena have been observed [5-9]. Considering the ER effect induced by an applied electric field, we are interested in finding out what happens if an electric field is applied to a granular flow.

Electric field was found to be useful to regulate and control the granular flows in many industrial processes, such as in the electrospouted bed [10], the electrofluidized bed [11] and the electrostatic valve [12]. Electric field may also be a powerful tool to gain understanding of the dynamics of granular flows [13, 14]. In this work we report our study on the behavior of granular flows in a vertical pipe by applying a horizontal ac electric field. Some interesting phenomena have been observed:

(1) the granular flow is retarded beyond a critical electric field V_1 when V increases. For $V < V_1$, the flow rate is unaffected by V and keeps a constant value Q_0. The flow rate abruptly drops at V_1 and then decreases with increasing V for $V > V_1$;

(2) in a reverse process, if we first apply an electric field of voltage $V > V_1$ to the flow, the granules become densely filled above the electrodes. We then decrease the voltage to a desired value V to measure the flow rate at that voltage, we find that the flow rate increases continuously as V decreases, and reaches Q_0 at V_2, which is lower than V_1; and

(3) if the electric field is high enough a granular arch is formed at the region of the electrodes. A computer simulation is performed and the results agree qualitatively with the measurements.

2 Experimental setup

The experimental setup is schemed in Fig. 1. Nickel spheres of an average diameter d = 0.25 mm is filled in a glass hopper placed on top of a glass pipe with inner diameter of 3mm and 100mm in length. The metallic particle is chosen for less electrostatic effect. Two parallel flat copper electrodes 4.4mm apart, 15 mm in height and 12 mm in width, are attached to the outside walls of glass pipe. A 50 Hz ac electric voltage V is applied

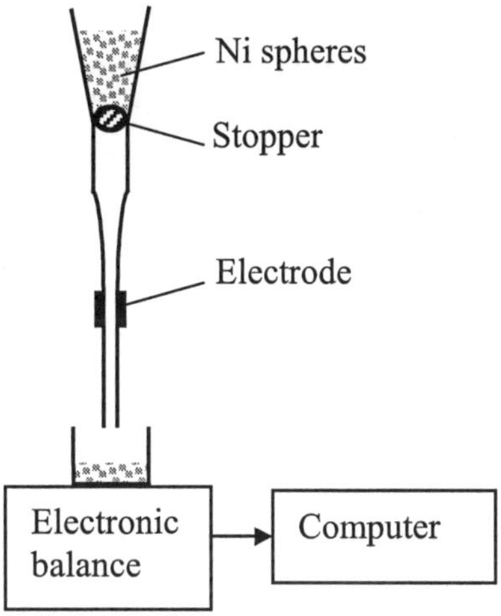

Figure 1. Scheme of experimental set up.

across the electrodes. The granular flow is initiated by pulling the stopper, which is inserted at the outlet of hopper. A weighing sensor with sensitivity of 0.02 g and recording rate of 0.02 s is applied at the bottom of the pipe. By measuring the granular mass at each desired electric voltage, the dependence of flow rate vs. voltage can be obtained.

3 Experimental results and computer simulation

When we apply an electric field E of 0 V/mm and 1200V/mm alternatively with an interval of several seconds, the mass M(t) measured as a function of time t shows a slope difference: the slope is higher at E = 0 and lower at E = 1200 V/mm as shown in Fig. 2. This phenomenon indicates that the granular flow can be retarded by electric field.

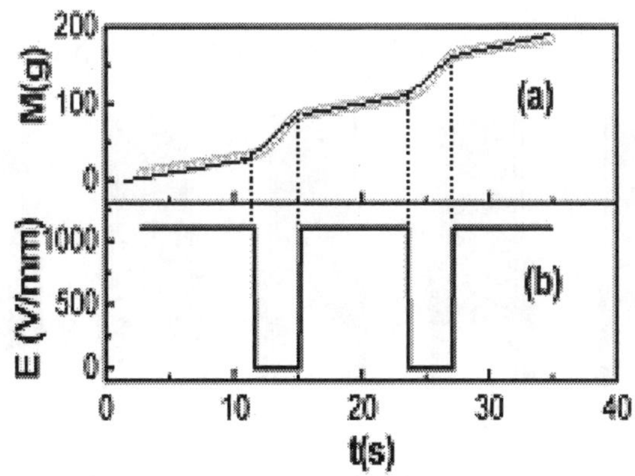

Figure 2. Relation of flow mass (a) vs. applied pulse electric filed (b).

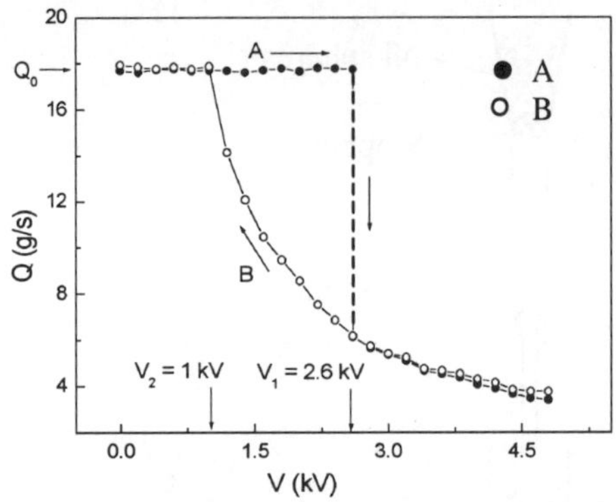

Figure 3. Flow rate Q vs. the applied voltage V. **A** and **B** indicate different processes. The arrows represent the direction of applied voltage.

If we change the applied voltage in steps, how does the voltage influence the granular flow? The following two processes are applied:

In the first process (denoted as A), we measure the flow rate at each desired voltage in a range from 0 V to 4.8 kV with an interval ΔV= 0.2 kV. In the second process (denoted as B), V is first fixed at a voltage 4.8 kV and then lowered to a desired voltage from 4.8 kV to 0 V with an interval of 0.2 kV. The flow rate Q vs. V is plotted in Fig.3. It can be seen that in process A, the granular flow keeps a constant value Q_0 with increasing V until $V=V_1$= 2.6 kV. Flow rate Q drops abruptly when we increase the voltage by a small amount (as small as 0.02 kV) higher than V_1. When we further increase the voltage, Q decreases gradually with increasing V. In process B when lowering the voltage, however, Q increases continuously without any sudden jump and reaches Q_0 at a voltage V_2 (1.0 kV). In the two processes flow rates follow the same Q(V) curve in the region between V_1 and 4.8 kV. The curve can be fitted approximately as $Q = CV^{-1}$, where C is a fitting parameter.

In the experiments we have observed that particles converge to the sidewall of the tube near the top edges of the electrodes as V increases. Particles near the sidewall thus move slower than that in the central part. Eventually when the voltage is high enough, at and above V_1, an unstable arch across the tube at a position near the top edge of the electrodes is formed as shown in Fig. 4. It can be seen that some particles in the arch fall

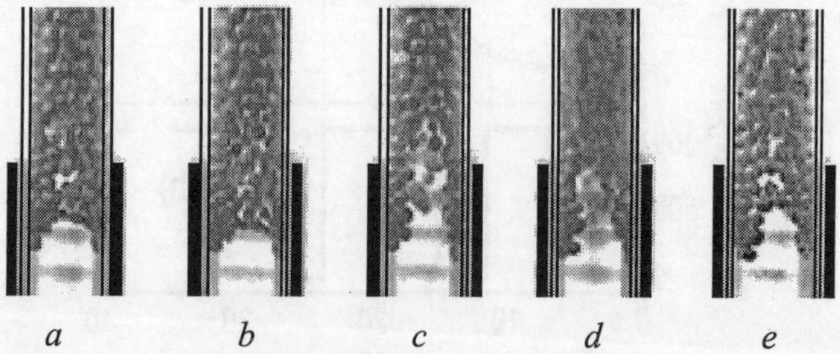

Figure 4. The unstable arch formed at high voltage. Pictures *a* to *e* were taken at a time sequence in an interval of 1/25 second.

randomly resulting a lower flow rate at high voltage.

To understand the mechanism behind, an analysis is proposed and a computer simulation is done.

We first consider the case of process A. The force acting on the polarized particle in an inhomogeneous electric field can be expressed as

$$\vec{f} = -\nabla(-\vec{p}\cdot\vec{E}) = 4\pi\varepsilon_0 R^3 \nabla E^2 .$$

When the electric field is high enough to affect the motion of the particles, i.e. the interaction is comparable to the kinetic energy of the falling particles, the particles will be dragged by the gradient of field and converge to the sidewall near the top of the electrodes as shown in Fig. 5. The arrows in Fig.5 represent the direction of the dragging force. This force against the sidewall induces a friction between particle and wall, which will reduce the velocity of particles along the wall. The attraction of polarized particles cause the particles moving closer and enlarges the local field between particles. The enhancement of local field between particles will further increase their mutual induction and cause particles to interact with each other and aggregate stronger. This positive-feedback effect, we believe, is the major contribution which causes the flow rate dropping dramatically at V_1. In fact there is a transition from dilute flow to dense flow occurred at V_1. When $V< V_1$ the remaining opening at the central area of converged particles in pipe is large enough for the particles passing through as a dilute state flow. The flow remains at a constant rate $Q = Q_0 = \rho vS$, where ρ and v are the average density and average velocity of particles, respectively, and S is the effective cross-sectional area. Although the v and S are reduced due to the electric field induced converging effect, the increase in ρ ia able to assure Q (= ρvS) to remain a constant Q_0 until V reaches a critical voltage V_1. At this voltage the density ρ reaches its maximum value, and it becomes dense flow for $V > V_1$.

In process B, the flow rate vs. voltage shows power law decay vs. V as described above, $Q=CV^{-\beta}$ and $\beta\approx 1$ in our experimental condition. Although the mechanism is still not yet clear[15], the rate of dense particle flow from a hopper can be expressed as $Q\approx aD^{5/2}$, where D is the diameter of outlet. Therefore, the electric field plays a role here as an effective opening in the case of dense flow, i.e. increasing voltage corresponds to decreasing the size of the opening.

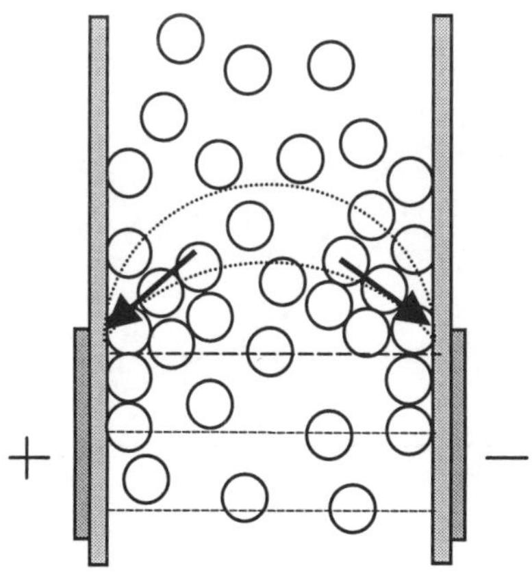

Figure 5. Sketch of the force acting on particles due to the field gradient in the region of electrodes. Dashed lines indicate electric lines in between electrodes.

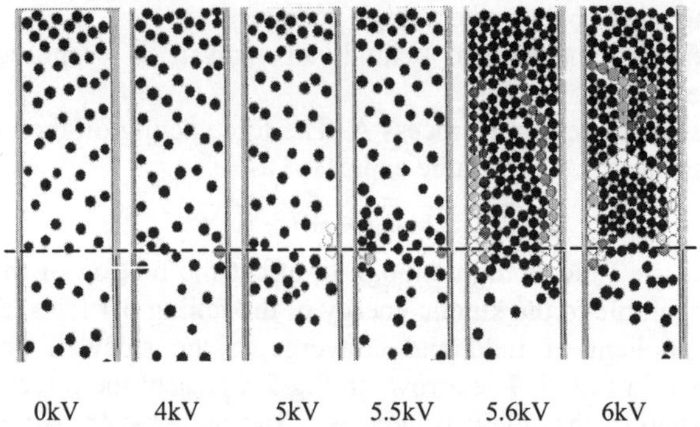

Figure 6. Two-dimensional simulation: particles in the pipe at different voltages. Dashed line indicates the top edge of electrodes.

Therefore, by applying the electric field higher than a critical voltage the initial dilute flow can be transferred to a dense flow, corresponding to a sudden drop of flow rate. This phenomenon may be similar to that of a first order phase transition. A quantitative explanation is still absent.

A two-dimensional model was proposed to simulate the granular flow under the electric field. The computer simulation is based on molecular dynamics model in which the interaction and friction between polarized particles and wall, the collision as well as the motion of particles under gravity were taken into account. The simulation results on the distribution of particles in the pipe and the flow rate Q vs. voltage V are plotted in Fig. 6 and Fig. 7 respectively. It can be seen that particles converge to sidewalls and slow the speed of particles above as voltage increases. There is a critical value $V=V_1\approx 5.6$ kV at which the particles become dense in the region above the electrodes. These pictures show that the granular flow is thin when $V< V_1$ while the flow is dense when $V\geq V_1$. The behavior of Q vs. V shown in Fig. 7 is consistent qualitatively with the experimental measurements as showed in Fig. 3.

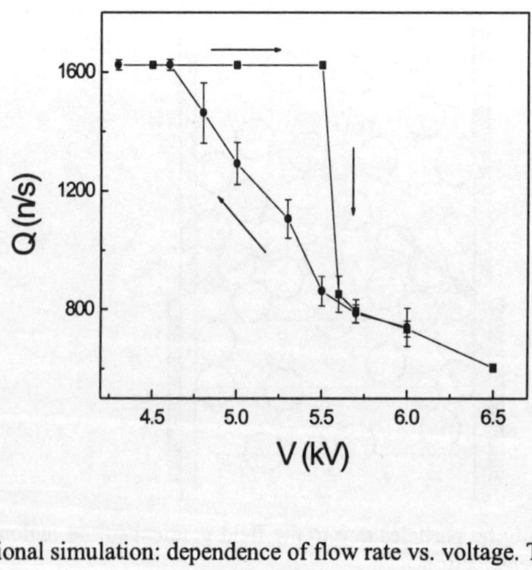

Figure 7. Two-dimensional simulation: dependence of flow rate vs. voltage. The arrows indicate the direction of the applied voltage.

4 Conclusions

By applying an electric field to the granular flow in a vertical pipe we have observed some novel phenomena. Applying an electric field to a dilute flow, we find that the flow rate is unaffected by V and remains as Q_0 till V reaches a critical value V_1. At $V = V_1$ the flow rate drops abruptly, and decreases with increasing V. In a reverse process, when V decreases from a value $V > V_1$, the flow rate increases continuously with decreasing V and reaches Q_0 at V_2. At V_1 the transition from thin flow to dense flow occurs, which is similar to a first order phase transition. The unstable arch is formed near the top edge of the electrodes as the electric field is high enough. A computer simulation is performed, and the result of which agrees qualitatively well with the measured behavior.

5 Acknowledgements

This research is supported by the State Key Program for Basic Research of China and the Chinese Natural Science Foundation.

References

1. H.M.Jaeger and S.R.Nagel, Science **255**, 1523 (1992).
2. H.M.Jaeger, S.R.Nagel, R.P.Behringer, Rev. Mod. Phys. **68**, 1259 (1996).
3. L.P. Kadanoff, Rev. Mod. Phys. **71**, 435 (1999).
4. P.G. de Gennes, Rev. Mod. Phys. **71**, S374 (1999).
5. T.Raafat, J. P. Hulin and H. J. Herrmann, Phys. Rev. E, **53,** 4345 (1996).
6. J. L. Aider, et al, Phys. Rev. E, **59,** 778 (1999).
7. S. Horikawa, et al, Physica A, **233**, 699 (1996).
8. A. Nakahara and T. Isoda, Phys. Rev. E, **55,** 4264 (1997).
9. O.Moriyama, N.Kuroiwa, M.Matsushita, & H. Hayakawa, Phys. Rev. Lett. **80**, 2833(1998).
10. C.M.Talbert, et al, IEEE Trans. Ind. Applicat., **IA-20**,1220 (1984).
11. P.W.Dietz and J.R.Melcher., Ind. Eng. Chem. Fundam., **17**, 28 (1978).
12. W.Balachandram., et al., IEEE Trans. Ind. Applicat., **33**,871 (1997).
13. W.Chen, M.Hou, K.Lu,Z.Jiang & L.Lam, Phys.Rev.E (in press)
14. W.Chen, K.Lu,Z.Jiang, M.Hou & L.Lam, *Powers and Grains*, Balkema,lisse p. 433, 2001.
15. R. M. Nedderman, "Static and Kinematics of Granular Materials", Cambridge University Press, 1992

THE CONTRIBUTION OF COLUMNAR STRUCTURE TO THE STRESS IN ER POLYMERIC BLENDS IN A SHEAR FLOW

P. RIHA

Institute of Hydrodynamics, Academy of Sciences, 166 12 Prague, Czech Republic
E-mail: riha@ih.cas.cz

J. TAKIMOTO, Y. MASUBUCHI, K. KOYAMA

Department of Materials Science and Engineering, Yamagata University, Yonezawa 992, Japan
E-mail: kkoyama@eie.yz.yamagata-u.ac.jp

The bridging of electrodes by the columns of highly viscous polymer is dominant feature of the response of most electro-rheological liquid polymeric blends to the imposed electric field. To link the columnar structure and its deformation to the macroscopic shear stress, the transitory process of structure reorganization is described as a stochastic process determining rheological properties of the blend. The shear stress predictions are compared to the experimental data for steady and start up shear flows. The data support the assumption that only viscous deformation prevails in shear flow.

1 Introduction

Electro-rheological (ER) liquid polymeric blends display changes in their rheological properties when subjected to the electric field transverse to the direction of deformation and flow [2-4]. The changes come from the induced columnar structure bridging the electrodes. To deform the structure and impose a flow, the specific amount of energy has to be supplied to disturb the mechanical stability of polymeric columns. When the bridging columns begin to break up, the induced structure changes its organization. That is, generally, from organization at rest when columns are bridging the electrodes to the destroyed structure and no bridging in the high intensity flows.

The stochastic formation of bridging columns is described in [3,4]. The rheological behavior of blends is linked with the columnar structure through the constitutive coefficients of the Maxwell-like constitutive equation describing the bulk blend behavior. The structure organization and fluid rheological behavior thus is interconnected by the plausible relationship. The objective of this paper is to analyze deformation of the columnar structure and link it by means of the probability of its breaking and the zero-shear viscosity with the macroscopic shear stress. For this purpose the kinetics of column formation is proposed and the approach based on the contribution of expectation of the column formation and breaking up to the macroscopic shear stress is used.

2 Theoretical

The formation of bridging columns between the plate electrodes can be described as a stochastic process $\phi = \phi(e^c, t)$ that is dependent on the stochastic variable e^c representing the threshold energy supply to disturb the bridging columns [3,4]. Generally, the induced columnar structure has different stability due to the different thickness and interfacial area between the columns and surrounding suspending fluid. Consequently, the columns have also different threshold energy when they begin to rupture and undergo a reversible process of breaking and reformation in flow.

The range of the stochastic process is a discrete set of states ϕ_n that take only integral values n since the process is linked to the number of columns bridging the electrodes. The governing master equation for the probability $p_n(t)$ that the state has value n in time t is proposed in [3,4],

$$\frac{dp_n}{dt} = k_g[N-(n-1)]p_{n-1} + k_r(n+1)p_{n+1} - [k_g(N-n)+k_r n]p_n, \tag{1}$$

where N denotes the upper bound of possible states. When the kinetic rates k_g and k_r are considered constant the above differential-difference equation can be solved in closed-form using the approach of the probability generating function $F(z,t)$ defined by

$$F(z,t) = \sum_{n=0}^{N} z^n p_n(t), \tag{2}$$

where z is an auxiliary variable [5]. Thus the master equation (1) can be written as an equation for $F(z,t)$ when multiplied by z^n and summed over all states n. The solution of that partial differential equation follows from the general solution given in [1],

$$F(z,t) = (1 + k_r/k_g)^{-N}\{z + k_r/k_g + (1-z)\exp[-(k_g + k_r)]\}^N, \tag{3}$$

as the particular case in which there are no broken columns initially.

The first moment of the stochastic process corresponds to its expectation value. Taking derivative of $F(z,t)$ with respect to z and allowing $z \to 1$, the first moment suitably reduced to represent the column density is,

$$\Phi(t) = N^{-1}\langle n\rangle_t = N^{-1}\sum_{n=0}^{N} np_n(t) = \lim_{z\to 1}\partial\left(F^{1/N}\right)/\partial z\,. \tag{4}$$

The expectation value is then

$$\Phi = \left(1 + k_r/k_g\right)^{-1}\left\{1 - \exp\left[-\left(k_g + k_r\right)t\right]\right\}. \tag{5}$$

The viscous shear stress related to deformation of the solvent and bridging columns confined between electrodes arises from the variation of conformations of polymer molecules as well as column organization. To determine this shear stress, the detailed calculations of the deformation of polymer molecules affected by the electric field should be carried out. This calculation is presently missing in the ER blend models. Without developing such extension of the theory, the shear stress should depend on viscous properties of the blend constituents and columnar structure. One simple way to indicate this dependence is to let $\sigma_{12} \propto \eta^E$ where η^E is a bulk shear viscosity at the particular shear rate D_{12} representing in all the above stated effects. The proportion of the columnar structure effect in η^E can be emphasized, for example, at very small deformation when $D_{12} \to 0$ and column organization corresponds nearly to the one at the quiescent state. Consequently, the number of bridging columns is maximal and the shear viscosity denoted η_0^E represents among others the effect of that column organization. When deformation is increased, the expectation that the column is breaking up in the time interval between t' and $t' + dt'$ is $k_r dt'$ and that it survives without breaking from time t' to time t is $\Phi(t',t)$. At the same time, the expectation that the column is forming in the time interval between t' and $t' + dt'$ is $k_g dt'$ and that it survives without forming from time t' to time t is $1 - \Phi(t',t)$. The contribution to the macroscopic shear stress from those columns that meet the above stipulations is then

$$d\sigma_{12} = \left\{- k_r dt'\Phi(t',t) + k_g dt'\left[1 - \Phi(t',t)\right]\right\}\eta_0^E D_{12}(t',t). \tag{6}$$

Introducing the expectation of column formation (5), the shear stress is,

$$\sigma_{12} = \int_{-\infty}^{t} k_g \exp\left[\left(k_g + k_r\right)\left(t' - t\right)\right]\eta_0^E D_{12}(t',t)dt', \tag{7}$$

when the effect of bridging at all past time t' is taken into account.

3 Experimental

The measurement of the rheological characteristics of ER fluids is done using the coaxial cylinders and parallel disks disposition of the rotational rheometer (MR-300V2, Rheology Co., Ltd., Japan). The disks and cylinders are arranged as electrodes for applying dc electric field.

For observation of the structure organization, the specially designed set-up is used. In this set-up, the fluid sample is confined in the gap (0.5 mm) between the horizontal ring-shaped outer electrode and disk-shaped rotating inner electrode (diameter 120 mm, shear rate range 0.1 - 10 sec^{-1}). The optical microscope and CCD camera is used for recording the observation in the vertical direction (sample height is 0.3 mm).

The polymeric blends used for the experiments are the blends of the urethane-modified polypropylene glycol (UPPG) and dimethyl siloxane (DMS). The preparation of the blends as well as the experimental procedure and the experimental results are described in details elsewhere [2]. In case of the stress growth tests, the shear-flow is imposed after formation of columnar structure in quiescent blend during 1 min. The tests are performed at 30°C.

4 Results

The representative photographs showing organization of the blend columnar structure at rest and under shearing are shown in Fig.1. When the bridging columns are deformed and broken up by shearing, they immediately reconnect with the opposite column ends and bridge again the electrodes.

The representative plot of the shear rate vs. shear stress and the shear stress growth are presented in Fig.2. The description of the experimental data for steady shear flow results from Eq. (7) in the form,

$$\sigma_{12} = \left(1 + k_r / k_g\right)^{-1} \eta_0^E D_{12} \,. \tag{8}$$

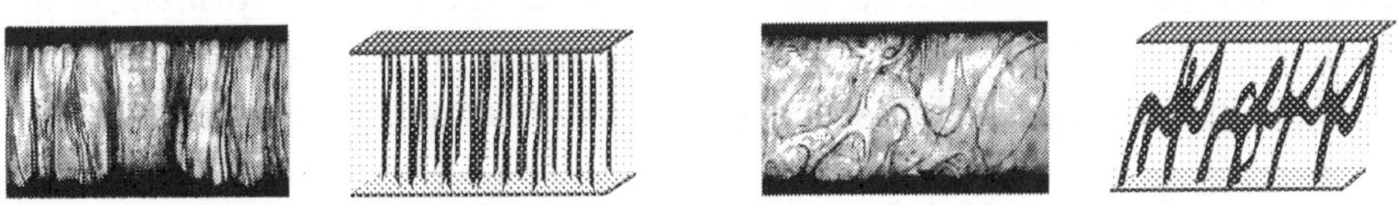

Figure 1. The photographs and sketches of the columnar structure between the electrodes at the quiescent state (on the left) and under shearing. The shear rate is 1 s^{-1} and the strength of the electric field 2 kV/mm [2].

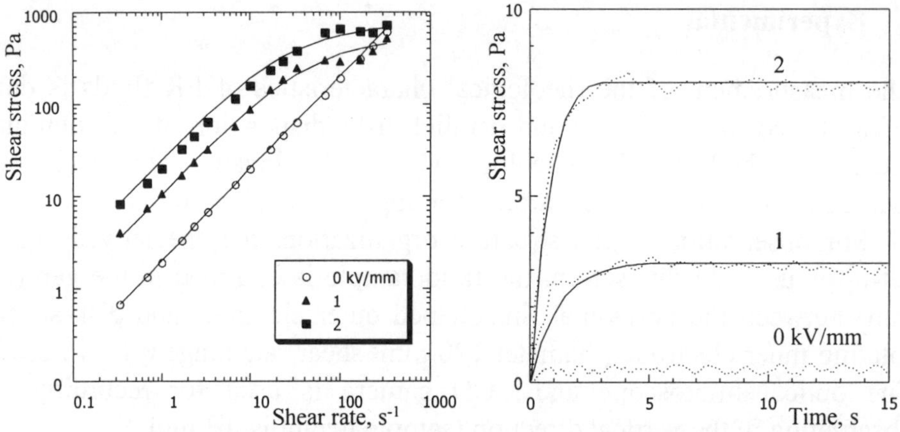

Figure 2. The shear rate dependence of shear stress (on the left) and the shear stress growth after onset of steady shear flow (shear rate D_{12}=0.3 s^{-1}) for UPPG (25 wt%)/DMS blend. The solid curves are given by Eqs. (8) and (9), respectively. The symbols denote the experimental data for corresponding strength of the electric field. The dashed curves represent the weighted fit of the experimental data.

This equation is exactly same as the corresponding equation following from the Maxwell-like rheological equation used previously as the plausible equation for description of rheological behavior of the polymeric blend [4]. The ratio of the kinetic rates was considered proportional to the shear rate, $k_r/k_g = aD_{12}$, and the parameters were determined by the nonlinear regression, $a = 0.02E^{0.8}$, $\eta_0^E = 1.9 + 8.7E^{3/2}$, where E denotes the strength of the electric field in kV/mm [4]. When the electric field does not affect the flow that is when E=0, the blend behaves like a Newtonian fluid, see Fig.2.

The description of shear stress growth after the onset of steady shear flow results from Eq. (7) in the form,

$$\sigma_{12} = \left(1 + k_r/k_g\right)^{-1}\left\{1 - \exp\left[-\left(k_g + k_r\right)t\right]\right\}\eta_0^E D_{12}. \tag{9}$$

The overall kinetic rate is considered also proportional to the shear rate, $k_g + k_r = bD_{12}$. The experimental data in Fig.2 provide, $b = 0.8E^{0.81}$.

According to Fig.2, the calculated results represent the experimental data well and suggest that the proposed concept of deformation and breaking of the bridging columns corresponds in principle to structural changes in ER fluids and its effect on the macroscopic shear stress.

5 Conclusions

The observation of the columnar structure formation as well as the description of rheological properties of investigated ER polymeric blends result the following evidences. The observed bridging of the electrodes by the more viscous constituent of the blend is dominant feature of the induced structure organization. The stochastic character of this feature is described and related to the macroscopic shear stress. The proposed description corresponds to the plausible Maxwell-like model used to describe the experimental data in [4] only in the case of steady shear flow. For unsteady shear flow, Eq. (7) embodies the kinetics of column formation while the phenomenological Maxwell-like model contains the notion of a “fading memory” to express the structure formation effect.

6 Acknowledgement

One of the authors (PR) gratefully acknowledges the financial support from the Grant Agency of the Academy of Sciences of the Czech Republic (grant No. IAA2060003).

References

1. Chesla S. E., Selvaraj P. and Zhu C., Measuring two-dimensional receptor-ligand binding kinetics by micropipette, *Biophys. J.* **75** (1998) pp. 1553-1572.
2. Kimura H., Aikawa K., Masubuchi Y., Takimoto J., Koyama K. and Uemura T., Positive and negative electro-rheological effect of liquid blends, *J. Non-Newtonian Fluid Mech.* **76** (1998) pp. 199-211.
3. Riha P., Kimura H., Aikawa K., Masubuchi Y., Takimoto J. and Koyama K., Stochastic aspects of bridging bonds of induced structure of electro-rheological fluids in shear flow, *Int. J. Mod. Phys. B* **13** (1999) pp.1798-1805.
4. Riha P., Kimura H., Aikawa K., Masubuchi Y., Takimoto J. and Koyama K., The shear-flow properties of electro-rheological liquid polymeric blends, *J. Non-Newtonian Fluid Mech.* **85** (1999) 249-256.
5. Van Kampen N. G. Stochastic processes in Physics and Chemistry (North-Holland, Amsterdam, 1981).

EFFECTS OF SHAPE AND SIZE OF DISPERSOID ON ELECTRORHEOLOGY

AKIKO KAWAI, KUNIO UCHIDA, AND FUMIKAZU IKAZAKI

Porous Materials Group Institute for Materials and Chemical Process

AIST Tsukuba Central 5 Higashi 1-1-1, Tsukuba, Ibaraki 305-8565, JAPAN
E-mail: akiko-kawai@aist.go.jp

The effects of shape and size of dispersoids on electrorheology were investigated. In order to study the effect of shape on electrorheology, three kinds of hydroxy-zinc complexes were used for dispersoids. Hydroxy-zinc complexes were $Zn_5(OH)_8Cl_2 \cdot H_2O$(1), $Zn_5(OH)_8(NO_3)_2 \cdot 2H_2O$(2), and $Zn_5(OH)_8(CH_3COO)_2 \cdot 2H_2O$(3). Shapes of (1) and (2) are plate. The shape of (3) is rod. The ER fluid containing (3) showed the lowest permittivity and the lowest ER effect. The ER phenomena containing dispersoids with different shapes were independent of their shapes and were explained by their dielectric properties. Zinc oxides prepared by the heat treatment of (1), (2), and (3) were used for studying the effect of size on electrorheology. The particle size influenced their dielectric property and influenced their electrorheology. The dielectric properties were responsible for the ER effect.

1 Introduction

Electrorheological (ER) fluid changes viscosity in the presence or absence of an external electric field. ER fluids are expected as vibratory control devices, clutches, and actuators. ER fluids are not yet industrially practical, because ER phenomena are not well understood and the design of ER fluid remains to be clarified.

In a dispersed ER system, dispersed particles polarized in an electric field are mutually connected by dipole-dipole interaction. We studied the relation between dielectric properties and ER effects of various ER fluids, finding that the large ER effect requires an appropriate polarization rate of the dispersoids corresponding to the shear rate and the effective permittivity $\Delta\varepsilon$'[1,2,3] defined as the permittivity difference between 100 and 100 kHz. Few papers discussed the effect of shape and size of the dispersoids on electrorheology. Sugimoto[4] suggested that dispersoids having sharp edge were not good as the dispersoids of ER fluid. In another report[2], which changed the particle size of the dispersoid, it was suggested that ER fluid containing larger size of dispersoid showed higher ER effect. However, the other parameter such as the amount of water on the dispersoid was not constant, so the ER effect was influenced not only by the size but also by the water content of the dispersoid. Several papers[5,6,7], which theoretically examined the effect of particle size on electrorheology, concluded that there is no effect of the particle size on the ER effect and the conclusion is contradictory to experimental results reported before[2,8].

In the present study, the effects of shape and size on electrorheology are respectively examined by using three kinds of layered hydroxy-zinc complexes with different shapes and by using zinc oxide with different sizes prepared by the heat treatment of the three layered hydroxy-zinc complexes as dispersoids.

2 Experiment

2-1. Sample:

Three kinds of layered hydroxy-zinc complexes with different shapes were prepared following the procedure of Kawai et al[9]. Three kinds of ZnO having different sizes were thermally converted from these hydroxy-zinc complexes at the temperature of 600, 450, and 300℃, for Cl-, NO_3-, and CH_3COO-type complexes, respectively. The heat treatment temperature, which was the end temperature of the conversion into ZnO, was determined by DTA-TG analysis. Hydroxy-zinc particles were dispersed into light silicone oil (Toshiba Silicone Co Ltd, TSF 437, 22 cSt at 25℃), because the layered zinc complexes were swollen in dimethylsilicone oil. ZnO particle was dispersed in dimethylsilicone oil (Toshiba Silicone Co Ltd, TSF 451-100, 100 cSt at 25℃).

The ratio of dispersoids to light silicone oil was 30%, and to dimethyl silicone oil was 10% by weight.

Crystal structures of particles were identified by X-ray powder diffraction (MAC Science Co. Ltd., XRD M03X-HF22). The interlayer water of the three hydroxy-zinc complexes was analyzed by TG-DTA. Scanning electron microscopy (Hitachi, SEM S-800) was used to observe their morphology. Particle size distribution was measured using laser diffraction (Horiba, LA-920).

2-2. Measurement:

Dielectric properties were measured from 20 Hz to 1 MHz by an LCR meter (HP 4284A, Hewlett-Packard Japan Ltd.). The ER effect was measured by a rotational viscometer (VT550, Haake). The fluid was held in a 1 mm gap between an inner rotating cylinder and outer cup and a DC voltage of 2.5 kV/mm was applied. The shear rate was changed from 0 to

700 s^{-1}.

3 Results and Discussion

3.1 Three Kinds of Hydroxy-Zinc Complexes

The XRD patterns of the synthesized three samples show the same patterns of hydroxy-zinc complexes reported before[9]. The amount of interlayer water was determined by TG-DTA and three kinds of hydroxy-zinc complexes were identified as $Zn_5(OH)_8Cl_2 \cdot H_2O$, $Zn_5(OH)_8(NO_3)_2 \cdot 2H_2O$, and $Zn_5(OH)_8(CH_3COO)_2 \cdot 2H_2O$, respectively. SEM photos of these complexes were shown in Fig. 1. These layered hydroxy-zinc complexes consist of $Zn(OH)_2$ layer and have anion between the layers. The shapes seem to be related with the interlayer anions.

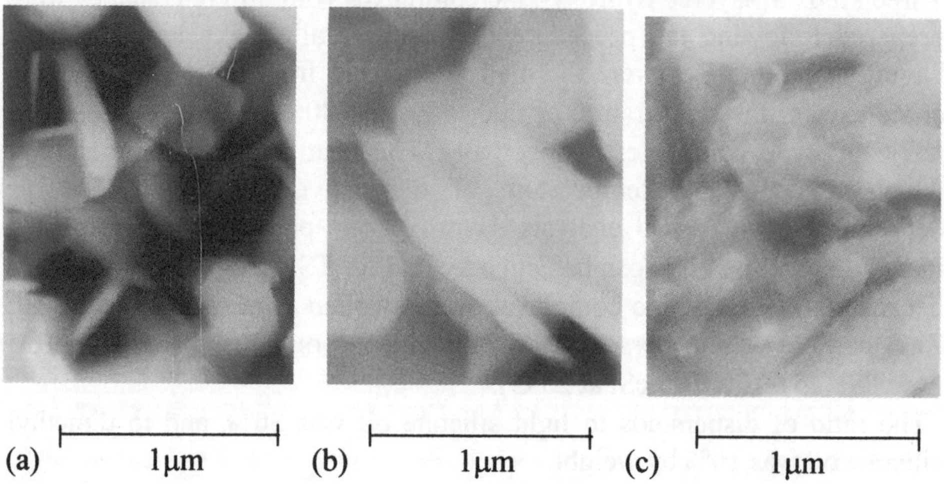

Figure 1. Scanning electron micrographs of (a) $Zn_5(OH)_8Cl_2 \cdot H_2O$, (b) $Zn_5(OH)_8(NO_3)_2 \cdot 2H_2O$, and (c) $Zn_5(OH)_8(CH_3COO)_2 \cdot 2H_2O$.

The SEM photos show that the Cl type and NO_3 type are plate-like, and CH_3COO type is rod-like particle. The ER effect of the fluids containing these hydroxy-zinc complexes in light silicone oil at 30 wt.% was shown in Fig. 2. In the shear rate of lower than 100 s^{-1}, peculiar behavior was observed in the CH_3COO type ER fluid. It can be due to the microstructure change of the system, significantly dependent on shear rate.

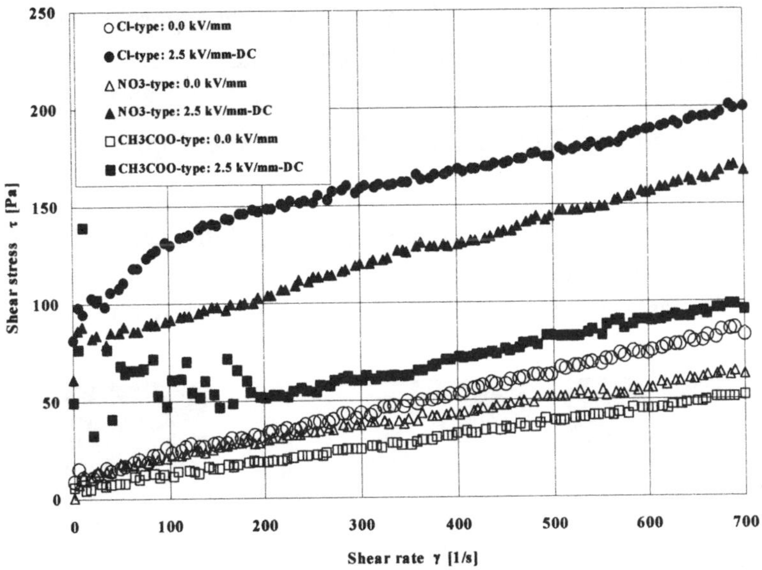

Figure 2. Flow curves of the ER fluid containing 30 wt % of $Zn_5(OH)_8Cl_2 \cdot H_2O$, $Zn_5(OH)_8(NO_3)_2 \cdot 2H_2O$, and $Zn_5(OH)_8(CH_3COO)_2 \cdot 2H_2O$ in light silicone oil

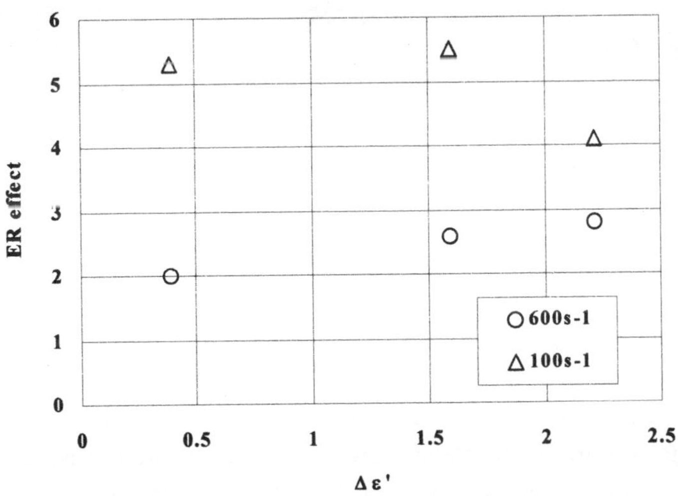

Figure 3. Relation between ER effect at 100 and 600 s^{-1} and Δε' of the three kinds of ER fluids

CH_3COO type hydroxy-zinc complexes have the lowest ER effect. The effective permittivity Δε', defined as the permittivity difference between 100 and 100 kHz, which indicate mainly the mobility of CH_3COO^- ions between layers, is the smallest of the three fluids. The relation between ER effect and Δε' of three kinds of fluids is shown in Figure 3. The effective permittivity Δε' is responsible for ER phenomena.

3.2 Three Kinds of Zinc Oxide Obtained From Different Starting Materials

Median diameters of ZnO from Cl-type, NO_3-type, and CH_3COO-type hydroxy-zinc complexes were 2.1 μm, 0.9 μm, and 2.2 μm, respectively, based on the laser diffraction measurement. Fig. 4 shows the morphology of these ZnO particles.

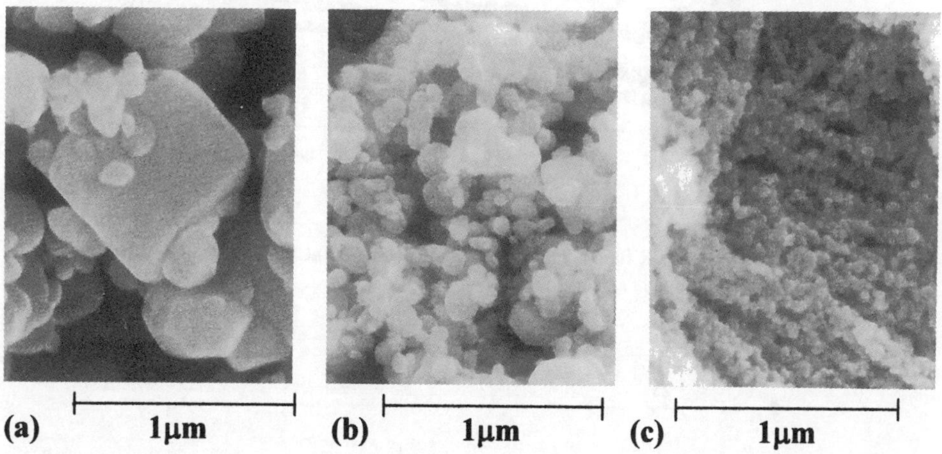

Figure 4. Scanning electron micrographs of ZnO powders obtained by the thermal treatment. The powders were obtained from (a) $Zn_5(OH)_8Cl_2 \cdot H_2O$, (b) $Zn_5(OH)_8(NO_3)_2 \cdot 2H_2O$, and (c) $Zn_5(OH)_8(CH_3COO)_2 \cdot 2H_2O$.

Laser diffraction results for these particles differed completely from those of SEM imaging, probably due to the aggregation of the particles. The particles are considered to be aggregated each other in the ER fluid, while primary particle sizes of these particles were fine.

The ER effect of the fluids containing ZnO in dimethyl silicone oil was shown in Fig. 5. One of the most different points of the dielectric property among these three fluids was Δε' of CH_3COO-type. Based on our idea[1,2,3], larger Δε' show higher ER effect. In fact, CH_3COO-type fluid shows the highest ER effect among the three. The dielectric property of Cl-type fluid is almost the same as that of NO_3-type fluid, the ER effects of Cl-type and NO_3-type fluid at high shear rate are almost same. Therefore the ER phenomena seems to be responsible to Δε'.

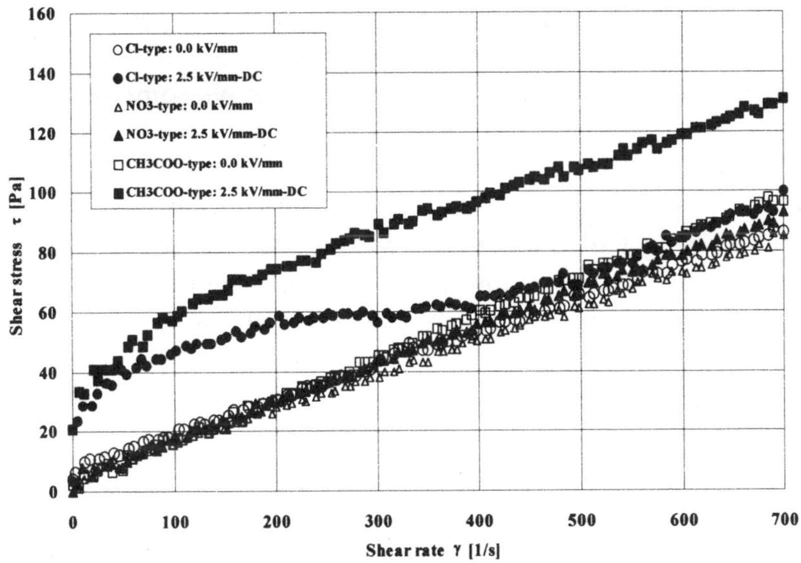

Figure 5. Flow curves of the ER fluid containing 10 wt. % of ZnO from Cl type, NO_3 type, and CH_3COO type hydroxy-zinc complex in dimethyl silicone oil

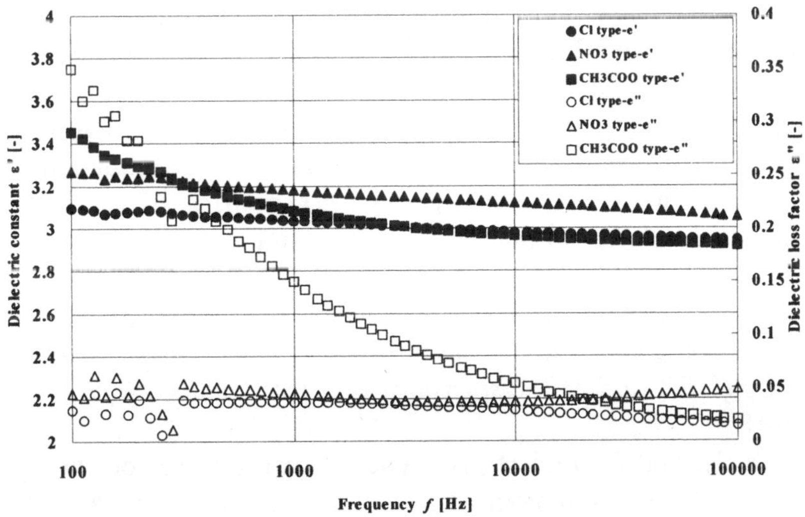

Figure 6. Dielectric properties of the ER fluids containing three kind of ZnO in dimethyl silicone oil.

But, the ER effects of Cl-type and NO_3-type fluid are different at low shear rate. In the case of different shapes of dispersoids, the ER phenomena at low shear rate are also peculiar. It is suggested that not only the dielectric property but also another factor influenced the electrorheology at low shear rate. At present time, the ER phenomena at low shear rate are influenced by the microstructure of the system, but further investigation is necessary for clarify the ER phenomena.

5. Conclusion

The effects of shape and size of the dispersoids on electrorheology in a DC field were investigated. Effective permittivity $\Delta\varepsilon'$ are responsible for the ER effect. The flow curve at low shear rate was peculiar, however, further investigation is necessary for the ER phenomena at low shear rate.

References

1. Kawai A., Ide Y., Inoue A., and Ikazaki F., Electrorheology of miscible blended liquid crystalline polymer: A dielectric property approach, *J. Chem. Phys.* **109** (1998) pp. 4587–4591.
2. Ikazaki F., Kawai A., Uchida K., Kawakami T., Edamura K., Sakurai K. Anzai H., and Asako Y., Mechanisms of electrorheology: the effect of the dielectric property, *J. Phys. D.* 31 (1998) pp. 336–347.
3. Kawai A. and Ikazaki F, Effect of three functional groups on electrorheology. *Int. J. Mod. Phys. B.* 15 (2001) pp. 1042–1049.
4. Sugimoto A., Ion Koukan Jushi Bunsankei niyoru Winslow kouka, *Nihon Kikai Gakkai Ronbunshu* 43 (1997) pp. 1075-1083.
5. Klingenberg D. J., Swol F. V., and Zukoski C. F., The small shear rate response of electrorheological suspensions. 1. Simulation in the point-dipole limit, *J. Chem. Phys.* 94(1991) pp.6160-61.
6. Takimoto J, Cluster formation in electrorheological fluids: Theory and computer simulation. *Nihon Reoroji Gakkaishi* **20** (1992) pp.95–100.
7. Mori M. J., Tsunamoto T. and Nakayama H., Computer simulation of electrorheological fluids in the binary system of particle size. *Proc.6th Int. Conf. ER and MR suspensions and their applications* (1998) pp. 180–185.
8. Shih Y. H. and Conrad H., Influence of particle-size on the dynamic strength of electrorheological fluids, *J. Mod. Phys. B.* 8 (1994) pp. 2835-2853.
9. Kawai A., Sugahara Y., Park I.Y, Kuroda K., and Kato C., Preparation of zinc oxide powders from 2-dimensional hydroxy-zinc complexes of basic zinc acetate, chloride, and nitrate, *Ceramic Transactions* 22 (1991) pp. 75-80.

ER PROPERTIES AND FLOW-INDUCED MICROSTRUCTURES OF AN ER FLUID BETWEEN TWO PARALLEL DISK ELECTRODES IN SQUEEZE FLOW MODE

M. NAKANO AND T. NAGATA

Department of Mechanical Systems Engineering, Yamagata University
4-3-16, Jonan, Yonezawa, Yamagata 992-8510, JAPAN
E-mail: nakano@mnaka.yz.yamagata-u.ac.jp

When an electrorheological fluid between two parallel disk electrodes is subjected to squeeze mode excitation, the microstructural flow behavior of the ER fluid has been investigated in close relation to the measured electrorheological properties. A dilute anhydrous ER fluid made of 10 vol% carbonaceous particles dispersed in silicone oil and two ITO glass disk electrodes are used to visualize the microstructural flow behavior. The squeeze tests are carried out on a mechanical testing machine in tensile, compression and oscillatory squeeze modes. The ER properties and the associated flow-induced microstructures are measured and visualized for different initial DC electric field strengths and initial gap h_i=0.5 mm between the electrodes in two different configurations: (a) the fluid is placed only between the electrodes (conf-1) and (b) the two electrodes are immersed in the ERF (conf-2).

1 Introduction

Electrorheological (ER) fluids consisting of electrically polarizable particles dispersed in dielectric oil exhibit dramatic changes in their rheological and mechanical properties like a Bingham fluid having yield stress under the application of an electric field [1]. These rheological changes are caused by the alignment of the particles into a fibrous structure in the direction of the field due to electrostatic forces, which is responsible for the increase in resistance against the flow of ER fluids. The response of the ER fluids to the electric field is fast, making them advantageous to be used in many control systems such as vibration control systems and robot control systems, as well as mechanical devices such as dampers, valves, clutches, brakes, engine mounts and so on [2, 3]. Fundamental flow modes in the application to these devices are classified into three, i.e., shear flow and squeeze flow modes where two electrodes are moveable relatively, and pressure flow mode where two electrodes are fixed [4]. The shear- and pressure-flow modes have been frequently utilized for the clutches, brakes and valves, and the ER properties and the flow behavior in these flow modes have been clarified by many researchers [5, 6, 7, 8]. While, the ER properties in the squeeze flow mode have been investigated by few research groups [9, 10], although there is a possibility that the ER fluid in this flow mode is employed in squeeze dampers featuring high damping forces. And also, the flow behavior of ER fluid in squeeze has never been observed so far.

In our previous study [10], the mechanical properties of an anhydrous ER fluid made of carbonaceous particles (30 vol%) have been determined in tensile, compression and oscillatory squeeze tests. It is of great importance to investigate the ER mechanical properties in close relation to the flow-induced microstructures of the ER fluid to understand the ER mechanism. In this study, the microstructural flow behavior and the ER properties of the ER fluid between two parallel disk electrodes have been investigated in tensile, compression and oscillatory squeeze tests in terms of the DC electric field strength, the number of oscillation cycles, and the ways to fill up ER fluid between two electrodes.

2 Experimental Apparatus and Procedure

In this study, a dilute anhydrous electrorheological fluid BA-1(Bridgestone) of black color made of carbonaceous particles (10 vol%, 3 ~5 μm) dispersed in silicone oil and two ITO glass electrodes are used to visualize the microstructural flow behavior. Figure 1 shows a test rig developed for measuring the mechanical properties and visualizing the micro flow behavior of the ER fluid, which is composed of two glass electrodes (**d** = ϕ60 mm) coated with ITO film between which the ER fluid is placed. The upper or lower electrode can move up and down in parallel to another electrode along the four guiding rods and ball bushes, while another one remains fixed. The mechanical properties and the flow behavior of the ER fluid are determined in tensile, compression and oscillatory squeeze tests. In tensile and compression tests, the test rig is installed to a mechanical testing machine (Autograph AG-50kNG, Shimadzu). The ER fluid placed between the electrodes is submitted to an elongational force in the tensile test as the upper electrode moves up, while is squeezed to flow out in the compression test as the upper electrode is forced to move down. The oscillatory squeeze test is carried out on an electromagnetic vibration exciter (VS-20-03, IMV), and the lower electrode is excited to move up and down sinusoidally at a constant exciting force, while the upper electrode is fixed, so the ER fluid is cyclically submitted to tensile and to compression. The strain amplitude is initially set as $|\varepsilon_0| = 0.05$. The sample is tested until 5000 cycles are completed.

In each test, the displacement **Δh** of the moving glass electrode and the force **F** acting on the upper electrode are measured on a laser displacement sensor and a load cell, respectively, and are transformed into the strain ($\varepsilon = \Delta h/h_i$, h_i: initial gap between two electrodes (=0.5mm)) and the stress ($\sigma = 4F/(\pi d^2)$) of the ER fluid to evaluate the ER

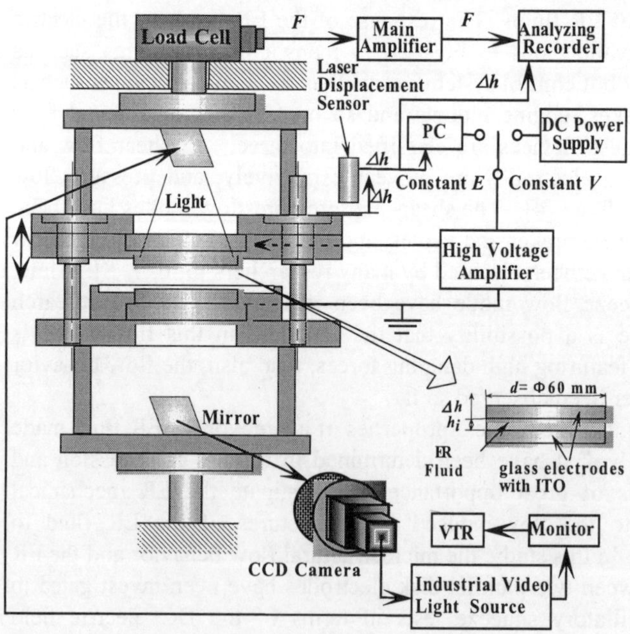

Figure 1. Test rig developed for measuring the mechanical properties and visualizing the micro-flow behavior of ER fluid in squeeze flow mode.

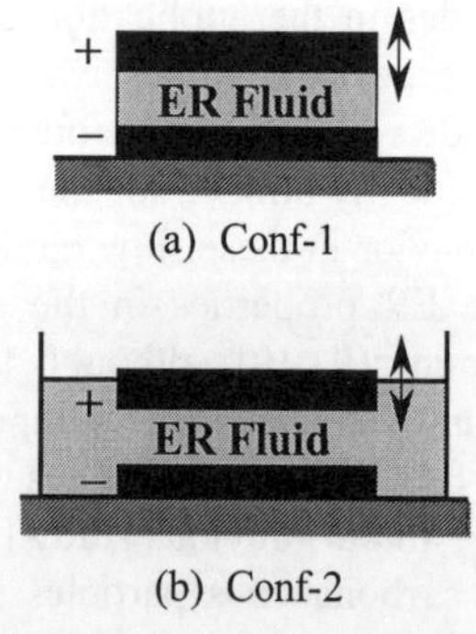

Figure 2. Two ways of filling up ER fluid between two electrodes.

properties. The tensile and compression tests are conducted at constant electric field strength (constant **E**) and at constant voltage (constant **V**). In the case of the constant **E**, the electric field strength is maintained constant by varying the input voltage as the gap is increasing (tensile test) or decreasing (compression test). In the case of the constant **V**, the applied voltage **V** is constant during all the experiments. So, the electric field strength varies with time **t** according to the equation: $\mathbf{E} = \mathbf{V} / (\mathbf{h_i} \pm \mathbf{v}\ \mathbf{t})$, where **v** is the velocity of the moving electrode. The oscillatory squeeze test is conducted at the constant **V**.

In each test, the ER fluid is filled up between two electrodes in the following two different ways, as shown in Fig.2:

(a) the fluid is placed only between the electrodes (Conf-1);

(b) two electrodes are immersed inside the ER fluid (Conf-2).

The flow-induced microstructures of the ER fluid between two electrodes are observed on a CCD camera through a reflecting mirror from the bottom side of the lower transparent glass electrode under a halogen light that enters from the upper side of the upper glass electrode in the tensile and compression tests, and do from the upper side of the upper transparent glass electrode in the oscillatory squeeze test.

3 Results and Discussion

3.1 Tensile Tests

Figure 3 shows the dependences of the initial DC electric field strength $\mathbf{E_0}$ on the tensile stress – strain curves for conf-1 and –2 under the constant E. For both conf-1 and –2, initially, the tensile stress increases proportionally with increasing strain, showing the elastic deformation of the ER fluid, and after that, the ER fluid presents a plastic region where the stress σ increases with the strain until it reaches a maximum value: the tensile strength σ_t and after the σ_t is attained the stress decreases with increasing strain, as presented in our previous paper [10]. The increase of the electric field produces the increase of the tensile strength, because of strengthening the fibrous structure of the particles. What is obvious on comparing Fig. 3(a) with Fig. 3(b) that in the case of conf-1, after the tensile strength is attained the stress remarkably decreases at $\mathbf{E_0}$ more than 1.5 kV/mm while it gradually decreases in the case of conf-2, and the tensile strength for the conf-2 is higher than that for the conf-1. This is happened because in the case of the conf-

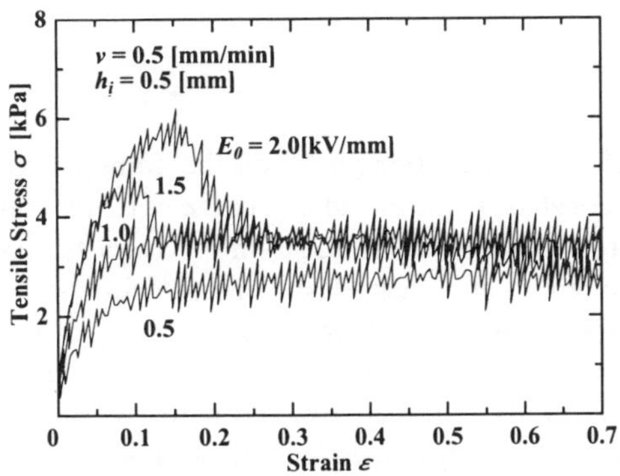

(a) Conf-1, Constant E

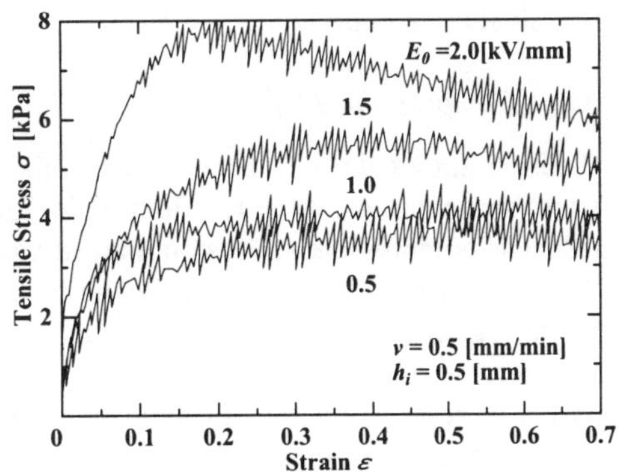

(b) Conf-2, Constant E

Figure 3. Stress-strain curves of ER fluid at various constant DC electric field strengths in tensile test.

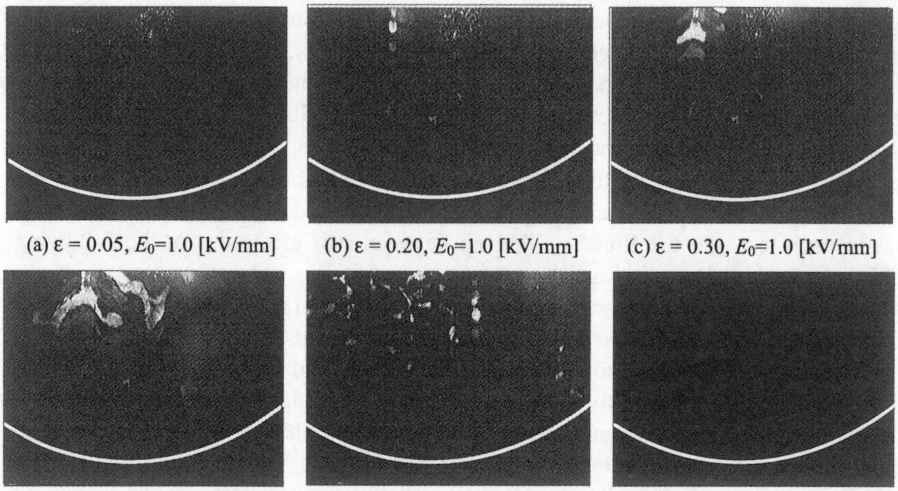

(a) ε = 0.05, E_0=1.0 [kV/mm] (b) ε = 0.20, E_0=1.0 [kV/mm] (c) ε = 0.30, E_0=1.0 [kV/mm]

(d) ε = 0.60, E_0=1.0 [kV/mm] (e) ε = 0.60, E_0=2.0 [kV/mm] (f) ε = 0.60, E_0=2.0 [kV/mm], Conf-2

Figure 4. Evolution of flow-induced microstructure of ER fluid between two electrodes with increasing strain and electric field strength in tensile test (Conf-1, Constant *E*).

2 the ER fluid outside of the electrode gap can enter into the gap as the upper electrode is being raised.

Figure 4 shows the evolution of the observed microstructures of the ER fluid between two electrodes with increasing strain and electric field strength under the constant **E** for the conf-1. A white circular arc drawn in each picture corresponds to the outline of the lower electrode. As seen in Fig.4(a)-(d), in the region of the strain beyond the tensile strength strain (**ε**=0.20), transparent areas like clacks, which might be caused by air penetration, are obviously observed around the center of the disk electrodes, and spread over the electrodes as the strain increases.

The increase of the electric field strength finely divides the transparent areas into many small segments as seen in Fig.4(e), representing the decrease of the contacting area of the ER fluid to the electrodes. While, in the conf-2, these transparent areas can not be observed as seen in Fig.4(f), because of the inflow of the ER fluid from outside of the gap. So, it is concluded that this air penetration weakens the resistance to the tensile motion of the ER fluid in conf-1.

3.2 *Compression Tests*

Figure 5 shows the influence of the initial electric field strength $\mathbf{E_0}$ on the compressive stress-strain curves of the ER fluid under the constant **V** for the conf-1 in the compression test. The compressive stress continuously increases with increasing strain for every $\mathbf{E_0}$. As the strain increases, the ER fluid exhibits a softening nonlinear elastic characteristic in an initial region of a small strain, and then it does a hardening nonlinear elasticity, showing the increase of the slope of the stress-strain curves with increasing strain. This is caused by the increase of the electric field strength **E (=V/h)** with increasing compressive strain. For higher electric fields, the elasticity of the ER fluid becomes harder, because the ER fluid becomes more resistant to flow. Comparing these results with those of tensile test, it can be concluded that the ER fluid is more resistant to compression than to tensile.

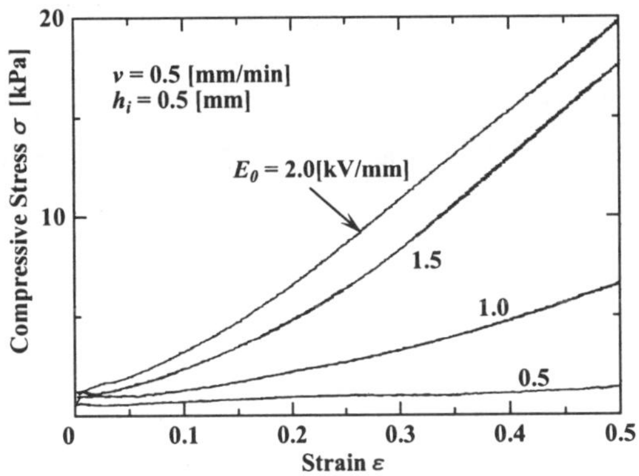

Figure 5. Stress-strain curves of ER fluid at various DC electric fields in compression test (Conf-1, Constant *V*)

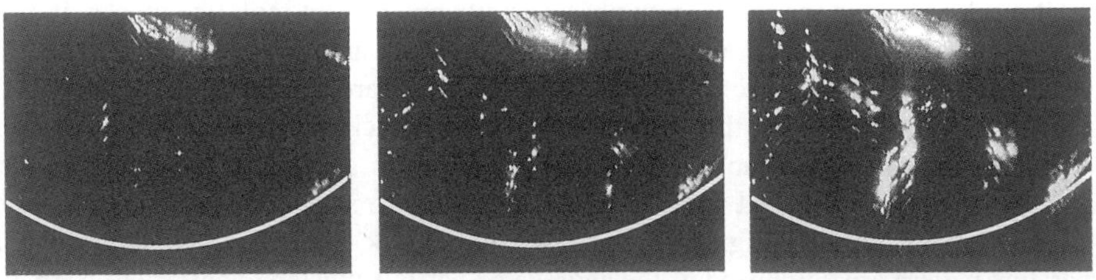

(a) ε = 0.1 (b) ε = 0.3 (c) ε = 0.5

Figure 6. Evolution of flow-induced microstructure of ER fluid between two electrodes with increasing strain in compression test (Conf-1, Constant *V*, E_0=1.5 [kV/mm])

The evolutions of the flow-induced microstructures of the ER fluid between two electrodes with increasing compressive strain are shown in Fig.6 in the case of the conf-1 and constant **V**. As the compressive strain increases, the transparent phases of silicone oil that can be observed like as lands appear and enlarge, representing the phase separation of the ER fluid that might produce particle aggregations, which are more resistant to compressive motion. This phase separation occurs for the initial electric field strength $\mathbf{E_0}$ more than 1.5 kV/mm, and become more remarkable as the electric field increases.

3.3 *Oscillatory Squeeze Tests*

The oscillatory squeeze tests are carried out at a constant maximum exciting force under the constant **V**. Figure 7(a) and (b) show the dependence of the hysteresis loops of stress-strain curve on the number of cycles **N** for conf-1 and -2, $\boldsymbol{f}$ = 5 Hz, $\mathbf{h_i}$ = 0.5 mm and $\mathbf{E_0}$ = 2.0 kV/mm. As **N** increases, the strain amplitude and the area inside the hysteresis loop decrease for conf-1, while the hysteresis loop for conf-2 does not change significantly. This decrease of the strain amplitude in conf-1 means the increase of the resistance to oscillatory squeezing motion. That is, the damping force of the ER fluid in conf-1 increases with the increase of **N**, because the lower electrode is excited at a constant maximum force.

Figure 8 shows the changes of the ER fluid microstructures induced by oscillatory squeeze flow with the increase of **N** at $\mathbf{E_0}$=2.0 kV/mm for the conf-1 and –2. For the conf-1 (Fig.8(a)), as the **N** increases, the ER fluid distributed uniformly at an initial stage gradually forms the phase separation structure, which is composed of the particle agglomera-

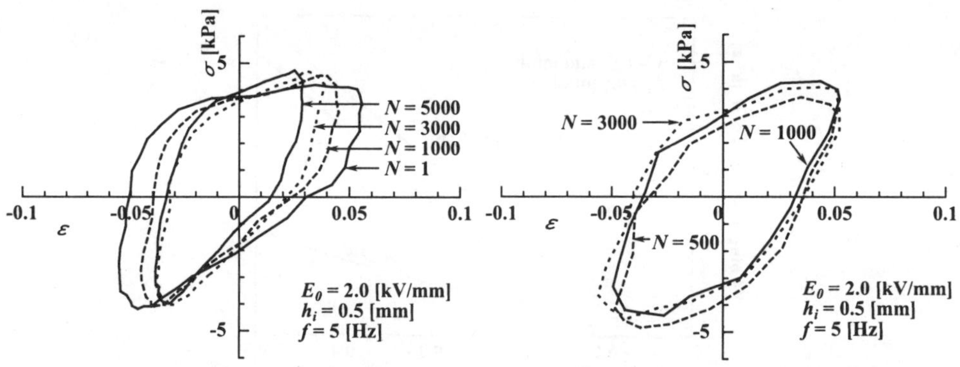

(a) Dependence of number of cycles N (Conf-1) (b) Dependence of number of cycles N (Conf-2)

Figure 7. Changes of hysteresis loops of stress vs. strain in oscillatory squeeze test.

tion of black areas and the transparent phases of silicone oil that look like cracks. It can be observed in these phase separation patterns that the black areas of the aggregated particle clusters are fixed to the electrodes and only the transparent phases of the silicone oil oscillates in large amplitude through the crack-like flow channels. This phase separation phenomena become remarkable and the particle clusters agglomerate densely as the **N** increases, so that the induced stress becomes large due to stronger agglomerated particle clusters. For the conf-2 (Fig.8(b)), the crack-like black phases in the induced phase separation pattern are composed of the aggregated particle clusters, contrary to the pattern for the conf-1, because the ER fluid outside of the electrode gap is possible to enter into the gap. As the electric field become stronger, the above-mentioned phase separation patterns appear and the transparent phases of silicon oil widely spread over the electrodes.

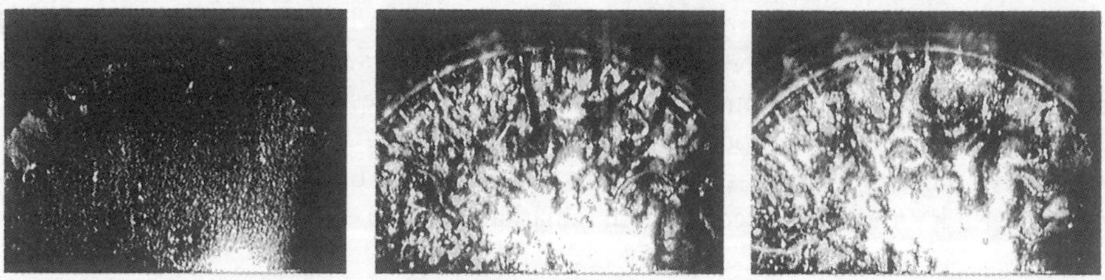

N=50 N=1000 N=5000

(a) Conf-1

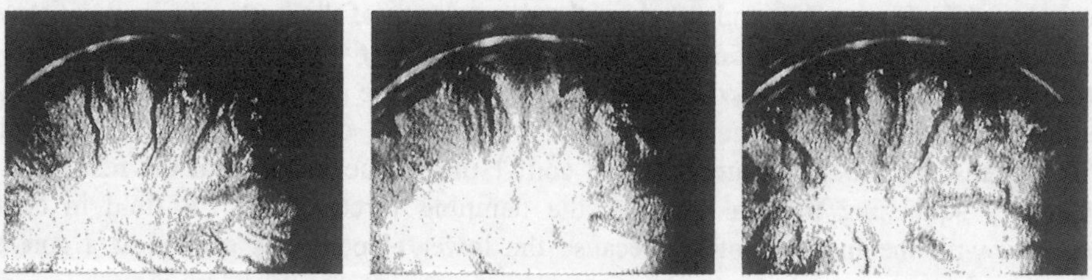

N=50 N=1000 N=5000

(b) Conf-2

Figure 8. Phase separation of ER fluid induced by oscillatory squeeze flow (E_0=2.0 [kV/mm], h_i=0.5 [mm], f=5 [Hz]).

4 Concluding Remarks

In this study, the microstructural flow behavior and the ER properties of an anhydrous electrorheological fluid made of carbonaceous particles dispersed in silicone oil have been investigated in tensile, compression and oscillatory squeeze tests.

In tensile test, the tensile strength was higher for conf-2 than for conf-1, which was caused by the penetration of air into the ER fluid between two electrodes and the associated formation of the ER fluid aggregations in case of the conf-1.

In compression test, at relatively high electric field strengths the ER fluid could be viewed as a phase separated composite in which the dispersed phase formed particle agglomerations, which played the role of reinforcing fibers, and the transparent phase of silicone oil appeared.

In oscillatory squeeze test for conf-1, as the number of oscillation cycles increased, the dispersed phase of particles concentrated to form the organic structure of the phase separation composed of many small cell-like agglomerations, indicating that the damping performance of the ER fluid decreased, and on the other hand, the shape of the hysteresis loop of the stress as a function of the strain for conf-2 did not change with the number of cycles.

References

1. Klass, D.L., and Martinek, T.W., Electroviscous fluids. I: Rheological properties, *J. Appl. Phys.*, **38-1** (1967) pp.67-74.
2. Nakano, M. and Koyama, K. eds., *Proceedings of the 6th International Conference on ER Fluids, MR Suspensions and their Applications*, World Scientific, (1998) p. 893.
3. Tao, R. ed., *Proceedings of the 7th International Conference on ER Fluids, MR Suspensions*, World Scientific, (2000) p. 848.
4. Nakano, M., Controllable damper using electrorheological fluids and new aspect of vibration control, *J. of the Society of Instrument and Control Engineers* (in Japanese), **34-9** (1995) pp.707-711.
5. Filisko, F.E., Henley, S. and Guist, G., Recent developments in the properties and composition of electrorheological fluids, *J. of Intelligent Material Systems and Structures*, **10-6** (1999) pp.476-480.
6. Aizawa, R., Vieira, S.L., Nakano, M. and Asako, Y., Hysteresis phenomenon in flow-curves of ER fluids containing sulfonated polymer particles, *Int. J. of Modern Physics B*, **15-6&7** (2001) pp.1070-1077.
7. Nakano, M., Aizawa, R. and Asako, Y., Steady and transient responses of electro-rheological suspension passing through a rectangular channel, *Int. J. of Modern Physics B*, **10-23&24** (1996) pp.2965-2972.
8. Nakano, M., Ito, K., Konno, M. and Aizawa, R., Dynamic viscoelasticity in oscillatory slit flow of an ER suspension containing sulfonated polymer particles, *Int. J. of Modern Physics B*, **13-14, 15&16** (1999) pp.1870-1877.
9. Sproston, J.L., Wahed, A.K.El. and Stanway, R., Electrorheological fluids in squeeze under AC and DC excitation, *Int. J. of Modern Physics B*, **13-14,15&16** (1999) pp.1861-1869.
10. Vieira, S.L., Nakano, M., Oke, R. and Nagata, T., Mechanical properties of an ER fluid in tensile, compression and oscillatory squeeze tests, *Int. J. of Modern Physics B*, **15-6&7** (2001) pp.714-722.

THE SIGNIFICANCE OF FLOW-MODIFIED PERMITTIVITY: A NEW MODEL AND COMPUTER SIMULATION OF ELECTRORHEOLOGY

YUE HU AND EE-YENN E. LIN

Physics Department, Wellesley College, Wellesley, MA 02481, USA
E-mail: yhu@wellesley.edu

UJITHA M. DASSANAYAKE

Physics Department, Brandeis University, Waltham, MA 02454, USA

We propose a model that takes into account the effect of flow-modified permittivity on electrorheology (ER). Due to dielectric relaxation, a shear flow causes the induced particle dipole moments in an ER fluid to tilt in a direction away from the direction of the applied DC electric field. Results from our computer simulation indicate that at high shear rates this misalignment (tilt angle) between the particle dipole moments and the applied electric field plays a crucial role in producing ER effects. By choosing particle-fluid dielectric and conductive mismatches to optimize the tilt angle, our simulation produces ER effects at much higher shear rates than those in earlier simulation work, even though there is no chain structure at these high shear rates. The increase in shear stress due to the applied electric field in our simulation is nearly constant over the wide range of shear rates examined, in qualitative agreement with experimental results. In addition, our model generates results that agree with earlier simulation work at low shear rates, where the particle dipole moments are essentially aligned with the field and the chain model is adequate.

1. Introduction

When a large electric field is applied to small particles suspended in an insulating fluid, the viscosity of the suspension can increase drastically. This is called the electrorheological (ER) effect. Since the discovery of the ER effect more than half a century ago,[1] the mechanism responsible for the effect has been the subject of much debate.[2] It is generally believed that the dielectric or conductive mismatch between the particles and the surrounding fluid causes the particles to become polarized in an external electric field. The polarized particles attract each other to form chain and column structures, and many researchers believe that the deformation of these chain structures is responsible for ER effects (the chain model).[3]

Critics of the chain model have pointed out that while chain structures cannot persist at high shear rates, nevertheless, ER effects are still appreciable.[4,5] Also, ER effects do not correlate directly with particle polarization, and fluids consisting of highly polarizable particles often show no ER effects.[6] Furthermore, fluids can lose their ER behavior if particles are dried thoroughly before mixing with fluids, even though the particle-fluid dielectric mismatch should be the same whether the particles are dry or moist.[5]

If particle polarization strength does not determine ER effects, then what does? Experimental evidence shows that the dielectric relaxation time seems to correlate strongly with ER effects and that there exists an optimal range of dielectric relaxation times; fluids with too long or short dielectric relaxation times do not make good ER fluids.[4]

In this paper, we consider the fundamental mechanism responsible for ER effects. We propose a model for ER effects that takes into account the effects of dielectric relaxation time and flow fields on particle polarization. Results of computer simulations based on this model are presented and analyzed. Finally, we argue that dielectric relaxation time is not necessarily, in itself, a quantity that predicts ER effects, and we suggest possible ways to maximize ER effects.

2. Model

As in many earlier theoretical studies, our model assumes that particles in ER fluids interact with each other as induced dipoles. The unique feature of our model, however, is that the strength and the direction of the dipole moments are modified by the shear flow of the fluids (flow-modified permittivity, or FMP). It is known that in an infinite laminar flow field of shear rate $\dot{\gamma}$, a spherical particle spins at an angular velocity $\Omega = \dot{\gamma}/2$.[7] The effect of particle spin on particle polarization has been studied extensively by H. Block and coworkers, both experimentally and theoretically.[8,9,10] They have shown that in a real ER fluid, where the assumption of a single particle in an infinite flow field is certainly not valid, particles still spin at an angular velocity equal to half the shear rate, even in the presence of an applied DC electric field as strong as 1 kV/mm.[11]

For particles of finite dielectric relaxation time τ_r, the spinning of a particle results in a component of the induced particle dipole moment perpendicular to the direction of the applied field, in addition to the usual parallel component. For a particle obeying Debye-type relaxation, the induced dipole moment in a DC field is[10]

$$\vec{\mu} = 4\pi\varepsilon_0\varepsilon_f R^3 E_a \vec{\alpha}, \tag{1}$$

where R is the particle radius, ε_0 the permittivity of the vacuum, ε_f the dielectric constant of the fluid, and E_a the strength of the applied electric field. The components of the dimensionless dipole moment $\vec{\alpha}$ in directions parallel and perpendicular to the applied field are:

$$\alpha_{\parallel} = \beta_\infty + (\beta_0 - \beta_\infty)/(1+\Omega^2\tau_r^2), \tag{2}$$

$$\alpha_{\perp} = (\beta_0 - \beta_\infty)\Omega\tau_r/(1+\Omega^2\tau_r^2). \tag{3}$$

Here the factors $\beta_0 = (\sigma_p - \sigma_f)/(\sigma_p + 2\sigma_f)$ and $\beta_\infty = (\varepsilon_p - \varepsilon_f)/(\varepsilon_p + 2\varepsilon_f)$ depend on particle-fluid conductive and dielectric mismatches.[12]

Fig.1 plots the tilt angle and the magnitude of the dimensionless dipole moment $\vec{\alpha}$

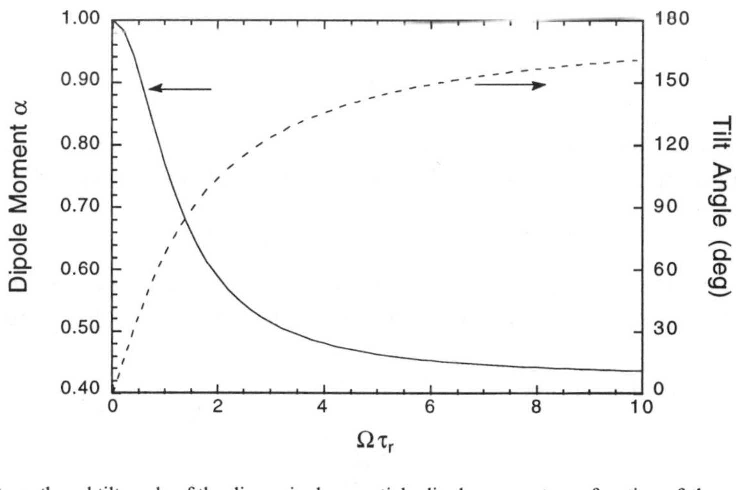

Fig. 1 Strength and tilt angle of the dimensionless particle dipole moment as a function of the product of the particle spinning rate and dielectric relaxation time, assuming $\varepsilon_f/\varepsilon_p = 10$, $\sigma_p \neq 0$, and $\sigma_f = 0$.

as a function of $\Omega\tau_r$, assuming a dielectric mismatch of $\varepsilon_f/\varepsilon_p = 10$, fluid conductivity $\sigma_f = 0$, and particle conductivity $\sigma_p \neq 0$. We can see that at $\Omega = 0$, the particle dipole moment is determined by the particle-fluid conductive mismatch, and the dipole is parallel to the applied electric field. As the shear rate increases, however, the magnitude of the dipole moment approaches a value determined by the dielectric mismatch, and the dipole moment and the applied field are no longer aligned.

We want to point out that this misalignment between the particle dipole moment and the applied electric field has been ignored or assumed to be unimportant in most earlier theoretical work on ER.[13,14,15] As will be seen later, we have found that it is precisely the tilt of the dipole that gives rise to ER effects at high shear rates.

3. Computer Simulation

Our simulation method is similar to that used by Klingenberg, *et al.*[13] The forces on each particle include: (1) electric dipole forces ($\vec{F}_{ele}$), (2) hard-core repulsive forces ($\vec{F}_{rep}$) all due to neighboring particles, and (3) the Stokes drag force from the fluid. For simplicity, we limit the motion of the particles to two dimensions, and we ignore mutual polarization and hydrodynamic interactions between particles. We also introduce the following scaling factors for force, length, and time: $F_0 = 3\pi\varepsilon_0\varepsilon_f \overline{R}^2 E_a^2/4$, $d_0 = 2\overline{R}$, and $t_0 = 3\pi\eta d_0^2/F_0$, where $\overline{R}$ is the mean particle radius and η the viscosity of the suspending fluid.

Traditionally, the ratio of the strength of the hydrodynamic interaction to that of the electrostatic interaction is specified by the Mason number $\mathrm{Mn} = \eta\dot{\gamma}/2\varepsilon_0\varepsilon_f\beta_\infty^2 E_a^2$. Our dimensionless shear rate here is proportional to the Mason number: $\dot{\gamma}^* = 32\beta_\infty^2\mathrm{Mn}$.

Initially, 45 particles are placed in a 12×12 box (Fig. 2(b)). The results reported here are averaged over 5 different initial configurations. The particle radii have a mean value of 0.5 and a standard deviation of 0.3. In dimensionless units, the equation of motion for the ith particle is

$$\frac{d\vec{r}_i^*}{dt^*} = \vec{v}_f^*\left(\vec{r}_i^*\right) + \frac{1}{d_i^*}\left(\vec{F}_{ele}^* + \vec{F}_{rep}^*\right), \tag{4}$$

where d_i^* is the diameter of the ith particle and $\vec{v}_f^*\left(\vec{r}_i^*\right) = \dot{\gamma}^* z_i^* \hat{x}$ is the shear field. We have also used asterisks to denote dimensionless quantities. The hard-core repulsive force between particles i and j is along the line connecting the centers of the two particles, and the form of the force is exponential with respect to the particle separation r_{ij}^*:[13]

$$\left|\vec{F}_{rep}^*\right| = 2\exp\left(-\left(r_{ij}^* - R_i^* - R_j^*\right)/0.01\right). \tag{5}$$

The simulation is carried out using an Euler method in the overdamped mode with shifted periodic boundary conditions. Each simulation is run for 5×10^6 steps, and the product of the time-step Δt^* and the shear rate γ^* is kept constant: $\Delta t^*\gamma^* = 2\times10^{-4}$. Shear stress is calculated using the equation[16]

$$\tau_{xz}^* = -\frac{1}{V^*}\sum_{i=1}^{N} F_{x,i}^* z_i^*, \tag{6}$$

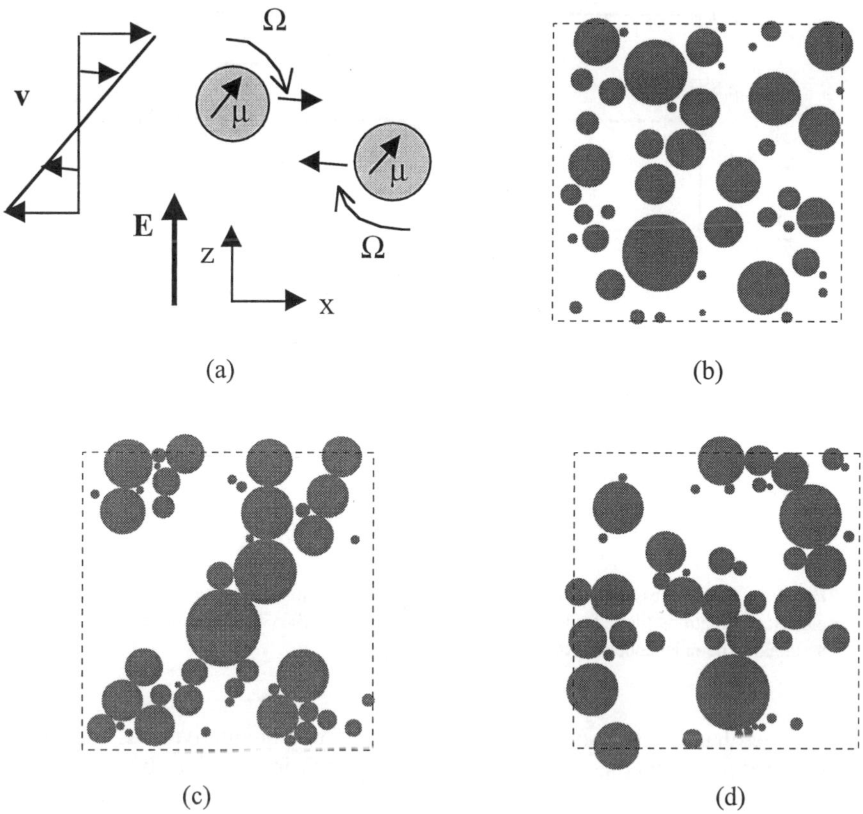

Fig. 2 (a) Two particles spinning in a shear field approach each other. The dipole moments are tilted due to FMP. (b) An initial configuration of 45 particles in a 12×12 box. (c) Chain structures are seen at a low shear rate of $\dot{\gamma}^* = 1$. (d) No chain structures are seen at a high shear rate of $\dot{\gamma}^* = 100$.

where V^* is the system volume, $F^*_{x,i}$ the x-component of the net (electric and repulsive) force on the ith particle, z^*_i the z-coordinate of the particle, and N the number of particles in the box. We also run simulations with no electric interactions to obtain the zero-field shear stress $\tau^*_{xz,0}$. The difference, $\Delta\tau^*_{xz} = \tau^*_{xz} - \tau^*_{xz,0}$, is an indictor of the strength of ER effects.

4. Simulation Results and Discussion

First we investigate the effect of the dipole tilt angle by assuming that dipoles have unit magnitude ($\alpha = 1$) but tilt at various angles relative to the direction of the applied electric field. Fig. 3 shows the shear stress as a function of tilt angle compared with $\tau^*_{xz,0}$ at shear rates $\dot{\gamma}^* = 1$ and $\dot{\gamma}^* = 100$. At the high shear rate $\dot{\gamma}^* = 100$, dipole-dipole interactions do not produce ER effects for tilt angles less than 100°, and instead negative ER effects are obtained. On the other hand, positive ER effects are achieved when the tilt

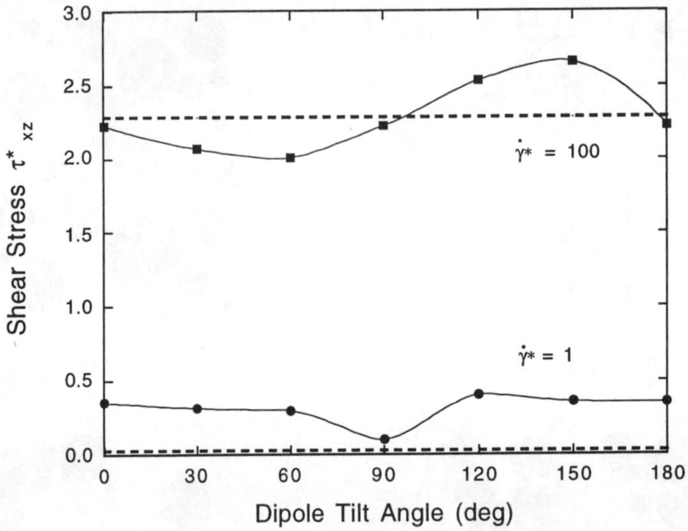

Fig. 3 Shear stress as a function of dipole tilt angle compared with shear stress without electric interactions (dotted lines) at shear rates $\dot{\gamma}^* = 1$ and 100. The magnitude of the dipole is assumed to be independent of the tilt angle.

angle is larger than 100°. These results can be understood qualitatively as follows: If two particles approach each other in the configuration shown in Fig. 2(a) but with the dipole tilt angle larger than 100°, the electrostatic forces between them are attractive, and the particles "ram" into each other harder compared to the situation where the electric interactions are absent. This results in a larger hard-core repulsive force and hence a larger overall shear stress. If the dipole moments are nearly parallel (or anti-parallel) to the electric field, the effect is the opposite. In this case, the electric interactions are repulsive as the two particles approach each other, and thus the particles are further apart than they would have been in the absence of electric interactions. Consequently, the hard-core repulsive forces are also weaker, resulting in a smaller overall shear stress. At the low shear rate $\dot{\gamma}^* = 1$, the effect of the tilt angle on shear stress is minimal, except for the extreme case when the dipole moments are perpendicular to the electric field (and parallel to the flow field).

To create appreciable ER effects at high shear rates, we choose the following parameters: $\varepsilon_f/\varepsilon_p = 10$, $\sigma_f = 0$, $\sigma_p \neq 0$, and $\tau_r^* = 0.2$. (This ensures that the dipole tilt angle is between 100°-160° when the shear rate is between 20-100.) The magnitude of the dipole moment is also reduced to about 0.4 at $\dot{\gamma}^* = 100$ (Fig. 1). From Fig. 4 we can see a nearly constant increase in the shear stress relative to $\tau^*_{xz,0}$ over the entire range of shear rates. If we take the shear stress at $\dot{\gamma}^* = 1$ as the yield stress, then the increase in shear stress at high shear rates ($\Delta\tau^*_{xz} = 0.12$) is comparable to the yield stress (0.34). If the dielectric relaxation time is too low ($\tau_r^* = 0.02$), there is a negative ER effect; if the dielectric relaxation time is too high ($\tau_r^* = 2$), there is no ER effect (Fig. 5).

We want to point out that the mechanism for ER effects at high shear rates is different from that at low shear rates. At low shear rates, chain structures are seen (Fig.

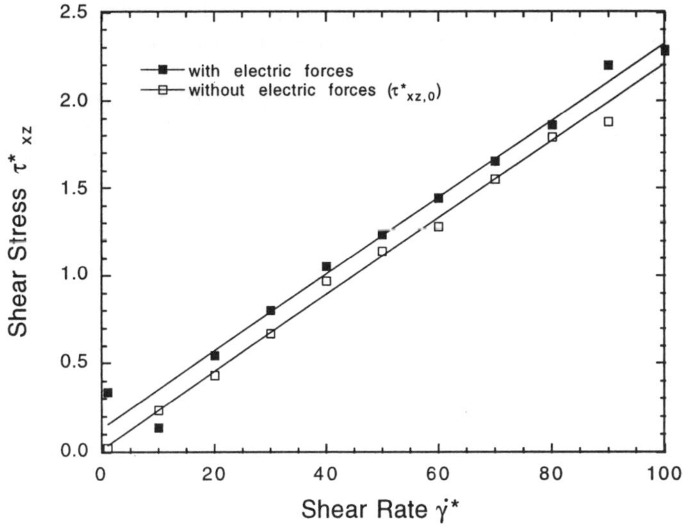

Fig. 4 Shear stress versus shear rate for $\tau_r^* = 0.2$. The lines are linear least-squares fits to the simulation data. The nearly constant increase in shear stress compared to $\tau_{xz,0}^*$ over the entire range of shear rates is in qualitative agreement with experiments.

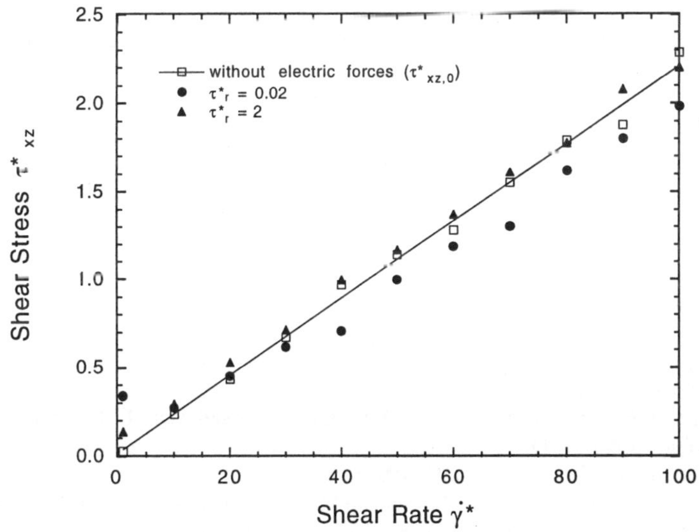

Fig. 5 Negative ER effects are seen if the dielectric relaxation time is too low ($\tau_r^* = 0.02$), and no ER effects are seen if the dielectric relaxation time is too high ($\tau_r^* = 2$). The solid line is a linear least-squares fit to $\tau_{xz,0}^*$.

2(c)), as in earlier simulations, and the shear stress results mostly from electrostatic interactions between particles in the stretched chains. The yield stress obtained in our simulation (0.34) is comparable to the value obtained by Bonnecaze, *et al.* (0.12),[14] although Bonnecaze, *et al.* included near-field electrostatic interactions and hydrodynamic interactions.

At high shear rates, however, there is no chain structure (Fig. 2(d)), and ER effects result from the combination of electrostatic interactions between titled dipole moments and the hard-core repulsive forces. We would like to stress that it is important to have particles of different sizes in order to obtain ER effects at high shear rates. If the particles are uniform in size, then particles settle into layered structures,[15] and a very low shear stress results.

The transition between high and low shear rates in our simulation occurs at $\dot{\gamma}^* \approx 20$ ($\text{Mn} \approx 0.6$). For the typical experimental parameters $E_a = 1000$ kV/mm and $\dot{\gamma} = 1000\ \text{s}^{-1}$, the Mason number is on the order of 10 for a typical ER fluid, and this certainly falls within the range of high shear rates in our simulation. We should point out that earlier simulations do not obtain ER effects at these high shear rates.

Finally, since the dielectric relaxation time τ_r^* by itself does not determine the dipole tilt angle, we argue that τ_r^* alone is not necessarily a good predictor of ER effects. Dielectric mismatch, conductive mismatch, and surface conductivity all contribute to determining the dipole tilt angle and thus all play roles in ER effects. Furthermore, in order to enhance ER effects, we recommend that ER devices be designed with the applied electric field not perpendicular to the flow field.

Acknowledgements

This work has been supported by the U.S. Department of Energy, a Brachman Hoffman Fellowship, and a Wellesley College Summer Research Fellowship. We also thank the Wellesley College Information Services for providing us with computers.

References

1. W. M. Winslow, *J. Appl. Phys.* **20**, 1137 (1949).
2. M. Parthasarathy and D. J. Klingenberg, *Mat. Sci. Eng.* **R17**, 57 (1996).
3. A. P. Gast and C. F. Zukoski, *Adv. Coll. Int. Sci.* **30**, 153 (1989).
4. H. Block and J. P. Kelly, *J. Phys. D: Appl. Phys.* **21**, 1661 (1988).
5. F. E. Filisko, in *Progress in Electrorheology*, ed. K. O. Havelka and F. E. Filisko (Plenum Press, New York, 1995), p.3.
6. K. O. Havelka, in *Progress in Electrorheology*, ed. K. O. Havelka and F. E. Filisko (Plenum Press, New York, 1995), p.43.
7. J. Hemp, *Proc. R. Soc. Lond.* **A434**, 297 (1991).
8. H. Block, K. M. W. Goodwin, E. M. Gregson, and S. M. Walker, *Nature* **275**, 632 (1978).
9. H. Block, J. P. Kelly, A. Qin, and T. Watson, *Langmuir* **6**, 6 (1990).
10. H. Block, E. Kluk, J. McConnell, and B. K. P. Scaife, *J. Coll. Int. Sci.* **101**, 320 (1984).
11. H. Block and P. Rattray, in *Progress in Electrorheology*, ed. K. O. Havelka and F. E. Filisko (Plenum Press, New York, 1995), p.19.
12. T. B. Jones, *Electromechanics of Particles* (Cambridge Univ. Press, New York, 1995).
13. D. J. Klingenberg, F. van Swol, and C. F. Zukoski, *J. Chem. Phys.* **94**, 6160 (1991).
14. R. T. Bonnecaze and J. F. Brady, *J. Chem. Phys.* **96**, 2183 (1992).
15. J. R. Melrose, *Mol. Phys.* **76**, 635 (1992).
16. M. Doi and S. F. Edwards, *The Theory of Polymer Dynamics* (Oxford Univ. Press, London, 1986).

ER EFFECT OF LIQUID CRYSTAL FLOWING BETWEEN TWO PARALLEL-PLATE ELECTRODES

TETSUHIRO TSUKIJI

Dept. of Mechanical Engineering, Sophia Univ., 7-1, Kioi-cho, Chiyoda-ku, Tokyo, 102-8554, Japan
E-mail: t-tukiji@hoffman.cc.sophia.ac.jp

SHINSUKE TANABE

Cisco Systems Co Ltd, Japan

Liquid crystal is one of homogeneous ER(Electro-rheological) fluids in some range of temperature. Transient responses of pressure drop are examined when liquid crystal flows between two parallel-plate electrodes for constant flow rates. When voltages are applied on the liquid crystal and removed, the pressure responses of the inlet of electrodes are measured with the pressure transducer. At the same time, liquid crystal between the transparent electrodes made of glass is visualized with the high-speed video camera to investigate the time history of the director of the liquid crystal. Outlet of the flow channel with two parallel-plate electrodes is atmosphere. Relation between the flow visualization results and the changes of pressure drop is investigated especially for transient period. In the present experiment the flow rates change from 0.001cc/sec(velocity is 1mm/sec) to 0.003cc/sec and the electric field intensity is from 0.2kV/mm to 1kV/mm. The gap of the electrodes is 0.2mm. The isotropic-nematic transition is 35.5℃ and smectic-nematic transition is 23.1℃. The open-loop test facility with the liquid crystal is set in a pyrostat to keep the temperature constant.

1 Introduction

The rheological property of the ER (Electro-Rheological)fluids can be controlled using the application of the electric fields. The ER fluids are well known as one of the functional or intelligent fluids[1]. Nowadays the welfare devices[2] and micro-machines[3] etc. are developed as the applications of the ER fluids. However, very little is known about the influence of the electric fields on the relation between the flow rate and the pressure, and the mechanism of the flow in the case of unsteady electric fields, for example the sinusoidal voltages and the voltages added or removed suddenly. Furthermore there has been only a little amount of information about the response of the flow to the voltages. Knowledge of the influence of the unsteady voltages on the flow characteristics is required for designing the fluid power applications using the ER fluids as a working fluid and expecting the dynamic characteristics of the applications. One of authors has measured the shear stress using rotational viscometer[4] and investigated relation between pressure drop and flow rate under application of unsteady electric fields in the case of ER suspensions [5].

In the present study the homogeneous ER fluid without the particles is used because it is easy to keep the dispersion stability. The flow channel made of transparent electrodes is designed to visualize the ER fluid inside the channel through transparent electrodes. The unsteady pressure drops are measured using the pressure transducer for the constant flow rates when the voltages are added and removed suddenly and at the same time the flow of the ER fluid is visualized using a high speed video camera. The present paper discusses the relation between the time variation of the pressure drop and the flow inside the

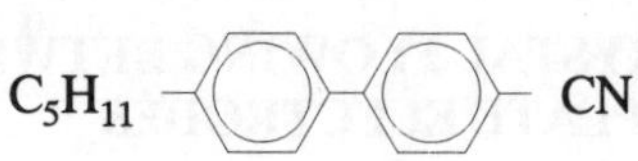

Figure 1. Molecular formula of the liquid crystal

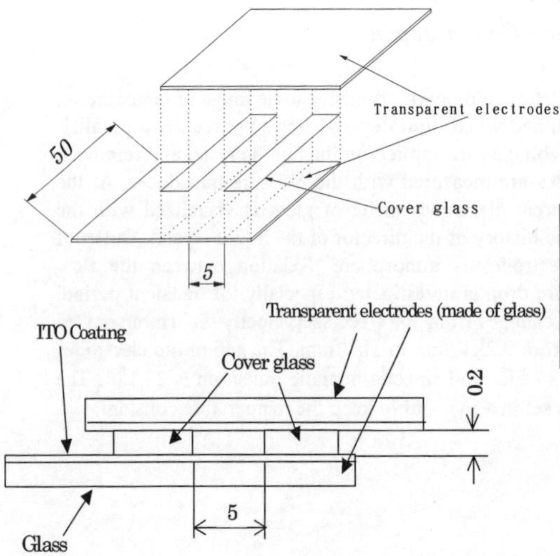

Figure 2. Parallel plate electrodes

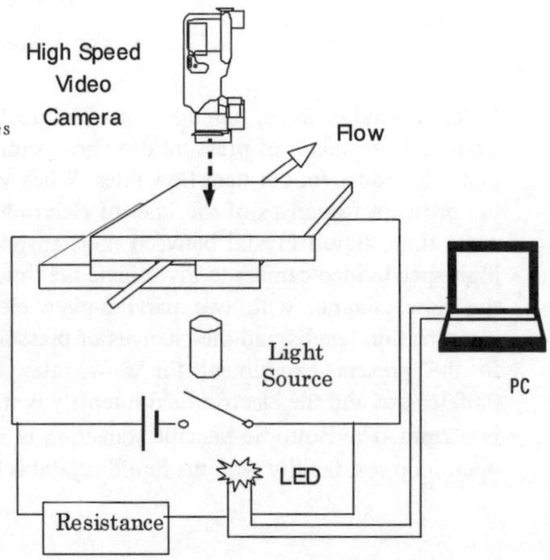

Figure 3. Electric circuit for LED

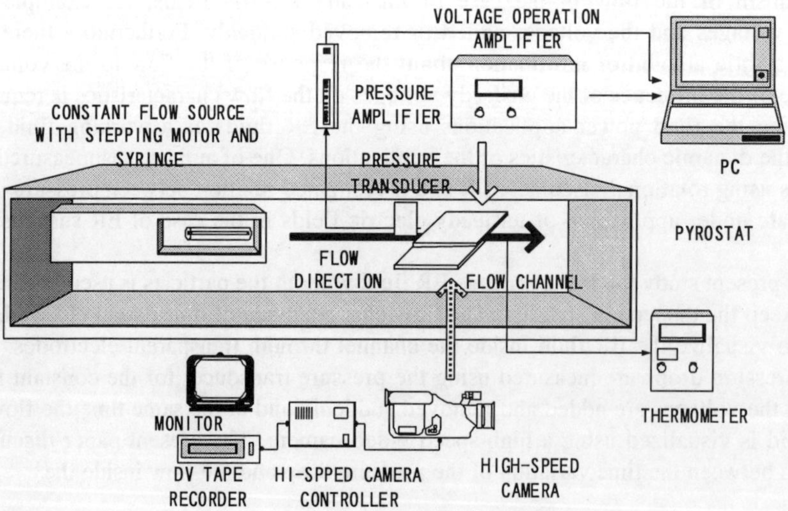

Figure 4. Experimental setup

electrodes. And the pressure response also is described. In the present experiment, nematic liquid crystal is used as one of the homogeneous ER fluids.

2 Experimental Setup and Methods

Molecular formula of liquid crystal used in the present experiment is shown in Fig.1. The liquid crystal was supplied by MERCK Japan Co., Ltd. The isotropic-nematic transition is 35.5℃ and smectic-nematic transition is 23.1℃.

Figure 2 shows the flow channel with two parallel-plate electrodes. This flow channel is made of transparent electrodes to visualize inside the flow channel through transparent electrodes. The electrode is made of glass coated with ITO material. The dimension of channel is 50×5×0.2mm. The electrodes are connected to a voltage operational amplifier.

The arrangement of the high speed video camera and the light source is shown in Fig.3. The electric circuit with the LED to add a light into an image of the flow during applying the voltages is also shown in this figure. The voltage is inputted into the personal computer when the voltages are applied.

The present experimental setup is shown in Fig.4. The liquid crystal is supplied to the two-parallel plate electrodes for the constant flow rates from the syringe whose piston is pushed by a stepping motor at a constant speed. The flow rates can be controlled by the speed of the stepping motor. The syringe is connected to the two electrodes using a glass tube. The inlet pressure of the electrodes is measured using the pressure transducer and the date are inputted to the personal computer every 1ms. The flow circuit with the syringe ,

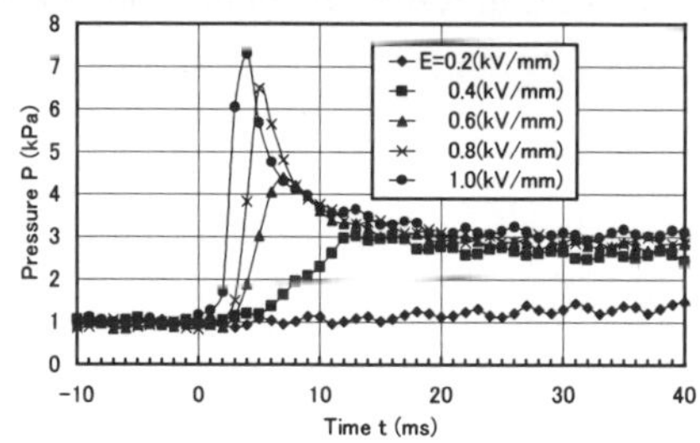

Figure 5. Pressure drops(Q=0.003cm^3/s)

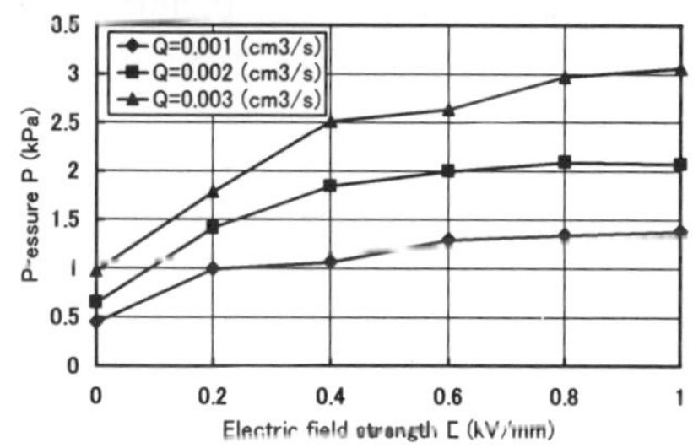

Figure 6. Pressure drops(Steady flow)

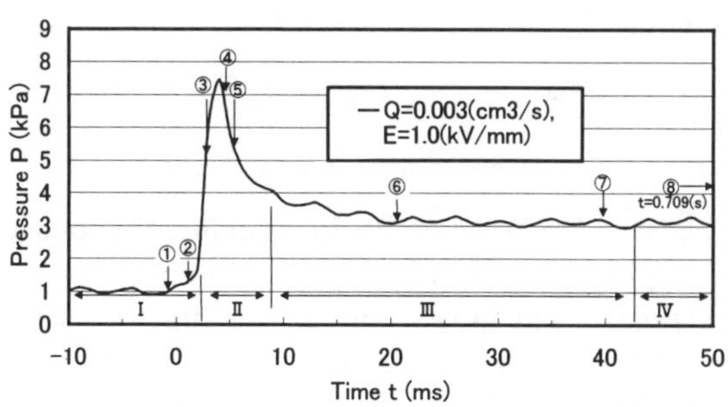

Figure 7. Pressure drops(Q=0.003cm^3/s,E=1kV/mm)

the glass tube and the electrodes is set in a pyrostat to keep the temperature constant. When the pressure is measured the flow of liquid crystal between the electrodes is visualized using the high speed video camera for 1125Frame/sec at the same time.

3 Experimental results and discussion

3.1 In the case of the voltage added suddenly

One of the typical results for the pressure responses is shown in Fig.5 when the voltage is applied suddenly. The temperature is 25±0.5℃ and the voltage is applied at t=0. When the electric field intensity E is greater than 0.6kV/mm, the pressure begins to increase 1～2ms after. The value of the pressure is maximum near t=3～6ms because of the compressibility[6] and it decreases gradually. It becomes almost constant for t=20～30ms. The pressure increases gradually after the application of the voltage and is almost constant at t=20～30ms for E=0.4kV/mm. When those electrodes are used as a valve in the fluid power applications, the surge pressure after application of the voltages has a bad influence on the applications. In the present experiment, the pressure can be controlled without the surge pressure using the electric field for E≒0.4kV/mm.

The average pressures for the steady flow are shown in Fig.6. The pressures are saturated with the electric field strength.

The time variations of the pressure drop for Q=0.003cm3/s, E=1.0kV/mm is shown in Fig.7. The pictures of the flow visualization results are omitted here and the

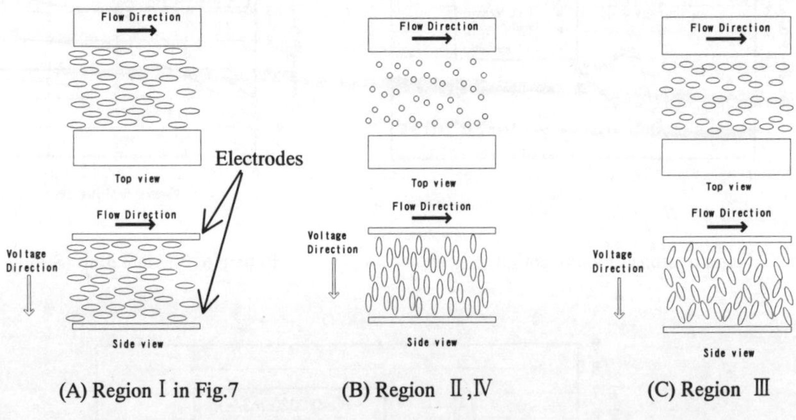

Figure 8. Directors in liquid crystal

results are explained. The liquid crystal flows before the voltage is applied at time①. The voltage is applied at t=0 however the flow is not changed. The picture begins to lighten from time③ and grows light at the times④. The time for 3～6ms is corresponding to that of the increases of the pressure in Fig.7. Furthermore the picture becomes dark at the times⑥ and ⑦, and it grows light again at the time⑧. The light source is placed at the opposite side of the camera as shown in Fig.3. If the light from the light source goes through the electrodes the picture is bright. If it is not easy to go through it is dark. In the case of the present nematic liquid crystal(p-type), when the electric field is applied, the director which is the average orientation of the molecular becomes the direction of the field. The picture become bright under the application of the electric field because the light goes through the electrodes. On the other hand it is dark without the voltage because the director is almost the flow direction. The orientations of the molecular are estimated from the flow visualization results and the predicted figures are shown in Fig.8. The figures of the upper side in Figs.(A)~(C) shows the top view from the camera and the bottom figures shows the front view from the clearance which is 0.2mm. The elliptic shapes show the molecular of the liquid crystal or the cluster of the molecular. We can expect that the director is the flow direction shown in Fig.8(A) for the region Ⅰ because the pictures are dark compared with that at the time⑧. We can also expect that the director is the direction of the electric field for the region Ⅱ after the application of the voltage because the pictures become bright. The pressure goes up for the region Ⅱ because the flow resistance increases due to the director. For the region Ⅲ, the pictures become dark. It is expected that the molecular of the liquid crystal are inclined as shown in Fig.8(C) because of the reaction of applying the voltage suddenly and the pressure decreases as the flow resistance. Furthermore the picture grows light again for the steady state in the region Ⅳ and the pressure become constant because of the director shown in Fig.8(B). Although the other structure of the molecular can be predicted for the region Ⅲ in Fig.8, one of the predicted structure is shown here. The flow resistance seems to be same in the regions Ⅲ and Ⅳ because of the same pressure drop although the difference of the director is expected in Figs.8(B) and (C).

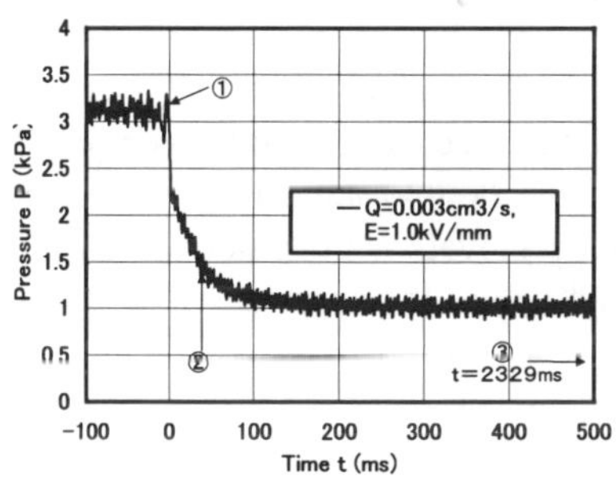

Figure 9. Pressure drop

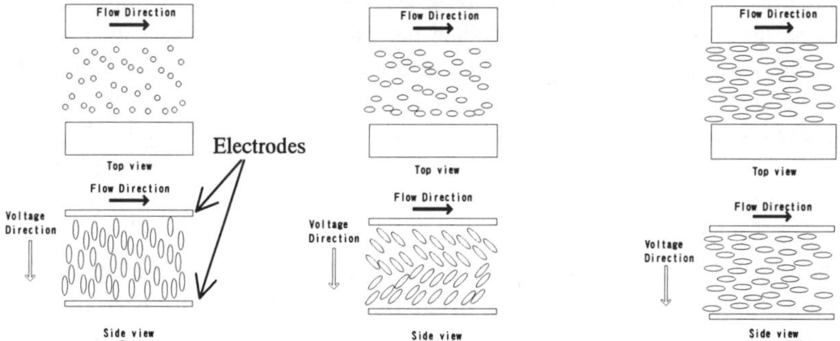

(A) time① in Fig.9 (B) time ② (C) time③

Figure10. Directors in liquid crystal

The relation between the time variation of the pressure and structure of the molecular can be explained to some extent as mentioned above. We obtained almost same results in the case of E≧0.6kV/mm. For E=0.4kV/mm, the director changes from Fig.8(A) to (B) after application of the field because the picture grows light gradually as increase of the pressure. We could not recognize the difference between the flow before and after the application of the field for E=0.2kV/mm.

3.2 In the case of the voltage removed suddenly

One of the typical result of the pressure drop is shown in Fig.9 for Q=0.003cm^3 and E=1kV/mm in the case of the voltage removed. The temperature is 25±0.5℃ and the voltage is removed at t=0. The visualized pictures are also omitted here and only the results are described. The pressure begin to decrease 1～2ms (which is confirmed from the expanded figure) after the voltage is removed. Furthermore the picture becomes dark at the time ② and it becomes more dark at the time ③. The pressure is almost constant near t=100～150ms, the time delay for the pressure to become constant is lager than that of the case applied the voltage suddenly. The response for the case applied the voltage is better than for the case removed. The predicted figures of the director are shown in Fig.10. The liquid crystal flows as shown in Fig.10(A) before the voltage is removed. As the picture is a little dark after the removal, the light is not easy to go through the electrodes because of the change of the director as shown in Fig.10(B). The flow resistance decreases due to the change of the director and the pressure also decreases. The picture becomes more dark after that and the director becomes the flow direction. The flow resistance becomes that without the voltage, and the pressure is almost constant. The other constructions of the molecular can be predicted in Fig.10(B), one of them is shown here from the predicted velocity profile between the electrodes. In the case of E=1.0kV/mm, Q=0.001,0.002cm^3/sec, the pressures decrease wholly compared with the case of Q=0.003cm^3/sec and the results of the flow visualization is almost same. The pressures become constant 100～150ms after removal. We can not recognize the difference of the visualized results between after and before the application of the field for E=0.2kV/mm because the electric field is small.

4 Conclusions

In the present study the nematic liquid crystal is used as one of homogeneous ER fluids to examine the flow characteristics. When the liquid crystal flows between two parallel-plate electrodes for constant flow rates the pressure drops were measured in both cases applied the electric field suddenly and removed it. At the same time the directors of the liquid crystal were predicted from the flow visualization results using the high speed video camera. Furthermore the time variations of the pressure drop are explained from the director. On the basis of the results obtained in the present investigation, the following conclusions can be drawn.

(1)In the case of applying the electric fields suddenly for E≧0.6kV/mm, the pressure increases 1～2ms after the application of the fields and the director becomes the direction of the field 3～6ms later. Therefore the flow resistance increases and the pressures have maximum values. After that the director is predicted to incline with the decrease of the pressure and the directors become the direction of the field again 40ms after the application. On the other hand the pressures become constant about 20～30ms after the application. The pressures increase gradually and become constant 20～30ms after for E=0.4kv/mm. The director becomes the direction of the field and constant about 40ms after.

(2)The pressures begin to decrease about 1～2ms after the removal of the electric fields and the values of the pressures become those without the fields 100～150ms after. Therefore the response for the application of the fields is better than that of the removal. The director becomes the flow direction gradually after the removal and constant 2～3s after.

5 Acknowledgements

We thank Mr. Sumio Syuutou for his assistance to design our experimental rig and MERCK Japan Co., Ltd. for supplying us the nematic liquid crystal. We also thank Prof. Yasufumi Ootubo, Prof.Takatsune Narumi and Prof. Shigeomi Chono for their advice.

References

1. For example, see, *Electro-Rheological Fluids and Magneto-Rheological Suspensions,* edited by Tao,R., World Scientific Publishing,. Singapore, (2000).
2. Inoue, A., Ryu,U., Miyabe,Y. and Nakagawa,A., Properties of ER fluid comprised of liquid crystalline polymer and it's application to welfare devices, *Preprint of the Japan Society of Mechanical Engineers,* No.00-1,Vol. IV (2000),pp.281-282(in Japanese)
3. Park,J., Yoshida,K. and Yokota,S., A micro fluid control system using homogeneous ER fluid, *Transactions of the Japan Society of Mechanical Engineers* **66**(648)C(2000),pp.2700-2706(in Japanese)
4. Tsukiji,T., Saito, S., Asako,Y. and Kawakami,T., Measurement of shear stress in ER suspensions, *Proc. of 4th JHPS International Symposium Fluid Power Tokyo'99,* pp.365-370.
5. Tsukiji,T. and Hori, K., Pressure response of ER fluids between two parallel-plate electrodes, *Electro-Rheological Fluids, Magneto-Rheological Suspensions and their Application,* edited by Nakano,M. and Koyama, K., World Scientific Publishing,. Singapore, (1998)pp.294-301.
6. Nakano,M. and Yonekawa, T., Pressure responses of ER fluid in a piston cylinder-ER valve system, *Electrorheological Fluids* edited by Tao,R.and Roy,G.D., World Scientific Publishing,. Singapore, (1993) pp.477-489.

EFFECT OF MAGNETIC HYSTERESIS OF THE SOLID PHASE ON THE RHEOLOGICAL PROPERTIES OF MR FLUIDS

J. DE VICENTE, J. D. G. DURÁN, A. V. DELGADO AND F. GONZÁLEZ-CABALLERO
Departamento de Física Aplicada. Facultad de Ciencias. Universidad de Granada, 18071 – Granada, SPAIN.
E-mail: jvicente@ugr.es

G. BOSSIS
Laboratoire de Physique de la Matière Condensée. Université de Nice-Sophia Antipolis. Parc Valrose, 06108 – Nice Cedex 2, FRANCE.
E-mail: bossis@unice.fr

An experimental investigation is described concerning the effect of the existence of a remanent magnetization of the dispersed particles on the rheological properties of magnetorheological fluids (MRF). Two MRF's were used: (1) solid phase: cobalt ferrite particles + silica gel (1.5 % w/w); liquid phase: silicone oil (viscosity 20 mPa·s); and (2) solid phase: carbonyl iron + silica gel; liquid phase: silicone oil. The cobalt ferrite particles were synthetized as monodisperse colloidal spheres with an average diameter of 850 nm. The dependence of the dimensionless shear stress (τ^*/ϕ) vs. Mason number (**Mn**) fails to scale when a "magnetorheological hysteresis procedure" is followed, specially for the higher volume fractions used ($\approx$ 7.5 %). The yield stress (τ_y) is first estimated from successive rheograms obtained decreasing the external field ($\mathbf{H_0}$) values for different ϕ. A more precise determination can be done by applying a stress ramp in the oscillatory regime. The critical stress amplitude (τ_c) needed to exceed the viscoelastic linear region (VLR) is obtained. It is found that both τ_y and τ_c strongly depend on the magnetic history of the sample. As expected, the previous results were not obtained in a classical MRF of carbonyl iron particles since they do not present magnetic hysteresis. We conclude that cobalt ferrite suspensions are an other kind of MRF which works at low fields (0 - 17.8 kA/m) with the opposite effect: decrease of the yield stress with the field. This property can be improved using particles with stronger remanent magnetization.

1 Introduction

Magnetorheological fluids (MRF) are non-aqueous suspensions of micron-sized magnetizable particles. They usually consist of a magnetically soft solid phase material. Iron has been a popular choice since it is the element with the highest saturation magnetization. Furthermore, it is easily obtained from iron pentacarbonyl through a relatively inexpensive chemical vapor deposition (CVD) process. Some properties of the magnetic dispersed phase have been studied: Felt *et al.* [1] studied the effect of particle size on the magnetorheological effect. Wu and Conrad [2] studied the role of surface roughness, and Foister [3] studied the effect of particle size distribution. Also, other soft magnetic materials have been studied [4].
Due to the increasing use of this kind of systems in many fields [5], it seems necessary to incorporate novel studies concerning the relation between their rheological and magnetic properties. The aim of the present work is to investigate the effect of the existence of a remanent magnetization in the

solid phase, after the application of an external magnetic field, on the rheological properties of MRF's. In order to do that, we will not only study their flow properties but also their dynamic oscillatory behavior.

This paper is organized as follows: Sec. 2 is devoted to the experimental part. Firstly, we describe the model system including its magnetic characterization. Then, the steady shear and dynamic oscillatory studies are described. Finally, in Sec. 3 the results are discussed.

2 Experimental

Cobalt ferrite particles were prepared following the coprecipitation method described elsewhere [6,7]. The method involves the precipitation of iron (II) and cobalt (II) hydroxides that, upon heating at 90°C in an oxidizing aqueous medium, yield the ferrite particles with a diameter of 850 ± 150 nm. The density of the particles was ascertained by picnometry and resulted 3.6 ± 0.1 $g \cdot cm^{-3}$. The MRF consisted in a suspension of these cobalt ferrite particles in silicone oil (47V20) with viscosity 0.02 Pa·s and density 0.95 $g \cdot cm^{-3}$ (Rhône-Poulenc, France). With the purpose of avoiding the sedimentation of the particles during the experiments, 1.5 % w/w of fine spherical particles of silica gel with 7 nm of diameter (Aerosil-300) was added to the suspensions.

The magnetization of the solids was measured as a function of the applied magnetic field $\mathbf{H_0}$ in a Manics DSM-8 magneto-susceptometer (France). The whole magnetic characterization can be found elsewhere [7].

Steady shear and dynamic oscillatory properties were investigated with a plate-plate (20 mm of diameter) controlled-stress rheometer (Haake RS150) at 25°C. The magnetic field was always oriented in perpendicular direction to the plates. The gap was fixed at 0.1 mm.

Figs. 1a and 1b schematically show the experimental procedures. In all cases studied, the suspensions were subjected to a steady preshear of 700 Pa during 60 s. The magnetic field was immediately applied at the end of the preshear, maintained for 480 s and suddenly removed. Seven experiments were performed for each sample, varing the external magnetic field according to the next sequence: 0, 29.7, 23.7, 17.8, 11.9, 5.9 and 0 kA/m. In the stress-amplitude sweep experiments the frequency was fixed to 1 Hz.

Also, frequency sweep experiments were performed in a similar way. For each magnetic field, the stress amplitude applied (τ_0) followed the sequence: 2, 3, 4, 5, 6, 7, 8, 9, 10, 12, 15, 18, 21, 24, 27, 30, 33 and 40 Pa. The curve corresponding to the maximum stress amplitude still in the "viscoelastic linear region" is only considered. The fact that the response to low enough excitations depends on its value has also been reported in the literature for ERF [8].

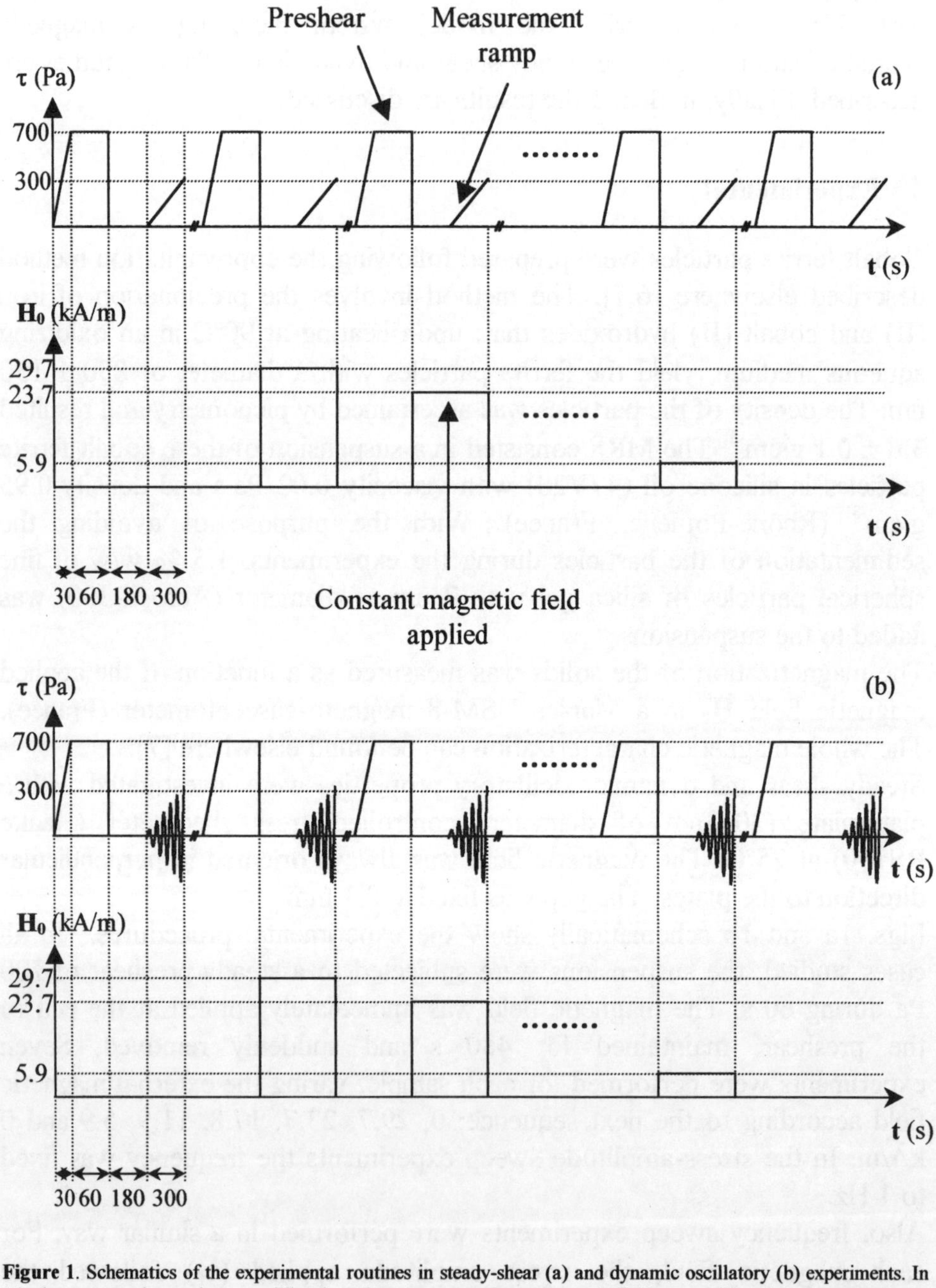

Figure 1. Schematics of the experimental routines in steady-shear (a) and dynamic oscillatory (b) experiments. In (a), the sequence is: preshear, equilibration, shear ramp, preshear, magnetic field on, equilibration, shear ramp, and so on. In (b) the steady shear ramps are substituted by an oscillatory shear, with increasing amplitude and frequency 1 Hz.

3 Results and discussions

Steady-state measurements performed as described above showed that the dimensionless shear-stress (τ^*), defined by

$$\tau^* = \frac{\tau}{\eta_c \dot{\gamma}} \tag{1}$$

normalized by the particle volume fraction (ϕ), depends linearly on the Mason number (**Mn**) in the range **Mn** = 10^{-4} - 10^{-1}, for the maximum magnetic field, and volume fractions between 0.5 and 7.5 %. Recall that **Mn** is given by

$$Mn = \frac{8\eta_c \dot{\gamma}}{\mu_0 \mu_{cr} \beta^2 H_0^2} \tag{2}$$

where η_c and μ_{cr} are the apparent viscosity and relative permeability of the continuous phase (silicone oil) respectively, $\beta = \frac{\mu_{pr} - \mu_{cr}}{\mu_{pr} + 2\mu_{cr}}$ is the magnetic contrast factor and μ_{pr} is the relative permeability of the particles. The results, not shown here for brevity, demonstrate a good agreement with the predictions of non-interacting chain models, no matter the hydrodynamic interaction considered [9]: a slope close to –1 is found in all cases when τ^*/ϕ is plotted vs. **Mn** in a log-log scale. For **Mn** below $\cong 10^{-4}$, a plateau is found for τ^*/ϕ, suggesting that the suspensions do not exhibit a pure Bingham behavior, other researchers found similar results [10].

It is also possible to perform a similar analysis in oscillatory measurements, by studying the viscoelastic behavior during a frequency sweep experiment analogous to the one described in Fig. 1b. for an amplitude stress sweep. In terms of the dynamic analogues of τ^* and **Mn**, the curves properly scale and have almost identical slopes ($-1 \pm 5\times10^{-4}$) (we do not show the results here for brevity).

However, the most important finding concerns the rheological behavior obtained for a high enough volume fraction ($\cong$ 7.5 %), by varying the external magnetic field in the sequence described in the Sec. 2. In this case, the curves (τ^* vs. **Mn**), both in stationary and dynamic conditions, no longer collapse although their separation decreases when the volume fraction decreases (inset in Fig. 2). Fig. 2 is a typical example of the steady flow curves in the cobalt ferrite suspensions.

Non-negligible thermodynamic forces are not a good explanation for the lack of scaling since for high fields the typical size of the aggregates must increase and hence the effect of brownian motion should be less significant.

We would expect larger viscosities for larger fields contrary to our experimental findings (Fig. 2).

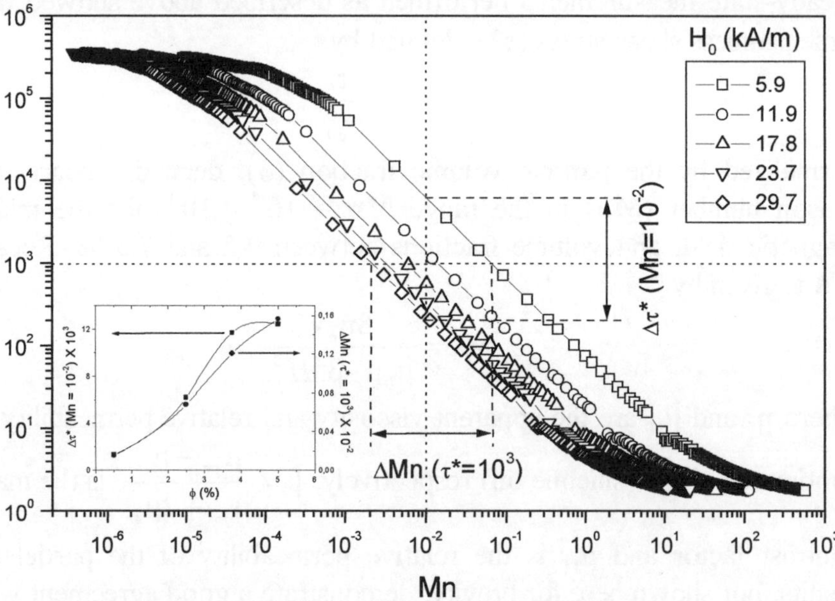

Figure 2. Typical flow curves of 2.5 % volume fraction cobalt ferrite suspensions for different external magnetic fields. Inset: width of the steady-state τ^* *vs.* **Mn** plots for cobalt ferrite suspensions of different volume fractions.

The possibility that we have two magnetic forces, one coming from the induced magnetization (at high fields), and the other from the permanent magnetization of the particles (at low fields) could explain the lack of scaling at low fields and the trend of the curves to collapse for $\mathbf{H_0} >\approx 20$ kA/m.

A stronger evidence of this interpretation comes from the analysis of the dependence of the yield stress (τ_y) on the external field. τ_y was calculated following the usual procedure of extrapolating the stress towards zero shear rate in a logaritmic scale. The results are shown in Fig. 3a. In our opinion, a more precise determination can be done by applying a stress ramp in the oscillatory regime (Fig. 1b). From dynamic measurements it is possible to obtain a similar behavior (Fig. 3b). We will call "critical stress" (τ_c) to the one corresponding to the sharp decrease in the curve **G'** (storage modulus) vs. τ_0 in an oscillatory amplitude sweep.

The increase in τ_y under the application of the magnetic field is easily explainable due to the structuration of the MRF. At high fields the model of multipolar interactions between polarizable particles is enough to predict the behavior observed. At medium fields, the chains formed by the permanent moments could be aligned with the external field. In this situation the yield stress will be low and the chains will break at low angle. At low fields, the chains are not oriented by the field but by the strain and they break at higher angle giving a higher yield stress [11].

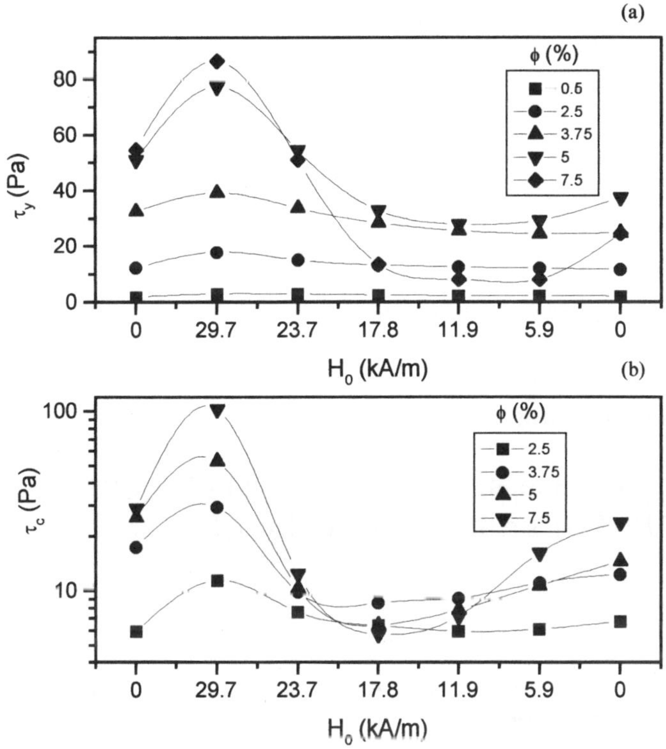

Figure 3. a) Yield stress as a function of the external magnetic field for different volume fractions. b) Same as a), but for critical stress.

To check this qualitative explanation we did similar experiences in carbonyl iron based MRF's. Following the same process as described in Fig 1a, the results obtained are coherent with our previous discussion (Fig. 4). The yield stress monotonically decreases towards its value at zero field, never surpassing it.

In conclusion, we have shown an other kind of MR fluid which works at low fields (between 0 and 17.8 kA/m) with the opposite effect: decrease of the yield stress with the field. This MRF can be improved by using particles with stronger remanent magnetization.

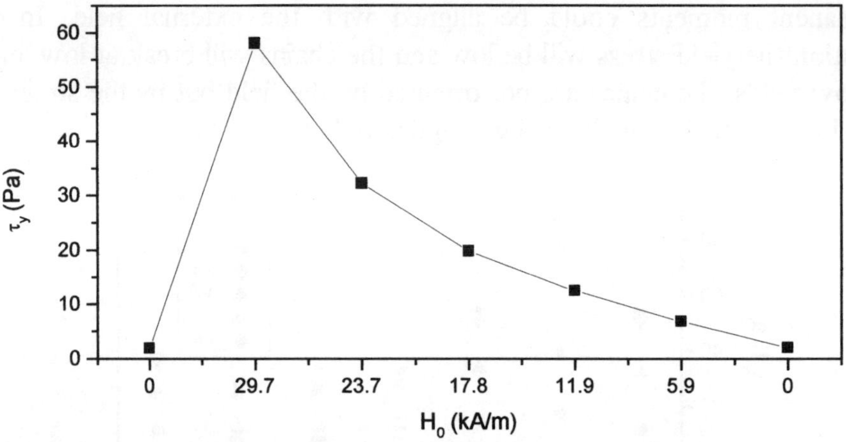

Figure 4. Yield stress of suspensions of carbonyl iron particles at 10 % volumen fraction, as a function of the applied magnetic field.

4 Acknowledgements

Financial support through projects MAT98-0940 (CICYT, Spain) and INTAS99-0510 (EU) is gratefully acknowledged.

References

1. Felt D. W., Hagenbüchle M., Liu J. and Richard J., *J. Intelligent Mater. Sys. Str.* **7** (1996) 589
2. Wu C. W. and Conrad H., *J. Appl. Phys.* **83** (1998) 3880
3. Foister R., U. S. Patent 5,667,715, 1997
4. Margida A. J., Weiss K. D. and Carlson J. D., in *Proc. 5th Int. Conf. on Electrorheological Fluids, Magnetorheological Suspensions and Associated Technology*, ed by W. A. Bullough, (World Scientific, Singapore, 1996) 544
5. Ginder J. M., U. S. Patent 5,549,837, 1996
6. Tamura H. and Matijević E., *J. Colloid Interface Sci.* **90** (1982) 100
7. de Vicente J., Delgado A. V., Plaza R. C., Durán J. D. G. and González-Caballero F., *Langmuir* **16** (2000) 7954
8. Otsubo Y. and Edamura K., *J. Rheol.* **38** (1994) 1721
9. Martin J. E. and Anderson R. E., *J. Chem. Phys.* **104** (1996) 4814
10. Volkova O., *Étude de la Rheologie de Suspensions de Particules Magnetiques*, Ph. D. Thesis, Université de Nice-Sophia Antipolis, 1998
11. Bossis G., Lemaire E., Volkova O. and Clercx H., *J. Rheol.* **41** (1997) 687

RHEOLOGICAL AND DIELECTRIC CHARACTERIZATION OF ELECTRORHEOLOGICAL FLUIDS COMPOSED OF BOTH DISPERSED PARTICLES AND LIQUID DROPS IN A DIELECTRIC MEDIUM

L. REJON*, L. GARCIA
Instituto de Investigaciones Eléctricas
Apartado Postal 1-475, 62001 Cuernavaca, Morelos, Mexico
lrejon@iie.org.mx

O. MANERO
Instituto de Investigaciones en Materiales, UNAM
Apartado Postal 70-360, Mexico, D.F., 04510, Mexico.
manero@servidor.unam.mx

The rheological and dielectric behavior of novel electrorheological suspensions composed of non-colloidal silica gel particles of irregular shape and oil-in-oil emulsions with different electrical conductivity (and permittivity) are studied in this work. The rheological behavior of the suspensions is examined in systems with various compositions of the oil phase under different intensities of constant DC electric fields. Results reveal that the electrorheological (ER) response of the oil-in-oil emulsion increases with the applied voltage but is affected negatively when the concentration of the oil with higher conductivity is increased. When a small volume fraction of silica gel is added, the ER response increases considerably and its magnitude is strongly dependent on the electric field. In this case, however, larger concentrations of the oil with the higher conductivity, in the presence of silica particles, increase the relative viscosity of the suspensions. The dielectric properties of the suspensions are examined to elucidate the relation between the dielectric relaxation process and the relative composition of liquid drop phases. The mechanisms responsible of the ER response are explained in terms of the relationship between the microstructure changes in the suspensions and their rheological and dielectric properties.

1 Introduction

Generally, ER suspensions are composed of polarizable particles suspended in a dielectric medium, and under the influence of high electric fields they show a rapid but reversible change in their rheological properties [1-2]. The ER effect is ascribed to the chain-like particle structure that presents high resistance to flow, inducing substantial increments in the suspension viscosity. These suspensions have been applied in dampers, shock absorbers, ER valves, mechanical and electrical interfaces. However, there are still significant problems to obtain the desired materials for particular devices, primarily due to the poor understanding of the materials structure and mechanisms responsible for the ER effect [3]. Other related problems include the lack of stability of the ER fluids (for instance, due to sedimentation) and sometimes it is necessary to apply high electric fields to obtain large yield stresses.*

ER fluid behavior is largely influenced by the structures formed in the presence of electric fields [4]. Moreover, the magnitude of the ER response is highly influenced by the shape, thickness, size and morphology of such structures. These are themselves dependent on the physical, chemical properties and morphology of the ER constituents. Recently, investigations on the rheological behavior of immiscible polymer blends or emulsions [5-7] and systems containing both dispersed particles and liquid drops in a

* Author to whom correspondence should be addressed: lrejon@iie.org.mx Phone + 52-7-3183811 ext 7289

continuous medium [8] have been carried out. In immiscible blends, the ER response is enhanced by the electric field but negatively affected by the flow field strength. Kimura et al [5] reported positive and negative ER response in a systems of urethane-modified propylene glycol (UPPG)/dimetylsiloxane (DMS) emulsions. They found that the ER response is associated with the domain composition without electric fields. The ER response takes place when the UPPG forms droplets, whereas it is absent when UPPG is the continuous phase. When an electric field is applied, the UPPG droplets elongate and bridge between the electrodes even in the presence of shear flow. The ER response is positive when the viscosity of the UPPG is larger that of the DMS and negative otherwise. Ha and Yang [9] reported the rheological behavior of oil-oil emulsions under dc fields when the dispersed phase is less conducting than the continuous phase. The viscosity of the emulsions was enhanced when the electric field predominates, but when the shear flow dominates, the electric field is rheologically less important. Finally, the negative ER responses were observed when the shear flow and electric field compete. Clearly, this behavior is related to different morphology and to the interfacial contribution that depends strongly on the conductivity ratio of the fluids. Since in liquid-liquid systems the ER response is due the formation of columns forming bridges between electrodes, a low yield stress is related to the weakness of the column [7]. Considering that the inclusion of polarizable solid particles in ER oil-oil emulsions may be more effective, Chin and Park [8] studied the ER behavior of a oil-oil emulsion with polyaniline particles. Results indicated that high yield stresses with improved ER response and reduced current density were obtained when the dispersed phase was constituted by liquid drops with higher conductivity and polyaniline particles. Unique microstructures were formed by the particles and deformed emulsion drops. The dependence of the yield stress on the electric field was strongly affected by the change in the relative composition of particulate and liquid drop phases.

The present work reports rheological measurements made on an oil-oil emulsion with polarizable solid particle under dc electric fields. Particular attention is given to the influence of the dielectric properties of the ER components on the structure and dielectric behavior of the ER systems. The relationship between the polarization mechanisms, structural arrangement and rheological behavior are discussed here in detail.

2 Experimental

2.1 Sample preparation

To prepare oil-in-oil emulsions with different conductivity, blends of silicon oil (Spectrum 1000) and chlorinated paraffin (Paraclor S-52, Charlotte Chemical Inc.) were made with volume ratios of 92/8 (E1) and 84/16 (E2), where the minor component formed dispersed drops. Table 1 contains the relevant properties of the liquid phases. The ER suspensions were prepared using silica gel particles of irregular shape (Merck 60, 0.015-0.040 mm) of approximately 22.1 mm in average size with a volume fraction of 0.02. The particles have a permittivity of 10.4 (at 60Hz) and a conductivity of 2×10^{-9} S/cm. The method to prepare the blend was reported previously [10].

Table 1. Properties of the liquids

Oil	Dielectric constant at 10 Hz	Conductivity (S/cm) at 1 kV	Density (g/cm^3)	Viscosity (Pa.s)
Silicon oil	2.4	7.7×10^{-15}	0.97	0.96
Paraclor S-52	7.8	1.1×10^{-11}	1.25	1.70

2.2 Rheological Characterization

Rheological measurements were made in an ARES (2KFRTN1) rheometer using parallel plates with a diameter of 50 mm and an electrode gap of 1-mm. DC voltages were supplied by a Trek High Voltage Amplifier model 609E-6. The method used was steady-state sweep at 25 °C.

2.3 Dielectric and electrical characterization

The procedure to measure the dielectric properties of the systems was recently reported [11]. The conductivities of the liquids were measured using a Guard-Ring Capacitors Type 2903 (Tettex AG Instruments) coupled with a HP 4339A High Resistance Meter. The DC voltage (1000 volts) was applied during the tests.

2.4 Structural arrangements observations

The sample cell and procedures to observe the change in the spatial arrangement of the silica particles and fluid drops with higher conductivity, at different electric fields, was previously reported [12].

3 Results and Discussion

3.1 Rheological properties

The shear viscosity of the emulsions is plotted with the shear rate in figures (1a) and (1b) for two concentrations of Paraclor S-52 (8, E1, and 16 vol %, E2) under various electric fields (0.0-2.5 kV/mm). The emulsions depict a non-Newtonian behavior attributed to the breaking of the microstructure of the dispersed phase as the shear rate increases. Two regions of different ER response were identified: the low shear-rate region, where the polarization forces predominate, and a high shear rate region where the flow strength dominates. At low shear rates, the E1 emulsion shows the ER response for electric fields higher that 0.5 kV/mm, but its structure is weak, since it is undoubtedly affected by the flow (figure 1a). Upon increasing the S-52 content in the emulsion, E2 (figure 1b) shows a lower ER response in the low shear rate region and a larger response for shear-rates higher than 1 s^{-1}. In the region where neither the flow nor the electric field dominate, the electro-hydrodynamics of neighboring drops become more complicated [9]. The observed variation of the viscosity in this range suggests a kinetic process of reformation-breakage of the dispersed phase structure under flow.

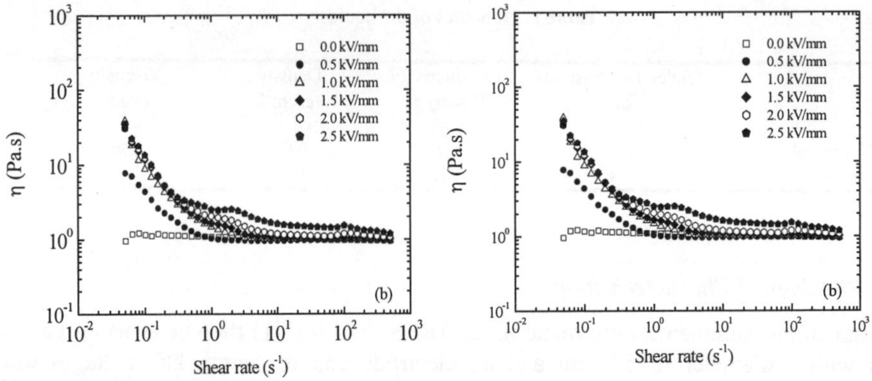

Figure 1. Viscosity as a function a shear rate for silicon oil/ Paraclor S-52 emulsions under various electric fields. Paraclor S-52 concentrations: (a) 8 vol %, E1 (b) 16 vol %, E2.

When a small amount ($\phi_{\text{particle}} = 0.02$) of polarizable solid particles is added to the emulsions, we observe an increment of the suspension viscosity over the entire range of shear rates and electric fields, as shown in figures 2(a) and 2(b). The largest increase in the viscosity is observed at low shear rates. The suspension with the E2 emulsion shows a viscosity enhancement at high shear rates. Moreover, the ER response in these suspensions is higher than the observed response in systems composed of silicon oil and silica particles (see figures 2c and 2d). The structures formed by the S-52 drops enhance the particle-particle interaction and form a more stable fibril structure. In the following section, attention is given to the effect of such structures on the behavior shown here.

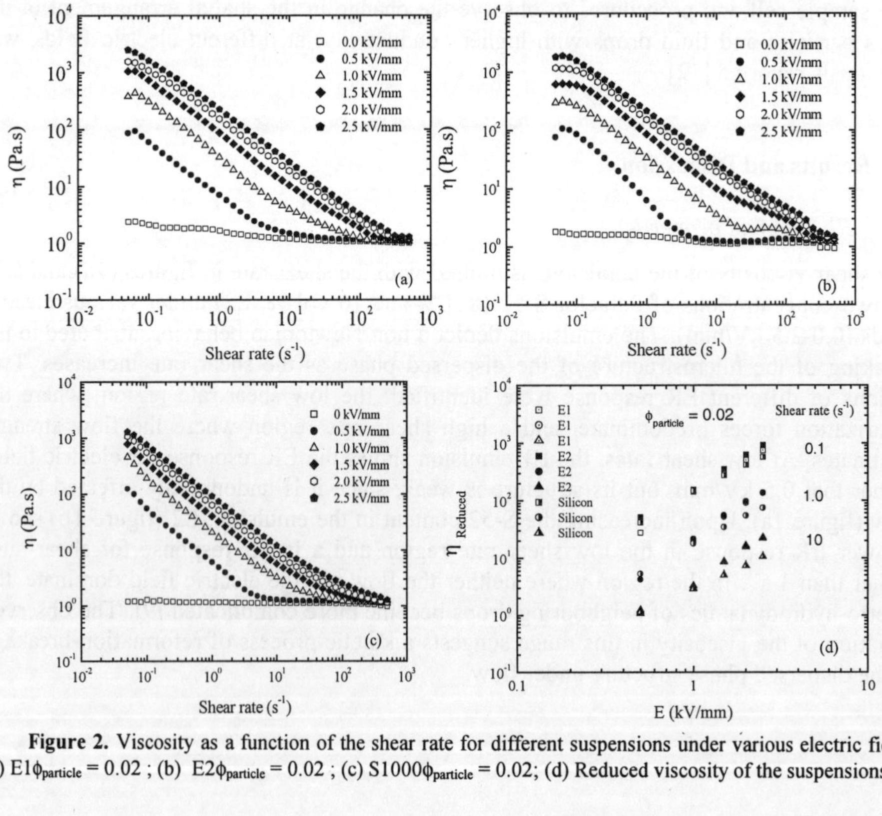

Figure 2. Viscosity as a function of the shear rate for different suspensions under various electric fields: (a) E1$\phi_{\text{particle}} = 0.02$; (b) E2$\phi_{\text{particle}} = 0.02$; (c) S1000$\phi_{\text{particle}} = 0.02$; (d) Reduced viscosity of the suspensions.

The suspensions show Bingham behavior with a yield stress over entire range of electric fields, as depicted in figure (3). The yield stress of the suspension composed of the silicon oil and silica particles varies as $E^{1.6}$ with the electric field. The departure from the theoretical squared-power dependence may be attributed to the distortion of the electric field around the particle due to its irregular shape and broad particle-size distribution. Similar behavior was reported in phosphate cellulose-based suspensions [13]. In the case of the suspension with emulsions E1 and E2, the yield stress varies as $E^{2.24}$ and $E^{2.05}$ respectively. In oil-solid particle systems, a power larger than two is found at high concentrations of the dispersed phase, attributed to the formation of cross-linked structures [14]. Similar behavior was observed in system based on silicon oil-Cereclor S45 emulsions and polyaniline particles [8], attributed to complex structural effects arising among the dispersed phases.

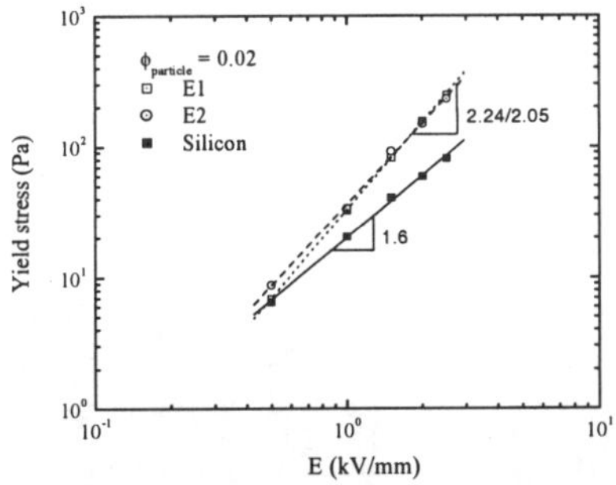

Figure 3. Yield stress as a function of the electric field strength for various silica suspensions.

3.2 *Structural arrangement*

In this section, microscopic observations of the structural arrangement in various ER systems are discussed. Figures 4(a) and 4(b) depict the structural features observed in emulsions E1 at 0.5 and 2.5 kV/mm respectively. The emulsions generate different types of time-dependent microstructures. The emulsion E1 presents a low response at low field whereas the E2 show the coalescence of the drops giving a better response. When the voltage is increased at 2.5 kV/mm, in the E1, the width of the elongated columnar structure is smaller that the width of the structures generated in emulsion E2, which may explain the higher resistance to break at larger shear rates observed in E2.

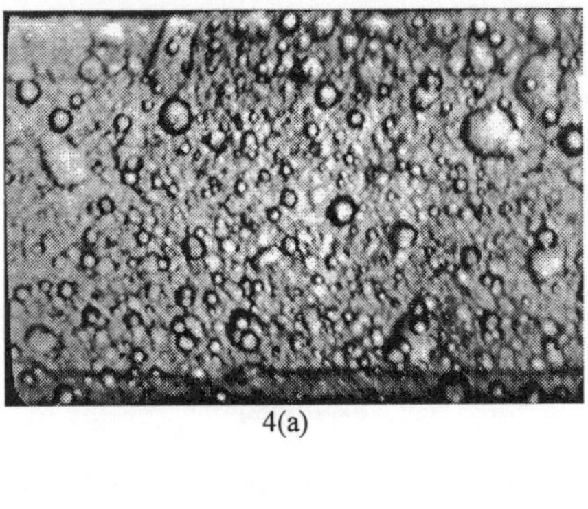

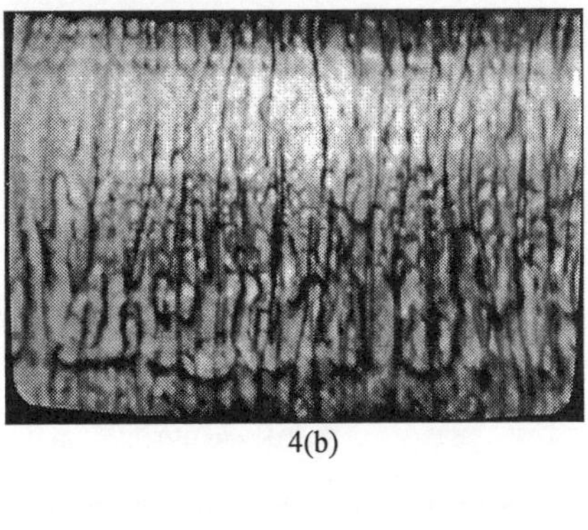

4(a) 4(b)

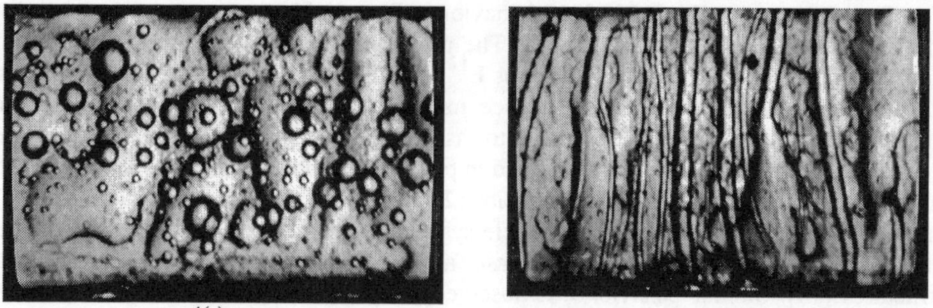

4(c) 4(d)

Figure 4. Structural arrangement observed in emulsions: E1 (a) 0.5 (b) 2.5 kV/mm and E2 (c) 0.5 (d) 2.5 kV/mm

Addition of solid particles to these emulsions (E1, E2), generate the structures depicted in figures 5 (a) and 5(b). The suspension based in emulsion E2 produces structures with wider columns, and hence a larger ER response. This reflects larger interactions between S-52 drops and silica particles.

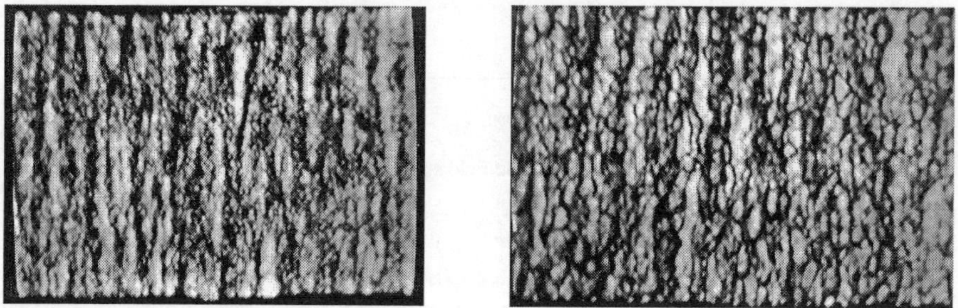

5(a) 5(b)

Figure 5. Structural arrangements in suspensions with silica particles ($\phi = 0.02$) and emulsions: (a) E1 and (b) E2

3.3 *Dielectric properties*

Figures 6 (a) and 6(b) show the variation of the permittivity and loss factor with frequency for emulsions (E1 and E2) and these with silica particles, respectively.

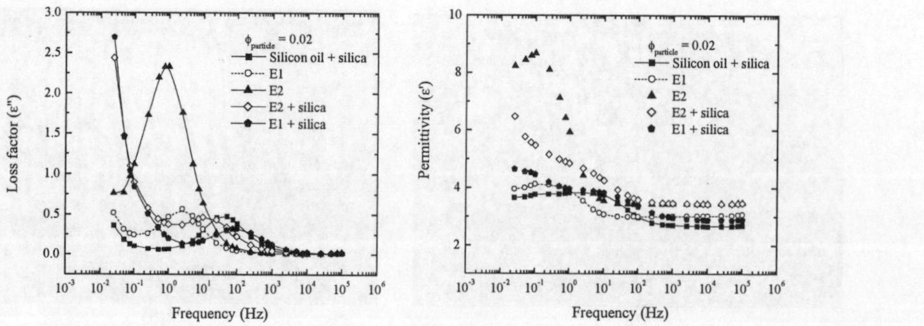

Figure 6. Dielectrical properties of suspensions with silica gel ($\phi = 0.02$) and emulsions: (a) E1 and (b) E2

The suspension shows an increment in the permittivity with the addition of silica particles. The maximum observed in the dielectric loss data reflects a Debye-type relaxation process, but it is likely that the relaxation peaks occurring at low frequencies may be ascribed to interfacial Maxwell-Wagner type polarization. Overall, the relaxation times are strongly dependent on the relative amounts of the dispersed phase composition. Emulsions E1 and E2 possess smaller relaxation times.

4 Conclusions

The dielectric response of the systems examined in this work suggests the presence of high polarization occurring in the interfacial region of the particles. Data suggest that the presence of polarizable particles promote the creation of fibrils with high resistance to flow (large ER response). Smaller relaxation times in the emulsions are then ascribed to weaker structures where the flow may be predominant, or where the ER response is small. These results are supported by microscopic observations of the structures formed in the systems.

5 Acknowledgements

The authors acknowledge the financial support from CONACYT (project 31123-U) and IIE (project 11794). Mr. Gustavo Regalado and Mr. Francisco Lopez are thanked for their work in the electrical evaluation of the materials and rheological characterizations.

6 References

1. M. Parthasarathy, D. J. Klingenberg, *Mater. Sci. and Eng.* **R17**, 57-103 (1996).
2. H. Block and J.P. Kelly, J. Phys. D: Appl. Phys. **21**, 1661-1677 (1988).
3. K. Havelka, TRIP, **3** No. 2, 66-69 (1995).
4. T.C. Halsey, Science **258**, 761-766 (992).
5. H. Kimura, K. Aikawa, Y. Masubuchi, J. Takimoto, K. Koyama and T. Uemura, J. Non-Newtonian Fluid Mech., **76**, 199-2211 (1998).
6. H. Orihara, Yasunari Hosoi, Kozo Tajiri, Yoshihiro Ishibashi, Masao Doi and Akio Inoue, J. Rheol., **43**, 125-136 (1999).
7. H. Orihara, Naoya Miwa, Masao Doi, and Akio Inoue, J. Rheol, **45**, 773-782 (2001).
8. B.D. Chin and O.O. Park, J. Rheol, **44**, 397-412 (2000).
9. Jong-Wook Ha and Seung-Man. Yang, J. Rheol, **44**, 235-256 (2000).
10. L. Marshall, F. Zukoski, J. W. Goodwin, J. Chem. Soc. Faraday Trans, **85**, 2785-2795 (1989).
11. L. Rejon, I. Castañeda-Aranda, O. Manero, Coll. and Surf. A: Physicochemical and Engineering Aspect, **182**, 93-107 (2001).
12. L. Rejon, M. A. Ponce, C. De la Luz, R. Nava and V.M. Castaño., J. Mat. Sci: Materials in Electronics, **7**, 433-436 (1996).
13. S.G. Kim, J.W. Kim, W.H. Jang, H..J. Choi and M.S. Jhon., Polymer, **42,** 5005-5012 (2001).

Y. Otsubo, M. Sekine, and S. Katayama, J. Rheol, **36**, 479-496 (1992)

OPTICAL ATTENUATION IN A MAGNETIC FLUID FILM UNDER PERPENDICULAR MAGNETIC FIELDS

CHIN-YIH HONG[+]

Department of Mechanical Engineering, Da-Yeh University, Changhwa 515, Taiwan

S.Y. YANG[*], H.E. HORNG

Department of Physics, National Taiwan Normal University, Taipei 116, Taiwan

H.C. YANG

Department of Physics, National Taiwan University, Taipei 106, Taiwan

The optical attenuation of light passing through a magnetic fluid film under a perpendicular magnetic field was investigated by measuring the optical transmission. A light with a wavelength of 1557 nm is normally incident to the magnetic fluid film, which is subjected to the magnetic fields perpendicular to the plane of the film. The experimental results show that the transmission I/I_o of a given magnetic fluid film decreases exponentially with the increasing thickness L of the film under a fixed magnetic field. By fitting the experimental $I/I_o(L)$ data to the equation $I/I_o = \exp(-L/\delta)$, the attenuation length δ can be obtained. It was found that the attenuation length could be reduced by raising the magnetic field strength, whereas the optical trnasmittance and attenuation length of magnetic fluids are independent of the sweep rate of the field. According to the experimental data, the optical attenuation length of the magnetic fluid ranges from tens to hundreds of micrometers.

1 Introduction

Due to its potential applications, optical phenomena of magnetic fluids subjected to an external magnetic field have been recently put under extensive study. One of the areas of research done on such phenomena concentrates on the optical transmission of magnetic fluid films under an external magnetic field. When a magnetic fluid thin film is subjected to an external magnetic field parallel to the plane of film, the particles in the film start to agglomerate and form chains along the direction of the field. Thus, the magnetic fluids become optically anisotropic,[1] and birefringence occurs as light rays pass normally through the thin film. This anisotropy may be the result of the shape anisotropy of magnetic particles,[2,3] the interaction between particles,[4,5] and/or the chain formation in the film.[6,7,8] Furthermore, the transmitted light intensity of this birefringence increases with an increase in both magnetic field strength and particle concentration, and it tends to saturate at higher fields.[9,10] On the other hand, the field is perpendicularly-applied to the film, magnetic particles agglomerate and form magnetic columns in the film. Some experimental results show that that the optical transmission of magnetic fluid films can be modulated by varying the field strength.[10] In addition to the influence of magnetic fields on the transmission of the magnetic fluid film, the temperature[11] and the wavelength effect[12,13] on the transmission of the magnetic fluid film were also reported. In this report, we investigate the effect of film thickness, of the strength of magnetic field and of the sweep rate of the field on the optical transmission of magnetic fluid films.

[+] Correspondence: e-mail: cyhong@mail.dyu.edu.tw, Tel. & Fax: 886-4-8529065

[*] Present address : Institute of Physics, Academia Sinica, Taipei 115, Taiwan

2 Experimental details

Fe_3O_4 magnetic fluid was prepared with the co-precipitation technique.[10] The saturated magnetization of the sample was 0.17 emu/g and the particle size was around 100Å. The fluid was then injected into glass cells of thickness 11.8 μm, 130 μm, 200 μm, 400 μm, 600 μm, and 1000 μm, to form magnetic fluid films. A pair of solenoids was used to provide a magnetic field perpendicular to the plane of magnetic fluid films. The magnetic fields generated by solenoids were uniform with deviation of less than 1 % in the field strength in the region of the film. In this work, the temperature was kept constant at 24.3 °C throughout each experiment by circulating cooling water around the solenoids. To measure the optical transmission of the magnetic fluid, a parallel light ray with a wavelength of 1557 nm was perpendicularly incident to the film, and the transmitted light, detected directly above the film, was guided by an optical fiber, to the spectroscopy to measure the transmitted intensity. Before the measurement of each data point, the magnetic field was charged up to the desired field strength and kept constant for one minute until the agglomeration of magnetic particles became stable. By dividing the measured transmitted intensity of the magnetic fluid film under a given field by that of the bare glass cell, the optical transmission of the film was obtained.

3 Results and discussion

When the magnetic fluid film is subjected to external magnetic field perpendicular to the plane of the film, magnetic particles in the film agglomerate and form particle columns. If the field strength exceeds some critical field strength, a hexagonal structural pattern (ordered structure) forms inside the film, with particle columns occupying its vertices. The areal density of the magnetic columns depends on the strength of magnetic field. From an optical point of view, since the magnetic column is opaque in comparison to that of surrounding fluid, it is reasonable to assume that the optically transparent area of the film reduces as that column areal density increases. Thus, the optical transmittance will change when the magnetic field strength varies. Furthermore, as an electromagnetic wave travels through the magnetic fluid, the absorption of the intensity increases as the traveling distance becomes longer and follows the exponential law of absorption. Therefore, the optical transmittance of an incident electromagnetic wave through a magnetic fluid film depends on film thickness as well.

Figure 1 gives the optical transmittance as functions of field strength with various film thickness. As expected, the transmittance is reduced when the field strength increases and finally levels off. Moreover, the thinner the film thickness, the higher the transmittance. Figure 2 plots the transmittance as functions of film thickness in log-linear scale. It shows that the curves are straight lines whose slopes depend on the field strength. The line with higher field strength has a steeper descent. Therefore, these experimental dates can be curve-fitted by using the relationship given by the exponential law of absorption, Transmittance = exp $(-L/\delta)$, to obtain the attenuation length δ of magnetic fluid film. The value of $-1/\delta$ represents the product of the constant ln(10), the conversion factor from logarithm to natural logarithm, and the slope of the straight line in Figure 2. Figure 3 gives the attenuation length of magnetic fluid films as a function of the field strength. The curve shows that at a lower field strength, the attenuation length has a higher value, which has a higher optical transmittance. It is consistent with the conclusion reached from the areal density of the magnetic columns under fields. The value of the δ varies from hundreds to tens micrometers.

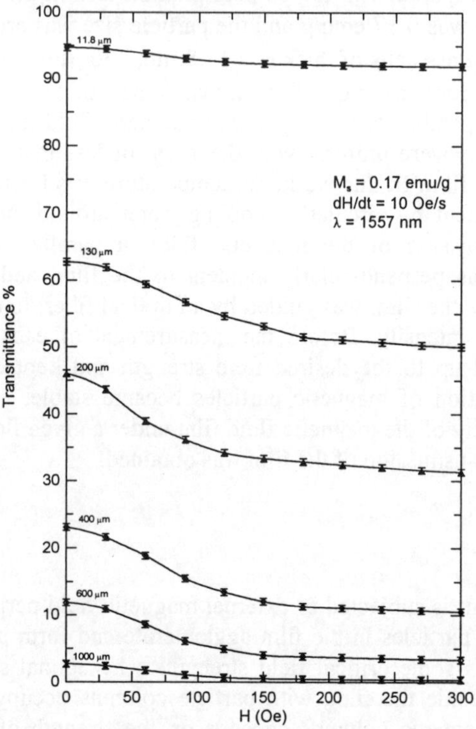

Fig.1 Optical transmittance of magnetic fluid films under perpendicular magnetic field as functions of field strength and film thickness. The conditions of the experiments are M_s = 0.17 emu/g, dH/dt = 10 Oe/s, and λ = 1557 nm.

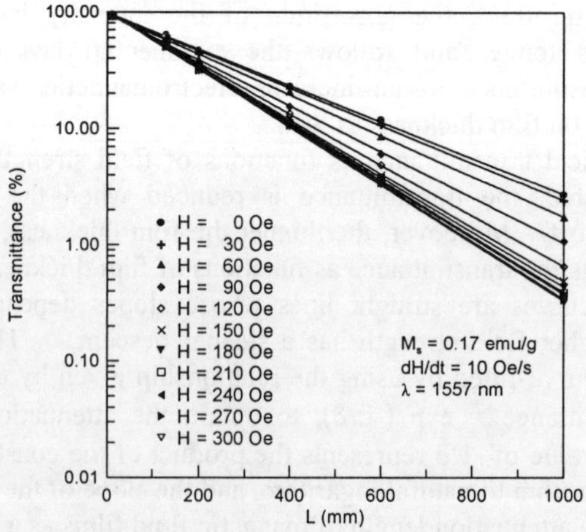

Fig.2 Plot of transmittance in log-linear scale as a function of film thickness at various field strengths. The relationship between transmittance and film thickness shows a linear behavior.

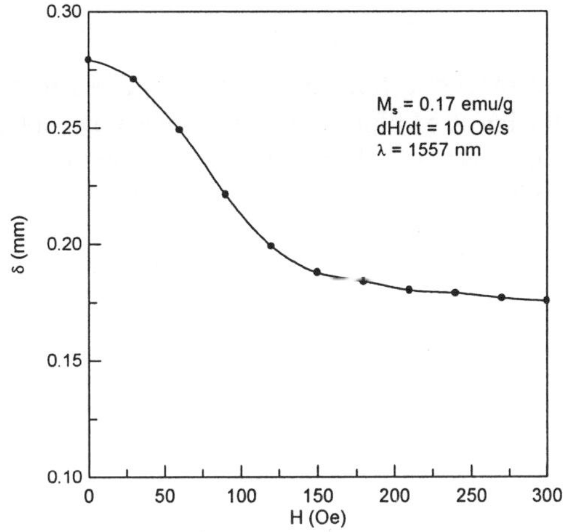

Fig. 3 Plot of attenuation length of the electromagnetic wave passing through magnetic fluid films subjected to a perpendicular field as a function of field strength.

The effect of the field sweep rate on the transmittance of magnetic fluid film is also investigated. Generally, the sweep rate affects the final state of the ordered structural pattern formed inside magnetic fluid films due to the time effect during structural formation for a given field strength, i.e., higher sweep rate resulting on smaller column size. However, if the structural pattern is still in a disordered state, the areal density of particle columns remains the same for a given field strength, regardless of the column size. Under our experimental conditions, a dis-ordered state of structural pattern in the film is to be expected. Thus, the optical transmittance of different sweep rates will have the same value for any given field strength. Figure 4 gives the optical transmittance of magnetic fluid film as functions of field strength for various sweep rates. Unsurprisingly, it demonstrates that the optical transmittance is indeed independent of the field sweep rate.

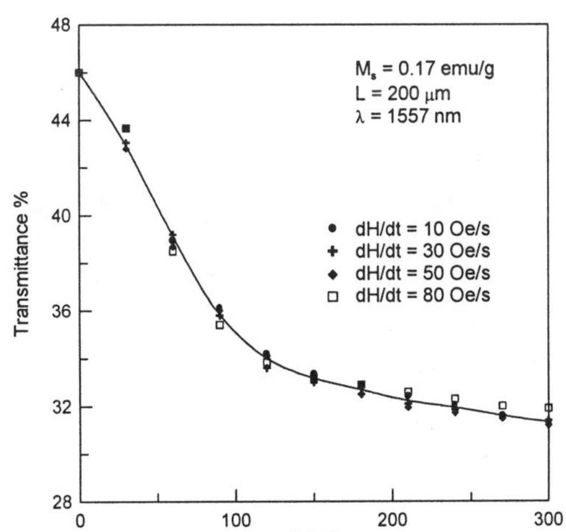

Fig. 4 Optical transmittance as functions of sweep rate of magnetic field. It shows the transmittance is independent of the sweep rate.

4 Conclusion

Optical transmittance of magnetic fluid thin films subjected to perpendicular magnetic fluids depends on the thickness of film, L, and is followed by exp(-L/δ) relationship. The attenuation length δ of the magnetic fluid films is dependent on the field strength similar to that of optical transmittance, which has a higher value at low field strength. On the other hand, these optical properties are independent of the field sweep rate.

5 Acknowledgements

This work was supported by the National Science Council under grant No. 89-2212-E-212-021.

References

1. P.C. Sholten, IEEE Trans. Magn. MAG-**16**, 221 (1980).
2. L. Sakhnini and J. Popplewell, J. Magn. Magn. Mater. **122**, 146 (1990).
3. M.F. da Silva and A. M. Figueiredo Neto, Phys. Rev. E., **48**, 4483, (1993).
4. L. Sakhnini and J. Popplewell, J. Magn. Magn. Mater. **122**, 142 (1990).
5. M. Xu and P. J. Ridler, J. Appl. Phys. **82**, 326 (1997).
6. S. Taketomi, Jpn. J. Appl. Phys. **22**, 1137 (1983).
7. Nihad A. Yusuf, Jpn. J. Appl. Phys. **27**, 2418 (1988).
8. Chin-Yih Hong, J. Appl. Phys. **85**, 5962 (1999).
9. Zhang Qi, Wang Jianhua and Zhu Hesun, J. Materials Science, **31**, 4807 (1996).
10. Chin-Yih Hong, J. Magn. Magn. Mater., **201**, 178(1999).
11. J.E. Martin, K.M. Hill, and C.P. Tigges, Phys. Rev. E, **59**, 5676(1999).
12. Tengda Du and Weili Luo, Appl. Phys. Lett., **72**, 272(1998).
13. N. Inaba, H. Miyajima, H. Takahashi, S. Taketomi, and S. Chikazumi, IEEE Trans. Magn., **25**, 3866(1989).

HEAT TRANSFER FROM AN ESF RADIAL PLATE CLUTCH SURFACE
Part 2 – Theory

VLADIMÍR ORAVSKÝ

Institute of Materials and Machine Mechanics of the SAS
Račianska 75, 831 02 Bratislava 3, Slovakia
E-mail: ummsorav@savba.sk

In the paper half of a radial electro-structured fluid (ESF) clutch with vertical axis, consisting of a lower stationary disc and upper rotating disc is examined for constant speed of rotation. The narrow gap between the discs is filled with a viscous liquid, so electro-rheological effect is not taken into consideration. Frictional tangential forces in the fluid generate heat which is transferred through the fluid and rotating disc (by conduction) into surrounding air (by convection).
A steady state model of heat transfer for the system in question is built up and solved to enable the establishment of heat transfer prediction to be achieved. Several novelties in approach and solution relative to existing ones are adopted and realized. For some particular cases theoretical results are compared with experimental values by means of a table and gaphically and a good agreement is achieved.
The paper is the second part of a double paper, the first part of which is focused more upon general philosophy and experimental side of the problem in question.

1 Introduction

A general introduction to the problem outlined in the abstract is given in Ref.[1] Here only matters relating to the theoretical side of the problem are tackled. After a short description of the system in question, assumptions made in its modelling are specified. Then a reduced so called **quasistatic** (**qs**) model (in steady state at constant speed of rotation) is presented.

The aim of the paper is to determine analytically and calculate qs **temperatures** (**T-s**) of the model and compare them with experimental results. Also to analyze influence of adjustable **parameters (Pm-s)** of the model upon qs T-s in order to validate a **heat transfer** (**hTr**) model of the **radial clutch** (**rC**).

In the paper, the most frequent and long terms are substituted by their abbreviations and acronyms, whereby plural is distinguished by additional suffix -s. Places of their first appearance are made conspicuous by a bold-type face. Complete list of abbreviations is at the end of the paper.

2 Description of the system and assumptions

A sketch of the system introduced in Ref.[1] is given in Fig. 1 with characteristic calculation Pm-s. It is the upper half of a rC with vertical axis consisting of a lower stationary disc and upper rotating disc. Fluid in the gap between the discs (Shell Tellus Oil 37 [6]) is taken as viscous fluid only, i.e. the electro-rheological effect is not considered. Notation in the Fig. 1 is adopted according to Ref.[4]: upper bars denote **dimensional** (**dim**) quantities and Pm-s. Their meaning is given in the attached nomeclature.

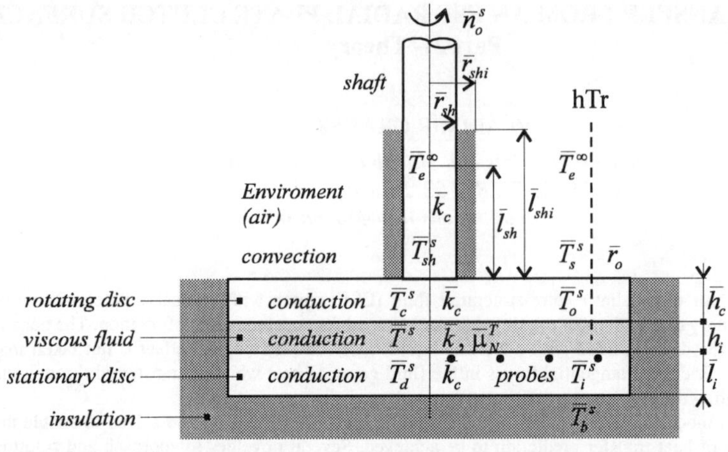

Fig. 1 Scheme of radial clutch (rC)

The most important **assumptions** adopted in the model are that bottom of the stationary disc and circumferential parts of rC are insulated (due to this **hTr** is only in the axial direction upwards), only qs state in rC is considered with constant revolutions $\overline{n}_o$ (therefore quantities and Pm-s have superscript *s*) and T-s at boundary surfaces $\overline{T}_i^s, \overline{T}_o^s, \overline{T}_s^s$ are supposed to be constant (but different) and unknown (but determinable in the course of solution). More detailed list of assumptions can be found in Ref. [4, 5]

An analytical model of rC based on the above assumptions is given in the next paragraph.

3 Quasistatic model of heat transfer and its solution

Full qs model of rC is described in Ref.[5] It contains 16 dim input Pm-s:

$$\overline{h}_c, \overline{h}_i, \overline{h}_{i\max}, \overline{r}_o, \overline{r}_{sh}, \overline{r}_{shi}, \overline{l}_{shi}, \overline{n}_o, \overline{n}_{o\max}, \overline{T}^a, \overline{T}^b, \overline{T}_e^\infty, \overline{k}, \overline{k}_c, \overline{h}_{eo}, \overline{\mu}_N^a \tag{1}$$

(their meaning see in Fig.1 and nomenclature) whereby two of them $\overline{h}_i$, $\overline{n}_o$ are adjustable in the experiment. Note that in general, revolutions $\overline{n}_o$ is a variable output quantity [4] but here, due to restriction to qs state only, it becomes an input Pm. Note also that $\overline{h}_{i\max}, \overline{n}_{o\max}, \overline{T}^a, \overline{T}^b$ are normalizing Pm-s and the room T $\overline{T}_e^\infty$ is given from experiment.

Values of all input Pm-s (1) for the special experimental stand in Fig.1 are given at the end of the paper.

There is no necessity to give **equations (Eq-s)** and solution of the qs model (distribution of T-s within rC) in detail. Note only that T within the stationary disc (Fig.1) proved to be constant [5]:

$$\overline{T}_d^s = \overline{T}_b^s = \overline{T}_i^s . \tag{2}$$

For the purpose of determination of overall qs hTr and unknown T-s: $\overline{T}_i^s, \overline{T}_o^s, \overline{T}_s^s$ it is enough to take into consideration **qs heat balance** of the following heat flows:

- overall qs heat flow generated in the fluid [5]: $\overline{q}_\tau^s = \frac{1}{2}\overline{A}_o\, \overline{\mu}_N^{s\phi}\, \overline{\omega}_o^2\, \overline{r}_o^2 / \overline{h}_i$, $\overline{A}_o = \pi\, \overline{r}_o^2$ (3)

  where $\overline{\mu}_N^{s\phi}$ is taken for the mean qs T within the gap $\overline{T}^{s\phi} = \left(\overline{T}_i^s + \overline{T}_o^s\right)/2$ (4)

- heat flow transferred by the fluid via conduction: $\bar{q}^{s} = \bar{A}_o\, \bar{k}\left(\bar{T}_i^{s} - \bar{T}_o^{s}\right)/\bar{h}_i$ (5)
- heat flow through the rotating disc via conduction: $\bar{q}_c^{s} = \bar{A}_o\, \bar{k}_c\left(\bar{T}_s^{s} - \bar{T}_o^{s}\right)/\bar{h}_c$ (6)

heat flow through the rotating shaft via conduction:

$$\bar{q}_{sh}^{s} = \bar{A}_{sh}\, \bar{k}_c\left(\bar{T}_s^{s} - \bar{T}_e^{\infty}\right)/\bar{l}_{sh}, \quad \bar{A}_{sh} = \pi\, \bar{r}_{sh}^{2} \quad (7)$$

Here the shaft length $\bar{l}_{sh}$ after which the shaft possesses a room T $\bar{T}_e^{\infty}$ is unknown but determinable from the requirement to fit agreement between the theory and experiment for one case only. This procedure is similar to calibration of a measuring apparatus before measurement and compensates to a certain degree indeterminateness and incompleteness of the model.

- heat flow transferred from rC into air via convection

$$\bar{q}_e^{s} = \left(\bar{A}_o - \bar{A}_{sh\,i}\right)\bar{h}_e^{s\phi}\left(\bar{T}_s^{s} - \bar{T}_e^{\infty}\right), \quad \bar{A}_{sh\,i} = \pi\, \bar{r}_{sh\,i}^{2}, \quad (8)$$

where overall qs hTr coefficient $\bar{h}_e^{s\phi}$ is taken for the mean outside qsT

$$\bar{T}_e^{s\phi} = \left(\bar{T}_s^{s} + \bar{T}_e^{\infty}\right)/2. \quad (9)$$

In view of qs heat flow balance it must hold: $\bar{q}_\tau^{s} \doteq \bar{q}^{s} \doteq \bar{q}_c^{s} \doteq \bar{q}_{sh}^{s} + \bar{q}_e^{s}$ (10)

Solution of the system of Eq-s (3)÷(10) is complicated due to the following difficulties:

viscosity of the fluid $\bar{\mu}_N^{s\phi}$ in Eq (3) is strongly T-dependent (with increasing T, viscosity nonlinearly monotonously decreases[6]). On the contrary to Ref.[2, 3], now another approximation by means of a hyperbolic form is preferentially used[5]:

$$\bar{\mu}_N^{s\phi} = \bar{\mu}_N^{a}\, f^{s\phi}, \qquad \bar{\mu}_N^{a} = \bar{\mu}_N^{T}\left(\bar{T}^{a}\right), \qquad f^{s\phi} = \frac{1 - aT^{s\phi}}{1 + bT^{s\phi}}, \quad (11)$$

where **nondimensional (ndim)** constants a, b have been identified for the fluid[6] in Ref.[5] and ndim mean qsT of the fluid $T^{s\phi}$ is given by the transformation:

$$T^{s\phi} = \frac{\bar{T}^{s\phi} - \bar{T}^{a}}{\bar{T}^{b} - \bar{T}^{a}} \quad \text{with the given range of T-s: } \bar{T}^{a} = 20\,°\text{C} \div \bar{T}^{b} = 40\,°\text{C}. \quad (12)$$

This enables not only considerable reduction of the errors of approximation[5] (relative to experimental input data[6]), but also to find a qs solution of the model without iteration in a compact analytical form.

1. The mean qs hTr coefficient of convection $\bar{h}_e^{s\phi}$ in Eq (8) is given by a nonlinear semi-empirical relation expressed through ndim Nusselt, Reynolds and Prandtl numbers[4, 8]. However, in the case in question it is possible to derive[5] that for the considered range of T-s (12), $\bar{h}_e^{s\phi}$ is fairly approximated by the formula:

$$\bar{h}_e^{s\phi} = \bar{h}_{eo}\sqrt{\pi.\bar{n}_o / 30}, \quad \bar{h}_{eo} = 2.328\,\text{W}\sqrt{\text{s}}/\text{m}^2\,°\text{C} \doteq \text{const} \quad \text{for } \bar{T}_e^{s}\,(20 \div 40)°\text{C} \quad (13)$$

Conditions (10) after manipulation with Eq-s (3), (5)÷(8), (11)÷(13) give the following **system of 3 ndim Eq-s** for unknown ndim T-s $\bar{T}_i^{s}$, $\bar{T}_o^{s}$, $\bar{T}_s^{s}$ [5]:

$$T_i^{s} - T_o^{s} = \mu_N\, n_o^{2}\, f^{s\phi}, \quad T_o^{s} - T_s^{s} = k\left(T_i^{s} - T_o^{s}\right)/h_i, \quad T_o^{s} - T_s^{s} = H\left(T_s^{s} - T_e\right) \quad (14)$$

where, in addition to nondimensionalization of T-s by means of the transformation similar to (12), the following **7 ndim Pm-s** are convenient to introduce:

$$\mu_N = \frac{1}{2}\frac{\overline{\mu}_N^a \,\overline{r}_o^2}{\overline{k}\left(\overline{T}^b - \overline{T}^a\right)}\left(\frac{\pi}{30}\overline{n}_{o\max}\right)^2, \quad n_o = \frac{\overline{n}_o}{\overline{n}_{o\max}}, \quad h_i = \frac{\overline{h}_i}{\overline{h}_{i\max}}, \quad k = \frac{\overline{k}\,\overline{h}_c}{\overline{k}_c\,\overline{h}_{i\max}} \tag{15}$$

$$T_e = \frac{\overline{T}_e^\infty - \overline{T}^a}{\overline{T}^b - \overline{T}^a}, \quad H_a = \left(\frac{\overline{r}_{sh}}{\overline{r}_o}\right)^2 \frac{\overline{h}_c}{\overline{l}_{sh}}, \quad H_b = \left(1 - \frac{\overline{r}_{shi}^2}{\overline{r}_o^2}\right)\frac{\overline{h}_{eo}\,\overline{h}_c}{\overline{k}_c}\sqrt{\frac{\pi}{30}\overline{n}_{o\max}} \tag{16}$$

and moreover **3 secondary ndim Pm-s**:

$$H = H_a + H_b\sqrt{n_o}, \quad K = k/h_i, \quad m_o = \mu_N\, n_o^2 \tag{17}$$

Note that explicit ndim Pm-s h_i, n_o have been introduced due to requirement of experiment, to follow and adjust them explicitly, and ndim Pm-s T_e or H_a are determined in the course of experiment or calculation.

Manipulation with Eq-s (14) gives the following solution: (18)

$$T_s^s = T_e + f^{s\phi}\, m_o\, K/H, \quad T_o^s = T_e + f^{s\phi}\, m_o\, K\left(1 + 1/H\right), \quad T_i^s = T_e + f^{s\phi}\, m_o\left(1 + K + K/H\right)$$

where in view of Eq-s (4), (11):

$$f^{s\phi} = \frac{1 - aT^{s\phi}}{1 + bT^{s\phi}}, \quad T^{s\phi} = \frac{1}{2}\left(T_i^s + T_o^s\right). \tag{19}$$

Therefore from the last two Eq-s (18) one derives:

$$T^{s\phi} = T_e + m\, f^{s\phi}, \quad \text{where} \quad \boxed{m = m_o\left(0.5 + K + K/H\right)} \tag{20}$$

is a complex ndim Pm containing all dim Pm-s (1) of the problem except the room T $\overline{T}_e^\infty$.

Combination of Eq-s (19), (20) leads to simple quadratic Eq:

$$bT^{s\phi 2} + \left(1 + am - bT_e\right)T^{s\phi} - \left(m + T_e\right) = 0 \tag{21}$$

from which only positive root is the **solution** of the problem in question:

$$\boxed{T^{s\phi} = \frac{1}{2b}\left[-\left(1 + am - bT_e\right) + \sqrt{\left(1 + am - bT_e\right)^2 + 4b\left(m + T_e\right)}\right]}. \tag{22}$$

Then unknown ndim T-s: T_s^s, T_o^s, T_i^s are found according to Eq (18). Corresponding dim T-s are calculated by a similar formula inverse to Eq (12):

$$\overline{T}_s^s = \overline{T}^a + T_s^s\left(\overline{T}^b - \overline{T}^a\right), \quad \overline{T}_o^s = \overline{T}^a + T_o^s\left(\overline{T}^b - \overline{T}^a\right), \quad \overline{T}_i^s = \overline{T}^a + T_i^s\left(\overline{T}^b - \overline{T}^a\right), \tag{23}$$

Comparison of $\overline{T}_i^s$ with the experimental value $\overline{T}_i^{s\exp}$ is evaluated by the absolute and relative difference:

$$\Delta\overline{T}_i^s = \overline{T}_i^s - \overline{T}_i^{s\exp}, \quad \delta\overline{T}_i^s = 100\,\Delta\overline{T}_i^s / \overline{T}_i^{s\exp}\ \%. \tag{24}$$

Algorithm of calculation as well as identification procedure of the unknown hTr coefficient H_a in (16) or the length $\overline{l}_{sh}$ can be found in detail in Ref. [5]

Following the above roughly described theoretical solution and calculation, let the numerical results be presented. For special input values, given at the end of the paper, the following ndim Pm-s (11), (15), (16) are arrived at:

$$a = 0.31, \ b = 0.94, \ \mu_N = 0.3289, \ k = 2.4559\ 10^{-2}, \ H_a = 0.0251, \ H_b = 3.5959\ 10^{-3}. \tag{25}$$

Then for experimental data given in Ref. [1], essential results of calculation are concentrated in the *Table 1*.

These results will be commented on and graphically evaluated in the next paragraph.

Table 1. Experimental and theoretical results

Gap width [mm]	Speed n_o [rpm]	AirTemp. T_e^{∞} [°C]	hTr coeff. (H_a=.0251) [1]	Viscosity μ_{Φ} [1]	Temp. of rot. disc Upper T_s^s [°C]	Lower T_o^s [°C]	Temp. of stationary disc (Lower Temp. of fluid) T_i^s [°C]	$T_i^{s\,exp}$	ΔT_i^s [°C]	δT_i^s [%]
0.5	500	25.4	0.0276	0.0527	27.27	27.33	28.38	28.1	+0.28	+1.00
		24.8		0.0542	26.73	26.78	27.86	28.3	-0.46	-1.54
	750	25.9	0.0282	0.1031	29.49	29.59	31.65	31.8	-0.15	-0.47
		26.0		0.1027	29.57	29.68	31.73	31.7	+0.03	+0.09
	1000	26.3	0.0287	0.1592	31.75	31.90	35.09	35.8	**-0.71**	**-1.99**
		26.4		0.1586	31.83	31.98	35.16	35.6	-0.44	-1.25
0.75	500	27.0	0.0276	0.0504	28.19	28.23	29.24	29.1	+0.14	+0.47
		27.0		0.0504	28.19	28.23	29.24	29.1	+0.14	+0.47
	750	26.5	0.0282	0.1059	28.96	29.03	31.14	31.6	-0.46	-1.44
		26.4		0.1063	28.87	28.94	31.06	31.4	-0.34	-1.07
	1000	24.2	0.0287	0.1853	28.43	28.55	32.26	32.3	-0.04	-0.14
		24.5		0.1832	28.68	28.80	32.46	31.7	**+0.76**	**+2.41**
1.0	500	27.6	0.0276	0.0497	28.48	28.51	29.50	29.0	+0.50	+1.73
		25.8		0.0542	26.76	26.79	27.87	27.3	+0.57	+2.10
	750	25.4	0.0282	0.1143	27.39	27.45	29.73	30.1	-0.37	-1.13
		27.2		0.1056	29.04	29.09	31.20	31.0	+0.20	+0.65
	1000	25.3	0.0287	0.1852	28.47	28.56	32.26	32.0	+0.26	+0.83
		25.2		0.1859	28.38	28.47	32.19	31.9	+0.29	+0.92

4 Evaluation of results

- experimental data are given in the shaded columns of the table,
- values $\overline{T}_s^s, \overline{T}_o^s, \overline{T}_i^s$ increase as expected. Differences between $\overline{T}_o^s$ and $\overline{T}_s^s$ within the rotating disc are small (less than 0.15 °C), but differences between $\overline{T}_i^s$ and $\overline{T}_o^s$ across the gap with the fluid are up to 3.72 °C. So assumption of uniform T of the fluid across the gap, frequently used (e.g. in Ref.[2, 3, 7]), has been proved untrue and inappropriate.
- comparison of theoretical and experimental results gives differences in the range (–0.71 ÷ +0.76) °C, resp. (–1.99 ÷ 2.41) % which is quite good agreement, peak-to-peak error is 1.47 °C, probable error for all cases is (0.0056 ± 0.4088) °C. On the contrary, in Ref.[2, 3] differences are up to 6 °C.
- extreme errors (in bold-type face) are nearly ± equal and the mean error is nearly zero (0.0056 °C). This proves that identification of the unknown hTr coefficient H_a is nearly optimum; for the shaft it has been identified by the value $H_a = 0.0251$. The corresponding length of the shaft after which it has external T (first gray column) is $\bar{l}_{sh} = 21.0\,\mathrm{mm}$. This is less than the length of insulation $\bar{l}_{shi} = 60.\,\mathrm{mm}$. This supports correctness of the theoretical calculation and fulfilment of the assumption $\bar{l}_{sh} \le \bar{l}_{shi}$,
- it is surprising that for the above value H_a, a substantial part of the total heat from rC is transferred through the shaft (given by ratio H_a/H) and much less through the surface of the rotating disc (given by ratio $H_b\sqrt{n_o}/H$): for 500 rpm it is only 9.19 %, for 750 rpm only 11.06 % and for 1000 rpm only 12.54 %. This proves that neglect of heat transferred through the shaft (7), which is quite usual[2, 3, 7], is inadmissible.

Comparison of theoretical and experimental T-s on stationary disc (undermost T of the fluid) is given in Fig. 2 for outPm-s given by (25). From Fig.2 one can again see and state a good agreement between theoretical and experimental results. One can also see nearly linear dependence of T of the stationary disc $\overline{T}_i^s$ upon external T $\overline{T}_e^\infty$, whereby influence of revolutions $\overline{n}_o$ is more conspicuous than influence of the gap width $\overline{h}_i$. The above proves also more detailed analysis of influence of the model Pm-s (15) upon results [5] which, due to lack of space, cannot be presented here.

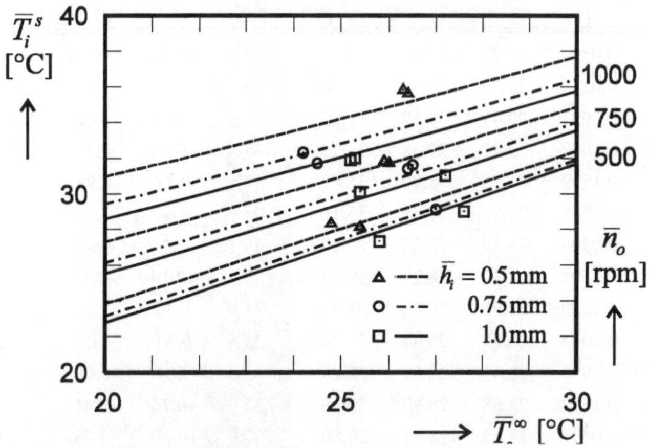

Fig. 2 Comparison of theoretical and experimental results

5 Conclusions

Thanks to novelties in approach and solution a good agreement between the theory and experiment has been achieved. This gives reliable basis for future prediction in practice.

Frequently used assumptions (simplifications) of a uniform temperature of the fluid within the gap and neglect of heat transferred through the shaft have shown to be untrue and inadmissible.

In future, a similar approach and methodology will be used for investigation of heat transfer in cylindrical clutches.

Acknowledgement

The author wishes to thank to the Royal Society of London for finacial support enabling his academic visit to the University of Sheffield. Special thanks are due to Messrs. W.A. Bullough and D.J.Ellam from this University who have started the problem in question and gave the author valuable input data and opportunity to take part in solving the problem theoretically. The author is also obliged to Mr. T. L. J. Divinec of Toronto, Canada whose financial support enabled him to attend the ERMR Conference in Nice.

References

1. D. J. Ellam, W. A. Bullough, and V. Oravský, Heat transfer from an ESF radial plate clutch surface, Part 1. *Proc. 8th Int. Conf. on Electro-rheological Fluids and Magneto-rheological Suspensions*, Nice, France, 2001 (World Scientific Publishing).
2. J. Kirk, Heat transfer through an ER radial clutch, Undergraduate thesis. University of Sheffield, 1999.
3. M. Nosek, Heat transfer through an ER radial clutch, Undergraduate thesis. University of Sheffield, 1998.

4. V. Oravský, Unified dynamic model of concentric and radial ER clutches including heat transfer, *Proc. 7th Int. Conf. on Electro-rheological Fluids and Magneto-rheological Suspensions*, ed. R. Tao (World Scientific, Singapore, 2000), pp.827–838.
5. V. Oravský, Quasistatic model of experimental device with a radial clutch and its solution, (in preparation for the *J. of Intel. Mater. Syst. and Structures*).
6. Shell Tellus Oils, UOCS/3, February 2000.
7. R. Smyth, K. H. Tau and W. A. Bullough, Heat transfer modelling of a cylindrical ER clutch, *J. Modern Physics B*, Vol. 8, 1994, pp. 3033–3050.
8. H. Y. Wong, Handbook of essential formulac and data on heat transfer for engineers, (Logman, London & New York, 1977).

Basic nomenclature

a, b $[-]$	ndim coefficients in an approximate substitution by a hyperbola ,
$\bar{h}_e$ $[\mathrm{W/m^2\,°C}]$	heat transfer coefficient of environment (air),
$\bar{h}_c, \bar{h}_i$ $[\mathrm{m}]$	width of rotating disc, width of gap with fluid,
$\bar{k}, \bar{k}_c$ $[\mathrm{W/m\,°C}]$	thermal conductivity of fluid, of material of discs and shaft,
$\bar{l}_{sh}, \bar{l}_{shi}$ $[\mathrm{m}]$	effective length of the shaft, length of the shaft insulation,
$\bar{n}_o$ $[\mathrm{rpm}]$	revolutions of the rotating disc,
$\bar{r}_o, \bar{r}_{sh}, \bar{r}_{shi}$ $[\mathrm{m}]$	radius of rotating disc, of the shaft and its insulation,
$\bar{T}^a, \bar{T}^b$ $[°\mathrm{C}]$	range temperatures of experiment and calculation,
$\bar{T}_e^{\infty}$ $[°\mathrm{C}]$	room temperature far from the experimental device,
$\bar{T}_i, \bar{T}_o, \bar{T}_e$ $[°\mathrm{C}]$	T of the stationary disc, lower and upper surface T of the rotating disc,
$\bar{\mu}_N$ $[\mathrm{Pa\,s}]$	viscosity of Newtonian fluid (Shell Oil Tellus 37),

Superscripts:		
	T	– temperature dependence,
	a, b, ϕ	– values corresponding to range T-s and mean T,
	s	– values at quasistatic state (with constant revolutions),
	$\bar{\ }$	– upper bar denotes dim quantities and Pm-s (ndim ones are denoted by the same symbol without upper bar).

Abbreviations

dim	dimensional	Pm	parameter
Eq	equation	qs	quasi-static
hTr	heat transfer	rC	radial clutch
ndim	nondimensional	T	temperature

Input data (for an experimental stand in Fig. 1) $a = 0.31$, $b = 0.94$

$\bar{h}_{eo} = 2.328\ \mathrm{W/m^2\,°C}$	$\bar{h}_c = 9.5\,\mathrm{mm}$	$\bar{h}_i = 0.5,\ 0.75,\ 1.0\,\mathrm{mm}$
$\bar{k} = 0.134\,\mathrm{W/m\,°C}$	$\bar{k}_c = 51.834\,\mathrm{W/m\,°C}$	$\bar{h}_{i\max} = 1.0\,\mathrm{mm}$
$\bar{n}_o = 500,\ 750,\ 1000\,\mathrm{rpm}$	$\bar{n}_{o\max} = 1000\,\mathrm{rpm}$	$\bar{l}_{shi} = 0.060\,\mathrm{m}$
$\bar{r}_o = 0.0425\,\mathrm{m}$	$\bar{r}_{sh} = 0.01\,\mathrm{m}$	$\bar{r}_{shi} = 0.01785\,\mathrm{m}$
$\bar{T}^a = 20\,°\mathrm{C}$	$\bar{T}^b = 40\,°\mathrm{C}$	$\bar{\mu}_N^a = 0.089\,\mathrm{Pa\,s}$

$\bar{T}_e^{\infty}$, $\bar{l}_{sh}$ given by experiment

THE UNIFORM MAGNETIC FIELD INFLUENCE ON THE MAGNETIC FLUID MENISCUS MOTION IN THE CYLINDRICAL CAPILLARY

V. BASHTOVOI, P. KUZHIR AND A. REKS

UNESCO Chair "Energy Conservation and Renewable Energies", Belarussian State Polytechnic Academy, 65 F. Skaryna Ave., Minsk, 220027, Belarus
E-mail: Bashv@itmo.by

G. BOSSIS, O. VOLKOVA

Laboratoire de Physique de la Matière Condensée, CNRS, U.M.R. 6622, Université de Nice-Sophia Antipolis, Parc Varlose – 06108, Nice, Cedex 2, France
E-mail: Bossis@unice.fr

An improvement of liophobic capillary-porous systems using magnetic fluids is proposed. The cycle of non-wetting liquid penetration and displacement is realized experimentally in the presence of the uniform magnetic field. Experimental investigations of the effect of the external uniform magnetic field on dynamics of capillary penetration of the Newtonian magnetic fluid into cylindrical capillaries at zero gravity and under gravity are presented. It is found that the pressure difference in the magnetic fluid between a meniscus and a free surface in a vessel increases in the field longitudinal to the capillary and decreases in the transverse one. In the longitudinal field, the velocity of penetration increases at zero gravity and does not vary under gravity. The transverse field slows down the process.

1 Introduction

Processes of capillary penetration and displacement of magnetic fluids have good prospects of their application in heat-mass transfer technique, apparatuses of chemical technology and mining engineering. The most interesting application of these processes can be found in devices operating on the base of liophobic capillary-porous systems.

These systems represent a capillary-porous body and a non-wetting liquid [1]. The latter penetrates into micro-pores of the body under high external pressure and is displaced under its capillary pressure. This penetration-displacement cycle was proposed to be used in some power devices, such as heat-engines, accumulators, dampers [2]. Non-wetting conditions are created by covering micro-pores with a thin molecular layer of organophobic material. That is rather complicated from the technology standpoint.

In order to except these difficulties, we propose to use two immiscible liquids, which wet pores in contact with the ambient air. In contact with each other, one of the liquids will be wetting and the other will be non-wetting if their surface tensions on the liquid-solid interface are different [3]. By analogy with systems of one liquid, while penetrating of the non-wetting liquid, the work is done under the system, and while displacing, the system does work returning the energy accumulated. Liophobic capillary-porous systems are purely dynamic ones and energy dissipation takes place. As for dissipation mechanisms, they can be the viscous one in volumes and on the interface of the liquids as well as viscous friction of the wetting perimeter. The latter determines the dynamic wetting angle of the system.

Using a magnetic fluid as one of the liquids makes possible to control capillary and dynamic processes in these systems by magnetic controlling capillary and viscous properties of magnetic fluids. Detailed investigations of magnetic fluid penetration and displacement are necessary for development of devices operating on the base of the principle described above.

Theoretical and experimental development of this problem is not completed yet. Investigations of statics and dynamics of capillary rise of the magnetic fluid under the non-uniform magnetic field have shown that a fluid column is drawn in the domain of a stronger field, and the main factor of the magnetic field effect is the volume magnetic force caused by the external magnetic field gradient [4,5]. As has been shown in [6], the effect of the external uniform magnetic field on statics of ascension into a cylindrical capillary is defined by effects of internal uniform and of induced non-uniform fields. The second effect is dominant in case of ascension from a thin layer of the magnetic fluid, the capillary rise height increases in the magnetic field longitudinal to the capillary and decreases in the transverse one as the square of the induction of the field.

Experimental studies of dynamics of penetration of wetting Newtonian magnetic fluid into a cylindrical capillary as well as results on experimental realization of penetration-displacement cycle are presented here. The velocity and the time of penetration are determined.

One of the mechanisms of the external uniform field effect on dynamics of magnetic fluid penetration is the same as the effect on statics of a fluid column in a capillary i.e. influence of the magnetic field on the pressure difference in the magnetic fluid. This pressure difference is regarded as that between free surfaces inside the capillary (meniscus) and outside it. Other mechanisms, such as the magneto-viscous effect, the field effect on wetting hysteresis, are not considered here.

2 Experiment

The experimental cell used for cycle realization and for investigation of magnetic fluid penetration at zero gravity is shown in Fig. 1a. A glass cylindrical vessel of diameter 25 mm with a vertically adjusted capillary both filled with the magnetic fluid are placed in a rectangular plexiglass box filled with aqueous solution of calcium chloride. Equal densities of liquids and their equal levels in the vessels (the layer depth of the magnetic fluid is 20 mm) provide conditions of hydroimponderability.

The experimental cell used for investigation of magnetic fluid penetration into a vertical capillary under gravity is shown in Fig. 1b. A glass cylindrical dish 120 mm in diameter is filled with the magnetic fluid, the layer depth is 8 mm. The capillary is stationary adjusted to the vertical position. The lower capillary end is immersed into the magnetic fluid.

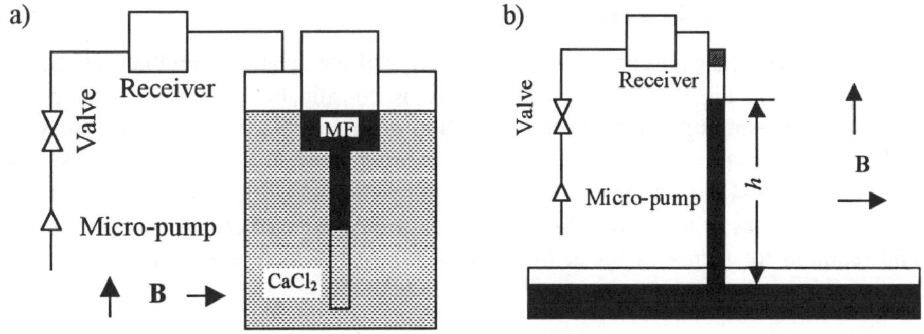

Fig. 1. Experimental cells for investigation of magnetic fluid penetration at zero gravity (a) and under gravity (b)

In contact with the ambient air as well as with solution of calcium chloride, the magnetic fluid wets capillary walls. The borosilicate glass cylindrical capillaries of diameter d = 0.27 mm or 0.80 mm and length l = 40 mm are used. The box of the first cell (Fig. 1a) and the capillary of the second one (Fig. 1b) are connected with the piston micro-pump. It produces pressure required to hold the meniscus at a given position in the capillary. This pressure determines the pressure difference Δp in the magnetic fluid and it is measured by the micro-manometer.

A method of cycle realization follows. A meniscus is initially situated on the capillary end immersed into the non-magnetic liquid. The pressure in this liquid is adjustable continuously from zero to the value such that the meniscus "breaks away" from the capillary end. This pressure exceeds slightly the pressure difference in wetting magnetic fluid. Under this pressure, the meniscus moves slowly along the capillary and quasi-stationary penetration of the non-wetting non-magnetic liquid takes place. The pressure is released drastically by the valve when the meniscus reaches the opposite capillary end. At this instant the meniscus changes direction of its way and returns to its initial position displacing the non-wetting liquid. The receiver smoothes pressure oscillations after valve opening.

While studying penetration dynamics of the wetting magnetic fluid under gravity (Fig. 1b), the meniscus is held at its initial position on the capillary end immersed into the magnetic fluid. Pressure is released drastically and the meniscus begins to move over the wetted surface of the capillary. The meniscus stops having reached the maximal capillary rise height h. Dynamics of the process is recorded by the video camera at 50 frames/sec and at exposure of 0.002 sec.

The uniform magnetic field longitudinal or transverse to a capillary (the induction B = 0 - 25 mT) is produced by Helmholtz coils. The magnetic fluid on the magnetite and kerosene base has density ρ = 1320 kg/m^3, surface tension σ= 0.029 N/m on the border with air and σ= 0.012 N/m on the border with solution of calcium chloride, wetting angle γ= 25 - 30^0 in contact with glass and air and γ= 45 - 50^0 in contact with glass and solution of calcium chloride, dynamic viscosity η_1 = 0.0054 Pa × sec, magnetization of saturation M_s = 38.2 kA/m. The magnetic fluid magnetization law is linear over the experimental range of the magnetic field, relative magnetic permeability is μ= 2.7. Evaluation of the magneto-viscous effect shows that the latter does not exceed 2% within the field range considered. Aqueous solution of calcium chloride has density ρ = 1320 kg/m^3, dynamic viscosity η_2 = 0.0022 Pa × sec.

3 Results

3.1. Penetration-displacement cycle

This cycle realized in the absence and in the presence of the magnetic field is presented in z-Δp coordinates in Fig. 2. Here z is the meniscus coordinate. The area inside closed curves of the cycle multiplied by the cross-section of the capillary presents the work expended on non-wetting liquid penetration and the work done in the process of displacement: $A = \Delta p \cdot (\pi d^2/4) \cdot l$. In the absence of the magnetic field, this work is given by $A = (\sigma \cdot \cos\gamma) \cdot (\pi dl)$. As is seen from Fig. 3, the pressure Δp exerted on the non-wetting liquid and required for meniscus break from the capillary end increases in the longitudinal field (by two times at $B = 25$ mT) and remains almost constant in the transverse field. That points to a possibility of magnetic control of processes in such liophobic systems.

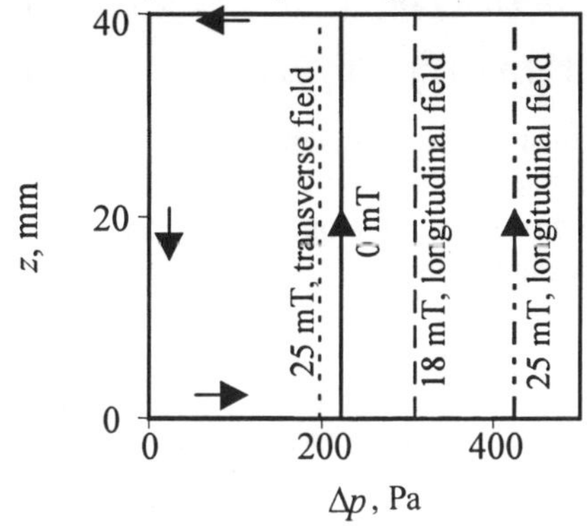

Fig. 2. The cycle of penetration and displacement of the non-wetting liquid contacting with wetting magnetic fluid

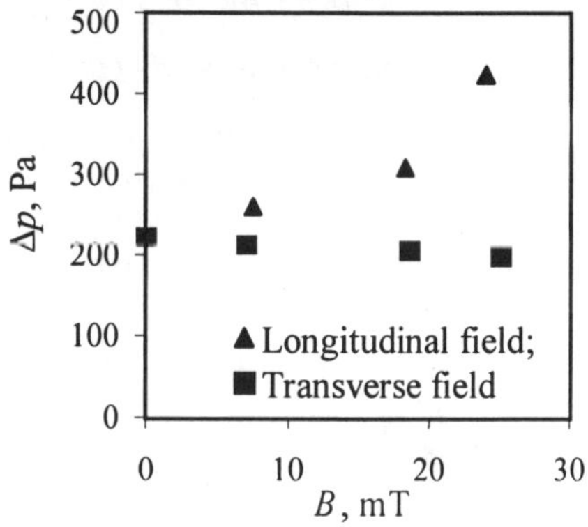

Fig. 3. Dependency of the pressure required for penetration of the non-wetting liquid into the capillary on the induction of the magnetic field

3.2 *Effect of field on dynamics of magnetic fluid penetration*

Penetration curves of the magnetic fluid (dependence of the meniscus position z on the time t) are presented in Fig. 4 for the case of zero gravity. Data point to the influence of the magnetic field on the penetration characteristics: the whole time of penetration T decreases in the longitudinal field (two times) and increases in the transverse one (by 50%) (Fig. 5). Accordingly, the velocity of penetration defined as $V = l / T$ increases in the longitudinal field and decreases in the transverse one.

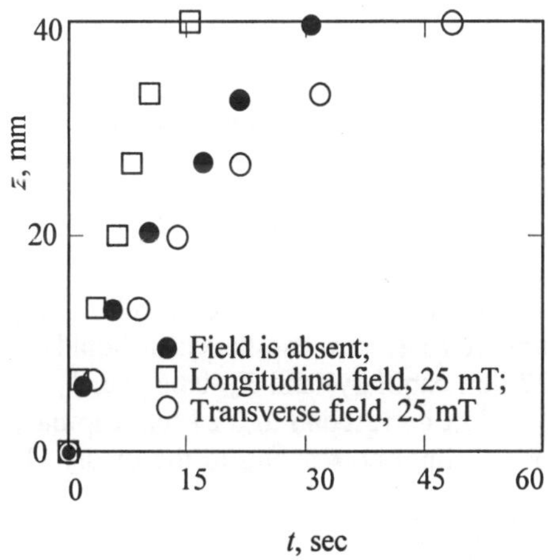

Fig. 4. Penetration curves of the magnetic fluid at zero gravity

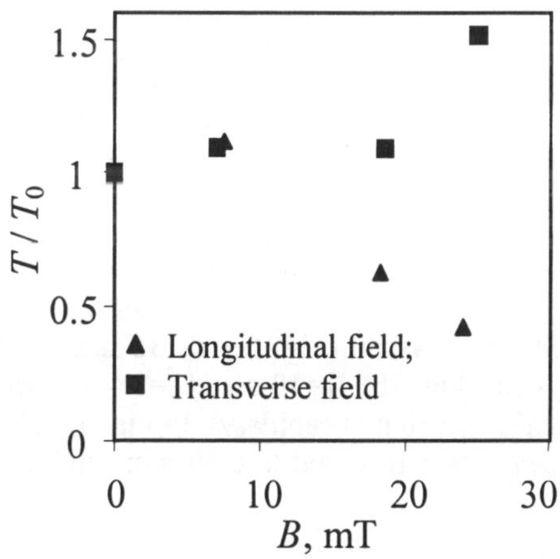

Fig. 5. Dependency of the dimensionless whole time of penetration on the induction of the magnetic field (zero gravity)

In case of capillary rise from a thin layer of the magnetic fluid under gravity, penetration curves have a saturation section with an asymptote h that is function of the magnetic field (Fig. 6). The maximal rise height h is given by the following semi-empirical expressions [6] for longitudinal and transverse fields respectively:

$$\rho gh = 4\sigma \cos \gamma / d + (\mu - 1)^2 / (2\mu) \cdot (B^2 / \mu_0), \tag{1}$$

$$\rho gh = 4\sigma \cos \gamma / d - (\mu - 1) / 2 \cdot [1 - 4 / (\mu + 1)^2] \cdot (B^2 / \mu_0), \tag{2}$$

where g is gravity acceleration, μ_0 is magnetic permeability of vacuum.

At $h > 0.5d$ these dependencies fit the experimental data well (Fig. 7) The experimental range of the vertical field is limited by the threshold value of the induction $B \approx 8.5$ mT such that the fluid surface in the vessel becomes unstable.

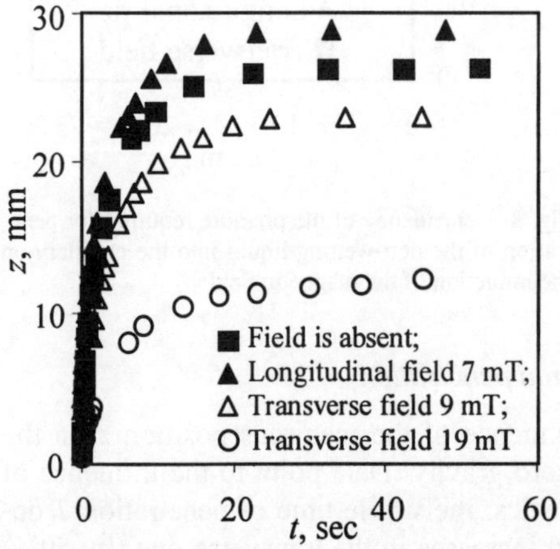

Fig. 6. Penetration curves of the magnetic fluid under gravity

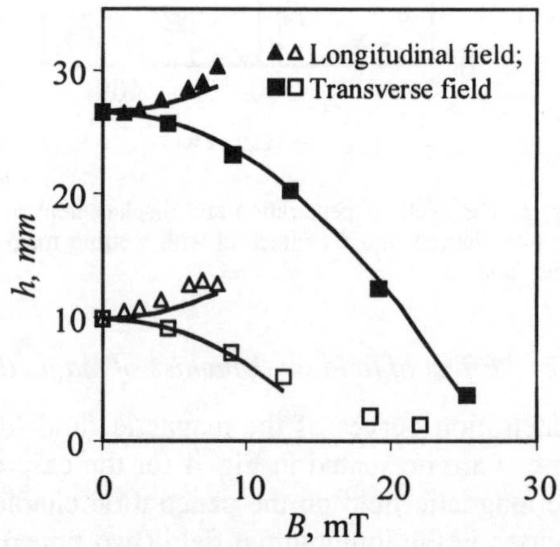

Fig. 7. Dependencies of the capillary rise height on the induction of the field

Since the whole time of the process is infinite, define the characteristic time T of penetration in terms of the period of meniscus passage of 90% of its maximal height. The characteristic velocity V of the meniscus will be $V = 0.9h / T$. It is found that the time T increases both in the longitudinal field (by 35% for 0.80 mm capillary at $B = 8$ mT) and in the transverse one (by 80% for 0.27 mm capillary at $B = 19$ mT) (Fig. 8). The velocity V decreases in the transverse field (by 75% for 0.27 mm capillary at $B = 19$ mT) (Fig. 9). Influence of the longitudinal field on the velocity V is not found.

In Fig. 5, 8, 9 T_0, V_0 are values of T and V respectively in the absence of field. In liquid-liquid system (Fig. 1a) T_0=32 sec, V_0=1.2 mm/sec (0.27 mm capillary); in liquid-gas system (Fig. 1b) T_0=10 sec, V_0=2.4 mm/sec (0.27 mm capillary) and T_0=0.66 sec, V_0=13 mm/sec (0.80 mm capillary). In Fig. 5, 7, 8, 9 solid dots correspond to 0.27 mm capillary, open dots correspond to 0.80 mm capillary. In Fig. 7 lines correspond to the theory (1), (2).

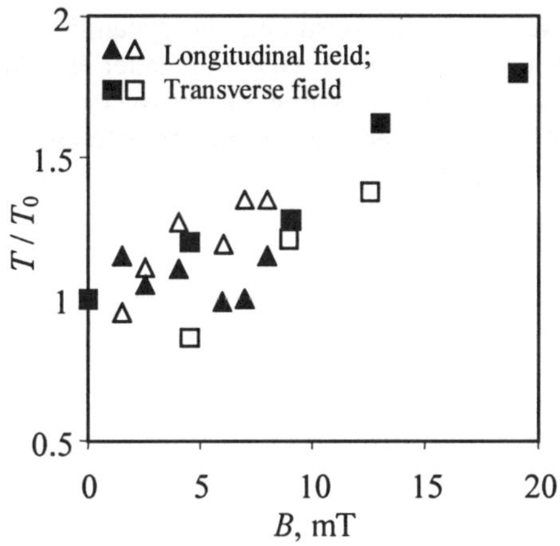

Fig. 8. Dependencies of the dimensionless characteristic time of penetration on the induction of the magnetic field (under gravity)

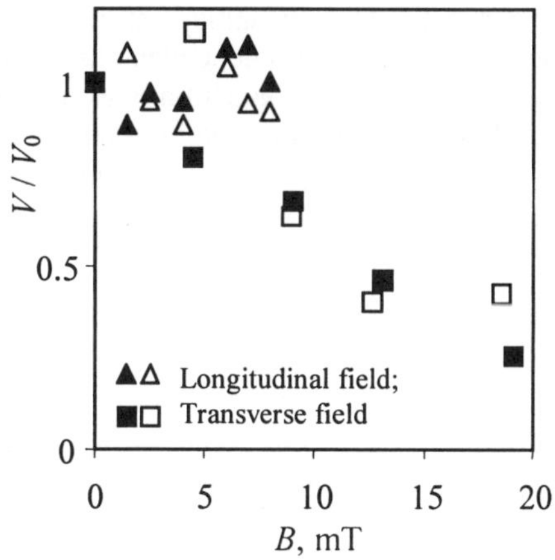

Fig. 9. Dependencies of the dimensionless characteristic velocity of penetration on the induction of the magnetic field (under gravity)

In the presence of the external uniform magnetic field, penetration of the Newtonian magnetic fluid is given by classical Washburn equation [7]. Assuming the pressure difference Δp in the magnetic fluid to be constant during penetration, this equation gives the following t vs. z dependencies:

$$t = 32 / (\Delta p d^2) \cdot [(\eta_1 - \eta_2) \cdot (z^2/2) + \eta_2 l z] \tag{3}$$

for the case of liquid-liquid system at zero gravity and

$$t = (32\eta_1 h) / (\rho g d^2) \cdot [-z/h - \mathrm{Ln}(1 - z/h)] \tag{4}$$

for the case of liquid gas system under gravity.

So, the expressions for the time of penetration will be:

$$T = 16 l^2 \cdot (\eta_1 + \eta_2) / (\Delta p d^2) \tag{5}$$

for the case of liquid-liquid system at zero gravity and

$$T = (44.8\eta_1 h) / (\rho g d^2) \tag{6}$$

for the case of liquid-gas system under gravity.

As is seen from analyses of (5), at zero gravity, the time of penetration should decrease in the longitudinal field and remain constant in the transverse one according to the dependence of the pressure difference Δp on the magnetic field. That is supported by the experiment in case of the longitudinal field (Δp gives double increase, T gives double decrease at $B = 25$ mT) but it is not absolutely true for the transverse field (Fig. 3, 5).

As for penetration under gravity, equation (6) gives linear dependence of the time T on the capillary rise height h and suggests that the velocity $V = 0.9h / T$ is independent of h. Hence, the characteristic time of penetration should increase in the longitudinal field and decrease in the transverse one. The characteristic velocity of penetration should remain constant at any field values and directions. That is qualitatively supported by the experiment in case of the longitudinal field. The experimental data show an opposite tendency in the transverse field (Fig. 8, 9).

Unexpected deceleration of penetration in the transverse field in both cases is probably connected with dynamic hysteresis of the wetting angle. Assuming the dynamic wetting angle to be field independent, taking into account relations between the latter and the velocity in Washburn equation does not explain this result. Possibly, the transverse magnetic field influences on the mechanism of wetting perimeter viscous friction, decreasing the dynamic wetting angle so that the pressure difference Δp at a given meniscus position decreases together with the velocity of the meniscus.

4 Conclusions

1. The proposed improvement of liophobic capillary-porous systems is 1) use of two immiscible wetting liquids with different interfacial tensions, 2) use of magnetic fluids for magnetic control of capillary and dynamic processes in these systems.
2. "Motive power" of magnetic fluid penetration is the pressure difference in the fluid between the meniscus and the free surface in the vessel. The pressure difference increases in the field longitudinal to the capillary and decreases in the transverse one. In the longitudinal field, the velocity of penetration increases at zero gravity and does not vary under gravity. In the transverse field, it decreases in both situations.

5 Acknowledgements

This work has been supported by the Foundation of Fundamental Researches of the Republic of Belarus and INTAS Fellowship Grant for Young Scientists. We are very grateful to V. Samoilov for making experimental setups and to O. Safonenko and L. Suloeva for provision us with magnetic fluids.

References

1. V.A. Yeroshenko in *Doklady AN USSR* (Reports of the Academy of Science of the Ukrainian SSR) **Ser. A, 10** (1990) p. 77 (Russ).
2. Ye.N. Serdun, A.G. Portianoy, A.P. Sorokin and G.A. Portianoy, *Teploenergetika* (Heat and power engineering) **12** (2000) p. 64 (Russ).
3. A. W. Adamson, *Physical Chemistry of Surfaces* (John Wiley and Sons, Inc., New York, 1976).
4. Yu.I. Dikansky, M.A. Bedzhanyan and O.V. Borisenko in *Fiziko-khimicheskie i Prikladnye Problemy Magnitnyh Zhidkostey* (Phisico-chemical and Applied Problems of Magnetic Fluids) (SGU, Stavropol, 1997), p. 28 (Russ.).
5. Ye.V. Dzerzhavina in *Tez. Dokl. 5 Vses. Konf. po Magn. Zhidk. 1* (Abstracts of Reports of the 5th All-Union Conference on Magnetic Fluids 1) (Plyos, 1988), p. 80 (Russ.).
6. V. Bashtovoi, P. Kuzhir and A. Reks, *J. Magn. Magn. Mater.* (2001) (to be published).
7. E.W. Washburn, *Phys. Rev.* **Ser.2, 17** (1921) p. 273.

RELAXATION THEORY FOR DYNAMIC ELECTRORHEOLOGICAL EFFECT

G. Q. GU

Information College, East China Normal University,
Shanghai, 200062, China
E-mail: guoguoqk@online.sh.cn

K. W. YU and P. M. HUI

Department of Physics, The Chinese University of Hong Kong,
Shatin, New Territories, Hong Kong
E-mail: kwyu@phy.cuhk.edu.hk, pmhui@phy.cuhk.edu.hk

We present a model to investigate the induced interaction between rotating particles in ER fluids. For particle undergoes uniform rotation or harmonic oscillation, the effects due to dipole relaxation are studied analytically. The induced interaction between particles which are rotating uniformly is derived. The result shows that the dynamic effect reduces the interaction between particles. A proof on the consistency of the relaxation equation with charge conservation upon rotation is also given.

1 Introduction

Recently, electrorheological (ER) fluids have attracted considerable attention both experimentally and theoretically due to their potential applications.[1,2] Typically, ER fluid consists of a suspension of particles with high dielectric constant (high conductivity) in a fluid of low dielectric constant (low conductivity). Under an applied external electric field, polarized charges are induced on the particle surface. The interaction between these induced charges leads to an anisotropic interaction between the particles and results in the ER effect. Under the influence of ER effect, the particles in an ER fluid first aggregate into chains and then the chains aggregate into columns.[3] The possibility of tuning the effect through an applied electric field and the rapid transitions between a fluid state and a solid state make this class of smart materials important for their potential industrial applications.

The interaction between particles provides the basis for investigating the rheological and transport properties of ER fluids. It is the starting point of many numerical simulations. In deriving the induced interactions between particles, the particles are usually assumed to be at rest.[3–5] In an ER fluid, however, particles may be driven into rotational motion due to, say, the influence of fluid flow or a dc field.[6,7] Rotational motion of the particles may lead to some dynamic rheological effects in suspensions.[8] It is thus important to study the effects of rotation of particle on the induced interaction between particles. In this work, we introduce a phenomenological theory for dynamic electrorheological effect based on the relaxation of the displaced polarized charges.

2 Relaxation of polarized charges

We assume that there is no free charge on the particles and the ER effect originates from the mismatch of dielectric constants between the particles and host. Let σ_{st} be the surface polarized charge distribution in the presence of an applied electric field when the particle is not in motion. We refer to σ_{st} as the static polarized charge distribution. Here, we propose a model to describe the dynamic ER effect with respect to particle rotations. The idea is that the polarized charge distribution is displaced constantly in space as the particle rotates. Momentarily at the rotated position, the displaced original polarized chage distribution deviates from the local equilibrium distribution. The distribution will tend to relax to an equilibrium distribution at a certain rate $1/\tau$ depending on the relaxation time τ, which is dependent on the material forming the particles. Mathematically, the time dependent surface polarized charge distribution $\sigma(\theta,\varphi,t)$ on a rotating particle satisfies the equation

$$\frac{\partial\sigma(\theta,\varphi,t)}{\partial t}+\frac{\partial\sigma(\theta,\varphi,t)}{\partial\theta}\frac{\partial\theta}{\partial t}+\frac{\partial\sigma(\theta,\varphi,t)}{\partial\varphi}\frac{\partial\varphi}{\partial t}=-\frac{1}{\tau}\left[\sigma(\theta,\varphi,t)-\sigma_{st}(\theta,\varphi)\right], \tag{1}$$

with the origin of the coordinates placed at the center of the rotating particle.

For the purpose of studying the aggregation of particles under an applied field, first we need to find the interaction between two particles. For simplicity we consider a two-particle system. In this case, the φ dependence drop off[9,10] and the relaxation equation becomes

$$\frac{\partial\sigma(\theta,t)}{\partial t}+\frac{\partial\sigma(\theta,t)}{\partial\theta}\frac{\partial\theta}{\partial t}=-\frac{1}{\tau}\left[\sigma(\theta,t)-\sigma_{st}(\theta)\right]. \tag{2}$$

For a particle undergoing uniform rotation, which can be realized by a simple shear, the relaxation equation is further simplified to

$$\frac{\partial\sigma(\theta,t)}{\partial t}+\omega\frac{\partial\sigma(\theta,t)}{\partial\theta}=-\frac{1}{\tau}\left[\sigma(\theta,t)-\sigma_{st}(\theta)\right]. \tag{3}$$

A general solution to the time-dependent surface polarized charge distribution on a rotating particle can be found via a Fourier analysis.

3 Charge conservation upon rotation

There is no charge flowing into or out of the particle surface and hence the charge on particle surface must be conserved. Here, we show that the relaxation equation (2) is consistent with the charge conservation. It is convenient to transform the angle θ to θ' through $\theta=\theta'+\frac{\pi}{2}$, i.e., by measuring the polar angle from a horizontal axis through the particle centre. The charge distribution on the particle surface can, in general, be decomposed into a sum of an odd and even functions:

$$\sigma\left(\theta'+\pi/2,t\right)=\sigma^{odd}(\theta',t)+\sigma^{even}(\theta',t), \tag{4}$$

$$\sigma_{st}\left(\theta'+\pi/2,t\right)=\sigma_{st}^{odd}(\theta',t)+\sigma_{st}^{even}(\theta',t). \tag{5}$$

The integrals of the odd functions σ^{odd} and σ_{st}^{odd} over the particle surface vanish, i.e.,

$$\int_{-\pi/2}^{\pi/2} \sigma^{odd}(\theta') \cos\theta' d\theta' = 0$$

$$\int_{-\pi/2}^{\pi/2} \sigma_{st}^{odd}(\theta') \cos\theta' d\theta' = 0$$

The total charge on the particle surface is thus

$$Q^{even}(t) = \int_{-\pi/2}^{\pi/2} \sigma^{even}(\theta', t) \cos\theta' d\theta' \tag{6}$$

The total charge $Q^{even}(t)$ on the particle surface satisfies the equation

$$\frac{\partial Q^{even}(t)}{\partial t} + \frac{1}{\tau} Q^{even}(t) = 0, \tag{7}$$

with the general solution

$$Q^{even}(t) = A \exp\left(-t/\tau\right).$$

The constant A is given by the initial condition that

$$Q^{even}(0) = Q_{st}^{even} = 0. \tag{8}$$

The total polarized charges add up to zero for a rotating particle as there is no free charge. As a result, the present formalism is consistent with charge conservation.

4 Single Particle: Uniform rotation and oscillatory motion

We consider the time evoluation of the induced dipole moment in a spherical particle. The applied field is chosen to be in the $\hat{z}$-direction. For an anticlockwise rotation along an axis in the $\hat{y}$-direction, the relaxation equation for a rotating particle is,[9]

$$\frac{d\vec{p}}{dt} = \vec{\omega} \times \vec{p} - \frac{1}{\tau}(\vec{p} - \vec{p}_0).$$

The equation can be readily solved by decomposing into components. The equation for the p_y gives $p_y = p_{y0} exp(-\alpha t)$ where $\alpha = \int_0^t (1/\tau) dt'$. Hence p_y vanishes in the long time limit. The $\hat{x}$ and $\hat{z}$ components are more conveniently handled by introducing the complex notation $\tilde{p} = p_x + i p_z$ with the initial condition $\tilde{p}_0 = i p_0$. The complex dipole moment satisfies the equation

$$\frac{d\tilde{p}}{dt} = \left(i\omega - \frac{1}{\tau}\right) \tilde{p} + \frac{\tilde{p}_0}{\tau},$$

which has the solution

$$\tilde{p} e^I - \tilde{p}_0 = \tilde{p}_0 \int_0^t \frac{e^I}{\tau} dt, \tag{9}$$

where I is an integration factor given by:

$$I = \int_0^t \left(\frac{1}{\tau} - i\omega(t) \right) dt. \tag{10}$$

For a particle in uniform rotation with angular velocity ω_0, the time-dependent dipole moment is given by

$$\tilde{p} = \frac{\tilde{p}_0}{1 - i\omega_0 \tau} \left(1 - i\omega_0 \tau e^{-t(1-i\omega_0\tau)/\tau} \right).$$

The steady state solution in the long time limit is thus

$$\tilde{p} = \frac{\tilde{p}_0}{1 - i\omega_0 \tau}.$$

Another interesting case is when the polarized particle undergoes an oscillatory motion $\theta(t) = \theta_0 \sin \omega' t$ and hence with an angular speed of $\omega(t) = \theta_0 \omega' \cos \omega' t$. Applying Eqs.(9) and (10), the dipole moment in the steady state is time-dependent and is given by

$$\tilde{p} = \frac{\tilde{p}_0}{1 - i\theta_0 \omega' \tau \cos \omega' t} = p_0 \frac{1 - i\theta_0 \omega' \tau \cos \omega' t}{1 + \theta_0^2 \omega'^2 \tau^2 \cos^2 \omega' t}. \tag{11}$$

Note that the dipole moment does not follow a simple harmonic oscillation. For a constant relaxation time τ, the time averaged dipole moment is given by

$$\frac{\langle p_x \rangle}{p_0} = 0 \quad \text{and} \quad \frac{\langle p_z \rangle}{p_0} = \frac{1}{\sqrt{1 + \theta_0^2 \omega'^2 \tau^2}}. \tag{12}$$

The latter result shows that oscillatory motion leads to a reduction in the ER effect. The reduction becomes increasingly significant for high oscillatory frequencies.

5 Interaction between two particles

An important quantity in the study of ER fluids is the induced interaction between two particles in the presence of an applied electric field. Previous studies usually assumed the particles to be stationary. Here we give the results for the interaction between a uniformly rotating particle and a stationary particle. Following a multiple-image method and a Rayleigh method previously developed[10–12], the leading term in the interaction turns out to take on the simple form of

$$\mathbf{F}_{rot}^{(1)} = \frac{1}{2} \left[1 + \frac{1}{1 + (\tau\omega)^2} \right] \mathbf{F}_{st}^{(1)}. \tag{13}$$

Here $\mathbf{F}_{st}^{(1)}$ is the leading term of the force between two stationary particles, which can be obtained via the Rayleigh method as[10]

$$\mathbf{F}_{st}^{(1)} = \frac{9\Omega_a}{8\pi a^3} \frac{(\epsilon_h - \epsilon_a)}{1 - \epsilon_{a,h}} \frac{(E_0)^2}{a\left[(T_1^a)^{-1} + 2P_2(\cos\theta'_{21}) \left(\frac{a}{R}\right)^3 \right]^2} \left[\left(\frac{a}{R}\right)^4 (3(\cos\theta'_{21})^2 - 1)\, \mathbf{e}_r + \sin(2\theta'_{21}) \mathbf{e}_\theta \right].$$

where a and Ω_a are the radius and volume of the particle respectively, and ϵ_a (ϵ_h) is the dielectric constant of the particles (host). The centers of the particles are separated by a distance R. The ratio of the dielectric constants is $\epsilon_{a,h} = \epsilon_a/\epsilon_h$, the factor $T_1^a = (1 - \epsilon_{a,h})/(2 + \epsilon_{a,h})$ and the θ'_{21} is the polar angle of the vector $\mathbf{R}_{12}$ joining the two particles with the origin taken to be at the center of particle 1.

The results in Eq.(13), which relates the angular velocity to the ratio of force on rotating particle to that on stationary particle, is useful to experimental measurements. On one hand, it gives the force on rotating particles in terms of the polarized relaxation time and the angular velocity of the rotating particles. On the other hand, it also gives a way to determine the polarized relaxation time of materials by measuring the reduction in the inter-particle force under different angular velocities.

6 Conclusion

We discussed the dynamic ER effects within the framework of a phenomenological theory based on the relaxation of the surface charge distribution. Analytic expressions for the induced dipole moment for a particle in uniform rotation and oscillatory motion are derived. In general, there is a reduction in the time averaged dipole moment as compared to that in a stationary particle. The reduction factor depends on the relaxation time, which in turn depends on the material forming the particles. For short relaxation time (fast relaxation processes), the reduction is small. While our theory is constructed with reference to the polarized charges, the same formalism can be applied to other processes concerning charge relaxation. For example, for the process leading to Quincke rotation,[7] the relaxation time is $\tau_M = (\varepsilon_a + 2\varepsilon_h)/(\sigma_a + 2\sigma_h)$, which is in the range of microseconds for typical ER fluids. Thus, while particle rotations affect the interaction between particles, the effect is expected to be small in existing ER fluids.

Acknowledgements

This work was supported in part by the Research Grants Council of the Hong Kong SAR Government under the grant CUHK4284/00P. G.Q.G. acknowledges the financial support from a Key Project of the National Natural Science Foundation of China under the grant 19834070. G.Q.G. and K.W.Y. are grateful to Dr. Z. W. Wang, who informed them his experiments on the dynamic ER effects of rotating particles and for many fruitful discussions.

References

1. R. Tao (eds.), Proceedings of the 7th International Conference on Electro-Rheological Fluids and Magneto-Rheological Suspension, published in *Int. J. Mod. Phys.* **B 15**, Nos.6 & 7 (2001).
2. H. Conrad, *MRS Bulletin* **23**, (8) 35 (1998).
3. D. J. Klingenberg and C. F. Zukoski, *Langmuir* **6**, 15 (1990); D. J. Klingenberg, *MRS Bulletin* **23**, (8) 30 (1998).
4. Z. W. Wang, Z. F. Lin and R. B. Tao, *Int. J. Mod. Phys.* **B 10**, 1153 (1996).

5. Z. W. Wang, Z. F. Lin and R. B. Tao, *J. Phys.* **D 30**, 1265 (1997).
6. A. J. C. Ladd, *J. Chem. Phys.* **88**, 5051 (1988).
7. L. Lobry and E. Lemaire, *J. Electrostat.* **47**, 61 (1999).
8. J. C. Bacri, R. Perzynski, M. I. Shliomis and G. I. Burde, *Phys. Rev. Lett.* **75**, 2128 (1995).
9. J. T. K. Wan, K. W. Yu and G. Q. Gu, *Phys. Rev.* **E 62**, 6846 (2000).
10. G. Q. Gu and P. M. Hui, *Int. J. Mod. Phys.* **B 15**, 1033 (2001).
11. J. T. K. Wan, G. Q. Gu and K. W. Yu, *Phys. Rev.* **E 63**, 52501 (2001).
12. G. Q. Gu and K. W. Yu, *Commun. Theor. Phys.* **27**, 359 (1997).

FLOW MODIFICATION INDUCED BY QUINCKE ROTATION IN A CAPILLARY

A. CEBERS
Institute of Physics, Salaspils, Latvia,

E. LEMAIRE, L. LOBRY
LPMC, CNRS Université de Nice-Sophia Antipolis, France.

When particles immersed in a semi-insulating liquid are submitted to a sufficiently high DC field, they can rotate spontaneously around any axis perpendicular to the field (Quincke rotation). Recently we have shown that due to Quincke effect the effective viscosity of a colloidal suspension could be reduced. When the suspension is submitted to a shear, the particles rotation is amplified by the electric torque and drives the suspending liquid. For a flow in a capillary, this effect manifests itself by an increase of the flow rate. We present the results of our experiments carried out with a rectangular cross section capillary. These results are compared with the direct determination of the apparent viscosity in a Couette flow rheometer.

1. Introduction

The possibility to decrease the apparent viscosity of a suspension by inducing particles rotation has been yet investigated in magnetic fluids.[1,2] Indeed if a rotating or an alternating magnetic field is applied to a suspension of rigid dipoles, the time dependent field is able to induce particles rotation. The magnetic torque can be either resistant or driving. The former case corresponds to the application of a static or low frequency field. The later is obtained when the field frequency is sufficiently high for the particles to rotate faster than the surrounding liquid. The rotation of the particles favours the liquid motion and thus decreases the apparent viscosity of the suspension. This phenomenon has been predicted and observed experimentally several years ago with a concentrated ferrofluid submitted to an alternating magnetic field. The authors mention a decrease of the apparent viscosity of about 25%,[3] they wonder if the interesting limit of a zero apparent viscosity is attainable. They finally state that, although there is no theoretical limit for the viscosity decrease, the required conditions to obtain zero apparent viscosity are incompatible with the stability of the ferrofluid against aggregation.[4] Nevertheless their result is of great interest for the active damping control since the viscosity can be tuned between 50 and 220 cP by changing the frequency of the applied magnetic field (H=2000 Oe) from 700 Hz to 0. In this paper, we propose another way to induce particle rotation. This technique consists of applying a DC electric field to a suspension of insulating particles dispersed in a slightly conducting fluid. When a non conducting particle immersed in a semi-insulating liquid is submitted to a sufficiently high amplitude DC field, it begins to rotate spontaneously around itself with an axis pointing in any direction perpendicular to DC field. This symmetry break is known as Quincke rotation and comes from the instability of the induced dipole whose direction is opposite to the electric field. The particle begins to rotate in order to flip its dipole in the field direction. The angular displacement of the dipole itself produces a mechanical torque on the particle which tends to increase the displacement further. As a consequence, in stationary state, the particle submitted to the DC electric field rotates at constant angular velocity which is given by the balance of the electric and viscous torque. In the first part, we propose equations for the polarisation relaxation which come from the fundamental equations of electrostatics. Then, introducing the electric torque in Navier Stokes equation, we predict the decrease of the viscosity induced by the DC electric field. We compare these predictions to experimental results obtained with a suspension of PMMA particles dispersed in a low conducting mixture of transformer oil and Ugilec (Elf Atochem) in Couette flow geometry. In the fourth part, we show the influence of Quincke rotation on the flow in a rectangular capillary.

2. Quincke rotation:

To derive the dipole moment of a rotating sphere in a DC electric field three methods are mainly used. Jones considers the sphere at rest in a rotating electric field and calculates the effective polarisation through the complex permittivities method.[5] Melcher deduces the polarisation from the charge distribution at the surface of the particle.[6] In this section, we propose another way to determine the dipole moment which is based on the resolution of the polarisation relaxation equations. These equations are similar to those of the magnetization relaxation in magnetic fluids.[7]

The total polarization of a particle can be divided in two parts: the instantaneous polarization $\vec{P}_\infty$ coming from the permittivity mismatch between the particle and the liquid, and the retarding polarization $\vec{P}$ related to the free charges accumulation on the particle surface. The instantaneous polarization is given by

$$\vec{P}_\infty = \chi_\infty \vec{E} \tag{1}$$

where χ_∞ is the high frequency susceptibility of the particle. For a sphere,

$$\chi_\infty = 4\pi\varepsilon_1 a^3 \frac{\varepsilon_2 - \varepsilon_1}{\varepsilon_2 + 2\varepsilon_1} \tag{2}$$

where a is the radius of the particle and ε_i, the permittivities (index 1 refers to the fluid and 2 to the particle).

$$\frac{d\vec{P}}{dt} = [\vec{\omega}_p \times \vec{P}] - \frac{1}{\tau}(\vec{P} - (\chi_0 - \chi_\infty)\vec{E}) \tag{3}$$

The first term of the right hand side is the rate of change of the polarization due to the particle rotation ($\vec{\omega}_p$ is the angular velocity), the second term comes from the relaxation of the polarization to the equilibrium state with the characteristic time $\tau=(\varepsilon_2+2\varepsilon_1)/(\gamma_2+2\gamma_1)$ (the so-called Maxwell-Wagner relaxation time) and χ_0 is the low frequency susceptibility :

$$\chi_0 = 4\pi\varepsilon_1 a^3 \frac{\gamma_2 - \gamma_1}{\gamma_2 + 2\gamma_1} \tag{4}$$

where γ_i are the conductivities.

Neglecting inertia, the dynamic of the particle is governed by the balance of the viscous and electrical torques :

$$0=-\alpha(\vec{\omega}_p-\vec{\Omega})+[\vec{P}\times\vec{E}] \tag{5}$$

where $\vec{\Omega}$ is the fluid angular velocity and α the rotational friction coefficient of the particle. For a spherical particle $\alpha = 8\pi\eta a^3$ where η is the viscosity of the liquid.
Let us first consider the stationary state of a spherical particle in a liquid at rest ($\vec{\Omega}=0$) submitted to an electric field in the z direction. Combining the relations (3) and (5), we obtain the following equation for the transversal component of the polarization P_y :

$$P_y\left(\frac{(\chi_0-\chi_\infty)\tau E^2}{\alpha}+\left(\frac{\tau E}{\alpha}\right)^2 P_y^2+1\right)=0 \tag{6}$$

Depending on the electric field strength and on the sign of $(\chi_0-\chi_\infty)$, equation (6) has either one or three solutions:
If $(\chi_0-\chi_\infty)>0$: $P_y=0$

If $(\chi_0-\chi_\infty)<0$: $P_y=0$ and $P_y=\pm\frac{\alpha}{\tau E}\sqrt{\frac{E^2}{E_C^2}-1}$ if $E>E_C=\sqrt{\frac{-\alpha}{\tau(\chi_0-\chi_\infty)}}$. (7)

Where E_c is the critical field below which the driving electric torque is always lower than the resistant viscous one, small though the angular velocity my be. For applied fields higher than E_c, the spontaneous rotation of the dielectric spheres is possible if $(\chi_0-\chi_\infty)<0$ what corresponds to the following condition : $\frac{\sigma_2}{\varepsilon_2}<\frac{\sigma_1}{\varepsilon_1}$.

Non-zero value of the transversal component of the polarization corresponds to the rotation of the particle with an angular velocity ω_p in the direction perpendicular both to the E field and to the transverse polarisation :

$$\omega_p=\pm\frac{1}{\tau}\sqrt{\frac{E^2}{E_C^2}-1} \tag{8}$$

It should be stressed that all the directions perpendicular to the E field are possible axes of rotation and finally the axis of rotation is chosen randomly in the plane perpendicular to the electric field. Otherwise, for a spherical particle the dipolar moment and the frictional coefficient have the same dependence on the radius, the angular velocity is independent on the size of the particle.
In our experimental conditions, we obtain $\tau\approx 1$ms and $E_c\approx 2$kV/mm. And, if a field of few E_c is applied, the particle spin rate will be about few hundreds of radians per second which is also the order of magnitude of the angular velocity of a particle in an ordinary shear flow. This observation suggests that Quincke rotation will significantly influence the rheological behaviour of a suspension.

3. "Negative" viscosity effect

Let us consider an electrorheological suspension in a simple shear flow. The particles rotate around themselves along the vorticity direction, if a DC electric field is applied in the velocity gradient direction, the spin rate of the particles will increase and become higher than the suspending liquid angular velocity. Consequently, the particles will drive the surrounding liquid and the apparent viscosity of the suspension will decrease. Tacking into account the electrical torque on the particles, the average shear stress in the suspension is given by :[8]

$$\sigma_{yz}=\eta_s\dot{\gamma}+\frac{1}{2}nP_yE \tag{9}$$

where η_s is the shear viscosity of the suspension without any applied electric field and n the number of particles per unit of volume. $\dot{\gamma}$ is the average shear rate in the suspension given by the relation :

$$\dot{\gamma}=2\left(\phi\omega_p+(1-\phi)\Omega\right) \tag{10}$$

Combining equations (3), (5) and (9), we obtain the rheological equation of state :

$$\begin{aligned}&\frac{1}{2}E^{*2}\dot{\gamma}^*+\left(\frac{2\eta_s}{n\alpha}\right)\left(1-E^{*2}(1-\phi)+\frac{1}{4}\dot{\gamma}^{*2}\right)\left(\frac{2\eta_s}{n\alpha}\right)(\sigma^*-\dot{\gamma}^*)-\dot{\gamma}^*(1-\phi)\left(\frac{2\eta_s}{n\alpha}\right)^2(\sigma^*-\dot{\gamma}^*)^2\\&+(1-\phi)\left(\frac{2\eta_s}{n\alpha}\right)^3(\sigma^*-\dot{\gamma}^*)^3=0\end{aligned} \tag{11}$$

where the variables $\dot{\gamma}^*$, σ^* and E^* are dimensionless : $\dot{\gamma}^*=\dot{\gamma}\tau$, $\sigma^*=\frac{\sigma\tau}{\eta_s}$ and $E^*=\frac{E}{E_c}$.

The resolution of this equation gives three solutions for the value of the apparent viscosity. Two of them are higher than the viscosity without any applied electric field; they correspond to an axis of rotation of the particles in the direction opposite to the vorticity. We have shown elsewhere that these solutions were unstable and that the rotation of the particles in the vorticity direction remained the only stable state.[9] The apparent viscosity of a suspension containing 15% of particles, all rotating in the vorticity direction, is drawn in figure 1 for different normalized electric field strengths. As expected the apparent viscosity is lower than the viscosity in absence of an electric field. It should also be stressed that for sufficiently high fields the viscosity can become negative. The field strength for which the negative viscosity appears is directly given by equation (11) when the applied shear stress tends to zero. This value, $E_{neg}=\sqrt{\frac{E_c^2}{1-\phi+\frac{n\alpha}{4\eta}}}$, is

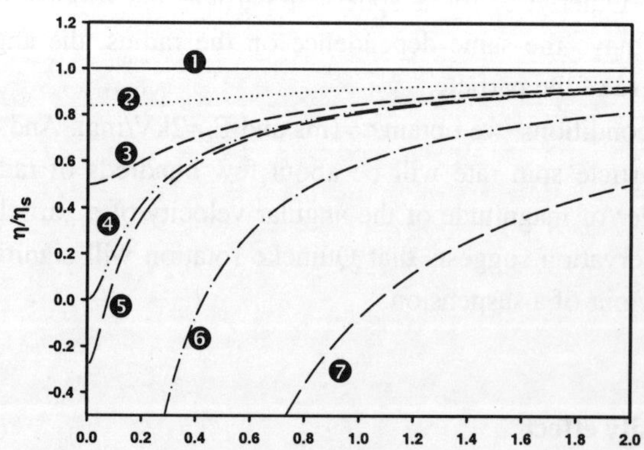

Fig. 1. Theoretical viscosity versus the shear rate for different values of the normalized electric field, E/E_c :
❶: 0, ❷: 0.7, ❸: 0.9, ❹: 0.98, ❺: 1, ❻: 1.5, ❼: 3

lower than E_c, the critical field at which an isolated particle begins to rotate. This observation can be explained by the influence of the vorticity creating by the arising collective rotation of the particles.

These theoretical predictions have been compared to experimental results obtained with a suspension of PMMA particles (diameter: 100 μ, ϕ=0.15, ε_2=2.6) in a mixture of transformer oil and Ugilec (Elf Atochem) (ε_1=5.6). The conductivity of the suspending liquid is controlled by adding AOT salts and is always much higher than that of the particles which can consequently be considered as insulating. The rheological measurements have been carried out with a controlled stress rheometer (Carrimed CSL 100) in concentric cylinders geometry. The gap between the cylinders is sufficiently small to approximate the flow by a simple shear flow ((R1-R2)/R2=0.08). The measurements have been performed decreasing the applied shear stress from its maximum value (~5Pa) to zero. The qualitative agreement is rather good, especially at high shear rates (Figure2). When the shear stress is decreased to low values, a discrepancy between experimental and theoretical behaviors appears. This may have different origins either practical or more fundamental. Indeed, the rheometer we used is not designed to measure so low viscosities ($\eta \leq 1$cP) and the measurements performed in that region are very inaccurate. From a more fundamental point of view, in the model, we have not taken into account the attractive dipolar interactions between particles which can be responsible for the formation of fibrillated structures. The formation of such structures remains possible even if the applied field is higher than E_c, the critical field at which an isolated particle begins to rotate. Indeed, we have shown that an anisotropic particle which can represent a chain of spherical particles, began to rotate for electric even more intense that the aspect ratio of the particle is large.[10] Consequently, if, because of dipolar interaction, the particles have formed chains, the suspension may behave like a classical ER suspension exhibiting a yield stress. Nevertheless, the Quincke rotation is a good way to reduce drastically the apparent viscosity of a suspension and, in the following section, we will be interested in the consequences of this viscosity reduction on the flow of the suspension through a capillary.

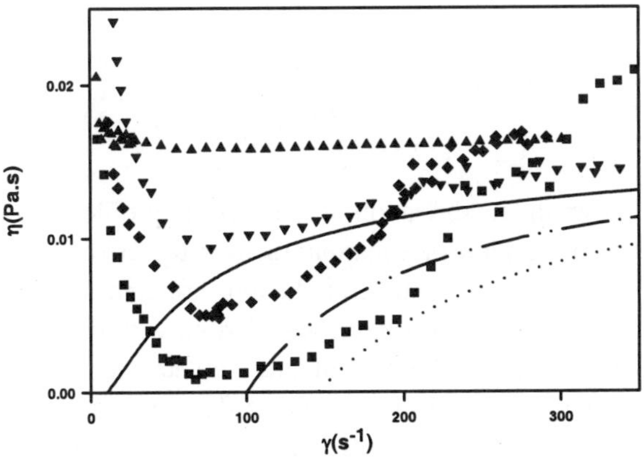

Fig. 2. Apparent viscosity of a suspension of PMMA particles in a low conducting liquid (γ=1.8 10^{-8}S/m) for E=0 (▲), E=1570V/m(▼,—), E=1960V/m (■,—..—) and E=2350V/m (◆, ….).

4. Capillary flow

We have studied the flow of the suspension in a rectangular channel when submitted to a DC electric field. Considering the dimensions of the capillary (length: L= 30 cm, height : h=750μm and width : l=1.5cm), the flow is assumed to be two dimensional ($\vec{v}=v_y\vec{e}_y$; $\vec{\nabla}v_y=\frac{\partial v_y}{\partial z}\vec{e}_z$). The electric field is applied along the shortest dimension of the channel and the pressure drop across the capillary is controlled by changing the altitude of the suspension reservoir. For various electric field strengths and pressure differences, the rate of flow, Q, is measured with an electronic balance at the outlet of the capillary.

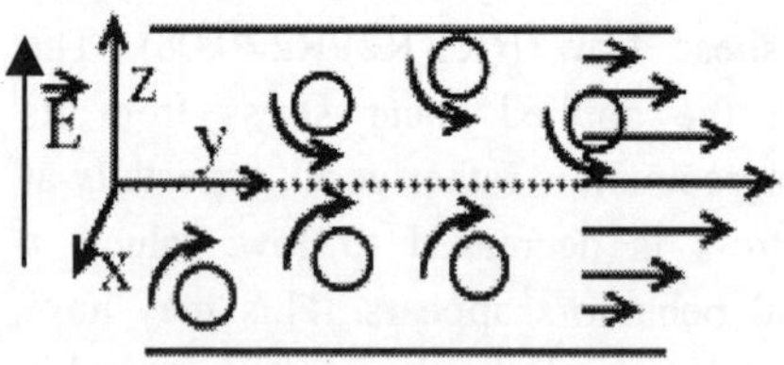

Fig. 3. Schematic representation of the rotation of the particles in the capillary

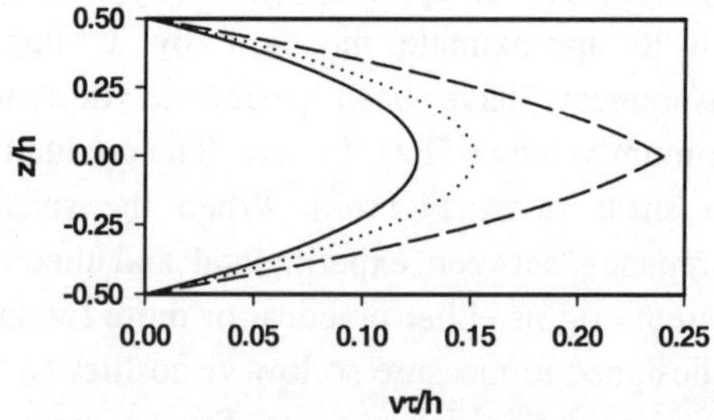

Fig. 4. Velocity profiles for different electric fields, E/E_c=0——— , E/E_c=0.8··········· , E/E_c=1.2 - - - - -

When the E field is applied, we observe a great increase of the rate of flow of the suspension. As in simple shear flow, we assume that all particles rotate with an axis along the vorticity direction and the rotation direction changes at the middle plane of the channel (figure 3). In the presence of the electric field, the behavior of the suspension is non newtonian (figure 1) and, knowing its rheological equation of state $\sigma = f(\dot{\gamma})$, we can deduce the velocity profile between the walls of the capillary. The equation of momentum can be expressed as:

$$\sigma = z\frac{\partial p}{\partial y} \tag{12}$$

and the integration of $\dot{\gamma} = \frac{\partial v_y}{\partial z}$ between z=h/2 at the wall and the coordinate z leads to :

$$v_y(z)=\int_z^{h/2} f\left(\varsigma\frac{\partial p}{\partial y}\right)d\varsigma \tag{13}$$

Figure 4 shows the modification of the velocity profile induced by the rotation of the particles. The integration of equation 13 over the cross section of the channel gives the rate of flow. These theoretical predictions are compared to the experimental results (figure 5). These experiments have been carried with the suspension of PMMA particles in the mixture of transformer oil and Ugilec with a conductivity of the suspending liquid slightly higher than the Couette flow experiment in order to enhance the effect of the electric field : γ=2.4 10^{-8} S/m. Figure 5 shows that the predictions overestimate the effect of Quincke rotation on the increase of the flow rate. This

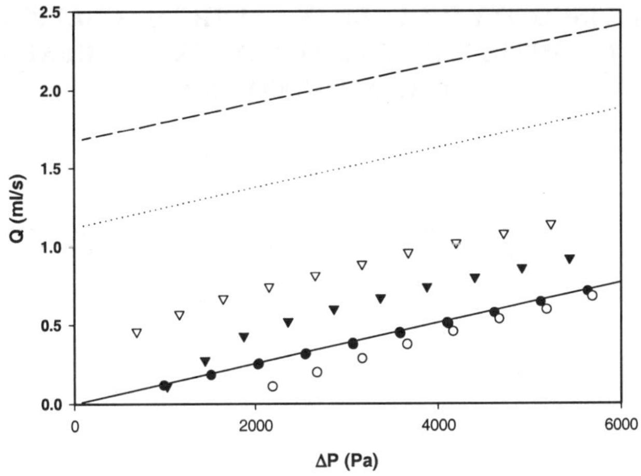

Fig. 5. Flow rate versus pressure drop for different electric fields: E=0 ●——, E=1.6kV/mm ○, E=2.1kV/mm ▼......, E=2.6 kV/mm ▽----- . The symbols correspond to experimental results and the lines to the theoretical predictions.

observation is in agreement with the results we have obtained in the Couette flow experiments where, at low shear rate, the measured apparent viscosity was much higher than the predicted one. We had explained this feature by the possible chain structuration of the particles. This argument could also explain the discrepancy observed in the present experiment especially since the shear stress remains low at the capillary center. In a future paper, we will show that the account of dipolar interaction between particles leads to a reduction of the flow rate.

References

1. M. I. Shliomis, Sov. Phys. JETP, **34**, 1291 (1972)
2. R. E. Rosensweig, Science, **271**, 614 (1996)
3. J.-C. Bacri, R. Perzynski, M. I. Shliomis, G. I. Burde, Phys. Rev. Lett., **75**, 2128 (1995)
4. M. I. Shliomis and K. I. Morozov, Phys. Fluids, **6**, 2855 (1994)
5. T. B. Jones, IEE Trans. Educ., IA **20**, 845 (1984)
6. J. R. Melcher, IEE Trans. Educ., **E 17**, 100 (1974)
7. A.Cebers, Mechanics of liquid and gas, **2**, 86 (1980) (in Russ.)
8. A.Cebers, E. Lemaire and L. Lobry, Magnetohydrodynamics, **36**, 347 (2000)
9. E. Lemaire and L. Lobry, J. Electrostatics, **47**, 61 (1999)
10. A.Cebers, E. Lemaire and L. Lobry, Phys. Rev. E, **63**, 016301 (2001)

NEGATIVE VISCOSITY INDUCED IN A FERROMAGNETIC COLLOIDAL SYSTEM SUBJECTED TO BOTH EXTERNAL SHEAR FLOW AND AC MAGNETIC FIELDS

HISAO MORIMOTO

Department of Mechanical Engineering, The University of Tokyo (JSPS Research Fellow)
7-3-1 Hongo, Bunkyo-ku, Tokyo, 113-8656, Japan

TORU MAEKAWA

Bio-Nano Electronics Research Centre, Toyo University
2100, Kujirai, Kawagoe, Saitama 350-8585, Japan

We investigate numerically the effect of an external ac magnetic field on the rheological characteristics of a ferromagnetic colloidal system which is subjected to a shear flow field. We carry out Brownian dynamics simulations, in which both translational and rotational motions of ferromagnetic particles are taken into account. The nondimensional parameters, which govern the problem, are the ratio of the magnetic dipole energy to thermal energy, λ, the ratio of the interactive energy between the magnetic field and the dipole moment to thermal energy, ξ, and the ratio of the shear rate to the viscosity, *Pe*, which is called the Peclet number. We analyse the dependence of the apparent viscosity on the frequency of the external magnetic field for $\lambda = 1.5$, $\xi = 2$ and $Pe = 1$. We find that when λ is as small as 1.5, in which case large clusters are not formed in the system, the apparent viscosity increases at low frequencies and becomes almost zero at high frequencies. The viscosity starts decreasing and becomes negative at intermediate frequencies.

1. Introduction

The rheological characteristics of ferromagnetic colloidal systems are more complicated than those of paramagnetic colloidal systems since a ferromagnetic particle is composed of a permanent dipole moment and therefore has a degree of freedom of rotation as well as translation, which generates asymmetric components in the stress tensor. The increase in the apparent viscosity caused by the asymmetric components of the stress tensor was estimated theoretically by Shliomis [1] and the estimated value agreed well with the experimental results [2]. It is also observed that negative viscosity and vortico-magnetic resonance are induced by the rotational motions of ferromagnetic particles which are subjected to both shear flow and ac magnetic fields [3,4]. However, the mechanism which causes negative viscosity has not yet been studied in detail.

In this paper, we focus on the apparent viscosity caused by the rotational motions of ferromagnetic colloidal particles which are subjected to both shear flow and ac magnetic fields. We develop a Brownian dynamics calculation method and investigate the effect of the frequency of the external ac magnetic field on the apparent viscosity of the ferromagnetic colloidal system.

2. Numerical modelling and calculation method

Ferromagnetic particles are dispersed in the solvent. Each particle is composed of a permanent magnetic dipole moment and covered with surfactant molecules (see Fig.1). The calculation system is shown in Fig.2. The system is two-dimensional and an external magnetic field is applied in the y-direction. Particles are subjected to a shear flow field. The interactions between particles are expressed by a dipole-dipole potential, van der Waals potential and repulsive potential caused by the surfactant-surfactant contact, which are summarized in Ref. [5].

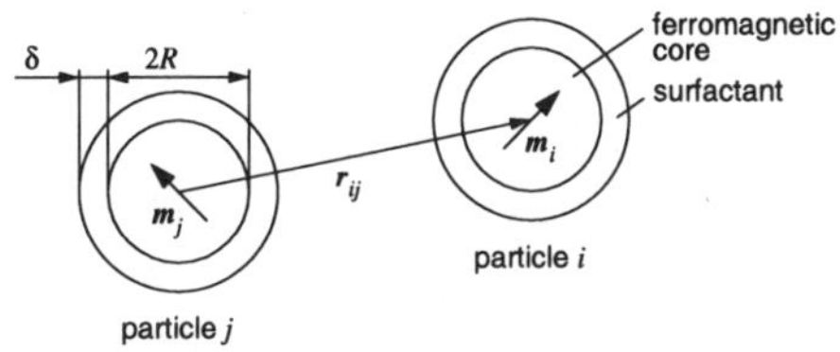

Fig. 1. Ferromagnetic particles.

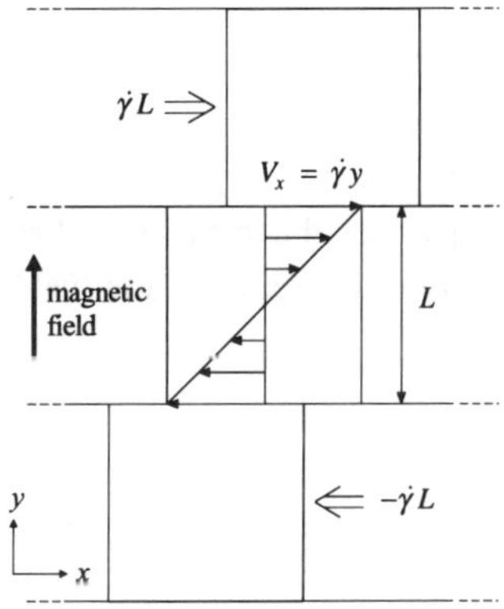

Fig. 2. Calculation system. Ferromagnetic particles are dispersed in the solvent. A shear flow is imposed in the x-direction and an external magnetic field is applied in the y-direction.

The translational and rotational motions are expressed by the following Langevin equations:

$$\frac{\partial \boldsymbol{v}}{\partial t} = -\frac{\zeta_t}{m}\left(\boldsymbol{v} - \boldsymbol{V}\right) + \frac{\boldsymbol{F}}{m} + \boldsymbol{P}, \tag{1}$$

$$\frac{\partial \boldsymbol{\omega}}{\partial t} = -\frac{\zeta_r}{I}\left(\boldsymbol{\omega} - \boldsymbol{\Omega}\right) + \frac{\boldsymbol{T}}{I} + \boldsymbol{\tau}, \tag{2}$$

where t, $\boldsymbol{v}$, $\boldsymbol{\omega}$, m and I are, respectively, the time, the velocity, the angular velocity, the mass and the moment of inertia of a particle. $\boldsymbol{F}$ and $\boldsymbol{T}$ are the force and torque acting on a particle through the interparticle potentials. $\boldsymbol{P}$ and $\boldsymbol{\tau}$ are the random force and torque caused by the solvent molecules. ζ_t and ζ_r are the friction coefficients. Here, we assume that Stokes law applies and the particles' motions do not induce any flow in the solvent. $\boldsymbol{V}$ and $\boldsymbol{\Omega}$ are the external velocity and angular velocity:

$$\boldsymbol{V} = \begin{pmatrix} \dot{\gamma} y \\ 0 \\ 0 \end{pmatrix}, \quad \boldsymbol{\Omega} = \frac{1}{2} curl\, \boldsymbol{V} = \begin{pmatrix} 0 \\ 0 \\ -\dot{\gamma}/2 \end{pmatrix}, \tag{3}$$

where $\dot{\gamma}$ is the shear rate. Moving boundary conditions are employed [6] and the Langevin equations are integrated following Ermak's procedure [7]. After the system has reached the steady-state in the absence of a shear flow field, a shear flow is imposed. The stress tensor, σ_{yx}, and the apparent viscosity, $\Delta\eta$, are expressed as follows:

$$\sigma_{yx} = -\frac{1}{V}\sum_i \left(y_i F_{xi} - \frac{T_z}{2} \right), \tag{4}$$

$$\Delta\eta = \sigma_{yx} / \dot{\gamma} = \Delta\eta_t + \Delta\eta_r . \tag{5}$$

The first and second terms on the right-hand side of Eq.(4) represent symmetric and asymmetric components, respectively. The apparent viscosity is induced by the symmetric and asymmetric components of the stress tensor, which are denoted by $\Delta\eta_t$ and $\Delta\eta_r$. The control parameters are the ratio of the magnetic dipole moment energy to thermal energy, λ, the ratio of the interactive energy between the magnetic dipole moment and the external magnetic field to thermal energy, ξ, and the Peclet number, *Pe*:

$$\lambda \equiv \frac{|\boldsymbol{m}|^2}{4\pi\mu_0 d^3 kT}, \quad \xi \equiv \frac{|\boldsymbol{m}|H}{kT}, \quad Pe \equiv \frac{3\pi\eta_0 (d+2\delta)^3 \dot{\gamma}}{kT}, \tag{6}$$

where $\boldsymbol{m}$, μ_0, d, δ, k, T, H and η_0 are, respectively, the magnetic dipole moment of a particle, the permeability, the diameter of a particle, the thickness of the surfactant layer, the Boltzmann constant, temperature, the intensity of the external magnetic field and the viscosity of the solvent.

3. Result and discussion

We carried out calculations using 121 particles. The nondimensional number density of particles, Nd^2/L^2, where N and L are the number of particles and the length of the calculation cell side, was fixed at 9.77×10^{-2}. The values of the control parameters are as follows: $\lambda = 1.5$ and $Pe = 1.0$. ξ was changed from 0 to 30 in the case of a dc magnetic field, whereas ξ was fixed at 2 in the case of an ac magnetic field. In such a small λ case, ferromagnetic particles do not form clusters. First, we investigate the relation between the apparent viscosity and the intensity of the external dc magnetic field. The dependence of the apparent viscosity, which is normalized by the solvent viscosity, on the intensity of the dc magnetic field is shown in Fig.3. The apparent viscosity caused by the asymmetric components of the stress tensor, $\Delta\eta_r$ (see Eqs.(4) and (5)), increases with an increase in ξ up to $\xi = 5$ and becomes constant for $\xi > 5$. From Eq.(2), the apparent viscosity, $\Delta\eta_r$, at the steady-state is estimated as follows:

$$\frac{\Delta\eta_r}{\eta_0} = 2\pi R^3 n^* \left(1 - \frac{\langle \omega_z \rangle}{\Omega_z} \right), \tag{7}$$

where R and n^* are, respectively, the radius of a particle and the number density of particles. $\langle\ \rangle$ represents the average with respect to the total number of particles. Note that, as we mentioned, clusters are not formed when λ is as small as 1.5. When the dc magnetic field is very strong, that is $\xi \gg 1$, $\langle \omega_z \rangle \to 0$, since the rotational motions are stopped by the magnetic field. The apparent viscosity, $\Delta\eta_r$, approaches the estimated value;

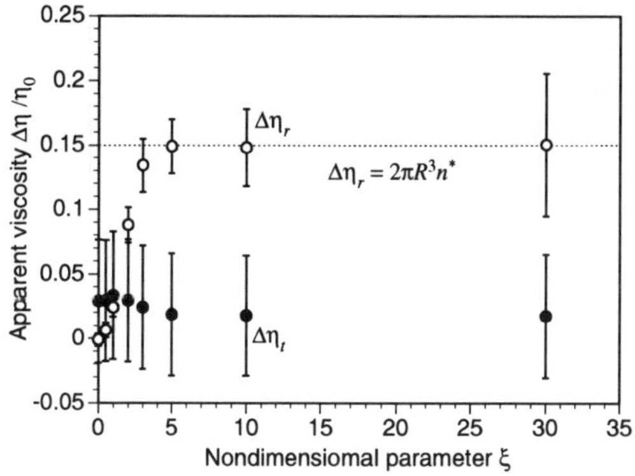

Fig. 3. Dependence of apparent viscosity on the intensity of a dc magnetic field. $\lambda = 1.5$ and $Pe = 1.0$.

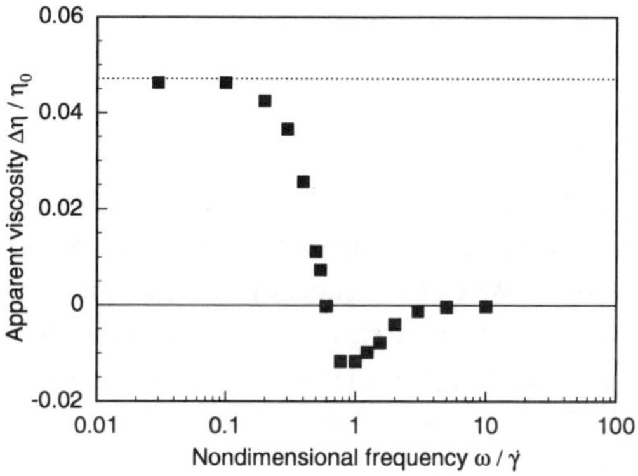

Fig. 4. Dependence of apparent viscosity on the frequency of an ac magnetic field. $\lambda = 1.5$, $\xi = 2.0$ and $Pe = 1.0$.

$\Delta\eta_r/\eta_0 = 2\pi R^3 n^*$, when $\xi > 5$ (see Fig.3). The apparent viscosity caused by the symmetric component of the stress tensor, $\Delta\eta_t$, on the other hand, is very small compared to $\Delta\eta_r$. Note that $\Delta\eta_t$ becomes large when chain clusters are formed through strong particle-particle interactions.

Now, we investigate the effect of the frequency of the external ac magnetic field on the apparent viscosity. The dependence of the apparent viscosity on the frequency is shown in Fig.4 where the angular frequency of the external magnetic field, which was normalized by the shear rate, $\omega/\dot{\gamma}$, is changed from 0.02 to 10 and ξ is fixed at 2. When the frequency is low, the apparent viscosity increases and approaches the constant value, 0.047, which is estimated from the $\Delta\eta_r - \xi$ curve in the dc magnetic field (see Fig.3). When the frequency is high, there is no effect of the magnetic field on the apparent viscosity since the particles' rotational motions cannot follow the magnetic field change. However, when the frequency is in the intermediate range, the apparent viscosity becomes negative. As these phenomena should be correlated with the rotational motions of the ferromagnetic particles, we investigated the time variations of the rotational angles of the particles, which is shown in Fig.5. The angles are averaged with respect to the total number of particles and the gradient of the curve represents the mean angular velocity of the particles. Basically, when the direction of the dipole is towards $x > 0$, the rotational motion caused by the external shear flow is

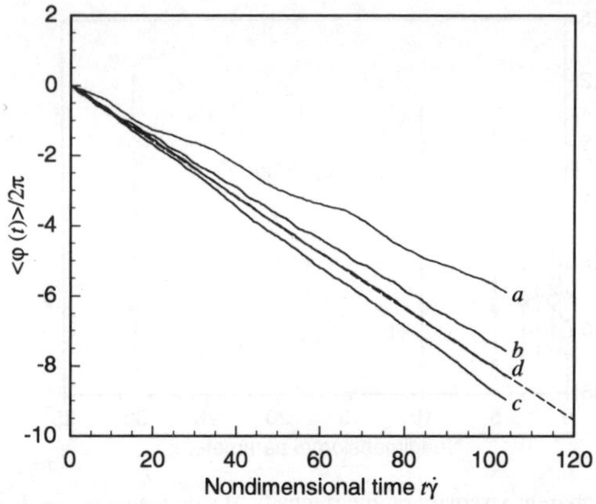

Fig. 5. Time variations of rotational angles of ferromagnetic particles. λ= 1.5, ξ =2.0 and Pe = 1.0. Solid lines *a*-*d* show the results for $\omega/\dot{\gamma}$ = 0.1, 0.5, 1.0 and 10, respectively. The slope of each curve gives the mean drift angular velocity of a particle and the slope of the broken line in the figure corresponds to the angular velocity of the solvent fluid.

decelerated by the magnetic field, whereas the rotational motion is accelerated by the magnetic field when the dipole is pointing to $x < 0$. When the frequency of the magnetic field is low ($\omega/\dot{\gamma} = 0.1$), the mean angular velocity of the particles is lower than the shear rate, which causes a high apparent viscosity. When the frequency of the magnetic field is high ($\omega/\dot{\gamma}$ = 10), the mean angular velocity of the particles coincides with the shear rate and therefore the increase in the apparent viscosity is zero. However, in the intermediate frequency range of the external magnetic field, the angular velocity of the particles becomes higher than the shear rate, which causes negative viscosity.

4. Conclusions

We studied the rheological characteristics of a ferromagnetic colloidal system, which is subjected to both shear flow and ac magnetic fields, numerically for a weak particle-particle interaction case. We investigated the effect of the external ac magnetic field on the apparent viscosity of the system and obtained the following results: (a) The apparent viscosity increases when the frequency of the magnetic field is low. (b) There is no effect of the magnetic field on the apparent viscosity when the frequency is high. (c) The apparent viscosity becomes low and negative in an intermediate frequency range. The mean angular velocity of particles exceeds the shear rate in the negative viscosity regions.

In the present analysis, we have focused on a weak particle-particle interaction case (λ = 1.5). We are going to study strong interaction cases, which have not yet been investigated either experimentally or numerically. The interaction between long chain clusters and the external ac magnetic field may cause different rheological characteristics.

Acknowledgements

This study has been supported by the High-Tech Research Centre Fund organized by the Ministry of Education, Science, Culture and Sports, Japan since 1996 and by the Toyo University Special Fund since 1999.

References

1. M.I. Shliomis, *Soviet Phys. JETP* **34**, 1291 (1972).
2. J.P. McTague, *J. Chem. Phys.* **51**, 133 (1969).
3. J.-C. Bacri, R. Perzynski, M.I Shliomis and G.I. Brude, *Phys. Rev. Lett.* **75**, 2128 (1995).
4. F. Gazeau, B.M. Heegaard, J.-C. Bacri, A. Cebers and R. Perzynski, *Europhys. Lett.* **35**, 609 (1996).
5. H. Morimoto and T. Maekawa, *J. Phys. A*, **33**, 247 (2000).
6. A. Lees and S.F. Edwards, *J. Phys. C*, **5**, 1921 (1972).
7. D.L. Ermak and H. Buckholz, *J. Comput. Phys.*, **35**, 169 (1980).

NONHOMOGENEOUS, NON-LINEAR DEFORMATION OF POLYMER GELS SWOLLEN WITH MAGNETO-RHEOLOGICAL SUSPENSIONS

DÉNES SZABÓ AND MIKLÓS ZRÍNYI

Department of Physical Chemistry, Budapest University of Technology and Economics
H-1521 Budapest, HUNGARY
E-mail: d_szabo@chem.bme.hu

Polymer gels swollen with a magneto-rheological suspension are highly elastic materials with considerable magnetic susceptibility. In this work the magnetic field induced deformation and motion of these magnetic polymer systems is discussed. We present a continuum material model by introducing magnetic equations into non-linear elasticity theory. The material properties of these magnetic rubber-like substances are characterized with a Langevin type magnetization and a neo-Hookean strain energy function. The non-linear character of the equations that describe the material properties and the nonhomogeneity of the deformation lead to a unique deformation mechanism. In order to demonstrate the characteristics of the magnetic field induced deformations we present numerical and finite element calculations and compare them with experimental results.

1 Introduction

Novel magnetic- and electric field sensitive polymeric materials have been developed recently. [1-3] In these polymer gels fine nanoparticles are embedded into a polymer network resulting in a highly elastic, field sensitive material. In this paper we deal with polymers containing magnetic particles. These so-called ferrogels are superparamagnetic with a considerable magnetic susceptibility of the order 0.01-0.1. Ferrogels are capable of undergo giant deformations (the strain can reach 1.5) in a magnetic field with ordinary strength.

Our goal is to understand and describe the deformation of ferrogels in magnetic field. To accomplish this we need to introduce magnetic equations into the non-linear elasticity theory. Magnetic interactions are not usual in elasticity problems. Since the field, and the force alike, might have a spatial distribution different material points experience force of different strength and direction, which leads to a complex, nonhomogeneous deformation. Not only the nature of the resulting deformation, but also the treatment of body forces is different from that of surface tractions.

In case of a highly elastic magnetic medium the non-linear character of both elastic and magnetic interactions lead to some very novel features of the deformation process. The non-linearity of the governing interactions and the spatial distribution of the force density make the treatment fairly complicated and result in deformations with some unique features as well as with special patterns never seen in deformations induced by surface tractions. In the following chapter we discuss the mechanism of magnetic field induced deformations. Then, we briefly outline the continuum material model for ferrogel, a rubber elastic, superparamagnetic medium. In order to compare the model with real experiments we show the solution of the equations of motion in a simple, one-dimensional situation and compare it with experimental results. Finally, one- and three-dimensional finite element calculations are shown in order to illustrate the real deformation of a ferrogel sample.

2 Deformation induced by magnetic field

The ferrogel is a polymer network swollen with a MR suspension. Most of the magnetic properties of ferrogels are similar to that of ferrofluids. They are superparamegnetic with a fast magnetic relaxation. The main difference is that the particles cannot move. Their motion is restricted by the polymer matrix. In uniform magnetic field, the moments of the particles rotate towards the direction of the field, but cannot form chains or alike. The ferrogel gets magnetised and its elastic modulus changes depending on the strength of the magnetising field. [4] It experiences no net force and does not get deformed. However, when it is placed into an inhomogeneous magnetic field, the particles experience forces. These magnetic forces acting on the individual particles generate a stress distribution, which becomes balanced by the network elasticity, owing to the strong adhesive interactions between chains and particles. Due to the cross-linking bridges and entanglements in the network, changes in molecular conformation can accumulate and lead to macroscopic deformation or – if the external field changes in time - motion. The magnetic field drives and controls the deformation or motion, and the momentary equilibrium shape is determined by the balance between magnetic and elastic interactions.

As an example to the nonhomogeneous deformation of ferrogels in magnetic field we present the deformation of a magnetic gel sheet in Fig 1. Picture 'A' and 'B' show the gel in undeformed and deformed states, respectively. The complexity and nonhomogeneity of the deformation is well seen on the transformation of the grid painted on the surface of the gel.

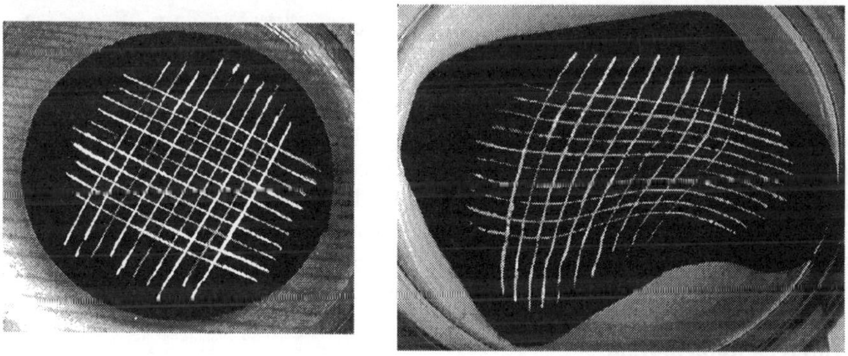

A B

Fig.1: Illustration of the complex nonhomogeneous deformation of a ferrogel sheet. Undeformed (A) and deformed (B) states. The magnetic field was induced by permanent magnets placed around the gel.

3 Mechanical model for ferrogels

The mechanical behaviour of the gel is assumed to be hyperelastic, thus exists a strain-energy function W for which the neo-Hooke law is a good approximation:

$$\mathrm{W} = \frac{G}{2}\left(\lambda_1^2 + \lambda_2^2 + \lambda_3^2 - 3\right) \tag{1}$$

where the material parameter G is the initial shear modulus. For incompressible materials, such as ferrogels, the equation of the constraint $\lambda_1\lambda_2\lambda_3 = 1$ is also needed.

The mechanical balance equations written with the nominal stress are

$$\operatorname{Div}\mathbf{t}(\mathbf{X}) + \mathbf{b}_o(\mathbf{X},t) = \mathbf{0}, \tag{2}$$

The vector $\mathbf{b}_o$ is the body-force written with the terms of the initial configuration. On a continuous magnetic material due to the applied magnetic field it can be written as

$$\mathbf{b}(\mathbf{x},t) = (\operatorname{grad}\mathbf{M}(\mathbf{x},t))\mathbf{H}(\mathbf{x},t). \tag{3}$$

The vector $\mathbf{H}(\mathbf{x},t)$ denotes the magnetic field strength and $\mathbf{M}(\mathbf{x},t)$ represents the magnetization of the material. The ferrogel shows superparamagnetic material behavior. Its magnetization follows the one-valued equation [5]

$$\mathbf{M}(\mathbf{x},t) = \Phi_m M_o \alpha(\xi)\frac{\mathbf{H}}{\mathrm{H}}, \tag{4}$$

with the Langevin function $\alpha(\xi)$. The parameter Φ_m stands for the volume fraction of the magnetic particles in the gel, M_o represents the saturation magnetization of the material of the particles, m is the magnetic moment of an individual particle and equals to $M_o V_m$ in the case of a mono-domain particle, where V_m is the volume of the particle. At moderate magnetic fields linearized relation can be assumed between the magnetization and magnetic field

$$\mathbf{M} = \mu_o \chi \mathbf{H}, \tag{5}$$

where μ_o represents the permeability of the vacuum. This simplification results in a simplified expression for the magnetic body-force, namely

$$\mathbf{b}(\mathbf{x},t) = \mu_o \chi (\operatorname{grad}\mathbf{H})\mathbf{H}. \tag{6}$$

Substitution the constitutive relation into the equilibrium equation (eq.2), we obtain a set of three partial differential equations of the second order for the coordinates of the material points. If the material is incompressible, the hydrostatic pressure field p must be also calculated from the incompressibility constraint. For time-dependent problems, boundary conditions and initial conditions for the displacement and velocity fields are necessary. For quasi-static analysis only boundary conditions should be defined.

4 Uniaxial deformation

Because of the relatively complex form of the Langevin-type magnetization and the magnetic force density it is not possible to achieve analytical solution for Eq.2 even in simple magnetic field distributions. In order to reveal the characteristics of the magnetic field induced deformations of a magnetic continuum it is worth to examine simple one-dimensional situations.

Let us consider a very simple physical situation Long and thin ferrogel cylinder is suspended in water vertically. The axis of the gel cylinder (z) is parallel with the magnetic field and its gradient. In this case, the deformation of the gel is uniaxial and can be considered as one-dimensional. In this situation eq.2 results in the following second-order, non-linear ordinary differential equation:

$$G\left(\frac{d^2u_z(Z)}{dZ^2} + \frac{2}{(du_z(Z)/dZ)^3}\frac{d^2u_z(Z)}{dZ^2}\right) + M(u_z(Z))\frac{dH(u_z(Z))}{dZ} = 0 \tag{7}$$

Here $u_z(z)$ means the displacement of the material point Z along the z axis. The relevant boundary conditions are: $u_z(0)=0$, and $t(Z_m)=0$. Here 0 and Z_m represents the position of the top and bottom end of the gel, respectively.

In our one-dimensional experiments the ferrogel cylinder was elongated by an inhomogeneous magnetic field induced by a solenoid-based electromagnet that was placed under the gel. The distribution of the field was determined along the axis of the gel by a teslameter.

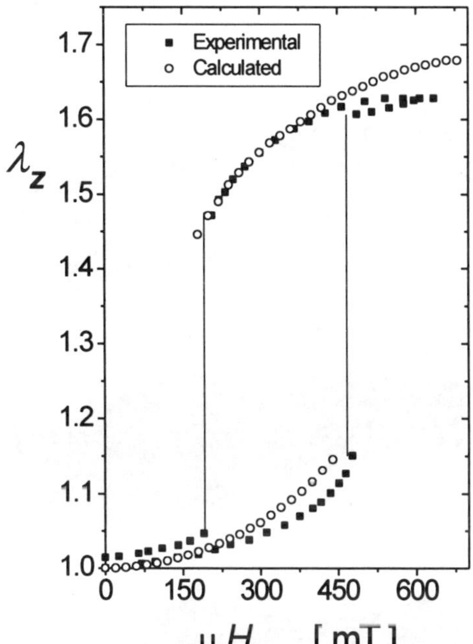

Fig.2: Uniaxial elongation of a ferrogel cylinder. The points represent the displacement of the bottom end of the gel. The blank points were calculated on the basis of Eq.7.

A typical result of the uniaxial experiments can be seen in Fig.2, where the displacement of the bottom end of the gel cylinder is plotted against the current intensity. It may be seen that a non-continuous change in the length appears within an infinitesimal change in the current intensity. While decreasing the current, another non-continuous size transformation occurs. The unique property of ferrogel's magnetoelasticity is the non-continuity, which is due to the non-linearity of the magnetic interactions. In Fig.2 we have also plotted the numerical solution of eq.7. The points fit quite well to the measured ones indicating the continuum model holds for ferrogels.

5 Finite element calculation

As mentioned previously, the analytical determination of the displacement field is generally extremely complicated; therefore a numerical procedure is necessary to calculate the deformation. Nowadays the finite element method (FEM) is prevalent in

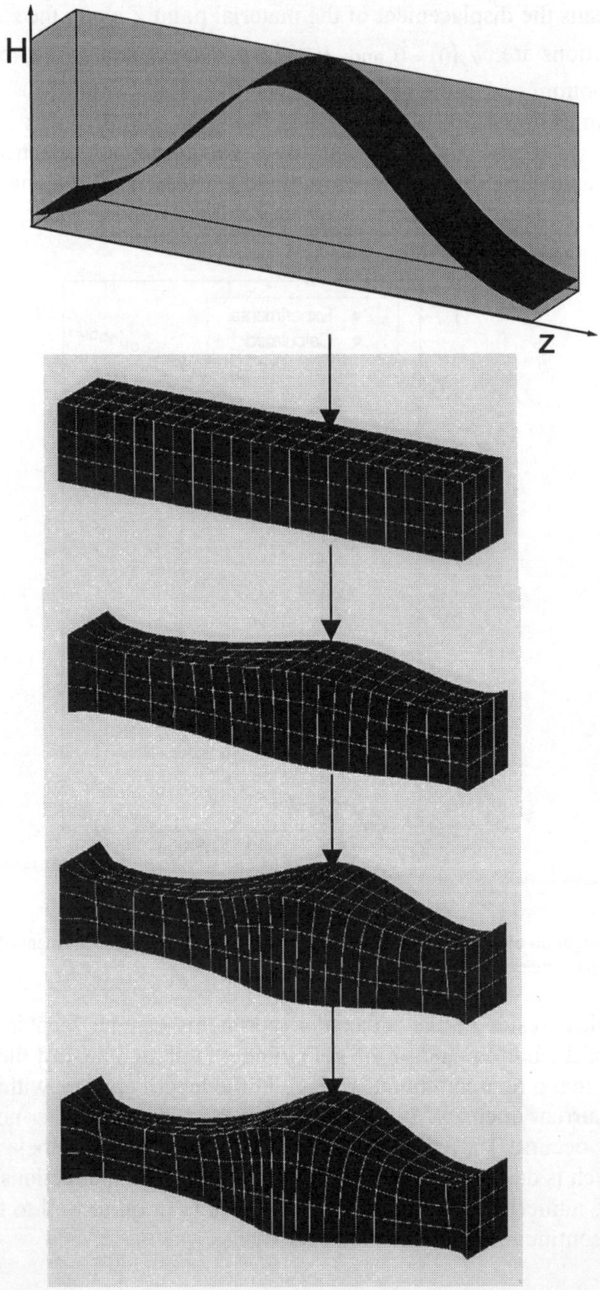

Fig.3: FEM calculation of the deformation of a ferrogel block in a uniaxial magnetic field. The field distribution is shown at the top. The arrows indicate the maximum field strength.

solid mechanics and provides a useful tool to study non-linear deformations in magnetic field. In what follows, we expose the results of our FEM procedure on ferrogels. With FEM fully 3D analysis can also be performed. In theory any arbitrary experimental situation can be modelled and analysed. As an illustrative example, in Fig.3 we present the deformation of a ferrogel block in a uniaxial field parallel with the gel axis. Looking at the set of pictures one might associate with the motion of a living worm. And indeed, the deformation and motion of ferrogels has a close relation with the one of simple living organisms. This is due to the nonhomogeneity of the magnetic field induced deformations.

6 Conclusion

To summarize our results, we can conclude that the above-presented mechanical model for magnetic continuums can successfully describe the deformation in static magnetic fields. With the aid of numerical or FEM calculations based on the model, it is possible to reveal the basic characteristics of the magnetic field induced deformations (i.e. non-continuity, nonhomogeneity) or predict the deformation of a highly elastic magnetic substance in an arbitrary magnetic field distribution. However, the effect of the field distribution on the resulting deformation, which is probably the most interesting point, has not analyzed yet. We leave this for future works. The model can also serve as a basis for designing ferrogel-based actuators or artificial muscles whose importance does not need to emphasize.

References:

1. M. Zrínyi, Trends in Polymer Science **5**, 277 (1997)
2. M Zrínyi, L. Barsi, D. Szabó, H.-G. Kilian, J. Chem. Phys. **106**, 5685 (1997)
3. K. Bohon, S. Krause, J.Polym.Sci.:Polym.Phys. **36**, 1091 (1998)
4. T. Mitsumata, K. Ikeda, J. P. Gong, Y. Osada, D. Szabó, M. Zrínyi, J. Appl. Phys. **85**, 1 (1999)
5. D.Szabó, I.Czakó-Nagy, M.Zrínyi,A.Vértes, Journal of Colloid and Interface Science **221**, 166 (2000)

STRUCTURE-ENHANCED YIELD SHEAR STRESS IN ELECTRORHEOLOGICAL FLUIDS

R. Tao, Y.C. Lan, and X. Xu
Department of Physics, Temple University, Philadelphia, PA 19122-6082, USA

A new technology, compression-assisted aggregation, is developed to enhance the strength of electrorheological (ER) fluids. The yield shear stress of ER fluids depends on the fluid microstructure. The unassisted electric-field induced ER structure mainly consists of single chains, whose weak points are at their ends. This new technology produces a structure consisting of robust thick columns with strong ends. As the weak points of the original ER structure are greatly enforced, the new structure makes ER fluids super-strong: At a moderate electric field and moderate pressure, the yield shear stress of ER fluids at 35% volume fraction exceeds 100 kPa, well above any requirement for major industrial applications.

1. introduction

Since the discovery of electrorheological (ER) fluids, there has been an expectation that ER fluids may revolutionize certain industrial sectors if they are strong enough [1,2]. Manufacture engineers estimated that a yield shear stress of 40 kPa is required for ER fluids to be suitable for major industrial applications [3]. However, a typical ER fluid now only has a yield shear stress around 5-10 kPa. While searching for strong ER materials remains active, the progress in this direction is slow and limited.

In this paper, we will report a completely different and novel approach that makes ER fluids super-strong. We do not seek for new materials, but employ a quick compression-assisted-aggregation process to force ER fluids to form a microstructure that is much stronger than the structure produced by unassisted aggregation [4]. After doing so, our ER fluid under a moderate electric field produces a yield shear stress exceeding 100 kPa, ten times more than the value without the compression-assisted aggregation. In addition, since this method is dealing with the fluid's microstructure, it is general and applicable to all kinds of ER fluids

2. The Weak Points of ER Microstructure

To develop strong ER fluids, we need to know the rheological properties of the induced ER structures. Especially we want to identify the weak points of these structures and find a way to strengthen them. Let us consider the single-chain structure, which is the basis of all other ER structures.

After the fluid is activated, chains are formed. The induced attractive force among the chain particles provides a capacity to resist an external stress. The yield shear stress of a chain is defined as the stress, below which a deformed chain can recover after the stress is removed, but above which the chain incurs non-recoverable damage even if the stress is released. The chain is originally along the field direction, spanning between the electrodes with particles touching each other. Upon application of a pair of shear forces at the chain's two end particles, which are of the same magnitude but in the opposite direction perpendicular to the field, the chain is strained and elongated, becomes slanted, and then breaks into two broken chains if the shear forces exceed a threshold. The breaking point should be the chain's weak point under the shear. The conventional wisdom assumes that the breaking point is in the middle of chain [5,6]. Why is it at the middle? The system is symmetric around the middle between the two poles; based on this symmetry, it is natural to guess that the chain could break at the middle. However, this is not the case. There is a symmetry breaking [7]. After a magnetic field is applied, the space is no longer isotropic. To form a long chain along the field direction is energetically favored in this space.

Let us assume that the shear process is quasi-static that the chain relaxes into the lowest energy state at each shearing step. Under such a deformation, the configuration with the minimum energy is no longer symmetric. The symmetric configuration becomes unstable. In particular, during the initial deformation, a gap emerges at one end. The gap widens as the strain increases. When the strain exceeds a critical value, the chain breaks. The breaking point is at either end, but not at the middle. In another words, the weak points of an ER chain under shear are at its two ends.

In order to demonstrate the above points, we apply the Green's function method to find the energy for various configurations [8]. The calculation includes the multipole contributions and the effect of electrodes. A comparison of the energies between different configurations enables us to identify the minimum-energy configuration. We thus find the deformation process and locate the breaking point.

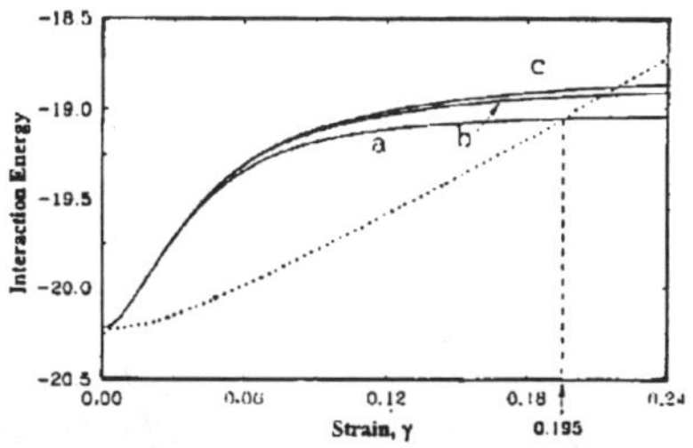

Fig.1 A comparison of the interaction energies for different broken chains: (a) the breaking point is between the first and second particles; (b) the breaking point is between the second and third particles; (c) the breaking point is at the middle. The dotted curve is for a slanted chain with only one gap between the first and second particles.

The energy calculation converges quickly as more multipole moments are included. When a pair of shear forces is applied to a chain through the electrode plates as in most ER applications, we assume that the pair of shear forces is applied to the first particle and the last particle in the chain. It is clear that when the sheared chain is slanted, it is also stretched. In order to locate the breaking point, we first examine where the gap or gaps should be due to this stretch. With various proposed configurations, we calculate the energies for all them and, afterwards, make a comparison. Some configurations are as follows: (1) gaps uniformly distributed along the chain, (2) two gaps, one is between the first and second particles and the other between the last and next last particles, (3) only one gap between the first and second particles, and (4) only one gap in the middle of the chain. Our calculation has found that when the chain has just one gap between the first and the second particles, the energy is the lowest. The configuration with one gap in the middle or the configuration with gaps uniformly distributed along the chain never has the lowest energy. This means that under the external shear force, the slanted chain first has one gap between the first and second particles. Obviously, this gap will get wider and wider if the shear strain becomes larger and larger. When the shear strain is big enough, the chain breaks. Where is the breaking point? In Fig.1, we plot the energy for the broken chains with several different possible broken points. When the broken point is between the first and second particles, the energy is the lowest among them. The broken chain with the broke point in the middle does not have the lowest energy. Therefore, when the chain breaks under the shear, the broken point must be at one of its ends. This implies that the weak points of an ER chain are at its two ends. When the shear strain is small, the chain becomes slanted and the gap between the first particle and second particle increases. When the shear strain exceeds a critical value, the chain breaks at one of its ends and the broken chain returns to the vertical position.

The above theoretical result has also been verified by experiments, which are depicted in Fig.2. The particles used in the experiment are glass beads coated with two layers: a metallic layer and an insulating TiO_2 surface layer. The particle diameter is about 40 μm. Silicon oil is the base liquid. An electric field close to 1kV/mm is applied across the two electrodes. The gap between the electrodes is about 2.8 mm. Therefore, there are about 70 particles in a single chain. Figure 2a shows that the chain shears slowly. Figure 2b shows that the chain breaks between the first particle and second particle when the shear strain exceeds a critical value. This experiment repeats many times at low shear rate and we always find that the chain break at one end, never at the middle. Therefore, it confirms the theoretical result that the weak points of an ER chain are at the chain's ends. In the experiment, we also increase the shear rate and find that the chain breaks at its one end even at the shear rate of 200 s^{-1}.

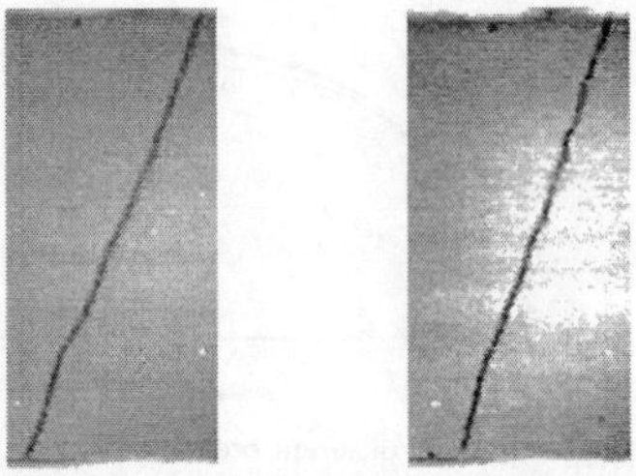

(a) (b)

Fig.2 (a) Under a shear force, the chain becomes slanted. (b) The chain breaks between the first and second particles when the shear strain exceeds the critical value.

While the above discussion is limited to the single-chain structure, the asymmetric behavior is general. Under an electric field, ER chains may slowly aggregate into thick columns, whose ideal structure is a body-centered tetragonal (bct) lattice [9]. Such columns developed from unassisted aggregation, in fact, are a bundle of a dozen chains [10]. Therefore, under a strong shear force, this bundle of a dozen chains may well break at one of its ends.

3. Superstrong ER Fluids

After understanding the microstructure of ER fluids and identifying the weak points, we are in a position to seek a new agile approach to produce super-strong ER fluids. Let us emphasize that the strength of ER fluids solely comes from the induced microstructure. Different microstructure delivers different strength. For example, both theoretical calculation [5] and experiment [8] have already found that a bct lattice structure has a higher yield shear stress than a single-chain structure. Unfortunately, the unassisted aggregation process, as discussed before, is slow, produces columns of a limited thickness, and is thus not very useful. We need a technology that can rapidly produce thick columns with strong and robust ends, i.e., thick columns with no weak points at their ends.

This technology is a compression-assisted-aggregation process. Immediately after an electric field is applied, we compress the ER fluid before a shear force is applied. The electric field produces chains in milliseconds. The compression pushes these chains to form thick columns with strong and robust ends. In addition, each column consists of at least a couple of hundred chains. Once the weak points of ER microstructure are strengthened and the columns are very thick, the ER fluids become super-strong.

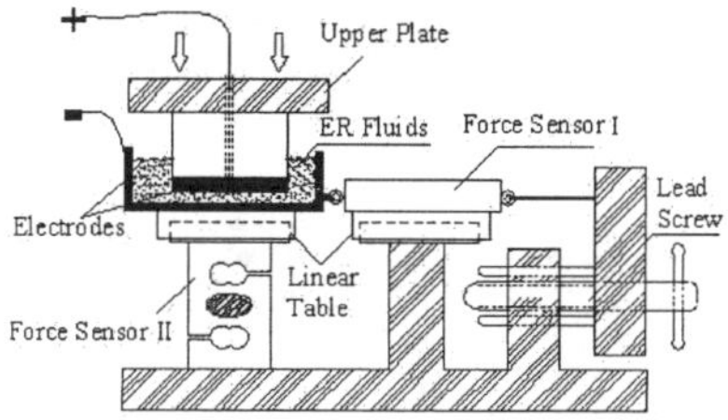

Fig.3 The experimental setup, in which the compression is along the field direction.

Our first experimental setup is in Fig. 3, in which one electrode is pushed down to compress the ER fluid along the field direction, similar to the experiment on MR fluids [11,12]. Our ER fluid consists of dielectric particles in pump oil. While the main results reported here are from the ER fluid with a volume fraction of 35%, we have tested ER fluids with volume fraction from 10% to 60% and found that our technology works well for all of them.

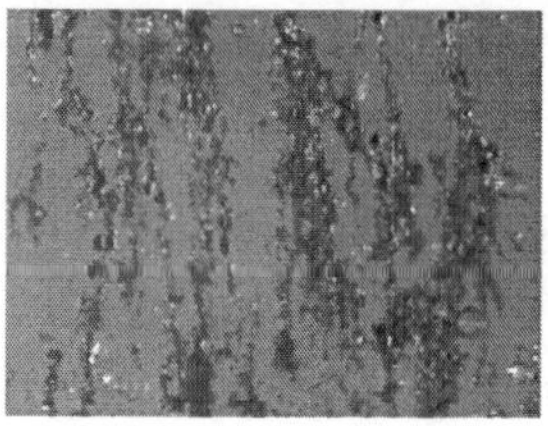

Fig. 4 Before the compression, the electric-field induced ER structure is dominated by single chains. There are some columns, but of limited thickness

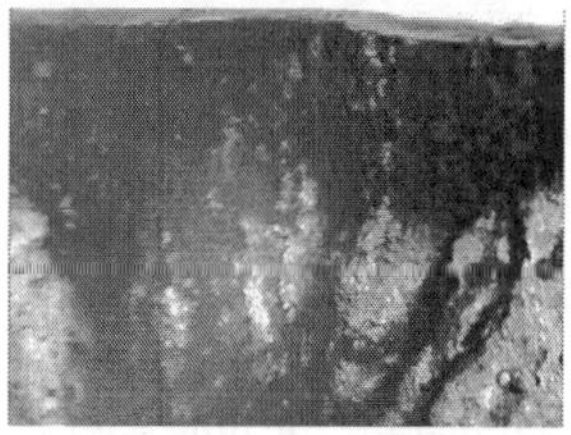

Fig.5 The compression process produces very thick columns. In addition, these columns have strong and robust ends. The original weak points are greatly enforced

The dielectric particles are well dispersed in the base liquid before an electric field is applied. When an electric field is applied, the ER particles quickly form chains along the field direction. It is clear from Fig. 4 that before the compression, the dominant ER structure is single-chain structure. There are some columns, but their thickness is very limited. As shown in Fig.5, after the compression, the aggregated columns are very thick. In addition, the ends of these columns are much thicker than their middle. Such structures are robust and strong because the original weak points in the structure are greatly enforced.

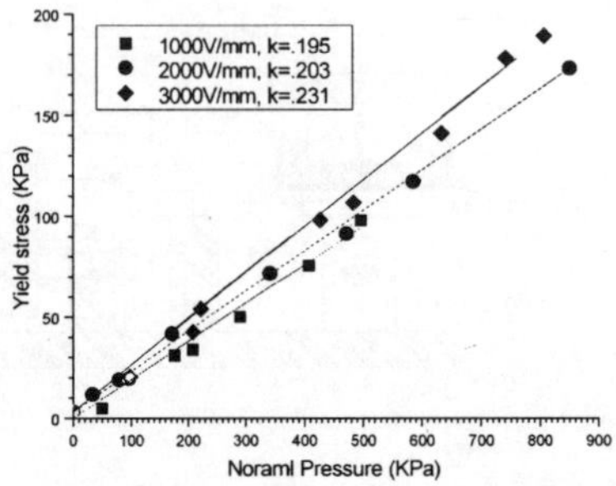

Fig.6 The yield shear stress increases linearly with the compression pressure. At 3 kV/mm and 800 kPa pressure, the yield shear stress is close to 200 kPa.

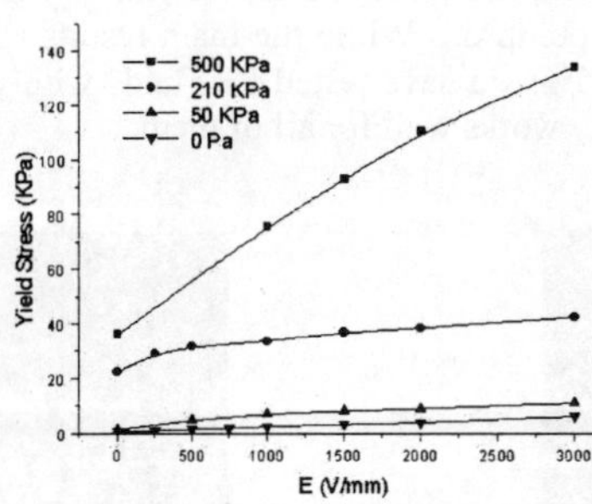

Fig.7 At fixed pressure, the yield shear stress increases with the electric field.

It is natural that the ER fluid with such a structure delivers very strong yield shear stress. As shown in Fig. 6, the yield shear stress increases linearly with the applied pressure. Without the compression-assisted aggregation, our ER fluid of 35% volume fraction has about 5 kPa yield shear stress at 3kV/mm. With the same electric field and a pressure of 800 kPa, the ER fluid has a yield shear stress close to 200 kPa, 40 times stronger than the original value. From Fig.6, we have an empirical expression

$$\tau_y(E) = \tau_0 + KP \qquad (1)$$

where P is the pressure and the slope K increases with the electric field. It is 0.231 at 3kV/mm, 0.203 at 2kV/mm, and 0.195 at 1 kV/mm.

In Fig. 7 we plot the relationship of the yield shear stress versus the electric field while the pressure is maintained as a constant. The strength of ER fluids is further enhanced if the electric field increases. We note that when the electric field is below 250V/mm, the yield shear stress has a deep drop. This implies that there is a minimum electric field required to maintain enough strength of the ER structure to sustain the pressure. When the electric field is below 250V/mm, the ER structure cannot hold the internal pressure difference and the structure may collapse and lead to sharp drop in the yield shear stress. When the electric field is off, the compressed ER fluid has a residual yield stress. This hysteresis indicates that the dielectric particles formed a solid structure under the compression-assisted aggregation and the solid structure remained after the external field was removed. However, this hysteresis is so weak that a light stir returns the ER fluid back to its liquid state immediately.

For ER fluids, application of the technology of compression-assisted aggregation is much easier and can have more variation than that for magnetorheological fluids [11]. As shown in Fig.8, the compression can be in the direction perpendicular to the electric field. In this setup, the two electrodes are on one plate and arranged alternatively. The compression is provided by pushing an insulate plate down. There is no motion to change the position between the two electrodes here. The transmission of force can be achieved between the electrode plate and the insulate plate. The experiment shows that such setup also substantially enhances the yield shear stress of ER fluids by the compression. For example, the structure-enhanced yield shear stress reaches 100 kPa at moderate electric field and moderate compression pressure. Equation (1) also applies to this case. The yield shear stress increases with the pressure linearly with a slope k=0.225 at 1kV/mm, close to the slope for the device in Fig. 3. The relationship between the yield shear stress and the applied electric field for the device in Fig. 8 is also similar to Fig.7.

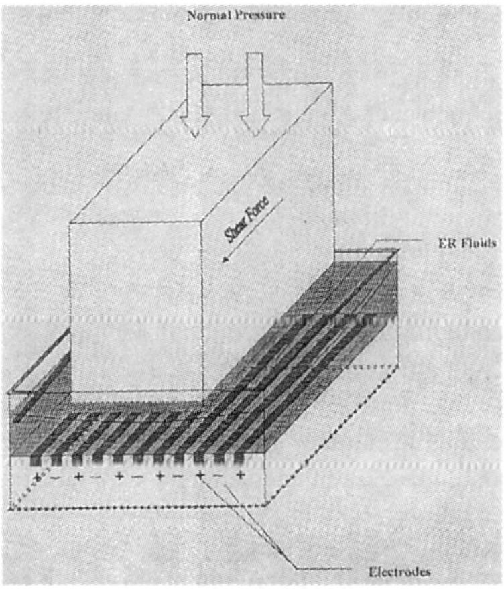

Fig. 8 The electrodes are inserted in the bottom plate. The top insulate plate is used to compress ER fluids.

In Fig. 8, the shear force is perpendicular to the electric field direction. We also tested the situation where the shear force is parallel to the electric field direction and have found that the results are almost the same. When the shear force is in an arbitrary direction, we can always decompose it into two components, perpendicular and parallel to the field. Then the yield shear stress is the vector summation of the responses from the above two directions. Our results thus indicate that for the device in Fig.8, the yield shear stress is almost the same for any direction.

There are many advantages for the device in Fig. 8. First, the electrodes in this device do not move in the compression, the electric field is thus stable and the chance of dielectric breakdown is considerably reduced. Second, the electric field is now limited near the plate, which contains the electrodes. The difficulties of insulation or grounding in ER applications are almost eliminated or greatly reduced here. The force is now transmitted between the top insulate plate and the plate containing both electrodes. As the top plate is not associated with the electrode, it is easy to modify it to meet the application requirement.

To conclude this paper, we provide a sketch of a new clutch (Fig.9) to illustrate the high potential of super-strong ER fluids in industrial applications. The bottom plate containing the electrodes may be connected to the wheel, while the top insulate plate is connected to the engine. ER fluids are filled between them. When the electric field is turned on and the top plate is pushed down, the strong enhanced yield shear stress of ER fluids makes the clutch engaged. When the pressure on the top plate is turned off or the electric field is turned off, the clutch is disengaged.

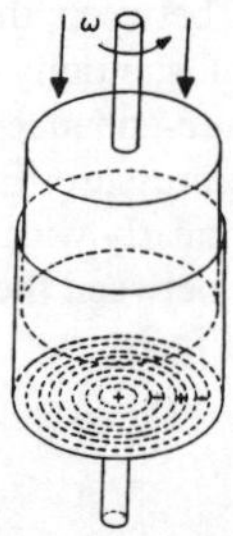

Fig. 9 A design of new clutch to transmit torque between two parallel plates.

Acknowledgement. This research is supported in part by a grant from NSF DMR-0075780.

References

1. W. M. Winslow, U. S. Patent No. 2 417 850 (1947); ibid, J. Appl. Phys. 20, 1137 (1949).
2. For example, see, *Electrorheological Fluids and Magneto-Rheological Suspensions* (edited by R. Tao, World Scientific Publishing, Singapore, 2000).
3. D. I. Hartsock, R. F. Novak, and G. J. Chaundy, J. of Rheology, **35**, 1305 (1991).
4. R. Tao, et al, "Method and apparatus for increasing and modulating the yield shear stress of electrorheological fluids", US Petent (pending).
5. G. L. Gulley and R. Tao, Phys. Rev. E **48**, 2744 (1993)
6. T. C. Halsey and J. E. Martin, Scientific American **10**, 58 (1993).
7. R. Tao, et al, International J. of Modern Physics B **15**, 918 (2001).
8. X. Tang, Y. Chen and H. Conrad, J. Intell. Mat. Sys. and Struct. **7**,517 (1996).
9. R. Tao and J. M. Sun, Phys. Rev. Lett. **67**, 398 (1991).
10. T. J. Chen, R. N. Zitter, and R. Tao, Phys. Rev. Lett. **68**, 2555 (1992).
11. X. Tang, X. Zhang, and R. Tao, J. of Appl. Phys. **87**, 2634 (2000).
12. R. Tao, J. of Physics: Condensed Matter (in press).

OSCILLATORY CONVECTION IN VISCOELASTIC, FERROMAGNETIC / DIELECTRIC LIQUIDS

P.G. SIDDHESHWAR

Department of Mathematics, Bangalore University, Central College Campus, Bangalore 560 001, India
E- mail : pgsiddheshwar@ hotmail.com

Oscillatory convection in viscoelastic ferromagnetic and dielectric liquids of the Rivlin-Ericksen, Maxwell and Oldroyd types is studied analytically by considering free-free , isothermal boundaries with idealized conditions on the magnetic / electric potential. The linear theory reveals the stabilizing nature of the strain retardation parameter and the destabilizing nature of the stress relaxation and magnetization / dielectric parameters. The Maxwell liquids are found to be more unstable than the one subscribing to the Oldroyd description whereas the Rivlin-Ericksen liquid is comparatively more stable. The results have implications in many non-isothermal applications of ferromagnetic and dielectric liquids especially in energy conversion devices.

1 Introduction

Ferromagnetic fluids, which are both thermally and magnetically responding, are now well known from the point of view of applications [2] and exhaustive literature on Rayleigh-Benard convection in Newtonian ferromagnetic liquids is presently available. Some recent works [1, 5] suggest that magnetic fluids demonstrate a non-Newtonian behaviour in the presence of a magnetic field. The situation in dielectric liquids is analogous to that in ferromagnetic liquids. The dielectric liquids have immense heat transfer applications [12]. Motivated by these works, Siddheshwar [6, 7] studied oscillatory convection in Oldroyd/Rivlin - Ericksen-Boussinesq-ferromagnetic fluids. In the present paper three viscoelastic descriptions for both ferromagnetic and dielectric liquids are considered for investigating a similar problem.

2 Mathematical formulation

We consider ferromagnetic/dielectric liquids of the Rivlin – Ericksen, Maxwell and Oldroyd types. Each fluid is discussed below using the notations of earlier related standard works.

2.1 Rivlin – Ericksen ferromagnetic liquid [4, 7]

We consider an infinite horizontal layer of viscoelastic ferromagnetic liquid of thickness d. The lower plane at Z = 0 and the upper one at Z = 1, where Z = z / d, are maintained at constant temperatures $T_0 + \Delta T$ and T_0 respectively. In addition to a temperature gradient, a vertical uniform DC magnetic field is also imposed across the layer. The fluid layer is initially quiescent. Let this state be slightly disturbed. Following the classical lines of linear stability theory as presented by Chandrasekhar [3], the non-dimensional equations governing small perturbation may be written as

$$\frac{1}{\mathbf{Pr}}\frac{\partial}{\partial \mathbf{t}}(\nabla^2 \mathbf{w}) - (\mathbf{R}+\mathbf{N})\nabla_1{}^2\theta + \mathbf{N}\frac{\partial}{\partial \mathbf{t}}(\nabla_1{}^2\phi) = \left(1+\frac{\mathbf{Q}}{\mathbf{Pr}}\frac{\partial}{\partial \mathbf{t}}\right)\nabla^4 \mathbf{w}, \quad (1)$$

$$\left(\frac{\partial}{\partial \mathbf{t}} - \nabla^2\right)\theta = \mathbf{w}, \quad (2)$$

$$\mathbf{M}_3\ \nabla_1^2\phi + \frac{\partial^2\phi}{\partial \mathbf{Z}^2} - \frac{\partial\theta}{\partial \mathbf{Z}} = 0, \quad (3)$$

where $\nabla_1^2 = \frac{\partial^2}{\partial \mathbf{X}^2} + \frac{\partial^2}{\partial \mathbf{Y}^2}$, $\nabla^2 = \nabla_1^2 + \frac{\partial^2}{\partial \mathbf{Z}^2}$, $\mathbf{M}_3$ is a magnetization parameter, **Pr**, **R** and $\mathbf{N}(=\mathbf{RM}_1)$ are the Prandtl, Rayleigh and magnetization Rayleigh numbers respectively [4], and **Q** characterizes viscoelasticity [8]. The quantity $\mathbf{M}_1$ also pertains to magnetization and the quantities **w**, **T**, ϕ denote vertical component of velocity, temperature and magnetic potential respectively. We now decompose the solution in terms of normal modes, so that

$$[\mathbf{w}, \theta, \phi] = [\mathbf{W(Z)}, \mathbf{T(Z)}, \mathbf{\Phi(Z)}] \exp\left[\sigma \mathbf{t} + i\,(\mathbf{a_X X} + \mathbf{a_Y Y})\right], \quad (4)$$

where σ is the time constant and $\mathbf{a_X}$ and $\mathbf{a_Y}$ are the horizontal wave numbers. Using (4) in (1) - (3), we get

$$\left[\left(1+\frac{\mathbf{Q}\sigma}{\mathbf{Pr}}\right)(\mathbf{D}^2-\mathbf{a}^2)^2 - \frac{\sigma}{\mathbf{Pr}}(\mathbf{D}^2-\mathbf{a}^2)\right]\mathbf{W}$$

$$-(\mathbf{R}+\mathbf{N})\ \mathbf{a}^2\ \mathbf{T} + \mathbf{N}\ \mathbf{a}^2\ \mathbf{D\Phi} = 0, \quad (5)$$

$$\left(\mathbf{D}^2 - \mathbf{a}^2 - \sigma\right)\mathbf{T} + \mathbf{W} = 0, \quad (6)$$

$$\left(\mathbf{D}^2 - \mathbf{M}_3\ \mathbf{a}^2\right)\mathbf{\Phi} - \mathbf{DT} = 0, \quad (7)$$

where $\mathbf{a}^2 = \mathbf{a_X}^2 + \mathbf{a_Y}^2$ and $\mathbf{D} = \frac{\mathbf{d}}{\mathbf{dZ}}$.

The boundary conditions appropriate for an analytical study are

$$\mathbf{W} = \mathbf{D^2W} = \mathbf{T} = \mathbf{D\Phi} = 0 \text{ at } \mathbf{Z} = 0, 1. \quad (8)$$

Past experience with problems of this kind suggests that the essential features of the problem can be extracted using these admittedly artificial conditions. Equations (5) - (8) constitute an eigenvalue system for the present problem. It is evident that when $\mathbf{Q} = 0$ the system reduces to that for an ordinary viscous Newtonian ferromagnetic liquid.

2.2 Rivlin – Ericksen dielectric liquid [8, 9]

The eigenvalue system for this liquid is

$$\left[\left(1+\frac{\mathbf{Q}\sigma}{\mathbf{Pr}}\right)(\mathbf{D}^2-\mathbf{a}^2)^2-\frac{\sigma}{\mathbf{Pr}}(\mathbf{D}^2-\mathbf{a}^2)\right]\mathbf{W}$$

$$-(\mathbf{R}+\mathbf{L})\,\mathbf{a}^2\ \mathbf{T}+\mathbf{L}\ \mathbf{a}^2\ \mathbf{D\Phi}=0, \tag{9}$$

$$\left(\mathbf{D}^2-\mathbf{a}^2-\sigma\right)\mathbf{T}+\mathbf{W}=0, \tag{10}$$

$$\left(\mathbf{D}^2-\mathbf{a}^2\right)\mathbf{\Phi}-\mathbf{DT}=0, \tag{11}$$

where $\mathbf{L}$ is the electrical Rayleigh number, $\mathbf{\Phi}$ is the electric potential and all the other quantities have the earlier meaning outlined in the context of ferromagnetic liquids. The boundary conditions (8) also apply here.

2.3 Oldroyd ferromagnetic liquid [4, 6]

The eigenvalue equations for this liquid are equations (6) and (7) together with the following equation that can be obtained from the conservation of linear momentum:

$$\left[(\mathbf{D}^2-\mathbf{a}^2)^2-\frac{\mathbf{G}\,\sigma}{\mathbf{Pr}}(\mathbf{D}^2-\mathbf{a}^2)\right]\mathbf{W}$$

$$-(\mathbf{R}+\mathbf{N})\,\mathbf{G}\ \mathbf{a}^2\ \mathbf{T}+\mathbf{N}\ \mathbf{G}\ \mathbf{a}^2\ \mathbf{D\Phi}=0, \tag{12}$$

where $\mathbf{G} = (1 + \Gamma\sigma)(1 + \Gamma\eta\,\sigma)^{-1}$ and the other quantities are as defined earlier. Amongst the undefined quantities, Γ and η are stress - relaxation and strain-retardation parameters.

2.4 Oldroyd dielectric liquid [9, 10]

The eigenvalue equations, in the notation of the earlier sections, for this liquid are equations (10) and (11) together with the following equation that can be obtained from the conservation of linear momentum:

$$\left[(\mathbf{D}^2-\mathbf{a}^2)^2-\frac{\mathbf{G}\,\sigma}{\mathbf{Pr}}(\mathbf{D}^2-\mathbf{a}^2)\right]\mathbf{W}$$

$$-(\mathbf{R}+\mathbf{L})\,\mathbf{G}\ \mathbf{a}^2\ \mathbf{T}+\mathbf{L}\ \mathbf{G}\ \mathbf{a}^2\ \mathbf{D\Phi}=0. \tag{13}$$

We note here that by taking $\eta = 0$ in equations (12) and (13) we get the results of Maxwell ferromagnetic/dielectric liquids.

3 Solution

Eliminating **T** and **Φ** between equations (5) - (7), we get

$$[(D^2 - a^2)(D^2 - M_3 a^2)(D^2 - a^2 - \sigma)\left(\left(1 + \frac{Q\sigma}{Pr}\right)(D^2 - a^2) - \frac{\sigma}{Pr}\right) + R a^2 (D^2 - M_3 a^2) - N a^4 M] W = 0. \tag{14}$$

Executing a similar procedure on equation (9), we get

$$[(D^2 - a^2)(D^2 - a^2)(D^2 - a^2 - \sigma)\left(\left(1 + \frac{Q\sigma}{Pr}\right)(D^2 - a^2) - \frac{\sigma}{Pr}\right) + R a^2 (D^2 - a^2) - L a^4 M] W = 0. \tag{15}$$

From the governing equations and equation (8), the following conditions may be obtained on **W** :

$$W = D^2 W = D^4 W = D^6 W = 0 \text{ at } Z = 0, 1. \tag{16}$$

Equations (14) and (16) indicate that the relevant solution for the lowest mode of **W** is given by

$$W = A \sin(\pi Z). \tag{17}$$

Substituting (17) into (14), we get the required eigenvalue relationship in the form:

$$R = \frac{\delta^2(\delta^2 + \sigma)\ (\delta^2\left(1 + \frac{Q\sigma}{Pr}\right) + \frac{\sigma}{Pr}) - \frac{N a^4 M_3}{\delta_1^2}}{a^2}. \tag{18}$$

To explore oscillatory convection we put $\sigma = i\omega$ in (18) to get

$$R = \frac{\delta^2}{a^2}\left[\delta^4 - \omega^2 Pr^{-1}(\delta^2 Q + 1) - \frac{N a^4 M_3}{\delta^2\ \delta_1^2}\right] + i\omega N_1, \tag{19}$$

where ω is the frequency and

$$N_1 = \frac{\delta^4}{a^2}\left[1 + (\delta^2 Q + 1)\, Pr^{-1}\right]. \tag{20}$$

Since $\mathbf{R}$ is a real quantity, either $\omega = 0 (\mathbf{N}_1 \neq 0)$ or $\omega \neq 0 \ (\mathbf{N}_1 = 0)$. But since $\mathbf{N}_1 \neq 0$, we take $\omega = 0$ and this gives

$$\mathbf{R} = \frac{\delta^6}{\mathbf{a}^2} - \frac{\mathbf{N}\mathbf{M}_3\mathbf{a}^2}{\delta_1^2}. \tag{21}$$

Similarly for Rivlin – Ericksen dielectric liquids, we get

$$\mathbf{R} = \frac{\delta^6}{\mathbf{a}^2} - \frac{\mathbf{L}\mathbf{a}^2}{\delta^2}. \tag{22}$$

We note here that stationary convection is predicted for Rivlin – Ericksen ferromagnetic/ dielectric liquids. To rectify this situation of physically improper prediction of stationary convection, Siddheshwar [7] used a hyperbolic heat equation in place of the parabolic heat equation of the present study. Oscillatory convection was rightly predicted by this study. The oscillatory Rayleigh number for Oldroyd – ferromagnetic / dielectric liquids can similarly be obtained as

$$\mathbf{R} = \frac{\delta^2(\delta^2 + \mathrm{i}\omega)\left[\dfrac{\delta^2(1+\Gamma\mathrm{i}\omega)(1-\Gamma\eta\mathrm{i}\omega)}{1+\Gamma^2\omega^2} + \dfrac{\mathrm{i}\omega}{\mathbf{Pr}}\right] - \dfrac{\mathbf{a}^4(\mathbf{N}\mathbf{M}_3\,\delta_{j1} + \mathbf{L}\,\delta_{j2})}{\delta_1^2}}{\mathbf{a}^2}. \tag{23}$$

In equation (23), $\mathbf{j} = 1$ gives the expression for a ferromagnetic fluid and $\mathbf{j} = 2$ that for a dielectric liquid. The imaginary part of equation (23) has to be zero as $\mathbf{R}$ is real. This yields the frequency of oscillations ω in the form

$$\omega^2 = -\frac{\mathbf{Pr}\,\Gamma\,(\eta - 1)\,\delta^2 + 1 + \mathbf{Pr}}{\Gamma^2\,(1 + \mathbf{Pr}\,\eta)}. \tag{24}$$

It is clear from equation (24) that for $\eta > 1$ oscillatory convection is impossible. Hence, in the computations we take values of η less than 1 which opens up the possibility of oscillatory convection. Evidently from equation (24) we see that the frequency ω is explicitly independent of the magnetization - related parameters but is implicitly dependent on them through the wave number.

4 Results and discussion

The paper deals with a linear stability analysis of Rayleigh – Benard convection in Rivlin – Ericksen, Maxwell and Oldroyd, ferromagnetic/dielectric liquids. Due to paucity of space we now discuss the results without seeking recourse to graphical displays.

The effect of magnetization on convection in viscoelastic fluids was found to be akin to its effect in Newtonian fluids. In the present paper we have two magnetization parameters $\mathbf{M_1}$ and $\mathbf{M_3}$. $\mathbf{M_1}$ is the ratio of magnetic to gravitational forces. $\mathbf{M_3}$ measures the departure of linearity in the magnetic equation of state. $\mathbf{M_3}$ equal to 1 corresponds to linearity and larger values imply non – linearity of magnetization. As in Finlayson [4], $\mathbf{M_1}$ and $\mathbf{M_3}$ have destabilizing effect on $\mathbf{R_c}$ of viscoelastic ferromagnetic fluids. This observation is true of all the three viscoelastic fluid descriptions considered in the paper. In the case of viscoelastic dielectric liquids it was found that $\mathbf{R_c}$ decreases with increase in value of electrical Rayleigh number. An identical qualitative result was reported by Stiles *et al.* [9]. It is apt to note here that **L** of the problem involving dielectric liquids plays a role similar to **N** of the corresponding problem in ferromagnetic liquids. In the case of all the three viscoleastic descriptions for ferromagnetic and dielectric liquids the computations delineate the fact that the stress relaxation parameter and strain retardation parameter have opposing influence on convection. The stress relaxation parameter Γ has a destabilizing influence whereas the strain retardation parameter η has a stabilizing influence. This result is true both in the presence and absence of external fields like magnetic/electric field. The Prandtl number **Pr** influences convection in viscoelastic ferromagnetic/dielectric liquids in the classical way of its influence in Newtonian fluids.

The critical value of wave number $\mathbf{a_c}$ and the frequency ω_c, which in turn give the critical Rayleigh number $\mathbf{R_c}$ for the onset of convection, were computed for various parameter ranges. It was observed that the effect of various parameters on $\mathbf{a_c}$ and ω_c in the case of viscoelastic fluids is much the same as that of Newtonian fluids. It is in place to note here that the parameter **Q** in Rivlin – Ericksen fluids plays the role of strain retardation parameter of Oldroyd fluids. With this observation we now summarize the following general conclusions:

(a) $\lim\limits_{\Gamma,\eta\to 0} \{\textbf{Results of present study}\} \to \{\textbf{Results of Finlayson [4]}\}$.

(b) For ferromagnetic/dielectric liquids the following result holds :

$$\lim_{\substack{M_1 (\text{or } L)\to 0 \\ \eta\to 0}} \{\textbf{Results of Oldroyd}\} \to \{\textbf{Results of Maxwell}\}.$$

(c) Comparing the results of Rivlin – Ericksen, Maxwell and Oldroyd ferromagnetic/dielectric liquids the following observation is true:

$$\mathbf{R_c^{Rivlin-Ericksen}} > \mathbf{R_c^{Oldroyd}} > \mathbf{R_c^{Maxwell}}.$$

5 Acknowledgements

The author is grateful to Prof. Georges Bossis, University of Nice – Sophia Antipolis, Nice, France for financial support and encouragement to attend ERMR 2001. The work was supported by the UGC – DSA programme, Department of Mathematics, Bangalore University, Bangalore -1, India.

References

1. Bacri J. C., Boilot J. C., Massart R., Malier L., Perzynzki R. and Zins D., Viscoelastic measurements in gels with ferrofluid particles, *Proc. of the Int. Symp. on Aerospace and Fluid Science* ed. by Kamiyama S. (Institute of Fluid Science, Tohoku Univ., Sendai, 1993) pp. 643-649.
2. Berkovsky B.M., Medvedev M. and Krakov M. S., Magnetic Fluids: Engineering Applications (Oxford University Press, Oxford, 1993).
3. Chandrasekhar S., Hydrodynamic and Hydromagnetic Stability (Oxford: Clarendon Press, 1961).
4. Finlayson B. A., Convective instability of ferromagnetic fluids, *J. Fluid Mech.* **40** (1970) pp. 753-767.
5. Kamiyama S., Koike K. and Wang Z., Rheological characteristics of magnetic fluids, *JSME Int. J.* **30** (1987) pp. 761-766.
6. Siddheshwar P. G., Oscillatory convection in a viscoelastic-Boussinesq-ferromagnetic fluid, *Ind. J. Engng. Matls. Sci.* **5** (1998) pp. 453-456.
7. Siddheshwar P. G., Rayleigh - Benard convection in a second – order ferromagnetic fluid with second sound, *Proc. 7th Asian Cong. Fluid Mech.* Ed. (Shenzhen, China, 1999) pp. 631-634.
8. Siddheshwar P. G. and Srikrishna C. V., Unsteady non-linear convection in a second – order fluid, *Int. J. Non – Linear Mech.* (2001) (In Press).
9. Stiles P. J., Lin F. and Blennerhassett P. J., Convective heat transfer through polarized dielectric liquids, *Phys. of Fluids.* **12** (1993) pp. 3273-3278.
10. Takashima M. and Ghosh A. K., Electrohydrodynamic instability in a viscoelastic liquid layer, *J. Phys. Soc. of Japan* **47** (1979) pp. 1717-1722.
11. Vest C. M. and Arpaci V. S., Overstability of a viscoelastic fluid layer heated from below, *J. Fluid Mech.* **36** (1969) pp. 613-617.
12. Zhakin A. I., Electroconvective heat transfer: Theory and applications, *Proc. of 5th Int. Conf. On Properties and Applications of Dielectric Materials, IEEE,* **1** (1997) pp. 94-100.

CHARACTERISTICS OF A YIELD STRESS SCALING FUNCTION FOR ELECTRORHEOLOGICAL FLUIDS

H. J. CHOI, J. W. KIM AND M. S. CHO

Department of Polymer Science and Engineering, Inha University, Inchon, 402-751, Korea

E-mail: hjchoi@inha.ac.kr

C. A. KIM AND M. S. JHON

Department of Chemical Engineering, Carnegie Mellon University, Pittsburgh, PA 15213-3890, USA

The electrorheological (ER) fluids exhibit a drastic change in rheological and electrical properties. Among these properties, yield stress is one of the critical evaluation parameters of the performance of ER devices. The published experimental data of yield dependence on the electric field strength and particle volume fraction are inconsistent due to the time dependence of material properties and measuring conditions. In this paper, we present a universal function, descriptive of the normalized yield stress, via scaling of the applied electric field strength. This scaling equation hybridizes both the polarization and conductivity models. Yield stress data for various ER fluids are collapsed onto a single curve for a broad range of electric field strengths, suggesting that the proposed scaling equation is adequate for predicting the ER property. Furthermore, the yield stresses, obtained from two different measuring techniques (static and dynamic methods), were also examined.

1 Introduction

The electrorheological (ER) effect refers to the abrupt, reversible change in viscosity for a given suspension of polarizable particles in an insulating oil by imposition of an electric field. ER fluids are considered to be one of the most promising "smart" materials [1,2].

Since their discovery, ER substances have been the subjects of intensive theoretical [3-5] and experimental research [6-10] due to their various potential applications. The change in viscosity resulting from reorientation of the dispersed particles, whose initially random distribution transforms into a fibrillated or layered structure, giving rise to higher viscosity. ER fluids under an externally applied electric field also exhibit Bingham fluid behavior (*e.g.*, yield stress) [11]. The Bingham fluid model has been the suitable rheological model for the steady shear response of many ER fluids and is described below [11].

$$\begin{aligned} \tau &= \tau_y + \eta\dot{\gamma} & \tau &> \tau_y \\ \dot{\gamma} &= 0 & \tau &\le \tau_y \end{aligned} \tag{1}$$

Here, τ is the shear stress, τ_y is the yield stress, $\dot{\gamma}$ is the shear rate, and η is shear viscosity. The dynamic yield stresses are obtained by extrapolating $\dot{\gamma}$ to zero in a controlled shear rate test. A strain applied perpendicular to the electric field distorts and rebuilds fibers between electrodes, and the dissipating energy originates from the deformation results in viscosity increase.

The yield stress is one of the critical design parameters in ER devices and has attracted considerable attention both experimentally [12-14] and theoretically [11,15]. However, many published experimental data of the yield stress dependence on the electric field strength and volume fraction of particles (ϕ) are inconsistent due to the hysteresis of the material systems and measuring conditions, including the actual measuring devices [16]. In this paper, we present a generalized scaling function for the normalized yield stress via scaling of the applied electric field strengths. We believe that this scheme will be useful in the data analyses for ER/magnetorheological (MR) fluid experimentalists. Several models have been proposed to describe the ER phenomena: the original model [16], the so-called "polarization model," attributed the attractive force between particles, to Maxwell-Wagner's interfacial polarization and employed point-dipole approximation. Within this framework, τ_y is represented as

$$\tau_y \propto \phi K_f E_0^{\ 2} f(\beta), \tag{2}$$

where E_0 is electric field strength, and $\beta = (K_p - K_f)/(K_p + 2K_f)$ is the dimensionless dielectric mismatch parameter. Here, K_p and K_f are the dielectric permitivities of the particle and the fluid, respectively. This polarization model agrees extremely well [11] with data for small ϕ and E_0. However, the yield stress data significantly deviates from Eq. (2) at high E_0 [15] and are better represented by the power law relationship $\tau_y \propto E_0^{\ m}$ (m<2) [17,18]. As the gap between the conducting particles in the fluid decreases, the electric response of the fluid becomes nonlinear through electrical breakdown at high electric field strengths. Davis [15] showed that over a range of E_0 greater than the "critical electric field strength (E_c)," the slope of E_0 *vs.* τ_y (on log paper) becomes 3/2, while the slope maintains 2 for small E_0. In this paper, we synthesized three different semiconducting polymer particles and then prepared ER fluid suspensions by dispersing them in silicone oil. The measured yield stresses of these ER fluids were then correlated via a universal yield stress function.

2 Experimental

2.1 Poly(p-phenylene) (PPP)

The poly(p-phenylene) PPP, which has been extensively studied as a conducting polymeric material, has recently been found to exhibit ER characteristics [19,20]. Under a nitrogen atmosphere and in the presence of aluminum chloride, water, and cupric chloride, which form a Lewis acid, benzene was converted to PPP. The stirring speed of approximately 400 ~ 500 rpm was maintained for 2 hrs at 45 °C. After polymerization, the mixture was washed using water and ethyl alcohol to remove the cupric chloride ($CuCl_2$) and aqueous acid. The PPP was then dried and milled. To control the conductivity of the PPP particles, ferric chloride ($FeCl_3$) solution was added as a dopant. Particles were doped with one of three different $FeCl_3$ concentrations in aqueous solution (2.5, 5.0 and 7.5 wt%) for 48 hrs. After doping, the PPP particles were filtered and then vacuum dried for 2 days to remove any trace of moisture. Depending on the doping degree, the samples were indexed to be PPP(0.0), PPP(2.5), PPP(5.0), and PPP(7.5) for undoped, doped with 2.5 wt% $FeCl_3$, 5.0 wt% $FeCl_3$, and 7.5 wt% $FeCl_3$ in aqueous solutions, respectively.

2.2 *Microencapsulated Polyaniline (MCPA)*

Because microencapsulation entraps core materials by circulating them with certain polymeric materials, we adopted this method to modify the polyaniline (PANI) particles [21]. As a core material, PANI was synthesized via chemical oxidation process. The polymerization temperature was kept at −5 °C, and the measured pH of the product before and during the study was approximately 5. As a next step, dried PANI particles were also used for the encapsulation, and prepolymer melamine formaldehyde (MF) was used as the shell material [22]. 25 g of a 20 % citric acid or poly(styrene sulfornic acid) and 100 g of distilled water was mixed. The pH of the solution was then controlled to 4 using 1M NaOH. Dried PANI particles were put into this solution. After stirring this mixed solution for 2 hrs at 40 ~ 45 °C in a water bath, MF solution was added while stirring. As the temperature rose to approximately 65 °C, the MF resins were cured, and PANI particles were encapsulated. After completion of the encapsulation reaction, the particles were filtered and washed with distilled water several times to remove the unreacted residuals. Finally, the products were dried in a vacuum oven at 25 °C. Microencapsulated polyaniline (MCPA) composites are denoted as MCPAi (i = 1, 2, 3). Here 1, 2, and 3 are related with thickness of MF resin. Because microencapsulation entraps core materials with polymers, we used this method to modify the PANI particles.

2.3 *Copolyaniline (COPA)*

The copolyaniline (COPA) was synthesized by using an aqueous solution of HCl, ammonium peroxysulfate ($(NH_4)_2S_2O_8$, oxidant or initiator), and feeding various mole fractions of *o*-ethoxyaniline monomer [23]. At first, PANI was synthesized through oxidant polymerization. A prechilled solution of ammonium peroxysulfate in 240 m*l* of 1M HCl was slowly dropped into a well-stirred reactor containing aniline or monomer mixture in 400 m*l* of 1M HCl. During the synthesis, the temperature was kept at 0 °C. The mole fractions of comonomer in the reaction system were controlled to be 0 %, 25 %, 50 %, 75 %, and 100 %, respectively. The speed of stirring was maintained to be about 400 ~ 500 rpm for 2 hrs. To obtain suitable particles for ER fluid, various procedures such as dedoping, milling, filtering, washing, and drying were followed after the synthesis. The pH of the product was adjusted to be 6.5 by adding either NaOH or HCl solution. The synthesized copolymer was then washed using distilled water to remove the initiator, the unreacted monomer, and the oligomer. It was further washed with ethanol and cyclohexane in order to make the surface of the synthesized particles to be hydrophobic. Finally, the synthesized particles were put into a vacuum oven for sufficient time to get rid of any trace of moisture.

2.4 *Rheological Mesurement*

All ER fluids were prepared by dispersing the particles in a silicone oil. The density and kinematic viscosity of the silicone oil are 0.95 g/m*l* and 30 cS at 25 °C, respectively.

Rheological properties were measured by a rotational Physica rheometer (PHYSICA MC 120, Stuttgart, Germany) with a Couette type geometry, a high-voltage generator, and an oil bath for temperature control. An electric field was applied for five minutes in order to obtain an equilibrium chain-like, or columnar, structure before applying the shear.

3 Results and Discussion

The relationship between τ_y and E_0 obtained from the PPP–based ER fluid is represented. As shown in Fig. 1(a), τ_y is enhanced with the doping degree. On a log plot, the data show

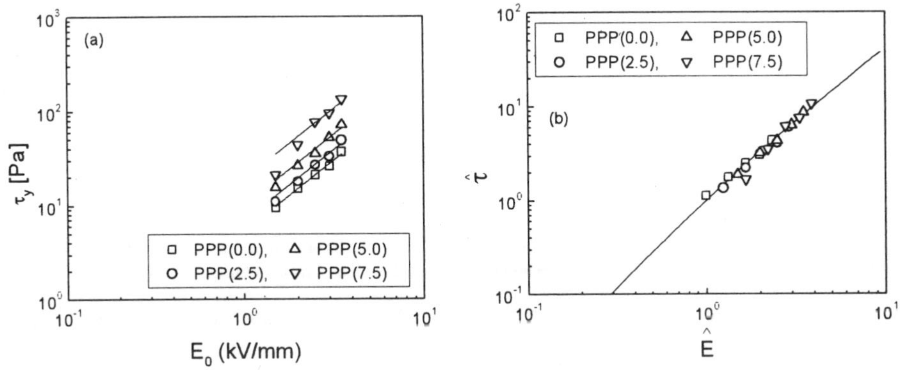

a linear relationship (slope of 1.5 in $\tau_y - E_0$ plot).

Figure 1. (a) Yield stress vs. electric field strength and (b) $\hat{\tau}$ vs. $\hat{E}$ for PPP(0.0), PPP(2.5), PPP(5.0), and PPP(7.5).

To present the yield stress data for the broad range of electric field strengths, Choi *et al.* [20,24] introduced the following simple yield stress equation:

$$\tau_y(E_0) = \alpha E_0^2 \left(\frac{\tanh\sqrt{E_0/E_c}}{\sqrt{E_0/E_c}} \right). \tag{3}$$

Here, α is a parameter which depends on the dielectric constant of the fluid and ϕ, and E_c is the critical electric field strength, which is proportional to the particle conductivity. Equation (3) has two limiting behaviors, *i.e.*, τ_y is proportional to E_0^2 at low E_0, and τ_y changes abruptly to $E_0^{3/2}$ at high E_0. To collapse the data onto a single curve, we normalized E_0 with E_c, and $\tau_y(E_o)$ with $\tau_y(E_c) = \alpha E_c^2 \tanh(1) = 0.762\,\alpha E_c^2$:

$$\hat{\tau} = 1.313\,\hat{E}^{3/2} \tanh\sqrt{\hat{E}}\,. \tag{4}$$

Here, $\hat{E} \equiv E_0/E_c$ and $\hat{\tau} \equiv \tau_y(E_0)/\tau_y(E_c)$. The PPP data are collapsed onto a single curve using Eq. (4), which is shown in Fig. 1(b). To obtain the best fit, we chose E_c to be 1.5 kV/mm for PPP(0.0), 1.2 for PPP(2.5), 1.0 for PPP(5.0), and 0.9 for PPP(7.5).

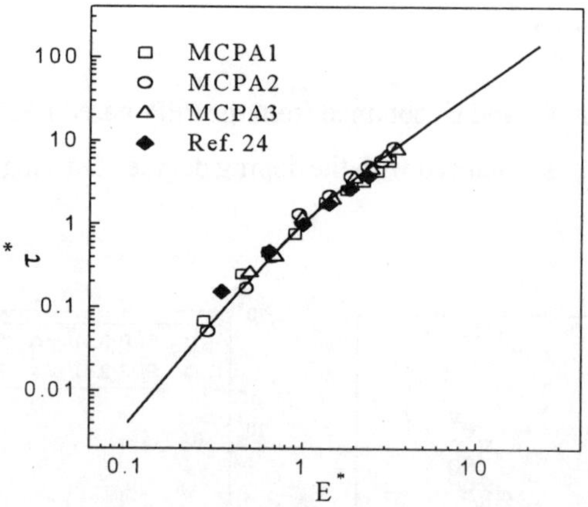

Figure 2. τ^* vs. E^* for 20 wt% suspensions of MCPA particles in silicone oil and 20 vol% suspension of 1.5 μm doubly coated particles in silicone oil (Data are taken from Ref. 24). The line is obtained from Eq. (6).

Furthermore, to improve the experimental data, we introduced one additional scaled parameter b:

$$\tau_y(E_0) = \alpha E_0^{\ 2}\left(\frac{\tanh (E_0/E_c)^{0.5+b}}{(E_0/E_c)^{0.5+b}}\right) \qquad (5)$$

To obtain a universal correlation with Eq. (5), we rescaled $\hat{\tau}$ and $\hat{E}$ with $\tau^* = \hat{\tau}\hat{E}^{4b}$ and $E^* = \hat{E}^{1+2b}$ as

$$\tau^* = 1.313E^{*1.5} \tanh E^{*0.5} . \qquad (6)$$

The data for the MCPA–based ER fluid were also found to collapse onto a single curve by using Eq. (4) However, to check the universality of the proposed scaling function, the data from the MCPA-based ER fluid was replotted using Eq. (6).

To examine the effect of the frequency, we also plotted the yield stress data measured at a 50 Hz ac field applied between the parallel plates by Tam *et al.* [25] using 1.5 μm doubly coated particles. As shown in Fig. 2 (set b = 0.28), the yield stress obtained under the ac field also fit to the scaling yield stress function [Eq. (6)] very well.

Thereby, we believe that Eq. (6) is a very useful expression in constructing the master yield stress curve for ER/MR fluids. Even though the proposed correlations (Eqs. (3) and (5)) are not derived from first principles, they are suitable for representing data. The beauty of these correlations is the fact that most of the ER/MR experimental data can be collapsed with two parameters (E_c and b).

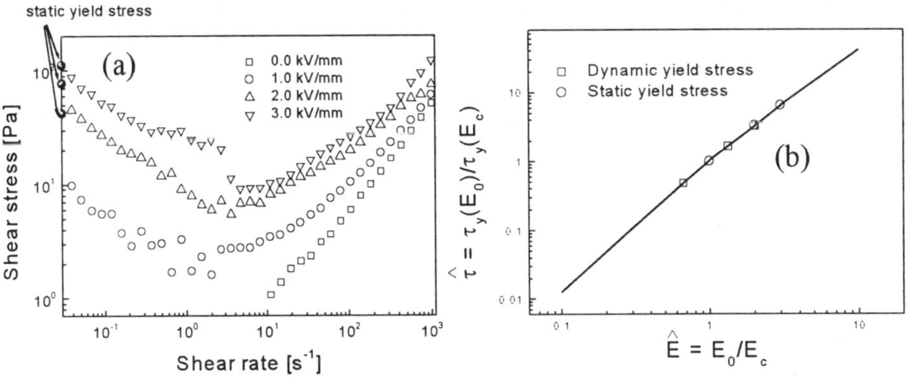

Figure 3. (a) yield stress obtained by CSR and CSS mode, (b) $\hat{\tau}$ *vs.* $\hat{E}$ for 20 wt% suspensions of COPA particles in silicone oil.

Furthermore, Figure 3 (a) represents the yield stress *vs.* shear rate for the COPA–based ER fluid at several different electric field strengths, obtained via two different measuring techniques; one is the static yield stress obtained from controlled shear stress (CSS), the other is the dynamic yield stress obtained from controlled shear rate (CSR). As shown in Fig. 3 (b), it was found that the proposed universal scaling function also fit very well for the data taken from both CSR and CSS modes.

This study was supported by research grants from the Korea Science and Engineering Foundation (KOSEF) through the Applied Rheology Center (ARC), an official KOSEF-created engineering research center (ERC) at Korea University, Seoul, Korea.

References

1. Halsey, T. C., Martin, J. E. and Adolf, D., Rheology of Electro-rheological Fluids, *Phys. Rev. Lett.* **68** (1992) pp. 1519-1522.
2. Tao R., Zhang J., Shiroyanagi Y., Tang X. and Zhang X., Electrorheological Fluids under Shear, *Int. J. Mod. Phys. B* **15** (2001) pp.918-929.
3. Filisko F.E. and Henley S. Parameters Affecting Lamellar Formations in ER Fluids: An Alternative Model for ER Activity, *Int. J. Mod. Phys. B* **15** (2001) pp.686-694.
4. Gonon P., Foulc J. N. and Atten P., A Conduction Model Describing Partile-Particle Interaction in the Case of Surface Conducting Particles, *Int. J. Mod. phys. B* **15** (2001) pp.704-713.
5. See H. and Saito T., Layered Model of Electrorheological Fluid under Flow, *Rheol. Acta* **35**(1996) pp.233-241.
6. Wen W. Ma H., Tam W. Y. and Sheng P., Frequency-induced Structure Variation in Electrorheological Fluids, *Appl. Phys. Lett.* **77** (2000) pp.3821-3823.
7. Bossis G., Abbo C., Cutillas S., Lacis S. and Métayer C., Electroactive and Electro structured Elastomers, *Int. J. Mod. Phys. B* **15** (2001) pp.564-573.

8. Bullough W. A., Atkin R. J., Urang S., Kum T, G., Musch C. and Rober T., Two Dimensional Flow of an ESF: Experiment, CFD, Bingham Plastic Analysis, *Int. J. Mod. Phys. B* **15** (2001) pp.745-757.
9. Choi H. J., Cho M. S. and To K., Electrorheological and Dielectric Characteristics of Semiconductive Polyaniline-Silicone Oil Suspensions, *Physica A* **254** (1998) pp.272-279.
10. Choi H. J., Kim J. W., Joo J. and Kim B. H., Synthesis and Electrorheology of Emulsion Intercalated PANI-clay Nanocomposites, *Synth. Met.* **121** (2001) pp.1325-1326
11. Parthasarathy M. and Klingenberg D. J., Electrorheology: Mechanisms and Models, *Mater. Sci. Eng* **R17** (1996) pp.57–103.
12. Choi H. J., Kim T. W., Cho M. S., Kim S. G. and Jhon M. S., Electrorheological Characterization of Polyaniline Dispersions, *Eur. Polym. J.* **33** (1997) pp.699-703.
13. Wen W., Wang N., Tam W. Y. and Sheng P., Magnetic Materials-based Electrorheological Fluids, *Appl. Phys. Lett.* **71** (1997) pp.2529–2531.
14. Kim J. W., Kim S. G., Choi H. J. and Jhon M. S., Synthesis and Electrorheological Properties of Polyaniline-Na^+-Monmorillonite Suspensions, *Macromol. Rapid Commun* **20** (1999) pp.450-452.
15. Davis L. C., Time-dependent and Nonlinear Effects in Electrorheological Fluids, *J. Appl. Phys.* **81** (1997) pp.1985–1991.
16. Conrad H., Wu C. W. and Tang X., Conductivity in Electrorheology, *Int. J. Mod. Phys.* B **13** (1999) pp. 1729-1738.
17. Choi H. J., Lee J. H. and Jhon M. S., Electrorheological Characterization of Semiconducting Polyaniline Suspensions, *Polym. Eng. Sci.* **39** (1999) pp.493-499.
18. Kim S. G., Choi H. J. and Jhon M. S., Preparation and Characterization of Phosphate Cellulose-Based Electrorheological Fluids, *Macromol. Chem. Phys.* **202** (2001) pp.521-526,
19. Choi H. J., Sim I. S. and Jhon M. S., Aqueous Ferric Chloride Doped Poly(p-phenylene) for Electrorheological Material, *J. Mater. Sci. Lett.* **19** (2000) pp. 1629-1631.
20. Lee Y. H., Kim C. A., Jang W. H., Choi H. J. and Jhon M. S., Synthesis and Electrorheological Characteristics of Micreoencapsulated Polyaniline Particles with Melamine-Formaldehyde Resins, *Polymer* **42** (2001) pp.8277-8283.
21. Choi H. J., Lee Y. H., Kim C. A. and Jhon M. S., Microencapsulated Polyaniline Particle for Electrorheological Material, *J. Mater. Sci. Lett.* **19** (2000), pp.533-535.
22. Choi H. J., Cho M. S., Kim J. W., Webber R. M. and Jhon M. S, Effect of Ionic and Nonionic Substituents on the Electrorheological Characteristics of Polyaniline Derivatives, *Int. J. Mod. Phys. B* **15** (2001) pp.988-995.
23. Sim I. S., Kim J. W., Choi H. J., Kim C. A. and Jhon M.S., Preparation and Electrorheological Characteristics of Poly(*p*-phenylene)-Based Suspensions, *Chem. Mater.* **13** (2001) pp.1243–1247.
24. Choi H. J., Cho M. S., Kim J.W., Kim C. A. and Jhon M. S., A Yield Stress Scaling Function for Electrorheological fluids, *Appl. Phys. Lett.* **78** (2001) pp.3806-3808.
25. Tam W. Y., Yi G. H., Wen W., Ma H., Loy M. T. and Sheng P., New Electrorheological Fluid: Theory and Experiment, *Phys. Rev. Lett.* **78** (1997) pp.2987–2990.

MULTISCALE ANALYSIS OF ELECTRORHEOLOGICAL FLUIDS

BOGDAN VERNESCU
Department of Mathematical Sciences, Worcester Polytechnic Institute, 100 Institute Road, Worcester, MA 01609, U.S.A
E-mail: vernescu@wpi.edu

We construct a microscale model for a rigid particle suspension in a viscous fluid that includes Maxwell electrostatic forces. Via homogenization techniques we characterize the properties the material exhibits at the macroscale. The change in the effective constitutive equations is due to the highly oscillating electrostatic forces. The material properties are determined by both hydrodynamic and electrostatic particle interactions.

1 Introduction

Electrorheological fluids are suspensions of small solid dielectric particles in a viscous non-conducting or weakly conducting liquid. Exposed to an electric field, the ER fluids undergo a rapid reversible transition that can dramatically increase their apparent viscosity, by a few orders of magnitude, simultaneously ordering their microstructure.

Experimental observations at microscopic level show that when an electric field is applied to an ER fluid, a rapid reversible transition of the material takes place in which the suspended particles align themselves into chains and then aggregate into fibrous column-like structures parallel to the electric field [1,2]. "Chains" have been experimentally observed using optical microscopy [3,4] and other structures have been observed more recently using confocal laser microscopy [5]. When the particles align themselves in this manner the apparent viscosity of the ER material increases significantly by several orders of magnitude and in a shearing experiment the material exhibits a dynamic yield stress which is proportional to the square of the electric field [1].

Through experiments [6,7,8,9,10] and others have observed that in a shear experiment the Bingham constitutive equation models the response of ER fluids quite reasonably. This was also confirmed by dynamic simulations[14,15]. The Bingham constitutive equation reduces in the case of simple shear flow to $\sigma = \tau_d + \mu_{pl}\dot{\gamma}$, where μ_{pl} is the plastic viscosity, $\dot{\gamma}$ is the shear rate, and τ_d is the Bingham or dynamic yield stress.

There are hundreds of particle-liquid combinations of ER fluids and a large range for the particle sizes (.1 to 100 μm in diameter) and particle volume fractions (10% - 40% [13]). A better understanding of the way the ER effect is influenced by the material parameters (such as volume fraction, specific surface area, permittivity, surfactants, hydration) is needed to design better suspensions.

In this paper we model the ER fluid, at the microscopic scale, as a suspension of rigid dielectric particles in a viscous non-conducting fluid. Electrostatic Maxwell forces are taken into account to account for the interparticle forces due to particle polarization. We use a multiscale method to upscale the microsopic information and derive the macroscopic behavior of an ER fluid. The method enables us to obtain, from laws of suspension mechanics and electrodynamics, the ER-fluid behavior as an asymptotic limit. In this way rigorous formulas for the effective properties are obtained facilitating the comparison between numerical simulations and experimental measurements for validating the significance of one or more microscopic phenomena.

2 Electrostatic polarization mechanism in anhydrous systems

We will start by analyzing the simplest model of an ER fluid. The particles and the fluid are considered to be pure dielectrics and the liquid is not hydrated. This models a large number of anhydrous systems [14,15,16,17] for which the ER effect is significant.

The starting point is the model for a suspension of rigid particles in a newtonian viscous fluid, coupled with the principles of electrodynamics. The electric field satisfies Maxwell's laws:

$$E = -\nabla\phi, \quad \mathrm{div}(P + E) = 0 \text{ in } \Omega_f \text{ and } \Omega_p \tag{1}$$

For simplicity both materials, the fluid medium Ω_f and the suspended particles Ω_p, are considered here linear dielectrics, so the electric polarization $P = \chi E$ with χ the electric susceptibility, $\eta = 1 + \chi$ the dielectric constant. Since $\eta = \eta_p$ in the particle and $\eta = \eta_f$ in the fluid, the Maxwell equations are coupled with the momentum equations.

The coupling between the momentum balance equations and Maxwell's equations is done through the Maxwell stress tensor. The material is subjected to electrical forces caused by the polarization of the particles. The electric force density can be expressed in the following form :

$$F_e = -\frac{1}{2}E^2\nabla\eta. \tag{2}$$

It is important to remark here that in the case of homogeneous dielectrics, the electric force is a measure supported on the surface of the particles. A more traditional expression is in terms of the Maxwell stress tensor [18] $\tau = (\tau_{ij})$, which, unlike the viscous stress, exists in both the fluid and solid regions:

$$F_e = \mathrm{div}\tau, \quad \text{where} \quad \tau_{ij} = \eta E_i E_j - \frac{1}{2}\eta E^2\delta_{ij} \tag{3}$$

At a point inside the fluid part of the material, the balance of momentum states:

$$\rho_f \frac{dv}{dt} = \mathrm{div}\sigma + \mathrm{div}\tau + f \tag{4}$$

where ρ_f is the constant fluid density, v is the velocity, f is the density of body forces and the viscous stress for the incompressible fluid is

$$\sigma_{ij} = -p\delta_{ij} + 2\mu D_{ij}(v), \quad \text{where } D_{ij}(v) = \frac{1}{2}\left(\frac{\partial v_i}{\partial x_j} + \frac{\partial v_j}{\partial x_i}\right) \tag{5}$$

with p the pressure and μ the viscosity of the fluid.

In each particle the velocity v is a rigid solid velocity field so $D_{ij}(v) = 0$ and the balance of momentum and balance of torques for each particle gives

$$\int_{\Omega_p} \rho_p \frac{dv}{dt} dV = \int_{\partial\Omega_p} \sigma n dS + \int_{\partial\Omega_p} [\tau n]\, ds + \int_{\Omega_p} f dV$$

$$\int_{\Omega_p} \rho_p \frac{dv}{dt} dV = \int_{\partial\Omega_p} \sigma n \times (x - x_G) dS + \int_{\partial\Omega_p} [\tau n] \times (x - x_G) dS + \int_{\Omega_p} f \times (x - x_G) dV \tag{6}$$

where ρ_p denotes the particle density, x_G is the center of mass of the particle, n is the unit normal outer to the particle and $[\cdot]$ denotes the jump on the particle surface.

The boundary conditions for the components of the electric field on the solid-fluid interface $\Gamma = \bar{\Omega}_p \cap \bar{\Omega}_f$ are $[E_\tau] = 0, \quad [\chi E \cdot n] = 0, \quad [v] = 0$ on Γ.

In the absence of the electric field, the behavior of suspensions is well understood. Suspensions [19,20,21] behave like viscous fluids with a modified, in general anisotropic, viscosity.

By contrast, ER fluids exhibit a different constitutive behavior when under the influence of an electric field. The Bingham-like behavior in shear experiments is a consequence of the Maxwell force term (2), which in the simplest case of constant dielectric permittivities η_f and η_p, is zero in the bulk of each phase and appears only on the fluid-solid interface.

3 Multiple scale analysis

We consider the suspension having a periodic structure, with Y denoting the rectangular periodicity box. Y is formed by solid particles part Y_p and the fluid part Y_f. We use double scale asymptotic expansions for the velocity, pressure and electric potential in the form:

$$\begin{aligned} &v^\epsilon(x) = v^0(x) + \epsilon v^1(x,y) + \mathcal{O}(\epsilon^2), \quad p^\epsilon(x) = p^0(x,y) + \epsilon p^1(x,y) + \mathcal{O}(\epsilon^2), \\ &\phi^\epsilon(x) = \phi^0(x) + \epsilon\phi^1(x,y) + \mathcal{O}(\epsilon^2) \end{aligned} \tag{7}$$

where the functions v^i, p^i, and ϕ^i are Y-periodic and where $y = x/\epsilon$.

From the Maxwell equations ϕ^1 can be expressed as

$$\phi^1(x,y) = \vec{\alpha}(y) \cdot \nabla_x \phi^0 + c_1 \tag{8}$$

where α_i is a Y-periodic function that satisfies a local problem:

$$-\text{div}_y(\eta(\nabla_y \alpha_i + e_i)) = 0 \text{ in } Y_f \cup Y_p, \ \ [\eta(\nabla_y \alpha_i + e_i)] \cdot n = 0, \ \ [\alpha_i] = 0 \text{ on } \Gamma. \tag{9}$$

with e_i the i-th vector of the Cartesian base and Γ the particles' surfaces.

The asymptotic two-scale expansion for the electric potential is thus

$$\phi^\epsilon(x) = \phi^0(x) + \epsilon \left(\alpha(\vec{y}) \cdot \nabla_x \phi^0 + c_1 \right) + \mathcal{O}(\epsilon^2). \tag{10}$$

The velocity corrector v^1 can be written as a sum of two terms corresponding to the viscous and electrostatic effect

$$v_k^1 = \chi_k^{ij} D_{ij} v^0 + \xi_k^{ij} \frac{\partial \phi^0}{\partial x_i} \frac{\partial \phi^0}{\partial x_j} + A_k(x), \tag{11}$$

where A is an additive constant vector with respect to y and χ^{ij} and ξ^{ij} satisfy the following local problems:

$$-2\mu \Delta_y \chi^{mn} = -\nabla_y \pi^{mn} \text{ in } Y_f, \;\; \nabla_y \cdot \chi^{mn} = 0 \text{ in } Y_f, \tag{12}$$

$$D_{ij}(\chi^{mn}) = \frac{1}{2}(\delta_{im}\delta_{jn} + \delta_{in}\delta_{jm}) \text{ in } Y_p, \;\; [\chi^{mn} \cdot n] = 0 \text{ on } \Gamma \tag{13}$$

and

$$-2\mu \Delta_y \xi^{mn} = D_{ij}(\eta A_{ijmn}) \text{ in } Y_f, \;\; \nabla_y \cdot \xi^{mn} = 0 \text{ in } Y_f, \;\; D_{ij}(\xi^{mn}) = 0 \text{ in } Y_p \tag{14}$$

Here the average on a set S was denoted by $\langle \cdot \rangle_S = \frac{1}{|S|} \int_S \cdot dA$ and $A_{ijkl} = A_{ik} A_{jl} - \frac{1}{2} A_{qk} A_{ql} \delta_{ij}$ with $A_{ik} = \delta_{ik} + \frac{\partial \alpha_k}{\partial y_i}$.

Therefore, the double scale expansion for $v^\epsilon(t,x)$ becomes

$$v^\epsilon(t,x) = v^0(t,x) + \epsilon \left(\chi_k^{ij} D_{ij} v^0 + \xi_k^{ij} \frac{\partial \phi^0}{\partial x_i} \frac{\partial \phi^0}{\partial x_j} + A_k(x) \right) + \mathcal{O}(\epsilon^2). \tag{15}$$

The macroscopic equation for the electric potential is

$$\text{div}_x \left(\eta^H \nabla_x \phi^0 \right) = 0 \text{ with } \eta_{ij}^H = \frac{1}{|Y|} \int_Y \eta \left(\delta_{ij} + \frac{\partial \alpha_i}{\partial y_j} \right) dy \tag{16}$$

The effective permittivity is determined thus using the solutions of the local Maxwell equation (9).

Next we establish the macroscopic flow equation

$$\frac{\partial v_i^0}{\partial t} + (v^0 \cdot \nabla_x) v_i^0 + \gamma_{ijkl}^H v_j^0 D_{kl}(v^0) = \frac{\partial \sigma_{ij}^H}{\partial x_j} + f_i \tag{17}$$

$$\sigma_{ij}^H = -p^0 \delta_{ij} + 2\mu_{ijkl}^H D_{lk}(v^0) + \tau_{ijkl}^H \frac{\partial \phi^0}{\partial x_k} \frac{\partial \phi^0}{\partial x_l} \tag{18}$$

where,

$$\mu_{ijkl}^H = \langle \mu(\delta_{il}\delta_{jk} + D_{ij_{(y)}} \chi^{lk}) \rangle_{Y_f} + \langle \mu \left(\delta_{ik}\delta_{ml} + D_{im_{(y)}} \chi^{lk} \right) y_j n_m \rangle_\Gamma \tag{19}$$

$$\gamma_{ijkl}^H = \left\langle \frac{\partial \chi_i^{kl}}{\partial y_j} \right\rangle_Y, \quad \tau_{ijkl}^H = \langle 2\eta A_{ijkl} + 2\mu D_{ij} \xi^{kl} \rangle_Y \tag{20}$$

We have to remark that the effective viscosity μ_{ijkl}^H in the macroscopic constitutive equation, (18), is the same as the one obtained for a suspension of solid particles in a newtonian fluid by several authors[19,20,22].

Thus the effective behavior of the ER-fluid is governed by (17), (16) and the incompressibility condition. We need to remark that the electric field has as effect the change of the constitutive equation (18). The macroscopic constitutive equation is not that of a newtonian fluid anymore. The additional term that appears in (17) is due to the electric forces and polarization effect.

4 Remarks

The first conclusion is that the transformed constitutive behavior is due to the particles' polarization. Indeed in a suspension with equal dielectric constants for the fluid and the solid no electrorheologic effect can be observed. If $\eta_f = \eta_p$ then the solution, to (9) is $\alpha = 0$ up to a constant. The homogenized coefficient then becomes $A_{ijkl} = \delta_{ik}\delta_{jl} - \frac{1}{2}\delta_{kl}\delta_{ij}$ and we obtain $\tau^0_{ij} = \eta\left(\frac{\partial\phi^0}{\partial x_i}\frac{\partial\phi^0}{\partial x_j} - \frac{1}{2}\frac{\partial\phi^0}{\partial x_k}\frac{\partial\phi^0}{\partial x_k}\delta_{ij}\right)$. The above tensor has no contribution in (17). Thus the electric field has no contribution in the local problem and therefore $\xi^{kl} = 0$. This shows that our model agrees with the experimental observations [23] that there is almost no electrorheological effect when the ratio of the dielectric constants of the two phases is close to 1 .

Next consider a shear experiment between two infinite electrodes with a uniform potential distribution. In this case the constitutive equation will be similar to the one for the Bingham fluid. We are thus able to determine a formula for the dynamic yield stress. We consider a uniform shearing of the ER fluid submitted to uniform electric field boundary conditions in a two dimensional slab, $0 < x_2 < 1$, by applying a shear force to the top electrode while the bottom remains fixed. The corresponding boundary conditions for such an experiment are $v^0(0) = 0,\ v^0(1) = \dot{\gamma}$ and $\phi^0(0) = 0,\ \phi^0(1) = \phi_0$. It is an easy check to see that the solution to (16) with the above boundary conditions is $\phi^0 = \phi_0 x_2$. Assuming the flow has a low Reynolds number, equation (17) reduces to

$$\frac{\partial}{\partial x_j}\left(-p^0\delta_{ij} + \mu^H_{ijkl}D_{lk}(v^0) + \tau^H_{ijkl}\frac{\partial\phi^0}{\partial x_k}\frac{\partial\phi^0}{\partial x_l}\right) = 0. \tag{21}$$

We also know that the limit flow is incompressible and so $\operatorname{div}(v^0) = 0$. Substituting the solution, $\phi^0 = \phi_0 x_2$, the flow equations admits a laminar solution and we obtain $v^0 = (\dot{\gamma}x_2, 0)$ and $p^0 = c$. The homogenized stress tensor reduces to

$$\sigma^H_{ij} = -p\delta_{ij} + \mu^H_{ijkl}D_{kl}(v^0) + \tau^H_{ij22}\phi_0^2 \quad \text{with} \quad D_{kl}(v^0) = \begin{pmatrix} 0 & \dot{\gamma}/2 \\ \dot{\gamma}/2 & 0 \end{pmatrix} \tag{22}$$

We can see that the stress tensor at the macroscopic scale has exactly the form of the Bingham constitutive equation for a shear experiment. Moreover, we see that the dynamic yield stress increases quadratically with the electric field. A similar conclusion is obtained by an averaging method in a seminal paper of Bonnecaze and Brady[11]. The homogenization method however is able to provide explicit formulas for the effective coefficients that enable us to compute the yield stress and compare it to the experimental data.

The upscaling method presented here can be extended to include the presence of ions in the fluid and the boundary layer effects.[24] These problems exhibit "size effects" at the macroscale, i.e. the macroscopic coefficients (viscosity, permittivity, yield stress) depend on the particle size. This property is similar to the one observed in composited with "imperfect boundaries".[25,26]

References

1. T. C. Halsey, *Science*, **258**, 761 (1992).
2. T. C. Halsey and J. E. Martin, *Scientific American*, 58 (1993).
3. J. D. Carlson, A. F. Sprecher and H. Conrad, *Electrorheological Fluids — Proc.Second Int. Conf. on ER Fluids* (Technomic, Pennsylvania, 1990).
4. Y. Shih, A. F. Sprecher and H. Conrad, *Int. J. Modern Phys. B*, **20**, 2877 (1994).
5. U. Dassanayake, S. Fraden, and A. vanBlaaderen, *J. Chem. Phys.* **112**, 3851 (2000).
6. H. Conrad., Y. Chen and A. F. Sprecher, *Int. J. Modern Phys. B*, **6**, 2575 (1992).
7. H. Conrad, A. F. Sprecher, Y. Choi and Y. Chen, *Journal of Rheology*, **35**, 1393 (1991).
8. Y. Otsubo and K. Edamura, *J. Rheology*, **38**, 1721 (1994).
9. Y. Otsubo and K. Edamura, *J. Colloid and Interface Science*, **172**, 530(1995).
10. W. S. Yen and P. J. Achorn, *Journal of Rheology*, **35**, 1375 (1991).
11. R. T. Bonnecaze and J. F. Brady, *Journal of Rheology*, **36**, 73(1992).
12. R. T. Bonnecaze and J. F. Brady, *Journal of Chemical Physics*, **96**, 2183 (1992).
13. Tolu-Honary, A. and Court, M. *Hydraulics & Pneumatics*, **January**, 133, (1994).
14. H. Block and J. P. Kelly, J.P., *J.Phys. D*, **21**, 1661 (1988).
15. T. C. Jordan and M. T. Shaw, *IEEE Trans. Elect. Insul.* **24**, 849 (1989).
16. C. J. Gow and C. F. Zukoski, *J. Coll. Interface Sci.* **136**, 175 (1990).
17. F. E. Filisko and L. H. Radzilowski, *J Rheol.* **34**, 539 (1990).
18. G. A. Maugin, Continuum Mechanics of Electromagnetic Solids, *Elsevier*, (1988).
19. Batchelor, G.K. The stress system in a suspension of force-free particles, *Journal of Fluid Mechanics*, **41** (1970), 545-570.
20. J. B. Keller, L. A. Rubenfeld and J. E. Molyneux, *Journal of Fluid Mechanics*, **30** (1967), 97-125.
21. E. J. Hinch and L. G. Leal, *J. Fluid Mech.*, **71**, 481 (1975).
22. T. Levy and E. Sanchez-Palencia, *Journal of Non-Newtonian Fluid Mechanics*, **13**, 63 (1983).
23. A.P. Gast and C.F. Zukoski, *Advances in Colloid and Interface Science*, **30**, 153 (1989).
24. B. Vernescu, forthcoming
25. R. Lipton and B. Vernescu, *J. Appl. Phys.*, **9**, 8964 (1996).
26. R. Lipton and B. Vernescu, *Math. Models & Methods in Appl. Sci.*, **5**, 1139 (1995).

Continuum modelling of electrorheological fluids

R. DROUOT
**Centre de Mécanique des Milieux Continus, Université de Versailles Saint-Quentin en Yvelines and Laboratoire de Modélisation Mécanique CNRS-Université Pierre et Marie Curie, 45 av. des Etats-Unis, 78035 Versailles Cedex, FRANCE*
Raymonde.Drouot@meca.uvsq.fr

G. NAPOLI*
napoli@lmm.jussieu.fr

G. RACINEUX*
Guillaume.Racineux@meca.uvsq.fr

A thermodynamical continuum modelling is proposed for electrorheological fluids. This thermodynamical approach tries to describe the response of an electrorheological fluid in the solid phase (which means with an applied adequate electric field) under mechanical solicitations. Thermodynamic formulations distinguish the contributions due to reversible and irreversible process. In an electrorheological fluid the microstructure generated by the application of an electric field introduces parameters whose the evolution of which influence the behavior. These parameters will be defined as internal variables, or hidden variables, in the formulation of thermodynamics. They are used for the description of dissipative effects. The choice and the number of pertinent parameters are crucial and need to be selected according to the nature of microscopic mechanism and experimental observations. The presence of fibrous structures in the E.R. fluid, when an electric field is applied, is an observed fact. In the theory of thermodynamics with internal variables, the introduction of internal variables doesn't modify, by hypothesis, the balance equations. Nevertheless, the experiments show the existence of hysteresis phenomenon due to the reorganization, at the microscopic level, of the particles that compose the fibers. In order to describe these phenomena, we introduce two internal variables following an idea of Kiryushin.

1. Introduction

Winslow[1] was the first to observe the electrorheological properties of fine electrically polarizable particles dispersed in a dielectric liquid. These fluids have been largely studied from different point of view. These electrorheological suspensions show dramatic change in the rheological properties when an electric field is applied. Many parameters influence the rheology of the material such as concentration, intensity of electric field etc... The complicated nature of these materials is due to a great number of microscopic physical process. Micromechanical analyses of electrorheological behavior have considerably clarified the structure of such systems (see for example Gast and Zukoski).[2]

We are interested in the case when the suspension under the action of an external electric field acquires dipole moment and these dipolar particles rapidly aggregate to form long chains. So it corresponds to certain range of values for concentration of particles suspensions. A macroscopic description of such materials needs the use of additional phenomenological parameters reflecting the internal structure. The standard general thermodynamics framework is used here to develop and identify a model for electrorheological fluid. The role (influence) of the internal structure will be described using the internal variables theory. The concept of internal variables has its origin in thermodynamics. Contemporary understanding has been recently reviewed by Maugin and Muschik[3] in the context of non equilibrium thermodynamics.

The main idea is that the thermodynamic state is determined not only by observable variables, but also by internal variables α or hidden variables. The formalism is based on continuum mechanics and on the study of dissipation potential from which the equation satisfied by internal variables are formed. This concept of hidden variables is developed for modeling of complex dissipative processes.

2. Internal Variables in Continuum Mechanics

Internal variables are considered as opposed to observable or controllable variables. Observable variables are inertial, governed by a balance law with the kinetic energy, internal variables do not possess inertia. The choice of internal variable depends on the level of observation and these variables can be a scalar, a vector or have different tensorial order. We briefly recall the mains "ideas" in this approach.[3] We note $\underline{X}$ the observable variables for system and α the internal variables. We suppose that :

(i) The balance laws associated with the observable variable are not modified (mass, momentum, moment of momentum, energy) ;
(ii) There is no reversible work associated to α variables.

The notion of internal variable is used for the description of dissipation mechanisms. For classical continuum theory, the Clausius-Duhem inequality governing a mechanical process is

$$-\rho\left(\dot{\psi} + \eta\,\dot{T}\right) + \underline{\underline{\sigma}} : \dot{\underline{\underline{\varepsilon}}} - \frac{1}{T}\,\underline{q}\cdot\text{grad}T \geq 0 \tag{1}$$

where ρ is the mass per unit volume, ψ is the Helmoltz free energy, $\underline{\underline{\sigma}}$ the Cauchy stress tensor, $\underline{\underline{\varepsilon}}$ the strain tensor, η the local entropy, T the temperature, $\underline{q}$ the heat flux vector, the upper dot indicates material time derivative.

As an example, if we assume the following decomposing of $\underline{\underline{\varepsilon}}$

$$\underline{\underline{\varepsilon}} = \underline{\underline{\varepsilon}}^{e} + \underline{\underline{\varepsilon}}^{p}, \tag{2}$$

where $\underline{\underline{\varepsilon}}^{e}$ is the elastic part or reversible part and $\underline{\underline{\varepsilon}}^{p}$ the anelastic part ; if we also assume the free energy function to be

$$\psi = \psi\left(\underline{\underline{\varepsilon}} - \underline{\underline{\varepsilon}}^{p}, T, \alpha\right), \tag{3}$$

the inequality (1) becomes

$$\underline{\underline{\sigma}} : \dot{\underline{\underline{\varepsilon}}}^p + A\,\dot{\alpha} - \frac{1}{T}\,\underline{q}.\underline{\mathrm{grad}T} \geq 0, \tag{4}$$

with

$$A = -\rho\frac{\partial\psi}{\partial\alpha}, \qquad \eta = -\frac{\partial\psi}{\partial T}, \qquad \underline{\underline{\sigma}} = \rho\frac{\partial\psi}{\partial\underline{\underline{\varepsilon}}^e}. \tag{5}$$

If we impose intrinsic dissipation and thermal dissipation to be separately positive, we obtain

$$\Phi_{int} = \underline{\underline{\sigma}} : \dot{\underline{\underline{\varepsilon}}}^p + A\,\dot{\alpha} \geq 0. \tag{6}$$

We recall that equilibrium is defined by

$$\dot{\underline{\underline{\varepsilon}}}^p = 0 \qquad \text{and} \qquad A = 0. \tag{7}$$

We postulate the existence of dissipation potential D so we have

$$\dot{\underline{\underline{\varepsilon}}}^p = \frac{\partial D}{\partial\underline{\underline{\sigma}}} \qquad \text{and} \qquad A = \frac{\partial D}{\partial\,\dot{\alpha}}. \tag{8}$$

The evolution equation for the internal variable α is obtain through the relation

$$A = -\rho\frac{\partial\psi}{\partial\alpha} = \frac{\partial D}{\partial\,\dot{\alpha}}. \tag{9}$$

3. Description of Electrorheological Fluid when an Electric Field is Applied

We are interested by the description of materials when an electric field is applied and then the material is in solid phase. The particles acquire dipole moments and they aggregate into chains of dipoles parallel to external electric field. We try to describe the behavior of the material when mechanical loading are considered. The media with fibrous structures will correspond to our suitable initial state of equilibrium. We shall emphasize the role played by the time evolution of the structure and its implication in dissipative effects.

We consider the solid phase of electrorheological fluid and use Lagrangian description. We restrict our study to non magnetizable media and suppose that only electric field is applied. The state variables are the Lagrangian strain tensor, polarization per unit volume and the temperature. The basic equations we need are those of electromechanics.[4]

In material frame we recall the electromechanical balance laws and the quasi-static approximations of Maxwell equations

$$\rho_0 = \rho J = \rho\det\left(\underline{\underline{F}}\right), \tag{10}$$

$$\rho_0\,\ddot{\underline{x}} = \underline{\mathrm{div}}\left(\underline{\underline{T}}\right) + J\left(\underline{f} + \underline{f}^{em}\right), \tag{11}$$

$$\rho_0\,\dot{e} = \underline{\underline{\Sigma}} : \dot{\underline{\underline{E}}} - {}^{L}\,\underline{E}.\dot{\underline{\Pi}} - \mathrm{div}\underline{q}\,, \tag{12}$$

$$ {}^{L}\underline{E} + \underline{E} = \underline{0}, \tag{13} $$

where the materials quantities are :

- ρ_0 is the mass density in the material frame,
- $x_i = x_i\left(X_k, t\right)$,
- $\underline{\underline{F}}$ is the deformation between its reference configuration K_R at time t_0 and its present configuration at time t,
- $\underline{\underline{E}} = \frac{1}{2}\left(\underline{\underline{F}}^T \underline{\underline{F}} - \underline{\underline{I}}\right)$,
- $\underline{\underline{\Sigma}}$ is Piola-Lagrange stress tensor,
- $\underline{\underline{T}} = \underline{\underline{\tilde{T}}} + {}^{L}\underline{E} \otimes \underline{\Pi}$ with $\underline{\underline{\tilde{T}}} = \underline{\underline{F}}\,\underline{\underline{\Sigma}}$,
- ${}^{L}\underline{E}$: local electric field in material frame,
- $\underline{\Pi}$: electric polarization per unit volume,
- $\underline{E}$: external electric field,
- $\underline{q}$: heat flux density vector.

All processes must satisfy the Clausius-Duhem inequality

$$ -\rho_0\left(\dot{\psi} + \eta\,\dot{T}\right) + \underline{\underline{\Sigma}} : \underline{\underline{\dot{E}}} - {}^{L}\underline{E}.\underline{\dot{\Pi}} - \frac{1}{T}\underline{q}.\mathrm{grad}T \geq 0. \tag{14} $$

The point of view developed here is identical to that of Y.M. Shkel[5], in continuum mechanics for the reversible part of the behavior, we propose a method for the description of dissipative phenomena. We will further consider only isothermal process and we consider the configuration K_i defined by

$$ \underline{\Pi} = \underline{P}^0\,,\quad \underline{E} = \underline{E}^0,\quad \underline{\underline{\mathrm{grad}}}\left(\underline{P}^0\right) = \underline{0},\quad \underline{\underline{\mathrm{grad}}}\left(\underline{E}^0\right) = \underline{0},\quad x_i = X_i \text{ and } \underline{E}^0 \,//\, \underline{P}^0. \tag{15} $$

Balance laws are all satisfied with

$$ \underline{\underline{T}}^0 = \underline{\underline{\tilde{T}}}^0 + {}^{L}\underline{E}^0 \otimes \underline{P}^0. \tag{16} $$

Without external mechanical loading we have a pre-stressed configuration due to electric field. Now we consider linearization of the balance laws about the configuration K_i and we introduce the perturbed quantities $\underline{u}$, $\underline{p}$, $\underline{e}$, ${}^{L}\underline{e}$ and $\underline{\underline{t}}$ in a classical way

$$ \underline{x} = \underline{X} + \underline{u}\,,\qquad \underline{\Pi} = \underline{P}^0 + \underline{p} - \underline{\underline{grad}}(\underline{u})\,\underline{P}^0\,,\qquad \underline{E} = \underline{E}^0 + \underline{e}\,, \tag{17} $$

$$ \underline{\underline{T}} = \underline{\underline{T}}^0 + \underline{\underline{t}},\qquad J = 1 + \mathrm{div}\underline{u}\,,\quad \text{and}\quad \underline{\underline{\varepsilon}} = \mathrm{sym}(\underline{\underline{\mathrm{grad}}}\,(\underline{u})). \tag{18} $$

The formalism is largely detailed by Maugin[4] and Eringen.[6] So we obtain the linear balance laws

$$ \underline{\mathrm{rot}}\,(\underline{e}) = \underline{0}, \tag{19} $$

$$ \mathrm{div}\left(\underline{e} + \underline{p} - \underline{\underline{\mathrm{grad}}}\,(\underline{u})\,\underline{P}^0\right) = 0\,, \tag{20} $$

$$ \rho = \rho^0\left(1 - \mathrm{div}\underline{u}\right), \tag{21} $$

$$\rho_0 \, \underline{\ddot{u}} = \underline{div} \, (\underline{\underline{t}}) + \underline{\underline{\text{grad}}} \, (\underline{e}) \, \underline{P}^0 \, , \tag{22}$$

$$\underline{e} + {}^L \underline{e} = \underline{0} \, , \tag{23}$$

The choice of thermodynamical potential gives the reversible behavior of the material; in this case the unit direction $\underline{\tau}^0 = \underline{P}^0 / |\underline{P}^0|$ gives initial anisotropy direction and occurs in the expression of invariants.

4. Description with Internal Variables - Evolution Equations

The fibers structures is characteristic of the material[7, 8] and following Kiryushin[9] we introduce two internal variables. This idea appears in some simple expression in thermodynamic[10]. In our model the fibers are considered as linear elastic rods with initial configuration defined by $\underline{\tau}^0$ direction. The linearized deformation of the rod around this configuration is defined by :

(i) ξ the average relative stretch of chain;
(ii) $\underline{\gamma}$ the average infinitesimal rotation of the section of the rod.

The additional parameters have different tensorial characters : $\underline{\gamma}$ is an axial vector associated to an anti-symmetric tensor with dimension (L^{-1}) and describes the bending of the rods while ξ is dimensionless and represents one component of the classical deformation tensor ; ξ will play a role in squeeze flow while $\underline{\gamma}$ will be significative in shear flow experiments. The internal variables appear in the expression of Helmotz free energy through invariants and we introduce the following partial derivatives

$$N = -\frac{\partial \psi}{\partial \xi}, \tag{24}$$

$$\underline{M} = -\frac{\partial \psi}{\partial \underline{\gamma}}. \tag{25}$$

As Kiryushin[9] we suppose that ξ and $\underline{\gamma}$ consist of two parts, one transported by the macroscopic deformation of the media and the other, noted with suffix $_{(i)}$, corresponding to microscopic restructuring, so that

$$\xi = \xi_{(i)} + \underline{\tau}^0 . \underline{\underline{\varepsilon}} \, \underline{\tau}^0, \tag{26}$$

$$\underline{\gamma} = \underline{\gamma}_{(i)} + \underline{\gamma}_{(c)}. \tag{27}$$

We introduce the dissipative parts of the stress tensor and the local electric field in the following manner

$$\underline{\underline{t}} = {}^D \underline{\underline{t}} + {}^R \underline{\underline{t}}, \quad {}^L \underline{e} = {}^{DL} \underline{e} + {}^{RL} \underline{e}, \tag{28}$$

and we obtain the expression of Clausius-Duhem inequality

$$\left({}^D \underline{\underline{t}} + N \left(\underline{\tau}^0 \otimes \underline{\tau}^0\right) + {}^{DL} \underline{e} \otimes \underline{P}^0\right) : \underline{\underline{\dot{\varepsilon}}}^0 - {}^{DL} \underline{e} . \left(\underline{\dot{p}} - \underline{\omega} \wedge \underline{P}^o\right) + N \, \dot{\xi}_{(i)} + M \, \underline{\dot{\gamma}} \geq 0 \tag{29}$$

$$\text{with } \underline{\omega} = \frac{1}{2}\underline{\text{rot}}\,(\dot{\underline{u}})\,. \tag{30}$$

Here the upper dot indicates partial time derivative. In absence of internal variable, we could have dissipation associated to state variables $\underline{\underline{\varepsilon}}$ and $\underline{P}$. The introduction of two hidden variables allows us some new terms. In classical mechanics, dissipative processes can be ordered in two mains categories :

(i) relaxation process (viscosity) ;
(ii) plasticity.

First of all let us consider the simple effect relaxation process associated to the new variable $\underline{\gamma}$ (it is not necessary of the good type). In frame of thermodynamics of irreversible processes, Onsager theory, we obtain complementary laws

$${}^{D}\underline{e} \wedge \underline{P}^0 = \alpha\underline{\omega} + \lambda\,\dot{\underline{\gamma}}\,, \tag{31}$$

$$\underline{M} = \mu\,\dot{\underline{\gamma}} + \lambda\underline{\omega}\,. \tag{32}$$

Evolution equation for $\underline{\gamma}$ follows from Eq. (24), (25), (31) and (32). The complexity of this equations depends on the choice of ψ. In first step, a quadratic dependence of ψ on $\underline{\gamma}$ will be taken so $\underline{M}$ will be linear in $\underline{\gamma}$.

The system is now closed and it is possible to look at some shear problem as those studied by E. Lemaire.[8] The resolution of the system of equations is not trivial and need to use sophisticated numerical method.

5. Conclusion

To summarize, the formalism of thermodynamics with hidden variables allows us to obtain in a systematic manner a modelling of some microscopic effect. Here we are dealing with the bending of fibers under mechanical loading and we think we will be able to describe the breaking of fibers.[11]

References

1. W.M. Winslow, *J. Appl. Phis.* **20**, 1137 (1949).
2. A.P. Gast and C.F. Zukoski, *Adv. Colloid Interface Sci.* **30**, 153 (1989).
3. G.A. Maugin and W. Muschik, *J. of Non Equili.Thermodynamics* **19**, 217 (1994).
4. G.A. Maugin and J. Pouget, *J. Acoustic. Soc AM.* **67**, 575 (1980).
5. Y.M. Shkel anf D.J. Klingenberg, *J. Rheol.* **43**, 1307 (1999).
6. A.C. Eringen and G.A. Maugin, *Electrodynamics of Continua I.(* ed. Springer-Verlay 1989).
7. W. Wen, D.W. Zheng and K.N. Tu, *J. Appl. Phys.* **85**, 530 (1999).
8. E. Lemaire, G. Bossis and Y. Gasseli, *Langmir* **8**, 385 (1992).
9. V.V. Kiryushin, *J. of Magnetism and Magnetic Mater.* **39**, 14 (1983).
10. T. Chen, X. Zhang, R.N. Zitter and R. Tao, *J. Appl. Phys.* **74**, 942 (1993).
11. E.M. Furst, A.P. Gast, *Phys. Review E.* **61**, 6, 6732 (2000).

STRESS RELAXATION OF MAGNETORHEOLOGICAL FLUIDS

W.H.LI*, G.CHEN, S.H.YEO AND H.DU

Centre for Mechanics of Micro-Systems, School of Mechanical and Production Engineering, Nanyang Technological University, Nanyang Avenue, Singapore 639798

In this paper, the experimental and modeling study and analysis of the stress relaxation characteristics of magnetorheological (MR) fluids under step shear strain are presented. The experiments are carried out using a rheometer with parallel-plate geometry. The applied strain varies from 0.01% to 100%, covering both the pre-yield and post-yield regimes. The effects of step strain, field strength, and temperature on the stress modulus are addressed. For small step strain ranges, the stress relaxation modulus $G(t,\gamma)$ is independent of step strain, where MR fluids behave as linear viscoelastic solids. For large step strain ranges, the stress relaxation modulus decreases gradually with increasing step strain. Moreover, the stress relaxation modulus $G(t, \gamma)$ was found to obey time-strain factorability. That is, $G(t, \gamma)$ can be represented as the product of a linear stress relaxation $G(t)$ and a strain-dependent damping function $h(\gamma)$. The linear stress relaxation modulus is represented as a three-parameter solid viscoelastic model, and the damping function $h(\gamma)$ has a sigmoidal form with two parameters. The comparison between the experimental results and the model-predicted values indicates that this model can accurately describe the relaxation behavior of MR fluids under step strains.

1 Introduction

MR fluids have found many potential applications because they exhibit dramatic changes in rheological properties when exposed to magnetic fields. Most of MR devices operate in dynamic conditions, where the MR fluids are subjected to complex shears, including steady shear, oscillatory shear, and instantaneous shear. It is crucial to understand clearly the MR rheology under such dynamic conditions because the mechanical behavior of MR fluids depends much on the working conditions [1].

Many studies have reported the rheological behavior of MR fluids under steady shear and oscillatory shear [2,3]. Under the steady shear, MR fluids operate in the post-yield regime. The studies have addressed how the shear rate, volume fraction, field strength, and temperature influence the rheological behaviors, including the yield stress and apparent viscosity [2]. Under the oscillatory shear, a few reports focused on study of linear-viscoelastic properties of MR suspension [3]. At small strain amplitudes, MR fluid behaves as a linear viscoelastic body, where the storage modulus shows an increasing trend with frequency but be independent of strain amplitude. At large strain amplitudes, MR fluids exhibit nonlinear viscoelastic or viscoplastic properties. The storage modulus depends not only on oscillatory frequency but also on strain amplitudes [3].

So far, little attention has been paid on study of stress relaxation of MR fluids under simple step shear, though the stress relaxation experiments can also provides useful information towards understanding of viscoelastic properties [4]. A step shear strain experiment is a direct means of measuring the strain-dependent rheology of materials, which is also a method to study viscoelastic properties of materials. In this paper, the step shear strain experiment is conducted to study stress relaxation of MR fluids. The applied strain varies from 0.01% to 100%, covering both the pre-yield and post-yield regimes. The

* Email: mwhli@ntu.edu.sg

obtained stress relaxation moduli at different strains are analyzed by time-strain factorability function. The whole results are quantitatively compared to the predictions of proposed five-parameter model.

2 Theoretical Background

In the following sections, the concept of a step strain method and the theory on time-strain factorability will be presented respectively.

2.1 *Simple Step Strain Method*

Conventionally, a step-strain method is used to measure the stress relaxation of polymer solutions and polymer melts [5]. An instantaneous strain, γ_0, is applied at time $t = 0$ followed by relaxing the resulting time-dependent stress $\tau(t, \gamma)$ over a period of time. The stress relaxation modulus, $G(t, \gamma)$, is defined as the ratio of resulting stress to the step strain:

$$G(t, \gamma) = \tau(t, \gamma) / \gamma_0 . \tag{1}$$

The basic mechanism of stress relaxation is viscous flow [5]. In amorphous viscoelastic materials, the stress is relaxed by molecules sliding past each other reducing applied internal forces over an extended period of time, and as $t \to \infty$, $G(t, \gamma) \to 0$. In solids, the relaxation time can be infinite, and $G(t, \gamma)$ approaches a constant finite value G_p.

2.2 *Time-Strain Factorability*

Nonlinear theory of viscoelastic materials [4], as applied to step shear, suggests that the stress relaxation modulus, $G(t, \gamma)$, obeys time-strain factorability. That is, the stress relaxation modulus is factorized into the products of two functions

$$G(t, \gamma) = G(t)\, h(\gamma), \tag{2}$$

where $G(t)$ is a linear relaxation modulus. And $h(\gamma)$ is a strain-dependent damping function describing the nonlinear rheological response. Therefore, the nonlinearity of the rheological behavior of MR fluids is only characterized by the damping function $h(\gamma)$. In particular, $h(\gamma)$ approaches unity at small strains and decreases monotonically as the strain increases. For example, at the strain $\gamma = 0$, $h(\gamma) = 1$, the material can be described by linear viscoelasticity theory.

Many forms of equations have been proposed for fitting the damping function [6]. In this paper, the damping function, $h(\gamma)$, containing a sigmoidal nature, is adopted and represented as

$$h(\gamma) = 1/(1+ a\gamma^b), \tag{3}$$

where a and b are two parameters to be determined based on experimental data.

3 Experiment

The MR sample, MRF-132LD, is supplied by the US Lord Corporation. This commercial product has the advantage of high yield stress, wide temperature adaptability, low contamination and so on.

The rheological measurements are carried out using a MR rheometer (Model: UDS 200) with parallel-disk geometry. The samples are placed in a MR cell with a constant gap of 1.0mm. The disk diameter is 20mm. The magnetic field is supplied by a built-in electromagnetic coil.

In each experiment, the suspension is firstly subjected to a steady shear at $\dot{\gamma}$= 300 s^{-1} for 2 minutes as an initial condition. The desired magnetic field is then applied by setting the coil current and wait for 1 minute, ensuring the MR suspension form into a stable structure. Finally, the step strain is applied to the sample, and the resulting stress relaxation modulus is recorded accordingly.

4 Results and Discussion

The stress relaxation tests are carried out under the strain-controlled mode. The applied strain varies from 0.01% to 100%. The evaluation of MR fluids is conducted at three temperatures of 20^0C, 40^0C, and 60^0C. The stress relaxation modulus, $G(t, \gamma)$, can be found from the stress relaxation test. The influences of step strain (γ_0), magnetic field strength (B), and temperature (T) on stress relaxation modulus can also be investigated.

4.1 Evolution of $G(t, \gamma)$ with Step Strain

Figure 1 shows the time dependence of stress relaxation at the field stress of 340 mT under various step strains ranging from 0.01% to 100%. For each applied step strain, the value of $G(t, \gamma)$ decreases quickly towards a plateau value G_p, and the relaxation time constant is within 1 second. At sufficiently small strains, the relaxation modulus has little variation, where MR fluid behaves as a linear viscoelastic body. From this figure, linearity can be obtained for strain values up to about 0.1%. That is the stress relaxation modulus, $G(t, \gamma)$, is independent of the applied strain at the small strain ranges $\gamma_0 \leq 0.1\%$. For strains higher than 0.01%, $G(t, \gamma)$ shows a sharply decreasing trend with strain, where MR fluids exhibit nonlinear viscoelastic properties. For example, the value of G_p at γ_0 = 100% is about three orders lower than that γ_0=0.1%.

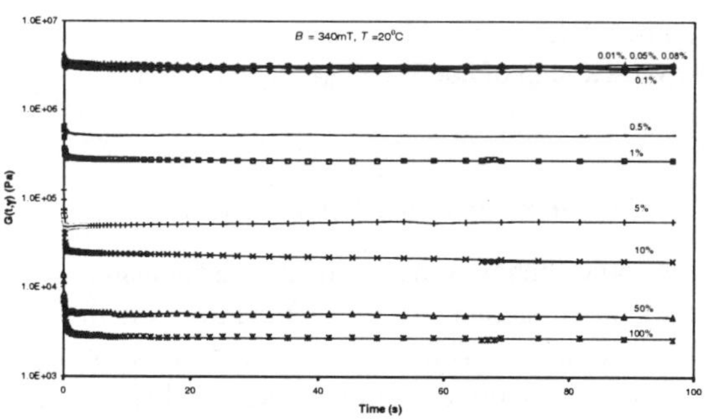

Figure 1. The stress relaxation modulus, $G(t, \gamma)$, under various step strains

4.2 *Damping Function* $h(\gamma)$

As shown in figure 1, the stress relaxation modulus $G(t, \gamma)$ are nearly parallel. Thus, they obey the time-strain factorability. This can be obtained by separating the shear stress modulus, $G(t, \gamma)$, into two components: a linear relaxation modulus $G(t)$ and a strain dependent function $h(\gamma)$. And the damping function is in the form of Eq. (3). In this section, the damping function will be determined after the two parameters, a and b, are identified.

In figure 1, the nonlinear stress relaxation, $G(t, \gamma)$, for large step strain $\gamma_0 > 0.1\%$ are vertically shifted to obtain the best superposition onto the linear data, i.e., $G(t)$ for $\gamma_0 \leq 0.1\%$. As shown in figure 2, the superposition is qualitatively good.
Referring to Eq. (3), the damping function can be expressed as the ratio of $G(t, \gamma)$ to the linear relaxation modulus $G(t)$:

$$h(\gamma) = G(t, \gamma) / G(t, 0.1\%). \tag{4}$$

Indeed, for large step strains, the damping function calculated from this ratio was found to be constant. As shown in figure 3, the experimental $h(\gamma)$ was found to be well fitted by the following equation:

$$h(\gamma) = 1/ (1+ 975\, \gamma^{0.99}). \tag{5}$$

Therefore, the damping function parameters are: $a = 975$, $b = 0.99$.

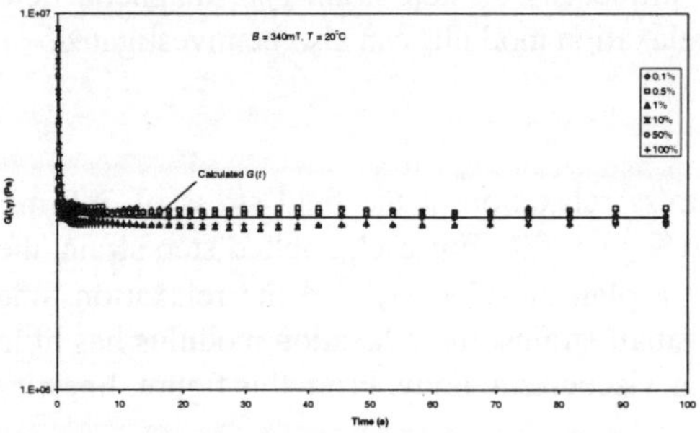

Figure 2. Examination of the time-strain factorability by vertical shift of nonlinear $G(t, \gamma)$ on linear $G(t, \gamma)$

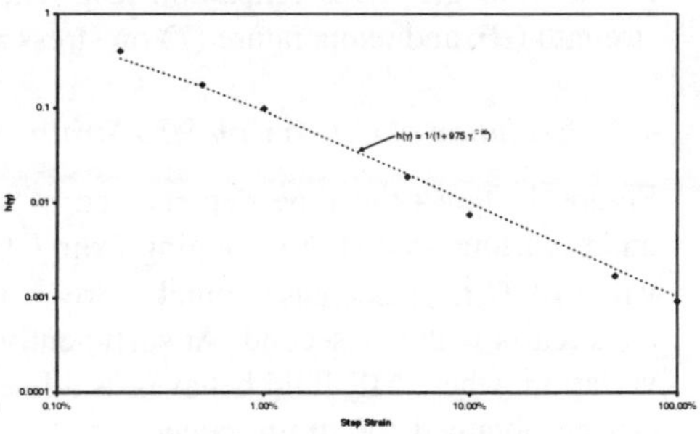

Figure 3. Damping function $h(\gamma)$ for experimental data and the best fit

4.3 *Linear Relaxation Modulus* $G(t)$

MR fluid behaves as a linear viscoelastic solid when subjected to sufficiently small step strains. Thus, a three-parameter viscoelastic model (standard solid) [5] is used to describe $G(t)$. As shown in figure 4, this model is formed by a spring K_2 in series with a Voigt model consisting of a spring K_1 in parallel with a dashpot C_1.

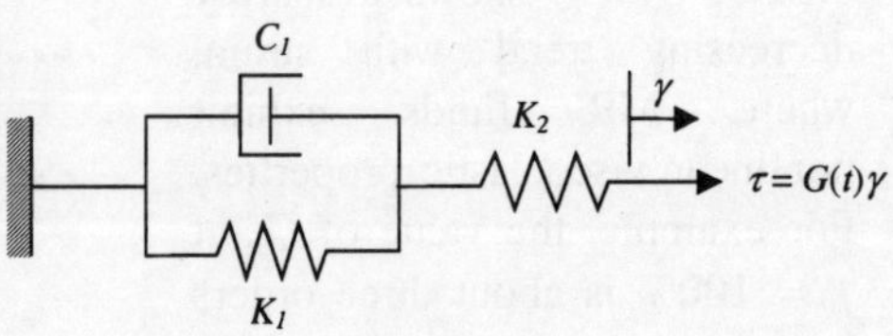

Figure 4. A three-parameter linear viscoelastic model

Based on this model, the stress relaxation modulus $G(t)$ is represented as

$$G(t) = K_2 - \frac{K_2^2}{K_1 + K_2}(1 - e^{-t/p_1}), \tag{6}$$

where $p_1 = C_1/(K_1 + K_2)$ is called the relaxation time constant.

The three parameters K_1, C_1, and K_2 depend on material properties. So they are all sensitive to magnetic field strength and temperature. The procedure for parameter identification is discussed as below.

For a specific field strength, eg. 340 mT, a sufficiently small step strain $\gamma_0 = 0.05\%$ is applied to the MR fluid, the stress relaxation modulus $G(t_i)$ is measured at different time instances t_i. Let $G_m(t_i)$ be the model-predicted value at time t_i as given by Eq. (6). The error in the model can then be presented by an objective function as

$$J = \sum_{i=1}^{N} (G(t_i) - G_m(t_i))^2 , \tag{7}$$

where N is the number of t_i when experimental data are recorded. The model parameters are then extracted by minimizing the objective function J. The three parameters to be determined are the linear viscoelastic model parameters K_1, C_1, and K_2. The objective function is minimized with respect to each of the three parameters. Thus, the condition is

$$\nabla J = \left[\frac{\partial J}{\partial K_1}, \frac{\partial J}{\partial C_1}, \frac{\partial J}{\partial K_2} \right] = 0. \tag{8}$$

To be physically meaningful, the model parameters must be positive, i.e. $K_1, C_1, K_2 > 0$.

The identified parameters against the four specified field strengths are shown in Table 1. It can be seen that all the three parameters increase steady with increasing field strength.

Table 1. Field strength dependence of parameters ($T = 20^0$C)

B (mT)	170	270	340	400
K_1	5.84E+06	7.70E+06	5.12E+07	8.86E+07
K_2	1.28E+06	2.10E+06	3.31E+06	4.20E+06
C_1	7.12E+05	4.90E+06	3.27E+07	7.42E+07

4.4 Prediction from Theory

Using the parameters estimated from the system identification as described in sections 4.2 and 4.3, the stress relaxation modulus is reconstructed and compared with the experimental data, as shown in figure 5. The solid lines represent the reconstructed curves for various step strains at the field strength of 340 mT. The plots indicate that the model can accurately describe the stress relaxation of MR fluids.

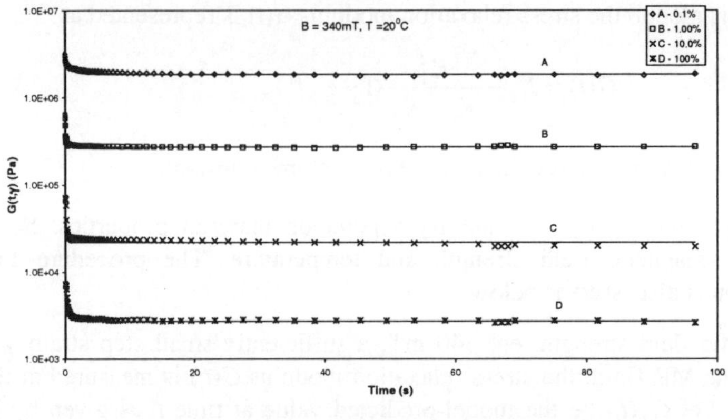

Figure 5. Comparison of stress relaxation modulus at various step strains (B=340mT, T=20^{0}C)

4.5 Field Dependence

Under a small step strain γ_0 = 0.05%, the stress relaxation at various field strengths is shown in figure 6. It can be seen that the stress relaxation modulus increases steady with increasing field strength. This result also indicates that the higher the field strength, the more elastic MR fluids behave.

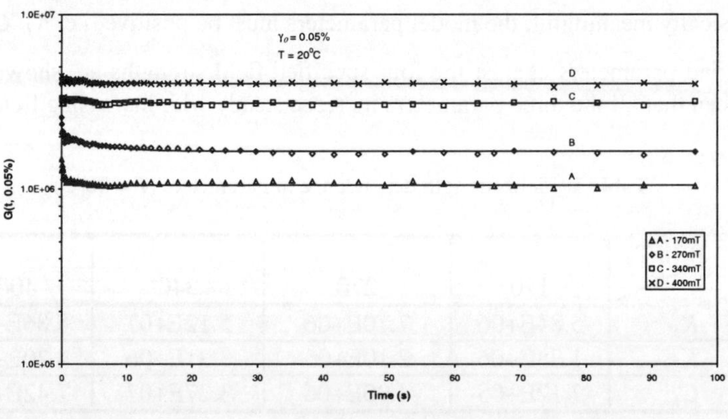

Figure 6. The stress relaxation under various field strengths, $\gamma_0 = 0.05\%$, $T = 20^0$C

4.6 Temperature Dependence

Figure 7 shows the stress relaxation modulus at three different temperatures of 20^{0}C, 40^{0}C, and 60^{0}C under a step strain of 0.5%. It is seen that the stress relaxation modulus decreases slightly with increasing temperature. The results indicate that MR fluids may thin more with increasing temperature.

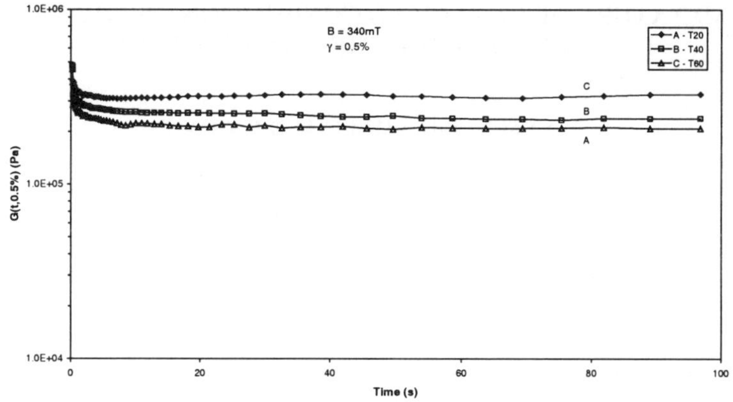

Figure 7. The stress relaxation modulus under various temperatures, $\gamma_0 = 0.5\%$, $T = 20^0$C

5 Conclusion

Experimental and modeling studies of the stress relaxation of MR fluids under step shear strain are presented. At small step strain ranges, the stress relaxation modulus is independent of the applied strain. MR fluid behaves as a linear viscoelastic solid. At large step strains, the stress relaxation modulus shows a gradually decreasing trend with step strain. Thus, the elastic component of MR fluids decreases with increasing step strain. The stress relaxation shows a sharp increasing trend with field strength; however, it decreases slightly with increasing temperature.

Moreover, a five-parameter model is proposed to describe the relaxation behavior. This model consists of time-dependent of stand solid model and a strain-dependent damping function. It has been shown that this model could accurately describe the relaxation behavior.

Acknowledgements

The authors wish to thank Prof. Faramarz Gordaninejad and Prof. Hideya NISHIYAMA for helpful comments.

References

1. J.D. Carlson, In 7th *Int. Conf. On ERFs, MR Suspensions and Associate Technology*, ed. R.Tao (World Scientific Singapore, 2000), p. 621.
2. D.W. Felt, M. Hagenbuchle, and J. Liu, *J.Intell.Mater.Syst.Struct.*, **7**, 589 (1996).
3. K.D.Weiss, J.D.Carlson, and D.A.Nixon, *J.Intell.Mater.Syst. Struct.*, **5**, 772 (1994).
4. M. Iza, and M. Bousmina, *J.Rheol.*, **44**, 1363, (2000).
5. J.D. Ferry, *Viscoelastic properties of polymers* (Wiley, New York, 1980).
6. P.R. Soskey, and H.H. Winter, *J.Rheol.*, **28**, 625 (1984).

ROLE AND NATURE OF HIGH FIELD CONDUCTION OF THE SUSPENDING LIQUID IN ELECTRORHEOLOGICAL FLUIDS.

P. ATTEN, J.-N. FOULC AND P. GONON

LEMD, CNRS and Joseph Fourier University, BP 166 – 38042 Grenoble Cedex 9, France
E-mail: atten@labs.polycnrs-gre.fr

We examine the interaction force between spheres of a slightly conducting material immersed in a dielectric liquid when subjected to a DC field. An approach is developed which refines the previous two-zone model retaining only the electrical conduction of the solid and liquid phases and taking into account the field enhanced dissociation of electrolytic impurities. Approximations on the shape of the equipotential surfaces inside the solid lead to a system of ordinary differential equations governing the distribution of the electrical potential along the sphere surface. Estimates are derived for the attraction force between spheres in contact, for the current and for the electric stress in the liquid lying between the spheres in the "contact zone". This field takes values in the range 200 - 300 V/μm which correspond to extremely high levels of salts dissociation. The different conduction phenomena in the liquid are then discussed and their role in limiting the interaction force is emphasized.

1 Introduction

The electrorheological (ER) effect arises from the electrically induced structuration of the particles in suspension in an insulating liquid [1]. The electrostatic forces are responsible for the attraction between solid particles and between particles and the electrodes [2]. In order to determine the attraction force between particles and, hence, to model the behaviour of ER fluids, the permittivity mismatch between particles and suspending liquid has long been retained [3,4] but, often, this is not adequate because the solid and liquid constituents are not perfectly insulating media. In the case of DC and low frequency AC applied fields, the distributions of electric field and surface charge are determined by the conduction properties of both the solid and liquid phases [5,6]. Indeed these distributions are intimately related with the passage of electrical current through the fibbers of contacting or nearly contacting particles.

A conduction model was developed which fairly well accounts for the attraction force between two centimetre scale half-spheres of a slightly conducting polymer immersed in a dielectric liquid when they are in contact [7,8] or at a very small distance [9]. This model is based on a very simple picture of the contacting spheres distinguishing an "electric contact" zone of radius δ (see Fig. 1) from an outer zone where the electric potential at the interface can be considered as constant. Here a refined treatment is proposed which validates the two-zone approximation and which leads to estimates of the electric stress in the liquid of the contacting zone.

2 Recall of the two zone conduction model

We consider two spheres of radius R of a solid material of conductivity σ_S much greater than that, σ_L, of the suspending insulating liquid ($\sigma_S >> \sigma_L$). When the distance s between the two spheres is high enough, each sphere roughly behaves as a conductor immersed in a much more insulating medium and takes the potential corresponding to the location of its centre [9]. When s tends to zero, the magnitude of the field between the spheres would increase dramatically [7]. Now, under high electric fields E, the conductivity of non-polar

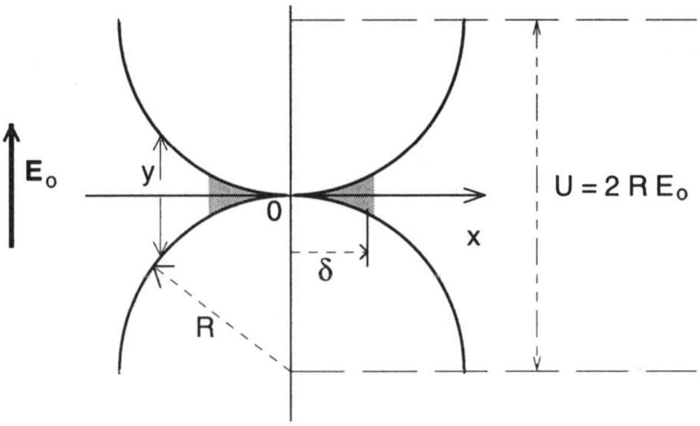

Fig. 1. Cross-section of two spheres in contact showing the electrical "contact zone" of radius δ.

liquids to a rough approximation increases exponentially with $E^{1/2}$. This field enhanced conduction practically results in a saturation of the field between the spheres because of the current flowing across the liquid and of the subsequent potential gradients in the solid spheres due to the constriction of the field lines close to the contact point.

In the case $\Gamma = \sigma_S / \sigma_L >> 1$, these basic remarks lead to the following two-zone picture (Fig. 1) : for $x > \delta$ the sphere surface is assumed to be approximately equipotential (the density of the current escaping from the sphere is low); for $x \leq \delta$, the field in the liquid saturates due to the enhanced conduction in the thin liquid layer. This picture makes it easy to calculate the attraction force. By integration of the axial component of the electrostatic pressure $\varepsilon_L E^2/2$ on the whole half-sphere surface, using some approximations, the following expression is obtained [7] :

$$F \cong \pi\varepsilon_L U^2 \left(\frac{R}{\delta}\right)^2 = 4\pi R^2 \, \varepsilon_L \, E_o^2 \left(\frac{R}{\delta}\right)^2 \tag{1}$$

The crucial point then is to determine the radius δ of the "electrical contact" zone. In the outer region ($x > \delta$), the resistance offered by the liquid is much greater than that of the solid phase. Very close to the contact point, conversely, the resistance of the very thin layer of liquid (the conductivity of which is enhanced by the field) is negligible compared with the one of the solid material where the current density takes rather high values. The conductance of the inner zone is, in practice, equal to that, C_S, of the solid spheres. A classical expression for the electrical contact gives $C_S = 2\sigma_S\delta$ [7]. By requiring that $C_S = C_L$ (C_L : conductance of the liquid in the outer region), an equation is obtained for δ [7]. Expressing C_S in terms of the current I_S flowing through the spheres leads to a very general expression relating the attraction force to $U = 2RE_0$ and I_S which fairly well accounts for the experiments on centimetre scale half-spheres [8].

The simplest conduction model assumes that the liquid has a conductivity associated with the presence of electrolytic impurities. Onsager [10] gave a treatment of the problem of field enhanced dissociation and showed that the rate of dissociation is such that :

$$k_D(E) = k_D(0) \; \mathrm{f}\left(\frac{E}{E_c}\right) \qquad \text{with} \qquad \mathrm{f}(x) = \frac{2\,\mathrm{I}_1\left(x^{1/2}\right)}{x^{1/2}} \tag{2}$$

where I_1 is a modified Bessel function and E_c is a characteristic field : $E_c = \pi\varepsilon_L (kT)^2/\mathrm{e}^3$ (kT : thermal energy, e : electron charge). For low values of the applied mean field E_o, δ is constant and the force $F \propto E_o^2$ [7]. For high field strengths, the field enhancement of conduction in the liquid plays an important role and, moreover, there is a saturation of the current density due to the thinness (here denoted d) of the liquid layer [8,11] :

$$j_{sat} = \frac{\sigma_L d}{2K\tau} \, \mathrm{f}\left(\frac{E}{E_c}\right) \tag{3}$$

Further approximations finally lead to the following expressions for δ and F [8] :

$$\frac{R}{\delta} \cong \left(\frac{E_c}{2E_o}\right)^{1/2} A(\Gamma, E_o, \delta) \qquad F \cong 2\pi R^2 \varepsilon_L E_c E_o A^2(\Gamma, E_o, \delta) \tag{4}$$

where $$A(\Gamma, E_o, \delta) = \ln\left[\beta \Gamma^2 \left(\frac{2E_o}{E_c}\right)^{9/4} \left(\frac{R}{\delta}\right)^{1/2}\right]$$

β being a constant depending on the properties of solid and liquid phases [8].

Fig. 2 shows the variations of the attraction force between two half spheres in contact immersed in a mineral oil (TF50 from Elf). For different values of the conductivity ratio $\Gamma = \sigma_S / \sigma_L$ we clearly observe a quadratic dependence of F on E_o for low applied field (and low values of F) and a roughly linear one for high field values [7,8,12].

3 Refined conduction model

Calculations based on approximations not so dramatic than the ones of the afore mentioned two zone picture were performed, which led to ordinary differential equations [13]. We derive here another equation for the case not previously considered of solid spheres in contact characterised by bulk conduction only. The first hypothesis in this model is to assume that the equipotential surfaces inside the spheres have a spherical shape and, to a first approximation, are normal to the sphere surface. The total current through an equipotential surface inside the solid sphere can then be estimated to be :

$$I = 2\pi \mathrm{R} \, \sigma_S \frac{\mathrm{d}V_s}{\mathrm{d}\theta} \frac{\sin^2\theta}{1+\sin\theta} \tag{5}$$

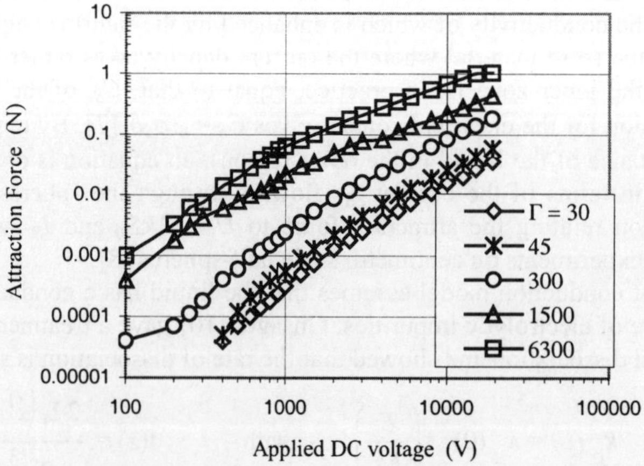

Fig. 2. Attraction force between two half spheres in contact (polyamid with conductivity σ_S = 1.7 10^{-8} S/m, radius R = 7 mm) immersed in a mineral oil as a function of the applied voltage U.

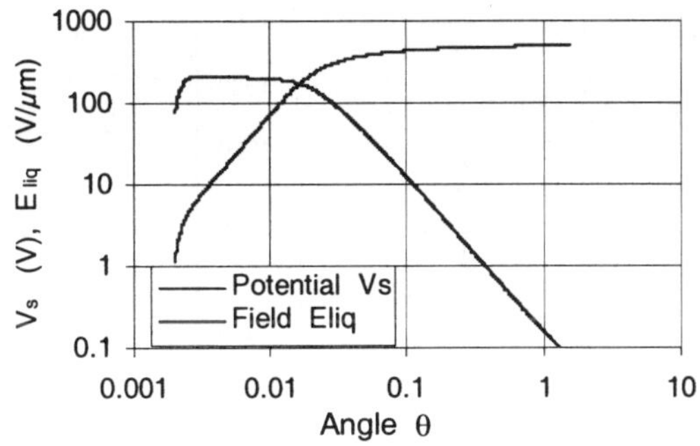

Fig. 3. Variations versus the angle θ of the potential V_s at the sphere surface and of the corresponding electric field E_{liq} in the liquid lying between the two spheres ($A = 11.6\ \mathrm{V}^{-1}$, $U = 500$ V).

where θ is the angle with the axis passing through the spheres centres and V_s is the distribution of electric potential at the sphere surface. As in the two-zone picture, the electric field in the liquid is taken as the ratio of V_s over distance to the symmetry plane :

$$E_{liq}\left(x = R\ \sin\theta\right) = \frac{V_s(\theta)}{R\left(1-\cos\theta\right)} \tag{6}$$

Expressing the conservation of electric current leads to the following equation :

$$\frac{\sin\theta}{\cos\theta}\frac{\mathrm{d}^2 V_s}{\mathrm{d}\theta^2} + \frac{2+\sin\theta}{1+\sin\theta}\frac{\mathrm{d}V_s}{\mathrm{d}\theta} - \frac{(1+\sin\theta)(1-\cos\theta)}{A}\,\mathrm{f}\left(\frac{V_s}{RE_c(1-\cos\theta)}\right) = 0 \tag{7}$$

where $A = \dfrac{K_L\,\varepsilon_L}{R^2\,\sigma_S}\,\Gamma^2$ and $\Gamma = \dfrac{\sigma_S}{\sigma_L}$, K_L and ε_L being the ions mobility and the permittivity respectively. Eq. (7) is solved by classical numerical methods and the solution fulfilling the boundary conditions is obtained by a shooting technique. Fig. 3 shows a typical solution exhibiting very different behaviours in two zones. In the outer zone the potential V_s has a very weak variation as a function of θ whereas, in the inner zone, the electric field E_{liq} across the liquid lying between the spheres is nearly constant. This shows that the two-zone picture is a very relevant approximation.

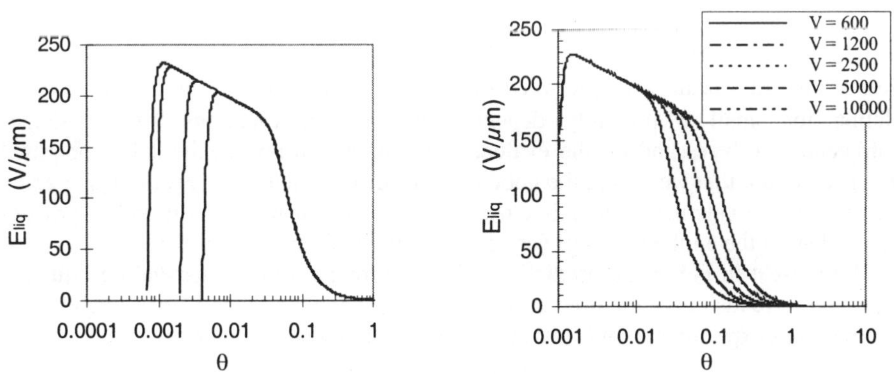

Fig. 4. Variations versus θ of the electric field E_{liq} in the liquid lying between the two spheres (mineral oil, $A = 11.6\ \mathrm{V}^{-1}$). a) Solutions for various θ_0 values ($U = 2500$ V); b) by varying U ($\theta_0 = 10^{-3}$).

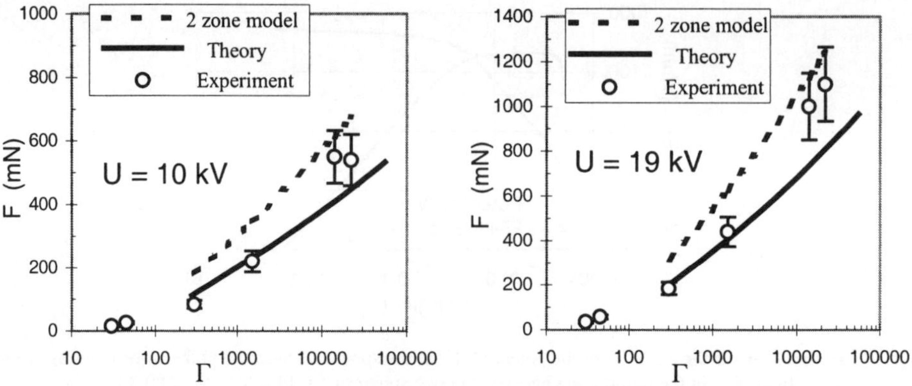

Fig. 5. Variations with the conductivity parameter Γ of the attraction force between two half-spheres of polyamid immersed in a mineral oil.

Equation (7) has a singularity at the angle $\theta = 0$. This difficulty can be disregarded because, from a practical viewpoint, the contact between the spheres always concerns a region of finite, although small, area. We can take as initial angle θ_0 a value of the order of 10^{-3} rad. Fig. 4a indicates that the choice of θ_0 and of the corresponding potential has but a negligibly small influence on the solution : very rapidly the solution reaches the asymptotic solution in the inner zone. Moreover Fig. 4b shows that this asymptotic behaviour is independent of the applied voltage U (it is controlled by the parameter A). Fig. 4b also shows that there is a regular shift (in logarithmic scale) of the steeply decreasing part of the $E_{liq}-\theta$ curves when doubling the applied voltage U. This corresponds closely to a power law with exponent 1/2 as predicted by relation (4).

The solution of (7) also gives the value of the attraction force F. Fig. 5 shows a comparison of these predictions with measured forces and prediction of the two zone approach [8,12]. It appears that the present derivation gives a lower bound whereas the previous estimation is an upper bound. Similar results are obtained for two other liquids.

4 Electric stress on the liquid in the inner zone

The interest of the present approach is to give rather good estimates for the electric stress in the thin liquid layer between the spheres. The variation law of the asymptotic solution for small θ probably depends on the approximations on the shape of equipotential surfaces and on the evaluation (5) of the current in the solid spheres. But there is no doubt that we obtain the good order of magnitude of E_{liq}. From Figs. 3 and 4, it is clear that the calculated field is close to 200 V/µm in mineral oil and is practically independent of the applied voltage (or equivalently of the mean field E_0).

We present in Fig. 6 the predictions obtained for 3 different suspending liquids. For comparison, we also give the predictions of the two-zone model and very rough estimates (referred to as experimental in Fig. 6) based on the following expression :

$$E_{liq} = E_\delta = UR/\delta^2 = F/(2\pi R^2 \varepsilon_L E_0) \tag{8}$$

where the value of δ is obtained from the measured force by using relation (1). We see that the field on the liquid is on the order of 200 V/µm for mineral oil and Jarylec® (a mixture of mono- and di-benzyl-toluene) and more than 260 V/µm for silicone oil.

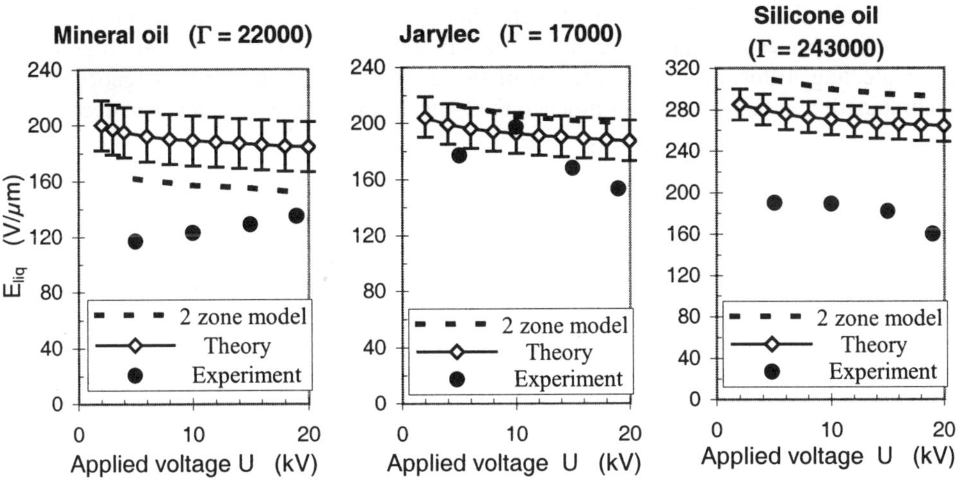

Fig. 6. Variations of the electric field E_{liq} in the liquid lying between the two spheres (inner zone) as a function of the voltage U applied across the two polyamid half-spheres (R = 7 mm).

5 Discussion

Three mechanisms of electrical conduction can contribute to the passage of current between two particles : i) the field enhanced dissociation of electrolytic impurities, ii) the injection of charge carriers from the solid/liquid interfaces, iii) the breakdown of the liquid. The calculated field values are only moderately lower than the ones characterising the breakdown of liquid layers of thickness about 10 µm (E_b ~ 400 V/µm). Indeed partial discharges were observed when working in silicone oil with half-spheres of radius R = 7 mm [12]. Hence the breakdown of the suspending liquid might appear as a phenomenon limiting the attraction force between particles and, therefore, the yield stress of electrorheological fluids. However it must be kept in mind that the breakdown field strongly depends on the distance. In practice, the breakdown phenomenon is triggered by electronic processes in the liquid phase. Electron avalanches are involved which require a potential difference across the liquid layer of at least 150 or 200 volts to initiate streamers (in gases the Paschen minimum voltage ~ 300-400 volts slightly depends on the gas nature). In ER fluids with particles of small enough size, the potential difference $U \leq 2RE_0$ between facing surfaces of particles will always take values lower than 150 V. Therefore with particles of size for example ≤ 10 µm, breakdown will not be a mechanism of non linear conduction of the liquid in the "electric contact" zone.

The very high values of E_{liq} also question the mechanism retained in the conduction model, which is based on bulk liquid conduction. Under electric fields in the range 200 to 250 V/µm, the enhancement factor f(E/E_c) takes values varying between 10^{10} and 10^{12}. Such huge values imply practically a quasi instantaneous total dissociation of the electrolytic impurities (or additives). A continuous current in the inner zone can be sustained only through the diffusion or convection into the liquid film in the inner zone, of salts coming from the liquid in the outer zone. In the experiments performed with centimetre scale half-spheres immersed in a rather big vessel (see results of Fig. 2 and in [8,12]), a convection associated with volume electric forces very likely maintains a steady non zero mean concentration of dissociable species. This explains the fairly good agreement between measured and predicted values of force and current.

For ER fluids, the problem is different because there is a consumption of the dissociable species in a short time [14]. Firstly in the liquid subjected to a very high field, by neglecting diffusion and convection of the dissociable species and by ignoring any recombination in this electrolysis process, the typical time for the concentration decay is :

$$\tau = (\varepsilon_L/\sigma_L)\,\left[\mathrm{f}\left(E/E_c\right)\right]^{-1} \tag{8}$$

and can take very small values << 1 s. Secondly the reservoir of dissociable species is the liquid of the outer zone which has a volume proportional to R^3 whereas the zone of very strong dissociation scales as R^2. Therefore the overall rate of salt consumption increases as the particles size decreases. We can conclude that the bulk conduction which plays a role in the case of big particles (~ 100 μm) very likely can be ignored in ER fluids with small size particles (~ 1 μm).

In the case of small particles, the dominant conduction process in the liquid film close to the contact area between particles is related with injection phenomena, i.e. creation of charge carriers at the interfaces solid/liquid. For metallic electrodes, injection is a universal phenomenon whose dependence on the electric field can be similar to field enhanced dissociation [15]. The problem for ER fluids is that the particles can be organic or inorganic and have ionic or semi-conducting conduction properties and nearly nothing is known about the electrochemical processes responsible for the exchange of charge at such solid/liquid interfaces. If we assume that the injected charge density is an increasing function of the electric field, we can expect a power law for the attraction force $F \propto E_0^{\alpha}$ with an exponent α (which ranges from 1 to 2) depending on the injection law [8]. Now the intensity of injection depends on the nature of the liquid and solid materials and, in particular, on the ionic or electronic character of the conduction in the solid phase. The couples of liquid and solid materials leading to very weak injections into the liquid should be the best suited for defining ER fluids of high performance. But extensive basic investigations are necessary in order to understand the mechanisms of injections and, hence, to select good materials.

References

1. Gast A. P. and Zukoski C. F., *Adv. Colloid Int. Sci.*, **30**, p. 153 (1989).
2. Jordan T. C. and Shaw M. T., *IEEE Trans. Elec. Insul.*, **24**, p. 849 (1989).
3. Block H. and Kelly J.P., *J. Phys. D : Appl. Phys.*, **21**, p. 1661 (1988).
4. Klingenberg D.J., Dierking D. and Zukovski C.F., *J. Chem. Soc. Farady Trans.*, **87**, p. 425 (1991).
5. Anderson R. A., *Proceed. 3rd Int. Conf. ER fluids*, ed. R. Tao (World Scientific, Singapore, 1992) p. 81.
6. Davis L. C., *J. Appl. Phys.*, 72, p. 1334 (1992).
7. Atten P., Foulc J.N. and Félici N., *Intern. J. Mod. Phys. B*, **8**, pp. 2731-45 (1994).
8. Boissy C., Atten P. and Foulc J.N., *Intern. J. Mod. Phys. B*, **10**, p. 2991 (1996).
9. Atten P., Boissy C. and Foulc J.N., *Proceed. 12th ICDL*, ed. C. Mazzetti (IEEE Cat. N 96CH35981) pp. 419-422 (1996).
10. Onsager L., *J. Chem. Phys.*, **2**, p. 599 (1934).
11. Atten P., Foulc J.-N. and Benqassmi H., "*Progress in Electrorheology*", K.O. Havelka et F.E. Filisko (Eds.), Plenum Press (New York), (1995) pp. 231-243.
12. Boissy C., *Ph D Thesis*, Université Joseph Fourier, Grenoble, France, (1996).
13. Tang X., Wu C. and Conrad H., *J. Appl. Phys.*, **78**, pp. 4183-88 (1995).
14. Felici N.J., *J. Electrostatics*, **40&41**, pp. 567-572 (1997).
15. Denat A., Gosse B. and Gosse J.P., *J. Electrostatics*, **11**, pp. 179-194 (1982).

A TWO-FLUID MODEL FOR ELECTRO- AND MAGNETORHEOLOGICAL SUSPENSIONS

KARL VON PFEIL, MICHAEL D. GRAHAM AND DANIEL J. KLINGENBERG

Department of Chemical Engineering, University of Wisconsin, 1415 Engineering Drive, Madison, WI 53706

JEFFREY F. MORRIS

School of Chemical Engineering, Georgia Institute of Technology, Atlanta, GA 30332

A two-fluid continuum model is developed to describe mass transport in electro- and magnetorheological suspensions. The particle flux is related to the field-induced stresses. Solutions of the resulting mass balance show column formation in the absence of flow, and stripe formation when a suspension is subjected simultaneously to an applied electric field and shear flow.

1 Introduction

Particle-level models and simulations of electrorheological (ER) and magnetorheological (MR) suspensions have been very valuable for understanding the relationships between particle properties, interactions and macroscopic suspension behavior.[1] Unfortunately, they are limited to relatively small numbers of particles, and hence are not capable of predicting the behavior of a suspension throughout an entire device. This issue is particularly important in situations where the field, temperature, or particle concentration are not uniform.

In this article we consider an alternative modeling approach, where ER suspensions are considered as continua. Specifically, we show how one can describe the evolution of the particle concentration without considering the precise positions of each particle. This is accomplished by solving a mass conservation equation for the particulate phase, coupled with an expression for the particle flux, which is obtained from a momentum balance. We find that by using this approach, one obtains suspension structures commonly observed experimentally. For electric fields applied to quiescent suspensions, this approach captures the formation of particulate columns; for electric fields applied to flowing suspensions, it captures the formation of particulate stripes oriented in the flow direction.

The analysis is illustrated below for ER suspensions. A similar analysis for MR suspensions would follow directly. In the following section we employ a continuum, two-fluid model to develop a conservation equation for the particle concentration. In Section 3, the particle flux is related to the particle contribution to the stress. Combining the flux expression with the conservation equation produces a convection-diffusion-like equation for the evolution of the particle concentration. Solutions of this equation for different situations are described in Section 4.

2 Two-fluid model: mass transport

Following Morris and Boulay,[2] we begin with the general mass conservation equation. Averaging over the particulate volume and dividing by the particle density yields the particle-phase conservation equation

$$\frac{\partial \phi}{\partial t} + \langle \boldsymbol{u} \rangle \cdot \nabla \phi = -\nabla \cdot \boldsymbol{j}, \tag{1}$$

where ϕ is the particle volume fraction, $\langle \boldsymbol{u} \rangle$ is the suspension average velocity, $\boldsymbol{j} \equiv \phi(\boldsymbol{U} - \langle \boldsymbol{u} \rangle)$ is the particle flux relative to the mean suspension motion, and $\boldsymbol{U}$ is the local average velocity of the particulate phase.

Averaging the general momentum balance over the particle phase, and restricting attention to low Reynolds numbers and neutrally buoyant, monodisperse spheres yields the relation for the particle flux

$$\boldsymbol{j} = \frac{2a^2}{9\eta_c} f(\phi) \nabla \cdot \boldsymbol{\sigma}^{(p)}, \tag{2}$$

where a is the particle radius, η_c is the viscosity of the suspending fluid, $f(\phi) = (1-\phi)^4$ is the sedimentation hindrance function, and $\boldsymbol{\sigma}^{(p)}$ is the particle contribution to the stress. Below we obtain the electrostatic part of $\boldsymbol{\sigma}^{(p)}$.

3 Field-induced mass transport

We consider the suspension to be composed of dielectric spheres immersed in a dielectric suspending fluid (conducting spheres in a weakly conducting fluid are treated analogously). The suspension structure is assumed to be isotropic and amorphous. Such a structure is reasonable for dilute quiescent or flowing suspensions prior to the application of an electric field. For such a suspension, the electrostatic stress is

$$\boldsymbol{\sigma}^{(p,E)} = \epsilon_0 \left[\left(\epsilon - \frac{1}{2} a_1 \right) \boldsymbol{E}\boldsymbol{E} - \frac{1}{2} (\epsilon + a_2) E^2 \boldsymbol{\delta} \right], \tag{3}$$

where $\epsilon_0 = 8.8542 \times 10^{-12}$ F/m is the permittivity of free space. A self-consistent mean-field analysis in the point-dipole limit yields a suspension dielectric constant[3]

$$\epsilon(\phi) = \epsilon_c \frac{1 + 2\beta\phi}{1 - \beta\phi}, \tag{4}$$

where ϵ_c is the dielectric constant of the suspending fluid, $\beta = (\epsilon_p - \epsilon_c)/(\epsilon_p + 2\epsilon_c)$, and ϵ_p is the dielectric constant of the particulate material. Assuming that the disperse and continuous phases themselves are not electrostrictive, the electrostriction coefficients of the suspension, a_1 and a_2, are[3] $a_1 = 0$ and

$$a_2(\phi) = \frac{(\epsilon(\phi) - \epsilon_c)(\epsilon(\phi) + 2\epsilon_c)}{3\epsilon_c}. \tag{5}$$

For the system considered here, this stress may be considered as entirely a particle contribution. This may be illustrated by taking the divergence of Eq. (3) and utilizing Maxwell's equations $\nabla \cdot \boldsymbol{D} = \rho^{(e)}$ ($\rho^{(e)}$ is the free charge density) and $\nabla \times \boldsymbol{E} = \boldsymbol{0}$ to give

$$\nabla \cdot \boldsymbol{\sigma}^{(p,E)} = \rho^{(e)} \boldsymbol{E} - \frac{1}{2} \epsilon_0 E^2 \nabla \epsilon - \frac{1}{2} \epsilon_0 \nabla \left(a_2 E^2 \right). \tag{6}$$

The terms on the right side are body forces. For the present analysis, $\rho^{(e)} = 0$, and the gradient terms can be nonzero only for nonzero gradients in the particle volume fraction; thus $\nabla \cdot \boldsymbol{\sigma}^{(p,E)}$ contains only particle contributions to the stress.

The particle stress will also contain a hydrodynamic contribution.[2] This contribution is neglected here in the interest of space, and to illustrate the effects of the electrostatic driving force. Hydrodynamic contributions will be addressed in a future publication.

Combining Eqs. (1), (2) and (6) yields a conservation equation for the evolution of the volume fraction when a field is applied to a flowing suspension. Here we consider only simple shear flow ($\langle \boldsymbol{u} \rangle = (\dot{\gamma} z, 0, 0)$) and limit attention to small changes in concentration from an initially uniform suspension. For an applied field $\boldsymbol{E} = (0, 0, E_0)$, the conservation equation becomes

$$\frac{\partial \phi'}{\partial t} + \dot{\gamma} z \frac{\partial \phi'}{\partial x} = -M \left(\frac{\partial^2 \phi'}{\partial x^2} + \frac{\partial^2 \phi'}{\partial y^2} - \kappa \frac{\partial^2 \phi'}{\partial z^2} \right), \tag{7}$$

where $\phi'(\boldsymbol{x}, t) \equiv \phi(\boldsymbol{x}, t) - \phi_0$ is the deviation of the volume fraction from its initially uniform value ϕ_0,

$$M = \frac{2a^2 \epsilon_0 \epsilon_c \beta^2 E_0^2 f(\phi_0) \phi_0}{3\eta_c (1 - \beta\phi_0)^3}, \tag{8}$$

and $\kappa = 2(1 - \beta\phi_0)/(1 + 2\beta\phi_0) > 0$. Equation (7) is similar to a convection-diffusion equation, with anisotropic diffusion and a *negative* diffusivity in the x and y directions. However, the particle phase motion has nothing to do with Brownian diffusion, but rather is caused by electrostatic forces. In fact, for typical ER suspensions, Brownian motion is negligible: for $a = 1 \times 10^{-5}$ m, $T = 298$ K, and $\eta_c = 0.1$ Pa·s, the Brownian diffusivity is $D_0 = kT/6\pi\eta_c a = 2.2 \times 10^{-16}$ m^2/s, whereas $M = 1.1 \times 10^{-9}$ m^2/s for $\epsilon_c = 2$, $\beta = 1$, $E_0 = 1 \times 10^6$ V/m, and $\phi_0 = 0.1$.

The negative apparent diffusivity is reminiscent of spinodal decomposition in phase separating systems.[4] Indeed, a uniform suspension in an electrostatic field is thermodynamically unstable. Consider the electrostatic contribution to the free energy, $\mathcal{F}^E = -\epsilon_0 \epsilon(\phi) E^2/2$.[5] Since $\partial^2 \epsilon/\partial \phi^2 > 0$ (e.g., Eq. (4)), $\partial^2 \mathcal{F}^E/\partial \phi^2 < 0$, which implies that the free energy of the suspension can be reduced by separating the system into dilute and concentrated phases.

Below we show that Eq. (7) predicts that the application of an electric field to a uniform suspension indeed causes the formation of more concentrated "phases." For a quiescent suspension, the field induces the formation of particle-rich columns oriented in the direction of the applied field; for sheared suspensions, the field induces the formation of particle-rich stripes oriented in the flow direction. Again, these structures are consistent with those commonly observed experimentally.

4 Results

4.1 Structure evolution in quiescent suspensions: column formation

Consider first the application of an electric field to an initially, uniform, quiescent suspension ($\dot{\gamma} = 0$). The electrodes are located at $z = 0$ and L. We seek a solution of Eq. (7) of the form

$$\phi'(\boldsymbol{x},t) = f(z)e^{i(k_x x + k_y y)}e^{\sigma t}, \tag{9}$$

which represents fluctuations with sinusoidal variation in the x and y directions. The z dependence, expressed through $f(z)$, is yet to be determined. σ represents the growth rate of the fluctuations. The full solution to Eq. (7) would be composed of an infinite series of many terms of the form of Eq. (9). Substitution of Eq. (9) into Eq. (7) yields

$$\kappa f''(z) + (k_x^2 + k_y^2 - \sigma/M)f(z) = 0. \tag{10}$$

Using the no-flux boundary conditions $f'(z) = 0$ at $z = 0, L$ yields

$$f(z) = \cos n\pi z/L \qquad n = 0, 1, \ldots, \tag{11}$$

and

$$\frac{\sigma L^2}{M} = (k_x L)^2 + (k_y L)^2 - \kappa n^2\pi^2. \tag{12}$$

Fluctuations will grow for all $\sigma > 0$; the fastest growing fluctuations (for any given k_x and k_y) will occur for $n = 0$, and grow with rate $\sigma_{\max}/M = k_x^2 + k_y^2 > 0$. This has several implications for the resulting suspension structures. First, since the fastest growing fluctuations occur for $n = 0$, the resulting concentration profile should not depend on z ($f(z) =$ constant). Second, the growth rate is symmetric with respect to k_x and k_y, suggesting that the structures will be cylindrical columns; that is, columns of particles oriented in the z direction, as commonly observed. Third, the maximum growth rate increases with k_x and k_y; in fact, the growth rate goes to infinity as k_x and $k_y \to \infty$. This implies that the dominant growing fluctuations will be thin columns. However, at some small length scale, this continuum approach will break down, suggesting that the fastest growing wavenumbers are limited to roughly the inverse of the particle diameter.

A finite difference numerical solution of the governing equation (Eq. (7)) is illustrated in Fig. 1. No-flux boundary conditions were applied at the electrodes ($f'(z) = 0$ at $z = 0, L$); periodic boundary conditions were applied at $x, y = 0, L$ to represent a system of infinite extent in the x and y directions. The volume fraction was initially uniform ($\phi_0 = 0.1$) with small, random initial fluctuations. The evolution of the concentration fluctuations are illustrated in Fig. 1, where positive fluctuations are indicated by dark cubes, and negative fluctuations are indicated by white cubes. The initial distribution and a distribution at a later time ($tM/L^2 = 0.05$) are shown. As time progresses, the concentration fluctuations form columnar structures, which are similar to the fibrous aggregates commonly observed in ER and MR suspensions.

The mechanism of column formation is revealed by considering the change in electrostatic free energy associated with concentration fluctuations, given by $\delta\mathcal{F}^E = -\epsilon_0 \int_V \delta\epsilon(\phi)\boldsymbol{E}\cdot\boldsymbol{E}_0 \, \mathrm{d}V/2$,[5] where $\delta\epsilon$ is the fluctuation in dielectric constant caused by the concentration fluctuation, $\boldsymbol{E}$ is the electric field, and $\boldsymbol{E}_0$ is the electric field prior to the fluctuation. Consider first a parallel concentration fluctuation of the form $\phi(x) = \phi_0 + A_\parallel \cos\beta_\parallel x$ with $A_\parallel \ll 1$. In this case the field is unaltered, and the change in the electrostatic free energy is $\delta\mathcal{F}^E = -\epsilon_0 E_0^2(d^2\epsilon/d\phi^2)A_\parallel^2 V/8 < 0$,

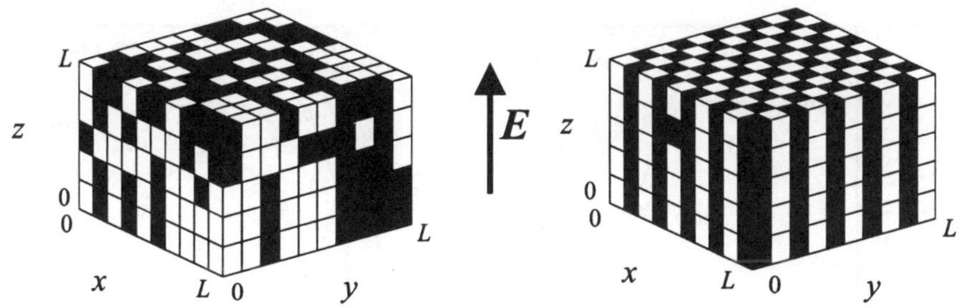

Figure 1. Initial structure, and structure after the application of an electric field to a quiescent suspension ($tM/L^2 = 0.05$). Dark: positive concentration fluctuation; light: negative concentration fluctuation.

implying that such fluctuations are unstable and will continue to grow. Next consider perpendicular fluctuations of the form $\phi(z) = \phi_0 + A_\perp \cos \beta_\perp z$. In this case the field is altered by the fluctuation, and the resulting change in free energy is $\delta\mathcal{F}^E = +\epsilon_0 E_0^2[(d\epsilon/d\phi)^2/\epsilon - (d^2\epsilon/d\phi^2)/2]A_\perp^2 V/4 > 0$, implying that such fluctuations are stable and will decay. Columnar fluctuations parallel to the applied field, which reduce the system free energy, will thus be favored.

4.2 Structure evolution in sheared suspensions: stripe formation

Now we consider the evolution of the concentration profile when a field is applied to a uniform suspension under shear flow. Again we seek a solution of the form of Eq. (9), which upon substitution into Eq. (7) yields the ordinary differential equation for $f(z)$

$$\kappa f''(z) + \left(k_x^2 + k_y^2 - \sigma/M - i\dot{\gamma}k_x z/M\right) f = 0. \tag{13}$$

Solution of this differential equation with boundary conditions $f'(z) = 0$ at $z = 0, L$ provides the eigenvalue σ as a function of the wavenumbers k_x and k_y. Again, positive values of $\sigma(k_x, k_y)$ indicate unstable or growing fluctuations, while negative values of $\sigma(k_x, k_y)$ indicate stable or decaying fluctuations.

Equation (13) was solved numerically to determine $\sigma(k_x, k_y)$. The results for $\dot{\gamma}L^2/M = 10^3$ and 10^4 are illustrated in Fig. 2, where contour plots of $\sigma L^2/M$ as a function of k_x and k_y are presented. The $\sigma = 0$ contours in Fig. 2(a) and (b) represent the limit of stability. Points (k_x, k_y) below the $\sigma = 0$ contour are stable (fluctuations decay), while points above the $\sigma = 0$ contour are unstable (fluctuations grow). The $\sigma = 0$ contours are roughly semicircles, which grow with increasing shear rate. Thus certain fluctuations tend to be stabilized by shear flow.

As discussed above the present model breaks down at large wavenumbers. This arises from neglecting the physical dimensions of the particles, as well as neglecting short range interactions between particles. Including these features will tend to stabilize large $|\boldsymbol{k}|$ fluctuations. Combining this information with that presented in Fig. 2 implies that at sufficiently large shear rates, the only unstable fluctuations will be those with small k_x values and all k_y below the large $|\boldsymbol{k}|$ cutoff (i.e., the region between the $\sigma = 0$ contour and the k_y axis). Within this region, the most unstable

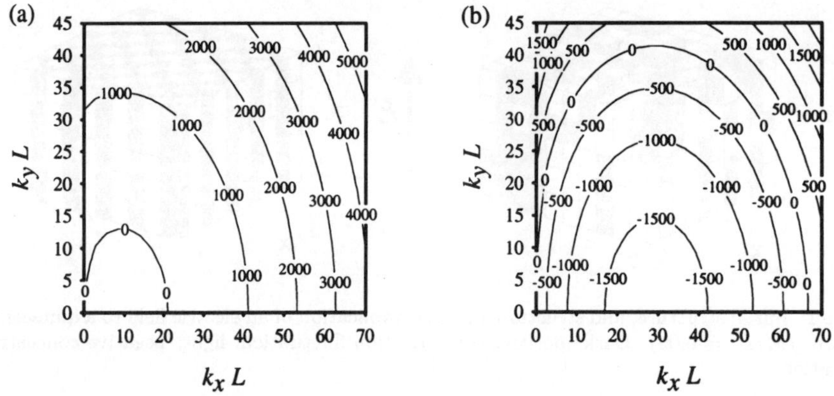

Figure 2. Contour plots of $\sigma L^2/M$ as a function of $k_x L$ and $k_y L$ for (a) $\dot{\gamma}L^2/M = 10^3$ and (b) $\dot{\gamma}L^2/M = 10^4$.

fluctuations will be those with $k_x = 0$ and nonzero k_y—that is, the dominant structure formed by that application of an electric field to a flowing suspension is predicted to be stripes of particle-rich regions oriented in the flow direction, as commonly observed experimentally.[6,7,8,9,10]

Equation (7) has also been solved by the finite difference method to illustrate the stripe formation process. A no flux boundary condition was applied at the electrodes ($f'(z) = 0$ at $z = 0, L$); periodic boundary conditions were applied at $x, y = 0, L$, to represent a system of infinite extent in the x and y directions. The evolution of concentration fluctuations is illustrated in Fig. 3 for $\dot{\gamma}L^2/M = 10^4$ and $\phi_0 = 0.1$. The initially random fluctuations eventually redistribute into stripes oriented in the flow direction, as described above. For the parameter values listed in the previous section, $\dot{\gamma}L^2/M = 10^4$ corresponds to a shear rate of $\dot{\gamma} = 10\ \mathrm{s}^{-1}$.

The mechanism of stripe formation in shear flow is related to the mechanism of column formation in quiescent suspensions. Fluctuations in the plane of shear are rotated and stretched by the shear flow, continually changing their orientation relative to the field. Fluctuations grow when parallel to the field, and are smoothed (decay) when perpendicular to the field. Because each fluid element will continually see both parallel and perpendicular fluctuations, the net effect is a uniform concentration in the plane of shear. However, since fluctuations in the vorticity direction are unaffected by shear and will continue to grow, the resulting structure will be sheets of higher concentration in the plane of shear (i.e., stripes oriented in the flow direction).

5 Conclusion

A two-fluid continuum model for mass transport in ER suspensions has been presented. The particle flux is related to the divergence of the particle contribution to the stress, which in turn is related to the suspension dielectric and electrostrictive

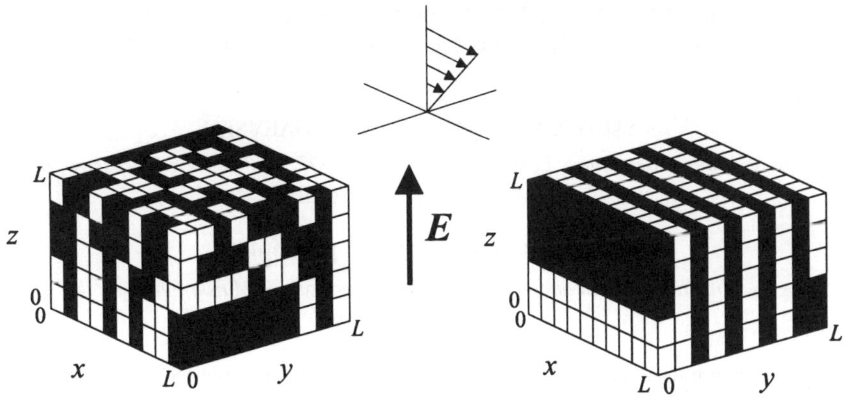

Figure 3. Initial structure, and structure after the application of an electric field to a flowing suspension ($\dot{\gamma}L^2/M = 10^4$, $tM/L^2 = 0.05$). Dark: positive concentration fluctuation; light: negative concentration fluctuation.

properties. Solutions of the resulting particle conservation equation capture common observations: column formation in quiescent suspensions, and stripe formation in sheared suspensions.

Future work will involve refining and improving this continuum approach. Hydrodynamic interactions must be included to probe the behavior when the concentration within the stripes gets large. Nonlocal contributions to the stress and flux should be included to correctly capture the large $|\boldsymbol{k}|$ behavior. Effects of particle interactions and structural anisotropy should be included to completely capture the structure evolution process in ER and MR suspensions.

References

1. M. Parthasarathy and D. J. Klingenberg, Mat. Sci. Eng. Rep. **R17**, 57 (1996).
2. J. F. Morris and F. Boulay, J. Rheol. **43**, 1213 (1999).
3. Y. M. Shkel and D. J. Klingenberg, J. Appl. Phys. **83**, 7834 (1998).
4. J. W. Cahn, J. Chem. Phys. **42**, 93 (1965).
5. J. A. Stratton, *Electromagnetic Theory* (McGraw-Hill, New York, 1941).
6. G. Bossis, E. Lemaire, J. Persello, and L. Petit, Prog. Coll. Polym. Sci. **89**, 1 (1992).
7. G. Bossis, Y. Grasselli, E. Lemaire, J. Persello, and L. Petit, Europhys. Lett. **25**, 335 (1994).
8. S. Henley and F. E. Filisko, J. Rheol. **43**, 1323 (1999).
9. O. Volkova, S. Cutillas, and G. Bossis, Phys. Rev. Lett. **82**, 233 (1999).
10. S. Vieira, L. B. Pompeo Neto, and A. C. F. Arruda, J. Rheol. **44**, 1139 (2000).

ER-FLUID RHEOLOGICAL PROPERTIES IN TERMS OF THE MULTIPARTICLE MODEL OF A COMPOSITE

YU. G. YANOVSKY AND V. E.ZGAEVSKII
Institute of Applied Mechanics, Leninsky Ave. 32A, 119991, GSP-1, V-334, Moscow, Russia
E-mail: iam@ipsun.ras.ru

Z. P. SHULMAN AND E. V. KOROBKO
Institute of Heat and Mass Transfer, P.Brovky, 15, Minsk, Belarus
E-mail: eva@itmo.by

The three-dimensional multi-particle well-ordered model could be considered as an analogy to a crystal body. We use this model for describing rheological properties of concentrated electrorheological fluids (ER fluids). According to this model, the particles of the suspension take their places at sites of a grid with specified type of symmetry and then an electric field is applied to the fluid. Taking into account hydrodynamic couple interaction of particles and forces of electrostatic interaction of particles polarized under the action of an external electric field and employing the mathematical apparatus of the microscopic theory of crystals, we construct the basic relationships for describing viscoelastic electrorheological properties of ER fluids.

1 Introduction

Electrorheological fluids (ER fluids) have the unique property of abrupt altering their effective viscosity when they are exposed to an electric field. The intensive use of ER fluids is stimulatory to their theoretical and experimental investigations.

We use the three-dimensional well-ordered multi-particle model of a composite (the crystal model) [6] for describing rheological properties of electrorheological fluids. Since hydrodynamic interaction between particles is allowed for, we have an opportunity to investigate fluids with high concentration of particles (more that 10% of volume in accordance with the classification introduced in papers [4,5]).

2 The model

An electrorheological fluid is a concentrated soliquid (suspension) consisting of dielectric particles in a viscous medium. Particles are all identical to each other. When no electric field all the particles are located at sites of a lattice with specified type of symmetry. The dielectric permeability of particles, ε_p, is higher than the dielectric permeability of the fluid ε_1. Let now the ER fluid be under the action of a homogeneous electric field of intensity $\vec{E}$. The presence of the electric fields results in polarization of particles. We shall approximate interaction between two particles by interaction between two equivalent dipoles (by analogy, it is possible to consider an interaction of charged particles). Besides electrostatic forces, viscose forces act on the particles. The latter forces may be associated with the displacement of particles in a viscose fluid due to the presence of a gradient of shear rate. Since the concentration of disperse particles is high enough, the hydrodynamic interaction between the particles should be allowed for. At that, we assume that the macroscopic gradient of shear rate is constant all over the whole region occupied by the ER fluid.

A representative element of the model is shown in Fig. 1. The ER fluid is exposed to an external electric field (the positive electrode, anode, is on the left). The ER fluid is under the action of gradients of shear rate. The left-hand electrode is immovable and the right-hand one moves parallel to its initial position at a specified velocity $\vec{U}$. Thus, each particle is under action of both electric and hydrodynamic forces.

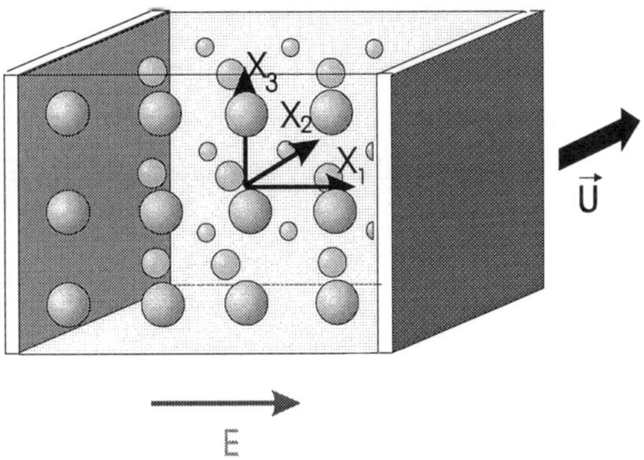

Figure 1. A representative element of the model.

3 Density of potential energy of electrostatic interaction

Consider two neighboring particles of the model. The potential energy of interaction of the particles can be represented in the form of the potential energy of interaction of equivalent dipoles [2]

$$W = \frac{(\bar{p}_1, \bar{p}_2)}{4\pi\varepsilon_0\varepsilon_1 r^3} - \frac{3(\bar{r}, \bar{p}_1)(\bar{r}, \bar{p}_2)}{4\pi\varepsilon_0\varepsilon_1 r^5} \tag{1}$$

where p_i, i = 1, 2, are the dipole moments for corresponding particles, $\bar{r}$ is the vector from the center of the chosen particle to the center of the nearest neighboring particle, $r = |\bar{r}|$ is the distance between the centers of the particles under consideration, and $\varepsilon_0 = 8.85 \cdot 10^{-5}$ F/m (SI) is a constant. The round brackets are used to denote the scalar product of vectors.

The dipole moment of a polarized sphere (a is the sphere diameter) used in the model can be written as

$$\bar{p} = \frac{\pi}{2}\varepsilon_0\varepsilon_1 \frac{(\varepsilon_p - \varepsilon_1)a^3}{(\varepsilon_p + 2\varepsilon_1)}\vec{E} \tag{2}$$

Since all the particles of the dispersed phase are identical, the dipole moments of equivalent dipoles are equal in magnitude and oriented in one direction. Substituting (2) into (1), we obtain

$$W = A\left[\frac{E^2}{r^3} - \frac{3(\bar{r}, E)^2}{r^5}\right] \tag{3}$$

where

$$A = \frac{\pi^2 \varepsilon_0^2 \varepsilon_1^2 (\varepsilon_p - \varepsilon_1)^2 a^6}{4(\varepsilon_p + 2\varepsilon_1)^2}$$

Let $\bar{u}$ be a small arbitrary displacement of one particle relative to another such that $u \ll r^0$, where r^0 is the initial distance between the centers of the particles; consequently, $\bar{r} = \bar{r}^0 + \bar{u}$. The potential energy of interaction of the dipoles reads

$$\begin{aligned} W = A\Bigg[& \frac{1}{r^{0^3}}(E^2 - 3\overset{1}{e_i}\overset{1}{e_k} E_i E_k) - \\ & - \frac{3}{r^{0^4}}\left(E^2 - 5\overset{1}{e_i}\overset{1}{e_k} E_i E_k\right)\overset{1}{e_n} u_n - \frac{6}{r^{0^4}}\overset{1}{e_k} E_k E_i u_i - \\ & - \frac{3}{2r^{0^5}}\left(E^2 - 5\overset{1}{e_i}\overset{1}{e_k} E_i E_k\right) u_n u_n + \\ & + \frac{15}{2r^{0^5}}\left(E^2 - 7\overset{1}{e_i}\overset{1}{e_k} E_i E_k\right)\overset{1}{e_n}\overset{1}{e_m} u_n u_m - \\ & - \frac{3}{r^{0^5}}\left(E_i u_i - 10\overset{1}{e_m} u_m \overset{1}{e_k} E_k\right) E_n u_n \Bigg] \end{aligned} \tag{4}$$

where $\overset{1}{e_i} = \dfrac{r_i^0}{r^0}$ are corresponding unit vectors. From here on, summation from 1 to 3 will be made over repeated index.

Each unit vector connects a pair of nearest particles. As it customary in the microscopic theory of crystals, we shall use the notation $\overset{1}{e_n}\left(\begin{smallmatrix} l & l' \\ s & s' \end{smallmatrix}\right)$ to designate such unit vector; here, l and l' are the numbers of elementary cells whilst s and s' are the numbers of particles in the cells. We shall use similar notation for displacements of particles relative to each other.

To obtain the density of electrostatic energy Φ, we should perform the summation of (4) with respect to nearest neighbors of a specified particle within the frameworks of an elementary cell and then divide the result by double volume of the elementary cell (we should divide by 2 because we consider here a pair interaction).

Let us now consider a particular case shown in Fig. 1. We assume that the electrostatic field $\bar{E}$ is directed along the X_1-axis and thus its projections are $E_i = E \cdot \delta_{i1}$, where δ_{i1} is the Kronecker delta (recall that $\delta_{i1}=1$ for $i=1$ and $\delta_{i1}=0$ for $i \neq 1$). The displacement is homogeneous and its gradient, $u_{i,k} = \dfrac{\partial u_i}{\partial x_k}$, is constant over the sample (here, u_i is the projection of a displacement of a particle when it displaced out from the initial position). From Fig. 1 it follows that $i = 2$ and $k = 1$.

The density of electrostatic energy of interaction of particles is

$$\Phi = \frac{A}{2V}\sum_{l's'}\left[-\frac{3}{2\overset{0}{r}{}^5}\left(E^2 - 5\overset{1}{e}{}_1^2\begin{pmatrix} l & l' \\ s & s' \end{pmatrix}E^2\right)u_2^2\begin{pmatrix} l & l' \\ s & s' \end{pmatrix} + \frac{15}{2\overset{0}{r}{}^5}\left(E^2 - 7\overset{1}{e}{}_1^2\begin{pmatrix} l & l' \\ s & s' \end{pmatrix}E^2\right)\overset{1}{e}{}_2^2\begin{pmatrix} l & l' \\ s & s' \end{pmatrix}u_2^2\begin{pmatrix} l & l' \\ s & s' \end{pmatrix}\right] \quad (5)$$

The formula (5) contains the terms that are quadratic with respect to displacements of particles. This fact is in good agreement with the results of the microscopic theory of ionic crystals. In the general case, the displacements of particles can be expressed via the displacement gradients

$$u_n = \overset{0}{r}\, u_{n,k}\,\overset{1}{e}_k \quad (6)$$

A density of work of external forces F that are the reason for the displacement gradient in a body $u_{2,1}$ is equal in magnitude to the electrostatic energy of the field (5) and opposite in sign to the field lines.

Substituting (6) into (5) we arrive at

$$F = \frac{3AE^2}{4V\overset{0}{r}{}^3}u_{2,1}^2\sum\left[\left(1 - 5\overset{1}{e}_1\overset{1}{e}_1\right)\overset{1}{e}_1\overset{1}{e}_1 - 5\left(1 - 7\overset{1}{e}_1\overset{1}{e}_1\right)\overset{1}{e}_1\overset{1}{e}_1\overset{1}{e}_2\overset{1}{e}_2\right] \quad (7)$$

From the direct calculations of (7) with involving the results of [6] performed for a cubic body-centered lattice it follows that

$$F = \frac{7\pi^2}{9}\cdot\frac{\varepsilon_0^2\varepsilon_p^2(\varepsilon_p - \varepsilon_l)^2 a^6}{(\varepsilon_p + 2\varepsilon_l)^2\cdot V\cdot \overset{0}{r}{}^3}E^2u_{2,1}^2 \quad (8)$$

4 Viscoelastic properties of ER fluids

Let σ_{21} be the force that should be applied to a unit area of the surface of a model to ensure the displacement gradient $u_{2,1}$. The force is directed along the X_2-axis and it applied to the plane element that is orthogonal to the X_1-axis. Then, using the conventional approach, we arrive at

$$\sigma_{21} = \frac{\partial F}{\partial u_{2,1}} = \frac{14\pi^2}{9}\cdot\frac{\varepsilon_0^2\varepsilon_p^2(\varepsilon_p - \varepsilon_l)^2 a^6}{(\varepsilon_p + 2\varepsilon_l)^2\cdot V\cdot \overset{0}{r}{}^3}E^2u_{2,1} \quad (9)$$

Taking into account the formula for the volume of an elementary cell $V = \frac{4}{3\sqrt{3}} r^{0^3}$ [6], we obtain

$$\frac{a^6}{V \cdot r^{0^3}} = \frac{\sqrt{3}}{4\pi^2}\varphi^2$$

where φ is the volume fraction of a filler. Substituting the last expression into the formula (9), we finally arrive at

$$\sigma_{21} = \frac{7}{6\sqrt{3}}\beta^2\varphi^2 E^2 u_{2,1} \tag{10}$$

where

$$\beta = \frac{\varepsilon_0\varepsilon_p(\varepsilon_p - \varepsilon_1)}{(\varepsilon_p + 2\varepsilon_1)}$$

According to [6], one can write out the viscose strain tensor

$$\sigma' = \frac{8}{3}Q_2\left(\frac{1}{3}k + 1\right)\dot{u}_{2,1}$$

where $Q_2 = \frac{3\sqrt{3}}{4}\pi\frac{d^2}{r^0 h}$, $k = \frac{3d^2}{2h}$, d - the radius of a region of effective contact of particles and h is the distance between the two nearest particles.

The quotients d/h and d/r^0 can be represented via the volume fraction φ and limiting volume fraction φ_{max} (the latter corresponds to the infinite viscosity of the medium). Taking into account [6] we obtain

$$\sigma'_{21} = 0.21\pi\Psi(\varphi)\eta\dot{u}_{2,1} \tag{11}$$

where

$$\psi(\varphi) = \left(\frac{\varphi}{\varphi_{max}}\right)^{4/3}\left[1 - \left(\frac{\varphi}{\varphi_{max}}\right)^{1/3}\right]^{-3}$$

An equation of motion of the fluid under consideration is a balance of external and internal forces applied to the system

$$\sigma_{21} + \sigma'_{21} = f_{21}(t) \tag{12}$$

where $f_{21}(t)$ is a time-dependent external specific force and σ_{21} and σ'_{21} are the stresses due to the internal elastic forces (10) and viscose forces (11), respectively. The force $f_{21}(t)$ is referred to the unit area, directed along the X_2-axis, and applied to the plane region that is perpendicular to the X_1-axis. Equation (12) it is valid for small displacement gradients only. Substituting (10) and (11) into (12), we arrive at the equation of motion

$$\frac{7}{3\sqrt{3}}\beta^2\varphi^2E^2u_{2,1}+0.21\pi\Psi(\varphi)\eta\dot{u}_{2,1}=f_{21}(t) \tag{13}$$

Equation (13) describes the behavior of a viscoelastic body in accordance with the Kelvin-Voigt-Mayer phenomenological model.

Let us assume that a time-independent force $\overset{0}{f}_{21}$ is applied to the body at time t = 0. In this case, the gradient $u_{2,1}$ varies with time as

$$u_{2,1}(t)=1.48\frac{\overset{0}{f}_{21}}{\beta^2\varphi^2E^2}\left(1-e^{-\frac{t}{\tau}}\right) \tag{14}$$

where $\tau=0.31\pi\frac{\psi(\varphi)\cdot\eta}{\beta^2\varphi^2E^2}$

If the load varies with time periodically, for example, $f=\overset{0}{f}_{21}e^{i\varpi t}$, where ϖ is the angular frequency we obtain

$$u_{2,1}(t)=1.48\frac{\overset{0}{f}_{21}}{\beta^2\varphi^2E^2}\cdot\frac{1}{1+i\varpi\tau}e^{i\varpi t} \tag{15}$$

Finally, we arrive at the complex dynamic shear modulus

$$G^*=0.67\beta^2E^2\varphi^2+i\varpi\cdot0.21\pi\cdot\Psi(\varphi)=G'+iG'' \tag{16}$$

6. Conclusions

It is shown that, in the frameworks of the approach proposed and under the assumptions made, the ER fluid can be treated as a viscoelastic body and consequently described on the basis of the linear theory of viscoelasticity. We obtain a complex shear modulus as a function of the electric field strength, volume fraction of filler, dielectric properties of compounds of the ER fluid, and viscosity of the fluid phase. The real part of the complex shear modulus is proportional to the electric field strength squared (this dependence of parabolic type is similar to the results of the papers [1,3]).

References

1. Martin J.E. and Anderson R.A., Chain model of electrorheology. *J. Chem. Phys.* **104** (1996) pp.4814-4827.
2. Panofsky W.K.H. and Phillips M. *Classical Electricity and Magnetism* (Addison – Wesley Publishing Company, Cambridge 42. Mass.).

3. Shkel Yu.M. and Klingenlerg D.J., A continuum approach to electrorheology. *J. Rheol.* **43** (5) (1999) pp.1307-1322.
4. Shulman Z.P., Korobko E.V. and Yanovskii Yu.G., The mechanism of the viscoelastic behaviour of electrorheological suspensions. *J. Non-Newtonian Fluid Mech.* **33** (1989) pp.181-196.
5. Yanovsky Yu.G., Korobko E.V. and Barancheeva V.V., Peculiarities of rheological behavior of electro-sensitive suspensions of various structural-rheological types. *Inzhenerno-Fizicheskii Zhurnal,* **57** (2) (1989) pp.219-224.
6. Zgaevskii V.E. and Yanovsky Yu.G., Calculation of effective viscosity of concentrated solid particle suspensions using a crystal model. *Composite Mechanics and Design* **2** (3) (1996) pp.113-144.

STRESS BEHAVIOR AND CONDUCTIVITY OF THE ELECTRO-RHEOLOGICAL SUSPENSIONS

KATSUFUMI TANAKA, KENSUKE ICHIZAWA, YOHEI ONISHI, ATSUSHI KUBONO AND RYUICHI AKIYAMA

Department of Polymer Science and Engineering, Kyoto Institute of Technology, Matsugasaki, Kyoto 606-8585, Japan

The steady stress under static electric fields and the conduction current passing through the electro-rheological (ER) suspensions with cation exchange resin particles were investigated simultaneously as a function of the shear rate. The shear thinning behavior was found for a suspension with a higher viscosity of the continuous phase. Further, a remarkable dip behavior was found for a dilute ER suspension. The apparent conductivity, which was calculated from the conduction current, probed sensitively the corresponding behavior to the stress responses.

1 Introduction

The flow behavior of a typical ER suspension is generally assumed to be the Bingham flow [1] with the yield stress (dynamic yield stress or induced stress) σ_y:

$$\sigma = \sigma_y + \eta_c \dot{\gamma}, \quad (1)$$

where σ is the stress under the electric field, η_c is the viscosity of the continuous phase of the suspension under no electric field, and $\dot{\gamma}$ is the shear rate. In this model, the yield stress is induced by the external electric field. It is thought that the yield stress is induced by the aggregated structure of polarized particles after application of the electric field. Further, the yield stress at a given strength of the electric field is generally assumed to be constant relative to the shear rate. However, the applicability or limitation of the Bingham flow should be clarified for the purpose of an appropriate design of a system using the ER suspensions in practical application.

In the present paper, some results of rheological behavior and conduction current are reported for the electro-rheological suspensions with various combinations of the viscosity of the continuous phase and the concentration of cation exchange resin particles. The shear thinning behavior is found for suspensions with higher viscosities of the continuous phase. It should be noted that a remarkable dip behavior is also found for a dilute ER suspension: The stress decreases as the shear rate increases, and it gradually increases again and approaches the stress under no electric field. The apparent conductivity, which is calculated from the conduction current, probes sensitively the corresponding behavior to the stress responses.

2 Experimental

Cation exchange resin particles [2,3] were suspended in silicone oils with various viscosities. The concentrations of the particles were 30 wt% for a typical concentration and 5 wt% for a dilute concentration. The viscosities of the silicone oils (η_c) were 10 Pas, and 50 Pas.

A rotational rheometer (Rheology Co., Ltd., MR-300V2E) was used with a fixture of a coaxial cylinder [2,3]. The DC electric field was applied to the ER suspensions, and the strength of the electric field was varied. The stress was calculated from the torque measured with the rheometer controlled by computer #1 using the software provided by the supplier. The shear rate was also calculated from the revolution angle measured with the rheometer as a function of time. The rheometer was calibrated using silicone oils with various viscosities.

In the present experiment, the analog outputs of the torque and the revolution angle from the rheometer were digitized by an analog-to-digital converter, and the data were stored as a function of time in another computer (computer #2). The current passing through the ER suspension was also digitized and stored in computer #2. The subtracted current passing through the ER suspension (ΔI), which was the difference between the current measured under the electric field and that under no electric field, was calculated as a function of time. The subtracted current was averaged over the period of each constant shear rate. The apparent conductivity (κ_{app}) was calculated using the averaged subtracted current ($\langle \Delta I \rangle$) in the following equation:

$$\kappa_{app} = \ln(R_2/R_1)(\langle \Delta I \rangle / V)/2\pi L, \quad (2)$$

where V is the applied voltage, R_1 (17.52 mm), R_2 (16.49 mm), and L (= 14.0 mm) are, respectively, the inner radius, the outer radius, and the immersed length of the coaxial cylinder of the rheometer. In the present experiment, the data for the torque and the revolution angle stored in computer #2 were used only for determining the period for a constant shear rate. Further details were reported elsewhere [3].

3 Results and Discussion

3.1 *Shear Thinning Behavior of the ER Suspensions with a Typical Concentration and Higher Viscosities of the Continuous Phase*

Figure 1 shows the flow curves plotted on a linear scale of an ER suspension with η_c = 50 Pas. The concentration of the particles was 30 wt%. The stress under 2.0 kV/mm shows the shear thinning behavior at higher shear rates. The stress deviates downward from the dotted line, which is assumed to be the Bingham flow, around the shear rate of 1 1/s although the onset of the shear thinning is less pronounced. (The shear thinning behavior was not found evidently on a logarithmic scale as reported previously [2].) Similar tendency of the deviation of the stress is also found in Fig.1 at lower strengths of the electric field. Therefore, the Bingham flow is not assumed generally to the flow of the ER suspension under higher shear rates. The shear thinning behavior was also found for the ER suspension with η_c = 10 Pas and the same concentration of the particles.

For the discussion further in detail, a general dimensionless parameter called Mason's number, M_n, is introduced because most important experimental conditions are included in the parameter [4]:

$$M_n = \eta_c\dot{\gamma}/2\varepsilon_c\varepsilon_0 (\beta E)^2, \quad (3)$$

$$\beta = (\kappa_p - \kappa_c)/(\kappa_p + 2\kappa_c), \quad (4)$$

where E is the strength of the electric field, ε_0 is permittivity of free space, κ_p is the conductivity of the particles, and ε_c and κ_c are, respectively, the dielectric constant and the conductivity of the continuous phase of the silicone oil [2,4-7].

In the present study, the conductivity of the particles is assumed to be much higher than that of the silicone oils. Therefore, β is also assumed to be unity. A typical value of M_n for a typical ER suspension is calculated to be 0.08 using η_c = 0.02 Pas, γ = 200 1/s, ε_c

= 2.70, and E = 1 kV/mm. When an ER suspension with the higher viscosity of the continuous phase and the lower strength of the electric field is investigated, the Mason's number becomes much larger, which corresponds to an experimental condition at higher shear rates. For instance, an important value to be M_n = 1 can be obtained for an ER suspension with η_c = 10 Pas, $\dot{\gamma}$ = 5 1/s, ε_c = 2.76 and E = 1 kV/mm. The condition corresponds to the shear rate one or two orders higher than the typical case. The Mason's number at which the onset of the shear thinning was found in Fig.1 is calculated (roughly) to be 0.3 using η_c = 50 Pas, $\dot{\gamma}$ = 1 1/s, ε_c = 2.76 and E = 2 kV/mm.

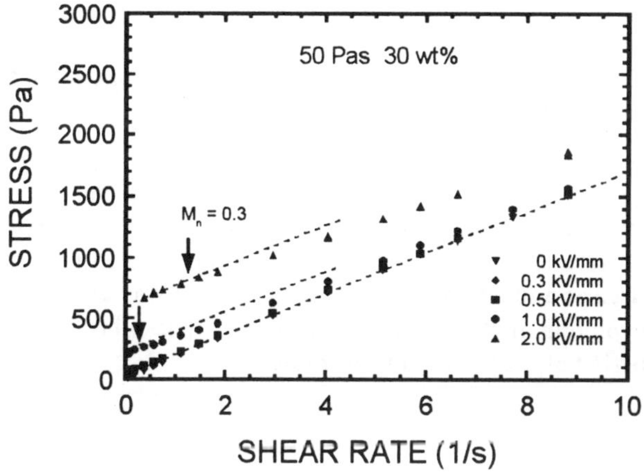

Figure 1. Flow curves of an ER suspension with η_c = 50 Pas. The concentration of particles was 30 wt%.

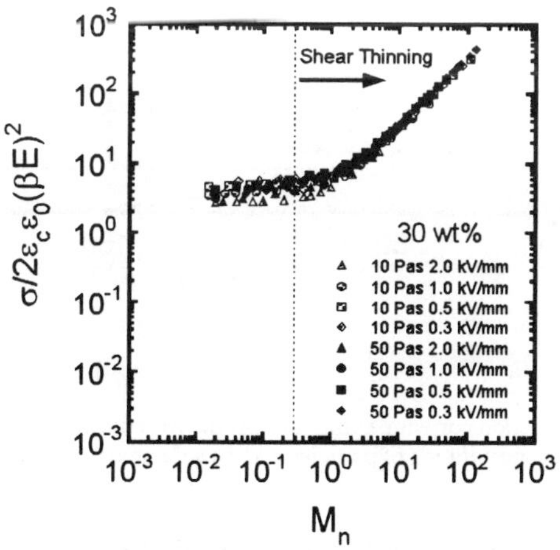

Figure 2. Reduced stress of the ER suspensions with η_c = 10 Pas and 50 Pas plotted against the Mason's number. The concentration of particles was 30 wt%.

Figure 2 shows the reduced stress of the ER suspensions plotted against the Mason's number. The stress was reduced by the factor $2\varepsilon_c\varepsilon_0(\beta E)^2$, which is proportional to the electric force (with the dimension of stress) acting between the particles. The Mason's number for the onset of the shear thinning around 0.3 is indicated by dotted line in the figure. Although the reduced stresses shows crudely a muster curve in Fig.2, the shear thinning behavior is not evident again on the logarithmic scale. On the other hand, the shear thinning would be sensitively probed by the conduction current passing through the ER suspension, because the shear thinning would be induced by the changes in the aggregated structure of the polarized particles in the ER suspension. Therefore, the conduction current was measured at the same time and the apparent conductivity was calculated for the discussion in detail.

3.2 Conduction Current and the Apparent Conductivity Corresponding to the Shear Thinning Behavior in the Stress

Figure 3 shows an example for changes in the conduction current passing through an ER suspension and the shear rate plotted against time. In the simultaneous measurement of rheological behavior and conduction current, the shear rate was decreased step by step from a higher shear rate, and the current as well as the revolution angle of the outer cylinder was measured as a function of time. The shear rate was calculated from the time derivative of the revolution angle of the outer cylinder. The subtracted current, ΔI, was averaged over the period of each constant shear rate.

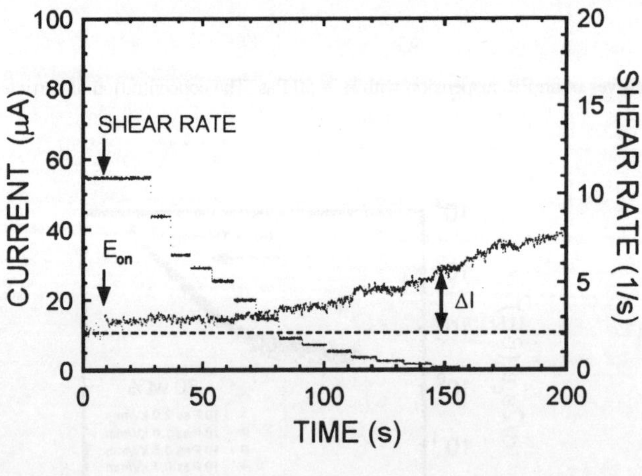

Figure 3. Changes in the conduction current passing through the ER suspension with η_c = 50 Pas and the shear rate plotted against time. The concentration of particles was 30 wt%, and the applied voltage was 1.0 kV.

Figure 4 shows the apparent conductivity plotted against the Mason's number for the ER suspensions with η_c = 10 Pas and 50 Pas. The concentration of the ER suspensions was 30 wt%. (The apparent conductivity at 0.3 kV/mm is not plotted in the figure because the conduction current was too small to evaluate the apparent conductivity.) The apparent conductivity is well reduced to a master curve plotted against the Mason's number. It decreases and shows a minimum around $M_n = 10$ as the Mason's number increases. (It seems that the apparent conductivity increases at the higher Mason's numbers.) The apparent conductivity is mostly constant at the lower limit of the Mason's number. Around $M_n = 0.3$, the slope of the decrease in the apparent conductivity changes, which is roughly related to the onset of the shear thinning behavior in the stress response.

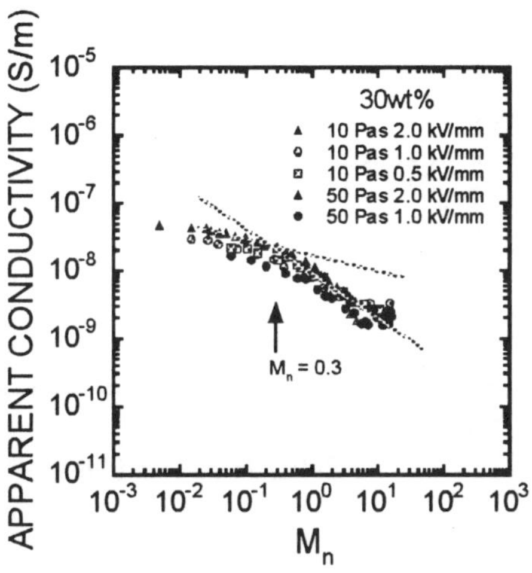

Figure 4. The apparent conductivity of the ER suspensions shown in Fig.2 plotted against the Mason's number.

The apparent conductivity is directly related to the current passing through the ER suspension. The current would flow along the aggregated chains of the polarized particles as conduction paths. Because the DC electric field is applied to the ER suspension, the decrease in the current at a given strength of the electric field would be caused by macroscopic fracture of the aggregated structure in the ER suspension. The dynamic equilibrium of the aggregated structure under shear deformation would be maintained at lower Mason's numbers. The fractured structure is recombined quickly to maintain the dynamic equilibrium, which corresponds to the relatively constant induced stress and the apparent conductivity. However, the aggregated structure would be macroscopically fractured at higher Mason's numbers. The shearing force from the silicone oil becomes comparable to the electric force bridging between the polarized particles, and the possibility for the recombination of the fractured structure becomes smaller as the Mason's number increases up to $M_n = 10$.

3.3 Dip Behavior and the Apparent Conductivity of a Dilute ER Suspension

Figure 5 shows the flow curves of a dilute ER suspension with $\eta_c = 50$ Pas. The concentration of the particles was 5 wt%. As shown in the figure, the stress at a given strength of the electric field shows a remarkable dip behavior at higher shear rates: The stress decreases as the shear rate increases above a critical shear rate, and it gradually increases again and approaches the stress under no electric field. The dip behavior is extremely remarkable for the dilute ER suspension. Further, the shear rate at the onset of the dip behavior is around 0.1 1/s at 2.0 kV/mm, and it is reduced to M_n = 0.03. Because the Mason's number at the onset of the dip behavior is one order smaller than that of the shear thinning behavior shown in Fig.1, the concentration of the particles plays a crucial role in the dip behavior of the ER suspension.

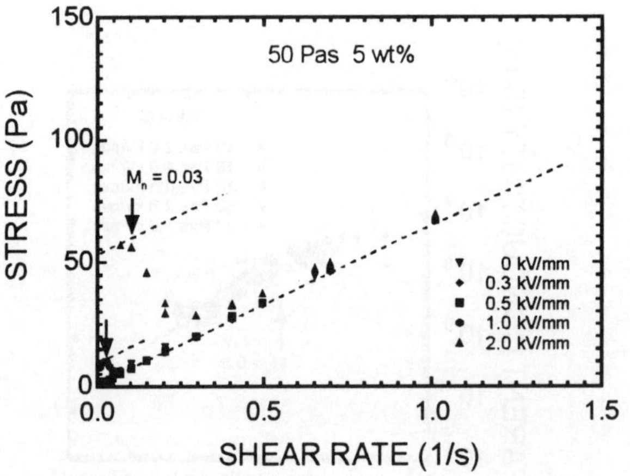

Figure 5. Flow curves of a dilute ER suspension with η_c = 50 Pas. The concentration of the particles was 5 wt%.

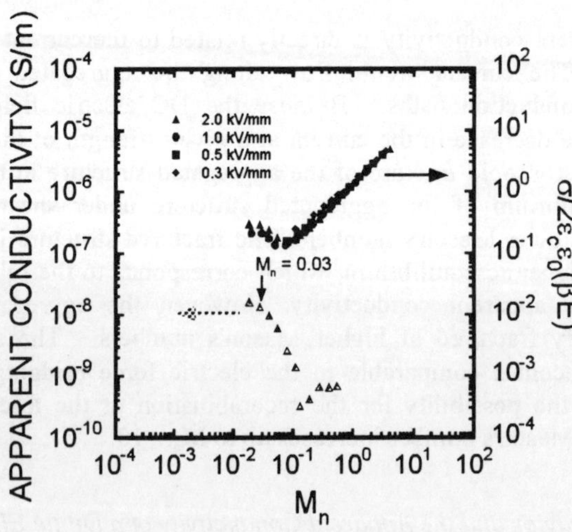

Figure 6. Reduced stress and the apparent conductivity of the dilute ER suspensions shown in Fig.5 plotted against the Mason's number.

Figure 6 shows the apparent conductivity and the reduced stress of the ER suspension shown in Fig.5 plotted against the Mason's number. (Only the apparent conductivity at 2.0 kV/mm is plotted in the figure.) In Fig.5, a muster curve is seen for the reduced stress. The Mason's number for the onset of the dip behavior is around 0.03, around which the apparent conductivity decreases steeply as the Mason's number increases. Further, the Mason's number for the minimum of the reduced stress is around 0.1. The apparent conductivity also shows the minimum around $M_n = 0.1$, and it gradually increases around $M_n = 0.3$. Around $M_n = 0.3$, the fractured clusters in the ER suspension would be recombined again, which may be related to the increase in the chance of collision among polarized particles [7]. The collision of the polarized particles with the electrodes would also play an important role in the increase in the apparent conductivity. The reasons for the increase in the apparent conductivity are still unclear at the present stage.

4 Conclusions

The steady stress and the current response were investigated at higher shear rates for the ER suspensions. The shear thinning was found for the ER suspensions with a typical concentration and higher viscosities of the continuous phase. Further, a remarkable dip behavior was also found for a dilute ER suspension. The apparent conductivity probed sensitively the corresponding behavior to the stress responses. The shear thinning and the dip behavior are closely related to the changes in the aggregated structure of the particles in the ER suspensions, and the concentration of the particles is one of the most important parameters to be considered.

References

1. Jordan, T. C. and Shaw, M. T., "Electrorheology", *IEEE Trans. Electric. Insulation* **24** (1989) pp.849-878.
2. Tanaka, K., Hashimoto, S., Takenouchi, T., Sugimoto, I., Kubono, A. and Akiyama, R., Shearing effects on the electrorheological response, In *Proc. of 7th International Conference on ER Fluids, MR Suspensions and Their Application*, eds. by R. Tao (World Scientific, Singapore, 2000) pp. 423-430. (Also published in *Int. J. Mod. Phys.* **13** (1999) pp. 1814-1821.)
3. Tanaka, K., Ichizawa, K., Akiyama, R. and Kubono, A., Simultaneous measurement of rheological behavior and conduction current for an electro-rheological suspension with cation exchange resin particles *J. Soc. Rheol., Jpn.* **29** (2001) in press.
4. Marshall, L. and Zukoski, C. F., Effects of electric fields on rheology of non-aqueous concentrated suspensions. *J. Chem. Soc., Faraday Trans.*, **85** (1989) pp. 2785-2795.
5. Davis, L. C., Polarization forces and conductivity effects in electrorheological fluids. *J. Appl. Phys.* **72** (1992) pp. 1334-1340.
6. Klingenberg, D. J., and Zukoski, C. F., Studies on the steady-shear behavior of electrorheological suspensions. *Langmuir* **6** (1990) pp. 15-24.
7. Tanaka, K., Nakamura, N., Takada, K., Iwaki, F., Kubono, A., Akiyama, R. and Kuramoto, N., Effects of chance of collision among particles on electrorheological response. In *Proc. of 6th International Conference on ER Fluids, MR Suspensions and Their Application*, eds. by K. Koyama and M. Nakano (World Scientific, Singapore, 1999) pp. 172-179. (Also published in *Int. J. Mod. Phys.* **15** (2001) pp. 930-937.)

DETERMINATION OF RHEOLOGICAL AND MAGNETIC PROPERTIES FOR MAGNETORHEOLOGICAL COMPOSITES VIA SHEAR MAGNETIZATION MEASUREMENTS

A.T. HORVATH

Dept of Mechanical Engineering, University of Wisconsin - Madison, 1513 University Ave, WI 53706, USA

D. J. KLINGENBERG

Dept of Chemical Engineering, University of Wisconsin - Madison, 1415 Engineering Drive, Madison, WI, 53706, USA

Y. M. SHKEL

Dept of Mechanical Engineering and Polymer Engineering Center, University of Wisconsin - Madison, 1513 University Ave, WI 53706, USA
E-mail: yshkel@engr.wisc.edu

Anisotropic solid-state magnetorheological (MR) composites are prepared by applying magnetic fields to isotropic suspensions in polymer melts and preserved by cross-linking the melts. The composites are studied through a continuum-level perspective at various magnetic field strengths under oscillatory shear. The minimum number of constraints are applied in the derivation of the magnetic and rheological invariants, which relate the measured macroscopic magnetostriction coefficients to the discrete parameters of any microscopic model. The field-induced stress is defined by magnetostriction coefficients, which are material parameters that describe the strain dependence of the magnetization tensor. A comparison of the independently measured rheological and magnetic properties allow for the determination of the magnetostriction coefficients.

1 Introduction

Ferromagnetic particles dispersed in a matrix (fluid or solid) exhibit useful magnetoactive properties. These materials change size and shape when exposed to an external magnetic field, and their rheological properties can be controlled with external magnetic fields. Such magnetoactive behavior has a wide variety of potential applications, from transducers to active damping devices. The objective of this article is to understand the response of magnetoactive composites to shear deformations, and to relate the magnetic and rheological properties of these materials. Here a phenomenological continuum representation of magnetoactivity, a microscopic model, and a novel experimental approach are employed for extracting the model parameters. We briefly describe a new experimental approach for measuring material properties that govern magnetoactivity.

Anisotropic solid-state MR composites are prepared by applying magnetic fields to isotropic MR suspensions in polymer melts; the resulting structure are preserved by cross-linking the melts. The composites are studied at various magnetic field strength under oscillatory shear. The field-induced stress is defined by magnetostriction coefficients, which are material parameters that describe the strain dependence of the magnetization tensor. The experimental setup includes a parallel plate rheometer providing oscillatory shear measurements and an inductance coil surrounding the composite sample. Sample oscillation in a dc magnetic field produces an e.m.f. due to shear-induced magnetization, which is measured by a lock-in amplifier. A comparison of the independently measured rheological and magnetic properties allow for the determination of the magnetostriction coefficients, which a thermodynamic analysis and a magnetic flux analysis predict to be the same. The minimum number of constraints are applied in the derivation of the magnetic and rheological invariants, which relate the measured macroscopic magnetostriction coefficients to the discrete parameters of any microscopic model. We will illustrate the relation between the macroscopic coefficients and the microstructure for an isolated chain model.

2 Phenomenology of Magnetostriction

2.1 Magnetorheology and Magnetostriction

In our previous papers[1,2] we demonstrate a thermodynamic approach to the derivation of the total stress tensor in polarizable anisotropic media. Following Landau and Lifshitz[3] (1984), we write the total stress in anisotropic media in terms of the free energy

$$\sigma_{ik} = F_0\delta_{ik} + \left.\frac{\partial F_0}{\partial u_{ik}}\right|_T - \frac{1}{2}\left.\frac{\partial B_l}{\partial u_{ik}}\right|_{T,\mathbf{H}} \cdot H_l - \frac{\mathbf{H}\cdot\mathbf{D}}{2}\delta_{ik} + \frac{H_iD_k + H_kD_i}{2}, \tag{1}$$

where $\sigma_{ik} = \sigma_{ik}^{mech.} + \sigma_{ik}^{mag.}$ is the mechanical (e.g., elastic) plus magnetostatic stress, $F_0(u_{ik}, T)$ is the free energy per unit volume without a field, and u_{ij} is the strain tensor [for infinitesimal deformations, $u_{ij} = (\partial U_i/\partial x_j + \partial U_j/\partial x_i)/2$ where $\mathbf{U}(\mathbf{x})$ is the displacement field]. The first two terms in Eq. (1) constitute the mechanical stress $\sigma_{ik}^{mech.}$, which can be defined as a total stress in the material with a given structure in the absence of a field ($\mathbf{H} = \mathbf{0}$). The third term arises from the change in magnetic properties of the material with deformation and is often called the *magnetostriction stress*. The last two terms represent the contribution of free and bound charges to the total stress and constitute the so-called Maxwell's stress tensor. The magnetic induction $\mathbf{B}$ in anisotropic media is

$$\mathbf{B} = \mu_0\boldsymbol{\mu}\cdot\mathbf{H}, \tag{2}$$

where $\mu_0 = 4\pi\times10^{-7}$ H/m is the magnetic permeability of free space and the second order permeability tensor, $\boldsymbol{\mu}$, relates the magnetic induction, $\mathbf{B}$, to the magnetic field, $\mathbf{H}$, in the material and is a function of the strain u_{ik}, temperature T, and field $\mathbf{H}$. Components of permeability tensor, $\boldsymbol{\mu}$, can be directly measured or predicted through modelling the structure of composite materials which may be comprised from a variety of magnetization mechanisms. One can see that magnetostriction part of the total stress tensor, $-\partial B_l/\partial u_{ik}\cdot H_l/2$, can be predicted if the deformation effect on the magnetic properties of the material is known. In the following section, we discuss approaches for experimental investigation of the deformation effect on material magnetization.

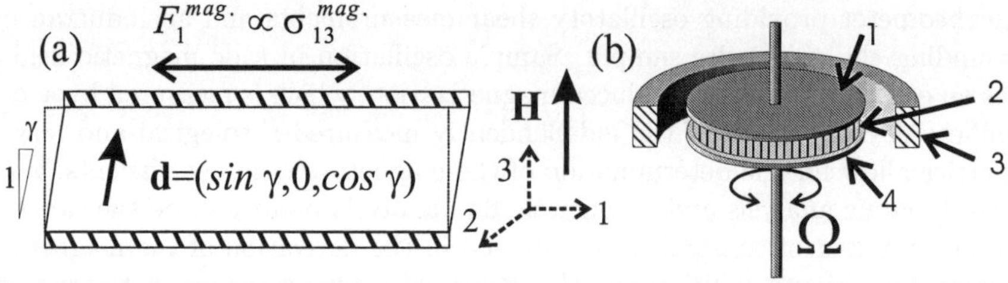

Figure 1. (a) Schematic diagrams of small amplitude shear deformation, and (b) experimental set-up: sample (2) between fixed (1) and moving (4) parallel disks; the measurement coil (3) is used to probe the change in permeability with deformation.

2.2 Magnetostriction of Uniaxial Anisotropic Suspensions

We consider the practically important type of anisotropy possessed by microstructures with a single distinct direction described by the unit vector **d**. Such uniaxial materials are labeled *transverse isotropic*. This type of anisotropy describes typical electro- and magnetorheological suspensions and elastomers, where particle columns or chains are oriented in a common direction, but where translational order is absent in the plane perpendicular to the columns. The magnetic properties of transverse isotropic materials differ in the **d** direction and in the plane perpendicular to **d**. These materials have a second order permeability tensor of the form

$$\mu_{ik} = \mu^{\perp}\delta_{ik} + (\mu^{\parallel} - \mu^{\perp})d_i d_k, \tag{3}$$

where $\mu^{\parallel}$ and $\mu^{\perp}$ are the relative permeability constants for probing fields applied parallel and perpendicular to **d**, respectively. In this case, the strain-dependent permeability tensor can be written[1,2]

$$\mu_{ik} = \mu_{ik}^0 + \alpha_1 u_{ik} + \alpha_2 u_{ll}\delta_{ik} + \alpha_3 u_{lm} d_l^0 d_m^0 \delta_{ik} + \alpha_4(u_{il}d_l^0 d_k^0 + u_{kl}d_l^0 d_i^0) \\ + \alpha_5 u_{ll} d_i^0 d_k^0 + \alpha_6 u_{lm} d_l^0 d_m^0 d_i^0 d_k^0 + (\mu^{\parallel} - \mu^{\perp})(\omega_{il}d_l^0 d_k^0 + \omega_{kl}d_l^0 d_i^0). \tag{4}$$

Eq. (4) introduces 6 phenomenological *magnetostriction coefficients* α_1–α_6. These material properties are scalar functions of thermodynamic variables, such as the temperature, the magnetic field magnitude, and the density. The permeability constants $\mu^{\parallel}$ and $\mu^{\perp}$ of the undeformed material are well-known material parameters, which can be measured by a variety of experimental methods and related to the material microstructure through a number of theoretical models. The parameters α_1 and α_2 were first introduced by Stratton[4] and Landau and Lifshitz[3] for isotropic materials. Microscopic models for predicting parameters α_3–α_6 have been discussed by Shkel and Klingenberg[2].

2.3 Example Solution for Field-Induced Stresses

Following our previous article,[2] we illustrate how the shear stress is related to magnetostriction parameters for the configuration illustrated schematically in Fig. 1 (a).

Consider a material with the anisotropy vector initially aligned with the magnetic field $\mathbf{H} = (0, 0, H)$, and subjected to a small simple shear deformation where the nonzero components of the strain tensor are $u_{13} = u_{31} = \gamma/2$. After deformation, the anisotropy vector is $\mathbf{d} = (\sin\gamma, 0, \cos\gamma)$, and the stress can be obtained by substitution this value into Eqs. (1),(2) and (4). To leading order in γ, the expression for the shear stress is[1,2]

$$\sigma_{13}^{mag.} = \mu_0 \frac{H^2}{2} \gamma \Big\{ (\mu^{\|} - \mu^{\perp}) - (\alpha_4 + \alpha_3 + \alpha_6) \Big\}, \tag{5}$$

Note that the phenomenological parameters are functions of the structure in the deformed state. For simple shear deformation, the leading order variation of the parameters with strain is $\mathcal{O}(\gamma^2)$, and thus the parameters in Eq. (5) can be evaluated for the undeformed structure.

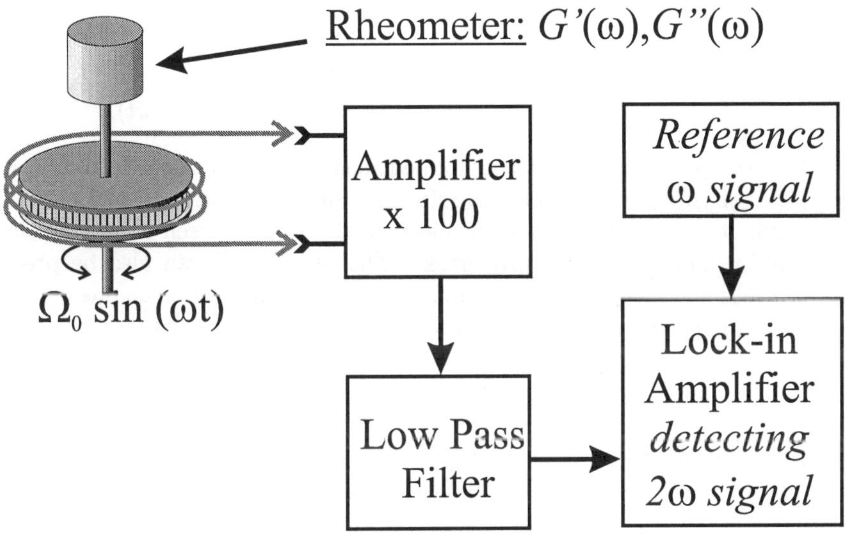

Figure 2. Dynamic potential testing of magnetostriction: Bias Magnetic field is applied across the sample. AC signal produced by sample deformation in the field was filtered and measured by Lock-in-Amplifier on double oscillatory frequency. Obtained magnetic e.m.f. data were compared with rheometric data.

3 Experimental Study of Magnetostriction

Consider a composite of ferromagnetic particles in a matrix contained between parallel disks as illustrated in Fig. 1(b). When the sample is simultaneously sheared and subjected to an external magnetic field, an e.m.f. is generated in the measurement coil. This e.m.f. is related to the deformation-dependent permeability tensor μ_{ik} defined by Eq. (4). For small oscillatory shear deformations $\gamma = \gamma_0 \sin(\omega t)$ ($\gamma_0 \sim \Omega_0 \ll 1$) of transverse isotropic materials (e.g., chain-like structures with

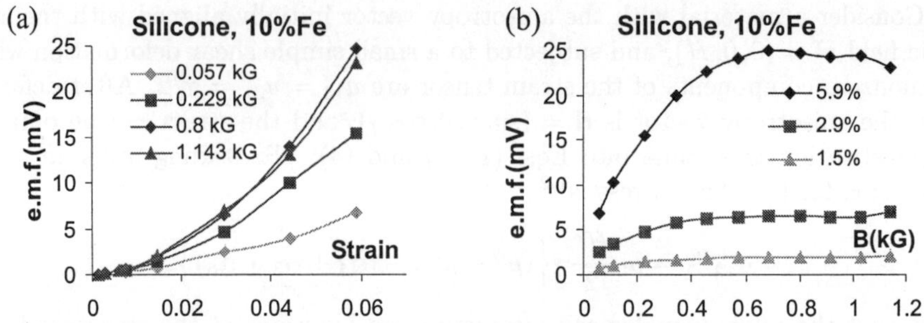

Figure 3. Preliminary results for the e.m.f. arising from the deformation of a magnetostrictive medium in a magnetic field: E.m.f. ($\mathcal{E}$) is quadratic with the strain amplitude and linear with the field strength for small fields. For large fields e.m.f. is saturated.

orientation **d**), the e.m.f. generated in the coil, $\mathcal{E}$, has the form

$$\mathcal{E} = 2\mu_0 \left[(\mu^{\parallel} - \mu^{\perp}) - (\alpha_3 + \alpha_4 + \alpha_6)\right] SH\gamma_0^2 \omega \sin(2\omega t),$$

where S is the plate area, H is the applied magnetic field strength and γ_0 is the shear strain amplitude. The shear modulus $G'(= \sigma_{13}^{mag.}/\gamma)$, defined by Eq. (5) has a similar form, which illustrates the intimate relationship between phenomenological magnetization and rheological properties. The response $\mathcal{E}$ can also be predicted from a microscopic model.[1,2] For structures composed of isolated chains of spheres, the microscopic prediction in the limit of linear magnetizable spheres is

$$\mathcal{E} = \mu_0 \mu_c 3\phi\beta \left[\frac{3\xi/2}{(1-\xi)^2} + \frac{\xi}{1-\xi} + \frac{\xi/2}{1+\xi/2}\right] SH\gamma_0^2 \omega \sin(2\omega t),$$

where μ_c is the relative permeability of the continuous phase, ϕ is the particle volume fraction, $\beta = (\mu_p - \mu_c)/(\mu_p + 2\mu_p)$ is the relative particle permeability, μ_p is the particle permeability, $\xi = \beta\zeta(3)(\sigma/\ell^0)^3/2$, where $\zeta(3) \approx 1.20206$, σ is the sphere diameter, and ℓ^0 is the sphere separation in a chain.

Both analyses yield the following predictions: $\mathcal{E}$ responds at twice the deformation frequency; $\mathcal{E}$ is linear in the oscillation frequency ω; $\mathcal{E}$ is quadratic in the strain amplitude; and $\mathcal{E}$ is proportional to the shear modulus.

3.1 Experimental Setup

Magnetostriction and magnetorheological characterizations of samples with aligned and random structures were conducted with a Bohlin VOR rheometer modified for work in a magnetic field. A disk shaped sample was tested with a parallel plate configuration. The rheometer provided oscillatory shear deformation in a magnetic field $(0 - 2T)$ applied across the sample using a custom-made electromagnet. The rheometer provided rheological and MR properties of samples at various frequencies, deformation magnitudes, and magnetic field strengths. In addition, the rheometer provided controlled oscillations for a magnetostriction study with *dynamic potential measurement* techniques. Magnetic measurements were conducted using customized

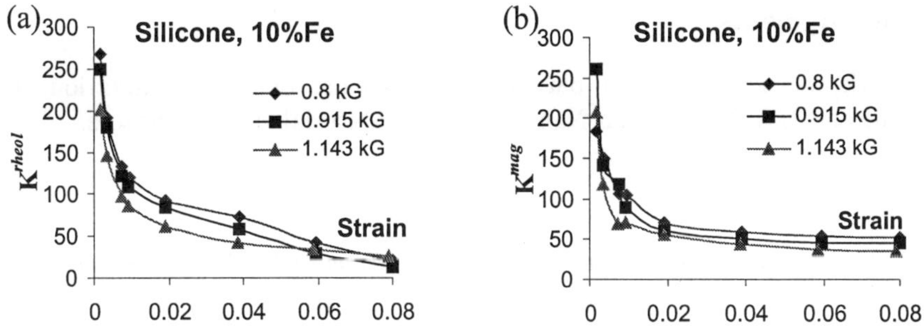

Figure 4. Rheological value of $K^{rheol} = 2\sigma_{13}/\mu_0 H^2 \gamma$ (a), and magnetostriction value of $K^{mag} = \mathcal{E}/2\mu_0 S H \gamma_0^2$ (b) yield same values.

dynamic microphones.[7] Amplified AC signals produced by the sample deformation were measured with a Stanford Research Systems Lock-in-Amplifier.

3.2 Experimental Results

Preliminary measurements were performed with a 10 vol.% suspension of iron particles in silicone rubber. The sample is a disk $20mm$ diameter, and thickness $\approx 2mm$. Data for the e.m.f. as a function of strain amplitude ($\omega = 2\pi rad/s$) for different field strengths are plotted in Fig. 3. Note that the data are consistent with the predictions listed above: the e.m.f. ($\mathcal{E}$) is quadratic in the strain amplitude and linear in the field strength for small fields. For large fields, the e.m.f. is saturates.

The combination of electrostriction parameters in Eqs. (3) and (5) can be expressed as $K = (\mu^{\parallel} - \mu^{\perp}) - (\alpha_3 + \alpha_4 + \alpha_6)$, where the parameter K can be determined from either rheological or magnetostriction measurements

$$K^{rheol} = \frac{2\sigma_{13}}{\mu_0 H^2 \gamma}, \quad K^{mag} = \frac{\mathcal{E}}{2\mu_0 S H \gamma_0^2}. \tag{6}$$

Fig. 4 presents K^{rheol} and K^{mag} values obtained for various field strengths—Fig. 4(a) gives the rheological value of K and Fig. 4(b) gives magnetostriction value of K corrected for the demagnetization factor due to the sample shape. One can see that both approaches yield the same values.

4 Conclusion

We have shown how the rheological response of MR composites can be related to macroscopically measurable magnetostriction properties. We are currently improving this new experimental technique to further test the phenomenological theory, and to quantitatively extract the phenomenological magnetostriction parameters.

Acknowledgments

This work was supported in part by the National Science Foundation (CTS-9502276) and UW-MRSEC on Nanostructured Materials and Interfaces.

References

1. Y. M. Shkel and D. J. Klingenberg, *Proceedings of the 7th International Conference on Electro-Rheological Fluids and Magneto-Rheological Suspensions, Honolulu, Hawaii, July 19-23*, 1999, pp.252-259, (World Scientific Publishing, 2000).
2. Y. M. Shkel and D. J. Klingenberg, J.Rheol. **45(5)**, 1322 (1999).
3. L. D. Landau and E. M. Lifshitz, *Electrodynamics of Continuous Media*, Pergamon, New York, 1984.
4. J.A. Stratton, *Electromagnetic Theory* (McGraw-Hill, New York, 1941).
5. Y. M. Shkel and N. J. Ferrier, Electrostriction Enhancement of Solid-State Capacitance Sensing, Submitted to IEEE/ASME Trans. on Mechatronics, 28 Jan. 2000.
6. Y. M. Shkel and D. J. Klingenberg, Electrostatic Stresses in Anisotropic Electroactive Composites: Tetragonal Lattice, in preparation, (2000).
7. P. Horowitz and W. Hill, *The Art of Electronics* (Cambridge University Press, 1998).

EVOLUTION OF CHAIN STRUCTURE OF ELECTRORHEOLOGICAL FLUIDS IN FLOW MODEL

X.P. ZHAO X.Y. GAO D.J.GAO

Institute of Electrorheological Technology 141#, Department of Applied Physics, Northwestern Polytechnical University, Xi'an,P.R. China, 710072

E-mail: xpzhao@nwpu.edu.cn

The movement of particles in electrorheological (ER) fluids is analyzed by means of molecular dynamic simulations. We found that the velocity profile of particles can be divided into two zones. One zone near electrodes where particles' velocity profiles change periodically like "breathing type" is called transition zone. The other in the middle of two electrodes where particles move smoothly like a plug is called "plug zone". In addition, the relationship between volume flow rate and relative pressure gradient is simulated out. Factors such as volume flow rate, critical electric field, critical pressure gradient and response time of shutting up were also analyzed respectively.

1 Introduction

Electrorheological (ER) fluids, which exhibit changes in rheological behavior in the presence of an external electric field, have been studied since Winslow in the 1940's [1]. All engineering applications of ER fluid are based on its the controllable mechanical response [2]. In general, the working behavior of actuators based on ER fluids is classified by three different mode: shear, flow and squeeze [3-7]. In the flow mode, which is called also Poiseulle flow, it is assumed that two electrodes are fixed.

Currently the leading research on poiseulle flow of ER fluid is confined to continuous medium hypothesis [8,9], which is based on the characteristic size and time of the investigated problem are both greater than the distance and collision time of particles respectively. Also the statistical average value of a large quantity of particles can be hardly affected by individual ones [10,11]. Actually, ER fluids can form chains and columns under external electric field. Tamura and Doi firstly used molecular dynamics to simulate Poiseuille flow of ER fluids. They treated the polarized particles as electric dipoles and got some results in accordance with experiment [12]. But they used a' mean field approximation' for the field, i.e. the field fluctuation were neglected and the each particle felt a space-averaged field [12]. Since immersed particles in ER fluid are μm-ordered in diameter and ER fluid will form asymmetrical chain-column structures in the presence of electric field. And based on the research of ER fluid's multi-chain interaction [13] in the past, we think it more reasonable to regard ER fluid as a scattering particle system in the effect of electric field. In this paper, molecular dynamic simulation is employed to establish a motion equation for every particle and to investigate the rheological behavior of the scattering particle system.

2 Model and simulations

First we investigate the fluid placed two parallel flat slabs between which the distance is h. Constant flow, which is caused by pressure gradient p' along the axis of x, is showed in **Fig.1** (a). The width and length of electrodes are b and L respectively, which are far greater than h. Therefore, the edge effect of electrodes could be neglected. Thus the discussed question can be regarded as a two-dimensional problem (**Fig. 1**(b)).

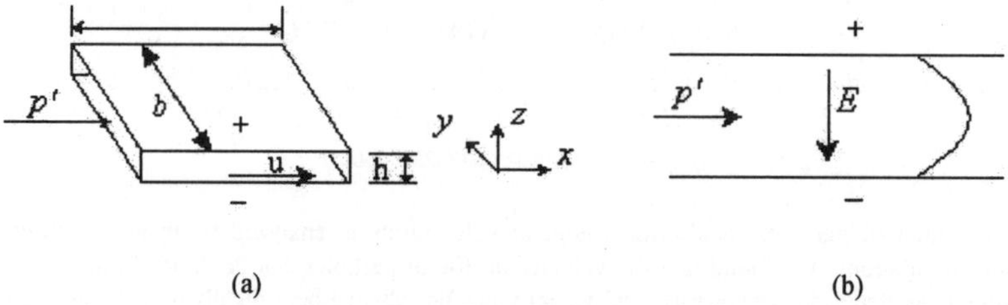

Figure 1. Sketch map of flow model.

The velocity distribution of section can be obtained from fluid mechanics [10,11], $u = \dfrac{p'}{2\mu} z(h-z)$. Here the parameter u is velocity distribution and μ is the fluid viscosity.

In the presence of electric field the interaction force among immersed particles of ER fluid is engendered, each particle follows its given motion equation. But when the volume of particles is relatively small, the fluid can still be regarded as continuous medium with the removal of electric field. In the effect of electric field the interaction force between two particles can be described as the following formula [14]:

$$\vec{F}_{ij}\left(\vec{R}_{ij}\right) = \frac{3p^2}{4\pi\varepsilon_0 K_f R_{ij}^4}\left[\left(3\cos^2\theta_{ij} - 1\right)\vec{e}_r - \sin\left(2\theta_{ij}\right)\vec{e}_\theta\right] \tag{1}$$

$$\vec{R}_{ij} = \vec{R}_i - \vec{R}_j \qquad 0 \le \theta_{ij} \le \pi/2$$

Here the parameter θ_{ij} is the angle between $\vec{R}_{ij}$ and the electric field line. The term of $\vec{e}_r$ is the unit vector, which parallels with $\vec{R}_{ij}$. The term of $\vec{e}_\theta$ is unit vector, which direction parallel with $\vec{e}_r \times \left(\vec{e}_r \times \vec{E}_0\right)$, namely, $\vec{e}_\theta = \dfrac{\vec{e}_r \times \left(\vec{e}_r \times \vec{E}_0\right)}{\left|\vec{e}_r \times \left(\vec{e}_r \times \vec{E}_0\right)\right|}$. The term of p is the electric polarization strength.

Short-distance repulsion force among particles decreases sharply with the increase of the distance. We use an exponential form in this paper as following description [15]:

$$\vec{F}_{ij}^{rep}\left(\vec{R}_{ij}\right) = -\frac{3p^2}{4\pi\varepsilon_0} \times \frac{1}{K_f R_{ij}^4} \exp\left[-100\left(\frac{R_{ij}}{\delta} - 1\right)\right]\vec{e}_r \tag{2}$$

There exists interaction between a particle and its mirror reflection dipole. So the total mirror reflection force that the particle receives is:

$$\vec{F}_i^{self} = \frac{\frac{3}{8}p^2}{4\pi\varepsilon_0 K_f}\left\{-\frac{1}{z_i^4} + \sum_{s=1}^{\infty}\left[\frac{1}{\left(z_i - sL_z\right)^4} - \frac{1}{\left(z_i + sL_z\right)^4}\right]\right\}\vec{e}_z \tag{3}$$

In general, the coefficient of inertia is about O(10^{-10}). Neglecting the inertia force and Brown force of particles, in the effect of pressure ingredient p' and electric field the motion equation is described as the following formula:

$$\sum_{j\neq i}\left[\vec{F}_{ij}\left(\vec{R}_{ij}\right)+\vec{F}_{ij}^{rep*}\left(\vec{R}_{ij}\right)\right]+\vec{F}_i^{self}-3\pi\delta\eta\left[\frac{d\vec{R}_i}{dt}-\frac{p'}{2\mu}z(h-z)\vec{e}_x\right]=0 \quad (4)$$

After eliminating its dimension, we will get:

$$\frac{d\vec{R}_i}{dt}=\sum_{j\neq i}\left[\vec{F}_{ij}^*\left(\vec{R}_{ij}^*\right)+\vec{F}_{ij}^{rep*}\left(\vec{R}_{ij}^*\right)\right]+\vec{F}_i^{self*}+S^*\frac{z_i^*\left(h^*-z_i^*\right)}{2}\vec{e}_x \quad (5)$$

$$i,j=1,2...M$$

Here the parameter M is the number of particles. δ ~Length, F_0 ~force, $3\pi\eta\delta^2/F_0$ ~ time, all these factors are non-dimension ones.

$$S^*=\frac{3\pi\eta\delta^2}{F_0}\cdot\frac{p'\delta}{\eta}=\frac{16}{\varepsilon_0 K_f(\beta E)^2}\cdot p'\delta \quad (6)$$

Here the parameter S^*, which is pressure gradient without dimension, is called relative pressure gradient,

Simulation unit cell $L_x \times L_z = 2\times 30$ is taken, and the number of particles in the total chain, which includes an entire chain in the effect of electric field, is 30. We take 5 chains periodically distributed left or right, to simulate infinite chains formed in the horizontal direction. Based on observing experiments[16], Ashurst[17] and Parthasarathy[18] et al supposed that there is a contact zone near electrodes in the simulation process. Here there is no slippage within particles and electrodes. So we think of the thickness of the contact zone as a radius of the particle, while the touch electrodes particlesare static.

Volume flux when $E=0$ is:

$$Q_0^*=\frac{S^*h^{*3}}{12}=\frac{\pi h^3 p'}{\frac{3}{4}\pi\varepsilon_0 K_f\delta^2(\beta E)^2} \quad (7)$$

The step length of time is 0.01, and applying four-order Runge-Kutta method equation group (5) can be worked out.

3 Simulation results and discussion

The dependence simulated of the volume flux on time is shown in **Fig.2** under different pressure gradients. **Fig.2** (a) is at the circumstance of the relative pressuring gradient (RPG) $S^*=0.001\sim 0.005$ and **Fig.2** (b) is $S^*=0.00025\sim 0.00075$ respectively. The volume flux continuously increases with the pressure gradient. **Fig.2** (b) shows that when $RPG<0.00075$ the volume flux decreases monotonously to zero, namely, flow is cut off. By contrast, **Fig.2**(a) indicates that when $RPG>0.001$ the volume flux does not decrease monotonously with time, but takes on somewhat periodical changes like impulse after its minimum. And the time intervals decrease with the pressure gradient.

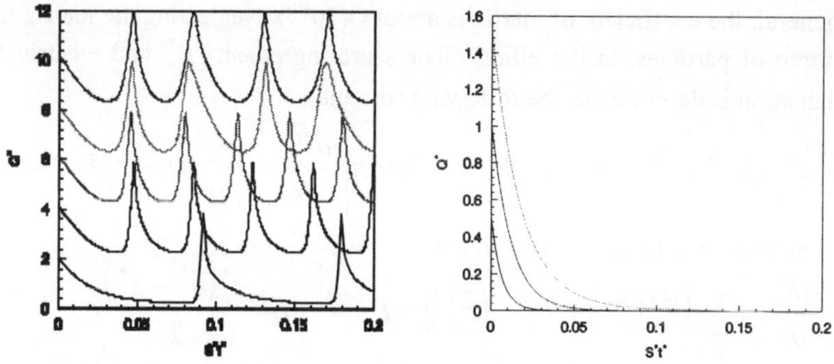

Figure 2. Sketch map of flow model. the dependence of volume flux on time, $h/\delta = 30$

(a) down to up $S^* = 1\times10^{-3}, 2\times10^{-3}, 3\times10^{-3}, 4\times10^{-3}, 5\times10^{-3}$ successively;

(b)down to up $S^* = 0.25\times10^{-3}, 0.50\times10^{-3}, 0.75\times10^{-3}$ successively.

3.1 The motion mode of particles

Fig. 3(b) shows the velocity sections at different time when the pressure gradient $S^* = 0.002$. As an example, we investigate it specially. The velocity section is showed as curve **a**, dot **A** in **Fig.2**(a). The velocity of particles decreases with time in the effect of electric field, and the velocity section changes monotonously with the envelope of curve **a** within "area I". The central particles, which slow up more rapidly than side particles, tend to be the same speed. As a result, the volume flux reduces gradually. It reaches curve **b** at the moment of $t^* = 18.5$, represented by dot **B** in **Fig.2** (a). Then the speed of particles close to two electrodes slabs changes sharply, increasing and then decreasing rapidly in a short time. During that period, the speed of the middle particles changes steadily. Curve **c** with the largest speed, represented by dot **C**, sees the greatest instant volume flux. Then the speed of every particle decreases continuously to reach curve **b** represented by dot **D** again. The process repeats itself like this.

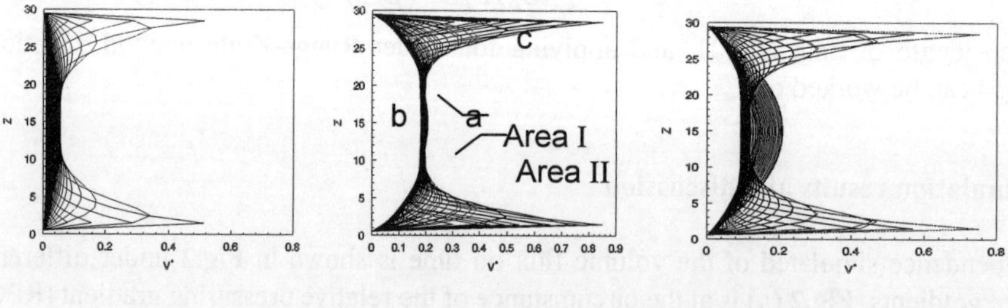

Figure 3. velocity section envelop area (area II),

$(a)S^* = 1\times10^{-3}$, $(b)S^* = 2\times10^{-3}$, $(c)S^* = 3\times10^{-3}$

The area II formed by velocity section in **Fig.3** could be called "velocity section envelope area". The velocity of the middle particles, which move forwards as a whole, is almost about the same and changes little with time. It is similar to the result from continuous medium mechanics method, so we call it "plug zone", in which the particles' speed is relatively steady. Another area is "transition zone" in which the speed of particles at both ends changes sharply in a certain time interval. The particles in transition zone exhibit transition motion mode like "breathing type", which is unlike steady motion of the particles in the middle zone. It is necessary to point out that continuous medium dynamics method takes no account of motion of individual particles, and only considers macroscopic motion characteristics of a large quantity of particles. So it can not explain the mutation motion like "glass ball effect".

3.2 Volume flow rate and its threshold value

The statistics averaging on the time period was conducted to instant volume flux, and averaging volume flux can be obtained as the formula (8) describes:

$$\overline{Q}^* = \frac{1}{nT}\int_0^{nT+t_0} Q^*(t^*)dt^* \tag{8}$$

Here the term of t_0 is any moment and T is a period of volume flux. Average volume flux represents macroscopic volume flux in the presence of electric field. Let $\alpha = \overline{Q}^* / Q_0^*$, then α is called volume flow rate, which is the rate of average volume flux with electric field to that without electric field. It changes between 0 and 1. It shows the control capacity of electric field over volume flux.

It is found that there is a pressure gradient threshold value S_0^*. When $S^* < S_0^*$, the volume flow rate $\alpha = 0$, that is to say, flow is completely shut off by electric field. According to the characteristic that the volume flow rate on RPG, an function can be made as the following expression:

$$\alpha = \begin{cases} \left(1 - \dfrac{c}{S^*}\right)^n \left(1 + \dfrac{c_1}{S^*}\right) & S^* > c \\ 0 & S^* \le c \end{cases} \tag{9}$$

Here the term of c is the threshold value of relative pressure gradient. The parameter c should be limited from 0.75×10^{-3} to 0.90×10^{-3}. In order to guarantee $\alpha < 1$, We can verify that when $n = 1$, $c_1 \le c$, and when $n = 2$, $c_1 \le 2c$ respectively.

After adjusting these parameters, we find out the below parameters, which can ensure volume flow rate and pressure gradient accord with the above-mentioned equation: $n = 1, c = 0.78\times10^{-3}, c_1 = c/2$.

Volume flow rate simulated accord well with expression (9), so we ensure the threshold value of volume flow rate $c = 0.78\times10^{-3}$ in the former expression. Applying molecules dynamic simulation only calculates out the given pressure gradient. It is difficult to give the specific value although the range of threshold value of volume flow rate can be pointed out. So expression (9) seems especially convenient and helpful. As expression (9) and (6) indicate, decreasing $p'\delta$, namely, decreasing the pressure gradient applied to particles, and increasing electric field strength all can reduce the volume flow rate. Since the non-dimension pressure gradient used in expression (9) does not reflect geometric factors between electrodes directly, we use the volume flux to replace it. The volume flux reflects the mutual effect of pressure gradient and geometric size, so the selection of the parameters in the approximate formula can be eliminated due to that of particle number in simulation system. The pressure gradient and volume flux are in one-to-one correspondence in a given simulation system, thus the threshold value $c_Q = \dfrac{ch^{*3}}{12} = 1.5853$ of volume flux can be obtained from that of pressure gradient. Here the term of h^* is the length between both electrodes subtracting that of both the contact zones at the end.

Two critical values can be obtained from critical volume flux.

(1) Critical electric field strength:

$$E_{cr} = \sqrt{\frac{\pi h^3 p'}{\frac{3}{4}\pi\varepsilon_0 K_f \delta^2 \beta^2 c_Q}} \tag{10}$$

As expression (10) indicates that the critical electric field strength increases with the pressure gradient for a given ER fluid material. On the condition that pressure gradient keeps constant, we can lessen the distance between two electrode slabs and employ dispersed particles with larger size to reduce critical electric field.

(2) Critical pressure gradient

$$P'_{cr} = \frac{c_Q}{\pi h^3 / 4F_0} = \frac{c_Q}{\pi h^3 \big/ \left(\frac{3}{4}\pi\varepsilon_0 K_f \delta^2 \beta^2 E_0^2\right)} \tag{11}$$

As seen from expression (11), critical pressure gradient is related to electric field strength, the distance between electrodes and the particles' size. Increasing electric field strength or particles' size appropriately both can increase the critical pressure gradient value.

3.3 *Corresponding time*

As seen from **Fig.2**, the time and the period decrease with the pressure gradient. The thing more concerned is how long the flow is cut off when the electric field strength is bigger than critical value. In contrast, the simulation to the circumstance of less distance between electrodes has been processed, here choose $L_x \times L_z = 2\times 10$ among which there is a whole long chain connected the electrodes, and the volume rate is guaranteed at 0.26. The corresponding time of different relative pressure gradient in the two simulation systems hardly changes with the relative pressure gradient in each simulation system. We can obtain from the non-dimension factor that the time cutting off the flow is relative to the fluid viscosity and electric field strength, but fewer relative to pressure gradient. The corresponding time increases with the fluid viscosity and decreases with the electric field strength.

4 Conclusion

The conclusion can be obtained as following descriptions by applying the molecule dynamics simulation to ER fluid under Poiseulle flow:

(1) The particles in middle area between two electrodes exhibit plug flow at the same speed in the effect of electric field. Those close to the contact zone flow in a transition form like breathing. The velocity of the plug zone, together with volume flux, decrease

with electric field. And there is a critical electric field strength or critical pressure gradient..

(2) Volume flow rate is related to pressure gradient, the space between electrodes, particles size and electric field strength. Increasing electric field strength and particles' size appropriately can decrease volume flow rate to increase critical pressure gradient.

(3)When RPG is smaller than the critical value, the response time that electric field cuts off flow of fluid is only related to the electric field strength, the fluid viscosity and the space between electrodes. But ii has almost slight rclation to pressure gradient. The response time decreases with electric field strength, increases with the fluid viscosity and the space between electrodes.

Acknowledgements

We would like to thank the National Natural Science Foundation of China under Grant No.59832090 and the Natural Science Foundation of China for Distinguished Young Scholar under Grant No.50025207 for their support for this work.

References

1. H.Block, J.P.Kelly, J.Phys.D: Appl. Phys, **21** (1988), 1661
2. K.D.Weiss, J.D.Carlson, D.A.Nixson, J.Intell.Mater.Syst.Struct, **5** (1994), 772
3. R. Stanway, J.L.Sproston, J.Dynamic Systems, Measurement and Control **116**(1994), 498
4. S.B.Choi, C.C.Cheong and G.W.Kim, Mechatronics 7 (1997), 53.
5. R.Stanway, J.L.Sproston., Mathematics and Control in Smart Structures Proceedings of SPIE – The International Society for Optical Engineering, 2715 (1996), 110.
6. G.M.Kamath, M.K.Hurt and N.M.Werely., J.Smart Materials and Structures 5 (1996), 576.
7. S.B.Choi, Y.T.Choi, et.al., Mechatronics 8 (1998),143.
8. K.R. Rajagopal, A.S. Wineman, *Acta Mechanica,* **91** (1992), 57.
9. R.J. Atkin, Xiao Shi, W.A. Bullough, *J. Rheol.,* **35** (1991), 1441.
10. Wu Wangyi, Fluid Mechanics (Beking University Press, Beijing, 1982) p.81~91
11. Chia-Shuan Yihm, Fluid Mechanics, McGraw-Hill Book Company 1969, p.1, 225.
12. H.Tamura, M. Doi, J. Physical Society of Japan **11** (1992), 3984.
13. X.Y. Gao, X.P. Zhao, C.Q. Zheng, *Journal of Phys.* ***D****: Appl. Phys.,* **31** (1998), 3397.
14. J.M. Sun, R. Tao, *Physical Review* ***E*****,** **52**(1),(1995), 813.
15. J.R. Melrose, *Phys. Rev. A,* **44** (1991), R4789.
16. D.J. Klingenberg, C.F. Zukoski, , Langmuir, **6** (1990),15.
17. W.T. Ashurst, W.G. Hoover, *Phys. Rev. Lett.*, **73** (1994), 205.
18. M. Parthasarathy, K.H. Ahn, B.M. Belongia and D.J. Klingenberg, Electrorheological fluids, edited by R. Tao and G.D. Roy, World scientific press, Singapore, 1994

PREDICTION OF THE DYNAMICS OF ELECTRORHEOLOGICAL FLUIDS

LUIZ B. POMPEO NETO AND ANTONIO CELSO F. ARRUDA

Deartment. of Petroleum Engineering, Faculty of Mechanical Engineering, UNICAMP
C.P. 6122, 13081-970 – Campinas – SP, Brazil
E-mail: pompeo@dep.fem.unicamp.br
arruda@dep.fem.unicamp.br

SHEILA L. VIEIRA

M.I.M.E. Department, Nitschke Hall 4005, University of Toledo
Toledo, OH 43606, USA
E-mail: svieira@eng.utoledo.edu

When Electrorheological fluids are subjected to electric fields (E) and/or shear fields $\dot{\gamma}$), that is, when they are under an electro-hydrodynamic (EHD) field, they present changes in their rheological properties. By making the apparent viscosity (which varies under EHD) as an indication of the ER structure state, and considering that it is composed of a dielectric polarizability component, an electric conductivity component and the original apparent viscosity of the suspension, it is possible to obtain an equation (derived from an equilibrium relation) that describes the dynamics of ER fluids. In this article we present an analysis of the factors that influence the dynamics of structure formation in ER fluids We concluded that the particle's electric conductivity proved to be fundamental for the understanding of the general behavior of the fluid, and that there is an optimum structuring level per unit applied field ratio ($\dot{\gamma}/E$), no matter whether the ER fluid is more characteristically dielectric or conductive. These conclusions explain the deviation of the resistance to shear of ER fluids from a quadratic relation (E^2), experimentally detected some time ago.

1 Introduction

Dispersions of electrically polarizable particles suspended in dielectric liquids have been termed electrorheological fluids (ERFs). What has been termed electrorheological effect, the increase in apparent viscosity of the suspension subjected to an electric field (E), would be a result of the alignment of the particles along the direction of E.

The complex behavior of the ERF under flow and subjected to an electric field, though, indicates that other factors might be playing a role in the structure formation within the suspension. In an effort to clear these factors Mason and collaborators [1] introduced the concept of electrohydrodynamic (EHD) drive, by means of which they derived expressions for stress components relative to the torques experienced by a particle suspended in dilute suspension. Besides the hydrodynamic torque and the torque electrostatically (induced due to the effective dielectric polarizability of the suspended particle), also to be considered is the torque due to the permanent electric dipole moment of the particle. This torque is associated to the electric conductivity of the particle, which is always operative and fundamental [2] to the understanding of the general behavior of the ERF when not negligible in relation to the dielectric polarization.

The torques with an electrostatic nature are proportional to the product of the dipole moment with the electric field strength to which the particle is subjected. As the *induced* dipole moment depends on the electric field strength, the torque electrostatically induced is proportional to E^2. On the other hand, as the *permanent* dipole moment is independent of E and, consequently, the torque due to the electric conductivity of the suspension is proportional to E (which is a way to express Ohm's law of electric conduction).

From an expression for the equilibrium of the three stress components (τ_h, τ_p, τ_i)[1] over the suspended particle, we proposed in a recent article [3] an empirical equation. This equation makes clear, to the macroscopic level, some of the effects caused by the dynamics of structure formation in ERFs, besides permitting to infer much of what may be happening down to the microscopic level.

2 Electrorheological Effect

If the analysis of the electrorheological (ER) effect is made by evaluating the relative increase in apparent viscosity, obtained as a function of the applied shear field ($\dot{\gamma}$) and electric field (E) [3,4], it will tend to result in a sigmoidal curve. One of the parts of this curve corresponds to the parabola, which is related to the structuring component due to the dielectric polarizability of the particles. The other part of the sigmoidal curve corresponds to the deviation of the quadratic trend as the structuring components start to dominate the structuring process due to either the electric conductivity or due to the shear of the suspension or due to both of the latter components.

As an illustration of such a distribution of the experimental data, we reproduce in figure 1 some of the results we obtained [3] for tests with dispersions of cornstarch particles suspended in 500 cSt silicone oil. The concentration of cornstarch and the humidity content of the samples were varied so as to produce variations in the dielectric polarizability (sample 2) and in the electric conductivity (sample 3) of a reference sample (sample 1)[2]. The samples were tested in a Haake rotational rheometer under two values of constant shear rate, as shown in Fig. 1. In this plot, α' and β'_m are defined as:

$$\alpha' = \frac{\eta_{ss}}{\eta_{E0}} \tag{1}$$

$$\beta'_m = \frac{1}{\dfrac{T_f}{2\pi}\sqrt{\dfrac{\dot{\gamma}}{E^2}}} \tag{2}$$

---

[1] τ_h is the hydrodynamic stress, τ_p is the stress due to the particles' permanent dipole moment per unit volume, and τ_i is the stress due to the dipole moment induced on the particle of dielectric constant K_1 suspended in a fluid with dielectric constant K_2.

[2] Sample 1: 10 w% starch particles containing 8.67 w% of humidity
Sample 2: 20 w% starch particles containing 8.67 w% of humidity
Sample 3: 10 w% starch particles containing 16 w% of humidity

where η_{SS} is the steady state value of the apparent viscosity of the ER fluid, η_{E0} is the initial value (with $E = 0$) of the apparent viscosity of the ER fluid, and T_f is taken as equal to the transient duration for the ER fluid subjected to a given shear rate and electric field strength.

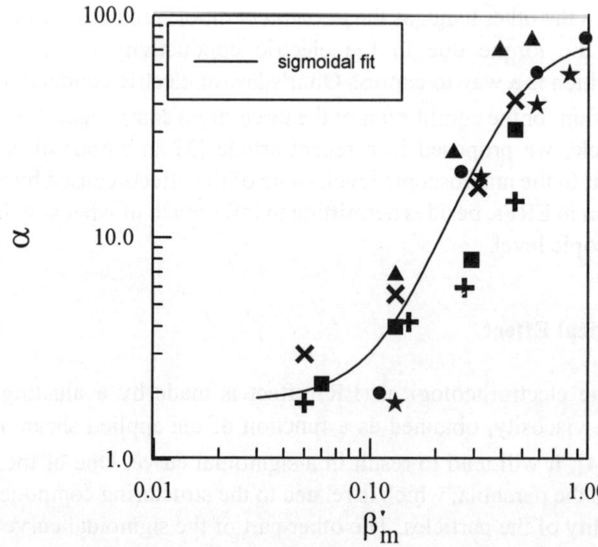

Figure 1. Reduction of experimental data from reference β] for $E = 1, 2, 3$ and 4 kV/mm – ● sample 1at $\dot{\gamma} = 50\ s^{-1}$; ■ sample 1 at $\dot{\gamma} = 100\ s^{-1}$; ▲sample 2 at $\dot{\gamma} = 50\ s^{-1}$; × 6 sample 2 at $\dot{\gamma} = 100\ s^{-1}$; ★ sample 3 at $\dot{\gamma} = 50\ s^{-1}$; + sample 3 at $\dot{\gamma} = 100\ s^{-1}$.

The transient duration was determined experimentally and proved to be consistently related to the level of EHD drive and to the electric properties of the ER fluid [3,4].

This character of ER suspensions, which manifests as an abrupt change in behavior [5] and which has been reported [6] as indicative of changes in quality of the microstructure [7] developing in typical ER fluids (fluids more characteristically dielectric with a certain degree of electric conductivity as ours [4]) under given EHD drive, should be additionally exemplified with the following references.

Yang and Shine [8] found that the steady state apparent viscosity increases in a sigmoidal fashion with the applied electric and shear fields during the characterization of liquid crystal polymer solutions, in which the solutes are known to rapidly orient in the direction of E due to their large permanent dipole moment in relation to electrically induced orientation processes:

Davis and Ginder [9] discuss the deviation of the yield stress from a quadratic relation with E for higher values of E in connection with a limit in the ER effect due to conductivity effects.

Kushid and Acrivos [10] obtained a diagram of state following a thermodynamic analysis of the aggregation process for particles with interparticle interactions varying from more characteristically conductive up to more characteristically dielectric. Transitions in the amount of aggregation due to particle and fluid electric properties and properties of the electric field are notable in their diagram of state.

Thus, as the sigmoidal behavior of the ER response continues valid given the combination of type of driving force dominating the process for fluids more characteristically conductive (with certain degree of dielectricity), it is possible to say that the constitutive relation for $\eta(\dot{\gamma}, E)$ that we proposed [3] has generalized validity. That constitutive relation predicts that, rather than an ultimate degree of structuring, an optimum degree of structuring is to be achieved. While the sigmoidal curve itself is not such constitutive relation, it is still useful for the assessment of the range ability of the ERF under the given EHD conditions.

3 Conclusions

In this article we present an analysis of the factors that influence the dynamics of structure formation in ER fluids. Besides the influences of the dielectric polarizability and of the initial apparent viscosity of the fluid, the influence of the particle's electric conductivity is analyzed. This latter property proved to be fundamental for the understanding of the general behavior of the fluid.

It was possible to confirm the change in the behavior of the response of the ER fluid as the range of the ratio ($\dot{\gamma} / E$) was extended, either already visualized as a qualitative change in the microstructure being formed in the suspension, or already detected as a deviation in the quadratic trend of the experimental data points $\eta(E)$. Thus, it is possible to gather observations in these senses cited throughout the literature and to predict that no matter whether the ER fluid is more characteristically dielectric or conductive, its response will display an optimum structuring level per unit applied field ratio ($\dot{\gamma} / E$).

References

1. Arp P. A., Foister R. T. and Mason S. G., Some Electrohydrodynamic Effects in Fluid Dispersions. *Adv. Colloid Interf. Sci.* **12** (1980) pp. 295-356.
2. Davis L. C., Time-dependent and nonlinear effects in electrorheological fluids. *J. Appl. Phys.* **81**(4) (1997) pp. 1985-1991.
3. Pompeo Neto L. B., Arruda A. C. F. and Vieira S. L., Evaluation of a Governing Equation for the Macroscopic Dynamics of Electrorheological Fluids. *J. Appl. Phys.* **89** (8) (2001) pp. 4657-4663.
4. Vieira S. L., Pompeo Neto L. B. and Arruda A. C. F., Transient Behavior of an Electrorheological Fluid in Shear Flow Mode. *J. Rheol.*, **44**(5) (2000) pp. 1139-1149.
5. Okagawa A., Cox R. G. and Mason S. G., Particle Behavior in Shear and Electric Fields. *J. Colloid Interface Sci.* **47**(2) (1974) pp. 536-567.
6. Wu C. W. and Conrad H., On the prediction of the shear stress of electrorheological suspensions. *J. Mater. Res.* **13**(11) (1998) pp. 3299-3303.
7. Tang X., Li W. H., Wang X. J. and Zhang P. Q., Structure Evolution of Electrorheological Fluids Under Flow Conditions, *Proc. 6th Intl. Conf. on ER Fluids, MR Suspensions and their Applications*, edited by M. Nakano and K. Koyama, World Scientific (1998) pp. 164-171.

8. Yang I-K. and Shine A. D., Electrorheology of a nematic poly(*n*-hexylisocyanate) solution. *J. Rheol.* **36**(6), (1992) pp. 1079-1104.
9. Davis L. C. and Ginder J. M., Electrostatic Forces in Electrorheological Fluids. *Progress in Electrorheology – Science and Technology of Electrorheological Materials*, edited by K. O. Havelka and F. E. Filisko, Plenum Press (1995) pp. 107-114.
10. Kushid B. M. and Acrivos A., Effects of Conductivity in Electric-Field-Induced Aggregation in Electrorheological Fluids. *Phys. Rev. E* **52** (1995) pp. 1669-1693.

PHYSICAL MODELING OF DAMPING CHARACTERISTICS OF MR FLUIDS

JEROME SARRAZIN, JEROME CLARACQ, JEAN-PIERRE MONTFORT

Université de Pau et des Pays de l'Adour, Lab. de Physique des Matériaux Industriels
BP 1155Pau, 64013, France

Received
Revised

The behavior of MR fluids involved in industrial devices is often modeled by analogical models which do no link the parameters with the physical properties of the fluid. From the equations of fluid mechanics, we give a physical meaning to parameters such as friction, elastic and mass coefficients used in the mentioned models. The predictions are checked with experimental data which mimic the behavior of devices such as antiseismic systems. The viscoelastic behavior of MR fluids is characterized by mechanical spectroscopy. The complex shear modulus $G^*(\omega)$ contains all the information about the restoring and viscous forces. For example, the relation between elastic modulus, yield stress and magnetic force is discussed in terms of strain amplitude.

1. Introduction

For describing the functioning of devices based on the properties of MR fluids, mechanical models are often used which associate dashpots, springs and additional masses[1,2]. The value of associated parameters is calculated by fitting the model with the experimental data and usually there is no straightforward relation with the physics of the system.

Here, we are aiming at giving a physical meaning to the mechanical parameters in the following situation : the MR fluid is behaving as a Bingham fluid ; it is submitted to a weak magnetic field and to an oscillating shear motion of small amplitude. The first part of the paper will be devoted to establish the analytical expression of the forces involved in two experimental devices; one of them is equipped with an additional spring[1], the second one mimics the behavior of base-isolation antiseismic systems[2]. The predictions will be compared with experimental data in the first case. The second part of the paper will focus on the viscoelastic parameters of the MR fluid.

2. MR fluids dampers

2.1. *Fluid mechanics approach*

The equipment used by Nishiyama et al. is composed of a flat plate immersed in a fixed rectangular vessel filled with a MR fluid (Fig. 1). The plate is connected to a vibrating exciter via a spring. The imposed magnetic field is perpendicular to the plate.

A simple mechanical model (Fig. 2) is able to describe the behavior of such a system. The fundamental principle of dynamics applied to the plate can be written as :

$$m\frac{d^2z}{dt^2}+k(z-z_g)=F(t) \tag{1}$$

where m is the mass of the plate, k the spring stiffness, $F(t)$ the drag force exerted by the fluid, $z(t)$ and $z_g(t$ the displacement of the plate and of the exciter, respectively.

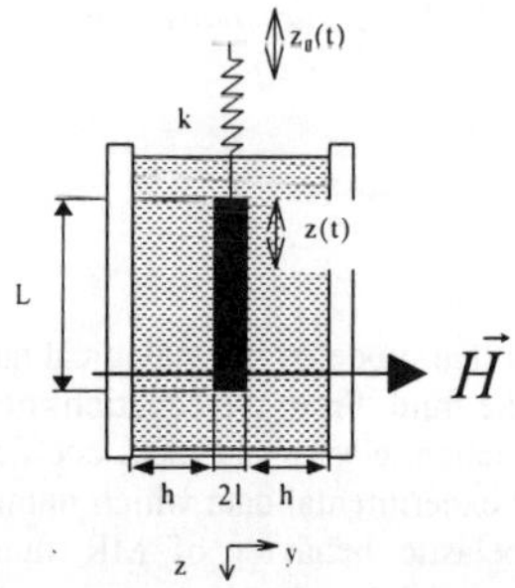

Fig. 1. Schematic description of the system

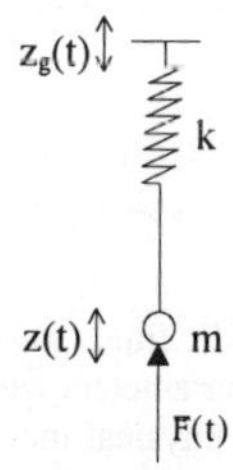

Fig. 2. Associated mechanical model

The equation of conservation in the z-direction is :

$$\rho\frac{\partial u}{\partial t}=\frac{\partial}{\partial y}(\tau_{yz}) \tag{2}$$

where ρ is the volumic mass, $u(y,t)$ is the velocity of the fluid and τ_{yz} the shear stress. Assuming a Bingham behavior of the fluid with a restoring force following Klingenberg approach[3], the shear stress can be written as :

$$\tau_{yz}(y,t)=\frac{9}{4}\mu_0\mu_m\beta^2H^2\phi\gamma(y,t)+\eta\frac{\partial u}{\partial y}=K\gamma(y,t)+\eta\frac{\partial u}{\partial y} \tag{3}$$

where γ stands for the strain and η is a viscosity coefficient. That coefficient can be seen as the suspending fluid viscosity and should vary with the magnetic field.
For an oscillatory flow, the total shear stress becomes :

$$\tau^*_{yz}(y,t)=K\gamma^*(y,t)+\eta\frac{\partial u^*}{\partial y}=\left(-i\frac{K}{\omega}+\eta\right)\frac{\partial u^*}{\partial y} \tag{4}$$

Therefore, equation (2) changes into :

$$u_0^{*}{}''+\frac{\omega^2\rho}{K+i\eta\omega}u_0^*=0\,. \tag{5}$$

The solution arises from the boundary conditions $u_0^*(l)=U_0$ and $u_0^*(l+h)=0$:

$$u_0^*(y)=U_0\frac{sh(\alpha(l+h-y))}{sh(\alpha h)}\quad\text{with}\quad\alpha^2=\frac{\omega^2\rho}{K+i\eta\omega} \tag{6}$$

The complex shear stress can thus be rewritten as:

$$\tau^*_{yz}(y,t)=-\left(\eta-i\frac{K}{\omega}\right)\alpha\frac{ch(\alpha(y-l-h))}{sh(\alpha h)}\frac{dz^*}{dt} \tag{7}$$

and then the drag force is :

$$F^*(t)=2S\tau^*_{yz}(l,t)=-2S\left(\eta-i\frac{K}{\omega}\right)\alpha\coth(\alpha h)\frac{dz^*}{dt}=-C^*\frac{dz^*}{dt} \tag{8}$$

where $S=L^2$ is the surface of the plate.

Therefore, Eq. 1 can be written as:

$$m\frac{d^2z^*}{dt^2}+k(z^*-z_g{}^*)=-C^*\frac{dz^*}{dt}. \tag{9}$$

It follows that the damping coefficient of the system $\zeta=\left|\frac{z_0^*}{z_{g0}}\right|$ is expressed by:

$$\zeta=\frac{k}{\left|k-m\omega^2+i\omega C^*\right|}=\frac{k}{\left|k-m\omega^2+2S(K+i\omega\eta)\alpha\coth(\alpha h)\right|}. \tag{10}$$

Consequently, the damping coefficient ζ depends on the physical characteristics of the MR fluid (η, ρ, ϕ, μ_0, μ_m) and on the magnetic field H through the coefficients $C'(\omega)$ and $C''(\omega)$. The comparison between our model and the data of Ref. 1 is presented in Fig. 3. The best fit is obtained for a viscosity of seven Pa.s which is higher than that deduced from the flow curve. The resonance frequency is not correct and can be adjusted by taking k=1250 N/m whereas the stiffness of the spring was given equal to 1500 N/m (Fig. 4).

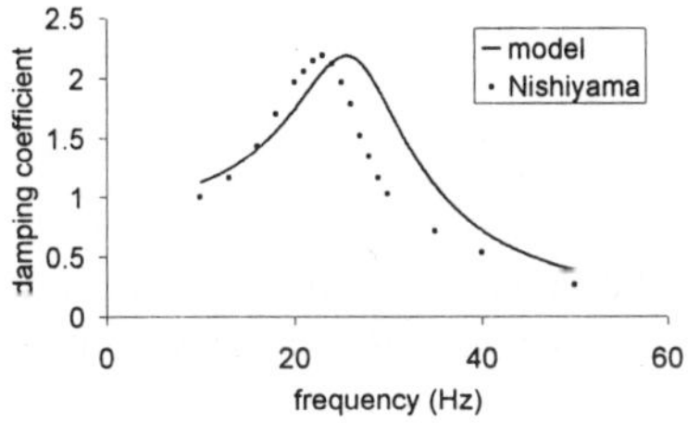

Fig. 3. Damping coefficient versus frequency

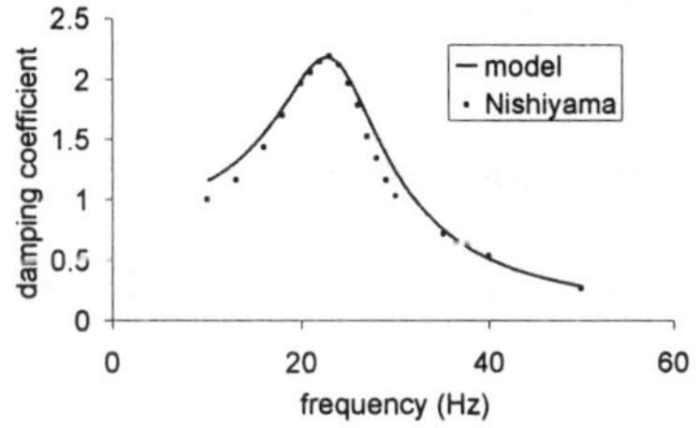

Fig. 4. Variation of ζ with the spring stiffness

Fig. 5 and 6 show how the damping coefficient is field and viscosity dependent. It is worth noting that the resonance frequency depends essentially on the elastic contribution (due to the restoring magnetic force) whereas the height of the peak mainly depends on the viscous contribution.

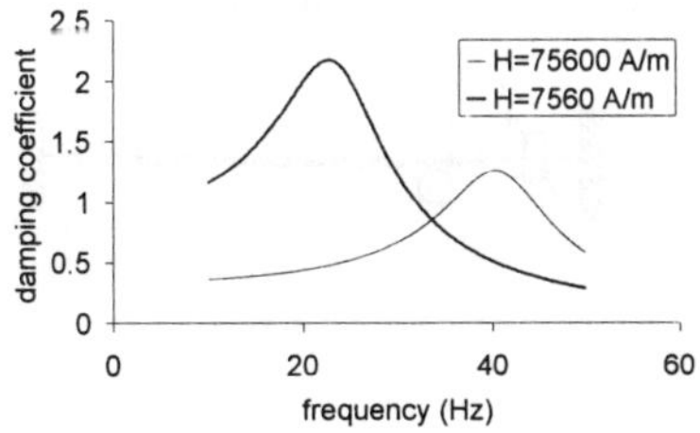

Fig. 5. Damping coefficient versus field strength

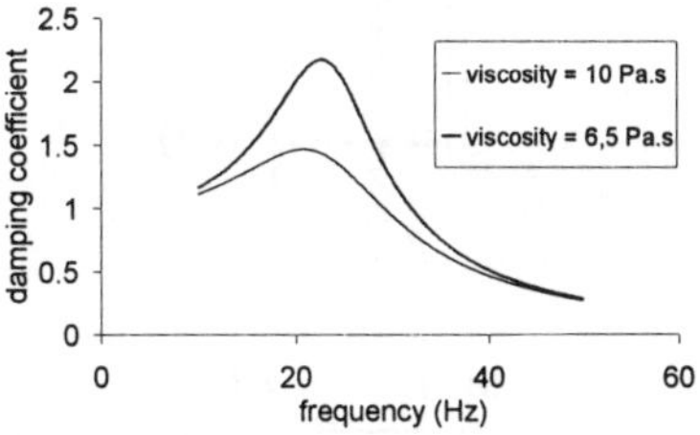

Fig. 6. Damping coefficient versus viscosity

The term $C^*(\omega)$ can be expanded when $|\alpha|h<<\pi$. That condition is fulfilled here, thus:

$$C^*(\omega)=2S\left(\frac{K}{i\omega h}+\frac{\eta}{h}+\frac{i\omega h\rho}{3}\right). \tag{11}$$

Therefore, the drag force becomes :

$$F^*(t) = -\frac{2SK}{h} z^*(t) - \frac{2S\eta}{h}\frac{dz^*}{dt} - \frac{2Sh\rho}{3}\frac{d^2z^*}{dt^2} \quad (12)$$

and Eq. (9) can be rewritten as:

$$m\frac{d^2z^*}{dt^2} + k(z^* - z_g{}^*) = -k'z^*(t) - C\frac{dz^*}{dt} - m_A\frac{d^2z^*}{dt^2}. \quad (13)$$

That equation is typical of analogical models which include springs, dashpots and additional masses. Here, the stiffness comes from the magnetic force, the friction coefficient is related to the viscosity of the fluid and the additional mass is due to the inertial term. Nevertheless, if we compare the strength of the various terms of Eq. (13) from Nishiyama's data, we obtain the following values : $K \approx 40$ Pa ; $\eta\omega \approx 940$ Pa ; $\rho h^2\omega^2/3 \approx 60$ Pa, at the resonance frequency. It appears that the viscosity term is dominant and changes with the magnetic field[1]. Moreover, the elastic term generated by the spring is $(1-1/\varsigma)kh/2S \approx 1360\,Pa$. It is by far higher than the contribution of the magnetic force. But, we will see in part 3 that Klingenberg expression does not fit the measured restoring forces which are much higher.

2.2. *Antiseismic system*

Let us now consider a similar system where the vessel is mobile and the spring acts on the vessel itself (Fig. 7). That is a first approach of suitable devices for the semi-active control of seismically loaded structures[2]. Let *M* be the mass of the structure and *F* the drag force exerted by the MR fluid onto the wall of the vessel.

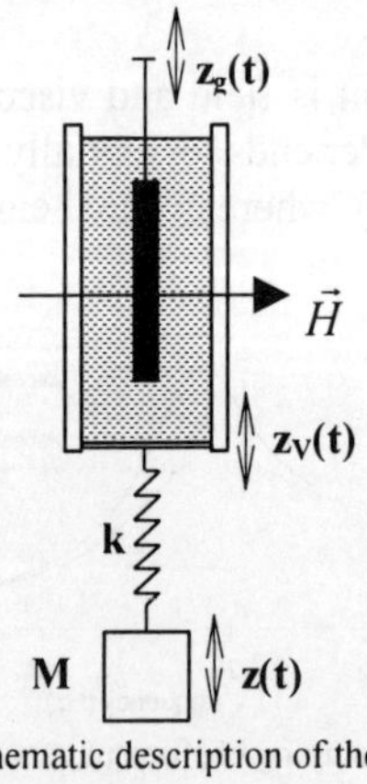

Fig. 7. Schematic description of the system

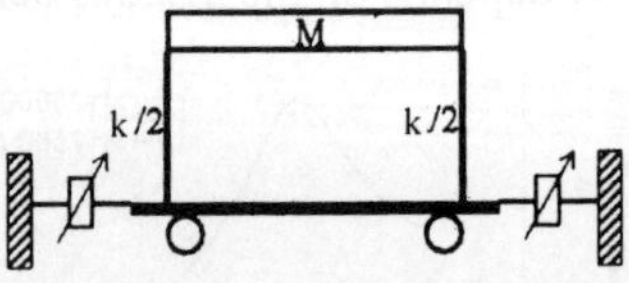

Fig. 8. Base isolation antiseismic system

The motion equation is :
$$M\frac{d^2z^*}{dt^2} = -k(z^* - z_v^*) = F^*(t) \quad (14)$$

The constitutive equations are the same as in part 1.1 but the boundary conditions differ, that is to say : $u^*(l,t) = U(t) = i\omega z_g^*(t)$ and $u^*(l+h,t) = \frac{dz_v^*}{dt} = i\omega z_v^*(t)$.

The velocity profile is now expressed by :

$$u^*(y,t)=\frac{i\omega}{sh(\alpha h)}\left[sh(\alpha(y-l))z_v^*(t)-sh(\alpha(y-l-h))z_g^*(t)\right] \tag{15}$$

which leads to the following drag force :

$$F(t)=\left(K+i\omega\eta\right)\frac{2S\alpha}{sh(\alpha h)}\left[ch(\alpha h)z_v^*(t)-z_g^*(t)\right]. \tag{16}$$

From Eq. (14), the expression of the damping coefficient is:

$$\zeta=\frac{k}{\left|(k-M\omega^2)\,ch(\alpha h)+\dfrac{M\omega^2 k}{2S\alpha(K+i\omega\eta)}sh(\alpha h)\right|} \quad \text{with } \alpha^2=\frac{\omega^2\rho}{K+i\eta\omega}. \tag{17}$$

Therefore, the damping curve can be obtained from data related to the loaded structure and also from the characteristics of the MR device.

3. Viscoelastic behavior of a MR fluid

It is necessary to gain more insight into the origin of the parameters we have introduced in the motion equations of the above systems. For doing so, we have investigated the viscoelastic properties of a MR fluid made of a silicone oil (zero-shear viscosity of 9.7 Pa.s) and iron particles of diameter eight microns. The volume fraction of the particles is 0.15 and the applied magnetic field is equal to 21 $kA.m^{-1}$. The sample is analyzed with two rotating rheometers: one is a stress-imposed device (Rheometric DSR) and the other one a strain-imposed rheometer (Rheometric ARES). A cone and plate geometry allows the sample to be sheared at a constant shear rate throughout the volume. The magnetic field is oriented along the rotating axis of the chamber. Three types of experiments have been conducted: with the DSR device, we performed stress sweeps and measured strain and strain rate (exp. A); with the ARES rheometer, we performed frequency sweeps at a given strain (exp.B) or strain sweeps at a given frequency (exp.C). Experiments A give the variations of the stress with the strain rate in order to check the assumption of Bingham behavior of the fluid. From experiments B and C, we measure the complex shear modulus of the fluid $G^*(\omega,\gamma)$, which includes both the elastic behavior (storage modulus G') and the viscous behavior (loss modulus G'').

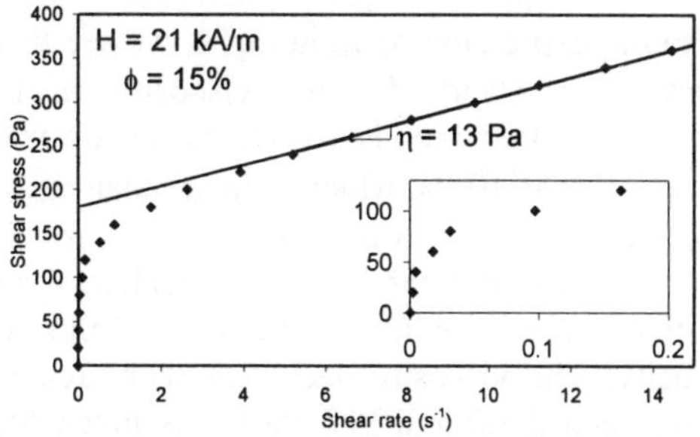

Fig. 8. Flow curve of the MR fluid under field (exp. A)

The flow curve of the sample (Fig. 8) shows the expected Bingham behavior with a yield stress τ_{yield} of about 180 Pa and a steady shear viscosity of 13 Pa.s. That value is higher than that of the pure oil and can be compared to the prediction of suspension models such as that given by Eilers[4]: $\eta_{susp.} = \eta_{oil}\,(1+5\phi/4(1-\phi/\phi_m))^2$ with Φ_m standing for the packing fraction. The above relation gives a value of viscosity close to 15 Pa.s which is consistent with the experimental value. But the notion of yield stress is very formal as shown in the insert of Fig. 8. The fluid undergoes a flow from the very beginning and therefore, Bingham equation only holds in the asymptotic part of the curve.

In addition, the complex shear modulus of the sample is very different with and without field (Fig.9). In a first approximation, at least at high frequencies, the fluid alone exhibits a newtonian behavior; that is to say G'' is proportional to ω, then the dynamic viscosity η' is constant. Its value of 50 Pa.s is higher than that deduced from experiment A.

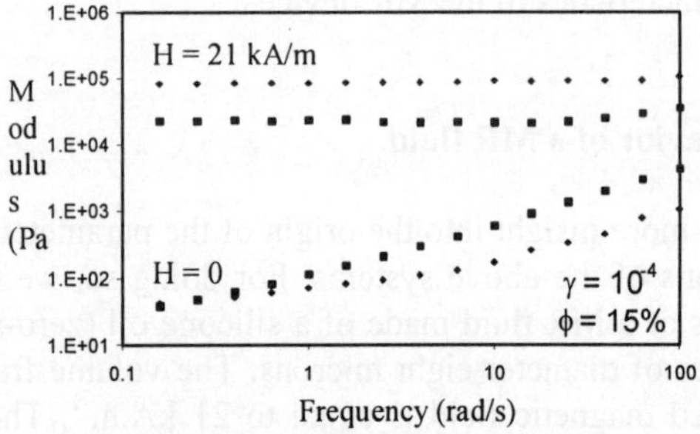

Fig. 9. Complex shear modulus of the "free" suspension (◊ G'; □ G'') and under field (◆ G'; ■ G'')-(exp. B)

That means that, at high shear rate, the columns of particles created by the magnetic field are destroyed and the fluid behaves as an homogeneous suspension. At low shear rates, the deviation from a newtonian behavior comes either from a sedimentation of the suspension or from a steric interaction between the particles. But, in that area, the values of G' and G'' are three orders of magnitude lower than those measured under magnetic field. Moreover, we observe a large plateau under field. That means that the behavior under field has to be related with the structuration of the fluid. The shear stress can be expressed in a complex form by:

$$\tau^* = G^*(\omega,\gamma)\,\gamma \quad = G'\gamma + i\,G''\gamma \tag{18}$$

If we compare it to the expression of τ_{yz} in Eq. 4, we see that G' plays the role of coefficient K and that G''/ω stands for the viscosity η. For our experiments, Klingenberg term K equals 200 Pa which is a value by far lower than the plateau value of G' and the elastic stress $G'\gamma$ is 10 Pa. which is lower than τ_{yield}. In fact, the insert of Fig. 8 confirms that the level of the stress is lower than τ_{yield} for $d\gamma/dt=\gamma\omega < 0.2\ s^{-1}$. But the very high value of G' indicates that the use of Klingenberg expression is not suitable for very low strains. Moreover, as G'' exhibits a plateau, it follows that one cannot use a constant value of viscosity. The viscosity decreases with frequency, then with shear rate, and the value – between 2000 and 300 Pa.s – is much higher than that of the homogeneous suspension. We will comment on the possible interpretation of such data and to the dependence in magnetic field strength and particle concentration in a next paper[5].

Another interesting feature is the impact of the strain strength on the viscoelastic behavior (Fig.10). At very low strains, the fluid behaves as a viscoelastic solid–$G'>G''$- and moves toward a liquid behavior at high shear strains – $G'<G''$. The transition occurs at the critical strain γ_{c2} One can expect to retrieve the values of G' and G'' of the free suspension at higher strains. In the range of strain explored here, the elastic component of the free suspension does not perturb the variation of G' which remains higher than K.

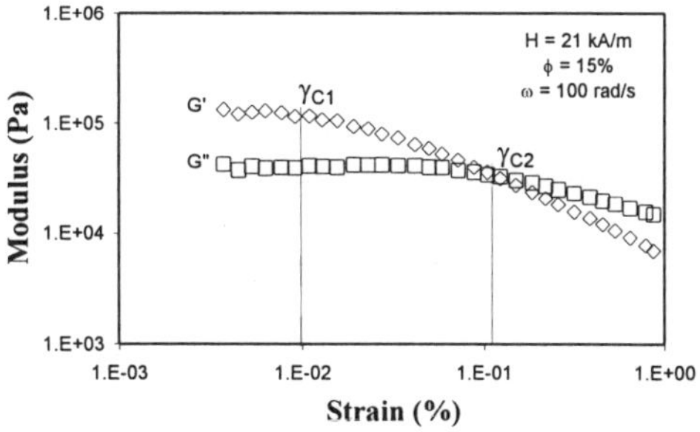

Fig. 10. Variations of the shear modulus with strain amplitude

4. Conclusion

For oscillating motions, the shear stress can be expressed in a complex form and the complex shear modulus $G^*(\omega)$ contains all the information about the elastic and viscous behavior of a MR fluid. The pre-yield and post-yield situations are described in the same way from a simple constitutive equation: $\tau^*=G^*\gamma^*$. Therefore, the mechanical response of damping systems - for example the damping coefficient ζ - should be expressed in terms of complex shear modulus whatever the strain. The dependence of G^* with magnetic field, particle concentration and strain will be further investigated and presented in a next paper[5].

References

1. H, Nishiyama, T. Oyama and T. Fujita, in *Physical Mechanisms – Proc. Int. Conf. on Electro-rheological Fluids and Magneto-rheological Suspensions*, ed. R. Tao (World Scientific, Singapore, 1999), p. 306.
2. G.J. Hiemenz and N.M. Wereley, in *Applications of Magnetorheological Fluids – Proc. Int. Conf. on Electro-rheological Fluids and Magneto-rheological Suspensions*, ed. R. Tao (World Scientific, Singapore, 1999), p. 657.
3. D.J. Klingenberg, *J. of Chem. and Phys.*, 425 (1991).
4. H. Eilers, *Kolloid-Z*, **97**, 313 (1969)
5. J. Claracq, J. Sarrazin and J. P. Montfort, in preparation for *Rheologica Acta*.

DYNAMICS OF MAGNETIC FLUID-PERMANENT MAGNET SYSTEM SUBJECTED TO VERTICAL VIBRATION

SEIICHI SUDO, KAZUTAKA ISE

Department of Mechanical Engineering,Iwaki Meisei University, Iino 5-5-1, Chuohdai, Iwaki 970-8551, Japan
E-mail: sudou@iwakimu.ac.jp

TOSHIAKI IKOHAGI

Institute of Fluid Science, Tohoku University, Katahira 2-1-1, Aoba-ku, Sendai 980-8577, Japan
E-mail: ikohagi@ifs.tohoku.ac.jp

This paper is concerned with the dynamic behavior of a magnetic fluid-permanent magnet system subjected to the vertical vibration. The surface response of magnetic fluid adsorbed to a disc permanent magnet is investigated experimentally. Experiment is performed on a vibration-testing system. The electrodynamic shaker is operated at a given frequency, displacement, and acceleration. The dynamic behavior of magnetic fluid surface is analyzed by a three-dimensional motion analysis system. The displacement of the permanent magnet is also measured with an optical displacement detector system. It is shown that the response amplitude depends on the excitation frequency at the constant excitation amplitude. The 1/2-subharmonic response on the surface of magnetic fluid is observed at higher frequencies. This response is observed as polygonal modes with large amplitude azimuthal waves. The response of the system such as jumping motion is also shown at the resonance frequency of the system.

1 Introduction

The study of dynamic behavior of magnetic fluids under an applied magnetic field is of fundamental interest and importance with respect to a variety of applications. Therefore, various phenomena of magnetic fluids have been conducted and reported[1]. Magnetic fluids are ultra stable colloidal systems containing single domain ferro or ferri magnetic particles[2]. The fluids respond readily to magnetic fields and have a higher saturation magnetization. In many applications of magnetic fluids, an understanding of the fundamental dynamics of magnetic fluid surface under various fields is essential. Zelazo and Melcher[3] performed experiments using rectangular containers partly filled with magnetic fluid and driven by a low-frequency transducer that vibrates the container in the horizontal plane. Bashtovoi and Rosensweig[4] found that subcritical standing waves are excited when a time-periodic field is superpose collinearly on a steady field with normal orientation to a pool of magnetic fluid. Authors have been studied the surface responses of a single magnetic fluid droplet on a horizontal solid nonmagnetic base subject to magnetic field and vertical vibration[5]. We also have been studied the interaction between the applied magnetic field and the magnetic fluid surface in an open circular cylindrical container in our previous papers[6,7].

However, the research data on the surface responses of a magnetic fluid adsorbed to a permanent magnet subject to forced vibration are insufficient, and there still remains a wide unexplored domain.

In the present paper, the dynamic behavior of free surface of magnetic fluid adsorbed to the permanent magnet on the table subjected to the vertical vibration is investigated with a three dimensional motion analysis system. Frequency characteristics of the magnetic fluid-permanent magnet system are also revealed.

2 Experimental Apparatus and Procedures

Figure 1 shows a schematic diagram of the experimental apparatus. The experimental apparatus consisted of the magnetic fluid-magnet system, vibration testing system (IMV Model VE3201), and measurement system. The magnetic fluid-permanent magnet system is composed of the disc permanent magnet and magnetic

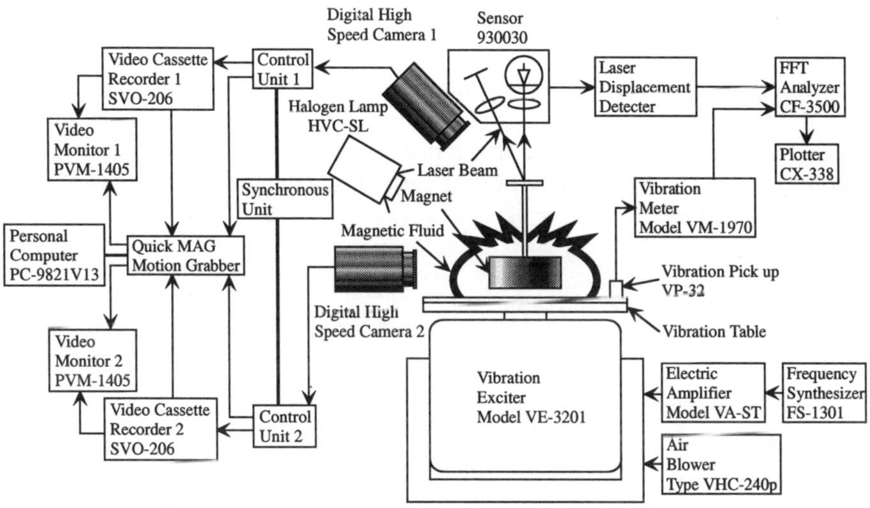

Figure 1. Schematic diagram of the experimental apparatus

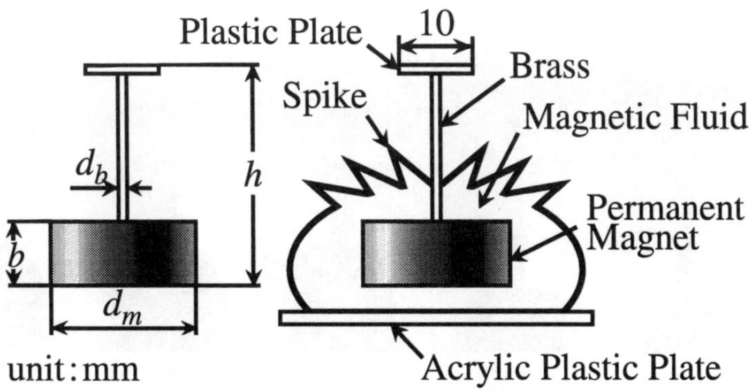

Figure 2. The magnetic fluid-permanent magnet system

fluid as shown in Fig.2. Table 1 shows the physical properties of ferrite magnet used in the experiment. The circular plastic stick with a small plate is bonded to the magnet for the measurement of magnet oscillation. Sample magnetic fluid used in the experiment was kerosene-based ferricolloid HC-50 (the saturation magnetization M_s = 33.42kA/m). The surface tension of the magnetic fluid is 0.0277 N/m (magnetic fluid/air at a temperature of 19°C). The side view photographs of magnetic fluid-permanent magnet systems are shown in Fig.3. The electrodynamic shaker is operated at a given frequency, displacement and acceleration within the range of maximum exciting force 4.1kN, maximum exciting acceleration 784m/s^2, maximum exciting amplitude 25mm, and maximum exciting frequency 3000Hz. The accelerometer is used to measure the vibration components of the vibrating table. The displacement of the magnetic fluid-permanent magnet system on the vibrating table is measured with an optical displacement detector system. The dynamic behavior of magnetic fluid surface is analyzed by a three-dimensional motion analysis system. The system of the three dimensional motion analysis is composed of two high speed video cameras, synchronize unit, two video cassette recorders, motion grabber, two video monitors, and a personal computer. The magnetic fluid-permanent magnet system is put on the vibrating table of the electrodynamic shaker. The excitation frequency is set as desired, and the vibrating table connecting to the exciter is vibrated vertically at an arbitrary level of input excitation acceleration.

Table 1. Physical properties of permanent magnet

Disk Magnet	d_m(mm)	b(mm)	d_b(mm)	h(mm)	B_0(mT)
Ferrite	15.0	4.0	3.0	35.2	90.0

Figure 3. Photographs of magnetic fluid-magnet systems

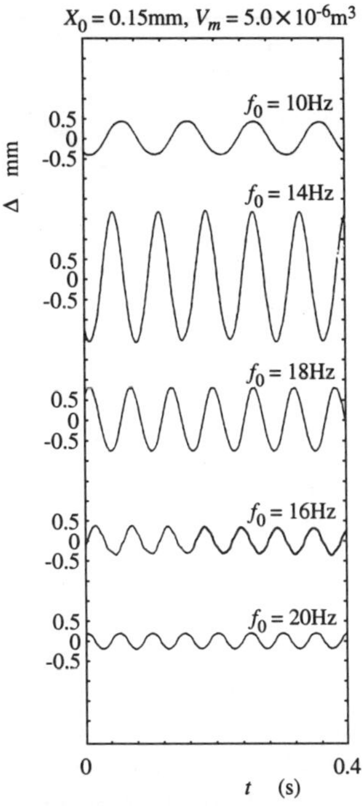

Figure 4. Output signals of magnet oscillation

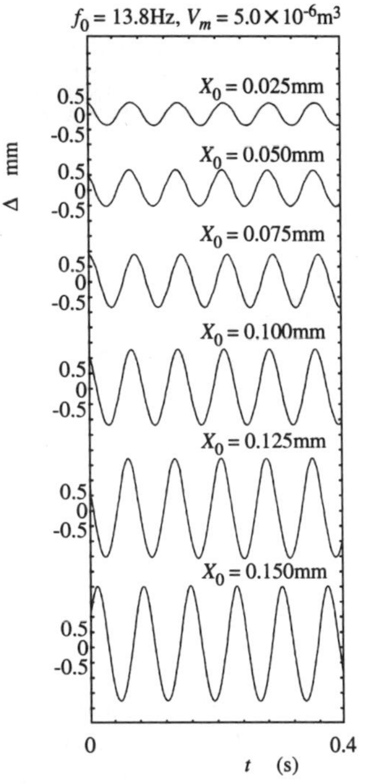

Figure 5. Output signals of magnet oscillation

3 Experimental Results and Discussion

3.1 Response of Permanent Magnet

The motion of magnetic fluid-permanent magnet system shown in Fig.2 is studied in this paragraph. The magnetic fluid-magnet system is vibrated vertically by the displacement

$$X = X_0 \cos 2\pi f_0 t \tag{1}$$

where X_0 is the excitation amplitude, f_0 is the excitation frequency, and t is the time. The displacement of the permanent magnet in the system is measured with the optical displacement detector shown in Fig.2. Figure 4 shows the responses of the magnet for various excitation frequencies at the constant excitation amplitude. It can be seen that the amplitude of magnet oscillation depends on the excitation frequency f_0. The response of magnet oscillation is harmonic. In this system, the amplitude of magnet oscillation is largest at f_0=13.8Hz. Figure 5 shows the responses of the magnet for various excitation amplitudes at the constant excitation frequency f_0=13.8Hz. The amplitude of magnet oscillation increases with the excitation amplitude.

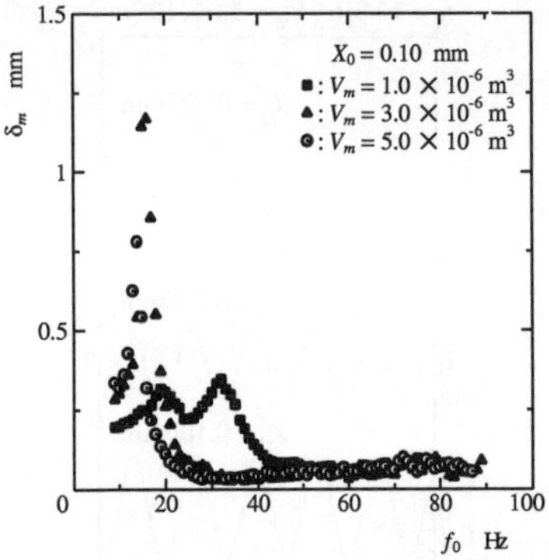

Figure 6. Frequency characteristics of the systems

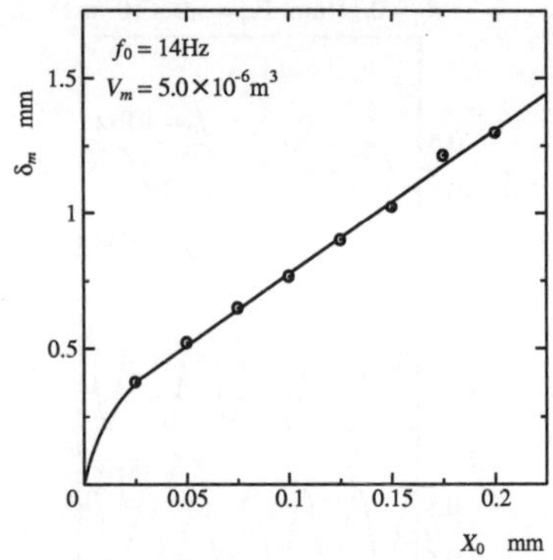

Figure 7. Relation between the response and excitation amplitudes

3.2 Frequency characteristics of the system

Figure6 shows the relation between the amplitude of magnet, δ_m and the excitation frequency f_0 with the constant excitation amplitude X_0 for the various volumes of magnetic fluid. It can be seen that frequency characteristics of the system show the clear peak at the certain frequency. The peak in the frequency response corresponds to the natural frequency of the system. The natural frequency of the system decreases with the volume of magnetic fluid adsorbed to the permanent manget. Figure7 shows the relation between the response amplitude of magnet, δ_m, and the excitation amplitude, X_0. The amplitude δ_m increases with the excitation amplitude X_0 almost linearly. The gradient $d\delta_m/dX_0$ depends on the excitation frequency. The gradient $d\delta_m/dX_0$ is heigher in the neighbourhood of the natural frequency of the system.

3.3 Spike response of magnetic fluid

As shown in Fig.2, the magnetic fluid spikes are formed at the upper part of the magnetic fluid bulk by the normal magnetic field instability[8]. These magnetic fluid spikes respond to the excitation in the elongation and contraction of their heights. The frequency of the spike pulsation for vertical excitation corresponds precisely to the exciting frequency. The shape oscillation of spikes dipends on the apparent gravitational acceleration $\bar{g}$. The critical level of magnetization for onset of the instability, which was derived by cowley & Rosensweig[8], is modified as follows:

$$M_c^2 = \frac{2}{\mu_0}\left[1 + \frac{1}{r_0}\right](\bar{g}\Delta\rho\sigma)^{\frac{1}{2}} \tag{2}$$

$$\bar{g} = g - X_0(2\pi f_0)^2 \cos 2\pi f_0 t \tag{3}$$

$$r_0 \equiv (\mu_c\mu_t/\mu_0^2)^{\frac{1}{2}} \tag{4}$$

where M_c is the critical magnetization, μ_0 is the permeability of the vacuum, $\Delta\rho$ is the density difference, σ is the surface tension, μ_c is the chord permeability, μ_t is the tangent permeability.

3.4 Shape oscillation of magnetic fluid bulk

Figure 8 shows photographs of a sequence of stages in the shape oscillation of magnetic fluid bulk with f_0 = 84Hz. Photographs are selected frames from high-speed pictures. In Fig.8, δ_t is the time interval for each frame, and picture numbers 1 to 6 reveal the process of shape oscillation which corresponds to the half period. It can be seen that the surface shape shows the heptagon oscillation. The surface

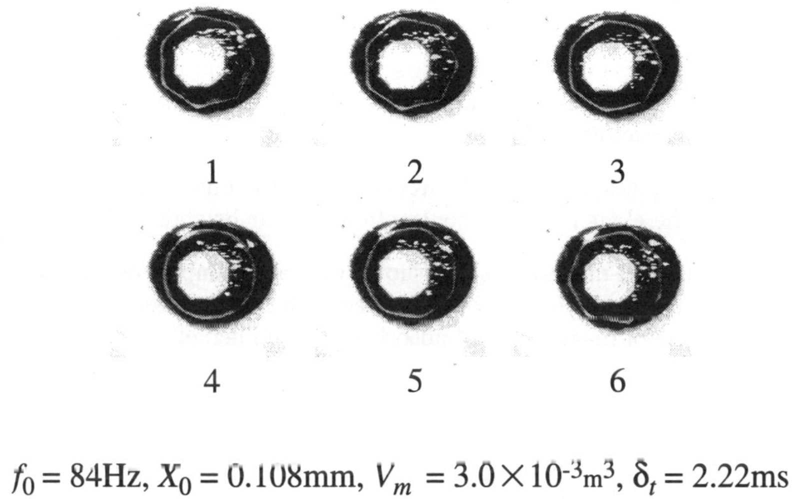

Figure 8. Photographs of surface shape oscillation

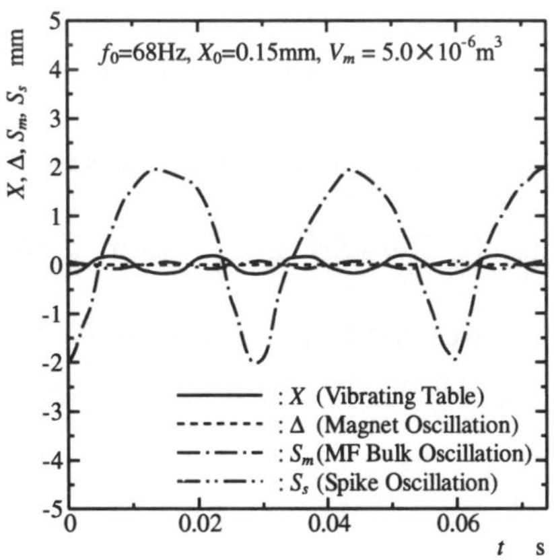

Figure 9. Dynamic behavior of magnetic fluid-magnet system

oscillation of magnetic fluid is the 1/2-subharmonic response. The amplitude of surface oscillation of the system response to the vertical vibration. In Fig.9, S_m (alternate long and short dash line) is the heptagon oscillation, S_s (two-dot chain line) is the spike oscillation, Δ (dotted line) is the permanent magnet oscillation, and X (continuous line) is the table oscillation. It can be seen that the amplitude of heptagon oscillation is sufficiently large compared with other oscillations and the frequency of the heptagon oscillation is exactly one-half that of the excitation.

4 Conculusions

The dynamic behavior of the magnetic fluid-permanent magnet system on the table subjected to the vertical vibration is investigated experimentally. The results obtained are summarized as follows.

1. The haptagon shape oscillation occurs to the magnetic fluid bulk at higher exciting frequencies. The frequency of the heptagon oscillation is exactly one-half that of the excitation.
2. Magnetic fluid spikes formed at the upper part of the fluid bulk respond to the excitation in the elongation and contraction of their heights.
3. The magnetic fluid-magnet system shows the large amplitude up-and-down motion at the natural frequency of the system. The natural frequency decreases with the volume of magentic fluid adsorbed to the magnet.

References

1. R. E. Rosensweig, *Ferrohydrodynamics*, Combridge Univ. Press (1985).
2. J. Popplewell, *Chem. Eng. Comm.* **67**, 243 (1988).
3. R. E. Zelazo and J. R. Melcher, *J. Fluid Mech.* **39**, 1 (1969).
4. V. G. Bashtovoi and R. E. Rosensweig, *J. Magn. Magn. Mater.* **122**, 234 (1993).
5. S. Sudo, M. Ohaba, K. Katagiri and H. Hashimoto, *J. Magn. Magn. Mater.* **122**, 248 (1993).
6. S. Sudo, M. Ohaba, K. Katagiri and H. Hashimoto, *Magnetohydrodynamics* **33**, 480 (1997).
7. S. Sudo, H. Nishiyama, K. Katagiri and J. Tani, *J. Intel. Mater. Systems Struc.* **10**, 498 (1999).
8. M. D. Cowley and R. E. Rosensweig, *J. Fluid Mech.* **30**, 671 (1967).

KINETICS OF ELECTRORHEOLOGICAL FLUID SPREADING OVER A HORIZONTAL SURFACE

V.I.BAYKOV and E.V.KOROBKO

A.V.Luikov Heat & Mass Transfer Institute, The National Academy of Sciences of Belarus
15 P.Brovka St.,Minsk 220072, Belarus

ABSTRACT

A Physical and Mathematical model of the process of fluid drop spreading is suggested which quantitatively describes all the stages of spreading kinetics on the basic of a unified approach. It has been established that gravitational forces exert a pronounced influence on the process of fluid drop spreading over a horizontal solid surface.

1. Introduction

Spreading of fluids over a horizontal surface plays a significant part in various physical and chemical events and technological processes. For investigations in this field it is convenient to consider a fluid drop spreading over a horizontal surface. However, theoretically it is rather difficult to describe kinetics of the spreading because it is necessary to allow for a number of factors that influence spreading of a real drop. In particular, of importance is understanding of dynamics of the processes that take place near solid, liquid and gaseous phases. At present considerable progress has been made towards this goal, a number of methods to construct a mathematical model of a spreading-out drop are suggested [1 - 7] and for some particular problems the relations between the current radius of a drop base and the time are obtained [2, 4, 5].

A general drawback of the developed models lies in the facts that firstly, they pertain to separate stages of spreading and, secondly, they fit experimental data only qualitatively and substantially fail to agree quantitatively.

In the present work kinetics of different stages of the spreading of a fluid drop over a solid surface is shown on the basis of a unified approach. Theoretical relations are obtained which are well consistent with experimental results.

2. Formulation of the Problem.

We will consider spreading of an electrorheological fluid (ERF) drop with mass m over a horizontal solid surface. Results of experiments [5, 8] have shown that in spreading the drop has the shape schematically depicted in Fig.1.

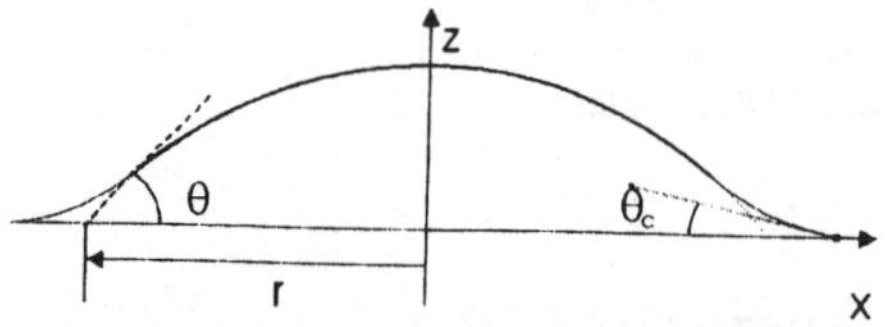

Fig.1 Scheme of ERF spreading

The large (central) part of the drop retains the form approaching a sphere segment with the sphere radius r and bounded by the convex surface with a contact angle Θ. On the drop periphery there is formed the zone of a dynamic meniscus having a concave surface with the contact angle approaching the equilibrium contact angle Θ_c. Such form of the external surface of the drop is indicative of the pressure redistribution in the main (central) part of the drop and in the dynamic meniscus zone, which is responsible for the spreading, and allows us to describe this process from a unified position.

We will make a number of simplifications. Instead of the process of spreading we may consider motion of the relatively solid surface of the zone of a dynamic fluid meniscus existing on the edge of the fluid drop. We assume that the fluid pressure at the dynamic meniscus inlet is equal to that in the central part of the drop

$$p = \frac{2\sigma}{r_1} \sin \Theta \tag{1}$$

where σ is the surface tension at the drop - gas boundary. The pressure at the end of the dynamic meniscus is assumed to be close to the equilibrium value

$$p_C = \frac{2\sigma}{r_2} \sin \Theta_C \tag{2}$$

Spreading is a quasi - stationary process, i.e. we neglect contribution of the local velocity derivative with respect to time as compared to that of the viscosity term. Inertia effects are also neglected.

The viscosity μ, the surface tension σ, and the equilibrium contact angle Θ_c of an ERF depend on an electric field intensity E: $\mu(E)$, $\sigma(E)$, $\Theta_c(E)$.

Motion of the fluid in the dynamic meniscus zone in the cylindrical coordinate axes, whose axies z is perpendicular to the solid surface and coincides with the axes of drop symmetry, is described, with the assumptions made above, by the equations

$$-\frac{\partial p}{\partial r} + \mu \frac{\partial^2 u}{\partial z^2} = 0; \qquad \frac{\partial p}{\partial z} = 0 \tag{3}$$

with the boundary conditions

$$u=0 \text{ at } z=0; \qquad \frac{\partial u}{\partial z}=0 \quad \text{at } z=h \tag{4}$$

From the second relation of system Eqs. (3) it stems that $P=P(r)$
Then having integrated the first equality of system Eqs. (3) with conditions (4), we arrive at

$$u=-\frac{h^2}{\mu}\frac{\partial p}{\partial r}\left(\frac{z}{h}-\frac{z^2}{2h^2}\right)$$

From this we obtain the following relations for the fluid flow rate in the dynamic meniscus zone

$$Q=\int_0^h 2\pi ru\partial z=-\frac{2\pi rh^3}{3\mu}\frac{\partial p}{\partial r}=const$$

At a result, the average velocity of the dynamic meniscus zone is

$$\bar{u}=\frac{dr}{dt}=\frac{Q}{S}=\frac{Q}{2\pi rh}=-\frac{h^2(r)}{3\mu}\frac{\partial p}{\partial r} \tag{5}$$

3. Solution of the Problem

Eq. (5) contains two unknown quantities, i.e. $h(r)$ and $\frac{\partial p}{\partial r}$. Since our interest is in average velocity $\bar{u}$, then with allowance for Eqs. (1), (2) we can write the following relation for the pressure gradient in the dynamic meniscus zone:

$$-\frac{\partial p}{\partial r}\approx\frac{p-p_C}{r_2-r_1}=\frac{2\sigma}{r_1(r_2-r_1)}\left(\sin\Theta-\frac{r_1}{r_2}\sin\Theta_C\right)$$

If we take into consideration that the dynamic meniscus zone r_2 - r_1 is much less than the main part of the drop r_1, then $r_1/r_2 \approx 1$ and, as a result

$$-\frac{\partial p}{\partial r}=\frac{2\sigma}{r_1(r_2-r_1)}(\sin\Theta-\sin\Theta_C)$$

Consequently, from Eq. (5) we obtain the relation for the average velocity of the dynamic meniscus

$$\bar{u} = \frac{dr_1}{dt} = \frac{2\sigma}{3\mu} \frac{h^2(r_1)}{r_1(r_2 - r_1)}(\sin\Theta - \sin\Theta_C) \tag{6}$$

Next, we will determine the drop height $h_1 = h(r_1)$ at the inlet of the dynamic meniscus zone. Since this part of the drop pertains to the main one, it is practically immobile and we can employ the equation for a static meniscus which for a concave surface is of the form

$$\frac{\partial^2 r_h}{\partial z^2}\left[1 + \left(\frac{\partial r_h}{\partial z}\right)^2\right]^{-3/2} = \frac{\rho g z}{\sigma} \tag{7}$$

(ρ is the fluid density, g is the free fall acceleration) with the boundary conditions

$$\frac{\partial r_h}{\partial z} = -ctg\Theta_C \text{ at } z = 0; \quad \frac{\partial r_h}{\partial z} = -ctg\Theta \text{ at } z = h_1$$

Single integration of this equation yields

$$\frac{\partial r_h}{\partial z}\left[1 + \left(\frac{\partial r_h}{\partial z}\right)\right]^{-\frac{1}{2}} = \frac{\rho g z^2}{2\sigma} + C$$

and with allowance for the boundary conditions we find

$$h_1 = \left(\frac{2\sigma}{\rho g}\right)^{1/2} (\cos\Theta_C - \cos\Theta)^{1/2} \tag{8}$$

And, finally, owing to the fact that the dimensions of the dynamic meniscus depend on one and the same inducing force, i.e. the pressure gradient, we can assume that

$$r_2 - r_1 = a h_1 \tag{9}$$

where a is the proportionality factor.

Substitution of Eqs. (8), (9) into formula (6) gives a kinetic equation that describes the process of fluid drop spreading over a horizontal surface

$$\frac{dr_1}{dt} = \frac{2}{3a}\ \frac{\sigma}{\mu}\left(\frac{2\sigma}{\rho g}\right)^{1/2}\ \frac{1}{r_1}(\sin\Theta - \sin\Theta_C)(\cos\Theta_C - \cos\Theta)^{1/2} \tag{10}$$

The contact angle Θ of the sphere segment is related with its base radius r_1 by the trigonometric expression

$$\frac{6V}{\pi r_1^{\,3}} = 3tg\frac{\Theta}{2} + tg^3\frac{\Theta}{2}, \tag{11}$$

where $V = \dfrac{m}{\rho}$ is the volume of the fluid drop.

4. Analysis of the Kinetic Equation of Spreading

Eqs. (10), (11) describe kinetics of the process of drop spreading both in the field of blunt and contact angles Θ and in the field of acute contact angles Θ. To make sure, preliminarily we perform dimensionless representation. We will introduce the dimensionless drop base radius $R = \dfrac{r_1}{r_0}$, with $r_0 = \left(\dfrac{3V}{2\pi}\right)^{1/3}$ being the base radius of the drop when it has the form of a hemisphere, and dimensionless time

$$\tau = t\frac{2}{3}\frac{\sigma}{\mu a}\left(\frac{2\sigma}{\rho g}\right)^{1/2}\left(\frac{2\pi}{3V}\right)^{2/3}$$

As a result, the system of Eqs. (10), (11) acquires the form

$$\begin{aligned} \frac{dR}{d\tau} &= \frac{1}{R}(\sin\Theta - \sin\Theta_C)(\cos\Theta_C - \cos\Theta)^{1/2} \\ \frac{4}{R^3} &= 3tg\frac{\Theta}{2} + tg^3\frac{\Theta}{2} \end{aligned} \tag{12}$$

Now we will analyze the system of Eqs. (12). First we will consider the case when the equilibrium contact angle Θ_c is small, i.e. $\Theta_c \approx 0$. Then for spreading of the drop in the field of blunt contact angles $\Theta = \pi - \alpha$ for small α we obtain the following relations from system (12)

$$\frac{dR}{d\tau}=\frac{\sqrt{2}\alpha}{R}, \qquad \frac{4}{R^3}=\left(\frac{\alpha}{2}\right)^3 \tag{13}$$

Integration of Eqs. (13) yields in the expression

$$R=\frac{2\sqrt{2}}{4^{1/3}}\tau \tag{14}$$

For the case of the drop spreading in the field of acute contact angles for small Θ we find from Eqs. (12)

$$\frac{dR}{d\tau}=\frac{1}{\sqrt{2}}\frac{\Theta^2}{R}, \qquad \frac{4}{R^3}=\frac{3}{2}\Theta \tag{15}$$

Solving this system with the initial condition $R = R_0$ at $\tau = 0$, we arrive at

$$R^8-R_0^8=\frac{512}{\sqrt{2}\cdot 9}\tau$$

The last relation for $R^8 \gg R_0^8$ allows us to write

$$R=\left(\frac{512}{\sqrt{2}\cdot 9}\tau\right)^{1/8} \tag{16}$$

Thus, from formulas (14), (16) it follows that kinetics of displacement of the drop perimeter changes from the dependence $R\sim\tau$ in the field of blunt contact angles to $R\sim\tau^{1/8}$ in the field of acute contact angles. This is well consistent with numerous experimental data [2, 5, 7, 8].

Now we will dwell on evaluation of the proportionality factor a. Proceeding from Eqs. (7), (8) we can assert that it depends on the fluid viscosity μ and the drop volume V. On the basis of the experimental data [5] on spreading of polydimethylsiloxane and polyphenylmethylsiloxane drops over glass, lawsan and aluminium surfaces we recommend the following relation to fit quantitatively results of the theorem and experiment $a=\left|\frac{\rho g}{\sigma}\right|$, where $|\cdots|$ indicates a numerical value.

In conclusion, we will consider the influence of the electric field intensity E on kinetics of spreading of an ERF drop. From the system of Eqs. (12) it follows that the functional dependence of the kinetics of the spreading in the field of blunt and acute contact angles Θ is mainly influenced by the dependence of the equilibrium contact angle $\Theta_c(E)$ on the field intensity. Figure 2 provides the plots of kinetics of an ERF drop spreading obtained by solving numerically system of Eqs.(12). It is pertinent to note that in the field of blunt contact angles $\Theta > \frac{\pi}{2}$ and in the field of acute contact angles $\Theta < \frac{\pi}{2}$ no ERF drop spreading occurs if $\sin\Theta < \sin\Theta_c(E)$. This result has been experimentally verifiedin [8].

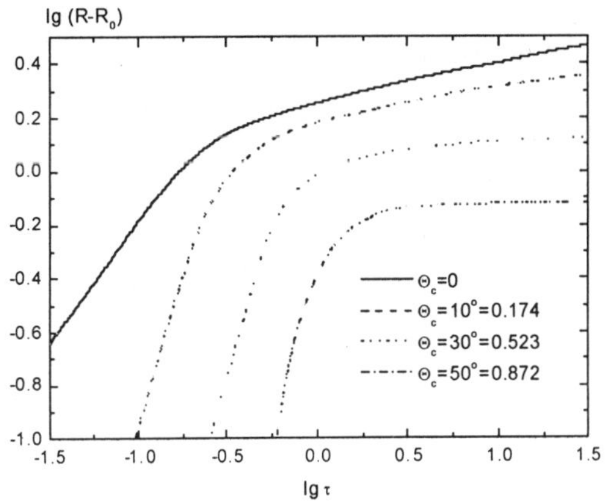

Fig. 2. Influence of an electric field on kinetics of drop spreading over a horizontale surface

References

1. E.A.Raud, B.D.Summ and B.D.Shchukin. *Dokl. Akad. Nauk SSSR.* **205**, No 5 (1972).
2. Yu.V.Sorokin, V.V.Khlynov and O.A.Esin. *Zh. Fiz. Khim.* **41**, No 7 (1967).
3. L.M.Hocking and A.D.Rivers. *J. Fluid Mech.* **121** (1982).
4. E.D.Shchukin, Yu.V.Gorynov, G.I.Den'schikova et al. *Kolloid. Zh.***25**, No 1 (1963).
5. A.A.Vavkushevsky, V.V.Arslanov, V.Yu.Stepanenko and V.A.Ogarev. *Kolloid. Zh.* **51**, No 3 (1989).
6. M.D.Lelah and A.Marmur. *Colloid and Interface Sci.* **82**, No 2 (1981).
7. B.D.Summ, E.A.Raud and E.D.Shchukin. *Dokl. Akad. Nauk SSSR.* **209**, No 1 (1973).
8. E.V.Korobko, R.G.Gorodkin, V.V.Melnichenko and L.W.Zhou. *Int. J. of Modern Phys.* **B10**, No 23 - 24 (1996).

MASS LOSSES OF MAGNETIZED RHEOLOGICAL MEDIUMS SUBJECTED TO MAGNETIC FIELD.

A. R. BAEV,
Institute of Applied Physics of National Academy of Science of Belarus
16, Academicheskaya, Minsk, 220072, Belarus
E-mail: konovalov@iaph.bas-net.by

N. P. MATOUSSEVITCH
Institute of Applied Physics of National Academy of Science of Belarus
16, Academicheskaya, Minsk, 220072, Belarus
E-mail: konovalov@iaph.bas-net.by

Theoretical analysis and experimental research of the film flowing of magnetorheologic fluids when infinitive solid plate is retrieving from the former mediums were carried out. We have got dependencies of fluids mass losses m vs: velocity of moving plate V ; magnitude of magnetic intensity, its gradient and angle φ between direction of intensity lines and normal vector to plane of plate. It is shown that theoretical and experimental function $m(\varphi)$ is anisotropy one and may change its magnitude more than 10 times. Real behavior and extreme characteristics of $m(\varphi)$ depends on rhelogiacal parameters of MRS and its magnetization. Obtained experimental data of dependence $m(V)$ are in qualitative and quantitative agreement with the modernized theory.

1 Introduction

The problem of mass losses of rheologic liquids, such as magnetic suspensions or magnetic fluids of high volume magnetic concentration (denote them MRS), controlled by external magnetic fields - is very actual for technic applications: to seal gap between rotating, oscillating or linearly moving part of apparatus, installation, to control its damping or oscillations [1], to ensure a reliable acoustic contact between the object to be tested and the acoustic emitter [2] and etc. Clearly, that hydrodynamic, magnetostatic and capillary phenomena which occur while solid plate is travels through fluid and pulling out film layer, are interesting to study the structure of thin MRS layers and to discover new film effects in applied magnetic field.

There are works, for example [2, 3], where hydrodynamic equations, describing the process of drawing out of a magnetic fluid film from confined volume by plate, travelling with speed V, does not take account of rheologic properties, demagnetizing factor of fluid volume and inertial terms. And this theory is correct for dimensionless speed $Ca=\eta V/\sigma<<1$ only, where η and σ - are viscosity and liquid surface tension. The former problem [4] was analyzed for non-magnetic Newtonian film flowing in wide range $Ca=10^{-3}$ - 10 and for different magnitude of hydrodynamic parameter $\gamma=\dfrac{\sigma(\eta^4 g)^{-1/3}}{\rho}$. Mass losses of non-magnetic rheologic fluids ($\gamma=0$) have been investigated in [5]. We are going to consider the problem of MRS losses, when magnetic field and the volume forces $\vec{f} = \mu_0 M \nabla H$ in fluid and the jump of magnetostatic pressure on the MRS boundary $\Delta P_{M^2} \sim M_n^2$ arise, where M_n - normal component of the MRS magnetization

2 Theory

The results of theoretical research of authors [4 - 5] are used to develop the model of MRF mass losses, which takes account of magnetic field interaction with the surface of film meniscus and of rheologic properties. We consider hydrodynamic and capillary phenomena which accompany the process of retrieving of non-magnetic solid plate from infinitive fluid volume, when V=const and non-homogeneous magnetic field applied. Let infinitive plane plate travels vertically in x direction through MRS volume. Applied magnetic field magnetizes fluid to saturation and magnetization $\overline{M}_S$=const, so that magnetic intensity $\overline{H}$ (x,y,z) lines and direction of $\overline{M}_S$ are parallel to plane $x=0$. The plane $x=0$ intersects surface of meniscus at $h=h_s$, where the velocity of fluid particle $u=0$. Quantity of MRS mass losses (or height its film h_0 on moving plate) is determined by interaction of the complex factors, such as magnetic field, speed V , fluid properties. According to [5] there are two part of the fluid volume: the first is range of dynamic meniscus where the velocity of flowing $u \geq 0$ (system of coordinate $x.y$), the second one - $u \leq 0$ – is proper to quasi-static part of the meniscus volume (system of coordinate $x_1.y$). x_1=0 is the plane of the infinitive fluid free surface when magnetic intensity $\overline{H}$ =0. Using the results of evaluation and approximation of inertia, viscous and capillary forces made in [4] and taking account of the effects of magnetic field interaction with magnetized fluid medium [5] the modified hydrodynamic equations and the boundary conditions for two parts of the meniscus volume are.

2.1 Dynamic meniscus:

$$\partial \tau / \partial u + \partial x[\rho g x + \Phi] / \partial x - \partial p / \partial x = 0, \quad \partial p / \partial y = 0 \qquad (1)$$

$$p - p_0 = -\sigma d^2h/dx^2 + \Delta P_{M^2} \text{ at } y=h, \quad u=V \quad y=0, \quad \tau=0 \quad y=h, \qquad (2)$$

$$|d^2h/dx^2| \, |dh/dx| \rightarrow 0 \quad y \rightarrow h_0 \quad \text{for } x \rightarrow \infty \qquad (3)$$

where $\Phi = \mu_0 \int_{\infty}^{x} M dH$ - is magnetostatic potential; ΔP_{M^2} =0,5μ_0 $(M_0^2 - M_n^2)$ - *is* an effective magnetostatic pressure difference between the coordinate on the meniscus surface $x_1=0$ and x_1, caused by the jump of the normal component of magnetization at the boundary MF-air.

2.2 *Quasi-static meniscus.*

There are space derivatives of the velocities and stresses much smaller than in zone of dynamic meniscus. Let neglect them in equation (4) and in edge condition (5) and suggest $|\partial H/\partial x_1| >> |\partial H/\partial y|$, and then the surface form of quasi-static meniscus is described by equations

$$\partial [\rho g x_1 + \Phi] /\partial x_1 + \partial p/\partial x_1 = 0, \quad \partial p/\partial y = 0 \qquad (4)$$

$$p - p_0 = -\sigma R_s^{-1} + \Delta P_{M^2} \quad \text{for } y = h, \qquad (5)$$

$$dh/dx_1 \to -\infty \quad \text{for } x \to l$$

it is suggested that meniscus surface adjoins to the non-moving plane surface, so that

$$dh/dx_1 = 0 \quad x = x_m \quad \text{for} \quad h = h_0 \qquad (6).$$

It is clearly that if the process of MRS losing is stationary, than

$$Q = \mu_0 \int_0^h u\,dy = \text{const} = V h_\infty, \qquad (7)$$

where h_∞ - is thickness of MRS layer, measured so far from $x_1=0$ that $u=V$ at $y = 0 - h$. Let the rheologic equation, describing MRS flowing in applied external field, has the order law: $\tau = k\,|\partial u/\partial y|^{m-1} |\partial u/\partial y$, where k and m are constants and

$$G_m = \partial \rho^{-1} \partial [\rho g x + \Phi + \Delta P_{M^2}]/\partial x = \text{const} \qquad (8)$$

Than on the base of (1 – 7) and using [4], it is possible to derive relations describing dependence of the MRS thickness vs. external forces, fluids properties and speed of plate for dynamic part of meniscus (9) and quasi-static one (10)

$$Ca_M^{-1} A_{11}(d^3L_M/dz^3) = 1 - A_{11}\, l_M^{-(2n+1)} [A_{12}(l_M-1)+A_{13}]^m \qquad (9)$$

$$Ca_M^{1/2} A_{21}\,(L_M^{-}1) = B - (4 - Ca_M A_{22} z_1^{\,2})^{1/2} + 0{,}5\ln[(1+A_{23}\, z_1^2)(1-A_{23}\, z_1^2)^{-1}]^{1/2} \qquad (10)$$

where $=h/h_0$, $z = x/h_0$, $z_1 = x/h_0$, $Ca_M = h_0^{1-m} V^m k/\sigma$, $A_{11} = \xi_M^{--(m+1)}$, $\xi_M = h_0(V^{-m} G_m/k)^{1/(m+1)}$, $A_{12} = 2+n^{-1}$, $A_{13} = \xi_M^{-(m+1)/m}$, $A_{21} = \xi_M^{(m+1)/2}$, $A_{22} = \xi_M^{-(m+1)}$, $A_{23} = Ca_M\, \xi_M^{m+1}/4$, B – is constant. Using the condition of continuity ("suiting") of the dynamic and quasi-static meniscus surfaces and calculation procedure [4] it is possible to find dependence of the MRS film thickness $\xi_M = \xi_M(Ca_M)$.

G

$\boldsymbol{G_m}$ **evaluation.** As follows from the theoretical analysis, for example [4], the meniscus surface form influences the MRS mass losses essentially in wide range of Ca_M and for h_0/x_m. <<1. Let analyze functions $\Phi(x_1)$ and $\Delta P_{M^2}(x_1)$. In accordance with conditions of magnetic field application, (mentioned above), it is possible to show that the more substantial ΔP_{M^2} and Φ variations are for $x_1 = 0 - x_m$. It is in this part of fluid volume the more intensive hydrodynamic and energy dissipation processes occur. Clearly, that there is impossible to solve the problem exactly because of complicated character of the the meniscus formation in applied magnetic field and its influence on hydrodynamic processes in MRS layer. That is why we simplify the problem according to equation (8) and assume, that

$$G_M = g + (\mu_0 \Delta H + \Delta P_{M^2})/\rho x_m = g + G_{M1} + G_{M2} = g + P_m/x_m = g + G_{M1}(1+\varepsilon) \qquad (11)$$

where ΔH - magnetic field difference between coordinate $x_1=0$ and x_m. in MRS; G_{M2} is caused by field of fluid magnetization; $G_{M1} = G_M$, when $g = G_{M2} = 0$. Using (11) and the former mentioned suggestions in regard of magnetic field. it is possible to show that $\vec{G}_M$ is function of the angle φ between $\overline{H}$ and the normal vector $\vec{N}$ to plane of the plate surface and

$$G_M \rho x_m \approx [1/2\mu_0(\overline{M}_S\vec{N})^2 + \mu_0 M_S \left\{ H_{x_1=0} - \left[(H_{1y} - (\overline{M}_S)_y + H_{y2})^2 + H_{x2}^2 + (H_{z2} + H_{z1})^2 \right]^{0,5} \right\}] \quad (12)$$

where $H_{1y} = H_y$ for $x_1 = x_m$. The method of surface "magnetic charges"[5] is used to find inducted magnetic intensity H_2 - by integration of magnetic field caused by magnetic charges ζ on the boundary surface MRF-air and MRF-plate. So that magnetic intensity inducted by effective "magnetic charges" on curved meniscus volume is $\sum_{i=1}^{i_0} I_i(x_m,\varphi)$ and G_M can be presented

$$G_M = G_{M1}\{2[(1+T_M^2)^{0,5} - 1]^{-1} + 1\}, \qquad (13)$$

where $G_M = G_{M1}$ if $M_S = 0$, $T_M^2 = 8\,\sigma\mu_0^{-2}\, G_{M1}\, \kappa_0^{-2} M_S^{-4}$; $\kappa_0 = \kappa_0(h_m, \varphi, \kappa_i, l_0)$, K_i - integral coefficients to characterize components of demagnetizing field.

If meniscus is alone (i_0=2) - there are two "charged" surfaces, if thin plate is retrieving through MRS - i_0=4. To simplify calculations, we suggest too that static meniscus surface ($0 < x_1 < x_m$) has curvature $\rho' = x_m^{-1}$ which coincides with the average meniscus curvature. $\tilde{\rho}' = \left(\int_0^{x_m} \frac{dx}{R} \right) / x_m \approx x_m^{-1}$ if G_M=const.

It follows from (11 - 13), that magnitude of an effective additive pressure gradient in the MRS, caused by meniscus demagnetizing factor, is close $G_{M2} \cong G_{M1}$ when

$$M_S \approx 32^{1/6} \left(\sigma |\nabla H_1|_x \right)^{1/3} \mu_0^{-1/3} k_0^{-2/3}$$

According [3] if $Ca_m \ll 1$ then $h_0 \sim (\tilde{\rho}')^{-1} \sim x_m$ and

$$\Omega = m_{\varphi=0} / m_{\varphi=\pi/2} \approx h_{0\varphi=\pi/2} / h_{0\varphi=0} \approx \left\{ \left[T_M^{-2} + 1 \right]^{0,5} - T_M^{-1} \right\}.$$

The results shown in fig. 1 are obtained using this method to calculate G_M

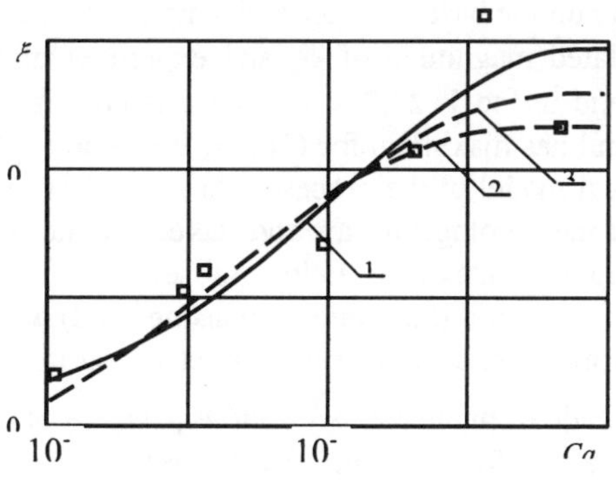

Fig. 1. The plot of $\xi_M(Ca_M)$. - Theory (1 - 3) and experiment ( ). n=1 (1); 0,84 (2); 0,56 (3). Sample - magnetic fluid.

3 Experimental methods.

To check the theory and to find the dependencies of mass loss of MRS m_o against the velocity of moving solid surface and magnitude and direction of magnetic field the special installation was designed. Elastic metallic strip is retrieving from the fluid volume with the controlled speed $(0,5 - 250)10^{-3}$ m/s. Device H331 is used to pull the strip If a specimen of magnetic fluid used the strip is moving in horizontal direction through its volume held between two parallelepiped magnets polarized in opposite direction. So, $g=0$ $|\nabla H_x|=1,76\times10^7$ A/m^2, $H=1,04\times10^5$ A/m and they are measured in vicinity of coordinate $x=0$ $y=0$ when $H=0$.
Influence of direction and magnitude of the homogeneous external field lines on suspension (and magnetic fluid) mass losses is studied for conditions when metallic strip moves through mediums in vertical direction ($g\neq0$) and fluid is in cylindrical container with diameter D=0,08 m. Fluid height is 0,015 m. As our previous experiments show the more accurate method of MF mass measurements is the method of weighting. So, the specific mass loss m_o and the height of the fluid layer h_∞ for $x \rightarrow \infty$ are determined: $m_o = (m_2 - m_1)/2$; $h_\infty = m_o/\rho$, where m_1 and m_2 – weights of strip before and after movement through MRS.
The source of controlled magnetic field is a pair of two wire coils. To check (qualitatively) magnetic field influence on magnetic fluid meniscus form the automatic system of visualization is used. It consist of special video-camera with computer. The discrimination of this system is 10^4 points per 1mm^2.
Two samples of fluids were used in experiment: magnetic fluid on transformer oil and magnetic volume concentration q =23,2 % ; a sample of magnetorheological suspension (q=8%) on the oil base. Rheologic properties of concentrated MF (q=23,2 %) in applied magnetic field were measured by A. Reks (BSPA Minsk) from which rheologic constants k and n were found. Calculated and mentioned above parameter γ_M for characteristic magnitude of the velocity gradient ϖ is close to zero.

4 Experimental results and discussion

Some basic results of research are illustrated by fig. 1 - 3. The laboratory findings show good qualitative agreement with the modified theory. Our more accurate calculation of the meniscus surface formation on the base of suggested simplified model, gives, as a rule, the difference between calculated magnitude of ξ_M and experimental data 10 - 15% when Ca_M from $(2 \div 3)\times 10^{-3}$ and up to $1\div2$ This difference is not higher 20 - 25% at whole experimental range Ca_M and has maximum for $Ca_M>1$. It can be explained by necessity to have more accurate physical model of the processes to study. For example, to use a more accurate approximation of the rheological law and take account of the peculiarity of the meniscus formation in applied magnetic field. As seen in fig.2, and follows from experiment, mass losses of magnetic fluid (and suspension used) are function of the angle between field lines and normal vector to plate surface retrieving from fluid.

It is shown that that magnitude of parameter $\Omega = m(\varphi)_{=0} / m_{\varphi=\pi/2}$ is increasing function vs. φ and can be 10 and more. As for function $m_d(H) = m/m_0$, it is decreasing one when $\pi/2 > \varphi^* > \varphi > 0$. And magnitude of φ^*, in turn, depends on Ca_m. If $\varphi = \pi/2$ - $dm_d/dH > 0$.

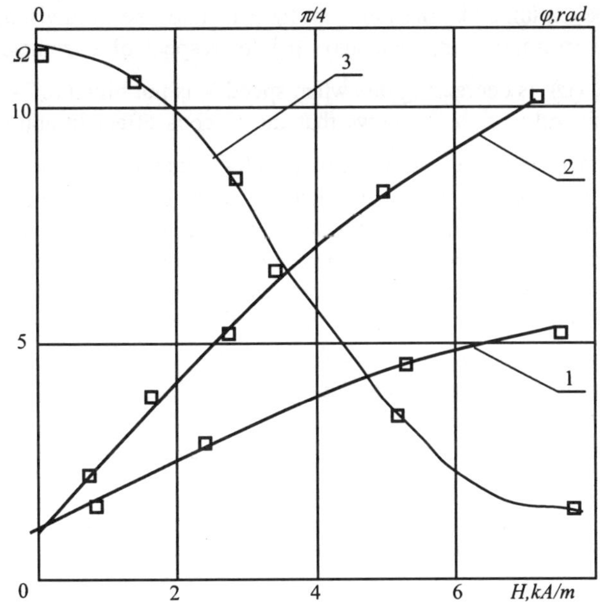

Fig. 2. The plot of $\Omega(H)=h_{0\varphi=\pi/2}/h_{0\varphi=\pi/2}$ (curves 1÷2) and $\Omega(\varphi)=h_0/h_{0g\neq 0}$ - 3.
V, 10^{-2} m/s =0,5 - 1, 3; 2,5 - 2.

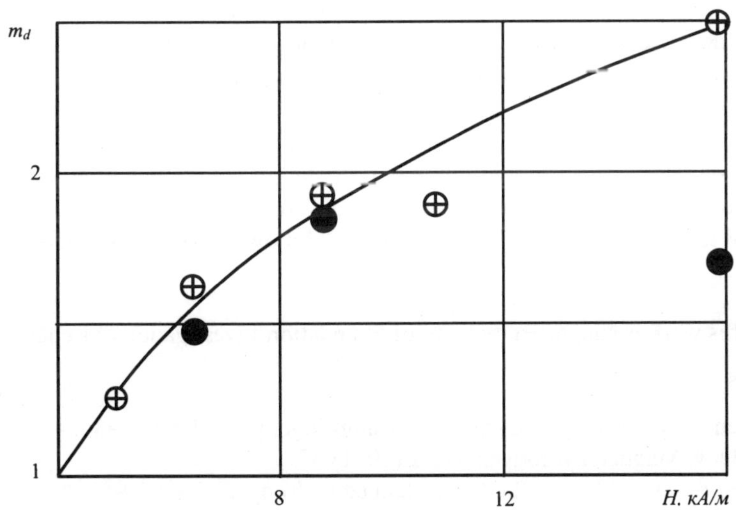

Fig. 3. The plot of $m_d(H)= m/m_0$. Sample – magnetorheological suspension.
V,10^{-4} m =1 (⊕); 10 (•).

This effect may be explained by predominant increasing of the fluid viscosity under action of magnetic field (primary homogeneous) [6]. In respect of ferromagnetic suspension used, function Ω (φ) is decreasing one when speed V is not high (10^{-4} - 10^{-3} m/s), but m_d increases predominantly vs. H. It shows that the viscous effect in applied field is more substantial than meniscus one. So we can conclude that the character of Ω (φ) and $\xi_M(Ca_M)$ behavior depends on concurrence of two effects: meniscus deformation, changing hydrodynamic resistance to film flow, while fluid is retrieving from the volume; variation of the viscosity properties under action of magnetic field.

These effects are to be taken into account while constructing of arrangements. As qualitative experiments show, a thickness of MF film on moving cylindrical objects is function like sinusoid against the angle ϕ - between direction of magnetic lines and direction of the object radius. Using dependence $\xi_M(Ca_M)$ it is possible: to use the developed method of "fluid film losses" to study the rheologic fluid properties; to measure speed and vector of acceleration of moving objects, measuring height of magnetizing fluid by electromagnetic or optic method.

5 Conclusions.

Using the results of the theoretical research [3, 4], we develop a new aspects of the film flowing of magnetized fluids with rheologic properties in applied magnetic field.

To check the proposed theory which takes into account of rheological properties and interaction of magnetic field with film meniscus the experimental study has been made.

The obtained research results have shown:

qualitative agreement between experimental data and theory in wide region of plate velocity moving;

substantial dependence of mass losses m v.s. angle φ between direction of magnetic field lines and the normal vector to moving solid surface. For example, for concentrate MF

$m^* = m_{\varphi=0} / m_{\varphi=\pi/2}$ can be 5 -10 and more. If the sample is MRS- $\boldsymbol{m^*<1}$.

It is proposed to use the method of retrieving film to investigate and check the rheological properties of magnetorheological suspensions and magnetic fluids. The obtained results can be used to measure cinematic characteristics of moving systems and another practical applications.

This work is supported by Fond of Foundation Investigation of Republic Belarus.

References

1. R Rosensweig Ferrogydrodynamics. - Cambridge: Univ. Press, 1985. --344p.
2. A.R. Baev, Magnetohydrodynamics 1 (1991) 39
3. V.M. Korovin and Yu. L. Raukher, Magnetohydrodynamics 1 (1987) 49
4. Z.P. Shulman, V. M. Buykove,. Izvestia Ac. Sci. USSR 4 (1987) 43.
5. V.M. Shulman, Rheology Izvestia Ac. Sci. USSR 4 (1987) 43.
6. V.M. Greenberg , Partial problems of electromagnetic and electrical field, (USSR, Moscow 1957) 43
7. B. V. Acta Physic. Chim., USSR, , 20 (1945)

TEMPERATURE DEPENDENCE OF MR FLUIDS

W.H.LI*, G.CHEN, S.H.YEO AND H.DU

Centre for Mechanics of Micro-Systems, School of Mechanical and Production Engineering, Nanyang Technological University, Nanyang Avenue, Singapore 639798

The rheological properties of MR fluids, MRF-132LD, are investigated under the steady shear and oscillatory shear for a range of operating temperatures from 20^0C to 60^0C. This was accomplished by using an advanced rheometer with the parallel-plate configuration. Under the steady shear, the Herschel-Bulkley model is used to model the rheology of the MR fluid. The corresponding parameters namely, τ_{yd} , K and n were determined at various temperatures, in an attempt to minimize the discrepancies between the experimental results and that predicted by the model. The results show that τ_{yd}, K and n all show decreasing trend with temperature. The results suggest that the MR fluid get "thinner" with increasing temperatures. Under the oscillatory shear, viscoelastic properties of the MR fluid were studied in the frequency sweep mode. The storage modulus, G', decreases with increasing temperatures in both the linear viscoelastic region and the nonlinear viscoelastic region. In addition, two critical frequencies, ω_{cr} and ω_m, were identified in the latter region. They were found to decrease with increasing temperatures. Finally, thermodynamics are used to explain temperature dependence of MR properties.

1 Introduction

Magnetorheological (MR) fluids have the promising ability to change their rheological properties in a rapid and reversible manner on the application of an external magnetic field. They can be used in various kinds of semi-active devices such as dampers, clutches, and brakes [1]. Any MR device in a dynamic application dissipates energy, which would remain within the mechanical system. The device dissipates the excess energy and converts it to heat, which would increases the working temperature. The temperature increase may lead to some problems, such as part failure of MR device [2]. Therefore, understanding the influence of temperature on MR effect is critical to accurately model and predict the properties of mechanisms that utilize MR fluids.

Very few reports can be found to investigate the temperature dependence of MR fluids. Some studies use MR fluids to control heat transfer through a channel [3]. The heat flux across the MR fluid region could be partially controlled by varying thermophysical properties of the fluid. These properties can be changed by the application of a magnetic field to the fluid. Other important attempts have been made by Gordaninejad and Breese to investigate heating of MR fluid dampers [4]. Their experimental and modeling results indicate the temperature effect on the damping force is significant.

In this paper, the influences of temperature on the MR rheology are investigated using a rheometer with parallel-plate geometry. The working temperature, regulated by a thermostat, varies from 20^0C to 60^0C. Rheological properties of MR fluids under steady shear and oscillatory shear at various temperatures were studied and analyzed.

* Email: mwhli@ntu.edu.sg

2 Experimental

The MR fluid, MRF-132LD, is used as the specimen in the experiment. The rheological properties of such specimen are experimentally investigated using a rheometer system (UDS 200) [5]. The rheometer system can operate in both stress-controlled mode and strain-controlled mode. This system consists of four main parts, namely: (a) the measuring drive system; (b) the electronics; (c) the user-interface software package; and (d) the thermostat. The measuring system is based on a highly dynamically motor-driven system without gears and without mechanical force transducer, consequently, it can guarantee a constant gap even for temperature tests causing thermal expansion of measuring systems. The test specimen is placed in a constant gap of 1.0mm, formed by two parallel disks whose diameters are both 20mm. A built-in coil incorporated into the system can generate the required magnetic field. The temperature of the specimen is controlled by a thermostat unit.

Prior to each experiment, the thermostat will be tuned to the required temperature. After half an hour, thermal equilibrium between the thermostat and MR cell will be reached where the experiments will proceed. In each experiment, the specimen is first subjected to a steady shear at $\dot{\gamma} = 300\,s^{-1}$ for 2 minutes as an initial condition. The desired magnetic field is then applied by setting the coil current and wait for 1 minute, ensuring the specimen form into a stable structure. For different purposes, two different working modes, including steady shear and oscillatory shear, are chosen accordingly.

3 Results and Discussion

3.1 *Rheology under Steady Shear*

For study of MR rheology under steady shear, the relationship between shear stress and shear rate should be experimentally obtained. Accordingly, the shear rate is being ramped logarithmically from 0.1 s^{-1} to 100 s^{-1} within 400 seconds, and 40 measuring points are recorded. Such procedure is repeated to measure MR rheology under steady shear at various temperatures. In our experiment, the tests are carried out at 6 different temperature levels: 10^0C, 20^0C, 30^0C, 40^0C, 50^0C, and 60^0C. The reason for choosing such narrow temperature ranges is due to the limitation of maximum coil current [5].

At a constant magnetic field strength $B = 400$mT, the experimental data of MR fluid under steady shear flow, i.e., τ vs. $\dot{\gamma}$, at various temperatures is shown in Figure 1. The yield stress seems to show a slightly decreasing trend with temperature.

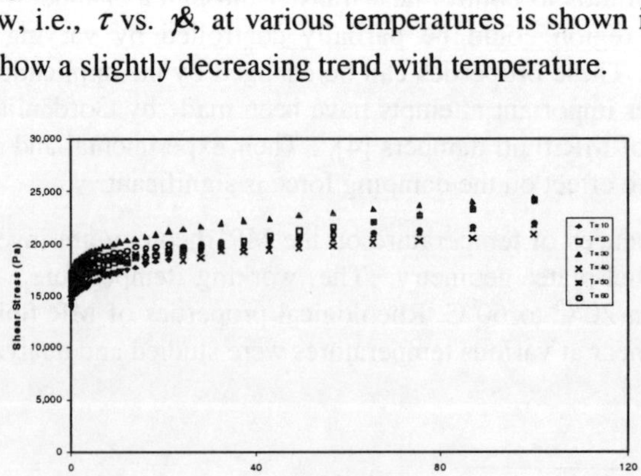

Figure 1. Plot of shear stress vs. shear rate at various temperatures, $B = 400$mT.

3.2 Herschel-Bulkley Model Study

The Bingham plastic model is widely used to model MR rheology in post-yield regime, with a constant plastic viscosity assumption. However, for cases where the MR fluid experiences post-yield shear thinning, the assumption of constant plastic viscosity may not be valid [6].

In order to accurately model the fluid behavior, the Herschel-Bulkley model is used. This model allows for a non-linear post-yield behavior.

The constitutive equation for one-dimensional Herschel-Bulkley model is

$$\begin{cases} \dot{\gamma} = 0, & |\tau| \le \tau_{yd} \\ \tau = \tau_{yd}\, sign(\dot{\gamma}) + K\dot{\gamma}^{n}, & |\tau| > \tau_{yd} \end{cases} \tag{1}$$

where τ is the shear stress, $\dot{\gamma}$ is the shear rate, τ_{yd} is the dynamic yield stress, K and n are two parameters that describe the magnitude and slope of the curve. The parameter K is sometimes referred to as the "consistency index", and has the units of $Pa \cdot s^{n}$, while the exponent n is dimensionless and is called the "behavior index". Both K and n depend on material properties and working condition such as temperature. For shear-thinning of MR suspensions, $0 < n < 1$.

Obviously, there are three parameters, i.e., τ_{yd}, K, and n, in this Herschel-Bulkley model. And these three parameters depend on the temperature, magnetic field, and other factors. Thus, the parameters are estimated for each temperature and at a particular magnetic field. The parameter estimation is done on the basis of minimizing the error between the model-predicted shear stress, given by Eq. (1), and the actual shear stress obtained from experimental measurements, as shown in Figure 1. Assume $\tau_m(\dot{\gamma})$ and $\tau_e(\dot{\gamma})$ are respectively the experimental data and model output values at the shear rate of $\dot{\gamma}$, then the error in the model is represented by the objective function J given by

$$J = \sum_{i=1}^{N} \left(\tau_{mi}(\dot{\gamma}) - \tau_{ei}(\dot{\gamma})\right)^2 \tag{2}$$

where N is the number of the experimental points.

The parameters are identified with MATLAB TOOLBOX for constrained optimization. The constraints for the three parameters are

$$\tau_{yd} > 0,\ K > 0,\ \text{and}\ 0 < n < 1 \tag{3}$$

The model parameters of τ_{yd}, K, and n as a function of temperature at the field strength 400 mT is shown in Figure 2. It can be seen from Figure 2a that the dynamic yield stress shows a slightly decreasing trend with temperature. The decrease in yield stress is about 11% from 10^0C to 60^0C at a shear rate of 0.1 s^{-1}. As shown in Figure 2b, the coefficient K also shows a steadily decreasing trend with temperature. The value of K ranges from 1423 $Pa \cdot s^{n}$ at 10^0C to 519 $Pa \cdot s^{n}$ at 60^0C for the field strength of 400 mT. These results imply that the MR fluid encounters the temperature thinning effect. Examination of the parameter n shows that it is also a function of temperature. As shown in Figure 2c, the value of n decreases gradually with increasing temperature. This indicates that the fluid thins more with an increase in temperature.

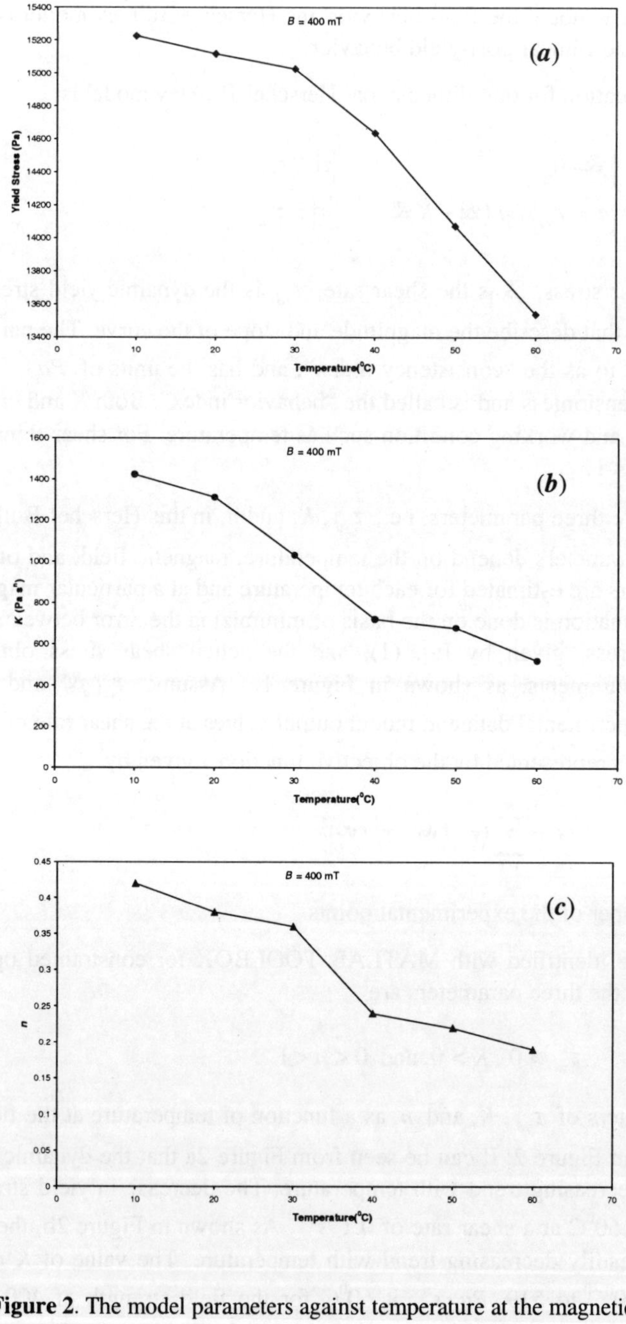

Figure 2. The model parameters against temperature at the magnetic field of 400mT (a) τ_{yd} vs. T; (b) K vs. T; and (c) n vs. T.

3.3 Viscoelastic Properties under Oscillatory Shear

With frequency sweep mode, the influence of temperature on the storage modulus (G') in both the linear and nonlinear regions is investigated and discussed.

The temperature dependence of G' under the conditions at the magnetic field strength B = 270 mT and the strain amplitude γ_0 = 0.05% are illustrated in Figure 3. The driving frequency is swept from 1 Hz to 100 Hz. The choice of this small strain amplitude is to guarantee that the MR fluid operates in the pre-yield region.

As shown in Figure 3, the decrease of G' with increase in temperature may be explained by the increase in thermal energy acquired by the particles forming the columnar structure. At higher temperatures, the particles receive higher energy and "vibrate" more rigorously. The tendency for these particles to break away from the ordered arrangement increases as a result, causing a decrease in the stiffness of the structure. Subsequently, this leads to the decrease in G', since it is an indicator of the stiffness exhibited by the columnar structure.

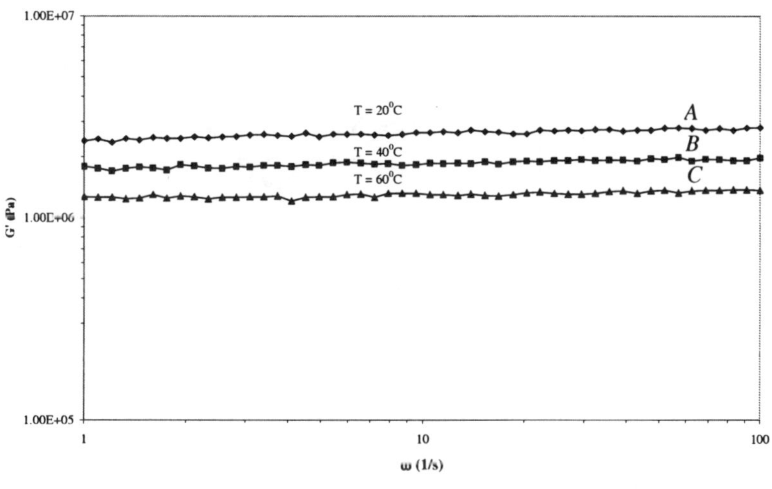

Figure 3. The plots of G' *vs.* ω at various temperatures
$\gamma_0 = 0.05\%$, $B = 270$mT, $T = 20^0$C, 40^0C, 60^0C

At a large strain amplitude γ_0 = 10%, the MR fluid behaves as nonlinear viscoelastic properties [7]. As shown in Figure 4, the frequency dependence of storage modulus at varying temperatures of 20^0C, 40^0C, and 60^0C are shown in Curves A, B, and C respectively. It can be observed that G' decreases with increasing temperature. This observation is similar to that made for the linear case, and the explanation provided for the linear case might be used to account for such behavior of the fluid. In addition, at a constant temperature of 20^0C as illustrated by Curve A in Figure 4, G' remains constant with increasing frequency until a critical frequency (ω_{cr}) is reached. Beyond the critical frequency, G' decreases as the frequency is further increased and reaches a minimum value at another critical frequency, ω_m. Further increase in frequency results in increasing G' values. The trend observed in G' may be explained by micro-structural variation of the MR fluid under different frequencies. At low frequencies of $\omega \le \omega_{cr}$, the microstructures of MR fluids have little variance because of very small shear rate, hence, G' remains constant at these small frequencies. Above the critical frequency, ω_{cr}, some of the columnar structures will break due to large shear rate, which accounts for the decrease in the elastic characteristic of the fluid. As a result, G' decreases with increasing frequency. At frequencies higher than ω_m, the microstructure of the MR fluid stop further changed and start reversing gradually, and so the interactions between the particles are not decreased considerably. Hence, G' increases steadily with frequency beyond ω_m.

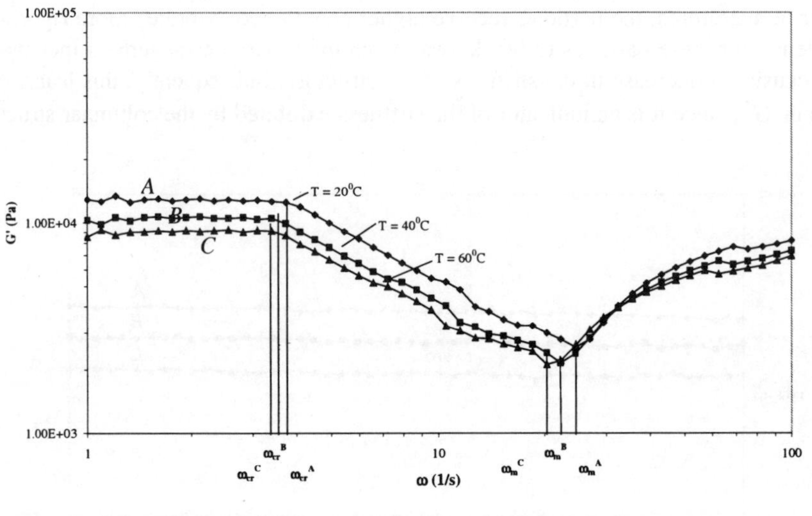

Figure 4. The plots of G' *vs.* ω at various temperatures
$\gamma_0 = 10\%$, $B = 270$mT, $T = 20^0$C, 40^0C, 60^0C

As shown in Figure 4, the respective critical frequencies at $T = 20^0$C, 40^0C, and 60^0C are indicated by ($\omega_{cr}^A, \omega_m^A$), ($\omega_{cr}^B, \omega_m^B$), and ($\omega_{cr}^C, \omega_m^C$), respectively. It can be seen that $\omega_{cr}^A > \omega_{cr}^B > \omega_{cr}^C$ and $\omega_m^A > \omega_m^B > \omega_m^C$. This result suggests that increasing temperature decrease the critical frequencies. This observation may be associated with the thermal energy possessed by the particles in the MR fluid at a given temperature.

At a given temperature, the suspended particles acquire an amount of thermal energy and tend to "vibrate and move randomly" about in the MR fluid. However, in the presence of a magnetic field, such tendency is overcome considerably and the suspended particles are "forced" to align into a columnar structure. With an increase in temperature, the thermal energy possessed by the particles increases and the tendency for them to break away from the ordered structure also increases. As a result, some of the chains formed may be broken. Therefore, the structure is expected to experience a drop in its stiffness, accompanied by the increase in temperature, and hence a reduction in G'.

4 Conclusion

Two different shear modes, *steady shear* and *oscillatory shear*, were used to experimentally study the effect of temperature. Both modes indicate that temperature effect is significant.

For the steady shear, the temperature dependence of MR rheology was experimentally investigated and analyzed with a Herschel-Bulkley model. Both the experimental and modeling results indicate that the dynamic yield stress decreases slightly with increasing temperature. MR fluids shows temperature-thinning effect.

For the oscillatory shear, the influences of temperature on viscoelastic properties of MR fluids were experimentally studied. The storage modulus shows a decreasing trend with temperature. In addition, a possible explanation of temperature effect based on thermodynamics was presented.

Acknowledgements

The study is supported by the Academic Research Fund, RG63/96, from Nanyang Technological University, Singapore. The authors wish to thank Mr. Terence Ong for helpful comments on this manuscript.

References

1. J.D. Carlson, In 7th *Int. Conf. On ERFs, MR Suspensions and Associate Technology*, ed. R.Tao (World Scientific Singapore, 2000), p. 621.
2. F. Gordaninejad, and S.P. Kelso, *J.Intell.Mater.Syst.Struct.*, **11**, 395 (2000).
3. S.A. Demchuk, *J.Mag.Mag.Mater.*, 312 (1993).
4. F. Gordaninejad, and D.G. Breese, *J.Intell.Mater.Syst.Struct.*, **10**, 634 (1999).
5. Perating Manual: *PHYSICA UDS 200* Paar Physica, Germany (1998).
6. X.J. Wang, and F. Gordaninejad, *J.Intell.Mater.Syst.Struct.*, **10**, 601 (1999).
7. K.D.Weiss, J.D.Carlson, and D.A.Nixon, *J.Intell.Mater.Syst.Struct.*, **5**, 772 (1994).

SIMULATION OF BIDISPERSE MAGNETORHEOLOGICAL FLUIDS

DAVID KITTIPOOMWONG AND DANIEL J. KLINGENBERG

Department of Chemical Engineering, University of Wisconsin, 1415 Engineering Drive, Madison, WI 53706

JOHN C. ULICNY

General Motors R&D Center, 30500 Mound Road, Warren, MI, 48090-9055

A method for simulating the steady-shear behavior of bidisperse, nonlinearly magnetizable MR suspensions is described. Results show that the yield stress of suspensions containing mixtures of large and small particles is larger than that of monodisperse suspensions, in agreement with previous experimental results.

1 Introduction

The objective of this study is to use particle-level simulations to predict the steady-shear rheological properties of bidisperse, nonlinearly magnetizable magnetorheological (MR) fluids. Particle motion is determined by the forces acting on the particles, which include magnetostatic, hydrodynamic, and short-range repulsive forces. The magnetostatic forces arise from magnetic dipole moments induced on the particles by the applied external field as well as the dipole fields created by the other particles. A self-consistent, mean-field calculation of these moments for arbitrary bidisperse suspensions is described in the following section. In Section 3, the simulation method is described. Results are discussed in Section 4. We find that the yield stress of suspensions containing mixtures of large and small particles is larger than that of monodisperse suspensions.

2 Self-Consistent Mean-Field Analysis of MR Fluids

Application of a magnetic field to a suspension induces moments on each sphere. Here we employ a self-consistent point-dipole approximation, where the moment on each sphere is proportional to the local field. This analysis follows that presented previously for chains of nonlinearly magnetizable spheres.[1]

Consider a collection of magnetizable spheres (bidisperse, diameters σ_I and σ_{II}, relative field-dependent permeability $\mu_p(H)$), immersed in a nonmagnetizable liquid, and subjected to a uniform magnetic field $\boldsymbol{H}_0 = H_0\boldsymbol{h}$ ($\boldsymbol{h}$ is a unit vector). The magnetic moment induced on sphere i, $\boldsymbol{m}_i$, is proportional to the *local field*, that is, the applied field plus the disturbance fields created by all the other spheres,

$$\boldsymbol{m}_i = \alpha_{m,i}(H_{\text{loc}})\boldsymbol{H}_{\text{loc}} = \alpha_{m,i}(H_{\text{loc}})\left(\boldsymbol{H}_0 - \sum_{i\neq j}\mathsf{T}(\boldsymbol{r}_{ij})\cdot\boldsymbol{m}_j\right), \tag{1}$$

where $\alpha_{m,i} = \pi\sigma_i^3\beta_i/2$ is the magnetizability of sphere i, $\beta_i = (\mu_p(H_{\text{loc}}) -$

$1)/(\mu_p(H_{\rm loc})+2)$ is the relative magnetizability, and

$$\mathsf{T}(\boldsymbol{r})=\frac{1}{4\pi}\left(\frac{\boldsymbol{\delta}}{r^3}-3\frac{\boldsymbol{rr}}{r^5}\right) \tag{2}$$

is the dipole interaction tensor, where $\boldsymbol{r}_{ij}\equiv\boldsymbol{r}_j-\boldsymbol{r}_i$ is the vector separation between the centers of spheres i and j.

In this analysis, we represent the field-dependent sphere permeability with the Frohlich-Kennelly equation,[1,2]

$$\mu_p(H_{\rm loc})=1+\frac{(\mu^0-1)M_s/[(1-\beta_i)H_{\rm loc}]}{(\mu^0-1)+M_s/[(1-\beta_i)H_{\rm loc}]}, \tag{3}$$

where μ^0 is permeability in the limit of vanishing field strength, M_s is the saturation magnetization, and $(1-\beta_i)H_{\rm loc}$ is the magnetic field strength within sphere i (for iron, $\mu^0\approx 10^3$ and $\mu_0 M_s\approx 2$ T, where $\mu_0=4\pi\times 10^{-7}$ N/A^2).

Resolving the above equations for all of the sphere moments is a formidable task, especially for nonlinearly magnetizable spheres. To simplify, we assume that each sphere of type i will have the same magnetic moment on average, and that the magnetic moment on each sphere is aligned with the applied field $\boldsymbol{H}_0$. These *mean-field* assumptions have been employed in a similar model for monodisperse electrorheological fluids.[3] Dotting Eq. 1 with $\boldsymbol{h}$, and ensemble averaging over all realizations yields the following equations for the moment magnitudes of the two different diameter particles,

$$m_I=\alpha_{m,I}H_0-\alpha_{m,I}T_h^{I-I}m_I-\alpha_{m,I}T_h^{I-II}m_{II}, \tag{4}$$

$$m_{II}=\alpha_{m,II}H_0-\alpha_{m,II}T_h^{II-I}m_I-\alpha_{m,II}T_h^{II-II}m_{II}, \tag{5}$$

where the notation

$$T_h^{A-B}\equiv\sum_{j\neq i}^{A-B}\langle\boldsymbol{h}\cdot\mathsf{T}\cdot\boldsymbol{h}\rangle \tag{6}$$

means the sum over all spheres j of type B, excluding sphere i at the origin and with the sphere at the origin being of type A, of the ensemble average of $\boldsymbol{h}\cdot\mathsf{T}(\boldsymbol{r}_{ij})\cdot\boldsymbol{h}$. Note that there are four distinct sums, depending on which type of sphere is at the origin, and which type of sphere is considered in the summation. The above two equations can be resolved for the moment magnitudes as

$$m_I=\alpha_{m,I}\lambda_I H_0,\qquad m_{II}=\alpha_{m,II}\lambda_{II}H_0, \tag{7}$$

where the local field enhancement factors (λ_I and λ_{II}) are defined

$$\lambda_I=\frac{1+\alpha_{m,II}(T_h^{II\text{-}II}-T_h^{I\text{-}II})}{1+\alpha_{m,I}T_h^{I\text{-}I}+\alpha_{m,II}T_h^{II\text{-}II}+\alpha_{m,I}\alpha_{m,II}(T_h^{I\text{-}I}T_h^{II\text{-}II}-T_h^{I\text{-}II}T_h^{II\text{-}I})}, \tag{8}$$

$$\lambda_{II}=\frac{1+\alpha_{m,I}(T_h^{I\text{-}I}-T_h^{II\text{-}I})}{1+\alpha_{m,I}T_h^{I\text{-}I}+\alpha_{m,II}T_h^{II\text{-}II}+\alpha_{m,I}\alpha_{m,II}(T_h^{I\text{-}I}T_h^{II\text{-}II}-T_h^{I\text{-}II}T_h^{II\text{-}I})}. \tag{9}$$

Unfortunately, the above summations are conditionally convergent. The conditional convergence can be avoided by employing the renormalization procedure proposed by Jeffery[4] and employed for ER fluids by Adriani and Gast.[3] Using this procedure to first order in particle volume fraction, the terms $T_h^{A\text{-}B}$ can be rewritten

$$T_h^{A\text{-}B} = -\frac{\rho_B}{3} - \frac{\rho_B}{4\pi}\int_{r>\frac{\sigma_A+\sigma_B}{2}}^{\infty} \frac{1}{r^3}(3\cos^2\theta - 1)(g^{A\text{-}B}(\mathbf{r}) - 1)d\mathbf{r} \qquad (10)$$

where $g^{A\text{-}B}(\boldsymbol{r})$ is the pair distribution function for spheres of type A and B, and ρ_B is number density of sphere type B. In the simulation, Eq. (10) is evaluated numerically over a sufficiently large volume such that the values converge. The integrals over $g^{A\text{-}B}(\boldsymbol{r})$ are converted to summations over particle pairs.

3 Dynamic Simulation of MR Fluids

An MR fluid is treated here as a neutrally-buoyant mixture of nonlinearly magnetizable spheres of two diameters (σ_I and σ_{II}) in a nonmagnetizable, Newtonian continuous phase. The motion of each particle is governed by Newton's equation of motion. Neglecting particle inertia, this results in a balance of forces on each sphere i, $\boldsymbol{F}_i^{\text{net}} = \boldsymbol{0}$, where $\boldsymbol{F}_i^{\text{net}}$ is the net force on sphere i. In these simulations, we consider three different types of forces acting on each particle: hydrodynamic, short-range repulsive, and magnetic forces. These forces are described below.

The magnetostatic force on sphere i caused by the presence of particle j in the uniform applied field $\boldsymbol{H}_0 = H_0\boldsymbol{e}_z$ is[1,6] $\boldsymbol{F}_{ij}^{\text{mag}} = \boldsymbol{m}_i \cdot \nabla \boldsymbol{B}_j(\boldsymbol{r}_i)$, where $\nabla \boldsymbol{B}_j(\boldsymbol{r}_i)$ is the gradient of the disturbance flux density created by sphere j, evaluated at the center of sphere i. Using the dipole moments obtained in the previous section (Eq. (7)), the force on sphere i caused by sphere j can be written

$$\boldsymbol{F}_{ij}^{\text{mag}} = F_0^{\text{mag}}(H_0)\left(\frac{r_{\text{min}}}{r_{ij}}\right)^4 \left[(3\cos^2\theta_{ij} - 1)\,\boldsymbol{e}_r + \sin 2\theta_{ij}\boldsymbol{e}_\theta\right], \qquad (11)$$

where $r_{\text{min}} = (\sigma_i + \sigma_j)/2$ is the minimum sphere separation, $r_{ij} = |\boldsymbol{r}_j - \boldsymbol{r}_i|$, $\boldsymbol{e}_r$ and $\boldsymbol{e}_\theta$ are the unit vectors at $\boldsymbol{r}_j$ of a spherical coordinate system centered at $\boldsymbol{r}_i$, and

$$F_0^{\text{mag}}(H_0) = \frac{3\pi}{16}\mu_0\beta_i\beta_j\lambda_i\lambda_j H_0^2 \left[16\frac{\sigma_i^3\sigma_j^3}{(\sigma_i + \sigma_j)^4}\right], \qquad (12)$$

is the magnitude of the magnetostatic force.

Particles also interact magnetically with the shearing surfaces. We assume these surfaces to act as constant magnetic potential surfaces. Thus the magnetic force exerted by a wall on sphere i is the sum of the forces exerted by the images of all other spheres j (including sphere i) on sphere i.

The hydrodynamic force on sphere i is taken as Stokes' drag,

$$\boldsymbol{F}_i^{\text{hyd}} = -3\pi\eta_c\sigma_i\left(\frac{d\boldsymbol{r}_i}{dt} - \boldsymbol{u}^\infty\right), \qquad (13)$$

where η_c is the viscosity of the continuous phase, $\boldsymbol{r}_i$ is the location of sphere i, and $\boldsymbol{u}^\infty$ is the ambient flow field. In these simulations, we consider either quiescent suspensions ($\boldsymbol{u}^\infty = \boldsymbol{0}$) or simple shear flow ($\boldsymbol{u}^\infty = (\dot\gamma z, 0, 0)$).

The repulsive force on sphere i due to particle j located at (r_{ij}, θ_{ij}) is modeled as $\boldsymbol{F}_{ij}^{\text{rep}} = -F_0\exp[-\kappa(r_{ij} - r_{\text{min}})/r_{\text{min}}]\boldsymbol{e}_r$ where $F_0 \equiv F_0^{\text{mag}}(H)$, and κ^{-1}

represents the range of the repulsive force. Particles near the shearing surfaces also experience short-range particle-wall repulsive forces, of the form $\boldsymbol{F}_i^{\text{wall}} = F_0 \exp[-\kappa(H_i - \sigma_i/2)/\sigma_i)]\boldsymbol{n}$, where $H_i = L_z/2 - |z_i|$ and $\boldsymbol{n} = [-z_i/|z_i|]\boldsymbol{e}_z$.

The governing equations are nondimensionalized by the following length, force and time scales: $L_s = \sigma_\ell$, $F_s = \pi\mu_0\sigma_\ell^2 M_s^2/48$, $t_s = 144\eta_c\sigma_\ell^2/\mu_0 M_s^2$, where σ_ℓ is the diameter of the largest sphere. The force scale is the asymptotic value of F_0^{mag} for the large spheres in the limit $H_0/M_s \to \infty$, where the particle magnetization saturates. The time scale is chosen to simplify the dimensionless equation of motion, which is written

$$\frac{d\boldsymbol{r}_i^*}{dt^*} = \frac{1}{\sigma_i^*}\left[\sum_{j\neq i}^{N} \boldsymbol{F}_{ij}^{\text{rep}*} + \boldsymbol{F}_i^{\text{wall}*} + \sum_{j\neq i}^{N'} \boldsymbol{F}_{ij}^{\text{mag}*}\right] + \boldsymbol{u}^{\infty*}, \tag{14}$$

where the asterisks denote dimensionless quantities, and the prime on the last term in the square brackets indicates that the sum extends over all spheres $j \neq i$, as well as the images of all the spheres.

Spheres in contact with the bounding surfaces ($L_z/2-|z_i| \leq 0.05\sigma_\ell$) are assumed to acquire the lateral velocity of the surface. These spheres obey slightly different equations of motion, described elsewhere.[7]

Simulations of a bidisperse sheared suspension is begun by randomly placing N_ℓ large spheres and N_s small spheres in the simulation cell of dimensions $L_x \times L_y \times L_z$. The volume fractions of the different type spheres are $\phi_\ell = \pi\rho_\ell\sigma_\ell^3/6$, $\phi_s = \pi\rho_s\sigma_s^3/6$, where $\rho_i = N_i/(L_x L_y L_z)$ is the number of spheres of type i per unit volume. The total volume fraction ϕ is the sum of the volume fractions of the components. Periodic boundaries are located at $x^* = \pm L_x^*/2$ and at $y^* = \pm L_y^*/2$. The shearing surfaces are located at $z^* = \pm L_z^*/2$. Equation (14) is integrated numerically using Euler's method with a time step Δt^* (which depends on the ratio of sphere diameters). Some runs were performed with different time step sizes to verify that the numerical integration is sufficiently accurate. Forces between particle pairs are evaluated within a cutoff radius of $r_c^* = L_z^*/2$.

Once the particle positions are known, properties of the suspension can be evaluated. Of particular interest in this study is the magnetic contribution to the shear stress, defined in dimensionless form as[7]

$$\tau_{xz}^* = -\frac{1}{V^*}\sum_i z_i^* F_{xi}^*, \tag{15}$$

where F_{xi}^* is the dimensionless x component of the nonhydrodynamic force on sphere i, z_i^* is the dimensionless z coordinate of the position of the ith sphere, and $V^* = L_x L_y L_z/\sigma_\ell^3$ is the volume of the simulation cell. This equation represents the shear stress of a particular configuration. The *dynamic yield stress*, τ_0^*, is calculated by averaging the shear stress above over many configurations at different shear rates, and then extrapolating these values to zero shear rate.[7]

4 Results and Discussion

The model described in the previous section was tested by comparing various predictions with known results. The simulations reproduced the values of the factors

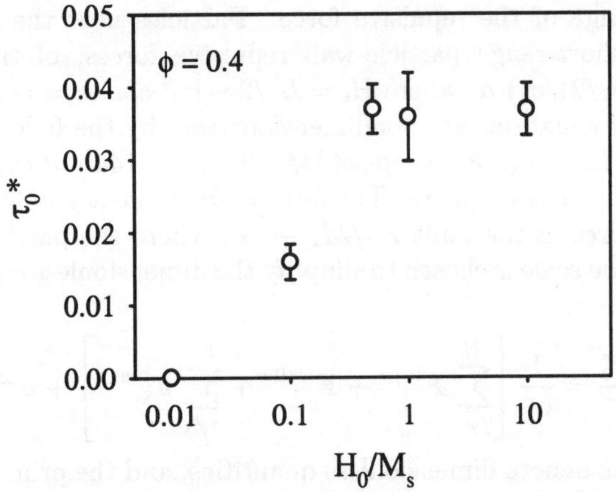

Figure 1. Yield stress as a function of magnetic field strength for a monodisperse suspension ($\phi = 0.4$).

λ_i reported previously for isolated chains of monodisperse spheres.[1] The simulations at large magnetic field strengths also agree with those reported previously,[7] as expected in the limit of complete saturation. Below we describe results for 3D simulations of nonlinearly magnetizable monodisperse and bidisperse MR fluids.

The dimensionless yield stress for monodisperse, 3D suspensions at $\phi = 0.4$ is plotted as a function of dimensionless magnetic field strength, H_0/M_s, in Fig. 1. The data appear to be consistent with the experimentally observed trend that the yield stress increases with applied magnetic field strength initially, and then saturates at large magnetic field strengths. The value of the yield stress at large field strengths, $\tau_0 \approx 7.7$ kPa may be compared to the experimental values reported by Rankin et al.[5] ($\tau_0 \approx 5.6$ kPa at $\phi = 0.1$). Assuming that the yield stress at saturation is linear in particle concentration, the simulated yield stresses are considerably smaller, but of the same order of magnitude as the experimentally measured values.

A snapshot of a bidisperse, 3D suspension ($\sigma_\ell/\sigma_s = 2$) after shearing to a strain $\gamma = 7$ ($\dot{\gamma}^* = 10^{-2}$) and relaxing is illustrated in Fig. 2. This simulation was performed with $\phi = 0.4, H_0/M_s = 1.0, L_x^* = L_y^* = L_z^* = 5$. For such concentrated suspensions in three dimensions, the fibrous columns are not as clear. However, the anisotropic nature is present, more clearly reflected by the voids that extend nearly from one solid surface to the other.

The dimensionless yield stresses of bidisperse suspensions are plotted as a function of sphere diameter ratio in Fig. 3 for $\phi = 0.4$, $\phi_\ell = \phi_s = 0.2$, and $H_0/M_s = 1$. The yield stress appears to increase with particle size ratio, which is supported further below.

The dimensionless yield stress for $\sigma_\ell/\sigma_s = 2$, $H_0/M_s = 1.0$, and $\phi = 0.4$ is plotted as a function of the fraction of large particles ($\phi_\ell/(\phi_\ell + \phi_s)$) in Fig. 4. The yield stress for mixtures of small and large particles appears to be larger than the

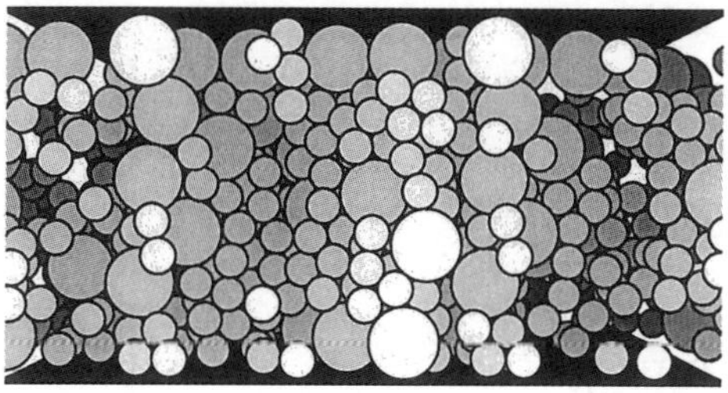

Figure 2. Structure of a bidisperse, 3D suspension ($\sigma_\ell/\sigma_s = 2$) after shearing to strain $\gamma = 7$ ($\dot{\gamma}^* = 10^{-2}$), and relaxing ($\phi = 0.4, H_0/M_s = 1.0$).

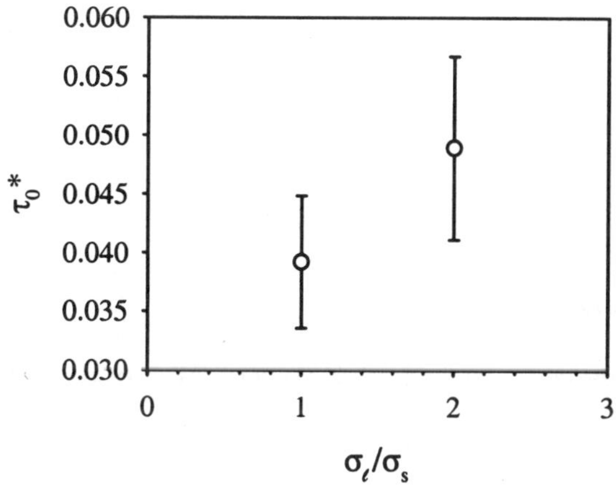

Figure 3. Dimensionless yield stress of bidisperse suspensions as a function of $\sigma_\ell/\sigma_s = 2$ ($\phi = 0.4$, $\phi_\ell = \phi_s = 0.2$, and $H_0/M_s = 1.0$).

value obtained for monodisperse suspensions. This is consistent with experimental results reported by Foister et al.[8]

5 Conclusions and Future Work

The model presented here for nonlinearly magnetizable, bidisperse MR fluids has been used to examine such features as the effects of sphere diameter ratio, magnetic field strength, and composition on the MR effect. We find that the field strength dependence of the yield stress is similar to that observed experimentally, and that the yield stress of suspensions of mixtures of small and large particles is larger than that of monodisperse suspensions.

The results presented here only examine a relatively small portion of parameter space. Future simulations will more completely examine the effects of such parameters as the sphere diameter ratio, σ_ℓ/σ_s, the total concentration, ϕ, the relative concentration of small and large spheres, ϕ_s/ϕ_ℓ, and the dimensionless applied

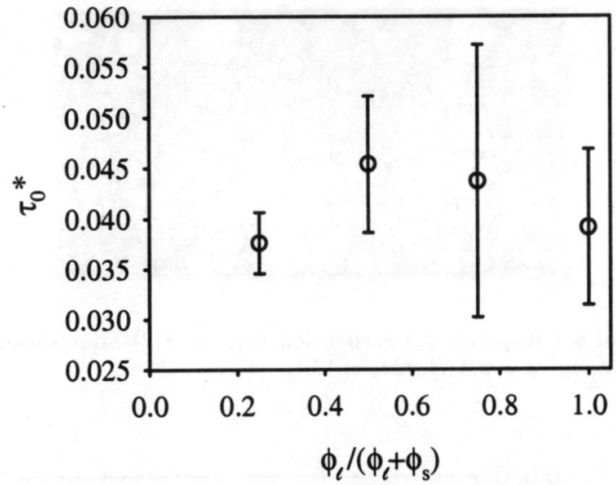

Figure 4. Dimensionless yield stress of bidisperse suspensions as a function of the fraction of large particles, $\phi_\ell/(\phi_\ell + \phi_s)$ ($\phi = 0.4$, $\sigma_\ell/\sigma_s = 2$, and $H_0/M_s = 1.0$).

magnetic field strength, H_0/M_s.

Acknowledgments

The authors acknowledge the support of General Motors Research.

References

1. Y. M. Shkel and D. J. Klingenberg, J. Rheol. **45**, 351 (2001).
2. R. M. Bozorth, *Ferromagnetism* (van Nostrand, New York, 1951).
3. P. M. Adriani and A. P. Gast, Phys. Fluids **31**, 2757 (1988).
4. D. J. Jeffrey, Proc. R. Soc. London, Ser. A **335**, 355 (1973).
5. P. J. Rankin, A. T. Horvath and D. J. Klingenberg, Rheol. Acta **38**, 471 (1999).
6. J. A. Stratton, *Electromagnetic Theory* (McGraw-Hill, New York, 1941).
7. D. J. Klingenberg, F. van Swol and C. F. Zukoski, J. Chem. Phys. **94**, 6160 (1991).
8. R. T. Foister, U.S. Patent No. 5,667,715 (1997).

A SINGLE-CHAIN MODEL FOR QUASI-STATIC SHEAR STRESS-STRAIN PROPERTIES OF MAGNETORHEOLOGICAL FLUIDS

ZEHUI JIANG
Department of Applied Physics, Harbin Institute of Technology, Harbin 150001, China

and

KUNQUAN LU
Institute of Physics, Chinese Academy of Sciences, Beijing 100080, China

Abstract

The quasi-static shear stress-strain property of magnetorheological fluids is analyzed, based on the calculation of the interaction energy of chains of equi-spaced magnetic spheres with the applied uniform magnetic field. First, the dipole moment of a chain of equi-spaced magnetic spheres imposed to a uniform magnetic field is determined, and simple expressions are given. Then, the interaction energy of a chain with the forcing field is determined, and the restoring force is calculated when the chain is slanted and elongated. In our analysis, only the contributions from the mutual interaction of the particles in chains are considered. Comparison with experimental results is made, and good agreement is obtained.

1. Introduction

Magnetorheological (MR) fluids typically consist of fine magnetizable particles suspended in nonmagnetic host fluids. Under action of the applied external magnetic field, the particles are magnetized and form chain-like structures with their axis parallel to the direction of the forcing field. When a shear field is applied perpendicular to the chains, they are slanted and elongated, and a restoring stress is aroused, which comes from the attractive magnetization interaction of the particles. To determine the shear stress-strain relation often is difficult because of the difficulty in handling the multipole interaction of the particles, which contributes significantly when the particles are in contact or approach touching. Consideration of the saturation or nonlinearity of the magnetization of the particles also makes a theoretical analysis difficult.

In practice, a MR fluid is commonly confined between two plates. It has been recognized that the nature (ferromagnetic or paramagnetic) of the plates also has a large effect on the yield stress [1]. For the ferromagnetic plates, the dipole moment of a particle will induces an image moment in the plate with almost the same strength because the permeability of the plate is quite larger compared to unity. In this case, a particle chain stretching the plate gap may be seen as an infinite one in view of the images induced in the plates.[2, 3]. For the paramagnetic plates, a particle chain has to be treated as a finite one because the amplitude of the image dipoles induced in the plate is neglectable. Here, we focus on the latter case and present a simple model for analysis the quasi-static shear stress-strain property of MR fluids. In this model, an attempt is made to check the effect of the size of the chains on the yield stress. Our results show that only when the relative magnetic permeability of the particles is high, the effect of chain length becomes obvious, and long chain may provide a yield stress larger by one order of magnitude than the chains containing two spheres.

2. Theory

First, we consider the magnetization of a linear chain of equi-spaced spheres subjected to a uniform magnetic field. The effect of the forcing field is to excite a dipole moment in each sphere. Due to the mutual induction of the spheres, the dipole moment is dependent on the configuration of the chain and the permeability of the particles [4, 5]. When the magnetization is in the linear or low-field regime, the determination of the magnetic dipole moment is analogous to that for its electrostatic counterpart. Here we adopt the theoretical model presented in references 6 and 7 for dielectric particle chains, and directly write down the expressions for the total longitudinal magnetic dipole moment $m_N^L\left(k_\mu, x\right)$ and the total transverse moment $m_N^L\left(k_\mu, x\right)$ of an N-particle chain,

$$m_N^L\left(k_\mu, x\right) = N^{\ln\alpha^L(k_\mu, x)/\ln 2} m_0 , \tag{1}$$

$$m_N^T\left(k_\mu, x\right) = N^{\ln\alpha^T(k_\mu, x)/\ln 2} m_0 , \tag{2}$$

$$m_0 = 4\pi\mu_f \frac{2k_\mu}{3-k_\mu} R^3 H_0 = 4\pi\mu_f \beta_\mu R^3 H_0 , \tag{3}$$

where m_0 is the dipole moment of single sphere with radium R in a uniform magnetic field H_0, the permeability of the sphere and the host fluid are, respectively, μ_p and μ_f, $k_\mu = \dfrac{\mu_p - \mu_f}{\mu_p + \mu_f}$, $\beta_\mu = \dfrac{\mu_p - \mu_f}{\mu_p + 2\mu_f}$. $\alpha^L\left(k_\mu, x\right)$ and $\alpha^T\left(k_\mu, x\right)$ are given by the following equations,

$$\alpha^L\left(k_\mu, x\right) = 2\left\{\sum_{n=0}^{\infty} k_\mu^n \left[\frac{\sinh^3\eta}{\sinh^3(n+1)\eta} + \frac{\sinh\eta\sinh^2(n\eta)}{\sinh^3(n+1)\eta}\right] - \left[\sum_{n=0}^{\infty} k_\mu^n \frac{\sinh\eta\sinh(n\eta)}{\sinh(n+1)\eta}\right]^2 \Bigg/ \sum_{n=0}^{\infty} k_\mu^n \frac{\sinh\eta}{\sinh(n+1)\eta}\right\}, \tag{4}$$

$$\alpha^T\left(k_\mu, x\right) = 2\sum_{n=0}^{\infty}\left(-k_\mu\right)^n \left[\frac{\sinh\eta}{\sinh(n+1)\eta}\right]^3, \tag{5}$$

where, r is the center-to-center separation of the spheres, and $x = r/2R$ is the reduced separation. η is defined by,

$$x = \cosh\eta = r/2R . \tag{6}$$

Generally, the interaction energy of the induced dipole moment of a particle chain with the applied magnetic field may be written as,

$$U_0 = -\frac{1}{2}\overline{m}\bullet\overline{H}_0$$

$$= -2\pi\mu_f\beta_\mu R^3 H_0^2 \left(N^{\ln\alpha^L(k_\mu,x)/\ln 2}\cos^2\theta + N^{\ln\alpha^T(k_\mu,x)/\ln 2}\sin^2\theta \right), \tag{7}$$

where θ is the angle between the chain axis and the direction of magnetic field H_0. Now we consider a single chain of N touching spheres aligned on the direction of an external field H_0, refer to Figure 1. The symbol h denotes the distance of the plate gap, $N = h/2R$. When an external stress along direction y is applied on this chain by moving the upper plate, the chain is deformed. Here we assume that the shear rate of the upper plate is quite small and neglectable. The relation between the shear angle θ and shear strain γ is,

$$\gamma = \tan\theta = y/h, \tag{8}$$

where y is the distance the upper plate moved. The reduced separation of the spheres in the deformed chain may be expressed as,

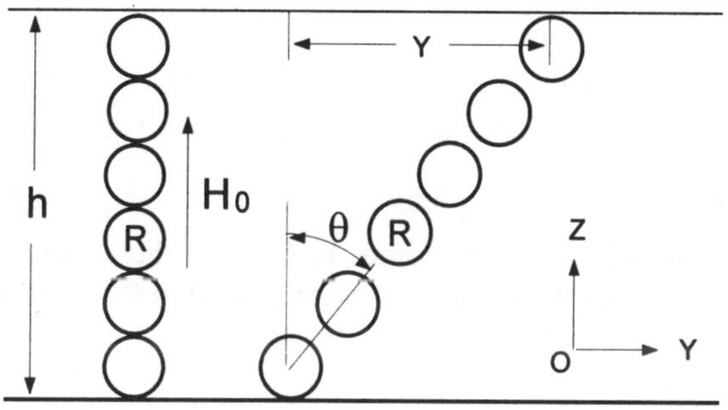

Figure 1. The deformation of a chain of initially touching spheres under shear.

$$x = \cosh\eta = 1/\cos\theta = \sqrt{1+\gamma^2}. \tag{9}$$

When the chain is strained by a quantity γ, the energy will generally decrease owing to the nonalignment of chained particles with the forcing magnetic field. This gives rise to a restoring force,

$$F_s = \frac{1}{h}\frac{\partial U_0}{\partial \gamma}$$

$$= -\pi\mu_f\beta_\mu R^2 H_0^2 \frac{\partial}{\partial\gamma}\left\{ N^{\ln\alpha^L(k_\mu,x)/\ln 2}\frac{1}{1+\gamma^2} + N^{\ln\alpha^T(k_\mu,x)/\ln 2}\frac{\gamma^2}{1+\gamma^2} \right\}. \tag{10}$$

Here we model a MR fluid consisting of single chains. If the volume fraction of the particles is not very high, each chain is far apart and the chain-chain interaction is weak and neglectable. The total energy of the fluids is the sum of that of each single chain,

$$U = N'U_0 = \frac{3\varphi Ah}{N4\pi R^3}U_0, \tag{11}$$

where $N' = \dfrac{3\varphi Ah}{N4\pi R^3}$ is the total number of the chains, and A is the area of the plate.

The total shear stress is then,

$$\tau = \frac{\partial U}{Ah\partial\gamma}$$

$$= -1.5\varphi\mu_f\beta_\mu H_0^2 \frac{\partial}{\partial\gamma}\left\{N^{\ln\alpha^L(k_\mu,x)/\ln 2-1}\cos^2\theta + N^{\ln\alpha^T(k_\mu,x)/\ln 2-1}\sin^2\theta\right\}$$

$$= -1.5\varphi\mu_f\beta_\mu H_0^2 \frac{\partial}{\partial\gamma}\left\{N^{\ln\alpha^L(k_\mu,x)/\ln 2-1}\frac{1}{1+\gamma^2} + N^{\ln\alpha^T(k_\mu,x)/\ln 2-1}\frac{\gamma^2}{1+\gamma^2}\right\}. \quad (12)$$

3. Discussion

To check equation (10), a comparison is made with previous experimental results of Lemaire *et al* [3]. In their experiments the restoring force for the chains containing 8 steel spheres is measured. The radius of spheres is 1 mm, and the magnetic permeability of them μ_p is determined by measuring the interaction force between two particles. For a field of 310 Oe, they got $\mu_p = 38\mu_0$, here correspondingly $k_\mu = 0.95$. Our calculation value for the yield stress is about two times larger than the experiment result, refer to Figure 2. The reason for this deviation may be attributed to the nonlinearity of magnetic material. [3, 5, 8, 9] Experiments [3] have shown that the magnetic permeability is sensitive to the field strength. Since in the touching areas of the particles the field is higher, local magnetization saturation may be induced and so the permeability is less than would be in linear medium. When we take the symbol $k_\mu = 0.88$, the calculation result is improved. On the other hand, this difference also may be attributed to the rupture mechanism. In experimental situation the distance of each couple of spheres is seldom separated equally.

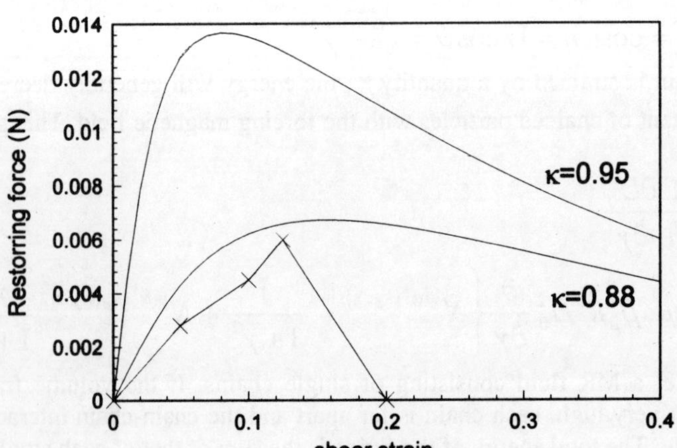

Figure 2. A comparison of our model (equation (10)) with previous experimental result. × — experimental data read from Fig.4 in reference 3.

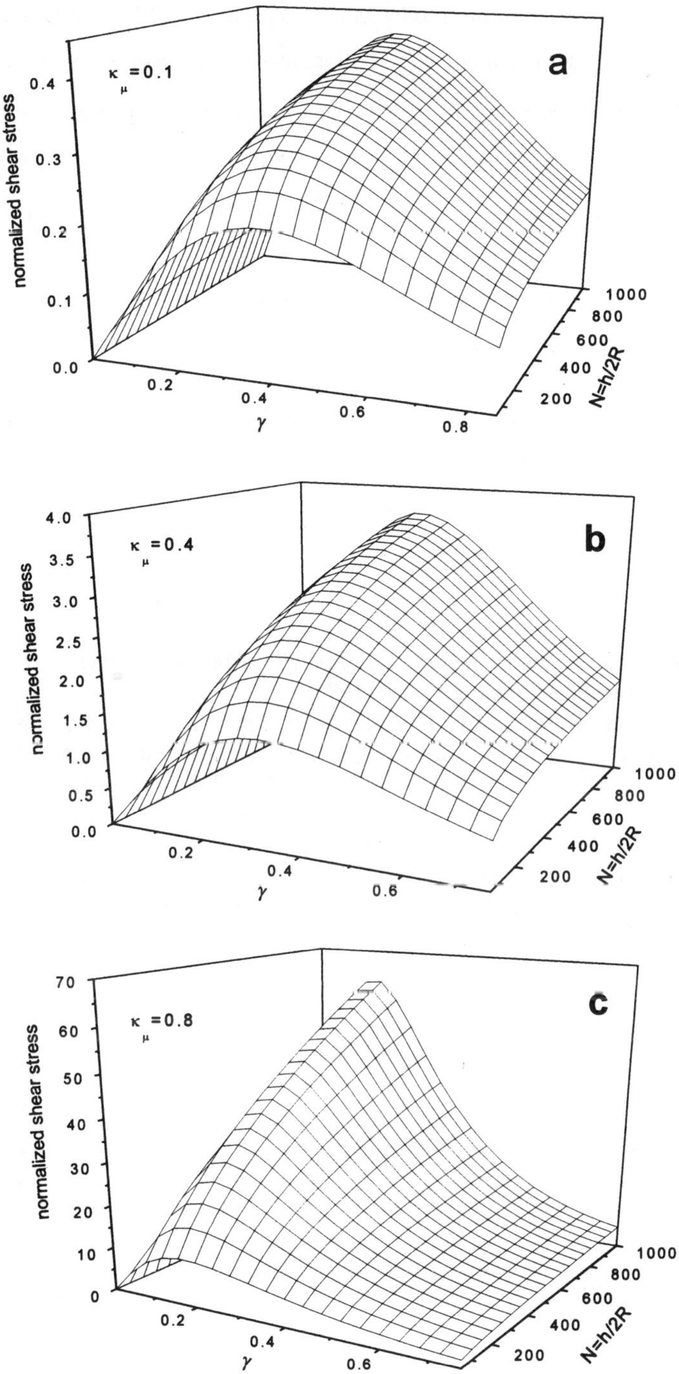

Figure 3. Normalized shear stress vs. γ and N for several values of symbol κ_μ.

In figure 3 the shear stress for our model MR fluid (equation (12)) as a function of γ and N for several values of symbol k_μ is illustrated. The shear stress is normalized to $1.5\varphi\mu_f\beta_\mu H_0^2$. It is seen clearly that the yield shear stress increases with the scaled magnetic permeability k_μ, and also increases with the chain length N when k_μ is kept at a constant value. Only when k_μ takes higher value, the effect of chain length becomes obvious, and long chain may provide a yield stress larger by one order of magnitude than the chains containing two spheres. We also note that the critical deformation γ_c depends weakly on the values of k_μ and decreases from about 0.3 to 0.1 when k_μ approaches 0.8 from 0.1.

In summery, the quasi-static shear stress-strain property of a model MR fluid confined between two nonmagnetic plates is studied theoretically, and the dependence of the stress on the magnetic permeability and chain length is discussed. The correctness of the obtained results should be further checked by comparison with other theories and experimental results.

4. References

1. E. Lemaire and G. Bossis, *J. Phys. D: Appl. Phys.* **24,** 1473 (1991).
2. R. E. Rosensweig, *J. Rheol.* **39**, 179 (1995).
3. E. Lemaire, G. Bossis, and O. Volkova, *in Proc. 5th Int. Conf. ERF, MRS, and Associated Tech.*, ed. W. A. Bullough (World Scientific, Singapore, 1996), p. 368.
4. E. J. Yarmechuk and J. F. Janak, *IEEE Trans.* **18**, 1268 (1982).
5. T. B. Jones, *Electromechanics of Particles*, (Cambridge Univ. Press, 1995).
6. Z. Jiang, X. He, J. Han, and S. Du, *Ann. Phys.* **266**, 244 (1998).
7. Z. Jiang, *et al.*, *J. Electrostatics*, **39**, 297 (1997).
8. T. B. Jones and B. Saha, *J. Appl. Phys.* **68**, 404 (1990).
9. C. Tan and T. B. Jones, *J. Appl. Phys.* **73**, 3593 (1993).

SURFACE SHEAR STRESS ENHANCEMENT UNDER MR FLUID DEFORMATION

S. GORODKIN[A),B)], N. ZHURAVSKI[A)]

[A)]*Heat and Mass Transfer Institute, 15 Brovka Street, Minsk 220072, Belarus*
E-mail: srgor@hmti.ac.by

[B)]*University of Rochester, Center for Opics Manufacturing, 240 East River Road, Rochester NY 14623, USA*
E-mail: sgor@lle.rochester.edu

W.KORDONSKI

QED Technologies Inc, 1040 University Ave., Rochester NY 14607, USA
E-mail: kordonski@qedmrf.com

The effect of a regular surface micro topography on a magnitude of surface shear stress in a normal magnetic field under MR fluid deformation between two magnetic plates has been experimentally investigated. It was revealed that the static yield stress was increased by a factor of 2.8 for radially grooved plates as against smooth ones. The effect was most conspicuous at small iron particles volume fractions and strong magnetic fields. It is assumed that grooves create a magneto mechanical barrier that prevents the slip of particle aggregates against the wall.

1 Introduction

The operation of hydraulic MR devices may be considered as a process based on magnetic field control of surface shear stress, τ_s, of an MR fluid. In the frame of commonly accepted Bingham model an applied magnetic field controls yield stress (and surface shear stress as a measured net value of this one) by generating a microstructure from ferromagnetic dispersed particles. A structure at rest is field-elongated chain of particles agglomerates. The structure is deformed under shear and MR fluid flow is accompanied either by the rupture of chains, their slip over a wall as a unit, or by a combination of these two processes. It is supposed that the stress is transmitted to the wall from the MR fluid by the particles contacting with wall. One of the ways to increase τ_s consists in enhancing the strength of a structural skeleton, e.g., by elevating the magnetic particles volume fractions[1] or by using the powders with their improved magnetic properties.[2] Similar methods are being employed with advantage but they have essential restrictions. It is a density and viscosity growth in the first case and a substantial rise in price in the second. According to the data of Tao et al,[3] a multiple increase surface shear stress is seen if the MR fluid is compressed in the field direction immediately after a magnetic field has been applied. However, this method much complicates the device design.

On the other hand, the efficiency of MR control depends on a degree of momentum transfer from the fluid to the channel wall and then to actuating elements of the device. At this stage the slipping over the wall may play a substantial role. It has been found[4-6] that using a magnetic wall with a rough surface can reduce the slip. By means of model experiments with small iron balls Tang et al.[7] demonstrated that the interacting force between adjacent particles in a chain is more substantial as against the one between the extreme particles and the wall. Therefore, the rupture of chains requires higher level of stresses than flow with the slip. It is considered[4,5,8] that the use of a magnetic wall enhances an attractive force of the structure against the surface, and the roughness augments the friction coefficient. As a result, the slip is decreased and τ_s is increased. For example, Tang and Conrad[6] found experimentally that a transition from a smooth surface (with arithmetical mean deviation of the profile above and below the reference line R_a=3.3 μm) to a rough sand paper treated one (R_a=22.9 μm) results in a 2-fold growth of τ_s under other conditions being equal.

In the present study an attempt was made to develop a method of enhancing the magnetic interaction between the structure and the wall, particularly to evaluate the effect of a regular surface micro topography on the effective surface shear stress.

2 Experimental

The plate-plate magnetorheometer described in detail before[9] was used for measurements of static yield. The schematic of the test instrument is shown in Fig.1.

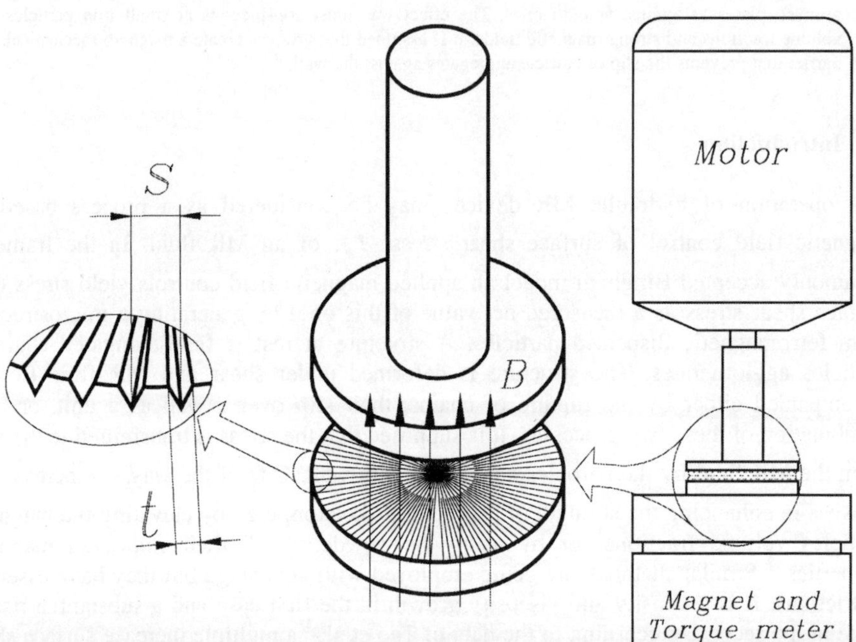

Figure 1. Schematic of measuring instrument

Removable circular plates with radius R of 10mm were made from low carbon steel. The upper rotating plate was attached to the motor shaft while the lower one was coupled to the torque sensor. The working surfaces of each pair of plates were either "smooth" (R_a~40μm) or regularly cut to form the grooves with different geometry and orientation. The plates to be tested are specified in the Table 1. Here, grooves depth b, the width t, and the spacing s (or s_R for radial grooves) are given in mm.

Table 1. Plates characterization.

Plates #	Grooves orientation	Cross section	b	t	$s(s_R)$	δ_R	δ_m
1	Smooth	-	-	-	-	0	0
2	Circumferential	Triangular	0.1	0.2	0.4	0.5	0.5
3	Radial	Triangular	0.05	0.1	0.5	0.05	0.1
4	Radial	Triangular	0.05	0.1	0.8	0.125	0.25
5	Radial	Triangular	0.1	0.2	0.8	0.25	0.5
6	Radial	Triangular	0.1	0.2	0.52	0.385	0.77
7	Radial	Triangular	0.2	0.4	0.89	0.45	0.9
8	Radial	Triangular	0.2	0.4	0.8	0.5	1
9	Radial	Rectangular	0.1	0.2	0.8	0.25	0.5

The relative spacing $\delta = t/s$ characterizes the scale of topography features. Since the spacing between radial grooves changes along the radius, the average value $\delta_m = \frac{1}{\pi R^2}\int_0^R \delta_r \cdot 2\pi r dr = 2\delta_R$, was used, where $\delta_R = t/s_R$ is the relative spacing at the edge of the plate (for circumferential grooves $\delta = \delta_R = \delta_m$ and $s = s_R$).

After loading the sample between the plates, the motor current was linearly increased up to the moment when the shaft with the upper plate began to move. The corresponding torque M_0 was recorded and yield stress was calculated as $\tau_s = 1.5M_0/\pi R^3$. The magnetic induction B was oriented as shown in the sketch and measured in the air gap by a Teslameter F4354/1. MR fluid samples were prepared manually by stirring EW carbonyl iron particles (BASF, Germany) with the main size of ~ 4 μm into 100 cP (at 25 ^{0}C) silicon oil. The following CI volume fraction range was used in experiments: $\phi = 0.05 - 0.4$. The gap between the plates was 0.45 mm, the induction was varied from residual one 0.125 T up to 0.65 T.

3 Results and Discussion

Figure 2 shows the surface shear stress for a fluid with lowest iron particle content vs the magnetic field induction for smooth plates (#1) and two pairs of the roughed plates (#2 and 3) having the radial triangular grooves with δ_m 0.5 and 0.25. The data taken from Tang and Conrad[6] measurements for similar concentration are also plotted on the graph. There is a good agreement between two "smooth cases", taking into account that the roughness of smooth walls in our experiment is more pronounced. Grooved plates significantly enhance stress. The effect is increased with field induction and depends on a grooves relative spacing.

The influence of topography scale on τ_s is displayed in Fig. 3. Seven plates (#1, 3-8) with radial triangular grooves and different grooves relative spacing were tested, and the normalized surface stress, i.e. the ratio τ_s/τ_{s1} (τ_{s1} is the stress obtained for smooth plates #1) was plotted as a function of δ_m for B=0.38 T and for three concentrations ϕ =0.05, 0.1 and 0.2. The curves have symmetrical extreme-type shape with sharp maximum at $\delta_m = 0.5$ and there is no noticeable effect when $\delta_m > 1$. The optimum groove spacing value has been selected for further measurements to find out how the geometry and orientation of grooves affect on the surface stress with variation of fluid concentration and field induction.

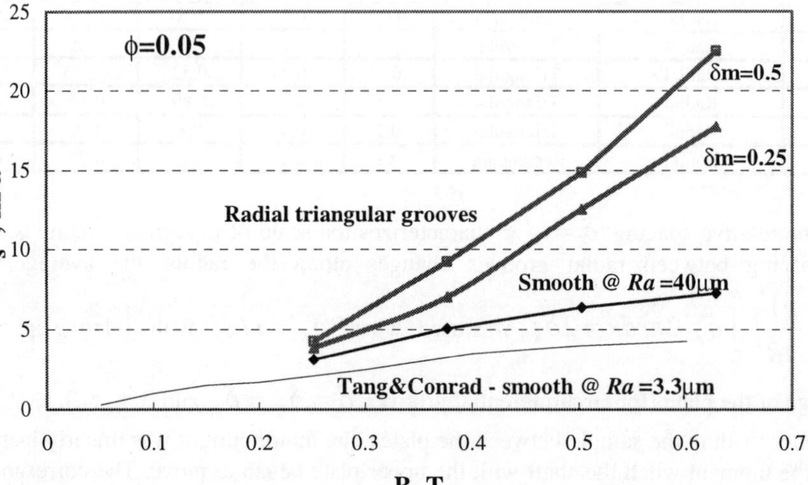

Figure 2. Surface shear stress vs the magnetic induction for smooth plates and grooved plates with two different relative spacing of radial triangular grooves. Tang and Conrad[6] data have plotted on the basis of their Fig.6

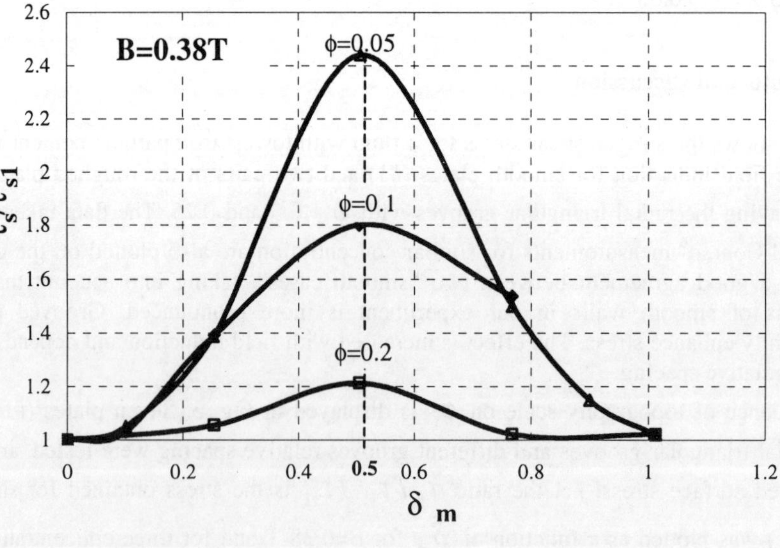

Figure 3. Normalized surface shear stress as a function of relative spacing for radial triangular grooves

To do this, plates # 2 with circumferential rectangular grooves and plates # 9 with radial rectangular grooves were additionally employed. Figure 4 demonstrates the influence of all listed above parameters on the ratio τ_s / τ_{s1}. Circumferential orientation of grooves only slightly changes the stress and there is no essential difference between triangular and rectangular shape. The ability of radial oriented grooves to enhance surface stress is monotonically increased with field induction and, on the contrary, it is dropped down with concentration. Abrupt decay of τ_s / τ_{s1} occurs when ϕ increases from 0.05 to 0.2 and then the effect of surface topography on the stress is significantly diminished at larger concentration.

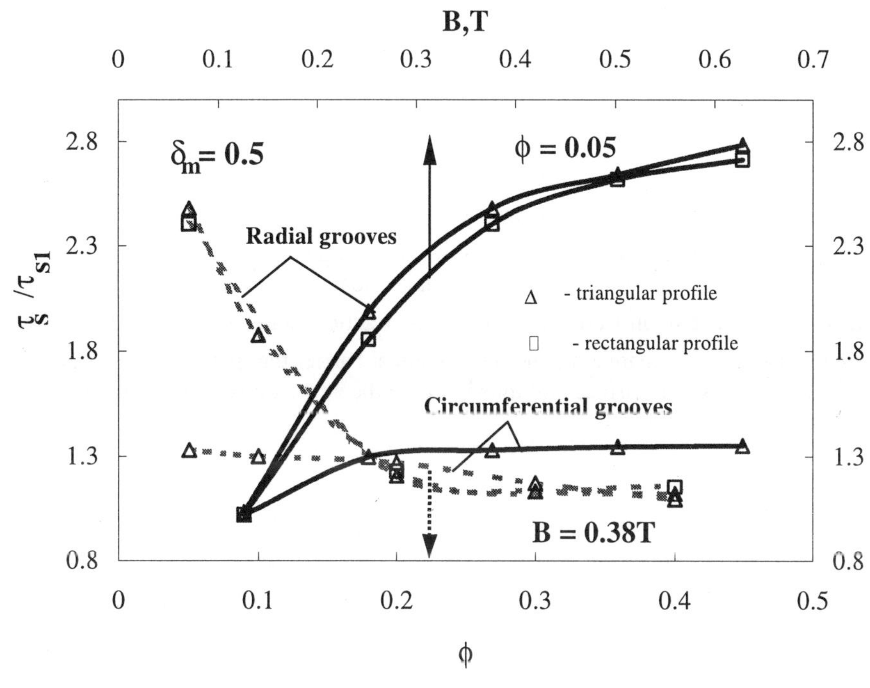

Figure 4. Normalized surface shear stress vs iron particles volume rate ϕ (at B=0.38 T, dotted lines) and field induction B (at ϕ=0.05, solid lines). Circumferential grooves have a triangular profile, radial ones have both triangular and rectangular profile.

The hypothesis to describe qualitatively the established facts is the following. According to current standpoint[4-8] when MR fluid is deformed the level of surface shear stress transmitted to the wall from the iron particles structure is defined not only by the structure toughness but also depends on the interaction of structure extreme particles with the wall. Both of those factors were changed in the experiments. The structure toughness was increased with particles concentration and field induction, while the structure/surface interaction was varied with surface topography. As one might expect, the normalized shear stress diminishes with concentration approaching the level of "smooth" surface. At the same time, the observed effect of surface topography sets the stage for contemplation. Most pronounced is effect of grooves orientation. When the regular surface topography is oriented across the flow, a mechanical barrier that inhibits the slip is created resulting in significant enhancement of the surface shear stress. With circumferential orientation, extreme particles are moved along the grooves and don't meet substantial resistance. Another factor to be considered in the slip problem is local magnetic field non-homogeneity produced by wall surface irregularities. Each groove contains sharp edges that produce an appreciable local magnetic field gradient, which acts as a magnetic "trap" that seizes particles and suppresses slipping phenomena as well. Thus, an effective complex magneto-mechanical barrier is formed resulting in wall slip reduction and so, in surface shear stress enhancement.

This speculation is supported by the fact that the grooves relative spacing δ_m, which defines the local magnetic field distribution, has a significant effect on surface shear stress enhancement, whereas influence of groove shape is not so essential.

4 Conclusion

A regular wall micro topography in the form of grooves with sharp edges substantially enhances surface shear stress under MR fluid deformation in magnetic field, which is normal to shear. The measured normalized yield stress τ_s / τ_{s1} (τ_{s1} is the stress obtained for "smooth" wall) has run to its 2.8 maximum value for the range of parameters has been used. The effect was most conspicuous for radial oriented grooves with a mean relative spacing $\delta_m = 0.5$, at small iron particles volume fractions and strong magnetic fields. It is assumed that grooves create a magneto mechanical barrier that prevents the slip of fluid microstructure against the surface and thus increase the stress transmitted to the wall from the MR fluid.

References

1. Kordonski et al., in *Proc. of the 6th Int. Conf. on ER Fluids, MR Suspensions and Their Applications*, ed. M.nakano and K.Koyama (World Scientific, Singapore, 1998), p. 535.
2. Margida et al, *Int. J. Mod. Phys.* **B 10**, 3335 (1996).
3. Tao et al., *J. Appl. Phys.* **87**, 2634 (2000).
4. Lemaire and Bossis, *J. Phys. D: Appl. Phys.* **24**, 1473 (1991).
5. Shorey et al., *Rev. Sci. Instrum.* **70**, 4200 (1999).
6. Tang and Conrad, *J. Rheol* **40**, 1167 (1996).
7. Tang et al., in *Proc. of the 5th Int. Conf. on ER Fluids, MR Suspensions and Their Applications*, , ed. W. Bullough (World Scientific, Singapore, 1996), p. 603.
8. Miyamoto and Ota, *Appl. Phys. Lett.* **64**, 1165 (1994).
9. Kordonski et al., in in *Proc. of the 7th Int. Conf. on ER Fluids, MR Suspensions and Their Applications*, ed. R.Tao (World Scientific, Singapore, 2000), p. 611.

COMPARATIVE INVESTIGATIONS ON ER FLUIDS WITH DIFFERENT POLARIZATION MECHANISMS

H. BÖSE AND A. TRENDLER

Fraunhofer-Institut für Silicatforschung, Neunerplatz 2, 97082 Würzburg, Germany
E-mail: boese@isc.fhg.de

The rheological and electric properties of various ER fluids in strong electric fields have been investigated simultaneously. From the measured data the dependences of shear stress and real and imaginary part of the high field permittivity on field strength, temperature and frequency have been derived. One group of ER fluids containing perovskite type particles obeys a polarization mechanism which is attributed to the local displacement of ions in the crystal structure of the particles. The contribution of conduction to ER activity is low and the electric properties of these ER fluids are nearly constant. The increase of shear stress with frequency is in accordance with the polarization model. In contrast, another group of ER fluids containing zeolite or ion conducting polymer particles shows a strong dependence of the complex permittivity on field strength and temperature demonstrating the pronounced nonlinearity of the electric properties. It is assumed that both conduction and polarization contribute to the ER activity in these systems. The relative extent of the contributions changes with external conditions like field strength and temperature.

1 Introduction

The electrorheological effect, i.e. the strong increase of the shear stress of suspensions in an electric field in the order of some kV/mm is generally attributed to the polarization of charge carriers in the particles and corresponding dipolar interactions which lead to structure formation between the electrodes. In the polarization model the permittivity of the particles has to be higher than that of the carrier liquid in order to generate attractive forces between the particles [1]. Recently a conduction model has been proposed which postulates a corresponding mismatch between the conductivities of the particles and the carrier liquid [2]. ER fluids which are dominated by their permittivities show an increase of shear stress with rising frequency of the electric field but for conductivity dominated ER fluids the frequency dependence is reverse [3]. This means that the first type has stronger ER activity at high frequencies whereas the latter works better at DC or AC with low frequency.

Different mechanisms of charge transport in the particles depending on their material properties with consequences on the electric properties of the corresponding ER fluids have been identified [4]. In perovskite particles like barium titanate a local polarization of ions in the electric field occurs. In contrast to this in zeolite particles the loosely bound cations can move through the crystal lattice. It has been shown that the current density of ER fluids based on zeolites strongly increases with temperature which is not observed for perovskite based ER fluids [4].

The purpose of this paper is a comparison of ER fluids based on particles with different charge transport mechanisms. A simultaneous analysis of their rheological and electric properties in strong electric fields under shear has been conducted in order to detect correlations between them. On the one hand various perovskite type particles with locally polarizable ions have been chosen for the investigations [4]. On the other hand two ER fluids with different types of zeolites [4] and one fluid with ion conducting polymer particles [5] have been used for comparison.

2 Experimental

The perovskite type powders barium titanate, strontium titanate and calcium titanate were dried at 140 °C before use. The powder (25 % by volume) was mixed with silicone oil (viscosity 10 mPas at 20 °C) using a high speed stirrer and dispersed by ultrasound. The ER fluids are designated $BaTiO_3$, $SrTiO_3$ and $CaTiO_3$. Correspondingly, zeolite particles with the approximated compositions $K_4Na_8Si_{12}Al_{12}O_{48}$ (KNaA) and $Na_{60}Si_{180}Al_{60}O_{480}$ (NaY) were dried at 250 °C in vacuum and mixed with silicone oil at the same concentration. Additionally a polymer particle ER fluid Rheobay 3565 (Bayer AG) containing ion conducting polyurethane (PU) particles in silicone oil was used for comparison.

The rheological measurements were performed with a rheometer Physica MC 100 with bob and cup geometry at a shear rate of 1000 s^{-1}. For the simultaneous electric investigations on the same sample the rheometer is equipped with a high voltage generator consisting of a function generator, a high voltage amplifier and an oscilloscope (see Fig. 1a) which yields the voltage and current signals (sine waveform) as well as the phase shift between them. The following measuring programs were performed:

- Variation of field strength at a fixed frequency of 100 Hz and various temperatures
- Variation of frequency at fixed field strength of 2 kV/mm and various temperatures

The measured electric properties are influenced by the measurement system (cables, connections, etc.). In order to calculate the sample properties corrections on the measured data have been performed using the results of measurements on the empty system without ER fluid. For the corrections the simple electric model of a curcuit with parallel capacitors and resistors for the measurement system and the sample shown in Fig. 1b has been applied. With the corrected figures of current and phase shift the dielectric properties of the sample (real and imaginary part of the permittivity, ε' and ε'', respectively) at high field strengths have been derived.

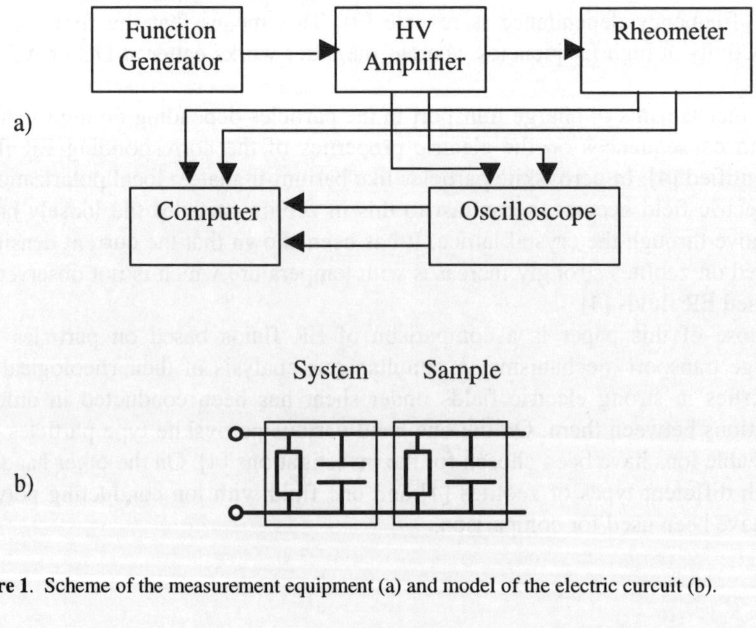

Figure 1. Scheme of the measurement equipment (a) and model of the electric curcuit (b).

3 Results and Discussion

3.1 Field strength depencence

The dependence of the shear stress on the field strength at 50 °C for the six ER fluids investigated is shown in Fig. 2a. At low field strengths of about 1 kV/mm the shear stress increase is stronger than linear for all ER fluids. The slower increase at high field strengths can be explained by the field dependent conductivity of the carrier liquid silicone oil between the particles rising in strong electric fields when the ER effect is dominated by the conductivities.

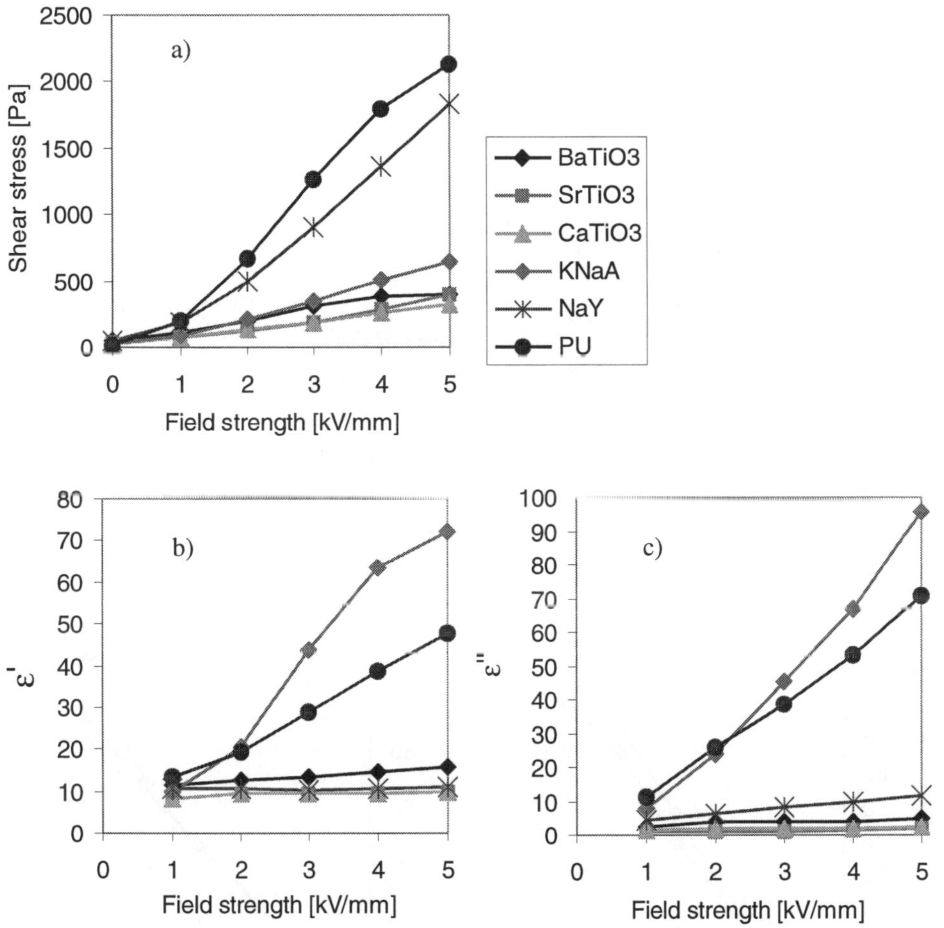

Figure 2. Field strength dependence of shear stress (a), ε' (b) and ε'' (c) at 50 °C and 100 Hz.

Fig. 2b reveals the dependence of the real part of permittivity ε' on the field strength. For all ER fluids the limit to low field strengths yields a value in the vicinity of 10 which is of the same order of magnitude as the permittivities determined from dielectric spectra at low voltages [4]. The permittivities of ER fluids containing perovskite or zeolite NaY particles in strong fields do not have a pronounced dependence on the field strength in contrast to the other ER fluids (KNaA and PU) which reveal strongly nonlinear permittivities.

The imaginary part of permittivity ε'' is also nearly constant in the whole range of field strengths for the perovskite ER fluids (see Fig 2c). Otherwise, the NaY ER fluid shows a weak increase of ε'' with field strength while ε'' of the other ER fluids (KNaA and PU) increases strongly. Since ε'' is a measure of the conduction properties of the samples this observation emphasizes the nonlinear conduction behaviour in ER fluids containing zeolite or PU particles. In case of ER fluids with perovskite particles ε'' is small demonstrating the low significance of conduction processes in these systems.

3.2 Temperature dependence

Most of the ER fluids investigated show a decrease of shear stress at higher temperatures. The exception is NaY with a maximum shear stress between 50 and 75 °C (see Fig. 3a).

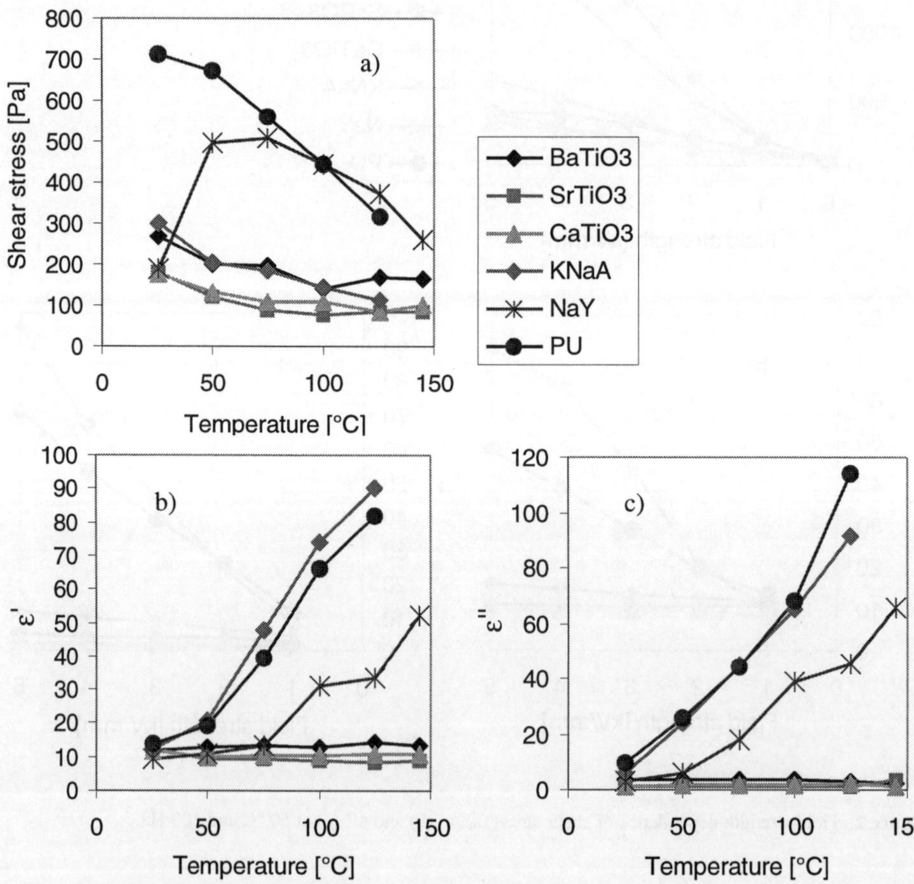

Figure 3. Temperature dependence of shear stress (a), ε' (b) and ε'' (c) at 2 kV/mm and 100 Hz.

The temperature dependence of the perovskite ER fluids is relatively weak and is mainly attributed to the viscosity change of silicone oil with temperature. The other ER fluids show a slightly stronger temperature dependence.

Drastic differences of the temperature dependence of ε' and ε'' occur between perovskites on the one hand and zeolites and PU on the other hand (see Fig. 3b, c). In the case of perovskites no change of ε' and ε'' can be observed in the whole range of temperatures considered. The absolute values of about 10 for ε' are similar to the permittivities in the low voltage limit. For the other ER fluids with zeolite or PU particles more or less strong increases of both parts of permittivity with temperature have been measured. The increase for NaY is less pronounced and starts at a higher temperature than for the other ER fluids (Fig. 3b, c).

3.3 Frequency dependence

Now first the results at ambient temperature of 25 °C and then at 100 °C are considered. The shear stress of the perovskite based ER fluids at 25 °C increases with the frequency to the upper limit of 1000 Hz. In contrast, the other ER fluids show a decrease in this frequency range (see Fig. 4a). In the range of low frequencies near 10 Hz the shear stress of the ER fluids KNaA and PU slightly increases. The shear stress values for the NaY based ER fluid are relatively low at 25 °C but drastically rise at 50 °C (Fig. 3a).

The shear stress results at high frequencies are consistent with the behaviour expected for ER fluids which are permittivity or conductivity dominated. Since the change of shear stress in the frequency range from 10 to 1000 Hz is small the frequency interval of investigation should be expanded but this is not possible due to experimental limitations. The dependence of ε' on frequency in Fig. 4b shows a small change for perovskite ER fluids but a stronger dependence for the other ER fluids at 25 °C.

At 100 °C the results are very different. For almost all ER fluids with a slight exception of NaY the shear stress increases with frequency (see Fig. 5a) which hints to a polarization mechanism at this temperature. On the other hand the frequency dependence of ε' in Fig. 5b shows only a weak change for perovskites but a strong decrease for the zeolite ER fluids and particularly for PU. ε'' could not be determined with sufficient accuracy under these conditions.

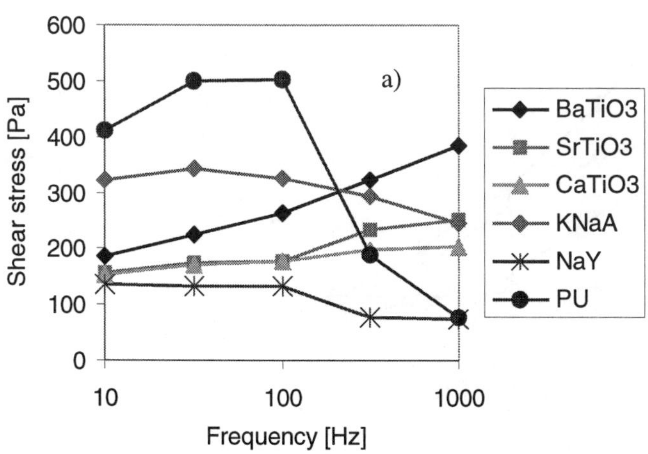

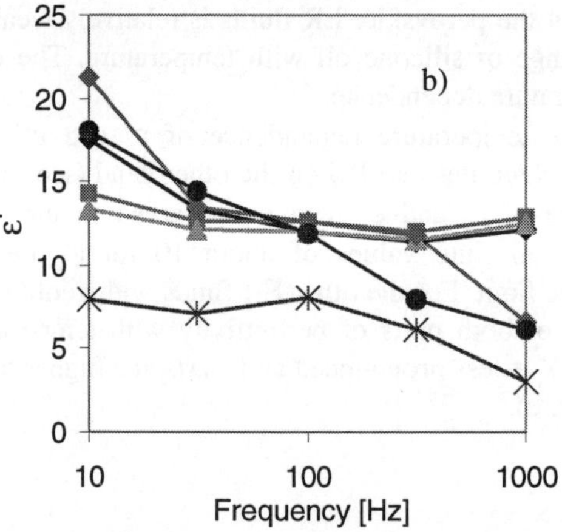

Figure 4. Frequency dependence of shear stress (a) and ε' (b) at 2 kV/mm and 25 °C.

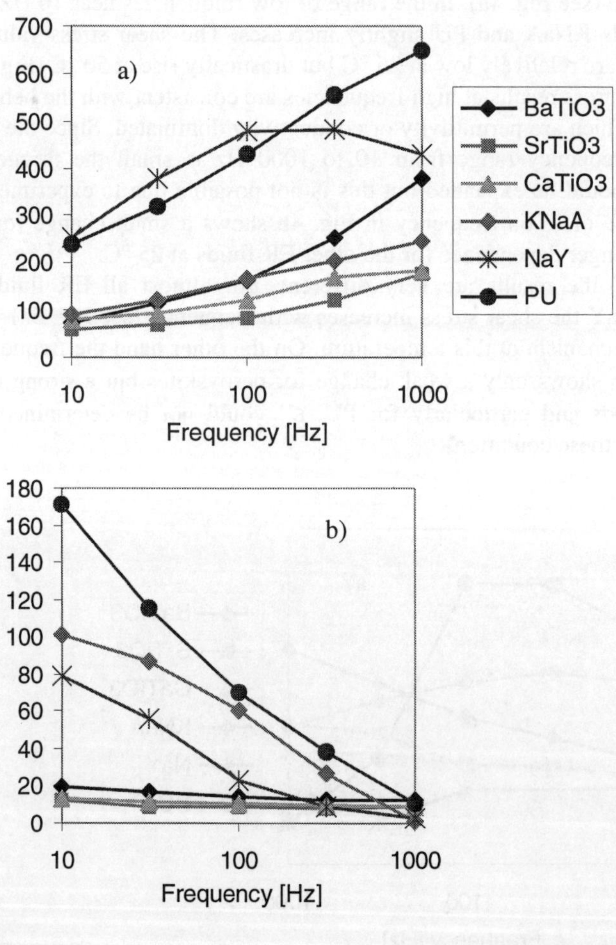

Figure 5. Frequency dependence of shear stress (a) and ε' (b) at 2 kV/mm and 100 °C.

The experimental results can be summarized in that way that perovskite ER fluids work with a polarization mechanism. For the other systems the mechanism is more complicated. Ion conduction seems to dominate the activity of the zeolite and PU based ER fluids but polarization by the permittivities is assumed to contribute. The relative contributions change not only with frequency but also with temperature and field strength. The electric properties of these systems are strongly nonlinear.

4 Conclusions

Simultaneous rheological and electric measurements have been performed on two groups of ER fluids containing different kinds of particles. In order to get more insight into the mechanisms of their ER activity the influences of field strength, temperature and frequency on shear stress and real and imaginary part of the permittivity, ε' and ε'', respectively, in strong electric fields have been investigated.

The ER fluids with perovskite type particles obey a polarization mechanism which is attributed to the local displacement of ions in the crystal structure of the particles. The contribution of conduction to ER activity is low. From the very weak influence of field strength and temperature on ε' and ε'' it can be derived that the electric properties of these ER fluids are nearly linear. The increase of shear stress with frequency is in accordance with theory.

In contrast to this the strong dependence of the dielectric properties of zeolite and PU based ER fluids on field strength and temperature demonstrates the pronounced nonlinearity of the electric properties of these systems. It is assumed that both conduction and polarization contribute to the ER activity. The relative extent of the contributions changes with external conditions like field strength and temperature.

5 Acknowledgements

Financial support for this work from the Deutsche Forschungsgemeinschaft is gratefully acknowledged by the authors.

References

1. Gast A. and Zukoski C., Electrorheological fluids as colloidal suspensions, *Adv. Colloid Int. Sci.* **30** (1989) pp. 153-202.
2. Conrad H., Properties and design of electrorheological suspensions, *MRS Bulletin* **23** (1998) pp. 35–42.
3. Klingenberg D. J., Particle polarization and nonlinear effects in electrorheological suspensions, *MRS Bulletin* **23** (1998) pp. 30–34.
4. Böse H. and Trendler A., Comparison of electrorheological and electric properties of ER fluids based on different materials, *Proc. 7th. Int. Conf. On Electrorheological Fluids and Magnetorheological Suspensions (R. Tao, Ed.), World Scientific, Singapore* (2000) pp. 80-87.
5. Bloodworth R., Electrorheological fluids based on polyurethane dispersions, *Proc. 4th. Int. Conf. On Electrorheological Fluids (R. Tao and G. D. Roy, Eds.), World Scientific, Singapore* (1994) pp. 67-83.

MAGNETORHEOLOGY OF A MILLIMETRIC STEEL SPHERES SUSPENSION

G.BOSSIS, O.VOLKOVA, S.LACIS

Laboratoire de physique de la Matiére Condensée, UMR 6622 Université de Nice Sophia-Antipolis, Parc Valrose, 06108 Nice, France E-mail: bossis@unice.fr

M.GUYOT

Laboratoire de Magnétisme et d'Optique,45 Avenue des Etats-Unis,78035 Versailles,France

The magnetorheology of a "suspension" of millimetric-size steel spheres in oil has been studied. After a precise measurement of the magnetic properties of the particles we show that the yield stress is well characterized and that it can be predicted by a finite element method and a non-affine model where the chains break in the middle. Some experiments in microgravity confirm that the sedimentation does not alter our results and also shows that the magnetic stress decreases with the shear rate.

1 Introduction

The prediction of the strength of an electrorheological or a magnetorheological suspension as a function of the applied field relies on the prediction of its yield stress. Many models, based on the calculation of interparticle forces, can give the right order of magnitude in some cases and fail in other ones. For instance a full three-dimensional model taking into account multipolar interactions up to the 50th order [1] is correct at low fileds but it will always predict a E^2 behavior which is not found experimentally in most cases due to the field limitation in the interparticle gap [2],[3]. In magnetorheological suspensions the situation can be easier to model since there is no charge transfer to deal with and calculations based on magnetic saturation of poles give the right order of magnitude for the yield stress [4],[5].In a magnetic material the permeability does not remain constant with the field and a precise estimation of the force between particles can only be obtained numerically by using finite elements[5],[6]. Even in this case some approximations are needed to avoid a full 3D calculation which would be very time consuming. The experimental situation is often not better as long as we want to characterize the particles (broad size distribution, existence of an oxide layer, difficulty to access to the bulk permeability of the particles, unknown structure, slippage on the walls[7]. In order to control most of the experimental parameters, we have used a suspension of monodisperse steel spheres with a diameter of one millimeter. Our aim in this paper is to examine the validity of the yield stress theories by comparing their predictions with the experimental results obtained on a reference "suspension". In a first section we shall describe our permeability measurements, then our yield stress measurements. As the sedimentation could be a factor of error in our experiments, we shall present in the last section some results obtained in microgravity.

2 Permeability of suspension

We have first measured the bulk permeability of the particles by using a rod of the same steel in an double coil system called hysteresimeter. As this steel presents a small hysteresis ,we have taken the average curve of magnetization. The curve is well fitted by a Frolisch- Kennelly law[8] with, as parameters, the initial permeability: $\mu_i=1+\chi_i$ and the saturation magnetization, Ms: $M=\chi_i H/(1+\frac{\chi_i}{M_s}H)$

A good agreement is obtained with M_s=1360 kA/m and μ_i=250. We have tested our finite element method and the consistency of these values by doing two other experiments one on the permeability of a chain of sphere and the other on the rupture force between two particles.

2.1 Permeability of a single chain of a steel spheres

First of all let us note that for such high permeabilities ($\mu_i=250$ for initial permeability and still $\mu_p=90$ for H=10kA/m), the Maxwell-Garnett theory for predicting the permeability, $\bar{\mu}$, of a suspension of spheres: $\frac{\bar{\mu}-1}{\bar{\mu}+2}=\Phi\cdot\beta$ with $\beta=\frac{\mu_p-1}{\mu_p+2}\approx 1$ gives a large uncertainty Furthermore the suspension is not isotropic as it would supposed to be in this model. In order to get the average permeability of the suspension, that we need to calculate the average field: $H=H_0/\bar{\mu}(H)$, we better have to start with the permeability of a chain of spheres. We shall consider 6 spheres because it is the number of particles between the two walls of the cell. We have placed 6 spheres in a plastic tube and measured their permeability.

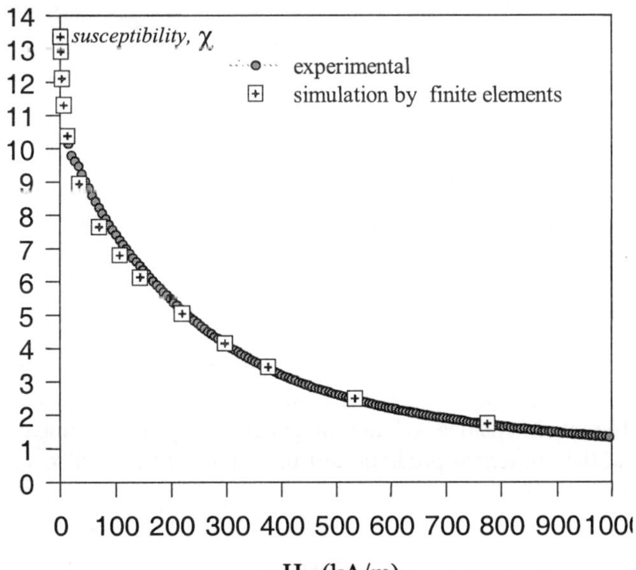

Fig.1 Comparison of the measured permeability of a chain of six spheres with the numerical predictions by finite elements using M_s=1360 kA/m and μ_i=250

The magnetization curve of this chain was obtained in the hysteresimeter and in the vibrating sample magnetometer (VSM). Figure 1 shows the susceptibility $\chi_{ch}=\mu_{ch}-1$ of a chain versus the magnetic field H_i inside the chain. The magnetic field inside and outside the chain are related by $H_i = \dfrac{H}{1+N_d \cdot \chi_{ch}(H_i)}$ with N_d=0.017 for the equivalent length/diameter cylinder [8]. The crosses are the predictions of a finite element calculation using as input the permeability of bulk steel fitted with the Frohlish-Kennely law. It is quite interesting to note that the initial susceptibility of a chain is an order of magnitude less than the one of a bulk steel. This is related to the fact that for a given average internal field H_i, the real field can be very large in the region of contact of pairs of particles inside the chain which greatly decreases the permeability. On one hand it is a good test of our numerical method and on the other hand it also allows to calculate the average permeability of the suspension by stating that the magnetic field H_{ch} acting on the chain and average field H in the suspension are equal. Then knowing the magnetic moment of a chain, $m_{ch}(H)$, the magnetization of the whole system of chains is: $M=\chi(H)H=N_{ch}\, m_{ch}/V = (3\Phi/2\, \pi\, a^2)\, m_{ch}$ where Φ is the volume fraction of the suspension. The average permeability of the suspension is represented in Fig.2a.

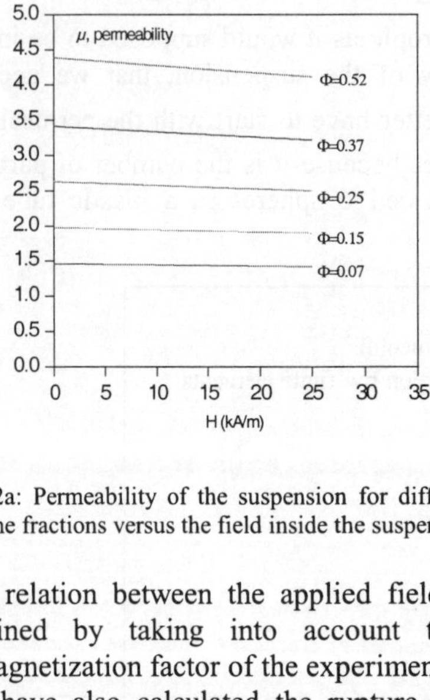

Fig. 2a: Permeability of the suspension for different volume fractions versus the field inside the suspension

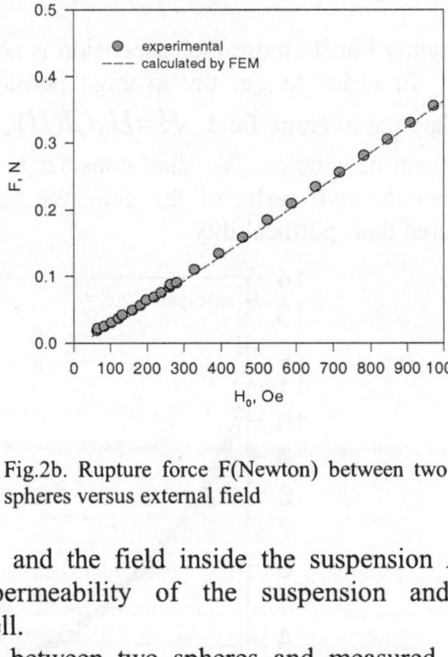

Fig.2b. Rupture force F(Newton) between two steel spheres versus external field

The relation between the applied field H_0 and the field inside the suspension H is obtained by taking into account the permeability of the suspension and the demagnetization factor of the experimental cell.
We have also calculated the rupture force between two spheres and measured it by attaching one sphere to a wire below the plate of an electronic balance and pulling on the other for different intensities of the magnetic field. The comparison between the experiment and the numerical prediction of the rupture force is also very good as can be seen in Fig. 2b.

3 Yield stress of the system of steel spheres

Two methods are often used for the experimental definition of the yield stress:
1) Measurements of the stress as a function of the strain. In this case we progressively increase the applied stress (or strain) and we record $\tau(\gamma)$ up to strains of a few unities. The stress corresponding to a plateau of the stress-strain plot will define the yield stress τ^y

2) We can also find the yield stress with the help of the flow curves. Here we obtain the applied shear stress as a function of the shear rate $\tau(\dot{\gamma})$. Then by the extrapolation of the curve towards $\dot{\gamma}\to 0$ we can obtain the yield stress τ^y.

The yield stress of a model suspension of steel spheres (*1mm* and *2mm* diameter) has been studied in the parallel plate geometry of the Carri-Med CSL-100 rheometer. In order to avoid any kind of slipping, a layer of spheres was glued on to each plate of the measuring cell. The number of spheres fixed on the plates has been adjusted as a function of each studied volume fraction. The two methods of yield stress measurements were used and they gave the same (within 5% of error interval) values of the yield stress. Figure 3 shows a characteristic flow curve obtained for a volume fraction Φ=15% in an external field H=31(kA/m). The dots connected by a line represent the values of the flow curve (shear stress versus shear rate) obtained by a slow increase of stress and the single big point corresponds to the yield stress obtained with the stress-strain measurements. We see that there is a good agreement between the extrapolation of the flow curve at zero shear rate and the stress-strain value. The hysteresis between the charge curve (increasing the stress) and the discharge curve (decreasing the stress) could be due to sedimentation since, when the structures are breaking, we can suppose that each sphere can fall a little bit. Actually the cell is transparent and we do not observe a sedimentation because the magnetic forces are much higher than the weight of the particles. This hysteresis is often observed even with colloidal suspensions and is due to a change of structure with the shear[8];we shall confirm this interpretation in the next section.

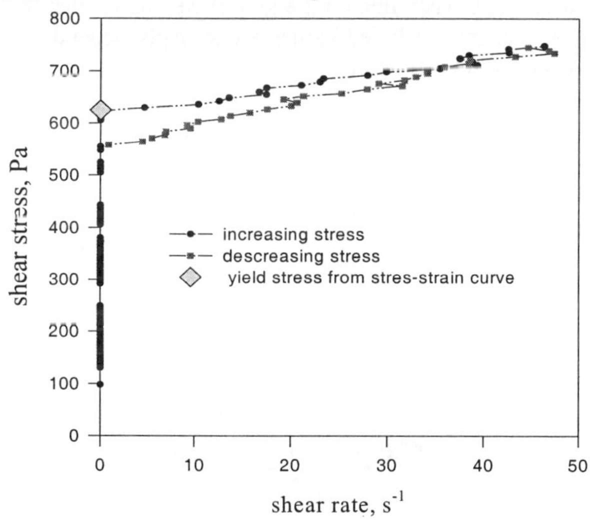

Figure 3. Flow curve for Φ=15% suspension and the yield stress found from stress-strain curve

Different volume fractions:Φ=0.07; 0.15; 0.25; 0.37; 0.52 were studied over a quite large range of applied fields.

Stress-strain measurements were also carried out in order to obtain the yield stress and simultaneously, the critical strain. The measurements were made in stress control mode and in the situation of equilibrium. That means that we set the stress and we measure the strain; then we increase the stress again only after the stabilization of the strain. It is quite easy to find the yield stress from the stress-strain curves. On the contrary defining the critical strain is more difficult since it changes with the position inside the cell: at the centre of the plates the strain is always zero, it increases with the distance from the centre of the plates and reaches its maximum at the extremity *(r=R)*. The other difficulty in the critical strain definition comes from the fact that the chains, which have been broken once, re-form themselves with the other pieces of broken chains. We have compared the experimental results with the finite elements calculations using the standard cubic network of chains. If we know the interparticle shearing force F_x between two particles, then we can find the shear stress as: $\tau=\frac{N_c}{S}F_x=\frac{\Phi}{(2/3)\pi a^2}F_x$ where Φ is the volume fraction of solid particles and N_c/S the number of chains per unit surface. The projection of the interparticle force on the shear direction increases with the strain, passes through a maximum and then falls down. The maximum of the force corresponds to the rupture and the corresponding stress is the yield stress. On the other hand the model is based on an initial structure made of a cubic lattice of simple chains which is affinely strained. The only input is the permeability of the bulk steel as a function of the field.

The test of the chain model is done by comparing the dimensionless calculated and measured yield stress, τ/μ_0H^2, (where H is the field inside the suspension) to the experimental one. This comparison is shown in Figure 4 for the volume fraction Φ=0.15 as a function of the internal average field *H*. The upper solid curve corresponds to the assumption that all the particles in the 6 particles chain are separating at the same time. On the contrary in the lower curve (long-dash line on the graph) we impose that the chain breaks only at its center. We see that the predicted values are strongly dependent on the assumption concerning the way the particles separate.

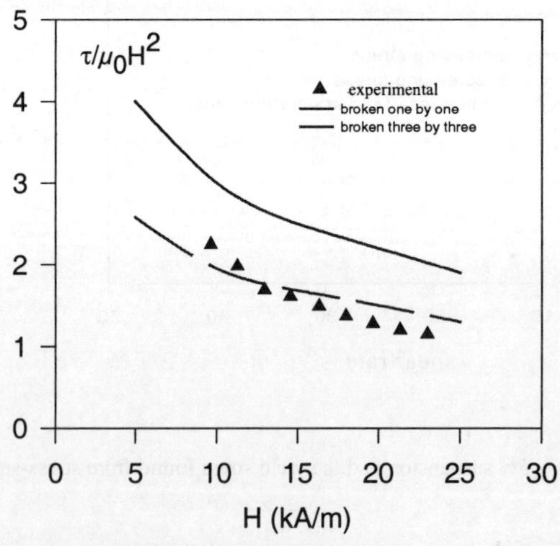

Figure 4. Normalised yield stress as a field function for Φ=15%

The experimental values are represented by the triangles and are normalized by the average field $H = H_0/\overline{\mu}(H)$. The agreement with the scheme where the chains break at the middle correspond to what we observe in the cell. Actually the affine motion of the chains gives an overestimation of the yield stress because, for a same strain the gaps are distributed among each pair of particles ,hence the distance between two particles does not decrease as quickly as in the case of one single gap.

In these experiments with millimetric spheres the liquid plays a role in thc viscous dissipation since the stress increases with the shear rate as can be seen in Fig.3. It is interesting to know if the hypothesis that the magnetic stress remains constant when we increase the shear rate (Bingham model) is a good one or not. We have checked this point by doing the same experiments in microgravity (parabolic flights with 20s of microgravity) on a gas of steel spheres.

4 Rheology of magnetic spheres in microgravity

The measurements were carried out with imposed ramp of velocity on the viscosimeter HAAKE-20 in the parallel plates geometry both in normal gravity and in microgravity for different values of the magnetic field. In the results plotted in Fig.5 we see that we have two regions. In the first part for shear rates between 0 and $200 s^{-1}$ we see a decrease of the stress when the shear rate is increasing. We can say that, once the chains have been broken the average distance between the particles is increasing and, as the magnetic interparticle force decreases with the distance, we observe this decrease of stress, which is not compensated by an increase of viscous dissipation since we have no liquid. Actually there is a small dissipation due to the inelasticity of the collisions between the steel spheres.

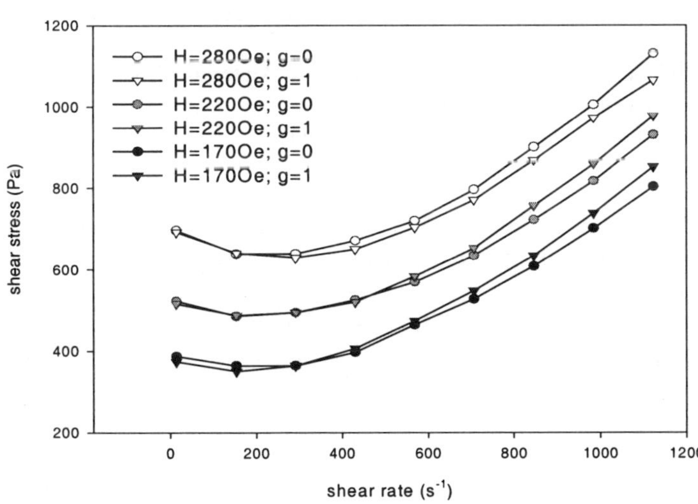

Fig.5. Comparison between the results obtained in microgravity (g=0) and in normal gravity (g=1)

This collisional dissipation dominates at higher shear rates and gives the strong increase of shear stress in the second part of the graph. We have verified that the sedimentation was not the cause of this anomalous behavior (decrease of stress when increasing the shear rate) by comparing experiments done in microgravity and in normal gravity. We can see that the difference between normal gravity and microgravity is very small below $400s^{-1}$ which confirms that sedimentation does not play a role in our experimental conditions, so we can trust the yield stress values we have measured as well as the existence of a true hysteresis as shown in Fig.3

6 Aknowledgements.

We gratefully acknowledge the CNES for their financial support and the possibility to do these experiments in microgravity.

References.

1. H.J.H Clercx, and G. Bossis "Many-body electrostatic interactions in electrorheological fluids" *Phys.Rev E* 48, 2721 (1993)
2. L.C.Davis "Ground state of an electrorheological fluid", Phys Rev A, **46**,R719(1992)
3. N.Felici,J.N.Foulc, and P. Atten, Electrorheological Fluids, edited by R.Tao and G.D.Roy (World Scientific,Singapore, p 139-(1994)
4. J.M. Ginder and L.C. Davis "Shear stresses in magnetorheological fluids:role of magnetic saturation", *Appl. Phys.Lett.*, **65**,3410 (1994);
5. J.M. Ginder , L.C. Davis and L.D.Elie. 1995 "Rheology of magnetorheological fluids:Models and measurements" Proc, 5th International Conf. Electro-Rheological Fluids, Magneto-rheological Suspensions and associated Technology, Univ. of Sheffield, UK, July 1995, 10-14
6. S.Lacis, O.Volkova, G.Bossis, in Modelling of Material Processing- Proc. Int. Colloq. (University of Latvia,Riga, p 122,1999)
7. O.Volkova, PhD Thesis, University of Nice, (1998)
8. D. Jiles, Introduction to magnetism and magnetic materials, Chapman&Hall, London, (1995)
9. 4. G.Bossis, E.Lemaire, O.Volkova, H Clercx, Yield stress in MR and ER fluids: a comparison between microscopic and macroscopic structural models, *J. of Rheology*, **41** 687 (1997)

AN EXPERIMENTAL INVESTIGATION ON MECHANICAL FATIGUE PROPERTY OF CHEMICAL STARCH-BASED ER FLUIDS

H. G. LEE, S. B. CHOI, J. H. KIM and C. C. CHEONG

Smart Structures and Systems Laboratory, Department of Mechanical Engineering
Inha University, Incheon 402-751, KOREA
E-mail: seungbok@inha.ac.kr

In this work, mechanical fatigue properties of a chemical starch-based ER fluid are empirically investigated. To do this, a flow-mode type electroviscometer, which has the same working mode as the cylindrical ER damper, is constructed. The linear reciprocating motion is obtained by driving a hydraulic power unit, and the number of fatigue cycle is measured by the limit switch. The field-dependent yield stress and response time to step input are evaluated as a function of the operating cycle. In addition, the vibration control performance of ER suspension for a passenger car is analyzed by incorporating the full-vehicle model with the yield stresses evaluated at different operating cycles.

1 Introduction

Electrorheological(ER) fluids belong to a colloidal suspension, which exhibit large reversible changes in their rheological behavior when subjected to external electric fields[1]. One salient property of the ER fluid itself is that it has a very fast response characteristic to the electric field and hence wide control bandwidth[2,3]. This inherent feature has triggered tremendous research activities in the development of various engineering applications. The successful development of a technology incorporating the ER fluid basically requires three principal ingredients : advanced ER fluid, reliable mechanism and efficacious control algorithm. In general, advanced ER fluids feature high yield shear stress at maximum electric field and low viscosity in the absence of the electric field. In addition to the effort to improve the field-dependent yield shear stress, researches on temperature sensitivity, mechanical fatigue, and sedimentation are need to be undertaken for the practical applicability of ER fluids. On the contrary to the studies on Bingham properties of ER fluids, the researches on the fatigue properties of ER fluids are considerably rare. A few researches on the variation of the yield shear stress[4], and the lubricant property[5] of the ER fluid have been undertaken. Recently, Park et al[6] investigated mechanical and electrical fatigue properties of arabic gum-based ER fluids.

This paper presents experimental results of mechanical property of chemical starch-based ER fluids. After establishing a flow-mode type electroviscometer, the linear reciprocating motion is generated by driving a hydraulic power unit. The number of fatigue cycles is measured by the limit switch and the test is undertaken up to one million cycles. The field-dependent yield stress and response characteristics to step input field are evaluated with respect to the operating cycle. For the completeness of this work, a full-vehicle model installed with ER suspension is adopted and its vibration control performance is analyzed by incorporating the yield stresses evaluated at different operating cycles.

2 Fatigue Test of ER Fluid

An experimental apparatus for the mechanical fatigue test of the ER fluid is constructed

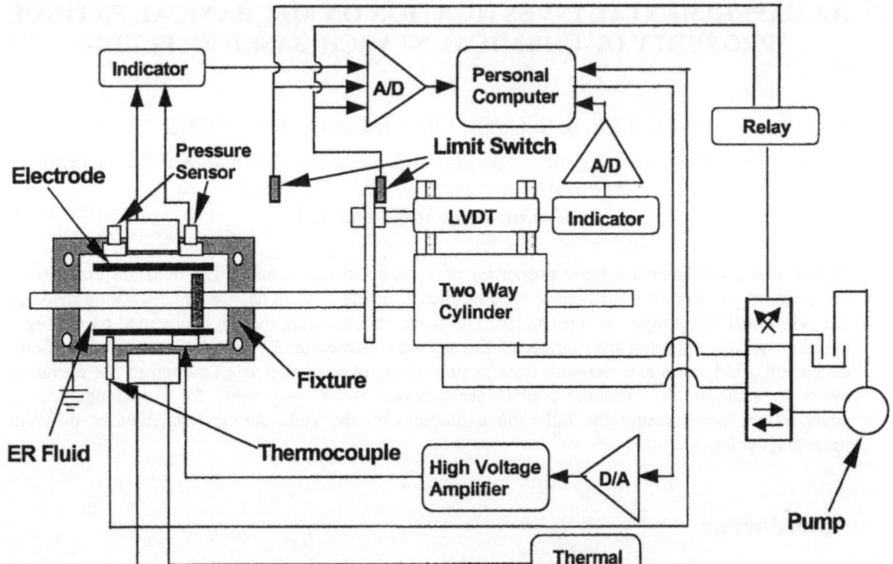

Figure 1. Experimental apparatus for fatigue test.

as shown in figure 1. The flow motion of the ER fluid filled in the cylinder housing is induced from the piston motion activated by driving a hydraulic power unit. The outer and inner cylinders are made of stainless steel so that they also serve as two electrodes. During the operating motion of the cylinder, the number of cycle is measured by the limit switch and directly fed back to the personal computer via A/D converter. In the mean time, the temperature of the ER fluid is maintained to be 25°C during whole operation controlling the thermal bath. The inner diameter of the cylinder is 38mm, and the electrode gap is 1mm. The excitation amplitude is set to be ±25mm, and the excitation frequency is fixed by 1.4Hz. The data are obtained at each 10,000 cycles until 100,000,

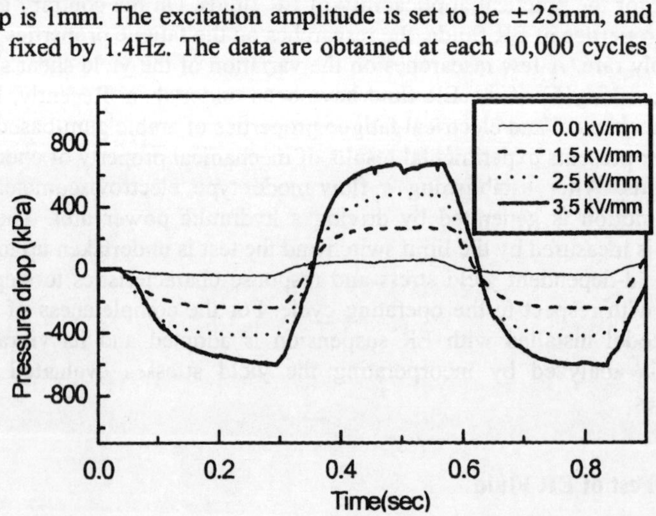

Figure 2. Pressure drop of the ER fluid.

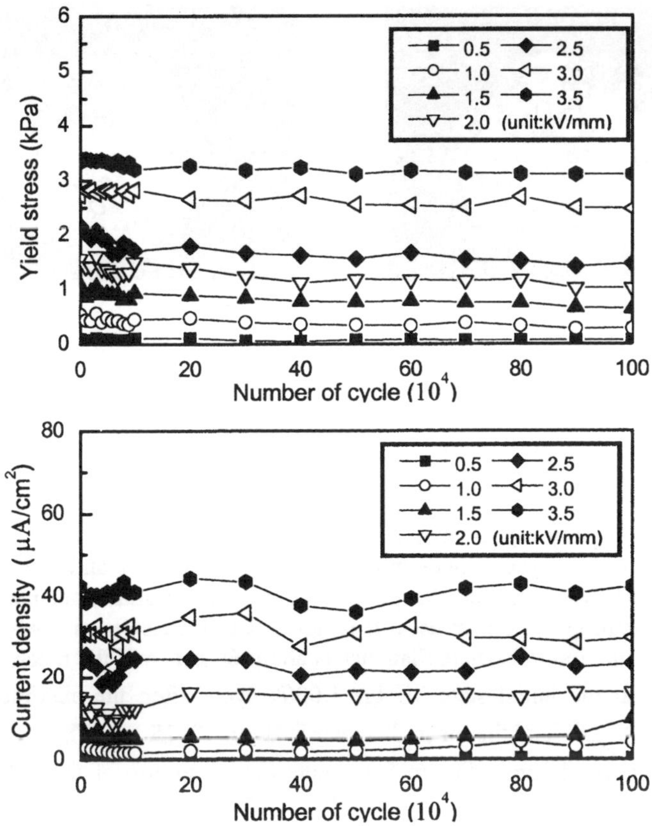

Figure 3. Bingham properties with respect to the operating cycle.

and at each 100,000 until one million cycles.

Figure 2 presents the measured pressure drop of the ER fluid at the first operating time. This is measured by two pressure sensors shown in figure 1. We clearly observe that the pressure drop is increased as the electric field increases. The maximum value of the pressure drop is then taken and converted to the yield stress. This conversion can be done by analyzing the flow-mode motion of the ER fluid[7]. On the other hand, the ER fluid employed in this work consists of chemical starch particle and silicon oil. The viscosity of the silicon oil is 10cS and the particle concentration is 50% by weight. It should be noted that the particles are dried for 168hours before using.

Figure 3 presents the measured yield stress and current density of the chemical starch-based ER fluid as a function of the operating cycle. It is clearly observed that the yield stress increases as the electric field increases. We can say from the result that the yield stress is not much degraded as the operating cycle increases. This result directly indicates that the chemical stach-based ER fluid exhibits favorable mechanical durability for a long time operation. It is also observed from the current density that the power required to obtain the yield stress is not increased with respect to the operating cycle. This is because the temperature of the ER fluid is controlled to be 25°C during whole operation. To more precisely observe the variation of the yield stress we distill the yield stress at two points ; initial state and after one million cycle state(we call it 1,000k after).

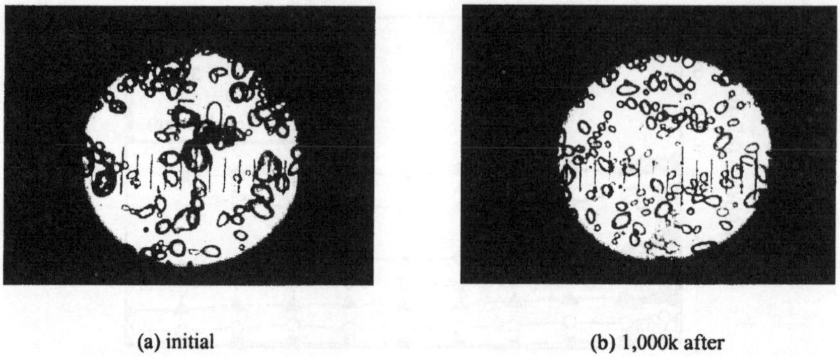

Figure 4. Microscopic photograph of the ER fluid.

The yield stress is expressed by $\tau_y = 590.4E^{1.39}$ Pa for the initial state, and $\tau_y = 263.8E^{1.98}$ Pa for the one million cycle state. Thus, we see that the yield stress at the initial state is slightly higher than that at the 1,000k after. In order to find out the reason, we take the microscope photographs of the ER particles and present them in figure 4. We clearly observe that the particle sizes at the initial state become much smaller as the operating cycle increases. It is noted that a division on a scale indicates 31 μm. Thus, the maximum particle diameter is about 60 μm and the mean diameter is about 20~25 μm at the initial state. At the 1,000k cycle after, the maximum particle diameter is about 35 μm and the mean diameter is about 10~15 μm. This may affect the degradation of the yield stress. Figure 5 presents time response of the ER fluid with respect to the operating cycle. The response time is obtained by applying step electric field and calculating the rising time when the pressure drop reaches to the steady state value[7]. We clearly see that the response time of the ER fluid has almost no variation with respect to the operating cycle. This implies that control software can be implemented in practice without any time delay.

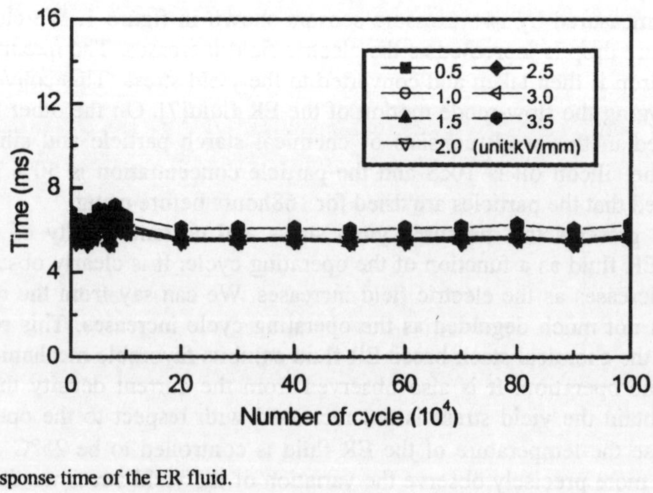

Figure 5. Response time of the ER fluid.

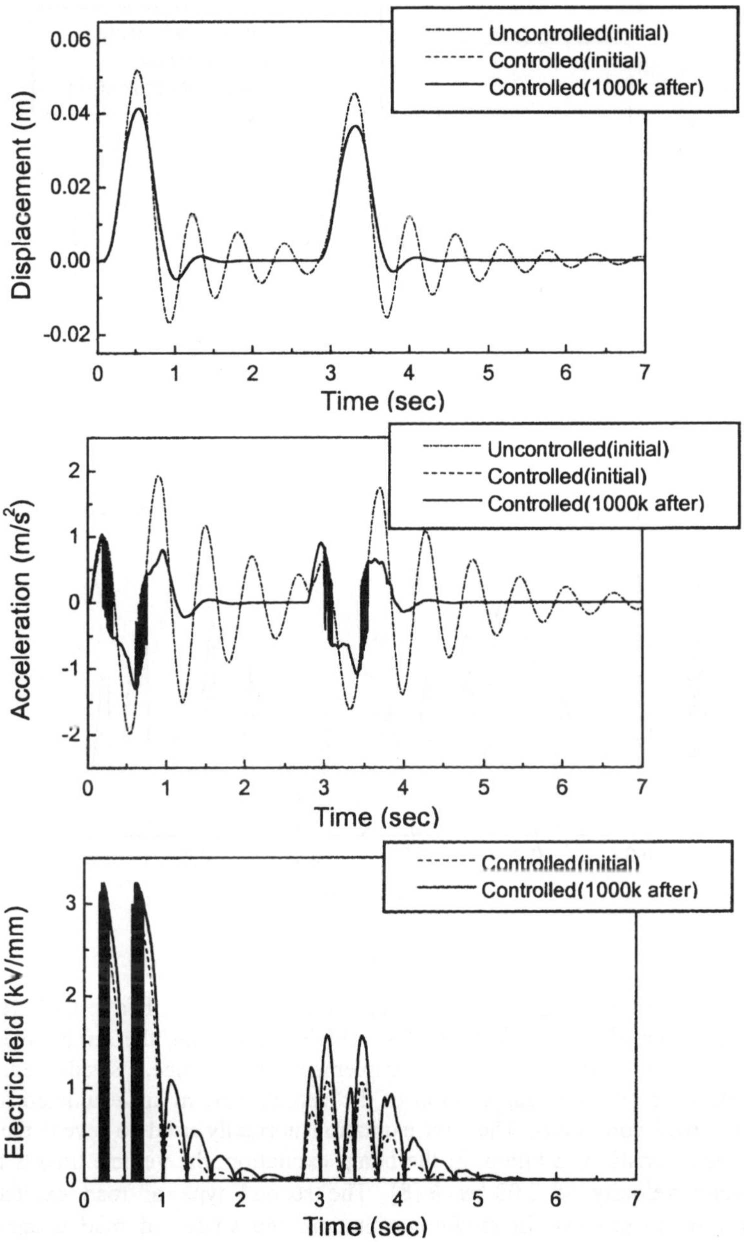

Figure 6. Bump response of the full-vehicle with ER suspension system.

3 Performance of Vehicle Suspension

The field-dependent yield stresses evaluated at two operating points ; initial state and 1,000k after, are incorporated with the full-vehicle system installed with ER suspension. Consequently, we analyze vibration control performance of the vehicle at two operating

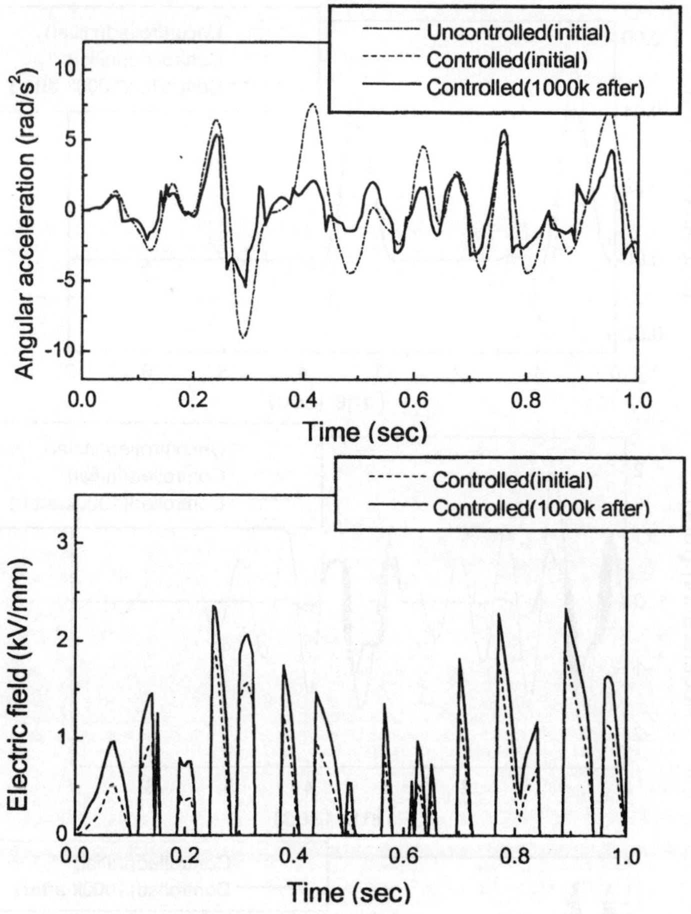

Figure 7. Random response of the full-vehicle with ER suspension system.

points. The full-vehicle model with ER suspension is well described in the reference[8]. Among many controller candidates for the vehicle suspension, the semiactive skyhook control algorithm, which is easy to implement in practice, is adopted. Control characteristics for vibration suppression of the vehicle system are evaluated under two types of the road conditions. The first excitation normally used to reveal the transmit response characteristic is a bump. In the bump excitation, the vehicle travels the bump with constant velocity of 3.08 km/h [8]. The second type of road excitation is a stationary random process. In random excitation, the values of road irregularity are chosen assuming that the vehicle travels on the paved road with the constant velocity of 72 km/h [8, 9].

Figure 6 presents the simulated bump responses of the full-vehicle model installed with ER suspension. We clearly see that the vibration magnitudes are substantially reduced by controlling the damping force of the ER suspension. It is also seen that there is almost no difference in the performance between the initial state and 1,000k after. However, we see from figure 6(c) that the control electric field applied to the ER suspension is slightly different. The magnitude of the input field at 1,000k after is a little higher than that at the initial state. This is easily expected from the yield stress at two operating points. Similar control characteristics are observed for the random excitation as shown in figure 7. These results directly indicate that control electric field to be applied to the ER suspension has a close relation to the yield stress. Therefore, an accurate evaluation of the yield stress at various conditions is very significant to accurately analyze control performances of ER devices.

4 Conclusion

Mechanical fatigue property of the chemical starch-based ER fluid was investigated under the flow-mode motion. It has been demonstrated via experiment that the field-dependent yield stress and current density have almost no variation with respect to the operating cycle. It was also shown that time response of the ER fluid was maintained to be constant regardless of the operating cycle. Finally, it has been analyzed that the magnitude of control electric field has a very close relation to the magnitude of the yield stress in the ER application devices.

Acknowledgement

This work was partially supported by National Research Laboratory(NRL) Program. This financial support is gratefully acknowledged.

References

1. Gandhi, M. V. and Thompson, B. S., Smart Materials and Structures. *Chapman and Hall, London* (1992)
2. Jordan, T. C. and Shaw, M. T., Electrorheology. *IEEE Transactions on Electrical Insulation* **24** (1998) pp. 849-879.
3. Block, H. and Kelly, J. P., Electro-rheology. *Journal of physics D : Apply Physics* **21** (1988) pp.1661-1677.
4. Xu, Y. Z. and Liang, R. F., Electrorheological Properties of Semiconducting Polymer-based Suspension. *Journal of Rheology* **35** (1991) pp.1355-1373.
5. Park, W. C., Choi, S. B., Cheong, C. C., Suh, M. S. and Yeo, M. S., Boundary Lubrication Characteristics of Silica-based Electro-rheological Fluids. *Journal of Intelligent Material Systems and Structures* **7** (1996) pp.511-516.
6. Park, W. C., Choi, S. B. and Suh, M. S., Material Characteristics of an ER Fluid and its Influence on Damping Forces of an ER Damper Part I : material characteristics. *Journal of Materials and Design* **20** (1999) pp.317-323.
7. Lee, H. G., Choi, S. B. Han, S. S. and Kim, J. H., Bingham and Response Characteristics of ER Fluids in Shear and Flow Modes. *Proceedings of the 7th International Conference on Electro-Rheological Fluids and Magneto-Rheological Suspension* (1999) pp.523-530.
8. Choi, S. B., Choi, Y. T. and Park, D. W., A Sliding Mode Control of a Full-car Electrorheological Suspension System via Hardware-in-the-loop Simulation. *ASME Journal of Dynamic Systems, Measurement and Control* **122** (2000) pp.114-121.
9. Nigam, N. C. and Narayanan, S., Applications of Random Vibrations. *Springer-Verlag, New York* (1994)

RHEOLOGICAL AND MAGNETORHEOLOGICAL BEHAVIOUR OF SOME MAGNETIC FLUIDS ON POLAR AND NONPOLAR CARRIER LIQUIDS

OANA BĂLĂU

Institute for Complex Fluids, University "Politehnica" of Timisoara, Bd. Mihai Viteazul no.1, 1900 Timisoara, Romania

DOINA BICA

Laboratory of magnetic fluids, Center for Fundamental and Advanced Technical Research, Romanian Academy – Timisoara Branch, Bd. Mihai Viteazul no.24, 1900 Timisoara, Romania

MARTINA KONERACKA, PETER KOPČANSKY

Institute for Experimental Physics, Slovak Academy of Sciences, Watsonova no.47, 04353 Kosice, Slovak Republic

DANIELA SUSAN – RESIGA

West University of Timisoara, Fac. of Physics, Bd. Vasile Parvan no.4, 1900 Timisoara, Romania

LADISLAU VÉKÁS[‡]

Laboratory of magnetic fluids, Center for Fundamental and Advanced Technical Research, Romanian Academy – Timisoara Branch, Bd. Mihai Viteazul no.24, 1900 Timisoara, Romania

Received October 24, 2001

Rheological and magnetorheological behaviour of monolayer and double layer sterically stabilized magnetic fluids, with transformer oil (UTR), dioctilsebacate (DOS), heptanol (Hept), pentanol (Pent) and water (W) as carrier liquids, were investigated.

The data for volumic concentration dependence of dynamic viscosity of high colloidal stability UTR, DOS, Hept and Pent samples are particularly well fitted by the formulas given by Vand (1948) and Chow (1994). The Chow type dependence proved its universal character as the viscosity data for dilution series of various magnetic fluids are well fitted by the same curve, regardless the nonpolar or polar charcater of the sample.

The magnetorheological effect measured for low and medium concentration water based magnetic fluids is much higher, due to agglomerate formation process, than the corresponding values obtained for the well stabilized UTR, DOS, Hept and Pent samples, even at very high volumic fraction of magnetic nanoparticles.

1. Introduction

Technical and biomedical applications of magnetic fluids depend mainly on their microstructural, physico-chemical, magnetic, rheological, magnetorheological properties and biocompatibility. Dimensional distribution, surfactant covering and

[‡] Corresponding author: Tel./Fax: +40-56-221547; e-mail: vekas@flumag2.mec.utt.ro

volume concentration of magnetic nanoparticles determine the above mentioned properties of magnetic fluids.

Indeed, the preparation procedure of magnetic nanoparticles and their mono-or double layer surface covering by surfactant molecules, depending on the properties of the carrier liquid, determine the degree of colloidal stability of the resulting magnetic fluid sample and consequently, its macroscopic behaviour under the influence of an applied magnetic field.

In what follows the rheological and magnetorheological behaviour of various concentration nonpolar and polar magnetic fluids will be presented and discussed.

2. Samples and Measuring Methods

Transformer oil (UTR), dioctylsebacate (DOS), pentanol (Pent), heptanol (Hept) and water (W) based magnetic fluids were prepared by specific procedures[1,2,3], depending on the physico-chemical properties of the carrier liquids, especially on their nonpolar or polar character. The chemical coprecipitation procedure was applied, i.e. the Fe_3O_4 nanoparticles were prepared by chemical condensation method. Stabilisation/dispersion of magnetic nanoparticles in the nonpolar UTR solvent was ensured by chemisorbed oleic acid monolayer, while in the case of the polar organic solvents (DOS, Pent, Hept) also a secondary physically adsorbed layer (dodecilbenzensulphonic acid (DBS) or a polymer) was neccesary.

In the case of water and physiological solution based, strongly polar magnetic fluids, two procedures were applied[3,4], using NH_4OH or NaOH for coprecipitation of Fe^{2+} and Fe^{3+} ions and oleic acid, lauric acid and dodecilbenzensulphonic acid for double layer sterical stabilization. The solid phase volume fraction attained was $\varphi_\rho=0.07$, corresponding to $M_o=300G$ nominal magnetization.

The physical volumic concentration φ_ρ of Fe_3O_4 solid nanoparticles in UTR, DOS, Pent, Hept samples attained $\varphi_\rho\approx0.2$, while the hydrodynamic volume fraction, φ_h, which includes also the surfactant layer (s), approaches $\varphi_h\cong0.6$, close to the highest possible volume fraction.

The rheological and magnetorheological behaviour of samples were investigated with an Ubbelohde-type capillary viscometer and with a Couette-type rotary rheometer (RHEOTEST-2). The experimental cells[5] have also magnetic field source, an electromagnet with plan-parallel pole pieces ($B_{max}\cong0.2T$) and respectively, a specially designed electromagnet ($B_{max}\cong0.1T$), for magnetorheological measurements. Details on the preparation procedures and complex characterization methods applied (e.g. electron microscopy, magnetic measurements, magnetooptics, small angle neutron scattering, ultracentrifugation, rheology, magnetorheology) for a large variety of nonpolar and polar magnetic fluid samples[1], may be found in a recent review[6].

3. Experimental Results and Discussion

3.1. *Physical volume fraction and temperature dependence of dynamic viscosity. Effective thickness of stabilization layer*

The physical volume concentration of particles (φ_ρ) and temperature (t) dependence of dynamic viscosity (η) was determined using the rotary rheometer for UTR, DOS, Pent and Hept samples, obtained by succesive dilution of the most concentrated initial magnetic fluid samples, all of them of high colloidal stability.

Dilution with pure solvent of the concentrated samples, up to very small volume fraction values (φ_ρ=0.02%), did not affect their colloidal stability, as it follows from detailed experimental investigations and theoretical analyses of the magnetisation curves for dilution series of concentrated magnetic fluids[7].

In Figs. 1,2 and 3 there are given the dependence of the relative viscosity, η/η_o, on physical volume fraction and temperature for two polar (DOS, Hept) samples and a nonpolar (UTR) sample, fitted to the formula of Vand[8]:

$$\eta/\eta_o = \exp\left[\left(2.5p\varphi_\rho + 2.7p^2\varphi_\rho^2\right)/(1-0.609p\varphi_\rho)\right], \tag{1}$$

where $p=\varphi_h/\varphi_\rho$ and η_o is the carrier viscosity.

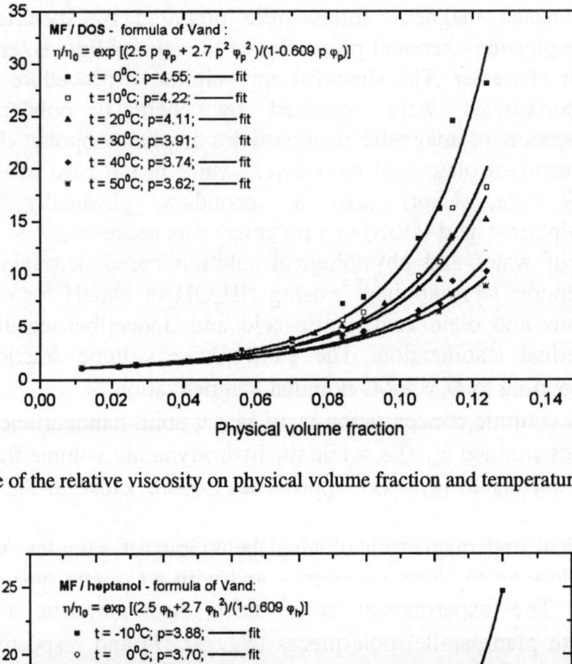

Fig. 1. Dependence of the relative viscosity on physical volume fraction and temperature for DOS samples

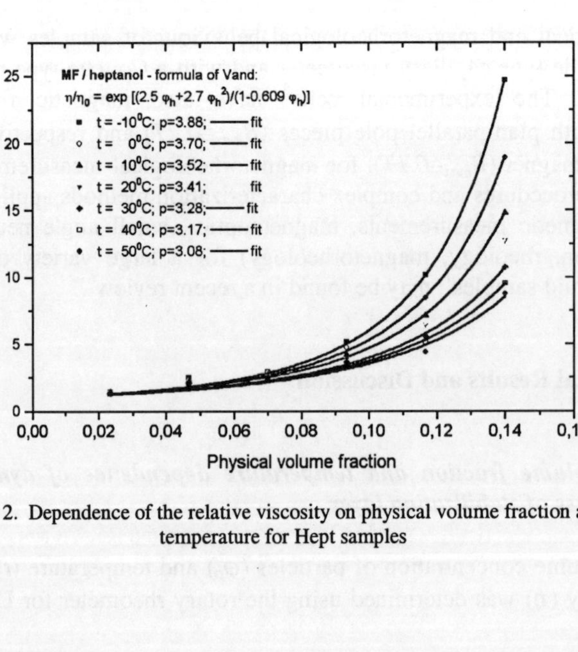

Fig. 2. Dependence of the relative viscosity on physical volume fraction and temperature for Hept samples

The fit parameter p is related to the effective stabilizant layer thickness:

$$\delta = \left(p^{1/3} - 1\right)\frac{d}{2}, \tag{2}$$

where d is the mean physical diameter of Fe_3O_4 nanoparticles, resulted from electron micrography analyses of samples (*d≅8.6nm* for UTR, DOS, Hept and Pent samples[9] and *d≅11.6nm* for W samples[4]). The fitted values of p indicate that the stabilizing layer has significantly greater thickness for the DOS samples, compared to the Hept and UTR samples, the UTR series having the most thin surfactant layer. Indeed, the DOS and Hept samples have Fe_3O_4 nanoparticles with double layer covering (oleic acid + polymer and DBS + DBS, respectively), while the UTR samples with chemisorbed monolayer (oleic acid) covering. As observed from the data outlined in Figs. 1,2,3, the effective thickness of the chemisorbed oleic acid monolayer is much less temperature sensitive (Fig. 3) compared to double layers (Figs. 1,2), where the second layer is physically (weakly) adsorbed.

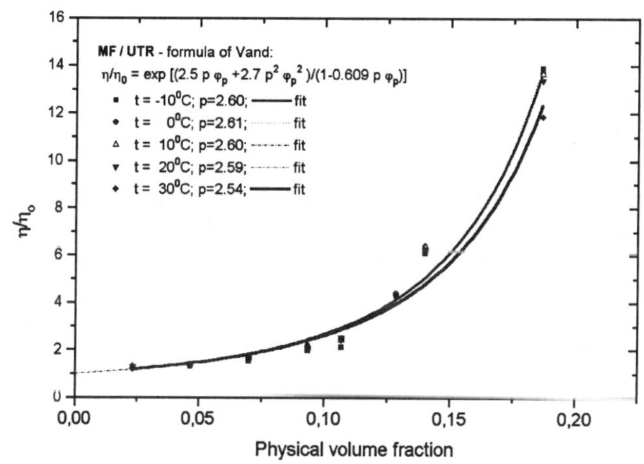

Fig. 3. Dependence of the relative viscosity on physical volume fraction and temperature for UTR samples

3.2. *Hydrodynamic volume fraction dependence of dynamic viscosity*

In previous analyses[5,10] it was shown that the hydrodynamic volume fraction dependence of the relative dynamic viscosity, $\eta/\eta_o = f(\varphi_h)$, is especially well described by the theoretical formula of Chow[12]:

$$\frac{\eta}{\eta_o} = \exp\left(\frac{2.5\varphi_h}{1-\varphi_h}\right) + \frac{A\varphi_h^2}{\left(1 - A\varphi_h^2\varphi_m\right)}, \tag{3}$$

were φ_m is the maximum hydrodynamic volume fraction, while A is the coupling coefficient (A_{theor} *=4.67* [12]). Good quality fits of Eq. (3) were obtained for UTR[10] and Pent[5] magnetic fluid series of samples.

Further investigations on other high stability magnetic fluids performed in this work evidence the ***universal character*** of the theoretical formula of Chow, Eq. (3). In Fig. 4 there are represented on the same curve the viscosity data, at different temperatures, for different concentration series of UTR, Pent, Hept and DOS magnetic fluid samples, fitted to Eq. (3).

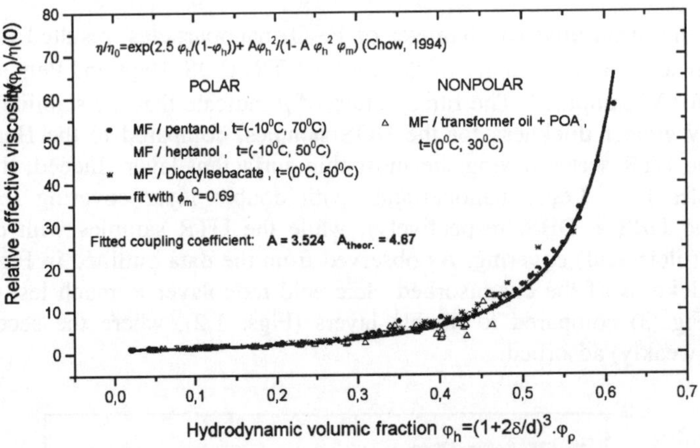

Fig. 4. Relative dynamic viscosity vs. hydrodynamic volume fraction for polar and nonpolar magnetic fluids samples

The maximum hydrodynamic volume fraction value φ_m=*0.69* used in the fit it was obtained by fitting the same data to the empirical formula of Quemada[5,13].

Eq. (3) does not take into account magnetic dipole-dipole interactions between particles nor the effect of agglomerate formation, therefore it is expected to fit well only for good stability magnetic fluids, independent of their nonpolar or polar character; deviations are expected at volume fraction values close to φ_m and at lower temperatures. Due to monolayer covering of particles, in case of nonpolar UTR samples at high concentrations, the distance between the magnetic core of particles could be even smaller than in the case of polar samples, which increases the role of dipolar interactions and produces larger deviations from the theoretical formula. The presence of agglomerates alters the interpretation of parameter p and fits as that of Fig. 4 are not possible along the way presented[5,10].

3.3. *Magnetorheological effect*

In the case of magnetic fluids the magnetic field induced increasse of effective viscosity (magnetorheological effect) is not a desired effect for most of applications, on the contrary to magnetorheological suspensions.

The existence of a large fraction of agglomerated particles, e.g. due to inefficient stabilization, significantly increase the magneto-rheological effect[11]. E.g., the use of technical grade oleic acid surfactant, instead of chemically pure one, drastically reduced the colloidal stability of a medium concentration (saturation magnetization M_o=*300 G*) decahydronaphtalene based magnetic fluid, whose effective viscosity increased approx. 1.8 times already in a moderate field (*B=0.06T*)[5].

On the contrary, for well stabilized samples, even of high magnetization (e.g. Pent with M_o=*830* G), the magnetorheological effect, under the same conditions, was very small, $\eta(B)/\eta(0) \cong 1.08$[13].

Water based magnetic fluids, of great interest for biomedical applications, represent a special case in what concerns stabilization mechanism, agglomerate formation and magnetorheological effect. The colloidal stability of water based magnetic fluids is reduced, even at medium or low volume fraction of magnetic nanoparticles, i. e. at correspondingly low saturation magnetization values *($M_o \leq 300G$).*

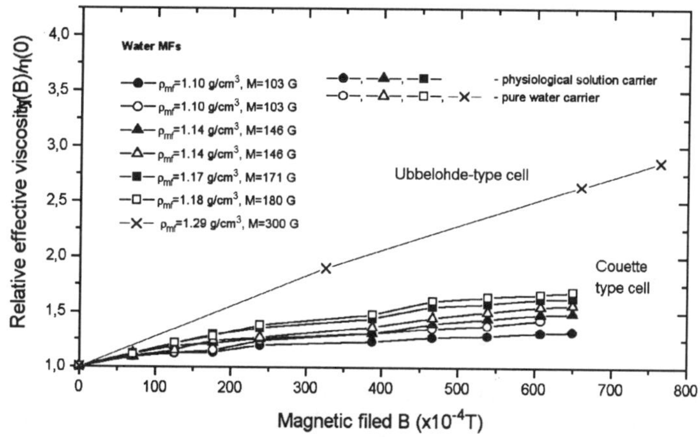

Fig. 5. Relative increase of the effective viscosity in magnetic field for pure water and physiological solution based magnetic fluids

In Fig. 5 the measured magnetorheological effect is compared for various concentration aqueous samples prepared. In the case of aquaeous magnetic fluids, intense agglomerate formation and phase separation occur, even at moderate field values. The size, shape and space configuration of condensed phase drops were evidenced by optical microscopy[14]. These are at the origin of the relatively large magnetorheological effect, up to $\eta(B)/\eta(0) \approx 3$ for the most concentrated sample, observed for water based magnetic fluids in moderate fields, below 0.1T.

4. Conclusions

The dependence of the relative dynamic viscsity η/η_o on the physical volume fraction φ_p of the magnetic particles is well fitted by the formula of Vand. For practically agglomerate free magnetic fluids the effective thickness of the stabilization layer and the corresponding hydrodynamic volume fraction values may be evaluated.

The theoretical formula of Chow fits well to dynamic viscosity data for various concentration magnetic fluids on nonpolar and polar carrier liquids, up to the highest value of the hydrodynamic volume fraction. The fitted value of the coupling coefficient is somewhat lower than the theoretically established value, mainly due to magnetic interactions between nanoparticles, of increasing significance at high packing.

Water based magnetic fluids, of special interest for biomedical applications, show a much larger increase of the effective viscosity, than that observed for the well stabilized concentrated magnetic fluids, already in moderate applied magnetic fields.

Acknowledgements

The support received through the NATO collaborative linkage grant No. LST. CLG. 977500 is gratefully acknowledged.

The researches performed were partly supported by the Romanian Academy through the grant No.87/2001. We are indebted to Professor Ioan Anton, member of the Romanian Academy, for his continuous advice in our work.

Special thanks are due to Professor George Bossis (University of Nice) for helpful discussions on water based magnetic fluids for biomedical applications. Dr. M. Liţă is thanked for electron micrography analyses performed.

References

1. Doina Bica, *Rom. Repts. Physics,* **47**, 265 (1995).
2. Doina Bica, L. Vékás, *Magn. Gidr. (Riga),* **2**, 194 (1994).
3. Doina Bica, L. Vékás, M. Raşa, „Preparation and magnetic properties of concentrated magnetic fluids on alcohol and water carrier liquids", *J. Magn. Magn. Mater.* (to be published)
4. M. Timko, Kopčansky, Martina Koneracka, A. Skumiel, M. Labowski, A. Jozefczak, Doina Bica, Oana Bălău, L. Vékás, C. Fanin, A. T. Giannitsis, „The comparative study of particle size distribution in water based magnetic fluid", (paper presented at CSMAG 2001, Kosice) (to be published).
5. L. Vékás, Doina Bica, I. Potencz, Daniela Gheorghe, Oana Bălău, M. Raşa, „Concentration and composition dependence of rheological and magnetorheological properties of some magnetic fluids", *Progr. Colloid Polym. Sci.,* **117**, 104 (2001).
6. L. Vékás, Doina Bica, M. Raşa, Oana Bălău, I. Potencz, Daniela Gheorghe, „Magnetric fluids – a special category of of nanomaterials. Preparation and complex characterization methods", *Micro and Nanostructures,* Edited by Dan Dascălu, Gheorghe Păun, Eugen Pincovschi, Vladimir Ţopa, Victor Voicu (Pub. House of the Romanian Academy, Bucharest 2001), 127.
7. M. Raşa, Doina Bica, A. Philipse, L. Vékás, „Dilution series approach for investigations of microstructural properties and particle interactions of high quality magnetic fluids", (submitted).
8. V. Vand, *J. Phys. Colloid. Chem.,* **52**, 277 (1948).
9. M. Raşa, Ph.D Thesis, West University Timişoara, Fac. Physics (1999).
10. L. Vékás, Doina Bica, Daniela Gheorghe, I. Potencz, M. Raşa, *J. Magn. Magn. Mater.,* **201**, 159 (1999).
11. S. Odenbach, K. Raj, „The influence of particle agglomerates on the magnetoviscous effect in ferrofluids", *Phys. Fluids,* (2001) (to appear).
12. T. S. Chow, *Phys. Rev. E,* **50**, 1274 (1994).
13. Daniela Gheorghe, PhD Thesis, West University Timişoara, Fac. Chemistry, Biology and Geography (2001).
14. V. Socoliuc, Doina Bica, „Experimental investigations of magnetic induced phase separation kinetics in aqueous ferrofluids", *Progr. Colloid Polym. Sci.,* **117**, 131 (2001).

AUTHOR INDEX

AUTHOR INDEX